Tolley's Health and Safety at Work Handbook 2005

Seventeenth Edition

Members of the LexisNexis Group worldwide

United Kingdom	LexisNexis UK, a Division of Reed Elsevier (UK) Ltd, 2 Addiscombe Road, CROYDON CR9 5AF
Argentina	LexisNexis Argentina, BUENOS AIRES
Australia	LexisNexis Butterworths, CHATSWOOD, New South Wales
Austria	LexisNexis Verlag ARD Orac GmbH & Co KG, VIENNA
Canada	LexisNexis Butterworths, MARKHAM, Ontario
Chile	LexisNexis Chile Ltda, SANTIAGO DE CHILE
Czech Republic	Nakladatelství Orac sro, PRAGUE
France	Editions du Juris-Classeur SA, PARIS
Germany	LexisNexis Deutschland GmbH, FRANKFURT and MUNSTER
Hong Kong	LexisNexis Butterworths, HONG KONG
Hungary	HVG-Orac, BUDAPEST
India	LexisNexis Butterworths, NEW DELHI
Ireland	Butterworths (Ireland) Ltd, DUBLIN
Italy	Giuffrè Editore, MILAN
Malaysia	Malayan Law Journal Sdn Bhd, KUALA LUMPUR
New Zealand	LexisNexis Butterworths, WELLINGTON
Poland	Wydawnictwo Prawnicze LexisNexis, WARSAW
Singapore	LexisNexis Butterworths, SINGAPORE
South Africa	LexisNexis Butterworths, Durban
Switzerland	Stämpfli Verlag AG, BERNE
USA	LexisNexis, DAYTON, Ohio

First published in 1988

A CIP Catalogue record for this book is available from the British Library.

ISBN 0 7545 2753 0

Typeset by Letterpart Ltd, Reigate, Surrey

Printed and bound in Great Britain by William Clowes Limited, Beccles and London

Visit LexisNexis UK at www.lexisnexis.co.uk

Tolley's
Health and Safety
at Work
Handbook
2005

List of Contributors

Lawrence Bamber, BSc, DIS, MASSE, FIRM, FIOSH, RSP,
Managing Director, Risk Solutions International

Mike Bateman, BSc, MIOSH, RSP,
Independent Health and Safety Consultant

Martin Bruffell, LLB,
Partner, Berrymans Lace Mawer

Jessica Burt, BA,
Solicitor, CMS Cameron McKenna

David Clarke, CIBiol,
Food Safety Consultant

Jill Cleaver, MSc, BSc (Hons),
Senior Consultant, Hu-Tech Associates Ltd

Nicola Coote, FIOSH, MIIRSM, MRSH, MCIPO, RSP,
Director of Personnel, Health and Safety Consultants Ltd

Dean Cross, BSc (Hons),
Consultant, Sypol Limited

Kitty Debenham, BA (Oxon),
Solicitor, Simmons & Simmons

Christopher Eskell, TD, LLB, MIMgt,
Partner, Bond Pearce

Janet Gaymer, MA (Oxon), LLM (Lond),
Partner, Simmons & Simmons

Margaret Hanson, BSc (Hons), MErgS,
Senior Consultant, Hu-Tech Associates Ltd

Hazel Harvey, MPhil, CCol, ASDC, MIOSH,
Head of Professional Affairs, IOSH

Dorothy Henderson, BA,
Partner, Travers Smith Braithwaite

Jo Hilliard, LLB,
Solicitor, Cartwrights

Nick Humphreys, LLB, LLM,
Solicitor, Barrister, Richards Butler

Ian Jerome, BSc,
Senior Consultant, Education and Training, Loss Prevention Council

Raj Lakha, BA, MBA, MIIRSM, MIOSH, MInstD, AIMC,
Director of Safety Solutions (UK) Ltd, Director of COASST International Ltd

David Leckie, LLB,
Barrister, Maclay Murray & Spens

Deborah Lloyd, BSc, LLB, LLM,
Solicitor, Ashurst Morris Crisp

Subash Ludhra, BSc, RSP, FIOSH, Dip Env Man,
Director Quality Assurance & Training, Initial Catering Services Ltd

Peter Neild, MPhil,
FCII, Berrymans Lace Mawer

Andrea Oates, BSc,
Health and Safety Researcher, Labour Research Department

Stephen Perry, LLB,
Solicitor, Bond Pearce

George Pitblado, I Eng, MI Plant E, MIIRSM, MWM Soc,
Partner, GMP Support Services Consultants

Hani Raafat, BSc, MSc, PhD, CEng, MIMechE, FIOSH, RSP,
Risk Management Consultant

Hamish Robertson, BSc (Hons),
Senior Consultant, Sypol Limited

Mark Rutter, BSc, PhD,
Environmental Advisor, CMS Cameron McKenna

Kajal Sharma, BSc,
Solicitor, CMS Cameron McKenna

Dr R C Slade,
College Safety Officer, King's College London

Mary Spear, BA (Cantab),
Solicitor, Cartwrights

Rebecca Stewart, BA,
Solicitor, Travers Smith Braithwaite

Roger Tompsett, BSc(Eng), MIOA,
Head of Acoustics, WS Atkins Noise & Vibration

Mark Tyler, MA, LLM, MIOSH,
Partner, CMS Cameron McKenna

George Ventris,
formerly secretary of the HSC's Construction Industry Advisory Committee (CONIAC) and a leader of HSE's Construction National Interest Group

Jeffrey Wale, LLB,
Partner, Berrymans Lace Mawer

Lawrence Waterman, FIOSH, RSP, MBIOH, ROH,
Chairman, Sypol Limited

Natalie Wood, BSc (Eng),
Solicitor, CMS Cameron McKenna

Health and Safety at Work Editorial Board

The following are all members of the newly appointed Health and Safety Editorial Board for the handbook. Each member brings with them a wealth of experience and knowledge of health and safety matters. The benefits to be gained from introducing this board are to ensure that this publication continues to provide practical, authoritative coverage of health and safety law in all aspects of industrial and office workplaces.

Lawrence Bamber

Lawrence Bamber is currently an Associate Consultant with Norwich Union Risk Services and is also a freelance OSH/ risk management consultant. He has worked in the OSH/ risk management field for over 30 years with General Accident, Stenhouse (Aon), as well as Norwich Union, and is well known for his training, consultancy and presentation skills. Lawrence is a past president of the Institution of Occupational Safety and Health (IOSH) and is the author of many papers and publications on OSH/ risk management.

Mike Bateman

Mike Bateman has spent over 30 years in the health and safety field, firstly as a HSE inspector, and then managing health and safety departments in the shipbuilding and aluminium industries. Since 1991 he has been running his own consultancy, attracting both large and small clients from a wide range of occupational areas. Mike has considerable practical experience of risk assessment and is the author of *Tolley's Risk Assessment Handbook*.

Professor Malcolm Harrington

Professor Malcolm Harrington, CBE, is Emiritus Professor of Occupational Medicine of the Faculty of Medicine and Dentistry, University of Birmingham. His major interests lie in epidemiology, occupational cancer, health care workers, and occupational health policy. In 1992, he was awarded the CBE, and in 2001 was awarded the lifetime achievement award by IOSH. During 1985–1996 Malcolm was Chairman of the Industrial Injuries Advisory Council. He is also the co-author of over 200 publications including research, review papers and chapters.

Mark Tyler

Mark Tyler is head of the Health, Safety and Products team at CMS Cameron McKenna. He holds corporate membership of IOSH, and is a member of the CBI Health and Safety Panel. Mark is also a co-author of the books *Product Safety* and *Safer by Design*. He has acted in a number of major cases which include the Southall and Ladbroke Grove Rail Inquiries, the Organophosphate group litigation, and the South Kensington Legionnaires' Disease outbreak.

Lawrence Waterman

Lawrence Waterman, FIOSH, RSP, MFOH, FRSA, has been a safety practitioner for over 25 years. Following a period in enforcement and in industry, he founded Sypol Limited, now one of the UK's leading health, safety and environmental consultancies. He is currently the Chairman of Sypol. Lawrence has a special interest in construction and in occupational health and hygiene which has been predominant in his career, but not to the exclusion of many other sectors and areas of work. He is a winner of the Gold Medal (jointly) for health and safety from the Institution of Civil Engineers and is Project Director of the Constructing Better Health occupational health pilot launched by the construction industry in October 2004. As of November 2004, Lawrence is President of IOSH.

Contents

List of Tables

List of Illustrations/Diagrams

Abbreviations and References

Many abbreviations occur only in one section of the work and are set out in full there. The following is a list of abbreviations used more frequently or throughout the work.

Legislation

ACoP	=	Approved Code of Practice
BPR	=	Biocidal Products Regulations
CAWR	=	Control of Asbestos at Work Regulations
CDM	=	Construction (Design and Management) Regulations
CHIP	=	Chemicals (Hazard Information and Packaging for Supply) Regulations
CHSW	=	Construction (Health, Safety and Welfare) Regulations
CLAW	=	Control of Lead at Work Regulations
COMAH	=	Control of Major Accident Hazards Regulation
COSHH	=	Control of Substances Hazardous to Health Regulations
CPA	=	Consumer Protection Act
DSEAR	=	Dangerous Substances and Explosive Atmospheres Regulations
EAWR	=	Electricity at Work Regulations
ECHR	=	European Convention on Human Rights
EPA	=	Environmental Protection Act
EPCA	=	Employment Protection (Consolidation) Act
FA	=	Factories Act
FSA	=	Food Safety Act
HRA	=	Human Rights Act
HSWA	=	Health and Safety at Work etc Act
LOLER	=	Lifting Operations and Lifting Equipment Regulations
OLA	=	Occupiers' Liability Act
OSRPA	=	Offices, Shops and Railway Premises Act
NONS	=	Notification of New Substances Regulations
POMSTER	=	Placing on the Market and Supervision of Transfers of Explosives Regulations
PUWER	=	Provision and Use of Work Equipment Regulations
Reg	=	Regulation
RIDDOR	=	Reporting of Injuries, Diseases and Dangerous Occurrences Regulations
s	=	Section
Sch	=	Schedule
SI	=	Statutory Instrument
SRSC	=	Safety Representatives and Safety Committees Regulations
SR&O	=	Statutory Rule and Order
TURERA	=	Trade Union Reform and Employment Rights Act 1993
UCTA	=	Unfair Contract Terms Act 1977

Case citations

AC	=	Appeal Cases
All ER	=	All England Reports
ALJR	=	Australian Law Journal Reports
Camp	=	Campbell Reports
CB, NS	=	Common Bench, New Series (ended 1865)

Ch	=	Chancery Reports
CL	=	Current Law
CLY	=	Current Law Year Book
CMLR	=	Community Law Reports
COD	=	Crown Office Digest
Con LR	=	Construction Law Reports
Cr App R	=	Criminal Appeal Reports
East	=	East's Term Reports
EG	=	Estates Gazette
Exch	=	Exchequer Reports
F & F	=	Foster & Finlayson (ended 1867)
H & C	=	Hurlstone & Norman
HSIB	=	Health and Safety Information Bulletin
ICR	=	Industrial Cases Reports
IRLR	=	Industrial Relations Law Reports
JP	=	Justice of the Peace and Local Government Review
KB/QB	=	Law Reports, King's (Queen's) Bench Division
KIR	=	Knight's Industrial Reports
LGR	=	Local Government Reports
LJKB	=	Law Journal Reports – New Series, Kings Bench
Lloyd's Rep	=	Lloyd's List Reports
M & W	=	Meeson & Welby
Med LR	=	Medical Law Reports
NLJ	=	New Law Journal
PIQR	=	Personal Injuries and Quantum Reports
RTR	=	Road Traffic Reports
SJ	=	Solicitors' Journal
SLT/SLT (Notes)	=	Scots Law Times/(Notes)
Taunt	=	Taunton Reports
TLR	=	Times Law Reports
WLR	=	Weekly Law Reports

Legal terminology

ECJ	=	European Court of Justice
HL	=	House of Lords
CA	=	Court of Appeal
EAT	=	Employment Appeal Tribunal
IT	=	Industrial Tribunal
J	=	Mr Justice, a junior judge, normally sitting in a court of first instance
LJ	=	Lord Justice, a senior judge, normally sitting in a court of appeal
plaintiff	=	the person presenting a claim in a civil action
defendant	=	the person against whom a claim is brought in a civil action
appellant	=	the person bringing an appeal in a civil action
respondent	=	the person against whom an appeal is brought in a civil action
tort	=	a species of civil action for injury or damage where the remedy or redress is an award of unliquidated damages
volenti non fit injuria	=	'to one who is willing no harm is done'; a complete defence in a civil action
novus actus interveniens	=	supervening act of third party
obiter dicta	=	words said by the way

ratio decidendi	=	principle of a case

Organisations

ACAS	=	Advisory, Conciliation and Arbitration Service
ACoP	=	Approved Code of Practice
ACTS	=	Advisory Committee on Toxic Substances
ASSE	=	American Society of Safety Engineers
BCC	=	British Chambers of Commerce
BHF	=	British Heart Foundation
BOHRF	=	British Occupational Health Research Foundation
BSC	=	British Safety Council
BSI	=	British Standards Institute
CBI	=	Confederation of British Industry
CCA	=	Centre for Corporate Accountability
CHAS	=	Chemical Health and Safety Assessment Scheme
CIA	=	Chemical Industries Association
CIEH	=	Chartered Institute of Environmental Health
CITB	=	Construction Industry Training Board
COIT	=	Central Office of the Industrial Tribunals
CORGI	=	Council for Registered Gas Installers
CPS	=	Crown Prosecution Service
CSCS	=	Construction Skills Certification Scheme
DEFRA	=	Department for Environment, Food and Rural Affairs
DWP	=	Department for Work and Pensions
EA	=	Environment Agency
EEF	=	Engineering Employers Federation
EMAS	=	Employment Medical Advisory Service
ENSHPO	=	European Network of Safety and Health Practitioners Organisation
EU	=	European Union
FPA	=	Fire Protection Association
HMIP	=	HM Inspectorate of Pollution
HMSO	=	Her Majesty's Stationery Office
HSC	=	Health and Safety Commission
HSE	=	Health and Safety Executive
ICOH	=	International Congress on Occupational Health
IEE	=	Institute of Electrical Engineers
IEMA	=	Institute of Environmental Management and Assessment
ILO	=	International Labour Office
INSHPO	=	International Network of Safety and Health Practitioners Organisations
IOSH	=	Institution of Occupational Safety and Health
JCT	=	Joint Contracts Tribunal
LSC	=	Learning and Skills Council
NEBOSH	=	National Examination Board in Occupational Safety and Health
NRPB	=	National Radiological Protection Board
OSHA	=	Occupational Safety and Health Administration (USA)
RoSPA	=	Royal Society for the Prevention of Accidents
RSSB	=	Rail Safety and Standards Board
SPA	=	Safety Passport Alliance
TUC	=	Trades Union Congress
WATCH	=	Working Group on the Assessment of Toxic Chemicals

Other terminology

ACM	=	Asbestos-containing Material
BRM	=	Business Risk Management
BS	=	British Standard
BSE	=	Bovine Spongiform Encephalepathy
CCTV	=	Closed Circuit Television
CHAN	=	Chemical Hazards Alert Notice
CRI	=	Corporate Responsibility Index
CSR	=	Corporate Social Responsibility
DSE	=	Display Screen Equipment
EHO	=	Environmental Health Officer
ELCI (ELI)	=	Employer's Liability (Compulsory) Insurance
EMF	=	Electromagnetic Field
EMS	=	Environmental Management System
FM	=	Facilities Management
FMD	=	Foot and Mouth Disease
FOPS	=	Falling Object Protective Structure
HAVS	=	Hand-arm Vibration Syndrome
IR	=	Infra-red (radiation) or Ionising Radiation
LEV	=	Local Exhaust Ventilation
LEL	=	Lower Explosive Limit
LOCAE	=	List of Classified and Authorised Explosives
LPG	=	Liquefied Petroleum Gas
MDF	=	Medium Density Fibreboard
MDHS	=	Methods for the Determination of Hazardous Substances
MEL	=	Maximum Exposure Limit
MEWP	=	Mobile Elevating Work Platform
MPE	=	Maximum Permissible Exposure
MSDS	=	Material Safety Data Sheet
NOEM	=	New or Expectant Mother
NVQ	=	National Vocational Qualification
OEL	=	Occupational Exposure Limit
OES	=	Occupational Exposure Standard
OHSAS	=	Occupational Health and Safety Assessment Series
OHSMS	=	Occupational Health and Safety Management Systems
PAH	=	Polycyclic Aromatic Hydrocarbons
PAT	=	Portable Appliance Testing
PCB	=	Polychlorinated Biphenyls
PLI	=	Public Liability Insurance
PPE	=	Personal Protective Equipment
ppm	=	parts per million
PTW	=	Permit to Work
RHS	=	Revitalising Health and Safety
ROPS	=	Roll-over Protective Structure
RPE	=	Respiratory Protective System
RSI	=	Repetitive Strain Injury
RSP	=	Registered Safety Practitioner
SHT	=	Securing Health Together
SWL	=	Safe Working Load
UEL	=	Upper Explosive Limit
ULD	=	Upper Limb Disorder
UV	=	Ultra-violet (radiation)
VCM	=	Vinyl Chloride Monomer
VDU	=	Visual Display Unit

WRULD	=	Work-related Upper Limb Disorder
WSA	=	Workers' Safety Advisors
YP	=	Young Person

Introduction

The handbook

1. Welcome to the 17th edition of *Tolley's Health and Safety at Work Handbook 2005*. When the first edition of the handbook was published, health and safety was still a matter for specialists – often professional safety advisers, HR directors and others who were required to 'deal with' health and safety on behalf of their organisations. The past year has seen a change in perspective highlighted by very public debates about risk and the nature of protection. Skilful media commentators such as Simon Jenkins and Jeremy Clarkson have questioned whether we are moving rapidly towards an over-protective, over-prescriptive approach to risk management. Participation in the debate on GM foods or the appropriateness of a particular train protection system requires an understanding of risk. In the boardroom this is linked to corporate governance and corporate social responsibility, on the shop floor it is associated with a radical rejection of 'compensation culture' in favour of a desire to return from work healthy and uninjured. This introduction sets the scene for health and safety and indicates ways in which the individual, focused chapters may be used to identify legal requirements and also characterise good practice with hints and tips on how this may be achieved and maintained.

This Introduction is in five main parts

2.
- drivers for health and safety performance (this starts at PARA 3);
- the changing world of work (this starts at PARA 6);
- recent trends in UK health and safety law and its enforcement (this starts at PARA 7);
- the structure of UK health and safety law (starting at PARA 19);
- the management of health and safety at work (starting at PARA 25).

Drivers for health and safety performance

Why do organisations seek to improve their performance?

3. This is an issue that has been subject to increasing research in the UK and around the world in recent years. There is also growing evidence of distinct differences between the perception of small businesses and larger companies. It is also worth noting that definitions of smaller businesses vary from less than 10 and less than 100 employees, to the Department of Trade and Industry's (DTi) category of small to medium-sized enterprises (SMEs), at less than 250 employees. In small companies, the attitudes and beliefs of the owners/managers are a dominant force whilst in larger corporations and public bodies the social norms are subject to the broader

forces of corporate governance. A combination of the following appear to be of greatest influence, but each overlaps with the other and may be present in different proportions and combinations:

- beliefs and attitudes of owners/managers in very small organisations;
- risk management and reputation risk;
- corporate governance;
- legal compliance.

Beliefs and attitudes of owner/managers in very small organisations

4. In the UK, across the European Union and in the United States of America, more than half of all workers are employed in organisations of less than 10 people. This represents a growing trend as the large industries of the most developed countries continue to decline and the new economy comprises services which are often provided by new organisations and by the self-employed. Identifying what motivates the leaders of such organisations is therefore important to governments and enforcing authorities for health and safety. It appears that the level of awareness of specific health and safety requirements embodied in regulations, is often poor and almost wholly dependent on suppliers, customers and peers. The suppliers to small businesses of goods and services are regarded as a reliable source of information. For example the labels on containers of materials and the accompanying safety data sheets are used by small businesses to establish working practices without, typically, any reference to the requirements of the *Control of Substances Hazardous to Health Regulations 2002* (*SI 2002 No 2677* as amended by *SI 2003 No 978*). For some small businesses which supply major clients, those larger organisations increasingly impose operating conditions and provide support which incorporates health and safety, as part of the management of a secure supply chain*. The last significant element of advice and guidance is from the informal contacts with other owners/managers. (* For a study into supply chain management for health and safety, see '*Managing risk – Adding value: How big firms manage contractual relations to reduce risk*' (1998), HSE Books.)

Whatever their source of information as to what they should be doing and how to do it, the owners/managers of small businesses who do act in this area seem to have a clear set of reasons for getting health and safety right. These motivators are:

- the *focus on employees* who are so well known because of daily contact that injury or ill health is treated almost as if a family member were affected (and in many cases family members do work in the small organisation, and children may attend the workplace in the school holidays, etc);
- the *focus on satisfying customer*s and responding to guidance from suppliers, a general approach to 'doing the right thing';
- a conviction amongst a minority that looking after worker health and safety and similar matters contributes to a *productive, healthier, happier workforce* and may have benefits such as controlling insurance premiums.

Why do larger organisations seek to improve their performance?

5. Most large organisations have been on a learning curve since the post-war period. At first there was an emphasis on managing the pure risk of getting health and safety

'wrong', and the impact that this could have on people (injuries), products (damage) and property (losses such as by fire). By the 1970s this effort had broadened to encompass environmental issues, and also the concept of business recovery – and as a result events which could lead to business disruption were being investigated to both prevent their occurrence but also to respond effectively if they did occur. By the 1980s risk management had matured into a discipline which also encompassed such matters as brand management and corporate reputation. This does not mean that large companies have eliminated such risks. The impact on companies involved in rail crashes or other high-profile health and safety failures is a testament to the correct evaluation that these are significant matters for corporate value – but most have implemented systems which are designed to identify and control such risks. Thus, protecting the reputation of a company as an efficient and well-managed organisation is one of the most significant drivers to improving health and safety performance.

Following scandals in the City of London in the late 1980s (such as Polly Peck and the Bank of Credit and Commerce International (BCCI)), and stimulated by press reports of 'Fat Cats' receiving unreasonable remuneration from poorly performing public companies, a series of committees reported on remedial steps – Cadbury, Greenbury and Hampel. The latter called for a new approach to corporate governance which included a component of risk management at the highest level. The Turnbull Report, which requires organisations to establish a risk management strategy, was adopted by the London Stock Exchange obliging listed companies (plcs) to publish an annual risk management report as part of their annual reporting arrangements. This has prompted many large businesses to adopt formal plan-do-check-act management systems to achieve the level of assurance required for directors to sign such reports. The approach to risk management has also dovetailed with an increasing interest in corporate social responsibility and the pressures on companies to be seen to look after their staff, the environment and their neighbouring communities. In the public sector, similar developments have been encouraged by the Audit Commission, and in the voluntary sector by the Charities Commission. (See '*Worth the risk*: *An introduction to risk management and its benefits*' (2001), Audit Commission; '*Accounting reporting by charities: Statement of recommended practice* (2000), SORP, which is effectively a mandatory equivalent of the Turnbull requirements.) Across the public and private sectors, therefore, the development of corporate governance is linked to and integrated with risk management – all organisations are asked to identify what could blow them off course, resulting in a diminution of services, a loss of profitability, the development of an unsustainable approach to their work. Health and safety is an important part of the matrix of risks identified in any such exercise, and the management and control of the health and safety risks has become part of what the board of directors or trustees have to achieve and assure. It was against this background that a guide was developed and published by the Health and Safety Executive (HSE) on the duties of directors (see also recent trends at PARA 7).

The third pressure for health and safety performance is the necessity to achieve legal compliance. Most people, including those who lead organisations, are social beings who accept that the rules they need to live and work by represent the framework for a civilised life. Even if an individual driver occasionally breaks the speed limit, he or she is unlikely to wish for unlimited speeds to be agreed for the street on which they live. For small organisations, there is an often-expressed view that they are very unlikely to be visited by an inspector, and there may be a low awareness of both the statutory requirements and often a misperception of the level of risks associated with their work. It is in the large organisation with a nervous board of directors and a management commitment to legal compliance that represents a significant stimulus. Identifying what regulations need to be complied with, those which are specifically applicable to the mix of work being undertaken, and then assuring that compliance, is a further push towards formal management systems.

The changing world of work

6. There is a wide range of evidence about the changes in the world of work. In the UK we have many more people working in call centres (about 4 per cent of the workforce) than in mining. This move towards service industries, as traditional manufacturing continues to decline in numbers of people employed, is matched by the increasing dominance of small businesses and the self-employed. Across the western world, it is more common for workers to be employed in companies with less than 10 staff than in companies with more than 250 employees. Flexible work patterns and the temporary nature of much employment, with hardly any young workers anticipating a 'job for life, has also altered the balance of health and safety issues with psychosocial risks giving rise to stress-related sickness absence overtaking musculo-skeletal injuries as the most common work-related illness. In the UK, the great increase in female employment and the pressure on an ageing workforce to stay at work longer is also changing the nature of the 'at risk' group which requires protection whilst at work. This listing of changes being experienced at work suggests that the law, which often moves slowly, the enforcement authorities, employers, employees and professionals safety advisers are each under pressure to adapt and change in turn.

HSC/E have sought to respond to this changing world of world through a strategy document designed to map out future priorities and actions over the years to 2010 and beyond (see '*A strategy for workplace health and safety in Great Britain to 2010 and beyond*' (MISC643)). The key lies in the language – a vision of health and safety as a cornerstone of a civilised society and a record which leads the world. It is notable that limited resources, the high level of working days lost to illness and accidents, and the fear that inhibits organisations from contacting HSE and local authorities, are seen as real issues to grapple with – as is the need to win hearts and minds and not grudging acceptance of health and safety arrangements. HSC has adopted a number of strategic themes to address these matters:

- Developing closer working partnerships, with a focus on: keeping people healthy and at work; improving education in risk management from an early age; dealing with the high levels of people receiving incapacity benefit by improving rehabilitation, and reducing sickness absence in the public sector. Occupational health is to be addressed mainly through strategic partnerships. It is worth noting that the *Constructing Better Health* initiative is being launched, managed and funded by representative organisations within the construction sector in partnership with HSE and the Department of Work and Pensions (DWP).
- Helping people benefit from effective safety management – through simplification, involving workers and providing accessible advice and support both directly and in partnership.
- Priorities – with less resources HSC has no choice but to select certain issues to concentrate on. Hence the reduction in HSE support for the European Week of Health and Safety (to be on Noise in 2005).
- Communications – to become more robust in defending the reputation of health and safety systems against both detractors and those who are over-zealous in their application.

The House of Commons Work and Pensions Select Committee reviewed the performance of HSC/E in 2004, taking evidence from many bodies including the

Institution of Occupational Safety and Health (IOSH). The Committee's report is critical of the failure to make progress on the *Revitalising* targets (see PARA 7 below), and is damning at the Government's slow progress in reforming the law on corporate killing. Calling for an increase in the number of inspectors in HSE's Field Operations Directorate, the Committee also questioned the reduction in the occupational health resource at a time that ill health had been adopted as a priority area. The report is supportive of HSC/E, but also makes many suggestions for improvement, including the need to address occupational road risk, to audit the performance of local authority regulators and to further encourage the involvement of the workforce.

The handbook reflects the continued changes in the world of work, and some themes are teased out in the remainder of this introduction.

Legal framework

Revitalising Health and Safety in the UK

7. Twenty-five years after the *Health and Safety at Work etc Act 1974* (*HSWA 1974*) came into force, reported accidents at work seemed to have been stuck on a plateau, albeit one lower than the levels reported when the Act was introduced. It was also recognised that ill health caused by work had not received enough attention. This prompted *Revitalising Health and Safety*, the Government's ten-year strategy statement published by the Health and Safety Commission (HSC) and launched by the Deputy Prime Minister in June 2000. The statement included targets for reducing incidence rates for fatal and major injury accidents and for work-related ill health by 10 per cent and 20 per cent respectively, and also for reducing the proportion of working days lost due to work-related injury and ill health.

The ten-point strategy for achieving these reductions included:

- positive engagement of small firms;
- motivation of employers through insurance incentives;
- cultivation of a more deeply engrained culture of self regulation;
- more partnerships with workers on health and safety issues;
- government leading by example, including promoting best practice through the supply chain;
- improved education on health and safety;
- introduction of a new occupational health strategy.

The last point was addressed in more detail in *Securing Health Together*, published by HSC a month later in July 2000. The target of healthy work, healthy at work and healthy for life was summarised in terms of workability. The occupational health strategy seeks, in addition to reducing the extent to which the workplace is a site for health harm, to improve access to the workplace for those who are ill or disabled by tackling rehabilitation and disability access with real vigour. Some three years after its launch, the strategy is being reworked to integrate with the Department of Work and Pensions (DWP) programme of health, work and recovery.

Since launching the two related documents there has been some limited progress. Pilot studies in occupational health have begun to provide the evidence required for creating an effective national occupational health support programme. The directors' guide on the responsibilities of the heads of companies was designed to embed health and safety at the highest levels of corporate life (see '*Directors' responsibilities for health and safety*' (2001), (IND(G)343), HSE). Although the strategy document refers to the significantly increased resources the Government has made available to HSC/HSE, since then there has been a squeeze on funding although the process is intended to be underpinned by enforcement. Proposals on corporate killing, and also the Government's stated intention to increase maximum fines, extend the range of offences which may be subject to custodial sentences and introduce other more innovative penalties have all been subject to delays, postponements and cancellations. However, HSE has started to 'name and shame' companies and individuals convicted of offences – the roll call can be accessed at www.hse-databases.co.uk/prosecutions/. Overall, as the Work and Pensions Select Committee pointed out, there is little realistic prospect of analysis in 2005 indicating that the 2004 (interim) *Revitalising* targets have been met.

Global legal framework

8. The International Labour Organisation (ILO) and the World Health Organisation (WHO) are both agencies of the United Nations. Each has a set of objectives embodied in programmes of action which reflect the international consensus that workers have the right to employment which does not damage their health or prejudice their safety. In the WHO *Occupational Health for All*, the statement in the introduction could not be more clear:

'According to the principles of the United Nations, WHO and ILO, every citizen of the world has a right to healthy and safe work and to a work environment that enables him or her to live a socially and economically productive life.'

[*Global Strategy on Occupational Health for All*, WHO Beijing Declaration 1994]

It is also worth noting that the founding definition of 'health' from WHO in 1948 is still that health is a state of complete physical, mental and social well-being and not merely the absence of disease or infirmity. This is, of course, much more positive than simply the prevention of harm, and can apply to the workplace as well. From the global perspective, each region can develop its own approach to health and safety at work. In many ways the most advanced region is the European Union. In addition, there are some health and safety regulations which derive directly from international agreements such as those which apply to marine safety, and to the transport of dangerous goods by road, rail and air.

The role of the European Union

9. For the past decade the European Union (EU) has provided the principal motor for changes in UK health and safety legislation. Article 118A in the Treaty of Rome 1957 gives health and safety prominence in the objectives of the EU. The Social Charter also contains a declaration on health and safety, although this has no legal force. Whilst the EU can issue its own regulations, it mainly operates through directives requiring Member States to pass their own legislation. The 'Framework Directive' adopted in 1989 as part of the creation of the Single Market contained many broad duties, including the requirement to assess risks and introduce appropriate control measures. Other directives have since been adopted or proposed, including many relating to technical standards and safety requirements for specific products. The proposed new EU Constitution is unlikely to do other than confirm the role of health and safety in the continued development towards an ever-closer union of nation states, soon to expand to 25 members. The new members are faced with some difficulties in bringing their economies into harmony with the established member states, and this applies to health and safety standards. One initiative

underway to foster such developments is a programme by IOSH for new member countries, providing a mechanism to support professional development and the sharing of advice and information.

Implementation of EU directives

10. The UK, along with other Member States, is required to implement EU directives by designated dates. The UK sits about halfway in the league-table for formal implementation of directives by the due date, and there is not at present an effective mechanism for evaluating implementation and enforcement. Implementation in the UK is normally overseen by HSC (and the HSENI in Northern Ireland), which as with its own regulatory initiatives, first circulates consultative documents incorporating the proposed regulations together with a related Approved Code of Practice (ACoP) and/or guidance to interested parties (trades unions, employers' organisations, professional bodies, local authorities etc). Following the consultation process, HSC submits a final version to the Secretary of State who then lays the regulations before Parliament. It is at this stage that a date for their coming into force is determined – usually that stipulated by the original EU directive.

Current EU initiatives

11. The process of EU directives and their local implementation peaked on 1 January 1993 when six sets of regulations came into operation on the same day. Since then, the pace of change has been much slower – a mixture of periodic updating of existing directives such as the Council Directive on Dangerous Substances which is amended every year, resulting in new CHIP Regulations covering the classification, labelling and packaging of dangerous substances, and new areas being covered, e.g. work is progressing on the Physical Agents Directive which will impact with new regulations on hand-arm vibration and whole body vibration. Few UK regulations are developed now, such as the proposed forthcoming regulations on improving arrangements for consultation of workers on health and safety matters, without them at least in part reflecting EU directives (as well as relevant Conventions such as that on human rights).

Balancing business with regulation

12. The latter stages of the Conservative Government of 1979–1997 saw a drive towards deregulation because of the perceived burden of legislation on business and particularly on small employers. The present Government was presumed by many to be automatically in favour of tougher health and safety laws and stricter enforcement. In practice, while HSC/HSE have been peripatetic between government departments with a period led by the Deputy Prime Minister and now a new home within DWP, the Government has displayed a high degree of reluctance to seek improvements in health and safety performance through new laws. As a result, when new regulations are considered, such as those for accident investigation (rejected in favour of new guidance) or the proposals for corporate manslaughter and for improved worker consultation – there is real pressure to establish that there will be direct benefits with little cost to modern British industry. The Work and Pensions Select Committee report challenged some of this in its report, but there is little sign of the Government changing its approach.

Health and Safety Executive

13. HSE has adopted eight Priority Programmes which are reflected in several ways, ie priorities for publications, for general resources and, above all, priorities for visiting inspectors on enforcement activities:

- Falls from height.
- Workplace transport.
- Musculoskeletal disorders.
- Work-related stress.
- Slips trips and falls.
- Construction sector.
- Agriculture sector.
- Healthcare sector.

Meanwhile, the Policy Section of HSE has been re-organised into a series of sections which combine health with traditional safety matters (the Health Directorate is now defunct). The current corporate plan for health and safety within HSE is instructive, i.e. dealing with display screens, continuing back awareness, seeking to reduce stress, (HSE is planning to publish new management standards for work organisations to reduce stress), targeting slips, trips and falls, improving the management of contractors and reducing work-related road risks. Each of the priority areas are reflected in a great deal of guidance available on the HSE website, and in particular the stress programme is led by the publication of management standards (defining what a 'good' organisation does, in order to be efficient, effective and with lower levels of stress imposed on its staff.)

Other influences

14. Both the Government and HSC/HSE are subject to pressure from public opinion and lobby groups. The aftermath of rail accidents at Southall, Paddington (Ladbroke Grove) and Hatfield and the Cullen Inquiry, together with the continuing campaign following the death of Simon Jones in his first day at work at Shoreham Docks, have kept the issue of corporate accountability in the public eye. This is reflected in the political calls for the law on corporate killing to be revised, for guidance to directors to be made mandatory, and for company reporting requirements to be amended to encompass health and safety performance.

Many civil actions for work-related injuries or ill health are supported by trades unions and these often bring issues into the public eye. Following the successful action by a social worker against Northumberland County Council, many more claims have been made for work-related stress (see *Walker v Northumberland County Council [1995] 1 All ER 737*). Compensation claims now total £2bn each year – 65 per cent through court claims and 35 per cent through industrial injuries benefit. Together with the estimated 40 million days lost each year through accidents and work-related ill health, the cost of getting health and safety wrong is obviously huge and growing.

Penalties and prosecutions

15. Penalties being imposed by the courts are steadily increasing. Great Western Trains pleaded guilty to a charge under *s 3* of the *HSWA 1974* following the Southall rail crash and were fined £1.5 million (plus £680,000 costs) – a record for a single

offence under health and safety legislation. In September 1994 six people were killed and seven seriously injured when a pedestrian walkway at the port of Ramsgate collapsed. After a twenty-five day trial early in 1997, two Swedish companies which had constructed and designed the walkway were fined £750,000 and £250,000 respectively, Lloyds Register of Shipping (which inspected the walkway) was fined £500,000 and Port Ramsgate, the client, was fined £200,000. With costs added to the total fines of £1.7 million, the bill to the defendants for the criminal proceedings alone came to a total of just under £2.5 million. The collapse in October 1994 of a rail tunnel being constructed at Heathrow did not result in any injuries but was described in court as one of the 'biggest near misses' in years. When the prosecution came to court in 1999, Balfour Beatty Civil Engineering were fined £700,000 plus £100,000 costs, with the tunnelling sub-contractor, Geoconsult GES MBH, fined £500,000 plus £100,000 costs.

Fines in excess of £100,000 for health and safety offences have now become relatively commonplace although both the Government and HSE believe that some courts continue to impose relatively lenient punishments. Guidelines on appropriate levels of penalties for breaches of health and safety legislation were laid down by the Court of Appeal in *R v F Howe & Son (Engineering) Ltd [1999] 2 All ER* and endorsed by the same court in *R v Rollco Screw and Rivet Co Ltd [1999] IRLR 439*. The first custodial sentence for a health and safety offence, one of three months, was imposed on Roy Hill by Bristol Crown Court, whilst Paul Evans received a nine-month term of imprisonment from Birmingham Crown Court in September 1998. Both penalties resulted from carrying out work involving asbestos without the necessary licence.

A recent case heard by the Court of Appeal will have a major influence on the level of fines imposed for breaches of health and safety legislation. In the case of *R v Colthrop Board Ltd, 31 January 2002 (Unreported)*, Lord Justice Gibbs giving judgment stated that although the remarks in a previous case that financial penalties in excess of £500,000 were reserved for cases of major public disasters, 'what is important is that companies in the position of the appellant can expect to receive financial penalties on a scale of up to at least half a million pounds for serious defaults and proportionately lesser sums if the limitation upon means or some lesser blame justifies it'.

This particular case did not involve a fatality. An employee suffered serious crushing injuries when he became entangled in a machine. A health and safety inspector had previously raised concerns about the machine. The Crown Court judge had imposed a total fine of £350,000, comprising of a fine of £200,000 for contravention of the *HSWA 1974, s 2(1)* and a further fine of £150,000 for contravention of the *Provision and Use of Work Equipment Regulations 1998* (*SI 1998 No 2306* as amended by *SI 2002 No 2174*).

The company lodged an appeal and argued that the fines imposed were excessive and were inconsistent with the scale of penalties to be derived from precedents in other cases and in particular the Court of Appeal. It also argued that the wrong approach had been taken to assessing its financial means for the purposes of sentencing after its business had ceased. When deciding whether the Crown Court judge had imposed a fine that was too high, Lord Justice Gibbs referred to other authorities, in particular *R v Friskies Petcare (UK) Ltd [2000] 2 Cr App Rep (S) 401*. He commented: 'we would not wish the sum of £500,00 to appear to be set in stone or to provide any sort of maximum limit for such cases. On the contrary, we anticipate that as time goes on and awareness of the importance of safety increases, the courts will uphold the sums of that amount and even in excess of them in serious cases, whether or not they involve what could be described as major public disasters'.

After the date of the accident the company had decided to stop trading. The sale of its assets realised £17 million. The Court of Appeal indicated that for a company that had ceased trading the asset value on the sale was the correct measure to use when assessing its means for sentencing purposes, and not its levels of profits while it had still traded (which would have resulted in lower assessment of means).

The Court of Appeal decided to reduce the total fine in this case to £200,000 (£100,000 for each offence). It took the view that despite the aggravating features, the company was of moderate size compared to others upon whom bigger fines had been imposed in the past. The case did not involve a fatality. The overall impact of the decision however will be more cases being referred to the Crown Courts for sentence and higher fines for corporate defendants.

Risk assessments

16. Much recent UK legislation has included a requirement for some type of risk assessment. The concept was introduced in the early 1980s in regulations applying to asbestos and lead but it came to wider prominence as a core requirement of the *Control of Substances Hazardous to Health Regulations 1988* (‘*COSHH*’). The group of Regulations which came into force on 1 January 1993 derived from EU directives continued this trend with four sets of regulations requiring an assessment of one kind or another. The most important of these is the general requirement for risk assessment contained in *Reg 3* of the *Management of Health and Safety at Work Regulations 1999* (*SI 1999 No 3242*) (‘the *Management Regulations*’). Other important regulations requiring more specific types of risk assessment relate to noise, manual handling operations, personal protective equipment, display screen equipment and fire precautions.

In practice a less formal type of risk assessment was already required by health and safety legislation – an assessment of the level of risk, the adequacy of existing precautions and the costs of additional precautions was necessary in order to determine what was ‘reasonably practicable’. Risk assessment lies at the heart of the original concept of self-regulation enshrined in the Robens Report 1992 – employers are required to demonstrate that they have identified relevant risks together with appropriate precautions and in most cases must have records available to prove this.

Supply-chain pressure

17. Since the introduction of *HSWA 1974*, many court decisions, most notably those involving *Swan Hunter Shipbuilders ([1984] IRLR 93–116)* and *Associated Octel ([1996] 1 WLR 1543)*, have emphasised that employers often have responsibilities for the activities of employees of other organisations. The structure of *HSWA 1974* and much subsidiary legislation is such that responsibilities overlap between employers rather than being neatly apportioned between them. Employers must do more than simply not turn a blind eye to the obvious health and safety failings of those with whom they come into contact: they must often take a pro-active interest in the health and safety standards of others.

This principle is exemplified by the Ramsgate prosecution (see PARA 15). Despite having engaged specialists to carry out the design, construction and checking of the passenger walkway, Port Ramsgate was still convicted of an offence. Mr Justice Clark stated that ‘the jury has found, in my judgement correctly, that an owner and operator of a port cannot simply sit back and do nothing and rely on others, however expert’.

In recent years a growing number of larger companies, local authorities and other public bodies have had increasingly formalised procedures for checking the health and safety standards of contractors wishing to work for them. This process has been accelerated by the demands of the *Construction (Design and Management) Regulations 1994* (*SI 1994 No 3140*) ('*CDM*') which require clients to satisfy themselves (via their planning supervisors) that potential principal contractors are capable of dealing with the health and safety issues associated with projects. The Regulations also place responsibilities on principal contractors in respect of sub-contractors. Consequently contractors are frequently required to provide details of their health and safety policies and generic risk assessments together with risk assessments and/or method statements for specific projects or activities. Many clients also take an extremely hands-on approach in policing the work of contractors on their premises. This has been turned into an aim of *Revitalising*, to make the Government a leader in procurement. This also chimes with current business developments, in that the supply chain used to be a little-regarded operation but now improving performance in an era of outsourcing is often focused on contractors. (See '*Managing risk – Adding value*'; '*Transforming the supply chain*' (2002), The Conference Board, New York.)

Recent changes in legislation

18.

- *Asbestos (Prohibitions) (Amendment) Regulations 2003 (SI 2003 No 1889);*
- *Biocidal Products (Amendment) Regulations 2003* (*SI 2003 No 429*);
- *Building (Amendment) Regulations 2003* (*SI 2003 No 2692*);
- *Carriage of Dangerous Goods and Use of Transportable Pressure Equipment Regulations 2004* (*SI 2004 No 568*);
- *Conduct of Employment Agencies and Employment Businesses Regulations 2003* (*SI 2003 No 3319*);
- *Control of Substances Hazardous to Health Regulations 2002* (*SI 2002 No 2677 as amended by SI 2003 No 978*);
- *Disability Discrimination Act 1995 (Amendment) Regulations 2003* (*SI 2003 No 1673*);
- *Employment Rights (Increase of Limits) Order 2003 (SI 2003 No 3038);*
- *Fishing Vessels (Working Time: Sea-fishermen) Regulations 2004 (SI 2004 No 1713);*
- *Health and Safety (Fees) Regulations 2004 (SI 2004 No 456);*
- *Pipelines Safety (Amendment) Regulations 2003 (SI 2003 No 2563);*
- *Regulation of Investigatory Powers (Communications Data) Order 2003 (SI 2003 No 3172);*
- *Working Time (Amendment) Regulations 2003* (*SI 2003 No 1684*).

The structure of UK health and safety law

19. Health and safety in the workplace involves two different branches of the law – criminal law (dealt with below at PARA 20) and civil law (referred to in PARA 24).

Criminal law

20. Criminal law is the process by which society, through the courts, punishes organisations or individuals for breaches of its rules. These rules, known as 'statutory

duties', are comprised in Acts passed by Parliament (e.g. *HSWA 1974*) or regulations which are made by government ministers using powers given to them by virtue of Acts (e.g. the *Manual Handling Operations Regulations 1992* (*SI 1992 No 2793* as amended by *SI 2002 No 2174*)).

Cases involving breaches of criminal law may be brought before the courts by the enforcement authorities which, in the case of health and safety law, are HSE and local authorities via their environmental health departments (see ENFORCEMENT). Magistrates' courts hear the vast majority of health and safety prosecutions although more serious cases can be heard by the Crown Courts. The maximum fine which can be imposed by the magistrates' courts is currently £20,000 for breaches of certain sections of *HSWA 1974* and £5,000 for most other offences.

Where cases are heard by the Crown Court there is no limit on the fines which can be imposed. PARA 15 details some significant fines resulting from recent cases. There are also a limited number of health and safety offences which can result in prison sentences of up to two years. These include:

- contravention of licensing requirements (e.g. for asbestos removal);
- explosives-related offences;
- contravention of an improvement or prohibition notice;
- contravention of a court remedy order.

As in all criminal prosecutions the case must be proved 'beyond all reasonable doubt'. There is a right of appeal to the Court of Appeal (Criminal Division) and eventually to the House of Lords or even the European courts, although in practice very few health and safety cases go to appeal. A death involving work activities might result in manslaughter charges which could lead to more severe penalties, although hitherto such cases have been rare. (PARA 7 refers to recent developments in relation to 'corporate killing' etc. and also to likely extensions to the range of penalties available for health and safety offences.)

The Health and Safety at Work etc Act 1974 (HSWA 1974)

21. *HSWA 1974* is the most important Act of Parliament relating to health and safety. It applies to everyone 'at work' – employers, self-employed and employees (with the exception of domestic servants in private households). It also protects the general public who may be affected by work activities. HSE has published a booklet, '*A guide to the Health and Safety at Work etc Act 1974*' (L1).

Some of the key sections of the Act are listed below.

Section 2 – Duties of employers

HSWA 1974, s 2(1) is the catch-all provision: 'It shall be the duty of every employer to ensure, so far as is reasonably practicable, the health, safety and welfare at work of all his employees.' See below for further discussion of the term 'reasonably practicable'.

Section 2(2) goes on to detail more specific requirements relating to:

- the provision and maintenance of plant and systems of work;
- the use, handling, storage and transport of articles and substances;
- the provision of information, instruction, training and supervision;

- places of work and means of access and egress;
- the working environment, facilities and welfare arrangements.

These are also qualified by the term 'reasonably practicable'.

Section 2(3) provides that an employer with five or more employees must prepare a written health and safety policy statement, together with the organisation and arrangements for carrying it out, and bring this to the notice of employees (see PARA 27).

Section 3 – Duties to others

HSWA 1974, s 3(1) provides: 'It shall be the duty of every employer to conduct his undertaking in such a way as to ensure, so far as is reasonably practicable, that persons not in his employment who may be affected thereby are not exposed to risks to their health or safety.'

Employers thus have duties to contractors (and their employees), visitors, customers, members of the emergency services, neighbours, passers-by and the public at large. This may extend to include trespassers, particularly if it is 'reasonably foreseeable' that they could be endangered, for example where high-risk workplaces are left unfenced.

Individuals who are self-employed are placed under a similar duty and must also take care of themselves. (If they have employees, they must comply with *section 2*).

Section 4 – Duties relating to premises

Under *HSWA 1974, s 4* persons in total or partial control of work premises (and plant or substances within them) must take 'reasonable' measures to ensure the health and safety of those who are not their employees. These responsibilities might be held by landlords or managing agents etc, even if they have no presence on the premises.

Section 6 – Duties of manufacturers, suppliers etc.

Those who design, manufacture, import, supply, erect or install any article, plant, machinery, equipment or appliances for use at work, or who manufacture, import or supply any substance for use at work, have duties under *s 6*.

Section 7 – Duties of employees

'It shall be the duty of every employee while at work:

(a) to take reasonable care for the health and safety of himself and of other persons who may be affected by his acts or omissions at work; and

(b) as regards any duty or requirement imposed on his employer or any other person by or under any of the relevant statutory provisions, to co-operate with him so far as is necessary to enable that duty or requirement to be complied with.'

[*HSWA 1974, s 7*]

Consequently employees must not do, or fail to do, anything which could endanger themselves or others. It should be noted that managers and supervisors also hold these duties as employees.

Section 8 – Interference and misuse

> 'No person shall intentionally or recklessly interfere with or misuse anything provided in the interests of health, safety or welfare in pursuance of any of the relevant statutory provisions.'

[*HSWA 1974, s 8*]

Section 9 – Duty not to charge

> 'No employer shall levy or permit to be levied on any employee of his any charge in respect of anything done or provided in pursuance of any specific requirement of the relevant statutory provisions.'

[*HSWA 1974, s 9*]

Levels of duty

22. Health and safety law contains different levels of duty:

Absolute

Absolute requirements must be complied with whatever the practicalities of the situation or the economic burden.

Practicable

The term 'practicable' means that measures must be possible in the light of current knowledge and invention.

Reasonably practicable

This term is contained in the main sections of *HSWA 1974* and many important regulations. It requires the risk to be weighed against the costs necessary to avert it (including time and trouble as well as financial cost). If, compared with the costs involved, the risk is small then the precautions need not be taken – it should be noted that such a comparison should be made before any incident has occurred. The burden of proof, however, rests on the person with the duty (usually the employer) – they must prove why something was not reasonably practicable at a particular point in time. The duty holder's ability to meet the cost is not a factor to be taken into account.

In effect, considering what is 'reasonably practicable' requires that a risk assessment be carried out. The existence of a well-documented and carefully considered risk assessment would go a long way towards supporting a case on what was or was not reasonably practicable. Neither risks nor costs remain the same forever and what is practicable or reasonably practicable will change with time – hence the need to keep risk assessments up to date.

Important regulations

23.

Construction (Design and Management) Regulations 1994 (SI 1994 No 3140)

The broad definition of 'construction work' used in the Regulations means that they apply to many medium-sized engineering and maintenance projects as well as to traditional construction activities and all demolition work. The Regulations provide for specific duties to be carried out by the 'client' (who must appoint a 'planning supervisor') and by the 'principal contractor' (who may in some cases also be the client). Key requirements of the Regulations are for the development and implementation of a formal 'health and safety plan' and the creation of a 'health and safety file' for the project.

Control of Substances Hazardous to Health Regulations 2002 (COSHH) (SI 2002 No 2677 as amended by SI 2003 No 978)

These Regulations require an assessment to be made of all substances hazardous to health in order to identify means of preventing or controlling exposure. There are also requirements for the proper use and maintenance of control measures and for workplace monitoring and health surveillance in certain circumstances.

Electricity at Work Regulations 1989 (SI 1989 No 635)

These Regulations contain requirements relating to the construction and maintenance of all electrical systems and work activities on or near such systems. They apply to all electrical equipment, from a battery-operated torch to a high-voltage transmission line.

Health and Safety (Consultation with Employees) Regulations 1996 (SI 1996 No 1513)

These Regulations extended the previous requirements (contained in the *Safety Representatives and Safety Committees Regulations 1977* (*SI 1977 No 500*) so that employers must now also consult workers not covered by trade union safety representatives.

Health and Safety (Display Screen Equipment) Regulations 1992 (SI 1992 No 2792 as amended by SI 2002 No 2174)

Where there is significant use of display screen equipment (DSE), employers must assess DSE workstations and offer 'users' eye and eyesight tests (which may necessitate provision of spectacles for DSE work).

Health and Safety (First-Aid) Regulations 1981 (SI 1981 No 917 as amended by SI 2002 No 2174)

Basic first-aid equipment controlled by an 'appointed person' must be provided for all workplaces. Higher risk activities or larger numbers of employees may require additional equipment and fully trained first-aiders.

Health and Safety (Safety Signs and Signals) Regulations 1996 (SI 1996 No 341 as amended by SI 2002 No 2174)

These Regulations require safety signs to be provided, where appropriate, for risks which cannot adequately be controlled by other means. Signs must be of the prescribed design and colours.

Health and Safety (Training for Employment) Regulations 1990 (SI 1990 No 1380)

Those receiving 'relevant training' (through training for employment schemes or work experience programmes) are treated as being 'at work' for the purposes of health and safety law. The provider of the 'relevant training' is deemed to be their employer – youth trainees and students on work experience placements therefore have the status of employees and must be protected accordingly.

Management of Health and Safety at Work Regulations 1999 (SI 1999 No 3242)

Employers and the self-employed are required to manage the health and safety aspects of their activities in a systematic and responsible way. The Regulations include requirements for risk assessment, the availability of competent health and safety advice and emergency procedures – several of these management issues are dealt with later in this Introduction (see PARA 25).

Manual Handling Operations Regulations 1992 (SI 1992 No 2793 as amended by SI 2002 No 2174)

Manual handling operations involving risk of injury must either be avoided or be assessed by the employer with steps taken to reduce the risk, so far as is reasonably practicable.

Noise at Work Regulations 1989 (SI 1989 No 1790)

Employers must carry out an assessment to determine the level of exposure to noise of their employees. The precautions required include noise reduction measures, provision of hearing protection and the establishment of hearing protection zones.

Personal Protective Equipment at Work Regulations 1992 (SI 1992 No 2966 as amended by SI 2002 No 2174)

Employers must assess the personal protective equipment (PPE) needs created by their work activities, provide the necessary PPE, and take reasonable steps to ensure its use.

Provision and Use of Work Equipment Regulations 1998 (PUWER 1998) (SI 1998 No 2306 as amended by SI 2002 No 2174)

These Regulations cover equipment safety, including the guarding of machinery. The definition of 'work equipment' also includes hand tools, vehicles, laboratory apparatus, lifting equipment, access equipment etc. The 1998 Regulations introduced additional requirements in respect of mobile work equipment and also replaced previous specific regulations relating to power presses, woodworking machines and abrasive wheels.

Reporting of Injuries, Diseases and Dangerous Occurrences Regulations 1995 (RIDDOR) (SI 1995 No 3163)

Fatal accidents, major injuries (as defined) and dangerous occurrences (as defined) must be reported immediately to the enforcing authority. Accidents involving four or more days' absence must be reported in writing within seven days.

Safety Representatives and Safety Committees Regulations 1977 (SI 1977 No 500)

Members of recognised trade unions may appoint safety representatives to represent them formally in consultations with their employer in respect of health and safety issues. The functions and rights of safety representatives are detailed in the Regulations. The employer must establish a safety committee if at least two representatives request this in writing.

Workplace (Health, Safety and Welfare) Regulations 1992 (SI 1992 No 3004 as amended by SI 2002 No 2174)

Physical working conditions, safe access for pedestrians and vehicles, and welfare provisions are covered by these Regulations.

Civil law

24. A civil action can be initiated by an employee who has suffered injury or damage to health caused by their work. This may be based upon the law of negligence, i.e. where the employer has been in breach of the duty of care which he owes to the employee. Being part of the common law, the law of negligence has evolved, and continues to evolve, by virtue of decisions in the courts – Parliament has had virtually no role to play in its development.

Civil actions may also be brought on the grounds of breach of statutory duty – it should be noted, however, that *HSWA 1974* and most of the provisions in the *Management of Health and Safety at Work Regulations 1999* (*SI 1999 No 3242*) do not confer a right of civil action, although the statutory duties owed by employers to employees under *HSWA 1974* have their equivalent obligations at common law.

Duty of care

Every member of society is under a 'duty of care', i.e. to take reasonable care to avoid acts or omissions which they can reasonably foresee are likely to injure their neighbour (anyone who ought reasonably to have been kept in mind). What is 'reasonable' will depend upon the circumstances.

Employers owe a duty of care not only to employees but also to such people as contractors, visitors, customers, and people on neighbouring property. In the case of the duty of care owed by employers to employees, it includes the duty to provide:

- safe premises;
- a safe system of work;
- safe plant, equipment and tools; and
- safe fellow workers.

Occupiers of premises are under statutory duties comprised in the *Occupiers' Liability Acts* of 1957 and 1984 (see OCCUPIERS' LIABILITY), and those suffering injury because of a defect in a product may sue the producer or importer under the *Consumer Protection Act 1987* (see PRODUCT SAFETY).

Vicarious liability

Employers are liable to persons injured by the wrongful acts of their employees, if such acts are committed in the course of their employment. Thus if an employee's careless driving of a forklift truck injures another employee (or a contractor or customer), the employer is likely to be liable. There is no vicarious liability if the act is not committed in the course of employment – thus the employer is not likely to be held liable if one employee assaults another.

Civil procedure

Civil actions must commence within three years from the time of knowledge of the cause of action. In an action for negligence, this will be the date on which the plaintiff knew or should have known that there was a significant injury and that it was caused by the employer's negligence. The plaintiff must be prepared to prove his case in the courts, but in practice most cases are settled out of court following negotiations between the plaintiff's legal representatives and the employer's insurers or their representatives. The *Employers' Liability (Compulsory Insurance) Act 1969* requires employers to be insured against such actions (see EMPLOYERS' LIABILITY INSURANCE), although some public bodies, for example local authorities, are exempt from the provisions of this Act.

As the result of recommendations made by Lord Woolf in his '*Access to Justice*' report of 1996, the *Civil Procedure Rules 1998* introduced widespread changes to civil procedure on 26 April 1999, affecting the progress of civil claims from their commencement to their conclusion. The rules involve a 'pre-action protocol' and govern the conduct of litigation in a way that is intended to limit delay. In most cases a single expert, medical or non-medical, will be instructed rather than each party using separate experts. Even if the case goes to court, the expert's report will usually be in writing, with both parties able to ask written questions of the expert and to see the replies. The new arrangements include a fast track system for personal injury claims up to a value of £15,000.

Damages

Damages are assessed under a number of headings including:

- loss of earnings (prior to trial);
- damage to clothing, property etc;
- pain and suffering (before and after trial);
- future loss of earnings;
- disfigurement;
- medical or nursing expenses; and
- inability to pursue personal or social interests or activities.

Defences

The plaintiff must prove breach of a statutory duty or of the duty of care on a balance of probabilities. However, a number of defences are available to the employer, including:

- *Contributory negligence*

  The employer may claim that the injured person was careless or reckless, for example, that he ignored clear safety rules or disobeyed instructions. Accidental errors are distinguished from a failure to take reasonable care. Damages will be reduced by the percentage of contributory negligence established, which will vary with the facts of each case.

- *Injuries not reasonably foreseeable*

  The employer may claim that the injuries were beyond normal expectation or control (an act of God). In cases of noise-induced hearing damage, mesothelioma (an asbestos-related cancer) or vibration-induced white finger, the courts have established dates after which a reasonable employer should have been aware of the relevant risks and taken precautions.

- *Voluntary assumption of risk*

  If an employee consents to take risks as part of the job, the employer may escape liability. However, this defence (*volenti non fit injuria*) cannot be used for cases involving breach of statutory duty – no one can contract out of their statutory obligations or be deprived of statutory protection.

Other civil actions

Other health and safety related situations may result in civil actions by employees. Employment protection legislation has recently been strengthened in relation to dismissals or redundancies resulting from health and safety activities (including refusal to work in situations of serious and imminent danger). Suspension or dismissal on maternity or medical grounds may also give a right of action. See EMPLOYMENT PROTECTION.

Management of health and safety at work

25. In 2003 the Institution of Occupational Safety and Health (IOSH) published its guidance to health and safety management systems, which stated in its introduction that:

> 'Work-related accidents and ill health can be prevented and wellbeing at work can be enhanced by organisations managing health and safety with the same degree of expertise and to the same standards as other core business activities.

[*'Systems in focus'* (2003), IOSH, Leicester]

The continuing development and adoption of formal occupational health and safety management systems (OHSMS) is a reflection of this conviction – such systems include those developed by HSE (*'Successful health and safety management'* (1997), (HSG65), 2nd edition), by the ILO (*'Guidelines on occupational health and safety management systems'*, (ILO-OSH 2001), ILO Geneva), and BSI (*'Guide to occupational health and safety management systems'*, (BS8800:1996), BSI London). Each system exhibits differences when compared to the others, but there is a similarity to the core elements of Plan-Do-Check-Act.

Figure 1: Plan-Do-Check-Act

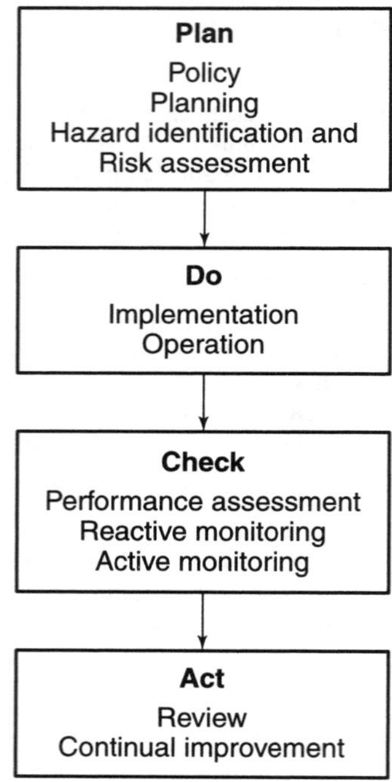

The costs of accidents and ill health

26. Various studies have been published to show the cost of accidents and ill health at work. Recent estimates of 40 million days lost to UK industry, of costs in the order of £20 bn, have a general impact but not at the company level. That is why a lot of work is going into developing business cases for taking health and safety more seriously. What is clear is that in large organisations, days lost represents in excess of 2 per cent of total costs. Staff take time off because they have been harmed, and the 'culture' permits and even encourages this. Large companies bear the cost, as do the individuals and their families. In small companies there is less absence, mainly because workers do not take the time off for equivalent ill effects – but this still impacts on productivity. And for small businesses, the loss of an employee for an extended period can be catastrophic. The UK has 6 times the number of people on long-term disability as in France, partly because we do not have ready access to rehabilitation and general occupational health services. This all represents work-in-progress but there are some lessons which can be learned:

- Where studies have been carried out, there is evidence that good health is good business.
- When staff are away from work ill, a return-to-work interview can provide invaluable information to prevent a recurrence.
- Failure to keep proper absence records hamstrings the ability to intervene positively.

- A culture which looks after employees suffers less stress and less absence than one which does not. High staff turnover as well as absence can be a lot more expensive than staff care and good work organisation.

The last full year of statistics (2003/04) shows an increase in the number of fatal injuries to workers by 4 per cent putting the figure up to 235, but this is largely the result of one of the worst ever work-related tragedies, when 23 cockle pickers were drowned in Morecambe Bay. This led to a private member's Bill being given government support for the licensing gangmasters. For direct employees, fatal accident rates fell from 183 to 168, and again the most common cause is falls from height, with being struck by a moving vehicle the cause of 44 deaths. Rates of injury are highest in agriculture and construction.

Health and safety policies

27. '*Successful health and safety management*' stated that 'accidents are caused by the absence of adequate management control' and stressed the importance of effective health and safety policies in establishing such control. *Section 2(3)* of the *HSWA 1974* requires employers to prepare in writing:

- a statement of their general policy with respect to the health and safety at work of their employees; and
- the organisation and arrangements for carrying out the policy.

It also requires the statement to be brought to the notice of all employees – employers with fewer than five employees are exempt from this requirement.

Policies are normally divided into three sections, to meet the three separate demands of *HSWA 1974*:

(i) The statement of intent

This involves a general statement of good intent, usually linked to a commitment to comply with relevant legislation. Many employers extend their policies so as to relate also to the health and safety of others affected by their activities. In order to demonstrate clearly that there is commitment at a high level, the statement should preferably be signed by the chairman, chief executive or someone in a similar position of seniority.

(ii) Organisational responsibilities

It is vitally important that the responsibilities for putting the good intentions into practice are clearly identified. In a small organisation this may be relatively simple but larger employers should identify the responsibilities held by those at different levels in the management structure. Whilst reference to employees' responsibilities may be included, it should be emphasised that the law requires the employer's organisation to be detailed in writing. Types of responsibilities to be covered in the policy might include:

- making adequate resources available to implement the policy;
- setting health and safety objectives;
- developing suitable procedures and safe systems;
- delegating specific responsibilities to others;
- monitoring the effectiveness of others in carrying out their responsibilities;

- monitoring standards within the workplace; and
- feeding concerns up through the organisation.

(iii) Arrangements

The policy need not contain all of the organisation's arrangements relating to health and safety but should contain information as to where they might be found, for example in a separate health and safety manual or within various procedural documents. Topics which may require detailed arrangements to be specified are:

- operational procedures relating to health and safety;
- training;
- personal protective equipment;
- health and safety inspection programmes;
- accident and incident investigation arrangements;
- fire and other emergency procedures;
- first aid;
- occupational health;
- control of contractors and visitors;
- consultation with employees; and
- audits of health and safety arrangements.

Employees must be aware of the policy and, in particular, must understand the arrangements which affect them and what their own responsibilities might be. They may be given their own copy (for example, within an employee handbook) or the policy might be displayed around the workplace. With regard to some arrangements detailed briefings may be necessary, for example as part of induction training.

Employers must revise their policies as often 'as may be appropriate'. Larger employers are likely to need to arrange for formal review and, where necessary, for revision to take place on a regular basis (e.g. by way of an ISO 9000 procedure). Dating of the policy document is an important part of this process.

Sources of health and safety advice

28. Within '*Successful health and safety management*', HSE emphasised the importance of establishing a positive health and safety culture within an organisation as a prerequisite of effective health and safety management. It referred to the 'four Cs' as key components in establishing such a culture: control, competence, communication and co-operation.

While competence in health and safety matters is relevant throughout any workforce, it is particularly important at management levels. The *Management of Health and Safety at Work Regulations 1999* (*SI 1999 No 3242*) have taken this concept further by requiring (in *Reg 7*) every employer to appoint one or more competent persons to assist him in complying with the law. The ACoP accompanying the Regulations states that the size and type of resource required will be relative to the size of the organisation and the risks present in its activities. Full-time or part-time specialists may be appointed, or use may be made of external consultants, although the 1999 Management Regulations state a preference for employees. Smaller employers may appoint themselves, provided that they are competent – the ACoP refers to competence as comprising both the possession of theoretical knowledge and the capacity to put it into practice in the work situation. An awareness of the limits of one's own knowledge and capabilities is also important, and it should be noted that the HSE was urged by the Work and Pensions Select Committee to become more prescriptive on the question of competence and issue clear guidance. It is possible that the establishment of individual chartered status for members of IOSH who meet the criteria, including a commitment to continuing professional development, will help to clarify matters.

Health and safety training is available from many different sources. The following organisations either provide training themselves or oversee training through accredited training centres.

- National Examination Board in Occupational Safety and Health (NEBOSH)

  tel: 0116 288 8858
- Institution of Occupational Safety and Health (IOSH)

  tel: 0116 257 3100
- Chartered Institution of Environmental Health (CIEH)

  tel: 0207 928 6006
- Royal Society for the Prevention of Accidents (RoSPA)

  tel: 0121 248 2000
- British Safety Council (BSC)

  tel: 0208 741 1231

There are many independent consultants who can provide advice and assistance on health and safety matters. Consultants are listed in the Yellow Pages and the Institution of Occupational Safety and Health (see above) maintains a consultants' register.

HSE itself can also be a valuable source of information and advice:

- HSE Books

  tel: 01787 881165

  fax: 01787 313995

  website: www.hse books.co.uk

  HSE has a huge range of priced publications and free leaflets, some of which are referred to elsewhere in this Introduction. '*The essentials of health and safety at work*' is a useful starting point for the small employer.
- HSE Infoline

  tel: 08701 545500

  Open Monday to Friday 8.30am to 5pm to provide information on workplace health and safety.
- HSE home page on the internet

  www.hse.gov.uk

  An online enquiry service can be accessed from the home page.

Of course, within this issue of *Tolley's Health and Safety at Work* you will find it easy to access much valuable information and practical advice, designed to keep you up-to-date with the law and all recent and relevant developments.

Using this book

29. In the following pages you will find a detailed compendium of information arranged in alphabetical order of subject, each prepared by an expert in the topic. It is possible to work through this material almost as a self-teaching aid in health and safety or to use each topic as a ready-reference when a question arises or a new company policy is being prepared in that subject area. The following is a general guide to the material.

Access Traffic Routes and Vehicles – this is one of HSE's current Priority Areas. If your workplace is visited by an inspector this is an aspect of your arrangements which is likely to be scrutinised. Many people are harmed from the poorly controlled and segregated movement of people and vehicles.

Accident Reporting and Investigation – more than a guide to the legal requirements, it is worth noting that HSC considered introducing a new legal duty to investigate. In the Management Cycle (Plan-Do-Check-Act) the reactive monitoring of performance is a critical source of information for improvement. Proper accident investigation is a critical function.

Asbestos – the new legal regime for asbestos in workplace buildings is coming into force. For the material responsible for more work-related deaths over the past 100 years than any other, it repays close attention.

Business Continuity – identifying threats to an organisation's ability to function, seeking to minimise and control the risks, mitigating the effects if the risks turn into reality, and recovering as rapidly as possible, are in a continuous chain of risk management effort. The proper management of occupational safety and health risks requires involvement with and an understanding of business continuity programmes.

Compensation for Work Injuries/Diseases – court cases lead to compensation of over £1.3bn each year (and rising), understanding what you need to do to comply with the Woolf requirements for handling claims is critical.

Construction and Building Operations – the UK's (and Europe's) most dangerous industry in terms of fatalities and people injured, the management of works requires close adherence to the rules to minimise the risks.

Control of Major Accident Hazards – the requirement to compile safety reports on major accident hazard sites and the annual revision of chemicals and quantities specified in the annex to the Seveso Directive mean that the regulatory regime needs to be carefully monitored and responded to by operators of such sites.

Dangerous Goods Systems – the definition, categorisation and safe handling of dangerous goods is an example of an international regime which has an impact at a local level.

Disaster and Emergency Management Systems (DEMS) – since the attacks including those on the New York World Trade Centre occurred, the world is perceived by many to be less stable and more dangerous, and this adds a further incentive to those seeking to develop effective disaster and emergency systems.

Display Screen Equipment – although compliance may seem technically straightforward, the very large numbers of workers who use this equipment for much of their working days and the potential harm and discomfort which can arise ensures that this remains a significant health and safety issue within many organisations.

Electricity – the Risk Control System (RCS) concept is a structured approach to specifying the management of electrical risk. The concept is now routinely used to identify the agenda which will assist HSE inspectors in the assessment of management systems associated with electrical safety.

Employers' Liability Insurance – the cost, and sometimes difficulty of obtaining, employers' liability insurance, has been a major talking point over the past year including ministerial meetings on the subject. Understanding your rights and obligations was never more critical.

Employment Protection – a useful overview to sit alongside the health and safety law, focusing on the interface between the two legal regimes.

Enforcement – when 'an Inspector calls', what can you expect and what do you need to do.

Environmental Management – environmental breaches far outweigh, in numbers of prosecutions and in total fines levied, health and safety breaches. Understanding the legal framework is essential.

Ergonomics – estimates vary, but about a third of all work-related sickness absence is due to musculo-skeletal problems (one of HSE's eight priority programmes), and this is just one aspect that ergonomics addresses.

Fire Prevention and Control – the proposed re-organisation of the Fire Service into the Fire and Emergency Service with emphasis on prevention is a reminder of the focus of workplace requirements.

First-Aid – one of the common factors in every workplace is the requirement to address this issue.

Food Safety and Standards – rarely out of the news, this remains one of the 'hot topics'.

Gas Safety – as prosecutions continue of landlords for failing to provide safe gas appliances, this remains an important topic for facilities managers.

Harassment in the Workplace – an element of the psychosocial issues which have come to the fore in recent years, and an example of the interface between health and safety and employment law.

Hazardous Substances in the Workplace – the new emphasis on occupational health has reinforced the necessity for dealing with hazardous substances effectively. Guidance on asthmagens, outbreaks of Legionnaires' Disease, are just some of the factors which need to be factored into an effective legal compliance programme.

Joint Consultation in Safety – Safety Representatives, Safety Committees, Collective Agreements and Work Councils – there are new plans to extend the legal rights for worker consultation, and in addition to an HSE initiative there are indications that the rules on Works Councils may be amended across the EU. The need to understand and implement current guidelines is clear.

Lifting Machinery and Equipment – the regulatory regime is clear, and the risks associated with lifting and handling operations necessitates careful compliance.

Lighting – one of the most significant causes of discomfort in indoor workplaces, and essential for safe working, the basic briefing will assist in providing advice on effective management.

Machinery Safety – one of the sets of Regulations introduced in 1992, the provision and use of work equipment has been updated since and remains a pillar of current effective safety management.

Managing Health and Safety – an understanding of the current thinking on management systems, what they offer, how they can be implemented and maintained, is vital to a modern organisation seeking to achieve both legal compliance and best practice.

Manual Handling – estimates vary, but about a third of all work-related sickness absence is due to musculo-skeletal problems (one of HSE's eight priority programmes).

Noise – with new regulations planned, stimulated by the EU Physical Agents Directive, and many thousands of workers adversely affected by exposures, understanding the current regulatory regime is essential. Noise-induced hearing loss remains the most common source of compensation claims for work-related harm.

Occupational Health and Diseases – this is one of the major drives from the Government and from HSE, encouraging employers to tackle the high cost (in financial and human terms) of work-related ill health and disability of the workforce.

Occupiers' Liability – a useful briefing on the duties of a building occupier.

Personal Protective Equipment – this may be at the bottom of the hierarchy of control, but it still represents a major element of effective protection against risk.

Product Safety – the interface between product safety and health and safety is a critical area to understand.

Risk Assessment – this remains the 'motor' of modern health and safety, but it needs to be done well. Significant risks ignored can lead to accidents and deaths, and therefore a balance needs to be struck.

Safe Systems of Work – driven by risk assessment, and as part of an overall approach to management, safe systems are the heart of modern health and safety.

Statements of Health and Safety Policy – an opportunity to set the tone for an organisation, to create a 'Mission Statement' declaring a real commitment to improving performance.

Stress and Violence in the Workplace – with the new Management Standards to minimise stress having been published by HSE, it is clear that reducing the adverse impact of excess stress and also incidents of violence and aggression within the workplace is both challenging and necessary.

Training and Competence in Occupational Safety and Health – this is a critical briefing on a topic which needs to be addressed by senior management. UK industry and commerce invests much in training, but unless it is effective the delegation of responsibility to junior staff will leave directors and senior managers vulnerable.

Ventilation – one of the key methods for the control of exposure to airborne contamination.

Vibration – the EU Physical Agents Directive has resulted in a new regulatory regime for hand-arm and whole-body vibration. Hand-arm vibration syndrome is the fastest growing foundation for work-related compensation claims.

Vulnerable Persons – people with disabilities, new or expectant mothers, young people, lone workers – the human rights and equal opportunities agenda is fully applicable to the development of risk management strategies and organisations need to address the vulnerabilities of individuals and groups.

Work at Heights – falls from heights, one of HSE's eight priority programmes, continue to represent a significant cause of workplace fatalities, and new Regulations are imminent from an EU Directive.

Working Time – this continues to be a troublesome area, and a clear strategy based on understanding the legal requirements is essential.

Workplaces – Health, Safety and Welfare – the management of the working environment represents one of the keys to effective improvements in health and safety performance.

Access, Traffic Routes and Vehicles

Introduction

A1001 Many hazardous situations arise from the regular daily flow of labour to and from the workplace, vehicles making deliveries and collecting items. Ensuring safe access and egress is therefore critical in reducing risks to employees and visitors to the premises. For this reason there is a duty on employers and factory occupiers to 'provide and maintain' safe access to and egress from a place of work both under statute and at common law. This is also part and parcel of compliance with the *Building Regulations 1991* (*SI 1991 No 2768*).

Access arrangements must include access facilities for disabled workers and visitors (i e people with physical and/or sensory impairments) (see further W11043 WORKPLACES – HEALTH, SAFETY AND WELFARE). Consideration must also be given to ensuring there are proper precautions to effect entry or exit to or from confined spaces, and in areas where a build-up of gas/combustible substances can be a real though not obvious danger (see A1022 below). Statutory requirements consist of general duties under the *Health and Safety at Work etc Act 1974* ('*HSWA 1974*'), which apply to all employers, and the more specific duties under the *Workplace* (*Health, Safety and Welfare*) *Regulations 1992* (*SI 1992 No 3004* as amended by *SI 2002 No 2174*).

The term 'access' is a comprehensive one and refers to just about anything that can reasonably be regarded as means of entrance/exit to a workplace, even if it is not the usual method of access/egress. Unreasonable means of access/egress would not be included, such as a dangerous short-cut, particularly if management has drawn a worker's attention to the danger. The fact that a worker is a trespasser has not prevented recovery of damages (*Westwood v The Post Office [1973] 3 All ER 184*).

Vehicle and pedestrian movements is one of the key areas on which the Health and Safety Executive is focusing its attention, as part of its strategy for *Revitalising Health and Safety*. As such, this area of health and safety is likely to be closely reviewed whenever an enforcement officer visits a premise on which vehicles and pedestrians are based.

Access to a fork lift truck qualified, for the purposes of the *Factories Act 1961, s 29*(*1*) (repealed as from 1 January 1996), the fork lift truck being a 'place' (*Gunnion v Roche Products Ltd, The Times, 4 November 1994*). The proper procedure is for the employer/factory occupier to designate points of access/egress for workers and see that they are safe, well-lit, maintained and (if necessary) manned and de-iced. Moreover, the statutory duties apply to access/egress points to any place where any employees have to work, and not merely their normal workplace.

This chapter summarises key statutory and common law duties in connection with:

- access and egress;
- vehicular traffic routes for internal traffic and deliveries;

- work vehicles and delivery vehicles – with particular emphasis on potentially hazardous activities involving such vehicles; and
- work in confined spaces, of necessity involving access and egress points.

Statutory duties concerning workplace access and egress – Workplace (Health, Safety and Welfare) Regulations 1992 (SI 1992 No 3004 as amended by SI 2002 No 2174)

A1002 Statutory duties centre around:

- the organisation of safe workplace transport systems;
- the suitability of traffic routes for vehicles and pedestrians; and
- the need to keep vehicles and pedestrians separate.

Organisation of safe workplace transport systems

A1003 Every workplace must (so far as is reasonably practicable) be so organised that pedestrians and vehicles can circulate in a safe manner. [*SI 1992 No 3004* (as amended by *SI 2002 No 2174*), *Reg 17*(*1*)].

Suitability of traffic routes

A1004 Traffic routes in a workplace must be suitable for the persons or vehicles using them, sufficient in number, in suitable positions and of sufficient size. [*SI 1992 No 3004* (as amended by *SI 2002 No 2174*), *Reg 17*(*2*)].

More particularly:

- pedestrians or vehicles must be able to use traffic routes without endangering those at work;
- there must be sufficient separation of traffic routes from doors, gates and pedestrian traffic routes, in the case of vehicles;
- where vehicles and pedestrians use the same traffic routes, there must be sufficient space between them; and
- where necessary, all traffic routes must be suitably indicated.

[*SI 1992 No 3004* (as amended by *SI 2002 No 2174*), *Reg 17*(*3*), (*4*)].

Compliance with these statutory duties involves provision of safe access for:

- vehicles, with attention being paid to design and layout of road systems, loading bays and parking spaces for employees and visitors; and
- pedestrians, so as to avoid their coming into contact with vehicles.

Traffic routes for vehicles

A1005 There should be sufficient traffic routes to allow vehicles to circulate safely and without difficulty. As for internal traffic, lines marked on roads/access routes in and between buildings should clearly indicate where vehicles are to pass e.g. fork lift trucks. Obstructions, such as limited headroom, are acceptable if clearly indicated. Temporary obstacles should be brought to the attention of drivers by warning signs or hazard cones or, alternatively, access prevented or restricted. Both internal and delivery traffic should be subject to sensible speed limits (e.g. 10 mph), which should be clearly displayed. Speed ramps (sleeping policemen), preceded by a warning sign or mark, are necessary, save for fork lift trucks, on workplace approaches. The traffic route should be wide enough to allow vehicles to pass and repass oncoming or parked traffic, and it may be advisable to introduce one way systems or parking restrictions. Traffic signs on roads, for example speed limit signs, must conform with those on public roads, whether or not the road is subject to the *Road Traffic Regulation Act 1984*.

Checklist – safe traffic routes

A1006 Safe traffic routes should:

- provide the safest route possible between places where vehicles have to call or deliver;
- be wide enough for the safe movement of the largest vehicle, including visiting vehicles (e.g. articulated lorries, ambulances etc.) and should allow vehicles to pass oncoming or parked vehicles safely. One way systems or parking restrictions are desirable;
- avoid vulnerable areas/items, such as fuel or chemical tanks or pipes, open or unprotected edges, and structures likely to collapse;
- incorporate safe areas for loading/unloading;
- avoid sharp or blind bends; if this is not possible, hazards should be indicated (e.g. blind corner);
- ensure that road/rail crossings are kept to a minimum and are clearly signed;
- ensure that entrances/gateways are wide enough; if necessary, to accommodate a second vehicle that may have stopped, without causing obstruction;
- set sensible speed limits, which are clearly signposted. Where necessary, ramps should be used to retard speed, and road humps or bollards to restrict the width of the road. These should be preceded by a warning sign or mark on the road;
- ensure that fork lift trucks should not have to pass over road humps, unless of a type capable of doing so;
- give prominent warning of limited headroom, both in advance and at an obstruction. Overhead electric cables or pipes containing flammable/hazardous chemicals should be shielded, i.e. using goal posts, height gauge posts or barriers;
- ensure that routes on open manoeuvring areas/yards are marked and signposted, and banksmen are employed to supervise the safe movement of vehicles;
- ensure that people at risk from exhaust fumes or material falling from vehicles are screened or protected; and
- restrict vehicle access where high-risk substances are stored (e.g. LPG) and where refuelling takes place.
- consider installation of refuge points (safe havens) where vehicles need to reverse into delivery areas or dead ends, or position barriers to prevent vehicles reversing from colliding into people.

Traffic routes for pedestrians

Checklist – safe traffic routes

A1007 In the case of pedestrians, the main object of the traffic route is to prevent their coming into contact with vehicles. Safe traffic routes should:

- provide separate routes/pavements for pedestrians, to keep them away from vehicles;
- where necessary, provide suitable barriers/guard rails at entrances/exits and at the corners of buildings;
- where traffic routes are used by both pedestrians and vehicles, be wide enough to allow vehicles to pass pedestrians safely;
- where pedestrian and vehicle routes cross, provide appropriate crossing points. These should be clearly marked and signposted. If necessary, barriers or rails should be provided to prevent pedestrians crossing at dangerous points and to direct them to designated crossing points;
- where traffic volume is high, traffic lights, bridges or subways should be used to control movement and ensure a smooth, safe flow;
- where crowds use or are likely to use roadways, e.g. at the end of a shift, stop vehicles from using them at such times;
- provide separate vehicle and pedestrian doors in premises, with vision panels on all doors;
- provide high visibility clothing for people permitted in delivery areas (e.g. bright jackets/overalls);
- where the public has access (e.g. at a farm or factory shop), public access points should be as near as possible to shops and separate from work activities;
- take into consideration needs for access of people with physical or sensory impairments.

(See also W11015 WORKPLACES.)

Vehicles

A1008 Vehicles account for a high percentage of deaths and injuries at work. In 1994, 77 people were killed, including six members of the public; there were 1,363 major injuries and 4,698 workers had to take three or more days off work as a result of vehicle injury. Many of these casualties occur whilst vehicles are reversing, though activities such as loading and unloading, sheeting and unsheeting, as well as cleaning, can similarly lead to injuries, especially where employees are struck by a falling load or a fall from a height on, say, a tanker or HGV. So, too, climbing and descending ladders on tankers during delivery and 'dipping' at petrol forecourts can be hazardous, access onto vehicles and egress being as important as design and construction. Tipping, too, has its dangers, with tipping vehicles, tipping trailers and tankers overturning in considerable numbers. Also, sheeting and unsheeting operations have led to sheeters slipping or losing their grip or falling whilst walking on top of loads or in consequence of ropes breaking; absence of, or inadequate, training being an additional factor in injuries involving work vehicles. This part of the chapter considers the general statutory requirements relating to vehicles at work, precautions in connection with potentially hazardous operations involving vehicles, as well as providing a checklist for vehicle safety. Specific construction and use requirements are not considered.

General statutory requirements

A1009 Both work and private vehicles come within the parameters of health and safety at work. Regarding work vehicles, employers have the direct responsibilities of provision and maintenance generally under *HSWA, s 2*, and, more specifically, under the *Provision and Use of Work Equipment Regulations 1998* (*SI 1998 No 2306* as amended by *SI 2002 No 2174*) (*PUWER*), vehicles qualifying as 'work equipment'. Regarding private vehicles, employers have much less control – at least, as far as design, construction and use are concerned – but should, nevertheless, endeavour to ensure regulated use via:

- restricted routes and access;
- provision of clearly signposted parking areas away from hazardous activities and operations; and
- enforcement of speed limits.

Work vehicles

A1010 Generally, work vehicles should be as safe, stable, efficient and roadworthy as private vehicles on public roads. As work equipment, they are subject to the controls of *PUWER*, which specifies provision, maintenance, access and safety provisions in the event of rolling or falling over, whilst employers must also ensure that drivers are suitably trained in conformity with the requirements of the *Management of Health and Safety at Work Regulations 1999* (*SI 1999 No 3242*).

Provision

A1011 All employers must ensure that vehicles:

- are constructed and adapted as to be suitable for its purpose;
- when selected, caters for risks to the health and safety of persons where the vehicles are to be used; and
- are only used for operations specified and under suitable conditions.

[*PUWER* (*SI 1998 No 2306* as amended by *SI 2002 No 2174*), *Reg 5*].

Compliance with these requirements on the part of operators of HGVs, fork lift trucks, dump trucks and mobile cranes presupposes conformity with the following checklist, namely:

(i) a high level of stability;

(ii) safe means of access and egress to and from the cab;

(iii) suitable and effective service and parking brakes;

(iv) windscreens with wipers and external mirrors giving optimum all-round visibility;

(v) a horn, vehicle lights, reflectors, reversing lights, reversing alarms;

(vi) suitable painting/markings so as to be conspicuous;

(vii) provision of a seat and seat belts;

(viii) guards on dangerous parts (e.g. power take-offs);

(ix) driver protection to prevent injury from overturning, and from falling objects or materials; and

(x) driver protection from adverse weather.

Maintenance

A1012 All employers must ensure that work equipment (including vehicles) is maintained in an efficient state, in efficient working order and in good repair. [*PUWER* (*SI 1998 No 2306* as amended by *SI 2002 No 2174*), *Reg 6*]. This combines the need for basic daily safety checks by the driver before using the vehicle, as well as preventive inspections and services carried out at regular intervals of time and/or mileage, in accordance with manufacturers recommendations. As regards basic daily safety checks, employers should provide drivers with a log book in which to record visual inspections undertaken and the findings of the following: brakes, tyres, steering, mirrors, windscreen washers and wipers, warning signals and specific safety systems (e.g. control interlocks).

Employers should see that drivers carry out the checks.

Training of drivers

A1013 All employers must:

- in entrusting tasks to employees, take into account their capabilities as regards health and safety; and
- ensure that employees are provided with adequate health and safety training on recruitment and exposure to new or increased risks.

[*Management of Health and Safety at Work Regulations 1999* (*SI 1999 No 3242*)].

In order to conform with these requirements, employers should ensure that, for general purposes, drivers of work vehicles are over 17 and have passed their driving test or, in the case of drivers of HGVs, that they are over 21 and have passed the HGV test. Moreover, to ensure continued competence, or to accommodate new risks at work or a changing work environment, employers should provide safety updates on an on-going basis as well as refresher training, and require approved drivers to report any conviction for a driving offence, whether or not involving a company vehicle. One method for ensuring this is to require all drivers to submit their licence annually so that a copy can be held in their personnel file.

Contractors and subcontractors

A1014 Similar assurances (see A1013 above) should be obtained from drivers of contractors and subcontractors visiting an employer's workplace. If they are not forthcoming, permission to work on site or in-house should be refused until either the contractor's vehicles comply with statutory requirement and/or his drivers are adequately trained. Training of contractor's drivers would normally be undertaken by contractors themselves, though site or in-house hazards, routes to be used etc. should be communicated to contractors by employers or occupiers. Contractors should be left in no doubt of the penalties involved for failure to conform with safe working practices – a useful way of ensuring enforcement on the part of contractors and subcontractors is to issue a licence.

Access to vehicles

A1015 In addition to *Regulation 5* of *PUWER* (see A1011 above), employers (and others having control, to any extent, of workplaces (see OCCUPIERS' LIABILITY)), who operate/use vehicles, are subject to:

- the fall prevention requirements of *Regulation 13* of the *Workplace* (*Health, Safety and Welfare*) *Regulations 1992* (*SI 1992 No 3004* as amended by *SI 2002 No 2174*) (see W9007 WORK AT HEIGHTS). As far as possible, compliance with this regulation would obviate the need for climbing on top of vehicles (by bottom-filling) and also require vehicle operators to ensure that loads are evenly distributed, packaged properly and secured in the interests of drivers going down slopes and up steep hills (see further A1018 below).
- the co-operation requirements of *Regulation 9* of the *Management of Health and Safety at Work Regulations 1999* (*SI 1999 No 3242*), specifying that where activities of different employers interact, different employers may need to co-operate with each other and co-ordinate preventive and protective measures. This regulation is particularly relevant for example, to tanker deliveries and 'dipping' at petrol forecourts as well as loading and/or unloading operations. For 'dipping' purposes or gaining top access to tankers, access should be by a ladder at the front or rear, such ladders being properly constructed, maintained and securely fixed; ideally, they should incline inwards towards the top. There should be a means of preventing people from falling whilst on top of the tanker. Failing this, employers of tanker drivers and forecourt owners should liaise on potential risks involved in tanker deliveries, e.g. the provision of suitable step-ladders on the part of the latter. As for carriage of goods and loading/unloading operations, consignors should ensure that goods are evenly distributed, properly packaged and secured.

Potentially hazardous operations

A1016 The following activities and operations are potentially hazardous in connection with vehicles.

Reversing

A1017 Approximately a quarter of all deaths at work are caused by reversing vehicles; in addition, negligent reversing can result in costly damage to premises, plant and goods. Where possible, workplace design should aspire to obviate the need for reversing by the incorporation of one-way traffic systems. Failing this, reversing areas should be clearly identified and marked, and non-essential personnel excluded from the area. Ideally, banksmen wearing high-visibility clothing should be in attendance to guide drivers through, and keep non-essential personnel and pedestrians away from, the reversing area. Refuge points, also known as safe havens, should be constructed where possible to enable an escape route or safe place for people to stay in the event of a vehicle reversing in an unsafe manner. Vehicles should be fitted with external side-mounted and rear-view mirrors – as, indeed, many now are. Closed-circuit television systems are also advisable for enabling drivers to see round 'blind spots' and corners.

Loading and unloading

A1018 Because employees can be seriously injured by falling loads or overturning vehicles, loading/unloading should not be carried out:

- near passing traffic, pedestrians and other employees;
- where there is a possibility of contact with overhead electric cables;
- on steep gradients;
- unless the load is spread evenly (racking will assist load stability);
- unless the vehicle has its brakes applied or is stabilised (similarly with trailers); or
- with the driver in the cab.

Tipping

A1019 Overturning of lorries and trailers is the main hazard associated with tipping. In order to minimise the potential for injuries, tipping operations should only occur:

- after drivers have consulted with site operators and checked that loads are evenly distributed;
- when non-essential personnel are not present;
- on level and stable ground away from power lines and pipework; and
- with the driver in the cab and the cab door closed.

Moreover, after discharge, drivers should ensure that the body of the vehicle is completely empty and should not drive the vehicle in an endeavour to free a stuck load.

Giving unauthorised lifts

A1020 Giving unauthorised lifts in work vehicles is both a criminal offence and can lead to employers being involved in civil liability. Thus, 'every employer shall ensure that work equipment is used only for operations for which, and under conditions for which, it is suitable'. [*PUWER* (*SI 1998 No 2306* as amended by *SI 2002 No 2174*), *Reg 5*].

Where a driver of a work vehicle gives employees and/or others unauthorised lifts, his employer could find himself prosecuted for breach of the above regulation, whilst the driver himself may be similarly prosecuted for breach of *HSWA, s 7*, as endangering co-employees and members of the public. In addition, although acting in an unauthorised manner and contrary to instructions, the employee may well involve his employer in vicarious liability for any subsequent injury to a co-employee and/or member of the public (see further *Rose v Plenty [1976] 1 All ER 97*).

Where instructions to employees not to give unauthorised lifts are clearly displayed in a work vehicle, but an employee nevertheless gives an unauthorised lift and a co-employee or member of the public is injured or killed as a result of the employee's negligent driving, it can be argued that the employee has exceeded the scope of his employment and so the employer is absolved from liability. (In *Twine v Bean's Express [1946] 1 All ER 202* a driver gave a lift to a third party who was killed in consequence of his negligent driving. There was a notice in the van prohibiting drivers from giving lifts. It was held that the employer was not liable, as the driver was acting outside the parameters of his employment when giving a lift. The injured passenger knew that the driver should not give lifts.)

Conversely, courts have taken the view that such conduct, on the part of drivers, does not circumscribe the scope of employment but rather constitutes performance of work in an unauthorised manner, so leaving the employer liable as the employee is doing what he is employed to do but doing it wrongly (see further *Rose v Plenty and Century Insurance Co v Northern Ireland Road Transport Board*). Yet again, if the passenger, having seen the prohibition on unauthorised lifts in the vehicle, nevertheless accepted a lift and was injured, arguably, if an adult rather than a minor, he has agreed to run the risk of negligent injury and so will forfeit the right to compensation.

Common law duty of care

A1021 At common law every employer owes all his employees a duty to provide and maintain safe means of access to and egress from places of work. Moreover, this duty extends to the workforce of another employer/contractor who happens to be working temporarily on the premises. Hence the common law duty covers all workplaces, out of doors as well as indoors, above ground or below, and extends to factories, mines, schools, universities, aircraft, ships, buses and even fire engines and appliances (*Cox v Angus [1981] ICR 683*, where a fireman injured in a cab was entitled to damages at common law).

Confined spaces

A1022 Accidents and fatalities such as drowning, poisoning by fumes or gassing, have happened as a result of working in confined spaces. See, for instance, the case of *Baker v T E Hopkins & Son Ltd [1959] 3 All ER 225* where a doctor was overcome by carbon monoxide fumes while going to rescue two workmen down a well – the defence of *volenti non fit injuria* failed). Normal safe practice is a formalised permit to work system or checklist tailored to a particular task and requiring appropriate and sufficient personal protective equipment. Hazards typical of this sort of operation are:

(*a*) atmospheric hazards – oxygen deficiency, enrichment (see *R v Swan Hunter Shipbuilders Ltd* at C8145 CONSTRUCTION AND BUILDING OPERATIONS), toxic gases (e.g. carbon monoxide), explosive atmospheres (e.g. methane in sewers);

(*b*) physical hazards – low entry headroom or low working headroom, protruding pipes, wet surfaces underfoot as well as any electrical or mechanical hazards;

(*c*) chemical hazards – concentration of toxic gas can quickly build up, where there is a combination of chemical cleaning substances and restricted air flow or movement.

In order to combat this variety of hazards peculiar to work in confined spaces, use of both gas detection equipment and suitable personal protective equipment are a prerequisite, since entry/exit paths are necessarily restricted.

Prior to entry, gas checks should test for (*a*) oxygen deficiency/enrichment, then (*b*) combustible gas and (*c*) toxic gas, by detection equipment being lowered into the space. This will determine the nature of personal protective equipment necessary. If gas is present in any quantity, the offending space should then be either naturally or mechanically ventilated. Where gas is present, entry should only take place in emergencies, subject to the correct respiratory protective equipment being worn. Assuming gas checks establish that there is no gaseous atmosphere, entry can then be made without use of respiratory equipment. Gas detection equipment should continue to be used whilst people are in the confined space so that any atmospheric change can subsequently be registered on the gas detection equipment. It is essential that an emergency plan is devised when gaining access to confined spaces, which includes effective two way communication between people inside and immediately outside the confined space, contact with emergency services and first aid, and fire prevention personnel on hand.

Statutory requirements

A1023 It should be noted that the *Confined Spaces Regulations 1997* (*SI 1997 No 1713*) came into force on 28 January 1998. These repeal *Factories Act 1961, s 30*, and impose requirements and prohibitions with respect to the health and safety of persons carrying out work in confined spaces.

A 'confined space' is defined in *Reg 1*(*2*) as 'any place, including any chamber, tank, vat, silo, pit, trench, pipe, sewer, flue, well or other similar space in which, by virtue of its enclosed nature, there arises a reasonably foreseeable specified risk'.

A 'specified risk' means a risk of:

(*a*) serious injury to any person at work arising from a fire or explosion;

(*b*) without prejudice to paragraph (*a*) –

  (i) the loss of consciousness of any person at work arising from an increase in body temperature;

  (ii) the loss of consciousness or asphyxiation of any person at work arising from gas, fume, vapour or the lack of oxygen;

(*c*) the drowning of any person at work arising from an increase in the level of a liquid; or

(*d*) the asphyxiation of any person at work arising from a free flowing solid or the inability to reach a respirable environment due to entrapment by a free flowing solid.

Regulation 4 prohibits a person from entering a confined space to carry out work for any purpose where it is reasonably practicable to carry out the work by other means.

If, however, a person is required to work in a confined space, a risk assessment must be undertaken to comply with the requirements of the *Management of Health and Safety at Work Regulations 1999* (*SI 1999 No 3242*), *Reg 3*. The risk assessment must be undertaken by a competent person and the outcome of the risk assessment process will then provide the basis for the development of a safe system of work. [*Confined Spaces Regulations 1997* (*SI 1997 No 1713*), *Reg 3*].

The risk assessment process should make use of all available information such as engineering drawings, working plans, soil or geological information and take into consideration factors such as the general condition of the confined space, work to be undertaken in the space to minimise hazards produced in the area, need for isolation of the space and the requirements for emergency rescue. In particular, information should be collected and assessed on the previous contents of the confined space, residues that still may be present, contamination that may arise from adjacent plant, processes, gas mains, surrounding soil, land or strata; oxygen level and physical dimensions of the space that may limit safe access and/or egress. The work to be undertaken should be assessed to determine if additional risks will be produced as a result of this work and systems developed to control these risks. All information collected should be recorded and a safe system of work developed for safe entry.

The main elements to consider when designing a safe system of work include the following:

- supervision,

- competence levels for personnel working in confined spaces,
- communications,
- testing/monitoring the atmosphere,
- gas purging,
- ventilation,
- removal of residues,
- isolation from gases, liquids and other flowing materials,
- isolation from mechanical and electrical equipment,
- selection and use of suitable equipment,
- personal protective equipment (PPE) and respiratory protective equipment (RPE),
- portable gas cylinders and internal combustion engines,
- gas supplied by pipes and hoses,
- access and egress,
- fire prevention,
- lighting,
- static electricity,
- smoking,
- emergencies and rescue,
- limited working time.

[*Confined Spaces Regulations 1997* (*SI 1997 No 1713*), *Reg 4*].

Regulation 6 provides for circumstances allowing the Health and Safety Executive to grant exemption certificates.

Accident Reporting and Investigation

Introduction

A3001 Employers (both onshore and offshore [*Reporting of Injuries, Diseases and Dangerous Occurrences Regulations 1995* (*RIDDOR*) (*SI 1995 No 3163*), *Reg 12*]) and other 'responsible persons' (see A3003 below) who have control over employees and work premises are required to notify and report to the relevant enforcing authority (see A3004 below) the following specified events occurring at work:

(*a*) accidents causing injuries, fatal and non-fatal, including:

  (i) acts of non-consensual physical violence committed at work, and

  (ii) acts of suicide occurring on, or in the course of, the operation of a railway, tramway, trolley or guided transport system

[*RIDDOR* (*SI 1995 No 3163*), *Reg 2*(*1*)];

(*b*) occupational diseases; and

(*c*) dangerous occurrences, even where no injury results.

The duty to report applies not only in the case of incidents involving employees, but also to visitors, customers and members of the public killed or injured by work activities [*RIDDOR* (*SI 1995 No 3163*), *Reg 3*(*1*)].

Employees also have certain obligations to report accidents to their employers.

Reporting of Injuries, Diseases and Dangerous Occurrences Regulations 1995 (RIDDOR) – (SI 1995 No 3163)

A3002 The *Reporting of Injuries, Diseases and Dangerous Occurrences Regulations 1995* cover:

(*a*) reportable work injuries (see A3007 below);

(*b*) reportable occupational diseases (see Appendix B at A3036 below) – now 47 in all;

(*c*) reportable dangerous occurrences (see A3009 and Appendix A at A3035 below) – now 83 in all;

(*d*) road accidents involving work (see A3019 below); and

(*e*) gas incidents (see A3020 below).

Records must be kept by employers and other 'responsible persons' (see A3005) of such injuries, diseases and dangerous occurrences for a minimum of three years from the date they were made. In addition, employers must also keep an Accident Book (Form BI510).

Two reporting routes now exist:

- Injuries and dangerous occurrences are reportable to the relevant enforcing authority on Form F2508 and diseases on Form F2508A (see A3012 and Appendices C and D at A3037, A3038 below); or
- Reports can be made by telephone or electronically to the national Incident Contact Centre (ICC) (see A3004).

As for notification of major accident hazards (see C10009 CONTROL OF MAJOR ACCIDENT HAZARDS), notification and reporting under RIDDOR is sufficient for the purposes of the *Control of Major Accident Hazards Regulations 1999* (*SI 1999 No 743*), *Reg 15*(*4*).

Relevant enforcing authority

A3003 Health and safety law is separately enforced either by the Health and Safety Executive (HSE), or by local authorities, through their environmental health departments, depending on the nature of the business activity in question. Under *RIDDOR*, notifications and reports should be directed to the authority responsible for the premises where the reportable event occurs (or in connection with which the work causing the event is being carried out). Details of HSE offices are listed in Appendix E to this chapter.

Incident Reporting Centre

A3004 For all incidents occurring after 1 April 2001, employers and others who are subject to the reporting requirements, can comply with their obligations by contacting the Incident Contact Centre (ICC). This will eliminate the need to identify the local agency or other local enforcing authority. The ICC will then pass the report on to them. Reporting to the local HSE office or local enforcement authority by phone and on the current statutory forms is still an option and this information will be forwarded to the ICC.

Incidents may be reported to the ICC through the following channels:

- By telephone: Telephone 0845 300 9923 (Monday to Friday, 8.30 a.m. to 5.00 p.m.)
- By facsimile: Fax 0845 300 9924
- By email: riddor@gov.uk
- By internet: www.riddor.gov.uk
- By post: Incident Contact Centre, Caerphilly Business Park, Caerphilly CF83 3GG

In taking advantage of these arrangements, the employer or other reporting person will not have a record of a statutory RIDDOR form, so the ICC will send out confirmation copies of reports which should be checked for accuracy and retained.

Persons responsible for notification and reporting

A3005 The person generally responsible for reporting injury-causing accidents, deaths or diseases is the employer. Failing that, the person having control of the work premises or activity will be the responsible person. [*RIDDOR* (*SI 1995 No 3163*), *Reg 2*(*1*)]. In certain cases these normal rules are displaced and there are specifically designated 'responsible persons', e.g. in the case of mines, quarries, offshore installations, vehicles, diving operations and pipelines (see Table 1 below).

Table 1

Persons generally responsible for reporting accidents

Death, major injury, over-3-day injury or specified occupational disease:	of an employee at work	that person's employer
	of a person receiving training for employment	the person whose undertaking makes immediate provision of the training
	of a self-employed person at work in premises under the control of someone else	the person for the time being having control of the premises in connection with the carrying on by him of any trade, business or undertaking
Specified major injury or condition, or over-3-day injury:	of a self-employed person at work in premises under his control	
Death, or specified major injury or condition:	of a person who is not himself at work (but is affected by the work of someone else), e.g. a member of the public, a shop customer, a resident of a nursing home	the person for the time being having control of the premises in connection with the carrying on by him of any trade, business or undertaking at which, or in connection with the work at which, the accident causing the injury happened

[*Reg 2*(*1*)(*b*), (*c*)].

Persons responsible for reporting accidents in specific locations

A mine	the mine manager
A quarry	the quarry owner
A closed tip	the owner of the mine or quarry with which the tip is associated
An offshore installation (except in the case of reportable diseases)	the duty holder
A dangerous occurrence at a pipeline	the owner of the pipeline

Persons responsible for reporting accidents in specific locations	
A dangerous occurrence at a well	the appointed person or, failing that, the concession owner
A diving operation (except in the case of reportable diseases)	the diving contractor
A vehicle	the vehicle operator
[*Reg 2*(*1*)(*a*)].	

In situations where the responsible person is difficult to identify because of the shared control of a site or operations, arrangements should be made to determine who will deal with *RIDDOR* reporting in line with the duty to co-operate and co-ordinate under the *Management of Health and Safety at Work Regulations 1999* (*SI 1999 No 3242*), *Reg 11*.

What is covered?

A3006 Covered by these Regulations are events involving:

(*a*) employees;

(*b*) self-employed persons;

(*c*) trainees ;

(*d*) any person, not an employee or trainee, on premises under the control of another, or who was otherwise involved in an accident (e.g. a visitor, customer, passenger or bystander) (see A3004 above).

Major injuries or conditions

A3007 Where any person dies or suffers a major injury as a result of, or in connection with, work, such an incident must be notified immediately and details formally reported. The person who dies or suffers injury need not be at work; it is enough if the death or injury arose from a work activity. For example, reporting requirements would apply to a shopper who fell and was injured on an escalator, so long as the injury was connected with the escalator (*Woking Borough Council v British Home Stores [1995] 93 LGR 396*); a member of the public overcome by fumes on a visit to a factory and who lost consciousness; a patient in a nursing home who fell over an electrical cable lying across the floor and was injured; or a pupil or student killed or injured in the course of his curricular work which was supervised by a lecturer or teacher.

Reportable major injuries and conditions are as follows:

(*a*) any fracture (other than to fingers, thumb or toes);

(*b*) any amputation;

(*c*) dislocation of shoulder, hip, knee or spine;

(*d*) loss of sight (whether temporary or permanent);

(*e*) a chemical or hot metal burn to the eye or any penetrating eye injury;

(*f*) any injury resulting from an electrical shock or electrical burn (including one caused by arcing or arcing products) leading to unconsciousness or requiring resuscitation or admittance to hospital for more than 24 hours;

(*g*) any other injury

(i) leading to hypothermia, heat-induced illness or unconsciousness,

(ii) requiring resuscitation, or

(iii) requiring admittance to hospital for more than 24 hours;

(*h*) loss of consciousness caused by asphyxia or by exposure to a harmful substance or biological agent;

(*j*) either

(i) acute illness requiring medical treatment, or

(ii) loss of consciousness

resulting from the absorption of any substance by inhalation, ingestion or through the skin;

(*k*) acute illness requiring medical treatment where there is reason to believe that this resulted from exposure to a biological agent or its toxins or infected material.

[*RIDDOR* (*SI 1995 No 3163*), *Reg 2*(*1*), *Sch 1*].

Injuries incapacitating for more than three consecutive days

A3008 Where a person at work is incapacitated for more than three consecutive days from their normal contractual work (excluding the day of the accident but including any days which would not have been working days) owing to an injury resulting from an accident at work (other than an injury reportable as a major injury listed in A3007 above), a report of the accident must be made direct to the ICC or sent in writing on Form F2508 to the enforcing authority as soon as is practicable and in any event within ten days of the accident. [*RIDDOR* (*SI 1995 No 3163*), *Reg 3*(*2*)].

Specified dangerous occurrences

A3009 83 types of reportable dangerous occurrences are specified in *RIDDOR* (*SI 1995 No 3163*), *Sch 2*, ranging from general dangerous occurrences (e.g. the collapse of a building or structure, the explosion of a pressure vessel, and accidental releases of significant quantities of dangerous substances) to more specific dangerous occurrences in mines, quarries, transport systems and offshore installations (see Appendix A at A3035 below).

Specified diseases

A3010 Where a worker suffers from an occupational disease related to a particular activity or process (as specified in *Schedule 3*), a report must be sent to the enforcing authority. [*RIDDOR* (*SI 1995 No 3163*), *Reg 5*(*1*)].

This duty arises only when an employer has received information in writing from a registered medical practitioner diagnosing one of the reportable diseases. (In the case of the self-employed, the duty is triggered regardless of whether or not the information is given to him in writing.) Many of these diseases are those for which disablement benefit is ordinarily prescribed (see O1053 OCCUPATIONAL HEALTH AND DISEASES). (In respect of offshore workers, such diseases tend to be communicable.)

Moreover, the *Industrial Diseases (Notification) Act 1981* and the *Registration of Births and Deaths Regulations 1987 (SI 1987 No 2088), Sch 2*, Form 14, require that particulars are to be included on the death certificate as to whether death might have been due to, or contributed to by, the deceased's employment. Such particulars are to be supplied by the doctor who attended the deceased during the last illness. (The full list of specified diseases is contained in Appendix B at A3036 below.)

Duty to notify/report

Duty to notify

A3011 'Responsible persons' must notify enforcing authorities (see A3003 and A3004 above) by the quickest means practicable (normally by telephone) of the following:

(*a*) death as a result of an accident arising out of, or in connection with, work;

(*b*) 'major injury' (see A3007 above) of a person at work as a result of an accident arising out of, or in connection with, work;

(*c*) injury suffered by a person not at work (e.g. a visitor, customer, client, passenger or bystander) as a result of an accident arising out of, or in connection with, work, where that person is taken from the accident site to a hospital for treatment;

(*d*) major injury suffered by a person not at work, as a result of an accident arising out of, or in connection with, work at a hospital;

(*e*) a dangerous occurrence (see A3009 above)

[*RIDDOR (SI 1995 No 3163), Reg 3(1)*];

(*f*) road injuries or deaths [*RIDDOR (SI 1995 No 3163), Reg 10(2)*] (see A3019 below); and

(*g*) gas incidents [*RIDDOR (SI 1995 No 3163), Reg 6(1)*] (see A3020 below).

Duty to report

A3012 Responsible persons must also formally report events causing death or major injury, dangerous occurrences and details associated with workers' occupational diseases. In cases of death, major injury and accidents leading to hospitalisation, the duty to report extends to visitors, bystanders and other non-employees. More particularly, within ten days of the incident, responsible persons must send a written report form to the relevant enforcing authority in relation to:

(*a*) all events which require notification (listed in (*a*) to (*g*) at A3011 above);

(*b*) the death of an employee if it occurs within a year following a reportable injury (whether or not reported under (*a*)) [*RIDDOR (SI 1995 No 3163), Reg 4*] (see A3018 below);

(*c*) incapacitation for work of a person at work for more than three consecutive days as a result of an injury caused by an accident at work [*RIDDOR (SI 1995 No 3163), Reg 3(2)*] (see A3008 above);

(*d*) specified occupational diseases relating to persons at work (see A3009 above and Appendix B Part I at A3036 below), and also specifically those suffered by workers on offshore installations (see Appendix B Part II at A3036 below) [*RIDDOR (SI 1995 No 3163), Reg 5(1)*], provided that, in both cases,

(i) the responsible person has received a written statement by a doctor diagnosing the specified disease, in the case of an employee, or

(ii) a self-employed person has been informed by a doctor that he is suffering from a specified disease.

[*RIDDOR* (*SI 1995 No 3163*), *Reg 5*(*2*)].

Exceptions to notification and reporting

A3013 There is no requirement to notify or report the injury or death of:

(*a*) a patient undergoing treatment in a hospital or a doctor's or dentist's surgery [*RIDDOR* (*SI 1995 No 3163*), *Reg 10*(*1*)]; or

(*b*) a member of the armed forces of the Crown [*RIDDOR* (*SI 1995 No 3163*), *Reg 10*(*3*)].

Duty to keep records

A3014 Records of injury-causing accidents, dangerous occurrences and specified diseases must be kept by responsible persons for at least three years. [*RIDDOR* (*SI 1995 No 3163*), *Reg 7*(*3*)].

Injuries and dangerous occurrences

A3015 In the case of injuries and dangerous occurrences, such records must contain:

(*a*) date and time of the accident or dangerous occurrence;

(*b*) if an accident is suffered by a person at work –

(i) full name,

(ii) occupation, and

(iii) nature of the injury;

(*c*) in the event of an accident suffered by a person not at work –

(i) full name,

(ii) status (e.g. passenger, customer, visitor or bystander), and

(iii) nature of injury;

(*d*) place where the accident or dangerous occurrence happened;

(*e*) a brief description of the circumstances;

(*f*) the date that the event was first reported to the enforcing authority;

(*g*) the method by which the event was reported.

[*RIDDOR* (*SI 1995 No 3163*), *Sch 4, Part I*].

Disease

A3016 In the case of specified diseases, such records must contain:

(*a*) date of diagnosis;

(*b*) name of the person affected;

(*c*) occupation of the person affected;

(*d*) name or nature of the disease;

(*e*) the date on which the disease was first reported to the enforcing authority; and

(*f*) the method by which disease was reported.

[*RIDDOR* (*SI 1995 No 3163*), *Sch 4, Part II*].

Action to be taken by employers and others when accidents occur at work

A3017

Figure 1

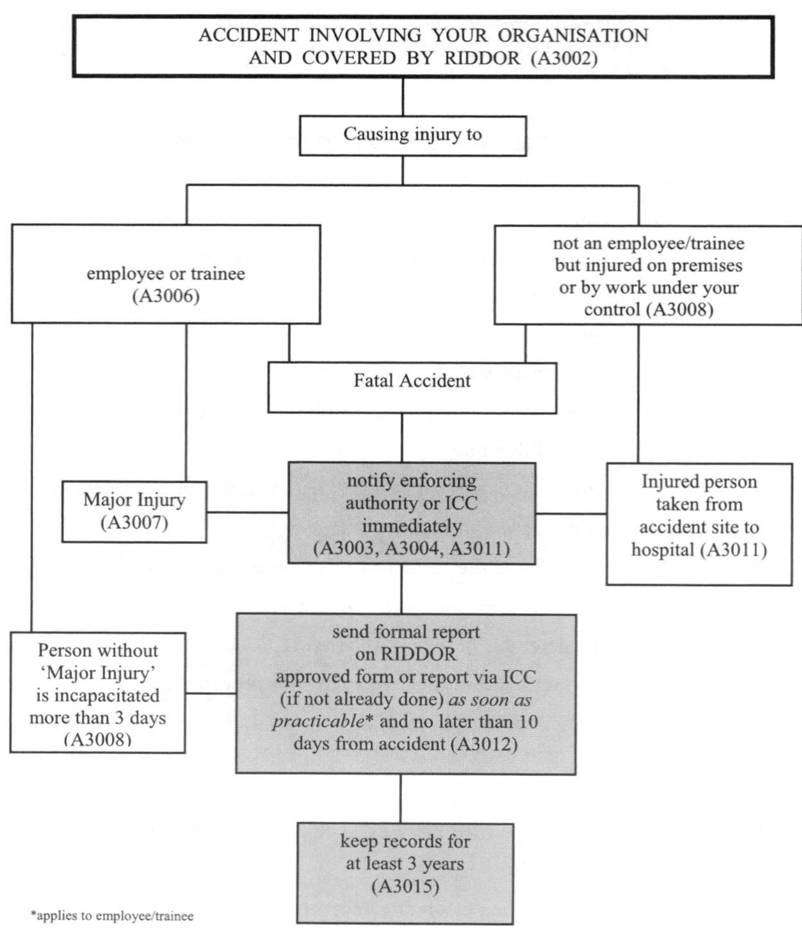

Reporting the death of an employee

A3018 Where an employee, as a result of an accident at work, has suffered a reportable injury/condition which is the cause of his death within one year of the date of the accident, the employer must inform the enforcing authority in writing of the death as soon as it comes to his attention, whether or not the accident has been reported under *Regulation 3*. [*RIDDOR* (*SI 1995 No 3163*), *Reg 4*].

Road accidents

A3019 Road accident deaths and injuries are notifiable and reportable if the death or major injury is caused by or connected with:

(*a*) exposure to any substance conveyed by road;

(*b*) loading or unloading vehicles;

(*c*) construction, demolition, alteration or maintenance activities on public roads; or

(*d*) an accident involving a train.

[*RIDDOR* (*SI 1995 No 3163*), *Reg 10*(*2*)].

The person injured, whether fatally or not, may or may not be engaged in the above-mentioned activities. Thus, an employee struck by a passing vehicle or a motorist injured by falling scaffolding, are covered. In addition, certain dangerous occurrences on public highways and private roads are covered (see Appendix A at A3035 below).

Duty to report gas incidents

Gas suppliers

A3020 Where:

(*a*) a conveyor of flammable gas through a fixed pipe distribution system, or

(*b*) a filler, importer or supplier (not by way of the retail trade) of a refillable container containing liquefied petroleum gas

receives notification of any death, injury or condition (see A3012 above) which has arisen out of, or in connection with, the gas supplied, filled or imported, he must immediately notify the HSE and then send a report of the incident within 14 days. [*RIDDOR* (*SI 1995 No 3163*), *Reg 6*(*1*)].

Gas fitters

A3021 Where an employer or self-employed person who is an approved gas fitter has information that a gas fitting, flue or ventilation is, or is likely to cause, death or major injury, he is required to report that fact to the HSE on a prescribed form within 14 days. [*RIDDOR* (*SI 1995 No 3163*), *Reg 6*(*2*)].

Penalties

A3022 Contravention of any of the provisions of *RIDDOR* is an offence. The maximum penalty on summary conviction is a fine of £5,000. Conviction on indictment in the Crown Court carries an unlimited fine.

Defence in proceedings for breach of Regulations

A3023 It is a defence under the Regulations for a person to prove that:

(*a*) he was not aware of the event requiring him to notify or send a report to the enforcing authority; and

(*b*) he had taken all reasonable steps to have all such events brought to his notice.

[*RIDDOR* (*SI 1995 No 3163*), *Reg 11*].

Obligations on employees

A3024 All employees have duties to inform their employers of serious and immediate dangers and recognisable short comings in safety arrangements.

A3025 There are separate reporting requirements on employees by the *Social Security* (*Claims and Payments*) *Regulations 1979* (*SI 1979 No 628*), *Reg 24*. Although not actually bound in with the separate scheme of health and safety legislation, these reporting requirements complement those which employers have under *RIDDOR*, and there is an overlap in the record keeping requirements for both sets of provisions – see further below.

An accident to which the employees' reporting requirements apply is one 'in respect of which benefit may be payable … '. [*Social Security* (*Claims and Payments*) *Regulations 1979* (*SI 1979 No 628*), *Reg 24*(*1*)]. Under *Regulation 24*, every 'employed earner' who suffers personal injury by an accident for the purposes of the Regulations is required by to give either *oral or written* notice to the employer in one of number of different ways. These ways include notice given to any supervisor or direct to a person designated by the employer. The method that has probably become the norm through custom and practice is by way of an entry of the appropriate particulars in a 'book kept specifically for these purposes' under the Regulations. These books are available from HSE Books as form BI 510. As well as containing sections for completing details of accidents etc. form BI 150 includes pages of instructions to employees and employers about their responsibilities. It is important only to use the latest (2003) version which has been updated so that it is compliant with the *Data Protection Act 1988*.

The entry in an Accident Book is to be made as soon as practicable after the happening of an accident by the employed earner or by some other person acting on his behalf. [*Social Security* (*Claims and Payments*) *Regulations 1979* (*SI 1979 No 628*), *Reg 24*(*3*]. Failure to so is an offence punishable by a fine. [*Social Security* (*Claims and Payments*) *Regulations 1979* (*SI 1979 No 628*), *Reg 31*].

Under *Reg 25* of the 1979 Regulations, once an accident is reported by the employee, the employer must take reasonable steps to investigate the circumstances and, if there appear to be any discrepancies between the circumstances reported and the findings in these investigations, these should be recorded.

Where the accident in question is one to which *RIDDOR* applies there is, as well as the usual reporting requirements (see A3014 above), a requirement to keep a record of the accident. [*RIDDOR* (*SI 1995 No 3163, Reg 7*]. The HSE guide to *RIDDOR* provides (at paragraph 86) that an employer may choose to utilise the form BI 150 Accident Book as this record, and the format of the form includes space for the employer to initial the report as being one reportable under *RIDDOR*.

Objectives of accident reporting

A3026 There should be an effective accident reporting and investigation system in all organisations. Accident reporting procedures should be clearly established in writing with individual reporting responsibilities specified. Staff should be trained in the system and disciplinary action may have to be taken where there is a failure to comply with it. Moreover, there is a case for all incidents, no matter how trivial they may seem and of the absence of any resulting injury being reported through the internal reporting procedures.

The capture of information about accidents and other incidents can be enhanced by the provision of a simple form used throughout the organisation.

Duty of disclosure of accident data

A3027 Employers are under a duty to disclose accident data to works safety representatives and, in the course of litigation, to legal representatives of persons claiming damages for death or personal injury.

Safety representatives

A3028 An employer must make available to safety representatives of both unionised and non-unionised workforces the information within the employer's knowledge necessary to enable them to fulfil their functions. [*Safety Representatives and Safety Committees Regulations 1977* (*SI 1977 No 500*), *Reg 7*(*2*); *Health and Safety* (*Consultation with Employees*) *Regulations 1996* (*SI 1996 No 1513*), *Reg 5*(*1*)]. The Approved Code of Practice in association with these Regulations states that such information should include information which the employer keeps relating to the occurrence of any accident, dangerous occurrence or notifiable industrial disease and any associated statistical records. (Code of Practice: Safety Representatives and Safety Committees 1976, para 6(c)). See JOINT CONSULTATION – SAFETY REPRESENTATIVES AND SAFETY COMMITTEES.

Legal representatives

A3029 In the course of litigation (and sometimes before proceedings have actually begun) obligations may arise to give disclosure of relevant documents concerning an accident – even if they are confidential. Legal representatives of an injured party would expect reasonable access to *RIDDOR* information. The decision in *Waugh v British Railways Board [1979] 2 All ER 1169* established that where an employer seeks to withhold on grounds of privilege a report made following an accident, he can only do so if its dominant purpose is related to actual or potential hostile legal proceedings. In *Waugh* a report was commissioned for two purposes following the death of an employee: (*a*) to recommend improvements in safety measures, and (*b*) to gather material for the employer's defence. It was held that the report was not privileged.

Accident Investigation

Rationale for undertaking investigations

A3030 Unlike the highly prescriptive legal requirements for notification under *RIDDOR* (*SI 1995 No 3163*), there are no specific legal obligations requiring the investigation of accidents or the production of accident reports. The HSC consulted on proposals

Figure 2: Internal Incident Notification Form

INTERNAL INCIDENT NOTIFICATION FORM

Time/date:	00.00 hrs	dd/mm/yy
Location:		
Description of incident:		
Person(s) injured:		
Nature of injury:		
Other persons involved:		
First aid administered:		
Accident Book completed:	Y/N	
Form completed by:		
Date:	dd/mm/yy	

for a statutory duty to undertake investigations into accidents, diseases and dangerous occurrences in 2001, but the outcome was a decision to develop new guidelines with a view to encouraging the implementation of best practice voluntarily.

Nevertheless a number of statutory requirements demand some level of investigation is carried out, if only to determine whether it is necessary to undertake a review of the adequacy of existing risk assessments for the relevant activity in order to comply to with *Regulation 3*(*3*) of the *Management of Health and Safety at Work Regulations 1999* (*SI 1999 No 3242*) (see R3017).

Accident investigation does however does however have wider rationales. The main drivers for it are:

(*a*) *learning lessons*: identifying and understanding the immediate and the underlying causes of accidents and identifying ways in which to prevent the recurrence of similar accidents.

(*b*) *reassurance and explanation*: particularly the victim of an accident, but also others connected with it often need to be reassured that appropriate action is being taken or that a satisfactory explanation has been given for what has happened.

(*c*) *monitoring of performance*: investigations produce data for the reactive monitoring of health and safety performance, measurement of results against an organisation's targets or for benchmarking, recording of essential information in databases ('corporate memory'), and the provision of management reports – ultimately to directors – on the management of operational risks.

(*d*) *providing information to other interested parties*: the information is likely to be needed with which to brief insurers, safety representatives, also those who are responsible for dealing with media. Tthere is also an expectation on the part of health and safety inspectors as well that an accident investigation will be produced, and disclosed to them once completed, and an absence of a formal report may be taken as an indicator of management failure.

(*e*) *disciplinary procedures*: where there is evidence of deliberate reckless failure to follow procedures an investigation may form part of the disciplinary process and evidence upon which the fairness of action perhaps ultimately resulting in dismissal might be judged.

(*f*) *allocation of blame/liability*: this will be part and parcel of any investigation undertaken by enforcement agencies and insurers, any information may need to be obtained and analysed by the organisation involved in the accident in preparation for what are inevitably adversarial proceedings.

Accident investigation literature often tends to be focussed on the risk management and accident prevention benefits that can accrue from the process, while treating issues relating to blame as subsidiary to other objectives. This can result in a number of difficulties and potential conflicts, as it is difficult in reality for accidents to be investigated in an entirely blame-free context. Accident investigations can – if there is not careful consideration given to legal and disciplinary implications – prejudice not just the organisation but also the position of individual employees and managers who may be subject to actual or implied criticism. It should also be borne in mind that the HSE's internal work instructions for inspectors steer inspectors towards directing or using an organisation's investigation process in order to achieve enforcement-related goals which include providing '*an early insight into the duty holder's thoughts regarding cause and blame enabling any potential defence or mitigation to any subsequent proceedings to be identified*'.

It is not possible to produce a single system of investigation which ultimately resolves all the potential conflicts between the different rationales and needs for the investigation. However, the processes do need to be operated in a way which takes into account these conflicts and seeks to minimise them by appropriately involving different parts of the organisation, insurers and legal representatives. Particular attention needs to be given to identifying the information which is obtained for advice which is given in situations where legal privilege may apply and where the wider dissemination of that material might cause prejudice in future legal proceedings.

The main elements of investigations

A3031 The following sections outline the main elements of an accident investigation process, and each stage of this process provides a framework within which more or less details and elaborate examination of relevant issues can be carried out depending on what is considered necessary or proportionate (see FIGURE 3). The process should not however be inflexible, and it may need to be adapted if for example there are parallel investigations taking place by enforcement authorities with a view to future prosecution.

Figure 3: Elements of the investigation process

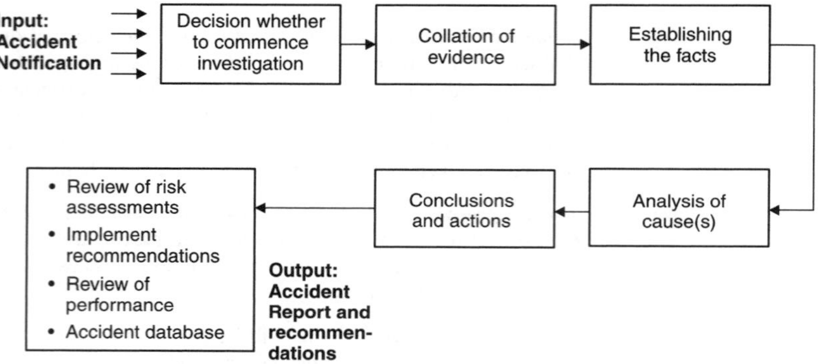

It is possible to devise ad hoc arrangements for each investigation as it arises, but ideally (and essentially in large organisations where there is a steady occurrence of accidents) there should be defined procedures. Responsibilities should be designated, and there should be basic but clear criteria as to the types of occurrence which should be investigated more thoroughly than is necessary merely for the purposes of carrying out RIDDOR notifications. Thought needs to be given to how an investigation team is to be constituted, setting terms of reference, and determining what the reporting lines are during the process and not just for the delivery of the accident report. An agreed format for the contents of accident reports is desirable so that consistent and comprehensive results are produced. Those with responsibility for the investigation process may require training in some areas such as interviewing skills, accident causation principles and report writing.

An HSE investigation or claim for personal injury will raise liability issues and opportunities for internal or external legal advice should be built into the process at each stage.

It is important that the process does not exist in isolation of what is happening elsewhere in the organisation. There may need to be liaison not just with legal advisors, but with others who may be dealing with necessary actions following an accident such as HR managers, insurance managers and of course line managers responsible for the activity in which the accident has occurred who may need immediate guidance on whether (or how) to recommence or continue the activities.

Collating evidence

A3032 Every situation will be different, but typically the evidence will comprise some or all of the following:

Conditions at the accident scene:

- photographs and videos;
- sketches and plans;
- measurements;
- records of weather/environmental conditions;
- samples of substances;
- equipment or retained parts;
- condition of safety equipment/devices;
- first-aid records;
- information obtained from a reconstruction of the accident;
- lists of eye-witnesses;
- identities of others involved;
- notes of comments made immediately after the accident;
- internal memos and emails dealing with the accident and aftermath;
- statements taken by managers/others;
- other electronic records (e.g. time of emergency calls, or computerised process data).

Safety documentation:

- accident book entries;
- RIDDOR records;
- relevant risk assessments;
- method statements/safe systems of work;
- training materials;
- training records;
- equipment operating instructions;
- equipment logs/maintenance records;
- personnel records of those involved in the accident;
- reports on relevant previous incidents;
- relevant health and safety audits, management reports, consultants' advice;
- relevant emergency procedures.

The most difficult part of this stage of the process is likely to be obtaining statements promptly from witnesses, either because they are distressed or because they are unco-operative (which may be for a variety of reasons including concern that they might be criticised). It is nevertheless important to obtain witnesses' evidence as soon as possible after an accident. The accuracy of their information is reduced the more time elapses, and recall can be affected by *post-event enhancement* –

the unconscious incorporation of ideas and perceptions based on discussing events with others in group setting and by *false confidence* – a hardening of a version of events from recalling and re-telling others or committing evidence to paper.

If witnesses are unco-operative or hostile it may serve no useful purpose to persevere in efforts to interview them. A period of delay can sometimes assist, in which it may be possible to resolve the underlying problems. Where the witness is an employee, ultimately the failure to co-operate with the employer may be treated as a disciplinary matter.

Another difficulty with interviewing witnesses can be that the HSE, police or other authorities may wish to restrict access to them – so that they can interview them first and take formal statements. There is no legal power to control witnesses in this way, but inspectors sometimes threaten employers with prosecution for obstructing them if they do not agree to postpone their interviews. It is therefore advisable to be open with the authorities about the interview process, and to offer inspectors or other officers an opportunity to attend interviews with witnesses about whom they have concerns.

<table>
<tr><td colspan="2">Interviewing Witnesses</td></tr>
<tr><td colspan="2">When carrying out an investigation, it is always necessary to interview the people involved in the accident. The following procedures should be adopted.</td></tr>
<tr><td colspan="2">Prepare for the interviews:
● gather information (facts not hearsay)
● visit the accident scene
● allow adequate time
● minimise the risk of interruptions</td></tr>
<tr><td colspan="2">The interview:
● ask what happened
● do not interrupt; be a good listener
● do not ask leading questions
● do no make assumptions
● do not lead the interviewee; some people will tell you what they think you want to hear!
● be considerate, not authoritarian or judgemental
● ask key questions in order to tease out the facts concerning the accident
● emphasise that the purpose of the investigation is primarily to prevent a recurrence. It is not a whistleblowing, blame apportionment, fault-finding exercise.</td></tr>
<tr><td>Do:
● put people at ease
● listen
● be sympathetic
● be firm</td><td>Don't:
● interrogate
● interrupt
● get into an argument
● point the finger</td></tr>
</table>

• be honest	• blame or find fault
• be tactful	• make promises
• be courteous	• describe what other witnesses have said
• be interested	• speculate about any facts or causes

(Adapted from *Tolley's Workplace Accident Handbook* (2003) ISBN 0 7545 2023 4)

Analysis of causes

A3033 This stage of the process is concerned with two inter-related questions: 'what happened?' and 'why did it happen?' These apparently simple questions can be subjected to highly sophisticated analytical techniques. (See for example 'Root causes analysis: Literature review' – HSE Contract Research Report 325/2001 at website: www.hse.gov.uk/researcher/crr_pdf/2001/crr01325.pdf, and 'Events and causal factors analysis' at website: www.tis.eh.doe.gov/analysis/trac/14/trac14.htm.) The following paragraphs summarise a basic approach, but those charged with accident investigations should have a fuller understanding of accident causation theory and knowledge of the background to *human factors* (see E17068).

Having collated the information about nature and circumstances of the accident the investigation team needs to decide what issues it needs to analyse. A useful starting point is often the organisation's standard template for accident investigation reports (see A3034 below) but not all the headings will necessarily be relevant in each case, and a degree of flexibility is needed so that the investigation does not become diverted by issues of low or no relevance.

Examples of the questions, which are typically explored, are:

- What was the chain of evens leading up to the accident?
- What planning of the activity had taken place?
- Were the risks known?
- Was the planning deficient?
- Were defective premises, equipment or maintenance a factor?
- Were the correct materials being used?
- Were those involved properly trained?
- Was the training adequate?
- Did people depart from instructions/procedures?
- Was supervision properly undertaken?
- Were the instructions/procedures adequate?
- Did organisational factors of the work play a role?, e.g. excessive hours, rushing to meet a deadline or target.
- Was there a failure to learn lessons from previous incidents or to follow-up previous warnings or complaints?
- Did the emergency procedures and provision of first-aid operate effectively?

It is not unusual for these or other questions to require additional evidence or for further queries to be raised with witnesses, and that the collating analysis stages may have to overlap.

Answers to these questions are used to inform the next stage of analysis which involves reaching an understanding as to the cause or causes of the accident, and determining whether the risk control measures that were in place are in need of improvement.

Causes of accidents can be complex and multi-factorial. It is often the unusual combination of a series of events which leads to injury or damage in what were hitherto routine workplace activities. Each event may be traced back to factors that preceded it, and so it has become common to analysis accidents in terms of three types of 'causes'.

1. 'Immediate' cause: an unsafe act (or omission) or an unsafe condition which leads directly to the injury or damage. (An example of an unsafe act is improper stacking of a load; an example of an unsafe condition is a machine with a missing guard). Occasionally the immediate cause of an accident may be a 'violation': a deliberate or reckless departure from correct procedures by an individual or group.

2. 'Underlying' cause: these precede the time of the accident and consist of management control factors (e.g. inadequate risk assessment or the absence of supervision), job factors (e.g. lack of competence or insufficient staffing), or environmental factors (such as working in harsh conditions).

3. 'Root' causes: a failing from which all the other causes have grown – usually a deficiency in the planning and organisation of management of health and safety.

Caution is needed in using this approach for a number of reasons. First, it may be difficult to separate and distinguish between several different causes as neatly in practice as it is in theory. Second, this approach does not give weight to the *gravity* of causal factors, which may be relevant in the wider employment or liability context. Third, and most importantly, there is a large element of subjectivity in determining what are the relevant underlying, and particular root, causes. The causal analysis therefore needs to be firmly based on the factual analysis, and the reasoning process explained at each stage by reference to the facts as they have found during the first stage of the analysis.

Preparing the report

A3034 There are no hard and fast rules about the structure or content of accident investigation reports. The essential features however are:

- description of circumstances of the accident;
- consideration of control measures;
- conclusions as to causes;
- recommendations for remedial action.

A basic template which can be used for reports is as follows.

Outline for Accident Investigation Report

Accident Investigation Report

1. Time and place of accident
2. Details of person(s) injured
3. Description of events leading to accident and injury
4. Working conditions/environment
5. Relevant equipment/substances
6. Planning and risk assessment
7. Adequacy of training
8. Instruction and supervision
9. Condition and use of PPE
10. Existence of similar risks elsewhere
11. Conclusions and cause(s) of accident
12. Recommendations for action

Names of investigation team:

....

Date of report: ....

Before finalising the report it is sometimes desirable to seek comments or corrections from those involved in the accident or connected with the background. Partly this is out of courtesy where individuals might perceive the conclusions as implied criticism, but it also provides a final opportunity to eliminate factual or analytical errors. It is also sensible that proposed recommendations for action are at least canvassed with the relevant managers who may become responsible for implementing them to ensure that they are not impractical or unrealistic. A draft may also be provided to the organisations' legal advisors to consider how it may affect liability issues and whether the conclusions are consistent with the findings of privileged investigations that have been, or are still being, carried out.

When the report is finalised there needs to be a clear and timely management plan of action to implement the recommendations.

The document itself will be disclosable in any subsequent legal proceedings and a copy may be requested by enforcing authorities. Delay or incompleteness in giving effect to the recommendations may expose the organisation to subsequent criticism.

Appendix A
List of Reportable Dangerous Occurrences

A3035 **PART I(A) – GENERAL**

Lifting machinery, etc.

1. The collapse of, the overturning of, or the failure of any load-bearing part of any—

   (a) lift or hoist;

   (b) crane or derrick;

   (c) mobile powered access platform;

   (d) access cradle or window-cleaning cradle;

   (e) excavator;

   (f) pile-driving frame or rig having an overall height, when operating, of more than 7 metres; or

   (g) fork lift truck.

Pressure systems

2. The failure of any closed vessel (including a boiler or boiler tube) or of any associated pipework, in which the internal pressure was above or below atmospheric pressure, where the failure has the potential to cause the death of any person.

Freight containers

3. (*a*) The failure of any freight container in any of its load-bearing parts while it is being raised, lowered or suspended.

   (*b*) In this paragraph, 'freight container' means a container as defined in regulation 2(1) of the Freight Containers (Safety Convention) Regulations 1984.

Overhead electric lines

4. Any unintentional incident in which plant or equipment either—

   (a) comes into contact with an uninsulated overhead electric line in which the voltage exceeds 200 volts; or

   (b) causes an electrical discharge from such an electric line by coming into close proximity to it.

Electrical short circuit

5. Electrical short circuit or overload attended by fire or explosion which results in the stoppage of the plant involved for more than 24 hours or which has the potential to cause the death of any person.

Explosives

6. (1) Any of the following incidents involving explosives—

(a) the unintentional explosion or ignition of explosives other than one—

(i) caused by the unintentional discharge of a weapon where, apart from that unintentional discharge, the weapon and explosives functioned as they were designed to do; or

(ii) where a fail-safe device or safe system of work functioned so as to prevent any person from being injured in consequence of the explosion or ignition;

(b) a misfire (other than one at a mine or quarry or inside a well or one involving a weapon) except where a fail-safe device or safe system of work functioned so as to prevent any person from being endangered in consequence of the misfire;

(c) the failure of the shots in any demolition operation to cause the intended extent of collapse or direction of fall of a building or structure;

(d) the projection of material (other than at a quarry) beyond the boundary of the site on which the explosives are being used or beyond the danger zone in circumstances such that any person was or might have been injured thereby;

(e) any injury to a person (other than at a mine or quarry or one otherwise reportable under these Regulations) involving first-aid or medical treatment resulting from the explosion or discharge of any explosives or detonator.

(2) In this paragraph 'explosives' means any explosive of a type which would, were it being transported, be assigned to class 1 within the meaning of the Classification and Labelling of Explosives Regulations 1983 and 'danger zone' means the area from which persons have been excluded or forbidden to enter to avoid being endangered by any explosion or ignition of explosives.

Biological agents

7. Any accident or incident which resulted or could have resulted in the release or escape of a biological agent likely to cause severe human infection or illness.

Malfunction of radiation generators, etc.

8. (1) Any incident in which—

(a) the malfunction of a radiation generator or its ancillary equipment used in fixed or mobile industrial radiography, the irradiation of food or the processing of products by irradiation, which causes it to fail to de-energise at the end of the intended exposure period; or

(b) the malfunction of equipment used in fixed or mobile industrial radiography or gamma irradiation causes a radioactive source to fail to return to its safe position by the normal means at the end of the intended exposure period.

(2) In this paragraph 'radiation generator' means any electrical equipment emitting ionising radiation and containing components operating at a potential difference of more than 5 kV.

Breathing apparatus

9. (1) Any incident in which breathing apparatus malfunctions—

(a) while in use, or

(b) during testing immediately prior to use in such a way that had the malfunction occurred while the apparatus was in use it would have posed a danger to the health or safety of the user.

(2) This paragraph shall not apply to breathing apparatus while it is being—

(a) used in a mine; or

(b) maintained or tested as part of a routine maintenance procedure.

Diving operations

10. Any of the following incidents in relation to a diving project—

(a) the failure or the endangering of—

(i) any lifting equipment associated with the diving operation, or

(ii) life support equipment, including control panels, hoses and breathing apparatus,

which puts a diver at risk;

(b) any damage to, or endangering of, the dive platform, or any failure of the dive platform to remain on station, which puts a diver at risk;

(c) the trapping of a diver;

(d) any explosion in the vicinity of a diver; or

(e) any uncontrolled ascent or any omitted decompression which puts a diver at risk.

Collapse of scaffolding

11. The complete or partial collapse of—

(a) any scaffold which is—

(i) more than 5 metres in height which results in a substantial part of the scaffold falling or overturning; or

(ii) erected over or adjacent to water in circumstances such that there would be a risk of drowning to a person falling from the scaffold into the water; or

(b) the suspension arrangements (including any outrigger) of any slung or suspended scaffold which causes a working platform or cradle to fall.

Train collisions

12. Any unintended collision of a train with any other train or vehicle, other than one reportable under Part IV of this Schedule, which caused, or might have caused, the death of, or major injury to, any person.

Wells

13. Any of the following incidents in relation to a well (other than a well sunk for the purpose of the abstraction of water)—

(a) a blow-out (that is to say an uncontrolled flow of well-fluids from a well);

(b) the coming into operation of a blow-out prevention or diversion system to control a flow from a well where normal control procedures fail;

(c) the detection of hydrogen sulphide in the course of operations at a well or in samples of well-fluids from a well where the presence of hydrogen sulphide in the reservoir being drawn on by the well was not anticipated by the responsible person before that detection;

(d) the taking of precautionary measures additional to any contained in the original drilling programme following failure to maintain a planned minimum separation distance between wells drilled from a particular installation; or

(e) the mechanical failure of any safety critical element of a well (and for this purpose the safety critical element of a well is any part of a well whose failure would cause or contribute to, or whose purpose is to prevent or limit the effect of, the unintentional release of fluids from a well or a reservoir being drawn on by a well).

Pipelines or pipeline works

14. The following incidents in respect of a pipeline or pipeline works—

(a) the uncontrolled or accidental escape of anything from, or inrush of anything into, a pipeline which has the potential to cause the death of, major injury or damage to the health of any person or which results in the pipeline being shut down for more than 24 hours;

(b) the unintentional ignition of anything in a pipeline or of anything which, immediately before it was ignited, was in a pipeline;

(c) any damage to any part of a pipeline which has the potential to cause the death of, major injury or damage to the health of any person or which results in the pipeline being shut down for more than 24 hours;

(d) any substantial and unintentional change in the position of a pipeline requiring immediate attention to safeguard the integrity or safety of a pipeline;

(e) any unintentional change in the subsoil or seabed in the vicinity of a pipeline which has the potential to affect the integrity or safety of a pipeline;

(f) any failure of any pipeline isolation device, equipment or system which has the potential to cause the death of, major injury or damage to the health of any person or which results in the pipeline being shut down for more than 24 hours; or

(g) any failure of equipment involved with pipeline works which has the potential to cause the death of, major injury or damage to the health of any person.

Fairground equipment

15. The following incidents on fairground equipment in use or under test—

(a) the failure of any load-bearing part;

(b) the failure of any part designed to support or restrain passengers; or

(c) the derailment or the unintended collision of cars or trains.

Carriage of dangerous substances by road

16. (1) Any incident involving a road tanker or tank container used for the carriage of dangerous goods in which—

(a) the road tanker or vehicle carrying the tank container overturns (including turning onto its side);

(b) the tank carrying the dangerous goods is seriously damaged;

(c) there is an uncontrolled release or escape of the dangerous goods being carried; or

(d) there is a fire involving the dangerous goods being carried.

17. (1) Any incident involving a vehicle used for the carriage of dangerous goods, other than a vehicle to which paragraph 16 applies, where there is –

(a) an uncontrolled release or escape of the dangerous goods being carried in such a quantity as to have the potential to cause the death of, or major injury to, any person; or

(b) a fire which involves the dangerous substance being carried.

17A. In paragraphs 16 and 17 above—

(a) 'road tanker' and 'tank container' have the same meanings as in Regulation 2(1) of the Carriage of Dangerous Goods (Classifications, Packaging and Labelling) and Use of Use of Transportable Pressure Receptacles Regulations 1996 ('the 1996 Regulations');

(b) 'carriage' has the same meaning as in Regulation 2(1) of the Carriage of Dangerous Goods by Road Regulations 1996; and

(c) 'dangerous goods' means any goods which fall within the definition of 'dangerous goods' in Regulation 2(1) of the 1996 Regulations, other than –

(i) explosives, or

(ii) radioactive material (other than that which is being carried in accordance with the condition specified in Schedules 1 to 4 of marginal 2704 to ADR), and in this sub-paragraph 'ADR' has the meaning assigned to it by Regulation 2(1) of the 1996 Regulations.

PART I(B) – DANGEROUS OCCURRENCES WHICH ARE REPORTABLE EXCEPT IN RELATION TO OFFSHORE WORKPLACES

Collapse of building or structure

18. Any unintended collapse or partial collapse of—

(a) any building or structure (whether above or below ground) under construction, reconstruction, alteration or demolition which involves a fall of more than 5 tonnes of material;

(b) any floor or wall of any building (whether above or below ground) used as a place of work; or

(c) any false-work.

Explosion or fire

19. An explosion or fire occurring in any plant or premises which results in the stoppage of that plant or as the case may be the suspension of normal work in those premises for more than 24 hours, where the explosion or fire was due to the ignition of any material.

Escape of flammable substances

20. (1) The sudden, uncontrolled release—

(a) inside a building —

(i) of 100 kilograms or more of a flammable liquid,

(ii) of 10 kilograms or more of a flammable liquid at a temperature above its normal boiling point, or

(iii) of 10 kilograms or more of a flammable gas; or

(b) in the open air, of 500 kilograms or more of any of the substances referred to in sub-paragraph (a) above.

(2) In this paragraph, 'flammable liquid' and 'flammable gas' mean respectively a liquid and a gas so classified in accordance with regulation 5(2), (3) or (5) of the Chemicals (Hazard Information and Packaging for Supply) Regulations 1994 (now revoked by SI 2002 No 1689).

Escape of substances

21. The accidental release or escape of any substance in a quantity sufficient to cause the death, major injury or any other damage to the health of any person.

PART II – DANGEROUS OCCURRENCES WHICH ARE REPORTABLE IN RELATION TO MINES

Fire or ignition of gas

22. The ignition, below ground, of any gas (other than gas in a safety lamp) or of any dust.

23. The accidental ignition of any gas in part of a firedamp drainage system on the surface or in an exhauster house.

24. The outbreak of any fire below ground.

25. An incident where any person in consequence of any smoke or any other indication that a fire may have broken out below ground has been caused to leave any place pursuant to either Regulation 11(1) of the Coal and Other Mines (Fire and Rescue) Regulations 1956 or section 79 of the Mines and Quarries Act 1954.

26. The outbreak of any fire on the surface which endangers the operation of any winding or haulage apparatus installed at a shaft or unwalkable outlet or of any mechanically operated apparatus for producing ventilation below ground.

Escape of gas

27. Any violent outburst of gas together with coal or other solid matter into the mine workings except when such outburst is caused intentionally.

Failure of plant or equipment

28. The breakage of any rope, chain, coupling, balance rope, guide rope, suspension gear or other gear used for or in connection with the carrying of persons through any shaft or staple shaft.

29. The breakage or unintentional uncoupling of any rope, chain, coupling, rope tensioning system or other gear used for or in connection with the transport of persons below ground, or breakage of any belt, rope or other gear used for or in connection with a belt conveyor designated by the mine manager as a man-riding conveyor.

30. An incident where any conveyance being used for the carriage of persons is overwound; or any conveyance not being so used is overwound and becomes detached from its winding rope; or any conveyance operated by means of the friction of a rope on a winding sheave is brought to rest by the apparatus provided in the headframe of the shaft or in the part of the shaft below the lowest landing for the time being in use, being apparatus provided for bringing the conveyance to rest in the event of its being overwound.

31. The stoppage of any ventilating apparatus (other than an auxiliary fan) which causes a substantial reduction in ventilation of the mine lasting for a period exceeding 30 minutes, except when for the purpose of planned maintenance.

32. The collapse of any headframe, winding engine house, fan house or storage bunker.

Breathing apparatus

33. At any mine an incident where—

(a) breathing apparatus or a smoke helmet or other apparatus serving the same purpose or a self-rescuer, while being used, fails to function safely or develops a defect likely to affect its safe working; or

(b) immediately after using and arising out of the use of breathing apparatus or a smoke helmet or other apparatus serving the same purpose or a self-rescuer, any person receives first-aid or medical treatment by reason of his unfitness or suspected unfitness at the mine.

Injury by explosion of blasting material etc.

34. An incident in which any person suffers an injury (not being a major injury or one reportable under regulation 3(2)) which results from an explosion or discharge of any blasting material or device within the meaning of section 69(4) of the Mines and Quarries Act 1954 for which he receives first-aid or medical treatment at the mine.

Use of emergency escape apparatus

35. An incident where any apparatus is used (other than for the purpose of training and practice) which has been provided at the mine in accordance with regulation 4 of the Mines (Safety of Exit) Regulations 1988 or where persons leave the mine when apparatus and equipment normally used by persons to leave the mine is unavailable.

Inrush of gas or water

36. Any inrush of noxious or flammable gas from old workings.

37. Any inrush of water or material which flows when wet from any source.

Insecure tip

38. Any movement of material or any fire or any other event which indicates that a tip to which Part I of the Mines and Quarries (Tips) Act 1969 applies, is or is likely to become insecure.

Locomotives

39. Any incident where an underground locomotive when not used for testing purposes is brought to rest by means other than its safety circuit protective devices or normal service brakes.

Falls of ground

40. Any fall of ground, not being part of the normal operations at a mine, which results from a failure of an underground support system and prevents persons travelling through the area affected by the fall or which otherwise exposes them to danger.

PART III – DANGEROUS OCCURRENCES WHICH ARE REPORTABLE IN RELATION TO QUARRIES

Collapse of storage bunkers

41. The collapse of any storage bunker.

Sinking of craft

42. The sinking of any water-borne craft or hovercraft.

Injuries

43. (1) An incident in which any person suffers an injury (not otherwise reportable under these Regulations) which results from an explosion or from the discharge of any explosives for which he receives first-aid or medical treatment at the quarry.

(2) In this paragraph, 'explosives' has the same meaning as in regulation 2(1) of the Quarries Regulations 1999.

Projection of substances outside quarry

44. Any incident in which any substance is ascertained to have been projected beyond a quarry boundary as a result of blasting operations in circumstances in which any person was or might have been endangered.

Misfires

45. Any misfire, as defined by regulation 2(1) of the Quarries Regulations 1999.

Insecure tips

46. Any event (including any movement of material or any fire) which indicates that a tip, to which the Quarries Regulations 1999 apply, is or is likely to become insecure.

Movement of slopes or faces

47. Any movement or failure of an excavated slope or face which—

(a) has the potential to cause the death of any person; or

(b) adversely affects any building, contiguous land, transport system, footpath, public utility or service, watercourse, reservoir or area of public access.

Explosions or fires in vehicles or plant

48. (1) Any explosion or fire occurring in any large vehicle or mobile plant which results in the stoppage of that vehicle or plant for more than 24 hours and which affects—

(a) any place where persons normally work; or

(b) the route of egress from such a place.

(2) In this paragraph, 'large vehicle or mobile plant' means—

(a) a dump truck having a load capacity of at least 50 tonnes; or

(b) an excavator having a bucket capacity of at least 5 cubic metres.

PART IV – DANGEROUS OCCURRENCES WHICH ARE REPORTABLE IN RESPECT OF RELEVANT TRANSPORT SYSTEMS

Accidents to passenger trains

49. Any collision in which a passenger train collides with another train.

50. Any case where a passenger train or any part of such a train unintentionally leaves the rails.

Accidents not involving passenger trains

51. Any collision between trains, other than one between a passenger train and another train, on a running line where any train sustains damage as a result of the collision, and any such collision in a siding which results in a running line being obstructed.

52. Any derailment, of a train other than a passenger train, on a running line, except a derailment which occurs during shunting operations and does not obstruct any other running line.

53. Any derailment, of a train other than a passenger train, in a siding which results in a running line being obstructed.

Accidents involving any kind of train

54. Any case of a train striking a buffer stop, other than in a siding, where damage is caused to the train.

55. Any case of a train striking any cattle or horse, whether or not damage is caused to the train, or striking any other animal if, in consequence, damage (including damage to the windows of the driver's cab but excluding other damage consisting solely in the breakage of glass) is caused to the train necessitating immediate temporary or permanent repair.

56. Any case of a train on a running line striking or being struck by any object which causes damage (including damage to the windows of the driver's cab but excluding other damage consisting solely in the breakage of glass) necessitating immediate temporary or permanent repair or which might have been liable to derail the train.

57. Any case of a train, other than one on a railway, striking or being struck by a road vehicle.

58. Any case of a passenger train, or any other train not fitted with continuous self-applying brakes, becoming unintentionally divided.

59. (1) Any of the following classes of accident which occurs or is discovered whilst the train is on a running line—

(a) the failure of an axle;

(b) the failure of a wheel or tyre, including a tyre loose on its wheel;

(c) the failure of a rope or the fastenings thereof or of the winding plant or equipment involved in working an incline;

(d) any fire, severe electrical arcing or fusing in or on any part of a passenger train or a train carrying dangerous goods;

(e) in the case of any train other than a passenger train, any severe electrical arcing or fusing, or any fire which was extinguished by a fire-fighting service; or

(f) any other failure of any part of a train which is likely to cause an accident to that or any other train or to kill or injure any person.

(2) In this paragraph 'dangerous goods' have the meaning assigned to it in Regulation 2(1) of the Carriage of Dangerous Goods (Classification, Packaging and Labelling) and Use of Transportable Pressure Receptacles Regulations 1996 (SI 1996 No 2092).

Accidents and incidents at level crossings

60. Any case of a train striking a road vehicle or gate at a level crossing.

61. Any case of a train running onto a level crossing when not authorised to do so.

62. A failure of the equipment at a level crossing which could endanger users of the road or path crossing the railway.

Accidents involving the permanent way and other works on or connected with a relevant transport system

63. The failure of a rail in a running line or of a rack rail, which results in—

(a) a complete fracture of the rail through its cross-section; or

(b) in a piece becoming detached from the rail which necessitates an immediate stoppage of traffic or the immediate imposition of a speed restriction lower than that currently in force.

64. A buckle of a running line which necessitates an immediate stoppage of traffic or the immediate imposition of a speed restriction lower than that currently in force.

65. Any case of an aircraft or a vehicle of any kind landing on, running onto or coming to rest foul of the line, or damaging the line, which causes damage which obstructs the line or which damages any railway equipment at a level crossing.

66. The runaway of an escalator, lift or passenger conveyor.

67. Any fire or severe arcing or fusing which seriously affects the functioning of signalling equipment.

68. Any fire affecting the permanent way or works of a relevant transport system which necessitates the suspension of services over any line, or the closure of any part of a station or signal box or other premises, for a period—

(a) in the case of a fire affecting any part of a relevant transport system below ground, of more than 30 minutes, and

(b) in any other case, of more than 1 hour.

69. Any other fire which causes damage which has the potential to affect the running of a relevant transport system.

Accidents involving failure of the works on or connected with a relevant transport system

70. (1) The following classes of accident where they are likely either to cause an accident to a train or to endanger any person—

(a) the failure of a tunnel, bridge, viaduct, culvert, station, or other structure or any part thereof including the fixed electrical equipment of an electrified relevant transport system;

(b) any failure in the signalling system which endangers or which has the potential to endanger the safe passage of trains other than a failure of a traffic light controlling the movement of vehicles on a road;

(c) a slip of a cutting or of an embankment;

(d) flooding of the permanent way;

(e) the striking of a bridge by a vessel or by a road vehicle or its load; or

(f) the failure of any other portion of the permanent way or works not specified above.

Incidents of serious congestion

71. Any case where planned procedures or arrangements have been activated in order to control risks arising from an incident of undue passenger congestion at a station unless that congestion has been relieved within a period of time allowed for by those procedures or arrangements.

Incidents of signals passed without authority

72. (1) Any case where a train, travelling on a running line or entering a running line from a siding, passes without authority a signal displaying a stop aspect unless—

(a) the stop aspect was not displayed in sufficient time for the driver to stop safely at the signal.

PART V – DANGEROUS OCCURRENCES WHICH ARE REPORTABLE IN RESPECT OF AN OFFSHORE WORKPLACE

Release of petroleum hydrocarbon

73. Any unintentional release of petroleum hydrocarbon on or from an offshore installation which—

(a) results in –

(i) a fire or explosion; or

(ii) the taking of action to prevent or limit the consequences of a potential fire or explosion; or

(b) has the potential to cause death or major injury to any person.

Fire or explosion

74. Any fire or explosion at an offshore installation, other than one to which paragraph 73 above applies, which results in the stoppage of plant or the suspension of normal work.

Release or escape of dangerous substances

75. The uncontrolled or unintentional release or escape of any substance (other than petroleum hydrocarbon) on or from an offshore installation which has the potential to cause the death of, major injury to or damage to the health of any person.

Collapses

76. Any unintended collapse of any offshore installation or any unintended collapse of any part thereof or any plant thereon which jeopardises the overall structural integrity of the installation.

Dangerous occurrences

77. Any of the following occurrences having the potential to cause death or major injury—

(a) the failure of equipment required to maintain a floating offshore installation on station;

(b) the dropping of any object on an offshore installation or on an attendant vessel or into the water adjacent to an installation or vessel; or

(c) damage to or on an offshore installation caused by adverse weather conditions.

Collisions

78. Any collision between a vessel or aircraft and an offshore installation which results in damage to the installation, the vessel or the aircraft.

79. Any occurrence with the potential for a collision between a vessel and an offshore installation where, had a collision occurred, it would have been liable to jeopardise the overall structural integrity of the offshore installation.

Subsidence or collapse of seabed

80. Any subsidence or local collapse of the seabed likely to affect the foundations of an offshore installation or the overall structural integrity of an offshore installation.

Loss of stability or buoyancy

81. Any incident involving loss of stability or buoyancy of a floating offshore installation.

Evacuation

82. Any evacuation (other than one arising out of an incident reportable under any other provision of these Regulations) of an offshore installation, in whole or part, in the interests of safety.

Falls into water

83. Any case of a person falling more than 2 metres into water (unless the fall results in death or injury required to be reported under sub-paragraphs (a)–(d) of regulation 3(1)).

[*The Reporting of Injuries, Diseases and Dangerous Occurrences Regulations 1995, Reg 2(1), Sch 2*].

Appendix B
List of Reportable Diseases

A3036 *PART I – OCCUPATIONAL DISEASES*

Column 1		Column 2
	Diseases	**Activities**
Conditions due to physical agents and the physical demands of work		
1.	Inflammation, ulceration or malignant disease of the skin due to ionising radiation.	
2.	Malignant disease of the bones due to ionising radiation.	} Work with ionising radiation.
3.	Blood dyscrasia due to ionising radiation.	
4	Cataract due to electromagnetic radiation.	Work involving exposure to electromagnetic radiation (including radiant heat).
5.	Decompression illness.	
6.	Barotrauma resulting in lung or other organ damage.	} Work involving breathing gases at increased pressure (including diving).
7.	Dysbaric osteonecrosis.	
8.	Cramp of the hand or forearm due to repetitive movements.	Work involving prolonged periods of handwriting, typing or other repetitive movements of the fingers, hand or arm.
9.	Subcutaneous cellulitis of the hand (beat hand).	Physically demanding work causing severe or prolonged friction or pressure on the hand.
10.	Bursitis or subcutaneous cellulitis arising at or about the knee due to severe or prolonged external friction or pressure at or about the knee (beat knee).	Physically demanding work causing severe or prolonged friction or pressure at or about the knee.
11.	Bursitis or subcutaneous cellulitis arising at or about the elbow due to severe or prolonged external friction or pressure at or about the elbow (beat elbow).	Physically demanding work causing severe or prolonged friction or pressure at or about the elbow.
12.	Traumatic inflammation of the tendons of the hand or forearm or of the associated tendon sheaths.	Physically demanding work, frequent or repeated movements, constrained postures or extremes of extension or flexion of the hand or wrist.

13.	Carpal tunnel syndrome.	Work involving the use of hand-held vibrating tools.
14.	Hand-arm vibration syndrome.	Work involving: (a) the use of chain saws, brush cutters or hand-held or hand-fed circular saws in forestry or wood-working; (b) the use of hand-held rotary tools in grinding material or in sanding or polishing metal; (c) the holding of material being ground or metal being sanded or polished by rotary tools; (d) the use of hand-held percussive metal-working tools or the holding of metal being worked upon by percussive tools in connection with riveting, caulking, chipping, hammering, fettling or swaging; (e) the use of hand-held powered percussive drills or hand-held powered percussive hammers in mining, quarrying or demolition, or on roads or footpaths (including road construction); or (f) the holding of material being worked upon by pounding machines in shoe manufacture.
Infections due to biological agents		
15.	Anthrax.	(a) Work involving handling infected animals, their products or packaging containing infected material; or (b) work on infected sites.
16.	Brucellosis.	Work involving contact with: (a) animals or their carcasses (including any parts thereof) infected by brucella or the untreated products of same; or (b) laboratory specimens or vaccines of or containing brucella.

17.	(a) Avian chlamydiosis.	Work involving contact with birds infected with chlamydia psittaci, or the remains or untreated products of such birds.
	(b) Ovine chlamydiosis.	Work involving contact with sheep infected with chlamydia psittaci or the remains or untreated products of such sheep.
18.	Hepatitis.	Work involving contact with: (a) human blood or human blood products; or (b) any source of viral hepatitis.
19.	Legionellosis.	Work on or near cooling systems which are located in the workplace and use water; or work on hot water service systems located in the workplace which are likely to be a source of contamination.
20.	Leptospirosis.	(a) Work in places which are or are liable to be infested by rats, fieldmice, voles or other small mammals; (b) work at dog kennels or involving the care or handling of dogs; or (c) work involving contact with bovine animals or their meat products or pigs or their meat products.
21.	Lyme disease.	Work involving exposure to ticks (including in particular work by forestry workers, rangers, dairy farmers, game keepers and other persons engaged in countryside management).
22.	Q fever.	Work involving contact with animals, their remains or their untreated products.
23.	Rabies.	Work involving handling or contact with infected animals.
24.	Streptococcus suis.	Work involving contact with pigs infected with streptococcus suis, or with the carcasses, products or residues of pigs so affected.
25.	Tetanus.	Work involving contact with soil likely to be contaminated by animals.

26.	Tuberculosis.	Work with persons, animals, human or animal remains or any other material which might be a source of infection.
27.	Any infection reliably attributable to the performance of the work specified in the entry opposite hereto.	Work with micro-organisms; work with live or dead human beings in the course of providing any treatment or service or in conducting any investigation involving exposure to blood or body fluids; work with animals or any potentially infected material derived from any of the above.
Conditions due to substances		
28.	Poisonings by any of the following: (a) acrylamide monomer; (b) arsenic or one of its compounds; (c) benzene or a homologue of benzene; (d) beryllium or one of its compounds; (e) cadmium or one of its compounds; (f) carbon disulphide; (g) diethylene dioxide (dioxan); (h) ethylene oxide; (i) lead or one of its compounds; (j) manganese or one of its compounds; (k) mercury or one of its compounds; (l) methyl bromide; (m) nitrochlorobenzene, or a nitro- or amino- or chloro-derivative of benzene or of a homologue of benzene; (n) oxides of nitrogen; (o) phosphorus or one of its compounds.	Any activity.

29.	Cancer of a bronchus or lung.	(a) Work in or about a building where nickel is produced by decomposition of a gaseous nickel compound or where any industrial process which is ancillary or incidental to that process is carried on; or (b) work involving exposure to bis(chloromethyl) ether or any electrolytic chromium processes (excluding passivation) which involve hexavalent chromium compounds, chromate production or zinc chromate pigment manufacture.
30.	Primary carcinoma of the lung where there is accompanying evidence of silicosis.	Any occupation in: (a) glass manufacture; (b) sandstone tunnelling or quarrying; (c) the pottery industry; (d) metal ore mining; (e) slate quarrying or slate production; (f) clay mining; (g) the use of siliceous materials as abrasives; (h) foundry work; (i) granite tunnelling or quarrying; or (j) stone cutting or masonry.
31.	Cancer of the urinary tract.	1. Work involving exposure to any of the following substances: (a) beta-naphthylamine or methylene-bis-orthochloroaniline; (b) diphenyl substituted by at least one nitro or primary amino group or by at least one nitro and primary amino group (including benzidine);

		(c) any of the substances mentioned in sub-paragraph (b) above if further ring substituted by halogeno, methyl or methoxy groups, but not by other groups; or
		(d) the salts of any of the substances mentioned in sub-paragraphs (a) to (c) above.
		2. The manufacture of auramine or magenta.
32.	Bladder cancer.	Work involving exposure to aluminium smelting using the Soderberg process.
33.	Angiosarcoma of the liver.	(a) Work in or about machinery or apparatus used for the polymerisation of vinyl chloride monomer, a process which, for the purposes of this sub-paragraph, comprises all operations up to and including the drying of the slurry produced by the polymerisation and the packaging of the dried product; or
		(b) work in a building or structure in which any part of the process referred to in the foregoing sub-paragraph takes place.
34.	Peripheral neuropathy.	Work involving the use or handling of or exposure to the fumes of or vapour containing n-hexane or methyl n-butyl ketone.
35.	Chrome ulceration of:	Work involving exposure to chromic acid or to any other chromium compound.
	(a) the nose or throat; or	
	(b) the skin of the hands or forearm.	
36.	Folliculitis.	
37.	Acne.	} Work involving exposure to mineral oil, tar, pitch or arsenic.
38.	Skin cancer.	

39.	Pneumoconiosis (excluding asbestosis).	1.(a) The mining, quarrying or working of silica rock or the working of dried quartzose sand, any dry deposit or residue of silica or any dry admixture containing such materials (including any activity in which any of the aforesaid operations are carried out incidentally to the mining or quarrying of other minerals or to the manufacture of articles containing crushed or ground silica rock); or (b) the handling of any of the materials specified in the foregoing sub-paragraph in or incidentally to any of the operations mentioned therein or substantial exposure to the dust arising from such operations. 2. The breaking, crushing or grinding of flint, the working or handling of broken, crushed or ground flint or materials containing such flint or substantial exposure to the dust arising from any of such operations. 3. Sand blasting by means of compressed air with the use of quartzose sand or crushed silica rock or flint or substantial exposure to the dust arising from such sand blasting. 4. Work in a foundry or the performance of, or substantial exposure to the dust arising from, any of the following operations: (a) the freeing of steel castings from adherent siliceous substance or; (b) the freeing of metal castings from adherent siliceous substance: (i) by blasting with an abrasive propelled by compressed air, steam or a wheel, or

(ii) by the use of power-driven tools.

5. The manufacture of china or earthenware (including sanitary earthenware, electrical earthenware and earthenware tiles) and any activity involving substantial exposure to the dust arising therefrom.

6. The grinding of mineral graphite or substantial exposure to the dust arising from such grinding.

7. The dressing of granite or any igneous rock by masons, the crushing of such materials or substantial exposure to the dust arising from such operations.

8. The use or preparation for use of an abrasive wheel or substantial exposure to the dust arising therefrom.

9.(a) Work underground in any mine in which one of the objects of the mining operations is the getting of any material;

(b) the working or handling above ground at any coal or tin mine of any materials extracted therefrom or any operation incidental thereto;

(c) the trimming of coal in any ship, barge, lighter, dock or harbour or at any wharf or quay; or

(d) the sawing, splitting or dressing of slate or any operation incidental thereto.

10. The manufacture or work incidental to the manufacture of carbon electrodes by an industrial undertaking for use in the electrolytic extraction of aluminium from aluminium oxide and any activity involving substantial exposure to the dust therefrom.

		11. Boiler scaling or substantial exposure to the dust arising therefrom.
40.	Byssinosis.	The spinning or manipulation of raw or waste cotton or flax or the weaving of cotton or flax, carried out in each case in a room in a factory, together with any other work carried out in such a room.
41.	Mesothelioma.	(a) The working or handling of asbestos or any admixture of asbestos;
42.	Lung cancer.	(b) the manufacture or repair of asbestos textiles or other articles containing or composed of asbestos;
43.	Asbestosis.	(c) the cleaning of any machinery or plant used in any of the foregoing operations and of any chambers, fixtures and appliances for the collection of asbestos dust; or
		(d) substantial exposure to the dust arising from any of the foregoing operations.
44.	Cancer of the nasal cavity or associated air sinuses.	1.(a) Work in or about a building where wooden furniture is manufactured;
		(b) work in a building used for the manufacture of footwear or components of footwear made wholly or partly of leather or fibre board; or
		(c) work at a place used wholly or mainly for the repair of footwear made wholly or partly of leather or fibre board.
		2. Work in or about a factory building where nickel is produced by decomposition of a gaseous nickel compound or in any process which is ancillary or incidental thereto.
45.	Occupational dermatitis.	Work involving exposure to any of the following agents:
		(a) epoxy resin systems;
		(b) formaldehyde and its resins;
		(c) metalworking fluids;

(d) chromate (hexavalent and derived from trivalent chromium);

(e) cement, plaster or concrete;

(f) acrylates and methacrylates;

(g) colophony (rosin) and its modified products;

(h) glutaraldehyde;

(i) mercaptobenzothiazole, thiurams, substituted paraphenylene-diamines and related rubber processing chemicals;

(j) biocides, anti-bacterials, preservatives or disinfectants;

(k) organic solvents;

(l) antibiotics and other pharmaceuticals and therapeutic agents;

(m) strong acids, strong alkalis, strong solutions (e.g. brine) and oxidising agents including domestic bleach or reducing agents;

(n) hairdressing products including in particular dyes, shampoos, bleaches and permanent waving solutions;

(o) soaps and detergents;

(p) plants and plant-derived material including in particular the daffodil, tulip and chrysanthemum families, the parsley family (carrots, parsnips, parsley and celery), garlic and onion, hardwoods and the pine family;

(q) fish, shell-fish or meat;

(r) sugar or flour; or

(s) any other known irritant or sensitising agent including in particular any chemical bearing the warning 'may cause sensitisation by skin contact' or 'irritating to the skin'.

46.	Extrinsic alveolitis (including farmer's lung).	Exposure to moulds, fungal spores or heterologous proteins during work in: (a) agriculture, horticulture, forestry, cultivation of edible fungi or malt-working; (b) loading, unloading or handling mouldy vegetable matter or edible fungi whilst same is being stored; (c) caring for or handling birds; or (d) handling bagasse.
47.	Occupational asthma.	Work involving exposure to any of the following agents: (a) isocyanates; (b) platinum salts; (c) fumes or dust arising from the manufacture, transport or use of hardening agents (including epoxy resin curing agents) based on phthalic anhydride, tetrachlorophthalic anhydride, trimellitic anhydride or triethylene-tetramine; (d) fumes arising from the use of rosin as a soldering flux; (e) proteolytic enzymes; (f) animals including insects and other arthropods used for the purposes of research or education or in laboratories; (g dusts arising from the sowing, cultivation, harvesting, drying, handling, milling, transport or storage of barley, oats, rye, wheat or maize or the handling, milling, transport or storage of meal or flour made therefrom; (h) antibiotics; (i) cimetidine; (j) wood dust; (k) ispaghula; (l) castor bean dust;

	(m) ipecacuanha;
	(n) azodicarbonamide;
	(o) animals including insects and other arthropods (whether in their larval forms or not) used for the purposes of pest control or fruit cultivation or the larval forms of animals used for the purposes of research or education or in laboratories;
	(p) glutaraldehyde;
	(q) persulphate salts or henna;
	(r) crustaceans or fish or products arising from these in the food processing industry;
	(s) reactive dyes;
	(t) soya bean;
	(u) tea dust;
	(v) green coffee bean dust;
	(w) fumes from stainless steel welding;
	(x) any other sensitising agent, including in particular any chemical bearing the warning 'may cause sensitisation by inhalation'.

PART II – DISEASES ADDITIONALLY REPORTABLE IN RESPECT OF OFFSHORE WORKPLACES

48. Chickenpox.
49. Cholera.
50. Diphtheria.
51. Dysentery (amoebic or bacillary).
52. Acute encephalitis.
53. Erysipelas.
54. Food poisoning.
55. Legionellosis.
56. Malaria.
57. Measles.
58. Meningitis.
59. Meningococcal septicaemia (without meningitis).
60. Mumps.

48. Chickenpox.
61. Paratyphoid fever.
62. Plague.
63. Acute poliomyelitis.
64. Rabies.
65. Rubella.
66. Scarlet fever.
67. Tetanus.
68. Tuberculosis.
69. Typhoid fever.
70. Typhus.
71. Viral haemorrhagic fevers.
72. Viral hepatitis.

[*The Reporting of Injuries, Diseases and Dangerous Occurrences Regulations 1995, Reg 5*(*1*)(*2*), *Sch 3*].

Appendix C

Prescribed Form for Reporting an Injury or Dangerous Occurrence

A3037

Health and Safety at Work etc Act 1974
The Reporting of Injuries, Diseases and Dangerous Occurrences Regulations 1995

HSE Health & Safety Executive

Report of an injury or dangerous occurrence

Filling in this form
This form must be filled in by an employer or other responsible person.

Part A

About you

1 What is your full name?

2 What is your job title?

3 What is your telephone number?

About your organisation

4 What is the name of your organisation?

5 What is its address and postcode?

6 What type of work does the organisation do?

Part B

About the incident

1 On what date did the incident happen?

/ /

2 At what time did the incident happen?
(Please use the 24-hour clock eg 0600)

3 Did the incident happen at the above address?

Yes ☐ Go to question 4

No ☐ Where did the incident happen?

☐ elsewhere in your organisation – give the name, address and postcode

☐ at someone else's premises – give the name, address and postcode

☐ in a public place – give details of where it happened

If you do not know the postcode, what is the name of the local authority?

4 In which department, or where on the premises, did the incident happen?

Part C

About the injured person

If you are reporting a dangerous occurrence, go to Part F.
If more than one person was injured in the same incident, please attach the details asked for in Part C and Part D for each injured person.

1 What is their full name?

2 What is their home address and postcode?

3 What is their home phone number?

4 How old are they?

5 Are they

☐ male?

☐ female?

6 What is their job title?

7 Was the injured person (tick only one box)

☐ one of your employees?

☐ on a training scheme? Give details:

☐ on work experience?

☐ employed by someone else? Give details of the employer:

☐ self-employed and at work?

☐ a member of the public?

Part D

About the injury

1 What was the injury? (eg fracture, laceration)

2 What part of the body was injured?

F2508 (01/96)

Continued overleaf

3 Was the injury (tick the one box that applies)

- ☐ a fatality?
- ☐ a major injury or condition? (see accompanying notes)
- ☐ an injury to an employee or self-employed person which prevented them doing their normal work for more than 3 days?
- ☐ an injury to a member of the public which meant they had to be taken from the scene of the accident to a hospital for treatment?

4 Did the injured person (tick all the boxes that apply)

- ☐ become unconscious?
- ☐ need resuscitation?
- ☐ remain in hospital for more than 24 hours?
- ☐ none of the above.

Part E

About the kind of accident

Please tick the one box that best describes what happened, then go to Part G.

- ☐ Contact with moving machinery or material being machined
- ☐ Hit by a moving, flying or falling object
- ☐ Hit by a moving vehicle
- ☐ Hit something fixed or stationary

- ☐ Injured while handling, lifting or carrying
- ☐ Slipped, tripped or fell on the same level
- ☐ Fell from a height

  How high was the fall?

  [metres]

- ☐ Trapped by something collapsing

- ☐ Drowned or asphyxiated
- ☐ Exposed to, or in contact with, a harmful substance
- ☐ Exposed to fire
- ☐ Exposed to an explosion

- ☐ Contact with electricity or an electrical discharge
- ☐ Injured by an animal
- ☐ Physically assaulted by a person

- ☐ Another kind of accident (describe it in Part G)

Part F

Dangerous occurrences

Enter the number of the dangerous occurrence you are reporting. (The numbers are given in the Regulations and in the notes which accompany this form)

[ ]

Part G

Describing what happened

Give as much detail as you can. For instance

- the name of any substance involved
- the name and type of any machine involved
- the events that led to the incident
- the part played by any people.

If it was a personal injury, give details of what the person was doing. Describe any action that has since been taken to prevent a similar incident. Use a separate piece of paper if you need to.

[ ]

Part H

Your signature

Signature

[ ]

Date

[/ /]

Where to send the form

Please send it to the Enforcing Authority for the place where it happened. If you do not know the Enforcing Authority, send it to the nearest HSE office.

For official use

Client number	Location number	Event number	
			☐ INV REP ☐ Y ☐ N

Appendix D

Prescribed Form for Reporting a Case of Disease

A3038

Health and Safety at Work etc Act 1974
The Reporting of Injuries, Diseases and Dangerous Occurrences Regulations 1995

Report of a case of disease

Filling in this form

This form must be filled in by an employer or other responsible person.

Part A

About you

1 What is your full name?

2 What is your job title?

3 What is your telephone number?

About your organisation

4 What is the name of your organisation?

5 What is its address and postcode?

6 Does the affected person usually work at this address?

Yes ☐ Go to question 7

No ☐ Where do they normally work?

7 What type of work does the organisation do?

Part B

About the affected person

1 What is their full name?

2 What is their date of birth?

/ /

3 What is their job title?

4 Are they

☐ male?

☐ female?

5 Is the affected person (tick one box)

☐ one of your employees?

☐ on a training scheme? Give details:

☐ on work experience?

☐ employed by someone else? Give details:

☐ other? Give details:

F2508A (01/96)

Continued overleaf

Part C

The disease you are reporting

1 Please give:

- the name of the disease, and the type of work it is associated with; or
- the name and number of the disease *(from Schedule 3 of the Regulations – see the accompanying notes).*

2 What is the date of the statement of the doctor who first diagnosed or confirmed the disease?

/ /

3 What is the name and address of the doctor?

Part D

Describing the work that led to the disease

Please describe any work done by the affected person which might have led to them getting the disease.

If the disease is thought to have been caused by exposure to an agent at work *(eg a specific chemical)* please say what that agent is.

Give any other information which is relevant.

Give your description here

Continue your description here

Part E

Your signature

Signature

Date

/ /

Where to send the form

Please send it to the Enforcing Authority for the place where the affected person works. If you do not know the Enforcing Authority, send it to the nearest HSE office.

For official use

Client number

Location number

Event number

☐ INV REP ☐ Y ☐ N

Appendix E
Health and Safety Executive Regional Offices

A3039 *London and South East Division*

Covers the counties of Kent, Surrey, East Sussex and West Sussex, and all London Boroughs.

St Dunstans House
201–211 Borough High Street
London SE1 1GZ
Telephone: 020 7556 2100
Fax: 020 7556 2200

3 East Grinstead House
London Road
East Grinstead RH19 1RR
Telephone: 01342 334 200
Fax: 01342 334 222

Home Counties Division

Covers the counties of Bedfordshire, Berkshire, Buckinghamshire, Cambridgeshire, Dorset, Essex (except London Boroughs in Essex), Hampshire, Hertfordshire, Isle of Wight, Norfolk, Suffolk and Wiltshire.

14 Cardiff Road
Luton LU1 1PP
Telephone: 01582 444 200
Fax: 01582 444 320

Priestley House
Priestley Road
Basingstoke RG24 9NW
Telephone: 01256 404 000
Fax: 01256 404 100

39 Baddow Road
Chelmsford CM2 0HL
Telephone: 01245 706 200
Fax: 01245 706 222

Midlands Division

Covers the counties of West Midlands, Leicestershire, Northamptonshire, Oxfordshire, Warwickshire, Derbyshire, Lincolnshire and Nottinghamshire.

McLaren Building
35 Dale End
Birmingham B4 7NP
Telephone: 0121 607 6200
Fax: 0121 607 6349

5th Floor
Belgrave House
1 Greyfriars
Northampton NN1 2BS
Telephone: 01604 738 300
Fax: 01604 738 333

1st Floor
The Pearson Building
55 Upper Parliament Street
Nottingham NG1 6AU
Telephone: 0115 971 2800
Fax: 0115 971 2802

Yorkshire and North East Division

Covers the counties and unitary authorities of Hartlepool, Middlesbrough, Redcar and Cleveland, Stockton-on-Tees, Durham, Hull, North Lincolnshire, North East Lincolnshire, East Riding, York, North Yorkshire, Northumberland, West Yorkshire, Tyne and Wear and the Metropolitan Boroughs of Barnsley, Doncaster, Rotherham and Sheffield.

Marshalls Mill
Marshall Street
Leeds LS11 9YJ
Telephone: 0113 283 4200
Fax: 0113 283 4296

Sovereign House
110 Queen Street
Sheffield S1 2ES
Telephone: 0114 291 2300
Fax: 0114 291 2379

Arden House
Regent Centre
Regent Farm Road
Gosforth
Newcastle upon Tyne NE3 3JN
Telephone: 0191 202 6200
Fax: 0191 202 6300

North West Division

Covers the counties of Cheshire, Cumbria, Greater Manchester, Lancashire and Merseyside.

Grove House
Skerton Road
Manchester M16 ORB
Telephone: 0161 952 8200
Fax: 0161 952 8222

Marshall House
Ringway
Preston PR1 2HS
Telephone: *see* Manchestert
Fax: 01772 836 222

Wales and West Division

Covers Wales, and the unitary authorities of Cornwall, Devon, Somerset, North West Somerset, Bath and North East Somerset, Bristol, South Gloucestershire, Gloucestershire, Hereford and Worcester, Shropshire and Staffordshire.

Government Buildings
Phase 1
Ty Glas
Llanishen
Cardiff CF14 5SH
Telephone: 029 2026 3000
Fax: 029 2026 3120

Inter City House
Mitchell Lane
Victoria Street
Bristol BS1 6AN
Telephone: 0117 988 6000
Fax: 0117 926 2998

The Marches House
Midway
Newcastle-under-Lyme
Staffordshire ST5 1DT
Telephone: 01782 602 300
Fax: 01782 602 400

Scotland

Covers all the Scottish unitary authorities and island councils.

Belford House
59 Belford Road
Edinburgh EH4 3UE
Telephone: 0131 247 2000
Fax: 0131 247 2121

375 West George Street
Glasgow G2 4LW
Telephone: 0141 275 3000
Fax: 0141 276 3100

Offshore Safety Division
Lord Cullen House
Fraser Place
Aberdeen AB25 3UB
Telephone: 01224 252 500
Fax: 01224 252 662

Asbestos

Introduction

A5001 Asbestos is the generic name for a group of naturally occurring fibrous minerals, metallic silicates, which have a wide range of industrial applications. The most common forms of asbestos are chrysotile (white), amosite (brown) and crocidolite (blue); others are fibrous actinolite, fibrous anthophyllite and fibrous tremolite. Mined in Canada, South Africa, Russia and elsewhere, the material has been manufactured into products which use its characteristics of heat and chemical resistance. Despite its contribution to fire protection, asbestos is now regarded as one of the most significant causes of occupational disease over the past 125 years. The use of asbestos has been extensive, but as a result of prohibitions in the UK on new uses of asbestos-containing materials, products such as brake linings and other friction products are now unlikely to be found in most workplaces. However, its widespread application in materials used in construction has led to many buildings containing asbestos, and the current focus of asbestos controls is largely on the management of this legacy.

Asbestos diseases

A5002 The diseases associated with asbestos only arise when asbestos fibres are inhaled and penetrate deep into the lung. Extremely small fibres are capable of bypassing the body's defence mechanisms to achieve this. These are therefore called respirable fibres and they are fibres with a diameter less than three micrometres. To illustrate how small these fibres are a human hair is approximately 50 micrometres in diameter. Asbestosis is a thickening of the wall of the alveoli, the tiny lung sacs where oxygen passes into the blood. Since the lung has a huge surface area one or two areas of such thickening would have little effect. However when millions of fibres are inhaled and reach the deep lung the reduction in gas exchange capacity caused by thickening of the alveoli walls becomes significant. The disease of asbestosis is recognised when the patient has a measurable decrease in lung capacity and even notices a shortness of breath. Signs of the disease can also be detected by x-ray. If exposure is extensive and prolonged the damage to the lungs can be severe and lead eventually to death. The progress of the disease is proportional to exposure, which in modern terms must be massive. It has occurred in industries where raw asbestos was handled in bulk, when raw asbestos was delivered in bales which were manually cut open and fed into hoppers for the manufacturing process. Workers often went home white with asbestos fibres adhering to their body and clothes. A condition associated with asbestosis was pleural plaques. This was areas of calcification or stiffening on the membrane on the outer surface of the lung. Asbestosis is a very similar disease to coal workers pneumoconiosis and since the level of exposure needed to develop these diseases no longer occurs in Britain today there are virtually no new cases appearing but workers exposed in the past are still alive, suffering from asbestosis and dying of it.

Mesothelioma is a cancer occurring on the outer membrane of the lung. It is normally a very rare disease and so its occurrence amongst asbestos workers was quickly recognised. Typically it is not detected in the early stages and therefore once diagnosed is advanced and usually progresses to death in months rather than years. However the latency period from initial exposure to onset of the disease appears to be long, in the region of decades. The risk of contracting mesothelioma is based on exposure. The greater the concentration of respirable asbestos fibres inhaled the greater is the risk of contracting the disease. In addition, the concentration of respirable asbestos fibres required to cause the disease is far lower than that associated with asbestosis. In the UK mesothelioma was at first associated with crocidolite or blue asbestos and the control limit (the maximum level to which workers could legally be exposed) for blue asbestos was tightened. Eventually it was recognised that all amphibole asbestos types were implicated. So amosite, brown asbestos, and crocidolite and the other amphibole asbestos types now have tighter control limits.

Lung cancer is now recognised as a disease associated with asbestos. The specific diagnosis of lung cancer being an asbestos-related disease is however occupational exposure. Since lung cancer occurring spontaneously, caused by smoking or caused by asbestos is indistinguishable it is the history of the patient which is used in the specific diagnosis. It is firmly established from epidemiology that there is an increased risk of lung cancer for smokers and that there is an increased risk of lung cancer for those exposed to asbestos. Indeed for smokers who are also exposed to asbestos it is estimated that the risk of lung cancer is multiplied rather than the two risks being additive.

When asbestos fibres are inhaled, they have a very long half life in the lungs. Their shape – long and thin – and robustness means that it is difficult for the usual lung clearance mechanisms to remove them from the lungs (in the manner in which dust particles are cleared), and they do not readily dissolve in lung fluid. Whilst the exact mechanisms which cause cancer remain obscure, the toughness and longevity of the fibres helps to explain why they are able to cause ill health many years after first exposure – although the abrasive and chemical effect of the fibres in very high lung concentrations can cause asbestosis on a shorter timescale this is largely an historical disease whilst the cancers continue to take their toll. If workers are exposed to airborne fibres, for example as a result of an incident in which asbestos has been inadvertently disturbed, there is no 'treatment' available to reduce the risk of future disease (except, of course to avoid inhaling cigarette smoke in future). However, it is worth noting that isolated exposures such as this represent a very (immeasurably) low level of risk. The significance of asbestos is that thousands of workers have been exposed on a daily basis to high levels, and that is why the last century saw an 'epidemic' of asbestos disease. This is not to argue that even single exposures to airborne asbestos should not be prevented, rather to encourage a recognition that the emphasis is on routine and continued exposure as this is the root of the problem.

Historical exposures – manufacturing

A5003 In the 1950s and 1960s the predominant concern arising from exposure to asbestos was for manufacturing workers contracting asbestosis, following exposure to massive concentrations of respirable fibres. These workers were involved in manufacturing asbestos insulating boards and panels, asbestos cement products and friction products such as clutch plates, break shoes and gaskets. Anecdotally, such workers would refer to working in 'snow storms' of the material, which because the fibres are soft and silky did not create any immediate discomfort, as would be the case with glass fibre which is immediately irritating to the eyes, throat and skin. In order to control this risk, and against the background of a well-funded lobbying campaign by the asbestos mining and manufacturing interests (a campaign which still functions, as the Canadian government's challenge through the World Trade Organisation to the

EU prohibitions on asbestos importation indicates) Asbestos Regulations were introduced imposing control limits on the maximum concentration of respirable asbestos fibres that workers could be exposed to. Where the concentrations could not be reduced below the control limits it was mandatory to provide suitable respirators. As the risk of lung cancer and mesothelioma was recognised and epidemiological data and analysis became available, the control limits for working with asbestos were gradually reduced. In addition, 'action levels' were adopted and defined in terms of concentrations of respirable airborne fibres measured over a twelve-week period. If the mean concentration of airborne fibres exceeded the action level then the area in the factory had to be designated an 'asbestos zone' and non-essential personnel excluded. If the airborne fibre concentration was likely to exceed the control limit then issuing and wearing respirators became mandatory.

In the *Control of Asbestos at Work Regulations 2002* (*SI 2002 No 2675*), the action levels are 72 fibre-hours per millilitre for chrysotile and 48 fibre-hours per millilitre for the other types of asbestos, such as amosite and crocidolite. This is equivalent to a continuous exposure to 0.15 fibres/ml for chrysotile and 0.1 fibres/ml for the other asbestos types. The control limits are 0.3 fibres/ml and 0.2 fibres/ml for other asbestos types, measured over four hours. There are also short-term control limits, measured over ten minutes, of 0.9 fibres/ml for chrysotile and 0.6 fibres/ml for other asbestos types. The Regulations also impose a general duty on all employers to assess the risk to the health of all employees before any work with asbestos commences and to put in place such control measures as are necessary to reduce the exposure to as low a level as reasonably practicable and, in any case, to below the control limit. Whilst these limits were introduced primarily for manufacturing, they did begin to have an effect on the management of asbestos installations in buildings, where everything from pipework lagging to floors, walls and ceilings could formed from asbestos products.

Asbestos prohibitions

A5004 The *Asbestos* (*Prohibition*) *Regulations 1992* (*SI 1992 No 3067*), as amended, prohibit all imported materials to which asbestos has been added as part of a deliberate manufacturing process, but excludes naturally-occurring minerals which may contain traces of asbestos. Many producers and importers of mineral products carry out comprehensive testing to prevent materials containing significant quantities of asbestos getting into the supply chain. Where asbestos is found the material should not be sold unless the amount is trivially small. Even in those cases suppliers must inform their customers that trace quantities of asbestos may occasionally be found in their products. This will allow 'high-energy' users (for example traces of asbestos in blast-cleaning materials would release airborne fibres) to take appropriate precautions or consider alternative materials. In addition to prohibitions on the sale of asbestos products, where they may be discovered in store rooms or warehouses, they may not be used and must be disposed of safely. Whilst there is no prohibition on asbestos which is already in place, new uses are effectively banned.

Asbestos removal

A5005 As the manufacture and use of asbestos declined, accelerated by the *Asbestos* (*Prohibition*) *Regulations 1992* (*SI 1992 No 3067*), the risk to the health of manufacturing workers also decreased. However, it was recognised that there was a new exposure group. These were the workers who were engaged to remove asbestos from buildings. Therefore, in the late 1980s, new Approved Codes of Practice (ACoPs) and Guidance Notes were introduced to specify how the *Control of Asbestos at Work Regulations 1987* (*SI 1987 No 2115*) (now revoked by *SI 2002 No 2675*) should be applied to the removal of asbestos from buildings. The *Asbestos (Licensing) Regulations 1983 (SI 1983 No 1649)* were also introduced to provide the enforcing authorities with information on which companies were involved in removing asbestos from buildings, where and when the work was being carried out. Guidance Notes recommended that after asbestos was removed from an area the area should be visibly free from dust and debris and that the airborne fibre concentration measured in the area should be below 0.01 fibres/ml. This is the detection limit of the optical microscopy techniques used to measure airborne fibres and is referred to as the 'clearance indicator'. The clearance indicator is simply a measure of cleanliness of an area and is not a risk based limit such as the control limit. Following the introduction of the *Control of Asbestos at Work Regulations 2002 (SI 2002 No 2675)*, new editions of the Approved Codes of Practice were published, including 'Work with asbestos insulation, asbestos coating and asbestos insulating board', (Fourth edition, 2002). A full list of asbestos legislation, Approved Codes of Practice and Guidance Notes are appended at the end of this chapter (see A5030).

Exposure of building maintenance workers

A5006 As the *Asbestos (Licensing) Regulations 1983 (SI 1983 No 1649)* were enforced, and the asbestos removal industry became more thoroughly regulated, the risk to the health of the workers in that industry began to be more effectively controlled. However, in the late 1990s it was recognised that there was still a very large group of workers whose exposure to asbestos was not adequately controlled by the existing asbestos control regime. Maintenance workers and those engaged in building refurbishment were still exposed to asbestos because many building owners did not know that asbestos was present and many employers of maintenance workers did not consider that they would be working with asbestos and so took no precautions. For these reasons, the *Control of Asbestos at Work Regulations 2002 (SI 2002 No 2675)* were introduced. These provisions require building owners and occupiers to identify where asbestos is present in every building and have a documented plan to control the risks to health presented by that asbestos. The aim is to reduce over time the number of asbestos-related deaths in the UK. Currently, past exposure to asbestos leads to 3,000 deaths per year, and the death rate is likely to rise further for another 10 years or so and is unlikely to decrease quickly as the latency period of asbestos-related cancer is between 15 and 60 years. Recent research has indicated that asbestos-related deaths should peak in a few years and then fall as a direct result of the control measures adopted.

Summary: despite lobbying by asbestos interests, the evidence of the health risks associated with asbestos has led to a sequential establishment of a statutory regime of control. The protection was first designed for manufacturing workers, was extended to asbestos removers and latterly has addressed the management of buildings which contain some of the thousands of tonnes of asbestos used in construction. The death rate from the diseases caused by inhalation of asbestos fibres in the past is still rising, but is predicted to peak and then start to fall in a few years as a consequence of the control measures.

Duties under the Control of Asbestos at Work Regulations 2002

A5007 The *Control of Asbestos at Work Regulations 2002 (SI 2002 No 2675)* (*'CAWR'*) place duties on every employer in respect of his employees, and make it clear that these duties, so far as is reasonably practicable, are owed also to any other person, whether at work or not, who may be affected by the work activity undertaken by the employer. The exceptions are in connection with *Regulation 9* of the 2002 Regulations where the duty to inform, instruct and train does to apply to non-employees unless they are in the premises where the work is being carried out, and *Regulation 21* where the duty to carry out medical surveillance and maintain health records for exposed workers only applies to the employer's own employees. The *CAWR* apply to the self-employed, applying the duties of both employer and employee (*Regulation 3(2)*).

After four years of debate, consultation and review, the long-awaited amendments to the regulations have been enacted under the *CAWR*. Not only are employers subject to health and safety duties, and the existing asbestos control regime, in the usual way but the *CAWR* now also apply to all companies which own, occupy, manage or have any contractual responsibilities for maintenance or repair of non-domestic premises. The new duties, therefore, apply to all business, commercial, industrial, educational, healthcare, military, administrative, retail and leisure properties including hotels and guest houses. Even in the absence of agreement, all parties which control access or maintenance of a building could still be liable. Duty holders are under a duty either to:

- manage the risk from asbestos-containing materials ('ACMs'); or
- co-operate with whoever manages that risk.

Affected premises

A5008 The duty applies not only to all non-domestic premises, but also applies to the common parts of commercial buildings, housing developments and leasehold flats. Given the widespread use of asbestos, particularly after the 1940s, many buildings still contain asbestos. An estimated 500,000 premises will be affected and a number of different parties with any measure of control will need to take steps to comply.

Identifying the duty holder

A5009 Persons with a primary responsibility for a building, such as the owner, sole occupier or part occupier, have duties under the *CAWR* (*SI 2002 No 2675*). Those with any responsibility for the maintenance of premises by virtue of a contract or tenancy also have a legal duty to manage the risks from asbestos and comply with the *CAWR*. If no such formal agreement exists, the owner, sub-lessor or managing agent with any de facto control of a workplace premise, or any employer or self-employed person who occupies commercial premises in which persons work is subject to the same duty.

Risk assessment

A5010 The duty holder is required to make a suitable and sufficient risk assessment as to whether asbestos is, or is liable to be present in the premises. It should be noted that wherever a duty such as this applies, compliance requires that the duty holder ensures that it has been carried out. In this case, this means ensuring that a suitable and sufficient assessment has been conducted, documented and is available and used to inform decision-making, advise incoming occupiers, maintenance workers and others. Commissioning others to carry out such work is acceptable practice, so long as the persons employed are competent and allocate sufficient resource to carry out the work properly.

This includes taking into account the age of the premises, building plans and any other relevant information. In carrying out inspections it must be presumed that materials contain asbestos unless there is strong evidence to the contrary. At this stage it is necessary simply to identify whether asbestos may be present and, therefore, determine whether a more thorough investigation is required. It is not, however, sufficient to conclude that a building is free of asbestos simply because it was built after 1990. That conclusion can only be reached if there is documentary evidence from a designer or architect who specified that no asbestos was to be used in the construction of the building. The most likely asbestos materials to be found are:

- Sprayed asbestos and asbestos loose packing – typically used as fire breaks in ceiling voids and to protect steelwork, fire protection in ducts, firebreaks, etc.
- Moulded or preformed lagging – generally used in thermal insulation of pipes and boilers.
- Insulating board – fire protection, thermal insulation, panelling, partitioning, ducts.
- Ceiling tiles.
- Millboard and paper products – insulating electrical products and facing wood fibreboard.
- Woven products – boiler seals and old fire blankets.
- Cement products in flat or corrugated sheets.
- Textured coatings.
- Bitumen roofing materials.
- Vinyl and thermoplastic floor tiles.

Records

A5011 If asbestos is present, an up-to-date record of the location and condition of asbestos-containing materials ('ACMs') or presumed ACMs in the premises must be made. This is often referred to as an 'asbestos register'. The asbestos register should record where asbestos is present in the building the form in which it is present, e.g. asbestos insulating board, pipe lagging, ceiling tiles, asbestos cement sheets, and the condition of the material. The health risk depends on the inhalation of respirable fibres, so that the condition of the asbestos, as recorded in the asbestos register, should reflect the extent to which fibres may be released from the ACM during normal occupancy and during any maintenance work. Since maintenance work will be required at some stage, and it is necessary to either determine the type of asbestos present before such work commences or assume that it is not solely chrysotile, it is often considered appropriate to identify the type of asbestos present in the ACM and record this in the asbestos register. This requires samples of suspected ACM to be taken for laboratory analysis in accordance with the Methods for the Determination of Hazardous Substances – 'Asbestos in bulk materials' (MDHS 77). From November 2004, all analyses of asbestos samples may only be undertaken by an accredited laboratory. At present, there is some concern that there are insufficient accredited laboratories to cope with the huge demand for surveys which the CAWR (*SI 2002 No 2675*) are producing. Since the asbestos register is compiled and maintained in order to quickly identify where asbestos is present before planned or emergency work is undertaken the most appropriate format for the asbestos register should be considered. An electronic format on a computer server will facilitate the maintenance of the register and ensure that an up-to-date version is available at several locations. However where maintenance work may be undertaken where there is no access to a computer system, or at weekends and outside normal office hours, a paper-based register may be more effective. It is a legal requirement that the information is maintained in a form which can be readily accessed by those who need it.

Management plan

A5012 The compilation of an asbestos register is only one part of the new duties. There must also be a documented plan to manage the health risks presented by the presence of ACMs. Where the health risk during normal occupancy is low, the plan may be to maintain the ACM in place and label it. There should then be written procedures for planned maintenance work and procedures for emergency work, and staff familiar with their duties in implementation. These procedures would normally include calling in licensed asbestos contractors to remove the ACM under controlled conditions before the maintenance work proceeded. If the presence of asbestos in the building represents a risk to health under normal occupancy conditions or it is envisaged that emergency situations would involve the removal of ACM, then the asbestos management plan may be to immediately encapsulate or remove ACMs for particular areas. This work requires the involvement of contactors licensed under the *Asbestos (Licensing) Regulations 1983 (SI 1983 No 1649)* working in accordance with the Approved Code of Practice 'Work with asbestos insulation, asbestos coating and asbestos insulating board (Fourth edition, 2002). Where asbestos removal operations are undertaken, it is good practice for the premises management to retain the services of a United Kingdom Accreditation Service (UKAS) accredited laboratory to both monitor and manage the asbestos removal work. The asbestos management plan should also contain the procedures for dealing with suspect ACM. Typically, such procedures would include the requirement to halt the work and call in a UKAS accredited laboratory to take samples and determine whether asbestos is present. The asbestos management plan should be reviewed regularly and the procedures and arrangements monitored to ensure that they are effective. It is important to test the procedures to ensure that they work. For instance, when routine maintenance is undertaken, checks that the tradesmen concerned were furnished with the appropriate information and that they followed the correct procedures should be made and recorded. In many commercial premises the effect of the *CAWR's* duty to manage may create more than one duty holder. Here it is necessary to examine the nature and extent of the repair and maintenance obligation owed by each party, or the level of control in determining their relative contributions to comply.

To ensure the effective compliance with the new duty to manage asbestos in non-domestic premises, there is a simple seven step programme to follow.

Step 1: What ACMs do you know about and how are you managing them

A5013 Are there likely to be ACMs on your premises? If there are none, no further action is required. If ACMs are present, or there is uncertainty, a review of current management arrangements is required. Check that:

- Maintenance and building work is effectively controlled, with a system to prevent disturbance of asbestos and unknown material.
- Maintenance and building staff know the rules about ACMs, and how to work safely on or adjacent to them.
- The condition of known ACMs or materials presumed to contain asbestos is known, there is a system to check for damage and deterioration.

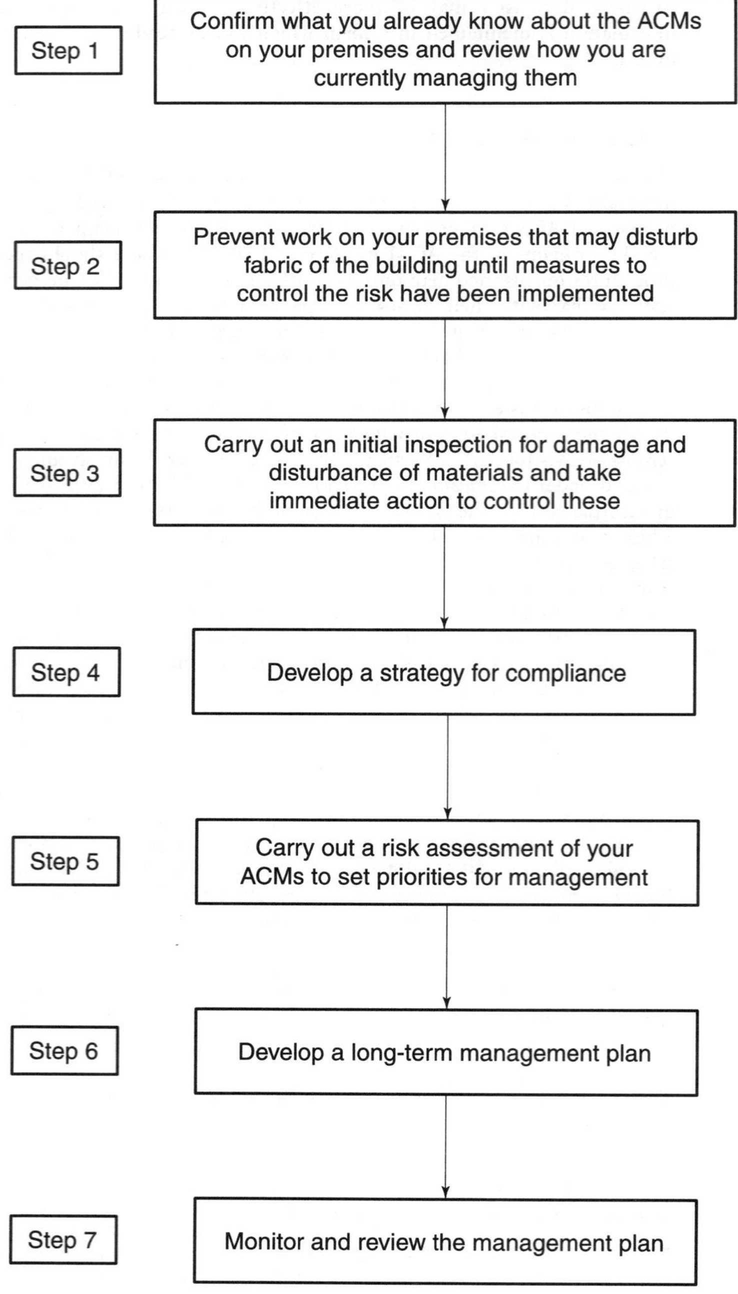
Step 1
Confirm what you already know about the ACMs on your premises and review how you are currently managing them
Step 2
Prevent work on your premises that may disturb fabric of the building until measures to control the risk have been implemented
Step 3
Carry out an initial inspection for damage and disturbance of materials and take immediate action to control these
Step 4
Develop a strategy for compliance
Step 5
Carry out a risk assessment of your ACMs to set priorities for management
Step 6
Develop a long-term management plan
Step 7
Monitor and review the management plan

- Damaged and deteriorated ACMs are repaired, removed or isolated to prevent exposure to released fibres.
- There is a management plan for recording findings on ACMs, monitoring their condition and taking action as necessary.
- Everyone, including appointed contractors, knows their roles and responsibilities for the management of ACMs on the premises.
- A responsible person has been appointed, is competent and resourced (premises or facilities manager, or for small businesses the owner/manager). Access to competent advice from a qualified safety practitioner or hygienist particularly for buildings which are known or suspected to contain significant quantities of ACMs.

Step 2: Control work on your premises

A5014 In every building, there needs to be a system which prevents anyone doing building or maintenance work, even running new telephone cables, without first checking whether ACMs may be disturbed (usually by reference to the Asbestos Register).

Step 3: Carry out an initial inspection

A5015 After stopping new activities which could damage or disturb ACMs, it is essential to check on the condition of all existing known or suspected ACMs, and take immediate action to prevent exposure to airborne fibres where such damage is identified. This check is looking for *any* damage to building materials. If any damaged materials are identified, they can be tested to see if they are ACMs, and kept isolated until the results are available and the material removed or repaired if asbestos is confirmed.

Step 4: Develop a strategy for compliance

A5016 After taking immediate control (steps 1–3), the remaining steps are designed to manage occupancy and building works. The first such step is to develop a formal strategy to manage the long-term risks:

- Commission a survey to establish the extent of ACMs.
- Document the management controls to minimise risk of damage to ACMs, and to identify and act on damage and deterioration effectively.

Step 5: Risk assessment and priorities

A5017 The combination of the intrinsic potential of the material to release fibres *and* the likelihood of disturbance leading to that release which represents the core of the risk assessment.

Step 6: Develop a long-term management plan

A5018 The plan should contain:

- How the location and condition of known or presumed ACMs is recorded.
- Priority assessments and an action plan.
- Decisions about management options and their rationale.
- Action timetable.

- Monitoring arrangements.
- Employees and their responsibilities.
- Contractor procedures including selection, appointment, information and monitoring.
- Training arrangements.
- Procedures and plan for implementation, including controlling work on the fabric of the building.
- Method for passing on information about ACMs to those who need it.
- Responsibility for the management plan and its updating.
- Procedure and timetable for review of the management plan.

Step 7: Monitor and review the management plan

A5019 Implementation of the management plan is monitored such that checks are routinely carried out on the ACMs, records are kept up-to-date. Incidents and accidents require careful investigation and lessons learned should be used to improve the plan and its implementation. For most premises it is recommended that the plan is subject to at least a 6 monthly thorough review.

Duties of third parties

A5020 The *CAWR* (*SI 2002 No 2675*) place any person under a duty to assist in the management of asbestos. Third parties will be required to co-operate with the duty holder so far as is necessary to enable the duty holder to comply with the *CAWR* by:

- making all information in relation to the premises available for inspection;
- providing them with information on the locations of asbestos, if known;
- allowing duty holders to identify asbestos or presumed asbestos for all parts of the premises repaired or maintained by the third party;
- assisting in the preparation and implementation of the asbestos management plan to control the health risk presented by ACMs;
- ensuring that everyone potentially at risk from ACMs receives information on the location and condition of the material so far as it is within the third party's control.

CAWR contractual and tenancy agreements

A5021 Organisations need to review the extent of their maintenance and repair obligations under any leases and be clear as to where the responsibility lies for asbestos control. Responsibility for dealing with any asbestos health risks, and therefore each party's contribution, should be considered in pre-contract lease negotiations and specific provisions should be made in the tenancy agreement. This will be especially important in determining who is responsible for common parts of the building where, for example, control over maintenance and repair activities in those parts could be shared.

Minimum requirements

A5022 It will be a defence to show that an employer has done everything in his power to prevent exposure to asbestos. He will need to be able to produce an up-to-date asbestos register in a suitable format and a documented asbestos management plan containing risk assessments, procedures and arrangements to manage the risk from exposure to asbestos during normal occupancy, planned maintenance and emergency situations. Evidence will also be required to demonstrate that the asbestos management plan has been implemented and that it is effective. The *CAWR* (*SI 2002 No 2675*) is accompanied by an ACoP 'The management of asbestos in non-domestic premises', which gives advice about ways to identify asbestos, the associated health risks and how to develop an asbestos management plan to manage those risks. The duty to manage asbestos took effect in May 2004 and employers should ensure familiarisation with the *CAWR* and to undertake surveys as soon as possible to locate whether there are any specific problem areas which should be dealt as a matter of urgency.

Asbestos building surveys

A5023 There are three types of asbestos surveys described in the Health and Safety Executive (HSE) publication MDHS 100 'Surveying, sampling and assessment of asbestos-containing materials'.

Type 1 survey

A5024 The Type I survey involves study of the building history, structure and plans to consider whether asbestos may be present. This is followed by a visual inspection to assess whether there are materials in place which have the appearance of ACMs. The output of a Type I survey may be the conclusion that the building is of a type which should not contain asbestos and that no suspect materials were seen, or that the building may contain asbestos and there were suspect materials which should be investigated. At this stage the building owner or occupier may assume that asbestos is present and draw up an asbestos management plan or commission a Type 2 survey.

Type 2 survey: risk assessments

A5025 A Type 2 survey involves a detailed inspection of the building with samples of suspect materials being taken for laboratory analysis. Such surveys should only be undertaken by organisations which hold UKAS accreditation for asbestos sampling with accreditation to ISO 17025 and comply with the provisions of EN45004 'General criteria for the operation of various types of bodies performing inspection'. After 20 November 2004 all organisations offering asbestos surveying services must hold this accreditation. The outcome of a Type 2 survey is a detailed report on the location of all ACMs within the building, in a form which can be readily understood and used by those who may be exposed. This usually means that the information is graphically displayed on a plan of the building. Photographs of the sampling site are also normally included to assist in the identification of the ACM position. The Type 2 survey report will also describe the nature of the material, e.g. asbestos insulating board, pipe lagging, ceiling tiles, asbestos rope, floor tiles, asbestos cement products etc. The report will also include the laboratory analysis of the type of asbestos present, e.g. amosite, crocidolite, chrysotile, etc. The condition of the ACM will be assessed in relation to the propensity for the release of respirable airborne fibres and so the consequent risk to health. This requires the judgement of a knowledgeable and experienced surveyor. This risk assessment will depend not only on the nature and condition of the ACM but also on its location and whether it is likely to be disturbed by normal occupancy or only during maintenance and refurbishment. This information is used to compile the asbestos register required in the *CAWR* (*SI 2002 No 2675*).

Type 3 surveys: asbestos removal specifications

A5026 The outcome of a Type 3 survey is used as part of the specifications given to licensed asbestos removal contractors to tender for asbestos removal contracts prior to refurbishment or building demolition. The Type 3 survey report contains details of the location and type of all ACMs and the type of asbestos present. However, the report also specifies the amount of each ACM present. This allows contractors to decide how the ACM should be removed and the extent of the operation and, therefore, provide a method statement for removal and an estimate of costs.

Asbestos waste

A5027 Most asbestos waste arising from asbestos removal operations, including asbestos cement is classified as a 'special waste' under the *Special Waste Regulations 1996* (*SI 1996 No 972*). Consequently, the waste must be transported in suitable containers and be adequately described in a consignment note in accordance with the *Environmental Protection* (*Duty of Care*) *Regulations 1991* (*SI 1991 No 2839*).

Asbestos works

A5028 If work is undertaken on an ACM – to repair and encapsulate it, or to remove it, then there is a simple sequence which is required:

1. Identification of the type of asbestos or assumption that it is asbestos but not solely chrysotile.
2. Plan of work – detailing the nature and probably duration of the work, the location, the work methods and controls (based on the risk assessment) and the equipment to be used to protect and decontaminate the workers and others nearby. The plan should also detail emergency procedures.
3. Notification – if a licensed contractor is employed, he may have this requirement as a condition of license.
4. Information, instruction and training as required for those who may be affected by the works and/or may be required to supervise etc.
5. Once works are underway, the licensed operator is required to comply with a range of technical duties for the containment of asbestos, the labelling of the containment area, the protection of the workers engaged in ACM work and their decontamination, and the handling of waste. During the works the efficacy of the containment and other controls should be checked by air testing carried out by a body independent of the contractor, should be accredited by UKAS.
6. On completion of ACM work within a contained area, the premises must be assessed to determine that they are clean and fit for normal occupation. Such testing, to be carried out by a body independent of the contractor, should be accredited by UKAS.

Licensing of asbestos works

A5029 A licensing system for asbestos works concerning insulation or coating (that is, effectively not applying to cement products, thermoplastic floor tiles and other lower risk materials) was introduced in 1983 by the *Asbestos (Licensing) Regulations* (*SI 1983 No 1649*, as amended by *SI 1998 No 3233*). A licence is required, except for:

- work where any person carrying it out will not spend more than one hour on the work in seven consecutive days; and
- the total time spent by all persons on the work does not exceed two hours;

or

- the employer or self-employed person is the occupier of the premises on which the work is carried out; and
- he has given HSE 28 days' notice in advance, specifying the type of work to be carried out and the location address;

or

- the work consists solely of air monitoring or collecting samples.

Where the licence does not apply (above) the employer must still comply with the *CAWR* (*SI 2002 No 2675*), including providing information, instruction and training, informing others, exercising effective control to minimise asbestos exposures and ensure that, as required, employees are under medical surveillance and health records are maintained.

Legislation, Approved Codes of Practice, and Guidance Notes

A5030 Legislation, Approved Codes of Practice and Guidance Notes dealing with asbestos for England and Wales include, but not exclusively, those listed below. Equivalent documentation exists for Scotland and Northern Ireland respectively.

All legislation, Approved Codes of Practice and Guidance Notes listed together with any subsequent amendments or revisions and any new relevant requirements should be considered before undertaking any work with asbestos or ACMs.

The following legislation, Approved Codes of Practice and Guidance Notes deal primarily with asbestos. Other legislation dealing with health and safety matters has not been listed here, although such legislation still applies to work with asbestos and should be considered at all times.

The following list was last revised in June 2004.

Legislation

- *Health and Safety at Work etc. Act 1974*;
- *Environmental Protection Act 1990*;
- *Environment Act 1995*;
- *Consumer Safety Act 1978*;
- *Asbestos (Licensing) Regulations 1983* (*SI 1983 No 1649*) and subsequent amendments (latest amendment *SI 1998 No 3233*);
- *Control of Asbestos at Work Regulations 2002* (*SI 2002 No 2675*);

- *Control of Asbestos in the Air Regulations 1990* (*SI 1990 No 556*);
- *Reporting of Injuries, Diseases and Dangerous Occurrence Regulations 1995* (*SI 1995 No 3163*);
- *Management of Health and Safety at Work Regulations 1999* (*SI 1999 No 3242*);
- *Workplace* (*Health, Safety and Welfare*) *Regulations 1992* (*SI 1992 No 3004*);
- *Personal Protective Equipment at Work Regulations 1992* (*SI 1992 No 2966*);
- *Provision and Use of Work Equipment Regulations 1998* (*SI 1998 No 2306*);
- *Construction* (*Design and Management*) *Regulations 1994* (*SI 1994 No 3140*);
- *Carriage of Dangerous Goods and Use of Transportable Pressure Equipment Regulations 2004* (*SI 2004 No 568*)
- *Carriage of Dangerous Goods by Rail Regulations 1996* (*SI 1996 No 2089*) (revoked in part by *SI 2004 No 568*);
- *Asbestos* (*Prohibitions*) *Regulations 1992* (*SI 1992 No 3067*);
- *Health and Safety* (*Emissions into the Atmosphere*) *Regulations 1983* (*SI 1983 No 943*);
- *Health and Safety* (*Fees*) *Regulations 2004* (*SI 2004 No 456*);
- *Environmental Protection* (*Duty of Care*) *Regulations 1991* (*SI 1991 No 2839*);
- *Waste Management Licensing Regulations 1994* (*SI 1994 No 1056*);
- *Controlled Waste Regulations 1992* (*SI 1992 No 588*);
- *Controlled Waste* (*Registration of Carriers and Seizure of Vehicles*) *Regulations 1991* (*SI 1991 No 1624*);
- *Environmental Protection* (*Prescribed Processes and Substances*) *Regulations 1991* (*SI 1991 No 472*) (as amended);
- *Asbestos Products* (*Safety*) *Regulations 1985* (*SI 1985 No 2042*).

Approved Codes of Practice

- L28 – 'Work with asbestos insulation, asbestos coating and asbestos insulating board' (Fourth edition, 2002);
- L27 – 'Work with asbestos which does not normally require a licence' (Fourth edition, 2002);
- L127 – 'The management of asbestos in non-domestic premises. Control of asbestos at Work Regulations 2002 (2002);
- 'A Guide to the Asbestos (Licensing) Regulations 1983' (1989);
- 'Waste management: The duty of care. A Code of Practice' (1991);
- 'Respiratory Protective Equipment: A practical guide for users' (1990);
- 'Respiratory Protective Equipment, legislative requirements and lists of HSE approved standards and type approved equipment' (Fourth edition, 1995).
- INDG288 – 'Selection of suitable respiratory protective equipment for work with asbestos' (1999).

Health and Safety Executive Guidance Notes

Environmental Hygiene Series

- EH 10 – 'Asbestos: Exposure limits and measurement of airborne dust concentrations' (revised 2001);
- EH 47 – 'The provision, use and maintenance of hygiene facilities for work with asbestos insulation, asbestos coating and asbestos insulating board' (revised 2002);
- EH 50 – 'Training operatives and supervisors for work with asbestos insulation and coatings' (March 1988);
- EH 51 – 'Enclosures provided for work with asbestos insulation, coatings and insulating board' (January 2000);
- EH 57 – 'The problems of asbestos removal at high temperatures' (December 1992).

Medical Series

- MS 5 – 'Lung Function' (July 1977);
- MS 6 – 'Chest x-rays in dust diseases' (September 1980);
- MS 13 – 'Asbestos' (revised, April 1999).

Methods for the Determination of Hazardous Substances Series

- MDHS 39/4 – 'Asbestos fibres in air, sampling and evaluation by Phase Contrast Microscopy (PCM) under the Control of Asbestos at Work Regulations' (November 1995), (Second Impression);
- MDHS 77 – 'Asbestos in bulk materials, sampling and identification by polarised light microscopy' (PLM) (June 1994);
- MDHS 100 – 'Surveying, sampling and assessment of asbestos-containing materials' (July 2001).

Health and Safety Guidance Series

- HSG 189/1 – 'Controlled asbestos stripping techniques for work requiring a licence' (Second edition, 1999);
- HSG 189/2 – 'Working with asbestos cement' (Second edition, 1999)
- HSG 53 – 'The selection use and maintenance of respiratory protective equipment' (1998);
- HSG 210 – 'Asbestos essentials task manual: Task guidance sheets for the building maintenance and allied trades' (2001);
- HSG 213 – 'Introduction to asbestos essentials: Comprehensive guidance on working with asbestos in the building maintenance and allied trades' (2001);
- HSG 227 – 'A comprehensive guide to managing asbestos in premises' (2002).

Business Continuity

What is business continuity planning?

B8001 Business continuity planning ('BCP') is not just about disaster recovery, crisis management, risk management or IT. It is a business risk management issue. It presents an opportunity to review the way an organisation performs its business processes, to improve procedures and practices, and increase resilience to interruption and loss.

To quote the Business Continuity Institute, the professional body for BCP:

> 'BCP is the act of anticipating incidents which will affect critical functions and processes for the organisation, and ensuring that it responds to any incident in a planned and rehearsed manner.'

The Turnbull Committee 'Guidance for Directors on Internal Controls' sets out an overall framework of best practice for business, based upon an assessment and control of its significant risks. For many companies, BCP will address some of these key risks and help them to achieve compliance.

Hence BCP as an activity has two primary objectives:

- minimise the risk of a disaster befalling the organisation; and
- maximise the ability of the organisation to recover from a disaster.

It differs from disaster recovery by recognising that approximately 80 per cent of disasters which organisations suffer are generated internally. As a result, BCP places more emphasis on managing out potential causes of disasters, than on recovering from those incidents which may not be avoided.

The BCP is therefore a combination of disaster/risk avoidance and the ability to recover.

Business continuity planning in context

B8002 Modern businesses cannot avoid all forms of corporate risk or potential damage. A realistic objective is to ensure the survival of an organisation by establishing a culture that will identify, assess and manage those risks that could cause it to suffer:

- inability to maintain customer services;
- damage to image, reputation or brand;
- failure to protect company assets;
- business control failure;
- failure to meet legal or regulatory requirements.

BCP therefore provides the strategic framework to achieve these objectives.

Why is business continuity planning of importance?

B8003 Very simply the answer to the above question is that disasters kill businesses.

In the UK, 80 per cent of companies who suffered a major disaster and did not have some form of BCP capability went into liquidation within eighteen months. A further 10 per cent suffered the same fate within five years.

Without BCP companies have only a one in ten chance of survival. Also, simply because:

- Customers expect continuity of supply in all circumstances.
- Shareholders expect directors/managers to be fully in control, and to be seen to be in control of any crisis.
- Employees and suppliers expect the business to protect their livelihoods.
- The company's reputation and brand is at risk without BCP.
- It is implicit in good corporate governance and demonstrates best practice in business management.

All the above needs to be taken into account against a business background in which:

- The pressures on business generally are changing and increasing.
- Technology is transforming and underpinning the business environment.
- Consolidation, restructuring, re-engineering, take-overs, acquisitions and mergers etc are now a fact of life.
- New risks and exposures are continually being created.

Business continuity drivers

B8004 The drivers or arguments that need to be understood by all involved in and concerned with BCP may be listed and grouped as follows:

- Commercial.
- Quality/Regulatory.
- Financial.
- Imported disaster.
- Public relations.

Commercial drivers

B8005 An organisation may operate in an intensively competitive market where any disruption to normal operation will be exploited by competitors. For example, when Perrier were forced into a total product withdrawal following a benzene contamination at source, the gaps on supermarket shelves were promptly filled by 30–40 competing brands.

The organisation may depend on a high degree of public confidence to generate/maintain sales, e.g. air transport industry. The issue for many organisations is not merely one of lost revenue, but also of lost customers, who may prove very difficult to get back. On occasions, it has been observed that companies who have suffered a disaster have spent three times their normal PR budget in an attempt to reassure customers that they are still around and worth doing business with.

Conversely, a visible and demonstrable BCP capability can be used to enhance and differentiate products and services.

Quality and regulatory drivers

B8006 If a company adheres to BS 7799 (the British Standard for information security management), there are specific requirements relating to the need for a BCP capability. If a company is BS EN ISO 9000 accredited, compliance with BS 7799 may well be an additional requirement.

Organisations who have implemented the Turnbull and Cadbury guidance on corporate governance will have agreed in principle to comply with all relevant ISO Standards. In addition, directors and officers have a duty to shareholders for the protection and wellbeing of their investments, hence the need for non-executive directors and an audit committee.

If the organisation operates under a licence agreement, a prolonged inability to operate could endanger the licence renewal. That inability to operate may also affect the ability to comply with statutory or regulatory requirements involving, for example, Inland Revenue, DTI, Stock Exchange, Customs and Excise, Companies House etc.

Financial drivers

B8007 Traditional financial measures of mitigating or cushioning against risk, such as insurance, are an important protective element within the overall organisational business risk management framework. Insurance on its own however has limitations which do not protect the complete business cycle.

Despite the fact that business interruption insurance is a fairly common component in most commercial insurance packages, cover is normally limited to two years from the event, and certain consequential losses, which may contain significant unquantifiable costs, will be excluded and hence non-recoverable.

For example, BI (business interruption) insurance does not cover the risk of lost customers, lost productivity or the attrition costs of key staff following a disaster. In some cases, this could be as much as 40 per cent of the total loss.

In some organisations, the disaster itself may be caused by, or made worse because of, loss of control over finances and cash flow, e.g. Barings, AIB.

The imported disaster

B8008 BCP is an approach which recognises that organisations do not operate in isolation of each other and that they rely on suppliers to continue to operate. Hence disasters elsewhere may have a knock-on effect within an organisation, and disasters within an organisation could adversely affect its customers.

With the concepts of single source supply and just-in-time delivery, there is an increasing need to undertake supplier audits to assess their BCP capability.

Also the risk of denial of access should be assessed, as in the case of the Manchester bomb explosion in 1996. Unaffected businesses were denied access by the emergency services for months after the explosion.

Natural disasters can also have the following knock-on effect:

Following the earthquake in Kobe, Japan, the world suffered a severe shortage of silicon chips (microprocessors). The price on the black market rose so dramatically, that it set off a spate of thefts worldwide. This led to any large office block thought to contain large numbers of PCs being hit.

The public relations disaster

B8009 Increasingly, companies strive to raise their profile in the market place and increase their visibility by promoting themselves as aggressively and publicly as possible. The public awareness they seek, and need, in order to sell their products and services is a window through which to view their operations.

Just as members of the public will be more aware of the products and services the company offer, so they will be more aware of any mistakes, accidents, disasters, poor decisions, poor performance etc. for which the company is ultimately responsible. The ill feeling that can be generated against the company by the general public can be so damaging that confidence in the company is totally undermined for a long period of time.

For example, Gerald Ratner jokingly referred to his products as 'cr*p' and later compounded the felony by suggesting that there was more value in a prawn sandwich than in some of his products. The ultimate consequence of his action was loss of sales and the subsequent break-up of his jewellery business. He lost control of his company and a number of employees lost their jobs through the resulting re-organisation.

Hoover's handling of the massive over-subscription to its 'Free Flights' promotion became a national scandal which caused the company to be pilloried almost weekly on prime time TV consumer programmes. Not only did several directors lose their jobs, but the Hoover crisis severely damaged the Hoover brand name so much that the company even considered changing its name away from one of the most powerful and successful brand names/images in history, hence the 'free flights fiasco!'

Thus, the ability of a company to avoid, or recover from, a disaster effectively can have a tremendous and positive PR impact, thus vastly improving stakeholders' (investors, public, customers, suppliers, employees etc.) confidence.

This area is covered in more detail in the section on crisis communication (B8038 below).

Convincing the board – the business case

B8010 A business case for BCP can be made by taking the following ten key points into account:

1. Find out how much time you have for your presentation and make sure you rehearse to make full use of what may be a limited time allocation.
2. Check the audience:
   - who will be there?
   - have non-executive directors previous experience of disaster and/or BCP?
3. Determine, in advance, if there are any regulatory influences that the board may be particularly sensitive to.
4. Achieve board buy-in/commitment to BCP by ensuring that they:

- understand the negative impact of business interruption;
- provide the necessary resources to reduce the risk of disaster;
- appreciate the need to prepare and plan.

5. Stress the benefits of BCP:
   - peace of mind/sleep at night;
   - risk reduction.
6. Be prepared for the 'It won't happen here' syndrome. Have some statistics (not too many) and some local or industry examples that the board may be aware of.
7. Spell out that the board is ultimately responsible for business risk management. N.B. Turnbull/Corporate governance.
8. Avoid being drawn into a detailed discussion about how the BCP capability is to be developed.
9. Avoid being drawn into a discussion on actual or potential disasters, unless they are specifically relevant.
10. Summarise and reinforce the need for BCP by highlighting the unique selling points from the above list.

The business continuity model

B8011 The model is a phased and active process consisting of five main steps:

- *Stage 1: Understanding your business*

  Business impact analysis ('BIA') and risk evaluation/assessment tools are used to: identify the critical business processes within the business; evaluate recovery priorities; and assess risks which could lead to business interruption and/or damage to the organisation's reputation.

- *Stage 2: Continuity strategies*

  This involves determining the selection of alternative strategies available to mitigate loss, assessing the relative merits of these against the business environment; and deciding which are likely to be the most effective in protecting and maintaining the critical business processes.

- *Stage 3: Plan development*

  The development of a proactive BCP plan is designed to improve the risk profile by upgrading optional procedures and practices; introducing alternative business strategies; and using risk financing measures; including insurance.

- *Stage 4: Plan implementation*

  The introduction of BCP capability is achieved via education and awareness of all stakeholders, including employees, customers, suppliers and shareholders. The end result is to establish a BCP culture throughout the organisation. The BCPs should be widely communicated and published to all concerned so that everyone is aware of rules, responsibilities, accountabilities, timescales and KPIs (key performance indicator).

- *Stage 5: Plan testing and maintenance*

  The plan is not the capability. The plan is the means by which the capability is realised.

  Failure to plan is planning to fail!

  BCP is therefore not a 'one-off' exercise. There is a need for ongoing plan testing, maintenance, audit and change in the management of the BCP and its processes. Regular communication and feedback on the results of testing and maintenance are imperative so as to ensure continual improvement of the BCP capability.

The process of developing a BCP capability is a cyclic and evolutionary one in order to take account of the ever-changing organisation.

Understanding your business (stage 1)

Risk evaluation

B8012 Risk evaluation is the process by which an organisation determines the pattern of risk that is unique to its operation.

Risk is used to describe the relationship between the impact and likelihood for a particular event or scenario:

Risk = Impact x Likelihood

Within the BCP framework, risk evaluation is the process which identifies those aspects of the company's operations, which are essential to the survival of the business, and assesses how well they are protected against serious disruption and/or potential disaster.

As in the field of occupational safety and health ('OSH'), risk evaluation is a combination of identification and assessment. Within BCP, there are three commonly used approaches to risk evaluation:

- scenario driven;
- component;
- critical business processes.

All approaches assume that all concerned fully understand what the company is in business for and what is vital for its survival.

Scenario driven approach

B8013 This approach involves the listing of all possible risk scenarios that can be thought of. Once the list has been compiled each scenario is considered in isolation with a view to planning specific counter-measures to tackle each risk scenario on an individual (and isolated) basis.

This approach has some serious limitations. It presupposes that risks exist in isolation and that they do not have a dynamic relationship with each other. This is clearly not the case. For example fire creates smoke, fumes and heat. The presence of smoke and fumes increases the risk of inhalation injury and may hinder evacuation or searching activities. If we then throw in the fact that the fumes are toxic or corrosive and also that they could adversely affect neighbouring sites, the single incidence of 'fire' has an exceptional number of associated risks.

Such an approach will have a high cost and also lead to complex planning implications.

Component approach

B8014 This approach examines each department or unit within the organisation in order to develop their individual risk profiles. Whilst this approach does eventually cover the whole organisation, it fails to prioritise amongst departments and hence does not determine those departments which are important/critical.

Also it sometimes misses linkages and dependencies between departments, e.g. the internal customer. The approach as a whole suffers from a lack of organisation-wide perspective, and hence it is difficult to establish which departments should be reinstated first.

Critical business approach

B8015 Business can be viewed as a set of linked processes. Some of these business processes, the critical ones, have a greater bearing on the core business or primary activity of the business. These are known as the critical business processes ('CBPs').

The fundamental question to be asked and understood is:

> 'What does my company actually do?'

Developing a model may help to clarify thinking and understanding in this regard. Reference to the company's vision, and mission statement and goals will also demonstrate where the organisation is focused. It is on mission critical activities – the critical business processes – where BCP should be focused.

There are four basic questions to be asked:

1. What is the business about?
2. When are we to achieve our goals?
3. Who is involved both internally and externally?
4. How are the goals to be achieved?

An organisation has many dependencies, both internally and externally, that support the mission critical business processes. These may include suppliers, customers, shareholders, key employees, IT systems, and manufacturing processes. It is important to identify these at an early stage and to involve people from the CBPs in the development of the BCP.

External influences on the CBPs may include government departments, regulators, competitors, trade bodies, and pressure groups. These must also be taken into account with the BCP framework.

In summary, the critical business approach:

- Needs a common understanding of the company's primary business objective – the focus or mission of the company. This is ultimately what the capability is being designed to protect and recover.
- Some BCPs are doomed to failure because the participants cannot agree to what the company's primary business objective actually is!
- Determines the CBPs by asking how important are they in achieving the business objective. Again agreement amongst the participants regarding the CBPs is vital in developing a relevant BCP.

- Develops a BCP capability to protect all identified CBPs whilst keeping focused on the primary business objective will prevent participants getting sidetracked or bogged down in detail.

Business impact analysis

B8016 Business impact analysis ('BIA') is the technique used to determine what the business stands to lose if its CBPs are disrupted.

BIA alerts the organisation to the likely cost of a business disruption and assists in the cost benefit analysis of risk reduction measures within the overall BCP capability. Indeed the approval and continuing support of your board for BCP may well hinge on the fact that the costs are justified in the light of potentially catastrophic business losses.

Quantification of business impact

B8017 It is important that all concerned in the BCP process understand that disasters are characterised by time and not event. What causes the disruption is less important to the survival of the organisation than how long the disruption lasts. BIA therefore should consider the impact of a disruption in any of the CBPs spread over time. Traditionally, the impact of a disruptive event was derived merely by determining how much revenue would be lost over the period during which the company was not operational:

Loss of revenue – *Out for 1 day* –> **Impact** = Annual Revenue/ No of working days/year

Loss of revenue – *Out for a year* –> **Impact** = Annual Revenue

On its own, however, loss of revenue does not give the full picture of the overall impact on the business. Other factors, which need to be taken into account, include:

- *Erosion of customer base*

  The organisation will lose customers if it cannot continue to service their requirements. Some will go simply because they perceive the service is unreliable and some will be forced to go for the sake of their own business continuity. Whatever the reason, it is generally estimated that it costs three times more to sell goods/services to a new customer than to an existing one. In the event of a prolonged disruption, the company could therefore have to treble its marketing, advertising, publicity and sales cost in order to restore its customer base.

- *Loss of key staff*

  Following a disaster, some companies have reported that up to 40 per cent of their workforce have left as a result. The loss of intellectual capital and the cost of recruitment can be considerable.

- *Loss of reputation*

  Most companies depend on their reputation as a reliable service provider to generate further sales. Also a great deal of resources, time and money, will have been spent over the years in cultivating a brand image linked to their reputation. Hence, any event which detracts from that image brings the company's reputation into question and not only jeopardises the previous investment but also seriously affects current and future sales. Such lack of confidence will also have a negative effect on the share price of publicly quoted companies.

- *Regulatory and contracted impact*

  The company may be subject to penalty clauses/payments if it cannot meet contracted obligations. It may also be barred from operating by an inability to comply with regulatory (e.g. prohibition notice requirements), reporting or licensing obligations.

- *Loss of credit-worthiness*

  The organisation may face increased costs of borrowing following a disruption to business, on the basis that it represents a greater investment risk to lenders as a result of the disaster.

Business impact Analysis: outputs

B8018 BIA enables the organisation to focus on the CBPs and their negative impact on the business, rather than to conduct a global, risk-specific analysis. The process also takes time sensitivity into account, thus enabling the recovery objectives to be agreed.

It is essential to rate the impact of the loss of CBPs on the business. Risk rating may be qualitative: high, medium, low or quantitative, i.e. 1–9 rating scale. Where possible, financial values should be placed on the business impact.

It is also important to involve key personnel who work in the CBPs in the BIA process, especially as the CBPs are cross-functional/cross-departmental, and agreement must be reached on the relative ratings. Agreement from the board should then be obtained to ensure that the output from the BIA, i.e. the plan is put into place.

In summary therefore:

- Ensure involvement of appropriate functions/departments.
- Determine the impact on the business of the loss of CBPs.
- Apply risk ratings, including time dependencies.
- Obtain broad approval for BIA outputs, i.e. the plan.

Continuity strategies (stage 2)

B8019 Having identified those areas where the organisation is most at risk, a decision has to be made as to what approach is to be taken to protect the operation. With the introduction of the Turnbull Guidance on internal control (corporate governance) this decision must be taken at board level.

Many possibilities exist and it is likely that any strategy adopted will comprise a number of these approaches. Whichever are chosen, there are certain considerations to bear in mind, as indicated below:

- Do nothing (risk retention) – in some instances, the board may consider the risk to be commercially acceptable.
- Change or end the process (risk avoidance) – deciding to alter existing procedures have to be done, bearing in mind the organisation's key focus/mission.
- Insurance (risk transfer) – provides financial recompense/support following a loss but does not provide protection for brand, image and reputation.

- Loss mitigation (risk reduction) – tangible loss control programme to eliminate or reduce risk.
- Business continuity planning (risk reduction) – an approach that seeks to improve organisational resilience to interruption, allowing for the recovery of key business systems and processes within the recovery time frame objective whilst maintaining the organisation's critical business processes.

Any strategy must recognise the internal and external dependencies of the organisation and must be accepted by all members of management involved in the CBPs. Hence, in summary, strategy selection is as follows:

- Identify possible BCP strategies.
- Assess suitability of alternative strategies against the outputs of the BIA.
- Propose cost/benefit analyses of the various strategies.
- Present recommendations to the board for decision and approval.

Developing a business continuity plan (stage 3)

B8020 As previously stated:

- Failing to plan is planning to fail.
- The plan is not the capability.

The BCP is the means by which the capability is realised and can be viewed as the route which takes the organisation from its present state (crisis/disaster) to its desired state (normal operation).

Too often organisations derive a false sense of security from the mere presence of a disaster recovery or BCP and lose sight of the fact that the plan is passive. It has no inherent capability to assist the organisation in any way until it is used. Its value to the organisation can only be assessed by how well it enabled the business to either avoid or recover from a disaster-threatening scenario, i.e. after the event.

The planning dilemma

B8021 *Question*: How does one include enough detail to provide meaningful instruction and guidance without including so much data that individuals or departments find the plan difficult to use?

There is no simple answer to this question since requirements will vary from business to business. However, via the outputs from the BIA, from the experience of individuals involved in the BCP process, and from regular testing and maintenance of the plan, a balance that suits the organisation will be reached through plan refinement.

Plan components

B8022 The objective of creating a BCP capability is firstly to design/engineer out as much of the operational risk from the identified CPBs as possible, i.e. risk avoidance.

Thereafter the BCP should provide a structural means of recovering those lost processes within a given timeframe.

Thus the plan will have two main components:

- *Proactive* – range of cost-justified risk reduction measures, procedural, physical, behavioural, processes, i.e. engineering.

- *Reactive* – disaster recovery plan.

Disaster recovery plan

B8023 The vast majority of disaster recovery plans will compose three stages:

1. *Emergency response – Immediate*

   This stage details the actions the company has predefined in immediate response to an emergency situation. Typically, it should include details of evacuation procedures, liaison with emergency services, how the BCP kicks in, notification procedures, damage assessment/ control/mitigation, deployment of first aid and other emergency resources.

2. *Fallback procedures – Short term*

   This stage details the process by which the company starts to cope with the aftermath of the disaster, i.e. the fallback position. This may involve a physical move to alternative premises, alternative working methods, or the commencement of contractual agreements with third parties to undertake one or more of the CBPs on behalf of the company. A list of potential suppliers of such services, together with contract details is a 'must' in this regard.

3. *Business resumption – Medium term*

   This stage details the process which determines when and how the organisation gets on the road back to normal operation. Typically it will contain logistical, commercial, and political criteria which will be used by the company to decide where the business will be resumed, which CBPs will be reinstated first and where resources will be allocated to get the operations back to normality as quickly as possible.

Hierarchical planning structure

B8024 In order to co-ordinate the recovery process, particularly when some of the decision-makers are not located at the site of the disaster, many companies make use of hierarchical management planning structure:

- *The command and control team*

  This team is usually staffed by senior managers/decision-makers, supported by a business continuity professional. Their objective is to analyse the information being fed to them by the damage assessment team and to co-ordinate the overall response of the company to the disaster. It is their job to ensure that the recovery is on track; that they respond quickly and effectively to any developments in order to achieve restoration; to allocate emergency response resources; and to handle crisis communications both, internal and external.

- *The damage assessment team*

  The damage assessment team is generally a multi-disciplinary group able to convey accurate information from the disaster site to the command and control team. They are also typically responsible for on-site OSH issues, site security, access control, liaison with emergency services, providing the command and control team with regular updates on restoration progress, and general tactical decision-making.

- *Departmental team(s)*

  At some point within the BCP, those departments involved in CBPs will be instructed by the command and control team to incorporate their individual recovery plans. This may involve moving to an alternate location or developing an alternative way of working. Regular progress reports will be provided to the command and control team to enable them to keep track of the overall recovery. In this way, slippage against expected recovery timescales might be quickly identified, thus allowing additional resources to be allocated to that department, if available.

Plan development process

B8025 The process for developing the BCP follows the criteria for project management methodology, increasingly employed by many organisations. It is essential that the BCP capability is developed in accordance with the following:

- Scope the development in terms of time and resources.
- Identify what skills are required to complete the recovery.
- Set milestones to demonstrate progress.
- Define reporting lines, roles, responsibilities, accountabilities.
- Establish a project tracking method.
- Set sign-off criteria.
- Move into maintenance phase.

Plan implementation (stage 4)

B8026 The implementation process for the BCP requires the achievement of certain milestones in the following chronological order.

Securing buy-in

B8027 Hopefully by this time, buy-in by the board will have been secured, using the techniques discussed above. It is vital, however, that this approval manifests itself in some way that allows it to be readily communicated throughout the organisation. A visible commitment to the BCP is required.

Policy

B8028 Having a clearly stated BCP policy is one good way of visibly demonstrating board level commitment. The policy can be issued under the signature of the MD/CEO as an endorsement of the BCP process. This then provides a mandate to those involved to get the BCP implemented throughout the organisation.

Project authorisation

B8029 The approval of the board should also include authorisation for the BCP project as a whole. This would generally be based on the information presented via the BIA, risk identification and risk evaluation exercises.

Individual responsibilities

B8030 All key individuals should, as part of their job, have clearly defined and agreed responsibilities, accountabilities, timescales, KPIs and action plans, so as to ensure smooth plan implementation. In this way individual personal performance is geared to successful BCP implementation and an ongoing BCP capability.

Phased roll-out

B8031 We have already determined that in order to provide the organisation with the most cost-beneficial BCP capability, there is a need to focus on the CBPs. In both the development and implementation stages, it is essential that the organisation stratifies its efforts to cover and protect those functions most directly involved in the core business (primary business objective) and the CBPs before moving on to less critical functions/areas.

This stratification will normally manifest itself as a phased roll-out of the BCP to the CBPs first, intermediate functions next, and non-business critical functions last, if at all.

Specialist services

B8032 Quite apart from the emergency services, not all the resources required to implement the plan will be available in-house or on-site.

The following list is by no means exhaustive but serves to illustrate further possible components of the BCP. It is imperative that contact details for all such support services are kept readily available (in at least two known locations) and up to date:

- Data recovery/IT back-up.
- Emergency telecommunications.
- Salvage/decontamination.
- Cleaning/restoration.
- Buildings/facilities management.
- Security.
- Counselling (post-traumatic stress disorder (PTSD)).

Plan testing and maintenance (stage 5)

B8033 Maintenance and testing are both vital components in ensuring that the BCP continues to support the disaster avoidance and recovery capability required by the organisation. Maintenance is essential to avoid the plan becoming unusable because it no longer reflects the organisation's structure, vision, mission or business priorities.

The objective of testing is to provide a high degree of assurance that the plan will work when it is needed, and also to highlight and rectify any shortcomings in the plan so that the BCP process is as streamlined as possible.

Maintenance should be an ongoing process throughout the lifetime of the BCP. Its primary focus is to change the plan to continuously reflect changes within the organisation. New business lines or processes, new reporting lines, reorganisations, changes of personnel, location, telephone numbers, user IDs etc all need to be reflected in the plan.

It is strongly advisable to allocate responsibilities for plan maintenance and testing to a group of individuals and make it part of their job descriptions, CYOs (current year objectives) and KPIs.

Testing strategies

The desk check/audit

B8034 Many plans fall into disrepair due to a lack of accuracy and, consequently, integrity in the detail of the plan itself. This situation seriously undermines the effectiveness of the plan and its ability to protect the organisation in a disaster situation. The desk check/audit is a type of testing which seeks to verify that the factual detail contained in the plan is both accurate and current. As well as confirming the factual correctness of the plan, it also provides an excellent indication of how well the maintenance function is being performed.

The walkthrough test

B8035 This test involves synthesising an incident or fictional set of circumstances, thereafter allowing participants to use the plan to resolve the described situation. Similar in concept to a war game, the rules of engagement are predefined and an external facilitator co-ordinates the introduction of extra information. The logic of the plan is tested conceptionally for robustness against a scenario designed to involve all sections of the plan.

Component testing

B8036 Component testing normally involves single functions or departments testing in isolation their own parts of the plan by enacting their emergency, fallback and resumption arrangements. The objective is to test the practicality of the plan and to derive an estimate of how effective the plan is in terms of logistics and recovery times. It has the merit of eventually covering the whole organisation but it does not adequately test the linkages between functions and departments, which may more fully represent the flow of activity within the organisation.

Full simulation

B8037 The organisation deliberately shuts down or denies access to CBPs, sometimes without warning, in order to get the most realistic feel for how groups and individuals react and perform in the event of a disaster.

Whilst this type of testing provides the highest possible degree of assurance that the plan works and the organisation can cope, it needs to be planned extremely carefully, otherwise it may well trigger off an alternative, real-life, disaster!

Normally a test of this type should not be undertaken until the organisation has successfully carried out the other types of test described with consistently favourable results.

Crisis communication and public relations

B8038 In a crisis, many decisions have to be taken within a very short timescale. Some of those decisions may determine whether the organisation fails or survives! This fact represents one of the most powerful arguments for BCP and hence should ensure that the organisation has a proactive, up-to-date plan to communicate with its internal and external audiences throughout the crisis.

Employees, suppliers, customers, shareholders, the public and the media all expect and need to be kept informed in order for the organisation to retain its reputation and to maintain confidence in the business. An inability to communicate with such audiences before, during and after a crisis can easily cause a public relations disaster, thereby compounding the original incident.

Examples of companies who did not, or chose not to, communicate include:

- Hoffman-La Roche: Seveso (July 1976).
- Eli Lilly: Opren Withdrawal (August 1982).
- Union Carbide: Bhopal (December 1984).
- Delta Airlines: Plane Crash (August 1985).
- Sandoz: Rhine Pollution (November 1986).

In all cases, the share price fell within a short time of the disaster and did not get back up to the market average for at least a year after the event.

The converse of the situation faced by these companies is one where the organisation does communicate and manages to secure excellent PR from a difficult situation for example:

- British Midland: Kegworth Air Crash.
- Commercial Union: City IRA Bomb:

  'We don't make a drama out of a crisis!'

In order to minimise the risk of negative PR and the damage it can do, many companies include a communications plan within their BCP. This also enables the company to be portrayed in a positive and responsible light.

The communication plan

B8039 The communications plan should ideally comprise three stages:

Stage 1: Pre-crisis – to minimise the risk of poor communications making the disaster worse

Stage 2: During crisis – to manage the volume and complexity of enquiries and information requirements

Stage 3: After crisis – to let shareholders know that we are back in business

Pre-crisis measures

B8040

- Understanding which media are influential in your market and your location and developing a working relationship with them prior to any incident will tend to make them more sympathetic to your version of events in the reporting of a crisis.
- Press statements are an excellent way of ensuring that there is a single source of authorised information being disseminated to the media. They provide a consistent means of updating stockholders. Having them prepared in advance, for most possible scenarios, will save the company valuable time in a crisis.
- In the event of a disaster, which generally excites the interest of the media, it is likely that directors/senior managers will be approached for comment in an interview format. The physical environment and the circumstances of the interview often conspire to generate a highly charged atmosphere, where the

untrained individual may fail to portray the organisation in a good light, hence missing the opportunity to use this very powerful medium of communication to good effect. Training and rehearsal can dramatically improve individual performance on camera and enhance the organisation's ability to harness the power of the media.

- It is therefore advisable for the company to identify and train not just directors and senior managers, but any individuals who have some detailed knowledge or experience of the organisation's critical business processes. They then become the company spokespersons if problems occur within their area of expertise.
- In the event of a disaster, organisations experience a phenomenon known as call deluge. Organisations can expect to receive up to ten times as many calls and enquiries following a disaster than on a normal day.

Under such a weight of demand, the risk of providing inaccurate or misleading information is vastly increased.

One way of managing call deluge is to set up a dedicated communications office to which are channelled all disaster-related calls. In some cases a separate crisis hotline telephone number should be given out and all received calls be dealt with by trained operators.

From this centre the enquiry can be satisfactorily answered and passed on to one of the company spokespersons, if considered necessary. This approach enables the company to maintain integrity and consistency in the information being released concerning the disaster and its ongoing effects and mitigation.

During crisis

B8041

- The communications office swings into action. It is advisable to include the setting up of a dedicated communications office within the testing and maintenance section of the BCP.
- Spokespersons needed for comment and/or interview should be readily available throughout the duration of the crisis. Ideally, they should be on site, or near to the site, working in tandem with the command and control team.
- Press/media statements should be released at regular and frequent intervals to update them on damage mitigation and recovery progress.
- Key stakeholders (e.g. customers) should be updated at regular and frequent intervals with regard to the recovery process and the estimated time for the reinstatement of their service.
- As the situation develops, all staff need to be kept informed, particularly if they are required to participate in the recovery process. In any event, all staff should be aware of the likely timescales for the business getting back into full operation.
- As the recovery from the disaster takes place, a crisis event log should be kept to enable significant events, actions or milestones to be recorded with the purpose of learning from them in the aftermath of the disaster, and possibly reviewing and amending the BCP in the light of unique experiences.

After the crisis

B8042

- Use the crisis events log to review and analyse how well the organisation performed towards its agreed recovery targets. Does the BCP need amending in the light of unique experiences during actual recovery?
- Identify those aspects which worked well and those where there is room for improvement to the BCP, in order to make things work more smoothly if ever there is a next time!
- Publicise a summary of performance to all stakeholders in order to demonstrate that the BCP has actually worked in practice. This will breed confidence in the organisation and its ability to rise above and proactively manage any future potential disaster scenarios.
- And finally consider a dedicated marketing campaign to make capital of how the crisis was managed in a positive manner, and to share the lessons learnt with others operating in similar environments. This will again demonstrate the organisation's capability to cope before, during, and after any type of disaster or crisis.

Compensation for Work Injuries/Diseases

Introduction

C6001 Compensation for work injuries, diseases and death is payable both under the social security system and in the form of damages for civil wrongs (torts). This chapter examines the two systems and the interaction between them.

The social security system is a form of public insurance, funded by employers/employees and taxpayers, and benefit is payable irrespective of liability on the part of an employer, i.e. 'no fault'. However, an employment relationship must be established before such benefits can be paid. The social security system was created by legislation, such as the various Social Security Acts, the *Social Security Administration Act 1992*, the *Social Security Contributions and Benefits Act 1992* ('*SSCBA 1992*'), the *Statutory Sick Pay Act 1994*, the *Social Security* (*Incapacity for Work*) *Act 1994*, the *Social Security* (*Recoupment of Benefits*) *Act 1997*, and other legislation.

Liability in tort depends on proof of negligence or breach of statutory duty against an employer. Employers must be insured for such liability (see EMPLOYERS' LIABILITY INSURANCE). The current law relating to work injuries and diseases is to be found in a variety of Acts, such as the *Law Reform* (*Personal Injuries*) *Act 1948*, the *Employers' Liability* (*Compulsory Insurance*) *Act 1969*, the *Employers' Liability* (*Defective Equipment*) *Act 1969*, the *Damages Act 1996*, numerous health and safety regulations, made under the *Health and Safety at Work etc. Act 1974*, and a wide body of case law.

Some useful guides worth referring to are:

- *Guidelines for the assessment of general damages in personal injury cases* (2002), 6th edition (ISBN 0 19 925795 7), issued by the Judicial Studies Board. This is a publication which aims to unify judicial approaches to awards of damages. These guidelines are not a legal document and a full examination of the law applicable to each case is required.
- *Industrial injuries handbook for adjudicating medical officers*, was published in 1997. It sets out medical examination procedures and offers guidance on legislation, including discussion of case law.
- *Quantum of damages*, Kemp and Kemp, 4th edition.

Social security benefits

C6002 Under the *Social Security* (*Incapacity for Work*) *Act 1994*, and the Regulations made under that Act, the higher rate of incapacity benefit is not payable until after 364 days. The effect that this has had on industrial injuries benefits, including disablement benefit, is that industrial benefits continue to be payable after 90 days, but the benefits which are payable before and at the same time, that is, any national insurance benefits, are less for the relevant period. The higher adult dependant's allowance is delayed for 364 days when the long-term incapacity benefit increases begin. Claimants whose industrial injuries benefit is lower than the rate of incapacity for work benefit and who are thus entitled to claim this benefit are also affected. (See C6015–C6017 below).

Entitlement to industrial injuries benefits requires compliance with national insurance contribution conditions for incapacity benefit. The incapacity for work tests under the *Social Security (Incapacity for Work) Act 1994* do not apply to industrial disablement benefits. The medical adjudication procedures for these benefits continue with some modifications as before. The incapacity for work tests will apply to sufferers from industrial disablement if they apply for incapacity benefit before the 90 qualifying days for industrial disablement benefits begin, or if they do not qualify for industrial disablement benefit under the percentage or other rules, or if incapacity benefit is payable at a higher rate than industrial disablement pension rates and incapacity benefit is chosen by the claimant.

Under the rules for claimants for incapacity benefit, the employee has to supply information and evidence of sickness and be prepared to submit to a medical examination to decide whether he is fit to work. If he has worked for more than eight weeks in the twenty-one weeks immediately preceding the first day of sickness, a test relevant to his incapacity to do work which he could reasonably be expected to do in the course of his occupation applies. This test criterion continues to the 197th day of incapacity. After that date, and for all claimants who do not qualify for the required period, an 'all work' test arises. Regulations provide for prescribed activities and a person's incapacity, by reason of a specific disease, or bodily or mental disablement, to perform these activities. These tests are detailed and are outside the scope of this publication. [*SSCBA 1992, ss 171(A)–(G)* as inserted by *Social Security (Incapacity for Work) Act 1994, ss 5–6*]. The *Social Security (Incapacity for Work) (General) Regulations 1995 (SI 1995 No 311)* (as amended by the *Social Security (Incapacity for Work) (General) Regulations 1996 (SI 1996 No 484)*), provide that certain people are deemed to be incapable of work, including those suffering from a severe condition as defined in the Regulations or those receiving certain regular treatment (such as chronic renal failure, hospital in-patients and those suffering from an infectious or contagious disease). 'Welfare to work beneficiaries' will also be treated as incapable of work in the circumstances prescribed by *Regulation 13A* of the 1995 Regulations.

A pregnant woman may also be deemed incapable of work if there is a serious risk to her health, or that of her unborn child, if she does not refrain from work, in the case of the 'own occupation' test; or in the case of the 'all work' test, if she does not refrain from work in any occupation. If she has no entitlement to maternity allowance or statutory maternity pay and the actual date of confinement has been certified, she is deemed to be incapable of work beginning with the first day of the sixth week before the expected week of confinement until the fourteenth day after the actual date of confinement and during this period she will therefore be entitled to claim incapacity benefit.

Whilst most actual work by a claimant disqualifies him from receiving this benefit on any day of such work, certain work does not stop benefit. [*Social Security (Incapacity for Work) (General) Regulations 1995 (SI 1995 No 311), Reg 17*]. Earnings from such work must not exceed £67.50 per week and in most cases must be for less than 16 hours per week. The exempt work is work done on the advice of a doctor which:

- helps to improve, or to prevent or delay deterioration in the disease or bodily or mental disablement which causes that person's incapacity for work; or
- is part of the treatment programme undertaken as a hospital in-patient or out-patient under medical supervision; or

- is done when the claimant is attending a sheltered workshop for people with disabilities.

Voluntary work and duties as a member of a disability appeal tribunal or the Disability Living Allowance Advisory Board are also exempt work.

Industrial injuries benefit

C6003 The benefits which are available to new claimants for industrial injuries benefits are:

- industrial injuries disablement pension;
- constant attendance allowance; and
- exceptionally severe disablement allowance.

Disabled person's tax credit may also be claimed in some circumstances (see C6016 below). Some older benefits continue to be payable to recipients who were receiving them when they were otherwise abolished, or whose entitlement arose before the relevant dates. (For details of these older benefits and more detailed social security benefits information, see *Tolley's Social Security and State Benefits Handbook*.)

Accident and personal injury provisions

C6004 The employed earner must have suffered personal injury caused after 4 July 1948 by an accident arising out of and in the course of his employment, being employed earner's employment. [*SSCBA 1992, s 94*]. The *Reporting of Injuries, Diseases and Dangerous Occurrences Regulations 1995* (*SI 1995 No 3163*) contain detailed provisions which require reports to be made on prescribed forms to the Health and Safety Executive after the occurrence of an accident or on receipt of a report from a registered medical practitioner of his diagnosis of a prescribed disease. A self-employed person may arrange for this report to be sent by someone else. Records must be kept for three years containing prescribed details of accidents, or the date of diagnosis of the disease, the occupation of the person affected and the nature of the disease. For more information on these Regulations see ACCIDENT REPORTING AND INVESTIGATION.

Industrial accident and disease records may be kept for three years on:

- a B510 Accident Book;
- photocopies of completed form F2508; or
- computerised records.

Personal injury caused by accident

C6005 'Personal injury' includes physical and mental impairment, a hurt to body or mind, which includes nervous disorders or shocks (R(I) 22/52; R(I) 22/59). The injury must have been caused by an accident. Although this is usually an unintended and unexpected occurrence, such as a fall, if a victim is injured by someone else, may also be considered to be an accident (*Trim Joint District School Board v Kelly [1914] AC 667*). A relevant accident may still have occurred where a series of accidents without separate definite times cause personal injury. An office worker was held to have suffered a series of accidents on each occasion she had been obliged to inhale her colleagues' tobacco smoke and this was held to have caused personal injury.

'Accident' must be distinguished from 'process', that is, bodily or mental derangement not ascribable to a particular event. Injuries to health caused by processes are not industrial injuries, unless they lead to prescribed industrial diseases (see C6010 below). 'There must come a time when the indefinite number of so-called accidents and the length of time over which they occur, take away the name of accident and substitute that of process' (*Roberts v Dorothea Slate Quarries Co Ltd (No 1) [1948] 2 All ER 201*). In *Chief Adjudication Officer v Faulds [2000] 1 WLR 1035*, the House of Lords, in disallowing a claim for damages for psychological injury suffered by a fireman, held there must be at least one identifiable accident that caused the injury. The fact that an employee might develop stress from a stressful occupation would not satisfy the definition of 'accident'.

Accident arising out of and in the course of employment

C6006 There is no need to show a cause for the relevant accident provided that the employee was working in the employer's premises at the time of the accident. An accident arising in the course of employment is presumed to have arisen out of that employment, in the absence of any evidence to the contrary.

What runs through all the case law is the common requirement giving rise to industrial injuries rights that the employee was doing something reasonably incidental to and within the scope of his employment, including extra-mural activities which the employee has agreed to do (R(I) 39/56). A male nurse who was injured in a football match watched by patients in the hospital grounds succeeded in a claim for benefit as this was reasonably incidental to and within the scope of his employment (R(I) 3/57), but a policeman who was injured whilst playing football for his force was unable to recover benefit despite the fact that his employers had encouraged him to play in the game (*R v National Insurance Commissioner, ex parte Michael [1977] 1 WLR 109*). Also, in *Faulkner v Chief Adjudication Officer [1994] PIQR P244*, a police officer who was injured whilst playing for a police football team was not entitled to industrial injuries benefit despite the benefit to the community resulting from his participation. He was not on duty at the time. Further guidance can be found in *Chief Adjudication Officer v Rhodes [1999] ICR 178* where it was held that the two main questions to be asked are:

- what are the employee's duties; and
- was he discharging them at the time of the accident.

Supplementary rules

C6007 Five statutory provisions establish rules under which the employee is deemed to be acting in the course of his employment duties. If the occurrence falls within these rules, the employee will be covered by industrial injuries benefit. These provisions are as follows:

- Illegal employment – if the employee was not lawfully employed, or his employment was actually void because of some contravention of employment legislation. [*SSCBA 1992, s 97*].
- Acting in breach of regulations, or orders of the employer – if the employee is not acting outside his authority under his employment duties, and an accident occurs while the employee is doing something for the purposes of, and in connection with, the employer's business [*SSCBA 1992, s 98*], he will be covered. For instance, a kitchen porter hung up his apron to dry in a recess near to the ovens where he was forbidden to go. He was injured when he fell into a shallow pit. The hanging up of the apron was for the purposes of his

employment, so the accident was deemed to have arisen out of, and in the course of, his employment duties (R(I) 6/55).

- Where a dock labourer was employed on loading a ship by the method of two slings, but he instead used a truck which he had not been authorised to use for this purpose, he was held to be acting in the course of his employment. (R(I) 1/70B). This contrasts with the earlier case of *R v D'Albuquerque, ex parte Bresnahan [1966] 1 Lloyd's Rep 69*, where a dock labourer was killed in an accident whilst driving a forklift truck, which he had no authority or permission to use to remove an obstruction. His widow was unable to recover industrial injuries benefit as her husband was held not to have been acting in the course of his employment.
- Travelling in an employer's transport – travel to and from work is not covered except where the employee is travelling in transport provided by the employer with his express or implied permission, whether or not the employee was bound to travel in this transport. [*SSCBA 1992, s 99*]. Outside this express provision, the employee will, in most cases, both be required to be on the employer's premises doing what he was authorised to do, unless his work takes him off the premises. A postman was able to recover benefit when he was bitten by a dog on the street, as his job required him to walk along streets (R(I) 10/57). If the journey is preparatory to the start of timed itinerant duties, for example, as a home help, there will be no entitlement to benefit in respect of injury sustained on the way to the first home, though if the employee has more discretion about his movements, he may be entitled to benefit on the way to his first call.
- An injury incurred while an employee is trying to prevent a danger to other people, or serious damage to property during an emergency. [*SSCBA 1992, s 100*].
- Accidents caused by another's misconduct, boisterousness or negligence (provided that the claimant did not directly induce or contribute to the accident by his own conduct), the behaviour of animals (including birds, fish and insects). If these cause an accident, or if a person is struck by lightning or by any object, respectively confer entitlement to industrial injuries benefit. [*SSCBA 1992, s 101*].

Relevant employment

C6008 'Employed earner's employment' includes all persons who are gainfully employed in Great Britain under a contract of service, or as an office holder, and who are subject to income tax under Schedule E. Self-employed people and private contractors are thus not entitled to industrial injuries benefits.

Certain classes of person are expressly included for the purposes of industrial injuries benefits. They include unpaid apprentices, members of fire brigades (or other rescue brigades), first-aid, salvage or air raid precautions parties, inspectors of mines, special constables, certain off-shore oil and gas workers and certain mariners and air crew. Most trainees on Government training schemes are excluded from the scheme. [*SSCBA 1992, ss 2, 95*].

If an industrial accident occurs outside Great Britain, industrial injuries benefit has since 1 October 1986 been payable when the employee returns to Great Britain, provided the employer is paying UK national insurance contributions, or if the claimant is a voluntary worker overseas and is himself paying UK contributions. Accidents which occur, or prescribed diseases which develop in other EU countries, are covered by common rules. An employee who is entitled to make a claim for industrial benefit in another EU member state should make his claim for benefit in that state, regardless of the country which will actually pay the benefit.

Certain types of employment are specifically excluded from cover. [*SSCBA 1992, s 95; Employed Earners' Employments for Industrial Injuries Purposes Regulations 1975* (*SI 1975 No 467*), *Regs 2–7* and *Sch 1, 2*].

Persons treated as employers

C6009 The *Employed Earners' Employments for Industrial Injuries Purposes Regulations 1975* (*SI 1975 No 467*), *Sch 3* also provide for cases where certain people who may not have a contract with the employee, or who may be an agency employer, to be the relevant employer for the person who has suffered the industrial injury, or who has developed a prescribed disease. An agency that supplies an office cleaner, or a typist, will be the relevant employer. For casual employees of clubs, the club will be the relevant employer.

Benefits for prescribed industrial diseases

C6010 The rules for certain prescribed diseases (namely deafness, asthma and asbestos related diseases), differ in some respects from the provisions affecting prescribed diseases outlined below. (See C6055–C6059 below.) The different rules applicable to diseases resulting from exposure to asbestos at work are dealt with for industrial injuries purposes by a special medical board and there is a separate state scheme for statutory compensation where one of these diseases develops and there is no remedy against an employer, or entitlement to state benefits (see C6061 below). (See leaflets NI 12, NI 207 and NI 272 for help with making a claim.) To obtain the right to industrial injuries benefits for all other prescribed diseases, the claimant must show:

- that he is suffering from a prescribed disease;
- that the disease is prescribed for his particular occupation (where a disease is prescribed for a general activity, for example, contact with certain substances, he must clearly show that this was more than to a minimal extent); and
- that he contracted the disease through engaging in the particular occupation (there is a presumption that if the disease is prescribed for a particular occupation, the disease was caused by it, in the absence of evidence to the contrary).

(See also OCCUPATIONAL HEALTH AND DISEASES.)

Claims are made on Form BI 100B and there is a right to, and it is advisable to, claim immediately after the disease starts. The 90-day waiting period for receipt of benefit applies as for accidents, as do the percentage disabilities and aggregated assessment rules, except in the case of loss of faculty resulting from diffuse mesothelioma when entitlement begins on the first day of the claim.

Assessments are made by two doctors who will decide on the percentage disability and how long the disability will last. Benefit will then be payable for the period stated in the assessment, but if the doctors are not sure of the period, benefit will be paid for a while with a further review. If the disease recurs during that period, there will be no need to make a further claim, but if the condition has worsened, the assessment may be reviewed. If there is a further attack after the period of the assessment, a fresh claim will have to be made, which will be subject to a further 90-day waiting period.

Industrial injuries disablement benefit

Entitlement and assessment

C6011 A person is entitled to an industrial injuries disablement pension if:

- he suffers from a prescribed industrial disease (see C6010 above);
- he suffers as a result of the relevant accident, a loss of physical or mental faculty such that the assessed extent of the resulting disablement amounts to not less than 14 per cent [*SSCBA 1992, S 103, Sch 6*]. See leaflet NI6 (July 1999) (update);
- 90 days (excluding Sundays) have elapsed since the date of the accident or onset of the prescribed disease or injury.

An assessment of the percentage disablement up to 100 per cent will be made by an adjudicating medical practitioner who, in the case of accidents, looks at the claimant's physical and mental condition, comparing him in those respects with those of a normal person of the same age and sex. No other factors are relevant. The assessment may cover a fixed period/or the life of the claimant. The degrees of disablement are laid down in a scale so that, for example, loss of one hand is normally 60 per cent and loss of both hands 100 per cent. Disfigurement is included even if this causes no bodily handicap.

Where the claimant suffered from a pre-existing disability before the happening of the industrial accident at the onset of the industrial disease, benefit will only be payable in respect of the industrial accident or disease itself, and the medical adjudicators will compare the original disability with the industrial disability for this purpose. Where one or more disabilities result from industrial accidents or diseases, the level of resulting disability may be aggregated, but not so as to exceed the 100 per cent disability and its corresponding rate of benefit. [*Social Security (General Benefit) Regulations 1982 (SI 1982 No 1408), Reg 11*].

A list of initial application forms is supplied in NI 6 (July 1999) Industrial Injuries Disablement Benefit. Once a claim for industrial injuries benefit has been made, the Secretary of State will send Form B1/76 to the claimant. This form asks for full details of the accident or disease. The claimant must then submit to examination by at least two adjudicating medical practitioners and must agree to follow any appropriate medical treatment. If he fails to comply with these requirements, he will be disqualified from receiving this benefit for six months. A decision in writing will be sent to the claimant. It is possible to appeal a refusal of industrial benefit; details of how to appeal will be sent with the decision if it is negative.

This benefit can only be paid 90 days after the date of the accident excluding Sundays. [*SSCBA 1992, s 103*].

Claims in accident cases

C6012 Claims should be made on Form BI 100A obtainable from the post office, job centre, or from the Benefits Agency, by claimants disabled by the accident for nine weeks. Claims should be made within six months of the accident to avoid loss of benefit, as the usual period for which benefit will be backdated is three months. [*Social Security (Claims and Payments) Regulations 1987 (SI 1987 No 1968), Reg 19, Sch 4, para 3*]. It is possible to apply to the DSS for a declaration from an adjudicating officer that an accident is covered by the scheme if a claimant suspects that an accident at work may have lasting effects which may not become apparent until some time later. If a positive finding is made, this will bind the DSS on any later claim for benefit.

[*Social Security Act 1998, s 29(4)*]. Form BI 95 should be used for making this application. (Claims in respect of prescribed diseases are mentioned at C6010 above; appeals at C6018 below.)

Rate of industrial injuries disablement benefit

C6013 Benefit is paid at one of two rates with the higher being paid to claimants aged over 18.

Constant attendance allowance and exceptionally severe disablement allowance

C6014 Constant attendance allowance is available if:

- the claimant is receiving industrial injuries disablement benefit based on 100 per cent disablement, or aggregated disablements that total 100 per cent or more;
- the claimant needs constant care and attention as a result of the effects of an industrial accident or disease.

Furthermore, if the carer of the claimant spends at least 35 hours a week looking after him and is of working age and is not earning more than £75 (before tax but after deduction of national insurance and other reasonable expenses) per week, invalid care allowance of £43.15 per week with allowances for the carer's own dependants (if any) may be granted to that person. Claims for industrial injuries benefit are made on Form BI 104. It is granted for a fixed period and may be renewed from time to time. There are four rates of payment:

- part-time (where full-time care is unnecessary);
- normal maximum rate (if the above conditions are fulfilled and full-time care is required);
- intermediate (if the claimant is exceptionally disabled and the degree of attendance required is greater, and the care necessary is greater than under the 'normal' classification, the benefit is limited to one and a half times the normal rate); and
- an exceptional rate (if the claimant is so exceptionally disabled as to be entirely dependent on full-time attendance for the necessities of life).

If the intermediate or exceptional rates are payable, an additional allowance (known as exceptionally severe disablement allowance) will be due if the condition is likely to be permanent. Constant attendance allowance may continue to be paid for up to four weeks if the claimant goes into hospital for free medical treatment.

Other sickness/disability benefits

C6015 However, most applicants for industrial injuries benefit will, as their incapacity results from accidents or diseases contracted during their employment, be able to claim statutory sick pay or incapacity benefit up to the date of commencement of industrial injuries benefit as 90 days is less than 28 weeks. The rate of statutory sick pay is higher than that for incapacity benefit for 28 weeks when they are then paid at the same rate. Incapacity benefit rates in the first 28 weeks apply to the self-employed and to the unemployed, or those with too few contributions. For the purposes of claiming incapacity benefit before being entitled to claim a disablement benefit, it should be noted that the presumption that a claimant for benefit for industrial or prescribed diseases has satisfied the contribution requirements for incapacity benefit (as was the case with sickness benefit), has been removed. An age addition is added to incapacity benefit where the claimant is under certain age limits. Claims for incapacity benefit for the first time claimant should be made on Form BI 202. Where entitlement to industrial injuries benefit is below the incapacity benefit rates, then that benefit may be claimed to top up industrial injuries benefit to the current rate of incapacity benefit.

It is possible to claim extra for a spouse or person looking after children if the spouse is over 60, or, if younger, child benefit is being paid, and the claimant was maintaining the family to at least the extent of the dependency benefit being claimed. None of these restrictive rules apply to industrial injuries benefit claims.

Income support and other state benefits such as housing benefit and council tax benefit may be available if the claimant with or without dependants does not have sufficient to live on.

Disability Tax Credit under Working Tax Credit

C6016 This benefit is for people aged 16 years or over who wish to work, but have a physical or mental disability which puts them at a disadvantage in securing a job under criteria set out in the regulations. The applicant must work for at least 16 hours a week to qualify and he receives a credit if he works for 30 hours or more a week. Self-employed people may qualify For a full analysis of the disability tax credit see: www.ircc.inrev.gov/taxcredits.

Payment

C6017 For the financial year 2003/2004 the weekly benefit rates payable for Statutory sick pay, Incapacity benefit, Industrial injuries disablement benefit, Constant attendance allowance and Exceptionally severe disablement are set by the *Social Security Benefits Up-rating Order 2003* (*SI 2003 No 526*).

Appeals

C6018 Appeals relating to all industrial injuries are made to an appeal tribunal and should be made within one month from the date that the decision maker sends the decision to the applicant.

Change of circumstances and financial effects of receipt of benefit

C6019

- *Hospital*

  If a claimant enters hospital, industrial injuries disablement pension continues to be payable, as does exceptionally severe disablement allowance. Constant attendance allowance will stop after four weeks. Statutory sick pay will be reduced after six weeks.

- *Taxation*

  Industrial injuries disablement benefit, constant attendance allowance, exceptionally severe disablement allowance and disabled person's tax credit are not taxable. Incapacity benefit is taxable for all claimants who have claimed after

13 April 1995. Those who were in receipt of invalidity benefit before that date are not liable to tax on it. Statutory sick pay is taxable under the PAYE system.

Benefit overlaps

C6020 Industrial injuries benefits may be taken at the same time as incapacity for work benefit with no reduction. Disability living allowance and attendance allowance may not be received at the same time as constant attendance allowance, but if the claimant should receive a higher rate of either of those benefits, constant attendance allowance will be topped-up to bring it up to the higher benefit rate. (For recoupment of benefit after awards of damages see C6044 below.)

Damages for occupational injuries and diseases

C6021 When a person is injured or killed at work in circumstances indicating negligence or breach of duty on the part of an employer, he may be entitled to an award of damages. Damages are assessed by judges in accordance with previously decided cases; very exceptionally they may be assessed by a jury. Damages are also categorised as general and special damages, according to whether they reflect pre-trial or post-trial losses. Calculation of damages is often made by reference to Kemp and Kemp, *The quantum of damages*, and to the Judicial Studies Board's document, *Guidelines for the assessment of general damages in personal injury cases* (2002) (6th edition).

Damages normally take the form of a lump sum; however, 'structured settlements', whereby accident victims are paid a variable sum for the rest of their lives, are now a viable alternative [*Damages Act 1996, s 2*] (see C6025 below). This may well involve greater reliance on actuarial evidence and a rate of return of interest provided by index-linked government securities.

Legal aid is currently the subject of extensive Government reform. The Legal Services Commission (LSC) replaced the Legal Aid Board in April 2000. Legal aid is no longer available in negligence personal injury cases. It is now only available in non-negligence personal injury claims, i.e. where the injury has been intentionally as opposed to negligently caused. There is one further exception, legal aid may be available where the personal injury claim forms a small part of another claim for which legal aid is available. Legal aid is still available for clinical negligence. The Access to Justice Act 1999 introduced the conditional fee system whereby solicitors represent clients on a 'no win no fee' basis. In the absence of legal aid, such a system will be the way in which most personal injury actions are brought.

Accident Line, which is part of a scheme run by solicitors who are members of a specialist panel of personal injuries lawyers, provides a free half-hour consultation for claimants who have suffered personal injuries, including industrial accidents. The telephone number is 0500 192 939. In addition, the Association of Personal Injuries Lawyers has a helpful website at: www.apil.com.

Basis of claim for damages for personal injury at work

C6022 The basis of an award of damages is that an injured employee should be entitled to recoup the loss which he has suffered in consequence of the injury/disease at work. 'The broad general principle which should govern the assessment in cases such as this is that the court should award the injured party such a sum of money as will put him in the same position as he would have been in if he had not sustained the injuries' (per Earl Jowitt in *British Transport Commission v Gourley [1956] AC 185*). The *Damages Act 1996*, provides that periodical payments may be made in some cases.

It is not necessary that a particular injury be foreseeable, although it normally would be (see further *Smith v Leech Brain & Co Ltd [1962] 2 QB 405*). Moreover, if the original injury has made the claimant susceptible to further injury (which would not otherwise have happened), damages will be awarded in respect of such further injuries, unless the injuries were due to the negligence of the claimant himself (*Wieland v Cyril Lord Carpets Ltd [1969] 3 All ER 1006*). In this case, a woman, who had earlier injured her neck, was fitted with a surgical collar. She later fell on some stairs, injuring herself, because her bifocal glasses had been dislodged slightly by the surgical collar. It was held that damages were payable in respect of this later injury by the perpetrator of the original act of negligence. Conversely, where, in spite of having suffered an injury owing to an employer's negligence, an employee contracts a disease which has no causal connection with the earlier injury, and the subsequent illness prevents the worker from working, any damages awarded in respect of the injury will stop at that point, since the supervening illness would have prevented (and, indeed, has prevented) the worker from going on working.

C6023 Listed below are the various losses for which the employee can expect to be compensated. Losses are classified as non-pecuniary and pecuniary.

- *Non-pecuniary losses*

  The principal non-pecuniary losses are:

  (i) pain and suffering prior to the trial;

  (ii) disability and loss of amenity (i.e. faculty) before the trial;

  (iii) pain and suffering in the future, whether permanent or temporary;

  (iv) disability and loss of amenity in the future, whether permanent or temporary;

  (v) bereavement.

  (Damages for loss of expectation of life were abolished by the *Administration of Justice Act 1982, s 1(1)* (see further 'loss of amenity' at C6034 below).)

  There are four main compensatable types of injury, namely:

  (i) maximum severity injuries (or hopeless cases), e.g. irreversible brain damage, quadraplegia;

  (ii) very serious injuries but not hopeless cases, e.g. severe head injuries/loss of sight in both eyes/injury to respiratory and/or excretory systems;

  (iii) serious injuries, e.g. loss of arm, hand, leg;

  (iv) less serious injuries, e.g. loss of a finger, thumb, toe etc.

  There is a scale of rates applicable to the range of disabilities accompanying injury to workers but it is nowhere as precise as the scale for social security disablement benefit. Damages for maximum severity cases can vary from several hundred thousand pounds to millions of pounds. In *Biesheuval v Birrell [1999] PIQR Q40*, the High Court awarded total damages of £9,200,000 (see C6024 below) to a student who was almost completely paralysed in all four limbs after a car crash. In *Dashiell v Luttitt [2000] 3 QR 4* a settlement of £5,000,000 was reached for brain damage sustained by a child aged 14 at the time of a school minibus crash. In *Capocci v Bloomsbury Health Authority (21 January 2000)* the High Court awarded £2,275,000 damages to a 13-year-old boy who had been asphyxiated at birth and as a result suffered cerebral palsy and other severe physical and mental handicaps. Less serious injuries attract lower damages – for example, in *Williams v Gloucestershire County Council (10 September 1999)* an out of court settlement of £3,639 was reached in a case where a 6-year-old lost the top of her little finger in an accident.

- *Pecuniary losses*

  These consist chiefly of:

  (i) loss of earnings prior to trial (i.e. special damages);

  (ii) expenses prior to the trial, e.g. medical expenses;

  (iii) loss of future earnings (see below);

  (iv) loss of earning capacity, i.e. the handicap on the open labour market following disability.

  In actions for pecuniary losses, employees can be required to disclose the general medical records of the whole of their medical history to the employer's medical advisers (*Dunn v British Coal Corporation [1993] ICR 591*).

General and special damages

C6024

- *General damages*

  General damages are awarded for loss of future earnings, earning capacity and loss of amenity. They are, therefore, awarded in respect of both pecuniary and non-pecuniary losses. An award of general damages normally consists of:

  (i) damages for loss of future earnings;

  (ii) pain and suffering (before and after the trial); and

  (iii) loss of amenity (including disfigurement).

- *Special damages*

  Special damages are awarded for itemised expenses and loss of earnings incurred prior to the trial. Unlike general damages, this amount is normally agreed between the parties' solicitors. When making an award, judges normally specify separately awards for general and special damages. A statement of special damage must be served with the statement of claim which should suffice to give the defendant a fair idea of the case he has to answer. More detailed information must be supplied after the exchange of medical and expert reports.

Example

An example of the way in which damages awards are assessed and broken down is the case of *Biesheuval v Birrell* (referred to in C6023 above) where the High Court awarded £9,200,000 damages consisting of:

Pain and suffering and loss of amenity	£137,000
Interest on general damages	£6,617
Past loss of earnings	£80,700

Pain and suffering and loss of amenity	£137,000
Interest on past loss of earnings	£14,929
Other special damages	£360,113
Interest on special damages	£54,516
Tax on interest	£41,215
Loss of future earnings	£3,700,000
Loss of pension rights	£67,491
Initial capital expenditure	£551,803
Recurring costs and future care	£4,267,000

Structured settlements

C6025 A structured settlement is an agreement for settling a claim or action for damages on terms that the award is made wholly or partly in the form of periodic payments. Such settlements are expected to become more usual in the case of larger awards of damages. Under the *Damages Act 1996*, these periodic payments must be payable in the form of an annuity for life, or for a specified period, and may be held on trust for the claimant if that should be necessary. Provision may be added to the settlements for increases, percentages or adjustments where the court or claimant's advisers secure these variations in his interest. Structured settlements in favour of claimants may be made for the duration of their life. Knowledge of the claimant's special needs is thus vital for the structure to be successful – it should be recognised that structured settlements will not be suitable for all cases. If the claimant dies while in receipt of periodic payments, they pass under his estate. Structured settlements may also be made in awards of damages in respect of fatal accidents. Tax-free annuities are payable directly to the claimant by the Life Office.

Structured settlement awards made by the Criminal Injuries Compensation Authority are also tax free.

When agreeing a settlement (whether structured or not), it is better to agree whether payments are net of repayable benefits. If a settlement offer is silent as to repayable benefits, then a deduction will have to be made in respect of them, possibly with unplanned results for the claimant.

Structured settlements are advisable where brain damage makes the injured person at risk and suggestible to pressure from relatives. They are also useful where the injured person has little experience or interest in investment, or dislikes the possibility of becoming dependent on the State or relatives should funds run out. Disadvantages are that annuities only last for the lifetime of the injured person. There is also a loss of flexibility to deal with changed circumstances and of the better return gained with the skilful investment of a lump sum.

Assessment of pecuniary losses

C6026 Assessing pecuniary losses, i.e. loss of future earnings, can be a difficult process. As was authoritatively said, 'If (the claimant) had not been injured, he would have had the prospect of earning a continuing income, it may be, for many years, but there can be no certainty as to what would have happened. In many cases the amount of that income may be doubtful, even if he had remained in good health, and there is always the possibility that he might have died or suffered from some incapacity at any time. The loss which he has suffered between the date of the accident and the date of the trial [i.e. special damages (see above)] may be certain, but his prospective loss is not. Yet damages must be assessed as a lump sum once and for all [see 'Provisional awards' at C6040 below], not only in respect of loss accrued before the trial but also in respect of a prospective loss' (per Lord Reid in *British Transport Commission v Gourley [1956] AC 185*). Moreover, if, at the time of injury, a worker earns at a particular rate, it is presumed that this will remain the same. If, therefore, he wishes to claim more, he must show that his earnings were going to rise, for example, in line with a likely increase in productivity – a probable rise in national productivity is not enough.

When the court assesses loss of earnings, the claimant has to mitigate his loss by taking work if he can. In *Larby v Thurgood [1993] ICR 66*, the defendant applied to the court to dismiss an action brought by a fireman who was severely injured in a road traffic accident and who had taken employment as a driver earning £6,000 per annum, unless he agreed to be interviewed by an employment consultant who would then give expert evidence on whether the claimant could have obtained better paid employment. This application was refused. Evidence of whether the claimant could have obtained better paid employment depended partly on medical evidence of his capabilities and the present and future state of the job market where he lived, which could be established by an employment consultant. His general suitability for employment, his willingness and motivation, were matters of fact for the judge; thus expert opinion was not required for that purpose.

A claimant may be earning practically as much as he was before his accident, but may be more at risk of losing his present job, of not achieving expected promotion, or of disadvantages in the labour market. Damages may be claimed for such prospective losses and are known as *Smith v Manchester* damages (after *Smith v Manchester City Council (1974) 118 SJ 597*).

Capitalisation of future losses

C6027 Loss of future earnings, often spanning many years ahead, is awarded normally as a once-and-for-all capital sum for the maintenance of the injured victim. The House of Lords considered the way in which lump sums should be calculated in the leading case of *Wells v Wells*; *Thomas v Brighton Health Authority [1999] 1 AC 345*. This also applies in Scotland (*McNulty v Marshall's Food Group Ltd [1999] SC 195*).

The award of damages is calculated on the basis of the present value of future losses – a sum less than the aggregate of prospective earnings because the final amount has to be discounted (or reduced) to give the present value of the future losses. Inflation is ignored when assessing future losses in the majority of cases (e.g. pension rights), since this was best left to prudent investment (*Lim Poh Choo v Camden and Islington Area Health Authority [1980] AC 174*). And where injury shortens the life of a worker, he can recover losses for the whole period for which he would have been working (net of income tax and social security contributions, which he would have had to pay), if his life had not been shortened by the accident. The present value of future losses can be gauged from actuarial or annuity tables. The net annual loss (based on rate of earnings at the time of trial) (the multiplicand) has to be multiplied by a suitable number of years (multiplier) which takes account of factors such as the claimant's life expectancy and the number of years that the disability or loss of earnings is expected to last. The multiplier normally ranges between 6 and 18, and is set out in actuarial tables. Both the multiplier and the multiplicand can vary for different periods; for example, where medical evidence shows that the need for care could increase or decrease over time – *Wells v Wells*; *Thomas v Brighton Health Authority [1998] 3 All ER 481*. In *McIlgrew v Devon County Council [1995] PIQR Q66*, the maximum multiplier of 18 was applied for permanent general losses, and the multiplier of 12 was applied to loss of earnings.

The multiplier is calculated on the assumption that the claimant will invest the lump sum prudently. Under *section 1(1)* of the *Damages Act 1996*, the Lord Chancellor may by order prescribe a rate of return which the courts must have regard to. Under this provision, the Lord Chancellor has set the rate of return at 2.5 per cent.

Example

In the case of a male worker, aged 30 at the date of trial, and earning £15,000 per year net, on a 2.5 per cent interest yield, the multiplier will be 22.80 (using Table 25 of the Ogden Tables). Hence, general damages will be about £342,000, assuming incapacity to work up to age 65.

In the case of a male worker, aged 50 at date of trial, earning £25,000 net, on a 2.5 per cent interest yield, the multiplier will be 12.06 (using Table 25 of the Ogden Tables). Hence, general damages will be about £301,500 assuming incapacity to work until age 65.

Deductions from awards under this head are made for the actual earnings of the injured claimant. Where the claimant takes a lighter, less well paid job, and thus suffers a loss of earnings, the courts have held that the fact that he gains more leisure through working shorter hours is not to be taken into account to reduce the amount of damages awarded for loss of earnings (*Potter v Arafa [1995] IRLR 316*).

Institutional care and home care

C6028 Expenses of medical treatment may be claimed, except where the victim is maintained at public expense in a hospital or a local authority financed nursing home, in which case any saving of income during his stay will be set off against the claim for pecuniary loss. Alterations to a home, purchase of a bungalow accessible to a wheelchair, adaptations to a car and equipment (such as lifting equipment), which are necessary for care, may be claimed. Nursing and care requiring constant or less attendance may be claimed whether or not the carer is a professional or voluntary carer. It was confirmed by the House of Lords in *Hunt v Severs [1994] AC 350*, that where an injured claimant is cared for by a voluntary carer, such as a member of his family, that damages could be recovered for this care, but the claimant should hold them in trust for the voluntary carer.

Where voluntary care is undertaken by relatives, compensation for the cost of this care is assessed as a percentage of the Crossroads rate agreed from time to time for community care by most local authorities; the actual percentage awarded being about two-thirds of that rate. A higher percentage of that rate will be awarded where the care being given is beyond the level of care normally provided by home helps (*Fairhurst v St Helens and Knowsley Health Authority [1995] PIQR 41*).

Damages for lost years

C6029 Damages may be payable up to retirement age for lost earnings resulting from the shortening of the claimant's life expectancy by reason of the injury or disease. Estimated costs of living expenses are deductible from these damages. Damages under this head may also be awarded to dependants if the victim has died.

Non-pecuniary losses (loss of amenity)

C6030 It is generally accepted by the courts that quantification of non-pecuniary losses is considerably more difficult than computing pecuniary losses. This becomes even more difficult where loss of sense of taste and smell are involved, or loss of reproductive or excretory organs. Unlike pecuniary losses, loss of amenity generally consists of two awards: an award for (i) actual loss of amenity and (ii) the impairment of the quality of life suffered in consequence (i.e the psychic loss).

Victims are generally conscious of their predicament, but in very serious cases they may not be, a distinction underlined in the leading case of *H West & Son Ltd v Shephard [1964] AC 326*. If a victim's injuries are of the maximum severity kind (e.g. tetraplegia) and he is conscious of his predicament, damages will be greater. However, where, as is often the case, the injuries shorten the life of an accident victim, damages for non-pecuniary losses will be reduced to take into account the fact of shortened life.

Types of non-pecuniary losses recoverable

C6031 The following are the non-pecuniary losses which are recoverable by way of damages:

- pain and suffering;
- loss of amenity;
- bereavement.

Pain and suffering

C6032 Pain and suffering refers principally to actual pain and suffering at the time of the injury and later. Since modern drugs can easily remove acute distress, actual pain and suffering is not likely to be great and so damages awarded will be relatively small.

Additionally, 'pain and suffering' includes 'mental distress' and related psychic conditions; more specifically (i) nervous shock, (ii) concomitant pain or illness following post-accident surgery and embarrassment or humiliation following disfigurement. Claustrophobia and fear are within the normal human emotional experience but even are not compensatable, unless amounting to a recognised psychiatric condition, such as post-traumatic stress syndrome (*Reilly v Merseyside Regional Health Authority [1995] 6 Med LR 246*).

In *Heil v Rankin and Another and joined appeals [2001] QB 272*, the Court of Appeal, reviewed the current awards for pain, suffering and loss of amenity. The court held that whilst payments under £10,000 should not increase, there would be a tapered increase for awards above that up to a maximum of 33 per cent for the highest level of damages.

Nervous shock

C6033 Nervous shock refers to actual and quantifiable damage to the nervous system, affecting nerves, glands and blood; and, although normally consequent upon earlier negligent physical injury, an action is nevertheless maintainable, even if shock is caused by property damage (*Attia v British Gas plc [1988] QB 304* where a house caught fire following a gas explosion. The claimant, who suffered nervous shock, was held entitled to damages).

Claimants fall into two categories, namely, (a) primary and (b) secondary victims, the former being directly involved in an accident and the latter are essentially spectators, bystanders or rescuers.

- *Primary victims*

  Primary victims can sue for damages for nervous shock/psychiatric injury, even if they have not suffered earlier physical injury (see *Page v Smith [1996] AC 155* in which the appellant, who was physically uninjured in a collision between his car and that of the respondent, had developed myalgic encephalomyelitis and chronic fatigue syndrome, which became permanent. It was held by the House of Lords that the respondent was liable for this condition in consequence of his negligent driving). Foreseeability of physical injury is sufficient to enable a claimant directly involved in an accident to recover damages for nervous shock. Thus, in the case of *Bourhill v Young [1943] AC 92* a pregnant woman, whilst getting off a tram, heard an accident some fifteen yards away between a motor cyclist and a car, in which the motor cyclist, driving negligently, was killed. In consequence, the claimant gave birth to a stillborn child. It was held on the facts of the case that the unknown motor cyclist could not have foreseen injury to the claimant who was unknown to him.

- *Secondary victims*

  As for secondary victims, defendants are taken to foresee the likelihood of nervous shock to rescuers attending to an injured person and to their close relatives, though the precise extent of nervous shock need not have been foreseen (*Brice v Brown [1984] 1 All ER 997*). Persons who witness distressing personal injuries, who are not related in either of these ways to the victim, are not only considered not to have been foreseen by the defendant, but are expected to be possessed of sufficient fortitude to be able to withstand the calamities of modern life. Only persons with a close tie with the victim or their rescuers who are within sight or sound of the accident or its immediate aftermath will be awarded damages for nervous shock as the law stands at the present. Husbands and wives will be presumed to have a sufficiently proximate tie whilst other relationships are considered on the evidence of the proximity of the relationship.

  In *Chadwick v British Railways Board [1967] 1 WLR 912*, following a serious railway accident, for which the defendant was held to be liable, a volunteer rescue worker suffered nervous shock and became psychoneurotic. As administratrix of the rescuer's estate, the claimant sued for nervous shock. It was held that (i) damages were recoverable for nervous shock, even though shock was not caused by fear for one's own safety or for that of one's children, (ii) the shock was foreseeable, and (iii) the defendant should have foreseen that volunteers might well offer to rescue and so owed them a duty of care.

  The class of persons who can sue for damages for nervous shock is limited, depending on proximity of the claimant's relationship with the deceased or injured person (*McLoughlin v O'Brian [1983] 1 AC 410* where the claimant's husband and three children were involved in a serious road accident, owing to the defendant's negligence. One child was killed and the husband and other two children were badly injured. At the time of the accident, the claimant was two miles away at home, being told of the accident by a neighbour and taken to the hospital, where she saw the injured members of her family and heard that her daughter had been killed. In consequence of hearing and seeing the results of the road accident, the claimant suffered severe and recurrent shock. It was held that she was entitled to damages for nervous shock as she was present in the immediate aftermath). This approach was confirmed by the House of Lords in *Alcock v Chief Constable of South Yorkshire Police [1992] 1 AC 310*, where it was stated that the class of persons to whom this duty of care was owed as being sufficiently proximate, was not limited to particular relationships such as husband and wife or parent and child, but was based on ties of love and affection, the closeness of which would need to be proved in each case, except that of spouse or parent, when such closeness would be assumed. Similarly, in *Hinz v Berry [1970] 2 QB 40* the appellant left her husband and children in a lay-by while she crossed over the road to pick bluebells. The respondent negligently drove his car into the rear of the car of the appellant. The appellant heard the crash and later saw her husband and children lying severely injured, the former fatally. She became ill from nervous shock and successfully sued the respondent for damages. On this basis, an employee who suffers nervous shock as a result of witnessing the death of a co-employee at work, will be unlikely to be able to claim damages against his employer for nervous shock, as being a 'bystander who happens to be an employee', as distinct from an active participant in rescue (*Robertson v Forth Bridge Joint Board [1995] IRLR 251* where an employee was blown off the Forth Bridge in a high gale and fell to his death; a co-employee who watched this was unable to sue for damages for nervous shock).

In *Hunter v British Coal Corporation [1999] QB 140*, a claimant who was 30 metres away from the scene of a fatal accident and was told of the victim's death 15 minutes later was unable to recover damages for the psychiatric injury he suffered because he felt responsible for the accident. It was held that he was neither physically nor temporarily close enough to the accident to be a primary victim.

In any event, reasonable fortitude, on the part of the claimant, will be assumed, thereby disqualifying claims on the part of hypersensitive persons (see *McFarlane v EE Caledonia [1994] 2 All ER 1* where the owner of a rig did not owe a duty of reasonable care to avoid causing psychiatric injury to a crew member of a rescue vessel who witnessed horrific scenes at the Piper Alpha disaster).

In *White v Chief Constable of South Yorkshire Police [1999] 1 All ER 1*, the House of Lords considered the question of psychiatric injury to police officers on duty during the Hillsborough football stadium disaster when 95 spectators were crushed to death. It was held that police officers were not entitled to recover damages against the Chief Constable for psychiatric injury suffered as a result of assisting with the aftermath of a disaster, either as employees or as rescuers. An employee who suffered psychiatric injury in the course of his employment had to prove liability under the general rules of negligence, including the rules restricting the recovery of damages for psychiatric injury.

Loss of amenity

C6034 Loss of amenity is a loss, permanent or temporary, of a bodily or mental function, coupled with gradual deterioration in health, e.g. loss of finger, eye, hand etc. Traditionally, there are three kinds of loss of amenity, ranging from maximum severity injury (quadraplegias and irreversible brain damage), multiple injuries (very severe injuries but not hopeless cases) to less severe injuries (i.e. loss of sight, hearing etc.). Damages reflect the actual amenity loss rather than the concomitant psychic loss, at least in hopeless cases. Though, if a claimant is aware that his life has been shortened, he will be compensated for this loss. The *Administration of Justice Act 1982, s 1(1)* states:

'(i) damages are not recoverable in respect of loss of expectation of life caused to the injured person by the injuries; but

(ii) if the injured person's expectation of life has been reduced by injuries, there shall be taken into account any pain and suffering caused or likely to be caused by awareness that his expectation of life has been reduced.'

Bereavement

C6035 A statutory sum of £10,000 is awardable for bereavement by the *Fatal Accidents Act 1976, s 1*A (as amended by the *Administration of Justice Act 1982, s 3(1)* and the *Damages for Bereavement (Variation of Sum) Order 1990 (SI 1990 No 2575)*). This sum is awardable at the suit of husband or wife, or of parents provided the deceased was under eighteen at the date of death if the deceased was legitimate; or of the deceased's mother, if the deceased was illegitimate. (In *Doleman v Deakin (1990) 87 (13) LSG 43* it was held that where an injury was sustained before the deceased's eighteenth birthday, but the deceased actually died after his eighteenth birthday, bereavement damages were not recoverable by his parents.)

Fatal injuries

C6036 Death at work can give rise to two types of action for damages:

- damages in respect of death itself, payable under the *Fatal Accidents Act 1976* (as amended); and
- damages in respect of liability which an employer would have incurred had the employee lived; here the action is said to 'survive' for the benefit of the deceased worker's estate, payable under the *Law Reform (Miscellaneous Provisions) Act 1934*.

Actions of both kinds are, in practice, brought by the deceased's dependants, though actions of the second kind technically survive for the benefit of the deceased's estate. Moreover, previously paid state benefits are not deductible from damages for fatal injuries. [*Social Security Act 1989, s 22(4)(c)*]. Nor are insurance moneys payable on death deductible, e.g. life assurance moneys. [*Fatal Accident Act 1976, s 4*].

Damages under the Fatal Accidents Act 1976

C6037 'If death is caused by any wrongful act, neglect or default which is such as would (if death had not ensued) have entitled the person injured to maintain an action and recover damages, the person who would have been liable if death had not ensued, shall be liable … for damages …'. [*Administration of Justice Act 1982, s 3(1)*].

Only dependants, which normally means the deceased's widow (or widower) and children and grandchildren can claim – the claim generally being brought by the bereaved spouse on behalf of him/herself and children. [*Administration of Justice Act 1982, s 1(2)*]. The basis of a successful claim is dependency, i.e. the claimant must show that he was, prior to the fatality, being maintained out of the income of the deceased. If, therefore, a widow had lived on her own private moneys prior to her husband's death, the claim will fail, as there is no dependency. The fact that both the deceased and his partner were at the time of the accident on state benefits is irrelevant to the question of loss in assessing damages – *Cox v Hockenhull [2000] 1 WLR 750*. Cointributory negligence on the part of the deceased will result in damages on the part of the dependants being reduced. [*Fatal Accidents Act 1976, s 5*].

Survival of actions

C6038 Actions for injury at work which the deceased worker might have had, had he lived, survive for the benefit of his estate, normally for the benefit of his widow. This is provided for in the *Law Reform (Miscellaneous Provisions) Act 1934*. Any damages paid or payable under one Act are 'set off' when damages are awarded under the other Act, as in practice actions in respect of deceased workers are brought simultaneously under both Acts.

Assessment of damages in fatal injuries cases

C6039 Damages in respect of a fatal injury are calculated by multiplying the net annual loss (i.e. earnings minus tax, social security contributions and deductions necessary for personal living (i.e. dependency)) by a suitable number of years' purchase. There is no deduction for things used jointly, such as a house or car. However, where a widow also works, this will reduce the dependency and she cannot claim a greater dependency in future on the ground that she and her deceased husband intended to have children.

It is possible to agree that fatal injury damages should be paid in the form of a structured settlement (see C6025 above).

Where both parents are dead as a result of negligence or a mother dies, dependency is assessed on the cost of supplying a nanny (*Watson v Willmott [1991] 1 QB 140*; *Cresswell v Eaton [1991] 1 WLR 1113*).

Provisional awards

C6040 Because medical prognosis can only estimate the chance of a victim's recovery, whether partial or total, or alternatively, deterioration or death, it is accepted that there is too much chance and uncertainty in the system of lump sum damages paid on a once-and-for-all basis. Serious deterioration denotes clear risk of deterioration beyond the norm that could be expected, ruling out pure speculation (*Willson v Ministry of Defence [1991] 1 All ER 638*). Similarly in the case of dependency awards under the Fatal Accidents Act 1976 it can never be known what the deceased's future would have been, yet courts are expected and called upon to make forecasts as to future income. To meet this problem it is provided that provisional awards may be made, e.g 'This section applies to an action for damages for personal injuries in which there is proved or admitted to be a chance that at some definite or indefinite time in the future the injured person will, as a result of the act or omission, which gave rise to the cause of action, develop some serious disease or suffer some serious deterioration in his physical or mental condition'. [*Administration of Justice Act 1982, s 6(1)*]. Moreover, 'Provision may be made by rules of the court for enabling the court to award the injured person:

- damages assessed on the assumption that the injured person will not develop the disease or suffer the deterioration in his condition; and
- further damages at a future date if he develops the disease or suffers the deterioration'.

[*Administration of Justice Act 1982, s 6(2)*].

A claim in respect of provisional damages must be included in the statement of claim to entitle the claimant to such damages. The disease or type of deterioration in respect of which any future applications may be made must be stated (the *Civil Procedure Rules 1998 (SI 1998 No 3132),Part 41 rule 2(2)*). The defendant may make a written offer if the statement of claim includes a claim for provisional damages, offering a specified sum on the basis that the claimant's condition will not deteriorate and agreeing to make an award of provisional damages in that sum.

Interim awards

C6041 In certain limited circumstances a claimant can apply to the court for an interim payment. This enables a claimant to recover part of the compensation to which he is entitled before the trial rather than waiting till the result of the trial is known – which may be some time away. This procedure is provided for in the *Civil Procedure Rules 1998* (*SI 1998 No 3132*), *Part 25*, but it only applies where the defendant is either:

- insured,
- a public authority, or
- a person whose resources are such as to enable him to make the interim payment.

Prior to making an interim payment a judge is under an obligation to take into account any effect the payment may have on whether there is a 'level playing field' for the hearing – see *Campbell v Mylchreest [1999] PIQR Q17*.

Compensation recovery applies to interim payments as well as to payments into court. Care needs to be taken when applying for interim payments to avoid putting the claimant at a disadvantage. Capital of over £8,000, which could include an interim award, removes entitlement to means tested benefits, particularly income support, and there are also reductions on a sliding scale in such benefits for any capital above £3,000. In the recent judgement of *Beattie v Department of Social Security [2001] EWCA Civ 498* the Court of Appeal held that payments from structured annuity funds constitutes 'income' for the purposes of the *Income Support* (*General*) *Regulations 1987* (*SI 1987 No 1967*).

Payments into court

C6042 A payment into court ('Part 36 payment') may be made in satisfaction of a claim even where liability is disputed. From the defendant's point of view, costs from the date of the Part 36 payment may be saved if the court does not order a higher payment of damages than that paid into court. The claimant may accept a Part 36 payment or Part 36 offer not less than 21 days before the start of the trial without needing the court's permission if he gives the defendant written notice of the acceptance not later than 21 days after the offer or payment was made. If the defendant's Part 36 offer or Part 36 payment is made less than 21 days before the start of the trial, or the claimant does not accept it within the specified period, then the court's permission is only required if liability for costs is not agreed between the parties. [*Civil Procedure Rules 1998* (*SI 1998 No 3132*), *Part 36 rule 11*].

The defendant may, instead of making a Part 36 payment, make a Part 36 offer (formerly known as a Calderbank letter) in which he sets out his terms for settling the action

Interest on damages

C6043 Damages constitute a judgment debt; such debt carries interest at 8 per cent (currently) up to date of payment. Courts have a discretion to award interest on any damages, total or partial, prior to date of payment (and this irrespective of whether part payment has already been made [*Administration of Justice Act 1982, s 15*]), though this does not apply in the case of damages for loss of earnings, since they are not yet due. Moreover, a claimant is entitled to interest at 2 per cent on damages relating to non-pecuniary losses (except bereavement), even though the actual damages themselves take into account inflation (*Wright v British Railways Board [1983] 2 AC 773*). Under *section 17* of the *Judgments Act 1838* (and *section 35A* of the *Supreme Court Act 1981* and *Schedule 1* to the *Administration of Justice Act 1982*), interest runs from the date of the damages judgment. Thus, where, as sometimes happens, there is a split trial, interest is payable from the date that the damages are quantified or recorded, rather than from the date (earlier) that liability is determined (*Thomas v Bunn, Wilson v Graham, Lea v British Aerospace plc [1991] 1 AC 362*). Moreover, interest at the recommended rate (of 8 per cent) is recoverable only after damages have been assessed, and not (earlier) when liability has been established (*Lindop v Goodwin Steel Castings Ltd, The Times, 19 June 1990*).

Awards of damages and recovery of state benefits

C6044 The *Social Security* (*Recovery of Benefits*) *Act 1997* and accompanying regulations made important changes to the rules for recoupment of benefit from compensation payments. One key change is that recoupment will not be taken from general damages for pain and suffering and for loss of amenity which it has been accepted should be paid in full. With respect to the other heads of compensation, namely loss of earnings, cost of care and compensation for loss of mobility, they are only to be subject to recoupment from specified benefits relevant to each of these heads of compensation. [*Social Security* (*Recovery of Benefits*) *Act 1997, s 8*].

Duties of the compensator

C6045 Before the compensator makes a compensation payment, he must apply to the Secretary of State for a certificate of recoverable benefits. [*Social Security* (*Recovery of Benefits*) *Act 1997, s 4*]. The Secretary of State must send a written acknowledgement of receipt of the application and must supply the certificate within four weeks of receipt of the application. [*Social Security* (*Recovery of Benefits*) *Act 1997, s 4*].

He must supply the following information with his application:

- full name and address of the injured person;
- his date of birth or national insurance number, if known;
- date of accident or injury when liability arose (or is alleged to have arisen);
- nature of the accident or disease;
- his payroll number (where known), if the injured person is employed under a contract of service and the period of five years during which benefits can be recouped includes a period prior to 1994 [*Social Security* (*Recovery of Benefits*) *Regulations 1997* (*SI 1997 No 2205*), *Reg 5*]; and
- the amount of statutory sick pay paid to the injured person for five years since the date when liability first arose, as well as any statutory sick pay before 1994, if the compensator is also the injured person's employer. The causes of his incapacity for work must also be stated.

The certificate of recoverable benefits will show the benefits which have been paid.

The compensator must pay the sum certified within 14 days of the date following the date of issue of the certificate of recoverable benefits. [*Social Security* (*Recovery of Benefits) Act 1997, s 6*]. If the compensator makes a compensation payment without having applied for a certificate, or fails to pay within the prescribed fourteen days, the Secretary of State may issue a demand for payment immediately. A county court execution may be issued against the compensator to recover the sum as though under a court order. It is wise for the compensator to check the benefits required to be set off against the heads of compensation payment so that he is sure that the correct reduced compensation is paid to the injured person. Adjustments of recoupable benefits and the issue of fresh certificates are possible.

Where the compensator makes a reduced compensation payment to the injured person, he must inform him that the payment has been reduced. Statements that compensation has been reduced to nil must be made in a specific form. Once the compensator has paid the Secretary of State the correct compensation recovery amount and has made the statement as required, he is treated as having discharged his liability. [*Social Security* (*Recovery of Benefits*) *Act 1997, s 9*].

Complications

Contributory negligence

C6046 Where damages have been reduced as the result of the claimant's contributory negligence, the reduction of compensation is ignored and recovery of benefits is set-off against the full compensation sum.

Structured settlements

C6047 The original sum agreed or awarded is subject to compensation recovery and for this purpose, the terms of the structured settlement are ignored and this original sum is treated as a single compensation payment. [*Social Security* (*Recovery of Benefits*) *Regulations 1997* (*SI 1997 No 2205*), *Reg 10*].

Complex cases

C6048 Where a lump sum payment has been made followed by a later lump sum, both payments are subject to recoupment of benefits where those benefits were recoupable. If the compensator has overpaid the Benefits Agency, he can seek a partial refund. [*Social Security* (*Recovery of Benefits*) *Regulations 1997* (*SI 1997 No 2205*), *Reg 9*].

Information provisions

C6049 Under *section 23* of the *Social Security* (*Recovery of Benefits*) *Act 1997*, anyone who is liable in respect of any accident, injury or disease must supply the Secretary of State with the following information within 14 days of the receipt of the claim against him:

- full name and address of the injured person;
- his date of birth or national insurance number, if known;
- date of accident or injury when liability arose (or is alleged to have arisen); and
- nature of the accident, or disease.

Where the injured person is employed under a contract of service, his employer should also supply the injured person's payroll number (if known) if the period of five years during which benefits may be recouped includes a period prior to 1994 and this is requested by the Secretary of State. [*Social Security (Recovery of Benefits) Regulations 1997 (SI 1997 No 2205), Regs 3, 5* and *6*].

If the Secretary of State requests prescribed information from the injured person, it must be supplied within 14 days of the date of the request. This information includes details of the name and address of the person accused of the default which led to the accident, injury or disease, the name and address of the maker of any compensation claim and a list of the benefits received from the date of the claim. If statutory sick pay was received by the injured person, the name and address of the employer who has paid statutory sick pay during the five-year period from the date of the claim or before 6 April 1994.

Appeals against certificates of recoverable benefits

C6050 Appeals against the certificate of recoverable benefits must be in writing and made not less than three months after the compensation payment was made. The appeal should be made to the Compensation Recovery Unit for a hearing before a tribunal. Leave to appeal to a Commissioner against the decision of the tribunal should be made not later than three months after notice of the tribunal's decision.

Treatment of deductible and non-deductible payments from awards of damages

C6051 There are three well established exceptions to deductibility of financial gains:

- recovery under an insurance policy to which the claimant has contributed all or part of the premiums paid on the policy;
- retirement pensions;
- charitable or ex-gratia payments prompted by sympathy for the claimant's misfortune.

Deductible financial gains

C6052 The courts, applying the principles outlined above, have ruled that the following financial gains are deductible:

- tax rebates where the employee has been absent from work as a result of his injuries (*Hartley v Sandholme Iron Co [1975] QB 600*);
- domestic cost of living expenses (estimated) must be set off against the cost of care (*Lim Poh Choo v Camden and Islington Area Health Authority [1980] AC 174*);
- estimated living expenses must be set off against loss of earnings (*Lim Poh Choo*, above);
- payment from a job release scheme must be set off against loss of earnings (*Crawley v Mercer, The Times, 9 March 1984*);
- statutory sick pay must be set off against loss of earnings (*Palfrey v GLC [1985] ICR 437*);
- sick pay provided under an insurance policy must be set off against loss of earnings not paid as a lump sum (*Hussain v New Taplow Paper Mills [1988] AC 514*);

- health insurance payment under an occupational pension plan paid before retirement must be set off against loss of earnings where no separate premium had been paid by the employee who had paid contributions to the pension scheme (*Page v Sheerness Steel plc [1996] PIQR Q26*);
- reduced earnings allowance (not a disability benefit) must be set off against loss of earnings (*Flanagan v Watts Blake Bearne & Co plc [1992] PIQR Q144*);
- payments under *section 5* of the *Administration of Justice Act 1982* of any saving to the person who has sustained personal injuries through maintenance at the public expense must be set off against loss of earnings.

Non-deductible financial gains against loss of wages

C6053 The following financial gains are non-deductible:

- accident insurance payments under a personal insurance policy taken out by the employee (*Bradburn v Great Western Railway Co (1874) LR 10 Exch 1*) (contributory);
- incapacity pension (from contributory insurance scheme) both before and after retirement age (*Longden v British Coal Corporation [1998] AC 653*). It should be noted that even though the incapacity pension was triggered by the accident, benefit flows from the prior contributions paid by the injured party (contributory);
- private retirement pensions (*Parry v Cleaver [1970] AC 1*; *Hewson v Downs [1970] 1 QB 73*; *Smoker v London Fire and Civil Defence Authority [1991] 2 AC 502*) (contributory or benevolent);
- redundancy payment unconnected with the accident or disease (*Mills v Hassal [1983] ICR 330*). Where the claimant was made redundant because he was unfit to take up employment in the same trade, however, the redundancy payment was deductible from damages for lost earnings (*Wilson v National Coal Board [1981] SLT 67*);
- ex-gratia payment by an employer (*Cunningham v Harrison [1973] QB 942*; *Bews v Scottish Hydro-Electric plc [1992] SLT 749*) (benevolent);
- ill health award and higher pension benefits provided by the employer (*Smoker v London Fire and Civil Defence Authority [1991] 2 AC 502*) (benevolent);
- moneys from a benevolent fund, paid through trustees (not directly to the injured person or dependant) in respect of injuries;
- charitable donations (but not where the tortfeasor is the donor).

The above points however now need to reconsidered in light of the case of *Gaca v Pirelli General plc (Judgment 26 March 2004)* ('*Gaca*'). In this case the Court of Appeal, overturned a county court decision and concluded that insurance monies received by an employee after an accident at work were no longer to be disregarded in the assessment of an employee's damages unless it was shown that the claimant had paid, or contributed to, the insurance premium, directly or indirectly.

In *Gaca*, Mr Gaca's employment with Pirelli had been terminated on the ground of ill health following a serious accident at work. Following termination, Mr Gaca received an ill health gratuity payment of £10,000 from his employers, and two payments totalling £122,787.12 out of a group insurance policy covering temporary total disablement and permanent total disability. The premiums for that policy had been paid by the employer.

Mr Gaca commenced a claim for damages for injuries against his employers, Pirelli. Although liability was conceded Pirelli contended that the proceeds of the insurance should be deducted from any award of damages. In a trial of a preliminary issue in the county court the recorder, following the Court of Appeal decision in *McCamley v Cammel Laird Shipbuilders Ltd [1990] 1 WLR 963* ('*McCamley*'), held that the proceeds of the insurance policy received by the claimant should not be deducted because they fell within the so-called 'benevolence exception' – the principle that charitable or benevolent payments should be disregarded in the assessment of damages in personal injury cases.

Pirelli appealed against that ruling and, by respondent's notice, Mr Gaca contended that the county court's decision could in any event be upheld on the alternative basis that the insurance monies received fell within the so-called 'insurance exception'. Under that, and following *Bradburn v The Great Western Railway Company (1874) LR Exch 1*, and a long line of subsequent authorities, insurance payments are not to be taken into account in the assessment of damages – a general proposition applicable whether or not the premiums had been paid by the claimant.

The Court of Appeal held that *McCamley* should no longer be followed because it had been wrongly decided. The instant case did not satisfy the criteria for the benevolence exception because:

- the payments were made by the tortfeasor; and
- the payment of benefits under the insurance policy was not equivalent or analogous to a payment made by a third party out of sympathy for the plight of a victim of an accident.

Nor did the instant case fall within the insurance exception. It could only do so if the claimant had paid to, or contributed to, the insurance premium directly or indirectly. As to what might constitute evidence of indirect payment or contribution, the Court of Appeal said that there was no direct English authority, though the issue had been discussed in a Canadian case, *Cunningham v Wheeler [1994] 113 DLR*. It was a matter which would have to be resolved as and when it came up for determination.

Accordingly, the insurance monies received had to be deducted from the damages. The decision of the county court had to be set aside.

Non-deductible financial gains against cost of care

C6054

- Loss of board and lodging expenses awarded despite their being provided voluntarily by parents where the claimant had formerly paid such expenses herself (*Liffen v Watson [1940] 1 KB 556*).

Social security and damages for prescribed diseases

C6055 The claims for social security benefits are similar to those for personal injury. Claims for damages are also similar, with more latitude for late court applications because of recurrence of, or worsening of, industrial diseases. Lists of prescribed diseases are contained in social security leaflet NI 2. It is advisable always to refer to the latest edition of the leaflet, as the list of diseases is subject to amendment from time to time. There are special provisions applicable to asthma and deafness resulting from industrial processes, and also for pneumoconiosis and similar diseases.

Occupational deafness (see DSS leaflet 207)

C6056 If the claimant's deafness was caused by an accident at work, then to qualify for disablement benefit his average hearing loss must have been at least 50 decibels in both ears due to damage of the inner ear. The claimant's disablement must be at least 20 per cent or more for him to qualify for disablement benefit (total deafness being 100 per cent). The claimant must have worked for at least the five years immediately before making his claim, or for at least ten years, in one of the listed jobs set out in the current edition of the leaflet. A claimant who has been refused benefit because the rules were not complied with will have to wait for three more years when he may qualify if he has worked in one of the listed occupations for five years. Leaflet NI 196 'Social Security Benefits Rates' shows the rates payable, relevant to each percentage of loss, for a weekly disablement pension.

The claim will be decided by a Medical Board, whose members will also decide on reviews if the claimant believes that his deafness has worsened. They will inform the claimant in detail of their decision. Awards will be made for a period of five years, and there will be a review at the end of that period with another hearing test to determine whether benefit should continue, or be reduced.

Claims must be made on Form BI 100 (OD) which may be obtained free from the Department of the Environment HS ORI Level 4, Caxton House, London SW1H 9NA (tel: 020 7273 5248), or from a post office and some borough libraries.

Asthma because of a job (see DSS leaflet NI 237)

C6057 As is well known, there has been an increase in the number of sufferers from asthma. To be able to claim disablement benefit because of asthma, a claimant must have worked for an employer for at least ten years and have been in contact with prescribed substances, and his disability must amount to at least 14 per cent. In the list of substances, item 23 includes 'any other sensitising agent encountered at work', so claims may be possible even if the offending sensitising substance is not yet on the list.

Claims must be made on Form BI 100 (OA) which may be obtained from the Department of the Environment (see address and telephone number above), or from a post office and some borough libraries.

Once a claim has been made, the claimant will be seen by a Medical Board whose members will decide whether he has asthma because of his job and, if so, how disabled he is by the asthma. They will also determine how long his disability is likely to last and the extent of his disability, and will then communicate all of this information to the claimant. If they consider that no change is likely, they may award benefit for life. If they expect that there will be changes, there will be a review at the end of the period for which benefit was awarded.

Repetitive strain injuries

C6058 Repetitive strain injuries are included in the *Social Security* (*Industrial Injuries*) (*Prescribed Diseases*) *Regulations 1985* (*SI 1985 No 967*), *Sch 1*, as item A4. (See OCCUPATIONAL HEALTH AND DISEASES.) The difficulties in establishing causation can be seen from the case of *Pickford v Imperial Chemical Industries [1998] 1 WLR 1189*, where the House of Lords reversed the decision of the Court of Appeal and upheld the trial judge's decision that a secretary was not entitled to damages for repetitive strain injury as she had failed to establish the cause of the injury and that the condition was not reasonably foreseeable. However, in many other cases damages for repetitive strain injury have been awarded – for example, *Ping v Esselte-Letraset Ltd [1992] PIQR P74* where nine claimants who worked at a printing factory were awarded damages ranging from £3,000 to £8,000. In *Fish v British Tissues (Sheffield County Court, 29 November 1994)*, damages of £57,482 were awarded to a factory packer for repetitive strain injuries to arms, hands and elbows.

Pneumoconiosis

C6059 Pneumoconiosis is compensatable under different heads; first, as a ground for disablement benefit, under the *Social Security Contributions and Benefits Act 1992* and, secondly, by way of claim made under the *Pneumoconiosis etc.* (*Workers' Compensation*) *Act 1979* – the latter being in addition to any disablement benefit previously paid. See the Department of Environment guidance on the *Pneumoconiosis etc.* (*Workers' Compensation*) *Act 1979* on the DETR website: www.environment.detr.gov.uk//pneumo/index.htm.

Pneumoconiosis/tuberculosis-pneumoconiosis/emphysema – disablement benefit

C6060 Where a person is suffering from pneumoconiosis accompanied by tuberculosis, then, for benefit purposes, tuberculosis is to be treated as pneumoconiosis. [*Social Security Contributions and Benefits Act 1992, s 110*]. This applies also to pneumoconiosis accompanied by emphysema or chronic bronchitis, provided that disablement from pneumoconiosis, or pneumoconiosis and tuberculosis, is assessed at, at least, 50 per cent. [*Social Security Contributions and Benefits Act 1992, s 110*]. However, a person suffering from byssinosis is not entitled to disablement, unless he is suffering from loss of faculty which is likely to be permanent. [*Social Security Contributions and Benefits Act 1992, s 110*].

Benefit may be payable if the claimant's disablement is assessed at 1 per cent or more, instead of the usual minimum of 14 per cent. If the assessment of disablement is 10 per cent or less (but at least 1 per cent), then the industrial injuries disablement pension will be payable at one-tenth of the 100 per cent rate.

Pneumoconiosis – payment of compensation under the Pneumoconiosis etc. (Workers' Compensation) Act 1979

C6061 Claims made under the *Pneumoconiosis etc.* (*Workers' Compensation*) *Act 1979* are in addition to any disablement benefit paid or payable under the *Social Security* (*Industrial Injuries*) (*Prescribed Diseases*) *Regulations 1985* (*SI 1985 No 967*). Indeed, whereas the latter consists of periodical payments, the former resemble damages awarded against an employer at common law and/or for breach of statutory duty, except that under the above Act fault (or negligence) need not be proved. This statutory compensation is available only if the employee is unable to claim compensation from any of his former employers (for example, if the employer has become insolvent). There are no requirements that the employee must have worked in a particular industry to be able to claim compensation under the scheme. Claimants are entitled to benefits from the date of claim. There is no ninety-day waiting period under the scheme; it is sufficient if the claimant suffers from the prescribed disease. Coal miners are excluded from the scheme because they have their own statutory scheme. Claims made under the *Pneumoconiosis etc.* (*Workers' Compensation*) *Act 1979* must be made in the manner set out in the *Pneumoconiosis etc.* (*Workers' Compensation*) (*Determination of Claims*) *Regulations 1985* (*SI 1985 No 1645*). The claim must be made (except in the case of a 'specified disease' – see C6062 below) within 12 months from the date on which disablement benefit (under the Social Security Regulations) first become payable; or if the claim is by a dependant, within 12 months from the date of the deceased's death. [*Pneumoconiosis etc.* (*Workers' Compensation*) (*Determination of Claims*) *Regulations 1985* (*SI 1985 No 1645*), *Reg 4*(*1*), (*2*)]. Awards for pneumoconiosis/byssinosis have slowly declined recently.

Claims for specified diseases

C6062 Claims for a 'specified disease', i.e:

- pneumoconiosis, including (i) silicosis, (ii) asbestosis and kaolinosis;
- byssinosis (caused by cotton or flax dust);
- diffuse mesothelioma;
- primary carcinoma of the lung coupled with evidence of (i) asbestosis and/or (ii) bilateral diffuse pleural thickening;

[*Pneumoconiosis etc.* (*Workers' Compensation*) (*Specified Diseases*) *Order 1985* (*SI 1985 No 2034*)],

must be made within 12 months from the date when disablement benefit first became payable, or, in the case of a dependant, within 12 months from the date of the deceased's death. [*Pneumoconiosis etc.* (*Workers' Compensation*) (*Determination of Claims*) *Regulations 1985* (*SI 1985 No 1645*), *Reg 4*(*3*)(*4*)]. These time periods can be extended at the discretion of the Secretary of State. Moreover, where a person has already made a claim and has been refused payment, he can apply for a reconsideration of determination, on the ground that there has been a material change of circumstances since determination was made, or that determination was made in ignorance of, or based on, a mistake as to material fact.

Construction and Building Operations

The situation on the development of health and safety law in construction

C8001 At the present time there is much discussion about health and safety in the construction industry and at the time of writing developments in legislation are awaited. It will be helpful to give the background to this concern.

The main kinds of accidents and ill-health in the construction industry are:

- *falls* – particularly through fragile roofs and roof lights, from ladders and from scaffolds and other work;
- *falling material and collapses*;
- *electrical accidents*;
- *mobile plant* – causing people to be struck by excavators, lift trucks, dumpers and other vehicles;
- *exposure to dusts* – including asbestos, and other hazardous substances;
- *lifting heavy and awkward loads*; and
- *exposure to high noise levels and vibration.*

In addition, contact dermatitis from exposure to wet cement among bricklayers and their labourers is widespread; perhaps as many as ten per cent of such workers leave the construction industry before retiring age because of it.

In the past there were many industries employing large numbers of people that presented such a range of serious risks. Today there are not many mines and steel works; foundries and heavy engineering works are relatively few; and docks operations no longer mix regiments of men with open holds vertical ladders, suspended loads and hordes of fork trucks. The construction industry however remains labour intensive with a high level of risk. Moreover, the shift towards subcontracting and (often bogus) self-employment has not helped the acceptance of responsibility by the employer. Generally, employers have followed the *Health and Safety at Work etc Act 1974* (‘*HSWA 1974*’) with its emphasis, expressed in terms derived from English common law, of the conscious management of risk. However, as the implications of the 1974 Act have become more apparent, with health and safety becoming a factor in business decision taking in general, the peculiar position of the construction industry has become even more evident. The assumption of the Act that the employer of the worker is wholly, or at least sufficiently, the master of the environment in which the worker is at risk is simply not the case in this industry. Decisions affecting the degree of risk to be managed by the employer (in this case the contractor) are taken by clients and their advisers, in particular, architects engineers and quantity surveyors.

The matter came to a head with the first appearance of the *Control of Substances Hazardous to Health Regulations* in 1988 and their preference for inherent reduction of risk through the choice of non or less toxic materials. The contractor's degree of freedom in this regard is in fact restricted and sometimes non-existent.

The making general of such kinds of preference across all types of risk by the Framework Directive 1989 further exposed the construction industry. (In point of fact this was actually implemented in *Regulation 4* and *Schedule 1* of the *Management of Health and Safety at Work Regulations 1999* (*SI 1999 No 3242*) ('*MHSWR 1999*') – the 1989 Directive had not been properly transcribed in the original Management Regulations 1992 ('*MHSWR 1992*')). What was then the EEC, soon followed the Framework Directive with the Temporary or Mobile Construction Sites Directive. This latter Directive set up a managerial framework for the following of the strategy of inherent risk reduction by all the relevant parties – the client's team as well as the contractor's, the latter retaining their individual responsibilities as employers of labour under the Framework Directive. The need to implement this Directive and so to attain fully the objectives of the Framework Directive in construction, and similar long standing concerns within the UK, led to the *Construction* (*Design and Management*) *Regulations 1994* (*SI 1994 No 3140* as amended by *SI 2000 No 2380*) ('*CDM Regulations*'). However, because of the defective *MHSWR 1992*, the regulatory system was not fully in place until 1999.

The UK concern with the health and safety performance of the construction industry centred on the figures for fatalities, which in this industry showed little sign of following the generally downward trend elsewhere. After a peak at the beginning of the 1990s, such a trend was however suggested and with the introduction of the *MHSWR 1992* (*SI 1992 No 2051*), followed during 1995 by the *CDM Regulations* and in 1996, by the *Construction* (*Health, Safety and Welfare*) *Regulations 1996* (*SI 1996 No 1592*) – there was hope that matters were at last under control. Unfortunately at the end of the decade fatalities started to rise and in February 2001 the Deputy Prime Minister held a summit meeting of the construction industry's representatives to decide what could be done. A range of initiatives by the various parties involved is under way following this summit. For its own part the Health and Safety Executive (HSE) has established a Construction Division of its Inspectorate and it accelerated the preparation of a second edition of the Approved Code of Practice to the *CDM Regulations* which appeared later in 2001. Preparation of revised construction regulations is in progress at the time of writing.

It is not only in relation to health and safety that the performance of the construction industry has been found wanting. In 1998 there appeared the report 'Rethinking Construction' prepared by a government-sponsored task force under the chairmanship of Sir John Egan, and in September 2002 its sequel 'Accelerating Change'. The emphasis of both reports is the need for a more cohesive approach to the procurement of construction projects, very much the purpose in the health and safety field of the *CDM Regulations*. However there has been dissatisfaction and disappointment in and around the construction industry at the way the *CDM Regulations* have operated and demands for their amendment have come, notably from contractors. Much of this has been against the role of the planning supervisor who is concerned in the main with the preparation of projects; whereas the rise in fatalities which was notable in 2000/2001 would seem rather to put the effectiveness of the principal contractor, the duty holder responsible for the site organisation, in question.

The Health and Safety Commission (HSC) has been considering, in the widest sense, what policy and legislative adjustments might be made to address the continuing concern. A discussion document was published in the Autumn of 2002 *Revitalising Health and Safety in Construction* and a formal consultative document on regulations which will revise the *CDM Regulations* will appear before the end of 2004. To understand what is at issue it will be useful briefly to consider how the *CDM Regulations* are supposed to work and how they implement the Temporary or Mobile Construction Sites Directive, for whatever the views of the interested parties, UK law must continue to fulfil that purpose.

C8002 The management system of the Temporary or Mobile Construction Sites Directive ('the Directive') assumes a form of contract where all parties are subject to supervision by or on behalf of the client. The duty to see that the strategy of risk reduction is followed is placed on the project supervisor, or if appropriate, the client – and there are co-ordinators for project preparation and execution to assist. In the French system, for instance, the 'client' is the client as specially defined for the purpose of construction procurement in the Civil Code and has well demarcated and understood responsibilities. If clients are unwilling, or not able to fulfil these directly, they can delegate them to an agent (unless they are a public body). In consequence there are many organisations, architectural practices for instance, who are able to support clients in the overall supervision of projects to the degree that the Directive assumes. In some forms of contract in fact, there is an overall supervisor (as mentioned in the Directive) below the client. But the point to note is that in all these forms of contract it is reasonable that the duty to see that the strategy is observed throughout the project is placed at, or close to, the very source of the project and its ultimate control.

With our different regime of procurement it was not possible simply to follow the wording of the Directive since it was necessary to work around the fundamental transfer of responsibility in most UK arrangements when the contractor is engaged. Because of this there is a lack of adequate institutional support in the UK for the client engaged in overall supervision of a project, and consequently a limit to what could be expected of clients in this regard. At least this was the thought behind the approach to health and safety supervision of projects when the *CDM Regulations* were made. Subsequently, as we shall see, HSC appears to be suggesting that clients might actually have a responsibility for overall supervision under the *HSWA 1974*.

C8003 To start with the more readily understandable part of the UK adaptation of the Directive, in project execution the *CDM Regulations* (*SI 1994 No 3140* as amended by *SI 2000 No 2380*) establish a principal contractor to co-ordinate and manage health and safety and place duties of co-operation on all other contractors. The principal contractor and other contractors have their own duties to follow the strategy under the *MHSWR 1999* (*SI 1999 No 3242*). Thus, on site the aims of the Directive should be achieved and there is a sense in which, for this aspect of the project, the principal contractor represents both the Directive's overall supervisor and the co-ordinator for the execution phase.

C8004 In project preparation, the duties of the client as understood here in the UK could be no more than those it was reasonable to expect of any business client in an environment where overall supervision of projects was little practised. And there was certainly no profession dedicated to, or commonly undertaking, such a role (the *CDM Regulations* allow a client to appoint an agent to take over even these restricted duties). However, there had to be some party on whom the duty to see that the strategy was observed as relevant (in full it contains matters that can only be addressed by the employer of labour) in design and planning. For this purpose the planning supervisor was created: a kind of substitute, Continental-style supervisor in relation to project preparation and also the co-ordinator for that phase. It is a grave error to describe the planning supervisor merely as the co-ordinator for preparation as reference to the *CDM Regulations* (*SI 1994 No 3140* as amended by *SI 2000 No 2380*), *Reg 14* (see C8030) will show.

The idea however, is not of the planning supervisor as some omniscient designer sitting in judgement on the actual designers: but of a party able to determine whether the design reflects a proper approach to the avoidance and reduction of risk (there being many actual designs that might reflect this in any particular case). To support this role a direct duty was placed on designers to observe the relevant parts of the strategy and critically, there is a duty on the client not to let work to start on site unless a health and safety plan has been sufficiently prepared. The planning supervisor is required *to be in a position* to give the client advice as to whether this is so, and effectively this gives to the planing supervisor the power of veto on the fulfilment of the project.

The health and safety plan, which is a requirement of the Directive, is a plan to manage the risks of the project as they emerge from the designers' attempts to reduce them (among the balance of all the design considerations). The plan must have been further developed (as possible up to that stage) by the principal contractor (relating to any other contractors who might have been engaged by then). This must been done in the light of all such contractors' own obligations to follow the entire strategy as employers of labour (assessing the risks as the necessary preliminary). Such a plan cannot merely be any plausible set of prescriptions and policies for the running of health and safety on site. A plan that addressed risks, or levels of risk, that should not exist, would sanction illegality.

In this way the client's power is enlisted to ensure the observance of the strategy. Since most clients will not be in a position to make the necessary judgements themselves, the position of the planning supervisor is as an important one and the status of this newly created role is thereby reinforced.

C8005 The duties of the client include those of appointing the planning supervisor and the principal contractor and a duty not to appoint any contractor or designer (or the planing supervisor) unless reasonably satisfied that the organisation or individual is competent and will allocate sufficient resources for the tasks they have. These latter duties are not in the Directive, but followed concern by major contracting firms that a responsible attitude to the duties of all relevant health and safety law would otherwise be undermined by what are commonly called 'cowboys'. Again, because most clients would not be able to make this kind of judgement unaided, the planning supervisor must be in a position to help the client.

Very unfortunately many of the firms or individuals initially taking up the planning supervisor role sought to fulfil it through lengthy questionnaires, asking the most pedestrian facts from contractors rather than concentrating, as they should have done, on the contractor's response to the health and safety specifics of the design. Further, many contractors did not take kindly to having their own ideas on planning criticised by what they regarded as amateurs. Most of this should have been no more than teething troubles; but a certain prejudice has set in and it has to be said that neither the second edition of the *CDM Regulations* Approved Code of Practice (ACoP) nor the discussion document *Revitalising Health and Safety in Construction* set out the critical role of the planning supervisor in forthright terms – and certainly the latter document does not attempt to show how the *CDM Regulations* (*SI 1994 No 3140* as amended by *SI 2000 No 2380*) are meant to work (in the manner attempted above) when it discusses what changes might be necessary.

The recognition of the importance of the planning supervisor must follow if there is recognition of the importance of risk reduction in design. Unfortunately this largely has not happened and the true prominence of the planning supervision remains too often obscured. There is evidence of widespread disregard of the duty to reduce risk and it appears that even nine years after the inception of the *CDM Regulations* there has been little progress towards integrating health and safety risk reduction in design into professional training though at last there is support for it in HSE designers' web pages; Construction Industry Council Technical Notes (available on the SiD website at: www.safetyindesign.org/index/htm); and particularly the HSE *E Learning Academy* (www.learning-hse.com/hse/frameset.html).

C8006 During the later part of 2004 it should become clear what amendments to the *CDM Regulations* (*SI 1994 No 3140* as amended by *SI 2000 No 2380*) the HSC have in mind. There is some indication in both the *Revitalising* document and in the 2001 ACoP that the Commission takes the view that clients have a more active role to play than is required by the *CDM Regulations* themselves. Thus, implying that the duties they might have under the *HSWA 1974, s 3* and the *MHSWR 1999* (*SI 1999 No 3242*), *Regs 3, 4*, in relation to others who might be affected by their businesses, extend to construction workers working for their contractors. The Commission even expresses the opinion that the duty to have advice on health and safety matters under the *MHSWR 1999* (*SI 1999 No 3242*), *Reg 7* establishes a need to have advice on construction health and safety, presumably from the time a project is set up. The point is discussed at C8028.

C8007 A further legal development is the changes to the regulations on working at heights, i.e. *Regulations 5* and *8* of the *Construction* (*Health, Safety and Welfare*) *Regulations 1996* (*SI 1996 No 1592*) ('*CHSW 1996*'). Specific regulations covering such risks were required across all industries by July 2004 under the Second Amendment to the Directive on the Use of Work Equipment. It is intended that there will be a single set of regulations applying to all industries and that the construction regulations referred to above will be repealed. However, following controversy over the application of new regulations to recreational climbing, implementation has slipped and the new regulations are not expected to appear until the end of 2004. Current HSE guidance will require amendment and it is likely that a new guidance document will be prepared for the construction industry.

C8008 Another matter of doubt about new legislation arises from the *Dangerous Substances and Explosive Atmospheres Regulations 2002* (*SI 2002 No 2776*). These Regulations replaced the *Highly Flammable Liquids and Liquefied Petroleum Gases Regulations 1972* (*SI 1972 No 917*). On the whole they do not appear to have a major impact on the construction industry. Indeed generally they do little beyond making more explicit what was already required under other legislation, particularly the *MHSWR 1999* (*SI 1999 No 3242*). However, the requirement on the zoning of workplaces where explosive atmospheres are likely to occur requires verification of by a person competent in the field of explosion protection before the workplace is used for the first time. The notion of zoning itself is a natural consequence of the risk assessment duty that these Regulations reinforce. In construction, however, where workplaces in the sense surely intended are moving and opening up all the time, verification appears problematical. Neither the Regulations themselves, the guidance available at the time of writing (HSE leaflet INDG370), nor the ACoP L180, convey the impression that typical construction situations were much in mind, when they were drafted, yet there is no exemption for the construction industry. If it might be thought that such atmospheres would be a rarity in construction, it should be noted that the description given in *Schedule 2* of the 2002 Regulations for the Zone 2 is that such an atmosphere is not likely to occur in normal operation but if it does will persist for a short period only.

It is understood that comprehensive advice on the 2002 Regulations is in preparation and that the needs of construction will be explicitly covered. Additional information sheets on explosive atmospheres in confined places in the construction industry and probably also in demolition are also expected. Progress has however been slow, though some general guidance is available on the HSE website at: www.hse.gov.uk/spd//content/dsear.htm. It should be noted that workplaces coming into use or modified after 30 June 2003 must comply with this and related requirements immediately, while there is a transitional period of three years allowed for workplaces in use before July 2003. Thus for the moment we appear to have a situation of doubt in the very industry (construction) where the verification of the zoning of workplaces might be supposed to have its earliest impact because workplaces there are in constant flux – and virtually all of them have now been opened since 30 June 2003.

Practical safety criteria

C8009 After the discussion of the legal framework for the management of construction health and safety, it will be helpful to look at some important matters in a more practical way. We should start with the strategy or *principles of prevention and protection* which are fundamental to the subject and enshrined in legislation via the *MHSWR 1999* (*SI 1999 No 3242*), *Reg 4* and *Sch 1* – in effect enhancing or directing the manner in which the general duties of employers under the *HSWA 1974, ss 2, 3* and *4* are discharged. As we have seen, the strategy is appropriately applied to other determining parties in construction projects under the *CDM Regulations* (*SI 1994 No 3140* as amended by *SI 2000 No 2380*).

C8010 Let us imagine a site on which a serious accident has just occurred: for example, a dumper truck lies at the bottom of an excavation. The layman will perhaps think that the obvious question to ask is 'was the truck being driven properly?'. But behind that question lie issues about the training and experience of the driver, the state of the vehicle, provision of roll-over protection, and the protection of the excavation's edge at that point. Then there is the question of the suitability of the type of vehicle for this particular operation (dumper trucks have their limitations) and the general system of work. Behind these considerations again lie further questions about the traffic system and how well had the need of the contractors to transport materials been thought through in the design. Could the layout of the buildings under construction have been adjusted to assist the flow of materials, or, on the other hand do they now present a constant risk? And perhaps finally, 'did we need such an excavation at all, or at this stage might the work have been phased otherwise?'.

The causation of an accident lies in the interaction of various factors, though it may be that one of them dominates. Perhaps after all, in the example above, the dangerous behaviour of a properly instructed and protected driver had more to do with his fate than anything else; but in general the remedy for the future lies in removing, reducing, or otherwise managing, the standing and underlying factors. Such factors may include a lack of training; an unsuitable system of work; or a design that cannot be realised without a burden of what should have been avoidable risk.

Considering the above example and the strategy as set out in *Schedule 1* of the *MHSWR 1999* (*SI 1999 No 3242*), we find that our order of questions is reversed. The pre-emptive management of risk starts at the beginning of the project with the envisioning of the need to transport materials: avoiding risk or combating (reducing) it at source. One does not design with indifference to the problems of contractors or with unrealistic expectations of a casual workforce.

Ladders

C8011 Before paying very necessary attention to the matters listed below, one must consider whether the job in question is the sort of job that should be done from a ladder, assuming that it has to be done, or so much of it done, at a height at all. (See the reference to the current legal requirements and their forthcoming replacement at W9007, W9008).

- Only ladders in a sound condition should be used.
- The 'one out four up' rule should be strictly adhered to in all situations (i.e. that the vertical height from the ground to the ladder's point of rest should be four times the distance between the base of the vertical dimension and the foot of the ladder).
- Ladders should be securely fixed near to their upper resting place or, where this is impracticable, 'footed' by an individual or securely fixed at the base to prevent slipping.
- Ladders should be inspected on a regular basis and a record of such inspections maintained.

(See W9023 WORK AT HEIGHTS for statutory requirements.)

Working platforms

C8012

- Working platforms should be adequately fenced by means of guard-rails and toe-boards.
- Platforms should be adequately covered with sound boards.
- Where mobile platforms are used, they should be stationed on a firm level base and, where possible, tied to the structure to prevent sideways movement. Wheel-locking devices should be provided and used.
- The following height to base ratios should be applied for all mobile working platforms:

  outdoor work – 3:1; indoor work – 3.5:1.

(See W9018 WORK AT HEIGHTS for statutory requirements.)

Materials

C8013

- Meticulous standards of housekeeping must be maintained on working platforms and other elevated working positions to prevent materials, tools and other items falling on to people working directly below. The correct positioning of toe-boards is most important here.
- Lifting operations should ensure correct hooking and slinging prior to raising, correct assembly of gin-wheels and a high degree of supervision.
- Catchment platforms or 'fans' should be installed to catch small items which may fall during construction, particularly where work is undertaken above a public thoroughfare.

Excavations

C8014 Excavations should be avoided where possible or phased or adjusted to reduce risk – though of course they will often be inherent in the project or be part of its essence.

The nature of the ground and the general environment must be considered in the risk assessment under *Regulation 3* of the *MHSWR 1999* (*SI 1999 No 3242*).

- Trenches should be adequately timbered with regard to the depth and width of the trench, the nature of the surrounding ground and the load imposed by subsoil.
- Excavated ground and building materials should be stored well away from the verge of any excavation.

Powered hand tools and machinery

C8015

- Electrically operated hand tools, such as drills, should comply with British Standard 2769: Series and, unless 'all insulated' or 'double insulated', must be effectively earthed.
- Portable tools should be cordless or should operate through reduced voltages, using 110 volt mains isolation transformers with the secondary winding centre tapped to earth. Temporary lighting arrangements should similarly use the 110 volt system.
- Power take-offs, cooling fans, belt drives and other items of moving machinery should be securely fenced to prevent workers coming into contact with them. All machinery should comply with the *Provision and Use of Work Equipment Regulations 1998* (*SI 1998 No 2306*) (see further MACHINERY SAFETY AND ELECTRICITY). The HSE booklet, '*Electrical safety on construction sites*', provides more detailed guidance in this area.
- The health risks associated with vibration should be assessed when buying or hiring hand-held power tools, and existing power tools should also be checked (see further NOISE AND VIBRATION).

It should be realised that the use of 110 volt systems is a means for complying with the duty in the *MHSWR 1999* (*SI 1999 No 3242*), *Sch 1*, to combat risks at source, since the risk of electrocution is removed leaving the smaller risk of burning. Cordless tools come close to avoiding electrical risks altogether.

Site transport

C8016 We have already touched on the need not to hinder *in the design* the flow of materials round the site.

- Employees should not travel on site transport, such as dumper trucks, and notices should be affixed to such vehicles to that effect.
- All vehicle movement and tipping operations on site should be supervised by a person outside the driver's cab.
- Site vehicles should be subject to regular maintenance, particular attention being paid to braking and reversing systems.
- Only competent and trained drivers should be allowed to drive site vehicles.
- Site roadways should be maintained in a sound condition, free from mud, debris, obstructions and large puddles. The verges of the roadway should be clearly defined and adequate lighting provided, particularly at tipping points, reversing and turning areas.
- Site speed limits should be established and clearly marked with signs corresponding to speed limit signs on public roads.

Demolition

C8017 The special position of demolition attracts the full requirements of the *CDM Regulations* (*SI 1994 No 3140* as amended by *SI 2000 No 2380*) (see C8023), and where all relevant requirements for workers to be properly trained and supervised needs particular careful attention.

- A pre-demolition survey should always be undertaken, making use of the original plans if available.
- Catching platforms should be installed not more than 6 m below the working level wherever there is a risk to the public.
- Employees should be provided with safety helmets incorporating chin straps, goggles, heavy duty gloves and safety boots with steel insoles. In certain cases, respiratory protection, safety belts or harnesses may also be necessary.
- Demolition should be undertaken, wherever possible, in the reverse order of erection.
- When using working platforms, all debris should be removed on a regular basis.
- Members of framed structures should be adequately supported and temporary props, bracing or guys installed to restrain remaining parts of the building.
- Employees should not work from the floor of a building which is currently being demolished.
- Where pulling arrangements, demolition ball, explosives or pusher arms are to be used, employees should be kept well away until these stages have been completed.
- Frequent inspections must be made of the demolition site to detect dangers which may have arisen following commencement of demolition.

Fire

C8018 It may be possible to ameliorate in design and planning circumstances where workers are at risk. The design may permit the early installation of staircases, and avoid the likelihood of the presence of incompatible trades.

- All sources of ignition should be carefully controlled, e.g. welding activities, the use of blow lamps, gas or liquid fuel fired appliances.
- All flammable materials, including waste materials, should be carefully stored away from the main construction activity.
- All employees should be aware of the fire warning system, training sessions being undertaken according to need.
- There should be sufficient access for fire brigade appliances in the event of fire.
- There should be adequate space between buildings, e.g. site huts, canteen, etc.
- High-risk buildings should be separated from low-risk buildings.
- Controlled areas, where smoking and the use of naked lights are forbidden, should be established and suitably marked with warning signs.

- An adequate supply of water should be available for fire brigade appliances and on-site fire-fighting.
- Fire wardens should be appointed to undertake routine site inspections, together with the operation of a fire patrol, particularly at night and weekends.

Some useful practical guidance includes:

- '*Fire safety in construction work*' (HSG168);
- HSE guidance (INDG370) on the *Dangerous Substances and Explosive Atmospheres Regulations 2002* (*SI 2002 No 2776*); and
- comprehensive official guidance is currently being prepared which is hoped will address the difficulty mentioned in C8008 above.

(See also FIRE PREVENTION AND CONTROL.)

Asbestos

C8019

- It should be assumed that buildings constructed or refurbished before the 1980s will contain asbestos-based materials.
- No work should be carried out which is likely to expose employees to asbestos unless an adequate assessment of exposure has been made.
- The area where work is to be carried out should be checked to identify the location, type and condition of any asbestos likely to be disturbed.

Two guidance booklets from the HSE, '*Introduction to Asbestos Essentials*' (aimed at anyone who is liable to control or carry out maintenance work with asbestos-containing materials) and '*Asbestos Essentials Task Manual*' (aimed at workers), provide information on ensuring that building maintenance work involving asbestos-containing materials is carried out safely and in accordance with the law.

Under the new *Control of Asbestos at Work Regulations 2002* (*SI 2002 No 2675*) there is a duty to manage asbestos in non-domestic buildings. From 21 May 2004 those having control of non-domestic premises must make an assessment as to whether asbestos is liable to be present and if so assess the risk and prepare a written plan describing the presence of the asbestos and the precautions to be observed. The information must be given to anyone liable to disturb the asbestos.

HSE has published '*A comprehensive guide to managing asbestos in premises*' (HSG227) and there is also an ACoP (L127) on this subject.

This new duty will assist contractors in protecting their staff and reinforce the duty on clients to give and pass forward information discovered about premises under the *CDM Regulations* (*SI 1994 No 3140* as amended by *SI 2000 No 2380*) (see C8068).

Also see further ASBESTOS.

Health and safety law

C8020 The *HSWA 1974*, the *MHSWR 1999* (*SI 1999 No 3242*) and a wide range of regulations on particular subjects apply in construction as in any other industry. Thus the use of cranes and lifting equipment is governed by the *Lifting Operations and Lifting Equipment Regulations 1998* (*SI 1998 No 2307* as amended by *SI 2002 No 2174*) and the *Control of Substances Hazardous to Health Regulations 2002* (*SI 2002 No 2677* as amended by *SI 2003 No 978*) deal with health risks. However the peculiarities of construction are such that there is a need for industry-based regulations. Such regulations apply to 'construction work', which is defined quite widely in the *CDM Regulations* (*SI 1994 No 3140* as amended by *SI 2000 No 2380*). The definition includes, for example, the installation and removal of services which are part of a structure and the installation and removal of fixed plant where there is a risk of falling two metres or more.

Generally, the term *construction work* covers the activities on all kinds of construction sites, from the smallest internal jobs to large-scale complex projects including:

- general building and construction work;
- refurbishment work;
- maintenance and repair work;
- engineering construction work;
- civil engineering work.

The exploration and extraction of mineral resources and the preparatory activities are excluded from the definition of construction work, which is extensively defined in the principal regulations. Reference to the full definitions should be made in any case of doubt or for litigation purposes. [*CDM Regulations* (*SI 1994 No 3140* as amended by *SI 2000 No 2380*), *Reg 2*(*1*)*; CHSW 1996* (*SI 1996 No 1592*), *Reg 2*(*1*)].

The industry-specific regulations are:

- the *Construction* (*Design and Management*) *Regulations 1994* (*SI 1994 No 3140* as amended by *SI 2002 No 2380*) (see C8023 below);
- the *Construction* (*Health, Safety and Welfare*) *Regulations 1996* (*SI 1996 No 1592*) (see C8041 below);
- the *Construction* (*Head Protection*) *Regulations 1989* (*SI 1989 No 2209 as amended*) (see C8108 below).

(The application of the *Work in Compressed Air Regulations 1996* (*SI 1996 No 1656*) (see C8092 below) is such that they will apply only when the *CDM Regulations* apply.)

C8021 The various enactments form a kind of matrix. Thus, where there are no particular requirements specified for the assessment of risk (as there are in the *COSHH Regulations* (*SI 2002 No 2677*) for instance), the requirements of *Regulation 3* of the *MHSWR 1999* (*SI 1999 No 3242*) will apply. Again, the *COSHH Regulations* apply only to the contractor. However, the duty contained in the *CDM Regulations* (*SI 1994 No 3140* as amended by *SI 2000 No 2380*) on the designer to pay adequate regard to the need to avoid or reduce risks preserves the strategic thrust of *COSHH* where the contractor might well be denied choice of the substances to be used. The general requirements of the *HSWA 1974* permeate the circumstances of the construction site.

In particular, under the *HSWA 1974* itself:

- contractors (as employers) owe health and safety duties to their employees [*HSWA 1974, s 2*];
- employees owe such duties to their employers (i.e. contractors/ subcontractors) [*HSWA 1974, s 7*];
- building owners (i.e contractors' employers) owe health and safety duties to subcontractors and their employees [*HSWA 1974, s 3*(*1*)] (also see *R v Associated Octel Co Ltd at* C8147); The current view of HSC as stated in the

second edition of the ACoP to the *CDM Regulations* is that this is broader than a mere duty arising from any operations of the owner's at or near the construction work (see C8028);

- self-employed contractors/workmen owe health and safety duties to other self-employed persons (e.g. building owners to main contractors), and to employees (of other organisations) [*HSWA 1974, s 3*(*2*)].

  Self-styled 'labour-only' subcontractors have been held to be employees of a large main contractor and so were entitled to the protection of some of the construction regulations (*Ferguson v John Dawson & Partners (Contractors) Ltd [1976] IRLR 346* where a nominated subcontractor was liable to a 'self-employed labour-only subcontractor' for breach of *Reg 28*(*1*) of the *Construction* (*Working Places*) *Regulations 1966* (now repealed), requiring provision of guard-rails and toe-boards at working platforms and places);
- building owners and occupiers have duties to employees of contractors and subcontractors. [*HSWA 1974, s 4*(*1*), (*2*)]. Thus, the *HSWA 1974, s 4*, states that anyone 'having control to any extent' of premises, must take reasonable care in respect of persons working there, who are *not* employees. This duty can be subject to indemnity clauses in a lease and/or contract (see further OCCUPIERS' LIABILITY), it is strict and must be carried out so far as is reasonably practicable, that is, subject to cost-effective constraints;
- although ostensibly applicable to all workplaces, the *Workplace* (*Health, Safety and Welfare*) *Regulations 1992* (*SI 1992 No 3004* as amended by *SI 2002 No 2174*) do *not* extend to building operations and works of engineering construction, unless some other activity is being carried on there (see further WORKPLACES – HEALTH, SAFETY AND WELFARE).

C8022 Further legislation significant in construction is listed below:

- *Provision and Use of Work Equipment Regulations 1998* (*SI 1998 No 2306*);
- *Control of Asbestos at Work Regulations 2002* (*SI 2002 No 2675*);
- *Asbestos* (*Licensing*) *Regulations 1983* (*SI 1983 No 1649*);
- *Manual Handling Operations Regulations 1992* (*SI 1992 No 2793* as amended by *SI 2002 No 2174*);
- *Personal Protective Equipment Regulations 1992* (*SI 1992 No 2966* as amended by *SI 2002 No 2174*);
- *Noise at Work Regulations 1989* (*SI 1989 No 1790*);
- *Confined Spaces Regulations 1997* (*SI 1997 No 1713*);
- *Lifting Operations and Lifting Equipment Regulations 1998* (*SI 1998 No 2307* as amended by *SI 2002 No 2174*);
- *Reporting of Injuries Diseases and Dangerous Occurrences Regulations 1995* (*SI 1995 No 3163*);
- *Workplaces* (*Health, Safety and Welfare*) *Regulations 1992* (*SI 1992 No 3004*);
- *Health and Safety* (*Enforcing Authority*) *Regulations 1998* (*SI 1998 No 494*);
- *Safety Representatives and Safety Committees Regulations 1977* (*SI 1977 No 500*);
- *Health and Safety* (*Consultation with Employees*) *Regulations 1996* (*SI 1996 No 1513*).

Construction (Design and Management) Regulations 1994 (SI 1994 No 3140 as amended by SI 2000 No 2380)

C8023 The *Construction (Design and Management) Regulations 1994 (SI 1994 No 3140* as amended by *SI 2000 No 2380*) ('*CDM Regulations*') implement (with minor exceptions) EU Directive 92/57/EEC on the Minimum Safety and Health Requirements at Temporary or Mobile Construction Sites. The Regulations are supported by an ACoP, '*Managing health and safety in construction*' (HSG224) (ISBN 0 7176 2139 1).

Generally, the *CDM Regulations* apply to construction work (as defined) carried out on a construction site which is notifiable to HSE, i.e. a construction project which:

- is scheduled to last for more than 30 days; or
- will involve more than 500 man-days of work; or
- includes any demolition work regardless of the size or duration of the work; or
- involves five or more workers being on site at any one time.

The Regulations always apply to construction design work

Notification of project

C8024 Notification of a construction project (where notifiable) (see 'Exclusions' at C8025 below) should be given to HSE in writing (ideally) as soon as practicable after appointment of the planning supervisor (see below), or (failing that), after the appointment of the principal contractor, but before construction work starts, specifying:

- date of forwarding;
- exact address of construction site;
- name and address of client(s);
- type of project;
- name and address of the planning supervisor;
- declaration of appointment by the planning supervisor;
- name and address of principal contractor;
- declaration of appointment of the principal contractor;
- date planned for start of the construction phase;
- planned duration of the construction phase;
- estimated maximum number of people at work on the construction site;
- planned number of contractors on construction site;
- name and address of any contractor(s) already chosen.

[*CDM* (*SI 1994 No 3140* as amended by *SI 2000 No 2380*), *Reg 3, Sch 1*].

Exclusions

C8025 The *CDM Regulations* are inapplicable (mainly) to construction work:

(*a*) where a client reasonably believes that:

(i) a project is not notifiable, and

(ii) no more than four people are working at any one time (except for demolition and dismantling),

[*CDM* (*SI 1994 No 3140* as amended by *SI 2000 No 2380*), *Reg 3*(*2*), (*3*)];

(*b*) of a minor nature where the local authority is the enforcing authority,

[*CDM* (*SI 1994 No 3140* as amended by *SI 2000 No 2380*), *Reg 3*(*4*)];

(*c*) carried out for a domestic client, unless (as a result of agreement/ arrangement with the developer):

(i) land is transferred to the client,

(ii) the developer undertakes to build on the land, or

(iii) after construction, the land will incorporate premises to be occupied by the client,

[*CDM* (*SI 1994 No 3140* as amended by *SI 2000 No 2380*), *Reg 3*(*8*)].

Objectives of the CDM Regulations

C8026 The *CDM Regulations* (*SI 1994 No 3140* as amended by *SI 2000 No 2380*) mirror the relationship between the Framework and Temporary or Mobile Construction Sites Directives. The purpose of the 1994 Regulations is to make the strategy of risk management, as set out in *Schedule 1* of the *MHSWR 1999* (*SI 1999 No 3242*) (*the principles of prevention and protection*) fully effective in construction where fundamental decisions that the strategy should regulate are taken apart from the actual employer of the labour. Further, the co-ordination of management of risk may be complicated by a multiplicity of firms on the same site – perhaps far beyond the practicable scope of the relevant co-ordination duties of the *MHSWR 1999* alone. C8001 above explains how the Regulations accomplish these purposes and implement the Temporary or Mobile Construction Sites Directive.

A health and safety plan (see C8033 below) is initiated by the planning supervisor and forms part of tendering documentation. The competent principal contractor and other contractors are selected on their response and proposed resource allocation (pricing) in health and safety terms. Before actual construction work begins, the plan will have been developed by the principal contractor in order to ensure that it is properly adjusted to contractors and site activities and accepted by the client (very probably on the advice of the planning supervisor). Its function is now to form the basis of the principal contractor's health and safety management of the project.

One of the things the plan must do is to convey information about the condition of the premises or site provided by the client, if need be, after making reasonable enquiries. This must be given to the planning supervisor not only for incorporating into the plan, but for use in its development – and to inform the designers so that they might mitigate or avoid any consequent risks. Since such information may well have its greatest effect at the earliest stages of the design, the early appointment of the planning supervisor to convey it at the beginning of the process is a necessity. It is an unfortunate fact the planning supervisors are often appointed too late in the project to have the influence and oversight that their duties require.

Although clients do not have a duty to monitor the performance of contractors under the *CDM Regulations* (though obvious shortcomings might suggest that a contractor was not competent or properly resourced which would put the original appointment in question), it will be useful for clients to specify that contractors comply with the health and safety plan. In turn, when contractors price compliance, they should advise clients of any risks not identified in the plan which appear in their assessments. Thus when the planning supervisor advises the client on the adequacy of the provision for health and safety, contracts should only be awarded to contractors/tenderers truly prepared to comply with health and safety requirements and standards.

This advice follows the carefully limited duties of the client under the *CDM Regulations*. However, as noted in C8006 above HSC appears now to be taking the view (see for instance the introduction to the second edition of the CDM ACoP) that clients do have some overriding duty under the *HSWA 1974, s 3* and the *MHSWR 1999* to monitor and even control the conduct of their contractors. This point is further discussed below.

Particular duties and responsibilities

Clients

C8027 Clients (including clients' agents) must, prior to construction work, and in respect of each project [*CDM Regulations* (*SI 1994 No 3140* as amended by *SI 2000 No 2380*), *Reg 4*(*1*)]:

(*a*) appoint:

(i) a competent planning supervisor [*CDM Regulations* (*SI 1994 No 3140* as amended by *SI 2000 No 2380*), *Reg 6*(*1*), (*3*)], and

(ii) a competent principal contractor (who must be a contractor) [*CDM Regulations* (*SI 1994 No 3140* as amended by *SI 2000 No 2380*), *Reg 6*(*1*)–(*3*)].

Such appointments can be terminated, changed or renewed (where necessary) to ensure that these roles are filled until construction is completed [*CDM* (*SI 1994 No 3140*), *Reg 6*(*5*)]. So long as they are competent to perform both roles, planning supervisors can also be principal contractors; and clients can be planning supervisors or principal contractors (or both) [*CDM Regulations* (*SI 1994 No 3140*), *Reg 6*(*6*)]. However, clients must not appoint planning supervisors, designers and contractors, unless satisfied as to their competence [*CDM Regulations* (*SI 1994 No 3140* as amended by *SI 2000 No 2380*), *Reg 8*(*1*)–(*3*)], and that all three have allocated, or will allocate, appropriate resources for the performance of their respective roles [*CDM Regulations* (*SI 1994 No 3140* as amended by *SI 2000 No 2380*), *Reg 9*];

(*b*) so far as is reasonably practicable, ensure that the construction phase of any project does not begin without preparation of a satisfactory health and safety plan [*CDM Regulations* (*SI 1994 No 3140* as amended by *SI 2000 No 2380*), *Reg 10*];

(*c*) as soon as is reasonably practicable but before commencement of work, ensure that the planning supervisor is provided with information relevant to the state of the premises on which construction work is to take place [*CDM Regulations* (*SI 1994 No 3140* as amended by *SI 2000 No 2380*), *Reg 11*];

(*d*) ensure that information in the health and safety file is kept available for inspection by any person, for the purposes of compliance with statutory requirements and prohibitions [*CDM Regulations* (*SI 1994 No 3140* as amended by *SI 2000 No 2380*), *Reg 12*(*1*)].

How far should the client go?

C8028 Paragraphs 13–16 of the second edition of the ACoP to the *CDM Regulations* (*SI 1994 No 3140* as amended by *SI 2000 No 2380*) (which appear in bold and therefore carry the strict ACoP status) give an outline of HSC's view of the practical extent of clients' duties. No distinction is made for different kinds of client. In so far as this is a summary of what is, or what will follow from, a client's duties as specified in the *CDM Regulations*, this would not create any problem. For instance what will follow from the client's duty to appoint the principal contractor, who if he is competent will certainly set up the monitoring and review (presumably) of the site work that paragraph 15(f) of the ACoP refers to. However the document says that clients who lack the knowledge or resources for carrying out the duties described, i.e. making arrangements to ensure that projects are properly managed at all stages, must appoint one or more competent people to help them. [*MHSWR 1999* (*SI 1999 No 3242*), *Reg 7*]. This manifestly is not a reference to appointing a planning supervisor to give a CDM client the defined advice such a client needs to fulfil obligations under those Regulations. Clearly then more is intended.

If *Regulation 7 of* the *MHSWR 1999* applies to the activities of the client (that is taking no account of any duties he might have concerning the interaction of his own staff and the construction workers should the circumstances of a project bring them together) – then what HSC is saying that *Regulation 3* of the *MHSWR 1999* on risk assessment must also apply to this role, as will *section 3 of* the *HSWA 1974*, since both share an operative reference to the *conduct* of the duty holder's *undertaking* (see *Regulation 3(1)(b) of* the *MHSWR 1999*). Taking this view then, the client in the purely commercial act of procuring a construction project is *conducting his undertaking* within the meaning of *section 3* of the *HSWA 1974*.

The CDM ACoP makes a similar assertion about designers in the purely intellectual act of design. There is reason to doubt this (see C8029). Here in relation to clients, the view that they must recruit advisers (among other *section 3* and *MHSWR 1999* duties), if they need to, to sustain the activist role stated in paragraph 13 is more plausible where clients are constantly procuring projects. And considering the practicalities, such clients will undoubtedly have competent staff or those who could readily become competent to hand, and will for sound business purposes keep a close watch in any event on what they are paying for. It is likely that such clients are already acting as their own planning supervisor. But let us consider a firm, perhaps of no great size, that knows next to nothing about construction health and safety and has made a perfectly adequate appointment of an adviser for its own purposes under the *MHSWR 1999* (*SI 1999 No 3242*), *Reg 7*. Should it have to engage a construction health and safety specialist and interest itself in construction health and safety monitoring, just for instance because it is opening new premises 50 miles away, something it has not done before and which, for all anyone knows, it will never do again?

The CDM duties of the client set out in the above example are well calculated to secure the management system described in paragraphs 14–16 of the ACoP. But they cannot *ensure* it, as paragraph 13 purports to require. It is perfectly possible for a circumspectly appointed principal contractor to fall down on the job and perhaps to have been showing signs of error for a while, at least to the trained and experienced eye, without any breach of the client's CDM duties. From time to time even the most reputable contractors fail. The CDM scheme is such that the principal contractor bears the responsibility for the health and safety management of the construction phase. Regular clients might assume some overriding responsibility and indeed be *appropriate* for such a duty in terms of the Continental thinking of the Temporary or Mobile Construction Sites Directive – but to demand such an active role of the occasional client seems unjust.

All clients need to be aware that this onerous understanding of the client's duties must determine the line the HSC's enforcement arm, HSE, will be taking. Whether or not they actually prosecute under *section 3* of the *HSWA 1974* as a consequence of the interpretation set out in the ACoP, inspectors are unlikely to take a conservative view of a client's undoubted duties under the *CDM Regulations* themselves. However, ACoP paragraphs 13–16 have, it is submitted, created unfortunate uncertainties, which need clarification. It is to be hoped that this might come with the publication of a Consultative Document promised before the end of 2003.

Designers

C8029 Designers must:

- advise clients as to their duties; and
- ensure that any design has regard to the need:
  - (i) to avoid foreseeable risks to the health and safety of any person involved in construction or cleaning work in or on the structure at any time, or anyone who may be affected by the work of such person (e.g. a member of the public);
  - (ii) to combat at source risks to the health and safety of any person at work involved in construction, cleaning work or any person who may be affected by such work;
  - (iii) to give priority to measures for protecting those involved in construction, cleaning and those who may be affected;
  - (iv) to ensure that the design includes adequate information about any aspect of the project or structure of materials which might affect construction or cleaning workers or those who may be affected by their work; and
  - (v) to co-operate with the planning supervisor (and any other designer preparing a design in connection with the project) for the purposes of compliance with statutory requirements and prohibitions.

[*CDM Regulations* (*SI 1994 No 3140* as amended by *SI 2000 No 2380*), *Reg 13*(*2*)].

The Regulations concerning the definition of and duties of designers were amended by the *Construction* (*Design and Management*) (*Amendment*) *Regulations 2000* (*SI 2000 No 2380*).

The amending Regulations were made as a result of a Court of Appeal decision in *R v Paul Wurth SA, The Times, 29 March 2000*, in which HSE had prosecuted Paul Wurth SA following a fatality during the installation of a conveyor at the British Steel Port Talbot works in 1997.

HSE had alleged that the company had failed in its duties as designers under *Regulation 13*(*2*) of the *CDM Regulations*. This places a duty on a designer to ensure that any design he prepares, and which he is aware will be used for the purposes of construction work, includes among its design considerations adequate regard to the need to avoid foreseeable risks to the health and safety of people as a result of construction work and subsequent cleaning work.

However, the Court of Appeal held that designers do not have responsibilities for designs which are prepared by others under their control, including their employees. This was contrary to the aim of the original provisions of the Regulations, and hence the need for the amendment.

Regulations 2 (concerning the definition of 'designer') and *3* (regarding a person preparing a design), and *12* (concerning the client's duty to ensure the health and safety file is available for inspection) and *13* (concerning requirements on designers), have all been amended. The amended Regulations ensure that legal duties on designers to build safety into a design, apply not only to a design prepared by them personally, but also to a design prepared by an employee or other person under their control.

The CDM duty to pay adequate regard to avoidance and or reduction of risk among the design considerations entails an implicit duty to assess those risks. Whether designers are themselves subject to a direct duty to assess under the *MHSWR 1999* (*SI 1999 No 3242*), *Reg 3*, in the act of design is doubtful, though the official view now appears to be that they are (see paragraph 115 of the CDM ACoP – compliance with the CDM designer duties will satisfy it, according to that document).

The duty to consider risk to those not in the designer's own employment relates to risks arising out of, or in connection with, the conduct by that designer of his undertaking (as stated in the *HSWA 1974, s 3*, which concerns an employer's general duty of care towards non-employees). If designers, whether corporate or self employed, had a duty under *section 3* which included the intellectual process of design, there would have been no need to enact the *HSWA 1974, s 6*. This particular section impacts on designers of industrial equipment (a class of person whose direct influence on the safety of workers was much more in mind in 1974 when the Act was made than that of architects civil engineers and quantity surveyors).

On the other hand, in their press release following the decision of the Court of Appeal in *R v Paul Wurth SA* (see above) which largely negated the CDM design duty as originally drafted, HSE stated that a *section 3* duty still applied to designers (by the reasoning referred to above in relation to risk assessment). Furthermore, a search of the enforcement pages of the HSE website reveals one or two successful prosecutions of construction designers for what appears to be an act of design under *section 3(1)* albeit at a magistrates' courts only. It is not known whether a defence of no case to answer was attempted in these instances, and interestingly there is evidence that in prosecutions relating to machinery (as opposed to plant erection etc. which might constitute construction work as defined), HSE may perhaps combine both *sections 3* and *6* of *HSWA 1974* charges in a way which discriminates between design and non-design aspects.

An additional issue arises from this opinion and the action of HSE: the possible different extent of a design duty under *section 3* of *HSWA 1974* and the carefully qualified one under *Regulation 13* of the *CDM Regulations* – and where this would leave the corresponding duty of the planning supervisor under *Regulation 14* of the same Regulations. The relevant magistrates' court prosecutions under *section 3* were, it seems, for failings of engineering design detail and thus very straightforward. Whether *section 3* would respect the subtitles of a design balance involving aesthetics etc in the way that *Regulation 13* does is not beyond doubt.

It may be for the higher courts to pronounce on the extent of a designer's *section 3* duty under *HSWA 1974* and *Regulation 3* duties under *MHSWR 1999*; but it is hard to see how the HSC/E view can sit comfortably with the existence of *section 6* of the 1974 Act.

As to the implicit duty to assess in the *CDM Regulations* (*SI 1994 No 3140* as amended by *SI 2002 No 2380*), *Reg 13*, the official guidance is in section 5 of the HSC/CONIAC document '*Designing for health and safety in construction*' (revised). Semi-official guidance, produced with the participation of HSE, is contained in the CIRIA Report 166 '*Work sector guidance for designers*'.

If readers are not clear of the practical bearing of the duty to reduce risks in design the discussion of the duty in the CDM ACoP gives a number of excellent examples where the designer can make a substantial contribution to the management of risk

Planning supervisors

C8030 Planning supervisors must:

- ensure that the project is notified to HSE (unless it is reasonably believed that the project is not notifiable (see 'Exclusions' at C8017 above));
- ensure that the design for a project includes:
  - (i) adequate regard to the need to avoid or reduce risk etc;
  - (ii) adequate information regarding structure and materials;
- ensure co-operation between designers, for the purposes of compliance with their duties as designers (under *CDM Regulations* (*SI 1994 No 3140* as amended by *SI 2000 No 2380*), *Reg 13*);
- be able to give adequate advice to:
  - (i) any client/contractor regarding competence of personnel/allocation of resources; and
  - (ii) any client regarding competence of a contractor and a contractor's allocation of resources as well as the health and safety plan;
- ensure preparation of a health and safety file, containing:
  - (i) information concerning aspects of the project, structure or materials that may affect health and safety; and
  - (ii) any other information which, foreseeably, will be necessary to ensure health and safety of persons involved in construction, cleaning and maintenance or demolition work;
- ensure that, on completion of the project, the health and safety file is delivered to the client,

  [*CDM Regulations* (*SI 1994 No 3140* as amended by *SI 2000 No 2380*), *Reg 14*]; and
- ensure that the health and safety plan is made available to every contractor before arrangements are made for them to manage or carry out construction work.

[*CDM Regulations* (*SI 1994 No 3140* as amended by *SI 2002 No 2380*), *Reg 15*(*1*), (*2*)].

We have already drawn attention to the critical importance of the planning supervisor. It is worth amplifying here the significance of the co-ordination that the planning supervisor has to bring about and the consequent demands on the expertise inherent in the role.

Although the primary duty is on designers to avoid and reduce risks, someone has to see that the approach to health and safety management in project preparation is coherent overall. In particular the planning supervisor needs to be satisfied that the designers have indeed paid adequate regard to avoiding or reducing risk among the design considerations. Unless planning supervisors have sufficient understanding of the possibilities and limitations of risk reduction in design (which entails familiarity with risk assessment), this duty cannot be fulfilled, for apart from anything else the planning supervisor will lack credibility in the necessary and perhaps difficult dialogue with the designers.

It is perhaps in relation to this duty to take such steps as are reasonable to ensure co-operation between designers (*Regulation 14(b)*), that the most characteristic role of the planning supervisor emerges in the area of risk assessment. Designers will range from the original architect (in the case of a building) to employees of mechanical and electrical contractors who take what amount to design decisions. The possibility of avoidable risk arising from the separate activities of these perhaps many and certainly various, parties is obvious. The planning supervisor has to identify hazards that might arise from incompatibilities of different aspects or stages of the design and must form a view of the resulting risk. He must also act as a catalyst between the designers to resolve the issues in a way that satisfies the *Regulation 13* duty (under the *CDM Regulations*) that each designer has.

Principal contractors

C8031 Principal contractors must:

- take reasonable steps to ensure co-operation between all contractors, for the purposes of compliance with statutory requirements and/or prohibitions;
- so far as is reasonably practicable, ensure that every contractor and every employee complies with the health and safety plan;
- take reasonable steps to ensure that only authorised persons are allowed where construction work is carried on;
- ensure that notification particulars (see C8024 above) are displayed prominently so that they can be read by construction personnel; and
- provide the planning supervisor promptly with any of the following information:

  (i) which is in possession of the principal contractor, or which the latter could ascertain by making reasonable enquiries of a contractor,

  (ii) which, reasonably, the planning supervisor would include in the health and safety file (in order to comply with his duties (under *Regulation 14* above)), and

  (iii) which is not in possession of the planning supervisor,

  [*CDM Regulations* (*SI 1994 No 3140* as amended by *SI 2002 No 2380*), *Reg 16*(*1*)].

  For the purposes of compliance with *Regulation 16*(*1*), principal contractors can:

  (i) give any necessary directions to any contractor, and

  (ii) include in the health and safety plan rules for the management of construction work reasonably required for health and safety management; such rules being in writing and brought to the attention of those affected,

  [*CDM Regulations* (*SI 1994 No 3140* as amended by *SI 2000 No 2380*), *Reg 16*(*2*), (*3*)];

- so far as is reasonably practicable, ensure that every contractor is provided with comprehensible information on health and safety risks to that contractor or to employees or other persons under that contractor's control;
- so far as is reasonably practicable, ensure that every contractor who is an employer provides his employees engaged in construction work with:
  - (i) information relating to:
    - — health and safety risks,
    - — protective measures,
    - — procedures to be followed in imminent danger and in danger areas, and
    - — persons appointed to implement those procedures, and
  - (ii) health and safety training, both on recruitment and/or exposure to new/increased risks; such training to be repeated periodically,

  [*CDM Regulations* (*SI 1994 No 3140* as amended by *SI 2000 No 2380*), *Reg 17*];
- ensure that:
  - (i) employees/self-employed personnel are able to discuss, and offer advice on, matters foreseeably affecting their health and safety, and
  - (ii) there are arrangements for the co-ordination of employees' views (or their representatives'),

  [*CDM Regulations* (*SI 1994 No 3140* as amended by *SI 2000 No 2308*), *Reg 18*].

There has been a view among contractors that the *CDM Regulations and MHSWR 1999* should be seen as quite separate from each other. This is seriously mistaken. The essential point to grasp about the duties of the principal contractor is that they exist to make the requirements of the *MHSWR 1999* fully effective in the complicated and usually difficult circumstances of most construction sites (when compared with the relatively stable conditions of, for example, a factory). With many contractors engaged the multiplicity of transactions required between the various parties would be impracticable without the central and authoritative position of the role. Practical advice on running a site for health and safety can be found in part 1 of '*Health and safety in construction*' (HSG150), and '*Successful health and safety management*' (HSG65) also contains useful guidance.

It should be noted that the duty on the principal contractor to follow the principles of prevention and protection, which secures the principal aim of the Temporary or Mobile Construction Sites Directive in project execution, lies in the *MHSWR 1999* (*SI 1999 No 3242*) by virtue of *Regulation 4 and Schedule 1* – and not in the *CDM Regulations* themselves.

Contractors

C8032 Contractors must:

- co-operate with the principal contractor, so that both can comply with their statutory duties;
- so far as is reasonably practicable, provide the principal contractor promptly with any information (including risk assessments for the purposes of the *MHSWR 1999* (*SI 1999 No 3242*), which might affect the health and safety

of construction workers, or those who might be affected by construction work, or which might justify a review of the health and safety plan;

- comply with directions given by the principal contractor, for the purposes of compliance by contractors;
- comply with rules applicable to them in the health and safety plan;
- provide the principal contractor promptly with information relating to deaths, injuries, conditions and dangerous occurrences notifiable under the *Reporting of Injuries, Diseases and Dangerous Occurrences Regulations 1995* (*SI 1995 No 3163*) (see further ACCIDENT REPORTING);
- provide the principal contractor promptly with any information which:
  - (i) is in the possession of the contractor, or which he could ascertain by reasonable enquiries, and
  - (ii) it is reasonable to suppose the principal contractor would provide to the planning supervisor, for the purposes of inclusion in the health and safety file, which is not in the possession of the planning supervisor or the principal contractor,

[*CDM Regulations* (*SI 1994 No 3140* as amended by *SI 2000 No 2380*), *Reg 19*(*1*)];

- not allow any employee to work on construction work, unless provided with:
  - (i) the name of the planning supervisor,
  - (ii) the name of the principal contractor, and
  - (iii) the contents of the health and safety plan relating to work being carried out by the employee.

  Self-employed personnel must also be provided with this information [*CDM Regulations* (*SI 1994 No 3140* as amended by *SI 2000 No 2380*), *Reg 19*(*2*)–(*4*)].

  Again the remarks in C8031 above about the inseparability of the *CDM Regulations* and *MHSWR 1999* apply. Practical guidance can be found in HSG150.

Health and safety plan

C8033 Paragraphs 229–247 of the CDM ACoP deal with the plan at its different stages.

The plan has three vital functions:

- To provide tendering contractors, particularly the principal contractor, with information on the issues to be managed from a health and safety aspect, and any other information that might be necessary to enable a competent contractor to judge the time and resources needed to perform the work in compliance with health and safety law.
- To become (as developed by the principal contractor) the basis for the effective health and safety management of the site.
- To provide a means of assurance that the principles of prevention and protection have been applied in the design and preparation of the project on behalf of the client and have been applied by the contractors in the early preparation of construction work. The plan significantly contributes to the means whereby the *CDM Regulations* (*SI 1994 No 3140* as amended by *SI 2000 No 2380*) secure this fundamental thrust of the Temporary or Mobile

Construction Sites Directive in legal and commercial circumstances which differ from those assumed in it (see C8031 above).

C8034 This third function (above) is often overlooked, but it works in this way:

- the plan is a plan to manage the risks remaining after the application of the appropriate principles in the design;
- before work on site can start, the plan must have been appropriately developed by the principal contractor – again in the light of the principles which will apply by virtue of all contractors' obligations under the *MHSWR 1999* (*SI 1999 No 3242*);
- the client permits the work to start on the basis that the plan as developed complies with *Regulation 15*(*4*) of the *CDM Regulations* (*SI 1994 No 3140* as amended by *SI 2000 No 2380*);
- *Regulation 15*(*4*) refers to the risks involved and it will be a nonsense if these risks ought to have been eliminated or reduced, for then the client would be accepting a plan to manage illegality.

C8035 The planning supervisor must see that the initial stage of the plan is prepared. [*CDM Regulations* (*SI 1994 No 3140* as amended by *SI 2000 No 2380*), *Reg 15*(*1*)]. It has become customary for the individual or firm actually to undertake this, having in any event, duties to obtain the necessary information [*CDM Regulations, Reg 15*(*3*)] and the status to receive that part of it provided at the earliest stage by the client [*CDM Regulations, Reg 11*].

The plan at this stage must be prepared in time to allow it to be provided to any tendering contractor. [*CDM Regulations* (*SI 1994 No 3140* as amended by *SI 2000 No 2380*), *Reg 15*(*2*)].

The plan as taken over by the principal contractor now becomes the operational plan dealing additionally with managerial, monitoring and welfare arrangements. [*CDM Regulations* (*SI 1994 No 3140* as amended by *SI 2000 No 2380*), *Reg 15*(*4*)].

The planning supervisor is required to be in a position to give the client adequate advice as to whether the principal contractor's plan as initially developed complies with *Regulation 15*(*4*). It might be thought that the third function of the plan referred to above is a somewhat theoretical matter – perhaps more a political demonstration of this country's compliance with the Temporary or Mobile Construction Sites Directive. It would be for the courts finally to say what the relevant regulations together achieve here; but consider the position of a planning supervisor after a serious accident. Let us say it occurred in circumstances of risk that might readily have been avoided or significantly reduced in design. In that event, the planning supervisor would have failed to ensure, so far as was reasonably practicable, that the design avoided the risk. He would have prepared or overseen the pre-tender plan on this incorrect basis; and finally, he would have revisited his errors in advising the client that work might proceed. It is evident that the requirement to be able to advise the client on the plan leaves no escape from the gravity of the planning supervisor's responsibility in risk reduction through design.

C8036 The health and safety plan will continue to evolve and provide a focus for the co-ordination of health and safety matters as the construction work progresses. The development of the plan will depend upon the principal and other contractors' risk assessments and what follows from their consequent application of the principles of prevention and protection.

C8037 Clients who intend for normal operations to continue at or near the construction work, need to assess the impact of the project on the health and safety of their staff and others, such as the construction workers who might be affected. The result of this assessment may radically affect the design and may well also give rise to matters that need to be conveyed in the plan itself – perhaps in the form of rules of conduct and pre conditions for the principal contractors management of the site. All such information will be essential for tendering contractors.

Health and safety file

C8038 The health and safety file is a permanent record containing information about the particulars and arrangements relating to the design, methods and materials, maintenance and other information relating to the construction. In practice, the 'file' amounts to a manual to alert those who will be responsible for the structure after construction on safety matters which must be managed after handover. The manual should contain appropriate information regarding maintenance, repair, renovation and demolition.

Prosecution – defence and civil liability

Defence

C8039 A defence is provided against prosecution where it can be shown that an employer or a self-employed person made all reasonable enquiries and reasonably believed either that the Regulations did not apply to the work in question or that he had been given the names of the planning supervisor and the principal contractor together with the relevant contents of the health and safety plan. [*CDM* (*SI 1994 No 3140* as amended by *SI 2000 No 2380*), *Reg 19*(*5*)].

Civil liability

C8040 Generally, breach of the *CDM Regulations* is not actionable, except as regards:

- the preparation of the health and safety plan, prior to construction work, under *Reg 10*; or
- allowing only authorised personnel onto premises where construction work is going on, under *Reg 16*(*1*)(*c*).

[*CDM* (*SI 1994 No 3140* as amended by *SI 2000 No 2380*), *Reg 21*].

Construction (Health, Safety and Welfare) Regulations 1996 (SI 1996 No 1592)

C8041 The *Construction* (*Health, Safety and Welfare*) *Regulations 1996* (*SI 1996 No 1592*) ('*CHSW 1996*') promote the health and safety of everyone carrying out 'construction work' (as defined, and see also C8020 above) but the Regulations do not apply to workplaces on construction sites which are set aside for non-construction purposes. However, *CHSW 1996* gives protection to other people who may be affected by the construction work.

A 'construction site' is defined in the Regulations as 'any place where the principal work activity being carried out is construction work'. [*CHSW 1996* (*SI 1996 No 1592*), *Reg 2*(*1*)].

For details of the requirements and standards of protection regarding falls from access equipment [*CHSW 1996* (*SI 1996 No 1592*), *Reg 6*], see:

- *Schedule 1:* Requirements for guardrails etc;
- *Schedule 2:* Requirements for working platforms;
- *Schedule 3:* Requirements for personal suspension equipment;
- *Schedule 4:* Requirements for means of arresting falls;
- *Schedule 5:* Requirements for ladders.

  For details of the requirements for welfare facilities [*CHSW 1996* (*SI 1996 No 1592*), *Reg 22*] see:

- *Schedule 6:* Welfare facilities.

The 1996 Regulations implement Annex IV of the Temporary or Mobile Construction Sites Directive, thereby extending to construction sites the health, safety and welfare requirements imposed on all other workplaces by the *Workplace* (*Health, Safety and Welfare*) *Regulations 1992* (*SI 1992 No 3004* as amended by *SI 2002 No 2174*). There are requirements governing operations in circumstances of risk that remain after the application of the principles of prevention and protection by designers, planners and contractors either under the *CDM Regulations* or *MHSWR 1999*.

The *CHSW 1996* lays duties principally on employers, self-employed contractors and employees, and, to a lesser extent, on persons in control of construction sites – a category included to prevent the avoidance of responsibility by commercial subterfuge. [*CHSW 1996* (*SI 1996 No 1592*), *Regs 4, 22, 29*(*2*)]. All contractors, big or small, are covered. Breach of duty, followed by conviction, will be visited with criminal sanction (normally payment of a fine). There is also strict civil liability where breach of the Regulations results in injury or damage (see ENFORCEMENT and INTRODUCTION). Characteristically, 'I find it necessary to make some general observations about the interpretation of regulations of this kind. They are addressed to practical people skilled in the particular trade or industry, and their primary purpose is to prevent accidents by prescribing appropriate precautions. Any failure to take prescribed precautions is a criminal offence. The right to compensation, which arises when an accident is caused by a breach is a secondary matter. The Regulations supplement, but in no way supersede the ordinary common law obligations of an employer to care for the safety of his men, and they ought not to be expected to cover every possible kind of danger' (*Gill v Donald Humberstone & Co Ltd [1963] 3 All ER 180*, as per Lord Reid).

Duties under the regulations

C8042 The following duties are laid on employers (and self-employed persons) in respect of employees involved in construction work and any persons under their control.

Safe place of work

C8043 Except in the case of a person making such place safe [*CHSW 1996* (*SI 1996 No 1592*), *Reg 5*(*4*)]:

- every place of work must be made and kept safe and free from health risks for any person at work there, so far as is reasonably practicable [*CHSW 1996* (*SI 1996 No 1592*), *Reg 5*(*2*)];
- so far as is reasonably practicable, suitable and sufficient safe access/egress, to/from, every place of work and any other place provided for use of a person at work, must be provided and properly maintained [*CHSW 1996* (*SI 1996 No 1592*), *Reg 5*(*1*)];

- so far as is reasonably practicable, suitable and sufficient steps must be taken to deny access to any place not complying with (*a*) and (*b*) above [*CHSW* (*SI 1996 No 1592*), *Reg 5*(*3*)]; and
- so far as is reasonably practicable, every place of work must have sufficient working space and be so arranged as to be suitable for any person working or likely to work there [*CHSW 1996* (*SI 1996 No 1592*), *Reg 5*(*5*)].

(For requirements relating to falls, fall-preventative equipment, scaffolds, personal suspension equipment, working platforms and ladders, set out in *Regs 6, 7* and *8*, see WORK AT HEIGHTS.)

Falls, materials falling and fragile roofing materials

The requirements of *CHSW 1996* (*SI 1996 No 1592*) on these subjects will be replaced from **xxxJuly** 2004 by a new set of regulations implementing the Second Amendment to the Directive on the Use of Work Equipment which will apply across all industries (see W9008). All the requirements of the present regulations on these subjects are discussed in W9014–W9027 and need not be repeated here.

Stability of structures

C8044 The 1996 Regulations require that:

- in order to prevent danger to any person, all practicable steps must be taken to ensure that any new or existing structure which may become unstable through construction work (including excavations) does not accidentally collapse [*CHSW 1996* (*SI 1996 No 1592*), *Reg 9*(*1*)];
- no part of a structure must be so loaded as to make it unsafe [*CHSW 1996* (*SI 1996 No 1592*), *Reg 9*(*2*)]; and
- any buttress, temporary support or structure (used to support a permanent structure) must be erected or dismantled under the surveillance of a competent person [*CHSW 1996* (*SI 1996 No 1592*), *Reg 9*(*3*)].

Demolition or dismantling

C8045 Suitable and sufficient steps must be taken to ensure that demolition or dismantling of structures is planned and executed so as to prevent, so far as practicable, danger to any person. Planning and execution of demolition must be carried out under the supervision of a competent person. [*CHSW 1996* (*SI 1996 No 1592*), *Reg 10*]. Note the special application of the substantive requirements of the *CDM Regulations* to demolition (see C8017 above).

Explosives

C8046 Explosive charges can only be used or fired if suitable and sufficient steps have been taken to ensure that no one is exposed to any risk of injury from the explosion or flying material. [*CHSW 1996* (*SI 1996 No 1592*), *Reg 11*].

Excavations

C8047 (In considering the following, please note that the risk assessment requirements under *Reg 3* of the *MHSWR 1999* (*SI 1999 No 3242*).)

The 1996 Regulations require that:

- to prevent danger to any person, all practicable steps must be taken to ensure that new or existing excavations which are temporarily unstable due to the carrying out of construction work (including other excavation work) do not collapse accidentally [*CHSW 1996* (*SI 1996 No 1592*), *Reg 12*(*1*)];
- so far as is reasonably practicable, suitable and sufficient steps must be taken to prevent any person being buried or trapped by a fall or dislodgement of material. [*CHSW 1996* (*SI 1996 No 1592*), *Reg 12*(*2*)]. In particular, as early as practicable in the course of the work, the excavation must be sufficiently supported so as to prevent, so far as reasonably practicable, the fall or dislodgement of material [*CHSW 1996* (*SI 1996 No 1592*), *Reg 12*(*3*)]. Suitable and sufficient supporting equipment must be provided [*CHSW 1996* (*SI 1996 No 1592*), *Reg 12*(*4*)]; and installation, alteration or dismantling must be carried out only under the supervision of a competent person [*CHSW 1996* (*SI 1996 No 1592*), *Reg 12*(*5*)];
- suitable and sufficient steps must be taken to prevent any person, vehicle, plant and equipment, accumulation of earth or other material, from falling into an excavation [*CHSW 1996* (*SI 1996 No 1592*), *Reg 12*(*6*)];
- where collapse of an excavation would endanger a person, no material, vehicle or plant and equipment must be placed or moved near any excavation [*CHSW 1996* (*SI 1996 No 1592*), *Reg 12*(*7*)]; and
- no excavation work must be carried out unless suitable and sufficient steps have been taken to identify and, so far as reasonably practicable, to prevent any risk of injury arising from any underground cable or service [*CHSW 1996* (*SI 1996 No 1592*), *Reg 12*(*8*)].

A HSE guidance booklet, '*Health and safety in excavations: Be safe and shore*', provides practical guidance on working in the ground without placing anyone at risk. It is intended for all those who may be involved with excavation works, e.g. clients, designers, planning supervisors, contractors, site managers and foremen. It aims to help identify the main failings which lead to injury and how to take the necessary measures, at the planning stage, to avoid them. It deals with:

- hazards, risks and control measures;
- planning, design and management;
- legal requirements; and
- sources of further information.

It covers pipe and cable laying, manhole construction, foundations, small retaining walls, and other structures where earthworks are required.

Cofferdams and caissons

C8048 The 1996 Regulations require that:

- every cofferdam or caisson must be:
  - (i) of suitable design and construction,
  - (ii) of suitable and sound material,
  - (iii) of sufficient strength and capacity for the purpose used, and
  - (iv) properly maintained,

[*CHSW 1996* (*SI 1996 No 1592*), *Reg 13*(*1*)]; and

- construction, installation, alteration or dismantling must be under the supervision of a competent person [*CHSW 1996* (*SI 1996 No 1592*), *Reg 13*(*2*)].

Prevention of drowning

C8049 The 1996 Regulations require that where:

- after falling, any person is liable to fall into water (or other liquid) with a risk of drowning, suitable and sufficient steps must be taken to:
  - (i) prevent, so far as is reasonably practicable, such person from falling,
  - (ii) minimise the risk of drowning, and
  - (iii) ensure the provision and maintenance of suitable rescue equipment,

  [*CHSW 1996* (*SI 1996 No 1592*), *Reg 14*(*1*)]; and
- where there are conveyances by water, there must be provision of safe transport to or from a place of work by water [*CHSW 1996* (*SI 1996 No 1592*), *Reg 14*(*2*)]. Vessels used for conveyance purposes must be:
  - (i) of suitable construction,
  - (ii) properly maintained,
  - (iii) under the control of a competent person, and
  - (iv) not overcrowded or overloaded,

  [*CHSW 1996* (*SI 1996 No 1592*), *Reg 14*(*3*)].

Traffic routes

C8050 Pedestrians and vehicles should be able to move safely and without health risks. To that end, traffic routes should be:

- suitable for persons and vehicles using them;
- sufficient in number;
- in suitable positions;
- of sufficient size; and
- indicated by suitable signs (see WORKPLACES – HEALTH, SAFETY AND WELFARE).

[*CHSW 1996* (*SI 1996 No 1592*), *Reg 15*(*1*), (*2*), (*6*)].

These requirements are not satisfied unless:

- pedestrians or vehicles can use a traffic route without causing danger to the health or safety of persons near it;
- any door or gate (intended for use by pedestrians) leading on to a traffic route for vehicles is so separated from that traffic route as to enable pedestrians to see any approaching vehicle or plant from a place of safety;
- there is sufficient separation between vehicles and pedestrians to ensure safety, or if that is not reasonably practicable that:
  - (i) other means are provided for the protection of pedestrians, and

(ii) effective arrangements are made for warning a person liable to be crushed or trapped by a vehicle of the vehicle's approach;

- a loading bay has at least one exit point for exclusive use of pedestrians; and
- where it is unsafe for pedestrians to use any gate intended primarily for vehicles, one or more doors for pedestrians is provided in the immediate vicinity of such gate (such door(s) being clearly marked and obstruction free).

[*CHSW 1996* (*SI 1996 No 1592*), *Reg 15*(*3*)].

Vehicles must not be driven on traffic routes unless, so far as is reasonably practicable, the route is free from obstruction and permits sufficient clearance. [*CHSW 1996* (*SI 1996 No 1592*), *Reg 15*(*4*)]. Where this is not reasonably practicable, drivers and persons riding on vehicles must be warned of any approaching obstruction or lack of clearance, e.g. by a sign. (For requirements relating to traffic routes generally, see ACCESS, TRAFFIC ROUTES AND VEHICLES.)

Vehicles

C8051 The 1996 Regulations require:

- unintended movement of vehicles (including mobile plant, locomotives and towed vehicles) must be either prevented or controlled [*CHSW 1996* (*SI 1996 No 1592*), *Reg 17*(*1*)];
- where persons may be endangered by vehicle movement, the person in effective control of the vehicle must warn any person at work of the risk of injury [*CHSW 1996* (*SI 1996 No 1592*), *Reg 17*(*2*)];
- construction work vehicles, when being driven, operated or towed, must be:

  (i) driven, operated or towed safely, and

  (ii) be so loaded as to be able to be driven, operated or towed safely,

  [*CHSW 1996* (*SI 1996 No 1592*), *Reg 17*(*3*)];
- no person must ride or be required or permitted to ride on any construction work vehicle other than in a safe place provided for that purpose [*CHSW 1996* (*SI 1996 No 1592*), *Reg 17*(*4*)] (for civil liability consequences of failure to do so, see ACCESS, TRAFFIC ROUTES AND VEHICLES);
- no person must remain or be required or permitted to remain on a vehicle during the loading or unloading of loose material unless a safe place is provided and maintained [*CHSW 1996* (*SI 1996 No 1592*), *Reg 17*(*5*)];
- excavating, handling and tipping vehicles must be prevented from:

  (i) falling into an excavation, pit or water, or

  (ii) overturning the edge of an embankment or earthwork,

  [*CHSW 1996* (*SI 1996 No 1592*), *Reg 17*(*6*)]; and
- in the case of rail vehicles, plant and equipment must be provided for replacing them on their track, or moving them, if derailed [*CHSW 1996* (*SI 1996 No 1592*), *Reg 17*(*7*)].

The HSE publication, '*The safe use of vehicles on construction sites*', provides guidance for clients, designers, contractors, managers, and workers involved with construction transport. It provides guidance on how to establish safe workplaces for vehicle operations, vehicle selection, inspection and maintenance, safe driving and work practices, and managing construction transport.

Doors and gates (not forming part of mobile plant and equipment)

C8052 The 1996 Regulations require that:

- where necessary to prevent risk of injury, any door, gate, hatch (including temporary ones) must incorporate (or be fitted with) suitable safety devices [*CHSW 1996* (*SI 1996 No 1592*), *Reg 16*(*1*)]; and
- compliance with the Regulations presupposes that:
  - (i) any sliding door, gate or hatch has a device to prevent it coming off its track during use,
  - (ii) any upward opening door, gate or hatch has a device to prevent it falling back,
  - (iii) any powered door, gate or hatch has suitable and effective features to prevent it trapping persons, and
  - (iv) any powered door, gate or hatch can be operated manually unless it opens automatically if the power fails,

  [*CHSW 1996* (*SI 1996 No 1592*), *Reg 16*(*2*)].

Prevention of fire risks

C8053 Comprehensive guidance on fire safety in construction can be found in '*Fire safety in construction work*' (HSG168). It should be noted that the Office of the Deputy Prime Minister has consulted on new regulations on fire safety that would, among other things, replace *Regulations 19–21* of *CHSW 1996*. It does not appear however that there would be any changes of substance in relation to fire precautions in the construction industry or their enforcement

The 1996 Regulations require that:

- so far as reasonably practicable, suitable and sufficient steps must be taken to prevent risk of injury from:
  - (*a*) fire or explosion,
  - (*b*) flooding, or
  - (*c*) substances liable to cause asphyxiation,

  [*CHSW 1996* (*SI 1996 No 1592*), *Reg 18*];

  In relation to fire or explosion it should be noted that where dangerous substances are used, for instance LPG oil based paints and sanding dust from wood, the *Dangerous Substances and Explosive Atmospheres Regulations 2002* (*SI 2002 No 2776*) ('*DSEAR*') will in effect amplify the general duty in (*a*) above [*CHSW 1996* (*SI 1996 No 1592*), *Reg 18*(*a*)]. In outline the *DSEAR* (which replace the *Highly Flammable Liquids and Liquefied Petroleum Gases Regulations 1972* (*SI 1972 No 917*)) require the following (which follow the familiar strategic pattern of the *MHSWR 1999*):

  - (i) assess risks of work with dangerous substances;
  - (ii) eliminate or reduce risks so far as is reasonably practicable;
  - (iii) provide equipment and procedures to deal with emergencies;
  - (iv) information and training for employees;

(v) classify places where explosive atmospheres might occur and mark zones as necessary.

These duties do not appear to add very much materially to what should have been in hand under the combination of the present regulations and the *MHSWR 1999* (*SI 1999 No 3242*). However, we have mentioned the apparent difficulty in relation to zoning of workplaces in construction where, on the face of it, the specified action might have to be taken to a quite disproportionate, degree. It is hoped that clarification in a guidance document specifically referring to construction will be available at or soon after the publication of this update.

- if a work activity gives rise to risk of fire, such activity must not be carried out unless the worker is suitably instructed to prevent risk [*CHSW 1996* (*SI 1996 No 1592*), *Reg 21*(*6*)];
- there must also be provision of:

  (*a*) suitable and sufficient fire-fighting equipment, suitably located (e.g. in emergency routes);

  (*b*) suitable and sufficient fire detectors and alarm systems, suitably located (e.g. in emergency routes);

  (*c*) training of all persons on site in the correct use of appliances,

  [*CHSW 1996* (*SI 1996 No 1592*), *Reg 21*(*5*)];
- fire-fighting equipment, detectors and alarm systems must be:

  (*a*) properly maintained, examined and tested [*CHSW 1996* (*SI 1996 No 1592*), *Reg 21*(*3*)];

  (*b*) indicated by suitable signs [*CHSW 1996* (*SI 1996 No 1592*), *Reg 21*(*7*)]; and

  (*c*) easily accessible, if not designed to come into use automatically [*CHSW 1996* (*SI 1996 No 1592*), *Reg 21*(*4*)].

Emergency routes and exits

C8054 The 1996 Regulations require:

- a sufficient number of suitable emergency routes and exits, indicated by suitable signs (see WORKPLACES – HEALTH, SAFETY AND WELFARE), must be provided to enable any person to reach a place of safety quickly in the event of danger. [*CHSW 1996* (*SI 1996 No 1592*), *Reg 19*(*1*)]. This should lead, as directly as possible, to an identified safe area [*CHSW 1996* (*SI 1996 No 1592*), *Reg 19*(*2*)]; and
- emergency routes (and traffic routes or doors thereto) must be kept clear and obstruction free and, if necessary, be provided with emergency lighting [*CHSW 1996* (*SI 1996 No 1592*), *Reg 19*(*3*)].

Emergency procedures

C8055 There must be prepared and implemented suitable and sufficient arrangements (to be tested by being put into effect at regular intervals) for dealing with any foreseeable emergency, including procedures for site evacuation. Moreover, persons likely to be affected must be acquainted with such arrangements. [*CHSW 1996* (*SI 1996 No 1592*), *Reg 20*(*1*), (*3*)].

Welfare facilities

C8056 Duties in connection with the provision of welfare facilities on construction sites are laid principally on those in control of sites (i.e. occupiers, see generally C8142 below, and OCCUPIERS' LIABILITY). The fact that occupation (or control) of construction sites can be, and frequently is, shared between the building owner and contractor(s) means that both have obligations to see that welfare facilities are provided. In practice, actual provision would be made by the contractor(s). Hence, although overall duties are imposed on occupiers, employers (and self-employed persons) must ensure that workers under their control are provided with welfare facilities.

Facilities to be provided are:

- suitable and sufficient sanitary conveniences at readily accessible places [*CHSW 1996* (*SI 1996 No 1592*), *Reg 22*(*3*)]. So far as is reasonably practicable, such conveniences must be:

  (i) adequately ventilated and lit,

  (ii) kept in a clean and orderly condition, with

  (iii) separate rooms containing sanitary conveniences provided for men and women (except where each convenience is in a separate room, the door of which can be secured from the inside),

  [*CHSW 1996* (*SI 1996 No 1592*), *Sch 6*];

- suitable and sufficient washing facilities (including, where necessary, showers) at readily accessible places. So far as is reasonably practicable, washing facilities must:

  (i) be provided in the immediate vicinity of sanitary conveniences (except showers), and in the vicinity of changing rooms (whether or not provided elsewhere),

  (ii) include a supply of clean hot and cold (or warm) water (ideally running water), and soap, towels etc.

  Rooms containing washing facilities must be:

  (iii) sufficiently ventilated and lit,

  (iv) kept in a clean and orderly condition, and

  (v) must have separate washing facilities provided for men and women (except for washing hands, forearms and face only) unless provided in a room the door of which can be secured from the inside and the facilities in each such room are intended for use by only one person at a time,

  [*CHSW 1996* (*SI 1996 No 1592*), *Sch 6*];

  It should be noted that in view of the prevalence of contact dermatitis among bricklayers etc. which may become so severe as to make the sufferer leave the industry, and the difficulty of taking the preferred precautions (i.e. elimination, control, and even the use of protective gloves) – the HSE policy is that sites where wet cement is used should not operate unless proper washing facilities are available.

- an adequate supply of wholesome drinking water at readily accessible places [*CHSW* (*SI 1996 No 1592*), *Reg 22*(*5*)]. So far as is reasonably practicable, every supply of drinking water must:

  (i) be conspicuously marked, and

(ii) be provided with a sufficient number of suitable cups or other drinking vessels, unless the supply is from a jet,

[*CHSW* (*SI 1996 No 1592*), *Sch 6*];

- suitable and sufficient accommodation for:

  (i) accommodating the clothing of any person at work which is not worn during working hours, and

  (ii) special clothing worn by a person at work but which is not taken home,

  [*CHSW 1996* (*SI 1996 No 1592*), *Reg 22*(*6*)].

  So far as is reasonably practicable, clothing accommodation should include facilities for drying clothing [*CHSW 1996* (*SI 1996 No 1592*), *Sch 6*];

- suitable and sufficient accommodation for changing clothing where:

  (i) a person has to wear special clothing at work, and

  (ii) that person cannot be expected to change elsewhere,

  [*CHSW 1996* (*SI 1996 No 1592*), *Reg 22*(*7*)].

  (See *Post Office v Footitt [2000] IRLR 243* in W11026.)

  Where necessary, facilities for changing clothing must be separate facilities for men and women [*CHSW 1996* (*SI 1996 No 1592*), *Sch 6*];

- suitable and sufficient rest facilities at readily accessible places [*CHSW 1996* (*SI 1996 No 1592*), *Reg 22*(*8*)]. So far as reasonably practicable, rest facilities must include:

  (i) suitable arrangements to protect non-smokers from discomfort caused by tobacco smoke, and where necessary,

  (ii) facilities for pregnant women or nursing mothers,

  [*CHSW* (*SI 1996 No 1592*), *Sch 6*].

  (For welfare facilities generally, see WORKPLACES – HEALTH, SAFETY AND WELFARE.)

Fresh air

C8057 Every workplace on a construction site must, so far as reasonably practicable, have a supply of fresh or purified air so as to ensure safety and absence of health risks; and plant used for supply purposes must, where necessary, contain effective devices for giving visible or audible warning of failure. [*CHSW* (*SI 1996 No 1592*), *Reg 23*].

(For ventilation requirements generally, see VENTILATION.)

Temperature

C8058 During working hours, so far as reasonably practicable, temperature at any indoor place of work must be reasonable, having regard to the purpose of the workplace. [*CHSW 1996* (*SI 1996 No 1592*), *Reg 24*(*1*)].

(For temperature requirements generally, see WORKPLACES – HEALTH, SAFETY AND WELFARE.)

Weather protection

C8059 Every place of work outdoors must, where necessary, be so arranged, so far as is reasonably practicable, as to provide protection from adverse weather. [*CHSW 1996* (*SI 1996 No 1592*), *Reg 24*(*2*)].

Lighting

C8060 Every place of work and traffic route must have suitable and sufficient lighting. In the case of artificial lighting where there would be a risk to a person's health or safety from failure of primary artificial lighting, suitable and sufficient secondary lighting must be provided. [*CHSW 1996* (*SI 1996 No 1592*), *Reg 25*(*1*), (*3*)].

(For lighting requirements generally see LIGHTING.)

Plant and equipment

C8061 The 1996 Regulations require that, so far as is reasonably practicable, all plant and equipment used for construction work must be safe and without health risks, and be:

- of good construction,
- of suitable and sound materials,
- of sufficient strength and suitability for its intended purpose, and
- so used and maintained that it remains safe and without health risks.

[*CHSW 1996* (*SI 1996 No 1592*), *Reg 27*].

Good order

C8062 So far as reasonably practicable, every part of a construction site must be kept in good order and every place of work in a reasonable state of cleanliness. Perimeters should be identified by suitable signs (see WORKPLACES – HEALTH, SAFETY AND WELFARE), and sites arranged so that their extent is readily identifiable. Moreover, no timber material with projecting nails must be used in work where nails might be dangerous or allowed to remain in a place where nails could be a source of danger. [*CHSW 1996* (*SI 1996 No 1592*), *Reg 26*].

Training

C8063 Any person who carries out construction work where training, technical knowledge or experience is necessary to reduce risk of injury, must either:

- possess such training, knowledge or experience, or
- be under the supervision of a person who so does.

[*CHSW 1996* (*SI 1996 No 1592*), *Reg 28*].

Inspection

C8064 The 1996 Regulations require that in the case of the following places of work:

- working platforms or personal suspension equipment,
- excavations, and
- cofferdams or caissons,

work can only be carried out if such place has been inspected by a competent person as follows:

(*a*) Working platforms and personal suspension equipment must be inspected by a competent person:

  (i) before being taken into use for the first time,

  (ii) after substantial addition, dismantling or other alteration,

  (iii) after any event likely to have affected their strength or stability, and

  (iv) at regular intervals not exceeding 7 days following the last inspection.

(*b*) Excavations must be inspected by a competent person:

  (i) before a person carries out work at the start of every shift,

  (ii) after any event likely to have affected the strength or stability of the excavation, and

  (iii) after an accidental fall of rock/earth.

(*c*) Cofferdams and caissons should be inspected by a competent person:

  (i) before the start of every shift, and

  (ii) after any event likely to have affected their strength or stability.

[*CHSW 1996* (*SI 1996 No 1592*), *Reg 29*(*1*), *Sch 7*].

Where, following inspection, a place of work or plant and materials is not safe, this must be communicated to those in control and such place of work must not be used until defects have been remedied. [*CHSW 1996* (*SI 1996 No 1592*), *Reg 29*(*3*), (*4*)].

In addition to the general duty to inspect places of work (above), in the case of scaffolds, excavations, cofferdams and caissons forming part of a place of work, employers must ensure that they are stable and of sound construction and that the requisite safeguards are in place before workers use such place of work for the first time. [*CHSW 1996* (*SI 1996 No 1592*), *Reg 29*(*2*)].

Reports

C8065 Following inspection, a report must be prepared before the end of the working period in which the inspection was completed, and presented within 24 hours to the person on whose behalf it was carried out. [*CHSW 1996* (*SI 1996 No 1592*), *Reg 30*(*1*), (*2*)]. The report (or a copy) must be kept at the site of such place of work and, after work there is completed, retained at the office of the person on whose behalf the inspection was made, for a minimum of three months. Such a report is available for inspection by an HSE inspector and, should he so require, extracts or copies must be sent to the inspector. [*CHSW 1996* (*SI 1996 No 1592*), *Reg 30*(*3*), (*4*)].

Reports are not necessary in the following cases:

- working platforms where persons are not liable to fall more than 2 metres; and
- mobile towers (unless remaining erect for 7 days or more).

[*CHSW 1996* (*SI 1996 No 1592*), *Reg 30*(*5*), (*6*)].

Enforcement

C8066 Penalties and defences in connection with the *CHSW 1996* (*SI 1996 No 1592*) are as for the *HSWA 1974* (see ENFORCEMENT).

Civil liability for breach of the regulations

C8067 Breach of the 1996 Regulations (and the *Work in Compressed Air Regulations 1996* (*SI 1996 No 1656*) and the *Construction* (*Head Protection*) *Regulations 1989* (*SI 1989 No 2209*) – see C8092 and C8108 below), resulting in injury to an employee, will give rise to civil liability, since safety regulations, even if silent regarding civil liability, are actionable (see further INTRODUCTION). A short guide to these Regulations, *'A guide to the Construction* (*Health, Safety and Welfare*) *Regulations 1996'* is available free from HSE. This lists the series of further guidance.

The Control of Asbestos at Work Regulations 2002 (SI 2002 No 2675)

C8068 Asbestos is the greatest cause of death in construction and a relatively high death rate is expected to remain for some years. The cause of any future danger in the industry will be encounters with asbestos or asbestos-containing materials in demolition, repair, conversion and refurbishment and it is imperative that chance unprepared encounters in these activities cease. *The Control of Asbestos at Work Regulations 2002* (*SI 2002 No 2675*) (see ASBESTOS) place a duty on those in charge of non-domestic premises to manage any asbestos that might be there. The consequent information, that increasingly should be available, will assist clients in their CDM duty to make reasonable enquiries about premises that are to be the scene of a project. The duties on employers will, of course, have to be fulfilled whether this requirement is successful in detecting asbestos in particular premises or not and contractors must continue to be suspicious and vigilant in dealing with any building dating from the era when asbestos was commonly a building or insulating material.

A fourth edition of the ACoP '*Work with asbestos which does not normally requiring a licence*' (L27) has been published. This will be relevant to the trades and operations likely to disturb asbestos that might be present in buildings and other structures.

Confined spaces and harmful atmospheres

C8069 The *Confined Spaces Regulations 1997* (*SI 1997 No 1713*) apply in all premises and work situations in Great Britain subject to the *HSWA 1974*, with the exception of diving operations and below ground in a mine.

The Regulations are supported by an ACoPand guidance – '*Safe work in confined spaces*' (L101) – which provides practical guidance with respect to the requirements of:

- the *Confined Spaces Regulations 1997* (*SI 1997 No 1713*) and the *HSWA 1974, ss 2, 4, 6* and *7*;
- the *Management of Health and Safety at Work Regulations 1999* (*SI 1999 No 3242*);
- the *Control of Substances Hazardous to Health Regulations 2002* (*SI 2002 No 2677* as amended by *SI 2003 No 978*) ('*COSHH Regulations*').

Note: the *COSHH Regulations* apply to all substances hazardous to health (other than lead or asbestos), such as toxic fume and injurious dust. The *Ionising Radiation Regulations 1999* (*SI 1999 No 3232*) may apply where radon gas can accumulate in confined spaces, such as sewers, and where industrial radiography is used to look at, for example, the integrity of welds in vessels;

- the *Personal Protective Equipment at Work Regulations 1992* (*SI 1992 No 2966* as amended by *SI 2002 No 2174*);
- the *Provision and Use of Work Equipment Regulations 1998* (*SI 1998 No 2306*).

Entry into confined spaces

C8070 Recent legal requirements on the zoning of explosive atmospheres (see C8008 above) should be noted. Confined spaces where flammable atmospheres may be present are not subject to any doubt as to the applicability of the precautions set out in the *Dangerous Substances and Explosive Atmospheres Regulations 2002* (*SI 2002 No 2776*).

A 'confined space' has two defining features. Firstly, it is a place which is substantially (though not always entirely) enclosed and, secondly, there will be a reasonably foreseeable risk of serious injury from hazardous substances or conditions within the space or nearby. [*SI 1997 No 1713, Reg 1*(*2*)].

Examples:

- ducts, vessels, culverts, tunnels, boreholes, bored piles, manholes, shafts, excavations, sumps, inspection pits, cofferdams, freight containers, building voids, some enclosed rooms (particularly plant rooms) and compartments within them, including some cellars, enclosures for the purpose of asbestos removal, and interiors of machines, plant or vehicles.

Some confined spaces are fairly easy to identify, for example closed tanks, vessels and sewers. Others are less obvious but may be equally dangerous, for example open-topped tanks and vats, closed and unventilated or inadequately ventilated rooms and silos, or constructions that become confined spaces during their manufacture. A confined space may not necessarily be enclosed on all sides.

Respiratory protective equipment

C8071 Where respiratory protective equipment (RPE) is provided or used in connection with confined space entry or for emergency or rescue, it should be suitable for the purpose for which it is intended, that is, correctly selected and matched both to the job and the wearer.

Where the intention is to provide emergency breathing apparatus to ensure safe egress or escape, or for self-rescue in case of emergency, the type commonly called an 'escape breathing apparatus' or 'self-rescuer' (escape set) may be suitable. These types are intended to allow time for the user to exit the hazard area. They are generally carried by the user or stationed inside the confined space, but are not used until needed. This equipment usually has a breathable supply of only short duration and provides limited protection to allow the user to move to a place of safety or refuge. This type of equipment is not suitable for normal work.

In some circumstances entry without the continuous wearing of breathing apparatus may be possible. Several conditions must be satisfied to allow such work including:

- a risk assessment must be done and a safe system of work in place including all required controls, and continuous ventilation;
- any airborne contamination must be of a generally non-toxic nature, or present in very low concentrations well below the relevant occupational exposure limits.

Duties and responsibilities

C8072 There have been several court cases involving accidents and fatalities in sewers and other confined spaces (see for instance *Baker v Hopkins & Son [1959] All ER 225*). More recently, in 1998 at Cardiff Crown Court, a record fine was imposed on Neath Port Talbot Council following the deaths of two employees (*R v Neath Port Talbot Council*). The judge, John Prosser, said that the accident should never have happened; the dangers of toxic gases associated with sewer work were well known. The case demonstrated the need for employers to carry out, with strict care, the undertaking of such work and that the difficulty and danger must not be underestimated.

Duties to comply with the *Confined Spaces Regulations* (*SI 1997 No 1713*), are placed on:

- employers in respect of work carried out by their own employees and work carried out by any other person (for example, contractors) in so far as that work is to any extent under the employers' control [*SI 1997 No 1713, Reg 3*(*1*)]; and
- the self-employed in respect of their own work and work carried out by any other person in so far as that work is to any extent under the control of the self-employed [*SI 1997 No 1713, Reg 3*(*2*)].

Duty to prevent entry

C8073 The principal duty imposed on employers is to prevent entering or working inside a confined space where it is reasonably practicable to undertake the work by other means. [*SI 1997 No 1713, Reg 4*].

The duty extends to others who are to any extent within the employers' control (such as contractors) and in many cases it will be necessary to modify working practices following a risk assessment of each requirement to enter a confined space.

Examples:

- modifying the confined space itself to avoid the need for entry, or to enable the work to be undertaken from outside;
- testing the atmosphere or sampling the contents of confined spaces from outside using appropriate long tools and probes.

Another case where it may be necessary to modify working practices is where employers or the self-employed have duties in relation to people at work who are not their employees – then the duty is to do what is 'reasonably practicable' in the circumstances. In many cases, the employer or self-employed will need to liaise and co-operate with other employers to agree the respective responsibilities in terms of the regulations and duties. It is also necessary to take all reasonably practicable steps to engage competent contractors.

Associated duties and responsibilities

C8074 In addition to the requirements of the *HSWA 1974*, other legislation imposes duties with regard to the design, construction and operation within confined spaces.

Some duties extend to erectors and installers of equipment and would include situations where plant and equipment unavoidably involved confined spaces. Where it is not possible to eliminate a confined space completely, procedures must be drawn up to minimise the need to enter such spaces both during normal use or working, and for cleaning and maintenance.

Regarding the type of PPE to be provided, this will depend on the identified hazards and the type of confined space. It may be necessary, for example, to include safety lines and harnesses, and suitable breathing apparatus.

Examples:

- the wearing of some respiratory protective equipment and personal protective equipment can contribute to heat stress;
- footwear and clothing may require insulating properties, e.g. to prevent softening of plastics that could lead to distortion of components such as visors, air hoses and crimped connections.

Risk assessment – the development of a safe system of work

C8075 The priority when carrying out a confined space risk assessment is to identify the measures needed so that entry into the confined space can be avoided. If it is not reasonably practicable to prevent work in a confined space the employer (or the self-employed) must assess the risks connected with persons entering or working in the space and also to others who could be affected by the work. The assessor(s) must understand the risks involved, be experienced and familiar with the relevant processes, plant and equipment and be competent to devise a safe system of working.

If, in the light of the risks identified, it cannot be considered reasonably practicable to carry out the work without entering the confined space, then it will be necessary to secure a safe system for working. The precautions required to create a safe system of work will depend on the nature of the confined space and the hazards identified during the risk assessment.

Use of a permit-to-work procedure

C8076 Not all work involving confined spaces requires the use of a permit-to-work system. For example, it is unlikely that a system would be needed where:

- the assessed risks are low and can be controlled easily; and
- the system of work is very simple; and
- it is known that other work activities being carried out cannot affect safe working in the confined space.

Although there is no set format for a permit system, it is often appropriate to include certain information relevant to all confined space working. In all cases, it is essential that a system is developed which ensures that:

- the people working in the confined space are aware of the hazards involved and the identity, nature and extent of the work to be carried out;

- there is a formal and methodological system of checks undertaken by competent people before the confined space is entered and which confirms that a safe system of work is in place;
- other people and their activities are not affected by the work or conditions in the confined space.

Isolation requirements, that is, the need to isolate the confined space to prevent dangers arising from outside, should also be included in the permit system. Permits are particularly appropriate if essential supplies and emergency services such as sprinkler systems, communications etc, are to be disconnected. The most effective isolation technique is to disconnect the confined space completely by removing a section of pipe or duct and fitting blanks. Other methods include the use of spectacle blinds and lockable valves.

Workforce involvement

C8077 Employees and their representatives should be consulted when assessing the risks connected with entering or working in a confined space. Particular attention is required where the work circumstances change frequently such as at construction sites or steel fabrications.

Model or generic risk assessments

C8078 Where a number of confined spaces (for example, sewers or manholes) are broadly the same in terms of the conditions and the activities being carried out, model risk assessments are permitted provided that the risks and measures to deal with them are the same. Any differences in particular cases which would alter the conclusions of the model risk assessment must be identified.

Planning an entry into a confined space

C8079 To satisfy the safe system requirement of the *Confined Spaces Regulations 1997* (*SI 1997 No 1713*), *Reg 4*, it is necessary to plan the work thoroughly and to organise various facilities and arrangements. For a large confined space and multiple entries, a logging or tally system may be necessary in order to check everyone in and out and to control duration of entry.

Competence for confined space working

C8080 The competent person carrying out the risk assessment for work in confined spaces will need to consider the suitability of individuals in view of the particular work to be done.

Examples:

- suitable build of individuals for exceptional constraints in the physical layout of the space (this may be necessary to protect both the individual and others who could be affected by the work to be done);
- medical fitness concerning claustrophobia or the wearing of breathing apparatus.

Procedures and written instructions

C8081 To be effective a safe system of work needs to be in writing, i.e. in the form of written instructions setting out the work to be done and the precautions to be taken. Each procedure should contain all appropriate precautions to be taken and in the correct sequence.

In particular, procedures for confined space working should include instructions and guidance for:

- *First aid:* the availability of appropriate first aid equipment for emergencies until professional medical help arrives.
- *First aiders*: the strategic positioning of trained personnel to deal with foreseeable injuries.
- *Limiting working time*: for example, when respiratory protective equipment is used, or when the work is to be carried out under extreme conditions of temperature and humidity.
- *Communications*: that is, the system of adequate arrangements to enable efficient communication between those working inside the confined space and others to summon help in case of emergency.
- *Engine driven equipment*: that is, the rules regarding the siting of such equipment which should be well away from the working area and downwind of any ventilator intakes.
- *Water surges*: especially the anticipation that sewers can be affected over long distances by water surges, for example following sudden heavy rainfall upstream of where the work is being carried out.
- *Toxic gas, fume or vapour*: procedures to ensure that work can be undertaken safely to include the availability of additional facilities and arrangements where residues may be trapped in sludge, scale or other deposits, brickwork, or behind loose linings, in liquid traps, joints in vessels, in pipe bends, or in other places where removal is difficult.
- *Testing/monitoring the atmosphere*: procedures for the regular testing for hazardous gas, fume or vapour or to check the concentration of oxygen before entry or re-entry into the confined space.
- *Gas purging*: the availability of suitable equipment to purge the gas or vapour from the confined space.
- *Ventilation requirements*: the provision of suitable ventilation equipment to replace oxygen levels in the space, and to dilute and remove gas, fume or vapour produced by the work.
- *Lighting*: procedures to ensure that the confined space is well lit by lighting equipment, including emergency lighting, which must be suitable for use in flammable or potentially explosive atmospheres.

  Generally all lighting to be used in confined spaces should be protected against knocks, for example, by a wire cage, and be waterproof. Where water is present in the space, suitable plug/socket connectors capable of withstanding wet or damp conditions should be used and protected by residual current devices (RCDs) suitable for protection against electric shock. The position of lighting may also be important, for example to give ample clearance for work or rescue to be carried out unobstructed.

Fire prevention and protection procedures

C8082 The presence of flammable substances and oxygen enrichment in a confined space creates a serious hazard to workers inside the space. There is also a risk of explosion from the ignition of airborne flammable contaminants. In addition, a fire or explosion can be caused by leaks from adjoining plant or processes and the use of unsuitable equipment.

Note: in the case of *R v Associated Octel Co Ltd [1996] 4 All ER 846* (see C8147 below) a contractor was badly burned when an explosion occurred in the confined space (a chemical storage tank) he was working in. The principal cause of the accident was unsuitable lighting which broke and ignited some acetone solvent contained in an old emulsion bucket.

There are many fire precautions necessary for safe working in confined spaces; some of the more important of these are outlined below:

- *Fire prevention measures*: procedures to ensure that no flammable or combustible materials are stored in confined spaces that have not been specifically created or allocated for that purpose. In any event, the quantity of the material should be kept to a minimum and stored in suitable fire-resistant containers.
- *Fire protection and fire-fighting equipment*: procedures to ensure the availability of appropriate fire-fighting equipment where the risk of fire has been identified. In some situations, a sprinkler system may be appropriate.
- *Smoking*: procedures to ensure the prohibition of all smoking within and around all confined spaces.
- *Static electricity*: procedures to ensure that the build-up of static in a confined space is minimised. It may be necessary to obtain specialist advice regarding insulating characteristics (for example, most plastics), steam or water jetting equipment, clothing containing cotton or wool, flowing liquids or solids such as sand.

Supervision and training

C8083 It is likely that the risk assessment will identify a level of risk requiring the appointment of a competent person to supervise the work and ensure that the precautions are adhered to. Competence for safe working in confined spaces requires adequate training – in addition, experience in the particular work involved is essential. Training standards must be appropriate to the task, and to the individuals' roles and responsibilities as indicated during the risk assessment.

Emergency arrangements and procedures

C8084 The arrangements for the rescue of persons in the event of an emergency must be suitable and sufficient and, where appropriate, include rescue and resuscitation equipment. The arrangements should be in place before any person enters or works in a confined space. [*SI 1997 No 1713, Reg 5*].

The arrangements must cover any situation requiring the recovery of a person from a confined space, for example incapacitation following a fall.

Size of openings to enable rescue from confined spaces

C8085 Experience has shown that the minimum size of an opening to allow access with full rescue facilities including self-contained breathing apparatus is 575 mm diameter. This size should normally be used for new plant, although the openings for some confined spaces may need to be larger depending on the circumstances, for example to take account of a fully equipped employee, or the nature of the opening.

Public emergency services

C8086 In some circumstances, for example where there are prolonged operations in confined spaces and the risks justify it, there may be advantage in prior notification to the local emergency services before the work is undertaken. In all cases, however, arrangements must be in place for the rapid notification of the emergency services should an accident occur. On arrival, the emergency services should be given all known information about the conditions and risks of entering and/or leaving the confined space before a rescue is attempted.

Training for emergencies and rescue

C8087 To be suitable and sufficient the arrangements for training site personnel for rescue and resuscitation should include consideration of:

- rescue and resuscitation equipment;
- raising the alarm and rescue;
- safeguarding the rescuers;
- fire safety;
- control of plant;
- first aid.

Regular refresher training in the emergency procedures is essential and practice drills including emergency rescues will help to check that the size of openings and entry procedures are satisfactory. The risk assessment may indicate that at least one person, dedicated to the rescue role, should be stationed outside the confined space to keep those inside in constant direct visual sight.

All members of rescue parties should be trained in the operation of appropriate fire extinguishers which should be strategically located at the confined space. In some situations, a sprinkler system may be appropriate. In all cases, in the event of a fire the local fire service should be called in case the fire cannot be contained or extinguished by first-aid measures.

The training syllabus should include the following, where appropriate:

- the likely causes of an emergency;
- rescue techniques and the use of rescue equipment, for example breathing apparatus, lifelines, and where necessary a knowledge of its construction and how it works;
- the checking procedures to be followed when donning and using breathing apparatus;
- the checking of correct functioning and/or testing of emergency equipment (for immediate use and to enable specific periodic maintenance checks);

- identifying defects and dealing with malfunctions and failures of equipment during use;
- works, site or other local emergency procedures including the initiation of an emergency response;
- instruction on how to shut down relevant plant as appropriate (this knowledge would be required by anyone likely to perform a rescue);
- resuscitation procedures and, where appropriate, the correct use of relevant ancillary equipment and any resuscitation equipment provided (if intended to be operated by those receiving emergency rescue training);
- emergency first aid and the use of the first aid equipment provided;
- liaison with local emergency services in the event of an incident, providing relevant information about conditions and risks, and providing appropriate space and facilities to enable the emergency services to carry out their tasks.

Rescue equipment

C8088 When safety harness and lines are provided, it is essential that proper facilities to secure the free end of the line are available. In most cases the line should be secured outside the entry to the confined space. Lifting equipment may be necessary and the harness should be of suitable construction, and made of suitable material to recognised standards capable of withstanding both the strain likely to be imposed, and attack from chemicals.

Maintenance of safety and rescue equipment

C8089 All equipment provided or intended to be used for the purposes of securing the health and safety of people in connection with confined space entry or for emergency or rescue, should be maintained in an efficient state, in efficient working order and in good repair. This should include periodic examination and testing as necessary. Some types of equipment, for example breathing apparatus, should be inspected each time before use.

Atmospheric monitoring equipment, and special ventilating or other equipment provided or used in connection with confined space entry, needs to be properly maintained by competent persons. It should be examined thoroughly, and where necessary calibrated and checked at intervals in accordance with recommendations accompanying the equipment or, if these are not specified, at such intervals determined from the risk assessment.

Records of the examination and tests of equipment should normally be kept for at least five years. The records may be in any suitable format and may consist of a suitable summary of the reports. Records need to be kept readily available for inspection by the employees, their representatives, or by inspectors appointed by the relevant enforcing authority or by employment medical advisers.

Equipment for use in explosive atmospheres

C8090 When selecting equipment for use in confined spaces where an explosive atmosphere may be present, the requirements of the EU-originated Regulations, *Equipment and Protective Systems Intended for Use in Potentially Explosive Atmospheres Regulations* (*SI 1996 No 192*), must be complied with. These Regulations apply to 'equipment' and 'protective systems' intended for use in potentially explosive atmospheres. Some of the terms used in the Regulations are defined below:

- *equipment* means machines, apparatus, fixed or mobile devices, control components and instrumentation thereof and detection or prevention systems which, separately or jointly, are intended for the generation, transfer, storage, measurement, control and conversion of energy or the processing of material and which are capable of causing an explosion through their own potential sources of ignition;
- *protective systems* means design units which are intended to halt incipient explosions immediately and/or to limit the effective range of explosion flames and explosion pressures; protective systems may be integrated into equipment or separately placed on the market for use as autonomous systems;
- *devices* means safety devices, controlling devices and regulating devices intended for use outside potentially explosive atmospheres but required for or contributing to the safe functioning of equipment and protective systems with respect to the risks of explosion;
- *explosive atmosphere* means the mixture with air, under atmospheric conditions, of flammable substances in the form of gases, vapours, mists or dusts in which, after ignition has occurred, combustion spreads to the entire unburned mixture.

Selection and use of equipment

C8091 All equipment must bear the approved CE mark properly fixed in accordance with the requirements of the 1996 Regulations (*SI 1996 No 192*).

Any equipment provided for use in a confined space needs to be suitable for the purpose. Where there is a risk of a flammable gas seeping into a confined space, which could be ignited by electrical sources (for example a portable hand lamp), specially protected electrical equipment must be used.

To be suitable the equipment should be selected on the basis of its intended use – proper earthing is essential to prevent static charge build-up; mechanical equipment may need to be secured against free rotation, as people may tread or lean on it.

Work in compressed air on construction sites – Work in Compressed Air Regulations 1996 (SI 1996 No 1656)

C8092 Replacing the *Work in Compressed Air Special Regulations 1958*, the *Work in Compressed Air Regulations 1996* (*SI 1996 No 1656*) reflect more modern decompression criteria and are more concerned with the long-term effects of rapid return to atmospheric pressure than the short-term effects which were addressed by earlier regulations. The Regulations require principal contractors to appoint competent compressed air contractors, and the compressed air contractors to appoint contract medical advisers. In addition, greater provision is required in connection with fire prevention and protection measures (including, in particular, emergency means of escape and rescue). Duties are laid on principal contractors, employers (including the self-employed – tunnellers to whom these Regulations are substantially addressed are generally self-employed), and employees.

It should be noted that due consideration must also have been given to any possibility of avoiding such a high risk work method altogether under the design duty of the *CDM Regulations*.

Principal contractor's duties

C8093 The principal contractor must appoint a compressed air contractor in respect of work in compressed air, who must be competent. A compressed air contractor may be the principal contractor himself, if competent. [*SI 1996 No 1656, Reg 5*(*1*), (*2*)].

Compressed air contractor's duties – notification

C8094 The 1996 Regulations require that:

- the compressed air contractor must not allow work in compressed air to be carried out unless written notice has been forwarded to HSE at least 14 days before commencement of work. Where this is not practicable, owing to an emergency, notice must be given as soon as practicable after the necessity for such work becomes known to the compressed air contractor, and, anyway, before work commences [*SI 1996 No 1656, Reg 6*(*1*), (*2*)];
- no person must work in compressed air unless written notice is forwarded to:
  - (i) the nearest suitably equipped hospital,
  - (ii) the local ambulance service,
  - (iii) the local fire service, and
  - (iv) other establishments in the vicinity with an operable medical lock,

  [*SI 1996 No 1656, Reg 6*(*3*), (*4*)];
- notification should be in writing and contain the following information:
  - (i) the fact that work in compressed air is being undertaken,
  - (ii) the location of the site,
  - (iii) date of commencement and anticipated completion of work,
  - (iv) name of compressed air contractor and a 24-hour contact telephone number,
  - (v) name, address and telephone number of the contract medical adviser,
  - (vi) intended pressure at which the work is to be undertaken,
  - (vii) anticipated pattern of work (e.g. shifts), and
  - (viii) number of workers likely to be in each shift,

  [*SI 1996 No 1656, Reg 6*(*4*), *Sch 1*].

Competent persons

C8095 The compressed air contractor must ensure that no person works in (or leaves) compressed air, except in accordance with a system of work which, so far as is reasonably practicable, is safe and without health risks. [*SI 1996 No 1656, Reg 7*(*1*)]. To this end, he must ensure that a sufficient number of 'competent persons' are immediately available on site to supervise execution of work in compressed air at all times and for up to 24 hours when work is being undertaken at or above a pressure of 0.7 bar. [*SI 1996 No 1656, Reg 7*(*2*)].

Plant and equipment

C8096 The compressed air contractor must ensure that all plant and equipment is:

- of a proper design and construction and of sufficient capacity;
- safe and without health risks and safely maintained; and
- where such plant and equipment is used for the purpose of containing air at a pressure greater than 0.15 bar, it is :
  - (i) examined and tested by a competent person and any faults rectified prior to use, and
  - (ii) re-examined and re-tested after modification or alteration.

[*SI 1996 No 1656, Reg 8*].

Compression and decompression procedures

C8097 The compressed air contractor must ensure that:

- compression or decompression is only carried out as per procedures approved by HSE;
- no worker is subjected to a pressure greater than 3.5 bar (except in emergencies);
- no worker is subjected to 'decanting' (i.e. rapid decompression in an airlock to atmospheric pressure followed promptly by rapid compression in an alternative airlock and subsequent decompression to atmospheric pressure – except in an emergency); and
- an adequate record is made of exposure in respect of times and pressures at which work in compressed air is carried out, and kept for a minimum of 40 years (including individual exposure records). Such records must be made available to the worker himself and his employer, the latter being required to keep the record for at least 40 years.

[*SI 1996 No 1656, Reg 11*].

Provision and maintenance of adequate medical facilities

C8098 The compressed air contractor must ensure provision and maintenance of adequate medical facilities (e.g. medical lock, recompression therapy) for the treatment of people working in compressed air and those who have worked in compressed air in the previous 24 hours. Where work is carried out at a pressure greater than 0.7 bar, facilities should include a medical lock; and where work is carried out at a pressure greater than 1.0 bar, a medical lock attendant should be present. [*SI 1996 No 1656, Reg 12*].

Emergencies

C8099 The compressed air contractor must ensure that no work in compressed air takes place in the absence of suitable and sufficient arrangements in the event of emergencies as follows:

- provision and maintenance of a sufficient number of suitable means of access;

- preparation of a suitable rescue plan which can be put into effect immediately (including the provision and maintenance of plant and equipment necessary to put the rescue plan into operation);
- provision and maintenance of suitable lighting;
- provision and maintenance of suitable means of raising the alarm; and
- in cases where an airlock is required, maintenance of the airlock, so that it is fit to receive persons in the event of emergency (with particular regard to air supply and temperature of the airlock).

[*SI 1996 No 1656, Reg 13*].

Fire precautions

C8100 The compressed air contractor must ensure provision of suitable and sufficient means for fighting fire and that any airlock or working chamber is maintained and operated so as to minimise the risk of fire, and must ensure the enforcement of the prohibition against smoking. [*SI 1996 No 1656, Reg 14*].

Information, instruction and training

C8101 The compressed air contractor must ensure provision of adequate information, instruction and training to employees, including, particularly, information relating to the risks arising from the work and the precautions to be observed. [*SI 1996 No 1656, Reg 15*].

Fitness for work

C8102 The compressed air contractor must ensure that no one works in compressed air where he has reason to believe that the worker is subject to a medical or physical condition likely to make him unfit or unsuitable for such work. [*SI 1996 No 1656, Reg 16*].

Prohibition against alcohol and drugs

C8103 The compressed air contractor must prohibit anyone from working in compressed air where he believes such worker to be under the influence of drink and/or drugs. [*SI 1996 No 1656, Reg 17*].

Employees' duties

C8104 Employees, including the self-employed, must:

- when required to do so, and at the cost of the employer (see OCCUPATIONAL HEALTH AND DISEASES), submit to medical surveillance procedures during working hours [*SI 1996 No 1656, Reg 10(6)*];
- avoid smoking or carrying smoking materials [*SI 1996 No 1656, Reg 14(2)*];
- avoid consumption of alcohol or drugs [*SI 1996 No 1656, Reg 17(2)*]; and
- wear a badge or label for 24 hours after leaving work in compressed air [*SI 1996 No 1656, Reg 19(2)*].

Contract medical adviser

C8105 Owing to the potentially serious dangers arising from pressure itself, or the construction work being done, appointment of a contract medical adviser is essential. Ideally, such person would be a doctor appointed by HSE, who can carry out statutory medical examinations on compressed air workers on site. His principal role is to actively monitor incidence of decompression illness during work. In the event of decompression illness arising, the contract medical adviser should advise the medical lock attendant regarding appropriate treatment. Both the contract medical adviser and medical lock attendant are responsible for collation and maintenance of exposure records and completion of the worker's health and exposure record, and, on completion of the contract, assist the compressed air contractor in the preservation of formal health and exposure records for the statutory 40-year period.

Competent persons

C8106 The phrase 'competent persons' can refer to:

- the engineer in charge;
- the compressor attendants;
- the lock attendants;
- the medical lock attendants (for work in compressed air over 1.0 bar); and
- the contract medical adviser.

Enforcement

C8107 In any proceedings for an offence consisting of a contravention of *Reg 14*(*3*) or *17*(*3*) (compressed air contractor's duty to ensure compliance with prohibitions against smoking, alcohol or drugs), it is a defence for any person to prove that he took all reasonable precautions and exercised all due diligence to avoid the commission of the offence. [*SI 1996 No 1656, Reg 20*].

Construction (Head Protection) Regulations 1989 (SI 1989 No 2209)

C8108 Head injuries account for nearly one-third of all construction fatalities, but fell significantly after the introduction of the *Construction* (*Head Protection*) *Regulations 1989* (*SI 1989 No 2209*). These Regulations specify requirements for head protection during construction work, including offshore operations , but not diving operations at work. They place duties on employers, persons in control of construction sites, self-employed persons and employees regarding the wearing of head protection. The purpose of head protection is to prevent/mitigate head injury caused by: falling/swinging objects, e.g. materials and/or crane hooks; and striking the head against something, as where there is insufficient headroom. Circumstances where head injury is not reasonably foreseeable on construction sites are limited, but it probably would not be required on/in:

- sites where buildings are completed and there is no risk of falling materials/objects;
- site offices, cabins, toilets, canteens or mess rooms;
- cabs of vehicles, cranes etc;
- work at ground level, e.g. road works.

Duties of employers

C8109 The following duties are laid on employers.

- *Provision/maintenance of head protection*

  Every employer (i.e. main contractor, subcontractor etc.) must provide each employee, while at work on building/construction operations, with suitable head protection, and keep it maintained/replaced (as recommended by the manufacturer). [*SI 1989 No 2209, Reg 3*(*1*), (*2*)].

  Moreover, head protection equipment must be kept in good condition and stored, when not in use, in a safe place, though not in direct sunlight or hot or humid conditions. It should be inspected regularly and have defective harness components replaced, and sweatbands regularly cleaned or replaced.

- *Ensuring head protection is worn*

  So far as reasonably practicable (for meaning, see E15039 ENFORCEMENT), every employer must ensure that each of his employees, whilst on construction work, wears suitable head protection, unless there is no foreseeable risk of injury to his head (other than by falling). [*SI 1989 No 2209, Reg 4*(*1*)].

  Moreover, every employer (or employee) who has control (for meaning, see below) over any other person engaged in construction work, must ensure, so far as is reasonably practicable, that such persons wear suitable head protection, unless there is no foreseeable risk of injury to the head, other than by falling. [*SI 1989 No 2209, Reg 4*(*2*)].

Persons in control of construction sites

C8110 For the purposes of the 1989 Regulations, the following persons may be deemed to be 'in control' of construction sites:

- main contractor;
- managing contractor;
- contractor bringing in subcontractors;
- contract manager;
- site manager;
- subcontractor;
- managers, including foremen, supervisors;
- engineers and surveyors;
- (sometimes) clients and architects with control over persons at work.

Procedures and rule making

C8111 Employers and others in control must:

- identify when/where head protection should be worn;
- inform site personnel procedurally when/where to wear head protection and post suitable safety signs to that effect;
- provide adequate supervision;
- check that head protection is, in fact, worn.

Supervision by those responsible for ensuring head protection is worn, is an on-going requirement, including monitoring helmet use at all times, starting early in the day and taking in arrivals on site.

Persons in control of construction works can (and should) make rules regulating the wearing of suitable head protection. Such rules must be in writing and be brought clearly to the attention of those involved. Such procedure is particularly useful to main/managing contractors on multi-contractor sites, and rules/regulations on head protection should form part of overall site safety procedures, such as construction phase safety plans in accordance with the *CDM Regulations* (*SI 1994 No 3140* as amended by *SI 2000 No 2380*) (see C8033 above).

Wearing suitable head protection – duty of employees

C8112 Employees must also make full and proper use of head protection and return it to the accommodation provided for it after use. [*SI 1989 No 2209, Reg 6* as amended by the *Personal Protective Equipment at Work Regulations 1992, Sch 2, para 24*]. They must also comply with the rules and regulations made for the wearing of head protection mentioned in C8109 above (see also PERSONAL PROTECTIVE EQUIPMENT). All employees, provided with suitable head protection, must take reasonable care of it and report any loss of it or obvious defect in it, to the employer etc. [*SI 1989 No 2209, Reg 7*].

Suitable head protection

C8113 Suitable head protection refers to an industrial safety helmet conforming to British Standard BS EN 397: 1995 'Industrial Safety Helmets' – Specification for construction and performance (or an equivalent standard). For work in confined spaces, 'bump caps' to BS EN 812: 1998 specification are more suitable.

Suitability of head gear involves the following factors: (*a*) fit, (*b*) comfort, (*c*) compatibility with work to be done, and (*d*) user choice.

- *Fit*

  Head protection should be of an appropriate shell size for the person who is to wear it, and have an easily adjustable headband, nape and chin strap. The range of size adjustment should be sufficient to accommodate thermal liners in cold weather.

- *Comfort*

  Head gear should be as comfortable as possible, including:

  (*a*) a flexible headband of adequate width and contoured vertically and horizontally to fit the forehead;

  (*b*) an absorbent, easily cleanable or replaceable sweatband;

  (*c*) textile cradle straps;

  (*d*) chin straps (when fitted) which:

  (i) fit round the ears,

  (ii) are compatible with any other personal protective equipment needed,

  (iii) are fitted with smooth, quick release buckles which do not dig into the skin,

  (iv) are made from non-irritant materials,

(v) are capable of being stowed on the helmet when not in use.

- *Compatibility with work to be done*

  Head gear should not impede work to be done. For instance, an industrial safety helmet with little or no peak is functional for a surveyor taking measurements, using a theodolite or to allow unrestricted upward vision for a scaffold erector. If a job involves work in windy conditions, at heights, or repeated bending or constantly looking upwards, a secure retention system is necessary. Flexible headbands and Y-shaped chin straps can help to secure the helmet on the head. If other personal protective equipment, such as ear defenders or eye protectors, are required, the design must allow them to be worn safely and in comfort.

- *User choice*

  In order to avoid possibly unpleasant industrial relations consequences or a possible action for unfair dismissal (see further EMPLOYMENT PROTECTION), it is sensible and advisable to allow the user to participate in selection of head gear.

Duties of self-employed personnel

C8114 Every self-employed person involved in construction/building operations, must:

- provide himself with suitable head protection and maintain/replace it, whenever necessary [*SI 1989 No 2209, Reg 3*(*2*)];
- ensure that any person over whom he has control, wears suitable head gear, unless there is no foreseeable risk of injury [*SI 1989 No 2209, Reg 4*(*2*)];
- give directions to any other self-employed person regarding wearing of suitable head gear [*SI 1989 No 2209, Reg 5*(*4*)];
- wear properly suitable head protection, unless there is no foreseeable risk of injury to the head, and make full and proper use of it and return it to the accommodation provided for it after use [*SI 1989 No 2209, Reg 6*(*2*)–(*4*)];
- where the presence of more than one risk to health or safety makes it necessary for him to wear or use simultaneously more than one item of personal protective equipment, see that such equipment is compatible and continues to be effective against the risk or risks in question [*Personal Protective Equipment at Work Regulations 1992* (*SI 1992 No 2966* as amended by *SI 2002 No 2174*), *Reg 5*(*2*)].

Exceptions

C8115 The following categories of workers on construction sites are exempt from the 1989 Regulations:

- divers actually diving or preparing to dive;
- Sikhs wearing turbans on construction sites [*Employment Act 1989, ss 11, 12*], though the Regulations do apply to Sikhs not normally wearing turbans at work. (The probability is that this exemption, under the *Employment Act 1989, s 11*, is now subordinate to the requirements of the *Construction* (*Head Protection*) *Regulations 1989* and the *HSWA 1974*, with the result that Sikhs working on construction sites will have to wear hard hats (*SS Dhanjal v British Steel plc (Case No 50740/91)*).)

Visitors on site

C8116 It is not necessary that visitors are provided with or, even less, wear head protection, under the 1989 Regulations. Nevertheless, in order to satisfy their general duty under the *HSWA 1974*, *s 4* and additionally to avoid any civil liability for injury at common law in an action for negligence and/or under the *Occupiers' Liability Act 1957* (see OCCUPIERS' LIABILITY), employers should provide visitors to the site with suitable head protection where there is a reasonably foreseeable likelihood of injury. This contention is further reinforced by the requirement for every employer to consider the risks inherent in his business which have the potential to harm his employees or any other persons who might be affected, and to take measures to remove or reduce such risks. [*MHSWR 1999* (*SI 1999 No 3242*), *Reg 3*].

Composition of construction products

C8117 Products must be suitable for construction works and works of civil engineering and can then carry the 'CE' mark. To that end, when incorporated into design and building, construction products should satisfy the following criteria, namely,

- mechanical resistance and stability;
- safety in case of fire;
- hygiene, health and the environment;
- safety in use;
- protection against noise; and
- energy economy and heat retention.

[*Construction Products Regulations 1991* (*SI 1991 No 1620*), *Reg 3, Sch 2* as amended by *SI 1994 No 3051*].

Manufacturers must show that their products conform to these specifications, if necessary, by submitting to third party testing (see also PERSONAL PROTECTIVE EQUIPMENT for products generally).

Protecting visitors and the public

C8118 HSE have issued revised guidance, HS(G)151 '*Protecting the public – your next move*' (June 1997), which provides practical advice on the measures to be taken to minimise risks to the public and others not directly involved in construction activities. The advice is aimed at preventing accidents and ill health and, to a limited extent, at reducing incidents of nuisance. The guidance does not cover deliberate illegal trespass or forced entry on to sites by protest groups or those intent on criminal activity.

In addition, reference should be made to the following related statutory provisions which have aspects relating to public safety during construction or building operations:

(*a*) Roads and streets

Highways Act 1980:

- *s 168* (building operations affecting public safety);
- *s 169* (the control of scaffolding on highways);
- *s 174* (the erection of barriers, signs and lighting etc.).

New Roads and Street Works Act 1991:

- *s 50* (lays down particular safety requirements for work in the street and, specifically, the measures to be taken to minimise inconvenience to the disabled).

Environmental Protection Act 1990:

- *s 79* as amended by the *Noise and Statutory Nuisance Act 1993* (noise or vibration emitted from buildings and from or caused by a vehicle, machinery or equipment in a street).

(*b*) Waste from building and demolition sites

Waste produced on construction sites is classed as controlled waste and as such must be controlled to comply with EU-based Directives:

Environmental Protection Act 1990:

- *ss 33–46* (deal with waste management and licensing control).

Controlled Waste (Registration of Carriers and Seizure of Vehicles) Regulations 1991 (*SI 1991 No 1624* as amended) (carriage of controlled waste by registered carriers only).

Waste Management Licensing Regulations 1994 (*SI 1994 No 1056* as amended) (registers, applications and waste regulation authorities for the recovery and disposal of waste).

Special Waste Regulations 1996 (*SI 1996 No 972* as amended) (hazardous properties of waste).

Identifying hazards and evaluating risks

C8119 Construction work is by its nature carried out by workers away from the home base and may not be open to direct management and supervision. The employer's general duty of care to ensure that employees are not put at risk by their work activities still applies. [*HSWA 1974, s 2*]. In addition, if the work activities impinge on others, such as another organisation or members of the public, a further duty of care applies. [*HSWA 1974, s 3*].

The *HSWA 1974, s 4* relates to the control of premises, rather than the control of undertakings. Persons who control premises used by people who are at work, but who are not their employees, need to ensure, so far as is reasonably practicable, that the premises, access to them and plant and substances used on them are safe and free from risks to health and safety. Site occupiers therefore share a duty of care with contractors (as both are employers) to ensure that all reasonably practicable precautions are taken to safeguard their own employees, other persons on site and the public.

Under the *Occupiers' Liability Act 1957* and the *Occupiers' Liability Act 1984* a duty of care is imposed on occupiers of existing premises regarding visitors. The duty of care extends to children – it should be noted that a child is regarded as being at greater risk than an adult (see O3010 OCCUPIERS' LIABILITY). The 1984 Act further extends the duty of an occupier to people other than lawful visitors, such as trespassers, to ensure that they are not injured whilst on the premises. This may involve making unauthorised access more difficult or putting up suitable warning signs regarding hazards on site.

Safety policies and written arrangements

C8120 The *HSWA 1974, s 2(3)* requires written safety arrangements only with regard to employee safety. However, by virtue of the *MHSWR 1999* (*SI 1999 No 3242*), *Reg 5* (health and safety arrangements) written arrangements are also required with regard to the protection of the public and other non-employees.

The *MHSWR 1999* refers expressly to occupiers' responsibility to co-operate and co-ordinate arrangements [*SI 1999 No 3242, Regs 11* and *12*] and to provide information and training on risks and precautions [*SI 1999 No 3242, Regs 10* and *13*]. These duties apply whether or not payment is involved, for example free surveys, estimates, measurements, maintenance and servicing under warranty, etc.

Measures to ensure the safety of visitors and members of the public must be one result of the risk assessment task as required by the *MHSWR 1999* (*SI 1999 No 3242*), *Reg 3* (see R3005 RISK ASSESSMENT). Chemical installations or other high risk manufacturing or chemical storage premises are also subject to specific duties towards the general public under the *Control of Major Accident Hazards Regulations 1999* (*SI 1999 No 743*) (see CONTROL OF MAJOR ACCIDENT HAZARDS).

Companies with cooling towers on site are also subject to specific controls ultimately designed to protect the public. In *R v Board of Trustees of the Science Museum, The Times, March 15 1993, CA* the prosecution had alleged that members of the public outside the Science Museum had been exposed to risks to their health from *legionella pneumophila*, because of inadequate maintenance of the museum's air conditioning system. The basis of their case had been that it is sufficient for the prosecution to show that there has been a risk to health.

The Court of Appeal decided that the word 'risks' in the *HSWA 1974, s 3(1)* implied the idea of potential danger. There was nothing in the subsection which narrowed this meaning. The *HSWA 1974* should be interpreted so as to make it effective in its role of protecting public health and safety. *Section 3(1)* was intended to be an absolute prohibition, subject to the defence of reasonable practicability.

Insurance cover

C8121 The *Employers' Liability (Compulsory Insurance) Act 1969* states that an employer's legal liability for death, disease or bodily injury suffered by employees as a consequence of employment must be insured by the employer for their mutual protection under the duty of care owed by the *HSWA 1974, s 2.* A copy of the current certificate of insurance (issued annually) must be displayed within all working premises. Other insurances, for example reflecting the risk of liability to non-employees under the duty of care imposed by the *HSWA 1974, s 3*, are likely to be essential even if not compulsory.

Site planning and layout

C8122 Risk assessment should decide how the site perimeters will be defined, what type of barriers and fencing will be most effective and where they should be placed.

For most sites the perimeter will be the geographical area within which the construction work will be carried out. Determining the perimeter is an important aspect of managing public risk. It must always be recognised that site perimeters need to be changed as the work progresses.

Under the *CDM Regulations* (*SI 1994 No 3140* as amended by *SI 2000 No 2380*), the duties of the principal contractor are wide and varied. Importantly, the *CDM Regulations* are excluded from use in civil proceedings, except for *Reg 10* which requires the client to ensure that an adequate health and safety plan has been prepared before the construction phase of the project commences, and *Reg 16(1)(c)* which requires the principal contractor to take appropriate steps to ensure that only authorised access is permitted to premises where construction work is continuing.

Similarly, where construction work is taking place on an occupied site, the client will impose existing security rules on everyone entering and leaving the site.

Under the *CDM Regulations*, other contractors involved in the construction work are required to co-operate with the principal contractor and comply with any directions or site rules. In addition, all contractors must provide appropriate information, including information relating to any injuries, diseases and dangerous occurrences.

Employers' liability for the actions and safety of the public and non-employees

C8123 Under the *HSWA 1974*, both employers and the self-employed have duties not only to their own workpeople but also to outside contractors, workers employed by them and to members of the public (whether within or outside the workplace), who may be affected by work activities. Undertakings must be conducted in such a way as to ensure, so far as is reasonably practicable, that they do not expose people who are not their employees to risks to their health and safety. The duty extends to, for example, risks to the public outside the workplace from fire or explosion, from falls of unsafely erected scaffolding or from the release of harmful substances into the atmosphere.

In general, the standard of protection required for visitors and others within a construction site will be similar to that given to employees. There may, however, be a need to apply different criteria to achieve these standards when assessing the risks to members of the public. For example, it will be necessary to consider that certain people, such as the very young or disabled, may be more vulnerable than others and that people visiting or passing a workplace may have less knowledge of the potential hazards and of how to avoid them.

The responsibilities of employers and the self-employed to non-employees will in certain circumstances extend to people entering workplaces without permission. This is apart from any liability under common law towards trespassers. The duty under the *HSWA 1974* to conduct the business in such a way as not to expose people to risks to health and safety implies taking certain precautions to deter people from unlawfully entering the workplace, for example by the provision of fences, barriers and notices warning of the danger. The duty towards 'unauthorised' people is qualified 'so far as is reasonably practicable', and on construction sites and other open-air workplaces, which have particular dangers as far as children are concerned, simply locking or guarding main doors and gates may not be adequate.

Duties of all people

C8124 The *HSWA 1974* imposes one duty on all people, both people at work and members of the public, including children: this is not intentionally to interfere with or misuse anything that has been provided in the interests of health, safety or welfare, whether it has been provided for the protection of employees or other people. The purpose of the provision is clearly to protect things intended to ensure people's safety, including fire escapes and fire extinguishers, perimeter fencing, warning notices for particular hazards, protective clothing, guards on machinery and special containers for dangerous substances.

Control measures

C8125 The general duties of protection owed to visitors apply equally to the emergency services who should be given a plan or map of the premises together with information on specific high risk areas where high voltage or dangerous chemicals may be present. In addition, to comply with occupier's liability legislation the duty of care towards visitors includes contractors.

Visitors

C8126 In general, visitors to a construction site should not be left unaccompanied and they should not, if possible, be taken into any hazardous areas. All visitors should be made to sign in on arrival and sign out on departure and, ideally, be given basic instructions on what to do in the event of an emergency. The main element of looking after visitors is to ensure that they are accompanied at all times, so that the host can lead them to safety in the event of fire or other emergency. Constant accompaniment of visitors will, of course, also improve security arrangements.

Contractors

C8127 Under the *MHSWR 1999* the occupier of the premises must ensure that contractors on site are provided with comprehensible information on:

- the risks to health and safety arising out of the activities on site; and
- the measures taken by the occupier to ensure compliance with statutory requirements.

[*SI 1999 No 3242, Reg 10*].

The ideal situation is where contractors can be provided with a completely separated area which can be designated as being under their control. Such an arrangement will normally only apply when the contractors are on site to undertake a major engineering or construction project. It will only be successful if a contractor is actually given full control of the area – and the main site occupier and his or her employees only enter the area when authorised by the contractor. However, the main site occupier will retain the key responsibility for safety matters – as illustrated in the case of a south coast town council which employed contractors to remove part of a damaged pier. The council had accepted by far the lowest quote and did not discuss the system of work. During the demolition, there was a huge explosion which removed the derelict part of the pier but also caused considerable damage to cars and buildings on the seafront. The contractor and town council were held jointly responsible but the council suffered the larger fine because it failed to employ reputable contractors and did not request a method statement from them.

General public

C8128 Measures to ensure the safety of the general public are usually more difficult and must be one result of the risk assessment task as required by the *MHSWR* (*SI 1999 No 3242*), *Reg 3*. Members of the public are owed a duty of care under the *HSWA 1974*, *ss 3* and *4*, to ensure that they are not put at risk by the employer's undertaking. For example, an employer engaged in construction work near a public place must ensure that risks are assessed and adequately controlled. The 1974 Act also imposes a duty on employees to co-operate with their employer on health and safety matters and not to do anything which puts others at risk.

As far as protecting the public is concerned, adequate control measures must be determined which should not rely on the use of protective equipment.

Risk assessments

C8129 Risk assessments are an integral part of the *CDM Regulations* (*SI 1994 No 3140* as amended by *SI 2000 No 2380*) and the *Confined Spaces Regulations 1997* (*SI 1997 No 1713*). Additionally, with respect to visitors and the general public, employers may be liable to pay compensation to people injured on their premises under the terms of the *Occupiers' Liability Act 1957.*

When carrying out a risk assessment to safeguard members of the public it is necessary to adopt a very wide approach and consider all the possible hazards and subsequent risks. This would include, but not be limited to, compiling data and control measures for:

- *chemical hazards*, e.g. mist, vapour, gas, smoke, dust, aerosol, fumes;
- *physical hazards*, e.g. noise, temperature, lighting, vibration, radiation (ionising and non-ionising), pressure;
- *biological hazards*, e.g. bacteria, parasites.

Public vulnerability

C8130 Experience has shown that members of the public are particularly vulnerable to those hazards and risks which are not readily identifiable by the normal senses of sight, smell or hearing. The problem may be exacerbated by disabilities and sensory impairments as well as by physical and mental conditions. These conditions would have been assessed for everyone inside the confines of the site but the employer must address the specific needs of all people who may be affected by the work when the hazards may extend outside the site boundaries. The needs of children and the elderly must always be given top priority.

Cooling towers

C8131 The public is particularly vulnerable to cooling towers, and employers must therefore ensure, so far as is reasonably practicable, that no one is put at risk from legionellosis as a result of work activities. Plant of this type includes hot and cold water services, air conditioning and industrial cooling systems, spas and whirlpool baths, humidifiers and air washers.

Designers, manufacturers, importers, suppliers and installers of such plant or water systems – and water treatment contractors – have a duty to ensure, so far as is reasonably practicable, that the plant or system is so designed and constructed that it will be without risks to health. Appropriate information must be provided to users, and tests carried out if required. [*HSWA 1974, s 6*].

The use of this type of plant must be notified to the local authority under the *Notification of Cooling Towers and Evaporative Condensers Regulations 1992* (*SI 1992 No 2225*) and the requirements of the *COSHH Regulations* (*SI 2002 No 2677*) must be complied with, particularly the provisions relating to:

- *risk assessment* – to include breakdowns, abnormal operation and the possibility of exposure of susceptible people (for example in hospitals) [*SI 2002 No 2677, Reg 6*]. The assessment must be reviewed at least once every five years or if there is a change in plant or operation;
- *prevention or control of exposure* [*SI 2002 No 2677, Regs 7*]. Where potential exposure to infection cannot be prevented there must be a written control scheme to minimise exposure;
- *health surveillance* where appropriate [*SI 2002 No 2677, Reg 11*];

- *information, instruction and training* [*SI 2002 No 2677, Reg 12*].

(See further '*Legionnaires disease: The control of legionella bacteria in water systems*' (L8) (ISBN 0 7176 1772 6), price £8.00, available from HSE books.

Dusts and fibres

C8132 Dust in the form of particulates suspended in air is generated from a number of construction work activities, including, but not limited to:

- cutting bricks, blocks, tiles, slabs etc.;
- sawing wood;
- mixing cement, plasters etc;
- grinding operations;
- blasting;
- demolition operations.

Fibres from asbestos demolitions are particularly harmful and strict precautions must be taken in case of an uncontrolled release of asbestos fibres from the workplace.

Environmental safety aspects during asbestos removal

C8133 Under the *Control of Asbestos at Work Regulations 2002* (*SI 2002 No 2675*), *Reg 3*, employers have duties not only to their own employees but also to, for example:

- visitors to the place where work with asbestos is being carried out;
- the occupier's employees if the work is done in someone else's premises;
- people in the neighbourhood who might be accidentally exposed to asbestos dust arising from the work.

Whenever two or more employers work with asbestos at the same time at one workplace they should co-operate in order to meet their separate responsibilities (for further detail see ASBESTOS).

Contractors undertaking work on materials containing asbestos were warned of the serious health hazards associated with such work when the *Control of Asbestos at Work Regulations* first came into force in 1988. The warning was reiterated in 1995 when a court case involving contractors cleaning an asbestos roof ruled that such contractors had a duty to take reasonable care and skill in the work including the taking of necessary precautions (*Barclays Bank plc v Fairclough Building Ltd (No 2) [1995] IRLR 605 CA*).

There have been numerous cases involving incidents during demolition and the removal of asbestos in the years following the above case and its warning. In a very significant case where the lives of children had been put at risk, HSE and the Environment Agency co-operated and raised a joint prosecution.

The case (*R v Rollco Screw and Rivet Co and Others [1998] HSE E198:98*) resulted in a defendant being jailed for nine months (the second custodial sentence for an asbestos offence). Five others and a company were ordered to pay fines and costs totalling £98,000. Birmingham Crown Court heard disturbing evidence of the casual attitude of the people carrying out the stripping of an asbestos roof and the subsequent disposal of blue, white and brown asbestos. The contractors were not licensed in accordance with the requirements of the *Asbestos (Licensing) Regulations 1983 (SI 1983 No 1649)*. The HSE inspector who conducted the investigation emphasised the importance of property owners and managers checking the qualifications of any person employed to carry out asbestos removal work. The inspector was especially critical of the clear attempt to gain financially from decisions taken which put the public at risk.

The QC for the Environment Agency gave evidence of nine contraventions of the *Environmental Protection Act 1990* (the keeping or disposal of controlled waste in such a way that pollution of the environment or harm to human health was likely) where approximately 300 bags of asbestos were dumped at different locations around the city. Some of the bags had been left open and others had burst, releasing asbestos fibres into the air. The court heard that children had been playing with loose asbestos material since it had been dumped recklessly and indiscriminately in a playground and in a supermarket car park as well as other locations.

Judge Charles Harris QC, in imposing the sentences, singled out one man for the custodial sentence and said that, unlike the others in the case, he knew the risks, he lied steadily and showed manifest dishonesty. He told him 'This was an act of the most astonishing criminal irresponsibility. You understood the nature of asbestos and yet you distributed it around Birmingham in places where people, including children, had easy access to it'. He described the co-defendants as being ignorant of the dangers of asbestos but their neglect had put others in serious danger.

Fumes, mists and vapours

C8134 Most dangers associated with fumes, mists and vapours from operations such as spreading adhesives, mixing and thinning paints and coatings and spraying are well known. In addition, care should be taken and appropriate precautions put in place for operations such as:

- cleaning operations;
- heat treatment processes;
- disturbance of sludge and scale from vessels;
- the release of gases from sewers and similar operations;
- emissions from extraction equipment, LEV systems and the venting of relief valves etc.;
- the use of aerosols;
- the use of pesticides.

Fumes from cutting, welding, soldering and brazing operations are particularly dangerous and require a thorough assessment.

Radiations and hazardous waves

C8135 The dangers associated from the use of equipment emitting radiations, radio-frequency waves and micro-waves are well documented and the precautions to be observed during the use of such equipment must be strictly adhered to at all times. Some operations where the public may be at risk from uncontrolled releases or discharges from the equipment or, in some cases, from natural causes are:

- infra-red and ultraviolet radiations during cutting and welding;
- ionising radiation from radiography;

- laser rays (for example the use of lasers during accurate alignment operations of machinery or structures);
- X-rays from high voltage sources (for example non-destructive testing operations);
- the presence of radon on some soils and rocks.

Fire and smoke inhalation

C8136 In addition to the dangers of fire spreading to areas and premises occupied by members of the public, it is necessary to consider the effects of smoke from burning refuse and discarded materials. Advice should be sought from appropriate specialists and approval granted before commencing any burning operations.

Noise control on construction sites

C8137 In addition to the *Noise at Work Regulations 1989* (*SI 1989 No 1790*) and HSE's supporting guidance to the Regulations ('*Reducing noise at work*' (L108), ISBN 0 7176 1511 1), there are other legislative requirements and codes designed to protect the public from noise on construction sites. Some of the most important are:

- *Control of Pollution Act 1974, s 71* – approval of codes and standards regarding noise control;
- *Control of Noise* (*Codes of Practice for Construction and Open Sites*) *Orders 1984* (*SI 1984 No 1992*) *and 1987* (*SI 1987 No 1730*); and
- *Construction Plant and Equipment* (*Harmonisation of Noise Emission Standards*) *Regulations 1988* (*SI 1988 No 36*1 as amended) – noise from plant used in or about building or civil engineering operations.

General construction operations

C8138 The public is also at risk, but perhaps to a lesser degree, from normal construction work activities. Policies and procedures should be in place to cover contingencies and incidents arising from:

- electricity;
- excavations;
- explosives;
- flooding;
- lifting operations;
- materials handling;
- mechanical plant and general construction equipment;
- overhead working;
- piling;
- scaffolding;
- underground services;
- vehicles;
- vibration.

References

C8139 There are numerous publications available as reference documents which provide advice on the measures to be taken to protect visitors and members of the public during construction activities. In particular, the following HSE publications and Approved Codes of Practice (ACoPs) may be useful when formulating safety policies and procedures.

Construction management

L21: Management of health and safety at work: Management of Health and Safety at Work Regulations 1999 – ACoP (ISBN 0 7176 2488 9)

HSG224: Managing health and safety in construction: Construction (Design and Management) Regulations 1994 – ACoP and guidance (ISBN 0 7176 2139 1)

HSG150: Health and safety in construction (revised)

Designing for health and safety in construction: A guide for designers on the Construction (Design and Management Regulations 1994 (revised)

HSG65: Successful health and safety management (revised)

HSR25: Memorandum of Guidance on the Electricity at Work Regulations 1989 (ISBN 0 7176 1602 9)

HSG168: Fire safety in construction work

L55: Preventing asthma at work (ISBN 0 7176 0661 9)

L64: Safety signs and signals: The Health and Safety (Safety Signs and Signals) Regulations 1996 –Guidance on regulations (revised) (ISBN 0 7176 0870 0)

L73: A guide to the Reporting of Injuries, Diseases and Dangerous Occurrences Regulations 1995 (2nd edition) (ISBN 0 7176 2431 5)

Dangerous substances

INDG370: Fire and explosion: How safe is your workplace? (a short guide to DSEAR)

COP2: Control of lead at work: In support of SI 1998 No 543 – ACoP (revised) (ISBN 0 7176 1506 5)

L5: General COSHH ACoP, Carcinogens ACoP and Biological Agents ACoP: Control of Substances Hazardous to Health Regulations 1999 – ACoP (revised) (ISBN 0 7176 2534 6)

L27: Work with asbestos which does not normally require a licence: Control of Asbestos at Work Regulations 2002 – ACoP and Guidance (revised) (ISBN 0 7176 2562 1)

L28: Work with asbestos insulation, asbestos coating and asbestos insulating board: Control of Asbestos at Work Regulations 2002 – ACoP and Guidance (revised) (ISBN 07176 2563 X)

L86: Control of substances hazardous to health in fumigation operations (ISBN 0 7176 1195 7)

L130: The compilation of safety data sheets: Chemicals (Hazard Information and Packaging for Supply) Regulations 2002 – ACoP (revised) (ISBN 0 7176 2371 8)

Explosives

L10: A guide to the Control of Explosives Regulations 1991 (ISBN 0 11 885670 7)

Gas

COP20: Standards of training in safe gas installation: ACoP (ISBN 0 7176 0603 1)

L56: Safety in the installation and use of gas systems and appliances: The Gas Safety (Installations and Use) Regulations 1998 – ACoP and guidance (revised) (ISBN 0 7176 1635 5)

L80: A guide to the Gas Safety (Management) Regulations 1996 (ISBN 0 7176 1159 0)

L81: Design, construction and installation of gas service pipes: Pipelines Safety Regulations 1996 – ACoP and guidance (ISBN 0 7176 1172 8)

LPG and petroleum spirit

COP6: Plastic containers with nominal capacities up to 5 litres for petroleum spirit: Requirements for testing and marking or labelling in support of SI 1982 No 830 (ISBN 0 11 883643 9)

Pesticides

L9: The safe use of pesticides for non-agricultural purposes: Control of Substances Hazardous to Health Regulations 1994 – ACoP (revised) (ISBN 0 7176 0542 6)

Pressure systems

L122: Safety of pressure systems: Pressure Systems Safety Regulations 2000 (ISBN 0 7176 1767 X)

L96: A guide to the Work in Compressed Air Regulations 1996 (ISBN 0 7176 1120 5)

Radiations

L121: Work with ionising radiation: Ionising Radiations Regulations 1999 – ACoP and guidance (ISBN 0 7176 1746 7)

Site vehicles

L117: Rider-operated lift trucks: Operator training (ISBN 0 7176 2455 2)

Responsibility for contractors and subcontractors

C8140 Most industrial/commercial organisations delegate corporate functions and duties, placed on them by statute, regulation and common law, to contractors and subcontractors. This practice is particularly common in the construction industry, where a main contractor, in order the more competently and expeditiously to discharge his contractual obligations towards his employer (or builder owner), sublets performance of parts of the contract, e.g. steel erection, to specialists.

Significantly, this practice of subletting performance of parts of the entire contract is regarded as sufficiently important in the building industry to justify the existence of a Standard Form of Building Contract (the JCT Standard Form). This means that the rights/obligations of all interested parties, namely, the employer, main contractors and subcontractors, both nominated and domestic, are specified in a formal jointly witnessed contract, known as the Joint Contracts Tribunal (JCT). When a dispute arises between any of the interested parties, for example, who is liable for an injury to an employee of a subcontractor, reference is made to the Conditions of Contract (or Subcontract).

If necessary, such a dispute will be decided by arbitration, since the contract provides for independent arbitration machinery in the form of the RIBA (Royal Institute of British Architects). RIBA arbitration does not exclude jurisdiction of the courts but, in practice, that is often the result, since arbitration is quicker and cheaper. In other words, the building industry has its own quasi-judicial internal disputes machinery and procedures. This is preferable to no machinery at all, since all interested parties know where they stand – at least, that is the theory.

Multiple occupation of construction sites – standard form work

C8141 Where, as is normal on large construction sites, standard form (JCT) building work is being carried out, multiplicity of occupation (or control) is not uncommon. Here control will be shared among building owner(s), main contractor(s) and subcontractor(s). It has been decided that the legal nature of the relationship between a building owner and main contractor is that of licensor and licensee (*Hounslow London Borough Council v Twickenham Garden Developments Ltd [1970] 3 All ER 326*).

This vests in the building owner some degree of control over the works e.g. if the contractor does not carry out and complete the works (in accordance with the requirements of Clause 2(1) of the JCT Standard Form Contract) the licence can, subject to certain exceptions, be terminated. Moreover, (given that some statutory duties can be modified) Clause 20(1) of the Conditions of Contract states that 'the contractor shall be liable for, and shall indemnify the employer against any expense, liability, loss or claim or proceedings whatsoever arising under any statute or at common law in respect of personal injury to or the death of any person whomsoever arising out of, or in the course of, or caused by the carrying out of the works, unless due to any act or neglect of the employer or of any person for whom the employer is responsible' (e.g. employee). The proviso to the clause 'unless due to any act or neglect of the employer' is limited solely to common law negligence and does not extend to statutory negligence.

In consequence, the building owner can insist that the main contractor(s) must take out insurance to meet that indemnity, and likewise the main contractor(s) can insist that the subcontractor(s) does the same (Clause 21.1 JCT Standard Form Contract). Thus, for the purposes of common law liability, based on occupation, persons injured on building sites can sue the building owner, main contractor and any subcontractor who may be responsible, the question of indemnity as between the liable parties being governed and determined by the terms of the JCT Contract.

Liability of employer/occupier in connection with contract work

C8142 In practice, two sorts of situation give rise to liability:

- injuries/diseases, or the risk of them, to employees of the contractor as a result of working on the occupier's premises; or, alternatively, injuries or the risk of them to the employer's own workforce as a result of the employer failing to acquaint the contractor's workforce with dangers, thereby endangering his own employees (see *R v Swan Hunter Shipbuilders Ltd* below);
- injuries/damage to members of the public, or pollution or nuisance to neighbouring landowners.

Such liability can be both criminal and civil (often strict).

Criminal liability

C8143 Criminal liability can arise under several statutes and at common law (e.g. where, owing to gross negligence, employers commit manslaughter, but in practice this is rare). Particularly relevant is the *HSWA 1974, ss 3(1), 4(2)*.

Health and Safety at Work etc Act 1974, s 3(1)

C8144 In particular, the *HSWA 1974, s 3(1)* states: 'It shall be the duty of every employer to conduct his undertaking in such a way as to ensure, so far as is reasonably practicable, that persons not in his employment who may be affected thereby are not thereby exposed to risks to their health or safety'.

The Swan Hunter case

C8145 An instructive case involving this section was *R v Swan Hunter Shipbuilders Ltd [1982] 1 All ER 264*. During construction of a ship at a shipbuilder's yard, subcontractors, who had no contract with the shipbuilders, were working on the ship while it was being fitted out. The shipbuilders were aware that, because of use of oxygen hoses with fuel gases in welding, there was a risk of fire due to the atmosphere in confined and poorly ventilated spaces in the ship becoming oxygen enriched. In regard to that danger they had provided information and instruction for their own employees by way of a book of rules which stipulated that at the end of the day's work, all oxygen hoses should be returned from the lower decks to an open deck, or, where impracticable, the hoses should be disconnected at the cylinder or manifold. This rule book was *not* distributed to the subcontractor's employees working on the ship, and an employee of the subcontractor failed to disconnect the oxygen hose and oxygen was discharged during the night. In consequence, on the next morning, when a welder working in the lower deck lit his welding torch, a fierce fire escaped.

The shipbuilders were charged with failing to provide/maintain a safe system of work, contrary to the *HSWA 1974, s 2(2)(a)*, and failing to provide such information/instruction as was necessary to ensure the health/safety of their employees, contrary to the *HSWA 1974, s 2(2)(c)*.

The shipbuilders were also charged with failing to conduct their undertaking in such a way as to ensure that persons not in their employment, who might be affected thereby, were not exposed to risks to health and safety, contrary to *s 3(1)*.

On appeal against conviction on all three counts, the Court of Appeal held that:

(i) duties imposed on an employer by the *HSWA 1974, ss 2 and 3* followed the common law duty of care of the main contractors to co-ordinate operations at a place of work so as to ensure not only the safety of his own employees but also that of the subcontractor's employees. The main contractor had to prove, on a balance of probabilities, that it was not reasonably practicable for him to carry out the duties under the *HSWA 1974, ss 2* and *3*;

(ii) the shipbuilders were under a duty, under the *HSWA 1974, s 2(2)(a)*, to provide/maintain a safe system of work for the subcontractor's employees, so far as was reasonably practicable, and provide them with information/ instruction so as to ensure their safety.

Accordingly, the main contractor, Swan Hunter Shipbuilders Ltd, was fined £3,000 and the subcontractor, Telemeters Ltd, £15,000 after eight men were trapped and killed on board HMS Glasgow.

Furthermore, if the main contractor fails, as in *Swan Hunter*, to comply with the *HSWA 1974, s 3(1)*, in his duties towards subcontracted labour, it is likely that he would be in breach of his duty towards his own employees, under *s 2* of the 1974 Act. Thus, anyone who is responsible for co-ordinating work has to ensure that reasonable safety precautions are taken for the workmen of a contractor or subcontractor.

The Rhone-Poulenc Rorer case

C8146 Responsibility for contractors' work has been considered in several cases since the *Swan Hunter Shipbuilders* case resulting in further interpretation of the *HSWA 1974, s 3* and the prosecution of employers. Decisions made in the courts have continued to emphasise the role of the main employer in organising and sharing responsibility for the safety of work carried out by contractors. One of these cases concerned a breach of the *Construction (Working Places) Regulations 1966, Reg 36(2)* (now revoked) in addition to the *HSWA 1974, s 3* – the case of *R v Rhone-Poulenc Rorer Ltd [1996] ICR 1054* concerned the provision of suitable means for preventing a fall through fragile materials.

Counsel for the prosecution alleged that some sort of physical safety device is required to fulfil an employer's duty to prevent employees from falling through fragile material: neither a system of work based on instruction nor a Code of Practice will suffice.

The court heard that an employee of a subcontractor was instructed to repair a roof light at the company's factory in Dagenham. The company provided one of its own employees to supervise the work. The subcontractor was told not to climb on to the roof. He did so, and fell through the roof light on to a concrete floor 28 feet below and was killed. Rhone-Poulenc was charged under the 1974 Act and the 1966 Regulations. The company was fined £7,500 in respect of the former and £2,500 in respect of the latter, with £55,000 costs.

The company appealed and argued that the trial judge had misdirected the jury in telling them that it was necessary to prove that it had been impracticable to comply with the Regulations, not merely that it was not reasonable to do so.

The Court of Appeal rejected this argument. Wright J stated that the requirement under *Reg 36(2)* was absolute and that employers had a duty to provide such suitable means as might be necessary for preventing, so far as was reasonably practicable, persons from falling through fragile material. This meant that some sort of physical device, for example a safety harness, was required, where in the circumstances guard rails or covering could not be supplied. Falls through fragile material could not be prevented by the provision of a supervisor, a body of instructions, or a code of practice.

In relation to the *HSWA 1974, s 3* it was held that because the occupier (Rhone-Poulenc Rorer Ltd) had been in breach of Regulations which applied to its own employees, it was also in breach of *s 3* so far as the contractor's employees were concerned.

The Associated Octel case

C8147 It was 1996 before the House of Lords had to deal with the interpretation of the *HSWA 1974, s 3*. In *R v Associated Octel Co Ltd [1996] 4 All ER 846* it was decided that the duty placed on the employer of contractors extended to persons not in his employment.

In 1990, during the course of an annual shutdown for planned maintenance, a specialist contractor, RGP, was called in for the purposes of cleaning and repairing a tank at Octel's chlorine plant at Ellesmere Port. Octel had effectively approved the system of work by virtue of having issued a permit-to-work that required the contractor's employee to enter the tank, taking lighting with him, and grind the internal surfaces of the tank and clean residues with an acetone solvent. In the event the light broke causing an explosion in which the contractor's employee was badly injured.

The HSE inspector who investigated the incident was critical of the system of work and identified aspects of the operations which were unsafe. The points raised included:

- the acetone was carried in an old emulsion bucket;
- Octel provided an unsuitable lamp from its own stores;
- there was inadequate ventilation;
- the precautions listed on the permit-to-work referred only to the grinding work and did not specify adequate precautions relating to work in a confined space.

The inspector referred to the fact that the site was a major hazard site and under the control of the *Control of Industrial Major Accident Hazard Regulations 1984* (*CIMAH*) (*SI 1984 No 1902 as amended*) (now replaced by the *Control of Major Accident Hazards Regulations 1999*) which required a safety case to be submitted and complied with. RGP had employees at Octel's site on a regular basis and had worked for Octel over a number of years. Previously, Octel had exercised a degree of control over RGP's work performance and had a high level of understanding of the dangers involved in tank cleaning operations.

The case argued in the Crown Court and the Court of Appeal was a complex one where the central issue concerned 'control' of the work of the contractor. Octel's counsel referred to the Robens Report of 1972, and cited, *inter alia*, as precedents *R v Board of Trustees of the Science Museum [1993] 3 All ER 853, R v Swan Hunter Shipbuilders Ltd [1982] 1 All ER 264, Mailer v Austin Rover Group Ltd [1989] 2 All ER 1087* and *RMC Roadstone Products Ltd v Jester [1994] 4 All ER 1037*. Octel argued consistently that there was no case to answer because it could not be shown that it was the conduct of Octel's undertaking that endangered the RGP employee. RGP is a specialist contractor and Octel claimed that it had no right to control the manner in which its independent contractor did its work.

Octel lost its argument before the trial judge, as well as before the Court of Appeal for different and complicated legal reasons. However, it was decided that the chemical business was Octel's undertaking, and the undertaking included having the tank cleaned, whether by its own employees or by contractors.

Lord Hoffmann delivered the judgment of the House of Lords, upholding the decisions of the Crown Court and the Court of Appeal. *The Times* (15 November 1996) reported the following points made by Lord Hoffmann:

- it was a question of fact in each case whether an activity which caused a risk to the health and safety of persons other than employees amounted to 'conduct of an undertaking';
- *section 3* of the 1974 Act was not concerned with vicarious liability, but imposed a duty upon the employer himself;

- if the employer engaged an independent contractor to do work forming part of the 'undertaking', then the employer had to stipulate whatever conditions were needed to avoid risks to health and safety;
- the question was simply whether the activity in question could be described as part of the employer's undertaking. Octel's undertaking was running a chemical plant and it was part of the conduct of that undertaking to have the factory cleaned by contractors;
- the tank was part of Octel's plant. The work formed part of a maintenance programme planned by the firm. The workers, although employed by an independent contractor, were almost permanently integrated into the firm's larger operations. In these circumstances, a properly instructed jury would undoubtedly have convicted.

The Port Ramsgate ferry walkway case

C8148 In this case a harbour operator, two foreign marine engineering companies (designers and builders) and Lloyd's Register of Shipping were all prosecuted and found guilty of failing to ensure the safety of passengers after the collapse of a walkway to a cross-channel ferry led to the deaths of six people and injured many more.

The case centred around the 'reasonably practicable' element of the *HSWA 1974, s 3(1)* and concerned the design, construction, installation and the supervision and approval inspection of the walkway, contracted by the Port of Ramsgate Ltd. The accident occurred only four months after the walkway was commissioned because of the failure of a weld joining one of the feet to an axle. The walkway collapsed and fell some 30 feet on to a floating pontoon.

The QC who was prosecuting on behalf of HSE gave evidence that the failure was the result of inaccurate stress calculations and also of inferior welding.

During the trial, many legal arguments took place as to who exactly was responsible. Port of Ramsgate Ltd was adamant that it had acted properly in placing contracts with appropriate, experienced contractors and that, by arranging for Lloyd's to manage the design, construction and installation, there was nothing else it could have done regarding the safety of its passengers. It argued strongly that there was no way of foreseeing the profound errors of judgement made by the various parties concerned.

The prosecution continued with its argument that there were measures that could have been taken to satisfy the 'reasonably practicable' element and that one of these might have been the inclusion of a simple fail-safe device such as safety chains.

The judge directed the jury to treat the question of the Port of Ramsgate's undertaking as a question of fact and left it for the jury to decide whether sufficient had been done to satisfy the grounds of 'reasonably practicability'. In the event, the jury found that all three companies were guilty of failing to satisfy the requirement. The judge imposed very high fines:

- Port of Ramsgate Ltd: £200,000;
- Lloyd's Register of Shipping: £500,000;
- Fartygsentreprenader AB and Others (Sweden): £1 million.

It should be noted that this was a highly complex case and its implications for further cases will be considerable.

Although the case was brought before the implementation of the *CDM Regulations*, it should be noted that these Regulations do not expressly require quality assurance checks to be carried out.

If the same type of case occurred offshore, the *Offshore Installations and Wells (Design and Construction) Regulations 1996* (*SI 1996 No 913*) would probably apply and in particular the requirement for 'verification' and the effect on the installation's safety case.

Civil liability

C8149 Normally employers are only liable for the negligent acts/omissions of their own employees or agents but there are certain important exceptions, such as where an employee of an occupier is injured by the contractor's negligence but overall control rests with the occupier.

The McDermid Nash case

C8150 An important case was heard in 1987 concerning a safe system of work, the delegation of the performance of that work and the competency of the person in charge of a situation where an individual, not being in the employ of that person, was injured.

During the case (*McDermid v Nash Dredging & Reclamation Co Ltd [1987] 2 All ER 878*) several points were made regarding the legal responsibilities of the various people concerned when contracting an employee of another employer:

- the employer's safety duties towards his employees are owed personally to those employees – they cannot be got rid of by delegation;
- an employer who arranges for work to be done by another person, whether under contract or any other relationship, where that other will use employees of the employer, is vicariously liable for acts of negligence of the contractor or other person;
- liability arises not because the contractor or other person is, for the time being, the employer of the employee in the legal sense but merely from the fact that the usual employer has entrusted the safety of his employee to that other person;
- the employer still remains liable for the proper performance and the legal duties of his employee;
- this continued liability does not rely on any contract between the employer and the other party. It arises whenever an employer entrusts the performance of safety duties to another by asking that other party to perform work on his behalf using his employees.

The UK subsidiary of a Dutch company secured a contract to dredge a fjord in Sweden. It was to be an enterprise where some 100 of the UK company's employees were to work on board Dutch vessels under the joint captaincy of a UK and a Dutch captain, each working alternate shifts. Mr McDermid's job was to keep the deck tidy and to tie up and untie the tug from its moorings when it went alongside the dredger.

While the Dutch captain was in charge, Mr McDermid's leg got caught in one of the ship's hawsers and was so badly injured it had to be amputated. The evidence showed that the Dutch captain had not set up a safe system to be sure that Mr McDermid was free of the ropes before applying power to move the tug away from its mooring.

Because of the difficulties of suing a Dutch company while operating in Swedish territorial waters, Mr McDermid sought damages from his UK employer, even though at the time he was working under the control of a foreign captain who was clearly not an employee of the UK company.

The court decided that there was no need to prove even a temporary relationship of employer and employee between the Dutch company and Mr McDermid. Under the common law of vicarious liability, the employer's safety duties are owed to the employee personally. They cannot be got rid of by delegation. Further, when an employer puts his employee into the hands of another he is entrusting the performance of his own safety duties to that other but the employer remains responsible.

Taking all the factors into account, the court ruled that Mr McDermid's employer was vicariously liable for the negligence of the Dutch tug captain and stated:

> 'It is clear therefore that if an employer delegates to another person, whether an employee or not, his personal duty to take reasonable care for the safety of his employees, the employer is liable for injury caused through the negligence of that person because it is in the eyes of the law his own negligence.'

The employer's appeal to the House of Lords was dismissed, and it was held that an employer's duty to his employee to exercise reasonable care to ensure that the system of work provided for him is safe, is 'personal' and 'non-delegable'. It is no defence for the employer to show that he delegated its performance to a person whom he reasonably believed to be competent to perform it.

It was also stated in court that the negligence of the Dutch captain was not casual but central. It involved abandoning a safe system and operating in its place a manifestly unsafe system.

Hazards left on highways

C8151 An occupier is not under a duty of care to check that a contractor removes hazards from a highway, and so is not liable for consequential injuries. In *Rowe v Herman and others, The Times, 9 June 1997*, the plaintiff (R) sought damages for injuries sustained by tripping over some metal plates left lying on the pavement by an independent contractor (L), which had been carrying out (and had finished) some building work for the occupier (H). H contended that he was not liable for L's negligence under the general principle that an employer is not liable for an independent contractor's negligence. There are two main exceptions to this basic principle – an employer is liable:

(i) where the work commissioned involved extra hazardous acts; and

(ii) where danger was created by work on a highway.

The county court held that although H was not liable for the hazard whilst work was being carried out, he became liable under the principle in (ii) once L had finished the work. H appealed.

The Court of Appeal allowed H's appeal. Previous cases, where the employer was held liable, concerned obstructions to the highway being caused as a result of work being carried out under statutory powers, such obstructions arising directly from the work which the employer was required to do (being integral to it). In this case H was not obliged to carry out the building work, and it was not an integral part of the work to obstruct the footway (the plates had merely been laid to prevent lorries from damaging the pavement). It was clear that if the accident had occurred whilst L was carrying out the work, H would not have been liable; it made no sense to suddenly place H under such a duty once L had left the site. Following the general principle that H had no control over the manner in which L carried out its work, so too H had no control over the way that L cleared up.

Control of Major Accident Hazards

Introduction

C10001 The *Control of Major Accident Hazards Regulations 1999* (*SI 1999 No 743*) implement the requirements of the 'Seveso II' Directive (96/82/EC) on the control of major accident hazards involving dangerous substances. Seveso II replaced the original Seveso Directive (82/501/EEC) which was implemented in Great Britain by the *Control of Industrial Major Accident Hazards Regulations 1984* (*CIMAH*) which in turn have been replaced by the *Control of Major Accident Hazards Regulations 1999* (*COMAH*). *COMAH* came into force on 1 April 1999. The provisions of Article 12 of Seveso II concerning land-use planning have been implemented in the *Planning* (*Control of Major Accident Hazards*) *Regulations 1999* (*SI 1999 No 981*).

The Health and Safety Executive (HSE) and the Environment Agency (in England and Wales) or the Scottish Protection Agency are jointly responsible as the competent authority (CA) for *COMAH*. The CA is the author of a report listing recent major accidents at industrial premises subject to *COMAH*. The report *'COMAH major accidents notified to the European Commission, England, Scotland and Wales 1999–2000'* details ten major accidents (one of which resulted in the death of an employee), and three near misses. The CA's concern at both the magnitude and frequency of these accidents is highlighted in the report together with its belief that a thorough implementation of *COMAH* will contribute significantly to future improvements in the safety of persons and the environment.

Major accidents in the EU since the introduction of the Seveso II Directive, and the results of studies on carcinogens and substances dangerous for the environment carried out by the European Commission, have resulted in a new Directive (2003/105/EC). Member States are required to implement the new Directive by 1 July 2005. In Great Britain the Health and Safety Commission (HSC) proposes to implement the Directive, and introduce other minor changes to *COMAH* requested by the CA, through a new set of Regulations, the *Control of Major Accident Hazards* (*Amendment*) *Regulations 2005*. The draft Regulations are contained within HSE Consultation Document 193. When the amendment regulations come into force the HSE guidance (see C10002 below) will be revised to take account of the changes and experience that has accrued over the last five years since *COMAH* came into force.

Control of Major Accident Hazards Regulations 1999 (SI 1999 No 743)

C10002 *COMAH* (*SI 1999 No 743*) gives effect to a safety regime for the prevention and mitigation of major accidents resulting from ultra-hazardous industrial activities. The emphasis is on controlling risks to both people and the environment through demonstrable safety management systems, which are integrated into the routine of business rather than dealt with as an add-on.

An occurrence is regarded as a major accident if:

- it results from uncontrolled developments (i.e. they are sudden, unexpected or unplanned) in the course of the operation of an establishment to which the Regulations apply; and
- it leads to serious danger to people or to the environment, on or off-site; and
- it involves one or more dangerous substances defined in the Regulations.

Major emissions, fires and explosions are the most typical major accidents.

The duties placed on operators by *COMAH* fall into two categories: 'lower tier' and 'top tier'. Lower tier duties fall upon all operators. Some operators are subject to additional, top-tier, duties.

An 'operator' is a person (including a company or partnership) who is in control of the operation of an establishment or installation. Where the establishment or installation is to be constructed or operated, the 'operator' is the person who proposes to control its operation – where that person is not known, then the operator is the person who has commissioned its design and construction.

As mentioned, the HSE and the Environment Agency (in England and Wales) or the Scottish Environment Protection Agency are jointly responsible as the competent authority (CA) for *COMAH*. A co-ordinated approach has been adopted with the HSE acting as the primary contact for operators. A charging regime has been introduced and the fees are payable by the operator to the HSE for work carried out by the HSE and the Environment Agency. Details of the current charging regime may be found on the HSE website at: www.hse.gov.uk. Under *COMAH* (*SI 1999 No 743*), *Reg 19*, the CA must organise an adequate system of inspections of all establishments, and then must prepare a report on the inspection.

The main HSE guidance to the Regulations is contained in *'Guide to the Control of Major Accident Hazards Regulations 1999'* (L111).

Application

C10003 *COMAH* (*SI 1999 No 743*) applies to an *establishment* where:

(*a*) dangerous substances are present; or

(*b*) their presence is anticipated; or

(*c*) it is reasonable to believe that they may be generated during the loss of control of an industrial chemical process.

'Loss of control' excludes expected, planned or permitted discharges.

'Industrial chemical process' means that premises with no such chemical process do not fall within the scope of the Regulations solely because of dangerous substances generated during an accident.

An 'establishment' means the whole area under the control of the same person where dangerous substances are present in one or more installations. Two or more areas under the control of the same person and separated only by a road, railway or inland waterway are to be treated as one whole area.

'Dangerous substances' are those which:

(*a*) are named at the appropriate threshold (see Table 1); or (*b*) fall within a generic category at the appropriate threshold (see Table 2).

There are two threshold levels: lower tier (Column 2 of the Tables) and top tier (Column 3).

Table 1

Column 1	*Column 2*	*Column 3*
Dangerous substances	*Quantity in tonnes*	
Ammonium nitrate (see note below)	350	2,500
Ammonium nitrate (see note below)	1,250	5,000
Arsenic pentoxide, arsenic (V) acid and/or salts	1	2
Arsenic trioxide, arsenious (III) acid and/or salts	0.1	0.1
Bromine	20	100
Chlorine	10	25
Nickel compounds in inhalable powder form (nickel monoxide, nickel dioxide, nickel sulphide, trinickel disulphide, dinickel trioxide)	1	1
Ethyleneimine	10	20
Fluorine	10	20
Formaldehyde (concentration=>90%)	5	50
Hydrogen	5	50
Hydrogen chloride (liquefied gas)	25	250
Lead alkyls	5	50
Liquefied extremely flammable gases (including LPG) and natural gas (whether liquefied or not)	50	200
Acetylene	5	50
Ethylene oxide	5	50
Propylene oxide	5	50
Methanol	500	5,000
4, 4-Methylenebis (2-chloraniline) and/or salts, in powder form	0.01	0.01
Methylisocyanate	0.15	0.15
Oxygen	200	2,000
Toluene diisocyanate	10	100
Carbonyl dichloride (phosgene)	0.3	0.75
Arsenic trihydride (arsine)	0.2	1

Phosphorus trihydride (phosphine)	0.2	1
Sulphur dichloride	1	1
Sulphur trioxide	15	75
Polychlorodibenzofurans and polychlorodibenzodioxins (including TCDD), calculated in TCDD equivalent	0.001	0.001
The following CARCINOGENS:	0.001	0.001
4-Aminobiphenyl and/or its salts, Benzidine and/or salts,		
Bis(chloromethyl) ether, Chloromethyl methyl ether,		
Dimethylcarbamoyl chloride, Dimethylnitrosomine,		
Hexamethylphosphoric triamide, 2-Naphthylamine and/or salts,		
1, 3 Propanesultone and 4-nitrodiphenyl		
Automotive petrol and other petroleum spirit	5,000	50,000

[*COMAH* (*SI 1999 No 743*), *Sch 1, Part 2*].

Ammonium nitrate:

(*a*) The 350/2500 quantities apply to ammonium nitrate and ammonium nitrate compounds in which the nitrogen content as a result of the ammonium nitrate is more than 28 per cent by weight (except for those compounds referred to in (*b*) below) and to aqueous ammonium nitrate solutions in which the concentration of ammonium nitrate is more than 90 per cent by weight.

(*b*) The 1250/5000 quantities apply to simple ammonium-nitrate based fertilisers which conform with the requirements of the *Fertilisers Regulations 1991* (*SI 1991 No 2197*) and to composite fertilisers in which the nitrogen content as a result of the ammonium nitrate is more than 28 per cent in weight (a composite fertiliser contains ammonium nitrate with phosphate or potash, or phosphate and potash).

[*COMAH* (*SI 1999 No 743*), Notes 1 and 2 of *Part 2* of *Sch 1*].

Substances are classified according to the *Chemicals* (*Hazard Information and Packaging for Supply*) *Regulations 1994* (*SI 1994 No 3247*), *Reg 5* (now revoked and replaced by *SI 2002 No 1689, Reg 4*).

Table 2		
Column 1	*Column 2*	Column 3
Categories of dangerous substances	*Quantity in tonnes*	
1. VERY TOXIC	5	20
2. TOXIC	50	200
3. OXIDISING	50	200
4. EXPLOSIVE (see C10004 below)	50	200

5. EXPLOSIVE (see C10004 below)	10	50
6. FLAMMABLE (as defined in Sch 1 Part 3)	5,000	50,000
7a. HIGHLY FLAMMABLE (as defined in Sch 1 Part 3)	50	200
7b. HIGHLY FLAMMABLE liquids (as defined in Sch 1 Part 3)	5,000	50,000
8. EXTREMELY FLAMMABLE (as defined in Sch 1 Part 3)	10	50
9. DANGEROUS FOR THE ENVIRONMENT in combination with		
risk phrases:		
R50: 'Very toxic to aquatic organisms'	200	500
R51: 'Toxic to aquatic organisms'; and R53: 'May cause long-term adverse effects in the aquatic environment'	500	2,000
10. ANY CLASSIFICATION not covered by those given		
above in combination with risk phrases:		
R14: 'Reacts violently with water' (including R14/15)	100	500
R29: 'in contact with water, liberates toxic gas'	50	200

[*COMAH* (*SI 1999 No 743*), *Sch 1, Part 3*].

Where a substance or group of substances named in Table 1 also falls within a category in Table 2, the qualifying quantities set out in Table 1 must be used.

Dangerous substances present at an establishment only in quantities not exceeding 2 per cent of the relevant qualifying quantity are ignored for the purposes of calculating the total quantity present, provided that their location is such that they cannot initiate a major accident elsewhere on site.

Explosives

C10004 Note 2 to *COMAH* (*SI 1999 No 743*), *Sch 1, Part 3*, defines an 'explosive' as:

(*a*) (i) a substance or preparation which creates the risk of an explosion by shock, friction, fire or other sources of ignition;

(ii) a pyrotechnic substance is a substance (or mixture of substances) designed to produce heat, light, sound, gas or smoke or a combination of such effects through non-detonating self-sustained exothermic chemical reactions; or

(iii) an explosive or pyrotechnic substance or preparation contained in objects;

(*b*) a substance or preparation which creates extreme risks of explosion by shock, friction, fire or other sources of ignition.

Aggregation

C10005 There is a rule for aggregation of dangerous substances set out in Note 4 to *COMAH* (*SI 1999 No 743*), *Sch 1, Part 3* . The rule will apply in the following circumstances:

(i) for substances and preparations appearing in Table 1 at quantities less than their individual qualifying quantity present with substances having the same classification from Table 2, and the addition of substances and preparations with the same classification from Table 2;

(ii) for the addition of categories 1, 2 and 9 present at an establishment together;

(iii) for the addition of categories 3, 4, 5, 6, 7a, 7b and 8, present at an establishment together.

The sub-threshold quantities of named dangerous substances or categories of dangerous substances are expressed as partial fractions of the threshold quantities in column 3 of Table 1 or of Table 2, and added together. When the total exceeds 1, top-tier duties apply.

If the total is less than or equal to 1, the sub-threshold quantities of named dangerous substances or categories of dangerous substances are expressed as partial fractions of the threshold quantities in column 2 of Table 1 or of Table 2, and added together. If the total exceeds 1, lower tier duties apply.

Definitive guidance on aggregation can be found in HSE's 'Guide to the Control of Major Accident Hazards Regulations 1999'.

Exclusions

C10006 *COMAH* (*SI 1999 No 743*) does not apply to:

- Ministry of Defence establishments;
- extractive industries exploring for, or exploiting, materials in mines and quarries;
- waste land-fill sites;
- transport related activities;
- substances at nuclear licensed sites which create a hazard from ionising radiation.

COMAH does apply to explosives and chemicals at nuclear installations.

Lower tier duties

General duty

C10007 Under *COMAH* (*SI 1999 No 743*), *Reg 4*, every operator is under a general duty to take *all measures necessary* to prevent major accidents and limit their consequences to persons and the environment.

Although the best practice is to completely eliminate a risk, the Regulations recognise that this is not always possible. Prevention should be considered in a hierarchy based on the principle of reducing risk to a level as low as is reasonably practicable, taking account of what is technically feasible and the balance between the costs and benefits of the measures taken. Operators must be able to demonstrate that they have adopted control measures which are adequate for the risks identified.

Where hazards are high, high standards will be expected by the enforcement agencies to ensure that risks are acceptably low.

Major accident prevention policy (MAPP)

C10008 Under *COMAH* (*SI 1999 No 743*), *Reg 5*(*1*), every operator must prepare and keep a document (MAPP) setting out his policy with respect to the prevention of major accidents.

The MAPP document must be in writing and must include sufficient particulars to demonstrate that the operator has established an appropriate safety management system. *COMAH* (*SI 1999 No 743*), *Sch 2,* sets out the principles to be taken into account when preparing a MAPP document and at *para 4* lists the following specific issues to be addressed by the safety management system:

(*a*) *organisation and personnel* – the roles and responsibilities of personnel involved in the management of major hazards at all levels in the organisation. The identification of training needs of such personnel and the provision of the training so identified. The involvement of employees and, where appropriate, sub-contractors;

(*b*) *identification and evaluation of major hazards* – adoption and implementation of procedures for systematically identifying major hazards arising from normal and abnormal operation and the assessment of their likelihood and severity;

(*c*) *operational control* – adoption and implementation of procedures and instructions for safe operation, including maintenance of plant, processes, equipment and temporary stoppages;

(*d*) *management of change* – adoption and implementation of procedures for planning modifications to, or the design of, new installations, processes or storage facilities;

(*e*) *planning for emergencies* – adoption and implementation of procedures to identify foreseeable emergencies by systematic analysis and to prepare, test and review emergency plans to respond to such emergencies;

(*f*) *monitoring performance* – adoption and implementation of procedures for the on-going assessment of compliance with the objectives set by the operator's major accident prevention policy and safety management system, and the mechanisms for investigation and taking corrective action in the case of non-compliance. The procedures should cover the operator's system for reporting major accidents or near misses, particularly those involving failure of protective measures, and their investigation and follow-up on the basis of lessons learnt;

(*g*) *audit and review* – adoption and implementation of procedures for periodic systematic assessment of the major accident prevention policy and the effectiveness and suitability of the safety management system; the documented review of performance of the policy and safety management system and its updating by senior management.

The MAPP is a concise but key document for operators which sets out the framework within which adequate identification, prevention/control and mitigation of major accident hazards is achieved. Its purpose is to compel operators to provide a statement of commitment to achieving high standards of major hazard control, together with an indication that there is a management system covering all the issues set out in (*a*)–(*g*) above. The guidance to the Regulations suggests that the essential questions which operators must ask themselves are:

- does the MAPP meet the requirements of the Regulations?
- will it deliver a high level of protection for people and the environment?
- are there management systems in place which achieve the objectives set out in the policy?
- are the policy, management systems, risk control systems and workplace precautions kept under review to ensure that they are implemented and that they are relevant?

Not only must the MAPP be kept up to date (*COMAH* (*SI 1999 No 743*), *Reg 5*(*4*)), but also the safety management system described in it must be put into operation (*COMAH* (*SI 1999 No 743*), *Reg 5*(*5*)).

Notifications

C10009 Notification requirements are set out in *COMAH* (*SI 1999 No 743*), *Reg 6*. 'Notify' means notify in writing.

Within a reasonable period of time before the start of construction of an establishment and before the start of the operation of an establishment, the operator must send to the CA a notification containing the information which is specified in *COMAH* (*SI 1999 No 743*), *Sch 3*. There is no need for the notification sent before start-up to contain any information which has already been included in the notification sent before the start of construction, if that information is still valid.

The operator of an existing establishment was required to send a notification containing the specified information by 3 February 2000, unless a report had been sent to the HSE in accordance with the *Control of Industrial Major Accident Hazards Regulations 1984*, *Reg 7*.

The following information is specified in *Sch 3* for inclusion in a notification:

(*a*) the name and address of the operator;

(*b*) the address of the establishment concerned;

(*c*) the name or position of the person in charge of the establishment;

(*d*) information sufficient to identify the dangerous substances or category of dangerous substances present;

(*e*) the quantity and physical form of the dangerous substances present;

(*f*) a description of the activity or proposed activity of the installation concerned;

(*g*) details of the elements of the immediate environment liable to cause a major accident or to aggravate the consequences thereof.

Notification is a continuing duty. *COMAH* (*SI 1999 No 743*), *Reg 6*(*4*) provides that an operator must notify the CA forthwith in the event of:

- any significant increase in the quantity of dangerous substances previously notified;
- any significant change in the nature or physical form of the dangerous substances previously notified, the processes employing them or any other information notified to the CA in respect of the establishment;

- *Regulation 7* ceasing to apply to the establishment as a result of a change in the quantity of dangerous substances present there; or
- permanent closure of an installation in the establishment.

Information that has been included in a safety report does not need to be notified.

Duty to report a major accident

C10010 Where a major accident has occurred at an establishment, the operator must immediately inform the CA of the accident [*COMAH* (*SI 1999 No 743*), *Reg 15*(*3*)].

This duty will be satisfied when the operator notifies a major accident to the HSE in accordance with the requirements of the *Reporting of Injuries, Diseases and Dangerous Occurrences Regulations 1995* (*SI 1995 No 3163*).

The CA must then conduct a thorough investigation into the accident [*COMAH* (*SI 1999 No 743*), *Reg 19*].

Top-tier duties

Safety report

C10011 Operators of top-tier sites are required to produce a safety report – its key requirement is that operators must show that they have taken all necessary measures for the prevention of major accidents and for limiting the consequences to people and the environment of any that do occur.

Safety reports are required before construction as well as before start-up. The HSE publication *'Preparing safety reports'* (HSG 190), gives practical and comprehensive guidance to site operators in the preparation of a *COMAH* safety report. *COMAH* (*SI 1999 No 743*), *Reg 7*(*1*) requires that within a reasonable period of time before the start of construction of an establishment the operator must send to the CA a report:

- containing information which is sufficient for the purpose specified in *COMAH* (*SI 1999 No 743*), *Sch 4, Part 1, para 3*(*a*) (see below); and
- comprising at least such of the information specified in *COMAH* (*SI 1999 No 743*), *Sch 4, Part 2* (see below) as is relevant for that purpose.

Within a reasonable period of time before the start of the operation of an establishment, the operator must send to the CA a report containing information which is sufficient for the purposes specified in *Part 1 of Sch 4*, and comprising at least the information specified in *Part 2 of Sch 4* [*COMAH* (*SI 1999 No 743*), *Reg 7*(*5*)].

The report sent before start-up is not required to contain information already contained in the report sent before the start of construction.

An operator must ensure that neither construction of the establishment, nor its operation, is started until he has received the CA's conclusions on the report. The CA must communicate its conclusions within a reasonable period of time of receiving a safety report [*COMAH* (*SI 1999 No 743*), *Reg 17*(*1*)(*a*)].

The operator of an existing establishment must send to the CA a report meeting the requirements of *Parts 1* and *2* of *Sch 4*. Ex-CIMAH top-tier sites that submitted CIMAH safety reports in parts may continue to submit COMAH reports in parts. If a CIMAH report, or any part of it, was up for its three-yearly review before 3 February 2000, a COMAH report was required before that date or such later date (no later than 3 February 2001) as agreed by the CA. If the review would have fallen after that date a COMAH report was required by 3 February 2001. In any other case the safety report was required by 3 February 2002.

PURPOSE OF SAFETY REPORTS

1. Demonstrating that a major accident prevention policy and a safety management system for implementing it have been put into effect in accordance with the information set out in *Sch 2;*

2. Demonstrating that major accident hazards have been identified and that the necessary measures have been taken to prevent such accidents and to limit their consequences to persons and the environment;

3. Demonstrating that adequate safety and reliability have been incorporated into the

   (a) design and construction, and

   (b) operation and maintenance,

   of any installation and equipment and infrastructure connected with its operation, and that they are linked to major accident hazards within the establishment;

4. Demonstrating that on-site emergency plans have been drawn up and supplying information to enable the off-site plan to be drawn up in order to take the necessary measures in the event of a major accident;

5. Providing sufficient information to the competent authority to enable decisions to be made in terms of the siting of new activities or developments around establishments.

   [*COMAH* (*SI 1999 No 743*), *Sch 4, Part 1*]

MINIMUM INFORMATION TO BE INCLUDED IN SAFETY REPORT

1. Information on the management system and on the organisation of the establishment with a view to major accident prevention.

   This information must contain the elements set out in *Sch 2.*

2. Presentation of the environment of the establishment:

   (a) a description of the site and its environment including the geographical location, meteorological, geographical, hydrographic conditions and, if necessary, its history;

   (b) identification of installations and other activities of the establishment which could present a major accident hazard;

   (c) a description of areas where a major accident may occur.

3. Description of installation:

   (a) a description of the main activities and products of the parts of the establishment which are important from the point of view of safety, sources of major accident risks and conditions under which such a major accident could happen, together with a description of proposed preventive measures;

   (b) description of processes, in particular the operating methods;

(c) description of dangerous substances:

(i) inventory of dangerous substances including –

— the identification of dangerous substances: chemical name, the number allocated to the substance by the Chemicals Abstract Service, name according to International Union of Pure and Applied Chemistry nomenclature;

— the maximum quantity of dangerous substances present;

(ii) physical, chemical, toxicological characteristics and indication of the hazards, both immediate and delayed, for people and the environment;

(iii) physical and chemical behaviour under normal conditions of use or under foreseeable accidental conditions.

4. Identification and accidental risks analysis and prevention methods:

(a) detailed description of the possible major accident scenarios and their probability or the conditions under which they occur including a summary of the events which may play a role in triggering each of these scenarios, the causes being internal or external to the installation;

(b) assessment of the extent and severity of the consequences of identified major accidents;

(c) description of technical parameters and equipment used for the safety of the installations.

5. Measures of protection and intervention to limit the consequences of an accident:

(a) description of the equipment installed in the plant to limit the consequences of major accidents;

(b) organisation of alert and intervention;

(c) description of mobilisable resources, internal or external;

(d) summary of elements described in sub-paragraphs (a), (b) and (c) necessary for drawing up the on-site emergency plan.

[*COMAH* (*SI 1999 No 743*), *Sch 4, Part 2*]

All or part of the information required to be included in a safety report can be so included by reference to information contained in another report or notification furnished by virtue of other statutory requirements. This should be done only where the information in the other document is up to date, and adequate in terms of scope and level of detail.

If an operator can demonstrate that particular dangerous substances are in a state incapable of creating a major accident hazard, the CA can limit the information required to be included in the safety report [*Reg 7*(*12*)].

An operator must provide the CA with such further information as it may reasonably request in writing following its examination of the safety report [*COMAH* (*SI 1999 No 743*), *Reg 7*(*13*)].

Review and revision of safety report

C10012 Where a safety report has been sent to the CA, the operator must review it:

- at least every five years;
- whenever a review is necessary because of new facts or to take account of new technical knowledge about safety matters; and
- whenever the operator makes a change to the safety management system, which could have significant repercussions.

Where it is necessary to revise the report as a result of the review, the operator must carry out the revision immediately and inform the CA of the details [*COMAH* (*SI 1999 No 743*), *Reg 8*(*1*)].

An operator must inform the CA when he has reviewed the safety report but not revised it [*COMAH* (*SI 1999 No 743*), *Reg 8*(*2*)].

Ex-*CIMAH* top-tier sites that have submitted *CIMAH* safety reports in parts must review each part within five years from the time when that part was sent to the CA.

When modifications which could have significant repercussions are proposed (to the establishment or an installation in it, to the process carried on there or the nature or quantity of dangerous substances present there), the operator must review the safety report and where necessary revise it, in advance of any such modification. Details of any revision must be notified to the CA.

On-site emergency plan

C10013 The HSE publication *'Emergency planning for major accidents'* (HSG 191) is essential reading.

COMAH (*SI 1999 No 743*), *Reg 9*(*1*), requires operators of top-tier establishments to prepare an on-site emergency plan. The plan must be adequate to secure the objectives specified in *Sch 5*, *Part 1*, of the Regulations namely:

- containing and controlling incidents so as to minimise the effects, and to limit damage to persons, the environment and property;
- implementing the measures necessary to protect persons and the environment from the effects of major accidents;
- communicating the necessary information to the public and to the emergency services and authorities concerned in the area;
- providing for the restoration and clean-up of the environment following a major accident.

The plan must contain the following information:

- names or positions of persons authorised to set emergency procedures in motion and the person in charge of and co-ordinating the on-site mitigatory action;
- name or position of the person with responsibility for liaison with the local authority responsible for preparing the off-site emergency plan (see C10014 below);
- for foreseeable conditions or events which could be significant in bringing about a major accident, a description of the action which should be taken to control the conditions or events and to limit their consequences, including a description of the safety equipment and the resources available;

- arrangements for limiting the risks to persons on site including how warnings are to be given and the actions persons are expected to take on receipt of a warning;
- arrangements for providing early warning of the incident to the local authority responsible for setting the off-site emergency plan in motion, the type of information which should be contained in an initial warning and the arrangements for the provision of more detailed information as it becomes available;
- arrangements for training staff in the duties they will be expected to perform, and where necessary co-ordinating this with the emergency services;
- arrangements for providing assistance with off-site mitigatory action.

[*COMAH* (*SI 1999 No 743*), *Sch 5, Part 2*].

Ex-*CIMAH* top-tier sites that were subject to *CIMAH* requirements for the preparation of an on-site emergency plan were required to prepare a *COMAH* emergency plan by 3 February 2001. Any other existing establishment was required to prepare an on-site emergency plan by 3 February 2002. New establishments must prepare such a plan before start-up.

When preparing an on-site emergency plan, the operator must consult:

- employees at the establishment;
- the Environment Agency (or the Scottish Environment Protection Agency);
- the emergency services;
- the health authority for the area where the establishment is situated; and
- the local authority, unless it has been exempted from the requirement to prepare an off-site emergency plan.

Off-site emergency plan

C10014 The local authority for the area where a top-tier establishment is located must prepare an adequate emergency plan for dealing with off-site consequences of possible major accidents. As with the on-site plan, it should be in writing.

The objectives set out in *COMAH* (*SI 1999 No 743*), *Sch 5, Part 1,* (see C10013 above) also apply to off-site emergency plans.

The plan must contain the following information:

- names or positions of persons authorised to set emergency procedures in motion and of persons authorised to take charge of and co-ordinate off-site action;
- arrangements for receiving early warning of incidents, and alert and call-out procedures;
- arrangements for co-ordinating resources necessary to implement the off-site emergency plan;
- arrangements for providing assistance with on-site mitigatory action;
- arrangements for off-site mitigatory action;
- arrangements for providing the public with specific information relating to the accident and the behaviour which it should adopt;

- arrangements for the provision of information to the emergency services of other Member States in the event of a major accident with possible trans-boundary consequences.

An operator must supply the local authority with the information necessary for the authority's purposes, plus any additional information reasonably requested in writing by the local authority.

In preparing the off-site emergency plan, the local authority must consult:

- the operator;
- the emergency services;
- the CA;
- each health authority for the area in the vicinity of the establishment; and
- such members of the public as it deems appropriate.

[*COMAH* (*SI 1999 No 743*), *Reg 10*(*6*)].

In the light of the safety report, the CA may exempt a local authority from the requirement to prepare an off-site emergency plan in respect of an establishment [*COMAH* (*SI 1999 No 743*), *Reg 10*(*7*)].

Reviewing, testing and implementing emergency plans

C10015 *COMAH* (*SI 1999 No 743*), *Reg 11,* requires that emergency plans are reviewed and, where necessary, revised, at least every three years. Reviewing is a key process for addressing the adequacy and effectiveness of the components of the emergency plan – it should take into account:

- changes occurring in the establishment to which the plan relates;
- any changes in the emergency services relevant to the operation of the plan;
- advances in technical knowledge;
- knowledge gained as a result of major accidents either on-site or elsewhere; and
- lessons learned during the testing of emergency plans.

There is a new requirement to test emergency plans at least every three years. Such tests will assist in the assessment of the accuracy, completeness and practicability of the plan: if the test reveals any deficiencies, the relevant plan must be revised. Agreement should be reached beforehand between the operator, the emergency services and the local authority on the scale and nature of the emergency plan testing to be carried out.

Where there have been any modifications or significant changes to the establishment, operators should not wait for the three-year review before reviewing the adequacy and accuracy of the emergency planning arrangements.

When a major accident occurs, the operator and local authority are under a duty to implement the on-site and off-site emergency plans [*COMAH* (*SI 1999 No 743*), *Reg 12*].

Local authority charges

C10016 A local authority may charge the operator a fee for performing its functions under *COMAH* (*SI 1999 No 743*), *Regs 10* and *11*, i.e. for preparing, reviewing and testing off-site emergency plans.

The charges can only cover costs that have been reasonably incurred. If a local authority has contracted out some of the work to another organisation, the authority may recover the costs of the contract from the operator, provided that they are reasonable.

In presenting a fee to an operator, the local authority should provide an itemised, detailed statement of work done and costs incurred.

Provision of information to the public

C10017 The operator must supply information on safety measures to people within an area without their having to request it. The area is notified to the operator by the CA as being one in which people are liable to be affected by a major accident occurring at the establishment. The minimum information to be supplied to the public is specified in *COMAH* (*SI 1999 No 743*), *Sch 6*:

INFORMATION TO BE SUPPLIED TO THE PUBLIC

- name of operator and address of the establishment;
- identification, by position held, of the person giving the information;
- confirmation that the establishment is subject to these Regulations and that the notification referred to in *Reg* 6 or the safety report has been submitted to the competent authority;
- an explanation in simple terms of the activity or activities undertaken at the establishment;
- the common names or, in the case of dangerous substances covered by *Sch 1, Part 3,* the generic names or the general danger classification of the substances and preparations involved at the establishment which could give rise to a major accident, with an indication of their principal dangerous characteristics;
- general information relating to the nature of the major accident hazards, including their potential effects on the population and the environment;
- adequate information on how the population concerned will be warned and kept informed in the event of a major accident;
- adequate information on the actions the population concerned should take, and on the behaviour they should adopt, in the event of a major accident;
- confirmation that the operator is required to make adequate arrangements on site, in particular liaison with the emergency services, to deal with major accidents and to minimise their effects;
- a reference to the off-site emergency plan for the establishment. This should include advice to co-operate with any instructions or requests from the emergency services at the time of an accident;

- details of where further relevant information can be obtained, unless making that information available would be contrary to the interests of national security or personal confidentiality or would prejudice to an unreasonable degree the commercial interests of any person.

Under *COMAH* (*SI 1999 No 743*), *Sch 8*, the CA must maintain a public register which will include:

- the information included in the notifications submitted by the operators under *COMAH* (*SI 1999 No 743*), *Reg 6*;
- top-tier operators' safety reports;
- the CA's conclusions of its examination of safety reports.

Powers to prohibit use

C10018 The CA is required to prohibit the operation or bringing into operation of any establishment or installation or any part of it where the measures taken by the operator for the prevention and mitigation of major accidents are seriously deficient [*COMAH* (*SI 1999 No 743*), *Reg 18*(*1*)].

The CA may prohibit the operation or bringing into operation of any establishment or installation or any part of it if the operator has failed to submit any notification, safety report or other information required under the Regulations within the required time. Where the CA proposes to exercise its prohibitory powers, it must serve on the operator a notice giving reasons for the prohibition and specifying the date when it is to take effect. A notice may specify measures to be taken. The CA may, in writing, withdraw any notice.

Enforcement

C10019 The Regulations are treated as if they are health and safety regulations for the purpose of the *Health and Safety at Work etc Act 1974*. The provisions as to offences of the 1974 Act, *ss 33–42*, apply. A failure by the CA to discharge a duty under the Regulations is not an offence [*COMAH* (*SI 1999 No 743*), *Reg 20*(*2*)], although the remedy of judicial review is available.

Radiation (Emergency Preparedness and Public Information) Regulations 2001 (SI 2001 No 2975)

C10020 The *Radiation* (*Emergency Preparedness and Public Information*) *Regulations 2001* (*SI 2001 No 2975*) revoke the *Public Information for Radiation Emergencies Regulations 1992* (*SI 1992 No 2997*) (subject to savings). The 2001 Regulations came into force on 20 September 2001.

The Regulations implement the emergency planning aspects of Council Directive 96/29/Euratom, which lays down basic safety standards for the protection of health of workers and the general public against the dangers arising from ionising radiation. The Regulations impose requirements on operators of premises where radioactive substances are present (in quantities exceeding specified thresholds). They also impose requirements on carriers transporting radioactive substances (in quantities exceeding specified thresholds) by rail or conveying them through public places, with the exception of carriers conveying radioactive substances by rail, road, inland waterway, sea or air or by means of a pipeline or similar means.

Essential guidance is contained in the HSE publication '*Guide to the Radiation* (*Emergency Preparedness and Public Information*) *Regulations 2001*' (L126).

The Regulations:

(*a*) impose a duty on the operator and the carrier to make an assessment as to hazard identification and risk evaluation and, where the assessment reveals a radiation risk, to take all reasonably practicable steps to prevent a radiation accident of limit the consequences should such an accident occur [*SI 2001 No 2975, Reg 4*];

(*b*) impose a duty on the operator and carrier to send the HSE a report of an assessment containing specified matters at specified times and empower the HSE to require a detailed assessment of such further particulars as may reasonably require [*SI 2001 No 2975, Reg 6* and *Sch 5* and *6*];

(*c*) impose a duty on the operator and the carrier to make a further assessment following a major change to the work with the ionising radiation or within three years of the date of the last assessment, unless there has been no change of circumstances which would affect the last report of the assessment, and send the HSE a report of that further assessment [*SI 2001 No 2975, Regs 5* and *6*];

(*d*) where an assessment reveals a reasonably foreseeable radiation emergency arising, impose a duty on the operator or the carrier (as the case may be) and, in the case of an operator, the local authority in whose area the premises in question are situated, to prepare emergency plans [*SI 2001 No 2975, Regs 7*, *8* and *9* and *Sch 7* and *8*];

(*e*) require operators, carriers and local authorities to review, revise and test emergency plans at suitable intervals not exceeding three years [*SI 2001 No 2975, Reg 10*];

(*f*) make provision as to consultation and co-operation by operators, carriers, employers and local authorities [*SI 2001 No 2975, Reg 11*];

(*g*) make provision as to charging by local authorities for performing their functions under the Regulations in relation to emergency plans [*SI 2001 No 2975, Reg 12*];

(*h*) in the event of the occurrence of a radiation emergency or of an event which could reasonably be expected to lead to such an emergency, make provision as to the implementation of emergency plans, and in the event of the occurrence of a radiation emergency, the making of both provisional and final assessments as to the circumstances of the emergency [*SI 2001 No 2975, Reg 13*];

(*i*) where an emergency plan provides for the possibility of an employee receiving an emergency exposure, impose a duty on the employer to undertake specified arrangements for employees who may be subject to exposures, such as dose assessments, medical surveillance and the determination of appropriate dose levels and impose further duties on employers in the event that an emergency plan is implemented [*SI 2001 No 2975, Reg 14*];

(*j*) impose requirements on operators and carriers, where an operator or carrier carries out work with ionising radiation which could give rise to a reasonably foreseeable radiation emergency, and on local authorities, where there has been a radiation emergency in their area, to supply specified information to the public [*SI 2001 No 2975, Regs 16* and *17* and *Sch 9* and *10*].

Public Information for Radiation Emergencies Regulations 1992 (SI 1992 No 2997)

C10021 The *Public Information for Radiation Emergencies Regulations 1992* (*SI 1992 No 2997*) were revoked by the *Radiation* (*Emergency Preparedness and Public Information*) *Regulations 2001* (*SI 2001 No 2975*). However, *Reg 3* of the 1992 Regulations continues in force to the extent that it applies in relation to the transport of radioactive substances by road, inland waterway, sea or air. Any other provisions of the 1992 Regulations continue in force so far as is necessary to give effect to *Reg 3*.

An employer (or self-employed person) must supply prior information where a radiation emergency is reasonably foreseeable from his undertaking in relation to the transport of radioactive substances by road, inland waterway, sea or air. He must:

(*a*) supply information to members of the public likely to be in the area in which they are liable to be affected by a radiation emergency arising from the undertaking with without their having to request it, concerning:

- (i) basic facts about radioactivity and its effects on persons/environment;
- (ii) the various types of radiation emergency covered and their consequences for people/environment;
- (iii) emergency measures to alert, protect and assist people in the event of a radiation emergency;
- (iv) action to be taken by people in the event of an emergency;
- (v) authority/authorities responsible for implementing emergency measures.

(*b*) make the information publicly available; and

(*c*) update the information at least every three years.

[*Public Information for Radiation Emergencies Regulations 1992* (*SI 1992 No 2997*), *Reg 3* and *Sch 2*]

Dangerous Goods Systems

Basic regulatory principles

The purpose of dangerous goods measures

D0401 Every day, organisations produce, package and transport dangerous goods. As well as industrial products, many household products also consist of 'common' dangerous goods used in numerous mundane applications. From the most common paints to the deadliest poisons, all these dangerous goods need to be carried, by sea, air, rail or, more often, by road.

In order to minimise risk, rules have been created to regulate this transport. The basic reason for transport control measures and systems is that *dangerous goods are dangerous*. This may seem to be an obvious statement, but it is necessary to remember that these products can cause serious harm or damage and that all reasonable precautions must be taken in order to control and minimise risks.

The aim of dangerous goods controls and regulations is to protect passengers, transport crews and equipment, property and the environment from any of the threats that may be posed by goods with the potential to cause danger within the transport systems.

Fundamentals of control systems

D0402 Most goods that are consigned for transport pose no specific risk to the transport system and are therefore not covered by the controls present in the dangerous goods regulations. Bulky loads, which might present a manual handling hazard, are not classified as hazardous for transportation and are exempt from all dangerous goods regulations. The standard requirements of the *Road Traffic Acts* and others will still need to be followed.

However, certain types of goods may pose a direct hazard to the transport system or the environment within which it operates, and the carriage of these goods needs to be regulated to ensure suitable protection. Such items may pose many different types of hazards, from flammable liquids to radioactive substances and corrosive agents to compressed gases. The dangers that these goods present will be the same whether they are carried by road, rail sea or air so they can be referred to as multi-modal hazardous goods. Certain types of hazardous goods are only regarded as dangerous for specific modes of transport, such as magnetised materials which are dangerous goods for air transport but are not regulated for road transport.

Substances and preparations are classified as workplace hazards by the *Chemicals (Hazard Information and Packaging) for Supply Regulations 2002* (*SI 2002 No 1689*) ('the *CHIP Regulations*') and are assigned certain categories of danger, such as toxic, corrosive, etc. These categories of danger are represented by black on orange hazard symbols, as indicated in FIGURE 1 below:

Substances classified as hazardous by the *CHIP Regulations* may also be classified as dangerous goods for transport, but this is not always the case. Some low to medium

Figure 1

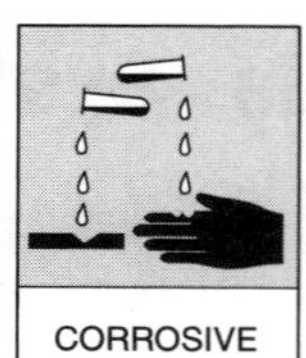

hazard materials are not considered to pose enough risk during transportation to warrant classification, the chance of exposure is lower during transport than use. Hazard classifications under the *CHIP Regulations* apply to materials that may show long-term chronic health effects as well as the results of acute exposure. A toxic paint that is hazardous to individuals who use it every day for a long period of time will not be classified as dangerous for transport if it poses no acute effects. However, some materials with chronic effects, such as dichloromethane (methylene chloride) *are* classified as dangerous goods. An in depth knowledge of the dangerous goods regulations is necessary in order to determine whether specific items are governed by the regulations.

Dangerous goods may be grouped into two categories. A very small number of substances have been judged to be too dangerous to transport under any circumstances as no same method of containment is practicable. Such items include some unstabilised peroxides, particularly sensitive explosives and spontaneously reactive substances. These goods are prohibited for carriage, and transport not allowed on any occasion. Prohibited items are quite rare for road transport, but are much more common for air transport due to the severe consequences of any loss of containment. By far the majority of dangerous goods may be accepted for carriage by road, subject to compliance with the transport of dangerous goods regulations.

The principle of the United Nations dangerous goods classification system

D0403 In order to maintain an international uniformity of approach to dangerous goods legislation, the United Nations has established guidelines on the regulation of dangerous goods. These guidelines are published as the 'Recommendations on the Transport of Dangerous Goods: Model Regulations' which is commonly referred to as the 'Orange Book'. These UN Recommendations have no legal force but are adopted worldwide as the basis of national legislation.

The Orange Book contains example procedural systems for the classification of types of danger, the identification of dangerous goods, containment system performance and use, documentation and training.

The UN classification system splits dangerous goods into nine hazard groups according to the kind of danger inherent to the item. Some of these classes are further broken down into smaller groupings, which are known as divisions. Sometimes items may show danger characteristics of more than one group. In this case, the material is assigned to the hazard class with the greatest risk, and the lesser risks are assigned as subsidiary hazard classes. Each class is allocated a diamond label to indicate the nature of the hazard on packaging and vehicles carrying dangerous goods. These are reproduced in FIGURE 2 below:

Figure 2

Class 1 Explosives

Class 3 Flammable Liquids

Class 3.2 Organic Peroxides

Class 4.1 Flammable Solids

Class 4.2 Spontaneously Combustible

Class 4.3 Water Reactive

Class 5.1 Oxidising Substances

Class 6.1 Toxic Substances

Class 6.2 Infectious Substances

Class 7 Radioactive Material

Class 8 Corrosive Substances

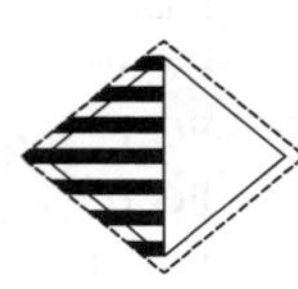

Class 9 Miscellaneous Items

In addition to indicating the kind of danger presented by a substance the classification system also gives qualitative guidance as to how dangerous that substance is in relation to similar types of dangerous goods. This degree of danger is indicated by the 'packing group', which is defined below:

PG I	=	Greatest danger
PG II	=	Medium danger
PG III	=	Least danger

The packing group is used to help determine the selection of acceptable packaging and threshold limits. The packing group system doesn't apply for Class 1,2,5.2, 6.2 or 7, as different specification criteria apply.

The legal framework

United Kingdom Carriage of Dangerous Goods Regulations

Applicability of Dangerous Goods Regulations

D0404 Although the UN Recommendations laid out in the 'Orange Book' are used as the basis for dangerous goods legislation, they have no actual legal authority. National controls are governed by Acts and Regulations established by Parliament which establish duties under national law. International agreements to allow cross border movement of dangerous goods are also authorised by specific national laws.

Within Europe, the carriage of dangerous goods is covered by the European Agreement concerning the International Carriage of Dangerous Goods ('ADR'). This agreement is maintained by the United Nations Economic Commission for Europe ('UNECE'), not by the European Union ('EU'). ADR covers not only the EU, but stretches as far afield as North Africa and the Middle East. The only EU requirement for the transportation of dangerous goods is that Member States' national regulations follow the ADR Agreement.

Interface of different modal controls and UK special concessions

D0405 While different modal regulations are all based upon UN Recommendations, separate specific requirements for road, rail, sea and air transport can cause the potential for conflict between various modes of transport. Differences between UK national requirements and the ADR requirements of international journeys can also lead to confusion about which requirements to follow. Luckily, the UK regulations generally follow the requirements of ADR and are flexible enough to accommodate such concerns.

UK regulations allow compliance with ADR for goods being transported within Britain as part of an international journey. A specific implication of international journeys ending or originating within the UK is that almost all such journeys will have to involve either travel by sea, air or channel tunnel. The requirements for sea or air transport, governed by the IMDG Code or IACO Technical Instructions are tougher than road transport requirements. To prevent having to label goods twice, compliance with UK national road regulations is not required as long as the goods have been correctly classified, packaged and labelled in accordance with applicable air and sea regulations.

Structure of UK Carriage of Dangerous Goods Regulations

D0406 The previous series of legislative documents governing dangerous goods consignment and road transport in Britain have recently been replaced by the *Carriage of Dangerous Goods and Use of Transportable Pressure Equipment Regulations 2004* (*SI 2004 No 568*) (*'CDGTPE Regulations'*) which came into force on May 10 2004.

These new Regulations act as one consolidated piece of legislation replacing the previous range of regulations. They also implement European Directive 2003/28/EC regarding the transport of dangerous goods by road and complete the implementation of Council Directive 1999/36/EC EC concerning transportable pressure equipment ('the Transportable Pressure Equipment Directive'). They directly reference the latest texts of the European Agreement concerning the International Carriage of Dangerous Goods by Road (ADR) (Current edition: 2003).

The main changes from the old rules include:

- New limited quantity and load thresholds for transporting dangerous goods.
- Packaging requirements now based on RID/ADR.
- The transport of diesel, gas oil and heating oil coming fully into scope.
- New reporting requirements.

The Regulations take account of national derogations agreed with the European Commission in respect of certain transport operations involving local journeys or the movement of dangerous goods in small quantities. They continue to use the mandatory Emergency Action Code system for placarding GB-registered vehicles on GB journeys instead of using the ADR placarding system. The Regulations will be supported by a number of letters of authorisation allowing time-limited relaxations to some requirements. Enforcement of the *CDGTPE Regulations* is undertaken by the Department of Transport, Local Government and the Regions rather than the Health and Safety Executive as used to be the case for the former regulations.

Revocations

D0407 The *CDGTPE Regulations* (*SI 2004 No 568*) completely revoke the following Regulations:

- the *Gas Cylinders* (*Pattern Approval*) *Regulations 1987* (*SI 1987 No 116*);
- the *Pressure Vessels* (*Verification*) *Regulations 1988* (*SI 1988 No 896*);
- the *Packaging of Explosives for Carriage Regulations 1991* (*SI 1991 No 2097*);
- the *Carriage of Dangerous Goods* (*Classification, Packaging and Labelling*) *and Use of Transportable Pressure Receptacles Regulations 1996* (*SI 1996 No 2092*);
- the *Carriage of Explosives by Road Regulations 1996* (*SI 1996 No 2093*);
- the *Carriage of Dangerous Goods by Road* (*Driver Training*) *Regulations 1996* (*SI 1996 No 2094*);
- the *Carriage of Dangerous Goods* (*Amendment*) *Regulations 1998* (*SI 1998 No 2885*);
- the *Carriage of Dangerous Goods* (*Amendment*) *Regulations 1999* (*SI 1999 No 303*);

- the *Transport of Dangerous Goods (Safety Advisers) Regulations 1999* (*SI 1999 No 257*);
- the *Transportable Pressure Vessels Regulations 2001* (*SI 2001 No 1426*);
- the *Carriage of Dangerous Goods and Transportable Pressure Vessels (Amendment) Regulations 2003* (*SI 2003 No 1431*).

The Regulations also partially revoke the following Regulation:

- the *Carriage of Dangerous Goods by Road Regulations 1996* (*SI 1996 No 2095*) ('*CDGRoad*').

  Regulations 1 to *4*, *7* to *19*, *21* to *29* and the *Schedules* are revoked. The remaining regulations empower certain approved documents to be used in conjunction with the new *CDGTPE Regulations* and enforce specific requirements regarding the unloading of petrol at petrol stations.

A number of legislative documents work in conjunction with the *CDGTPE Regulations* to provide complete legislative control of dangerous goods consignment and road transport in Britain. These are detailed below:

- the *Radioactive Material (Road Transport)Regulations 2002* (*SI 2002 No 1093*) ('*RAMRoad*').
- the *Radioactive Material (Road Transport)(Amendment) Regulations 2003* (*SI 2003 No 1867*) ('*RAMRoad Amendment*').
- the Dangerous Goods Emergency Action Code List ('DGEACL') (ISBN 0 11 341275 4).

These Regulations (except for *RAMRoad*) are empowered by the *Health and Safety at Work etc. Act 1974* and so follow the same format and type of presentation. Apart from items of Class 7 (additional measures covered by *RAMRoad*), the Regulations and supporting legal documents apply to all classes of dangerous goods. In each case, the appropriate regulation establishes the legal duties to ensure compliance and detailed schedules provide additional information to back up the regulations.

The Approved List

D0408 The former Approved Carriage List ('ACL') has been replaced by the ADR 'Approved List' listed in Table A of chapter 3.2 of ADR. This contains lists of all categories of dangerous goods authorised for road transport. If a substance is listed by name, or its hazardous properties described then it is subject to control and must be packaged, marked and labelled, its transport must be undertaken, the driver must be trained and a Dangerous Goods Safety Advisor ('DGSA') appointed in accordance with *CDGTPE* (*SI 2004 No 568*). Low quantities of dangerous goods in small receptacles or restricted load sixes limited in may be exempt from some or all of the above regulations under limited quantities concessions.

Each entry in the Approved List provides information to be used by the consignor to fulfil classification, identification, package selection, labelling and documentation completion duties. For example:

UNNO	*Proper Shipping Name*	*Class*	*CLS Code*	*Packing Group*	*Labels*	*Limited Quantities*	*Packaging*		
							Packaging instructions	Special packing provisions	Special provisions
1106	Amylamine	3	FC	II	3 + 8	LQ4	P001 IBC02		MP19

UN Portable Tank		*ADR Tank*		*Vehicle for tank carriage*	*Transport category*	*Special provisions for carriage*				*HIN*
Instructions	Special provisions	Tank code	Special provisions			Packages	Bulk	Loading, unloading and handling	Operation	
T7	TP1	L4BH	TE1	FL	2				S2 S20	38

UNNO = United Nations Number
CLASS = Hazard Class
CLS Code = Hazard Classification Code
HIN = Hazard Identification Number

It is essential to check the implication of any classification codes in the table above as these can significantly affect consignor duties. These codes are described in chapter 3.2 of ADR. The carrier of the dangerous goods can use the Approved List to confirm the accuracy of the consignor's transport documentation, to determine the competence required by the driver carrying the load and to decide what vehicle markings are appropriate for the journey.

Controls of radioactive materials follow a significantly different pattern and refer to further sections of ADR and are discussed separately.

Consignment duties

D0409 Procedures for classification, identification, package selection and warning marks and labels are based upon the ADR system and are contained within *Part 2* of *CDGTPE* (*SI 2004 No 568*) (*Regulations 9* to *25*). A range of concessions for limited quantities are available, which are discussed later. Classification requires cross references to be made to the schedules in the back of the Regulations as well as to the relevant chapters within ADR, which provide detailed information on classification procedures.

Dangerous goods classification and identification requires use of both *CDGTPE* and the ACL.

For package selection decisions and marking and labelling duties, *CDGTPE* along with its schedules and both the Approved List and Requirements must be utilised.

Carriage duties

D0410 The consignor is required to create a 'transport document', which is handed to the carrier as an indication that they will probably have to comply with the remaining duties found in the *CDGTPE* (*SI 2004 No 568*) *Regulations*.

There are load size thresholds that may affect duties – if exceeded the vehicle operator and driver must comply with the duties laid down in the Regulations.

The Regulations also lay down requirements to:

- the suitability of vehicles for particular loads;
- loading;
- unloading and stowage procedures;
- the carriage of safety equipment;
- the need for safe working systems;
- emergency response actions; and
- parking and supervision duties.

The ADR Agreement

Structure of the ADR Agreement

D0411 Volume I of this two volume document identifies hazard classes and allocates dangerous materials into these classes by UN number. It provides information to be used by the consignor to classify his hazardous goods for transportation in preparation for a journey. Volume II contains the duties to be followed by the carrier of the goods such as specific packing instructions, labelling requirements, package testing information, placarding and vehicle suitability/driver training arrangements.

Volume I comprises of an introduction, which is a brief summary of the duties, the legal text of the ADR Agreement and Annex A, the general provisions and provisions concerning dangerous substances and articles which outline the duties of the consignor of the goods.

The general provisions (chapter 1) detail the scope and availability of ADR, along with basic definitions, general obligations and transitional measures. Chapter 2 of Annex A then details general provisions and principles of hazard classification along with specific class provisions and approved test methods for classification and allocation of dangerous goods to a particular class and packing group. The final part (chapter 3) of volume I is a comprehensive dangerous goods list which details hazard class, packing group and other pertinent information for items falling into each specific UN number category. Its scope is the same as the ACL for UK regulations, but in addition includes details for explosives and radioactive materials (Classes 1 and 7).

Volume II begins with the concluding parts of chapter 3, detailing special provisions and exemptions for certain materials where appropriate. Chapter 4 describes packing and tank provisions, which covers general provisions along with special provisions for portable tanks, intermediate bulk containers ('IBCs') and other large vessels. Special provisions for explosives, gases, organic peroxides, self reactive substances and radioactive materials are also laid out. Chapter 5 describes consignment procedures including general provisions, marking and labelling requirements for packages, tanks and vehicles along with documentation requirements. Chapter 6 describes detailed requirements for the construction and testing of packages, IBC's, gas cylinders and tanks, etc and the resultant UN mark for approved packaging/receptacles. Chapter 7 details provisions concerning conditions of carriage, loading, unloading and handling of dangerous goods such as general conditions, mixed loading prohibitions, handling and stowage, cleaning vehicles after unloading and any additional provisions that may apply to specific items or goods. Annex B outlines provisions concerning transport equipment and transport operations and details requirements for vehicle crews, equipment, operation and documentation. Such matters as the provision of fire fighting equipment and spillage kits are covered, along with driver training requirements and other miscellaneous requirements to be complied with by the vehicle crew.

The ADR chapters are broken down into subchapters and subsections in a logical manner that allows easy reference to information throughout both volumes of the agreement.

Transport of radioactive materials

General principles of transport of radioactive material

D0412 The International Atomic Energy Agency ('IAEA') is an agency of the United Nations which addresses the whole field of radioactive material control and not just the transportation of such substances.

The IAEA produces recommendations for the effective control of radioactive materials. The Transport Regulations are published in a collection of volumes entitled the Safety Series, and sixth in the series is the 'Regulations for the Safe Transport of Radioactive Material', referred to as Safety Series No. 6.

In 1996 'Regulations for the Safe Transport of Radioactive Material No. ST-1' was published in the IAEA Safety Series as a successor to Safety Series No. 6.

The ADR Agreement and the *RAMRoad Regulations* (*SI 2002 No 1093*) have adopted and adapted Safety Series No.6 and incorporated the information into the procedural duties of both systems, rendering the recommendations that they convey as legal duties. Therefore, the safety series books are not operational documents as the relevant directions from the series can be identified in the appropriate transport controls.

UK Radioactive Transport Regulations

D0413 The *Radioactive Material (Road Transport) Regulations 2002 (SI 2002 No 1093)* ('*RAMRoad*'), control the movement of Class 7 items on road vehicles in Great Britain. *RAMRoad* is under the control of the Department of Transport, Local Government and the Regions, Radioactive Materials Transport Branch.

RAMRoad consists of 75 regulations and does not begin with an arrangement of regulations section. It also provides information on how to carry out the duties allocated through 14 schedules.

The ADR radioactive material system

D0414 General consignor duties are outlined in chapter 1.7 of the ADR Agreement which provides the general requirements for transporting radioactive materials. Rather than the ECE which governs the requirements for all other classes of goods, the standards used are based on the IAEA Regulations for the Safe Transport of Radioactive Material (ST-1). These focus upon the need for the containment of radioactive contents, control of external radiation levels, prevention of criticality and heat damage. Chapter 6.4 of the ADR Agreement covers the requirements for the construction, testing and approval of packages and materials for containing radioactive materials.

In addition to radioactive and fissile properties, any subsidiary hazards of dangerous goods items such as flammability, corrosivity, etc, must be taken into account in the documentation, packaging, labelling, marking, placarding, stowage, segregation and carriage requirements to ensure compliance with all the requirements of ADR.

General requirements

Classification of dangerous goods

Basic principles of classification

D0415 The purpose of classification is twofold; firstly, to determine which goods are dangerous during transport; secondly, to show what kind(s) of danger are to be found in a particular article or substance.

The United Nations Committee of Experts on the Transport of Dangerous Goods developed the system for testing and classification. Various modal and national experts have then modified this to produce the most appropriate controls for their method of transport.

All modes base their controls upon the UN system of 9 classes, follow the UN system of testing for the various classes and follow UN packaging group criteria.

The nine classes divide dangerous substances into the different types of hazard they present. Some of these classes are further broken down into smaller groupings, which are known as divisions.

Class/Division		*Typical technical reference*
1	Explosives	Pyrotechnic; non-detonating self-sustaining exothermic chemical reaction.
2	Gases	Vapour pressure above 300kPa at 50°C; gaseous at 20°C at 101.3kPa.
3	Flammable liquids	Initial boiling point; flashpoint.
4.4	Flammable solids	Burning time in seconds per 100mm; rate of detonation or deflagration; self-reactive.
4.2	Spontaneously combustible	Pyrophoric substance; self-heating substance.
4.3	Water reactive	Water reactive; rate of evolution of flammable gas; spontaneous ignition.
5.1	Oxidising substances	Increased burning rate.
5.2	Organic peroxide	Assigned only by competent authority.
6.1	Toxic	LD_{50} oral; LD_{50} dermal; LC_{50} inhalation.
6.2	Infectious substances	Genetically modified; biological products; diagnostic specimens.
7	Radioactive materials	Specific activity more than 70kBq/kg.
8	Corrosive	Full thickness destruction of intact skin tissue.
9	Miscellaneous	Allocated by regulatory authority.

The packing group indicates the degree of danger present by a substance, of which there are 3 levels.

UN Packaging Group	**Meaning**
PG1	HIGH danger
PG2	MEDIUM danger
PG3	LOW danger

The packing group plays a wide role – from the determination of acceptable packaging and the definition of threshold limits, to stowage and segregation requirements. The packing group system does not apply to Class 1, 2, 5.2, 6.2 or 7, as different specification criteria and special packaging requirements apply.

UK classification procedures

D0416 Classification in the UK is based on the UN system of nine classes, the procedures for which are detailed in ADR. The ADR Approved List contains a large range of substances that have been previously classified and assigned packaging groups.

ADR allocates a substance that does not meet the criteria of any other hazard class, but is regarded as an 'aquatic pollutant', to Class 9 as an 'environmentally hazardous substance'. Solutions and mixtures with a concentration of 25 per cent or more of such a substance, should be also be classified as such.

ADR classification procedures

D0417 These largely conform to the UN system, including the hazard class system and the use of packing groups.

The ADR classification system differs from the former UK system in a few ways, as follows:

- Class 2 (gases) – the former separate divisions 2.1, 2.2 and 2.3 no longer exist; gases are split into nine groups according to their hazardous properties as follows:

  A Asphyxiant

  O Oxidising

  F Flammable

  T Toxic

  TF Toxic and flammable

  TC Toxic and corrosive

  TO Toxic and oxidising

  TFC Toxic, flammable and corrosive

  TOC Toxic, oxidising and corrosive

- Class 3 (flammable liquids) – includes gas oil, diesel fume or heating oil (light), with a flashpoint in excess of 61°C.

The classification and packing group of a substance can be determined by looking up the name or UN number in the Approved List or alphabetically in Table B of chapter 3.2.

Identification of dangerous goods

Basic UN principles

D0418 In an emergency the emergency services, vehicle drivers, other transport staff and the general public need to know what dangers they face from certain substances. This creates a need for a product reference that can be quickly interpreted for an effective response.

The UN have devised 2 such references:

- Proper Shipping Names ('PSNs') – These are recommended names for a wide range of commonly moved substances, listed in the UN Orange Book. However, they are often complex chemical names and the need for the identification to be achieved on a worldwide basis led to the development of a second system.
- UN Numbers – These four digit identification numbers are directly linked to the PSN.

UK/ADR identification

D0419 UK regulations are in line with the UN Recommendations and require the use of the same PSN and UN Numbers.

For consignment and transport, the PSN and corresponding UN number can be found in the ADR Approved List.

The Approved List is ordered by UN Number and may be cross referenced using the PSN.

A solution or mixture that contains a dangerous substance and one or more non-dangerous substances is identified using the PSN for the listed one, with the addition of the word 'solution' or 'mixture' e.g. methanol solution.

This does not apply where:

- The solution or mixture is identified in the Approved List.
- The entry in the Approved List only applies to the pure substance.
- The class, physical state or packing group is different to that of the pure substance.
- There is a significant change in the measures to be taken in an emergency.

If this occurs, then the solution or mixture is identified under the appropriate generic or not otherwise specified (n.o.s.) entry in the Approved List. The index contains a range of generic names that cover substances belonging to particular families (e.g. Alcohols n.o.s. and Ethers n.o.s.) and also involves more descriptive names that identify their danger (e.g. Flammable n.o.s. and Corrosive n.o.s.).

Dangerous goods containment

Principles of containment

D0420 All packages used for the transport of dangerous goods have to meet three criteria. The selected package must:

- meet the general requirements established by the regulation;
- be authorised to UN specification standards;
- meet the detailed directions within the applicable regulation for the packaging of the product.

All packages that meet these criteria are available for use to carry the product in the particular transport system.

This means that for most substances and modes of transport a wide range of package types and materials are available. However, in the worst-case scenario only one system of containment will be authorised for transport use.

UN specification sets the standards for packages used in the transportation of dangerous substances, in the UK and by ADR or RID. The UN system requires packaging design and materials to have proven their competence by passing practical tests. Firstly the packs are weathered to their most vulnerable condition, if this is likely to be a factor in their performance. They are then subjected to a number of physical tests such as being dropped, held in a stack and placed under pressure demands to simulate transport situations. Approved packages are given markings that must be clear, legible and readily visible.

Each of the regulatory systems has a basic set of standards that must be met by every package ever used. The requirements are wide ranging and largely overshadowed by the UN specifications, but there are some duties not repeated elsewhere such as the compatibility of the product and packaging material, the provision of ullage and the inspection of all packs before transport. Each system also carries a section detailing the significance of the packaging code, for example:

UN 1A1/X/550/095/GB/0103

UN = UN packaging symbol

1A1 = Type of packaging and material

X = Packing Group competence and maximum density of liquids

550 = Hydraulic pressure test result (kPa)

95 = Year of manufacture

GB = Authorising country

0103 = Packaging ID number

Figure 3: An example of a UN mark as it may appear on a fibreboard (cardboard) package

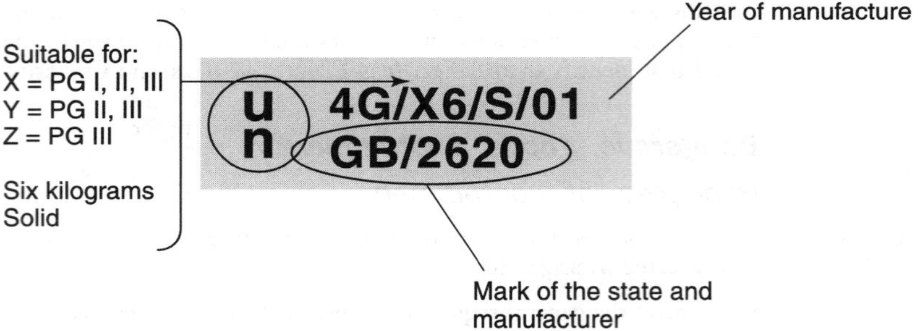

UK/ADR selection

D0421 Within the ADR Agreement, The General Packaging Conditions and the detail of UN specification are in chapter 4. Class specific provisions are detailed in Part 2 of ADR. The particular packaging requirements of each package type are identified in specific subchapters within Part 4.

Each packing instruction identified in the Approved List describes a range of packing options for substances depending upon their Packing Groups. These must be carefully examined in order to determine any volume or mass limits that may apply to a particular package type.

Intermediate bulk container and tank systems

D0422 Intermediate bulk containers ('IBCs') are designed for mechanical handling and have a general maximum capacity of 3m^3 (3,000 litres). The selection procedures are similar to those for packages:

- they must meet a set of general requirements;

- be authorised to UN specification; and
- meet with the detailed directions in the applicable regulation for the product concerned.

IBCs that are acceptable packaging options in the ADR Agreement are listed in Volume II, chapter 4.1.

There are many types of tank used in the transport of dangerous goods, but they all are all built to tightly controlled criteria. A range of descriptions are used to identify the different types, unfortunately, terminology is not consistent across the various transport modes. In each modal control, definitions will be found indicating what kind of equipment is being referred to by any particular name.

The various controls offer two avenues of control of tank traffic:

- Identification of how tanks must be built and equipped to meet the demands of the regulation.
- Description of the type tank and ancillary equipment that must be used for the transport of a particular substance.

Chapter 4.2 to 4.5 of ADR contains constructional and equipment details for the various types of ADR approved tanks, which detail requirements for portable tanks, fixed tanks, demountable tanks, tank containers, tank swap bodies, reinforced plastic tanks, welded tanks and vacuum waste tanks.

Marking, labelling and placarding

Basic principles

D0423 To ensure that everyone involved in handling or moving dangerous goods are aware of the hazards they present, information warning of the dangers are required to be clearly displayed on the outside of packages, tanks, vehicles, etc. This is also vital for rapid and effective action from the emergency response teams faced with dangerous goods incidents.

Terminology is not yet fully harmonised and a distinction is drawn between packages and transport units:

- *Package*: drums, jerricans, boxes, gas cylinders, IBCs, etc.
- *Transport unit*: road freight or road tanker vehicles, rail wagons or rail tanks, multi modal freight containers and tank containers.

Modal control requirements are all based on the UN Recommendations, the Orange Book, which identify the need for:

- Marking and labelling on packages.
- Placarding and marking of transport units.

Marking and labelling

D0424 *Marking* on packages is the application of the complete Proper Shipping Name ('PSN') and UN Number, in a readily visible and legible location on the outside of all packages.

Labelling is the application of diamond shaped warning labels representing both the class of danger and any subsidiary risks. Class labels are distinguished by the appropriate class number in the bottom corner, with the upper half containing a pictorial symbol representing the danger. Text indicating the nature of the hazard

Figure 4

S28555/ 2

Shipping Label for Fisher Scientific

DICHLOROMETHANE

UN 1593

Technical Name:

Fisher Item: D/1856/17
2.5L DICHLOROMETHANE HPLC

Supplied by:
Fisher Scientific UK Ltd., Bishop Meadow Road
Loughborough, Leics, LE11 5RG

Chemical Class – 6.1 Poisonous
Sub Class 1 –
Sub Class 2 –

e.g. 'Corrosive' may also be shown in the lower half – this is mandatory for Class 7, radioactive materials. Division numbers must also be shown on Class 1 and Class 5 labels.

The standard size for such labels is 100mm x 100mm, with a border line of 5mm. Symbols, text and numbers must be in black, except those containing the Class 8, corrosive, number and any associated text, which must be in white.

Limited quantity concessions of dangerous goods are not subject to the standard marking and labelling duties. The different modal regulations for these are not yet aligned and therefore the application of these varies depending on the type of journey.

Placarding and marking

D0425 Transport units must bear the appropriate placards and markings to warn of the dangers of the goods being transported. *Placards* are large versions of the warning labels required on packages, the size for which must be at least 250mm x 2,500mm with a border line of 12.5mm.

For certain journeys specific *markings* are required, the type, number and location of these vary depending on the transport mode and the nature of the journey –

national or international. For example, the relevant UN number must be displayed on the outside of tank transport units. An elevated temperature mark is required for those units transporting liquids at temperatures of 100°C or more, or solids at 250°C or more. An emergency action code ('EAC'), which provides guidance for the emergency services in the event of an incident involving the load, along with other markings may also be needed for tank or bulk load transport, depending on the type of journey.

UK and ADR requirements

D0426 Every package must be marked with the PSN, UN Number and class risk label. The marks and labels required for certain products may be established through the relevant substance entries in the first six columns of the Approved List. Special packing provisions apply variations to the standard marking and labelling, these must be checked before determining the requirements of the package.

The ADR requirements for marking and labelling are the same ones applicable to rail (RID) journeys. These requirements are stated in chapter 5.2 'Marking and Danger Labels on Packages', in each hazard class section in Volume I. It requires every package to be marked with the PSN and labelled with the appropriate class and subsidiary risk labels. Labels are referred to by the label model number (e.g. label model no.3 is the label required for flammable liquids of Class 3) and are described in chapter 5.2.

Documentation requirements

D0427 Documentation relating to a dangerous goods shipment describes important aspects of the cargo and its containment as well as indicating its point of origin and eventual destination. This information is required to ensure that all individuals involved in the transport process are fully aware of the dangers of the material that they are handling and employ suitable equipment and procedures to minimise any risks involved. The documentation also provides vital information in a concise form to emergency personnel. Description of packages etc should be as clear as possible to enable rapid identification, and substance descriptions must conform exactly to the requirements of the regulations in order to minimise misunderstanding and errors during an emergency.

UK transport documentation

D0428 With the exception of goods carried under limited quantities exemptions, the consignor must provide each shipment with documentation detailing the designation of the goods (the proper shipping name, along with the designation 'waste' if applicable) and the class of each item of goods preceded by the word 'Class' or the classification. The UN Number, preceded by the letters 'UN' and the transport category must also be stated. The mass/volume of each package along with the number of packages, or the total mass/volume of each transport category along with the consignor and consignee names and addresses is also required. The consignor must also provide any other information necessary for the vehicle operator to comply with their transport documentation requirements. This information along with the emergency action code, the prescribed temperature and a container packing certificate, will provide emergency information to the driver and is referred to as 'instructions in writing'. This is most commonly done by the provision of a Tremcard. In addition, transport documentation for explosives must detail the total number of packages carried, the total *net* mass of explosives carried and the name and address of the vehicle operator.

Under the provisions of the *CDGTPE Regulations* (*SI 2004 No 568*), the driver of the vehicle must ensure that any documentation relevant to the load being carried is kept readily available on the vehicle, and must produce the documentation on request to any authorised person. The driver must also ensure that any documentation relating to dangerous goods *not* being carried is removed from the vehicle or is secured in a closed container and clearly marked to show that such information inside does not relate to the load currently on board. Once a journey is complete, the vehicle operator must maintain copies of transport documentation (but not emergency information) for three months.

ADR documentation

D0429 The ADR Agreement imposes specific duties relating to the provision and carriage of documentation when goods are moved by roads on international journeys. Similar to UK requirements the consignor, vehicle operator and driver all have duties under the provisions of ADR.

The consignor has a duty to provide a transport document (information on the goods consigned), a declaration of compliance with ADR, written emergency instructions (actions to be implemented in the event of an accident) and a container packing certificate (only needed if the goods are loaded onto a freight container exceeding $3m^3$ capacity) to the vehicle operator.

It is the vehicle operator's responsibility to ensure that the appropriate documents are carried. These could include:

- the transport document and declaration of compliance;
- a container packing certificate, a vehicle approval certificate;
- the driver training certificate;
- the written emergency instructions; and
- a journey authorisation permit.

It is the driver's duty to ensure that the relevant documents are kept readily available and that any irrelevant documents are kept separate and clearly identified to avoid confusion.

Compliance thresholds/limited quantities

Principles of thresholds

D0430 Often dangerous goods are transported in such small receptacles that there is no significant risk of them causing a serious incident. Such items are subject to either simplified dangerous goods controls, or in certain circumstances are exempt from the scope of the regulations. This means that even if the substance is still dangerous, the quantity is so small that the package presents no particular danger for transport and is not considered to be a regulated package.

The definition of what is deemed to be a small receptacle varies greatly from class to class and packing group to packing group. Careful attention to detail is required to enable consignors to take advantage of limited quantity exemptions. Limited quantity concessions are available for shipments placed in combination packs, and there is usually a maximum weight specified for the package as well as a maximum size for the receptacles within the package.

The transport category of the dangerous goods determines the receptacle thresholds for limited quantity concessions. The transport categories are allocated on the basis

Figure 5

Limited quantity – concessionary package. Full scope of Regulations does not apply

Not limited quantity – regulated load. Full scope of Regulations applies

of packing groups and class specific allocation in accordance with ADR 1.1.3.6 and entries for each specific material listed in column 15 of the Approved List:

Transport category		*Maximum mass/ volume threshold per transport unit*
0	Specific materials listed in Table 1.1.3.6.3 of ADR and in the Approved List	0
1	Packing Group I goods (unless otherwise specified) Specific materials listed in Table 1.1.3.6.3 of ADR and in the Approved List	20 litre/ kilogram
2	Packing Group II goods (unless otherwise specified) Specific materials listed in Table 1.1.3.6.3 of ADR and in the Approved List	333 litre/ kilogram

3	Packing Group III goods (unless otherwise specified) Specific materials listed in Table 1.1.3.6.3 of ADR and in the Approved List and any other dangerous goods not listed in other categories.	1000 litre/ kilogram
4	Specific materials listed in Table 1.1.3.6.3 of ADR and in the Approved List. These materials are effectively exempt from ADR controls	Unlimited

Exemptions and concessions

D0431 Within ADR there are further concessions that confer exemption from the associated duties. To qualify for exemption, which extends to cover both the carrier of the goods as well as the consignor, both the receptacle and the package size maximum limits must be complied with. Details of such concessions are listed in column 7 of the Approved List. However, there are no concessions for Classes 1, 4.2, 6.2 or 7.

There are certain marking requirements applicable to any package that has limited quantity status: a package containing a single dangerous substance must display the appropriate PSN, preceded by the letters 'UN', a package containing dangerous substances with different PSNs must display the PSN for each of the substances, or the letter 'LQ'. Packages consigned as limited quantity and suitably marked are not subject to any other ADR duties.

Training duties

D0432

- *Classification principles*

  If appropriate safety standards are to be established and maintained it is essential that adequate training be provided to all personnel involved in the consignment and transport of dangerous goods. All such training must be commensurate with individual responsibility and is a requirement of the UN Recommendations.

- *Regulatory training requirements*

  Formal training for drivers of road vehicles carrying dangerous goods is a requirement of both ADR and the *CDGTPE Regulations* (*SI 2004 No 568*). In addition there is a requirement on employers to provide adequate training for all employees loading, unloading and handling dangerous goods, and records to be kept by both the employer and employee training records must be kept for a minimum of 5 years.

Specific requirements

Vehicle requirements

D0433 The vehicle and its ancillary equipment is an integral part of the containment system, which must be effective in order to allow the goods to be carried in a safe manner.

UK and ADR vehicle and equipment requirements

D0434 A range of requirements relating to the vehicle and its load carrying competence are laid down in the ADR Agreement (Annex B 'Provisions Concerning Transport Equipment and Transport Operations'). The information is presented at two levels:

a set of duties that underpin all decision making on vehicle provision, equipment and use; and another set of directions related to each danger class and in some cases to specific item numbers within a class.

Annex B contains a number of general requirements regarding the physical competence of the vehicle. Chapter 8.5 deals with any extra provisions or variations particular to a class, or a certain substance within a class. The degree of supplementary information varies between classes and covers a range of issues. For example, Class 1 has the greatest level of detail, ranging from vehicle construction requirements, to bodywork and exhaust systems, where as Class 4 provisions cover aspects such as temperature control and the use of closed or sheeted vehicles. Chapters 8.5 and 9 should therefore always be checked to confirm any additional requirements.

The *CDGTPE Regulations* (*SI 2004 No 568*) require a check to be carried out on the vehicle, its equipment and the driver's documentation before loading the vehicle, to ensure that the journey can be undertaken in a compliant manner.

Placarding and marking of different types of vehicle

D0435 The operator and driver of the vehicle must ensure that any placards and markings are displayed as required during the journey, that they are kept in a clean and visible state and removed or covered when not in use.

UK requirements

D0436 Vehicle placarding and marking is the one major area where the new *CDGTPE Regulations* (*SI 2004 No 568*) differ substantially from ADR requirements. The former system has been retained as an authorised national concession by the UK and incorporated into the current legislation. For road journeys in the UK, combinations of placards and markings are required for different transport units, such as:

- Vehicles carrying packages (including IBCs) or freight containers loaded with packages.
- Road tanker vehicles (i.e. fixed tanks).
- Vehicles carrying tank containers.
- Vehicles carrying bulk loads
- Vehicles carrying bulk loads in freight containers.

Some examples of UK vehicle labelling are given below in FIGURE 6.

Placards are the warning labels that represent the danger class of the load.

Markings consist of clearly visible rectangular orange-coloured plates, which can incorporate such items as the appropriate UN Number, the Emergency Action Code (EAC) and an emergency telephone number, where specialist advice may be obtained at any time during the carriage of the goods. Dangerous goods that are moved in tanks or as bulk loads may also display the appropriate class danger placard on a combined hazard-warning panel. Alternatively they can all be displayed separately according to the requirements under ADR.

Regulation 55 and *Schedule 9* of *CGDTPE* (*SI 2004 No 568*) contain the requirements relating to the type, number and location of placards and markings for the transport of dangerous goods. The details for a particular load can be ascertained

Figure 6

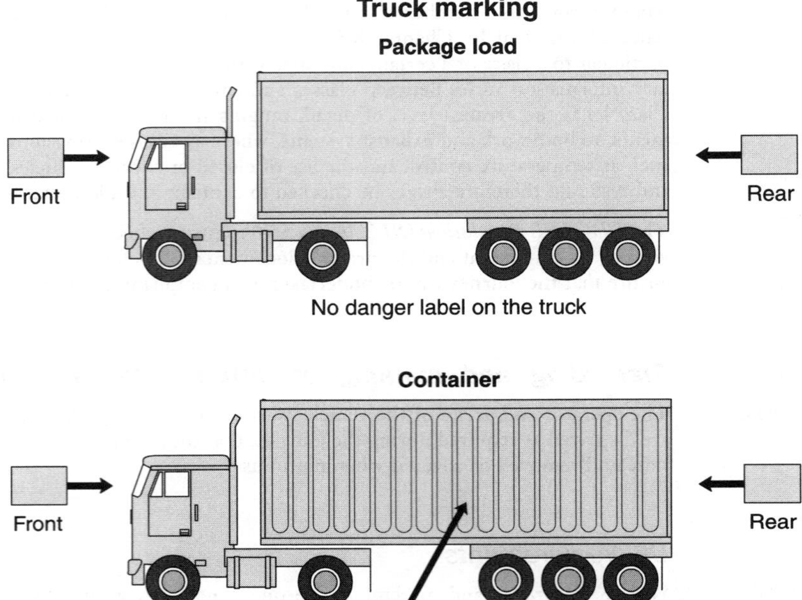

through the relevant entries in the Approved List, except for the EAC which may be found within the DGEACL. The EAC's effectively replace the ADR HIN codes for UK only journeys.

There are separate guidelines for those classes of dangerous substances not covered in *CDGTPE*. Those relevant to Class 7 items, are described in *Regulation 48* and *Schedule 6* of *RAMRoad* (*SI 2002 No 1093*).

ADR requirements

D0437 The requirements for ADR journeys and UK journeys are very closely aligned, in that all transport units must be placarded and marked.

Different placards and markings are required for the following transport units:

- Vehicles carrying packages (including IBCs).
- Vehicles carrying freight containers loaded with packages.
- Road tanker vehicles (i.e. fixed tanks).
- Vehicles carrying tank containers.
- Vehicles carrying a demountable tank(s).

- Vehicles carrying bulk loads.
- Vehicles carrying bulk loads in freight containers.
- Battery-vehicles (transport units with a manifold permanently attached to a frame, permanently fixed to the unit).

An example of ADR vehicle labelling is given below in FIGURE 7.

Figure 7

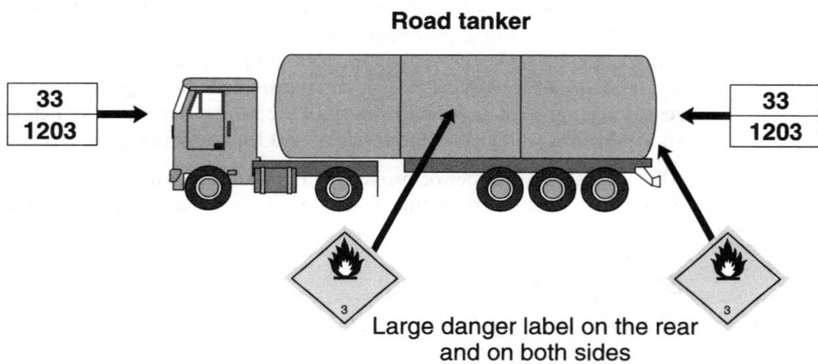

Placards are the warning labels that represent the danger class of the load. In the ADR Agreement they are referred to as 'labels'.

Markings consist of clearly visible rectangular orange-coloured plates, which must be securely attached to the unit. If the goods are being transported in tanks or as bulk loads the markings must incorporate the appropriate SIN and hazard identification number ('HIN' or Kemler code), which identifies the dangers of the load being carried. This is explained in Appendix B.5, which also contains three tables for ascertaining the SIN and HIN details for particular loads and confirming the placarding requirements.

The nature, number and location of the placards and markings can be located in chapter 5.3, once the class, item number and danger letter of the substance being transported has been established.

Regulated loads

D0438 There are some concessions that ease the transport controls for vehicles that are carrying only small quantities of dangerous substances. These Limited Quantity ('LQ') exemptions and their conditions of applicability are detailed by reference to column 7 in the Approved List.

Principles of concessions

D0439 Some requirements of the *CDGRoad Regulations* (*SI 1996 No 2095*) apply when any quantity of dangerous goods are being carried, e.g. to adequately segregate any toxic or infectious substances from food. The application of most requirements depends on whether the goods are being transported in a tank/bulk or in packages. If it is being transported in a tank or in bulk then the provisions of *CDG Road* apply. The Approved List provides full details of such requirements.

Filling, loading and stowage

Filling and loading requirements

D0440 The *CDGTPE Regulations* (*SI 2004 No 568*) contain duties to ensure the safe loading and unloading of vehicles.

A duty on all persons, to ensure that the manner in which goods are loaded, stowed or unloaded is not liable ' to create a significant risk or significantly increase any existing risk to the health and safety of any person'. This duty gives provisions for particular classes of dangerous goods, for example, where flammable goods are being transported by tank, the chassis must be properly earthed and flow rates controlled to prevent static discharges. It also provides more general health and safety requirements, such as prohibiting smoking in the vicinity of vehicles whilst loading and unloading and also cleaning tanks and freight containers after use.

ADR contains a number of provisions to ensure the safety of loading and unloading operations – which are in line. Class specific provisions can be found in chapter 2.

Stowage requirements

D0441 The *CDGTPE Regulations* (*SI 2004 No 568*) provide particular stowage requirements for certain dangerous goods. These requirements are detailed in chapter 7 of ADR and the subsections provide information on the various classes of dangerous goods, e.g. flammable solids must be loaded in a fashion that ensures free air circulation and the maintenance of a uniform temperature. A general duty to ensure that all goods are suitably stowed and secured on vehicles is also established.

Segregation requirements

D0442 From time to time a road traffic accident may occur, a tank valve may fail, a fire may break out. These incidents could result in two substances reacting dangerously if they came into direct contact with each other. Segregation ensures that acceptable levels of safety are maintained in the transport system, in case one of these events does occur.

Dangerous goods must not be packed together in the same outer packaging with dangerous or other goods if there is a possibility that they will react dangerously together to cause: combustion and/or the evolution of considerable heat; the evolution of flammable, toxic or asphyxiant gases; or the formation of corrosive or unstable substances. Toxic or infectious substances must not be carried on the same vehicle as food, unless it can be 'effectively separated'. It is therefore left to the operator to assess whether the loading arrangements are satisfactory. The circumstances under which the mixed packing of substances in a particular class is permitted with either other items from the same class, or with products from other classes, is clearly laid out in chapter 5.1 of ADR.

Operational procedures

D0443 The vehicle crew (driver, assistant(s), and attendant(s)) is the primary channel of safety during road journeys involving the carriage of dangerous goods. The working procedures, duties, training and instruction of the crew are both a safeguard against the likelihood of an incident occurring involving their hazardous load and a primary means of minimising the consequences of any such incident.

UK duties and equipment

D0444 The *CDGTPE Regulations* (*SI 2004 No 568*) contain various duties, many of which must be carried out by the vehicle crew.

The driver must:

- Ensure that any relevant transport documentation is kept readily available.
- Ensure that any relevant placards and markings required for the journey are displayed and removed when not relevant.
- If the load exceeds a specified limit then the special supervision and parking requirements must be satisfied.
- Ensure that any segregation, stowage or load safety instructions are followed.
- Ensure that the crew complies with any emergency information relating to the goods in the event of an accident and also ensure that the appropriate emergency services are contacted.
- Ensure that steps are taken to minimise the risk of fire or explosion.
- Ensure that the engine is shut off during loading and unloading.

The vehicle transporting dangerous goods must be equipped with a safety kit appropriate to the measures detailed in the 'emergency information', e.g. where toxic gases are being carried, the crew must be provided with suitable respiratory equipment.

The duties relating to the vehicle crew and its equipment are contained within Annex B of ADR.

Vehicle equipment requirements on dangerous goods journeys are:

- At least one wheel scotch.
- Two standing warning signs.
- A suitable warning vest or clothing.
- One intrinsically safe handlamp for each crewmember.
- Any other general or load-specific safety kit listed on the written 'emergency instructions'.

The duties laid down for the vehicle crew are contained in chapter 7.5 'Special Provisions Concerning Loading, Unloading and Handling'. These duties state that:

- If the load exceeds a specified limit then the vehicle must be supervised at all times unless it is in a secure factory or depot site.
- Crew members are the only personnel permitted to be carried on dangerous goods journeys.
- Crew members must know how to operate their fire extinguishers.
- Any relevant transport documentation is kept readily available.

Emergency response requirements

D0445 One function of the Dangerous Goods Safety Advisor ('DGSA') is to monitor the implementation of proper emergency procedures in the event of any accident or incident that may affect the safety during the transport of dangerous goods.

The *CDGTPE Regulations* (*SI 2004 No 568*) require that emergency information, usually in the form of a Tremcard, 'shall comprise details of the measures to be taken by the driver in the event of an accident or an emergency and other safety information regarding the goods being carried … '. This information covers areas such as inherent dangers, measures to be taken if someone comes in to contact with the substance, measures to deal with a spillage and measures to be taken in the event of a fire. Additional requirements for specific types of material are given in chapter 7.5.11 of ADR.

Environmental protection

D0446 The *CDGTPE Regulations* (*SI 2004 No 568*) do not define a requirement for the Dangerous Goods Safety Advisor to be involved in environmental pollution activities. However, they should be aware of the relevant regulations or documents that deal with environmental measures.

The written emergency instructions should contain information on the potential for environmental damage and guidance to the immediate protective measures. In addition, those substances with known environmental effects should be labelled accordingly.

Road transport training

D0447 When carrying dangerous goods, special training and certification requirements apply to many drivers. Drivers must hold a vocational training certificate ('VTC'), issued by a competent authority established by *Part 3* of *CDGTPE* (*SI 2004 No 568*). These can only be obtained by attending a government approved training course and passing the examinations. There are some exemptions to these requirements and they are listed in the limited quantities exemptions. The training requirements of the ADR are similar to those of the UK requirements, in that the driver must hold a VTC in align with a European driving license.

Disaster and Emergency Management Systems (DEMS)

Historical Development of Disaster and Emergency Management

D6001 The events which took place in New York on 11 September 2001 have marked a paradigm shift in the thinking, planning and perception of 'man-made disasters'. Throughout the 1990s the Federal Emergency Management Agency (FEMA) of the USA (a Competent Authority dealing with natural and man-made disasters) was being criticised by Congress and the media for being too resource consuming and inefficient. After '911' it was being hailed for its expertise in emergency response management. Suddenly 'corporate America' and the democratic world began to take more seriously the importance of 'Disaster and Emergency Management', 'Civil Defence Management', 'Civil Protection Management' and 'Business Continuity'. In the United Kingdom since June 2001 the 'civil contingencies' function has been transferred from the Home Office to the Cabinet Office, giving the Prime Minister's Office a greater strategic oversight of national 'Major Incidents'.

The UK experience broadly indicates that Disaster and Emergency Management (DEM) has evolved in three phases:

Disaster and Emergency Management (DEM) has evolved in three phases:

- *Phase 1: pre-CIMAH 1984* (*Control of Industrial Major Accident Hazards Regulations 1984* (*SI 1984 No 1902*))

  DEM was largely confined to local government, the emergency services, large organisations and civil protection agencies (environmental, military etc.). This phase is marked by large scale macro plans and detailed, sequential procedures to be followed in the event of a 'Disaster' or 'Emergency'. It is exemplified in the guidance prevalent on 'incidents' involving radioactivity, available from central government, the United Nations, Red Cross and others.

- *Phase 2: Liberalisation Phase*

  With the advent of decentralisation of industry and the growth of privatisation, organisations (large or otherwise) began to focus on 'procedures' and 'Business Continuity Planning'. British Telecom exemplified this with the establishment of its Disaster Recovery Unit and the offering of its expertise to customers at large. 'Major Incidents' such as Chernobyl, Piper Alpha, Kings Cross, The Marchioness and Challenger to name a few, which hit the world headlines in the 1980s also focused planners minds on effective DEM not just isolated procedures.

- *Phase 3: Holistic Phase*

  This has two dimensions, first the role of Europe. The impact of the *Framework and Daughter Directives*, as transposed into the '*6 Pack Regulations*' of 1992, introduced for the first time explicit and strict duties on organisations in general to plan for 'serious and imminent dangers'. The European Commission also began to take the risks of trans-national 'Disasters' more seriously, as highlighted by the greater role given to the Civil Protection Unit, in DGXI of the European Commission. Second, this phase sees domestic UK legislation becoming more cognisant of Disaster and Emergency issues (*Environment Act 1995* makes provisions for environmental emergencies; *Control of Major Accident Hazards Regulations 1999* (*COMAH*) (*SI 1999 No 743*) replaces *CIMAH 1984*; also the need for effective planning when carrying dangerous goods via road legislation, *Carriage of Dangerous Goods by Road Regulations 1996* (*SI 1996 No 2095* partially revoked by *SI 2004 No 568*)).

The most noticeable index of how DEM has changed is the availability of information and templates on 'Disaster planning' and 'emergency planning' to organisations of all sizes. There is also a greater media focus on such issues, with the BBC for instance having dedicated web resources on 'Disasters'.

Origin of Disaster and Emergency Management as a modern discipline

D6002 DEM is largely a fusion of four branches of knowledge:

- *Occupational Safety and Health (OSH)*

  This concerns 'internal' or work-based causes of systems failures, with major impact on life and property. It has been exemplified by Heinrich in his publication *Unsafe Acts and Unsafe Conditions* in the early part of the 1900s, Bird and Loftus with their 'management failing' explanations of accidents and incidents and Turner with his 'incubation' explanation of Man-Made Disasters in the 1970s. OSH academics have led the forefront in terms of analysing the branch and root causes of major industrial disasters.

- *Security Management*

  In the 1970s, security threats in the UK such as bomb explosions, terrorism and electronic surveillance failures have given insights to causation and the motivation behind man-made emergencies. Also, guidance from the Home Office as well as the emergency services has enabled a practical understanding of how to cope with emergency situations.

- *Business Management*

  The late 1980s saw a shift in the academic paradigm in economics and business management from 'static' or closed business planning – where businesses were told to make the assumption of *cetaris paribus*, that is assume all things are constant with the business acting as if it was the only one in the market place – to 'dynamic' or open planning. The latter sees uncertainty and risk being factored into decision making models. This influenced the development of 'business continuity management', developing strategies when the business faces major corporate uncertainty and crises as well as 'contingency planning'.

- *Insurance*

  The fourth significant influence comes from insurance and loss control. The occurrence of Disasters and Accidents has involved loss adjusters and actuarial personnel. The former have developed methods of analysing the basic and underlying causes of an event, whilst the latter have been developing statistical methods for calculating the chance of failure and the risk premiums needed to indemnify that failure.

  DEM uses both quantitative (statistics, quantified risk assessments, hazard analysis techniques, questionnaires, computer simulation etc.) and qualitative methods (inspections, audits, case studies etc.).

Definitions

D6003 Stan Kaplan (*The Words of Risk Analysis*, 1997, Risk Analysis, Vol 17, No 4) stated that 50 per cent of the problems with communication are due to individuals using the same words with different meanings. The remaining 50 per cent are due to individuals using different words with the same meaning. In DEM this is a basic problem. Some authors have argued distinguishing definitions has no practical value and can be ' ... highly undesirable to try to control how others use them ...' (*Managing the Global Consequences of a Disaster,* Richard Read, Paper at the 2nd International Disaster and Emergency Readiness Conference, The Hague, October 1999, page 130 of IDER Papers). Nevertheless there is much confusion over the meaning given to core terms, so that basic definitions can assist in avoiding the Kaplan dilemma. Legislation or approved codes have not provided definitions of 'Disaster' or 'Emergency' for instance. *The New Oxford Dictionary of English*, OUP, (1999) provides the following primary meanings:

(*a*) Catastrophe: 'an event causing great and often sudden damage or suffering: a disaster', page 287.

(*b*) Crisis: 'a time of intense difficulty or danger', page 435.

(*c*) Disaster: 'a sudden event, such as an accident or a natural catastrophe, that causes great damage or loss of life', page 524.

A catastrophe and a crisis are types of Disaster, namely more severe.

(*d*) Emergency: ' a serious, unexpected and often dangerous situation requiring immediate action', page 603.

Whilst both Disasters and Emergencies can be sudden, the former has a macro, large scale impact whilst the latter requires an immediate response.

(*e*) Accident: '1. an unfortunate incident that happens unexpectedly and unintentionally, typically resulting in damage or injury,...2. an event that happens by chance or that is without apparent or deliberate cause', page 10.

It should be noted that the *COMAH Regulations* (*SI 1999 No 743*) have introduced the term 'Major Accident'; this is an event due to:

(i) unexpected, sudden, unplanned developments in the course of the operation;

(ii) leads to serious danger to people and the environment both on site at the place of work and off site;

(iii) the event involves at least one dangerous substance as defined by *COMAH.*

Given the potential for both human and property loss, and given the *COMAH* requirements for emergency On-Site and Off-site Plans, there seems to be little practical difference between a Major Accident and an Emergency. Both are *response based* concepts. However, if the legislators wished a Major Accident to be different from an Emergency then they would have either said so or implied so. One can regard a Major Accident as a type of Emergency situation.

(*f*) Incident: ' an event or occurrence', page 923.

The *Reporting of Injuries, Diseases and Dangerous Occurrences Regulations 1995* (*RIDDOR*) (*SI 1995 No 3163*), do not define an accident or incident, but gives a classification of the types.

(*g*) Major Incident

In *Dealing with Disaster* (Home Office, Third Edition currently being revised), a 'Major Incident' is the only term explicitly defined by the Home Office: A 'Major Incident' is any Emergency that requires the implementation of special arrangements by one or more of the emergency services, the NHS or the local authority for:

(i) the initial treatment, rescue and transport of a large number of casualties;

(ii) the involvement either directly or indirectly of large numbers of people;

(iii) the handling of a large number of enquiries likely to be generated both from the public and the news media, usually to the police;

(iv) the need for the large scale combined resources of two or more of the emergency services;

(v) the mobilisation and organisation of the emergency services and supporting organisations, e.g. local authority, to cater for the threat of death, serious injury or homelessness to a large number of people, page 43.

This definition is accepted by the police, fire service, local government and broadly the NHS (they also have a specific definition of Major Incident).

Thus the term 'Major Incident' is a broad phrase encompassing an array of events. 'Accident' has not been included as a type of Major Incident given the definition of the latter requiring 'major' mobilisation of human and physical resources which in most Accidents is not necessarily the case.

As an example:

(1) two trains missing each other would be an 'incident' (near miss);

(2) an employee or member of the public being injured on a train – this would be an 'Accident', irrespective of the type of injury or fatality (following *RIDDOR*);

(3) an event at a *COMAH* site where dangerous substances ignite causing damage to the plant, injury to personnel and emissions into the local community, would be a 'Major Accident' under *COMAH*;

(4) the immediate event after the collision and the response needed to the chaos – this would be an 'Emergency'. For example, the response by the emergency services to Ladbroke Grove;

(5) if two trains collide causing multiple fatalities and immediate property and environmental damage, that would be referred to as a 'Disaster'. For example Ladbroke Grove Rail Crash in 1999;

(6) if the collision, with the multiple fatalities and property damage is difficult to access, manage and control, this would be a 'crisis'. Ladbroke Grove fell short from being a 'crisis' in contrast to the Clapham Junction in 1988 where a triple train crash caused major access and logistical problems;

(7) if the event generated major environmental, public and social harm that has 'longer term' implications, over and above the immediate human and property loss, this would be a 'catastrophe' For example, Kings Cross Underground Fire (1987), Chernobyl (1986), Piper Alpha (1988) to name a few that had wider consequences over and above the immediate impact.

Events (1)–(7) would be Major Incidents.

Figure 1 summarises the essential differences between the above events.

Figure 1: Classification of Incidents, Accidents and Major Incident types

				'MAJOR INCIDENTS'				
	Event:	Incident	Accident	Major Accident	Emergency	Disaster	Crisis	Catastrophe
Characteristic:				Types of emergency		Types of disaster		
1. MPL	Low							Very High
2. RISK:								
Severity	Near miss etc							Multiple Fatalities
Consequence	Minor							Major
3. NUMBERS	1–5							100 +
4. SOCIO-LEGAL IMPACT	No change likely							New Laws or Guidance
5. TIME	Short							Longer Impact
6. COST:								
Individual	Short Term							Irreparable
Social	None Usually							Irreparable
Environment	None Usually							Irreparable

Key:

MPL	=	Maximum Potential Loss (economic and property loss)
Risk	=	Severity x Consequence (Severity refers to the quantum of harm generated by the event whilst Consequence measures the scale of impact)
Numbers	=	The number of individuals affected
Socio-Legal Impact	=	The impact on social attitudes and legislation/guidance as a result of the event
Time	=	The length of the event
Cost	=	The loss suffered by the individual, society or nature

Such classifications are important, firstly from a philosophical perspective one needs to know how they differ, and secondly from a planning perspective as the resource allocation will accordingly differ. Thirdly, from a response perspective, the response to an incident differs from a Disaster, with the organisation needing to define and clarify when an event is an incident and not a Disaster.

The Need for Effective Disaster and Emergency Management

D6004 There are several reasons why DEM is needed: legal reasons (see D6005–D6022); insurance reasons (see D6023); corporate reasons (see D6024); societal reasons (see D6025); environmental reasons (see D6026); and humanitarian reasons (see D6027).

Legal reasons

D6005 There is an array of specific emergency-related legislation – much of which is currently under review, which may culminate in an Emergency Planning Bill. Existing legislation includes the following.

The Civil Defence Act 1948

D6006 This Act requires the UK Government to make arrangements to deal effectively to protect the population if and when a hostile attack takes place by a 'foreign power'. The first 'modern' *Civil Defence Act* was enacted in 1937, in response to threats from nationalists groups around the world during the era of the Empire. Under *s 1* of the *Civil Defence Act 1948*, a strict duty exists such that the responsible Minister considers the resources needed to protect civil society. Such resources include:

- training, equipment, management of suitable civil defence forces;
- providing the emergency services with training in civil defence;
- ensuring that suitable persons from the public can be trained to assist the civil defence forces/emergency services;
- ensuring that adequate commodities and rations are available if a civil defence situation arose;
- allowing alterations to be made to buildings or indeed temporary shelter to be constructed at short notice.

The Act is primarily a strategic control document such that Government has identified the source of the threat from a foreign power and has competent persons and suitable resources to protect its population. In addition the Act, highlights the links between national and local government and other competent authorities, the need for strategic planning, the importance of regular communication. The Act also provides a definition of 'civil defence' (in the broadest sense, protecting the public and the homeland).

The Civil Defence (Grant) Regulations 1953 (SI 1953 No 1777)

D6007 These Regulations were made under the *Civil Defence Act 1948, s 3*, and allow for grants to be paid to local authorities to assist with the costs of civil defence/war planning. The Regulations specify what are acceptable and legitimate costs that central Government will reimburse to a local authority when 'exceptional' events arise. Also a 'police authority' and/or 'fire and ambulance service' can make claims under the Regulations where specific and measurable resources are deployed for civil defence (which ordinarily would not have been).

It should be noted that these Regulations in particular have provided the impetus for local government in England and Wales to establish 'emergency planning units'. These units coordinate, control and oversee the effective preparation of Disaster Management and Emergency Response in a defined area. Their remit can range from natural catastrophe (e.g. floods) through to a population being exposed to chemical biological nuclear and radiological (CBNR) risks. The units' resources can be audited and expenses/costs/losses incurred measured.

The Local Government Act 1972, s 138

D6008 *Section 138* of the *Local Government Act 1972* allows local authorities to make grants and loans available to those affected by Disasters and Emergencies. This could be to necessary business services, infrastructure organisations (electric, water etc) or it could be the public, again under 'exceptional conditions'.

The Civil Protection in Peacetime Act 1986

D6009 The *Civil Protection in Peacetime Act 1986* is 'An Act to enable local authorities to use their civil defence resources in connection with Emergencies and Disasters unconnected with any form of hostile attack by a foreign power' (Preamble Note to the Act). The 1986 Act allows local government to use civil defence resources to mitigate the effects of any Disaster whether it is natural or man-made, whether 'accidental' or 'internally hostile'. It can be noted that the greater the perceived threat of attack, the more likely local government is to deploy resources. In the 1990s as the perceived threat from the Eastern European block was diminishing, the deployment of resources also was reduced. Since '911' the resource allocation by local government to the identification of persons, social groups and technologies that can cause civil harm has increased.

The Emergency Powers Act 1920

D6010 For a 'state of emergency' to be proclaimed under the *Emergency Powers Act 1920* (as amended by the *Emergency Powers Act 1964*) there must be (or about to be) some interference in the 'essentials of life' e.g. food, water, fuel, light or transport. The proclamation of a state of emergency has a 'shelf life' of one month, although additional proclamations can be issued during or by the month end. It is important to note that Parliament must be notified by the Secretary of State. Regulations can be made under 'Order of Council' when a state of emergency exists. These Orders cannot impose compulsory military service or industrial conscription and neither can they prohibit the right to strike by recognised trade unions. The Regulations must be approved by both Houses of Parliament.

The Civil Defence (General Local Authority Functions) Regulations 1993 (SI 1993 No 1812)

D6011 These Regulations became law in August 1993. Local government is not under a duty to provide 'protected emergency centres' for war time preparation and planning. It was a requirement that county and regional councils have a Main and Standby Centre, whilst, district councils provide one Emergency Centre.

However, county and regional councils are required to:

- Create, assess, review, update and exercise Plans.
- Ensure county and district council staff/other relevant persons are trained.
- Action the Plans created (not just a 'paper exercise').

- To work with and consult other neighbouring authorities.

District councils are required to:

- Ensure that relevant information is given to the county council.
- Provide the necessary support in the making/revising of Plans.
- Assist where necessary in preparations.
- Assist in implementation of new Plans.
- To ensure staff are adequately trained.

Unitary authorities are also covered and have similar functions to county and regional councils.

It should be noted that Scotland, Wales and Northern Ireland have additional and specific legislation e.g. the *Civil Defence (Scotland) Regulations 2001 (SI 2001 No 139)*.

The Health and Safety at Work etc Act 1974

D6012 The intentions of the *Health and Safety at Work etc Act 1974 (HSWA 1974)* are captured by s *1(1)(a)–(d)* which states that the Act is concerned with 'securing the health, safety and welfare of persons at work', protecting the public, effective control over dangerous substances/explosives and the effective control of emission of noxious substances into the atmosphere. Implicit at least is the intention that this Act will influence actions that contribute to Disaster and Emergency situations, whether industrial, chemical or environmental. The Act does give expressed powers to the Health and Safety Commission to investigate and to hold inquiries in relation to '... any accident, occurrence, situation ...'. [*HSWA 1974, s 14(1)*].

The Management of Health and Safety at Work Regulations 1999 (MHSWR) (SI 1999 No 3242)

D6013 These Regulations provide the main detail applicable to all organisations, for providing 'procedures for serious and imminent danger and for danger areas' [*MHSWR (SI 1999 No 3242), Reg 8*] and 'contacts with external services'. [*MHSWR (SI 1999 No 3242), Reg 9*]. In relation to industrial and chemical based incidents, Accidents and Major Incidents, the Regulations require the existence and implementation of safety procedures (internal) as well as links with emergency and para-emergency services (external). This contrasts with the 1992 and 1994 amended version of the same Regulations. *Regulation 1* (Citation, commencement and interpretation) does not define 'procedure for serious and imminent danger', although the ACOP L21 to the *MHSWR 1999 (SI 1999 No 3242)* exemplifies such situations, 'e.g. a fire, or for the police and emergency services an outbreak of public disorder' (page 20).

Regulation 8(1) says that a strict duty exists on all employers to:

(*a*) 'establish and where necessary give effect to appropriate procedures to be followed in the event of serious and imminent danger to persons at work in his undertaking'.

(*b*) 'nominate a sufficient number of competent persons to implement those procedures in so far as they relate to the evacuation from premises of persons at work in his undertaking'.

(*c*) 'ensure that none of his employees has access to any area occupied by him to which it is necessary to restrict access on grounds of health and safety unless the employee concerned has received adequate health and safety instruction'.

Regulation 8(2) says 'Without prejudice to the generality of *paragraph (1)(a)*, the procedures referred to in that sub-paragraph shall:

(*a*) so far as is practicable, require any persons at work who are exposed to serious and imminent danger to be informed of the nature of the hazard and of the steps taken or to be taken to protect them from it;

(*b*) enable the persons concerned (if necessary by taking appropriate steps in the absence of guidance or instruction and in the light of their knowledge and the technical means at their disposal) to stop work and immediately proceed to a place of safety in the event of their being exposed to serious, imminent and unavoidable danger; and

(*c*) save in exceptional cases for reasons duly substantiated (which cases and reasons shall be specified in those procedures), require the persons concerned to be prevented from resuming work in any situation where there is still a serious and imminent danger'.

Regulation 8(3) says 'A person shall be regarded as competent for the purposes of paragraph (1)(b) where he has sufficient training and experience or knowledge and other qualities to enable him properly to implement the evacuation procedures referred to in that sub-paragraph'.

Regulation 8 can be summarised:

- Procedures need to be sequential, logical, documented and clear (clarity of procedures).
- Procedures need to be justified and authorised (legitimise procedures).
- A 'hierarchy of procedural control' seems to be advocated by the Regulation i.e:
  - give information to those potentially affected by serious and imminent dangers;
  - take actions or steps to protect such people;
  - stop work activity if necessary to reduce danger;
  - prevent the resuming of work if necessary until the danger has been reduced or eliminated.
- Appoint competent persons preferably from within the organisation who will assist in any evacuations. It is implied that the role and responsibility of such persons needs to be clearly demarcated.
- Generally prohibit access to dangerous areas (site management). If access to a 'danger area' is required i.e. a place which has an unacceptable level of risk but must be accessed by the employee, then appropriate measures must be taken as specified by other legislation (see below).

In addition, the Regulations imply:

- Risk Assessments will identify foreseeable events that may need to be covered by procedures. Such assessments may also identify 'additional risks' that need additional procedures. Thus the Risk Assessment, as shall be discussed below, is a vital tool to keep procedures in tune with current generic and specific risks.
- Procedures need to be dynamic – reflecting the fact that events can occur suddenly.
- There may be a need to co-ordinate procedures where workplaces are shared.
- Procedures should also reflect other legislative requirements (see below).

Regulation 9 is a new addition to the *MHSWR* (*SI 1999 No 3242*). *Regulation 9* reads ' Every employer shall ensure that any necessary contacts with external services are arranged, particularly as regards first-aid, emergency medical care and rescue work'. It can be inferred that 'necessary contacts with external services' does not only relate to the emergency services but the organisation needs to identify both private sector and voluntary organisations that can be called on for assistance. The organisation needs to identify contact names, addresses and contact numbers of such bodies and develop relations with them.

The Control of Major Accident Hazards Regulations 1999 (COMAH) (SI 1999 No 743) (see C10002 for full details of COMAH 1999).

D6014 In summary:

The General Rule

After the Seveso Disaster in 1976, the European Union (EU) initiated an EU Directive called 'Seveso I'. Whilst no human was killed at Seveso the immense impact geographically and commercially in a key agricultural region meant the Italian Government and the European Commission were keen to ensure adequate safety and emergency systems were installed at 'high risk' production processes. At Seveso, the plant was manufacturing TriChloroPhenol (TCP) and due to the requirement to switch off batch production during the weekend, the operator had not distilled the chemical. He had assumed the temperature of the substance would drop to a tolerable level until he returned on Monday. However, latent heat transfer resulted in an exothermic runaway reaction causing the change in chemistry of the substance to a dioxin. Seveso is referred to as a boiling liquid evaporating vapour explosion (BLEVE).

When an EU Directive is introduced, Member States are obliged to transpose this instrument into domestic legislation within four years. In 1984 the UK implemented the EU Directive as the *Control of Major Incident Hazards Regulations* (*CIMAH*). Likewise other Member States implemented the Directive. *CIMAH* was more of a 'why' document i.e. why a petro-chemical or manufacturing process needs to prevent a Seveso type Disaster. After Chernobyl and Bhopal, the EU realised it needed a more forceful document that showed the 'what' and the 'how' to prevent major industrial accidents. This resulted in the Seveso II Directive, which the UK transposed as the *Control of Major Accident Hazards Regulations 1999* (*COMAH*) (*SI 1999 No 743*). *COMAH* replaced *CIMAH* and indeed *COMAH* has been implemented in most of the 15 EU Member States and is used as a 'benchmark' globally (e.g. by the Government's of Brazil, India, and the Caribbean States etc).

COMAH applies to those establishments that keep on site any substance that is specified in *Sch 1* of *COMAH* (*SI 1999 No 743*). For example, keeping a certain amount of ammonium nitrate or oxygen on site. However, there are 'Top-Tier' and 'Lower-Tier' Threshold levels. If an establishment keeps an amount greater than or equal to an Upper Threshold Level, they are called Top-Tier. For instance, if an establishment keeps at least 2,500 tonnes of ammonium nitrate, then it is Top-Tier. If an establishment keeps a lower quantity than this, (greater than or equal to a lower threshold) then it is a Lower-Tier establishment. For example, the Lower-Tier for ammonium nitrate is 350 tonnes. Therefore, knowing the amount of substance being stocked is critical to knowing what tier the establishment falls under.

It should be noted that even if the quantity of substance being held is below the Threshold Level, the establishment could be still subject to *COMAH* so long as the specified substance could be produced, assuming there is loss of control of the chemical process. This implies dangerous substances are produced through reaction with the establishment being unable to control the production of such amounts. This could happen if there is production failure or through unexpected chemical reactions.

Top-Tier establishments:

(i) Must produce both On-Site and Off-Site Emergency Plans. The former deals with risks that occur on the establishment and is produced by the operator. The latter deal with risks to the immediate social and physical environment due to the On-Site Emergency. It is prepared by the local authority. Both Plans should link together and be co-ordinated.

(ii) Top-tier establishments that exceed Top-Tier Thresholds are required to provide information to the public of known and foreseeable risks, as well as the controls adopted. They would also have to provide a Safety Report, which shows the actions taken by the operator to prevent Accidents and mitigate harm to the environment. The establishment must be able to prove that they are operating the plant in accordance with the safety report.

(iii) Operators of existing *CIMAH* sites which became Top-Tier establishments on 1 April 1999 are given until February 2001 to prepare On-Site Emergency Plans and supply information to the local authority. Whilst other establishments that became Top-Tier sites on 1 April 1999 (e.g. those now encompassed by *COMAH* but not *CIMAH* (*SI 1984 No 1902*)), have until 3 February 2002 to prepare the On-Site Emergency Plan and supply the information to the local authority. However, for those establishments that became top-tier after 1 April 1999, they have a duty to prepare the On-Site Emergency Plan before starting any operation.

(iv) The local authority has six months to prepare an Off-Site Emergency Plan. The six months can start from the time they receive all necessary information to prepare the Plan. However, the period can be extended to nine months. It is recommended that arrangements are made nevertheless by the operator until the Off-Site Plan is available.

(v) Emergency Plans should be reviewed and revised at least every three years.

Lower-Tier establishments:

(i) Whilst under no duty to produce Emergency Plans, they are required however to develop Emergency Arrangements in the Major Accident Prevention Policy (MAPP).

(ii) The MAPP should specify safety and emergency management response systems in place in the event of a Major Accident and the procedures for identifying foreseeable emergency situations.

Other establishments

COMAH does not apply to Ministry of Defence establishments, transport related activity, extractive industries exploiting/exploring in mines and quarries, waste land-fill sites nor to nuclear licensed sites that may have substances which generate ionising radiation. However, *COMAH* will apply to both chemicals and explosives at nuclear installations.

Breach of COMAH

A breach of *COMAH Regulations* carries the same penalties as those under *ss 33 – 42* of the *Health and Safety at Work etc Act 1974*, for example £20,000 fine per offence on summary conviction and unlimited fine and/or up to two years imprisonment on indictment.

The Reporting of Injuries, Diseases and Dangerous Occurrences Regulations 1995 (RIDDOR) (SI 1995 No 3163)

D6015 *RIDDOR* (*SI 1995 No 3163*) will apply to 'Accidents' through to 'Major Incidents'. *RIDDOR* is a reporting requirement that must be followed by the employer or 'responsible person' in notifying the enforcing authority by the quickest means practicable, when there is an event resulting in a reportable injury, reportable occupational disease, reportable dangerous occurrence, gas incidents, road incidents involving work and fatal or non-fatal injuries for example. In addition, work injuries lasting for three days or more are reported to the enforcing authority. (see A3001 ACCIDENT REPORTING AND INVESTIGATION for further details).

The Health and Safety (First-Aid) Regulations 1981 (SI 1981 No 917)

D6016 These Regulations do not explicitly mention first-aid Arrangements necessary in the event of an Emergency or Disaster. However, they will apply in all types of Accidents and Major Incidents as defined above. The Regulations are a general statement of best practice, of ensuring the existence of adequate medical equipment, competent and trained first-aiders or appointed persons and information on first-aid facilities and equipment to staff/others. The Regulations should be read as best practice to be followed in the event of an Accident or Major Incident.

The Construction (Health, Safety and Welfare) Regulations 1996 (SI 1996 No 1592)

D6017 *Regulation 20* requires employers to make emergency arrangements to cope with foreseeable emergency situations. It is primarily aimed at CDM sites where the Health and Safety Plans are operational. The Plans should factor in emergency arrangements and procedures in the event of a Major Incident on site.

The Confined Spaces Regulations 1997 (SI 1997 No 1713)

D6018 *Regulation 5* imposes duties on employers and others to make arrangements for emergencies in 'confined spaces'. In summary:

- 'Suitable and sufficient arrangements' need to be made for rescue of persons working in confined spaces, with no person accessing a confined space until such arrangements have been developed. [*SI 1997 No 1713, Reg 5*(*1*)].
- So far as is reasonably practicable, the risks to persons required to put the arrangements into operation must also be considered as should resuscitation equipment in the event of it being required. [*SI 1997 No 1713, Reg 5*(*2*)].
- There is also a duty to act, to operationalise the arrangements when an emergency results. [*SI 1997 No 1713, Reg 5*(*3*)].

Question: What are suitable and sufficient arrangements for rescue and resuscitation? This should include appropriate equipment, rescue procedures, warning systems that an emergency exists, fire safety systems, first-aid, control of access and egress, liaison with the emergency services, training and competence of rescuers etc.

The Carriage of Dangerous Goods by Road Regulations 1996 (SI 1996 No 2095 partially revoked by SI 2004 No 568)

D6019 A duty is imposed on the operator of a 'container', 'vehicle' or 'tank' to provide information to other operators engaged/contracted, regarding the handling of emergencies or accident situations. Such information includes:

- hazardousness of the goods being carried and controls needed to make safe such hazards;
- actions required if a person is exposed/makes contact with the goods being carried;
- actions required to avert a fire and the equipment that should or should not be used to fight the fire;
- the handling of a breakage/spillage; and
- any additional information that should be given that can assist the operators or others.

The Carriage of Dangerous Goods and Use of Transportable Pressure Equipment Regulations 2004 (SI 2004 No 568)

D6020 The *Carriage of Dangerous Goods and Use of Transportable Pressure Equipment Regulations 2004* (*SI 2004 No 568*) came into force on 10 May 2004 and are relevant to the carriage of dangerous goods by road and rail. The Regulations:

- consolidate over a dozen sets of regulations;
- implement EC Directives 2003/28/EC, 2003/29/EC and complete the implementation of Directive 1999/36/EC;
- directly reference the 2003 texts of the international agreements – RID for rail and ADR for road – that govern the land carriage of dangerous goods.

(See DANGEROUS GOODS SYSTEMS.)

The Environment Act 1995

D6021 Whilst the Act does not explicitly deal with DEM issues, its main objective is the mitigation or prevention of such events. For example, *ss 14–18* creates the flood defence committees to co-ordinate the prediction, consequence and response needed in the event of floods.

'Legislating the Criminal Code: Involuntary Manslaughter', Law Commission (Law Com No 237)

D6022 Disasters such as Piper Alpha, Kings Cross, Clapham Junction, Herald of Free Enterprise or Ladbroke Grove for example have seen multiple fatalities, allegations of managerial failure and negligence contributing to the Disasters. However, health and safety law has not permitted the successful prosecution and imprisonment of directors and senior managers because of the difficulty in establishing that the individuals in charge were 'the embodiment of the company'. All of the major Disasters have involved large and complex organisations with many management layers, so who is the 'embodiment of the company' or the 'directing mind'? The Law Commission has drafted an *Involuntary Homocide Bill 1995*, which is still being considered by the UK Government. The main intention of the Bill is stated in the preamble, to ' Create new offences of reckless killing, killing by gross carelessness and corporate killing to replace the offence of manslaughter in cases where death is caused without the intention of causing death or serious injury'. The intention is to have a codified system of law as regards these three offences. Thus:

(*a*) *Reckless killing*

This offence would be committed if the answers are 'yes' or 'affirmative' to all the following:

(i) a person through their conduct causes the death of another;

(ii) a person is aware that a risk exists and that their conduct whether causing that risk or acting on that risk will cause death or serious injury; and

(iii) it is unreasonable for the person to take a risk, knowing the circumstances prevalent.

(*b*) *Killing by gross carelessness*

This offence would arise if the answers are 'yes' or 'affirmative' to all the following:

(i) a person by their conduct causes the death of another;

(ii) a reasonable person in the position of the person under question realises that there is a risk that the person's conduct will cause death or serious injury;

(iii) the person 'is capable of appreciating' the risk at that time in question; and

(iv) either the person's conduct 'falls far below' what could be reasonably expected under those circumstances or, 'he or she intends by his conduct to cause some injury, or is aware of, and unreasonably takes, the risk that it may do so, *and* the conduct causing (or intended to cause) the injury constitutes an offence' (page 127/128). The former focuses on a breach of a duty whilst the second of this 'either or' option seems to require the elements of a crime, the *'action'* (action of 'unreasonably take' and 'conduct causing') and the '*intent*' ('intends by his conduct').

Both reckless killing and killing by gross carelessness are:

(1) forms of 'individual manslaughter';

(2) alternative verdicts to 'murder';

(3) alternatives to each other for the court to consider;

(4) both applicable to workplace and other situations, for example be available to the court in cases of road deaths or serious injury (although the Law Commission do not recommend a change to the existing offence of 'causing death by bad driving');

(5) reckless killing would carry a maximum sentence of life imprisonment whilst the Commission do not recommend a maximum for killing by gross carelessness.

(*c*) *Corporate killing*

This contrasts to the two proposed individual manslaughter offences above. The Law Commission says on page 128/129:

(i) that there should be a special offence of corporate killing, broadly corresponding to the individual offence of killing by gross carelessness;

(ii) that (like the individual offence) the corporate offence should be committed only where the defendant's conduct in causing the death falls far below what could reasonably be expected;

(iii) that (unlike the individual offence) the corporate offence should not require that the risk be obvious, or that the defendant be capable of appreciating the risk; and

(iv) that for the purposes of the corporate offence, a death should be regarded as having been caused by the conduct of a corporation if it is caused by a failure, in the way in which the corporation's activities are managed or organised, to ensure the health and safety of persons employed in or affected by those activities.

Therefore:

(1) Corporate killing as an offence relies less on the 'intent' component and more on the actions/conduct and safety systems in place (or otherwise). There is also no reference to 'foreseeability of risk', with the Commission arguing it is difficult to apply this concept to a corporation vis-à-vis an individual.

(2) The Law Commission stress the role of 'management failure' by the corporation as a cause of death even though the 'immediate cause' could be the act or omission of some individual such as an employee.

(3) Corporate killing applies to 'incorporated bodies' only e.g. limited and public limited companies and would not apply to 'sole traders'. Thus, the offence of corporate killing can only be by an incorporated body.

(4) The Law Commission recommended that this offence should not require consent to bring a private prosecution for corporate killing.

(5) The Law Commission recommended that this offence be triable by indictment only.

(6) The Law Commission is suggesting that on conviction of corporate killing the court should be empowered (upon the suggestion of the prosecution or the HSE) to recommend improvements to management systems to ensure death or serious injury at work does not arise again.

(7) Assuming that the jury cannot convict under the proposed offences, it should be able to consider conviction under *s 2* or *3* of the *Health and Safety at Work etc Act 1974*.

(*d*) *Unlawful act manslaughter*

The Law Commission recommends the abolition of unlawful act manslaughter as it currently stands. This refers to where a person causes death whilst committing a criminal act, with this act carrying a risk of *some* harm to the other person. The Commission says '.. we consider that it is wrong in principle that a person should be convicted for causing *death* when the gravest risk apparently inherent in his conduct was the risk of causing *some injury*'.

Insurance reasons

D6023 There are also insurance based reasons for effective DEM:

- given the positive correlation between risk and premium, the existence of DEM indicates hazard and risk control, consequently the premium ought to be less;
- if insurers are not satisfied or dissatisfied with the DEM system in place then they may not insure the operation, in turn increasing corporate risk as well as reducing corporate credibility. It may also prohibit the operation from tendering for contracts.

Corporate reasons

D6024

- The corporate experience of companies like P & O European Ferries (Dover) Ltd, indicates that proactive DEM systems would have saved the company considerable money, publicity and reputation. In March 1987, the *Herald of Free Enterprise* the roll-on roll-off car ferry left Zeebrugge for Dover, thereafter it sunk resulting in 187 deaths. The case against the company and the five senior corporate officers collapsed for reasons cited above ('embodiment of the company' and 'controlling mind', see D6022 above).
- Failures to have sound DEM systems can also have dire financial consequences as Disasters as disparate as Piper Alpha to The Challenger Space-shuttle indicate. In the former case, Occidental Petroleum were generating 10 per cent of all the UK's North Sea output from Piper Alpha. Lord Cullen, who led the inquiry said 'The safety policy and procedures were in place: the practice was deficient'. Piper Alpha showed that the company had ineffective emergency response procedures resulting in persons being trapped, dying of smoke inhalation or jumping into the cold North Sea (something the procedures prohibited but a significant number of those that did jump into the sea survived). Such a Disaster had consequences for Occidental's financial reputation, share value, ability to attract investment and growth potential. In contrast, in the case of The Challenger Disaster in January 1986, the o-rings that held the two segments of the rocket boosters, which carried the fuel to propel the Challenger into space, fell apart resulting in the Challenger exploding shortly after taking off. This shocked the USA public, resulting in delays and questions as to the viability of NASAs programmes. The failure of the o-rings was well known with the engineers down grading the risk from 'high to 'acceptable'.

Societal reasons

D6025 Society expects its organisations to plan and prepare for the worst case scenarios and when this doesn't happen, society seeks the closure, forefeiture or expulsion of the organisation from society e.g. the legal and political costs to Union Carbide in India with its mis-management of its plant in Bhopal in India. The company has been banned from operating in India.

Environmental reasons

D6026 Failure to prepare and plan for Disasters and Emergencies will also have environmental costs, no better illustrated than the Chernobyl Disaster in April 1986, with its impact not only on the population of Russia but as far as Wales in the United Kingdom. The aim was to test if the power reduction of a turbine generator would be sufficient given the use of some suitable voltage generator, to power an emergency core cooling system (ECCS) for a few minutes whilst the stand-by diesel generator became operational. The power fell to 7 per cent of full power with a benchmark of 20 per cent being critical for that make of RBMK reactor. Consequently, the reactor exploded with radioactive fall out polluting the physical environment.

Humanitarian reasons

D6027 The human cost for not planning for worst case outcomes is the most significant. All the major Disasters documented in the media resulted in multiple fatalities – of all demographic and socio-economic groups. After the Ladbroke Grove Disaster in October 1999, when two trains collided resulting in 31 fatalities and 160 being critically injured, the public and the media made a significant outcry for change. This culminated in John Prescott, the Secretary of State for the Environment, Transport and the Regions establishing a Committee headed by Lord Cullen (of the Piper Alpha Inquiry), to investigate 'safety culture' on the railways.

Disaster and Emergency Management Systems (DEMS)

D6028 A DEMS is outlined in Figure 2 below:

Figure 2: Disaster and Emergency Management Systems (DEMS)

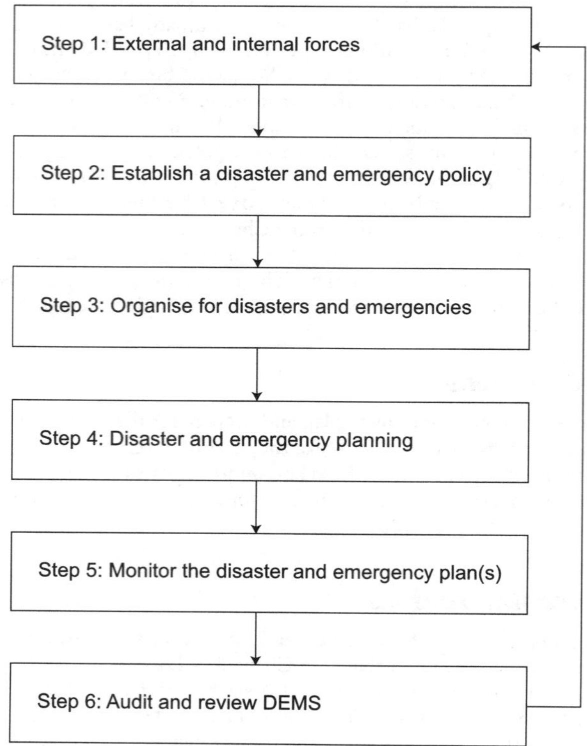

External and Internal Factors

D6029 The starting point with DEMS is understanding the variables that can influence or affect DEMS. These are both internal to the organisation and wider societal variables. It would be an erroneous assumption for the organisation to make, if it believed that Disaster and Emergency Planning can be in isolation of wider societal variables; these variables need to be understood and factored into any DEMS. The larger the organisation and the more hazardous its operation, the more it needs to provide specific detail.

External Factors influencing DEMS

D6030 *The Natural Environment*

The organisation needs to identify physical variables that can influence its reaction to a Disaster or Emergency. The organisation needs to identify:

(i) Seasonal weather conditions – including weather type(s) and temperature range. Effective rescue can be hampered by failing to know, note and record such variables. A simple chart, identifying in user-friendly terms on a monthly basis the weather conditions can assist the reader of any DEM Plans.

(ii) Geography around the site – a description of the physical environment such as terrain, rivers or sea, soil type, geology as well as longitude and latitude positioning of the site. These should be brief descriptions unless the site is remote and could have difficulty being accessed in an Emergency.

(iii) Time and distance – of the site from major emergency and accident centres, fire brigades, police etc. Both long and short routes need to be determined. Longitude and latitude co-ordinates should be identified.

Societal Factors

These include:

(i) Demographics – the organisation needs to identify the age range, gender type, socio-economic structure of the immediate vicinity around the site. A Major Incident with impact on the local community requires the organisation to calculate risks to the community. *COMAH* Off-Site Emergency Plans indicates why planning is not just an On-Site 'in these four walls' activity.

(ii) Social attitudes – the organisation needs to investigate briefly and be aware of attitudes (reactions and responses) of the local town or city and the country to Disasters and Emergencies. Union Carbide made the assumption that the population and authorities in Madhya Pradesh, where Bhopal is located, would react like the communities in the USA. The response rates, awareness levels and information available to different communities differs.

(iii) Perception of risk – how does the local community view the site or operation? Is the risk 'tolerable' and 'acceptable' to them for having the operation in their community? Both Chernobyl and Bhopal highlight that economic necessity can alter the perception of risk in contrast to the statistical level of risk.

(iv) History of community response – in brief the organisation needs to determine how many Major Incidents there have been in the past and how effectively have the community assisted (emergency services and volunteers). Such support will be critical in a Disaster.

(v) The built environment – a brief description of the urban environment, namely the street layout, urban concentration level, population density, access/egress to railways, motorways are useful. These issues could be covered by the inclusion of a map of the area.

Government and political factors

Relevant factors are:

(i) The policy of national government to Disasters and Emergencies. Are they proactive in advising organisations, what information and guidance do they give? The organisation needs to identify its local or regional office of the HSE, the Environment Agency or a similar body and make contact with officials and seek early input into any planning.

(ii) Local government – their Plans, information and guidance. DEM's need to be aware of any local government restrictions on managing Major Incidents. In the case of *COMAH* sites, they will need to involve the local authority in preparing Off-Site Emergency Plans.

(iii) Committees and agencies – there are many legal and quasi-legal bodies created by legislation which have guidance, information and templates on response. For example the flooding committees mentioned above under the *Environment Act 1995*.

(iv) Emergency and medical services – making contact, obtaining addresses/contact numbers and knowing the efficiency of the emergency services/accident and emergency are critical actions.

Legal factors

The organisation needs to know the legal constraints it has to operate within, not only to comply with the law but also to follow the best advice contained in legislation as a means of preventing Disaster and Emergency.

Sources of information issues

These include:

(i) The local and national media – will not only report any Incidents but can be critical in relaying messages. Any planning requires the identification of local and national newspapers (their addresses/contact numbers), local/national radio details and any public sector media (through local government).

(ii) Local library – they will in turn have considerable contact details and networks with other libraries so can be an efficient medium to relay urgent information. Making contact with the local librarian and identifying such contact details in any planing will be required.

(iii) Business associations – Chambers of Commerce, Business Links, Training and Enterprise Councils need to be identified for the same reasons cited for local libraries. In the event of a Major Incident they can convey and provide information on a local rescue or occupational health organisation.

(iv) Voluntary organisations – such as Red Cross, St. Johns Ambulance should be identified and noted.

The organisation needs to build information networks locally and nationally. The experience of major Disasters such as Chernobyl or Kings Cross indicate that letting others know of the Incident is not a shameful or embarrassing matter – rather it can warn others and halt others from attending the site.

Technological factors

These concern what technology and equipment exists in the local community; what doesn't; lifting equipment, rescue, computers, monitoring equipment measuring equipment etc.

Commercial factors

These include:

(i) Insurance – liaise with insurers from the beginning and ensure that all Plans are drawn to their attention and if possible approved by them. This can avoid difficulties of any claim that may arise due to a Disaster or Emergency as well as ensure that best advice from the insurer has been factored into the Plans.

(ii) Customer and supplier response – larger organisations need to identify the responses from customers and suppliers, their willingness to work with the organisation in the event of a worst case scenario and possible alternatives. Suppliers need to be identified. Issues of customer convenience and loyalty also need to be addressed.

(iii) Attitudes of the bank – liquidity issues will arise when a Major Incident strikes. Most organisations do not and cannot afford to make financial provisions for such eventualities. Therefore, if the operation is highly hazardous, establishing financial facilities before hand with the bank is necessary to ensure availability of liquid cash.

(iv) Strength of the economic sector and economy – larger organisations that are significant players in a sector or the economy need to be cognisant of the 'multiplier effect' that damage to their operation can do to the community and suppliers, employees and others.

Internal Factors affecting DEMS

D6031 DEMS also need to be constructed after accounting for various internal or organisational variables.

Resource factors

Cash flow and budgetary planning whether annual or a longer period needs to account for resource availability for Major Incidents. This includes physical, human and financial resources. This is wider than banking facilities and encompasses equipment, trained personnel and contingency funds.

Design and architecture

Issues include – is the workplace physically/structurally capable of withholding a Major Incident? What are the main design and architectural risks? Also layout, access/egress, emergency routes, adequacy of space for vehicles etc.

Corporate culture and practice

The organisation needs to identify its own collective behaviour, attitude, strengths and weaknesses to cope with a Major Incident. Being a 'large organisation' does not mean it is a 'coping organisation'; there is also the delay in response that is associated with hierarchical structures. In this case, the organisation needs to consider small 'matrix' or project team cell in the hierarchy dedicated to incident response.

Individual's perceived behaviour

(*a*) The *Hale and Hale Model* is an attempt to explain how individuals internalise and digest perceived information of danger; make decisions/choices according to the cost/benefit associated with each decision/choice; and the actual actions they take as well as any reaction*s* that result from their actions. In short, these five variables need to be understood in the organisation setting and a picture built up of the behavioural response of an individual.

(*b*) The *Glendon & Hale Model* is a macro model of how the organisation (behaving like a system), being dynamic and fluid, with objectives and indeed limitations (systems boundary) can shape behaviour and in turn influence Human Error. Following Rasmussen, Human Error can be Skill Based (failing to perform an action correctly), Rule Based (have not learned the sequences to avoid harm) or Knowledge Based (breaching rules or best practice). If all three error types are committed then the danger level in the system is also greater. The model is a focus on how wider systems can contribute to Human Error and how that error can permeate into the organisation. The inference is the organisation needs to clearly define and communicate its intention and objectives and continuously monitor individual response.

(*c*) *Reducing Error and Influencing Behaviour*, HSG 48, Health and Safety Executive, 1999 identifies the role of 'Human Error' and 'Human Factors' in Major Incident causation. The HSE says 'a human error is an action or decision which was *not intended*, which involved a deviation from an accepted standard, and which led to an undesirable outcome' (page 13). Following Rasmussen they classify four types:

(i) *slips* (unintended action);

(ii) *lapses* (short term memory failure) with slips and lapses being skill based;

(iii) *mistakes* (incorrect decision) which are rule based; and

(iv) *violations* (deliberate breach of rules) which are knowledge based.

Different types of Human Errors contribute in different ways to Major Incidents the HSE say and exemplify. Organisations need to identify from reported incidents the main types of Human Errors and why they are resulting and the negative harm generated. (Note that in the previous version of HSG 48 called *Human Factors in Industrial Safety*, five types of Human Errors were defined;

(i) *misperception* (tunnel vision, excluding wider factors from one's senses e.g. the belief that smoking is safe on the underground or that a 'smouldering' is not a significant fire as in the case of Kings Cross in 1987);

(ii) *mistaken action* (doing something under the false belief it is correct e.g. the pilot switching off the good engine under the belief he was switching off the one with the fire, so both are off, hence the crash landing in the Kegworth Disaster);

(iii) *mistaken priority* (a clash of objectives, such as safety and finance as implied in the Herald of Free Enterprise Disaster in 1987);

(iv) *lapse of attention* (short term memory failure, not concentrating on a task e.g. turning on a valve under repair as in Piper Alpha in 1988); and

(v) *wilfulness* (intentionally breaching rules e.g Lyme Bay Disaster when the principal director received a two year imprisonment for the death of teenagers at a leisure centre under his control)).

The HSE says that understanding Human Factors is a means of reducing Human Error potential. Human Factors is a combination of understanding the Person's behaviour, the Job they do (ergonomics) and the wider organisational system. Major Incidents can result if the organisation does not analyse and understand these three variables.

Information systems at work

The types of information systems, their effectiveness, and accuracy need to be identified. Thus telephone, fax, e-mail, cellular phone, telex etc need to be assessed for performance and efficacy during a worst case scenario.

Establish a Disaster and Emergency Policy

D6032 The DEM Policy is a concise document that highlights the corporate intent to cope with and manage a Major Incident. It will have three parts: Statement of Policy for managing Major Incidents (see D6033); Arrangements for Major Incident management (see D6034); and Command and Control charge of Arrangements (see D6035).

Statement of policy for managing major incidents

D6033 This should be a short (maximum 1 page) missionary and visionary statement covering the following:

- Senior management commitment to be responsible for the co-ordination of Major Incident response.
- To comply with the law, namely:
  - the protection of the health, safety and welfare of employees, visitors, the public and contractors;
  - to comply with the duties under *Regs 8* and *9* of the *Management of Health and Safety at Work Regulations 1999* (*SI 1999 No 3242*) (see D6013);
  - to comply with any other legislation that may be applicable to the organisation.
- To make suitable arrangements to cope with a Major Incident and to be proactive and efficient in the implementation process.
- That this Statement applies to all levels of the organisation and all relevant sites in the country of jurisdiction.
- The commitment of human, physical and financial resources to prevent and manage Major Incidents.
- To consult with affected parties (employees, the local authority and others if needed).
- To review the Statement .
- To communicate the Statement.

- Signed and dated by the most senior corporate officer.

The Arrangements for Major Incident management

D6034
- This refers to what the organisation has done, is doing and will do in the event of a Major Incident and *how* it will react in those circumstances. The Arrangements are a legal requirement under *Regs 5, 8* and *9* of the *Management of Health and Safety at Work Regulations 1999* (*SI 1999 No 3242*).
- Arrangements should be realistic and achievable. They should focus on major actions to be taken and issues to be addressed rather than being a 'shopping list'. The Arrangements will have to be verified (in particular for *COMAH* sites).
- Arrangements could be under the following headings with explanations under each. To repeat, the larger the organisation and the more complex the hazard facing it, the more detailed the Arrangements need to be. For example:

  (i) Medical assistance – including first-aid availability, first-aiders, links with accident and emergency at the medical centre, other specialists that could be called upon, rules on treating injured persons, specialised medical equipment and its availability etc.

  (ii) Facilities management – the location, site plans and accessibility to the main facilities (gas, electricity, water, substances etc.), rendering safe such facilities, availability of water supply in-house and within the perimeter of the site etc.

  (iii) Equipment to cope – identification of safety equipment available and/or accessible, location of such equipment, types (personal protective equipment, lifting, moving, working at height equipment etc.).

  (iv) Monitoring equipment – measuring, monitoring and recording devices needed, including basic items such as measuring tapes, paper, pens, tape recorders, intercom and loud-speakers.

  (v) Safe systems – procedures to access site, working safely by employees and contractors under Major Incident conditions (what can and cannot be done), hazard/risk assessments of dangers being confronted etc., risks to certain groups and procedures needed for rescue (disabled, young persons, children, pregnant women, elderly persons).

  (vi) Public safety – ensuring non access to Major Incident site by the public (in particular children, trespassers, the media and those with criminal intent), warning systems to the public etc.

  (vii) Contractor safety – guidance and information to contractors at the Major Incident on working safely.

  (viii) Information arrangements – the supply of information to staff, the media and others (insurers, enforcers) to inform them of the events. Where will the information be supplied from, when will it be done and updates?

  (ix) The media – managing the media, confining them to an area, handling pressure from them, what to say and what not to say etc.

  (x) Insurers/loss adjusters – notifying them and working with them at the earliest opportunity.

(xi) Enforcers – notifying them of the Major Incident, working with them including several types e.g. HSE, Environment Agency (or Scottish equivalent) as well as local authority (environmental health, planning, building control for instance).

(xii) Evidence and reporting arrangement – to cover strict rules on removal or evidence by employees or others, role and power of enforcers, incident reporting e.g under *RIDDOR* (*SI 1995 No 3163*) etc.

(xiii) The emergency services – working with the police, fire, ambulance/ NHS, and other specialists (Red Cross, Search & Rescue), rules of engagement, issues of information supply and communication with these services.

(xiv) Specialist arrangements for specific Major Incidents such as bomb explosions – issues of contacting the police, ordnance disposal, access and egress, rescue and search, economic and human impact to name a few issues were most evident during the London Dockland and Manchester Bombings in the 1990s.

(xv) Human aspects – removing, storing and naming dead bodies or seriously injured persons during the incident. Informing the next-of-kin, issues of religious and cultural respect. Issues of counselling support and person-to-person support during the incident.

This is not an exhaustive list. The arrangements should not repeat those in the Safety Policy, rather the latter can be abbreviated and attached as a schedule to the above, so that the reader can have access succinctly to specific OSH arrangements such as fire safety, occupational health, safe systems at work, dangerous substances for instance.

Command and Control Chart of Arrangements

D6035 This highlights who is responsible for the effective management of the Major Incident.

- It should be a graphical representation preferably in a hierarchical format, clearly delineating the division of labour between personnel in the organisation and the emergency services/others.
- The chart should display three broad levels of command and control, namely Strategic, Operational and Tactical. The first relates to the person(s) in overall charge of the Major Incident. Will this be the person who signed the Statement of Policy for Major Incidents or will it be another (disaster and emergency advisor, safety officer, others)? This person will make major decisions. The second relates to co-ordinators of teams. Operational level personnel need to have the above arrangements assigned to them in clear terms. The third refers to those at the front end of the Major Incident, for instance first-aiders.

  It is most important to note that internal Command and Control of Arrangements does not mean *overall* command and control of the Major Incident. This can (will be) vested with the appropriate emergency service, normally the police or the fire authority in the UK. In the event of any conflict of decisions, the external body such as the police will have the final veto. Therefore, the Chart and the Arrangements must reflect this variable.
- The chart should list on a separate page names/addresses/emergency phone, fax, e-mail, cellular numbers of those identified on the chart. It should also

list the numbers for the emergency services as well as others (Red Cross, specialist search and rescue, loss adjusters, enforcing body).

- The chart should also clearly ratify a principle of command and control, as to who would be 'In-Charge 1', 'In-Charge 2', if the original person became unavailable.
- The chart and the list of numbers should also be accompanied by a set of 'rules of engagement' in short 'bullet points' to remind personnel of the importance of command and control e.g. safety, obedience, communication, accuracy, humanity for instance.

Summary

D6036

- The three parts of the DEM Policy need to be in one document. Any detailed procedures can be separately documented ('Disaster and emergency procedures') and indeed could be an extensive source of information. However, unlike the Safety Policy and any accompanying Safety Manual, the same volume of information cannot apply to the DEM Policy. For obvious reasons it must be concise, clearly written, very practical and without complex cross-referencing.
- The DEM Policy must 'fit' with the Safety Policy, there can be no conflict so the safety officer and the DEM officer need to cross check and liaise. The DEM Policy must also fit with the broader corporate/business policy of the organisation.
- The DEM Policy must be proactive and reactive. The former concerned with preventing/mitigating loss and the latter concerned with managing the Major Incident when it does arise in a swift and least harmful manner.
- The DEM Policy needs to be reviewed 'regularly'. This could be when there is 'significant change' to the organisation, or as a part of an audit (semi-annual or annual).
- It must be remembered that the DEM Policy is a 'live' document so that it must be accessible and up-to-date.
- Although accessibility is important, the Policy should also have controlled circulation to core personnel only (for example those identified in the chart, the legal department). If the Policy was to be accessed by individuals wishing to harm the organisations, this will enable such persons to pre-empt and reduce the efficacy of the Policy.
- Finally and most crucially, the core contents of the Policy need to be communicated to all staff and others (contractors, temporary employees, possibly local authority).

Organise for Disasters and Emergencies

D6037 Once the establishment has accounted for External and Internal factors and has produced a DEM Policy taking account of such factors, it is then necessary to ensure personnel and others are aware of the issues raised in the DEM policy. The '4' C's approach of HSG 65 provides a logical framework to generate this (*Successful Health & Safety Management*, HSE): communication (see D6038); co-operation (see D6039); competence (see D6040); and control (see D6041).

Establish effective communication

D6038 Communication is a process of transmission, reception and feedback of information, whether that information is verbal, written, pictorial or intimated. Effective communication of the DEM Policy therefore is not a matter of circulating copies but requires the following:

Transmission

- The whole policy should not be circulated as it will mean little to employees and others. Rather an abridged version, possibly in booklet format or as an addition to any OSH documentation supplied will make more sense. Such copies must be clear, user friendly, non-technical as possible, be aware of the end user's capabilities to digest the information, be logical/sequential in explanation and use pictorial representation as much as possible.
- Being aware of the audience is central to the effective communication of the DEM Policy. The audience is not one entity but will consist of:
  - — direct employees;
  - — temporary employees;
  - — contractors;
  - — the media;
  - — the enforcers (*COMAH* sites);
  - — the local authority (*COMAH* sites);
  - — insurers;
  - — emergency services (*COMAH* sites);
  - — the public (*COMAH* sites).

  This does not necessarily mean separate copies for each of these entities but the abridged copy will need to satisfy the needs of all such groups.
- Transmission should start from the board, through to departmental heads, and disseminated downwards and across.

Reception

- What format will the end-user receive the abridged copy in (hard copy, electronic on disc, via e-mail, etc)?
- When will the copy be circulated – upon induction, upon training, ad hoc ?

Feedback

- Will the end-user have the opportunity to raise questions, make suggestions, be critical if they spot inconsistencies in the DEM Policy?
- There should also be 'tool box talks' and other general awareness programmes to inform individuals of the DEM Policy. This could be combined with general OSH programmes or wider personnel programmes, so that a holistic approach is presented.
- The importance of feedback is that the policy becomes owned by all individuals, which in turn is the single most important factor in successful pre-planning to prevent Major Incidents.

In general it may be useful to retain copies of the abridged and the full policy with other safety documentation in an 'in-house' company library. For smaller organisations, this could be one or two folders on a shelf through to a dedicated room for larger organisations. Second, the abridged copy could be pasted onto an intranet site.

Co-operation

D6039 Co-operation is concerned with collaborating, working together to achieve the shared goal and objectives:

- Firstly, co-operation between strategic, operational and tactical level management is critical. This reflects the chart in the DEM Policy, as discussed above. This could be consolidated as a part of a broader corporate meeting or preferably dedicated time to cover OSH and fatal incident issues. This could be a semi-annual event, with a dedicated day allotted for all grades of management to interface. This is not the same as a board level discussion or a management discussion.
- To give responsibility to either the Safety Committee or the safety group to co-ordinate review, debate and assessment of the DEM Policy or to fuse this function within a broader business/corporate review committee. The former has advantages as it is safety dedicated whilst the latter would integrate DEM Policy issues into the wider business debate.
- Involvement of Safety Representatives/Representatives of Employee Safety (ROES), is both a legal requirement as well as inclusive safety management. These persons can be central 'nodes' in linking 'management' and the 'workers' together. In the UK, there has been an increasing realisation that trained safety representatives are a knowledgeable resource with many being trained to IOSH/NEBOSH standard (as with the AEEU (electrical engineers) trade union, with their National Academy of Safety and Health, NASH).
- Co-operation also needs to extend to contractors. The person responsible for interfacing with contractors needs to up-date them and make them aware of the DEM Policy and seek their support and suggestions. It may also be valuable to invite contractors to OSH/fatal incident awareness days or the general committee meetings as observers.
- Co-operation is also needed between the organisation and external agencies such as the local authority, insurers, enforcers, media etc. This can be achieved through providing abridged copies of minutes or a 'newsletter' (1–2 sides of A4) distributed semi-annually informing such bodies of the DEM Policy and any changes as well as other OSH issues. *COMAH* sites will have to demonstrate as a legal requirement that Plans and policies are up-to-date and effective.

Competence

D6040 Competence is a process of acquiring knowledge, skill and experience to enhance both individual and corporate response. Thus, competence is about enhancing and achieving standards (set by the organisation or others).

- Competence is important, so that certain key persons are trained to understand the DEMS process. The above issues and their link to OSH in particular require personnel that can assimilate, digest and convey the above issues. Training does not necessarily mean formal or academic training but can be vocational or in-house. Neither does it mean the organisation expending vast sums but can be a part of a wider in-house OSH awareness programme (e.g. 1 day per quarter of a year).

- For larger organisations, they may be able to recruit 'competent persons' to advice on OSH and fatal incident matters.
- In short, all employees need to brought up to a minimal standard. Piper Alpha showed that whilst Occidental Petroleum had detailed procedures, the employees generally did not fully understand them nor had they the minimal understanding of Major Incident evacuation. The organisation had failed to impart knowledge, skill and experience sufficient to cope with fires and explosions on off-shore sites.

Control

D6041 Control refers to establishing parameters, constraints and limits on the behaviour and action. This ensures that on the one hand an effective DEM policy exists and on the other, personnel will act and react in a co-ordinated and responsible manner. Controls can be achieved via for example:

- Contractual means – as a term of a contract of employment that instruction and direction on OSH and fatal incident matters must be followed by individuals.
- Corporate means – the organisation continuously makes individuals aware of following rules and best practice.
- Behavioural means – by establishing clear rules, training, supply of information, leading-through-example, showing top-level management commitment etc.
- Supervisory means – ensuring that supervisors monitor employee safety attitudes and risk perceptions.

The cumulative effect of the 4 C's should be a positive and proactive culture in which not only OSH issues but the DEM Policy issues are understood. Factors that can mitigate or prevent a positive and proactive culture developing include lack of management commitment, lack of awareness of requirements, poor attitudes to work and life, misperception of the risk facing the organisation, lack of resources or the unwillingness to commit resources, fatalistic beliefs etc.

Disaster and Emergency Planning

D6042 'Planning' is a process of identifying a clear goal and objectives and pursuing the best means to achieve that goal/objectives. Although 'Disaster' and 'Emergency' are two separate but related terms, in the case of planning the two need to be viewed together. This is because in practice one cannot divorce the serious event (the Disaster) from the response to that serious event (Emergency).

Disaster and Emergency Planning can be viewed in three broad stages:

- stage 1: before the event; (see D6043)
- stage 2: factors to consider during the event; (see D6050)
- stage 3: after the event. (see D6051)

Stage 1: before the event

D6043 Once the DEM Policy has been established and a culture created where the Policy has been understood and positively received, next one needs to be establish a 'state-of-preparedness', that is addressing issues, speculating on scenarios and developing support services when the Major Incident does strike. One can summarise this stage into three sections, with a special section on *COMAH* and the specific legal requirements for *COMAH* sites.

The Risk Assessment of Major Incident potential and consequent Contingency Planning

D6044 What is the probability of the Major Incident resulting? What would be the severity? What type of Major Incident would it be? Figure 3 depicts a 'Major Incident Matrix':

Figure 3: Major Incident Matrix

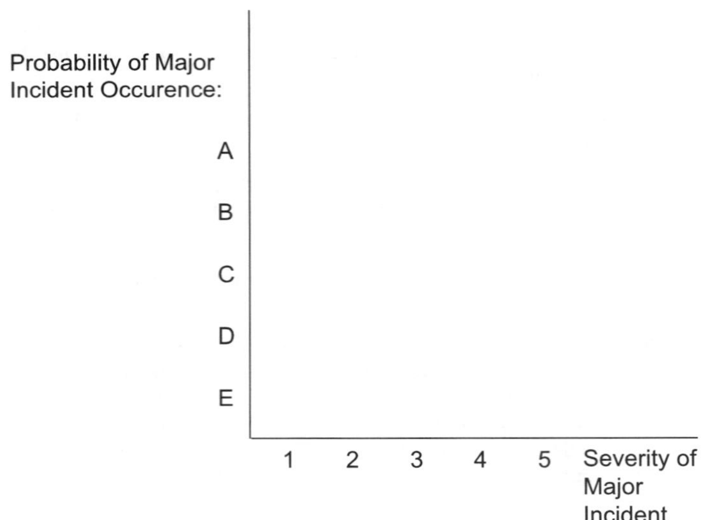

Where:

A=	Certainty of Occurrence (Probability = 1)
B=	Highly Probable
C=	0.5 Probability of Occurrence
D=	Low Probability
E=	Most Unlikely

And:

1=	Serious Injuries
2=	Multiple Fatalities

3= Multiple Fatalities and Serious Economic/Property Damage

4= Fatal Environmental , Bio-Sphere and Social Harm

5= Socio-Physical Catastrophe

The planning responses therefore will differ according to the cell one finds oneself in. For example, cell 1E requires the least complex state of preparedness in terms of resources, detail of planning and urgency. Whilst 5A is a major societal event that requires macro and holistic responses, in which case the organisation's sole efforts at planning are futile. The organisation needs to assess in which cell the Major Incidents it will confront, will mainly fall in and accordingly develop Plans. A conglomerate or an 'exposed operation' such as oil/gas refinery work may require planning at all 25 levels. This is 'Contingency Planning' – an analysis of the alternative outcomes and providing adequate responses to those outcomes.

During the concerns over the 'Millennium Bug', HSBC Bank provided some general advice on Contingency Planning (*Tackling the Millennium Bug: A final check on preparations including Contingency Planning*, HSBC, 1999). They suggested a 10-point strategy. 'A Contingency Plan is a carefully considered set of alternatives to your usual business processes' (page 18). This advice is generic and can be used more broadly for everyday planning, as follows.

(1) *Identify* – assume things will go wrong – focus on where the probability and severity would be of failure.

(2) *Prioritise* the processes into three categories:

- (*a*) Essential for business continuity.
- (*b*) Important for business continuity.
- (*c*) Non-essential for business continuity.

(3) *Analyse* e.g:

- (*a*) Identify the effects of risks on operation(s).
- (*b*) Assess 'domino effects'.
- (*c*) Grade risk using a suitable risk assessment approach.

(4) Develop solutions e.g:

- (*a*) For each operation develop a solution/control measure to the risk.
- (*b*) Encourage participation – involve suitable and critical stakeholders
- (*c*) Ensure new risks are not created.

(5) *Costs* – carry out a 'cost benefit analysis'.

(6) Formalise documents e.g:

- (*a*) Write and keep any assessments and inspections.
- (*b*) Top management involvement and 'joined up business'.
- (*c*) Develop an approach to implementation.
- (*d*) Regular training to be conducted to ensure improved competence.
- (*e*) Identify and deploy resources needed (if available).
- (*f*) Communications to be clear and simple – before, during and after.
- (*g*) Contractual issues to be dealt with so that contractors, customers and insurers are kept informed.

(7) *Coordinate* – managers and team to ensure implementation of contingency plan.

(8) *Test* e.g. controlled testing of the contingency plan(s) (see below).

(9) *Update* e.g:

(*a*) Update the plan.

(*b*) Quality assurance and control of the plan.

(*c*) Accuracy – ensure that the organisation's lawyer and insurer has sight of the plan.

(10) Communicate e.g.

(*a*) Plan signed off by top management.

(*b*) Communicate salient features to employees.

Appendices can be included containing any relevant information.

Decisional Planning

D6045 Planning before the event also involves co-ordinating different levels of decision making. Figure 4 depicts a Decisional Matrix:

Figure 4: Decisional Matrix

	Production Level	1. INPUT	2. PROCESS	3. OUTPUT
Management Level				
A. STRATEGIC				
B. OPERATIONAL				
C. TACTICAL				

Where:

A= Strategic or Board/Shareholder/Controller Level (Gold Level)

B= Operational or Departmental Level (Silver Level)

C= Tactical or Factory/Office Level (Bronze Level)

And:

1= Inputs or those resources that make production possible, thus raw materials, people, information, machinery and financial resources.

2= Process or the manner in which the inputs are arranged or combined to enable production (method of production)

3= Output or the product or service that is generated.

Decisional Planning requires Major Incident risks to be identified for each cell. Essentially this is assessing where in the production process and at which management level can problems arise that can lead to Major Incidents. This matrix is not 'closed' but is 'open' and 'dynamic', which means External Factors, which were discussed above as well as Internal Factors need to be accounted for. For example, in a commercial operation at board level, they need to consider:

- INPUTS: purchasing policy of any raw materials used (safety etc.), recruitment of stable staff, adequacy of resources to enable safe production, known and foreseeable risks of inputs used, reliability of supply, commitment to environmental safety and protection etc.
- PROCESS: external factors and their impact on production, production safety commitment, commitment to researching/investigating in the market for safer production processes etc.
- OUTPUT: boardroom commitment to quality assurance, environmental safety of the service and/or product produced, customer care and social responsibility etc.

The end result should be, in each cell major risks to production and their impact on a Major Incident needs to be identified. This need not be a major, time consuming exercise, rather it can be determined through a 'brain-storming' session or each management level fills out the matrix during their regular meeting and then it is jointly co-ordinated in a short (2–5 page) document.

Testing

D6046 This is considered to be a crucial aspect of pre-planning by all the emergency services in the UK. Testing is an objective rehearsal to examine the state of preparedness and to determine if the Policy and the Plan will perform as expected.

The Home Office guidance, *Dealing with Disaster* (1998, Third Edition) identifies Training and Exercising as types of Testing. Training is more personnel focused, aiming to assess how much the human resource knows about the Policy and Plan and means of enhancing the knowledge, skill and experience of that resource. Exercising is a broader approach, examining all aspects of the Policy and Plan, not just the human response but also physical and organisational capability to deal with the Major Incident.

Training

D6047 Carrying out a Training Needs Analysis (TNA) to determine the quantity (in terms of time) and quality of training needed is essential. It may be that personnel and others adequately understand the Policy and Plan, therefore the training response should be proportionate. Excessive training is a motivational threat to the interest and enthusiasm of the person as is under-training. The HSE in HSG 65, *Successful Health & Safety Management* suggest that the TNA consists of issues such as:

- Is the training 'necessary'? In other words, is there an alternative way of ensuring competence and awareness of the Policy and Plan, rather than just training? Could circulating the information be better? What about regular 'tool box talks'? It is important to weigh the costs and benefits of each option.
- Is it 'needed'? Will the training meet Personal, Job or Organisational needs thereby enhancing awareness and appreciation of the policy and plan ?
- What will be the intended learning objectives of the training? Will it be 'this is a course on the content of the DEM Policy and Plan'?
- What type of training will be offered? Class-room, on-site or both? What are the costs and benefits associated with each?
- Management of the training – where will the training be conducted? Who will deliver it? When will it be delivered – day or evening? What training aids will be provided – if any?

- How does one measure the effectiveness of such training ?

Exercising

D6048 The Home Office cite three types of exercising:

Seminar Exercises

— a broad, brain-storming session, assessing and analysing the efficacy of the Policy and Plan

— seminars need to be 'inclusive', that is bring staff and others together in order to co-ordinate strategic, operational and tactical issues.

Table-Top Exercises

This is an attempt to identify visually using a model of the production or office site and surrounding areas, what types of problems could arise (access/ egress, crowd control, logistics etc.).

Live Exercises

This is a rehearsal of the Major Incident, actively and pro-actively testing the responses of the individual, organisation and possibly the community (emergency services, the media etc.). London Underground, the railway sector commonly, the civil aviation sector, tend to annually assume that a Major Incident has occurred and the consequential conditions are created.

Synthetic Simulation

Whilst this is not mentioned in Home Office or other emergency service guidance, this fourth approach is becoming a most popular option. It applies the techniques used in flight simulation and military simulation, to Major Incidents. Thus, using computers one can model the site and operation and in 3-D format move around the screen. This then enables various eye-points around the site, achieved by moving the mouse. Some advanced systems enable 'computer generated entities' to be included into the data-set. For example, a collision can be simulated outside the production plant or the rate of noxious substance release modelled and calculated across the local community.

There are costs and benefits associated with each of the four approaches. The need is for developing exercise budgets which enable all four to be used in proportion and to compliment each other.

Training and exercising are both necessary to ensure that human resources understand the policy and plan, as well as to ensuring that problems can be identified and competent responses developed.

It should be noted that the guidance to the *COMAH Regulations* (*SI 1999 No 743*) does not distinguish between 'training' or 'exercising'. It identifies drills, which test a specific aspect of the Emergency Plan (e.g. fire drills), seminar exercises, walk through exercises, which involves visiting the site, table-top exercises, control post exercises, which assess the physical and geographical posting of the personnel and emergency services during a Major Incident and live exercises.

COMAH sites

D6049 *COMAH* sites, in addition to the above, also need to ensure that On-Site and Off-Site Emergency Plans are compliant with guidance and are realistic and achievable.

On-Site Emergency Plan

- A 'Site Main Controller' needs to be appointed. This is a strategic level person, who oversees the Major Incident (refer above to organisation chart under Policy). This person must:
  - oversee and take control over the event;
  - control the activities of site incident controller, that is operational and tactical level co-ordinators;
  - must contact and confirm that the emergency services have been contacted;
  - must confirm that On-Site Plan has been initiated;
  - mobilise key personnel, identified in the policy and pre-planning;
  - continuously review and assess the Major Incident;
  - authorise the evacuation process;
  - authorise closing down of plant and work equipment;
  - ensure casualties are being cared for and notify their relatives, as well as listing missing persons;
  - monitor weather conditions;
  - liaise with the HSE, local authority or the Environment Agency;
  - account for all personnel and contractors;
  - traffic management, site access/egress etc. must be strategically controlled;
  - record keeping of all decisions made, for later assessment and inquiry;
  - provide for welfare facilities (food, clothing etc.) of personnel;
  - liaise with the media;
  - compliance with the law – non-removing of evidence etc;
  - control of affected areas after the Major Incident.
- On-Site Emergency Control Centre (ECC)
  - This is a room or area near the Major Incident site, where the command and control decisions will be made.
  - The ECC will also co-ordinate liaison with the media and the public.
  - The ECC should contain: telecommunication equipment, cellular communication equipment, detailed site plans and maps, technical drawings of the operation and its shut-off points, list of critical/hazardous substances and waste on site, location of all safety equipment, location of fire safety equipment and access/egress points etc.
- The On-Site Emergency Plan must contain at least the following (*Sch 5, Part II* of *COMAH* (*SI 1999 No 743*)):

- name/position of person that can activate the Emergency Plan and that person (unless being the same person) who will co-ordinate/ manage and mitigate the impact of the Major Incident;
- name/position of the person who will liaise with the local authority regarding the Off-Site Emergency Plan;
- description of the foreseeable factors that can increase (or mitigate) the severity and consequence of the event, thereby either classifying it as a Major Accident or worse;
- Arrangements for mitigating risks to persons on-site. Warning systems and rules that such persons must follow need to be listed;
- Arrangements for warning the local authority, so that they can activate the Off-Site Emergency Plan, as well as ensuring the local authority receives any other relevant information;
- Arrangements for training staff and effective liaison by them with the emergency services;
- Arrangements for providing assistance with 'off-site mitigating action' such as special equipment that off-site personnel can use, media liaison etc.

Off-Site Emergency Plan

- *Schedule 5, Part III* of *COMAH* (*SI 1999 No 743*) says that this Plan must contain at least:
  - name/position of person that can activate the Emergency Plan and the person (unless the same person), who will co-ordinate the off-site actions. This should be senior, strategic level management that will co-ordinate links with the local authority and the public;
  - Arrangements for receiving 'early warning of incident'. This can be via on-site team, technology etc. In addition, there must be Arrangements for alerting others of the event and procedures for call-out of the emergency services, specialist assistance, other search and rescue bodies;
  - Arrangements for co-ordinating resources so that the Off-Site Emergency Plan can be effectively implemented. This should identify those organisations that can assist in the Off-Site Emergency, how each of these organisations will be alerted, how personnel of the organisation affected will recognise and identify the emergency services and specialist services and vice-versa, channels of communication between personnel and those other emergency services, specialist organisations, meeting place, off-site needs to be identified, for direct communication between personnel and others, access to the operation to use equipment;
  - Arrangements for providing on-site assistance, such as from the fire brigade;
  - Arrangements for off-site occurrences and the mitigating actions needed to protect the public and the environment;
  - Arrangements for providing information to the public and the appropriate behaviour expected from them. In addition, details of how the media could be utilised to disseminate information to the community and emergency organisations;

  - Arrangements for the provision of information to member states, when there is a 'reasonable likelihood' of impact on such neighbouring states.

- Off-Site Emergency Control Centre (ECC)
  - This parallels the on-site ECC and is at a safe distance from the site. It is a command and control centre for all off-site liaison.
  - It must also link up with the on-site ECC to ensure that Plans are in unison.

Thus, under *COMAH*, the On-Site and Off-Site Emergency Plans are required to reflect, account for and provide details of the Arrangements and Command and Control variables identified in stage 2 of the DEMS process. The difference is that under Emergency Plans, these variables are specific and must be practical, whilst in the Disaster and Emergency Policy of DEMS, these variables will also include wider issues not necessarily a legal requirement under COMAH but significant to Major Incident management. However, not all sites (majority of sites) are *COMAH* regulated.

Stage 2: during the event

D6050

- *Activate Disaster and Emergency Policy and pre-planning*

  Major Incident response and management ought to be clear and effective if one has followed the previous stages in the DEMS; accounted for External and Internal factors and uncertainties, developed a Chain of Command as well as Arrangements to cope, Organised the operation to cope with a Major Incident and finally pre-planned if the event actually arises. In the main, activating the Arrangements made is the central action (refer above).

- *Co-ordination*

  This is the central action to ensuring efforts are in unison. This includes both liasing with external bodies such as the emergency services, media and internally at strategic, operational and tactical levels (as discussed above). The police will normally have overall command and control, so their guidance must be followed.

- *Code of conduct*

  A code of conduct is a set of 'golden rules' that staff and contractors need to follow if the Major Incident arises. This includes:

  (i) following instruction from those with command and control responsibilities;

  (ii) using self initiative;

  (iii) facilities management (gas, electricity, water) – making safe, when and when not to switch on or off;

  (iv) medical assistance – liaison with health service, first-aiders, when not to administer and when to;

  (v) rules of evacuation;

  (vi) site access and egress;

  (vii) site security;

  (viii) work and rescue equipment – its safe use and logistics of use;

(ix) record keeping of the Major Incident (audio-visual, verbal and written);

(x) resources needed for effective Major Incident management;

(xi) media and public relations management;

(xii) issues of care and compassion of injured persons.

The code of conduct is a reflection of the Disaster and Emergency Policy and the pre-planning. It should be reinforced verbally and in writing (1 page of A4) and reiterated to the Major Incident team before and during the Major Incident. Whilst this seems bureaucratic, it must be noted that if the core team (internal and external) fail to follow best practice and safe-guard themselves, this increases the risk factor of the Major Incident and could even lead to a double tragedy. A five minute or so reiteration is a minor time cost.

COMAH sites will have to be aware of these issues also as well as activating the On-Site and Off-Site Emergency Plans.

Stage 3: after the event

D6051 Major Incident Planning does not cease as soon as the Major incident is physically over. There are continuous issues over time that need to be addressed (see below).

I. Immediate term (immediately after the major incident and within a few days)

D6052 The following paragraphs explain what steps should be taken.

Statutory investigation

D6053 This will involve the HSE, local authority, Environment Agency/SEPA for instance. Under statute, these bodies have powers and a duty to investigate a Major Incident.

- The organisation investigated must fully co-operate with these bodies and afford any assistance they require.
- There must be no removal or tampering with anything at the Major Incident site (or Off-Site in *COMAH* cases). There may be forensic or other data gathering required by these bodies.
- There must be provision for all documentation and information to be provided to these bodies.

Business Continuity

D6054 This has to be planned for even whilst the investigation by authorised inspectors is being carried out.

'Business Continuity' is a planning exercise to ensure critical facilities, processes and functions are operational and available during and immediately after the Major Incident thereby enabling the organisation to function commercially and socially.

There are many approaches to Business Continuity. Two are highlighted below – first, a 'generic approach', secondly the advice from the Business Continuity Institute (BCI). In addition, there are approaches from the Department of Trade and Industry, Disaster Recovery Institute of the USA, Australian National Audit Office as well as guidance contained in BS/ISO/ICE 17799:2000 (which is an information technology standard).

Generic approach

Business Continuity is a forecasting process of recovery, assessment and ensuring the adequacy of resources for the organisation to continue its operation.

(*a*) Recovery

Recovery is a state of regaining or salvaging assets that otherwise would have been permanently lost. In the event of a Major Incident, recovery phase needs to focus on:

(i) Human resource recovery

— ensuring personnel are safely evacuated;

— others such as lawful visitors and trespassers are evacuated;

— all persons in general are removed in the quickest and most practicable means.

(ii) Information resource recovery

— essential documents and software;

— private and confidential documents.

(iii) Physical resource recovery

— primary electrical and mechanical facilities such as power supply, water and gas services;

— essential work equipment, if possible and moveable.

(iv) Financial resource recovery

(v) Valuable assets recovery (if practicable)

The organisation also needs to rate the recovery potential – how possible is recovery? This could be rated from certain (1) through to not possible (5). This calculation requires different recovery responses; when the Major Incident strikes, those on the scene need to decide if the rating is 1 or say 5. Accordingly, the response will vary – if 5, then there is little purpose risking life and resources to salvage assets.

(*b*) Assessment

Upon the immediate recovery, the organisation must ask, how much damage (deterioration, infliction of harm or erosion of value) has been inflicted to the operation by the Major Incident? Damage Assessment is a three stage activity:

(i) Identify the type of damage

— human resource damage: physical and/or behavioural;

— informational resource damage: primary, secondary and tertiary documentation;

— physical resource damage: work equipment, property, facilities, environment;

— financial resource damage: cash, art, etc.

(ii) Identify severity of damage

This can be a qualitative scale which says 'low' or 'high' to the age or a quantitative scale – the damage is rated on some scale e.g. from 0 – 5. The severity needs to be forecasted for human, informational, physical and financial resources.

— Human resources could be scaled from serious injury (1) through to fatality and multiple fatality (5).

— Informational: from minor harm (1) through to destroyed (5).

— Physical: from reparable (1) to unsalvageable (5).

— Financial: from no impact on cash flow (1) to financial ruin (5).

(iii) Consequence of damage

How much does the damage affect the chance of the operation being resumed immediately or in the next few days? The longer it will take to resume, the worse has been the consequence. Consequence could be rated in quantitative terms also, such as a '1' for 'resume immediate' so the harm has been minimal through to say '5', or 'resumption will take months'. The consequence needs to be examined in each case:

— Human: how long will it take people to get back to work?

— Informational: how long will it take for manual and electronic systems to be operational?

— Physical: how long will it take for necessary facilities to be operational?

— Financial: how long will it take for cash flow to become positive or to access financial facilities?

Severity x Consequence will give an index of forecasted potential Loss, which can then be assessed as a spreadsheet over time. Loss adjustors use various detailed statistical models based upon such generic principles.

(*c*) Adequacy of resources

Once the organisation has forecasted the potential for recovery and hypothesised about damage assessment, next it needs to ensure that the organisation will have adequate resources to carry on operating, in light of the resource losses identified by the assessment. Adequacy needs to consider:

(i) Human resources:

— adequacy of competent and trained personnel at strategic, operational and tactical levels of the organisation;

— availability of key advisory support services (lawyer, accountant, etc.);

— if the Major Incident is classified as a 'Crisis', then there is a chance some of the key personnel are not available. Pre-planning therefore requires liaison with recruitment and selection specialists.

(ii) Physical resources:

— telephone, fax, e-mail, cellular connectivity and reliability;

- furniture and fittings;
- stationary;
- work equipment (including computers/type-writers, filing cabinets);
- working stock;
- working space;
- vehicles;
- safety equipment (personal protective equipment);
- etc.

(iii) Informational resources:

- legal documents (organisation's certificates of incorporation, insurance liability certificates etc.);
- personnel documents (PAYE, NIC, personnel records);
- financial documents (availability of bank books);
- sales/marketing documents, this is at the heart of the operation and there will be a need to develop databases, contact potential customers and re-establish commercial functionality.

(iv) Financial resources:

- adequacy of working capital.

Business Continuity Institute's approach

In *Business Continuity Management: A Strategy for Business Survival* by the BCI, they say ''BCM is not just about disaster recovery, crisis management, risk management or about IT. It is a business issue. It presents you with an opportunity to review the way your organisation performs its processes, to improve procedures and practices and increase resilience to interruption and loss. Business Continuity Management is the act of anticipating incidents which will affect critical functions and processes for the organisation and ensuring that it responds to any incident in a planned and rehearsed manner'.

Business Continuity Management (BCM) is the process of proactively identifying, anticipating, resourcing and planning to ensure that business operations can continue after the Disaster. In contrast, Business Continuity Planning (BCP) is a stage in the BCM process. It is the tactical and operational document that will be used when Disaster strikes. The BCI defines BCM as 'an approach that seeks to improve organisational resilience to interruption, allowing for the recovery of key business and systems processes within the recovery time frame objective, whilst maintaining the organisation's critical functions'.

The BCI present a 5 step approach to BCM.

- STEP 1: 'Understand your business'

Ensure that the main strengths, weaknesses, opportunities and threats are identified. Also ensure that key 'threats' and 'risks' are identified. Further, the support of the board/strategic management team (SMT) is necessary for the necessary resource allocations.

- STEP 2: *'Consider Continuity Strategies'* when the worst case scenario strikes e.g.

  (*a*) *Do nothing* – it maybe that no resource deployment or salvage strategy is required as the risks of significant loss are quite low.

  (*b*) *Changing or ending the process* – the board/SMT may consider that production or service processes/procedures need to be modified.

  (*c*) *Insurance* – should it be used? What are the costs?

  (*d*) *Loss Mitigation* – what actual/tangible actions can be taken to reduce risks?

  (*e*) *Business Continuity Planning* – is a more detailed plan needed given the severity and long term nature of the risk/threat?

- STEP 3: 'Develop the Response'

  If BCP is needed then develop the plan. The BCI suggest the following sections to a BCP:

  — Section 1: General Introduction and Overview

    (i) Objectives

    (ii) Responsibilities

    (iii) Exercising

    (iv) Maintenance

  — Section 2: Plan Invocation

    (i) Disaster declaration

    (ii) Damage assessment

    (iii) Continuity actions and procedures

    (iv) Team organisation and responsibilities

    (v) Emergency (Crisis) Operations Centre

  — Section 3: Communications

    (i) Who should be informed

    (ii) Contacts

    (iii) Key messages

  — Section 4: Suppliers

    (i) List of recovery suppliers

    (ii) Details of contract provision

- STEP 4: *'Establish a Continuity Culture'* – through education, training, information and awareness.

- STEP 5: *'Exercise, Plan Maintenance and Audit'*

  (*a*) Exercise the Plan frequently with participation of key stakeholders.

  (*b*) Plan Maintenance accounting for internal and external changes.

  (*c*) Use Auditing techniques for 'gap analysis' purposes.

Additional matters to consider

D6055 Also under the 'immediate term', in addition to the above, one needs to consider:

- *Insurers/loss adjusters*

  Assuming that the insurance contract covers direct and consequential damage from a Major Incident (and not all will), the organisation needs to notify the insurer and ensure all paperwork is completed promptly. Most insurance contracts stipulate a time limit by which the paperwork has to be lodged with the insurer; the Major Incident will divert attention to other issues, so ensuring that a person is appointed to activate this insurance task is vital (this should be the organisation's lawyer).

  The insurer in turn will notify their loss adjusters to investigate the basic and underlying causes of the event. Again, full disclosure and co-operation are implied insurance contractual requirements. Although, the organisation must check all documents that the loss adjusters completes, to ensure they are accurate and cover all aspects of the event.

- *The police*

  In the suspicion that criminal neglect played a role in the event, then the police will need to interview all core board members, senior management and others.

- *Building contractors*

  The organisation needs to plan for building contractors to visit the site and make it safe and secure. This may have to be done forthwith after the event, even if insurance issues have not been resolved. Therefore, adequacy of resources, as discussed above becomes vital.

- *Visitors*

  Major incidents also lead to the public and media making visits. The arrangements made to handle such groups must extend to after the event.

- *Counselling support*

  Counselling support to affected employees and possibly contractors is not just a personnel management requirement which shows 'caring management' but increasingly a legal duty of the organisation to provide such support. Issues of 'post traumatic stress', 'nervous shock' and 'bereavement' means that the organisation has common law obligations to offer medical and psychological support. This should involve a medical practitioner and an occupational nurse. The insurance policy can be extended before the event to cover the cost of such services.

II. Short Term (a week onwards after the event)

D6056

- *Investigation and inquiry*

  This can be both a statutory inquiry (although most are called within days) and/or the in-house investigation of the event and lessons to be learnt. Issues to consider include:

  — Basic causes: was it a fire, bomb, an explosion or a natural peril?

  — Underlying causes: what led up to such a peril occurring? Examine the managerial, personnel, legal, technical, organisational and natural factors that could have caused the event.

— Costs and losses involved.

— Lessons for the future.

— Did the Disaster and Emergency Plan operate as expected? Were there any failings? What improvements are required? There needs to be complete de-briefing and examination of the entire process involving internal personnel and external agencies.

The organisation should weigh up the possibility of external persons carrying out this exercise or whether in-house staff are objective and dispassionate enough to assess what went wrong.

- *Visit by enforcers*

  The organisation also needs to be prepared for further visits from the enforcers and the possibility of statutory enforcement notices being served to either regulate or prohibit the activity. Multiple notices are possible, from health and safety officers, fire authority, Environment Agency/SEPA, building control or planning officers. This will affect the operation, production process and have economic implications. This ought not to occur if the organisation has taken due care to pre-plan for the Major Incident and had continuous safety monitoring of the operation. Notices will be served if there is a failure to make safe the site.

- *Coping with speculation*

  The public, the media and employees will speculate about causation and there is the risk of adverse publicity. Public relations is a central activity. For legal and moral reasons, it is best practice to disclose all known facts unless the statutory investigation prohibits otherwise.

III. Medium term (a month plus)

D6057 The 'normalisation process' will begin. The organisation needs to carry on the operation, be prepared for further visits from enforcers, loss adjusters and re-assess its corporate/financial health.

IV. Long term (six month onwards)

D6058 *Systems review*

— A review of the impact of the Major Incident and whether the organisation is recovering from it.

— Impact on reputation.

— Legal threats.

— Any positive outcomes – learning from mistakes, improving technical know-how, wider industrial benefits from knowing the chain reaction of events etc.

V. Longer term (one year onwards)

D6059 The organisation's memory and experience needs to be included in:

- in-house training programmes;
- factored into systems and procedures;
- review of entire *espirit de corps* and corporate philosophy.

Disaster and Emergency Planning is an extensive exercise, being dynamic and accounting for a diverse array of phenomena as industrial, man-made, environmental, socio-technical, radiological and natural events.

Monitor the Disaster and Emergency Plan(s)

D6060 Monitoring is a process assessing and evaluating the value, efficiency and robustness of the Disaster and Emergency Plan(s). This involves looking at all three stages as discussed above (before, during and after the event stages) and not just the core document, 'the plan'. It is comprehensive and holistic in its questioning of the entire planning process.

Monitoring can be classified as proactive or reactive. The former attempts to identify problems with the plan(s) before the advent of a Major Incident. It is a case of continuously comparing the plan(s) with even minor incidents and loop-holes identified in any training/exercising sessions. Reactive monitoring occurs after a Major Incident or occurrences that could have led up-to a Major Incident, thereby reflecting back and assessing if the plan(s) need improvement. Both types are important.

Figure 5: Monitoring Matrix

Stages in Planning	**1. Before the Major Incident**	**2. During the Major Incident**	**3. After the Major Incident**
Monitoring Types			
A. PRO-ACTIVE	• Risk Assessments	• Inspections	• Inquiries
	• Testing via Training or Exercises	• Live Interviews	• Systems Review
	• Major Incident Assessment	• Feedback	• Counselling Reports
	• Facilities Inspections		
B. REACTIVE	• Incident Statistics	• Critical Assessments	• Brainstorming
	• Incident Reports	• Incident Levels	• Loss Assessments
	• Warnings	• Audio-Visual Assessment	• Enforcement
	• Notices		

The above cells are not strictly mutually exclusive; many techniques are both proactive and reactive. A third dimension is added in *COMAH* cases, that of, On-Site and Off-Site Emergency Plans (types of planning).

Cell A1 (proactive before the Major Incident)

- Risk Assessments and Hazard Analysis techniques will identify significant hazards and their risk level. Monitoring such risk is an index of danger, which in turn is a variable in the type and potential of the Major Incident, outlined in Figure 1. The HSE's *Five Steps approach* or their *Quantified Risk Assessment* methodology provides outlines of assessing risk. Hazard techniques include Hazard and Operability Studies, HAZANS, Fault and Event Tree Analysis etc.

- Testing will identify any problems or concerns with the Disaster and emergency plan. For example, a live exercise or a synthetic simulation could identify factors the plan has not considered or which may not be practicable if the Major Incident was to arise. Thus, enabling questioning, critical appraisal and comments should not be perceived as a threat or being awkward with the plan(s) but can provide vital information.
- Major Incident Assessment, which is a periodical overview of the plan(s), every quarter or semi-annual by both internal and external persons. This could identify areas of concern. This Assessment compares the plan(s) with the potential threat – can the former cope with the threat? Threats change as technology and know-how changes, so such Assessments become another vital source of information.
- Facilities inspections of gas, electricity, water, building structure, equipment available/not available etc. can highlight issues of physical resourcing and adequacy of such resourcing.

Cell B1 (reactive after the Major incident)

- Incident statistics will show the type of incidences, the type of injury, when and where it occurred. This enables the organisation to hypothesise/build a picture of the potential and severity of a bigger incident. One cannot divorce occupational health and safety data from 'Disaster and Emergency Management'.
- Analysing incident reports should enable issues of causation to be assessed. What type of occurrence could trigger a Major Incident? Identifying and developing a pattern of causes will enable one to assess if the plan(s) account for such causes.
- Warnings from employees, contractors, enforcers, public and others of possible and serious problems are to be treated with seriousness. All such warnings are to be analysed and a common pattern and trend spotted.
- Any notices served by enforcers will identify failings in the operation and the remedial actions required. These can be factored into the plan(s).

Cell A2 (proactive during the Major Incident)

- Inspections will be made even as the event occurs. Inspections can range from the stability of the structure through to how personnel behaved and coped. As these inspections are made, the command and control team needs to evaluate if any aspect of the plan(s), which is a 'live document' need immediate changes.
- Live Interviews with internal personnel and emergency services' personnel will enable a continuous appraisal of any difficulty with procedures, arrangements and instructions that emanate from the plan(s). Again, these can lead to immediate changes to the plan(s).
- Feedback is a proactive technique of requesting regular, interval information on and off-site. This enables a picture to be constructed of what could happen next; trying to anticipate the next sequence and if the plan(s) can cope with it.

Cell B2 (reactive during the Major Incident)

- Critical Assessments are carried out after some unexpected occurrence, which causes uncertainty and may even threaten the efficacy of the plan(s). The Critical Assessment is by the command and control team as a whole. Why did this happen? Why did we not account for it in the plan(s)?
- Incident levels – in particular if serious injury or fatalities are increasing, then at a moral or philosophical level one needs to ask if the plan(s) have been overwhelmed by reality. All forms of planning, including the statutory *COMAH* planning must not be viewed with rigidity. If the plan(s) are failing, it is better to re-appraise and re-plan. *COMAH* (*SI 1999 No 743*) does not overtly allow for this, although it stresses flexibility and continuous appraisal of the event. In such a case, there has to be quick and clear decision-making, with consequential command and control, as well as immediate communication of this 'alternative plan'. Training and exercising sessions need to factor in this dimension and equip people with decisional techniques.
- Audio-visual assessments can be a dramatic means of understanding the actual event. This can be video or photographic footage shot by the incident personnel or from the emergency services. This enables monitoring of the extent, potential and actual threat from the Major Incident.

A3 (proactive after the Major Incident)

- Inquiries are proactive, even though the event has happened, the inquiry (whether internal or external) will identify strengths and weaknesses in the plan(s), which can lead to future improvements in planning.
- Systems review, is an overhaul of the entire reaction and holistic experience of the organisation to the trauma of a Major Incident. This involves developing future coping strategies for personnel and issues of how well did the organisation respond? Were there adequate resources in place to cope?
- Counselling reports will identify the experiences, perceptual and cognitive issues that affected personnel and others. This can provide probably the most significant information on behavioural response of the command and control team and those that were injured. Which in turn can be factored into training and exercise programmes, which itself will lead to personnel skill improvements.

Cell B3 (reactive after the Major Incident)

- Brainstorming is an open-ended, participative and indeed critical analysis of what went wrong and what was right with the plan(s). Brainstorming should also be inclusive, involving emergency services and possibly enforcers as well, so that their guidance is factored in.
- The loss adjusters report will be a vital document as to the chain reaction that lead to the event and the consequences that followed. For legal reasons, their findings may have to be applied before insurance cover is available.
- Enforcement notices and enforcers reports will contain recommendations, which need to be viewed as lessons for the future.

Audit and Review

D6061 The final stage of the DEMS process is Audit and Review. An audit is a comprehensive and holistic examination of the entire DEMS process (Figure 2). An Audit will identify stages in this process that need improvement. A Review is an act of 'zooming in' into that particular stage and carrying out those improvements.

Major Incident Auditing

D6062

- Major Incident Auditing can be qualitative or quantitative. The former adopts a 'yes' or 'no' response format to questions. The latter asks the auditor to rate the issue being examined from say 0–5.
- The Audit must be comprehensive, assessing every aspect of the DEMS process. This means that the Audit will take time to be completed. The Audit is not some 'inspection' which is more random focusing upon a hazard rather than the complete system.
- Audits essentially benchmark (compare and contrast) performance. This can be against the DEMS process identified above or against legislation (e.g. *COMAH* (*SI 1999 No 743*)). The benchmarking could also be against another site or wider industry standards.
- Should Audits be carried out in-house or rely on external consultants? There are costs and benefits associated with both, with no definitive answer. The *Management of Health and Safety at Work Regulations 1999* (*SI 1999 No 3242*) in the UK, emphasises the need to develop and use in-house expertise in relation to general OSH issues, with a reliance on external specialists as a last resort. This may be interpreted as best practice for DEM.
- Audits can be annual or semi-annual. The more complex the operation and risk it poses, the greater the need for semi-annual audits.
- Audits should be proactive, that is learning from the weaknesses in the DEMS process and reducing or eliminating such weaknesses for the future.
- Finally, the results of the Audit need to be fed-back into the DEMS process and all affected persons informed of any changes and risk management issues arising.
- Audits are holistic (assess anything associated with Major Incidents), systemic (assess the entire DEMS process, i.e. the 'system') and systematic (that is logical and sequential in analysis).

Review

D6063

(1) The Review of any specific problems needs to be actioned by the organisation. The consultant will identify the areas of concern and make recommendations but the final discussion and implementation lies with the organisation. This needs to be led by senior officials in the organisation.

(2) Reviews are by definition 'diagnostic', meaning the organisation needs to look at causation and cure of the failure in any part of the DEMS process.

(3) Budgeting both in time and resource terms is critical in the Review, as it will require management and external agency involvement.

(4) A Review can be carried out at the same time as an audit. A Review can also be a legal requirement, as with *Reg 11* of *COMAH* (*SI 1999 No 743*), which requires a Review and where necessary a revision of the On-Site and Off-Site Emergency Plans for Top-Tier establishments.

Case studies

D6064 A summary of four Major Incidents follows (off-shore, rail underground, air aviation and land based chemical plant).

Piper Alpha

D6065

- Offshore Disaster in 1988. 167 workers died in North Sea.
- Major explosion and fire on offshore platform. Piper Alpha involved four rigs linked together. Gas, crude oil and compressed gases were drilled.
- Piper Alpha involved various levels of work along the platform – mining at a certain level, accommodation above that and the helicopter pad at the very top.
- At 22.00 hours whilst 62 worked night shifts and 226 were on the platform, an explosion occurred then a fireball swept the platform. Thereafter a number of small explosions occurred.
- The water systems and emergency systems failed to respond.
- Three may day calls were sent out and personnel assembled at deck d.
- The radio and lighting systems then failed.
- At 22.20 p.m. there was a rupture of the gas riser on another rig connected to Piper Alpha.
- This was followed by explosions and ignition of gas and crude.
- At 22.50 p.m. another explosion occurred followed by the structure collapsing.
- The formal inquiry found both technical and organisational failings. The maintenance error that led to the leak was due to lack of training and poor maintenance procedures.
- There was a breakdown in the communication systems and the permit to work system at shift changeover.
- There were inefficient safety procedures.
- Human error therefore existed at various levels.

Moorgate

D6066

- Underground rail accident in 1974.
- A passenger train carrying 300, overshot the platform at a speed to be around 30 – 40 mph.
- The warning light was knocked down on the track and hit hydraulic buffers.
- The lead car and hit the tunnel roof with the second carriage hitting this.
- 42 people were killed plus the driver and 74 injured.
- Driver was inexperienced and guard had little training.
- Brakes had to be checked visually.
- Driver should have slowed train down via manual controls.

- Safety systems were criticised – poor use of modern technology to slow down the train and the design and layout of impact equipment.

Kegworth

D6067

- A Boeing 737 had taken off to Belfast. A known fault to the right hand engine had been logged and corrected.
- The air conditioning on a 737 is driven from the right hand engine in most cases.
- During flight, the pilot spotted vibrations, excess of smoke and fumes; he throttles the right engine back. The left engine throttles back automatically at the same time.
- It is said that there was a warning light showing fire in the right engine.
- The pilot seeks to land at the East Midlands Airport near Derby.
- The signal was incorrect and really the problem was with the left engine. The correct functioning had been turned off. Thus, there was one faulty engine and one switched off. The plane could have landed with one engine.
- The pilot landed on the M1 near Kegworth village.
- Kegworth demonstrates importance of cognitive factors and ergonomic design.

Flixborough

D6068

- Chemical plant destroyed.
- A part (reaction vessel) is removed and in its place a bent pipe is inserted. The pipe is of inferior material and no problem assessment was carried out to assess the impact the pipe would have.
- The process was restarted and the pipe ruptured, releasing flammable vapour clouds that in turn ignited.
- Explosion resulted and other pipes were ruptured. Fires resulted in the complex.
- Poor maintenance and monitoring were prime factors in disaster.

Conclusion

D6069 All organisations and societies need to prepare for worst case scenarios. DEMS provide a logical framework to understand the main stages in effective preparation. The larger the organisation, the more detailed and analytical the preparation needs to be. Finally, DEMS is a live and open system requiring continuous monitoring.

Sources of information

D6070 These include:

Legislation

- *COMAH* sites and indeed non-*COMAH* sites will find it useful in relation to *COMAH* (*SI 1999 No 743*) as a legal or managerial benchmark.

Guidance notes

- *Emergency Planning for Major Accidents: Control of Major Accident Hazards Regulations 1999* , HSG 191 is joint Guidance from the HSE, Environment Agency and the Scottish equivalent , SEPA. This interprets *COMAH* in a user-friendly manner and recommends implementation approaches.
- *Dealing with Disaster* by the Home Office (Third Edition) is a broad but useful outline of the managerial issues involved in planning. It is aimed at the Police, Fire Service and voluntary bodies.
- *Planning for Major Incidents: The NHS Guidance*, published by the Department of Health, NHS Executive (1998 Edition) is similar in approach to the Home Office Guidance. This is also available at http://www.open.gov.uk/doh/epcu/epcu/index.htm
- There are many other useful guidance documents that can be searched for at the following web sites:
  - http://www.open.gov.uk/hse

    This will enable one to search for specific guidance, case studies and documents.
  - epc.ho@gtnet.gov.uk

    The Home Office Emergency Planning College can be contacted at this web site where further assistance can be obtained.
  - http://www.environment-agency.gov.uk

    The Environment Agency will have details of specific natural environment or flood related guidance;

The European Commission

DG XI is the department responsible for environment, nuclear safety and civil protection. It can be accessed via the EC web site http://www.europa.eu.int

International Agencies

The United Nations web site is http://www.un.org and then search for 'humanitarian affairs'.

Professional bodies e.g:

- International Institute of Risk & Safety Management, telephone 020 8741 0835.
- Institution of Occupational Safety & Health, telephone 0116 257 3100.
- Business Continuity Institute, telephone 0161 237 1007.
- Fire Protection Association, telephone 020 7902 5306.
- Loss Prevention Council, telephone 020 8207 2345.
- Society of Industrial and Emergency Safety Officers, telephone 01642 816281.

Please note that since the reorganisation of 'civil contingencies' in the UK Government, the Home Office is no longer responsible for strategic 'Emergency Planning' rather it is the Cabinet Office. Consequently, the guidance and information sources will change and readers are requested to visit the Cabinet Office web site on a regular basis.

Display Screen Equipment

Introduction

D8201 The *Health and Safety (Display Screen Equipment) Regulations 1992 (SI 1992 No 2792)* ('the *DSE Regulations*'), together with the other 'six pack' Regulations, came into operation on 1 January 1993. Their aim was to combat upper limb pains and discomfort, eye and eyesight effects, together with general fatigue and stress associated with work at display screen equipment ('DSE'). The Health and Safety Executive states in the guidance booklet to the Regulations (*Work with display screen equipment* (L26)) that they do not consider there are any radiation risks from DSE or special problems for pregnant women. The Regulations require assessments to be carried out to identify any health and safety risks at workstations used by 'users' or 'operators' (as defined). 'Users' have the right to an eye and eyesight test and any spectacles (or contact lenses) found to be necessary for their DSE work must be provided by their employer. Several minor amendments were made to the Regulations by the *Health and Safety (Miscellaneous Amendments) Regulations 2002 (SI 2002 2174)*. The latest HSE guidance booklet (L26) incorporates these amendments.

The Regulations summarised

Definitions

D8202 *Regulation 1* of the *DSE Regulations (SI 1992 No 2792)* contains a number of important definitions which are summarised below:

- *Display screen equipment*

  Any alphanumeric or graphic display screen, regardless of the display process involved.

- *Workstation*

  An assembly comprising display screen equipment, any optional accessories, any disk drive, telephone, modem, printer, document holder, work chair, work desk, work surface or other peripheral item, and the immediate environment around.

- *User*

  An employee who habitually uses DSE as a significant part of his normal work.

- *Operator*

  A self-employed person who habitually uses DSE as a significant part of his normal work.

The HSE guidance states that the main factors to consider in determining whether a person is a 'user' or 'operator' are:

- use of DSE normally for continuous or near-continuous spells of an hour or more;
- use of DSE more or less daily;
- the need to transfer information quickly to or from the DSE;

Other factors are:

- high levels of attention and concentration are required;
- high dependence on DSE with little choice about using it;
- special training or skills are needed to use the DSE.

Detailed examples are given in the HSE guidance booklet providing pen-portraits of definite users, possible users and those who are definitely not users. Homeworkers, teleworkers and agency workers all come within the scope of the Regulations, if they fulfill the definition of user or operator – see D8213 below.

Exclusions

D8203 The *DSE Regulations* (*SI 1992 No 2792*) do not apply to:

- drivers' cabs or control cabs for vehicles or machinery;
- DSE on board a means of transport;
- DSE mainly intended for public use;
- portable systems not in prolonged use;
- calculators, cash registers or other equipment with small displays;
- window typewriters.

However, the *Health and Safety at Work etc. Act 1974* and regulations made under it, such as the *Workplace* (*Health, Safety and Welfare*) *Regulations 1992* (*SI 1992 No 3004*), still apply to the use of such equipment.

Assessment of workstations and reduction of risk

D8204 *Regulation 2* of the *DSE Regulations* (*SI 1992 No 2792*) requires employers to perform a 'suitable and sufficient' analysis of all workstations which are:

- used for their purposes by 'users';
- provided by them and used for their purpose by 'operators';

to assess the health and safety risks in consequence of that use. They must then reduce the risks identified to the lowest level reasonably practicable. As for other types of assessments, an assessment must be reviewed if there is reason to suspect it is no longer valid or there have been significant changes. Further guidance on assessment, including an assessment checklist (at D8216), is provided later in the chapter.

Requirements for workstations

D8205 *Regulation 3* of the *DSE Regulations* (*SI 1992 No 2792*) states that all workstations must meet the requirements laid down in the *Schedule* to the Regulations. The requirement must relate to a component present in the workstation concerned and be relevant in relation to the health, safety and welfare of workers. For example there is no need to provide a document holder (referred to in the *Schedule*) if there is little or no inputting from documents and some individuals with back complaints may benefit from a fixed back rest or a special chair without a back rest. The HSE guidance booklet provides further examples of where the requirements of the *Schedule* may not be appropriate.

Daily work routine of users

D8206 Employers are required by *Regulation 4* of the *DSE Regulations* (*SI 1992 No 2792*) to plan the activities of 'users' so that their DSE work is periodically interrupted by breaks or changes of activity. Breaks should be taken before the onset of fatigue and preferably away from the screen. It is best if users are given some discretion in planning their work and are able to arrange breaks informally rather than having formal breaks at regular intervals. Some software tools provide a means of ensuring that users take regular breaks but HSE guidance draws attention to the limitations of such software.

Eyes and eyesight

D8207 *Regulation 5* of the *DSE Regulations* (*SI 1992 No 2792*) gives 'users' the right (at their employer's expense) to an appropriate eye and eyesight test by a competent person before becoming a 'user' and at regular intervals thereafter. (Some employers are re-testing at two yearly to five yearly intervals – in some cases varying the period depending on the age of the employee. Professional advice should be sought in cases of doubt). 'Users' are also entitled to tests on experiencing visual difficulties which may reasonably be considered to be caused by DSE work. Such tests are an entitlement for 'users' and are not compulsory.

Tests are normally carried out by opticians and involve a test of vision and an examination of the eye. Where companies have vision screening facilities, 'users' may opt for a screening test to see if a full eye test is needed. If the eye test shows a need for spectacles (other than the 'user's' normal spectacles) the basic cost of these must be met by the employer, but employees must pay for extras, e.g. designer frames or tinted lenses.

Provision of training

D8208 'Users' are required by *Regulation 6* of the *DSE Regulations* (*SI 1992 No 2792*) to be provided with adequate health and safety training in the use of any workstation upon which they may be required to work. Training may also be required where workstations are substantially modified. The training should include:

- the causes of DSE-related problems, e.g. poor posture, screen reflections;
- the user's role in detecting and recognising risks;
- the importance of comfortable posture and postural change;
- equipment adjustment mechanisms, e.g. chairs, contrast, brightness;
- use and arrangement of workstation components;
- the need for regular screen cleaning;
- the need to take breaks and changes of activity;
- arrangements for reporting problems with workstations, or ill health symptoms;
- information about the Regulations (especially eyesight tests and breaks);

- the user's role in assessments.

The HSE has published a leaflet (*Working with VDUs* (INDG36)) which provides a useful reference for training purposes.

Provision of information

D8209 *Regulation 7* of the *DSE Regulations* (*SI 1992 No 2792*) requires employers to ensure that operators and users at work within their undertaking are provided with adequate information. The table below shows the responsibility of the 'host' employer in this respect (and also gives a good guide to his responsibilities generally under the Regulations).

The 'host' employer must inform on:	*Regulation*:	*Own users*:	*Other users* (i.e. agency staff):	*Operators* (self-employed):
DSE and workstation risks		YES	YES	YES
Risk assessment and reduction measures	2 and 3	YES	YES	YES
Breaks and activity changes	4	YES	YES	NO
Eye and eyesight tests	5	YES	NO	NO
Initial training	6(1)	YES	NO	NO
Training when workstation substantially modified	6(2)	YES	YES	NO

DSE workstation assessments

Decide who will carry out the assessments

D8210 As with other types of risk assessments, DSE workstation assessments within an organisation may be carried out by an individual or by members of an assessment team. Those responsible for making assessments should have received appropriate training so that they are familiar with the requirements of the *DSE Regulations* (*SI 1992 No 2792*) and they should have the ability to :

- identify hazards (including less obvious ones) and assess risks from the workstation and the kind of DSE work being done;
- use additional sources of information or expertise as appropriate (recognising their own limitations);
- draw valid and reliable conclusions;
- make a clear record and communicate the findings to those who need to take action;
- recognise their own limitations so that further expertise can be utilised where necessary.

They may be health and safety specialists, IT managers, facilities managers or line managers, and there will often be benefits in involving employees' safety representatives in the assessment process.

Identify the 'users'

D8211 An important first step is to identify the users' (together with any 'operators') of DSE within the organisation. The definitions in *Regulation 1* of the *DSE Regulations* (*SI 1992 No 2792*) refer to habitual use of DSE. The HSE guidance booklet (L26) provides considerable advice on the factors which must be taken into account, of which time spent using DSE is the most significant. Some organisations have adopted a rule of thumb that anyone spending more than 50 per cent of their time in DSE work is a DSE 'user'. However, the HSE guidance indicates that a less simplistic approach should be taken.

The assessment checklist provided at D8216 includes reference to the other factors which the HSE states should be taken into account in deciding whether an individual is a 'user' or an 'operator'. It should be noted that employers have duties to assess workstations used for the purposes of their undertaking by *all* 'users' or 'operators'. This includes 'users' employed by others (e.g. agency-employed staff), 'operators' (e.g. self-employed draughtsmen or journalists), peripatetic staff (e.g. journalists, sales staff, careers advisors) and homeworkers or teleworkers (see D8213 below).

In practice, most employers do not find it too difficult to decide who their 'users' and 'operators' are. Many have taken the approach of assessing the workstations of those individuals where doubt exists and, if necessary, making a final decision then. More time and expense can often be wasted debating a few borderline cases than would be involved in including them in the definition.

Decide on the assessment approach

D8212 In some situations, especially smaller workplaces, it is practicable to carry out a personal assessment of each individual 'user's' or 'operator's' workstation, using a simple checklist (a completed example of such a checklist is provided at D8216). However, in larger organisations the number of DSE workstations may be so great that this approach becomes impracticable (particularly where there are regular office rearrangements taking place) or would require an unnecessarily large input of resources. In such cases the issuing of a self-assessment checklist to identified 'users' will often be more appropriate. A sample checklist, together with guidance to 'users' on its completion, is provided at D8216. The HSE guidance booklet (L26) provides a checklist covering similar topics. Providing that workers are given the necessary training and guidance in how to use such checklists, this approach is perfectly acceptable. However, it must be supported by other actions such as:

- an inspection of areas where DSE workstations are situated (evaluating general issues such as lighting, blinds, housekeeping, desk space and the standards of chairs and DSE hardware);
- providing employees with the option of an assessment of their individual workstation by a specialist;
- responding promptly to problems identified in the completed checklists.

It will often be appropriate to require staff to carry out a workstation self-assessment every year or two, and particularly after office rearrangements.

Homeworkers and teleworkers

D8213 Homeworkers and teleworkers (working away from their employer's premises) are subject to the *DSE Regulations* (*SI 1992 No 2792*), whether or not their workstation is provided by their employer. They will be subject to the normal risks from DSE work, some of which may be increased because of their social isolation, the absence of supervision and the practical difficulties of carrying out risk assessments. Risks can be reduced by:

- training such staff to carry out workstation self-assessments (see above);
- requiring them to carry out a self-assessment of their main workstation;
- encouraging them to carry out ad hoc assessments of temporary workstations, e.g. hotel rooms;
- emphasising the importance of ensuring good posture and taking adequate breaks;
- providing clear communication routes for reporting equipment defects and possible health problems;
- responding promptly and effectively to reports of defects or problems.

A free HSE guidance leaflet (*Homeworking: Guidance for employers and employees on health and safety* (INDG226)) provides general guidance on risk assessment in the home environment and the employer's responsibilities for home electrical systems and equipment.

Observations at the workstation

D8214 Where workstations are to be assessed on an individual basis the assessment checklist questions are intended to identify the principal factors that the assessor(s) need to look out for (as illustrated in the completed example).

Some of the more common problems identified are related to:

- Posture:
  - height of screen;
  - height of seat or position of backrest;
  - position of keyboard or keyboard technique;
  - need for footrest or document holder.
- Vision:
  - angle of screen;
  - position of lights or need for diffusers;
  - need for blinds to control sunlight;
  - adjustment of brightness or contrast controls.

Discussions with 'users' and 'operators'

D8215 One of the dangers of self-assessments is that a minority of employees will blame their DSE work for a variety of unrelated problems. An individual assessment provides an opportunity for a two-way dialogue between the assessor(s) and the 'user' and allows the assessor to evaluate whether stated problems are related to deficiencies in the workstation.

Common causes of problems are:

- Back, shoulders, neck:
  - the height or position of screen;

  - positioning of the seat, including the backrest;
  - the need for a footrest or document holder.
- Hands, wrists, arms:
  - position of keyboard or mouse;
  - keyboard technique.
- Tired eyes or headaches:
  - failure to take breaks or change activities;
  - reflected light (artificial or sunlight).
- Discussions with 'users' can also reveal other important pieces of information, e,g:
  - there are problems with sunlight at certain times of day or periods of the year;
  - the 'user' does not know how to adjust their chair or brightness/contrast controls;
  - the 'user's' chair is broken and incapable of being adjusted;
  - the 'user' has never been offered an eye test.

Assessment records

D8216 As for other types of assessment, there is no standard format for DSE workstation assessment records. The completed sample assessment checklist and the self-assessment checklist (both below), together with the guidance on its completion are offered as examples of record formats that have been found successful in practice.

The HSE guidance booklet states that records may be stored in electronic as well as paper form. Self-assessment checklists would particularly lend themselves to being completed and submitted electronically. No guidance is provided on how long records should be kept. Prudent employers may prefer to retain them indefinitely, bearing in mind that civil claims for alleged DSE-related conditions may be submitted many years after it is claimed the condition was first initiated.

Display Screen Equipment Workstation Assessment			
USER'S NAME: *Christine Jones*			LOCATION: *Sales*
FACTOR			COMMENT
1	**WORK PATTERNS**		
1.1	Most time spent per day at DSE		*5 to 6 hours*
1.2	Average time per day at DSE		*4 hours*
1.3	Number of days per week at DSE		*5*
1.4	Longest spell without break		*2 hours*
1.5	Can breaks be taken?		*Yes – at Christine's discretion*
1.6	Concentration important?		*Accuracy is important*
1.7	Speed of operation important?		*Sometimes*
'USER' STATUS CONFIRMED			*Yes*
2	**PROBLEMS EXPERIENCED** Has the user significant experience of problems with:		
2.1	Back, shoulders or neck		*Regular pains in shoulder and neck*
2.2	Hands wrists or arms		*Occasional aching wrists*
2.3	Tired eyes or headaches		*Sometimes*
2.4	Suitability of software		*Suitable for all tasks*
2.5	Other problems		*None*
3	**LIGHTING/ENVIRONMENT**		
3.1	Artificial lighting:	Adequate to see documents	*Yes*
		Any reflection or glare problems	*None – fittings recessed and diffused*
3.2	Sunlight:	Any reflection or glare problems	*On winter mornings from window behind*
		Suitable blinds available (if necessary)	*Good vertical blinds provided*
3.3	Noise:	Hindering to communication	*No*
		Distracting or stressful	*Occasionally distracting*
3.4	Temperature and ventilation:	Satisfactory in summer and winter	*Office hot and stuffy in summer*
4	**SCREEN**		
4.1	Set at suitable height		*Screen height too low*
4.2	Stable image with clear characters		*Good. Able to vary colours*
4.3	Brightness and contrast adjustable		*Both have adjustable controls*
4.4	Swivels and tilts easily		*Yes*
4.5	Cleaning materials available		*In stationery cupboard*

5	**KEYBOARD**	
5.1	Separate and tiltable	*Yes*
5.2	Sufficient space in front	*Too near edge of desk (wrists bent)*
5.3	Keys clearly visible	*Yes*
6	**DESK AND CHAIR**	
6.1	Desk size adequate	*Satisfactory*
6.2	Sufficient leg room	*Materials being stored in desk well*
6.3	Desk surface low reflectance	*Yes – wooden surface*
6.4	Suitable document holder (if required)	*Not provided*
6.5	Chair comfortable and stable	*Yes*
6.6	Chair height adjustable	*Yes*
6.7	Back adjustable (height and tilt)	*Christine did not know how to adjust*
6.8	Footrest available (if required)	*Not required*

OTHER COMMENTS

Christine was advised to take short breaks more regularly.
Some lengthy jobs without breaks seem to be the cause of her tired eyes and headaches.
Repositioning of the screen and provision of a document holder should overcome the shoulder and neck problems.
Christine was shown how to adjust her chair back.

No.	Actions required	Responsibility
3.2	Close blinds when sunlight bright.	*C Jones*
3.4	Free up office windows (seized up by paint).	*Facilities Manager*
4.1	Provide screen stand and keep top of screen at eye level.	*Office Manager/C Jones*
5.2	Keep space in front of keyboard – wrists horizontal.	*C Jones*
6.2	Remove items from desk well.	*C Jones*
6.4	Provide document holder.	*Office Manager*
6.7	Ensure induction includes chair adjustment mechanisms.	*Training Department*

Assessor's name: *B Wright* Signature: *B Wright* Date: *24/5/04*

PROGRESS WITH ACTIONS

All recommendations acted upon although Christine still needs to remember to take regular breaks on lengthy jobs. Pains in shoulder and neck have ceased and Christine only very occasionally experiences headaches and tired eyes.

B. Wright 12/7/04

Planned date for assessment review: July 2006

DISPLAY SCREEN EQUIPMENT WORKSTATION SELF ASSESSMENT CHECKLIST		
USER'S NAME:		LOCATION:
1	LIGHTING AND WORK ENVIRONMENT	COMMENTS
1.1	Is artificial lighting adequate?	
1.2	Does it cause any reflection or glare problems?	
1.3	Any reflection or glare problems from sunlight?	
1.4	Are suitable blinds available (if necessary)?	
1.5	Are temperature and ventilation satisfactory in summer and winter?	
2	SCREEN AND KEYBOARD	
2.1	Is your screen set at a suitable height?	
2.2	Stable image with clear characters?	
2.3	Brightness and contrast adjustable?	
2.4	Screen swivels and tilts easily?	
2.5	Cleaning materials available?	
2.6	Is your keyboard tiltable?	
2.7	Have you sufficient space in front of it?	
3	DESK AND CHAIR	
3.1	Is your desk size adequate?	
3.2	Is there sufficient leg room under it?	
3.3	Do you have a suitable document holder (if required)?	
3.4	Is your chair comfortable and stable?	
3.5	Can you adjust your seat height?	
3.6	Can you adjust the height and tilt of your chair back?	
3.7	Do you have a footrest (if required)?	
4	HAVE YOU HAD SIGNIFICANT EXPERIENCE OF PROBLEMS WITH:	
4.1	Your back, shoulders or neck?	
4.2	Your hands, wrists or arms?	

4.3	Tired eyes or headaches?	
4.4	The suitability of the software you use?	
4.5	Other problems?	
ANY OTHER PROBLEMS OR COMMENTS?		
WOULD YOU LIKE A FURTHER ASSESSMENT OF YOUR WORKSTATION?		Yes/No
Signature:		Date:
COMPLETED FORMS MUST BE SENT TO THE HUMAN RESOURCES DEPARTMENT		

GUIDANCE ON COMPLETING THE DSE WORKSTATION SELF-ASSESSMENT

Lighting and work environment

1.1 Artificial lighting should be adequate to see all the documents you work with.

1.2 Recessed lights with diffusers shouldn't cause problems. Lights suspended from ceilings might.

1.3 There may be problems in the early morning or afternoon, especially in winter when the sun is low.

1.4 Blinds provided should be effective in eliminating glare from the sun.

1.5 Strong sunlight may create significant thermal gain at times.

Screen and keyboard

2.1 The top of your screen should normally be level with your eyes when you are sitting in a comfortable position.

2.2 There should be little or no flicker on your screen.

2.3 You should know where the brightness and contrast controls are.

2.4 The screen should swivel and tilt so that you can avoid reflections.

2.5 You should know where to get cleaning items for your screen (and keyboard, if necessary).

2.6 Small legs at the back of your keyboard should allow you to adjust its angle.

2.7 Space in front the keyboard allows you to keep your wrists horizontal and to rest your hands and wrists when not keying in.

Desk and chair

3.1 Your desk should have sufficient space to allow you to have your screen and keyboard in a comfortable position and accommodate documents, document holder, phone etc.

3.2 There should be enough space under the desk to allow you to move your legs freely.

3.3 If you are inputting from documents, using a document holder helps avoid frequent neck movements.

3.4 Chairs with castors must have at least five (four is very unstable).

3.5 You should be able to adjust your seat height to work in a comfortable position (arms approximately horizontal and eyes level with the top of the screen).

3.6 The angle and height of your back support should be adjustable so that it provides a comfortable working position.

3.7 DSE users who are shorter may need a footrest to help them keep comfortable when sitting at the right height for their keyboard and screen (see 3.5).

Possible problems

4.1 Problems with back, shoulders or neck might indicate your screen is at the wrong height, an incorrectly adjusted chair or need for a document holder.

4.2 Problems with hands, wrists or arms might indicate incorrect positioning of the keyboard or a poor keying technique. Hands should not be bent up at the wrist and a soft touch should be used on the keyboard, not overstretching the fingers.

4.3 Tired eyes or headaches could indicate problems with lighting, glare or reflections.

They may also indicate the need to take regular breaks away from the screen. Persistent problems might need an eye test – contact Human Resources to request one.

4.4 The software should be suitable for the work you have to do.

IF YOU HAVE A PROBLEM DISCUSS IT WITH YOUR LINE MANAGER. REFER UNRESOLVED PROBLEMS OR REQUESTS FOR FURTHER ASSESSMENTS TO THE HUMAN RESOURCES DEPARTMENT.

Special situations

D8217 The HSE guidance booklet (L26) provides further guidance on several aspects of DSE work which have developed over the last decade as information technology has changed.

Shared workstations

D8218 'Hot desking' arrangements mean that many workstations are shared, sometimes by workers of widely differing sizes. Assessments of such workstations should take into account aspects such as:

- whether chair adjustments can accommodate all the workers involved;
- the availability of footrests;
- adequacy of leg room for taller workers;
- arrangements for adjusting the heights of screens (e.g. stands).

In some situations the use of desks with adjustable heights may be appropriate.

Work with portable DSE

D8219 Work with laptop computers as well as mobile phones or personal organisers which can be used to compose, edit or view text is becoming increasingly common. Where mobile phones or personal organisers are used in this way for prolonged periods they are subject to the *DSE Regulations* (*SI 1992 No 2792*), as is all work with laptops.

Mobile phones or personal organisers should not generally be used for sufficiently long periods to require a workstation assessment but work involving laptops is likely to be of much longer duration. The revised HSE booklet (L26) contains detailed guidance on the selection and use of laptop computers.

The design of laptops is such that postural problems are much more likely to result and there will be a far greater need to ensure that sufficient breaks are taken. Where the laptop is used in the office, the risks can be reduced considerably by the use of a docking station and the HSE booklet also provides guidance on this. Alternative means of reducing risks include using laptops with detachable or height adjustable screens and a facility for attaching an external mouse, keyboard or keypad. Other risks associated with the use of portable equipment (e g manual handling issues and possible theft involving assault) should also be considered.

Pointing devices

D8220 Much work at DSE workstations involves the use of a mouse, trackball or similar device to move the cursor around the screen and carry out operations. The HSE booklet (L26) provides guidance on:

- *Choice of pointing devices*

  Suitability of the device for:

  - the environment (space, position, dust, vibration);
  - the individual (right or left handed, physical limitations, existing upper limb disorder);
  - the task (some devices are better than others in respect of speed or accuracy).

- *Use of pointing devices*

  Issues to be considered include:

  — positioning (close to the midline of the user's body, not out to one side);

  — work surface (particularly its height and degree of support for the arm);

  — mousemats (smooth, large enough, without sharp edges);

  — software settings (suitable for the individual user);

  — task organisation (point device use mixed with other activities);

  — training (how to set up and use devices);

  — cleaning and maintenance.

The booklet also contains specific guidance on touch screens and speech interfaces.

After the assessment

Review and implementation of recommendations

D8221 The general guidance provided in the chapter on RISK ASSESSMENT is equally applicable to recommendations made as a result of DSE workstation assessments. Some issues of general importance will need to be reviewed with others, particularly if there are significant cost implications or practical problems. Responsibility for implementing recommendations should be allocated clearly and all recommendations followed up. It should be noted that the sample assessment checklist includes a space which can be used at a follow-up of an assessment to describe 'Progress with actions'.

Assessment review/re-assessments

D8222 Changes to the layout of office accommodation often take place with bewildering rapidity. Some of these changes have significant implications for DSE workstations, others do not. To carry out a re-assessment every time the office layout is changed would be an inefficient use of resources. A better approach is to make both management and DSE 'users' aware that a review or reassessment should be requested from the assessor or assessment team whenever significant changes take place. The use of self-assessments is particularly relevant in such situations. This can be supported by observations by the assessor(s) of changes which have occurred or are in progress.

Because of the frequency of changes a periodic review of DSE workstation assessments would be beneficial. The HSE does not provide any guidance on this aspect. Reviews at two-yearly intervals may be appropriate for larger workplaces where changes are regularly taking place whilst five yearly intervals would probably be suitable for smaller workplaces where the layout of workstations is more static. Significant changes in DSE equipment, software, furniture, lighting etc. would obviously justify a review of relevant workstation assessments.

Electricity

Introduction

E3001 When used properly electricity is a safe, convenient and efficient source of energy for heat, light and power. However, the most recent comprehensive analysis of electrical incidents (which was published in 1997) indicated that between 1990 and 1995, 533 people were killed in electrical incidents in Great Britain. In 1995 alone there were 121 serious electrical fires costing more than £45 million in insurance claims. These are just two facts that can be gleaned from '*Electrical incidents in Great Britain statistical summary*', published by the Health and Safety Executive ('HSE') and available from HSE Books, PO Box 1999, Sudbury CO10 6FS. The publication also gives information on a wide range of electrical incidents, including examples of some of the most commonly recurring in a variety of locations, such as offices, farms, construction sites and the home. By recording the statistics and nature of the incidents occurring, the HSE aims to promote a greater awareness and understanding of the causes of electrical accidents so that those responsible for electrical safety are more fully aware of the risks involved and consequently, accidents and deaths can hopefully be prevented. Further information about electrical safety can be obtained from the HSE Safety Policy Directorate website.

Another particular issue usually resulting in a serious injury or fatality is the inadvertent contact with overhead electric power lines. It is estimated that one-third of such contacts are fatal. The HSE has published guidance aimed at those working near overhead lines. The guidance contains separate sections dealing with agriculture and horticulture, arboriculture and forestry, construction, railways and other transport systems with overhead conductors and how the risks arise. See '*Guidance Note GS6*', revised in 1997, available from HSE Books, and '*Mechanical harvesting*' (2003) (AFAG 603 04/03 C100) and the range of safety guidance notes produced by the Arboriculture Industry Advisory Committee.

Given that electrical accidents occurring at work often result in severe injuries or damage, high standards in relation to electrical installations and the use of electrical plant and apparatus are essential and adequate systems of control and maintenance are required. Furthermore, there is a need for employers to ensure that those who work on or use such installations, plant and apparatus are sufficiently competent and that the workers are suitably trained, instructed and supervised.

Electrical hazards

E3002 Electrical hazards may arise from bad design, construction or installation of relevant equipment, as a result of inadequate standards of protection or maintenance, or from inappropriate usage of the equipment or, indeed, its misuse. Such hazards can lead to electric shock or electric burns to the individual concerned, or can result in damaged equipment, an explosion or even a general fire.

An electric shock is the result of an electric current flowing through a part of the body. It can affect the nervous system and bodily organs and functions. The value of the current and the time it flows through the body are the two critical factors that determine the effect on the body. The heart is particularly susceptible to a condition known as ventricular fibrillation from currents as low as 50 milliamps flowing for a few seconds. No accidental current should be allowed to pass through the body but the risk of any effects, should it occur, should be kept to a minimum by ensuring that the current passing will be as small as possible and that it will pass for as short a time as possible.

One effect of a shock may be a rapid movement away from the source, which might lead to a further incident such as a severe knock or a fall. When an individual is working above ground level, a fall could be fatal. Additionally, extensive and deep burns, at both the point of entry and at the point of exit, can result from a current passing through the body.

To assist in ensuring that people respond safely and appropriately to an incident involving (or potentially involving) electricity, the HSE has issued a poster which should be displayed in relevant work environments. The poster 'Electricity Shock: First Aid Procedures' takes into account advice from the European Resuscitation Council and provides basic advice on how to break the contact between an electrical source and a casualty and how to implement resuscitation. Employers should ensure that such posters are placed in the appropriate work environment.

Preventative action

E3003 Preventive action against shock and burns includes the following:

- inspection of all electrical equipment, particularly portable hand-held tools;
- checking suitable equipment is installed for circuit protection;
- testing of equipment installed for circuit protection;
- regular inspection of equipment to minimise the risks of shocks to personnel;
- avoidance of work near live conductors;
- the use of proper systems and methods of working;
- ensuring that those using the equipment or involved with it are competent so to do.

Precautions against electrical dangers must be taken in the light of legal requirements, relevant standards and codes of practice. The HSE has published guidance relating to ensuring safe working practices when working with or near to electricity entitled '*Electricity at work: Safe working practices*', (HSG85) (March 2003), (ISBN 0 717621642). In the event that the working environment exposes people to electrical dangers, the guidance should be considered and where appropriate, its recommendations implemented.

One additional point relating to prevention is ensuring the good design and construction of electrical equipment. In some instances, this may be assumed if the equipment complies with recognised standards and is marked in accordance with legal requirements. Further, it is important to note that good installation, protection and maintenance require competent staff or contractors to be employed and the correct operation and use of equipment will also depend on competence, achieved through adequate training, instruction and supervision.

Legal requirements

E3004 Although electricity is not specifically mentioned in every case, the legal requirements relating to the safe working with electricity, the safety of electrical installations and the use of electricity are covered in the following:

- the *Health and Safety at Work Act etc 1974*;
- the *Management of Health and Safety at Work Regulations 1999* (*SI 1999 No 3242*);
- the *Construction* (*Design and Management*) *Regulations 1994* (*SI 1994 No 3140*);
- the *Electricity at Work Regulations 1989* (*SI 1989 No 635*);
- the *Provision and Use of Work Equipment Regulations 1998* (*SI 1998 No 2306*);
- the *Equipment and Protective Systems Intended for Use in Potentially Explosive Atmospheres Regulations 1996* (*SI 1996 No 192*).

Each of the requirements impacts on how work with and near to electricity should be undertaken. Additionally, it is important to recall that the common law is also relevant when considering both the duties owed and the legal implications and consequences arising from an incident. Where an incident occurs, an employer may well find itself on the wrong side of the law both in terms of a criminal offence and also a civil claim for damages or loss.

General duties under the Health and Safety at Work etc Act 1974

E3005 The *Health and Safety at Work etc Act 1974* ('*HSWA 1974*') provides a comprehensive legal framework for occupational health and safety. Although the Act does not expressly refer to electricity, many of its general requirements, for example, safe methods of working, training and supervision, are relevant to electricity and its use. The *HSWA 1974* has been used for prosecutions issued/taken by the HSE with respect to electricity incidents including deaths.

The Management of Health and Safety at Work Regulations 1999

E3006 As with the *HSWA 1974*, the *Management of Health and Safety at Work Regulations 1999* (*SI 1999 No 3242*) do not specifically refer to electricity. The assessment of risks required by these Regulations, however, certainly extends to working with electricity and importantly, the principle of prevention identified in the Regulations should also be borne in mind. The general requirements of these Regulations are echoed in the *Electricity at Work Regulations 1989* (*SI 1989 No 635*).

The Construction (Design and Management) Regulations 1994 (as amended by the Construction (Design and Management) (Amendment) Regulations 2000)

E3007 Although not expressly mentioned, the use of electricity and the installation of electrical equipment certainly comes within the responsibilities of such persons as the designer, the client and the main contractor. The HSE refers to electricity in its guidance, '*Health and safety in construction*', (HSG150), which was updated and reissued to include information previously contained in '*Health and safety for small construction sites*', (HSG130). The HSE guidance (HSG150) sets out various matters which should be considered when construction works are being planned. Depending on the nature of the project there will be other issues to consider such as dangers from underground services, arc welding (see HSE publications '*Avoiding danger from underground services*', (ISBN 0 7176 1744 0) and '*Electrical safety in arc welding*', (ISBN 0 7176 0704 6)), and/or working with overhead cables (see *Guidance Note GS6*). (See also C8015 in CONSTRUCTION AND BUILDING OPERATIONS.)

The Electricity at Work Regulations 1989

E3008 The *Electricity at Work Regulations 1989* (*SI 1989 No 635* as amended) are aimed at the users rather than the suppliers or manufacturers of electrical equipment. The Regulations apply to all places of work, including factories, shops, offices, laboratories and educational establishments. In accordance with the *HSWA 1974*, they lay down the principles of electrical safety in general terms and also raise particular issues to be addressed. Further, it is necessary to bear these principles in mind prior to and during any work with electricity. Details of the design, selection, erection, inspection and testing of electrical installations have been published by the Institution of Electrical Engineers ('IEE') in '*IEE Wiring Regulations*' (currently 16th edition) (as amended in February 2001). (This can be ordered from Publications Sales, IEE, PO Box 96, Stevenage, Hertfordshire SG1 2SD (tel: 01483 767 328) or can be ordered on-line at: www.iee.org. Additionally, the IEE website contains information on proposed changes and consultation relating to the standards).

The Regulations refer to the duties to prevent danger or injury i.e. the prevention of danger amounts to the avoidance of risk of injury. The Regulations also provide for different levels of duty, ranging from reasonably practicable to an absolute duty. In the case of an alleged breach of an absolute duty, it is a defence that reasonable measures have been taken and all due diligence observed. Importantly, the HSE has issued a '*Memorandum of Guidance on the Electricity at Work Regulations 1989*', (ISBN 0 7176 1602 9). The Memorandum includes reference to the poster (referred to at E3002) and recommends that the poster is displayed in areas where workers are at a greater than average risk of electric shock, including supply industries, electricity generation, transmission and utilisation and also companies carrying out electrical testing.

There are numerous HSE (and also DTI) guidance notes and publications and also British Standards documents which are of relevance to working with electricity. These documents should be considered when undertaking any work relating to electrical systems depending on for example, the nature of the work being undertaken and the system or component being worked on. A number of these guidance notes are referred to below however when undertaking any work impacting on electrical systems care should be taken to check that all relevant (and current) information is considered with respect to the particular risks associated with the work to be undertaken.

Safe system of work – general obligations

E3009 There is an overriding need to provide and maintain a proper safe system of work in connection with work on electrical systems, irrespective of whether they are alive or have been made dead. [*Electricity at Work Regulations 1989* (*SI 1989 No 635*), *Regs 4, 13, 14*].

The HSE's leaflet, '*Electrical safety and you*', (IND(G)23(L)), is especially aimed at small firms and is helpful in describing the main hazards. The leaflet gives simple guidance on risk assessment and outlines the basic measures common to all industries required to assist in controlling the risks associated with using electricity at work. It also directs readers to more specific guidance produced by the HSE and other organisations.

In terms of one of the associated obligations, employers should use safety signs where there is a significant risk to health and safety that has not been avoided or controlled. This frequently involves the use of the 'Danger electricity' warning sign. These requirements are specified in the *Health and Safety* (*Safety Signs and Signals*) *Regulations* (*SI 1996 No 341*).

Design, construction and maintenance of electrical systems – general requirement

E3010 All systems must be constructed and maintained so as to prevent danger, so far as is reasonably practicable. Construction includes design of the system, selection of equipment used in it and installation.

Although the Regulations are user-orientated and do not impose duties on manufacturers and designers of electrical equipment, the *Provision and Use of Work Equipment Regulations 1998* (*SI 1998 No 2306*) provide that every employer must ensure that work equipment is so constructed or adapted as to be suitable for the purpose for which it is used or provided. (Obligations would also arise under the *HSWA 1974*.) This means that the equipment supplied must be suitable for the work to be undertaken and that therefore any electrical use associated with the equipment must also be suitable for the purpose for which the equipment is supplied. Generally any system complying with the current *IEE Wiring Regulations* and the relevant HSE guidance will go a long way to satisfying the *Electricity at Work Regulations 1989* (*SI 1989 No 635*). In general, the relevant HSE guidance may include '*Safety in electrical testing at work*' (INDG354) (2002) (ISBN 0717622967), and '*Electrical switchgear and safety: A concise guide for users*' (INDG372) (2003) (ISBN 0717621871).

Portable electrical equipment

E3011 Since many accidents occur when portable tools are being used, the HSE has emphasised three stages of inspection and testing.

The first stage is a frequent visual inspection by the user which includes checking that the cable sheath is not damaged; the plug is not damaged; that there are no inadequate joints in the cable; that the sheath of the cable is securely attached to the plug and equipment on entry to both; that the equipment has not been used for work for which it is not suited, causing it, for example, to become wet or contaminated; that there is no damage to the external casing of the equipment and that there is no evidence of overheating or burns.

The second stage involves a more formal regular visual inspection by a competent person and might include the checking of connections within the plug and equipment; that the correct fuse is being used in the plug, and that there is no indication of any overheating or burning.

The third stage comprises a regular inspection and testing of the equipment by a competent person.

There are two levels of competency: (i) where the person is not skilled in electrical work and uses a simple pass/fail type of portable appliance tester (PAT); and (ii) where more sophisticated electrical skills are used and the readings on the instruments used need interpretation. (See HSE publication '*Maintaining portable and transportable electrical equipment*', (HSG107) (1994) and '*Maintaining portable electrical equipment in offices and other low-risk environments*', (INDG236 11/99 C250) (1996).)

Strength and capability of electrical equipment

E3012 No electrical equipment should be used where the strength and capability of the particular piece of equipment may be exceeded in such a way as may give rise to danger. [*Electricity at Work Regulations 1989* (*SI 1989 No 635*), *Reg 5*]. This is an absolute requirement and therefore must be complied with, irrespective of whether risk or injury is foreseeable. *Regulation 29* of the 1989 Regulations provides a defence if it can be shown that reasonable steps were taken and all due diligence was observed to avoid a breach. Before electrical equipment is put into use, it must therefore be properly selected and adequately rated for the work to be carried out.

In terms of more general duties, the *Electrical Equipment* (*Safety*) *Regulations* (*SI 1994 No 3260* as amended) relate to laws concerning certain electrical equipment designed for use within certain voltage limits. The purpose behind the standard requirements (which arise from an EC Directive) is essentially directed to consumer protection. The Regulations require that electrical equipment is safe and that it is constructed in accordance with good engineering practice with the affixing of CE marking to electrical equipment and a written declaration of conformity being provided. Under the Regulations, the HSE may make arrangements for the enforcement of these Regulations in relation to equipment for use in the workplace under the *HSWA 1974*. It is a defence to proceedings brought under these Regulations to show that their requirements were satisfied in relation to the matter at hand. The DTI has issued two publications of relevance to this area which can be found on the DTI website at: www.dti.gov.uk and which are called '*Compliance cost assessment: the Electrical Equipment* (*Safety*) *Regulations 1994*' and '*Electrical equipment:* (*implementing the low voltage directive*)*: guidance notes on UK regulations*'.

Siting of equipment in adverse or hazardous environments

E3013 The *Electricity at Work Regulations 1989* (*SI 1989 No 635*), *Reg 6* provide that where it is reasonably foreseeable that electrical equipment is going to be exposed to:

- mechanical damage;
- the effects of weather, natural hazards, temperature or pressure;
- the effects of wet, dirty, dusty or corrosive conditions; or
- any flammable or explosive substances, including dusts, vapours and gases,

the equipment must be so constructed or protected so as to prevent, so far as reasonably practicable, danger from exposure of the equipment to the compromising situation.

This is aimed at conditions both indoors and outdoors, and includes the weather-proofing of switchboards housing electrical equipment.

Insulation, protection and placement of conductors

E3014 The *Electricity at Work Regulations 1989* (*SI 1989 No 635*), *Reg 7* provide that all conductors in a system giving rise to danger must either:

- be suitably covered with insulating material and protected, so far as is reasonably practicable to prevent danger; or
- have such precautions taken as will prevent danger, including being suitably placed.

The purpose of this requirement is to prevent danger from conductors in a system that can give rise to danger, such as an electric shock, by resorting to permanently safeguarding the live conductors. Where it is not possible to insulate fully, such as an electric overhead travelling crane, this requirement can be satisfied by live conductors being out of reach and therefore in a safe position. If the conductors intermittently come within reach, perhaps when a ladder is used, then a safe system of work should be used to limit or control such access e.g. a permit to work system may be of value in such situations.

Earthing or other suitable precautions

E3015 Precautions must be taken, either by earthing or other suitable means, to prevent danger arising when it is reasonably foreseeable that any conductor (other than a circuit conductor) which may become charged, does become charged. This might occur when a system is misused or there is a fault in the system. The *Electricity at Work Regulations 1989* (*SI 1989 No 635*), *Reg 29* provide a defence if it can be shown that reasonable steps have been taken and all due diligence observed to avoid a breach.

The usual precautionary measures include earthing any conductive parts that can be touched, i.e. the connection of such parts to the earth. This will include metal-cased equipment that can be touched. One way of earthing such equipment, along with pipes for water, gas or oil, is to connect all the equipment together. Such cross-bonding or equipotential bonding avoids the risk of dangerous voltages running through different exposed metal items. To be effective, the bonding conductors must be capable of carrying any fault current for the time it flows and the time will be dependent on the fuse or other protective system used. Another way of reducing the risks of danger when using electrical equipment is to use a residual current device (RCD) designed to operate rapidly if a small leakage current flows.

Reducing voltage will reduce the risk from a shock, e.g. 110-volt centre-tapped transformers are frequently used to do this on construction sites. Reduced voltage is particularly appropriate when working inside metal containers, such as boilers, with portable tools.

Removing the path to earth and working in an earth-free area, is another suitable precautionary measure. This approach means that even if the electricity source is earth-referenced, there can be no current path back to earth from the earth-free area and therefore no shock through an individual by a current to earth. This type of system is often used for testing electrical equipment.

Integrity of referenced connectors

E3016 If a circuit conductor is connected to earth or to any other reference point, nothing which might reasonably be expected to give rise to danger, by breaking the electrical continuity or introducing high impedance, must be placed in that conductor unless suitable precautions are taken to prevent that danger. [*Electricity at Work Regulations 1989* (*SI 1989 No 635*), *Regs 9* and *10*].

This requirement is especially important in the case of three-phase supplies, where the neutral conductor is connected to earth at source in the distribution system, so that phase voltages are not adversely affected by unbalanced loading. This does not mean that certain electrical devices, like joints or bolted links, cannot be connected in referential circuit conductors as long as suitable precautions have been taken to ensure that no danger is caused from their use or from their installation or removal. Fuses, thyristors, transistors and the like must not be installed in this way as they could give rise to danger if they become open circuit.

Every joint or connection in a system must be both mechanically and electrically suitable for use. [*Electricity at Work Regulations 1989* (*SI 1989 No 635*), *Reg 10*]. This requirement includes the connections to plugs, sockets and other means of joining or connecting conductors, whether these connections are permanent or temporary. *Regulation 29* of the 1989 Regulations provides a defence if it can be shown that reasonable steps have been taken and all due diligence observed to avoid a breach.

Excess current protection

E3017 There is also a requirement that efficient measures must be provided for protecting every part of a system from excess of current as may be necessary to prevent danger. [*Electricity at Work Regulations 1989* (*SI 1989 No 635*), *Reg 11*]. This duty is absolute, but the defence under *Regulation 29* is available. The provision recognises that faults may occur in electrical systems and requires that protective devices, such as fuses or circuit breakers, are installed to ensure that all parts of an electrical system are safeguarded from the consequences of fault conditions.

The main fault conditions are (i) overloads, (ii) short circuits, and (iii) earth faults. In all cases the protective device aims to detect the abnormal current flowing and then to interrupt the fault current before the danger causes damage or injury. The *IEE Wiring Regulations* give detailed guidance on selection and rating of protective devices.

Cutting off supply and isolation of electrical equipment

E3018 Suitable means must exist for cutting off the electrical supply to any electrical equipment and for the isolation of any electrical equipment. [*Electricity at Work Regulations 1989* (*SI 1989 No 635*), *Reg 12*]. This will include means of identifying circuits.

Isolation means the disconnection and separation of the electrical equipment from every source of electrical energy in such a way that this disconnection and separation is secure.

Precautions for work on equipment made dead

E3019 Adequate precautions must be taken in respect of electrical equipment which has been made dead in order to ensure that, while work is being carried out on or near that equipment, there is no danger of the equipment becoming electrically charged during the work.

Several accidents occur each year because of work on a de-energised system which inadvertently is still live or becomes live whilst the work is being carried out. A safe system of work must therefore be used. This can include the following:

- isolation from all points of supply;
- securing each point of isolation, for example by locking off;
- earthing the equipment that is being worked upon;
- testing and thereby verifying that the equipment is dead before working on it;

- creating a safe working zone only accessible to authorised persons;
- safeguarding from other live conductors in proximity, for example by screening; and
- issuing a permit to work.

Work on or near live conductors

E3020 The *Electricity at Work Regulations 1989* (*SI 1989 No 635*), *Reg 14* provide that no person must carry out work on or so near to any live conductor (other than one suitably covered with insulating material to prevent danger) that danger may arise, unless:

- it is unreasonable for it to be dead; and
- it is reasonable for him to be at work on or near it, while it is live; and
- suitable precautions, including provision of suitable protective equipment, are taken to prevent injury.

There are limited circumstances where live working is permitted, such as where it is not practicable to carry out work with the equipment dead, for example during testing; or where making the equipment dead might endanger other users of the equipment. This requirement imposes an absolute duty not to work on live electrical equipment unless the circumstances justify it. Such circumstances will need to be well documented together with the measures and precautions to be taken to prevent injury during the work. A written company policy specifying the criteria for live working and the precautions to be taken should be maintained.

Precautions

E3021 Live work should only be done by competent employees (see E3023 below) who are in possession of adequate information and experience about the nature of the work and the system. Appropriate insulated tools, equipment and protective clothing, e.g. rubber gloves or rubber mats, should be used as well as screens. Such work should be done with another competent person present if this would minimise the risk of injury. In addition, access to the work area should be restricted and earth-free work areas established.

Working space, access and lighting

E3022 To prevent injury, adequate working space, adequate means of access and adequate lighting must be provided at all electrical equipment on which or near which work is being done in circumstances that might give rise to danger. [*Electricity at Work Regulations 1989* (*SI 1989 No 635*), *Reg 15*]. (For access and lighting provisions, see ACCESS, TRAFFIC ROUTES AND VEHICLES and LIGHTING respectively.)

Competent person

E3023 No person shall be engaged in any work activity where technical knowledge or experience is necessary to prevent danger or injury, unless he possesses such knowledge or experience, or is under such degree of supervision, as may be appropriate having regard to the nature of the work.

Any supervision on electrical work, particularly live electrical work, must be by a suitably competent person.

Employees should retain records and information relating to the selection and appointment of people used to undertake work on electrical systems. This could include experience statements, references and qualification details.

The Provision and Use of Work Equipment Regulations 1998

E3024 Electricity is not specifically mentioned in the *Provision and Use of Work Equipment Regulations* (*SI 1998 No 2306*), but *Regulation 4* provides that 'work equipment shall be so constructed or adapted as to be suitable for the purpose for which it is used or provided'. Consideration should be given to these Regulations and the '*Safe use of work equipment: Provision and Use of Work Equipment Regulations 1988*' (Approved Code of Practice and Guidance L22 (1988)). See also M1004 in MACHINERY SAFETY.)

Use of electrical equipment in (potentially) explosive atmospheres

E3025 The *Equipment and Protective Systems Intended for Use in Potentially Explosive Atmospheres Regulations 1996* (*SI 1996 No 192*) now govern the use of electrical and mechanical equipment for use in potentially explosive atmospheres (e.g. whether above or below ground or on offshore installations) and the relevant devices and components. The 1996 Regulations revoked the *Electrical Equipment for Explosive Atmospheres* (*Certification*) *Regulations 1990* (*SI 1990 No 13* as amended) from 1 July 2003 – the date which reflected the transition period for the relevant EU Directive.

Certain electrical equipment and systems are exempt, e.g. medical devices, equipment for domestic use and personal protective equipment. [*SI 1996 No 192, Sch 5*]. The Regulations also do not apply to equipment and protective systems where the explosion results exclusively from the presence of explosives substances (in this regard see the *Dangerous and Explosives Atmospheres Regulations 2002* (*SI 2002 No 2766*) and H2144 in HAZARDOUS SUBSTANCES IN THE WORKPLACE).

Manufacturers of electrical and mechanical equipment, components systems and devices, are under a duty to ensure that all such equipment etc that is put onto the market or into service complies with the necessary health and safety requirements (e.g. relating to potential ignition sources/hazards arising from external effects) [*SI 1996 No 192, Sch 3*], and the appropriate conformity assessment procedures which include the manufacturer providing a declaration of conformity [*SI 1996 No 192, Regs 6,8* and *10*]. Suppliers of such equipment etc. must also see that it is safe to put into circulation, though not in the case of products put into circulation before 1 March 1996 or previously supplied within the EU. [*SI 1996 No 192, Reg 7*]. There is also a general duty imposed on a 'responsible person' (which includes manufacturers and their representatives) to provide a written attestation indicating that the equipment conforms to the relevant requirements, how the equipment is best incorporated into other systems and that it has been issued by the manufacturer or its authorised representative. [*SI 1996 No 192, Reg 8*].

The conformity assessment procedures apply to the particular equipment group and the category for the particular equipment. These groups and categories are set out in the schedules to the Regulations and largely reflect the relevant EU Directive. Conformity assessment procedures (for which fees are payable) are to be determined by notified bodies, with personnel appointed, if necessary, by the Secretary of State. [*SI 1996 No 192, Regs 11* and *13*]. The DTI has issued guidelines for organisations seeking notified body status with regard to undertaking the testing, inspection and certification of relevant equipment and systems for use in potentially explosive atmospheres (see DTI publication URN 99/583 and '*Equipment and protective systems intended for use in potentially explosive atmospheres*' (February 2002) (second edition) (URN 02/6091)).

As mentioned above, the *Electrical Equipment for Explosive Atmospheres (Certification) Regulations 1990* were revoked and the requirements relating to certification bodies were effectively replaced with the system set out in the Regulations.

Where an enforcement officer has reasonable grounds for suspecting that the CE marking is not properly affixed, a notice requiring certain information from the responsible person must be served prior to prosecution occurring. Breach of the Regulations is an offence, leading to a maximum period of imprisonment or a fine. [*SI 1996 No 192, Regs 15, 16* and *17*]. The Regulations provide that it is a defence that all due diligence was taken and that the defendant took all reasonable steps to avoid the commission of the offence. [*SI 1996 No 192, Reg 18*].

Product liability

E3026 Electricity is a product for the purposes of the *Consumer Protection Act 1987*. Consequently, where a defect in an electrical installation or system results in an injury, damage and/or death, liability is strict (see PRODUCT SAFETY).

Non-statutory standards and codes

E3027 The Institution of Electrical Engineers ('IEE') has been producing the IEE Wiring Regulations since 1882 – they are now in their sixteenth edition (which has been subject to amendment in 2001). The 16th edition includes requirements for design, installation, inspection, testing and maintenance of electrical installations in or about buildings generally. Although they have no statutory force, they provide a good indication of standard industrial practice for the purposes of the *Electricity Regulations 1989*.

The National Inspection Council for Electrical Installations Contracting ('NICEIC') enrols contractors whose work is of an approved standard. NICEIC surveys work to check that it complies with IEE Wiring Regulations.

The British Standards Institution ('BSI'), has issued many British Standards and codes of practice for electrical equipment and practice. Such standards and codes are subject to revision and supplementation and it will always be important to ensure that the most current standard is being considered. The standards range from insulation, earthing terminals and electrical connections of small equipment, to a complex set of precautions and specialised electrical equipment, based on the IEC concepts of flame-proofing, intrinsic safety and other types of protection for electrical equipment in flammable atmospheres.

A few of the BSI codes relate to earthing, street lighting, electrical equipment for industrial use and for office machines and the distribution of electricity on construction and building sites.

The British Approvals Service for Electrical Equipment in Flammable Atmospheres ('BASEEFA'), linked with the HSE, is the official UK body for testing and certificating electrical apparatus for use in hazardous atmospheres, to IEC standards as accepted by CENELEC and BSI.

The British Electrical and Allied Manufacturers' Association ('BEAMA'), issues a specialised range of standards and codes drawn up in consultation with users and others.

The British Electrical Approvals Board for Household Equipment ('BEAB'), gives its seal of approval to such domestic type equipment as satisfies design and safety standards.

The HSE has published a guidance note, PM 82, '*The selection, installation and maintenance of electrical equipment for use in and around buildings containing explosives*', to assist those responsible for design selection, installation, operation and maintenance of electrical equipment (including mobile mechanical handling equipment) used at premises where explosives are manufactured, handled or stored. It aims to prevent fires and explosions due to electrical causes. The main topics covered by the guidance note include:

- site supplies;
- area and building zoning and categorisation;
- selection and siting of equipment;
- lighting protection;
- radio frequency ignition hazards;
- portable equipment;
- fork-lift trucks;
- maintenance and testing of equipment and systems.

The important point to note is that when working with electricity consideration must be given to the relevant standards and guidance notes and standards in order to ensure that the most up-to-date information is utilised when assessing and managing the risks involved with working with electricity.

Electricity (Standards of Performance) Regulations 1993

E3028 The *Electricity* (*Standards of Performance*) *Regulations 1993* (*SI 1993 No 1193*) were amended a number of times and revoked by the *Electricity* (*Standards of Performance*) *Regulations 2001* (*SI 2001 No 3265*) which provide for, amongst other things, compensation to be payable by electricity suppliers for interruptions to supply and breach of any other performance standards. There are a number of requirements for information on each public electricity supplier's performance standards and for each of them to establish a complaints procedure.

Electricity Safety, Quality and Continuity Regulations 2002

E3029 In 2001, 10 people died and nearly 350 were injured after coming into contact with electricity supply networks. To help tackle this issue, the *Electricity Safety, Quality and Continuity Regulations 2002* (*SI 2002 No 2665*) came into force on 31 January 2003.

The rules define the safety responsibilities of generators, distributors, suppliers and meter operators, and require electricity companies to:

- warn the public of the dangers which can arise from activities carried out near overhead electric lines; and
- classify the risk of interference, vandalism, or unauthorised access for each of their substations and overhead line circuits. Information relevant to public safety will need to be maintained in a risk register.

In addition:

- all poles supporting bare-low voltage conductors must have safety signs;
- all newly installed underground cables, whether low or high voltage, must be installed in ducts or overlaid with warning tiles or tapes (currently this requirement only covers high voltage cables); and
- certain safety and protective equipment dating back to old networks will be phased out.

Conclusion

E3030 As indicated by the above discussion, there are a wide range of provisions, both general and specific, that relate to working with electricity and electrical installations and the use of electrical plant and apparatus. There are certain absolute duties which should be borne in mind when working with electricity, e.g. live working. Further, as noted there are particular situations which are recognised as giving rise to an increased risk such as overhead cables and/or construction work and in these situations, it would be prudent to ensure appropriate measures are taken to manage the risks.

Employers' Liability Insurance

Introduction

E13001 Most employers carrying on business in Great Britain have for 30 years been under a statutory duty to take out insurance against claims for injuries/diseases brought against them by all but a few employees. The minimum cover must be £5m (see E13011 below) and it is a criminal offence to fail to have such insurance (see E13014 below). A prudent employer would also be well advised to review all forms of business insurance cover and have insurance in place to cover excluded employees (see E13010 below), claims by employees not related to injury or disease and public liability insurance against claims by the public and other workers. In *Mattis v Pollock (t/a Flamingo's Nightclub) [2003] EWCA Civ 887* an employer was held liable to a customer when his nightclub bouncer returned from home to exact revenge on a customer with a knife! Prior to the statutory requirement most responsible employers were already insuring against these risks and generally, until recent years, there have been no significant problems, other than the insolvency of a major employers' liability insurer (see E13024 below).

A significant problem has developed because of major increases in premiums during recent years. The principal cause of this has been insurers seeking to recover from offering premiums at, what turned out to be, well below the cost of claims over a considerable number of years. In 2002 the amount paid in claims was 117 per cent relative to premium income. The outcry from employers forced the Government to investigate. It produced its First Stage Report on 3 June 2003 and its Second Stage Report (which is also the final report) (ref ELCI 2) on 3 December 2003. It identified employers' liability insurance as being important for three reasons:

- to provide an incentive to reduce accidents at work and improve health and safety;
- to properly compensate employees who are injured through their employer's negligence; and
- to protect businesses by spreading the risk of workplace accidents and which might otherwise ruin them financially.

The first reason is perhaps a slightly surprising one when from the report it is clear that premiums are not sufficiently risk based. Taken with the third reason the premium of the safety conscious employer will still be influenced by the uncaring or negligent employer. This is one of the downsides of any form of compulsory insurance. One of the positive outcomes has been a British Insurance Brokers Association (BIBA) voluntary code of practice on renewal notifications effective from 1 November 2003. Amongst other matters the report promises greater levels of enforcement.

(Readers are advised to refer to the following useful websites for further advice/information: www.dwp.gov.uk; www.abi.org.uk; www.fsa.gov.uk; www.businesslink.gov.uk and www.hse.gov.uk/consult/condocs/strategycd.pdf)

When insurance is obtained a certificate is issued. The employer must keep a copy prominently displayed in the workplace so that employees can see it. It is a criminal offence to fail to take out such insurance and/or to fail to display a certificate (see E13013–E13015 below); however, such a failure does not give rise to any civil liability on the part of a company director in England and Wales. In *Richardson v Pitt-Stanley [1995] 1 All ER 460* the plaintiff suffered a serious injury to his hand in an accident at work. He obtained judgment against his employer, a limited liability company, for breach of the *Factories Act 1961, s 14(1)* (failure to fence dangerous parts of machinery). Before damages were assessed, the company went into liquidation and there were no assets remaining to satisfy the plaintiff's judgment. The company had also failed to insure against liability for injury sustained by employees in the course of their employment, as required by *section 1* of the *Employers' Liability (Compulsory Insurance) Act 1969* ('the 1969 Act'). *Section 5* of the 1969 Act makes such a failure a criminal offence. The plaintiff then sued the directors and secretary of the company who, he alleged, had committed an offence under *section 5* of the 1969 Act, claiming as damages a sum equal to the sum which he would have recovered against the company, had it been properly insured. His action failed. It was held that the 1969 Act did not create a civil as well as criminal liability. In Scotland however the Sheriff Principal in *Quinn v McGinty 1999 SLT (Sh Ct) 27*, on similar facts did not follow the Court of Appeal in *Richardson v Pitt-Stanley*.

The law is contained in the 1969 Act and in the *Employers' Liability (Compulsory Insurance) Regulations 1998 (SI 1998 No 2573)*. In addition, the requirements of the 1969 Act extend to offshore installations but do not extend to injuries suffered by employees when carried on or in a vehicle, or entering or getting onto or alighting from a vehicle, where such injury is caused by, or arises out of use, by the employer, of a vehicle on the road. [*SI 1998 No 2573, Reg 9 and Sch 2, para 14*]. Such employees would normally be covered under the *Road Traffic Act 1988, s 145* as amended by the *Motor Vehicles (Compulsory Insurance) Regulations 1992 (SI 1992 No 3036)*. As from 1 July 1994, liability for injury to an employee whilst in a motor vehicle has been that of the employer's motor insurers.

Employees suffering from industrial diseases, which often take years to become apparent, can track down their employer's insurers thanks to a voluntary Code of Practice launched in November 1999. The Code was drawn up by the Department of the Environment, Transport and the Regions, The Association of British Insurers, and the Non-Marine Association at Lloyds, and sets out the procedures and standards of service required of insurers. The Code does not have statutory authority and is a voluntary code that commits member insurers to make a thorough search when requested of all records which exist. The Code obliges insurers to keep records of all policies issued for 60 years. There is no charge levied for conducting a search. For the period 1 November 1999 to 31 October 2000, 1,062 enquiries were circulated to its member insurers with a 24.6 per cent success rate. Copies of the Code and enquiry forms can be obtained from the Association of British Insurers, Employers Liability Enquiry Unit, 51 Gresham Street, London EC2V 7HQ, tel: 020 7216 7456; fax: 020 7367 8612.

Purpose of compulsory employers' liability insurance

E13002 The purpose of compulsory employers' liability insurance is to ensure that employers are covered for any legal liability to pay damages to employees who suffer bodily injury and/or disease during the course of employment and as a result of employment. It is the liability of the employer towards his employees which has to be covered; there is no question of compulsory insurance extending to employees, since employers are under no statutory or common law duty to insure employees against risk of injury, or even to advise on the desirability of insurance; it is their potential legal liability to employees which must be insured against (see E13008 below). Such liability is normally based on negligence, though not necessarily personal negligence on the part of the employer. Moreover, case law suggests that employers' liability is becoming stricter. The rule that employers must 'take their victims as they find them' underlines the need for long-tail cover because the employer may find himself liable for injuries/diseases which 'trigger off' or exacerbate existing conditions.

An employers' liability policy is a legal liability policy. Hence, if there is no legal liability on the part of an employer, no insurance moneys will be paid out. Moreover, if the employee's action against the employer cannot succeed, the action for damages cannot be brought against an employer's insurer (*Bradley v Eagle Star Insurance Co Ltd [1989] 1 All ER 961* where the employer company had been wound up and dissolved before the employer's liability to the injured employee had been established) (see below for the transfer of an employer's indemnity policy to an employee). The effect of this decision has been reversed by the *Companies Act 1989, s 141*, amending the *Companies Act 1985* (*CA 1985*), *s 651* which allows the revival of a dissolved company within two years of its dissolution for the purpose of legal claims and, in personal injuries cases, the revival can take place at any time subject to the existing limitation of action rules contained in the *Limitation Act 1980*. For example, in the case of *Re Workvale Ltd (No 2) [1992] 2 All ER 627*, the court exercised its discretion under *section 33* of the *Limitation Act 1980* to allow a personal injuries claim to proceed after the three-year limitation period had expired. This meant that the company could also be revived under the provisions of *CA 1985, s 651*(*5*) and (*6*) (as amended). Thus, proceedings under the *Third Parties* (*Rights against Insurers*) *Act 1930, s 1*(*1*)(*b*) may be brought in this manner.

If the employer becomes bankrupt or if a company becomes insolvent, the employer's right to an indemnity from his insurers is transferred to the employee who may then keep the sums recovered with priority to his employer's creditors. This is only so if the employer has made his claim to this indemnity by trial, arbitration or agreement before he is made bankrupt or insolvent (*Bradley v Eagle Star Insurance Co Ltd [1989] 1 All ER 961*). The employee must also claim within the statutory limitation period from the date of his injury (see E13007 below for subrogation rights generally).

The policy protects an employer from third party claims; an employee as such is not covered since he normally incurs no liability. Although offering wide cover an employers' liability policy does not give cover to third party non-employees (e.g. independent contractors and members of the public). Such liability is covered by a public liability policy which, though advisable, is not compulsory.

This section examines:

- the general law relating to contracts of insurance (see E13003–E13006 below);
- the insurer's right of recovery (i.e. subrogation) (see E13007 below);
- the duty to take out employers' liability insurance (see E13008–E13012 below);
- issue and display of certificates of insurance (see E13013 below);
- penalties (see E13014, E13015 below);
- scope and cover of policy (see E13016–E13019 below);
- 'prohibition' of certain terms (see E13020 below);
- trade endorsement for certain types of work (see E13021, E13022 below).
- insolvent employers' liability insurers (see E13024 below).

General law relating to insurance contracts

E13003 Insurance is a contract. When a person wishes to insure, for example, himself, his house, his liability towards his employees, valuable personal property or even loss of profits, he (the proposer) fills in a proposal form for insurance, at the same time making certain facts known to the insurer about what is to be insured. On the basis of the information disclosed in the proposal form, the insurer will decide whether to accept the risk or at what rate to fix the premium. If the insurer elects to accept the risk, a contract of insurance is then drawn up in the form of an insurance policy. (Incidentally, it seems to matter little whether the negotiations leading up to contract took place between the insured (proposer) and the insurance company or between the insured and a broker, since the broker is often regarded as the agent of one or the other, generally of the proposer (*Newsholme Brothers v Road Transport & General Insurance Co Ltd [1929] 2 KB 356)*.) However, a lot depends on the facts. If he is authorised to complete blank proposal forms, he may well be the agent of the insurer.

Extent of duty of disclosure

E13004 A proposer must disclose to the insurer all material facts within his actual knowledge. This does not extend to disclosure of facts which he could not reasonably be expected to know. 'The duty is a duty to disclose, and you cannot disclose what you do not know. The obligation to disclose, therefore, necessarily depends on the knowledge you possess. This, however, must not be misunderstood. The proposer's opinion of the materiality of that knowledge is of no moment. If a reasonable man would have recognised that the knowledge in question was material to disclose, it is no excuse that you did not recognise it. But the question always is – Was the knowledge you possessed such that you ought to have disclosed it?' (*Joel v Law Union and Crown Insurance Co [1908] 2 KB 863* per Fletcher Moulton LJ).

The knowledge of those who represent the directing mind and will of a company and who control what it does, e.g. directors and officers, is likely to be identified as the company's knowledge whether or not those individuals are responsible for arranging the insurance cover in question (*PCW Syndicates v PCW Reinsurers [1996] 1 Lloyd's Rep 241*).

An element of consumer protection, in favour of insureds, was introduced into insurance contracts by the Statement of General Insurance Practice 1986, a form of self-regulation applicable to many but not to all insurers. This has consequences for the duty of disclosure, proposal forms (E13005 below), renewals and claims. In particular, with regard to the last element (claims), an insurer should not refuse to indemnify on the grounds of:

(*a*) non-disclosure of a material fact which a policyholder could not reasonably be expected to have disclosed; or

(*b*) misrepresentation (unless it is a deliberate non-disclosure of, or negligence regarding a material fact). Innocent misrepresentation is not a ground for avoidance of payment.

The trend towards greater consumer protection in (*inter alia*) insurance contracts is reflected in the *Unfair Terms in Consumer Contracts Regulations 1999* (*SI 1999 No 2083*). These Regulations apply in the case of 'standard form' (or non-individually negotiated) contracts. [*SI 1999 No 2083, Reg 5*]. A contractual term in such a contract shall:

(i) be regarded as 'unfair' if contrary to the requirement of 'good faith' it 'causes a significant imbalance in the parties' rights and obligations arising under the contract, to the insured's detriment' [*SI 1999 No 2083, Reg 5(1), Sch 2*];

(ii) always be regarded as having been individually negotiated where it has been drafted in advance and the consumer has not been able to influence the substance of the term [*SI 1999 No 2083, Reg 5(2)*];

(iii) require a written contract term to be expressed in plain, intelligible language. If there is doubt about the meaning of terminology, a construction in favour of the insured will prevail [*SI 1999 No 2083, Reg 7*]; and

(iv) where the insurer claims that a term was individually negotiated, he must prove it [*SI 1999 No 2083, Reg 5(4)*].

Complaints (other than ones which are frivolous or vexatious) or which are to be handled by a qualified body [*SI 1999 No 2083, Sch 1*] relating to 'unfair terms' in standard form contracts, are considered by the Director General of Fair Trading, who may prevent their continued use. [*SI 1999 No 2083, Reg 12*]. See also P9056 PRODUCT SAFETY.

Filling in proposal form

E13005 Generally only failure to make disclosure of relevant facts will allow an insurer subsequently to invalidate the policy and refuse to compensate for the loss. The test of whether a fact was or was not relevant is whether its omission would have influenced a prudent insurer in deciding whether to accept the risk, or at what rate to fix the premium.

The arm of *uberrima fides* (i.e. the utmost good faith) is a long one. If, when filling in a proposal form, a statement made by the proposer is at that time true, but is false in relation to other facts which are not stated, or becomes false before issue of the insurance policy, this entitles the insurer to refuse to indemnify. In *Condogianis v Guardian Assurance Co Ltd [1921] 2 AC 125* a proposal form for fire cover contained the following question: 'Has proponent ever been a claimant on a fire insurance company in respect of the property now proposed, or any other property? If so, state when and name of company'. The proposer answered 'Yes', '1917', 'Ocean'. This answer was literally true, since he had claimed against the Ocean Insurance Co in respect of a burning car. However, he had failed to say that in 1912 he had made another claim against another insurance company in respect of another burning car. It was held that the answer was not a true one and the policy was, therefore, invalidated.

Loss mitigation

E13006 There is an implied term in most insurance contracts that the insured will take all reasonable steps to mitigate loss caused by one or more of the insured perils. Thus, in the case of burglary cover of commercial premises, this could extend to provision of security patrols, the fitting of burglar alarm devices and guard dogs. In the case of employers' liability, it will extend to appointment or use of services of an accredited safety officer and/or occupational hygienist, either permanently or temporarily, particularly in light off the *Management of Health and Safety at Work Regulations 1999* (*SI 1999 No 3242*), to oversee, for example, application of the *COSHH Regulations 1999* (*SI 1999 No 437*). Again, in the case of fire cover, steps to mitigate the extent of the loss on the part of the insured, might well extend to regular visits by the local fire authority and/or advice on storage of products and materials by reputable risk management consultants. Indeed, it is compliance with this implied duty in insurance contracts that accounts for the growth of the practice of risk management, and good housekeeping on the part of more and more companies.

Subrogation

E13007 Subrogation enables an insurer to make certain that the insured recovers no more than exact replacement of loss (i.e. indemnity). 'It [the doctrine of subrogation] was introduced in favour of the underwriters, in order to prevent their having to pay more than a full indemnity, not on the ground that the underwriters were sureties, for they are not so always, although their rights are sometimes similar to those of sureties, but in order to prevent the assured recovering more than a full indemnity' (*Castellain v Preston (1883) 11 QBD 380 per Brett LJ*). Subrogation does not extend to accident insurance moneys, whereby the insured (normally self-employed) is promised a fixed sum in the event of injury or illness (*Bradburn v Great Western Railway Co (1874) LR 10 Exch 1* where the appellant was injured whilst travelling on a train, owing to the negligence of the respondent. He had earlier bought personal accident insurance to cover him for the possibility of injury on the train. It was held that he was entitled to both damages for negligence *and* insurance moneys payable under the policy (see further COMPENSATION FOR WORK INJURIES/ DISEASES)). The right of subrogation does not arise until the insurer has paid the insured in respect of his loss, and has been invoked infrequently in employers' liability cases. In *Morris v Ford Motor Co Ltd [1973] 2 All ER 1084* the Ford Motor Co had subcontracted cleaning at one of their plants to the X company, for which the appellant worked. Whilst engaged on this work at the plant, the appellant was injured owing to the negligence of an employee whilst driving a forklift truck. The appellant claimed damages from the respondent company for the negligence of their employee, on the grounds of vicarious liability. X company had, however, entered into a contract of indemnity with the respondent company, agreeing to indemnify the company for all losses or claims for injury arising out of the cleaning operations. Although accepting that they were bound by the terms of this contract of indemnity, the X company argued that they should be subrogated against the negligent Ford employee, on the ground that the employee had carried out his work negligently. It was held that the agreement by the British Insurance Association that they would not sue an employee of an insured employer in respect of injury caused to a co-employee, unless there was either (*a*) collusion and/or (*b*) wilful misconduct on the part of the employee, was binding and that the X company could not recoup its loss from the negligent employee.

Duty of employer to take out and maintain insurance

E13008 'Every employer carrying on business in Great Britain shall insure, and maintain insurance against liability for bodily injury or disease sustained by his employees, and arising out of and in the course of their employment in Great Britain in that business.' [*Employers' Liability* (*Compulsory Insurance*) *Act 1969, s 1*(*1*)].

Such insurance must be provided under one or more 'approved policies'. An 'approved policy' is a policy of insurance not subject to any conditions or exceptions prohibited by regulations (see E13020 below). [*Employers' Liability* (*Compulsory Insurance*) *Act 1969, s 1*(*3*)]. This now includes insurance with an approved EU insurer. [*Insurance Companies* (*Amendment*) *Regulations 1992* (*SI 1992 No 2890*)].

There is no duty under the *1969 Act* to warn or insure the employee against risks of employment outside Great Britain (*Reid v Rush Tompkins Group plc [1989] 3 All ER*

228) although the *1998 Regulations* require the employer to insure employees employed on or from offshore installations or associated structures – see E13009 below.

Employees covered by the Act

E13009 Cover is required in respect of liability to employees who:

(*a*) are ordinarily resident in Great Britain; or

(*b*) though not ordinarily resident in Great Britain, are present in Great Britain in the course of employment here for a continuous period of not less than 14 days; or

(*c*) though not ordinarily resident in the United Kingdom, have been employed on or from an offshore installation or associated structure for a continuous period of not less than seven days.

[*Employers' Liability* (*Compulsory Insurance*) *Regulations 1998* (*SI 1998 No 2573*), *Reg 1*(*2*)].

Employees not covered by the Act

E13010 An employer is not required to insure against liability to an employee who is (*a*) a spouse, (*b*) father, (*c*) mother, (*d*) son, (*e*) daughter, (*f*) other close relative. [*Employers' Liability* (*Compulsory Insurance*) *Act 1969, s 2*(*2*)(*a*)]. Those who are not ordinarily resident in the UK are not covered by the Act except as above. Nor are employees working abroad covered. Such employees can sue under English law in limited circumstances (*Johnson v Coventry Churchill International Ltd [1992] 3 All ER 14* where an employee, working in Germany for an English manpower leasing company, was injured when he fell through a rotten plank. He was unable to sue his employer under German law; although he was working in Germany, it was held that England was the country with the most significant relationship with the claim because he had made the contract in England, his employers had covered him with personal liability insurance and he therefore expected them to compensate him through these insurers for any personal injury sustained in Germany).

Degree of cover necessary

E13011 The amount for which an employer is required to insure and maintain insurance is £5 million in respect of claims relating to any one or more of his employees, arising out of any one occurrence. [*Employers' Liability* (*Compulsory Insurance*) *Regulations 1998* (*SI 1998 No 2573*), *Reg 3*(*1*)].

Between 1 January 1972 (when the *Employer's Liability* (*Compulsory Insurance*) *Act 1969* came into force) and 1994, insurers, in practice, provided unlimited cover under employers' liability policies. As from 1 January 1995, as a result of payments made in respect of claims exceeding the amount of premiums received during the period 1989–1993, unlimited liability was withdrawn, but most insurers continued to offer a minimum of £10 million indemnity for onshore work. A consultative document issued by the Department of the Environment, Transport and the Regions entitled The Draft Employers' Liability (Compulsory Insurance) General Regulations [C4857 September 1997] (hereafter referred to as 'the 1997 consultative document') which preceded the *1998 Regulations* assumed that this practice would continue.

Where a company has subsidiaries, there will be sufficient compliance if a company insures/maintains insurance for itself *and* on behalf of its subsidiaries for £5 million in respect of claims affecting any one or more of its own employees and any one or more employees of its subsidiaries arising out of any one occurrence. [*Employers' Liability* (*Compulsory Insurance*) *Regulations 1998* (*SI 1998 No 2573*), *Reg 3*].

Insurers and the courts have interpreted the legislation to mean that all injuries resulting from one incident (e.g. an explosion) are treated as one occurrence, and each individual case of gradually occurring injury or disease is treated as an individual occurrence – the only exception being a situation where a sudden and immediate outbreak of a disease amongst the workforce is clearly attributable to an identifiable incident (e.g. the escape of a biological agent). The introduction to the 1997 consultative document suggested that this interpretation might be challenged and set out a possible alternative regulation to be used instead of what is now *Reg 3* of the *1998 Regulations* if clarification was felt necessary. This was not adopted and no new regulations are currently proposed and therefore it would appear that the Government are now satisfied that the position is clear.

Exempted employers

E13012 The following employers are exempt from the duty to take out and maintain insurance:

(*a*) nationalised industries;

(*b*) any body holding a Government department certificate that any claim which it cannot pay itself will be paid out of moneys provided by Parliament;

(*c*) any Passenger Transport Executive and its subsidiaries, London Regional Transport and its subsidiaries;

(*d*) statutory water undertakers and certain water boards;

(*e*) the Commission for the New Towns;

(*f*) health service bodies, National Health Service Trusts;

(*g*) probation and after-care committees, magistrates' court committees, and any voluntary management committee of an approved bail or approved probation hostel;

(*h*) governments of foreign states or commonwealth countries and some other specialised employers;

(*j*) Railtrack Group plc and its subsidiaries (the exemption ceasing when it is no longer owned by the Crown);

(*k*) the Qualifications & Curriculum Authority.

There are other types of employer specified in the Regulations, but these are the main exceptions. [*Employers' Liability Compulsory Insurance Act 1969, s 3; Employers' Liability* (*Compulsory Insurance*) *Regulations 1998* (*SI 1998 No 2573*), *Sch 2*].

Issue, display and retention of certificates of insurance

E13013 The insurer must issue the employer with a certificate of insurance, which must be issued not later than 30 days after the date on which insurance was commenced or renewed. [*Employers' Liability* (*Compulsory Insurance*) *Act 1969, s 4*(*1*)*; Employers' Liability* (*Compulsory Insurance*) *Regulations 1998* (*SI 1998 No 2573*), *Reg 4*]. Where there are one or more contracts of insurance which jointly provide insurance cover of not less than £5 million, the certificate issued by any individual insurer must specify both the amount in excess of which insurance cover is provided by the individual policy, and the maximum amount of that cover. [*Employers' Liability* (*Compulsory Insurance*) *Regulations 1998* (*SI 1998 No 2573*), *Reg 4*(*3*)].

A copy or copies of the certificate must be displayed at each place of business where there are any employees entitled to be covered by the insurance policy and the copy certificate(s) must be placed where employees can easily see and read it and be reasonably protected from being defaced or damaged. [*Employers' Liability* (*Compulsory Insurance*) *Regulations 1998* (*SI 1998 No 2573*), *Reg 5*]. The exception is where an employee is employed on or from an offshore installation or associated structure, when the employer must produce, at the request of that employee and within ten days from such request, a copy of the certificate. [*Employers' Liability* (*Compulsory Insurance*) *Regulations 1998* (*SI 1998 No 2573*), *Reg 5*(*4*)].

An employee must, if a notice has been served on him by the Health and Safety Executive, produce a copy of the policy to the officers specified in the notice and he must permit inspection of the policy by an inspector authorised by the Secretary of State to inspect the policy. [*Employers' Liability* (*Compulsory Insurance*) *Regulations 1998* (*SI 1998 No 2573*), *Regs 7, 8*].

A change introduced by the *1998 Regulations* is that employers are now required by law to retain any certificate of employers' liability insurance (or a copy) for a period of 40 years beginning on the date on which the insurance to which it relates commences or is renewed [*Employers' Liability* (*Compulsory Insurance*) *Regulations 1998* (*SI 1998 No 2573*), *Reg 4*(*4*)]. Companies may retain the copy in any eye-readable form in any one of the ways authorised by the *Companies Act 1985, ss 722 and 723* [*Employers' Liability* (*Compulsory Insurance*) *Regulations 1998* (*SI 1998 No 2573*), *Reg 4*(*5*)].

Penalties

Failure to insure or maintain insurance

E13014 Failure by an employer to effect and maintain insurance for any day on which it is required is a criminal offence, carrying a maximum penalty on conviction of £2,500. [*Criminal Justice Act 1982, s 37*(*2*)].

Failure to display a certificate of insurance

E13015 Failure on the part of an employer to display a certificate of insurance in a prominent position in the workplace is a criminal offence, carrying a maximum penalty on conviction of £1,000. [*Criminal Justice Act 1982, s 37*(*2*)].

In the 1997 consultative document, the Government suggests that penalties should be increased to become the same as those under the *Health and Safety at Work etc Act 1974* (*HSWA 1974*), i.e. fines of £20,000 in a magistrates' court and unlimited in the Crown Court. To implement this change will require primary legislation and currently it would appear that the Government has no plans to introduce such legislation.

Cover provided by a typical policy

Persons

E13016 Cover is limited to protection of employees. Independent contractors are not covered; liability to them should be covered by a public liability policy. Directors who are employed under a contract of employment are covered, but directors paid by fees who do not work full-time in the business are generally not regarded as 'employees'. Liability to them would normally be covered by a public liability policy. Similarly, since the judicial tendency is to construe 'labour-only' subcontractors in the construction industry as 'employees' (see CONSTRUCTION AND BUILDING OPERATIONS), employers' liability policies often contain the following endorsement: 'An employee shall also mean any labour master, and persons supplied by him, any person employed by labour-only subcontractors, any self-employed person, or any person hired from any public authority, company, firm or individual, while working for the insured in connection with the business'. The public liability policy should then be amended to exclude the insured's liability to 'employees' so designated.

In the 1997 consultative document, the Government points out that the issue as to what constitutes 'an employee' cannot be completely resolved without primary legislation, but proposes to issue guidance on interpretation, although it was proposed that guidance on interpretation would be issued. Such guidance has not been issued and therefore presumably the Government are now satisfied that the position is clear.

Scope of cover

E13017 The policy provides for payment of:

(*a*) costs and expenses of litigation, incurred with the insurer's consent, in defence of a claim against the insured (i.e. civil liability);

(*b*) solicitor's fees, incurred with the insurer's consent, for representation of the insured at proceedings in any court of summary jurisdiction (e.g. magistrates' court or Crown Court), coroner's inquest, or a fatal accident inquiry (i.e. criminal proceedings), arising out of an accident resulting in injury to an employee. It does *not* cover payment of a fine imposed by a criminal court.

The policy will often contain an excess negotiated between the insurer and employer, i.e. a provision that the employer pay the first £x of any claim. For the purposes of the *1969 Act*, any condition in a contract of insurance which requires a relevant employee to pay, or an insured employer to pay the relevant employee, the first amount of any claim or any aggregation of claims, is prohibited. Agreements will still be permitted which provide that the insurer will pay the claim in full and may then seek some reimbursement from the employer. [*Employers' Liability* (*Compulsory Insurance*) *Regulations 1998* (*SI 1998 No 2573*), *Reg 2*].

Geographical limits

E13018 Cover is normally limited to Great Britain, Northern Ireland, the Channel Islands and the Isle of Man, in respect of employees normally resident in any of the above, who sustain injury whilst working in those areas. Cover is also provided for such employees who are injured whilst temporarily working abroad, so long as the action for damages is brought in a court of law of Great Britain, Northern Ireland, the Channel Islands or the Isle of Man – though even this proviso is omitted from some policies.

Employers must also have employers' liability insurance in respect of employees who, though not ordinarily resident in the United Kingdom, have been employed on or from an offshore installation or associated structure for a continuous period of not less than seven days; or who, though not ordinarily resident in Great Britain, are present in Great Britain in the course of employment for not less than fourteen days. [*Employers' Liability* (*Compulsory Insurance*) *Regulations 1998* (*SI 1998 No 2573*), *Reg 1*(*2*)].

Conditions which must be satisfied

E13019 (*a*) Cover only relates to bodily injury or disease; it does not extend to employee's property. This latter cover is provided by an employers' public liability policy.

(*b*) Injury must arise out of and during the course of employment. If injury does not so arise, cover is normally provided by a public liability policy.

(*c*) Bodily injury must be caused during the period of insurance. Normally with injury-causing accidents there is no problem, since injury follows on from the accident almost immediately. Certain occupational diseases, however, may not manifest themselves until much later, e.g. asbestosis, mesothelioma, pneumoconiosis, deafness. Here legal liability takes place when the disease manifests itself, or is 'discovered'. Moreover, at least as far as occupational deafness is concerned, liability between employers can be apportioned, giving rise to contribution between insurers (see further NOISE AT WORK and VIBRATION).

(*d*) Claims must be notified by the insured to the insurer as soon as possible, or as stipulated by the policy.

The *Employers' Liability* (*Compulsory Insurance*) *Regulations 1998* (*SI 1998 No 2573*), *Reg 2* does not fetter the freedom of underwriters to apply certain conditions in connection with intrinsically hazardous work; for instance, exclusion of liability for accidents arising out of demolition work, or in connection with use of explosives.

'Prohibition' of certain conditions

E13020 All liability policies contain conditions with which the insured must comply if the insurer is to 'progress' his claim, e.g. notification of claims. Failure to comply with such condition(s) could jeopardise cover under the policy: the insured would be legally liable but without insurance protection. In the case of an employers' liability policy, an insurer might seek to avoid liability under the policy if the condition requiring the insured to take reasonable care to prevent injuries to employees, and/or comply with the provisions of any relevant statutes/statutory instruments (e.g. *HSWA 1974; Ionising Radiations Regulations 1999* (*SI 1999 No 3232*)), or to keep records, was not complied with.

The object of the 1969 Act was to ensure that an employer who had a claim brought against him would be able to pay the employee any damages awarded. Regulations made under the Act, therefore, seek to prevent insurers from avoiding their liability by relying on breach of a policy condition, by way of 'prohibiting' certain conditions in policies taken out under the Act. More particularly, insurers cannot avoid liability in the following circumstances:

(*a*) some specified thing being done or being omitted to be done after the happening of the event giving rise to a claim (e.g. omission to notify the insurer of a claim within a stipulated time) [*Employers' Liability* (*Compulsory Insurance*) *Regulations 1998* (*SI 1998 No 2573*), *Reg 2*(*1*)(*a*)];

(*b*) failure on the part of the policy-holder to take reasonable care to protect his employees against the risk of bodily injury or disease in the course of employment [*Employers' Liability* (*Compulsory Insurance*) *Regulations 1998* (*SI 1998 No 2573, Reg 2*(*1*)(*b*)]. As to the meaning of 'reasonable care' or 'reasonable precaution' here, 'It is eminently reasonable for employers to entrust ... tasks to a skilled and trusted foreman on whose competence they have every reason to rely'. (*Woolfall and Rimmer Ltd v Moyle and Another [1941] 3 All ER 304*). The prohibition is therefore, not broken by a negligent act on the part of a competent foreman selected by the employer. Where, however, an employer acted wilfully (in causing injury) and not merely negligently (though this would be rare), the insurer could presumably refuse to pay (*Hartley v Provincial Insurance Co Ltd [1957] Lloyd's Rep 121* where the insured employer had not taken steps to ensure that a stockbar was securely fenced for the purposes of the *Factories Act 1937, s 14(3)* in spite of repeated warnings from the factory inspector, with the result that an employee was scalped whilst working at a lathe. It was held that the insurer was justified in refusing to indemnify the employer who was in breach of statutory duty and so liable for damages). This was confirmed in *Aluminium Wire and Cable Co Ltd v Allstate Insurance Co Ltd [1985] 2 Lloyd's Rep 280*;

(*c*) failure on the part of the policy-holder to comply with statutory requirements for the protection of employees against the risk of injury [*Employers' Liability (Compulsory Insurance) Regulations 1998 (SI 1998 No 2573, Reg 2(1)(c)*] – the reasoning in *Hartley v Provincial Insurance Co Ltd* (see (*b*) above), that wilful breach may not be covered, probably applies here too;

(*d*) failure on the part of the policy-holder to keep specified records and make such information available to the insurer [*Employers' Liability (Compulsory Insurance) Regulations 1998 (SI 1998 No 2573, Reg 2(1)(d)*] (e.g. accident book or accounts relating to employees' wages and salaries (see ACCIDENT REPORTING AND INVESTIGATION));

(*e*) by means of the use of an excess in policies [*Employers' Liability (Compulsory Insurance) Regulations 1998 (SI 1998 No 2573, Reg 2(2)*] – see E13017 above

Trade endorsements for certain types of work

E13021 There are no policy exceptions to the standard employers' liability cover. Trade endorsements, however, are used frequently in underwriting employers' liability risks, and there is nothing in the *1969 Act* to prevent insurers from applying their normal underwriting principles and applying trade endorsements where they consider it necessary, i.e. they will amend their standard policy form to exclude certain risks. Thus, there may be specific exclusions of liability arising out of types of work, such as demolition, or the use of mechanically driven woodworking machinery, or work above certain heights, unless the appropriate rate of premium is paid. This does mean that there are still circumstances where an employee will not obtain compensation from his employer based on the employer's insurance cover.

Measure of risk and assessment of premium

E13022 Certain trades or businesses are known to be more dangerous than others. For most trades or businesses insurers have their own rate for the risk, expressed as a rate per cent on wages (other than for clerical, managerial or non-manual employees for whom a very low rate applies). This rate is used as a guide and is altered upwards or downwards depending upon:

(*a*) previous history of claims and cost of settlement;

(*b*) size of wage roll;

(*c*) whether certain risks are not to be covered, e.g. the premium will be lower if the insured elects to exclude from the policy certain risks, such as the use of power driven woodworking machinery;

(*d*) the insured's attitude towards safety.

Many insurers survey premises with the object of improving the risk and minimising the incidence of accidents and diseases. This is an essential part of their service, and they often work in conjunction with the insured's own safety staff.

Extension of cover

E13023 In addition to employers' liability insurance, it is becoming increasingly common for companies to buy insurance in respect of directors' personal liability. Indeed, in the United States, some directors refuse to take up appointments in the absence of such insurance being forthcoming.

Insolvent employers' liability insurers

E13024 The problems that arise following the insolvency of employer's liability insurer gained particular prominence following the much publicised liquidation of Chester Street Holdings Limited ('Chester Street') on 9 January 2001. It was initially feared that individuals whose claims pre-dated 1972 (the date when employers' liability insurance became compulsory) would be at risk of not receiving compensation. The situation has, however, largely been clarified following the establishment of the Financial Services Compensation Scheme ('the FSCS') which was created under the *Financial Services and Markets Act 2000* and acts as a 'safety net' for customers of finance sector companies who are unable to play claims against authorised companies. The FSCS came into effect on 1 December 2001. Under the FSCS policy holders (save for any period of cover when the policy holder was a nationalised industry which is specifically excluded from the provisions of the scheme) are eligible for protection if they are insured by an authorised insurance company under a contract of insurance issued in the UK, Channel Islands or Isle of Man. The FSCS scheme pays 100 per cent compensation for post-1972 compulsory employers' liability insurance claims. If the employer still exists and is solvent then the employer initially pays the compensation to the claimant although is able to recoup 100 per cent of the compensation outlay from the FSCS. The FSCS pays 90 per cent compensation for pre-1972 claims providing that the employer is not still in existence and solvent. If however the employer is still in existence – the employer must meet its liability to the claimant and is not compensated by the FSCS.

Special provisions apply to Chester Street for pre-1972 claims where:

- liability was established and quantified before insolvency on 9 January 2001, 90 per cent compensation is payable. Such claims are administered under the *Financial Services and Markets Act 2002* using the guidelines established by the Policyholders Protection Board;
- liability was agreed and quantified between 9 January and 30 November 2001. Such claims are subject to the requirements of a scheme set up by the Association of British Insurers who pay 90 per cent of the award.

Responsibility for claims handled by the Policyholder Protection Board (PPB) were transferred to the FSCS on 1 December 2001. PPB rules continue to apply to any claim arising from the insolvency of an insurance company where it occurred before 1 December 2001.

Further information can be obtained from FSCS, 7th Floor, Lloyds Chambers, 1 Portsoken Street, London E1 8BN, tel: 020 7892 7300.

Employment Protection

Introduction

E14001 The nature of employment protection is continually changing. At a domestic (national) level, the last decade has seen the extension of certain employment protection rights to workers as well as employees. For example, both the *Public Interest Disclosure Act 1998* and the *Working Time Regulations 1998* (*SI 1998 No 1833*) apply to 'workers' which is a wider category than persons categorised as employees under traditional English employment law analysis.

A number of new individual rights have been also introduced including the right to be accompanied at a disciplinary or grievance hearing and the right to request flexible working. Collective rights have also been enhanced with trade unions being given the right to statutory recognition in certain circumstances.

Europe has also continued to have a significant impact on the regulation of employment relations. Directives issued by Europe have required the introduction of legislation prohibiting discrimination on the grounds of sexual orientation and religion or belief, legislation requiring employers with more than 50 employees to set up national information and consultation procedures if requested by employees and will, in the future, require the prohibition of age discrimination. Further changes are also in the pipeline, perhaps the most important of which is the requirement for employers and employees to follow statutory dismissal, disciplinary and grievance procedures which will come into force on 1 October 2004.

The relationship between an employer and an employee is a contractual one and as such must have all the elements of a legally binding contract to render it enforceable. In strict contractual terms an offer is made by the employer which is then accepted by the employee. As in the case of the offer, the acceptance may be oral, in writing, or by conduct, for example by the employee turning up for work. The consideration on the employer's part is the promise to pay wages and on the employee's part to provide his services for the employer. Once the employer's offer has been accepted, the contract comes into existence and both parties are bound by any terms contained within it (*Taylor v Furness, Withy & Co Ltd (1969) 6 KIR 488*).

The contractual analysis of the employment relationship is not entirely satisfactory in explaining the relationship between employee and employer. To fit the contract model, various elements comprising the reality of the employment relationship become part of the contract by implication.

An employment contract is unlike many other contracts, because many of the terms will not have been individually negotiated by the parties. The contract will contain the express terms that the parties have agreed – most commonly hours, pay, job description – and there will be a variety of other terms which will be implied into the contract from other sources and which the parties have not agreed. Many of these are relevant to health and safety. If any of the express or implied terms in the contract are breached, the innocent party will have certain remedies. The fact that various employee rights, particularly in relation to health and safety, are implied into the contractual terms and conditions is important for this reason.

In addition to terms implied into the contract by the common law, statute (such as the *Employment Rights Act 1996*) has created additional employment protection rights for employees, including some specific rights in relation to health and safety. These are in addition to detailed rights and duties arising from health and safety legislation and regulations which are discussed elsewhere. An employer will often lay down health and safety rules and procedures. While the law allows an employer the ultimate sanction of dismissal as a method of ensuring that safety rules are observed, such dismissals should be lawful, that is generally with notice, and should be fair. Furthermore, statute has created specific protection from victimisation for employees who are protecting themselves or others against perceived health and safety risks. All of these provisions are the subject of this section.

As a specific health and safety protection measure, the *Health and Safety at Work etc Act 1974, s 2* (*HSWA 1974*) lays a general duty on all employers to ensure, so far as is reasonably practicable (for the meaning of this expression, see E15039 ENFORCEMENT), the health, safety and welfare of all their employees. An employer is also under a duty to consult about health and safety matters. In addition, a number of codes of practice have been issued under the *HSWA 1974* by the Health and Safety Executive, for example relating to safety representatives and allowing them time off to train.

Sources of contractual terms

Express terms

E14002 These are the terms agreed by the parties themselves and may be oral or in writing. Normally the courts will uphold the express terms in the contract because these are what the parties have agreed. However, if the term is ambiguous the court may be called upon to interpret the ambiguity, for example what the parties meant by 'reasonable overtime'.

Generally the express terms cause no legal problems and the parties can insert such terms into the contract as they wish. There are, however, a number of restrictions which include:

(*a*) An employer cannot restrict his liability for the death or personal injury of his employees caused by his negligence. Further, he can only restrict liability for damage to his employee's property if such a restriction is reasonable (*Unfair Contract Terms Act 1977, s 2*).

(*b*) The terms in the contract cannot infringe discrimination legislation including:

- the *Equal Pay Act 1970* and the *Sex Discrimination Acts 1975* and *1986*;
- the *Race Relations Act 1976*;
- the *Disability Discrimination Act 1995*;
- the *Employment Equality* (*Sexual Orientation*) *Regulations 2003* (*SI 2003 No 1661*);
- the *Employment Equality* (*Religion or Belief*) *Regulations 2003* (*SI 2003 No 1660*).

(*c*) The terms in the contract cannot infringe the *Part-time Workers (Prevention of Less Favourable Treatment) Regulations 2000* (*SI 2000 No 1551*). The Regulations provide that unless justified on objective grounds, a part-time worker has the right not to be treated less favourably than a comparable full-time worker on the ground that the worker is a part-timer in relation to the terms of the contract.

(*d*) The terms in the contract cannot infringe the *Fixed-term Workers (Prevention of Less Favourable Treatment) Regulations 2002* (*SI 2002 No 2034*). The Regulations provide that unless justified on objective grounds, a fixed-term worker has the right not to be treated less favourably than a comparable full-time worker on the ground that the worker is a fixed-term worker.

(*e*) The employer cannot have a notice provision which gives the employee less than the statutory minimum notice guaranteed by the *Employment Rights Act 1996, s 86*.

(*f*) Some judges have suggested that any express terms regarding hours are subject to the employer's duty to ensure his employee's safety and must be read subject to this, so that a term requiring an employee to work 100 hours a week will not be enforceable (see for example Stuart-Smith LJ in *Johnstone v Bloomsbury Health Authority [1991] IRLR 118*). More specifically, the provisions of the *Working Time Regulations 1998* (*SI 1998 No 1833*) affect the contractual term in relation to working hours. The 1998 Regulations, save in the case of specified exemptions, set a maximum working week of 48 hours averaged over a 17 week reference period. In addition, the Regulations provide for an obligatory daily rest period, weekly rest, rest breaks, limits on night work and minimum annual leave and otherwise regulate working time. In *Barber v RJB Mining UK Ltd [1999] IRLR 308*, the court decided that the maximum imposed on weekly working time by the Regulations was part of the employees' contract. This decision gives some protection to employees who refuse to work beyond the statutorily stated maximum. It is possible under the 1998 Regulations for a worker to agree with his or her employer in writing to opt-out of the 48 hour working week, subject to complying with certain requirements.

(*g*) The terms in the contract cannot infringe the *Maternity and Parental Leave etc Regulations 1999* (*SI 1999 No 3312*), which entitle employees with one year's continuous service who have responsibility for a child born after 15 December 1999, to be absent from work for up to 13 weeks' parental leave.

(*h*) Confidentiality provisions should be subject to an employee's right to make a protected disclosure in accordance with the provisions set out in the *Employment Rights Act 1996*.

Whilst the *Sex Discrimination Acts 1975* and *1986* removed some of the restrictions on women and their employment generally and health and safety specifically, some restrictions/prohibitions on certain types of employment by women, in the interests of health and safety at work, still remain.

Under the *Control of Lead at Work Regulations 2002* (*SI 2002 No 2676*) an employer is prohibited from employing women of reproductive capacity or young people in particular activities relating to lead processes as follows:

(*a*) In the lead smelting and refining process:

- (i) handling, treating, sintering, smelting or refining any material containing 5 per cent or more of lead; or
- (ii) cleaning where any of the above activities have taken place.

(*b*) In the lead acid manufacturing process:

(i) manipulating lead oxides;

(ii) mixing or pasting;

(iii) melting or casting;

(iv) trimming, abrading or cutting of pasted plates; or

(v) cleaning where any of the above activities have taken place.

Common law implied duties – all contracts

E14003 Both the employer and employee owe duties towards each other. These are duties implied into every contract of employment and should be distinguished from the implied terms discussed below which are implied into a particular individual contract. Although there are a number of different duties, three are of major importance in relation to health and safety:

(*a*) the duty on the part of the employee to obey lawful reasonable orders and to perform his work with reasonable care and skill; and

(*b*) the duty on the part of the employer to ensure his employee's safety.

The duty to obey lawful reasonable orders ensures that the employer's safety rules can be enforced and, as it is a contractual duty, breach will allow the employer to invoke certain sanctions against the employee, the ultimate of which may be dismissal.

The same is true of the duty to perform his work with reasonable care and skill. Should the employee be in breach of this duty and place his or others' safety at risk, the employer may impose sanctions against him including dismissal.

The imposition of the employer's duty is to complement the statutory provisions. Statutes such as the *HSWA 1974* provide sanctions against the employer should he fail to comply with the legislation or any regulations made thereunder. The common law duty provides the employee with a remedy should the duty be broken, either in the form of compensation if he is injured, or, potentially, with a claim of unfair dismissal. The employer's duty to ensure his employees' safety is one of the most important aspects of the employment relationship. At least one judge has argued that it is so important that any express term must be read subject to it (see *Johnstone v Bloomsbury Health Authority [1991] IRLR 118* at E14002 above). Breach of this duty can lead to the employee resigning and claiming constructive dismissal (see below). In *Walton & Morse v Dorrington [1997] IRLR 488*, an employee claimed that she had been constructively dismissed (unfairly) because her employer had breached the implied term of her contract of employment that it would provide, so far as reasonably practicable, a suitable working environment. The employee had been forced to work in a smoke-filled environment for a prolonged period of time and her employer did not take appropriate steps to redress the problem when she raised the issue. The Employment Appeal Tribunal agreed that she had been constructively dismissed because the employer had breached its duty to her. Further, the employer's duty to ensure his employee's safety has been held to cover stressful environments resulting in injury to the employee. In *Walker v Northumberland County Council [1995] IRLR 35*, the High Court held that an employer was liable for damages on the basis that they owed their employee a duty not to cause him psychiatric damage by the volume and/or character of work that he was required to undertake. In *Fraser v The State Hospitals Board for Scotland (11 January 2000) (2000 Rep LR 94)*, the Court of Session held that there was no reason to qualify an employer's duty to take reasonable care for the safety of employees so as to restrict the nature of the injury suffered to a physical one in circumstances where the employee claimed damages for psychological damage as a result of disciplinary measures. The claim failed on the basis of lack of foreseeability.

Common law implied terms – individual contracts

E14004 The court may imply terms into the contract when a situation arises which was not anticipated by the parties at the time they negotiated the express terms. As such, the court is 'filling in the gaps' left by the parties' own negotiations. The courts use two tests to see if a term should be implied, (i) the 'business efficacy' test (*The Moorcock (1889) 14 PD 64*) or (ii) the 'officious bystander' or 'oh of course' test (*Shirlaw v Southern Foundries Ltd [1939] 2 KB 206*). Once the court has decided, by virtue of one of these tests, that a term should be implied, it will use the concept of reasonableness to decide the content of the term. Often this will involve looking at how the parties have worked the contract in the past. For example, if the contract does not contain a mobility clause, but the employee has always worked on different sites, the court will normally imply a mobility clause into the contract (*Courtaulds Northern Spinning Ltd v Sibson [1988] IRLR 305*). (See also *Aparau v Iceland Frozen Foods plc [1996] IRLR 119*, where the Employment Appeal Tribunal refused to imply a mobility clause on the basis that there were other ways of achieving the necessary flexibility.) In relation to dismissal, a term has been implied that, except in the case of summary dismissal, the employer will not terminate the employment contract while the employee is incapacitated where the effect would be to deprive the employee of permanent health insurance benefits (*Aspden v Webbs Poultry and Meat Group (Holdings) Ltd [1996] IRLR 521)*. Terms can also be implied by the conduct of the parties or by custom and practice in a particular industry or area. The test for this is relatively difficult to fulfil – the term must be notorious and certain and, in effect, everyone in the industry/enterprise must know that it is part of the contract. Arguments based on custom and practice come into play in relation to issues such as statutory holidays and redundancy policies.

Collective agreements

E14005 Collective agreements are negotiated between an employer or employer's association and a trade union or unions. This means that they are not contracts between an employer and his individual employees because the employee was not one of the negotiating parties. Some terms of the collective agreement will be procedural and will govern the relationship between the employer and the union; some, on the other hand, will impact on the relationship between the employer and each individual employee, for example a collectively bargained pay increase. As the employee is not a party to the collective agreement, the only way he can enforce a term which is relevant to him is if the particular term from the collective agreement has become a term of his individual employment contract. Procedural provisions, policy and more general aspirations are not suitable for incorporation into an individual contract of employment. Incorporation is important because the collective agreement is not a legally binding contract between the employer and the union (*Trade Union and Labour Relations* (*Consolidation*) *Act 1992, s 179*(*1*)) and thus needs to be a term of an employment contract to make it legally enforceable.

The two main ways that a term from a collective agreement becomes a term of an employment contract is by express or implied incorporation. Until recently, implied incorporation was the most common and was complex. It generally required the employee to be a member of the union which negotiated the agreement, to have knowledge of the agreement and of the existence of the term, and to have conducted himself in such a way as to indicate that he accepted the term from the collective agreement as a term of his contract. A decision of the Employment Appeal Tribunal case, *Healy & Others v Corporation of London (24 June 1999) (unreported)*, illustrates that habitual acceptance of the benefits of a collective agreement does not, in itself, lead to the conclusion that the terms of that collective agreement have become contractually binding on an individual employee. There can be many reasons for an individual to accept the benefits of collective bargaining which do not amount to an acceptance that the underlying agreement forms part of his or her contract.

Express incorporation meant that the employee had expressly agreed (normally in his contract) that any term collectively agreed would become part of his contract. This used to be unusual, but with the change made to the statutory statement which must be given to all employees (see below) employees must be told of collective agreements which apply to them, and this has been held as expressly incorporating those agreements into the contract.

Statutory statement of terms and conditions

E14006 By the *Employment Rights Act 1996, s 1* every employee no later than two months after starting employment, must receive a statement of his basic terms and conditions. The statement must contain:

(*a*) the names of the employer and employee;

(*b*) the date the employment began;

(*c*) the date the employee's continuous employment began;

(*d*) the scale or rate of remuneration and how it is calculated;

(*e*) the intervals when remuneration is paid;

(*f*) terms and conditions relating to hours;

(*g*) terms and conditions relating to holidays;

(*h*) terms relating to sick pay (if any);

(*i*) terms and conditions relating to pensions;

(*j*) notice requirements;

(*k*) job description;

(*l*) title of the job;

(*m*) if the job is not permanent, the period of employment;

(*n*) place of work, or if various the address of the employer;

(*o*) any collective agreements which affect terms and conditions and, if the employer is not a party to the agreements, the persons with whom they were made;

(*p*) if the employee is required to work outside the UK for more than one month, the period he will be required to work, the currency in which he will be paid, any additional benefits paid to him and any terms and conditions relating to his return to the UK.

The terms in (*a*), (*b*), (*c*), (*d*), (*e*), (*f*), (*g*), (*k*), (*l*) and (*n*) must all be contained in a single document. In relation to pensions and sick pay the employer may refer the employee to a reasonably accessible document, and in respect of notice the employer can refer the employee to a reasonably accessible collective agreement or to the *Employment Rights Act 1996, s 86* which contains provisions relating to minimum notice periods.

In addition, if the employer employs more than twenty employees, he must give them written details of any disciplinary and grievance procedures which apply to them. If the employer employs fewer than twenty employees, he must let them know to which person they can take a grievance to– there is no requirement for him to give details of the disciplinary procedures. This exemption for small employers is, however, to be removed from 1 October 2004.

There is no duty on an employer to give details of any disciplinary or grievance procedures relating to health and safety. Given the employer's duties under the *HSWA 1974, s 2*, however, and given the law relating to unfair dismissal, it is good industrial relations practice to ensure that all employees know of all the disciplinary procedures which could be invoked against them. The written statement must also provide (either by instalments or in one single document) details of whether the employment is contracted out of the State Second Pension.

Works rules

E14007 Works rules may or may not be part of the contract. If they are part of the contract and thus contractual terms, they can be altered only by mutual agreement, that is the employee must agree to any change. It is unusual, however, for such rules to be contractual – to be so, there would have to be some reference to them within the contract and an intention that they are terms of the contract. The more usual position with regard to the employer's rules was stated in *Secretary of State for Employment v ASLEF (No 2) [1972] 2 QB 455* where Lord Denning said that they were merely instructions from an employer to an employee. This means that they are non-contractual and the employer can alter the rules without the consent of the employees. The fact that they are not contractual does not mean that they cannot be enforced against an employee. All employees have a duty to obey lawful, reasonable orders (see E14003 above) and thus failing to comply with the rules will be a breach of this duty and therefore a breach of contract. The only requirement that the law stipulates is that the order must be lawful and reasonable and it is unlikely that an order to comply with any health and safety rules would infringe these requirements.

Disciplinary and grievance procedures

E14008 It has already been noted that the employer must give details of grievance procedures to all employees. Failing to do so could lead to the employee resigning and claiming constructive dismissal (*W A Goold (Pearmak) Ltd v McConnell [1995] IRLR 516*). In addition, if the employer employs more than 20 employees he must give details of the disciplinary grievance procedures to those employees. This requirement will also apply to those employers with less than 20 employees from 1 October 2004. Many employers adopt the ACAS Code of Practice on Disciplinary Procedures. This gives guidelines as to the sanctions which can be employed for breaches of the employer's rules. The Code incorporates the statutory right to be accompanied by a fellow employee or by a trade union representative during certain grievance and disciplinary proceedings under the *Employment Relations Act 1999, ss 10–15.* This right is enforceable in the employment tribunal and compensation is payable for any failure. It is worth noting that fellow employees are under no duty to perform the role of accompanying individual.

Subject to the above, employers can currently establish their own disciplinary procedures. Such procedures may become part of the contract. If, for example, the employer gives the employee a copy of the procedures with the contract, and the contract refers to the procedures and the employee signs for receipt of the contract and the procedures, it is likely that they will be contractual. Employers may prefer their disciplinary procedures not to be contractual. If the procedures are contractual, any employee will be able to claim that his or her contract has been breached if they are not followed. This possibility also applies to those employees who have been employed for less than the one year qualifying period required to bring a claim for unfair dismissal. In response to a claim of breach of contract, a court may award damages against the employer. These damages are based on an assessment of the time for which, if the procedure had been followed, the employee's employment would have continued.

From 1 October 2004, where an employer wishes to dismiss an employee for any reason or take action (other than suspension or the issue of warnings) on grounds of capability or conduct, the employer must follow a statutory dismissal and disciplinary procedure (DDP). A modified DDP applies to immediate 'gross misconduct' dismissals. If an employer fails to follow the statutory DDP, a tribunal will be able to increase the compensation awarded to an employee by up to 50 per cent, although this cannot take an award above the statutory cap (see below at E14022). It will be automatically unfair for an employer to dismiss an employee without completing the relevant DDP. Employees will receive a minimum of four-weeks' pay as compensation.

From 1 October 2004, an employee or ex-employee who has a grievance which could form the basis of a tribunal complaint will not be able to bring a tribunal claim based on that grievance unless they have raised the grievance with their employer under the statutory grievance procedure (GP). Having raised a grievance under the statutory GP, an employee must wait 28 days before lodging a tribunal application otherwise the application will be refused. If an employee attempts to lodge a tribunal claim without having raised the matter as a grievance and waited 28 days, the tribunal claim will be rejected but the employee will be automatically granted an extension of three months which is added to the original time limit of three months.

If an employee raises a grievance under the statutory GP and the employer fails to follow the statutory GP:

(*a*) the admissibility criteria will not apply; and

(*b*) the tribunal will be able to increase any award it makes by up to 50 per cent.

The principles underlying both the statutory disciplinary and grievance procedures are:

- There should be a letter setting out the problem.
- There should be time provided for the recipient to consider and understand it.
- The employer and employee should meet to discuss the problem. The employee has a right to be accompanied at this meeting.
- The employer's decision should be notified to the employee.
- An appeal should be held if the issue has not been settled.

ACAS has also issued a draft Revised Code of Practice on Disciplinary and Grievance Procedures which take into account the new statutory rules. The revised draft Code of Practice will also come into effect on 1 October 2004.

Employee employment protection rights

Right not to suffer a detriment in health and safety cases

E14009 By the *Employment Rights Act 1996, s 44* every employee has the right not to be subjected to a detriment, by any act or any failure to act, by his employer on the grounds that:

(*a*) having been designated by the employer to carry out activities in connection with preventing or reducing risks to health and safety at work, the employee carried out (or proposed to carry out) any such activities;

(*b*) being a representative of workers on matters of health and safety at work or a member of a safety committee –

(i) in accordance with arrangements established under or by virtue of any enactment; or

(ii) by reason of being acknowledged as such by the employer;

the employee performed (or proposed to perform) any functions as such a representative or a member of such committee;

(*ba*) the employee took part (or proposed to take part) in consultation with the employer pursuant to the *Health and Safety* (*Consultation with Employees*) *Regulations 1996* (*SI 1996 No 1513*) or in an election of representatives of employee safety within the meaning of those Regulations (whether as a candidate or otherwise);

(*c*) being an employee at a place where –

(i) there was no such representative or safety committee; or

(ii) there was such a representative or safety committee but it was not reasonably practicable for the employee to raise the matter by those means;

he brought to his employer's attention, by reasonable means, circumstances connected with his work which he reasonably believed were harmful or potentially harmful to health or safety;

(*d*) in circumstances of danger which the employee reasonably believed to be serious and imminent and which he could not reasonably have been expected to avert, he left (or proposed to leave) or (while the danger persisted) refused to return to his place of work or any dangerous part of his place of work; or

(*e*) in circumstances of danger which the employee reasonably believed to be serious and imminent, he took (or proposed to take) appropriate steps to protect himself or other persons from the danger.

In considering whether the steps the employee took or proposed to take under (*e*) were reasonable, the court must have regard to all the circumstances including the employee's knowledge and the facilities and advice available to him (*Employment Rights Act 1996, s 44*(*2*)). In *Kerr v Nathan's Wastesavers Ltd (1995) IDS Brief 548*, however, the Employment Appeal Tribunal stressed that tribunals should not place too onerous a duty on the employee to make enquiries to determine if his belief is reasonable. Danger under (*d*) does not necessarily have to arise from the circumstances of the workplace, but can include the risk of attack by a fellow employee (*Harvest Press Ltd v McCaffrey [1999] IRLR 778*). Danger under (*e*) can include danger to others as well as the employee himself (*Mosiak v City Restaurants (UK) Ltd [1999] IRLR 180*). Various actions by the employer can constitute a detriment to the employee (such as disciplining the employee). Likewise, a failure to act on the part of the employer can also constitute a detriment (for example not sending the employee on a training course). Furthermore, the section is not restricted to the health and safety of the employee or his colleagues. In *Barton v Wandsworth Council (1995) IDS Brief 549* a tribunal ruled that an employee had been unlawfully disciplined when he voiced concerns over the safety of patients due to what he considered to be the lack of ability of newly introduced escorts. This shows that the legal protection is triggered in relation to any health and safety issue and includes cases where the employee voices concerns, and is not limited to only those circumstances where the employee commits more positive action. The *Employment Rights Act 1996, s 44(3)*, however, provides that an employee is not to be regarded as subjected to a detriment if the employer can show that the steps the employee took or proposed to take were so negligent that any reasonable employer would have treated him in the same manner. Furthermore, if the detriment suffered by the employee is dismissal, there is special protection under *s 100* (see below).

If the employee should suffer a detriment within the terms of *s 44* he may present a complaint to an employment tribunal (*Employment Rights Act 1996, s 48*). The complaint must be presented within three months of the act (or failure to act) complained of, or, if there is a series of acts, within three months of the date of the last act. The tribunal has a discretion to waive this time limit if it was not reasonably practicable for the employee to present his complaint in time. If the tribunal finds the complaint well founded, it must make a declaration to that effect and may make an award of compensation to the employee, the amount of compensation being what the tribunal regards as just and equitable in all the circumstances (*Employment Rights Act 1996, s 49(2)*). The amount of compensation shall take into account any expenses reasonably incurred by the employee in consequence of the employer's action and any loss of benefit caused by the employer's action. Compensation can be reduced because of the employee's contributory conduct.

The protection from being dismissed or subjected to a detriment on the health and safety grounds specified in *s 44* and *s 100* of the *Employment Rights Act 1996* has been reinforced by a new, more general protection for whistleblowers.

The Public Interest Disclosure Act 1998 provides protection to workers who make disqualifying disclosures about health and safety matters, as well as about criminal acts, failure to comply with legal obligations, miscarriages of justice, damage to the environment and deliberate concealment of any of these matters. Under the Act, which inserts new sections into the *Employment Rights Act 1996*, a worker who makes a 'qualifying disclosure' which he reasonably believes shows one of these matters, may be protected against dismissal or being subjected to a detriment.

Disclosures are only protected if they are made to appropriate persons – which often means that the employer must to be approached in the first instance. There are other possibilities available under the Act: disclosure to a legal adviser, disclosure to a prescribed person (e.g. the FSA or the Commissioners of the Inland Revenue), and a more general category for disclosures made provided that certain specified conditions are met.

In terms of health and safety risks, protection for qualifying disclosures is not limited to cases of imminent or serious danger – it can apply where the health and safety of any individual has been, is being or is likely to be, endangered. In all cases the worker must have a reasonable belief and make the disclosure in good faith (except in the case of disclosure to a legal adviser).

If a disclosure is protected, and an employee is subjected to any detriment or dismissed as a result, it is unlawful. A dismissal in these circumstances is deemed to be automatically unfair and the tribunal is not required to consider whether or not the employer's actions were reasonable. There is no minimum qualifying period for entitlement to make an unfair dismissal claim for this reason, no upper age limit applies and there is no limit on the compensation available to whistleblowers who are unfairly dismissed because they have made a protected disclosure.

Dismissal on health and safety grounds

E14010 In addition to the normal protection against dismissal (see below), where an employee is dismissed (or selected for redundancy) and the reason or principal reason for the dismissal is one of the grounds listed in the *Employment Rights Act 1996, s 44*, the dismissal will be automatically unfair. The only defence available to an employer applies to a dismissal taken by the employee to protect himself or others from danger that the employee reasonably believed was serious and imminent (*Employment Rights Act 1996, ss 44*(*1*)(*e*) and *100*(*1*)(*e*)). The employer can escape a finding of unfair dismissal if he can show that the actions taken or proposed by the employee were so negligent that any reasonable employer would have dismissed the individual concerned. In respect of a dismissal falling within *s 100*, the normal qualifying period of employment does not apply nor does the upper age limit (*Employment Rights Act 1996, ss 108*(*3*)(*c*) and *109*(*2*)(*c*)). Thus an employee who has only been employed for a few weeks or who is over the normal retirement age for the job can claim unfair dismissal for a breach of *s 100*. A dismissal which is not automatically unfair under *s 100* may nevertheless be unfair under the general reasonableness test under *s 98*.

There is no limit on the amount of compensation that may be awarded for an unfair dismissal on health and safety grounds.

Dismissal for assertion of a statutory right

E14011 By the *Employment Rights Act 1996, s 104*, an employee is deemed to be unfairly dismissed where the reason or principal reason for that dismissal was that the employee –

(*a*) brought proceedings against an employer to enforce a right of his which is a relevant statutory right, or

(*b*) alleged that the employer had infringed a right of his which is a relevant statutory right.

It is immaterial whether or not the employee has the right or whether or not the right has been infringed, as long as the employee made it clear to the employer what the right claimed to have been infringed was and the employee's claim is made in good faith. A statutory right for the purposes of the section is any right under the *Employment Rights Act 1996* in respect of which remedy for infringement is by way of complaint to an employment tribunal, a right under *s 86* of the 1996 Act (minimum notice requirements), rights in relation to trade union activities under the *Trade Union and Labour Relations* (*Consolidation*) *Act 1992* and the rights conferred by the *Working Time Regulations 1998* (*SI 1998 No 1833*).

This is an important right for employees. If, for example, after the employee has successfully claimed compensation from his employer for a breach of s *44* he is dismissed, the dismissal will be automatically unfair under *s 104*. Again, if the employer unlawfully demotes or suspends without pay as a disciplinary sanction for breach of health and safety rules, and after proceedings against him for an unlawful deduction from wages the employer dismisses the employee, this will be unfair under *s 104*. As with dismissal in health and safety cases under *s 100*, the normal qualifying period of employment does not apply – neither does the upper age limit.

Enforcement of safety rules by the employer

The rules

E14012 Given the statutory duty on the employer, under the *HSWA 1974, s 2*, to have a written statement of health and safety policy, and the common law duty on the employer to ensure his employees' safety, the employer should lay down contractual health and safety rules, breach of which will lead to disciplinary action against the employee. These rules must be communicated to the employee and be clear and unambiguous so that the employee knows exactly what he can and cannot do.

The employer's disciplinary rules will often classify misconduct, e.g. as minor misconduct, serious misconduct and gross misconduct. It is unlikely that a tribunal would uphold as fair a dismissal for minor misconduct. It will underline the importance of health and safety rules if their breach is deemed to be serious or gross misconduct. The tribunal will, however, look at all the circumstances of the case – it does not automatically follow, therefore, if an employer has stated that a breach of a particular rule will be gross misconduct, that a tribunal will find a resultant dismissal fair.

The procedures

E14013 Once an employer has laid down his rules, he must ensure that he has adequate procedures (which should be non-contractual) to deal with a breach. The procedures used by an employer are scrutinised by a tribunal in any unfair dismissal claim and past cases indicate that many employers have lost such claims due to inadequate procedures. As discussed below, an employer in an unfair dismissal claim must show the tribunal that he acted reasonably. This concentrates on the fairness of the employer's actions and not on the fairness to the individual employee (*Polkey v A E Dayton Services Ltd [1987] IRLR 503*). This means that an employer cannot argue that a breach of procedures has made no difference to the final outcome and that he would have dismissed the employee even if he had adhered to his procedure. Breach of procedures themselves by an employer is likely to render a dismissal unfair regardless of which rule was broken.

Many employers adopt the ACAS procedures. Essentially any disciplinary procedure should contain three elements: an investigation, a hearing and an appeal and should observe the principles of natural justice.

(a) Investigation

The law requires that the employer has a genuine belief in the employee's 'guilt', and that the belief is based on reasonable grounds after a reasonable investigation (*British Home Stores v Burchell [1978] IRLR 379*). If the employer suspends the employee during the investigation, this suspension should be with pay and in accordance with the disciplinary procedure. An investigation is important because it may reveal defects in the training of the employee, or reveal that the employee was not told of the rules, or that another employee was responsible for the breach. In all of these cases, disciplinary action against the suspended employee will be unfair. Any investigation should be as thorough as possible and documented. It should also take place as soon as possible since memories fade quickly and this is particularly important if other employees are to be questioned as witnesses. Likewise, taking too long to start an investigation may lead the employee to think that no action will be taken and to then discipline him may itself be unfair. No disciplinary action should be taken until a careful investigation has been concluded.

(b) Hearing

Once the employer has investigated, he must conduct a hearing to make a decision as to the sanction, if any, he will impose. To act fairly, the employer must comply with the rules of a fair hearing. These are:

(i) The employee must know the case against him to enable him to answer the complaint. This also means that the employee should be given sufficient time before the hearing with copies of relevant documents to enable him to prepare his case.

(ii) The employee should have an opportunity to put his side of the case, i.e. the employer should listen to the employee's side of the story and allow the employee to put forward any mitigating circumstances.

(iii) The employee must be allowed to be accompanied at the hearing by a fellow employee or a trade union representative of his choice (*Employment Relations Act 1999*).

(iv) The hearing should be unbiased, i.e. the person chairing the hearing should come to it with an open mind and not have prejudged the issue.

(v) The employee should be provided with an explanation as to why any sanctions are imposed.

(vi) The employee should be informed of his right to appeal (and the way in which he should go about it) to a higher level of management which has not been involved in the first hearing. If the employee fails to exercise his right of appeal, however, he will not have failed to mitigate his loss, if ultimately a tribunal finds that he has been unfairly dismissed and thus his compensation will not be reduced (*William Muir (Bond 9) Ltd v Lamb [1985] IRLR 95*). Failing to allow an employee to exercise a right of appeal will almost certainly render any dismissal unfair (*West Midlands Co-operative Society Ltd v Tipton [1986] IRLR 112*).

(c) Appeal

In an unfair dismissal case a tribunal is required to consider the reasonableness of the employer's action taking into account the resources of the employer and the size of the employer's undertaking. This means that in the case of all but very small undertakings, the tribunal will expect the employer to have provided an appeal for the employee. All the rules of a fair hearing equally apply to an appeal. Only an appeal which is a complete rehearing of the case (rather than merely a review of the written notes of the disciplinary hearing) can rectify procedural flaws committed earlier on in the procedure (*Jones v Sainsbury's Supermarkets Ltd (2000) (unreported)*. An appeal, however, cannot endorse the sanction imposed by the earlier hearing for a different reason, unless the employee has had notice of the new reason and has been given an opportunity to put forward his argument in respect of it.

With effect from 1 October 2004, employers must also ensure that their procedure complies with the statutory dismissal and disciplinary procedure (see E14008 above).

Sanctions other than dismissal

E14014 There are a variety of sanctions apart from dismissal that an employer may impose. It is important however that the 'punishment fits the crime'. The imposition of too harsh a sanction may entitle the employee to resign and claim constructive dismissal (see below).

(a) Warnings

The ACAS Code recommends three warnings in cases of normal misconduct before dismissing: the first oral, the second written and a final written warning stating that a repetition will result in dismissal. These are only guidelines, however, and it clearly depends on the circumstances of the case. A minor breach of a health and safety rule, for example, may justify a final written warning given the potential seriousness and consequences of breaches of such rules. The ACAS Code urges that, apart from gross misconduct, no employee should be dismissed for a first breach of discipline, although, again, breaches of health and safety rules have been held to be gross misconduct. The Code also recommends that warnings should remain on the employee's record for a definite period of time (six to twelve months). Once this time has expired, the warnings will be ignored when looking to see if the procedure has been followed in later cases of misconduct, but can be considered when the employer is looking at the employee's work record to decide what sanction to impose.

(b) Fines or deductions

The employer must have contractual authority or the written permission of the employee before he can make a deduction from the employee's wages as a disciplinary sanction. Deducting without such authority is a breach of the *Employment Rights Act 1996, s 13* and gives the employee the right to sue for recovery in the employment tribunal. It will also lead to a potential constructive dismissal claim.

(c) Suspension without pay

Any suspension without pay will have the same consequences as a fine or deduction if there is no contractual authority or written authorisation from the employee to impose such a sanction.

(d) Demotion

Most demotions will involve a reduction in pay, and thus without written or contractual authority to demote the employer will be in breach of the *Employment Rights Act 1996, s 13* and liable to a constructive dismissal claim.

Suitable alternative work

E14015 Where an employee is suspended from work on maternity grounds, the employer must offer available suitable alternative work. Alternative work will only be suitable if:

(*a*) the work is of a kind which is both suitable in relation to the employee and appropriate for the employee to do in the circumstances; and

(*b*) the terms and conditions applicable for performing the work are not substantially less favourable than corresponding terms and conditions applicable for performing the employee's usual work (*Employment Rights Act 1996, s 67*).

If an employer fails to offer suitable alternative work, the employee may bring a claim before an employment tribunal which can award 'just and equitable' compensation. Such complaint must normally be lodged within three months of the first day of the suspension (*Employment Rights Act 1996, s 70*(*4*)).

Remuneration on suspension from work

E14016 An employee who is suspended on medical grounds is entitled to normal remuneration for up to 26 weeks. An employee who is suspended on maternity grounds, if no suitable alternative work is available, is entitled to normal remuneration for the duration of the suspension. However, in either case, if the employee unreasonably refuses an offer of suitable alternative work, no remuneration is payable for the period during which the offer applies. An employee may bring a complaint to an employment tribunal if an employer fails to pay the whole or any part of the remuneration to which the employee is entitled (*Employment Rights Act 1996, ss 64, 68, 70*(*1*)). (See *British Airways Ltd v Moore [2000] IRLR 296*, in which the Employment Appeal Tribunal upheld a purser's claim to a flying allowance on the basis that suitable work must be on terms and conditions not substantially less favourable).

Dismissal

E14017 Dismissal is the ultimate sanction that an employer can impose for breach of health and safety rules. All employees are protected against wrongful dismissal at common law, but, in addition, some employees have protection against unfair dismissal. The protection against unfair dismissal comes from statute (the *Employment Rights Act 1996*) and therefore the employee must satisfy any qualifying criteria laid down by the statute before he can claim. Given that the protection against wrongful and unfair dismissal rest alongside each other, an employee may claim for both, although he will not be compensated twice. Wrongful dismissal is based on a breach of contract by the employer and compensation will be in the form of damages for that breach – that is, the damage the employee has suffered because the employer did not comply with the contract. Unfair dismissal, on the other hand, is statute based and is not dependent on a breach of contract by the employer. Compensation for such dismissal is based on a formula within the statute (see E14022 below).

Wrongful dismissal

E14018 A dismissal at common law is where the employer unilaterally terminates the employment relationship with or without notice. A wrongful dismissal is where the employer terminates the contract in breach, for example, by giving no notice or shorter notice than is required by the employee's contract and the employee's conduct does not justify this. An employer is entitled to dismiss without notice only if the employee has committed gross misconduct. In all other circumstances the employer must give contractual notice to end the relationship, or pay wages in lieu of notice. However, this does require qualification. Firstly, the law decides what is gross misconduct and not the employer. Just because the employer has stated that certain actions are gross misconduct does not mean that the law will regard it as such. Only very serious misconduct is regarded by the law as gross, such as refusing to obey lawful and reasonable orders, gross neglect, theft. Secondly, contractual notice periods are subject to the statutory minimum notice provisions contained in the *Employment Rights Act 1996, s 86*. Any attempt by the contract to give less than the statutory minimum notice is void. These periods apply to all employees who have been employed for one month or more and are:

(*a*) not less than one week if the employee has been employed for less than two years;

(*b*) after the employee has been employed for two years, one week for each year of service, up to a statutory maximum of 12 weeks.

Where an employer terminates the contract and pays the employee in lieu of notice, in the absence of an express right to do so, this will be a technical breach of contract. The employee can waive his right to notice or accept wages in lieu of notice. If the contract gives notice periods which are longer than the statutory minimum, the contractual notice prevails. Therefore, if the employer has an employee who has been employed for six years, and the employer sacks him with four weeks' notice, the employee can sue for a further two weeks' wages in the employment tribunal. Finally, if the employer fundamentally alters the terms of the employee's contract, without his consent, in reality the employer is terminating (repudiating) the original contract and substituting a new one. The employee should therefore be given the correct notice before the change comes into effect. A unilateral change by the employer to a fundamental term of the contract (followed by resignation by the employee) will amount to constructive dismissal (i.e. repudiation) and compensation in the form of damages for breach of contract. The employee must take all reasonable steps to mitigate his loss by seeking other employment.

Unfair dismissal

Dismissal

E14019 While all employees are protected against wrongful dismissal, generally employees must be employed for one year or more before they gain protection against unfair dismissal. In certain circumstances, however, an employee is protected immediately and does not need a year of employment. One of these is dismissal on certain health and safety grounds discussed above.

Once an employee is protected against unfair dismissal the *Employment Rights Act 1996, s 95* recognises three situations which the law regards as dismissal. These are:

(*a*) the employer terminating the contract (with or without notice);

(*b*) a fixed term contract which expires and is not renewed;

(*c*) the employee resigning in circumstances in which he is entitled to do so without notice because of the employer's conduct – a constructive dismissal.

In the first situation the employer is unilaterally ending the relationship. Even if the employer gives the correct amount of notice so that the dismissal is lawful, it does not necessarily follow that the dismissal will be fair.

The second situation needs no explanation. If a fixed term contract has come to an end and is not renewed, this is, in effect, the employer deciding to end the relationship. As from 25 October 1999, it is no longer possible for employees to validly waive unfair dismissal rights in fixed term contracts (*Employment Relations Act 1999*).

The third situation, constructive dismissal, is much more complex. On the face of it the employee has resigned. However if the reason for his resignation is the employer's conduct, then the law treats the resignation as an employer termination. The action on the part of the employer which entitles the employee to resign and claim constructive dismissal is a repudiatory breach of contract. In other words, the employer has committed a breach which goes to the root of the contract and has, therefore, repudiated it. This means that not all breaches by the employer are constructive dismissals but that serious breaches may be. It is also important to recognise that, as discussed above, the terms of the contract may include those which have not been expressly agreed by the parties and therefore rules, disciplinary procedures, terms collectively bargained, and all the implied duties discussed in E14003 above, may all be contractual terms. Breach of the health and safety duties owed to all employees is likely to give rise to a constructive dismissal claim (*Day v T Pickles Farms Ltd [1999] IRLR 217*). In addition, the law requires as an implied term of the contract that both the employer and employee treat each other with mutual respect and do nothing to destroy the trust and confidence each has in the other. Breach of this duty may give rise to a constructive dismissal claim. In one case, a demotion imposed as a disciplinary sanction was held to be excessive by the Employment Appeal Tribunal. Its very excessiveness was a breach of the duty of mutual respect which entitled the employee to resign and claim constructive dismissal. It has also been held that an employer's disclosure in a reference on behalf of an employee of complaints against the employee, before first giving the individual an opportunity to explain, was a fundamental breach of the implied term of mutual trust and confidence which amounted to constructive (since the employee resigned) and unfair dismissal (*TSB Bank v Harris [2000] IRLR 157*). In *Reed v Stedman [1999] IRLR 299*, the Employment Appeal Tribunal held that in a case where the employer was aware of an employee's deteriorating health, and the employee concerned had complained to colleagues at work about harassment, it was encumbent on the employer to investigate and their failure to do so was enough to justify a finding of breach of trust and confidence and thus constructive dismissal.

Where an employee with one or more year's continuous service (unless one of the specified reasons apply which render a dismissal automatically unfair) is constructively dismissed, he or she will also have the right to claim unfair dismissal. Obviously, the employee must resign before he can make a claim for unfair dismissal. In the majority of cases the repudiatory breach by the employer is a fundamental alteration of the contractual terms (for example hours). In this situation, the employer still wishes to continue the relationship, albeit on different terms. The employee has two choices: he can resign or he can continue to work under the new terms. If the employee continues to work and accepts the changed terms, the contract is mutually varied and no action will lie, provided that the employee was given the correct notice before the change was implemented. If the employee resigns, however, he will have been dismissed. In *Walton & Morse v Dorrington [1997] IRLR 488*, the employee waited to find alternative employment before she resigned. The Employment Appeal Tribunal decided that, in her circumstances, this was a reasonable thing to have done and agreed that she had not accepted her employer's breach of its duty to her and had been constructively dismissed (see E14003 above).

Reasons for dismissal

E14020 *Section 98* of the *Employment Rights Act 1996* gives five potentially fair reasons for dismissal. These are:

(*a*) capability or qualifications;

(*b*) conduct;

(*c*) redundancy;

(*d*) contravention of statute;

(*e*) some other substantial reason.

Dismissal on health and safety grounds could potentially fall within most of these reasons. It should, however, be remembered that, where an employee is dismissed in circumstances where continued employment involves a risk to the employee's health and safety, the employer may nevertheless face claims of unfair dismissal. Before terminating employment, an employer should consider all the circumstances of the case and assess the risk involved and take measures which are reasonably necessary to eliminate the risk.

Illness may make it unsafe to employ the employee; breach of health and safety rules will normally fall under misconduct; to continue to employ the employee may contravene health and safety legislation or it may be that the employer has had to reorganise his business on health and safety grounds and the employee is refusing to accept the change. This latter situation could be potentially fair under 'some other substantial reason'. (However, an employer may be liable under the *Disability Discrimination Act 1995*.)

Reasonableness

E14021 Merely having a fair reason to dismiss does not mean that the dismissal is fair. *Section 98(4)* of the *Employment Rights Act 1996* requires the tribunal in any unfair dismissal case to consider whether the employer acted reasonably in all the circumstances (including the size and administrative resources of the employer's undertaking). This means that the tribunal will look at two things – (i) was the treatment of the employee procedurally fair, and (ii) was dismissal a reasonable sanction in relation to the employee's actions and all the circumstances of the case.

Procedures have already been discussed at E14013 above. If the employer has complied with his procedures, he will not be found to have acted procedurally unfairly unless the procedures themselves are unfair. This is unlikely if the employer is following the ACAS Code.

In respect of the fairness of the decision, the tribunal should consider whether the employer's decision to dismiss fell within the band of reasonable responses to the employee's conduct which a reasonable employer could adopt. The tribunal will look at three things – (i) has the employer acted consistently, (ii) has he taken the employee's past work record into account, and (iii) has the employer looked for alternative employment. The latter aspect is of major importance in relation to redundancy, incapability due to illness or dismissal because of a contravention of legislation, but will not usually be relevant in dismissals for misconduct. It is important to note in addition that special obligations apply to an employer in the case of a disabled employee within the meaning of the *Disability Discrimination Act 1995*.

When looking at consistency, the tribunal will look for evidence that the employer has treated the same misconduct the same way in the past. If the employer has treated past breaches of health and safety rules leniently it will be unfair to suddenly dismiss for the same breach, unless he has made it clear to the employees that his attitude has changed and breaches will be dealt with more severely in the future. Employees have to know the potential disciplinary consequences for breaches of the rules, and if the employer has never dismissed in the past he is misleading employees unless he tells them that things have changed. The law, however, only requires an employer to be consistent between cases which are the same. This is where a consideration of the employee's past work record is important. It is not inconsistent to give a long-standing employee with a clean record a final warning for a breach of health and safety rules and to dismiss another shorter-serving employee with a series of warnings behind him, as long as both employees know that the penalty for breach of the rules could be dismissal. The cases are not the same. It would, however, be unfair if both the employees had the same type of work record and length of service and only one was dismissed, and dismissal had never been imposed as a sanction for that type of breach in the past. In order for a misconduct dismissal to be fair, the employer must have had a reasonable belief in the guilt of the employee of the misconduct in question on the basis of a reasonable investigation.

Remedies for unfair dismissal

(a) Reinstatement

E14022 The first remedy that the tribunal is required to consider is reinstatement of the employee. When doing so the tribunal must take into account whether the employee wishes to be reinstated, whether it is practicable for the employer to reinstate him and, if the employee's conduct contributed to or caused his dismissal, whether it is just to reinstate him. In order to resist an order for reinstatement, an employer must provide evidence to show that it is not practicable because the implied term of mutual trust and confidence between employer and employee has broken down (*IPC Magazines Ltd v Clements, EAT/456/99; Gentle & Ors v Perkins Group Ltd, EAT/670/99*). Reinstatement means that the employee must return to his old job with no loss of benefits. If reinstatement is ordered and the employer refuses to comply with the order, or only partially complies, compensation will be increased. Reinstatement, however, is rarely ordered by tribunals.

(b) Re-engagement

If the tribunal does not consider that reinstatement is practicable, it must consider whether to order the employer to re-engage the employee. In making its decision the tribunal looks at the same factors as when it considers reinstatement. Re-engagement is an order requiring the employer to re-employ the employee on terms which are as favourable as those he enjoyed before his dismissal, but it does not require the employer to give the employee the same job back. Failure on the part of the employer to comply with an order of re-engagement will lead to increased compensation, although tribunals rarely make re-engagement orders.

(c) Compensation

Compensation falls under a variety of different heads. In an unfair dismissal case the employee will receive:

Basic award: This is based on his age, years of service and salary –

(i) one and a half weeks' pay for each year of service over the age of 41;

(ii) one week's pay for each year of service between 41 and 22;

(iii) half a week's pay for each year of service below the age of 22.

This is subject to a statutory maximum of £270 a week, and a maximum of twenty years' service. Compensation is reduced by one-twelfth for each month the employee works during his 64th year. Where the employee is unfairly dismissed for health and safety reasons under the *Employment Rights Act 1996, s 100*, the minimum basic award is £3,600 (at present).

Compensatory award: This is payable in addition to the basic award to compensate the employee for loss of future earnings, benefits etc, which are in excess of the basic award. As with the basic award the compensatory award can be reduced for contributory conduct. The present maximum compensatory award is £55,000.

Additional award: If the employer fails to comply with a reinstatement or re-engagement order, the tribunal may make an additional award. This will be between 26 and 52 weeks' pay (subject to a maximum of £270 per week).

Note: There is no limit on:

- the compensatory award for employees who are dismissed for health and safety reasons (see E14010);
- awards of compensation in sex discrimination and equal pay cases (and interest can be included in such awards);
- the amount of the compensation that a tribunal may award in a case of discrimination under the *Disability Discrimination Act 1995*.

In cases of sex and disability discrimination, compensation may include compensation for injury to feelings.

Health and safety duties in relation to women at work

Sex discrimination

E14023 Since health and safety issues may give rise to sex discrimination claims under the *Sex Discrimination Act 1975* (*SDA 1975*), it is appropriate to examine what particular considerations an employer needs to bear in mind in its relations with female employees.

The steps necessary to be taken by an employer, in order to comply with his duties under *HSWA 1974, s 2*, may differ for women.

New or expectant mothers are particularly vulnerable to adverse or indifferent working conditions. Indeed, most employers have probably taken measures to guard against risks to new and expectant mothers, in accordance with their general duties under *HSWA 1974, s 2*, and the *Management of Health and Safety at Work Regulations 1999* (*SI 1999 No 3242*). In addition, employers are required to protect new and expectant mothers in their employment from certain specified risks if it is reasonable to do so, and to carry out a risk assessment of such hazards. If the employer cannot avoid the risk(s), he must alter the working conditions of the employee concerned or the hours of work, offer suitable alternative work and, if no suitable alternative work is available, suspend the employee on full pay (see E14024 below).

Although *SDA 1975* prohibits discrimination on grounds of sex, *s 51(1)* provides that any action taken to comply with certain existing health and safety legislation (e.g. *HSWA 1974*) will *not* amount to unlawful discrimination. In *Page v Freight Hire (Tank Haulage) Ltd [1981] IRLR 13*, the complainant was an HGV driver. The employer, acting on the instructions of the manufacturer of the chemical dimethylformamide (DMF), refused to allow her to transport the chemical which was potentially harmful to women of child bearing age. She brought a complaint of sex discrimination. It was held that the fact that the discriminatory action was taken in the interests of safety did not of itself provide a defence to a complaint of unlawful discrimination. However, the employer was protected by *s 51(1)* of *SDA 1975* because the action taken was necessary to comply with the employer's duty under *HSWA 1974*.

Pregnant workers, new and breastfeeding mothers

E14024 *The Management of Health and Safety at Work Regulations 1999* (*SI 1999 No 3242*) impose a duty on employers to protect new or expectant mothers from any process or working conditions or certain physical, chemical and biological risks at work (see E14025 below). The phrase 'new or expectant mother' is defined as a worker who is pregnant, who has given birth within the previous six months, or who is breastfeeding. 'Given birth' is defined as having delivered a living child or, after 24 weeks of pregnancy, a stillborn child.

Risk assessment

E14025 The 1999 Regulations require employers to carry out an assessment of the specific risks posed to the health and safety of pregnant women and new mothers in the workplace and then to take steps to ensure that those risks are avoided. Risks include those to the unborn child or child of a woman who is still breastfeeding – not just risks to the mother.

An interesting development in relation to this requirement is the case of *Day v T Pickles Farms Ltd [1999] IRLR 217*, where the employee suffered nausea when pregnant as a result of the smell of food at her workplace. As a result of the nausea, she was unable to work and was eventually dismissed after a prolonged absence.

The Employment Appeal Tribunal found that she had not been constructively dismissed. However, the Employment Appeal Tribunal held that the obligation to carry out a risk assessment which considers possible risks to the health and safety of a pregnant female employee is relevant from the moment an employer employs a woman of childbearing age. The question of whether the applicant had been subjected to a detriment was remitted to the employment tribunal.

The Health and Safety Executive has published a booklet entitled '*New and expectant mothers at work – A guide for employers*'. The booklet provides guidance on what employers need to do to comply with the legislation. The booklet includes a list of the known risks to new and expectant mothers and suggests methods of avoidance.

The main risks to be avoided are as follows:

- Physical agents:
  - shocks/vibrations/movement (including travelling and other physical burdens),
  - handling of loads entailing risks,
  - noise,
  - non-ionising radiation,
  - extremes of heat and cold.
- Biological agents:
  - such as listeria, rubella and chicken pox virus, toxoplasma, cytomegalovirus, hepatitis B and HIV.
- Chemical agents:
  - such as mercury, antimiotic drugs, carbon monoxide, chemical agents of known and percutaneous absorption and chemicals listed under various Directives.
- Working conditions:
  - such as mining work and work with display screen equipment (VDUs).

Where a risk has been identified following the assessment, affected employees or their representatives should be informed of the risk and the preventive measures to be adopted. The assessment should be kept under review.

In particular, employers must consider removing the hazard or seek to prevent exposure to it. If a risk remains after preventive action has been taken, the employer must take the following course of action:

(i) temporarily adjust her working conditions or hours of work (*Management of Health and Safety at Work Regulations 1999* (*SI 1999 No 3242*), *Reg 16*(*2*)).

If it is not reasonable to do so or would not avoid the risk:

(ii) offer suitable alternative work (*Employment Rights Act 1996, s 67*).

If neither of the above options is viable:

(iii) suspend her on full pay for as long as necessary to protect her health and safety or that of her child (*Management of Health and Safety at Work Regulations 1999* (*SI 1999 No 3242*), *Regs 16*(*2*) and *16*(*3*)*; Employment Rights Act 1996, s 67*).

Appendix 1 of the booklet lists aspects of pregnancy such as morning sickness, varicose veins, increasing size etc. that may affect work and which employers may take into account in considering working arrangements for pregnant and breastfeeding workers. These are merely suggestions and not requirements of the law.

Night work by new or expectant mother

E14026 Where a new or expectant mother works at night and has been issued with a certificate from a registered doctor or midwife stating that night work would affect her health and safety, the employer must first offer her suitable alternative daytime work, and suspend her as detailed at E14015 above if no suitable alternative employment can be found (*Management of Health and Safety at Work Regulations 1999* (*SI 1999 No 3242*), *Reg 17*).

Notification

E14027 An employer is not required to alter a woman's working conditions or hours of work or suspend her from work under *Management of Health and Safety at Work Regulations 1999* (*SI 1999 No 3242*), *Reg 16*(*2*) or (*3*) until she notifies him in writing that she is pregnant, has given birth within the previous six months or is breastfeeding. Additionally, the suspension or amended working conditions do not have to be maintained if the employee fails to produce a medical certificate confirming her pregnancy in writing within a reasonable time if the employer requests her to do so. The same applies once the employer knows that the employee is no longer a new or expectant mother or if the employer cannot establish whether she remains so (*Management of Health and Safety at Work Regulations 1999* (*SI 1999 No 3242*), *Reg 18*). However, an employer has a general duty under *HSWA 1974* and the *Management of Health and Safety at Work Regulations 1999* to take steps to protect the health and safety of a new or expectant mother, even if she has not given written notification of her condition.

Maternity leave

E14028 All pregnant workers have a right to 26 weeks' ordinary maternity leave, regardless of length of service and number of hours worked. Employees who have, at the beginning of the 14th week before the expected week of childbirth, being continuously employed for a period of not less than 26 weeks may be entitled to additional maternity leave. The ordinary maternity leave period starts from:

(*a*) the notified date of commencement; or

(*b*) the first day of absence because of pregnancy or childbirth after the beginning of the fourth week before the expected week of confinement; or

(*c*) the date of childbirth,

which ever is the earlier, and continues for 26 weeks or until the end of the compulsory leave period if later. Additional maternity leave continues until the end of the period of 26 weeks from the day after the last day or ordinary maternity leave.

If an employee is prohibited from working for a specified period after childbirth by virtue of a legislative requirement (e.g. under the *Public Health Act 1936, s 205*), her maternity leave period must continue until the expiry of that later period. The *Employment Rights Act 1996, s 72(1)*, provide that an employee entitled to maternity leave should not work or be permitted to work by her employer during the period of two weeks beginning with the date of childbirth.

Enforcement

Introduction

E15001 In the last decade there has been a movement away from a purely legalistic approach towards health and safety at work to one concerned with loss prevention, asset protection, accountability and consultation with the workforce, a situation without parallel under previous protective legislation. An effective system of enforcement is still, however, essential if workplaces are to be kept safe and accidents prevented. Prior to the *Health and Safety at Work etc Act 1974* (*HSWA 1974*), the principal sanction against breach of a statutory requirement was prosecution. This preoccupation with criminal proceedings was criticised by the Robens Committee (para 142) as being largely ineffective in securing the most important end result, namely that the breach should be remedied as soon as possible. The *HSWA 1974*, therefore, has given Health and Safety Executive (HSE) inspectors a range of enforcement powers which do not necessarily depend on prosecution for their efficacy. Most important of these are the powers to serve improvement and prohibition notices. In 2002/03 there were 5,159 prohibition notices and 8,104 improvement notices issued by the HSE. This is an increase of approximately 10 per cent on the number of notices issued the previous year. Contravention of an improvement/prohibition notice carries with it, on summary conviction, a maximum fine of £20,000 or, alternatively, six months' imprisonment and on conviction on indictment an unlimited fine or two years' imprisonment (see E15036 below). In particular, a prohibition notice may be served by an inspector where he believes there to be a risk of serious personal injury, regardless of whether any offence has actually been committed. Equally important, though less obvious perhaps, HSE inspectors and environmental health officers, the two principal enforcement authorities, can use powers given to them by *HSWA 1974*, i.e. serve improvement and prohibition notices, in order to enforce remaining pre-*HSWA* statutory requirements, e.g. duties under the *Factories Act 1961* and *Offices, Shops and Railway Premises Act 1963* (*OSRPA 1963*), since these qualify as 'relevant statutory provisions' (see E15012 below).

More recently, the *Environmental Protection Act 1990* and the *Environment Act 1995* have conferred similar notice-serving powers on the Environment Agency. The *Radioactive Material* (*Road Transport*) *Act 1991* has given transport inspectors powers to detain, search and generally 'quarantine' vehicles carrying radioactive packages in breach of that Act and the *Radioactive Substances Act 1993* similarly empowers inspectors in respect of premises containing radioactive substances and mobile radioactive apparatus.

Such enforcement powers apart, breach of a statutory requirement is still a criminal offence. In 2002/03 there were 1,688 prosecutions for health and safety offences of which 1,260 led to conviction. Persons committing a breach, or permitting one to occur, should be in no doubt that they stand to be prosecuted. It is not just the employer who is liable to prosecution: employees' and junior and middle management and even visitors to the workplace can also be prosecuted, either in tandem with the employer or alone. If the employer is a company or local authority, the company, its directors and/or officers, as well as councillors may be charged with an offence (see E15040 below).

Prosecutions and other enforcement procedures are the responsibility of the appropriate 'enforcing authority' (see E15007–E15011 below). In addition, where an employee is injured or killed as a result of negligence or breach of a statutory requirement, a civil action may be brought against the employer for damages.

This section deals with the following aspects of enforcement:

— The role of the Health and Safety Commission and the Health and Safety Executive (see E15002–E15006).

— The 'enforcing authorities' (see E15007–E15011 below).

— The 'relevant statutory provisions' which can be enforced under *HSWA 1974* (see E15012 below).

Part A: Enforcement Powers of Inspectors'

— Improvement and prohibition notices (see E15013–E15018 below).

— Appeals against improvement and prohibition notices (see E15019, E15020 below).

— Grounds for appeal against a notice (see E15021–E15025 below).

— Inspectors' investigation powers (see E15026 below).

— Inspectors' powers of search and seizure (see E15027 below).

— Indemnification by enforcing authority (see E15028 below).

— Public register of notices (see E15029 below).

Part B: Offences and Penalties

— Prosecution for contravention of the relevant statutory provisions (see E15030, E15031 below).

— Main offences and penalties (see E15032–E15038 below).

— Offences committed by particular types of persons, including the Crown (see E15039–E15045 below).

— Sentencing guidelines (see E15043 below).

The role of the Health and Safety Commission and Executive

E15002 Under the *HSWA 1974* the administration and enforcement of health and safety law is carried out by the Health and Safety Commission (HSC) and its executive arm the Health and Safety Executive (HSE). The HSC consists of a chairman and not fewer than six, nor more than nine, other members. Of those members three represent employers, three employees and the remainder are from bodies such as local authorities and professional organisations. [*HSWA 1974, s 10(2), (3)*]. The HSE is the executive arm of the HSC and consists of a director general appointed by the HSE, a deputy director general and head of operations. Both the HSC and

HSE are independent of any government department, but owe direct responsibilities to the relevant Secretary of State (the Secretary of State for the Environment, Transport and Regions).

The functions of the HSC

E15003 The functions of the HSC include promoting the general aims of the *HSWA 1974* and other health and safety legislation generally, drawing up proposals for regulations, replacing/updating existing law and regulations, preparing approved codes of practice and ensuring arrangements are in place for research and training. [*HSWA 1974, s 11*(*2*)].

The HSC is given consequential powers to make agreements with other government departments, established advisory committees etc for the purpose of discharging its functions. [*HSWA 1974, s 13*]. It may also delegate the exercise of certain of its functions to the HSE [*HSWA 1974, s 11*(*4*)(*a*)], for example, it may ask the HSE to carry out research or provide publicity or educational materials.

The HSC can also ask the HSE to conduct a special investigation and to report to them on their findings or, with the consent of the Secretary of State, direct that an inquiry be held into any accident, occurrence or matter. [*HSWA 1974, s 14*].

Preparation by the HSE of draft regulations and issue of approved codes of practice

E15004 One of the most important functions of the HSC is to prepare draft regulations for the Secretary of State, including new regulations seeking to implement EU directives relating to health and safety at work. This normally involves preparation of a consultative document and draft regulations which are considered by industry and employee bodies, trade associations and other interested bodies, such as local authorities and educational establishments. Draft regulations are submitted to the Secretary of State for approval after consultation. Although the power to make regulations rests with the Secretary of State, he has a duty to consult the HSC before doing so.

Such regulations can:

(*a*) repeal or modify any of the 'relevant statutory provisions' (see E15012 for the definition of this term);

(*b*) exclude or modify any of the 'existing statutory provisions' e.g. the *Employers' Health and Safety Policy Statements* (*Exception*) *Regulations 1975* (*SI 1975 No 1584*), exempt employers employing fewer than *five* employees from issuing a written company safety policy, as required by *HSWA 1974, s 2*(*3*);

(*c*) make a specified authority responsible for the enforcement of any of the relevant statutory provisions.

[*HSWA 1974, s 15*(*3*)(*c*)].

Although the HSC does not have power to make regulations, it can approve and issue of codes of practice providing practical guidance on the requirements imposed by health and safety legislation or regulations. [*HSWA 1974, s 16*]. In all cases the HSC must consult with any appropriate government departments and obtain the consent of the Secretary of State before the issue of Approved Codes of Practice ('ACOPs').

Functions of the Health and Safety Executive

E15005 The main function of the HSE is to enforce health and safety legislation unless responsibility lies with another enforcing body, such as a local authority. [*HSWA 1974, s 18*(*1*)]. The HSE's functions are principally performed by its inspectors and include carrying out routine inspections of premises, investigating workplace accidents, dangerous occurrences or cases of ill health and providing advice to companies and individuals on the legal requirements under health and safety legislation. The HSE also publish guidance documents (see E15006 below), provide an information service, carry out research and license or approve certain hazardous operations, such as nuclear site licensing and accepting offshore installation safety cases.

The *Health and Safety* (*Fees*) *Regulations 2004* (*SI 2004 No*) also provide that the HSE may, in certain limited circumstances, charge for their services. In particular, fees are payable for (*inter alia*):

(*a*) an approval under mines and quarries legislation;

(*b*) an approval of certain respiratory equipment;

(*c*) an approval under the *Agriculture* (*Tractor Cabs*) *Regulations 1974* (*SI 1974 No 2034* as amended by *SI 1990 No 1075*);

(*d*) an approval under the *Freight Containers* (*Safety Convention*) *Regulations 1984* (*SI 1984 No 1890*);

(*e*) a licence under the *Asbestos* (*Licensing*) *Regulations 1983* (*SI 1983 No 1649* as amended by *SI 1998 No 3233*));

(*f*) examination or surveillance by an employment medical adviser;

(*g*) medical surveillance by an employment medical adviser under the *Control of Lead at Work Regulations 2002* (*SI 1998 No 2676*);

(*h*) an approval or reassessment of approval of dosimetry services and for type approval of apparatus under the I*onising Radiations Regulations 1999* (*SI 1999 No 3232*);

(*i*) an application for an explosives licence under *Part IX* of the *Dangerous Substances in Harbour Regulations 1987* (*SI 1987 No 37*);

(*j*) a vocational training certificate under the *Carriage of Dangerous Goods by Road* (*Driver Training*) *Regulations 1996* (*SI 1996 No 2094*);

(*k*) an approval under the *Carriage of Dangerous Goods by Road* (*Driver Training*) *Regulations 1996* (*SI 1996 No 2094*);

(*l*) a vocational training certificate under the *Transport of Dangerous Goods* (*Safety Advisers*) *Regulations 1999* (*SI 1999 No 257*);

(*m*) notifications and applications under the *Genetically Modified Organisms* (*Contained Use*) *Regulations 2000* (*SI 2000 No 2831*);

(*n*) notifications and applications under the *Notification of New Substances Regulations 1993* (*SI 1993 No 3050* as amended by *SI 2001 No 1055*);

(*o*) an approval under the *Health and Safety* (*First-Aid*) *Regulations 1981* (*SI 1981 No 917* as amended *by SI 2002 No 2174*);

(*p*) an approval under the *Offshore Installations and Pipeline Works* (*First-Aid*) *Regulations 1989* (*SI 1989 No 1671*);

(*q*) enforcement provisions in relation to a railway safety case prepared pursuant to the *Railways* (*Safety Case*) *Regulations 2000* (*SI 2000 No 2688* as amended by *SI 2001 No 3291*);

(*r*) advice and assessments in relation to gas safety cases prepared pursuant to the *Gas Safety* (*Management*) *Regulations 1996* (*SI 1996 No 551*).

HSE publications

E15006 The HSE (through HSE Books) also publish a series of guidance notes/advisory literature for employers, local authorities, trade unions etc on most aspects of health and safety of concern to industry. These are divided into five main areas i.e. Chemical Safety (CS), Environmental Hygiene (EH), General Series (GS), Medical Series (MS), and Plant or Machinery (PM).

In addition there are the Health and Safety (Guidance) Series (HS(G)); the Health and Safety (Regulations) Series (HS(R)); Legal Series (L); Best Practicable Means Leaflets (BPM); Emission Test Methods (ETM); Health and Safety Commission Leaflets (HSC); Health and Safety Executive Leaflets (HSE); Industry General Leaflets (IND(G)); Industry Safety Leaflets (IND(S)); similarly, Methods for the Determination of Hazardous Substances (MDHS); Toxicity Reviews (TR); Occasional Papers; and Agricultural Safety Leaflets (AS).

Enforcing authorities

E15007 Whilst the HSE is the central body entrusted with the enforcement of health and safety legislation, in any given case enforcement powers rest with the body which is expressed by statute to be the 'enforcing authority'. Here the general rule is that the 'enforcing authority', in the case of industrial premises, is the HSE and, in the case of commercial premises within its area, the local authority (the enforcing authority in over a million premises), except that the HSE cannot enforce provisions in respect of its own premises, and similarly, local authorities' premises are inspected by the HSE. Each 'enforcing authority' is empowered to appoint suitably qualified persons as inspectors for the purpose of carrying into effect the 'relevant statutory provisions' within the authority's field of responsibility. [*HSWA 1974, s 19*(*1*)]. Inspectors so appointed can exercise any of the enforcement powers conferred by *HSWA 1974* (see E15013–E15018 below) and bring prosecutions (see E15030, E15031 below).

The appropriate 'enforcing authority'

E15008 The general rule is that the HSE is the enforcing authority, except to the extent that:

(*a*) regulations specify that the local authority is the enforcing authority instead; the regulations that so specify are the *Health and Safety* (*Enforcing Authority*) *Regulations 1998* (*SI 1998 No 494*); or;

(*b*) one of the 'relevant statutory provisions' specifies that some other body is responsible for the enforcement of a particular requirement.

[*HSWA 1974, s 18*(*1*), (*7*)(*a*)].

Activities for which the HSE is the enforcing authority

E15009

The HSE is specifically the enforcing authority in respect of the following activities (even though the main activity on the premises is one for which a local authority is usually the enforcing authority in accordance with the *Health and Safety (Enforcing Authority) Regulations 1998(SI 1998 No 494), Sch 1* (see E15010 below):

1. Any activity in a mine or quarry;
2. Fairground activity;
3. Broadcasting, recording, filming, or video recording and any activity in premises occupied by a radio, television or film undertaking where such work is carried on;
4. The following work carried out by independent contractors:
   (*a*) certain construction work;
   (*b*) installation, maintenance or repair of gas systems or work in connection with a gas fitting;
   (*c*) installation, maintenance or repair of electricity systems;
   (*d*) most work with ionising radiations;
5. Use of ionising radiations for medical exposure;
6. Any activity in radiography premises where work with ionising radiations is carried out;
7. Agricultural activities, including agricultural shows (but excluding garden centres);
8. Any activity on board a sea-going ship;
9. Ski slope, ski lift, ski tow or cable car activities;
10. Fish, maggot and game breeding (but not in a zoo);
11. The operation of a railway;
12. Any activity in relation to a pipeline;
13. Enforcement of *HSWA 1974, s 6* (duties of manufacturers/suppliers of industrial products).

[*Health and Safety (Enforcing Authority) Regulations 1998 (SI 1998 No 494), Reg 4(4)(b)* and *Sch 2*].

The HSE is the enforcing authority against the following, and for any premises they occupy, including parts of the premises occupied by others providing services for them. (This is so even though the main activity is listed in *Sch 1* of the 1998 Regulations, see E15010 below):

14. County councils;
15. Local authorities;
16. Parish or community councils;
17. Police authorities;
18. Fire authorities;

19. International HQ's and defence organisations and visiting forces;
20. United Kingdom Atomic Energy Authority (UKAEA);
21. The Crown (except where premises are occupied by the HSE itself).

[*Health and Safety* (*Enforcing Authority*) *Regulations 1998* (*SI 1998 No 494*), *Reg 4*(*1*)–(*3*)].

The HSE is also the enforcing authority for the premises set out below, even if occupied by more than one occupier:

22. Airport land;
23. The Channel Tunnel system;
24. Offshore installations;
25. Building/construction sites;
26. University, polytechnic, college, school etc. campuses;
27. Hospitals;

[*Health and Safety at Work* (*Enforcing Authority*) *Regulations 1998* (*SI 1998 No 494*), *Reg 3*(*5*)].

and for the following:

28. Common parts of domestic premises (*Reg 3*(*1*));
29. Certain areas within an airport (*Reg 3*(*4*)(*b*));
30. Common parts of railway stations/termini/goods yards (*Reg 3*(*6*)).

Activities for which local authorities are the enforcing authorities

E15010 Where the main activity carried on in non-domestic premises is one of the following, the local authority is the enforcing authority (i.e. the relevant county, district or borough council):

1. Sale or storage of goods for retail/wholesale distribution (including sale and fitting of motor car tyres, exhausts, windscreens or sunroofs), except:
   (*a*) at container depots where the main activity is the storage of goods which are of transit to or from dock premises, an airport or railway;
   (*b*) where the main activity is the sale or storage for wholesale distribution of dangerous substances;
   (*c*) where the main activity is the sale or storage of water or sewage or their by-products or natural or town gas.
2. Display or demonstration of goods at an exhibition, being offered or advertised for sale.
3. Office activities.
4. Catering services.

5. Provision of permanent or temporary residential accommodation, including sites for caravans or campers.
6. Consumer services provided in a shop, except:
   (*a*) dry cleaning;
   (*b*) radio/television repairs.
7. Cleaning (wet or dry) in coin-operated units in laundrettes etc.
8. Baths, saunas, solariums, massage parlours, premises for hair transplant, skin piercing, manicuring or other cosmetic services and therapeutic treatments, except where supervised by a doctor, dentist, physiotherapist, osteopath or chiropractor.
9. Practice or presentation of arts, sports, games, entertainment or other cultural/recreational activities, save where the main activity is the exhibition of a cave to the public.
10. Hiring out of pleasure craft for use on inland waters.
11. Care, treatment, accommodation or exhibition of animals, birds or other creatures, except where the main activity is:
   (*a*) horse breeding/horse training at stables;
   (*b*) agricultural activity;
   (*c*) veterinary surgery.
12. Undertaking, but not embalming or coffin making.
13. Church worship/religious meetings.
14. Provision of care parking facilities within an airport.
15. Childcare, playgroup or nursery facilities.

[*Health and Safety* (*Enforcing Authority*) *Regulations 1998* (*SI 1998 No 494*), *Reg 3*(*1*) *and Sch 1*].

Transfer of responsibility between the HSE and local authorities

E15011 Enforcement can be transferred (though not in the case of Crown premises), by prior agreement, from the HSE to the local authority and vice versa. The Health and Safety Commission is also empowered to effect such a transfer, without the necessity of such agreement. In either case, parties who are affected by such transfer must be notified. [*Health and Safety* (*Enforcing Authority*) *Regulations 1998* (*SI 1998 No 494*), *Reg 5*]. Transfer is effective even though the above procedure is not followed (i.e. the authority changes when the main activity changes) (*Hadley v Hancox (1987) 85 LGR 402*, decided under the previous Regulations).

Where there is uncertainty, these Regulations also allow responsibility to be assigned by the HSE and the local authority jointly to either body. [*Health and Safety* (*Enforcing Authority*) *Regulations 1998* (*SI 1998 No 494*), *Reg 6*(*1*)].

'Relevant statutory provisions' covered by the Health and Safety at Work etc Act 1974

E15012 The enforcement powers conferred by *HSWA 1974* extend to any of the 'relevant statutory provisions'. These comprise:

(*a*) the provisions of *HSWA 1974, Part I* (i.e. *ss 1–53*); and

(*b*) any health and safety regulations passed under *HSWA 1974*, e.g. the *Ionising Radiations Regulations 1985* (*SI 1985 No 1333*), the *Management of Health and Safety at Work Regulations 1999* (*SI 1999 No 3242*) and the *Workplace* (*Health, Safety and Welfare*) *Regulations 1992* (*SI 1992 No 3004*); and

(*c*) the 'existing statutory provisions', i.e. all enactments specified in *HSWA 1974, Sch 1*, including any regulations etc. made under them, so long as they continue to have effect; that is, the *Explosives Acts 1875–1923*, the *Mines and Quarries Act 1954*, the *Factories Act 1961*, the *Public Health Act 1961*, the *Offices, Shops and Railway Premises Act 1963* and (by dint of the *Offshore Safety Act 1992*) the *Mineral Workings* (*Offshore Installations*) *Act 1971* [*HSWA 1974, s 53*(*1*)].

Part A: Enforcement Powers of Inspectors

Improvement and prohibition notices

E15013 It was recommended by the Robens Committee that 'inspectors' should have the power, without reference to the courts, to issue a formal improvement notice to an employer requiring him to remedy particular faults or to institute a specified programme of work within a stated time limit.' (Cmnd 5034, para 269). 'The improvement notice would be the inspector's main sanction. In addition, an alternative and stronger power should be available to the inspector for use where he considers the case for remedial action to be particularly serious. In such cases he should be able to issue a prohibition notice.' (Cmnd 5034, para 276). *HSWA 1974* put these recommendations into effect.

Improvement notices

E15014 An inspector may serve an improvement notice if he is of the opinion that a person:

(*a*) is contravening one or more of the 'relevant statutory provisions' (see E15012 above); or

(*b*) has contravened one or more of those provisions in circumstances that make it likely that the contravention will continue or be repeated.

[*HSWA 1974, s 21*].

In the improvement notice the inspector must:

(i) state that he is of the opinion in (*a*) and (*b*) above; and

(ii) specify the provision(s) in his opinion contravened; and

(iii) give particulars of the reasons for his opinion; and

(iv) specify a period of time within which the person is required to remedy the contravention (or the matters occasioning such contravention).

[*HSWA 1974, s 21*].

The period specified in the notice within which the requirement must be carried out (see (iv) above) must be at least 21 days – this being the period within which an appeal may be lodged with an Employment tribunal (see E15019 below). [*HSWA 1974, s 21*]. In order to be validly served on a company, an improvement notice relating to the company's actions as an employer must be served at the registered office of the company, not elsewhere. Service at premises occupied by the company will only be valid if the notice relates to a contravention by the company in the capacity of occupier (*HSE v George Tancocks Garage (Exeter) [1993] COD 284*).

Although formal procedures requiring inspectors to give prior written notice of their intention to serve an improvement notice have been withdrawn, the HSC has announced (*Press Release C11:98, 31 March 1998*) that informal consultation between inspectors and employers regarding the issue of such notices will continue. Inspectors are expected to discuss with the employer concerned the alleged breaches of the law and any remedial action required in an attempt to resolve any points of difference before issuing a notice.

Failure to comply with an improvement notice can have serious penal consequences (see E15036 below).

Prohibition notices

E15015 If an inspector is of the opinion that, with regard to any activities to which *s 22(1)* applies (see below), the activities involve or will involve a risk of serious personal injury, he may serve on that person a notice (a prohibition notice). [*HSWA 1974, s 22(2)*].

It is incumbent on an inspector to show, on a balance of probabilities, that there is a risk to health and safety (*Readmans Ltd v Leeds CC [1992] COD 419* where an environmental health officer served a prohibition notice on the appellant regarding shopping trolleys with child seats on them, following an accident involving an eleven-month-old child. The appellant alleged that the Employment tribunal had wrongly placed the burden of proof on them, to show that the trolleys were not dangerous. It was held by the High Court (allowing the appeal), that it was for the inspector to prove that there was a health and/or safety risk).

Prohibition notices differ from improvement notices in two important ways:

(*a*) with prohibition notices, it is not necessary that an inspector believes that a provision of *HSWA 1974* or any other statutory provision is being or has been contravened;

(*b*) prohibition notices are served in *anticipation* of danger.

The *HSWA 1974, s 22* applies where, in the inspector's opinion, there is a hazardous activity or state of affairs generally. It is irrelevant that the hazard or danger is not mentioned in *HSWA 1974*; it can exist by virtue of other legislation, or even in the absence of any relevant statutory duty. In this way notices are used to enforce the later statutory requirements of *HSWA 1974* and the earlier requirements of the *Factories Act 1961* and other protective occupational legislation.

A prohibition notice must:

(*a*) state that the inspector is of the opinion stated immediately above;

(*b*) specify the matters which create the risk in question;

(*c*) where there is actual or anticipatory breach of provisions and regulations, state that the inspector is of the opinion that this is so and give reasons;

(*d*) direct that the activities referred to in the notice must not be carried out on, by or under the control of the person on whom the notice is served, unless the matters referred to in (*b*) above have been remedied.

[*HSWA 1974, s 22(3)*].

Failure to comply with a prohibition notice can have serious penal consequences (see E15028 below).

Differences between improvement and prohibition notices

E15016 Unlike an improvement notice, where time is allowed in which to correct a defect or offending state of affairs, a prohibition notice can take effect immediately.

A direction contained in a prohibition notice shall take effect:

(*a*) at the end of the period specified in the notice; or

(*b*) if the notice so declares, immediately.

[*HSWA 1974, s 22(4)* as substituted by *Consumer Protection Act 1987, Sch 3*].

Risk of injury need not be imminent, even if the notice is to take immediate effect (*Tesco Stores Ltd v Kippax COIT No 7605–6/90*).

An improvement notice gives a person upon whom it is served time to correct the defect or offending situation. A prohibition notice, which is a direction to stop the work activity in question rather than put it right, can take effect immediately on issue; alternatively, it may allow time for certain modifications to take place (i.e. deferred prohibition notice). Both types of notice will generally contain a schedule of work which the inspector will require to be carried out. If the nature of the work to be carried out is vague, the validity of the notice is not affected. If there is an appeal, an Employment tribunal may, within its powers to modify a notice, rephrase the schedule in more specific terms (*Chrysler (UK) Ltd v McCarthy [1978] ICR 939*).

Effect of non-compliance with notice

E15017 If, after expiry of the period specified in the notice, or in the event of an appeal, expiry of any additional time allowed for compliance by the Employment tribunal, an applicant does not comply with the notice or modified notice, he can be prosecuted. If convicted of contravening a prohibition notice, he may be imprisoned. [*HSWA 1974, s 33(1)(g), (2A)(b)*]. In *R v Kerr; R v Barker (1996) (unreported)* the directors of a company were each jailed for four months after allowing a machine which was subject to a prohibition notice – following an accident in which an employee lost an arm – to continue to be operated.

Service of notice coupled with prosecution

E15018 Where an inspector serves a notice, he may at the same time decide to prosecute for the substantive offence specified in the notice. The fact that a notice has been served is not relevant to the prosecution. Nevertheless, an inspector will not normally commence proceedings until after the expiry of 21 days, i.e. until he is satisfied that there is to be no appeal against the notice or until the Employment tribunal has heard the appeal and affirmed the notice, since it would be inconsistent if conviction by the magistrates were followed by cancellation of the notice by the Employment tribunal. The fact that an Employment tribunal has upheld a notice is not binding on a magistrates' court hearing a prosecution under the statutory provision of which the notice alleged a contravention; it is necessary for the prosecution to prove all the elements in the offence (see E15031 below).

Employment tribunals are mainly concerned with hearing unfair dismissal claims by employees'; only a tiny proportion of cases heard by them relate specifically to health and safety. Moreover, they are not empowered to determine breaches of criminal legislation.

Appeals against improvement and prohibition notices

E15019 A person on whom either type of notice is served may appeal to an Employment tribunal within 21 days from the date of service of the notice. The Employment tribunal may extend this time where it is satisfied, on application made in writing (either before or after expiry of the 21-day period), that it was not reasonably practicable for the appeal to be brought within the 21-day period. On appeal the Employment tribunal may either affirm or cancel the notice and, if it affirms it, may do so with modifications in the form of additions, omissions or amendments. [*HSWA 1974, ss 24*(*2*), *82*(*1*)(*c*); *Employment Tribunals* (*Constitution and Rules of Procedure*) *Regulations 2001* (*SI 2001 No 1171*), *Sch 5*].

Effect of appeal

E15020 Where an appeal is brought against a notice, the lodging of an appeal automatically suspends operation of an improvement notice, but a prohibition notice will continue to apply unless there is a direction to the contrary from the Employment tribunal. Thus:

(*a*) in the case of an improvement notice, the appeal has the effect of suspending the operation of the notice [*HSWA 1974, s 24*(*3*)(*a*)];

(*b*) in the case of a prohibition notice, the appeal only suspends the operation of the notice in the following circumstances:

   (i) if the Employment tribunal so directs, on the application of the appellant; and

   (ii) the suspension is then effective from the time when the Employment tribunal so directs.

[*HSWA 1974, s 24*(*3*)].

Grounds for appeal

E15021 The main grounds for appeal are:

(*a*) the inspector wrongly interpreted the law (see E15022 below);

(*b*) the inspector exceeded his powers, though not necessarily intentionally, under an Act or regulation (see E15023 below);

(*c*) breach of law is admitted but the proposed solution is not 'practicable' or not 'reasonably practicable', or that there was no 'best practicable means' other than that used where 'best practicable means' is also a defence to a charge of statutory nuisance (depending on the terminology of the particular statute) (see E15024 below);

(*d*) breach of law is admitted but the breach is so insignificant that the notice should be cancelled (see E15025 below).

The merits of appealing the issue or terms of an enforcement notice should always be considered carefully as, in practice, this may be the employer's only opportunity to dispute the reasonableness of the requirements imposed. Failure to comply with a notice is a strict liability offence and if a prosecution is brought it is not open to the employer to argue that he did everything 'reasonably practicable' to comply with the notice (*Deary v Mansion Hide Upholstery Ltd [1983] ICR 610*, see E15031 below).

The delay involved in lodging an appeal can have important practical consequences. Months might elapse before an appeal against a notice can be heard by the Employment tribunal. For this reason particularly, and in view of the fact that notice of appeal suspends operation of an improvement notice, some recipients opt for appealing, since at the time of service of notice they may not be in a position to meet the requirements of the notice; whereas, three or four months later, the position may have changed or a compromise arrangement reached with the inspector. If, however, the sole reason for appealing is to gain time and nothing else, the Employment tribunal is unlikely to be sympathetic and costs could be awarded against the unsuccessful appellant, though this is rare. [*Industrial Tribunals* (*Constitution and Rules of Procedure*) *Regulations 1993*, *Reg 8*(*4*), *Sch 4*, which states that: 'a Employment tribunal may make an Order that a party shall pay to another party either a specified sum in respect of the costs of or in connection with an appeal incurred by that other party or, in default of agreement, the taxed amount of those costs'.]

Inspector's wrong interpretation of the law

E15022 It is doubtful whether many cases have, or indeed would, succeed on this ground. Where regulations impose a strict duty (for example the duty under *Reg 6*(*1*) of the *Provision and Use of Work Equipment Regulations 1992* (*SI 1992 No 2932* as amended by *SI 2002 No 2174*)) to provide work equipment in good repair) there is no scope for argument by the employer. However, where the statute provides a defence, for example, it requires the duty to be carried out 'so far as reasonably practicable' (discussed more fully at E15031 below) or it provides for a due diligence defence, there is some scope to argue that the inspectors' interpretation of the law is incorrect. For example, in *Canterbury City Council v Howletts and Port Lympne Estates Limited (The Times, 13 December 1996)*, a prohibition notice was served on Howletts Zoo following the death of a keeper while he was cleaning the tigers' enclosure. It was Howletts' policy to allow their animals to roam freely; the local authority argued that the zoo's keepers could have carried out their tasks in the tigers' enclosure with the animals secured. The High Court affirmed the Employment tribunal's decision to set aside the notice, holding that *HSWA 1974, s 2* was not intended to render illegal certain working practices simply because they were dangerous.

Employers should note that contesting a prohibition notice on the ground that there has been no legal contravention is pointless, since valid service of a prohibition notice does not depend on legal contravention (*Roberts v Day, COIT No 1/133, Case No 3053/77*).

Inspector exceeded powers under statute

E15023 It can happen that an inspector exceeds his powers under statute by reason of misinterpretation of the statute or regulation (*Deeley v Effer, COIT No 1/72, Case No 25354/77*). This case involved the requirement that 'all floors, steps, stairs, passages and gangways must, so far as is reasonably practicable, be kept free from obstruction and from any substance likely to cause persons to slip'. [*OSRPA 1963, s 16*(*1*)] (now replaced by equivalent duties under the *Workplace* (*Health, Safety and Welfare*) *Regulations 1992* (*SI 1992 No 3004*). The inspector considered that employees were endangered by baskets of wares in the shop entrance. The Employment tribunal ruled that the notice had to be cancelled, since the only persons endangered were members of the public, and *OSRPA 1963* was concerned with dangers to employees.

Proposed solution not practicable

E15024 The position in a case where, although breach of the law is admitted, the proposed solution is not considered practicable, depends upon the nature of the obligation. The duty may be strict, or have to be carried out so far as practicable or, alternatively, so far as reasonably practicable. In the first two situations cost of compliance is irrelevant; in the latter case, where a requirement has to be carried out 'so far as reasonably practicable', cost effectiveness is an important factor but has to be weighed against the risks to health and safety involved in failing to implement the remedial measures identified in the enforcement notice. Where there is a real danger of serious injury the cost of complying with the notice is unlikely to be decisive. Thus in a leading case, the appellant was served with an improvement notice requiring secure fencing on transmission machinery. An appeal was lodged on the ground that the proposed modifications were too costly (£1,900). It was argued that because of the intelligence and integrity of the operators a safety screen costing £200 would be adequate. The Employment tribunal dismissed the appeal: the risk justified the cost (*Belhaven Brewery Co Ltd v McLean [1975] IRLR 370*).

The cost of complying with a notice is likely to carry less weight in the case of a prohibition notice than an improvement notice, as there must be 'a risk of serious personal injury' for a prohibition notice to be served (*Nico Manufacturing Co Ltd v Hendry [1975] IRLR 225*, where the company argued that a prohibition notice in respect of the worn state of their power presses should be cancelled on the ground that it would result in a 'serious loss of production' and endanger the jobs of several employees'. The Employment tribunal dismissed this argument, having decided that using the machinery in its worn condition could cause serious danger to operators). Similarly, an undertaking by a company to take additional safety precautions against the risk of injury from unsafe plant until new equipment was installed was not sufficient (*Grovehurst Energy Ltd v Strawson (HM Inspector) COIT No 5035/90*).

Where cost is a factor this is not to be confused with the current financial position of the company. A company's financial position is irrelevant to the question whether an Employment tribunal should affirm an enforcement notice. Thus in *Harrison (Newcastle-under-Lyme) Ltd v Ramsay (HM Inspector) [1976] IRLR 135*, a notice requiring cleaning, preparation and painting of walls had to be complied with even though the company was on an economy drive.

Employment tribunals have power under *HSWA 1974, s 24* to alter or extend time limits attaching to improvement and prohibition notices (*D J M and AJ Campion v Hughes (HM Inspector of Factories) [1975] IRLR 291*, where even though there was an imminent risk of serious personal injury, a further four months were allowed for the erection of fire escapes as it was not practicable to carry out the remedial works within the time limits set). Extensions of time in which to comply with notices are most commonly granted where the costs of the improvements and modifications required by the notice are significant.

Breach of law is insignificant

E15025 Employment tribunals will rarely cancel a notice which concerns breach of an absolute duty where the breach is admitted but the appellant argues that the breach is trivial: *South Surbiton Co-operative Society Ltd v Wilcox [1975] IRLR 292*, where a notice had been issued in respect of a cracked wash-hand basin, being a breach of an absolute duty under the *Offices, Shops and Railway Premises Act 1963*. It was argued by the appellant that, in view of their excellent record of cleanliness, there was no need for officials to visit the premises. The appeal was dismissed.

Inspectors investigation powers

E15026 Inspectors have wide ranging powers under *HSWA 1974* to investigate suspected health and safety offences. These include powers to:

(*a*) enter and search premises;

(*b*) direct that the premises or anything on them be left undisturbed for so long as is reasonably necessary for the purpose of the investigation;

(*c*) take measurements, photographs and recordings;

(*d*) take samples of articles or substances found in the premises and of the atmosphere in or in the vicinity of the premises;

(*e*) dismantle or test any article which appears to have caused or be likely to cause danger;

(*f*) detain items for testing or for use as evidence;

(*g*) interview any person;

(*h*) require the production and inspection of any documents and to take copies; and

(*i*) require the provision of facilities and assistance for the purpose of carrying out the investigation.

[*HSWA 1974, s 20*]

Under these powers inspectors may require interviewees to answer such questions as they think fit and sign a declaration that those answers are true. [*HSWA 1974, s 20*(*1*)(*j*)]. However, evidence given in this way is inadmissible in any proceedings subsequently taken against the person giving the statement or his or her spouse. [*HSWA 1974, s 20*(*7*)]. Where prosecution of an individual is contemplated the inspector will, therefore, usually exercise his evidence-gathering powers under the *Police and Criminal Evidence Act 1984* (*PACE*). Evidence given in this way is admissible against that person in later proceedings. Interviews conducted under *PACE* are subject to strict legal controls, for example, interviewees must be cautioned before the interview takes place, they have certain 'rights to silence' (although these were qualified by the *Criminal Justice and Public Order Act 1994*), and there are rules relating to the recording of the interview.

Inspectors' powers of search and seizure in case of imminent danger

E15027 Where an inspector has reasonable cause to believe that there are on premises 'articles or substances ('substance' includes solids, liquids and gases – *HSWA 1974, s 53*(*1*)) which give rise to imminent risk of serious personal injury', he can:

(*a*) seize them; and

(*b*) cause them to be rendered harmless (by destruction or otherwise).

[*HSWA 1974, s 25*(*1*)].

Enforcing authorities are given similar powers regarding environmental pollution, under the *Environment Act 1995, ss 108, 109.*

Before an article forming 'part of a batch of similar articles', or a substance is rendered harmless, an inspector must, if practicable, take a sample and give to a responsible person, at the premises where the article or substance was found, a portion which has been marked in such a way as to be identifiable. [*HSWA 1974, s 25*(*2*)]. After the article or substance has been rendered harmless, the inspector must sign a prepared report and give a copy of the report to:

(i) a responsible person (e.g. safety officer); and

(ii) the owner of the premises, unless he happens to be the 'responsible person'. (See A3005 ACCIDENT REPORTING AND INVESTIGATION for the meaning of this term.)

[*HSWA 1974, s 25*(*3*)].

Analogous powers are given to transport inspectors under the *Radioactive Material* (*Road Transport*) *Act 1991, s 5* in respect of vehicles carrying radioactive packages.

A customs officer may assist the enforcing authority or the inspector in his enforcement duties under the *HSWA 1974* by seizing any imported article or substance. He may then detain it for not more than 2 working days and may disclose information about it to the enforcing authorities or inspectors. [*HSWA 1974, ss 25A, 27A inserted by Consumer Protection Act 1987, Sch 3*].

Indemnification by enforcing authorities

E15028 Where an inspector has an action brought against him in respect of an act done in the execution or purported execution of any of the 'relevant statutory provisions' (see E15012 above), and is ordered to pay damages and costs (or expenses) in circumstances where he is not legally entitled to require the enforcing authority which appointed him to indemnify him, he may be able to take advantage of *HSWA 1974, s 26*. By virtue of that provision, the authority nevertheless has the power to indemnify the inspector against all or part of such damages where the authority is satisfied that the inspector honestly believed:

(*a*) that the act complained of was within his powers; and

(*b*) that his duty as an inspector required or entitled him to do it.

In practice there will be very few circumstances where an inspector is held liable for advice given or enforcement action taken as part of his statutory duties. The Court of Appeal has ruled (in *Harris v Evans and Another [1998] 3 All ER 522*) that an enforcing authority giving advice which leads to the issue of enforcement notices does not owe a duty of care to the owner of the premises affected by the notice and a claim for economic loss arising from such allegedly negligent advice cannot therefore arise. The court said that if enforcing authorities were to be exposed to liability in negligence at the suit of owners whose businesses are adversely affected by their decisions it would have a detrimental effect on the performance by inspectors of their statutory duties. The *HSWA 1974* contains its own statutory remedies against errors by inspectors and the court was not prepared to add to those measures. The court did, however, suggest that a possible exception might arise if a requirement imposed by the inspector introduced a new risk or danger which resulted in physical damage or economic loss.

Public register of improvement and prohibition notices

E15029 Improvement (though not in the case of the *Fire Precautions Act 1971*) and prohibition notices relating to public safety matters have to be entered in a public register as follows:

(*a*) within 14 days following the date on which notice is served in cases where there is no right of appeal;

(*b*) within 14 days following the day on which the time limit expired, in cases where there is a right of appeal but no appeal has been lodged within the statutory 21 days.

(*c*) within 14 days following the day when the appeal is disposed of, in cases where an appeal is brought.

[*Environment and Safety Information Act 1988, s 3*].

Notices which impose requirements or prohibitions solely for the protection of persons at work are not included in the register. [*Environment and Safety Information Act 1988, s 2*(*3*)].

In addition, registers must be kept of notices served by:

(i) fire authorities, under the *Schedule* to the *Environment and Safety Information Act 1988* for the purpose of *s 10* of the *Fire Precautions Act 1971* (not improvement notices);

(ii) local authorities, under the *Schedule* to the *Environment and Safety Information Act 1988* for the purpose of *s 10* of the *Safety of Sports Grounds Act 1975*;

(iii) responsible authorities (as defined by the *Environment and Safety Information Act 1988, s 2*(*2*)) and the Minister of Agriculture, Fisheries and Food under *s 2* of the *Environment and Safety Information Act 1988* for the purpose of *s 19* of the *Food and Environment Protection Act 1985*;

(iv) enforcing authorities, under *s 20* of the *Environmental Protection Act 1990*; and

(v) enforcing authorities, under the *Radioactive Substances Act 1993*.

These registers are open to inspection by the public free of charge at reasonable hours and, on request and payment of a reasonable fee, copies can be obtained from the relevant authority. [*Environment and Safety Information Act 1988, s 1*]. Such records can also be kept on computer.

Part B: Offences and Penalties

Prosecution for breach of the 'relevant statutory provisions'

E15030 Prosecutions can follow non-compliance with an improvement or prohibition notice, but equally inspectors' will sometimes prosecute without serving a notice. Service of notices remains the most usual method of enforcement. In 2002/03 approximately 13,000 enforcement notices were issued by the HSE compared with about 1,600 prosecutions for health and safety offences. Prosecutions are most commonly brought after a workplace accident or dangerous incident (such as a fire or explosion). Investigations by the enforcing authorities may take many months to complete and, in complex cases, it is not unusual for prosecutions to be commenced up to a year after the original incident. Prosecutions normally take place before the magistrates but there is increasing pressure for more serious cases, involving work-related fatalities or major injuries, to be prosecuted on indictment in the Crown Court, where increased penalties are available. The determining factor behind prosecution on indictment is the gravity of the offence. Whilst the decision whether to prosecute, issue an enforcement notice, or simply to give advice lies within the discretion of the inspector concerned, the enforcing authorities aim to pursue a policy which is open, consistent and proportionate to the risks in deciding what enforcement action to take (see the HSC's *Enforcement Policy Statement* of *January 2002*, and the *Enforcement Concordat* published by the Better Regulation Taskforce in April 1999 which has been adopted by the HSE and most local authorities).

In practice the enforcing authorities usually only investigate the most serious workplace accidents; it is HSE policy to investigate all work-related deaths which are reported to HSE, although the Health and Safety Statistics 2002/03, published by the HSE, shows a 15 per cent decrease in the total number of prosecutions brought by the HSE in the period under review. The greatest number of cases brought was in the manufacturing sector (37 per cent), followed closely by construction (36 per cent). In the same period the HSE prosecuted a total of 22 managers and directors, of which 11 led to prosecutions.

Burden of proof

E15031 Throughout criminal law, the burden of proof of guilt is on the prosecution to show that the accused committed the particular offence (*Woolmington v DPP [1935] AC 463*). The burden is a great deal heavier than in civil law, requiring proof of guilt beyond a reasonable doubt as distinct from on a balance of probabilities. While not eliminating the need for the prosecution to establish general proof of guilt, *HSWA 1974, s 40* makes the task of the prosecution easier by transferring the onus of proof to the accused for one element of certain offences. *Section 40* states that in any proceedings for an offence consisting of a failure to comply with a duty or requirement to do something so far as is practicable, or so far as reasonably practicable, or to use the best practicable means to do something, the onus is on the accused to prove (as the case may be) that it was not practicable, or not reasonably practicable to do more than was in fact done to satisfy the duty or requirement, or that there was no better practicable means than was in fact used to satisfy the duty or requirement. However, *s 40* does not apply to an offence created by *s 33(1)(g)* of the *HSWA 1974* – failing to comply with an improvement notice.

In *Deary v Mansion Hide Upholstery Ltd [1983] ICR 610*, an improvement notice was served on the defendant company requiring it to provide the fire resistant storage for polyurethane foam. The company did not comply, nor did it appeal. It was irrelevant that the company had complied with the notice so far as 'reasonably practicable' as that was not a requirement of the offence charged. A similar burden of proof exists under the *Environmental Protection Act 1990.*

Human rights legislation in the UK could have significant implications on the issue of the burden of proof in criminal cases. The case of *R v Lambert [2001] 3 WLR 206*, considered the effect of Article 6(2) of the European Convention on Human Rights ('ECHR') which requires that only an evidential burden can be placed on a defendant. In *R v Lambert* the defendant appealed against his conviction on the grounds that the trial judge's direction that the legal burden of proof was transferred to the defendant was contrary to Article 6(2). The appeal was dismissed by the Court of Appeal and the House of Lords on the grounds that the defendant could not rely on the *Human Rights Act 1998* ('*HRA 1998*') in relation to a conviction which predated the coming into force of the Act. However, in that case the defendant was challenging the trial judge's direction and not the authority which had brought the proceedings. Accordingly, the appeal brought by the defendant was not a proceeding by or instigated by a public authority so he was not entitled to claim retrospectivity under *s 22(4)* of the *HRA 1998*.

R v Lambert highlighted the potential for courts to review transfer of burden of proof cases. As a consequence, a recent preliminary Crown Court ruling in the case of *HSE v Klockner Moeller Limited* on 17 May 2002 has challenged the legitimacy of the transfer of burden provisions under the *HSWA 1974, s 40*. In *that case* the judge ruled, following submissions on how he ought to direct a jury at trial as to the effect of *s 40* that it should be interpreted such that the burden on the defence on the issue of reasonable practicability is an evidential one requiring only that the defence 'give sufficient evidence to raise the issue'. The effect of the ruling is to reverse the burden of proof required under *s 40* so that once the defence has met the evidential burden the prosecution must prove guilt beyond reasonable doubt. The present position is that the prosecution must have sought leave to appeal this challenge to the meaning of *s 40*. If the Crown Court's interpretation of *s 40* is upheld the onus of proof requirements on defendants for certain elements of offences committed under the *HSWA 1974*, will be considerably reduced.

Main offences and penalties

E15032 Health and safety offences are either (a) triable summarily (i.e. without jury before the magistrates), or (b) triable summarily and on indictment (i.e. triable either way), or (c) triable only on indictment. Most health and safety offences, however, fall into categories (*a*) and (*b*). The main offences falling into these two categories are set out below.

Summary only offences

E15033

(*a*) contravening a requirement imposed under *HSWA 1974, s 14* (power of the HSC to order an investigation);

(*b*) contravening a requirement imposed by an inspector under *HSWA 1974, s 20*;

(*c*) preventing or attempting to prevent a person from appearing before an inspector, or from answering his questions;

(*d*) intentionally obstructing an inspector or customs officer in the exercise of his powers;

(*e*) falsely pretending to be an inspector.

'Either way' offences

E15034

(*a*) failure to carry out one or more of the general duties of *HSWA 1974, ss 2–7*;

(*b*) contravening either:

   (i) *HSWA 1974, s 8* – intentionally or recklessly interfering with anything provided for safety;

   (ii) *HSWA 1974, s 9* – levying payment for anything that an employer must by law provide in the interests of health and safety (e.g. personal protective clothing);

(*c*) contravening any health and safety regulations;

(*d*) contravening a requirement imposed by an inspector under *HSWA 1974, s 25* (power to seize and destroy articles and substances);

(*e*) contravening a requirement of a prohibition or improvement notice;

(*f*) intentionally or recklessly making false statements, where the statement is made:

(i) to comply with a requirement to furnish information; or

(ii) to obtain the issue of a document;

(*g*) intentionally making a false entry in a register book, notice etc. which is required to be kept;

(*h*) failing to comply with a remedial court order made under *HSWA 1974, s 42*.

[*HSWA 1974, s 33*(*1*)].

In England and Wales there is no time limit for bringing prosecutions for offences, except for offences tried summarily in the magistrates' courts – where the time limit is 6 months from the date the complaint was laid. [*Magistrates' Courts Act 1980, s 127*(*1*)]. In Scotland the 6-month time limit extends to 'either way' offences tried summarily. The period may be extended in the case of special reports, coroners' court hearings or in cases of death generally. [*HSWA 1974, s 34*(*1*)] There is no time limit for commencing hearings in the Crown Court.

Summary trial or trial on indictment

E15035 Most offences triable either way are tried summarily. However an increasing number of serious offences are being referred to the Crown Court, which has increased sentencing powers, for trial on indictment. The Court of Appeal has advised magistrates to exercise caution in accepting jurisdiction in health and safety cases where the offence may require a penalty greater than they can impose or where death or serious injury has resulted from the offence (see *R v F Howe & Son (Engineering) Ltd* at E15043 below). The defendant may also refuse to consent to summary trial and opt for trial on indictment.

Penalties for health and safety offences

E15036 Penalties tend to relate to the three main categories of offences characterising breach of health and safety legislation, namely:

(*a*) breaches of *ss 2–6* of the *HSWA 1974* – serious offences:

(i) summary conviction – a maximum £20,000 fine;

(ii) conviction on indictment – an unlimited fine (but no imprisonment);

(*b*) breaches of improvement or prohibition orders, or orders under *s 42* of the *HSWA 1974* to remedy the cause of the offence – serious offences:

(i) summary conviction – a maximum £20,000 fine, or imprisonment for up to six months or both;

(ii) conviction on indictment – an unlimited fine or imprisonment for up to two years or both;

(*c*) most other offences, including breaches of health and safety regulations:

(i) summary conviction – a maximum fine of £5,000;

(ii) conviction on indictment – an unlimited fine. For licence breaches and explosives offences, the court may also order up to two years imprisonment (see, for example, *R v Hill (1996) (unreported)*, where a demolition contractor was jailed for three months for breach of asbestos licensing regulations).

[*HSWA 1974, s 33(1A), (2)* and *(2A)* as inserted by the *Offshore Safety Act 1992*, and *HSWA 1974, s 33(3)*].

In December 1999, the Lord Chancellor announced that the Government intends to legislate to increase the penalties available for health and safety offences as soon as parliamentary time allows. A private members bill, the *Health and Safety at Work (Offences) Bill*, was introduced in December 1999, but to date it has not been progressed. The Bill, if enacted, would raise the maximum fine which can be imposed in the Magistrates' Court for contraventions of requirements under health and safety regulations from the present £5,000 limit to match the £20,000 maximum that can be imposed for breaching the general duties under the *HSWA 1974, ss 2–7*. It would also make imprisonment an option for most health and safety offences in both the higher and lower courts.

Following the Court of Appeal decision in the case of *R v Colthorp Board Mill Ltd (January 2002) (unreported)*, an increase in the level of fines imposed for breaches of health and safety legislation is anticipated.

In his judgment, Lord Justice Gibbs referring to the authorities, in particular *R v F Howe & Son (Engineering) Ltd [1999] 2 All ER 249* and *R v Friskies Petcare UK Ltd (2000) 2 Cr App R (S) 401*, stated that although the remarks in a previous case that financial penalties in excess of £500,000 were reserved for cases of major public disaster, 'we would not wish the sum of £500,000 to appear to be set in stone or to provide any sort of maximum limit for such cases. On the contrary, we anticipate that as time goes on and awareness of the importance of safety increases, that courts will uphold sums of that amount and even in excess of them in serious cases, whether or not they involve what could be described as major public disasters'. The overall effect of the decision is expected to result in more cases being referred to the Crown Courts for sentence and correspondingly higher fines for corporate defendants. Notably, however, the latest statistics from the HSE for the year 2002/03 show that the average fine per case has decreased by 21 per cent from the previous year 2001/02 – although some of this drop is a result of seeing larger higher court fines in the year 2001/02 than in 2002/03.

Defences

E15037 Although no general defences are specified in the *Health and Safety at Work etc Act 1974*, some regulations passed under the Act carry the defence of 'due diligence' (for example, the *Control of Substances Hazardous to Health Regulations 2002* (*SI 2002 No 2677* as amended by *SI 2003 No 978*)).

Manslaughter

E15038 In cases of workplace death, manslaughter charges may also be brought if there is sufficient evidence. The decision to prosecute rests with the Crown Prosecution Service, not the enforcing authorities under health and safety legislation. Manslaughter convictions linked to breaches of safety legislation are rare but are becoming more common. There have been a series of individual prosecutions for manslaughter of directors or managers responsible for workplace deaths and, in

1994, the first successful prosecution for corporate manslaughter involved OLL Ltd, the activity centre responsible for organising the Lyme Bay canoeing trip in which four teenagers died.

Offences committed by particular types of persons

Corporate offences – delegation of duties to junior staff

E15039 Companies cannot avoid liability for breach of general duties under *ss 2–6* of the *HSWA 1974* by arguing that the senior management and the 'directing mind' of the company had taken all reasonable precautions, and that responsibility for the offence lay with a more junior employee or agent who was at fault. The *HSWA 1974* generally imposes strict criminal liabilities on employers and others (subject to the employer being able to establish that all reasonably practicable precautions had been taken) and it is not open to corporate employers to seek to avoid liability by arguing that their general duties have been delegated to someone lower down the corporate tree. In *R v British Steel plc [1995] IRLR 310*, British Steel were prosecuted under *HSWA 1974, s 3* after a fatal accident to a subcontractor who was carrying out construction work under the supervision of a British Steel engineer. British Steel argued that it was not responsible under *s 3* for the actions of the supervising engineer as the engineer was not part of the 'directing mind' of the company and all reasonable precautions to ensure the safety of the work had been taken by senior management. The Court of Appeal dismissed this argument. A similar decision was reached in *R v Gateway Foodmarkets Ltd, [1997] IRLR 189*, which concerned a breach of *HSWA 1974, s 2* arising out of a fatal accident to an employee who fell through a trap door in the floor of a lift control room. The accident occurred while the store manager was manually attempting to rectify an electrical fault in the lift in accordance with a local practice which was not authorised by Gateway's head office. The Court of Appeal held that the failure at store manager level was attributable to the employer.

However, it does not follow that an employer will automatically be held criminally responsible for an isolated act of negligence by an employee performing work on its behalf. This is because it may still be possible for the employer to establish that it has done everything reasonably practicable in the conduct of its undertaking to ensure that employees and third parties are not exposed to risks to their health and safety by virtue of the way it has conducted its business (see *R v Nelson Group Services (Maintenance) Limited [1999] IRLR 646, Court of Appeal*).

The definition of 'reasonably practicable' was authoritatively laid down in *Edwards v National Coal Board [1949] 1 All ER 743* where it was said that: 'Reasonably practicable' is a narrower term than 'physically possible', and seems to imply that a computation must be made by the owner in which the quantum of risk is placed on one scale and the sacrifice involved in the measures necessary for averting the risk (whether in money, time or trouble) is placed in the other, and that, if it be known that there is a gross disproportion between them – the risk being insignificant in relation to the sacrifice – the defendants discharge the onus on them.'

In a series of decisions the Court of Appeal have concluded that the words 'so far as reasonably practicable' provide a limited defence to what are otherwise absolute obligations on employers and other duty holders (*R v British Steel Plc [1995] IRLR 310*, approved by the House of Lords in *R v Associated Octel Co Ltd [1996] 1 WLR 1543*). Although the circumstances where such a defence may be established are likely to be rare – it involves doing more than simply exercising reasonable care – the Court of Appeal in *R v Nelson Group Services (Maintenance) Limited* made clear that an isolated act of negligence by an employee performing work on behalf of the company does not preclude that employer from establishing a defence that it has done everything reasonably practicable. The court said: *'It is not necessary for the adequate protection of the public that the employer should be held criminally liable even for an isolated act of negligence by the employee performing the work. Such persons are themselves liable to criminal sanctions under the Act and under the Regulations. Moreover it is a sufficient obligation to place on the employer in order to protect the public to require the employer to show that everything reasonably practicable has been done to see that a person doing the work has the appropriate skill and instruction, has had laid down for him safe systems of doing the work, has been subject to adequate supervision, and has been provided with safe plant equipment for the proper performance of the work.'*

The question of what is reasonably practicable is a question of fact which must be determined in the light of the circumstances of each case. The burden of proving, on the balance of probabilities, that all reasonably practicable steps have been taken rests with the employer (see also E15024).

Similarly, where there is a defence of due diligence to the legislation allegedly breached, the company may be able to avoid liability if the members of senior management responsible for actual control of the company's operations have exercised due diligence and the failure occurs because of the actions of a junior member of staff: *Tesco Supermarkets Ltd v Nattrass [1972] AC 153*, a House of Lords decision on the wording of the *Trade Descriptions Act 1968*. The case was considered by the Court of Appeal in *R v British Steel plc [1994] IRLR 540* who distinguished it from the factual circumstances before them on the basis that the *Tesco* decision involved application of a due diligence defence which was not part of *HSWA 1974, s 3*.

Offences of directors or other officers of a company

E15040 Where an offence is committed by a body corporate, senior persons in the hierarchy of the company may also be individually liable. Thus, where the offence was committed with the consent or connivance of, or was attributable to any neglect on the part of a director or officer of the company, that person is himself guilty of an offence and liable to be punished accordingly. Those who may be so liable are:

(*a*) any functional director;

(*b*) a manager (which does not include an *employee* in charge of a shop while the manager is away on a week's holiday (*R v Boal [1992] 1 QB 591*, concerning *s 23* of the *Fire Precautions Act 1971* – identical terminology to *HSWA 1974, s 37*));

(*c*) a company secretary;

(*d*) another similar officer of the company;

(*e*) anyone purporting to act as any of the above.

[*HSWA 1974, s 37*(*1*)].

It is not sufficient that the company through its 'directing mind' (its board of directors) has committed an offence – there must be some degree of personal culpability in the form of proof of consent, connivance or neglect by the individual concerned. Evidence of this sort can be difficult to obtain and prosecutions under *s 37*(*1*) have, in the past, been rarely compared with prosecutions of companies (although they are increasing in number). In 2000/01, the HSE prosecuted individuals on 55 separate charges, of which 45 led to conviction. This included 36 charges against directors and managers, of which 31 led to conviction.

Directors, managers and company secretaries can be personally liable for ensuring that corporate safety duties are performed throughout the company (for example, a failure to maintain a safe system of work can give rise to personal liability). Liability may also arise as a result of a failure to perform an obligation placed on individuals by their employment contracts and job descriptions – for example, obligations imposed under a safety policy – not just in relation to duties imposed by law. In the case of *Armour v Skeen (Procurator Fiscal, Glasgow) [1977] IRLR 310*, an employee fell to his death whilst repairing a road bridge over the River Clyde. The appellant, who was the Director of Roads, was held to be under a duty to supervise the safety of council workmen. He had not prepared a written safety policy for roadwork, despite a written request that he do so, and was found to have breached *HSWA 1974, s 37(1)*.

Similar duties exist under the *Environmental Protection Act 1990, s 157* and the *Environment Act 1995, s 95(2)–(4)*. Directors convicted of a breach of *HSWA 1974, s 37* may also be disqualified, for up to two years, from being a director of a company, under the provisions of the *Company Directors Disqualification Act 1986, s 2(1)* as having committed an indictable offence connected with (*inter alia*) the management of a company. In *R v Chapman (1992) (unreported)*, a director of a quarrying company was disqualified and fined £5,000 for contravening a prohibition notice on an unsafe quarry where there had been several fatalities and major injuries.

It is worth bearing in mind the recent developments with respect to a possible offence of corporate killing being introduced.

The Home Secretary, David Blunkett, confirmed on 20 May 2003 that the Government will publish a draft Bill on corporate manslaughter. A timetable for legislation and further details will be announced in Autumn 2003.

The Government has been looking carefully at this complex area of law and consulting with companies, business leaders and trade unions.

The legislation will be targeted at companies themselves, which is the area of weakness in the current law. No new burdens will be placed on companies which already comply fully with health and safety legislation. The criminal liability of individual directors will not be targeted by the proposals.

Reforming the laws on corporate manslaughter is part of the Government's wider agenda to modernise the criminal justice system – putting victims at the heart, protecting the public and ensuring that justice is done.

The Government has carried out a Regulatory Impact Assessment ('RIA'), the results of which are still being assessed but the preliminary indications are that the costs of a change in the law will not be large. The Government will take the RIA results into account in finalising its proposals to make the law on corporate manslaughter more effective and improve the lot of victims.

Directors' insurance

E15041 Companies can now buy insurance in order to protect directors. [*Companies Act 1985, s 310(3)*]. Moreover, directors need not contribute towards premiums, as they had to previously. Such insurance, which must be mentioned in the Annual Report and Accounts, may (subject to the terms of the policy), protect directors against:

(*a*) civil liability for claims made against them in breach of directorial duties, e.g. by shareholders when directors acted in breach of their duty of care to the company, as well as legal costs and expenses incurred in defence or settlement of such claims;

(*b*) legal costs and expenses involved in defending criminal actions (e.g. breach of *HSWA 1974, s 37*), but not the fine or other penalty incurred, it being illegal to insure against payment of penalties.

Companies can also indemnify directors against such costs and damages, but only where judgement is ultimately given in the individual's favour or he is acquitted.

Offences due to the act of another person

E15042 *Section 36(1)* of *HSWA 1974* makes clear that although provision is separately made for the prosecution of less senior corporate staff, e.g. safety officers, works managers, this does not prevent a further prosecution against the company itself. The section states that where an offence under *HSWA 1974* is due to the act or default of some other person, then:

(*a*) that other person is guilty of an offence; and

(*b*) a second person can be charged and convicted, whether or not proceedings are taken against the first-mentioned person.

Where the enforcing authorities rely on *HSWA 1974, s 36*, this must be made clear to the defendant. In *West Cumberland By Products Ltd v DPP, The Times, 12 November 1987*, the conviction of a company operating a road haulage business for breach of regulations relating to the transport of dangerous substances was set aside as the offence charged related to the obligations of the driver of the vehicle and, in prosecuting the operating company, reliance was not placed on *HSWA 1974, s 36*.

Sentencing Guidelines

E15043 The level of fine imposed for health and safety offences will ultimately depend on the facts and circumstances of the case, including the gravity of the offence, whether the breach resulted in death or serious injury, and any mitigating evidence the defendant is able to put forward (including details of its means and ability to pay any fine imposed). In the past there have been wide variations in the sentences handed down by different courts for similar breaches of the legislation and there has also been concern at the general low level of fines imposed for health and safety offences. In *R v F Howe & Son (Engineering) Limited [1999] 2 All ER 249*, the Court of Appeal sought to address these concerns by laying down guidelines to assist magistrates and judges in sentencing health and safety offences. The case concerned an appeal by the company against fines totalling £48,000 and an order for costs of £7,500 imposed in respect of four health and safety offences arising from a fatal accident to one of the company's employees who was electrocuted while using an electric vacuum machine to clean a floor at the company's premises. In reducing the level of fine imposed to reflect the company's limited financial resources, the Court of Appeal laid down the following general guidelines:

Sentencing Guidelines

General Principles

The level of fine should reflect:

- the gravity of the offence and the standard of the defendant's conduct;
- the degree of risk and extent of danger;
- the extent of the breach – an isolated incident may attract a lower fine than a continuing unsafe state of affairs;
- the defendant's resources and the effect of the fine on its business.

Aggravating Factors

- failure to heed warnings;
- if the defendant deliberately flouts safety legislation for financial reasons;
- if the offence results in a fatality.

Mitigating factors

- prompt admission of liability and guilty plea;
- steps taken to remedy deficiencies;
- a good safety record.

In applying these factors, the Court of Appeal made clear that every case needs to be considered on its own facts. It declined to lay down a tariff for particular offences or to link the level of fine directly to the defendant's turnover or net profit. However, it emphasised the importance of the defendant's means, as well as the gravity of the offence, in determining the appropriate level of fine, stating that this should be large enough to impress on both the management and shareholders of the defendant company the importance of providing a safe working environment. Although there might be cases where the offences are so serious that the defendant company ought not to be in business, in general the fine 'should not be so large as to emperil the earnings of employees or create a risk of bankruptcy'. In essence, the courts must answer two questions in determining the appropriate level of fine:

1. What financial penalty does the offence merit?
2. What penalty can the defendant reasonably be ordered to pay?

The Court of Appeal specifically made clear that the size and resources of the defendant company and its ability to provide safety measures or to employ in-house safety advisers are not a mitigating factor: the legislation imposes the same standard of care irrespective of the size of the organisation.

It also emphasised that if the offence results in a fatality that is a serious aggravating factor: 'the penalty should reflect public disquiet at the unnecessary loss of life.' In general, cases involving death or serious injury should be dealt with by the Crown Court, which has increased sentencing powers: the judgement in *Howe* made clear that 'magistrates should always think carefully before accepting jurisdiction' in such cases.

In *R v Rollco Screw & Rivet Co Ltd [1999] IRLR 439*, the Court of Appeal approved the principles laid down in *Howe* and indicated that the fine imposed should make clear that there is a personal responsibility on directors for their company's health and safety arrangements. It did, however, acknowledge that caution was necessary in the case of smaller companies, where the directors were also shareholders, to avoid imposing a fine which amounted to double punishment of the individuals concerned.

The court went on to suggest that, in appropriate circumstances, corporate defendants may be ordered to pay fines and costs by instalments over many years. In reducing Rollco's total payment period to 5 years and 7 months, the court held that there was no maximum period for payment of fines and costs. Although there might be good reason to limit the period of payment of fines and costs by a personal defendant, who may suffer anxiety because of his continuing financial obligations, the same considerations do not apply to a corporate defendant. The Court of Appeal indicated that, in proper circumstances, it might be appropriate to order payment of fines and costs by a corporate defendant over a 'substantially longer period' than would be appropriate in the case of an individual.

Publication of convictions for health and safety offences and enforcement notices

E15044 In an attempt to improve compliance with health and safety legislation, the HSE are also pursuing an active policy of naming companies that breach the legislation. Details of firms and individuals convicted of health and safety offences are published by the HSE in an annual report that can be accessed at their website: www.hse.gov.uk/action/content/off00–01.pdf. The HSE 'name and shame' website also publicises all prosecution cases taken by HSE since April 1999 which resulted in a conviction as well as, more recently, a register of improvement and prohibition notices issued by the HSE. This information can be accessed at: www.hse-databases.co.uk/notices/. In addition a local authority annual report names duty holders and individuals convicted of health and safety offences in Great Britain following investigations by their inspectors. This report can also be accessed on the HSE website.

Position of the Crown

E15045 The general duties of *HSWA 1974* bind the Crown. [*HSWA 1974, s 48*(*1*)]. (For the position under the *Factories Act 1961*, see W11031 WORKPLACES – HEALTH, SAFETY AND WELFARE.) However, improvement and prohibition notices cannot be served on the Crown, nor can the Crown be prosecuted [*HSWA 1974, s 48*(*1*)], although Crown employees' can be prosecuted for breaches of *HSWA 1974* [*HSWA 1974, s 48*(*2*)]. Non-statutory procedures are in place for the issue of Crown improvement and prohibition notices, and for the censure of Crown bodies in circumstances in which a prosecution would otherwise have been brought. In 2002/03, two censures, 14 improvement notices and 1 prohibition notice were issued against Crown bodies. Crown immunity is no longer enjoyed by health authorities, nor premises used by health authorities (defined as Crown premises) including hospitals (whether NHS hospitals or NHS trusts or private hospitals). [*National Health Service and Community Care Act 1990, s 60*]. Health authorities are also subject to the *Food Safety Act 1990*. Most Crown premises can be inspected by authorised officers in the same way as privately run concerns, though prosecution against the Crown is not possible. [*Food Safety Act 1990, s 54*(*2*)].

Environmental Management

Introduction

E16001 The management of environmental performance is now well accepted as a critical issue for organisations in both the public and private sectors.

Broadly speaking, environmental management refers to the controls implemented by an organisation to minimise the adverse environmental impacts of its operations. Historically, environmental management tends to have been driven by a complex and interacting array of external pressures, to which business somewhat reluctantly responded. However, in many cases a positive policy towards environment issues can lead to opportunities to improve business performance in general.

As the benefits become more apparent, business is now becoming more proactive in its approach to environmental management. External pressures are still important influences, but increasingly they tend to shape the nature and scope of business' environmental management practices, rather than triggering them in the first place. The emergence of co-operative and constructive stakeholder dialogue as an element of corporate environmental management is an encouraging indication that business recognises the value of effective and proactive environmental management. Furthermore, the ongoing development of new and increasingly innovative approaches to environmental management reflects the fact that there is commercial value to be gained from continuous improvement.

Examples of sources of pressure on business to adopt more sustainable management practices include:

- corporate social responsibility (CSR) expectations;
- management information needs;
- employees;
- legislation;
- market mechanisms;
- the financial community;
- the supply chain;
- community and environmental groups;
- environmental crises;
- the business community;
- customers; and
- competitor initiatives.

Internal drivers

Corporate social responsibility (CSR)

E16002 Environmental management is an integral part of effective corporate social responsibility ('CSR'). The concept of CSR has changed substantially during the last decade, reflecting the changed conditions in which business operates. The removal of trade barriers, the subsequent growth and political influence of trans-national companies, the opening of previously restricted markets and a general reduction in corporate taxation rates have contributed to radical changes in the way business operates, including increasing the extent to which it controls its own performance. Consequently, the notion of CSR has also changed, with its increasing recognition by the media and its rise up the boardroom agenda. One reason for this has been a tendency for protest groups to target companies' wider impacts on global society as well as over environmental impact issues. In addition, society is looking less to government to control the social and environmental impacts of business activity, and instead is seeing business as being accountable for those impacts. Good corporate governance is no longer merely a reflection of responsible fiscal performance. As a result, the mandate of corporate directors and managers is expanding and they are recognising the need to operate in a more transparent and inclusive manner. Proactive environmental management is a key aspect of that.

In the UK, the Government is embracing the concept of CSR. The Secretary of State for the Environment, Margaret Becket, has stated publicly that the efficient use of resources should be taken seriously by business as CSR is becoming a mainstream consideration, with the market rewarding responsible behaviour and with the companies measuring and managing their environmental impacts often making substantial savings on their operating costs. She also identified the boundaries to CSR, where the market does not reward companies that comply with voluntary standards, and suggested that fiscal measures introduced by the Government will continue to move these boundaries in favour of responsible environmental management.

In November 2002, a report published by the European Business Campaign as part of its ongoing campaign aimed at stimulating the uptake of CSR into core business, gave for the first time an overview of CSR throughout Europe. It concluded that CSR is being taken increasingly seriously throughout Europe and produced a 'CSR matrix' to illustrate CSR activity in 17 European countries. Using seven indicators, the matrix demonstrated that the Netherlands and the UK have a strong CSR profile and that southern and eastern European countries were also making good progress.

A recent initiative by Business in the Community (BITC), which consists of 700 member companies, is the development of a benchmark CSR index based on an overall score achieved for strategy and management practices in several social and environmental areas. The index covers 122 companies, 53 of which are in the FTSE100.

Management information needs

E16003 Effective business management relies on timely and reliable information on the multitude of factors that influence it. As managers' understanding of the relationship between environmental performance and business performance increases, so too does their requirement for information pertaining to environmental performance. Such information helps to increase their control over those factors.

This reflects the growing acceptance of the sustainability concept, which recognises that long-term business success requires environmental, social and economic factors to be balanced. While there is no clear guidance or agreement on how such a balance should be achieved, it is clear that it must be based on appropriate information on all three primary elements. Thus the recognition of the relevance of environmental performance to business performance, the value of controlling it and the need for expanded management information is an increasingly important driver of environmental management practices. Structured environmental management systems not only provide a means of controlling environmental performance *per se*, but also allow more informed strategic and operational decisions to be made by management.

Employees

E16004 Employees have a potentially strong influence over the environmental management practices of an organisation. The desire to minimise staff turnover means that companies are becoming more responsive to employee enquiries and suggestions regarding corporate environmental performance. Conversely, companies wishing to attract high calibre recruits are increasingly realising the importance of maintaining a strong and positive corporate image, which is often dependent on environmental performance, amongst a number of other things.

External drivers

Legislation

E16005 UK companies are influenced by a range of international treaties, conventions and protocols; European regulations and directives; and domestic legislation. The latter may be a tool for implementing European directives, or they may have been enacted independently of any requirement of the European Union ('EU').

Enforcement of the various legal instruments within England and Wales is primarily the responsibility of the Environment Agency, although local authorities also play a role. In Scotland, responsibility lies with the Scottish Environment Protection Agency.

Good corporate environmental management requires a thorough understanding of the legal requirements imposed on a company, in addition to evidence that the company has made reasonable attempts to ensure ongoing compliance with them, either through technological, procedural and/or administrative mechanisms. Many published corporate environmental policies now commit to going 'beyond compliance'. That is, legislative compliance is increasingly being seen as a minimum standard.

Other initiatives are forcing companies that have not voluntarily responded, to give more systematic consideration to environmental management and performance in making strategic and operational business decisions.

The Company Law Review, launched by the Department of Trade and Industry (DTI) in March 1998, which has produced a number of consultation papers, has been acknowledged as the most fundamental review of company law in 150 years. Among the proposals in the Company Law Review is the establishment of an institution of a Companies Commission. One of the Commission's roles would be to monitor the operation of the Combined Code and determine how or whether it should be amended or extended to unlisted companies. A White Paper was issued in July 2002 as the first part of the Government's response. This includes proposals to improve governance to encourage and support responsible business. Auditors would have a statutory right to ask for company information from employees and certain contractors, and directors would be obliged to volunteer information. Failure to provide information to auditors would incur penalties of up to 2 years in jail and unlimited fines. If implemented, these measures would increase greatly the requirement for disclosure of environmental information, which would in turn force companies to improve their environmental management practices.

The White Paper concluded that although the basic goal for directors should be the success of the company in the collective best interests of shareholders, directors should also recognise the company's need to foster relationships with its employees, customers and suppliers, in order to maintain its business reputation. It added that the company 's impact on the community and the working environment should also be considered.

Draft Regulations requiring quoted companies to produce an annual Operating and Financial Review ('OFR') were published in May 2004. From January 2005, around 1,000 of the largest UK companies will have to publish an OFR outlining the performance and future direction of the business. This will need to include consideration of the current and future impacts of environment issues including compliance with environmental legislation. *The Combined Code Principles of Good Governance and Code of Best Practice*, published in June 1998, lays out a set of principles with detailed code provisions on implementation. The listing rules require the annual report and accounts of each listed company incorporated in the UK to include a narrative statement as to how it has applied the principles and whether or not the company has complied with the provisions. Principle D.2 states that the company should maintain a sound system of internal controls to safeguard shareholders' investment and the company's assets. Such controls could include environmental issues. Under the provisions, the directors are required, at least annually, to conduct a review of the effectiveness of internal controls and to report to shareholders that they have done so. If the company does not have an internal audit function it should periodically review the need for one.

The European Commission's Green Paper on CSR, published in July 2001, defined CSR in terms of both company internal management and its impact on society and argued for a European framework for CSR. It covered areas such as social reporting and codes of practice and could eventually lead to legislation within this area. The Green Paper was followed in July 2002 by a Communication on a new strategy for CSR in which the European Commission defined CSR as 'company behaviour, which integrates social and environmental concerns into their operations over and above legal requirements'. Although it stated that CSR should not be mandatory, it did propose a new social and environmental role for business, and it also set up a European multi-stakeholder forum to establish principles for codes of conduct for CSR. A report on the work of the multi-stakeholder forum is due to be published in 2004.

Market mechanisms

E16006 While the command and control approach embodied in environmental legislation and regulations represents a significant pressure on business, policy makers have begun to examine new tools to encourage better management of environmental performance.

A range of measures is beginning to emerge designed to influence the economics of polluting activities. These so-called market-based or economic instruments impose costs on pollution-causing activities and provide incentives for companies to look for ways of minimising environmental damage. Such instruments being discussed or implemented include:

Climate Change Levy

The climate change levy ('CCL'), which came into effect in April 2001, is a tax on the business use of fossil fuels. It is designed to encourage energy conservation and a switch to cleaner fuels and renewable energy sources such as wind and solar power. All revenues are recycled back to business through a 0.3 per cent cut in employers' National Insurance contributions and additional support for energy-efficiency measures and energy-saving technologies.

Landfill Tax

The Landfill Tax was introduced in October 1996 to encourage companies to reduce their volume of waste produced. In 1999, the Government made a commitment to increase this by £1 per tonne a year for five years until 2004. The rate of tax paid by landfill operators from 1 April 2003 was £14 per tonne. It was confirmed in the 2003 Budget that the rate will increase by £3 per tonne in 2005–06 and thereafter by at least £3 per tonne per year until it reaches £35 a tonne by 2011 at the latest. There was to be further consultation on options to ensure that the tax will be revenue-neutral to business. There were no plans to increase the £2 per tonne rate of landfill tax applying to inactive or inert waste in the near future.

Aggregates Levy

An aggregates levy of £1.60 per tonne began in April 2002, with the revenues raised returned to business and the local communities affected by quarrying, through a 0.1 per cent cut in employers' National Insurance contributions and a new sustainability fund. It is intended that the levy will help to ensure that the environmental impact of aggregate extraction is reflected in the price. This is aimed at encouraging more efficient use of aggregates and the development of alternatives including waste glass, tyres and recycled construction and demolition waste.

Vehicle and Fuel Duty

Lower levels of duty on cleaner fuels and changes to company car taxation and vehicle excise duty are designed to reduce pollution, in particular carbon dioxide, from vehicles.

Emissions Trading

Companies that reduce their emissions below the quota can sell the unused part of their quota to other firms, thus providing an incentive to improve emissions performance. Such a scheme is included in the Kyoto Protocol, adopted in 1997 and recently ratified by the EU and Japan, which sets out formal targets for cuts in greenhouse gas emissions by developed countries. Article 17 of the Protocol allows developed countries that reduce their emissions by more than their assigned target to gain credits, which can be sold to other developed countries. In the UK, a voluntary domestic emissions trading scheme for the six greenhouse gases included in the Kyoto Protocol began in April 2002. It is one of the first national greenhouse gases trading schemes in the world. The EU adopted a Directive for an EU-wide trading scheme to be implemented from 2005 to 2008. This differs from the UK scheme in that it only covers, carbon dioxide, it is mandatory and is restricted to a limited number of sectors.

The financial community

E16007 The emerging realisation of the link between environmental performance and business performance has encouraged investors, shareholders and insurers to develop a direct interest in the environmental performance of companies. Investors are becoming increasingly concerned that companies which fail to manage their social and environmental exposure will suffer. The financial community is therefore increasingly seeking information on how environmental issues will potentially impact on the long-term viability of the companies in which they have a commercial interest. In particular, they are concerned about the extent to which environmental risks are being controlled. Brand reputation accounts for an increasing proportion of stock market valuations. Therefore, they want reassurance that companies are not in breach of legal requirements with the consequent threats of fines, damage to reputation and the need for unanticipated expenditure. They need to be sure that assets, in the form of plant, equipment, property and brand value, against which they have lent money, are correctly valued. Raw materials, by-products and end products may need to be replaced, modified and/or discontinued, which may require provisions or contingent liabilities to be included in the corporate accounts. Such a situation may arise as a result of substances being phased out (for example, legal controls on CFC production), or it may reflect changing market attitudes such that the demand for less environmentally damaging products begins to decline. Additional research and development costs are likely to be associated with such changes.

Arguably the most striking example of progress is the rapid growth in socially responsible investment ('SRI'). The Dow Jones sustainability indexes and FTSE4Good are two examples of indexes designed to track stocks for the purpose of SRI. They include only those companies with the strongest records of corporate social and environmental performance. Morley, the UK Fund Management Company, has launched a sustainability matrix, which rates FTSE 100 companies according to social and environmental criteria. Business sustainability is rated A–E, depending on factors such as renewable energy use, healthcare and education. The aim of grading companies in this way is to provide clear and transparent analysis of their social and environmental policies and to encourage improvements, to protect and increase shareholder value and to continue to raise awareness of CSR.

In its sixth report published in February 2002, Business in the Environment (BiE), the business led campaign for corporate environmental responsibility, highlighted increasing concern among city investors over the environmental risks to which large companies are exposed. It also showed that most FTSE 100 companies are gaining competitive advantage by moving ahead of legislation in managing their environmental impacts.

Hermes, the independent fund manager has published 'The Hermes Principles', setting out ten investment principles to address what owners should expect from UK public companies and what these companies expect from their owners. The Hermes Principles are designed to encourage companies to communicate clearly the plans they are pursuing and the likely financial and wider consequences of those plans. Principles 9 and 10 deal with social, ethical and environmental issues. These principles call for companies to manage effectively relationships with their employees, suppliers and customers and say that they should behave ethically and have regard for the environment and society as a whole. In addition, they require that companies should support voluntary and statutory measures designed to minimise the externalisation of costs to the detriment of society.

At a recent AGM of the oil company ExxonMobil, already subjected to a boycott campaign by environmental pressure groups because of it opposition to the Kyoto

Protocol, a resolution urging the company to invest in renewable energy received a significant share of the vote. Investors and some mainstream financial analysts were concerned that the companies' environmental stance threatened shareholder value.

Consideration of environmental risks is now an established component of acquisitions, mergers, flotations, buyouts or divestments. Management of companies involved in any of these deals must be able to demonstrate that environmental liabilities do not constitute an unacceptable risk for investors or insurers.

Supply chain

E16008 Most businesses are both purchasers and suppliers of a range of goods and services. Introducing environmental criteria into procurement decision-making processes emphasises the importance of issues other than price and quality in purchasing goods or services. Examples of environmental criteria are selecting materials, components or products that were manufactured using relatively less energy than alternatives, or that require relatively less energy in operations. Another approach is to use environmental criteria as specifications for a particular material, component or product, such as conforming to an eco-label. The UK Government has recently introduced procedures for procurement under its *Greening Government* initiative which outlines how environmental issues can be taken into account for all buying decisions. As part of a plan to ban hazardous chemicals from all their products in the next few years, B&Q and Homebase are pressing suppliers to start substituting chemical substances of concern with safer alternatives. This means that one company can directly influence the environmental performance of another. As a result, many companies are taking active steps to control their environmental performance throughout the supply chain. For example, it is quite common for purchasing companies to require their suppliers to demonstrate ongoing compliance with formal environmental management standards such as ISO 14001 or EMAS, and to provide information on the environmental performance of the supplier and the products and services it provides. Regardless of the approach that is used, the criteria should be relevant to each procurement decision and should be designed to enable a business to directly influence the attributes and performance of its products at other stages of their life cycles.

The EU Eco-label award scheme has been in operation since 1993, when the first product groups were established, and was comprehensively revised in 2000. An Eco-label (the 'Flower symbol') is awarded to products possessing characteristics that contribute to improvements in environment protection. The main objective of the scheme is to encourage business to market greener products by providing information to allow consumers to make informed environmental choices when purchasing. There has been a steady increase in both the number of applications from manufacturers and the number of eco-labelled products marketed in the last few years. By July 2002, 118 licenses for the use of the logo had been granted for several hundred products across the EU. The eco-label scheme (as laid down in Regulation (EC) No 1980/2000) is part of a wider approach on Integrated Product Policy (IPP) within the EU Action Programme. The Commission has published a Green Paper on IPP that will be a key innovative element of future environmental policy and sustainable consumption and production.

Community and environment group pressure

E16009 The majority of public pressure on companies to improve their environmental performance is initiated by local communities which experience the direct effects of pollution. Most companies recognise the importance and value of working co-operatively with local communities. In fact many have established community liaison panels, which comprise representatives of the local community. Such panels interact with the company on a regular basis and provide input on environmental and other community issues.

Environmental pressure groups, often supported by a high level of legal and technical expertise, are also influential. Previously the relationship between environmental pressure groups and business was characterised by mutual mistrust and reactive criticism. The emergence of the concept of 'stakeholder engagement', whereby companies take a more inclusive approach to business management, means that environmental pressure groups can be expected to have more direct access to companies and management in the future. In fact their opinions are already actively sought by many organisations, which recognise the importance of constructive dialogue. While companies will not necessarily implement all suggestions made by external pressure groups, the trend towards more timely and constructive dialogue is likely to continue.

A good example of the potential for corporate reputation damage that could occur when a company fails to engage with environmental groups is Greenpeace's Stop Esso campaign. This campaign targeted Esso's parent company ExxonMobil worldwide, for its negative stance on climate change and lack of investment in renewable energy sources, by calling for a boycott of the company's products. Greenpeace claimed that by September 2002, its boycott campaign had resulted in around one million motorists, a quarter of the number of regular buyers of Esso petrol, boycotting its filling stations. Due largely to this campaign, Deutsche bank warned in October 2002 that ExxonMobil was being labelled as 'environmental enemy number one' and that this posed a significant risk to its business.

Similarly, BP has come under pressure from the Worldwide Fund for Nature (WWF) which said in January 2003 that it was selling all its shares in the company. Although the value of the shares was insignificant in financial terms, it is likely to impact on the environmental reputation of BP. This was followed by an announcement by one of the UK's leading ethical investment funds, Henderson Global Investors, that it intended to sell several million pounds worth of BP shares.

The evolution of the internet and other sophisticated communication media has significantly increased public awareness of and access to information about corporate environmental performance, as well as increasing the speed with which such information can be transferred and responded to. Under a new EU Directive for implementing the first objective of the Aarhus Convention concerning public access to information, the public will shortly be given stronger rights to access information on companies' impacts on the environment. This is also expected to cover health and safety, which broadens the scope of the rules since many industrial projects can potentially affect public health. However, legally classified and commercially confidential information would still be excluded. The result of this is that pressure on companies to improve environmental performance has increased and has spread into the international arena. Nowadays, companies are finding it necessary to develop and adopt consistent environmental and social performance standards in all markets in which they operate, since their performance is increasingly subject to international scrutiny.

Environmental crises

E16010 In many cases, high profile environmental crises are a catalyst for improved environmental management. There are two main types:

(*a*) crises that are generated as a result of the actions of an individual company, the effects of which are generally experienced at a local scale;

(*b*) crises that are generated by collective action or by natural forces, the effects of which are often experienced at the national or international scale.

Individual companies that have been associated with environmentally damaging events such as oil spills generally find themselves exposed to intense pressure to improve their environmental management practices in the immediate future. The need to correct the damage to reputation caused by environmental crises is a further incentive to respond quickly. Obviously the costs of responding to pressures arising from catastrophic environmental incidents can be extremely high, and most companies seek to avoid those by incorporating environmental issues into their corporate risk management programmes.

Global or national environmental crises tend to emerge more gradually, and with considerably more debate about accountability and appropriate responses. Nevertheless, there are a number of examples of global environmental issues that have facilitated more systematic and intensive environmental management practices than may otherwise have occurred in the same time period. The most obvious examples are the depletion of the ozone layer and climate change. At a national level, issues such as water shortages, soil erosion and regional air pollution have triggered the adoption of improved environmental management practices on an extensive scale.

Business community

E16011 Trade associations and business groups such as the Confederation of British Industry (CBI) and the International Chamber of Commerce (ICC) have played a leading and effective role in encouraging businesses to adopt environmental management practices. Increasingly the importance of doing this within a sustainable development framework is being accepted. Sustainable development is generally recognised as the inter-relationship and inter-dependence of the three core elements of economic, environmental and social consideration, although it is a concept that is open to wide interpretation. However, it is generally understood to mean achieving a better quality of life with effective environmental protection. The most widely used definition is contained in the 1987 United Nations Brundtland report to the World Commission on Environment and Development. This definition states that it is 'development which meets the needs of the present without compromising the ability of future generations to meet their own needs.'

Thousands of companies worldwide have been helped by the ICC voluntary business initiative 'Business Charter for Sustainable Development'. Launched in 1991, as a tool to help companies tackle the challenges and opportunities of the environmental issues that emerged in the 1980s and early 1990s, the Charter has established a set of 16 principles to guide company strategies and operations towards sustainable development. Its principles for environmental management have provided a global alignment of business to common objectives and have helped thousands of companies worldwide establish the foundation on which to build their own integrated environmental management systems. The Charter highlights such areas as employee and customer education, research facilities and operations, contractors and suppliers and emergency preparedness. It requires organisations to support the transfer of technology, be open to concerns expressed by the public and employees and carry out regular environmental reviews and report progress.

In 1995 the World Industry Council on the Environment (WICE) and the Business Council for Sustainable Development (BCSD) merged to become the World Business Council for Sustainable Development (WBCSD). In 2003, it consisted of a coalition of some 165 international companies from 30 countries committed to the principles of sustainable development. WBCSD seeks to promote the effective implementation of these principles through a combination of advocacy, research, education, knowledge-sharing and policy development.

The Responsible Care programme is an example of an international initiative from a specific industry sector. The programme has been adopted by around 41 national chemical industry associations, including the Chemical Industries Association ('CIA') in the UK. All of those associations have made acceptance of Responsible Care requirements compulsory for individual member companies. Responsible Care is designed to promote continuous improvement, not only in environmental management, but also health and safety. The companies must also adopt a policy of openness by releasing information about their activities. Adherence to the principles and objectives of Responsible Care is a condition of membership of CIA.

In the UK, the Prince of Wales Business Leaders Forum (PWBLF), which incorporates Business in the Environment (BiE), was formed in 1990 and currently has 50 international member companies. It has played a major role in raising awareness of the relevance of environmental management and sustainable development to business, and promoting practical tools for improving performance.

Individual companies have also contributed to the development of improved environmental performance standards. As the relevance of good environmental management to overall business performance has become more apparent, progressive companies have voluntarily adopted a number of innovative and unique approaches to corporate environmental management. This has had the effect of constantly moving the frontiers of 'acceptable' environmental management practices and created substantial peer pressure which in turn has encouraged other companies to adopt similar or even more effective practices.

Customers

E16012 There is some evidence of the role of customers in the uptake by companies of a sound stance on environmental and other ethical issues. The Co-operative Bank is a well-established proponent of active ethical and environmental policies under which some customers are denied banking. It has attributed one fifth of its 2001–2002 pre-tax profits directly to these policies. Surveys for the bank showed that 14 per cent of account holders regarded ethical issues as the most important of a list of factors when choosing it, whilst 26 per cent viewed them as an important factor. The bank has increased its share of the current account market by 30 per cent since 1996. However, some competitors still feel little pressure from customers on these issues, preferring instead to work with industry to reduce environmental risks.

Further evidence of the importance of responsible environmental and social policies can be found in the Government's second annual CSR report. Published in May 2002, it claims that around half of consumers have identified that CSR is important when choosing to buy, and 20% will boycott or select products on these grounds.

Environmental management guidelines

Standards for environmental management

E16013 Various guidelines exist for responding to those many pressures to minimise damage to the environment and health. Effective environmental management, like quality management or financial management, requires *inter alia*:

(*a*) the setting of objectives and performance measures;

(*b*) the definition and allocation of responsibilities for implementing the various components of environmental management;

(*c*) the measurement, monitoring and reporting of information on performance;

(*d*) a process for ensuring feedback on systems and procedures so that the necessary changes can be actioned.

There are two main instruments which influence current approaches to environmental management. The first is the EU Eco-Management and Audit Scheme (EMAS) Regulation. The second is the International Standards Organisation Series of Environmental Management Standards.

The Eco-Management and Audit Scheme ('EMAS') Regulation

E16014 This voluntary scheme came into force in July 1993 and has been open for participation by companies in all Member States since April 1995.

EMAS has three main aims:

(*a*) the establishment and implementation of environmental policies, programmes and management systems.

(*b*) the systematic, objective and periodic evaluation of these measures.

(*c*) the provision of information to the public on environmental performance.

The Regulation sets out a number of elements of systematic environmental management. Sites that can demonstrate ongoing compliance with those requirements to an independent assessor have the right to be registered under EMAS.

Following the adoption of a company policy, an initial environmental review is made to identify the potential impacts of a site's operations, and an internal environmental protection system must be established. The system must include specific objectives for environmental performance and procedures for implementing them. The system, and the results of the initial environmental review, must be described in an initial environmental statement. The statement must be validated by an accredited external organisation before being submitted to nominated national authorities in individual Member States for registration of the site under the scheme.

There are a number of key points here:

(*a*) the first stage in developing environmental management is to carry out a thorough review of impacts on the environment;

(*b*) setting up an environmental management system is a prerequisite of registration under the scheme;

(*c*) external validation of the environmental statement is intended to ensure consistency in environmental management systems;

(*d*) the description of the environmental management system within the statement will be on the public record.

Once a site has been registered under the scheme, it will require regular audits to review the effectiveness of the environmental management system as well as giving information on environmental impacts of the site. Here it is sufficient to note that development of procedures for internal auditing is a crucial part of an environmental management system. In addition to the audit, the preparation and external validation of an environmental statement submitted to the competent authority for continued registration and made public, are elements of an on-going procedure.

In order to attract more registrations and make the scheme more competitive, EMAS was revised by Regulation (EC) No 761/2001, with the revisions taking effect in April 2001. The scheme is now open to all sectors of the economy, including financial companies, transport and local and public bodies. In addition, registered organisations can use an official logo to publicise their participation in EMAS as well as gaining regulatory benefits. The new scheme also encourages more involvement by employees in implementation and strengthens the role of the environmental statement to improve the transparency of organisations and their stakeholders.

By the end of 2002, 3,780 organisations had registered under EMAS throughout Europe, including 76 sites in the UK. This was only two more than at the end of the previous quarter, suggesting that the revised Regulation has not stimulated interest. It is thought that organisations still see less value in EMAS compared with ISO 14001, for which there were 20,400 certified organisations at the end of 2002, as it does not extend beyond Europe and ISO is more simply achieved.

International standards on environmental management (the ISO 14000 series)

E16015 The International Standards Organisation is continuing to develop a series of standards for various aspects of environmental management. All are designed to assist organisations with implementing more effective environmental management systems. Table 1 outlines the various standards and guidelines within the series.

The most high profile of these standards is ISO 14001, published in June 1996, which sets out the characteristics for a certifiable environmental management system ('EMS'). ISO 14001 was based on the British Standard on Environmental Management Systems (BS 7750), although the latter has been superseded by the international standard.

The format of ISO 14001 reflects the procedures and manuals approach of the ISO 9000 quality management systems series. In practice this means that organisations which operate to the requirements of ISO 9000 can extend their management systems to incorporate the environmental management standard, although the existence of a certified quality management system is not a prerequisite for ISO 14001.

ISO 14001 requires an organisation to develop an environmental policy which provides the foundation for the rest of the system. The standard includes guidance on the development and implementation of other elements of an EMS, which is ultimately designed to allow an organisation to manage those environmental aspects over which it has control, and over which it can be expected to have an influence. ISO 14001 does not itself stipulate environmental performance criteria.

Applications for ISO 14000 are growing at a considerable rate. The period 2000–2001 showed the largest increase since the introduction of the standard in 1995. At the end of 2001, there were at least 36,765 ISO 14000 certificates issued to organisations in 112 countries. The majority of these were issued in Europe and the Far East.

Table 1: The ISO 14000 series	
ISO 14001: 1996	Environmental management systems – Specifications with guidance for use
ISO 14004: 1996	Environmental management systems – General guidelines on principles, systems and supporting techniques

ISO 14010: 1996	Guidelines for environmental auditing – General principles
ISO 14011: 1996	Guidelines for environmental auditing – Audit procedures – Auditing of environmental management systems
ISO 14012: 1996	Guidelines for environmental auditing – Qualification criteria for environmental auditors
ISO 14040: 1997	Environmental management – Life cycle assessment – Principles and framework
ISO 14041: 1998	Environmental management – Life cycle assessment – Goal and scope definition and inventory analysis
ISO 14050: 1998	Environmental management – Vocabulary
ISO/FDIS 14020	Environmental labels and declarations – General principles
ISO/DIS 14021	Environmental labels and declarations – Environmental labelling – Self-declared environmental claims – Terms and definitions
ISO/DIS 14024	Environmental labels and declarations – Environmental labelling TYPE 1 – Guiding principles and procedures
ISO/FDIS 14031	Environmental performance evaluation – Guidelines
ISO/DIS 14042	Environmental management – Life cycle assessment – Impact assessment

First published in 1996, two of the key environmental management standards, ISO 14001 and ISO 14004 are being updated to reflect current thinking and practice with the final revisions to be published in 2004. Consultation drafts have been issued and these focus on issues related to the compatibility of these international standards with ISO 9001:2000 and clarification of the existing text to assist in interpretation. The revisions are aimed at helping users better understand and implement the standards, while at the same time enhancing environmental protection.

Differences between ISO 14001 and EMAS

E16016 The key differences between ISO 14001 and EMAS are:

(*a*) ISO 14001 can be applied on a company-wide basis, whereas EMAS registration is only granted at individual site level or for particular services;

(*b*) ISO 14001 does not currently include a requirement for public reporting of environmental performance information, whereas sites registered under EMAS must produce an independently validated, publicly available environmental statement;

(*c*) ISO 14001 does not require a register of environmental effects or legislation;

(*d*) the level of control of contractors and suppliers required in EMAS is not matched in ISO 14001, which stipulates only that required procedures are communicated to them.

The new revised EMAS allows integration of an existing ISO 14001 certification, to allow a smoother transition and avoid duplication when upgrading from ISO 14001 to EMAS.

Implementing environmental management

Practical requirements

E16017 Both EMAS and ISO 14001 set a pattern for companies wishing to develop environmental management systems. The key steps in implementing such systems involve:

(*a*) conducting an initial review, designed to establish the current situation with respect to legislative requirements, potential environmental impacts and existing environmental management controls;

(*b*) developing an environmental policy which will provide the basis of environmental management practices as well as informing day to day operational decisions;

(*c*) establishing specific objectives and performance improvement targets;

(*d*) developing a programme to implement the objectives and establish operational control over environmental performance;

(*e*) ensuring information systems are adequate to provide management with complete, reliable and timely information; and

(*f*) auditing the system to compare intended performance with actual performance.

Review

E16018 In order to be able to actively manage its interactions with the environment, an organisation needs to understand the relationship between its business processes and its environmental performance. An environmental review should therefore be conducted, which clarifies how various business activities could potentially affect, or be affected by, the quality of different components of the environment (air, land, water and the use of natural resources). This will allow an organisation to understand which of its activities have the greatest potential impact on the environment. Usually, but not always, these will relate to procurement and/or manufacturing processes. It is also important to consider which elements of environmental management and performance have the greatest potential impact on business performance, for example, high profile environmental prosecutions can have a significant impact on corporate reputation and brand value.

The initial and subsequent review should also consider the current and likely future legislative requirements that the company must comply with. It should consider the overall organisational strategy, any other relevant corporate policies, customer specifications and community expectations which could influence environmental management practices. The review also offers the opportunity to establish a baseline of actual management organisation, systems and procedures, its compliance record and the range of initiatives already in place to improve performance.

Ideally, an environmental review should involve input from a range of stakeholders, both internal and external to the organisation. This promotes a wider perspective on potential environmental impacts, and ensures that the resulting policies and programmes to be developed by the organisation, reflects a comprehensive range of issues and risks. Consequently, the chances of unidentified and therefore uncontrolled risks emerging will be minimised.

Policy

E16019 The results of the initial review should inform the development of a written environmental policy. The policy directs and underpins the remainder of the EMS, and represents a statement of intent with regard to environmental performance standards and priorities.

The policy should be endorsed by the highest level of management in the company, and should be communicated to all stakeholders.

Objectives

E16020 It is important that the policy be supported by objectives that are both measurable and achievable. They should be cascaded throughout an organisation, so that at each level there are defined targets for each function to assist in the achievement of the objectives. A key part of ensuring continuous improvement in environmental performance is to review and update objectives in the light of progress and changing regulations and standards.

Objectives should be developed in conjunction with the groups and individuals who will have responsibility for achieving them. They should also be clearly linked to the overall business strategy and as far as possible with operational objectives. This ensures that environmental management is viewed as relevant and integral to business performance, rather than being seen as an isolated initiative.

The process of objective setting also needs to consider the most appropriate performance measures for tracking progress towards the ultimate objective. For example, if an objective is to reduce waste by 20 per cent over five years, a number of parameters could be used to reflect different aspects of the organisation's waste reduction efforts towards that, including volumes of waste recycled, efficiency with which raw materials are converted to product, and proportion of production staff that have received waste management training.

ISO 14031 provides useful guidance on the principles to be applied in the selection of appropriate environmental performance indicators. Performance in whichever parameters are selected should be measured regularly to enable corrective actions to be taken in a timely manner. As far as possible, performance measures and the achievement of quantified targets should be linked to existing appraisal systems for business units or individuals.

The Institution of Chemical Engineers (IChemE) has recently launched its own system to enable companies to quantify their sustainable development performance. Key environmental indicators used are resource use, emissions, effluent and waste. The system calculates the environmental burden of each product by using a weighted calculation of its acidification, global warming, damage to human health, ozone depletion and photochemical ozone formation potential. The National Chemicals Industries Association (CIA) has pledged to recommend that its members should use the system.

Responsibilities

E16021 Allocation of responsibilities is vital for successful environmental management. Its implementation will typically involve changes in management systems and operations, training and awareness of personnel at all levels and in marketing and public relations.

A wide range of business functions will therefore need to be involved in developing and implementing environmental management systems. The commitment of senior personnel to introduce sound environmental management throughout the organisation, and to communicate it to all staff is vital.

Companies have adopted a range of organisational approaches as part of their environmental management systems. In some cases there is a single specialist function with responsibility for monitoring and auditing the system. An alternative is to have a central environmental function with only an advisory role which can also undertake verification of the internal audits carried out by other divisions or departments. The approach needs to be adapted to the culture and structure of the organisation, but whatever system is adopted, there are a number of crucial elements:

(*a*) access to expertise in assessing environmental impacts and developing solutions;

(*b*) a degree of independence in the auditing function;

(*c*) a clear accountability for meeting environmental management objectives;

(*d*) adequate information systems to help those responsible for evaluating performance against objectives, to identify problem areas and to ensure that action is taken to solve them.

Training and communications

E16022 Both ISO 14001 and EMAS include training and communications as a key requirement of the EMS. In particular, they focus on ensuring that employees at all levels, in addition to contractors and other business partners, are aware of company policy and objectives; of how their own work activities impact on the environment and the benefits of improved performance; what they need to do in their jobs to help meet the company's environmental objectives; and the risks to the organisation of failing to carry out standard operating procedures.

This can involve a significant investment for companies, but if integrated with existing training modules and reinforced regularly, many hours of essential training can be achieved. An important benefit of training is that it can be a fertile ground for new ideas to minimise adverse environmental impacts.

Communication with, for example, regulators, investors, public bodies and local communities is an important part of good environment management and requires a preparedness to be open, honest and informative. It also requires clear procedures for liaising with external groups.

Operational controls

E16023 The operational control elements of an EMS define its scope and essentially set out a basis for effective day-to-day management of environmental performance. Typically, such controls would include:

(*a*) a register of relevant legislation and corporate policies that must be complied with;

(*b*) a plan of action for ensuring that the policy is met, objectives are achieved and environmental management is continuously improved. This sets out the various initiatives to be proactively implemented and milestones to be achieved, and could be considered a 'road map' for guiding environmental performance;

(*c*) a compilation of operating procedures which define the limits of acceptable and unacceptable practices within a company and which incorporate consideration of the environmental interactions identified in the review and the objectives and targets that were defined subsequently;

(*d*) an emergency response plan to be implemented in the event of a sudden, unexpected and potentially catastrophic event which could potentially influence a company's environmental performance in an adverse manner;

(*e*) a programme for monitoring, measuring, recording and reviewing environmental performance. This should incorporate a mechanism for regularly reporting back to senior management, since they are the key enablers of the EMS and because they retain ultimate responsibility of business performance.

These operational controls, and indeed all elements of an EMS, should be documented.

Information management and public reporting

E16024 The critical factor in an EMS is the quality and timeliness of information to internal and external users. The importance of providing performance information to external users is increasing as expectations of greater transparency and CSR continue to grow. Such expectations have been supported in the UK by strong encouragement from Government for companies to voluntarily and publicly report on their environmental performance. The Chairman of the Environment Agency has stated recently that annual environmental reporting should be compulsory for all FTSE listed companies. He said that they should include information on how they identify and manage environmental risks, and report environmental performance against published targets. Despite the Prime Minister's November 2000 challenge to start reporting, fewer than 80 of the FTSE top 350 companies in the UK currently mention the environment in annual reporting, which is considerably lower than most other EU countries. In France and Denmark all publicly quoted firms are required by law to disclose data on environmental impact in annual financial reports. Public environmental reporting is encouraged, but not mandated, by ISO 14001, although EMAS has always included a requirement for the preparation of an environmental statement as a condition of registration under the scheme.

Regardless of whether they have adopted EMAS, ISO 14001, or neither, many companies have a statutory duty to report on some aspects of their environmental performance to demonstrate legislative compliance. Some of that information is publicly available. An increasing number of companies are choosing to provide information on their environmental performance, either in their annual report and accounts, or in a stand-alone document. In most cases, the information provided extends well beyond a demonstration of legislative compliance, and leans towards a general overview of all significant aspects of environmental management and performance. Furthermore, many companies, recognising the importance of sustainability, are beginning to measure and report on their performance in terms of social impact.

As stakeholders and communication media become more sophisticated, and our understanding of environmental interactions increases, so information requirements become more complex. This means that management information systems must incorporate database management, modelling, measuring, monitoring and flexible reporting. The trend also means that environmental reports that have been produced solely as a means of improving public relations and to defuse external pressures are becoming less acceptable to many stakeholders. The main reasons for this appear to be that they are seldom generated in response to internal management information needs, which in turn means that they tend to focus on statements of management intent, qualitative claims and descriptive anecdotes rather than actual performance. They are therefore less likely to include detailed and verifiable information.

Instead, companies are recognising the value of maintaining a constructive and open dialogue with stakeholders, and published reports are being used as part of those efforts. The Association of Chartered Certified Accountants ('ACCA') has held annual awards for environmental reporting for the last decade, over which time it has identified a steady increase in both the quantity and quality of environmental reports. Furthermore the initial tendency of large manufacturing companies to report has spread to small and medium sized enterprises, the public sector and service industries such as financial services. The structure of the awards was changed in 2001 to take account of the increasing awareness of the environmental, social and economic impacts of business. Under the title 'The ACCA Awards for Sustainability Reporting', the new scheme includes three different award categories: the ACCA UK Environmental Reporting Awards, the ACCA Social Reporting Awards and a new category, the ACCA Sustainability Reporting Awards.

Institutional investors needing reassurance that companies are aware of their environmental risks, are also driving companies into providing more environmental information. Morley Fund Management announced in 2001 that unless FTSE 100 companies publish an annual environmental report, it would vote against the adoption of annual reports and accounts. The Association of British Insurers ('ABI') issued new investment guidelines in October 2001 after consultation with a wide range of fund managers, corporate executives and NGOs. These aimed to improve disclosure by companies of their approach to external social, ethical and environmental risks and single out environmental risks as among the most significant CSR-related risks faced by companies. The guidelines set out the business case for CSR and recommend that CSR reporting should be integrated into annual financial reports to allow for independent verification. By setting out what institutional investors will expect to see disclosed in the annual reports of companies in which they hold stakes, the guidelines will increase transparency and allow shareholders to engage with companies where they consider significant risks have not been assessed adequately.

The introduction in July 2000 of a requirement for occupational pension funds to state the extent to which they consider environmental, social and ethical factors in investment decisions, has also increased the pressure for environmental reporting amongst companies. However, a survey of the SRI policies of the UK's top 100 occupational pension funds, published by Friends of the Earth in August 2001, concluded that although most funds now include SRI in their investment strategy, many impart little force on fund managers to ensure ethical standards. Furthermore, it found that most funds still have no means for monitoring whether trustees and managers are meeting their stated ethical policies.

While no standards on environmental reporting currently exist, a number of sources of guidance are available. Current 'best practice' environmental reporting in the UK is dictated by the United Nations Environment Program (UNEP)/SustainAbility benchmark study into corporate environmental reporting, the results of which have been published annually since 1996. The Global Reporting Initiative ('GRI') was established in 1997 in an attempt to develop globally applicable guidelines for reporting on the economic, environmental and social performance of business, governments and NGOs. It incorporates the participation of corporations, NGOs, accountancy organisations, business associations and other stakeholders from around the world. The GRI's Sustainability Reporting Guidelines, released in June 2000, provided an expanded model for voluntary non-financial reporting and reflect the move towards broader sustainability reporting. New guidelines published in 2002, update and expand the original guidelines, which have been used by hundreds of firms worldwide. They include a core set of indicators, which, for the most part, are applicable to organisations in all sectors of commerce and include a list of expanded performance indicators aimed at placing a greater emphasis on social and economic issues, rather than just concentrating on environmental issues. The guidelines also insist that statements on a company's vision and strategy regarding sustainable development, as well as specific company information on governance, should be addressed before a report can be described as in accordance with GRI guidelines. The GRI is now in the process of developing sector-specific supplements to the guidance, with the first tranche including the financial services, tour operator, automotive and telecommunication sectors.

Guidelines on corporate environmental reporting were issued and sent to the heads of the 350 largest firms in the UK by the Government in November 2001. They were developed in collaboration with consultants, academics and business, and aimed to set out how to produce an environmental report by drawing on existing reporting schemes, including the GRI, and outlines ten basic elements that should feature in a report. The guidelines were accompanied by the results of a survey on the costs and benefits of environmental reporting across different sectors and company size. The survey was sent out to 109 companies, 38 of which were FTSE 100 companies. The content quality of the reports was found to be generally very high and the annual costs of environmental reporting ranged widely from £6,500 to £535,000, with an average cost per firm of £92,716. Four companies that published only on the Internet showed a significant reduction in reporting costs.

The WBCSD has produced its own guide to corporate sustainability reporting. The guidance is intended to help companies to identify the relevant information to be included in order to mitigate corporate reputational and financial risks and to gain competitive advantage. There are also recommendations on how to fulfil the information needs of the financial community. In addition, the organisation has also set up a web-based portal to help companies report on their sustainable development activities. The portal provides examples of reporting practices from a variety of companies in a range of business sectors and is designed to generate ideas on what information to include in a report, rather than to prescribe rigid practices.

The European Commission has recently adopted a draft Recommendation providing guidelines on the information relating to environmental expenditures, liabilities and risks that companies should publish in their annual accounts and reports. Among those proposed is experience in implementing environment protection measures. It also advocates that where relevant, this should contain quantitative measures in areas such as emissions and consumption of water, energy and materials. Unlike a Directive, the Recommendation will not be binding.

In an attempt to increase the credibility of published reports and identify opportunities for improving management information systems, a number of companies are seeking independent third party assurance on the reliability, completeness and likely accuracy of information contained in their reports. However, progress on developing standards for independent auditing of environmental reports has been relatively slow. One promising initiative is the global assurance standard for corporate public reporting on social, environmental and economic performance, launched by the

British Organisation Account Ability ('AA') in March 2003. The AA 1000 Assurance Standard was developed after a two-year worldwide consultation involving hundreds of organisations, including the investment community, NGOs and business. It is openly accessible on a non-commercial basis and has been piloted by a number of leading companies in the UK. The Standard will also place new demands on external auditors and verifiers of CSR and environmental reports. They will be required to demonstrate independence and impartiality by publicly disclosing commercial relationships with their clients, as well as proving their competency and commenting where a report has omitted information that could be important to stakeholders. The standard is being supported by several organisations including the GRI.

Auditing and review

E16025 Auditing of environmental management systems, whether conducted by internal or external parties, is a means of identifying potential risk areas and can assist in identifying actions and system improvements required to facilitate ongoing system and performance improvement.

Benefits of environmental management

Effective environmental management

E16026 Effective environmental management will involve changes across all business functions. It requires commitment from senior management and is likely to need additional human and financial resources initially. However, it can also offer significant benefits to businesses. These include:

(*a*) avoidance of liability and risk. Good environmental management allows businesses to choose when and how to invest in better environmental performance, rather than reacting at the last minute to new legislation or consumer pressures. Unforeseen problems will be minimised, prosecution and litigation avoided;

(*b*) gaining competitive advantage. A business with sound environmental management is more likely to make a good impact on its customers. The business will be better placed to identify and respond rapidly to opportunities for new products and services, to take advantage of 'green' markets and also respond to the increasing demand for information on supplier environmental performance;

(*c*) achievement of a better profile with investors, employees and the public. Increasingly investors and their advisers are avoiding companies with a poor environmental record. The environmental performance of businesses is an increasing concern for existing staff and potential recruits. Some businesses are finding that a good environmental record helps to boost their public image;

(*d*) cost savings from better management of resources and reduction of wastes through attention to recovering, reusing and recycling; and reduced bills from more careful use of energy;

(*e*) an improved basis for corporate decision-making. Effective environmental management can provide valuable information to corporate decision-makers by expanding the basis of such decisions beyond financial considerations. Companies that understand the interactions between their business activities and environmental performance are in a strong position for integrating environmental management into their business, thereby incorporating key elements of the principles of sustainable development.

Effective environmental management can turn environmental issues from an area of threat and cost to one of profit and opportunity. As standards for environmental management systems are adopted and are applied widely, the question will increasingly become, as with quality management, can a company afford not to adopt environmental management? The external pressures on organisations to improve environmental performance are unlikely to abate. Environmental management systems can help companies respond to the pressures in a timely and cost-effective way.

The future

E16027 Environmental management is now well-established as an essential component of effective business management. Environmental management systems have been widely adopted and in most cases these have facilitated demonstrable and ongoing improvements in environmental performance. The involvement of a range of stakeholders in corporate environmental management is no longer the exception; the constructive contribution that they make is actively sought. The publication of environmental reports is also increasing and has become relatively common among leading companies; most include a mechanism for obtaining feedback from external stakeholders, and an increasing number are independently assured in a similar way to annual financial reports and accounts.

However, there is already a noticeable trend towards sustainability management, whereby companies are attempting to systematically balance economic, environmental and social considerations in business strategies and operations. The strong inter-relationships and inter-dependence between these three elements of sustainable development make it imperative for companies to move towards a more integrated approach to managing them.

In late 1998, the UK Government announced a number of national 'sustainability indicators' to be used to assess the effectiveness of Government policies and initiatives. The UK Strategy for Sustainable Development has four main objectives: social progress that meets the needs of everyone; prudent use of natural resources; effective protection of the environment; and maintenance of high and stable levels of economic growth. Pressures on business to follow this will increase, and the efforts of a number of organisations including WBCSD, UNEP, the GRI consortium and organisations seeking to implement their guidelines will help to respond to those pressures.

Continued development of this approach will be greatly influenced by shareholders, market analysts and financial institutions recognising the value of non-financial performance information as a basis for evaluating management competence and predicting business performance. Consequently they can increasingly be expected to insist on changes to the information presented to them by companies, thereby generating a radical shift in the traditional business paradigm.

As a direct consequence of the 2002 Johannesburg World Summit on Sustainable Development , the Government announced the launch of a UK strategy for Sustainable Consumption and Production (SCP). Containing the steps to take forward commitments made in Johannesburg, it will set out a framework for future action by Government and business. The strategy will be linked to existing and proposed policies and concentrate on waste and energy.

Therefore, while environmental management will remain a critical issue for business to address, it will increasingly become integrated with other aspects of business management, reflecting a growing acceptance of the importance of sustainability.

Ergonomics

Introduction

E17001 Ergonomics is concerned with the fit between people and the things they use. People vary enormously in many attributes and abilities such as height, strength, visual ability, ability to handle information and so on. Ergonomics applies scientific information about human abilities, attributes and limitations to ensure that the tools or equipment used, tasks undertaken, workstation, environment and work organisation are designed to suit people. Ergonomics adopts a people-centred approach, aiming to fit the work to the person, rather than forcing the person to adapt to poorly designed equipment, furniture, environments, or work systems. Correct application of ergonomics will produce a work system which optimises human performance and minimises the risk to workers' health and safety. Ergonomics can be applied to any environment or system (work, leisure, travel, home etc) with which people interact.

The word 'ergonomics' comes from the Greek words 'ergos' meaning 'work' and 'nomos' meaning 'natural law'. Historically the term 'ergonomics' has been used in the UK and Europe, while 'human factors' has been used in North America, although this distinction is now blurring. Essentially the terms 'ergonomics' and 'human factors' are synonymous, although in some quarters a distinction has been drawn between physical workplace design (which is referred to as 'ergonomics') and system design and human behaviour (which is referred to as 'human factors'). This chapter does not make that distinction and the term 'ergonomics' is used to cover all aspects of the design of equipment, environment or system in relation to people.

The benefits of applying ergonomics

E17002 The benefits of ergonomically designed equipment, furniture, workplaces and tasks are as follows:

- *Improved safety* as people are less likely to make a mistake if equipment is designed taking account of abilities (e.g. if characters on displays are a suitable size and colour that can be read accurately).
- *Reduced ill health and sickness absence* as equipment and workstations are designed to fit people (e.g. seats are comfortable), and tasks are designed to take account of abilities (e.g. work rates and weights handled are within the person's capabilities).
- *Reduced fatigue and stress* as tools, equipment and systems are easier to use.
- *Increased efficiency, performance, productivity and quality of product* (e.g. if tools are designed to fit the hands of the users they will be easier to hold, users will be more comfortable, and may work more efficiently and make fewer mistakes).
- *Increased job satisfaction* as tasks are easier to perform.

Although people are highly adaptable and resourceful and are often able to use poorly designed equipment, systems etc, their use can lead to stress, errors, fatigue and injury. The consequences of not applying ergonomics can be enormous, potentially leading to human suffering through physical discomfort and disability, stress, and accidents. It may also lead to reduced efficiency and productivity.

Ill health resulting from poorly designed furniture, tasks, lifting and handling etc is also extremely costly. Musculoskeletal injuries are the most significant occupational ill health outcome in the UK. There were an estimated 1.2 million people reporting a work-related musculoskeletal disorder in 1995, resulting in approximately 9.9 million days of sickness absence. Manual handling injuries account for a significant proportion of these; more than a third of reportable accidents in 1996/7 were due to handling activities.

Not applying ergonomics can also be one factor that contributes to stress. Stress arises as a result of a mismatch between the demands of the job or situation and the abilities of the individual. Stress results in a significant amount of ill health, with an estimated 500,000 people reporting work related stress/depression, resulting in approximately 6.5 million working days lost in 1995.

Health effects of stress include headaches, indigestion, disturbed sleep and fatigue, changes in appetite, increased alcohol consumption, smoking or drug taking, loss of concentration, irritability, and loss of self esteem. Work-related stress may be caused by poor communication, lack of appropriate training, high workload, lack of control over work, inadequate feedback about work, or repetitive or boring work, i.e. if tasks, jobs and systems are not designed with full consideration of the users. Individuals vary in their response to these factors, but the employer has a responsibility to reduce the risks by designing work appropriately.

As well as increasing the potential for musculoskeletal injuries and increased stress, poor design which results in increased fatigue and error can contribute to accidents and system failures. Inadequate attention to human factors issues has been cited as a contributing element in many catastrophic failures (e.g. Ladbrook Grove rail disaster, Three Mile Island disaster, Kegworth air disaster – see DISASTER AND EMERGENCY MANAGEMENT SYSTEMS (DEMS)). These usually arise as a consequence of a series of errors which are a result of inappropriate design and management. The application of ergonomics can help to reduce these problems. More information on human error is contained in a HSE publication *Reducing error and influencing behaviour* (HSG48) (see E17068 below).

Core disciplines

E17003 Ergonomics adopts a multi-disciplinary approach to an issue and thus draws on a number of other key disciplines, using this integrated knowledge to obtain a holistic view of the work system. The core disciplines include anatomy (the structure of the body); physiology (the function and capabilities of the body); biomechanics (the effect of movement and forces on the body); psychology (the performance of the mind and mental capability including perception, memory, reasoning, concentration etc); and anthropometry (the physical dimensions of the body).

The following text outlines:

- an overview of the ergonomic approach (E17004);
- design guidelines and principles relating to different aspects of the system (E17005–E17043);
- musculoskeletal disorders (E17044–E17050);

- tools which can be used to assess work and assist with design (E17051–E17058);
- outline of legislation which promotes ergonomic design (E17059–E17068).

The ergonomic approach

E17004 The ergonomic approach considers all aspects of the person's interaction with the task, workstation, environment and work system so that these can be designed to fit the abilities of the users and maximise the ease of use. The model below can be used to illustrate the framework of this approach.

Figure 1: Ergonomic Approach

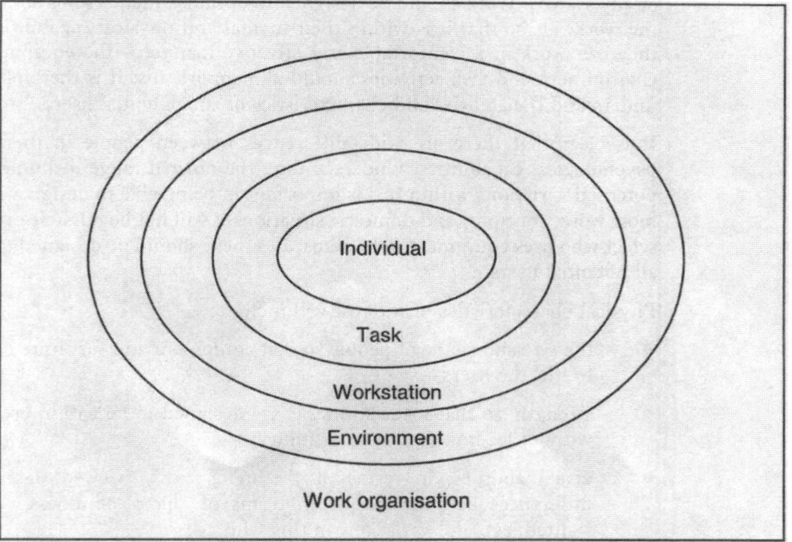

Ergonomics takes account of the physical and mental capabilities and needs of the potential users/workers/population (e.g. the size, strength and ability to handle information), before considering the task and the context that it will be undertaken in. By establishing the physical and mental capabilities of the user group, certain criteria or constraints can be established for the task. For example, most people can accurately remember up to seven numbers in short-term memory. A picking task requiring order numbers to be remembered should therefore be designed so that the order number is seven digits or less.

Likewise, the workstation should be designed in relation to the physical requirements of the person and the operational requirements of the task. For example, the users' height and reach will be relevant to specifying the dimensions of the workstation; the sequence of use of tools will be relevant to their arrangement on the workstation.

The work environment needs to be designed in relation to the person, task and workstation requirements. For example, the lighting and noise should be within acceptable limits for human work but also appropriate to the requirements of the task. (See also LIGHTING, and NOISE AT WORK and VIBRATION.)

The work system is the organisation within which these elements sit and which ultimately affect the whole work. This includes the way in which the person, task and workstation are organised, as well as the requirements and constraints that affect the way work is conducted. It should complement and not constrain the effectiveness of the person, task, workstation and environment interactions.

Ergonomic criterion for these elements are discussed in more detail in paragraphs E17005–E17043 below.

Designing for people

E17005 When adopting an ergonomic approach, the person is the central point of the design of any system. Work should be designed to optimise the person's ability to complete the work, i.e. so that it is within their mental and physical capabilities, and so that they can work in a comfortable and effective manner – the equipment, furniture, environment and task activities should all support this. It is therefore important to understand the abilities and characteristics of the potential users.

It is clear that there are wide differences between people in their physical and psychological capabilities. Understanding the normal range of human abilities and potential variations within this is important in being able to design appropriately. In most work, transport and domestic situations it will not be possible or appropriate to select who uses equipment or systems, and these should be designed to be useable by all potential users.

Physical characteristics of interest will include:

- the size and shape of people so that equipment and furniture can be designed to 'fit' the users;
- strength, so that loads and tasks are designed to be within users' capabilities without leading to fatigue or injury;
- visual abilities, in terms of the items to be viewed, taking account of differences between people in terms of colour blindness, long and short sightedness, and variations in this with age;
- hearing abilities, such that instructions and warnings can be clearly heard.

Psychological characteristics of interest will include:

- short and long term memory;
- uptake and processing of information;
- reaction time;
- motivation;
- concentration;
- perception.

These physical and psychological characteristics will vary between people and will depend on individual factors such as age, gender, training and skills, health and fitness, and previous injury or disability. For any individual these may also vary over time (e.g. visual ability tends to decrease with age).

Posture

E17006 In order to work comfortably and reduce the risk of musculoskeletal injury or discomfort, people should be able to adopt 'neutral' postures. A neutral posture is one in which the joints of the body are at about the midpoint of their comfortable range of movement and with the muscles relaxed (e.g. arms hanging by the side of the body). Deviations from this place a strain on the muscles and soft tissue.

The trunk is in a neutral posture when the natural 'S' curve of the spine is maintained – as when standing. The lower part of the spine (lumbar region) flattens when in a sitting posture due to rotation of the pelvis.

The wrist is in a neutral posture when the muscles are relaxed and the hand is in line with the forearm. The arm is in a neutral posture when relaxed by the side of the body.

Figure 2: The 'S' Shaped Curve of the Spine (S Pheasant Bodyspace: Anthropometry, ergonomics and the design of work (1996) p 69 fig. 4.1 reproduced with permission of Taylor and Francis)

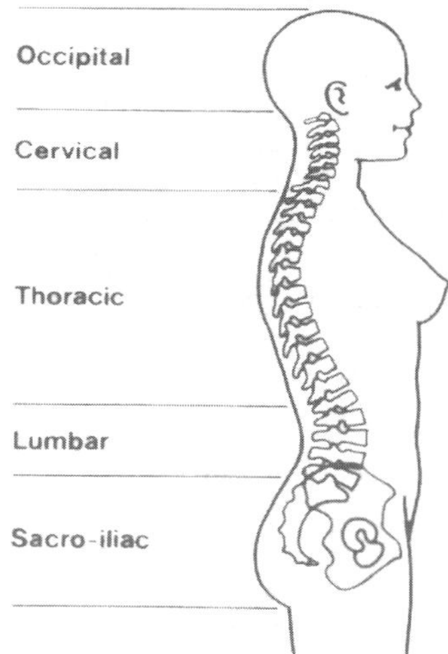

Depending on the degree of deviation and the duration of maintaining the posture, awkward postures (i.e. when the posture deviates from the neutral) and static postures (i.e. postures in which there is no movement for a period of time), can both lead to fatigue, discomfort, strain and possibly injury. Awkward postures such as bending, twisting and stretching, reaching the arms behind the shoulders, and reaching above shoulder height or below knee height should be avoided as far as possible; this can be achieved through careful positioning and adjustment of the equipment and furniture used. Introducing movement and changes in posture will help alleviate the discomfort which can arise from static postures.

Anthropometry

E17007 Anthropometry (measurements of body dimensions) can be used to ensure that people will be able to reach, use and operate tools, equipment and workstations, by structuring and positioning items appropriately (e.g. so they are within easy reach, so they adjust sufficiently, so they allow or prevent access etc).

The size of the body varies with age, gender and ethnicity. Tables of anthropometric data are available which give data for many different body dimensions, representing different population groups (see S Pheasant *Peoplesize* (*Open Ergonomics*)*; Anthropometrics: An introduction for schools and colleges* (1984) BSI p 7310). Variations in body size usually follow a normal Gaussian distribution, with most people falling within a central value for a given body dimension with fewer people tailing off at either side. It is common to design to take account of 90 per cent of the population, e.g. to design for those who are taller than the smallest 5 per cent of the population, and smaller than the tallest 5 per cent of the population (i.e. to design within the fifth to ninety-fifth percentile data). However, in some situations it is necessary to design for the extremes of the population. For example, the height of a door should take account of the tallest potential user; guarding on a machine should take account of the finger width of the smallest potential user.

Population stereotypes

E17008 Population stereotypes are cues and expectations which provide information regarding the meaning, state or operation of items. We have certain expectations concerning colours, directions of movement, shapes, sounds, and abbreviations. For example, a red tap indicates that it supplies hot water; turning a control dial clockwise turns the power up. Designs that take account of population stereotypes, will result in increased accuracy and faster reaction times. Population stereotypes can vary by culture and often by industry.

Examples of these stereotypes are shown below.

Population stereotypes in the UK
Colour • Red = hot, danger, warning, stop • Green = safe, go • Yellow = caution • Blue = cold, information *Shape of signs* • Round = enforceable • Triangle = caution/attention • Square = advisory *Direction* • Clockwise = on, tightens (screws), increases • Anti-clockwise = off, loosens, decreases • Upwards = switches off, increases • Downwards = switches on, decreases

Allocating tasks to people or machines

E17009 In most systems people interact with machines. When designing a system, tasks can be allocated to either the person or the machine. Decisions on whether a task should be undertaken by a person or by a machine will depend on an understanding of their relative capabilities. Some general guidelines on the relative capabilities of people and machines are shown below:

Figure 3: Guidance for allocating functions (adapted from Sanders and McCormick, *Human Factors in Engineering and Design* (1987) McGraw Hill)

Person's Capabilities	
• Sensing stimuli – both at low levels and over background interference	• Sensing stimuli outside range of human sensitivity
• Recognising patterns of stimuli	• Monitoring for prescribed events, especially infrequent ones
• Sensing unexpected occurrences	• Storing coded information quickly and in large quantities
• Decision making	• Retrieving coded information quickly and accurately
• Information recall	• Calculating
• Creating alternatives	• Exerting considerable force
• Interpreting subjective data	• High speed operations
• Adapting responses	• Repetitive tasks
	• Sustained accuracy and reliability
	• Counting or measuring
	• Simultaneous operations

In allocating function between people and machines economic or social factors may also need to be considered beside performance. In some situations the 'best' performance may not be required and contact with people may be preferable to dealing with machines. Also, in order to optimise job design, retaining some 'interesting' tasks for people may be more important than mechanising.

Designing the task

E17010 The task is the collection and sequence of activities and events that allow the work to be completed. The task should be designed to be within the capabilities of the person, but must take account of the constraints of the work process, environment or system. Mismatches between task requirements and individuals' capabilities increases the potential for human error.

Task design will influence the activities of the person e.g. the postures adopted, amount of force applied, repetition, and duration of the activities. These may have an influence on performance outcome or on health (e.g. musculoskeletal disorders). Issues to consider include:

- *Duration of the task* – long durations of work without a break may lead to reduced performance, and increased discomfort. Decline in concentration usually becomes evident after 30 minutes; physical discomfort increases if static postures are maintained or movements are repeated without a break.
- *Breaks and changes in activity* – infrequent breaks can lead to reduced performance (e.g. on a vigilance task such as inspection) and increased discomfort. Regular changes in posture and movement help reduce physical discomfort and injury. Prolonged periods looking at a display or screen without a break can also lead to visual discomfort. Varying the tasks that are undertaken will also help to reduce boredom.
- *Amount of repetition of movements required* – highly repetitive work which involves the same muscle groups and movements can lead to physical discomfort and boredom.
- *Frequency of the task* – frequently performed tasks may lead to boredom and increased physical discomfort (as the same movements are made). Conversely, there may be training and information needs for infrequently performed tasks.
- *Pacing* (if the person is required to work at a speed set by a machine or process) – this will dictate the speed of work and amount of recovery the person has between tasks or operations. This can have an impact on discomfort and performance. A comfortable work rate should be established.
- *Work rate* – excessively busy periods or quite periods can lead to stress and discomfort. There should not be benefits for completing the task early as this can lead to people rushing the task.
- *Complexity of the task* – a number of sub-tasks usually are required to complete the operation, and more than one operator may be required to assist in the operation. Tasks can be divided so that operators each undertake one sub-routine or a more complex combination of tasks. Jobs should be designed to facilitate development of skills, interest, variety and commitment. Several sub-tasks may be undertaken by one operator, but to reduce boredom and to prevent overload and the risk of musculoskeletal injury these should vary in the demands they place on the physical and psychological capabilities of the workers.
- *Awkward or static postures required* – these may be dictated by task requirements, or poor tool or workstation design. These can contribute to musculoskeletal disorders and should be avoided.
- *Force required to undertake the task* – application of frequent or excessive force, particularly in relation to the capabilities of the body part applying the force, can lead to discomfort. It may be appropriate to mechanise the task or use a tool or equipment to assist in generating force if required.
- *Sequence of use of tools/equipment* – appropriate layout of equipment and tools will facilitate ease of use, so that, for example, those used most frequently are positioned closest to the user.
- *Levels of concentration and attention required* – the level of arousal (the state of consciousness) helps to maintain attention and affects performance. Low levels of arousal are found in undemanding jobs, where activities are either very mundane or very infrequent, e.g. production line packing, night security work. High levels of arousal can be found in stressful and demanding jobs, e.g. air traffic control and ticket inspection. Both high and low levels of arousal can lead to poor performance. Tasks should be designed to maintain

the vigilance and arousal necessary for the task. Methods of facilitating concentration and vigilance include reducing the task time and introducing frequent breaks; providing music or background noise for routine tasks; and exaggerate the size or colour of the stimulus.

- *Information and training required for the task* – some tasks will require personnel to be trained in how to undertake them. For others, simple information or instructions may be adequate. The need for training and information will depend on the complexity of the task, and the experience and abilities of potential users.

Task design will be influenced by the equipment and workstation design, the organisation of the work, and the abilities of the individuals. Application of anthropometric and biomechanical knowledge will help ensure the physical elements of the task are within the physical capabilities of the person. Understanding the psychological abilities of the person such as their memory and concentration will help to design a task within limits acceptable to the person. The provision of any information or instruction should also account for the way in which people process information.

In general it is those who are undertaking the task who have the most knowledge about the needs of the work system. Involving workers in evaluation and re-design is an important component in ensuring the design meets their needs and those of the system, and that any changes made are accepted by workers.

Design of equipment

E17011 Equipment should be designed and selected for those who will use it, taking account of the task to be completed and the conditions under which it is used.

Hand tool design

E17012 Tools enable the use of the hand to be extended, e.g. to extend its ability to apply force, manipulate items, make precise movements and so on. Appropriate design of hand tools can facilitate the task, and reduce the risk of musculoskeletal injury.

Hand tools should be designed to allow neutral arm, wrist and finger postures in operation. The hand is in a neutral posture with the hand in line with the wrist, the thumb facing upwards and the palm facing inwards. This can be seen when the shoulder is relaxed and the arm is allowed to hang relaxed by the side of the body. The maximum grip strength can be applied when the wrist is in a neutral posture; deviations from this (ulnar or radial i.e. side to side, flexion or extension i.e. up or down) will reduce the amount of force that can be applied, with the hand able to apply the least force in a flexed posture. Working with the hand in a neutral posture also minimises the strain placed on joints and other soft tissue, which can be compressed or experience friction in awkward postures.

Angling either the work piece or the tool handle can help reduce the amount of wrist deviation required when using a hand tool; however, the potential for variation in orientation of the tool on the work piece may limit the benefit of this.

Key points in hand tool design are:

1. Handles should be long enough to fit the whole hand i.e. at least 100 mm, although a length of 120 mm is preferable.

2. In general, larger handles decrease the amount of muscular activity required when using the tool. A handle thickness of approximately 40mm is generally recommended. Slightly thicker handles may be beneficial when applying torque (e.g. for screwdrivers).
3. Tools with two handles which require a hand span (e.g. pliers, scissors) should have a span of approximately 60 mm. If the tool is used repetitively, an automatic spring opener will reduce the strain on the weaker finger extensor muscles (used to open the hand).
4. Ideally the handle surface should be compressible, textured and non-conductive. It should be free from sharp edges; avoid ridges or finger contouring on the handle as this can place pressure on the soft tissue of the hand. Avoid cold surfaces (e.g. metal), particularly if the tool is powered by compressed air.
5. An excessively smooth handle surface requires the user to grip the handle more tightly, increasing the amount of force required, and this should be avoided. If the hands are sweaty or if the user is wearing gloves they will also have to grip the handle more tightly.
6. It should be possible to use the tool in either hand; if this is not possible, specific tools for left handed workers should be provided.
7. Power-assistance – power assisted tools can greatly increase the speed of task completion, remove a large degree of force exertion from the operator and reduce some awkward postures such as rotating the wrist. However, the user may experience vibration from the tool. Pneumatically powered tools can blow cold air exhaust over the hands, increasing the risk of discomfort and musculoskeletal problems such as Hand Arm Vibration Syndrome (see OCCUPATIONAL HEALTH AND DISEASES). If power tools are used, ensure they are low-vibration and well maintained to reduce vibration transmitted to the hand and arm. Mounting power tools in a jig can reduce vibration transmission.
8. If force has to be applied through the tool, where possible use power tools rather than tools requiring manual application of force. However, power tools often weigh more than manually operated tools.
9. Weight – frequently used tools should weigh as little as possible; a maximum of 0.5kg is recommended. The distribution of weight in the tool should be even with the centre of gravity as close to the hand as possible. Counterbalances may be required to support heavier tools or those with an off-set centre of gravity.

Design of controls

E17013 Displays and controls are the mechanisms by which people and machines interact. The person gives instructions to the machine through the controls (e.g. knobs, switches, levers, buttons). Controls can be discrete, having a set number of conditions (e.g. on/off/standby); or continuous where the condition varies along a scale, (e.g. volume). Controls should be designed to facilitate the changes that they allow and the amount of effort required to operate them. The following factors should be considered:

- The amount of force required to operate the control (and the amount of resistance to prevent accidental operation). Fingers and hands should be used for quick, precise movements; arms and feet for operations requiring force.

- The size of the control should be appropriate for the force required, and body part required to activate it.
- The location of controls should facilitate their use. Hand-operated controls should be easily reached and grasped, between elbow and shoulder height. Controls should be sufficiently far apart to allow space for the fingertips. Sequence and frequency of use, and importance may also dictate the location of controls.
- Appropriate feedback should be provided to indicate activation of the control.
- Controls should be coded by controls by colour, shape, texture or size (to allow identification by touch).

Design of displays

E17014 The state of the machine is relayed to the person through a display. Displays should be designed to enable users to easily and accurately assess information from the machine. The following principles should be followed:

1. Qualitative displays showing a small number of conditions should be used for discrete information (e.g. on/off). Quantitative displays should be used to present numerical or continuous information, e.g. temperature, speed etc. Qualitative displays may be lights or words etc. Quantitative displays may be scales, counters etc.
2. Display scales should be clear, unobscured and concise. Scale intervals should increase left to right, bottom to top, preferably increasing in units of 10s, 100s, 1000s, etc. as appropriate.
3. Displays should conform to population stereotypes in terms of colours used (e.g. red = danger)
4. Ensure labelling, if used, is clear.
5. Limit the number of warning lights to avoid confusion and aid identification.
6. On dials with pointers, avoid parallax (the difference in scale reading depending on the viewing angle) by keeping the dial and pointer close together.
7. Text should be a suitable size and font to allow easy reading. If this is not appropriate, B and 8, O and 0 can be confused, and if resolution is not adequate, F and P can also be confused.

Grouping of controls and displays

E17015 The result of activating a control is often indicated in a display. The relationship between displays and controls should be clear; this can be achieved through location, arrangement, text, shapes, colours and responsiveness. Grouping of controls and displays can improve their association with their function and facilitate their ease of use. Controls and displays can be grouped according to sequence of use (e.g. start, run, finish) or by function (e.g. keeping all controls concerned with lights together) or by frequency of use, such that those most frequently used are within convenient reach and within the immediate viewing arc. Locating controls by importance of use can also aid their operation (e.g. emergency controls should be placed within convenient reach).

Text

E17016 Text used as information instruction or labelling should be clear and concise. Bold, italics, large fonts, underlining and colour can all be used to draw attention to or highlight information but should be used sparingly to preserve the meaning when they are used. Lower case should be used for phrases or sentences as UPPERCASE DECREASES THE DIFFERENTIATION BETWEEN LETTERS AND IS SLOWER TO READ.

Positive instructions should be used, not negative or double negatives. For example: 'When the alarm sounds turn the machine off' – not – 'When the alarm sounds do not leave the machine on'.

Software design

E17017 The field of Human Computer Interaction (HCI) is a specialist area; in this context suffice to say that software should be designed to make the use of a system intuitive and accessible (e.g. through the arrangement of icons, use of colours and text, structure of menus and commands).

Workstation design

E17018 The workstation (the area where the task is conducted) should be designed so that it is appropriate for the person using it. The workstation may be a desk or work surface, but may equally be at a conveyor belt or a driver's cab. Attention needs to be given to the physical dimensions of the workstation, and the arrangement of any necessary equipment/tools/components in it to allow good posture and acceptable movements.

The issues discussed below cover the main points that should be considered in the design of workstations.

Worksurface height

E17019 Many tasks will involve workers using a work surface. The height of the worksurface should permit a relaxed, upright posture and facilitate the use of equipment at the workstation. The surface height (whether designed for sitting or standing tasks) should be such as to avoid users stooping or reaching to its surface. In most situations the height of the worksurface is fixed. For seated tasks a height adjustable chair should be provided so that the working height can be set appropriately. A footrest may be required to support the feet when sitting at a comfortable height.

An appropriate worksurface height is dependent on the work that is done at that worksurface. As a general rule, the top of the worksurface should be level with the user's elbow height (the user may be sitting or standing, as required at the workstation). The height of the worksurface may need to be reduced to take account of the thickness of the item being worked on so that this is at an appropriate height.

There may be a range of users using the same workstation, and unless easy height adjustment is provided, it will not be possible to obtain the optimum height for all potential users. For a fixed height workstation some compromise will have to be made, based on the task requirements and the likely users.

The following working heights are recommended for different tasks, in relation to the user's height (S Pheasant *Ergonomics, work and health* (1991) Taylor and Francis).

Task	**Height**
Manipulative, requiring force and precision	50 mm – 100 mm BELOW elbow height
Delicate (including writing)	50 mm – 100 mm ABOVE elbow height
Heavy-requiring downward pressure	100 mm – 250 mm BELOW elbow height
Lifting and handling	Hands between knuckle and elbow height
Two-handed pushing/pulling	Hands just below elbow height
Hand-operated controls	Located between elbow and shoulder height

The variation in recommended working heights for different tasks arises from the aim to minimise the effort required to perform the tasks, by optimising the posture and muscle groups used.

Using this data, as an example, tasks such as component assembly where pneumatic screwdrivers are used should be conducted at between 100 mm and 250 mm below elbow height so that downward pressure can be applied. If the unit being assembled stands 50 mm above the worksurface, the worksurface will need to be 150 mm – 300 mm below elbow height. This assumes that all work at the workstation is of this nature. Workstations that are also used for writing tasks, for example, should be split-level, providing a writing surface at or just above elbow level. If a fixed height workstation and height adjustable chair are provided, users should adjust the height of the chair depending on the task. For example, VDU users who also write at their workstation will find it more comfortable to sit higher when they are keying and lower when they are writing.

Height adjustable workstations can be beneficial, particularly for standing tasks and where a range of different users may work at the same workstation. Where height adjustable equipment is provided users should be trained in how to adjust it and how to identify an appropriate working height.

Workstation characteristics

E17020 Other characteristics recommended for the workstation include:

- The workstation should be a suitable size such that all equipment can be located and arranged conveniently for the person to complete the task.
- The workstation surface properties should not present a risk to the user – there should be no sharp edges or unprotected hot/cold surfaces. A front edge of 90° can cause discomfort if the arms are rested or pivoted on it.
- Surrounding workstations or other items in the environment, e.g. columns or posts, should not constrain the user's posture and ability to get into and from the workstation easily.
- There should be sufficient legroom underneath the worksurface to allow the user to be able to sit comfortably, sufficiently close to the workstation, without their posture being constrained.

Workstation layout

E17021 To facilitate a good working posture, the equipment and items used to complete the tasks should be within convenient reach of the user, so they do not have to stretch or lean. The zone of convenient reach and the normal working area can be used to establish a suitable layout of the workstation.

Zone of convenient reach

E17022 The zone of convenient reach is defined by the area from the shoulder to the fingertips with the arm outstretched (upward, downward and to the sides), which can be reached without any undue exertion. Items required for the task (e.g. control buttons, handles, work surfaces, tools) should be placed within this zone. The extent of this area obviously depends on the length of the arm and its arc. The fifth percentile arm length (i.e. 95 per cent of the population will be able to reach items if placed within this area) is 735 mm for British men and 655 mm for British women when reaching forwards (S Pheasant *Peoplesize* (*Open Ergonomics*). If designing for both male and female users the shorter dimension (i.e. 655 mm) should be used.

Figure 4: Zone of Convenient Reach (E Grandjean Fitting the Task to the Man (1988) p 51 fig. 42 reproduced with the permission of Taylor and Francis)

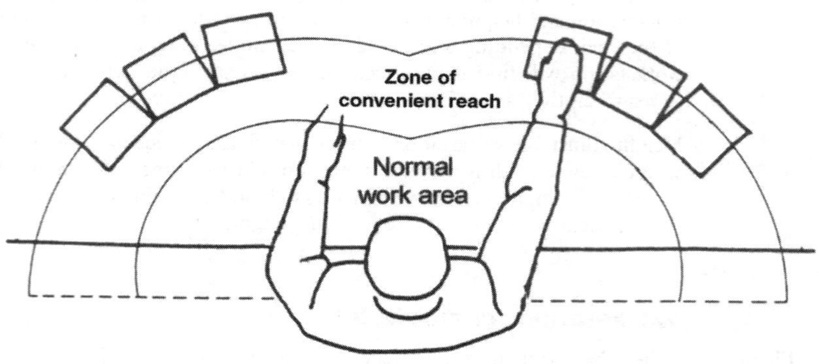

Normal working area

E17023 The normal working area is defined by the comfortable sweep of the forearm with the elbow bent at 90° and the upper arm in line with the trunk. This provides an area that can be reached without extension of the arm at the shoulder and requiring no trunk movement. The extent of the area depends on the length of the arm from the elbow to the fingertips. The fifth percentile forearm length is 443 mm for men and 402 mm for women (S Pheasant *Peoplesize* (*Open Ergonomics*). Again, if designing for both male and female users the shorter dimension (i.e. 402 mm) should be used.

Arrangement of items on workstation

E17024 Frequently used items and critical items (e.g. emergency controls) should be placed in the normal working area. Occasionally used items can be sited within the zone of convenient reach. Items can be grouped by function or sequence of use, or importance.

Sufficient space should be provided at the workstation to allow a flexible arrangement of all the equipment that is required for the task. Items should be arranged so operators do not have to reach across their body e.g. the phone should be placed on the left hand side if it is answered with the left hand.

Visual considerations

E17025 The following factors should be taken into consideration.

Viewing angle

E17026 Items can be viewed both by movement of the eyes, and by movements of the head, neck and trunk.

Awkward neck and trunk postures may be required if the viewed item is low, high or to one side and this may cause discomfort. Items to be viewed should be positioned such that no twisting or bending of the neck, head or trunk is required to see them, and so that the eye muscles are in a neutral position.

The neutral position of the eye is generally taken as about 15° below the horizontal line of sight (i.e. 15° below level with the eyes). The eyes can comfortably move about 30–45° below the horizontal line of sight before the head must be inclined; an upward gaze of approximately 15° is achievable before the head must be tilted backwards to view further. An upward gaze is fatiguing when sustained for any length of time, as the muscles controlling upward movement of the eyes are not as strong as the muscles controlling downwards movement. The most comfortable eye position is an arc of 30° from the horizontal position of the eyes downward and this is the acceptable viewing angle.

Acceptable Viewing Angle

E17027 Items outside about a 60° arc in front of the person will involve neck twist to view. Items to be viewed should not be positioned outside this area.

Viewing distance

E17028 The required viewing distance generally depends on the person's visual ability, the size of the item being viewed and the lighting levels. A viewing distance of between 500 mm and 750 mm is likely to be suitable for most items, but a shorter viewing distance may be required for fine work.

Sitting versus standing

E17029 There are biomechanical benefits to both sitting and standing. Sitting can help to reduce the development of fatigue in the lower limbs, provided the chair provides adequate support and the feet are also supported. Tasks that require generally static postures (with limited or no trunk movement) will benefit from being undertaken while sitting. However, inappropriate seating and prolonged sitting can lead to discomfort and fatigue, and may prevent good posture (e.g. armrests may prevent

Figure 5: Viewing Angle (S Pheasant Ergonomics, Work and Health (1991) reproduced with permission of Palgrave Macmillan)

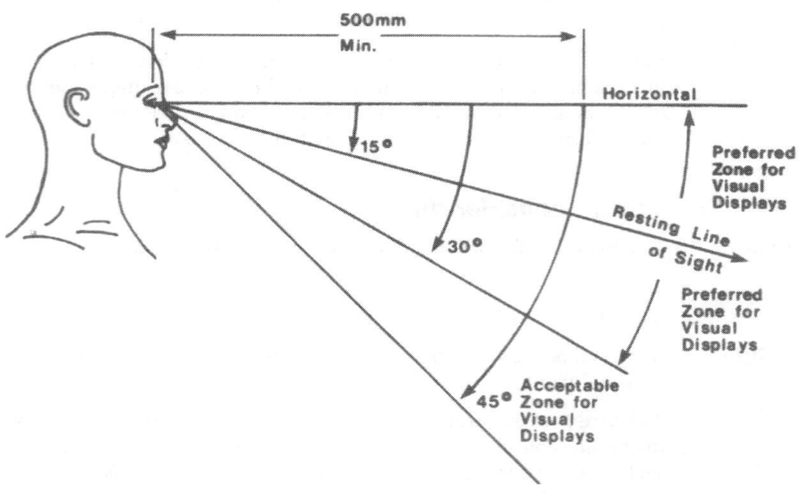

the chair being brought sufficiently close to the workstation) (see WORKPLACES – HEALTH, SAFETY AND WELFARE W11008).

Standing is beneficial if the task requires a range of body movement, or handling of heavy items. However, prolonged standing, particularly with little walking, can result in workers experiencing tired or swollen legs and lower back discomfort. The *Workplace (Health, Safety and Welfare) Regulations 1992* (*SI 1992 No 3004* as amended by *SI 2002 No 2174*) require a suitable seat be provided for each person at work whose work (or a substantial part of it) can or must be done seated. Whether sitting or standing, workers should be able to vary their posture and should be able to take short, frequent breaks away from the workstation.

Sit/stand stools can be used to allow work to be completed at standing height while removing some of the strain in the legs and back associated with standing.

Anti-fatigue matting can be used at standing workstations to cushion the feet, and this is especially beneficial on concrete floors as these are particularly fatiguing.

Physical work environment

E17030 The work environment (the temperature, humidity, lighting and noise and general layout of the place of work) should also be designed to suit the person and the work. In general, the effects of the environment can impact on performance, subjective comfort, perception, attitudes, safety and health.

Lighting

E17031 Appropriate levels of lighting are necessary to maintain acceptable levels of quality and performance, and to help avoid visual discomfort. Poor lighting can lead to visual fatigue (symptoms include red or sore eyes, blurred vision, headaches).

Inappropriate lighting can also force users to adopt poor postures in order to view items more easily, and this can also lead to discomfort.

Issues to consider in designing visual environments are the brightness of the light sources, their position, the evenness of the light distribution, the contrast between the different work items, and the potential for glare/reflections from other items in the work area.

The amount of illuminance required should be determined by the task demands e.g. visually demanding tasks such as inspection, or fine and precise work require a higher level of light than that required in walkways. Insufficient lighting can result in eye strain and discomfort, while too much light, or a bright light source within the visual field will cause glare and result in visual discomfort and reduced visual performance. Where light levels are already high it may be appropriate to improve the legibility of the source document or item being viewed (e.g. increasing size and contrast) rather than further increasing the lighting levels.

The location of lights will also effect the visual performance and comfort – light sources within the visual field may cause glare (which may be visually disabling or causes discomfort). Excessive differences in lighting levels within the work areas should be avoided. Surfaces of a highly reflective material should be covered or replaced to reduce glare. Directional lighting may be appropriate for some inspection tasks, but diffused light is generally more satisfactory as it reduces the amount of glare and shadows produced.

Light may come from a natural source (from windows) or an artificial source (electric lights). It is important to remember that lighting levels may vary by the time of day or year and therefore appropriate adjustments may need to be made e.g. provision of local lighting (i.e. lamps), or blinds.

The furniture and equipment in the room should be arranged so that the sources of light do not cause light to fall directly onto reflecting surfaces or into the eyes (e.g. sit at right angles to a window rather than facing or with the back to the window).

Colours should be chosen to provide contrast to aid detection but not to exaggerate glare or reflections. Colours with good contrast should be used for displays (e.g. black and white, blue and yellow etc). Where lighting levels are high, dark characters on a light background are generally easier to read.

In general it is appropriate to allow individuals to have control over the lighting levels in their area, as individuals vary in what they find comfortable. In addition, having control over the work environment can help reduce psychosocial stress.

Increased lighting levels may be required for those with poorer eyesight (e.g. older people), although this should be done with care as older people are generally more susceptible to glare.

For further details see LIGHTING.

Additional guidance is provided in the HSE Guidance Note *Lighting at Work* (HSG38).

Noise

E17032 Noise can have an effect both on hearing ability (hearing loss) and on performance. Too much or too little noise can adversely affect concentration. Where work is monotonous, noise can be introduced to provide a stimulus; but where noise is excessive it should be reduced to facilitate concentration. Noise levels should not exceed approximately 55dB(A) for tasks requiring concentration.

Noise can be used effectively as a warning or alarm, although excessive use of noise for warnings can be detrimental to their effectiveness. Hearing generally deteriorates with age; this, and the differences in hearing ability between people, should be taken into account when designing critical alarms.

There are several harmful effects of noise which include:

1. Physical damage to the eardrum and ossicles induced by excessively high noises e.g. explosives.
2. Hearing damage due to exposure to high levels of noise. Damage can be divided into temporary threshold shift (reversible) or noise-induced hearing loss (irreversible).
3. Annoyance and stress – which can lead to reduced concentration. Noises that are annoying are generally unexpected, infrequent, high frequency and those over which the listener has no control.
4. Hindering communication – which can lead to an increased rate of accidents as well as stress.

Where excessive noise is present (over 80dB(A)) methods to eliminate, reduce or control the risk (e.g. baffling and hearing protection) are required. Hearing protection should only be used as a last resort, with other methods of noise control being explored first, as hearing protection (muffs or plugs) is often found to be uncomfortable, incompatible with other forms of PPE, or the task requirements and may therefore not be worn.

For further information see NOISE AT WORK and VIBRATION.

Vibration

E17033 Exposure to vibration is associated with musculoskeletal disorders, other health problems, and reduced performance. There are two main categories of vibration transmission to the body: whole body and hand/arm (see N3027).

Whole body vibration usually occurs via the feet or seat when the person is working on/in a vibrating environment (e.g. transport systems). Low frequencies of vibration (such as those from motor vehicles) can lead to chest pains, difficulties in breathing, back pain and impaired vision.

Hand-arm vibration affects only the hands and arms and can be experienced when using power tools, such as pneumatic screwdrivers and chainsaws. Exposure of the hands and arms to vibration is a known risk factor in the development of Hand Arm Vibration Syndrome (HAVS), a condition which affects the nerves and blood supply to the hand resulting in tingling or numbness, impairment or loss of function, and blanching of the fingers, particularly on exposure to the cold. It is particularly higher frequencies of vibration (such as those associated with hand-held tools) that have been associated with muscle, joint and bone disorders affecting the hand and arm. Transmission of vibration has been found to increase as the gripping force increases and where tight gloves are worn.

Vibration, as with noise, should be tackled at its source. Larger machines can be isolated or the vibrations insulated, e.g. by introducing better mountings on the machine. Regular maintenance of machines can also reduce vibration. Relocation of the machine and insulation of the source should be considered. Floors, seats and handgrips can all be fitted with damping material. Hand tools should be designed to reduce the vibration created, the weight and amount of force required, so that tools do not have to be gripped tightly. Where this cannot be achieved, exposure time should be limited, and the hands kept warm.

For further information see NOISE AT WORK and VIBRATION.

Thermal environment

E17034 The internal body temperature needs to be maintained within a relatively narrow range for health, and the ambient conditions which facilitate this are also relatively narrow. Deviations from comfortable temperatures can lead to reduced performance, fatigue, discomfort, ill health (hypo and hyperthermia as the body temperature rises or drops below those necessary for health); in extreme circumstances this can be fatal.

Body temperature, and therefore comfort, is dependent on the ambient conditions (air temperature, humidity, any radiant heating, and air velocity), the activity being undertaken (and therefore the heat produced within the body), and the clothing worn. Individuals' response to the thermal environment vary and some people are more affected by high or low temperature than others; it is most satisfactory if employees are able to control the environment to suit themselves (e.g. opening windows etc), although this can be difficult in large, open plan offices.

As far as possible, environments should be designed and controlled to allow a comfortable working temperature (e.g. through heating, air conditioning etc). Where this is not possible (e.g. due to a work process or outdoor working), appropriate clothing can be used as a control measure, or in extreme situations it may be necessary to limit exposure time.

Low temperatures reduce manual dexterity and where this is required the hands should be kept warm (e.g. through appropriate gloves). Manual handling injuries are more likely in extreme thermal environments and account should be taken of this when planning tasks.

Guidelines for control of employees' work in hot/humid/cold environments are shown below.

1. Maintain temperatures between 20–24°C for sedentary work and 13–20°C for physical work.
2. Maintain humidity at between 40 and 70 per cent.
3. Reduce air velocity (draughts) to less than $0.1ms^{-1}$ in moderate and cold environments. Increasing air velocity can be an effective control measure in hot environments, provided the air temperature is not close to or above body temperature when increased air velocity will increase the thermal load on the body.
4. Decrease physical work load in hot environments.
5. Schedule rest pauses or rotate personnel around more strenuous tasks in hot conditions.
6. In hot environments, allow rest periods to be taken in cooler environments
7. Schedule outdoor work to avoid periods of very high or low temperature.
8. Permit gradual acclimatisation to hot or cold environments (7–10 days)
9. Maintain hydration by consuming sufficient drinking water, particularly in hot environments.
10. Avoid long exposure periods in cold environments.
11. Avoid rest for long periods in cold environments.
12. Provide protection from wind and foul weather when working outside.

Physical hazards

E17035 Other physical hazards may be present in the workplace (e.g. slipping, falling, tripping hazards). Ideally these should be eliminated at source, through appropriate design (e.g. removing the need for an operator to enter a potentially dangerous area). Where hazards are still present, guarding, safety barriers, good housekeeping, appropriate lighting, and good workplace layout will reduce the possibility of all types of accident. Raising awareness through education and training can be effective, although it should not be relied on as a first line of control. Personal protective equipment (PPE) should always be seen as the last resort control measure, due to issues related to fit, compatibility, impact on performance, loading on the body, etc which may mean that it is not worn correctly, or possibly at all.

Job design and work organisation

E17036 The design and organisation of work can have a significant impact on health. The demands of the job should be matched to the skills and abilities of the individuals. Excessive demands or, conversely, under utilisation of skills can both lead to a decrease in performance. The design of the job or work task is important for obtaining the maximum performance from the person for the minimum effort, and also in terms of facilitating a level of job satisfaction. Factors to consider include job rotation, job enlargement and job enrichment, as well as job scheduling, job demands, and shift patterns.

Job rotation

E17037 Job rotation, whereby people are moved from one task to another, allows changes in posture, movement, and mental demands. This may lead to the acquisition of additional skills and may help the person identify more with the completed product or service. Job rotation can also help to overcome boredom, particularly in repetitive jobs, and help to prevent fatigue, loss of concentration and deterioration in performance. To be effective as a control measure for musculoskeletal discomfort people should rotate to tasks that are significantly different in terms of postures, movements and forces required.

Job enlargement

E17038 Job enlargement whereby the number and range of tasks undertaken by the individual is increased, allows them to become more flexible, develop other skills and have greater variety in their tasks.

Job enrichment

E17039 Jobs can be enriched by incorporating motivating or growth factors, such as increased responsibility and involvement, opportunities for advancement and a greater sense of achievement. This should provide the workers with greater autonomy over the planning, execution and control of work.

Work scheduling

E17040 Ideally workers should have control over how the work is scheduled. As far as possible work should be self-paced rather than machine-paced. Machine pacing may be too fast or too slow for an individual, both of which can lead to problems (lack of recovery time for muscles and soft tissue; frustration etc). Machine pacing has been implicated as being a contributory risk factor in upper limb disorders, as well as stress and stress-related illnesses.

Rest breaks help alleviate both mental and physical fatigue. Allowing the worker to select when and how often they take breaks is preferable to fixed breaks, although there is some evidence that people work until fatigue occurs rather than stopping before this point. Therefore, under some circumstances (e.g. tasks that require high concentration or those requiring physical effort) it may be preferable to give fixed rest breaks before the person is likely to become fatigued.

Breaks should be taken away from the workstation, not only because this allows for physical movement and a change of scene but also because it may facilitate reducing exposure to hazards such as noise. Short, frequent breaks are better than long, less frequent breaks.

Shift-work

E17041 It is well recognised that shift-work affects health and safety at work. In particular, where shifts involve people changing their eating and sleeping patterns, symptoms such as irritability, depression, tiredness and gastrointestinal problems (e.g. indigestion, loss of appetite and constipation) can occur. These health problems are largely due to the disturbance of the body's physiological cycles. Although many shift-workers adapt to the disruptions, it is widely recognised that productivity levels are lower on night shifts than day shifts.

Night shifts should therefore be avoided if at all possible. Younger and older workers (those below 25 and above 50) may be more vulnerable to ill health effects from shift work. Those with digestive disorders, sleep problems or other health problems may find shift work aggravates these. Shift rotations should be short and followed by at least 24 hours' rest.

Job support

E17042 Appropriate design of jobs and the structure of the organisation and the way it operates are important design criteria. The following points are recommended for providing adequate support in a job.

- Provide opportunity for learning and problem solving, within the individual's competence.
- Provide opportunity for development and flexibility in ways that are relevant to the individual.
- Enable people to contribute to decisions affecting their jobs and their objectives.
- Ensure that the goals and other people's expectations are clear and provide a degree of challenge.
- Provide adequate resources (e.g. training, information, equipment, materials).
- Provide adequate supervision, support and assistance from contact with others.
- Provide feedback on performance and communication of objectives.

Organisational support

E17043 The following factors are considered to be necessary to provide an acceptable background against which people can work:

- Industrial relations policies and procedures should be agreed and understood and issues handled in accordance with these arrangements.
- Payment systems should be seen as fair and reflect the full contribution of individuals and groups.
- Other personnel policies and practices should be fair and adequate.
- Physical surroundings and the health and safety provision should be satisfactory.

Musculoskeletal disorders

E17044 The term 'musculoskeletal disorders' (MSDs) refers to problems affecting the muscles, tendons, ligaments, nerves, joints or other soft tissues of the body. These disorders most commonly affect the back, neck and upper limbs. Musculoskeletal disorders are the most common work related condition affecting the general population in Britain; they account for more than half of all self-reported occupational ill health (1.2 million cases in 1995). Of these, 44 per cent affected the back, 32 per cent the upper limbs or neck, and 9 per cent the lower limbs. An estimated 9.9 million working days were lost in Britain due to MSDs in 1995. The cost to employers of MSDs affecting the neck and upper limbs was estimated to be at least £200 million (HSE, 1995).

Musculoskeletal disorders give rise to clinical effects in the individual, i.e. symptoms (e.g. pain, numbness, tingling) and signs (e.g. changes in the appearance of the limb which may be identified on medical examination). They usually also result in some functional changes (e.g. reduced ability to use the part of the body affected and restrictions to movement or strength). They can also have a general effect, such that there may be a resultant reduction in general health or quality of life (e.g. through restriction of activities).

MSDs are frequently work related, but not necessarily caused by work. MSDs have been the subject of much civil litigation in recent years and a significant number of personal injury cases have been successfully brought against employers (See OCCUPATIONAL HEALTH AND DISEASES.). The employer's duty of care to their employees with respect to upper limb disorders is now well established in the civil courts. This civil law duty runs parallel to the employer's statutory responsibility under health and safety legislation.

There is currently no specific legislation relating to musculoskeletal disorders. However, the *Manual Handling Operation Regulations 1992* (*SI 1992 No 2793* as amended by *SI 2002 No 2174*), and the *Health and Safety* (*Display Screen Equipment*) *Regulations 1992* (*SI 1992 No 2792* as amended by *SI 2002 No 2174*) were introduced with the aim of reducing musculoskeletal disorders associated with these activities. There are also general duties under the *Health and Safety at Work etc Act 1974* and the *Management of Health and Safety at Work Regulations 1999* (*SI 1999 No 3242*). (For more details on legislation see E17059–E17068.)

Activities outside of work may also present similar types of risk of MSDs to those experienced in work activities. However, non-work activities are usually not as repetitive, forceful or prolonged as work tasks, and the individual is likely to have more control over whether the activities are undertaken and how long for.

Back disorders

E17045 Back discomfort is extremely common in adults. Over 80 per cent of adults will experience an episode of back pain during their life, but the vast majority of these (90 per cent) will recover within two weeks, although recurrence of back pain is common. Back pain is rarely caused by a single injury, but is generally cumulative in nature. However, a minor incident may be seen as the final straw which triggers discomfort.

Causes of back discomfort include manual handling activities; awkward postures such as twisting, leaning, bending, stretching and prolonged static postures; poor seat design (including car seats); low physical fitness; and exposure to whole body vibration.

Upper limb disorders

E17046 Upper limb disorders (ULDs) are a sub-category of MSDs and the term refers to injuries occurring in the upper limbs (fingers, hands, wrists, arms, shoulders and neck). It covers strains, sprains, injury and discomfort, including conditions such as tenosynovitis and carpal tunnel syndrome as well as non-specific disorders. The cause of some of these disorders can be related specifically to the work undertaken and become known as work-related upper limb disorders (WRULDs). These disorders are usually cumulative in nature, arising from a series of micro-traumas.

The main physical risk factors in the development of upper limb disorders are application of force, repetitive movements and awkward or static postures. It is usually the interaction of at least two of these risk factors that leads to disorders. Duration of exposure to these factors is obviously also significant, with long durations of exposure increasing the risk of injury.

1. *Force* – application of excessive force in relation to the capabilities of the upper limb muscle group, will place a strain on the muscles involved and may lead to injury. Different muscle groups are able to apply different levels of force (e.g. shoulder muscles can apply more force than finger muscles).

2. *Repetition* – frequent, repetitive movements require the same muscles to contract and relax over and over again. If this activity is very rapid or continues for a long period of time the muscle can become fatigued and this can lead to discomfort.

3. *Posture* – poor postures (both awkward postures and static postures) can lead to injury and discomfort by placing the muscles and joints under unnecessary strain.

4. *Vibration* – exposure to hand/arm vibration is a recognised risk factor in the development of Hand Arm Vibration Syndrome (HAVS).

WRULDs are not confined to particular activities or industries, but are widespread throughout the workforce. Some jobs have particularly been associated with WRULDs, and these involve recognised risk factors (force, repetition, awkward/static postures, and long durations of exposure to these factors). Jobs such as assembly line work, construction work, garment machinists, meat and poultry processors, and display screen equipment work (particularly intensive data entry tasks) pose a risk of WRULDs, but this list of tasks is not exhaustive and is provided for illustration only.

Common ULDs are listed in O1037.

For further advice and guidance, see the following publications produced by HSE:

- *The law on VDUs: An easy guide* (HSG90) is aimed at small businesses, and contains illustrated, practical advice on avoiding risk from using ordinary office computers.
- *Work with display screen equipment* (L26) discusses the same issues as above but in full technical and legal detail and is aimed at large firms and health and safety professionals.
- *Aching arms (or RSI) in small businesses* (INDG171(rev1)) is a free leaflet aimed at reducing RSI due to work activities other than those caused by using DSE. It offers advice for identifying risk factors such as using force, repetitive movements, or poor posture, and gives practical ideas and tips for preventing RSI.

Lower limb discomfort

E17047 Lower limb disorders are less common and often less debilitating than other musculoskeletal disorders. However, lower limb discomfort can be caused by prolonged standing, particularly on concrete floors; inappropriate (particularly hard soled), heavy or ill-fitting footwear; operation of foot controls; and a lack of adequate foot support if sitting.

Psychosocial risk factors

E17048 As well as the physical factors that can lead to musculoskeletal disorders, there is strong evidence of the role of work related psychosocial factors (i.e. the worker's psychological response to work and workplace conditions) in the development of these disorders. Relevant factors include the design, organisation and management of work, the context of the work (overall social environment) and the content of the work (the specific impact of job factors).

Specifically, repetitive, monotonous tasks, excessive or undemanding workloads, lack of control over the task or organisation of the workplace, working in isolation, poor communication, and lack of involvement in decision making have been implicated as psychosocial risk factors.

It is thought that many of the effects of these psychosocial factors occur via stress-related processes which result in biochemical and physiological changes which can result in discomfort. Some can also have a direct impact on working practices and behaviours, e.g. work pressure may mean workers do not adjust the workstation to suit themselves at the start of the shift, or may forego rest breaks.

Individual differences

E17049 For biological reasons some people may be more likely to develop a musculoskeletal injury. These include new employees who may need time to develop the necessary skills/rate of work; those returning from holiday or sickness absence; older/younger workers; new/expectant mothers; those with particular health conditions. Account should also be taken of differences in body size which may require awkward postures/reaches.

Prevention of WRMSDs

E17050 The greatest risk reduction benefits will be achieved by tackling both physical and psychosocial risk factors in the workplace. These factors are best identified and tackled through consultation with the workforce. Appropriate design of tasks, workstations, tools as outlined in E17005–E17043 can help to prevent injury to the musculoskeletal system, by ensuring work is designed to be within the physical capabilities of people. Tasks or equipment may need to be modified for those returning to work following an absence related to a work-related musculoskeletal disorder.

Ergonomics tools

E17051 The following techniques can be used to ensure thorough and comprehensive investigation of all parts of the work so that these can be designed appropriately.

Risk assessments

E17052 Ergonomic risk assessments can identify the risk of injury, accident or reduced performance due to ergonomic deficiencies in the work or tasks. Any assessment should be systematic and comprehensive, accounting for all elements of the work including irregular activities. Checklists can be used to ensure all elements within the task are covered, i.e. the person, the work task, the workstation, the work environment and work system. One example of a simple ergonomics checklist is provided in J Dul and Weerdmeester *Ergonomics for beginners* (2001) Taylor and Francis. The particular questions that may be asked may relate specifically to the work situation present.

All hazards should be identified and recorded and an assessment of the risk posed by the hazard should be made. All hazards must be addressed, eliminated where possible or reduced as far as practicable.

A risk assessment checklist concerning the risks of manual handling is provided in the *Manual Handling Operations Regulations 1992* (*SI 1992 No 2793* as amended by *SI 2002 No 2174*). A list of factors to consider in the assessment of the risk of injury through DSE work is included in the *Health and Safety* (*Display Screen Equipment*) *Regulations 1992* (*SI 1992 No 2792* as amended by *SI 2002 No 2174*), although this is not presented as a risk assessment checklist. The revised HSE guidance on *Upper limb disorders in the workplace* (HSG60 rev) contains a thorough risk assessment checklist for considering the risks associated with tasks.

Several checklists and assessment tools exist to facilitate in the assessment of working postures and risk and include:

- the Rapid Upper Limb Assessment tool (RULA)

  See L McAtamney and E N Corlett 'RULA: A survey method for the investigation of work-related upper limb disorders' (1993) *Applied Ergonomics* 24(2) pp 1991–1999;

- the Quick Exposure Checklist (QEC)

  See G Li and P Buckle 'Evaluating change in exposure to risk for musculoskeletal disorders: A practical tool' (1999) HSE Contract Research Report 251/1999;

- the Rapid Entire Body Assessment tool (REBA)

  See S Hignett and L McAtamney 'Rapid Entire Body Assessment (REBA)' (2000) *Applied Ergonomics* 31, pages 201–205.

See also RISK ASSESSMENT.

Task analysis

E17053 A task analysis provides information about the sequence and inter-relationship of activities undertaken in a task. It can be used to help in the evaluation of existing tasks, or the design of new ones, by identifying the demands on the individual, connections between activities, unplanned or unusual activities. From this tasks can be allocated to different individuals (or the decision may be taken to mechanise), difficulties with equipment, tasks, workstations can be identified and equipment and training needs can be identified.

Workflow analysis

E17054 In a workflow analysis the movements of an individual at the workstation (e.g. to the equipment that is used) or within a work area are assessed. This helps to identify the pattern of activities and the routes within the workplace, to ensure they are optimised. Equipment can be arranged to ensure that items used together or in sequence are placed appropriately.

Frequency analysis

E17055 Information on the frequency with which tools and equipment are used can be important in positioning items or controls. The frequency of use of tools or adoption of postures can be recorded regularly using a frequency count.

User trials

E17055 Mock ups of prototype equipment or workstations, and trials of new furniture can be useful ways of assessing the suitability of new equipment. A representative sample of the user population should be used to evaluate the design. Rating scales can be used to allow people to score their opinion of a new item in terms of comfort, ease of use (clarity, adjustment etc) and so on. Rating scales should have a mid-point (neutral); either 5 or 7 point rating scales are generally adequately discriminating.

Data collection

E17056 Questionnaires and (formal or informal) interviews can also be useful means of collecting information and obtaining users' views of equipment, tasks or workplaces. Open (free response) or closed (series of options) questions can be used depending on the information that is required. Anonymity may be required in some situations (e.g. asking about health issues).

Discomfort surveys can be conducted to gain an understanding of the extent of any discomfort or disorders that are experienced by the workforce. In these, individuals report any discomfort they experience at the end of their shift, according to the part of the body affected, and rate its severity (e.g. on a scale of 1–5). The responses of individuals from a work area can be collated, and this combined data can be used to identify in which parts of the body operators are experiencing discomfort and also to identify any trends in relation to the work tasks or work areas. This information can also be used as a baseline against which the benefit of any interventions can be evaluated. It is a useful tool to use in relation to risk assessment as it can assist in the interpretation of risks, and prioritisation of risk reduction measures. An example of a body map is shown in figure 6 below.

Alternative ways of using the body map are for groups of employees to apply stickers to a body map to indicate any discomfort experience, thus building up a pattern of discomfort.

Figure 6: Body Map

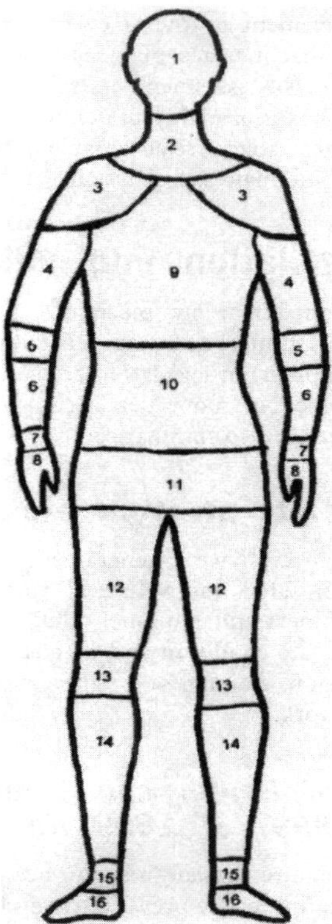

Accurate assessment of tasks requires good observational skills and sufficient time to assess the task adequately thoroughly. Observers should ensure that a representative sample of the workforce or tasks undertaken are observed. Irregularly undertaken tasks (e.g. maintenance) should not be overlooked in assessments. In some tasks, movements may be rapid, or only last for a short period. It can be useful therefore to collect video material of tasks to allow subsequent analysis.

Using existing data

E17057 In some cases data will be available which may indicate ergonomic issues. Productivity, accident and ill health data may all be used to identify any particular issues, and the extent of the problem. An accident book may also provide useful information on problems experienced. Data can be analysed by work area or type of injury in order to help prioritise areas for risk reduction measures.

Regular communication (e.g. through safety meetings) can also provide a useful route for identifying problem tasks or equipment.

Involving employees

E17058 A number of people can contribute to the design of the workplace or task. These may include planners, procurers, human resources as well as those who undertake the task itself. Involvement of the relevant people in any proposed changes is important to ensure that it is acceptable. In particular, involvement of those who undertake the task in risk assessment, redesign and evaluation is a key element in ensuring that the task is thoroughly understood, any redesign is suitable, and that any new equipment or changes are acceptable to the users. This can be achieved for example through questionnaires, discussions, and trials of new equipment.

Relevant legislation and guidance

E17059 Ergonomics is mentioned in several pieces of legislation, although there is no single specific piece of legislation concerning this subject. Requirements for ergonomic design (explicit or implicit) in legislation are summarised below. Specific pieces of legislation are discussed in more detail elsewhere in this publication. Selected relevant HSE guidance is also summarised.

Health and Safety at Work etc Act 1974

E17060 Under this Act employers have a general duty to ensure (as far as is reasonably practicable) the health, safety and welfare at work of their employees. This includes the provision of machinery and equipment that is without risks to health, the duty to keep workplaces in a safe condition and without risks to health, and to provide the information, instruction, training and supervision necessary to ensure employees' health and safety at work.

Management of Health and Safety at Work Regulations 1999 (SI 1999 No 3242)

E17061 These Regulations require that any risks to health from work have to be assessed; employers must take appropriate preventative and protective measures, and employees must be informed of the risks and preventative measures taken.

Further information on these Regulations is contained in RISK ASSESSMENT.

Manual Handling Operations Regulations 1992 (SI 1992 No 2793 as amended by SI 2002 No 2174)

E17062 These Regulations take an ergonomic approach to handling tasks. They apply to all manual handling tasks, and place a duty on employers to avoid hazardous manual handling activities as far as is reasonably practicable. Where this is not practical employers should assess the risk of injury, and then take appropriate steps to reduce the risk of injury as far as possible through appropriate design of the task, load, and environment, taking account of the capabilities of the individual.

Further information on these Regulations is contained in MANUAL HANDLING.

Health and Safety (Display Screen Equipment) Regulations 1992 (SI 1992 No 2792 as amended by SI 2002 No 2174)

E17063 Particular ergonomic issues relate to work at display screen equipment (DSE) i.e. VDUs. The *Health and Safety* (*Display Screen Equipment*) *Regulations 1992* set out particular duties on the employer to assess the health and safety risks associated with DSE use, and to reduce the risks identified. The health risks particularly associated with DSE use include musculoskeletal disorders, eyestrain and stress. Specific requirements for the display screen equipment, workstation, software and task design are set out in the *Schedule* to these Regulations. These requirements ensure that it is possible to make certain adjustments to the equipment and furniture. It is advisable as part of the assessment to ensure that equipment is positioned in such a way for the user as to prevent awkward postures (e.g. twist to view screen), and that the adjustments are suitable for the user (e.g. seat height adjustment is adequate).

Laptops computers pose a particular problem as it is not possible to independently adjust the distance between the screen and the keyboard, and this may result in visual and musculoskeletal discomfort. Laptops may be used in environments that do not allow adjustment to the workstation (e.g. trains, hotel rooms). There are also manual handling issues related to the transport of laptops, and can be concern over security and theft. These risks should be considered in a risk assessment.

Although there is currently no guidance from HSE on laptop use, it would be good practice to:

- Provide a docking station for the laptop when office based. Alternatively, provide a separate keyboard and mouse, and view the laptop screen (preferable); or a separate screen and use the laptop keyboard. If viewed, the height of the laptop screen may need to be raised so it is a suitable height.
- Provide information on adapting a work area to facilitate a comfortable posture (e.g. sitting in the passenger seat when working in a vehicle, using pillows to raise the height of the chair in a hotel).
- Encourage good working practices (e.g. regular breaks are particularly important because of the potential postural problems arising from laptop use).
- Provide a rucksack, or suitable carrying case, for easy transport of laptops. Rucksacks enable the weight to be carried close to the spine, reducing the amount of asymmetric loading and resultant stress on the spine and muscles that can occur if the bag is carried on one shoulder.
- Provide additional power supplies at regularly used locations, so these do not have to be carried.

If users are required to work at home, suitable furniture and equipment should be provided, and a risk assessment should be undertaken to ensure the workstation is suitable and the user can achieve a comfortable posture.

Provision and Use of Work Equipment Regulations 1998 (SI 1998 No 2306 as amended by SI 2002 No 2174)

E17064 The Regulations require employers to ensure that work equipment is suitable for the purpose and safe to use for the work so it does not pose any health and safety risk. General duties include: 'In selecting work equipment, every employer shall have regard to the working conditions and to the risks to the health and safety of persons which exist in the premises or undertaking in which that work equipment is to be used and any additional risk posed by the use of that work equipment.'

Further information on these Regulations is contained in MACHINERY SAFETY M1004.

Personal Protective Equipment at Work Regulations 1992 (SI 1992 No 2966 as amended by SI 2002 No 2174)

E17065 These Regulations place a duty on employers to ensure that suitable personal protective equipment (PPE) is provided to employees who may be exposed to a risk to their health and safety while at work, in circumstances where such risks cannot be adequately controlled by other means. PPE should take into account the ergonomic requirements of the wearer, and be capable of fitting them correctly.

PPE is designed to protect the wearer from a hazard; however, it may introduce other risks, such as reduced vision and hearing which may restrict ability to detect warnings. PPE may also restrict movement or the person's ability to perform a task (e.g. through reduced dexterity when wearing gloves). Some chemical protective clothing (particularly water vapour impermeable garments) can contribute to heat strain as the wearer has limited potential to evaporate sweat from the body. In selecting PPE the compatibility of different forms of PPE should be considered: it is often difficult to wear a hard hat with hearing defenders, or a chemical protective suit with a hard hat, although some integrated forms of PPE are available.

Reporting of Injuries, Diseases and Dangerous Occurrences Regulations 1995 (RIDDOR) (SI 1995 No 3163)

E17066 Under the RIDDOR Regulations certain work-related accidents, diseases and dangerous occurrences have to be reported to the enforcing authorities. In terms of musculoskeletal disorders (which may arise due to poor ergonomics and inappropriate design), RIDDOR specifies that 'cramps in the hands or forearm due to repetitive movements' must be reported if they arise from 'work involving prolonged periods of handwriting, typing or other repetitive movements of the fingers, hands or arms.'

Further information on these Regulations is contained in ACCIDENT REPORTING AND INVESTIGATION A3002.

Upper limb disorders in the workplace (HSG60 (rev))

E17067 This revised guidance on upper limb disorders in the workplace provides a very useful and comprehensive approach to tackling these issues. It models a management approach, outlining seven stages in addressing these issues. These seven stages are:

1. Understand the issues and commit to action on ULDs.
2. Create the right organisational environment.
3. Assess the risks of ULDs in the workplace.
4. Reduce the risk of ULDs.
5. Educate and inform the workforce ULDs.
6. Manage any episodes of ULDs.
7. Carry out regular checks on the programme effectiveness.

These stages are not necessarily sequential, and different stages may well interact, such as educating the workforce concerning ULDs may be part of creating the right organisational environment where these issues are taken seriously and tackled supportively.

The guidance also includes a two-stage risk assessment checklist. The first stage is a screening tool to help identify tasks where there may be a risk of injury and to assist in prioritising assessments. It contains five sections asking a small number of questions concerning any signs and symptoms of ULDs, repetition, working postures, application of force and exposure to vibration. More detailed risk assessment worksheets are also included. These are divided into eight sections concerning repetition, posture of the fingers, hands and wrist, posture of the arms and shoulders, posture of the head and neck, force, working environment, psychosocial factors, and individual differences. The risk assessment form encourages the identification of appropriate risk reduction measures for the risks identified.

The guidance also contains useful suggestions for reducing the risk of injury, medical aspects of ULDs and case studies illustrating how organisations have successfully tackled these issues.

Reducing error and influencing behaviour (HSG48)

E17068 This useful guidance promotes the consideration of human factors as a key factor in effective health and safety management. It provides practical advice on identifying, assessing and controlling risks arising from humans' interaction with the working environment.

Fire Prevention and Control

Elements of fire

F5001 Fire is a chemical reaction resulting in heat and light. For a fire to occur the following need to be combined in the correct proportions:

- oxygen (supplied by the air around us which contains about 21 per cent oxygen);
- fuel (combustible or flammable substances either solids, liquids or gases);
- source of heat energy (ignition sources such as open flame, hot surfaces, overheated electrical components).

For a fire to burn the fuel must be in a gaseous or vapour form. This means that solids and some liquids such as oils will need to be heated up until sufficient vapour is given off to burn. The form that the fuel takes will also influence the ease with which it ignites and the speed with which it burns. Generally, finely divided materials are easier to ignite and burn more quickly than fuels in a solid form.

Once initiated, a fire will continue to burn as a result of the reinvestment of energy during the process provided sufficient fuel vapour, heat and oxygen are available and the process is not interrupted.

The 'triangle' of fire symbolises the fire process. Each side of the triangle represents fuel, heat or oxygen. Take any one side away and the fire will be extinguished or more importantly, stop all three sides from coming together and a fire can be prevented.

The first part of this chapter (F5001–F5029) considers generally the practical aspects of fire prevention and control. It should be borne in mind that *special risks* involving flammable or toxic liquids, metal fires or other hazards should be separately assessed for loss prevention and control techniques. For precautions against fire hazards from flammable liquids, see the HSE Guidance documents HS(G)51 and HS(G)176. Control of these hazards is among the areas dealt with under the *Dangerous Substances and Explosives Atmospheres Regulations 2002* (*SI 2002 No 2776*).

It is the responsibility of management to consider how safe is safe: that is, to balance the costs of improvement against the financial consequences of fire. Considerable improvement can often be made immediately at little or no cost. Other recommendations which may require a financial appraisal must be related to loss effect values. In certain cases, however, due to high loss effect, special protection may be needed almost regardless of cost.

Modern developments in fire prevention and protection can now provide a solution to most risk management problems within economic acceptability. It must be pointed out, however, that it is a waste of time and money installing protective equipment unless it is designed to be functional, the purpose of such equipment is understood and accepted by all personnel and the equipment is adequately inspected and maintained. The reasons for providing such equipment should, therefore, be fully covered in any fire-training course. Fire routines should also be amended as necessary to ensure that full advantage is taken of any new measures implemented.

Common causes of fires

F5002 The following, in no particular order of significance, are the most common causes of fire in the workplace:

- wilful fire raising and arson;
- careless disposal of cigarettes, matches;
- combustible material left near to sources of heat;
- accumulation of easily ignitable rubbish or paper;
- inadvertence on the part of contractors or maintenance workers, usually involving hot work;
- electrical equipment left on inadvertently when not in use;
- misuse of portable heaters;
- obstructing ventilation of heaters, machinery or office equipment;
- inadequate cleaning of work areas;
- inadequate supervision of cooking activities.

Fire classification

F5003 There are five classes of fire which are related to the fuel involved and the method of extinction, as follows.

- *Class A*

  This relates to fires generally involving solid organic materials, such as coal, wood, paper and natural fibres, in which the combustion takes place with the formation of glowing embers. Extinction is achieved through the application of water in jet or spray form. Water extinguishes the fire by removing or limiting the heat by cooling.

- *Class B*

  This relates to fires involving:

  (i) liquids, which can be separated into those liquids which mix with water, e.g. acetone, acetic acid and methanol; and those which do not mix with water, e.g. waxes, fats, petrol and hydrocarbon solvents; and

  (ii) liquefiable solids, e.g. animal fats, solid waxes, certain plastics.

  Foam, carbon dioxide and dry powder can be used on all these types of fire. However, some types of foam break down on contact with water-miscible liquids and thus special alcohol-resistant foam is needed for large volumes of such liquids. Water must not be used on fats, petrol, etc. With foam, carbon dioxide and dry powder, extinction is principally achieved by limiting or removing the oxygen by smothering.

- *Class C*

  This relates to fires involving gases and should only be extinguished by shutting off the gas supply if it is safe to do so. Burning gas should not be extinguished as to do so without isolating the supply may result in a build-up of unburnt gas which might explode.

- *Class D*

  This relates to fires involving combustible metals, such as aluminium or magnesium. These fires burn with very high temperatures, and their extinction can only be achieved by the use of special powders. These powders form a crust over the surface isolating the burning metal from the surroundings. It is extremely dangerous to use water on burning metals. Special training is required where the use of extinguishers on combustible metals is concerned.

- *Class F*

  This relates to fires involving cooking oils and fats. Extinguishers using special wet chemical extinguishing agents, foam or dry powder can be assessed for their effectiveness for this type of fire using test fires defined in BS 7937: 2000. Extinction is achieved primarily by limiting or removing the oxygen by smothering and in some cases by a degree of cooling. It is extremely dangerous to use water on fires involving cooking oils and fats.

The table below classifies fires which can be controlled by portable fire appliances (see also BS EN 2 and BS EN 3).

Suitability of extinguishing types

Class of fire	*Description*	*Appropriate extinguisher*
A	Solid materials, usually organic, with glowing embers	Water (foam, dry powder or CO2 will work but may be less effective than water due to a lack of cooling properties).
B	Liquids and liquefiable solids:	Foam, CO2, dry powder
	• miscible with water e.g. acetone, methanol	Alcohol-resistant foam, CO2, dry powder and the skilled use of water spray
	• immiscible with water e.g. petrol, benzene, fats, waxes	Foam, dry powder, CO2
F	• cooking oils and fats	Type F, special wet chemical extinguisher

Electrical fires

F5004 Fires involving electrical apparatus must always be tackled by first isolating the electricity supply and then by the use of carbon dioxide or dry powder, both of which are non-conducting extinguishing mediums. Dry powder and carbon dioxide extinguish the fire by limiting or removing the oxygen by smothering.

EU Regulations, which apply in all EU member countries including the UK, prohibit the use of halon as an extinguishant in most applications after 31 December 2003. Portable extinguishers and fixed systems using halon are required, by these regulations, to have been decommissioned by this date.

Fire extinction – active fire protection measures

F5005 Extinction of a fire is achieved by one or more of the following:

- *starvation* – this is achieved through a reduction in the concentration of the fuel. It can be effected by:
  - (i) removing the fuel from the fire;
  - (ii) isolating the fire from the fuel source; and
  - (iii) reducing the bulk or quantity of fuel present;
- *smothering* – this brings about a reduction in the concentration of oxygen available to support combustion. It is achieved by preventing the inward flow of more oxygen to the fire, or by adding an inert gas to the burning mixture;
- *cooling* – this is the most common means of fire-fighting, using water. The addition of water to a fire results in vaporisation of some of the water to steam, which means that a substantial proportion of the heat is not being returned to the fuel to maintain combustion. Eventually, insufficient heat is added to the fuel and continuous ignition ceases. Water in spray form is more efficient for this purpose as the spray droplets absorb heat more rapidly than water in the form of a jet.

Property risk

F5006 Fire safety precautions needed for the protection of life are dealt with by legislation, as detailed in F5030–F5077.

In order to ensure the survival of a business in the event of fire, property protection must be considered and a business fire risk assessment carried out. The involvement of a company's insurer is essential as insurers have considerable experience in this field. The business fire risk assessment follows similar stages to the life safety risk assessment (see F5053) but assesses the importance of each area to the function of the business and how vulnerable they are to fire.

For example, an area where essential records or documents are stored is likely to have a serious effect should a fire occur. Essential equipment, plant or stock which, if destroyed or severely damaged by fire, might be difficult to replace or have a serious effect on production would require special consideration, and often high fire protection requirements to minimise such an effect.

These risks should be determined by management; they need to be identified, considered and evaluated. A report should be produced by each departmental head, outlining areas which may require special consideration. Such a report should also include protection of essential drawings, records and other essential documents.

A typical area of high loss effect would be the communication equipment room. The loss of this equipment could have a serious and immediate effect upon communications generally. Fire separation (to keep a fire out) is therefore considered essential, and automatic fire suppression by a self-contained extinguishing system should be strongly recommended.

Essential data should be duplicated and the copy stored in a safe area which preferably is off-site. Paper records can be converted to electronic format and similarly stored.

Passive and active fire protection

F5007 Passive fire protection is where part of the structure of a building is inherently fire resistant. Most buildings are divided into fire-resisting compartments. These serve two functions. One is to limit the spread of fire and can be a property protection measure as well as life safety. The other is to protect escape routes by making the escape route a fire-resisting compartment. All internal fire escape stairways are fire-resisting compartments.

In day-to-day work these compartments are most obvious where doorways pass through the compartment walls. The doors in these openings are nearly always self-closing fire resisting door sets. It is essential that these doors are not obstructed and are allowed to self-close freely at all times. Any glazing in such doors must also be fire resistant and if damaged must be repaired to the appropriate standard. (Guidance on the most common types of fire door is given in BS 8214: 1990 *Code of practice for fire door assemblies with non-metallic leaves.*)

The breaching of fire resistant walls with services such as pipes, ducts and cables should be avoided. Where there is no alternative, it is important that all the openings are suitably protected to the same rating as the fire wall. Ducts should be fitted with fire dampers while openings around services should be suitably sealed or fire stopped. (Guidance on the fire safety aspects of the design and construction of air handling duct work is given in BS 5588–9: 1999 *Fire precautions in the design, construction and use of buildings: Code of practice for ventilation and air conditioning ductwork.*)

Active fire protection involves systems that are activated when a fire occurs, for example automatic fire detection or automatic sprinkler systems. As such systems are only required in an emergency it is essential that they are designed, installed, tested, inspected and maintained in accordance with acceptable standards and good practice.

Fire procedures and portable equipment

Fire procedures

F5008 The need for effective and easily understood fire procedures cannot be over-emphasised. It may be necessary to provide a fire procedure manual, so arranged that it can be used for overall fire defence arrangements, and sectioned for use in individual departments or for special risks.

It is essential that three separate procedures are considered:

- procedure during normal working hours;
- procedure during restricted manning on shifts;

- procedure when only security staff are on the premises.

All procedures should take into consideration absence of personnel due to sickness, leave, etc. The fire brigade should be called immediately if any fire occurs, irrespective of the size of the fire. Any delay in calling the fire brigade must be added to the delay before the fire brigade's actual arrival, which will be related to the traffic conditions or the local appliances already attending another fire.

A person should be given the responsibility for ensuring that pre-planned action is carried out when a fire occurs. Large fires often result from a delayed call, which may be due not to delayed discovery but to wrong action being taken in the early stages following discovery of a fire. A pre-planned fire routine is essential for fire safety. The fire brigade, when called, should be met on arrival by a designated person available to guide them directly to the area of the fire. It is essential that all fire routines, when finalised, be made known to the fire brigade.

Fire equipment

F5009 There have been a number of cases where a person using an extinguisher has been seriously injured. Investigations have shown that either the wrong type of extinguisher was supplied or the operator had no training in the correct use of the extinguisher. The need for training staff cannot be over-emphasised, especially in areas of special risk, oil dipping tanks, furnace areas, highly flammable liquids, gas or cylinder fires etc.

The following recommendations are given in order to allow an evaluation of an existing problem and may need to be related to process risks:

- it is essential that persons be trained in the use of extinguishers, especially in areas where special risks require a specific type of extinguisher to be provided;
- any person employed to work, who is requested to deal with a fire, should be clearly instructed that at no time should that person jeopardise their own safety or the safety of others;
- persons who may be wearing overalls contaminated with oil, grease, paint or solvents should not be instructed to attack a fire. Such contaminated materials may vaporise due to heat from the fire, and ignite.

Types of fire extinguisher

F5010 The type of extinguisher provided should be suitable for the risk involved, adequately maintained and appropriate records kept of all inspections, tests etc. All fire extinguishers should be fitted on wall brackets or located in purpose built floor stands. It has been found that if this is not done, extinguishers are removed or knocked over and damaged. Extinguishers should be sited near exits or on the route to an exit.

Water extinguishers

F5011 This type of extinguisher is suitable for ordinary combustible fires, for example wood and paper, but is not suitable for flammable liquid fires. Spray type water extinguishers are recommended. Water extinguishers should be labelled 'not to be used on fires involving live electricity'.

Foam extinguishers

F5012 These are suitable for small liquid spill fires or small oil tank fires where it is possible for the foam to form a blanket over the surface of the flammable liquids involved. Foam extinguishers may not extinguish a flammable liquid fire on a vertical plane. Alcohols miscible with water, when on fire, will break down ordinary foam and should be considered a special risk. Alcohol resistant foams are available.

Dry powder extinguishers

F5013 This type of extinguisher will deal effectively with flammable liquid fires and is recommended, as it is capable of quick knock-down of a fire. The size of the extinguisher is important and it must be capable of dealing effectively with the possible size of the spill fire which may occur, with some extinguishant in reserve. The recommended minimum size is a 9 kg trigger-controlled extinguisher. (Dry powder extinguishers are also safe on fires involving electrical equipment.) Some dry powders have been tested and listed as suitable for use on normal combustible material (Class A) fires.

BCF extinguishers

F5014 The extinguishing medium is a halon. Manufacture of this class of chemicals is no longer permitted due to its adverse effect on the environment. New extinguishers of this type are no longer available although some old ones may still be in service. Recharging of existing extinguishers is no longer practised and replacement by dry powder or carbon dioxide extinguishers is recommended. EU Regulations prohibit the use of halon as an extinguishant in most applications (aircraft and military applications will be exempt) by the end of 2003. Any halon extinguishers in normal service should be decommissioned and disposed of safely by 31 December 2003.

Carbon dioxide extinguishers

F5015 For fires involving electrical equipment, carbon dioxide extinguishers are recommended. Carbon dioxide (CO_2) extinguishers are quite heavy and may be at high pressure. A minimum size of 2 kg is recommended. CO_2 is not recommended for flammable liquid fires, except for small fires. Training in the use of CO_2 extinguishers is essential. Particular care should be taken not to hold the discharge horn on a CO_2 extinguisher as the surface gets very cold and can cause cold burns to unprotected skin. As carbon dioxide is an asphxyiant it should not be used in confined unventilated spaces.

Colour coding and distribution of portable fire extinguishers

F5016 All newly certified fire extinguishers for use throughout the EU are coloured red in compliance with BS EN 3. Manufacturers are allowed under BS 7863 to affix different coloured areas on or above the operation instructions label. Existing extinguishers (complying with BS 5423) need not be replaced until they have served their useful life. The colour coded panels help identify the extinguisher type for example red for water, blue for dry powder and black for CO_2.

Extinguishers are distributed in accordance with BS 5306–8: 2000 *Code of practice for selection and installation of portable fire extinguishers*, based on an extinguishers rating (capability) rather than by the type and size of the unit.

Fire alarms in 'certificated' premises

F5017 A manually operated fire alarm system is required in any 'certificated' premises and in some cases an automatic fire detection system will be required (see F5039). Both these needs will be indicated on the fire certificate. The need for a fire detection and alarm system may also result from a fire risk assessment. The systems should comply with the requirements of BS 5839–Part1: 2002 *Fire detection and alarm systems for buildings: Code of practice for system design, installation, commissioning and maintenance.*

Fire alarm systems should be tested once a week

Good 'housekeeping'

F5018 The need for good 'housekeeping' cannot be over-emphasised. Poor housekeeping is the greatest single cause of fire. A carelessly discarded cigarette end, especially into a container of combustible waste or amongst combustible storage, often results in fire. The risk is higher in an area which is infrequently used. The following are essential guidelines:

- smoking should, preferably, be prohibited but where smoking is permitted, suitable deep metal ashtrays should be provided. Ashtrays should not be emptied into combustible waste unless the waste is to be removed from the building immediately. It is recommended that smoking ceases before close of work so that if smouldering occurs this will be detected before staff leave the premises;
- combustible waste and contaminated rags should be kept in separate metal bins with close fitting metal lids;
- cleaners should, preferably, be employed in the evenings when work ceases. This will ensure that combustible rubbish is removed from the building to a place of safety before the premises are left unoccupied;
- rubbish should not be kept in the building overnight, or stored in close proximity to the building;
- smoking should be prohibited in places which are infrequently used, e.g. stationery stores, oil stores, or telecommunications intake rooms. Suitable 'smoking prohibited' notices should be displayed throughout such areas – notices should comply with BS 5499 and the *Health and Safety* (*Safety Signs and Signals*) *Regulations 1996* (*SI 1996 No 341*);
- where 'no smoking' is enforced due to legal requirements (for example, areas where flammable liquids are used or stored) or in areas of high risk or high loss effect, it is recommended that the notice read 'Smoking prohibited – dismissal offence';
- materials should not be stored on cupboard tops, and all filing cabinets should be properly closed, and locked if possible, at the end of the day.

Pre-planning of fire prevention

F5019 A pre-planned approach to fire prevention and control is essential. Fire spreads extremely fast, and the temperature can rise to 1,000°C in only one minute. Smoke can be flammable and toxic. The essential factor is to re-evaluate the risk, identify areas of high loss effect or high risk, and plan accordingly to meet requirements and legal responsibilities.

Means of escape

F5020 Means of escape routes consist of two main stages:

- travel within rooms;
- travel from rooms to the entrance to a stairway or corridor protected against the effects of fire or directly to a final exit where a protected route is not provided.

  In particular, for offices:

  (i) total travel distance between any point in a building and the nearest final exit or access to a protected route should not exceed:

  — 25 m, if there is only one exit, or

  — 45 m, if more than one exit;

  (ii) two or more exits are necessary:

  — from a room in which more than 60 people work, or

  — if any point in the room is more than 12 m from the nearest exit;

  (iii) minimum widths of escape routes (doors, corridors, stairs):

  — doors from rooms should have a minimum width of 750 mm for up to 100 people and 1.05 m for up to 200 people;

  — main corridors should be at least 1.05 m wide;

  — office corridors should be divided by fire-resisting doors if they are longer than 45 m;

  (iv) stairways should be at least 1.05 m wide and in a fire-resistant enclosure;

  (v) one stairway may be acceptable in a building of up to four storeys only. Depending on the floor area the stairs will need to be protected against the affects of fire including a lobby approach;

  (vi) escape doors should never be locked. If, for security reasons, they have to be secured against external entry, panic bolts or similar fastenings (complying with BS EN 1125) should be fitted;

  (vii) fire exit notices should be affixed to or above fire escape doors;

  (viii) corridors and stairways, which are a means of escape and form part of a protected route should have a half-hour fire resistance and preferably be constructed from brick or concrete with a non-combustible surface;

  (ix) fire alarms should be audible throughout the building. In large or multi-storey buildings such alarms will normally be electrically operated with suitable backup power supplies. In very small buildings a bell, gong or air horn may be sufficient;

  (x) it should not normally be necessary for a person to travel more than 30 m to the nearest alarm point.

In any certificated premises the fire certificate must be available on demand. The following are essential:

- all doors affording means of escape in case of fire should be maintained easily and readily available for use at all times that persons are on the premises;

- all doors not in continuous use, affording a means of escape in case of fire, should be clearly indicated;
- sliding doors should also clearly indicate the direction of opening and, where possible, not be used on escape routes;
- doors should be adequately maintained and should not be locked or fastened in such a way that they cannot be easily and immediately opened by persons leaving the premises. Moreover, all gangways and escape routes should be kept clear at all times.

Unsatisfactory means of escape

F5021 The following are unsatisfactory means of escape and should not be used in the event of fire:

- lifts;
- portable ladders;
- spiral staircases;
- escalators;
- lowering lines.

Fire drill

F5022 As well as being a statutory requirement in procedures relating to 'events of serious and imminent danger' (under the *Management of Health and Safety at Work Regulations 1999* (*SI 1999 No 3242*), *Reg 8*(*1*)) – as a matter of good housekeeping (in addition to being probably required as well by the fire certificate (see F5039, 'Contents of a fire certificate')), employers should acquaint the workforce with the arrangements for fire drill. This consists of putting up a notice in a prominent place stating the action employees should take on:

- hearing the alarm, or
- discovering the fire.

Ideally, employees should receive regular fire drills, even though normal working is interrupted. Indeed, fire alarms should be sounded weekly so that employees may familiarise themselves with the sound, and evacuation drills should be carried out at least annually. Trained employees should be designated as fire wardens and carry out head counts on evacuation, as well as acting as last man out and generally advising and shepherding the public. In addition, selected employees should be trained in the proper use of fire extinguishers. Moreover, periodical visits by the local fire authority should be encouraged by employers, since this provides a valuable source of practical information on fire fighting, fire protection and training.

Typical fire drill notice

F5023

When the fire alarm sounds:

1. Switch off electrical equipment and leave room, closing doors behind you.
2. Walk quickly along escape route to open air.
3. Do not use lifts.

4. Report to fire warden at assembly point.
5. Do not re-enter building.

When you discover a fire:

1. Raise alarm (normally by operating a break glass call point).
2. Leave the room, closing doors behind you.
3. Leave the building by escape route.
4. Report to fire warden at assembly point.
5. Do not re-enter building.

Fires on construction sites

F5024 Each year there are numerous fires, of a major kind, on construction sites and in buildings undergoing refurbishment. For that reason the *Joint Code of Practice on the Protection from Fire of Construction Sites and Buildings Undergoing Renovation'* (published by the Fire Protection Association) proposes that the main contractor should appoint a *site fire safety co-ordinator*, responsible for assessing the degree of fire risk and for formulating and regularly updating the *site fire safety plan*; and he should liaise with the co-ordinator for the design phase (see F5025). The site fire safety plan should detail:

- organisation of and responsibilities for fire safety;
- general site precautions, fire detection and warning alarms;
- requirements for a hot work permit system;
- site accommodation;
- fire escape and communications system (including evacuation plan and procedures for calling the fire brigade);
- fire brigade access, facilities and co-ordination;
- fire drill and training;
- effective security measures to minimise the risk of arson;
- materials storage and waste control system.

Role of site fire safety co-ordinator

F5025 The site fire safety co-ordinator must:

- ensure that all procedures, precautionary measures and safety standards (as specified in the site fire safety plan) are clearly understood and complied with by all those on the project site;
- ensure establishment of hot work permit systems;
- carry out weekly checks of fire fighting equipment and test all alarm and detection devices;
- conduct weekly inspections of escape routes, fire brigade access, fire fighting facilities and work areas;
- liaise with local fire brigade for site inspections;
- liaise with security personnel;

- keep a written record of all checks, inspections, tests and fire drill procedures;
- monitor arrangements/procedures for calling the fire brigade;
- during the alarm, oversee safe evacuation of site, ensuring that all staff/ visitors report to assembly points;
- promote a safe working environment.

Emergency procedures

F5026 The following emergency procedures should be implemented, where necessary:

- establish a means of warning of fire, e.g. handbells, whistles etc.
- display written emergency procedures in prominent locations and give copies to all employees;
- maintain clear access to site and buildings;
- alert security personnel to unlock gates/doors in the event of an alarm;
- install clear signs in prominent positions, indicating locations of fire access routes, escape routes and positions of dry riser inlets and fire extinguishers.

Designing out fire

F5027 Construction works should be designed and sequenced to accommodate:

- permanent fire escape stairs, including compartment walls;
- fire compartments in buildings under construction, including installation of fire doors;
- fire protective materials to structural steelwork;
- planned fire fighting shafts duly commissioned and maintained;
- lightning conductors;
- automatic fire detection systems;
- automatic sprinkler and other fixed fire fighting installations.

Moreover, adequate water supplies should be available and hydrants suitably marked and kept clear of obstruction.

Other fire precautions on site

F5028 Portable fire extinguishers can represent the difference between a conflagration and a fire kept under control. Therefore, personnel should be trained in the use of portable fire fighting equipment and adequate numbers of suitable types of portable extinguishers should be available. They should be located in conspicuous positions near exits on each floor. In the open, they should be 500 mm above ground bearing the sign 'Fire Point' and be protected from both work activities and adverse weather conditions. In addition, all mechanically-propelled site plant should carry an appropriate fire extinguisher, and extinguishers, hydrants and fire protection equipment should be maintained and regularly inspected by the site fire safety co-ordinator.

Plant on construction sites also constitutes a potential danger. All internal combustion engines of powered equipment, therefore, should be positioned in the open air or in a well-ventilated non-combustible enclosure. They should be separated from working areas and sited so that exhaust pipes/gases are kept clear of combustible materials. Moreover, fuel tanks should not be filled whilst engines are running and compressors should be housed singly away from other plant in separate enclosures.

Consequences of failure to comply with code

F5029 Non-compliance with the provisions of this code could well result in insurance ceasing to be available or being withdrawn, thereby constituting a breach of a Standard Form contract (see C8141 CONSTRUCTION AND BUILDING OPERATIONS). Where fire damage is caused to property by the negligence of employees of a subcontractor, then, in accordance with Clause 6.2 of the JCT Contract, to the effect that the contractor is liable for 'injury or damage to property', the contractor is liable and not the employer (or building owner); even though, under Clause 6.3B, the employer is required to insure against loss or damage to existing structures, to the work in progress, and to all unfixed materials and goods intended for, delivered to or placed on the works (*National Trust v Haden Young Ltd, The Times, 11 August 1994*). The practical implementation of fire precautions on construction sites forms part of the overall construction health and safety plan (see C8033 CONSTRUCTION AND BUILDING OPERATIONS).

Fire and fire precautions – legislation

F5030 Fire safety legislation in the UK is concerned with the health and safety of the occupant's buildings. This will in most cases also include those, such as firefighters, who may need to enter a building on fire as part of their professional duties. There are two main sets of regulations that control this area: (*a*) the imposition of controls on the design and construction of buildings and (*b*) requirements leading to the safe management of buildings that are in use.

The provision of fire safety in new buildings relates to the physical provisions at the time of construction and are controlled through the Building Regulations in England and Wales, by Building Standards (Scotland) Regulations in Scotland and Building Regulations (Northern Ireland).

Buildings in use are currently controlled by a range of statutes relating to fire certification, workplace fire risk assessment and licensing of premises for consumption of alcohol, entertainment and similar licensed activities. These statutes generally relate to physical provisions and management requirements.

This section deals, almost exclusively, with the safe management of the workplace. The principle legislation relating to fire safety in the workplace is the *Fire Precautions Act 1971*. This Act is concerned with the fire safety provisions and management of designated premises. Designated premises include factories, offices, shops and railway premises as well as hotels and boarding houses. In addition to the 1971 Act the *Fire Precautions* (*Workplace*) *Regulations 1997* (*SI 1997 No 1840*) as amended by the *Fire Precautions* (*Workplace*) (*Amendment*) *Regulations 1999* (*SI 1999 No 1877*) extend control of fire safety to nearly all other workplaces.

The current position – overview

F5031 Currently statute law relating to fire and fire precautions is extensive and complex. The legislation includes:

- the *Fire Precautions Act 1971* ('*FPA 1971*') (and regulations and orders made thereunder);
- the *Fire Safety and Safety of Places of Sport Act 1987*;

- the *Fire Precautions (Workplace) Regulations 1997 (SI 1997 No 1840)* as amended by the *Fire Precautions (Workplace) (Amendment) Regulations 1999 (SI 1999 No 1877)*;
- the *Health and Safety at Work etc Act 1974* ('*HSWA 1974*') (in the form of the *Fire Certificates (Special Premises) Regulations 1976 (SI 1976 No 2003)* (see F5060 below));
- the Petroleum Acts (and regulations made thereunder);
- the *Public Health Acts 1936–1961*;
- the *Building Act 1984* and the Building Regulations made thereunder;
- the *Fire Services Act 1947* and the *Fires Prevention (Metropolis) Act 1774*;
- as well as certain regulations, made under the *Factories Act 1961*.

This complex situation has been recognised by the UK Government and it is proposed to rationalise UK fire safety legislation using a regulatory reform order under the *Regulatory Reform Act 2001*. Consultation has been carried out during 2003 and it is expected that the Reform Act will find its way onto the statute books during 2004.

Many of the existing regulations will be rationalised. At least 50 items of legislation have been identified in the consultation document as needing repeal or amendment. The *Petroleum Spirit (Consolidation) Act 1928* which imposes requirements on storage and transportation of petroleum spirit is being reviewed and has been amended by the *Dangerous Substances and Explosive Atmospheres Regulations 2002 (SI 2002 No 2776)* ('*DSEAR*'), but is expected to remain for the time being and not be changed by the reform order. The fire safety aspects of the *Building Act 1984* and the current *Building Regulations 2000 (SI 2000 No 2531)* made under that Act are also expected to be affected but only in the areas that refer to the *FPA 1971*.

Much of the regulation concerned with dangerous and hazardous materials under the *HSWA 1974* has already been rationalised under the *DSEAR*. This includes the repeal of the *Highly Flammable Liquids and Liquefied Petroleum Gases Regulations 1972 (SI 1972 No 917)*.

At the time of writing, the exact content of the proposed Regulatory Reform Order has not been confirmed. However it is expected that the *FPA 1971* will be repealed in favour of risk assessment based legislation similar to the existing *Fire Precautions (Workplace) Regulations* of 1997 and 1999.

In view of this, many fire authorities are reluctant to issue new fire certificates under the *FPA 1971* and are concentrating on ensuring that adequate fire risk assessments have been carried out under the *Fire Precautions (Workplace) Regulations* for those premises that they consider high risk such as schools, hospitals and hotels. During all or part of the currency of this chapter the *FPA 1971* will still be in place. This Act imposes requirements relating to fire precautions upon occupiers of premises (whether or not employers) where there is a fire risk.

Readers should also be aware of the Fire and Rescue Services Bill published by the Office of the Deputy Prime Minister on 12 January 2004.

The Bill replaces the *Fire Services Act 1947* with a new statutory framework, which is designed to drive forward the Government's plans for a more preventative and risk-based approach, as set out in the White Paper, *Our Fire and Rescue Service*. It is expected to become law later this year (2004).

Under the Bill, the community fire safety work, which most fire and rescue authorities currently undertake, will be placed on a statutory footing. The wider role of the service beyond firefighting is also recognised in the Bill, which will create a duty on the service to respond to non-fire incidents, such as road traffic accidents, major flooding and terrorist incidents.

The Bill also places the Fire and Rescue National Framework (which calls on fire authorities to set up regional management boards) on a statutory footing. And it gives the Secretary of State powers to procure equipment and services for fire authorities and to direct authorities during specific emergencies – a move designed to ensure a co-ordinated and strategic response.

The Bill also:

- imposes duties on authorities and water undertakers to ensure an adequate supply of water for firefighting activities;
- updates existing powers to allow fire authorities to enter into reinforcement schemes with other authorities to provide mutual assistance;
- maintains the existing ability for fire authorities to charge for particular services. The Bill will continue to exclude the possibility of charging for firefighting;
- includes reserve powers to set up new bodies to negotiate pay and conditions of service. These powers could be used if the current review of National Joint Council arrangements does not result in satisfactory new mechanisms;
- devolves responsibility for the fire and rescue service in Wales to the National Assembly for Wales;
- abolishes the Central Fire Brigades Advisory Council, which has now been replaced with the more effective and flexible Practitioners' and Business and Community forums.

Specific fire precautions regulations

F5032 Until the placement of the Regulatory Reform Order (See F5031), the main pieces of legislation relating to fire safety in workplaces in the UK are the *Fire Precautions (Workplace) Regulations 1997 (SI 1997 No 1840)* as amended by the *Fire Precautions (Workplace) (Amendment) Regulations 1999 (SI 1999 No 1877),* jointly referred to as 'the Workplace Regulations', and the *FPA 1971*. The key to operation of the *FPA 1971* is certification – that is, if premises were to be put to certain uses, they would require a valid fire certificate. The Workplace Regulations require all employers to conduct a fire risk assessment of the workplaces for which they are responsible. However, application must still be made for a fire certificate for premises designated under the *FPA 1971*, i.e. hotels and boarding houses, offices, factories and railway premises.

Similarly until the Regulatory Reform Order is in place fire certificates will also be required for 'special premises' (see below.)

The *FPA 1971* puts responsibilities on the owners or occupiers of properties whereas the responsibilities under the Workplace Regulations fall upon the employer.

Thus virtually all places of employment are now caught by the Workplace Regulations.

The Workplace Regulations do not allow the contents of a fire certificate to be such that people complying with the certificate will contravene these Regulations. The Workplace Regulations allow fire authorities to modify certificates accordingly.

Similarly, safety certificates issued under the *Safety of Sports Grounds Act 1975* or the *Fire Safety and Safety of Places of Sport Act 1987* are not allowed to require anyone to do anything that would cause them to contravene the Workplace Regulations.

The principal Regulations made under the *Fire Precautions Act 1971, s 12* (the regulation-enabling section) are:

- the *Fire Precautions (Factories, Offices, Shops and Railway Premises) Order 1989 (SI 1989 No 76)*;
- the *Fire Precautions (Hotels and Boarding Houses) Order 1972 (SIs 1972 Nos 238, 382)*;
- the *Fire Certificates (Special Premises) Regulations 1976 (SI 1976 No 2003)*; and
- the *Fire Precautions (Workplace) Regulations 1997 (SI 1997 No 1840)*.

(For offences and penalties in connection with the 'fire regulations', see F5040 below.)

Fire safety signs must be provided and maintained where indicated by a risk assessment to ensure that occupants leave safely in event of fire. [*Health and Safety (Safety Signs and Signals) Regulations 1996 (SI 1996 No 341), Reg 4*].

Fire certification

F5033 The prime tool of the *FPA 1971* was the requirement for 'designated' premises to be 'fire-certificated' – normally by the local fire authority, though in the case of exceptionally hazardous industrial premises (i.e. special premises) by the Health and Safety Executive (HSE). To date, two designation orders have been made affecting factories, offices, shops and railway premises *and* hotels and boarding houses (see further F5034 below). A fire certificate must be produced on demand to a fire officer, for inspection purposes, or a copy of it. [*FPA 1971, s 19(1)(c)*]. (For offences/penalties, see F5040 below.)

Fire certificates specify:

- use/uses of premises;
- means of escape in case of fire;
- how means of escape can be safely and effectively used;
- alarms and fire warning systems; and
- fire-fighting apparatus to be provided in the building.

Moreover, at its discretion, the fire authority can, additionally, impose requirements relating to:

- maintenance of means of escape and fire-fighting equipment;
- staff training; and
- restrictions on the number of people within the building.

Prior to the issue of a fire certificate, premises must be inspected by the fire authority, though failure to do so or inadequate inspection or advice cannot involve the fire authority in liability for negligence (see F5035 below).

Fire certification – what you must do

F5034 If your premises are a factory, office, shop or railway premises and fall under any of the following categories you must apply to your local fire authority for a fire certificate. (A shop includes restaurants. pubs and 'take away' outlets as these are considered as retail premises.)

(*a*) (i) where more than 20 people are at work, or

(ii) more than 10 people are at work elsewhere than on the ground floor,

(in shops, factories, offices and railway premises);

(*b*) in buildings in multiple occupation containing two or more individual factory, office, shop or railway premises, when the aggregate of people at work exceeds the same totals;

(*c*) in factories where explosive or highly flammable materials are stored, or used in or under the premises, unless, in the opinion of the fire authority, there is no serious risk to employees.

[*Fire Precautions* (*Factories, Offices, Shops and Railway Premises*) *Order 1989* (*SI 1989 No 76*), *Reg 5*].

In light of the amendment of the *FPA 1971* by the *Fire Safety and Safety of Places of Sport Act 1987*, some relaxation of earlier requirements is now possible (see F5041 below).

Applications for local authority fire certificate

F5035 Applications for a fire certificate (see F5034 above) must be made to the local fire authority on the correct form (see F5036 below), obtainable from each fire authority. Plans may be required and the premises will be inspected before issue of a certificate. If the fire authority is not satisfied as to existing arrangements, it will specify the steps to be taken before a certificate is issued, and notify the occupier or owner that they will not issue a certificate until such steps are taken within the specified time. [*FPA 1971, s 5*]. (Hospitals, NHS trusts and private hospitals must conform to the requirements of FIRECODE, which is a suite of documents produced by NHS Estates, making recommendations on fire safety in hospitals.) For the position of hospitals, see generally E15045 ENFORCEMENT and F5049 below.)

In view of the expected and imminent repeal of the *FPA 1971* many fire authorities are reluctant to issue new fire certificates under the *FPA 1971* and are concentrating on ensuring that adequate fire risk assessments have been carried out, under the Workplace Regulations.

Form for application for fire certificate

F5036 Applications for a fire certificate must be made on the form prescribed by the *Fire Precautions* (*Application for Certificate*) *Regulations 1989* (*SI 1989 No 77*).

Who should apply?

F5037 Application should normally be made by the occupier in the case of factories, offices and shops. In the following cases however it must be made by the owner or owners:

- premises consisting of part of a building, all parts of which are owned by the same person (i.e. multi-occupancy/single ownership situations);
- premises consisting of part of a building, the different parts being owned by different persons (i.e. multi-occupancy/plural ownership situations).

[*FPA 1971, s 5, Sch 2 Part II*].

FIRE PRECAUTIONS ACT 1971, s.5

APPLICATION FOR A FIRE CERTIFICATE

For Official Use Only

To the Chief Executive of the Fire Authority*

* In the case of Crown premises, substitute H.M. Inspector of Fire Services.

I hereby apply for a fire certificate in respect of the premises of which details are given below. I make the application as, or on behalf of, the occupier/owner of the premises.

Signature ....

Name: Mr/Mrs/Miss ....
(in block capitals)

If signing on behalf of a company or some other person, state the capacity in which signing ....

Address..

Telephone number.. Date ....

To be completed by the Applicant:–

1. Postal address of the premises ....

....

....

....

2. Name and address of the owner of the premises

Name ....

Address ....

....

....

....

(In the case of premises in plural ownership the names and addresses of all owners should be given.)

3. Details of the premises

(If the fire certificate is to cover the use of two or more sets of premises in the same building, details of each set of premises should be given on a separate sheet.)

(a) Name of occupier ....

(and any trading name, if different) ....

....

(b) Use(s) to which premises put ....

(c) Floor(s) in building on which premises situated (e.g. basement(s), ground floor, first floor etc.) ....

....

(d) Number of persons employed to work in the premises ....

(e) Maximum number of persons at work or it is proposed will work in the premises at any one time (including employees, self-employed persons and trainees)–

(i) below the ground floor of the building ....

(ii) on the ground floor of the building ....

(iii) on the first floor of the building ....

(iv) in the whole of the premises ....

(f) Maximum number of persons other than persons at work likely to be in the premises at any one time ....

(g) Number of persons (including staff, guests and other residents) for whom sleeping accommodation is provided in the premises–

(i) below the ground floor of the building ....

(ii) above the first floor of the building ....

(iii) in the whole of the premises ....

---

4. If the premises consist of part only of a building, the uses to which the other parts of the building are put (on a floor by floor basis):

....

....

....

---

5. (a) Total number of floors (excluding basements) in the building in which the premises are situated ....

(b) Total number of basements in that building ....

---

6. Approximate date of construction of the premises ....

---

7. Nature and quantity of any explosive or highly flammable materials stored or used in or under the premises

Materials	*Maximum quantity stored*	*Method of storage*	*Maximum quantity liable to be exposed at any one time*

(Continue on a separate sheet if necessary)

8. Details of fire-fighting equipment available for use in the premises

Nature of equipment	*Number Provided*	*Where installed*	*Is the equipment regularly maintained?*
(a) Hosereels			Yes/No
(b) Portable fire extinguishers			Yes/No
(c) Others			Yes/No
(specify types e.g. sand/water buckets. fire blanket)			

(Continue on a separate sheet if necessary)

Categories of fire risk premises

F5038 Where there is in the opinion of the fire authority a serious risk to persons from fire on premises, unless steps are taken to minimise that risk, and the fire authority thinks that a particular use of premises should be either prohibited or restricted, it may apply to the court for an order prohibiting or restricting that use of premises until remedial steps are taken. [*FPA 1971, s 10*(*2*)]. This requirement does not impose a duty on a fire authority to advise an occupier not to reopen; hence, if he reopens and suffers loss from fire, the fire authority will not be liable

The Workplace Regulations 1997 amend the *FPA 1971*, *ss 10* to *10B,* to include tents and other movable structures and places of work in the open air.

Contents of a fire certificate

F5039 A fire certificate specifies:

- the particular use or uses of premises which it covers;
- the means of escape in the case of fire (as per plan);
- the means for securing that the means of escape can be safely and effectively used at all relevant times (e.g. direction signs/emergency lighting/fire or smoke stop doors);
- the means for fighting fire for use by persons in the building;
- the means for giving warnings in the case of fire;
- in the case of any factory, particulars as to any explosives or highly flammable materials stored or used on the premises.

[*FPA 1971, s 6*(*1*)].

In addition, a fire certificate may require:

- maintenance of the means of escape and their freedom from obstruction;
- maintenance of other fire precautions set out in the certificate;
- training of employees on the premises as to what to do in the event of fire and keeping of suitable records of such training;
- limitation of number of persons who at any one time may be on the premises;

- any other relevant fire precautions.

[*FPA 1971, s 6*(*2*)].

Offences/penalties for breach of the Fire Precautions Act 1971

F5040 Failure to have/exhibit a valid fire certificate, or breach of any condition(s) in such certificate is an offence, committed by the holder, carrying:

- on summary conviction, a maximum fine at level 3 on the standard scale*; and
- on conviction on indictment, an indefinite fine or up to two years' imprisonment (or both).

[*FPA 1971, s 7*(*1*), (*5*) as amended by the *Criminal Justice Act 1991, s 17*(*1*)].

An occupier also commits an offence if the fire authority or inspectorate is not informed of a proposed structural or material alteration. [*FPA 1971, s 8*]. A breach carries the same penalties as for breach of *s 7* (above). Other offences include obstructing a fire inspector [*FPA 1971, s 19*(*6*)] and forging/falsifying a fire certificate [*FPA 1971, s 22*(*1*)].

*The standard scale of fines is defined in the *Criminal Justice Act 1982* and may be increased from time to time.

Exemptions from the need to have a fire certificate

F5041 A fire authority can grant exemption from certification requirements in the case of 'designated use' premises Exemption from certification requirements can be granted, either on application for a fire certificate, or at any time during the currency of a fire certificate. The usual criterion for exemption is that the fire authority deems the premises concerned to be low risk.

It is not necessary formally to apply for exemption as the fire authority will offer an exemption, if they see fit after inspecting the premises. Any exemption certificate must specify the greatest number of persons who can safely be in the premises at any one time. Such exemptions can be withdrawn by the fire authority without an inspection but must give notice of withdrawal.

Change of conditions affecting premises for which exemption is granted

F5042 If, while an exemption is in force an occupier proposes to carry out material changes in the premises, the occupier must inform the fire authority of the proposed changes. Not to do so is to commit an offence. This applies where it is proposed:

- to make an extension of, or structural alteration to the premises which would affect the means of escape from the premises; or
- to make an alteration in the internal arrangement of the premises, or in the furniture or equipment, which would affect the means of escape from the premises; or
- to keep explosive or highly flammable materials under, in or on the premises, in a quantity or aggregate quantity greater than the prescribed maximum; or
- to make use of the premises which involves there being a greater number of persons on the premises than specified in the exemption certificate.

Offences/penalties – exempted premises

F5043 The penalties for being found guilty of committing the offences (explained in F5042 above) are liable to the same penalties as described in F5040.

Duties on occupiers of exempted premises

F5044 Premises which are exempt from fire certification must be provided with:

- means of escape in the case of fire; and
- means of fighting fire.

A code of practice has been published ('*Code of practice for fire precautions in factories, offices, shops and railway premises not required to have a fire certificate*' (1993), Home Office and Scottish Home and Health Department, ISBN 0 11 340904 4). This means that an occupier who followed the recommendations of the code will normally be considered to have complied with these requirements.

Definition of 'escape'

F5045 In relation to premises 'escape' means escape from them to some place of safety beyond the building.

Improvement notices

F5046 Use of improvement notices, which have proved effective in general health and safety law for upgrading standards of health and safety at the workplace, has been duplicated in fire precautions law. Thus, where a fire authority is of the opinion that the duty to provide:

- means of escape; and
- means of fire-fighting;

has been breached, they can serve on the occupier an improvement notice, specifying, particularly by reference to a code of practice (see F5044 above), what measures are necessary to remedy the breach. They can also require the occupier to carry out this remedial work within three weeks. The occupier has a right to appeal. [*FPA 1971, s 9D*]. Service of such notice need not be recorded in a public register (see further E15029 ENFORCEMENT).

Building Regulations

F5047 Where the *Building Regulations 2000* (*SI 2000 No 2531* as amended by *SI 2001 No 3335, SI 2002 No 440 and SI 2002 No 2871*) apply to premises, with respect to requirements as to means of escape in case of fire, the fire authority cannot serve an improvement notice requiring structural or other alterations. However, a notice can be served if the fire authority is satisfied that the means of escape in case of fire are inadequate, by reason of matters/circumstances of which particulars were not required by the Regulations. [*FPA 1971, s 9D(3)*].

The *Building Regulations 2000* (as amended) include requirements to make 'appropriate' provisions for the early warning of fire and 'reasonable facilities to assist firefighters in the protection of life'. This brings these Regulations into line with the Workplace Regulations 1997. The amendments are reflected in the accompanying approved document, i.e. '*Approved Document B: Fire Safety*' (2000 edition) that should be read in conjunction with '*Approved Document B: Fire safety (2002*

Amendments)' available from the Stationary Office or the Office of the Deputy Prime Minister, Building Regulations Department.

The amendments incorporate changes and discussion on the replacement of references to British Standards by harmonised BS EN standards, the application of the CE marking to building products and the implications of the Construction Products Directive.

The following requirements are specified in respect of new premises:

- The building must be designed/constructed so that there are *means of escape* in case of fire, to a place of safety outside.

Internal fire spread

- To inhibit internal fire spread, internal linings must:
  - (i) adequately resist flame spread over surfaces;
  - (ii) if ignited, have a reasonable rate of heat release.
- The building must be designed and constructed so that, in the event of fire, its stability will be maintained for a reasonable period; and a common wall should be able to resist fire spread between the buildings.
- The building must be designed and constructed so that unseen fire/smoke spread within concealed spaces in its fabric and structure, is inhibited.

External fire spread

- External walls shall adequately resist fire spread over walls and from one building to another.
- A roof should be able to resist fire spread over the roof and from one building to another.
- The building must be designed and constructed to provide facilities to firefighters and enable fire appliances to gain access.

Means of warning and escape

- The building shall be designed and constructed so that there are appropriate provisions for the early warning of fire, and appropriate means of escape in case of fire from the building to a place of safety outside the building capable of being safely and effectively used at all material times.

Access and facilities for the fire service

- The building shall be designed and constructed so as to provide reasonable facilities to assist firefighters in the protection of life and reasonable provision made within the site of the building to enable fire appliances to gain access.

Appeals against improvement notices

F5048 An appeal must be lodged against an improvement notice within 21 days from the date of service. Moreover, unlike prohibition notices, the effect of an appeal is to suspend operation of the improvement notice. Presumably, a ground of appeal would be that occupiers of 'low risk' premises have achieved satisfactory fire safety standards by means other than those specified in a code of practice.

If the appeal fails, the occupier must carry out the remedial work specified in the notice. Failure to do so carries with it similar penalties as explained in F5040 above.

Prohibition notices

F5049 In places of work generally, health and safety inspectors can serve prohibition notices requiring a hazardous activity to cease in cases where there is thought to be a serious risk of personal injury. In the case of fire hazards where there is thought to be a serious risk of injury to persons from fire, the fire authority is in a similar way empowered to serve on the occupier a prohibition notice. The effect of such a notice (which need not affect all the premises) will be to prohibit use of the premises or activity until the risk is removed. Prohibition notices can be placed on premises that are:

- providing sleeping accommodation;
- providing treatment/care;
- for the purposes of entertainment, recreation or instruction, or for a club, society or association;
- for teaching, training or research;
- providing access to members of the public, whether for payment or otherwise;
- places of work.

This includes hospitals, factories and places of public worship, but not private dwellings. [*Fire Safety and Safety of Places of Sport Act 1987, s 13*]. Thus, a fire authority could restrict/prohibit the use of any part of a hospital, factory or place of religious worship, presenting a serious fire risk to persons – premises which are also subject to the *Building Act 1984* (see W11044 WORKPLACES – HEALTH, SAFETY AND WELFARE). In particular, after consultation with the fire authority, if a local authority is not satisfied with the means of escape in case of fire, it can serve a notice requiring the owner to carry out remedial work in residential premises of all kinds, including inns, hotels and nursing homes and certain commercial premises with sleeping accommodation above them, whether a fire certificate is in force or not. [*Building Act 1984, s 72*].

The fire authority is most likely to serve prohibition notices in cases where it considers that means of escape are inadequate or could be improved. As with prohibition notices served generally in respect of workplaces, it can be immediate or deferred. Occupiers must lodge an appeal within 21 days.

Entry of such notices must appear in a public register. [*Environment and Safety Information Act 1988, ss 1, 3* and *Sch*].

Offences/penalties

F5050 A person found guilty of contravening a prohibition notice is liable to the same penalties as described in F5040 above.

The Workplace Regulations

Premises to which the Workplace Regulations apply

F5051 By way of implementation of the EU Framework and Workplace Directives in relation to fire safety, the *Fire Precautions (Workplace) Regulations 1997 (SI 1997 No 1840)* pursuant to the *FPA 1971, s 12*, extend fire precautions requirements to most places of work. This includes some workplaces currently designated and certificated under the *Factories Act 1961* and the *Offices, Shops and Railway Premises Act 1963*.

The *Management of Health and Safety at Work Regulations 1999* (*SI 1999 No 3242*) amend some aspects of the Workplace Regulations, mostly where references to the Management Regulations 1992 have been changed. Also *Regulation 3* of the Management Regulations 1999 requires employers, among other things, to carry out a fire risk assessment to identify measures needed to comply with the provisions of the *Fire Precautions* (*Workplace*) *Regulations 1997* (as amended).

The *Fire Precautions* (*Workplace*) (*Amendment*) *Regulations 1999* (*SI 1999 No 1877*) make amendments to the Workplace Regulations 1997. The main changes are to extend the scope of the Regulations by removing some of the excepted categories of workplace that appeared in the Workplace Regulations 1997. These amendments have been brought in to address concerns expressed by the EU that the 1997 Regulations did not fully implement the Framework and Workplace Directives (see above) and to resolve the resulting legislative overlaps.

Workplaces to which the Workplace Regulations do not apply

F5052 These include the following:

- Workplaces used only by the self-employed.
- Private dwellings.
- Mine shafts and mine galleries, other than surface buildings.
- Construction sites (any workplace to which the *Construction* (*Health, Safety and Welfare*) *Regulations 1996* (*SI 1996 No 1592*) apply).
- Ships within the meaning of the *Docks Regulations 1988* (*SI 1988 No 1655*) (including those under construction or repair by persons other than the crew).
- Means of transport used outside the workplace and workplaces which are in or on a means of transport.
- Agricultural or forestry land situated away from the undertaking's main buildings.
- Offshore installations (workplaces to which the *Offshore Installations and Pipeline Works* (*Management and Administration*) *Regulations 1995* (*SI 1995 No 738*) apply).

[*Fire Precautions* (*Workplace*) *Regulations 1997* (*SI 1997 No 1840*), *Reg 3*(*5*) as amended by *Fire Precautions* (*Workplace*) (*Amendment*) *Regulations 1999* (*SI 1999 No 1877*), *Reg 5*(*c*)].

All other workplaces, including those subject to certification under the *FPA 1971* are covered by the Workplace Amendment Regulations 1999.

Fire risk assessment

F5053 The Workplace Regulations require all employees to assess the risk the fire poses to their employees and others who may be present in the workplace.

The fire risk assessment may involve six stages:

Stage 1 – identifying the fire hazards;

Stage 2 – identifying the people at risk;

Stage 3 – removing or reducing the hazards;

Stage 4 – assigning a risk category;

Stage 5 – deciding if existing arrangements are satisfactory or need improvement;

Stage 6 – recording the findings. This is a statutory requirement if more than five people are employed.

Further, detailed guidance is available in '*The Fire Protection Association Library of Fire Safety, Volume 5 – Fire risk management in the workplace*' published by the Fire Protection Association and in '*Fire Safety. An employer's guide*' published by the Stationary Office and HSE books.

Where there are five or more employees, the fire risk assessment must be recorded and retained.

The assessment should be reviewed periodically and when any significant changes occur to, for example, layout, management or staffing.

Fire-fighting and fire detection

F5054 Where necessary (whether due to the features of a workplace, the activity carried on there, any hazard present there or any other relevant circumstances, i.e. as a result of a fire risk assessment) in order to safeguard the safety of employees in case of fire:

(*a*) a workplace must, to the extent that is appropriate, be equipped with appropriate fire-fighting equipment and with fire detectors and alarms; and

(*b*) any non-automatic fire-fighting equipment so provided must be easily accessible, simple to use and indicated by signs,

and for the purposes of sub-paragraph (*a*) what is appropriate is to be determined by the dimensions and use of the building housing the workplace, the equipment it contains, the physical and chemical properties of the substances likely to be present and the maximum number of people that may be present at any one time i.e. as a result of a fire risk assessment. [*SI 1997 No 1840* (as amended), *Reg 4*(*1*)].

Where necessary in order to safeguard the safety of his employees in case of fire, an employer must:

- take measures for fire-fighting in the workplace, adapted to the nature of the activities carried on there and the size of his undertaking and of the workplace concerned and taking into account persons other than his employees who may be present;
- nominate employees to implement those measures and ensure that the number of such employees, their training and the equipment available to them are adequate, taking into account the size of, and the specific hazards involved in, the workplace concerned; and
- arrange any necessary contacts with external emergency services, particularly as regards rescue work and fire-fighting. [*SI 1997 No 1840* (as amended), *Reg 4*(*2*)].

Emergency routes and exits

F5055 Routes to emergency exits from a workplace and the exits themselves must be kept clear at all times, where this is necessary for safeguarding the safety of employees in case of fire. [*SI 1997 No 1840* (as amended), *Reg 5*(*1*)].

The following requirements must be complied with in respect of a workplace where necessary (whether due to the features of the workplace, the activity carried on there, any hazard present there or any other relevant circumstances) in order to safeguard the safety of employees in case of fire:

- emergency routes and exits must lead as directly as possible to a place of safety;
- in the event of danger, it must be possible for employees to evacuate the workplace quickly and as safely as possible;
- the number, distribution and dimensions of emergency routes and exits must be adequate having regard to the use, equipment and dimensions of the workplace and the maximum number of persons that may be present there at any one time;
- emergency doors must open in the direction of escape;
- sliding or revolving doors must not be used for exits specifically intended as emergency exits;
- emergency doors must not be so locked or fastened that they cannot be easily and immediately opened by any person who may require to use them in an emergency;
- emergency routes and exits must be indicated by signs; and
- emergency routes and exits requiring illumination must be provided with emergency lighting of adequate intensity in the case of failure of their normal lighting. [*SI 1997 No 1840* (as amended), *Reg 5(2)*].

Maintenance

F5056 The workplace and any equipment and devices provided in respect of the workplace under *Regs 4* and *5* of the 1997 Regulations, must be subject to a suitable system of maintenance and be maintained in an efficient state, in efficient working order and in good repair, where this is necessary for safeguarding the safety of employees in case of fire. [*SI 1997 No 1840* (as amended), *Reg 6*].

Enforcement

F5057 The duty of enforcing the workplace fire precautions legislation falls on fire authorities, who may appoint inspectors for this purpose. [*SI 1997 No 1840* (as amended), *Reg 10*].

A fire authority can issue an enforcement notice in respect of a serious breach. The enforcement notice must notify the person who is served with it that the fire authority is of the opinion that he is in breach of the workplace fire precautions legislation and that such breach is putting one or more employees at serious risk. The notice must also:

- specify the steps required to remedy the breach;
- require those steps to be taken within a given time; and
- provide details of the appeals procedure relating to enforcement notices.

[*SI 1997 No 1840* (as amended), *Reg 13*].

Offences and penalties

F5058 A person is guilty of an offence if:

- being under a requirement to do so, he fails to comply with any provision of the workplace fire precautions legislation;
- that failure places one or more employees at serious risk (i.e. subject to a risk of death or serious injury which is likely to materialise) in case of fire; and
- that failure is intentional or is due to his being reckless as to whether he complies or not.

[*SI 1997 No 1840* (as amended), *Reg 11*(*1*)].

Any person guilty of an offence under *Reg 11*(*1*) is liable:

- on summary conviction, to a fine; or
- on conviction on indictment, to a fine, or to imprisonment for a term not exceeding two years, or both.

[*SI 1997 No 1840* (as amended), *Reg 11*(*2*)].

A person is not guilty of an offence under *Reg 11*(*1*) in respect of any failure to comply with the workplace fire precautions legislation which is subject to an enforcement notice. [*SI 1997 No 1840* (as amended), *Reg 11*(*3*)].

Application to the Crown

F5059 The workplace fire precautions regulations do not extend to premises used solely by the armed forces, but otherwise apply to premises which are Crown occupied and Crown owned. *Section 10* of the *FPA 1971* only binds the Crown in so far as it applies to premises and workplaces owned by the Crown but not occupied by the Crown. [*SI 1997 No 1840* (as amended), *Reg 18*].

Fire Certificates (Special Premises) Regulations 1976 (SI 1976 No 2003)

F5060 It is expected that these Regulations will be revoked by the Regulatory Reform Order referred to in F5031. Until this Order enters the statute book the *Fire Certificates* (*Special Premises*) *Regulations 1976* (*SI 1976 No 2003*) remain in force. Under these Regulations, in the case of premises containing hazardous materials or processes (i.e. special premises), a fire certificate must be obtained from HSE. This applies even if only a small number of persons are employed there. Exemption from this requirement may be granted where the Regulations are inappropriate or not reasonably practicable.

When a certificate has been issued by HSE the occupier of those premises must post a notice in those premises, stating:

- that the certificate has been issued; and
- the places where it (or a copy) can be inspected easily by any person who might be affected by its provisions; and
- the date of the posting of the notice.

[*Fire Certificates* (*Special Premises*) *Regulations 1976*, *Reg 5*(*5*), (*6*)].

These conditions do not override those applicable to a licence for the storage of petroleum spirit under the *Petroleum Spirit* (*Consolidation*) *Act 1928* apart from the special provision for harbours.

'Special premises' for which a fire certificate is required from HSE

F5061 The 'special premises' for which a fire certificate is required from HSE are set out for reference in the table below.

'Special premises' for which a fire certificate is required by HSE

1 Any premises at which are carried on any manufacturing processes in which the total quantity of any highly flammable liquid under pressure greater than atmospheric pressure and above its boiling point at atmospheric pressure may exceed 50 tonnes.

2 Any premises at which is carried on the manufacturing of expanded cellular plastics and at which the quantities manufactured are normally of, or in excess of, 50 tonnes per week.

3 Any premises at which there is stored, or there are facilities provided for the storage of, liquefied petroleum gas in quantities of, or in excess of, 100 tonnes except where the liquefied petroleum gas is kept for use at the premises either as a fuel, or for the production of an atmosphere for the heat-treatment of metals.

4 Any premises at which there is stored, or there are facilities provided for the storage of, liquefied natural gas in quantities of, or in excess of, 100 tonnes except where the liquefied natural gas is kept solely for use at the premises as a fuel.

5 Any premises at which there is stored, or there are facilities provided for the storage of, any liquefied flammable gas consisting predominantly of methyl acetylene in quantities of, or in excess of, 100 tonnes except where the liquefied flammable gas is kept solely for use at the premises as a fuel.

6 Any premises at which oxygen is manufactured and at which there are stored, or there are facilities provided for the storage of, quantities of liquid oxygen of, or in excess of, 135 tonnes.

7 Any premises at which there are stored, or there are facilities provided for the storage of, quantities of chlorine of, or in excess of, 50 tonnes except when the chlorine is kept solely for the purpose of water purification.

8 Any premises at which artificial fertilisers are manufactured and at which there are stored, or there are facilities provided for the storage of, quantities of ammonia of, or in excess of, 250 tonnes.

9 Any premises at which there are in process, manufacture, use or storage at any one time, or there are facilities provided for such processing, manufacture, use or storage of, quantities of any of the materials listed below in, or in excess of, the quantities specified —

Phosgene	5 tonnes
Ethylene oxide	20 tonnes
Carbon disulphide	50 tonnes
Acrylonitrile	50 tonnes
Hydrogen cyanide	50 tonnes
Ethylene	100 tonnes

	Propylene	100 tonnes
	Any highly flammable liquid not otherwise specified	4,000 tonnes

10 Explosives, factories or magazines which are required to be licensed under the Explosives Act 1875.

11 Any building on the surface at any mine within the meaning of the Mines and Quarries Act 1954.

12 Any premises in which there is comprised —

(*a*) any undertaking on a site for which a licence is required in accordance with section 1 of the Nuclear Installations Act 1965 or for which a permit is required in accordance with section 2 of that Act; or

(*b*) any undertaking which would, except for the fact that it is carried on by the United Kingdom Atomic Energy Authority, or by, or on behalf of, the Crown, be required to have a licence or permit in accordance with the provisions mentioned in sub-paragraph (a) above.

13 Any premises containing any machine or apparatus in which charged particles can be accelerated by the equivalent of a voltage of not less than 50 megavolts except where the premises are used as a hospital.

14 Premises to which Regulation 26 of the Ionising Radiations Regulations 1985 (SI 1985 No 1333) applies.

15 Any building, or part of a building, which either —

(*a*) is constructed for temporary occupation for the purposes of building operations or works of engineering construction; or

(*b*) is in existence at the first commencement there of any further such operations or works

and which is used for any process of work ancillary to any such operations or works (but see F5062 below).

[*Fire Certificates (Special Premises) Regulations 1976, Sch 1 Part I*].

Application of Fire Certificates (Special Premises) Regulations to temporary buildings used for building operations or construction work

F5062 A fire certificate is required in the case of buildings constructed for temporary occupation for the purposes of building operations or works of engineering construction (see CONSTRUCTION AND BUILDING OPERATIONS). This is a requirement of the *Fire Certificates (Special Premises) Regulations 1976 (SI 1976 No 2003)*, and application for such a certificate should be made to HSE. Buildings that are already in existence for such purposes when such operations or works begin are included in this requirement. An exemption is available, however, if:

- fewer than 20 persons are employed at any one time, or fewer than 10 elsewhere than on the ground floor; and
- the nine conditions set out in *Sch 1 Pt II* of the 1976 Regulations are satisfied.

[*Fire Certificates (Special Premises) Regulations 1976 (SI 1976 No 2003, Reg 3(1), Sch 1, para 15*].

In all events suitable means of escape in case of fire, adequate fire-fighting equipment, exit doorways that can be easily and immediately opened from inside, unobstructed passageways as well as distinctive and conspicuous marking of fire exits must be provided.

Other dangerous processes

F5063 Until the *DSEAR* (*SI 2002 No 2776*) came into force, fire prevention measures for certain particularly dangerous processes were controlled by specific regulations. Most of these have been repealed or revoked and hazards are now controlled by the requirements of the *DSEAR*. Some of the regulations replaced by *DSEAR* are:

(*a*) the *Celluloid* (*Manufacture, etc.*) *Regulations 1921* (*SR & O 1921 No 1825*) – applying to the manufacture, manipulation and storage of celluloid and the disposal of celluloid waste;

(*b*) the *Manufacture of Cinematograph Film Regulations 1928* (*SR & O 1928 No 82*) – applicable to the manufacture, repair, manipulation or use of cinematograph film;

(*c*) the *Cinematograph Film Stripping Regulations 1939* (*SR & O 1939 No 571*) – applicable to the stripping, drying or storing of cinematograph film;

(*d*) the *Highly Flammable Liquids and Liquefied Petroleum Gases Regulations 1972* (*SI 1972 No 917*) – applicable to premises containing highly flammable liquids and liquefied petroleum gases;

(*e*) the *Magnesium* (*Grinding of Castings and Other Articles*) *Special Regulations 1946* (*SR & O 1946 No 2017*) – prohibition on smoking, open lights and fires;

(*f*) the *Factories* (*Testing of Aircraft Engines & Accessories*) *Special Regulations 1952* (*SI 1952 No 1689*) – applicable to the leakage or escape of petroleum spirit;

Some regulations with specific fire safety requirements that are still on the statute book are given below. However most of these are expected to be amended or replaced under the Regulatory Reform Order (see F5031 above).

- the *Offshore Installations* (*Prevention of Fire and Explosion, and Emergency Response*) *Regulations 1995* (*SI 1995 No 743*) – to prevent and minimise the effects of fire and explosion on offshore installation;
- the *Construction* (*Health, Safety and Welfare*) *Regulations 1996* (*SI 1996 No 1592*), *Reg 21*; and
- the *Work in Compressed Air Regulations 1996* (*SI 1996 No 1656*), *Reg 14*.
- the *Electricity at Work Regulations 1989* (*SI 1989 No 635*), *Reg 6*(*d*) – electrical equipment which may reasonably foreseeably be exposed to any flammable or explosive substance, must be constructed or protected so as to prevent danger from exposure;
- the *Dangerous Substances in Harbour Areas Regulations 1987* (*SI 1987 No 37*) – applicable to risks of fire and explosion in harbour areas. A fire certificate is not required for these premises.

Application for fire certificate for 'special premises'

F5064 In order to obtain a fire certificate for 'special premises' (see F5061 for definition), the occupier or owner (see F5037) must apply to the relevant HSE inspectorate, e.g.

the factory inspectorate in respect of factory premises. No form is prescribed but certain particulars must be provided, including:

- the address and description of the premises;
- the nature of the processes carried on or to be carried on there;
- nature and approximate quantities of any explosive or highly flammable substance kept or to be kept on the premises;
- the maximum number of persons likely to be present on the premises;
- the name and address of the occupier.

Plans may be required to be deposited and premises will be inspected, and the occupier may well have to make improvements to the fire precautions before a certificate is issued. [*Fire Certificates (Special Premises) Regulations 1976 (SI 1976 No 2003, Reg 4*].

Appeals relating to the issue of a fire certificate

F5065 An appeal may be made to the magistrates' court by an applicant for a fire certificate in relation to:

- a requirement specified by a fire authority or HSE; or
- the fire authority or HSE refusing to issue a certificate; or
- the contents of a certificate.

An appeal must be lodged within 21 days of the date of notice from the fire authority. [*Fire Certificates (Special Premises) Regulations 1976 (SI 1976 No 2003, Reg 12*].

Duty to keep fire certificate on the premises

F5066 Every fire certificate must be kept on the premises to which it relates (and preferably displayed), as long as it is in force. [*FPA 1971, s 6(8)*]. Failure to comply with this subsection is an offence. [*FPA 1971, s 7(6)*].

Notification of alteration to premises

F5067 Any proposed structural alterations or material internal alteration to premises for which a fire certificate has been issued, must first be notified to the local fire authority or inspectorate (as appropriate). Similarly, any proposed material alteration to equipment in the premises must be so notified. Not to do so is to commit an offence (see F5040 above). The premises may be inspected by staff from the relevant enforcement agency at any reasonable time while the certificate is in force. [*FPA 1971, s 8(1), (2)*]. A material alteration is one that may affect means of escape or other fire safety provisions such as removing a hose reel or changing the direction of an escape route.

Penalties

F5068 A person found guilty of contravening the notification requirements is liable to the same penalties as described in F5040 above.

Licensed premises and fire safety

F5069 Some premises must have a current licence in order to operate. A licence will be refused or not renewed by the local magistrates if the local fire authority is not satisfied as to the fire precautions necessary in the premises. If people are employed on these premises the Workplace Regulations will also apply (see F5051). Many of these licensing regulations are expected to be repealed or revoked under the Regulatory Reform Order (see F5031).

Cinemas

F5070 Safety in cinemas is controlled by the *Cinematograph (Safety) Regulations 1955 (SI 1955 No 1129)* (as subsequently amended). Cinemas must be provided with:

- adequate, clearly marked exits, so placed as to afford safe means of exit;
- doors which are easily and fully openable outwards;
- passages and stairways kept free from obstruction;

[*Cinematograph (Safety) Regulations 1955 (SI 1955 No 1129), Reg 2*];

- suitable and properly maintained fire appliances;
- proper instruction of licensee and staff on fire precautions;
- treatment of curtains so that they will not readily catch fire;
- use of non-flammable substances for cleaning film or projectors;

[*Cinematograph (Safety) Regulations 1955 (SI 1955 No 1129), Reg 5*];

- prohibition on smoking in certain parts of the premises [*Cinematograph (Safety) Regulations 1955 (SI 1955 No 1129), Reg 6*];
- appropriate siting of heating appliances [*Cinematograph (Safety) Regulations 1955 (SI 1955 No 1129, Reg 24*].

Theatres

F5071 Under *s 12(1)* of the *Theatres Act 1968* premises used for the public performance of a play must be licensed. The conditions for obtaining or having a licence renewed include:

- compliance with rules relating to safety of persons in the theatre, and particularly, staff fire drills, provision of fire-fighting equipment, maintenance of a safety curtain and communication with the fire service;
- gangways and seating correctly arranged and free from obstruction, doors and exits, marking and method of opening, lighting arrangements;
- scenery and draperies must be non-flammable and there must be controls over smoking and overcrowding.

Gaming houses (casinos, bingo halls etc.)

F5072 The issue or retention of a licence to operate depends *inter alia* on compliance with fire requirements [*Gaming Act 1968*].

Premises for music, dancing etc.

F5073 The issue or retention of a licence to operate premises for public music or entertainment depends on compliance with the fire requirements. [*Local Government (Miscellaneous Provisions) Act 1982, Sch 1*].

Similarly, in the case of premises used for private music/dancing, e.g. dancing schools, there must be compliance with the fire requirements. [*Private Places of Entertainment (Licensing) Act 1967*].

Schools

F5074 In the case of local authority controlled schools, including special schools, the 'health and safety of their occupants, and in particular, their safe escape in the event of fire, must be reasonably assured', with particular reference to the design, construction, limitation of surface flame spread and fire resistance of structure and materials therein. [*Standards for School Premises Regulations 1972 (SI 1972 No 2051)*].

As far as a school is a place of work, the *Fire Precautions (Workplace) Regulations 1997 (SI 1997 No 1840)* also apply.

Children's and community homes

F5075 Both these local authority controlled establishments must carry out fire drills and practices and, in addition, consult with the fire authorities. [*Children's Homes Regulations 1991 (SI 1991 No 1506)*; *Community Homes Regulations 1972 (SI 1972 No 319)*].

As far as a children's or community home is a place of work, the *Fire Precautions (Workplace) Regulations 1997 (SI 1997 No 1840)* also apply.

Residential and nursing homes

F5076 Similar requirements apply in the case of:

- residential homes [*National Assistance (Conduct of Homes) Regulations 1962 (SI 1962 No 2000)*]; and
- nursing homes [*Nursing Homes and Mental Nursing Homes Regulations 1981 (SI 1981 No 932)* (as amended)].

In particular, satisfactory arrangements must be made for the evacuation of patients and staff in the event of fire.

As far as a residential or nursing home is a place of work, the *Fire Precautions (Workplace) Regulations 1997 (SI 1997 No 1840)* also apply.

Crown premises

F5077 As with other health and safety duties and regulations, generally speaking, statutory fire duties and fire regulations apply to the Crown. [*FPA 1971, s 40*; *Fire Precautions (Workplace) Regulations 1997 (SI 1997 No 1840* as amended), *Reg 18*]. However owing to Crown immunity in law, proceedings cannot be enforced against the Crown (see further E15045 ENFORCEMENT). This has the effect that Crown premises, that is, government buildings such as the Treasury and the Foreign Office as well as royal palaces, are required to be fire-certificated but, if they fail to apply for certification, they cannot, like other occupiers, be prosecuted. Secondly, failure to acquire fire-certificate status could prejudice the safety of fire-fighters called upon to combat fires in Crown premises. Moreover, fire certificates are not required in (*a*) prisons, (*b*) special hospitals for the mentally incapacitated and (*c*) premises occupied exclusively by the armed forces. [*FPA 1971, s 40*(*2*)].

Civil liability at common law and under the Fires Prevention (Metropolis) Act 1774 for fire damage

F5078 Only *sections 83* and *86* of the *Fires Prevention* (*Metropolis*) *Act 1774* still exist. The Law Commission has proposed that *section 83* be repealed and the Reform Order is expected to remove *section 86*. Until these proposals become law these sections are still in force and there effect is summarised as follows.

Both sections only apply to England and Wales. *Section 83* allows an insurance company to insist that, if there is any hint of fraud, the money paid out by it in respect to fire damage can only be used to rebuild or repair the damaged property. *Section 86* prevents any action being taken against a person who owns property that accidentally burns down.

Liability of occupier

F5079 An occupier of premises where fire breaks out can be liable to:

- lawful visitors to the premises injured by the fire or falling debris (and is also liable to unlawful visitors, i.e. trespassers, as the principle of 'common humanity', enunciated in *Herrington v British Railways Board [1972] 1 All ER 749* applies as does the *Occupiers' Liability Act 1984*, see OCCUPIERS' LIABILITY);
- firemen injured during fire-fighting operations; and
- adjoining occupiers.

In order to ensure therefore that occupiers and others involved may minimise their liability, insurance cover, though not compulsory, is highly desirable. The basic principles relating to fire cover are considered below (see F5080).

Fire insurance

F5080 Insurance is intended to provide the insured with cover in the event of an accident that is not predictable. This applies to fire insurance and is reflected in the terms of insurance policies. Thus if a fire occurs in a premises where it is thought that the insured might be involved in deliberately starting the fire the insurance company may refuse to pay out under the policy. However if the insured is innocent of any involvement, even if the fire was started deliberately, the insurer will pay out as arson is not predictable in the context of this issue. Where a high fire risk has been identified and adequate measures are not taken to manage the risk there may be a case for non-payment on the grounds of negligence. In practice insurance plays a very important role in the management of many risks in industry, including fire, and adequate appropriate fire cover is considered an essential part of modern business risk management.

First-Aid

Introduction

F7001 People can and do suffer injury or fall ill at work. This may or may not be as a result of work related activity. However, it is important that they receive immediate attention.

The *Health and Safety (First-Aid) Regulations 1981* (*SI 1981 No 917* as amended by *SI 2002 No 2174*)), which came into operation on 1 July 1982, require employers to have facilities for the provision of first-aid in their place of work.

Medical treatment should be provided at the scene promptly, efficiently and effectively before the arrival of any medical teams that may have been called. First-aid can save lives and can prevent minor injuries from becoming major ones. Employers are responsible for making arrangements for the immediate management of any illness or injury suffered by a person at work. First-aid at work covers the management of first-aid in the workplace – it does not include treating ill or injured people at work with medicines.

However, the Regulations do not prevent specially trained staff taking action beyond the initial management of the injured or ill at work.

The employer's duty to make provision for first-aid

F7002 Employers must provide or ensure that there is equipment and facilities provided that are adequate and appropriate for rendering first-aid if any of their employees are injured or become ill at work. Employers must also ensure that there are an adequate and appropriate number of suitable persons who:

- are trained and have such qualifications as the Health and Safety Executive may approve for the time being; and
- have additional training, if any, as may be appropriate in the circumstances.

Where such a suitable person is absent in temporary and exceptional circumstances, an employer can appoint a person, or ensure that a person is appointed:

- to take responsibility for first-aid in situations relating to an injured or ill employee who needs help from a medical practitioner or nurse;
- to ensure that equipment and facilities are adequate and appropriate in the circumstances.

With regard to any period of absence of the first-aider, consideration must be given to:

- the nature of the undertaking;
- the number of employees at work; and
- the location of the establishment.

The assessment of first-aid needs

F7003 The employer should assess his first-aid needs and requirements as appropriate to the circumstances of his workplace. The employer's principal aim must be the reduction of the effects of injury and illness at the place of work. Adequate and appropriate first-aid personnel and facilities should be available for rendering assistance to persons with common injuries or illnesses and those likely to arise from specific hazards at work. Similarly, there must be adequate facilities for summoning an ambulance or other professional medical help.

The extent of first-aid provision in a particular workplace that an employer must make depends upon the circumstances of that workplace. There are no fixed levels of first-aid – the employer must assess what personnel and facilities are appropriate and adequate. Employers with access to advice from occupational health services may wish to use these resources for the purposes of conducting such assessment, and then take the advice given as to what first-aid provision would be deemed appropriate.

In workplaces that employ:

- qualified medical doctors registered with the General Medical Council; or
- nurses whose names are registered in Part 1, 2, 10 or 11 of the Single Professional Register maintained by the United Kingdom Central Council for Nursing, Midwifery and Health Visiting,

the employer may consider that there is no need to appoint first-aiders.

There is no requirement for the results of the first-aid risk assessment to be recorded in writing, although it may nevertheless be a useful exercise for the employer – for he may subsequently be asked to demonstrate that the first-aid provision is adequate and appropriate for the workplace.

When assessing first-aid needs, employers must consider the following:

- the hazards and risks in the workplace;
- the size of the organisation;
- the history of accidents in the organisation;
- the nature and distribution of the workforce;
- the distance from the workplace to emergency medical services;
- travelling, distant and lone workers' needs and requirements;
- employees working on shared or multi-occupied sites;
- annual leave and other absences of first-aiders and appointed persons.

The hazards and risks in the workplace

F7004 The *Management of Health and Safety at Work Regulations 1999* (*SI 1999 No 3242*) require employers to make a suitable and sufficient assessment of the risks to health and safety at work of their employees. The assessment must be designed to identify the measures required for controlling or preventing any risks to the workforce: highlighting what types of accidents or injuries are most likely to occur will help employers address such key questions as the appropriate nature, quantity and location of first-aid personnel and facilities.

Where the risk assessment conducted by an employer identifies a low risk to health and safety, employers may only need to provide (i) a first-aid container clearly identified and suitably stocked, and (ii) an appointed person to look after first-aid arrangements and resources and to take control in emergencies.

Where risks to health and safety are greater, employers may need to ensure the following:

- the provision of an adequate number of first-aiders so that first-aid can be given immediately;
- the training of first-aiders to deal with specific risks or hazards;
- informing the local emergency services in writing of the risks and hazards on the site where hazardous substances or processes are in use;
- the provision of a first-aid room(s).

Employers will need to consider the different risks in each part of their company or organisation. Where an organisation occupies a large building with different processes being performed in different parts of the premises, each area's risks must be assessed separately. It would not be appropriate to conduct a generic assessment of needs to cover a variety of activities – the parts of the building with higher risks will need greater first-aid provision than those with lower risk.

The size of the company

F7005 In general, the amount of first-aid provision that is required will increase according to the number of employees involved. Employers should be aware, however, that in some organisations there may be few employees but the risks to their health and safety might be high – and, as a result, their first-aid needs will also be high.

The history of incidents and accidents

F7006 When assessing first-aid needs, employers might find it useful to collate data on accidents that have occurred in the past, and then analyse, for example, the numbers and types of accidents, their frequency and consequences. Organisations with large premises should refer to such information when determining the first-aid equipment, facilities and personnel that are required to cover specific areas.

The character and dispersion of the workforce

F7007 The employer should bear in mind that the size of the premises can affect the time it might take a first-aider to reach an incident. If there are a number of buildings on the site, or the building in question comprises several storeys, the most suitable arrangement might be for each building or floor to be provided with its own first-aiders. Where employees work in shifts or in self-contained areas, first-aid arrangements may need to be tailored to reflect that fact.

Employees who are potentially at higher risk, such as trainees, young workers or people with disabilities, will need to be given special consideration.

The distance from the workplace to emergency medical services

F7008 Where a workplace is far from emergency medical services, it may be necessary to make special arrangements for ensuring that appropriate transport can be provided for taking an injured person to the emergency medical services.

In every case where the place of work is remote, the very least that an employer should do is to give written details to the local emergency services of the layout of the workplace, plus any other relevant information, such as information on specific hazards.

The needs and requirements of travelling, distant and lone workers

F7009 Employers are responsible for meeting the needs of their employees whilst they are working away from the main company premises.

When assessing the needs of staff who travel long distances or who are constantly mobile, consideration should be given to the question of whether they ought to be provided with a personal first-aid kit.

Organisations with staff working in remote areas must make special arrangements for those employees in respect of communications, special training and emergency transport.

Personnel on sites that are shared or multi-occupied

F7010 Employers with personnel working on shared or multi-occupied sites can agree to have one employer on the site who is solely responsible for providing first-aid cover for all the workers. It is strongly recommended that this agreement is written to avoid confusion and misunderstandings between the employers. It will highlight the risks and hazards of each company on the site, and will make sure that the shared provision is suitable and sufficient. After the employers have agreed the arrangement, the personnel must be informed accordingly.

When employees are contracted out to other companies their employer must ensure they have access to first-aid facilities and equipment. The user employer bears the responsibility for providing such facilities and equipment.

First-aiders on annual leave or absent from the workplace

F7011 Adequate provision of first-aid must be available at all times. Employers should therefore ensure that the arrangements they make for the provision of first-aid at the workplace are adequate to cover for any annual leave of their first-aiders or appointed persons. Such arrangements must also be able to cover for any unplanned or unusual absences from the workplace of first-aiders or appointed persons.

Other important points

F7012 Employers are not obliged by these regulations to provide first-aid facilities for members of the public. Although the regulations are aimed at employees, including trainees, many organisations like health authorities, schools and colleges, places of entertainment, fairgrounds and shops do make first-aid provision for persons other than their employees. Other legislation deals with public safety and first-aid facilities for the public – for example, the *Road Traffic Act 1960* regulates first-aid provision on buses and coaches.

Where employers extend their first-aid provision to cover more than merely their employees, the provision for the employees must not be diminished and should not fall below the standard required by these regulations.

The compulsory element of employers' liability insurance does not cover litigation resulting from first-aid given to non-employees. It is advised that employers check their public liability insurance policy on this issue.

Reassessing first-aid needs

F7013 In order to ensure that first-aid provision continues to be adequate and appropriate, employers should from time to time review the first-aid needs in the workplace, particularly when changes have been made to working practices.

Duty of the employer to inform employees of first-aid arrangements

F7014 Employers are under a duty to inform employees of the arrangements that have been made for first-aid in their workplace. This may be achieved by:

- Distributing guidance to all employees which highlights the key issues in the first-aid arrangements, such as listing the names of all first-aiders and describing where first-aid resources are located.
- Nominating key employees to ensure that the guidance is kept up to date and is distributed to all staff, and to act also as information officers for first-aid in the workplace.
- Internal memos can be used as a method of keeping the personnel informed of any changes in the first-aid arrangements.
- Displaying announcements up on notice boards informing employees of the first-aid arrangements and of any changes to those arrangements.
- Providing new employees with the information as part of their induction package.

  Any person with a reading or language problem must be given the information in a way that they can understand.

First-aid and the self-employed

F7015 The self-employed should provide, or ensure that there is provided, such equipment, if any, as is appropriate in the circumstances to enable them to render first-aid to themselves whilst at work. The self-employed who work in low-risk areas, for example at home, are required merely to make first-aid provision appropriate to a domestic environment.

When self-employed people work together on the same site, they are each responsible for their own first-aid arrangements. If they wish to collaborate on first-aid provision, they may agree a joint arrangement to cover all personnel on that particular site.

Number of first-aiders

F7016 Sufficient numbers of first-aiders should be located strategically on the premises to allow for the administration of first-aid quickly when the occasion arises. The assessment of first-aid needs may have helped to highlight the extent to which there is a need for first-aiders. The table below gives suggested numbers of first-aiders or appointed persons who should be available at the workplace. The suggested numbers are not a legal requirement – they are merely for guidance.

There are no hard and fast rules on numbers – employers will have to make a judgement, taking into account all the circumstances of their organisation. If the company is a long way from a medical facility or there are shift workers on site or the premises cover a large area, the numbers of first-aid personnel set out below may not be sufficient – the employer may have to make provision for a greater number of first-aiders to be on site.

Selection of first-aiders

F7017 First-aiders must be reliable and of good disposition. Not only should they have good communication skills, but they must also possess the ability and aptitude for acquiring new knowledge and skills and must be able to handle physical and stressful emergency incidents and procedures. Their position in the company should be such that they are able to leave their place of work immediately to respond to an emergency.

Table 1

First-aid personnel

Category of risk	*Numbers employed at any location*	*Suggested number of first-aid personnel*
Lower risk – e.g. shops, offices, libraries	Fewer than 50	At least one appointed person
	50–100	At least one first-aider
	More than 100	One additional first-aider for every 100 employed
Medium risk – e.g. light engineering and assembly work, food processing, warehousing	Fewer than 20	At least one appointed person
	20–100	At least one first-aider for every 50 employed (or part thereof)
	More than 100	One additional first-aider for every 100 employed
Higher risk – e.g. most construction, slaughterhouse, chemical manufacturer, extensive work with dangerous machinery or sharp instruments	Fewer than 5	At least one appointed person
	5–50	At least one first-aider
	More than 50	One additional first-aider for every 50 employed

	Where there are hazards for which additional first-aid skills are necessary	In addition, at least one first-aider trained in the specific emergency action

The training and qualifications of first-aid personnel

F7018 Individuals nominated to be first-aiders must complete a programme of competence-based training in first-aid at work run by an organisation approved by the HSE. See APPENDIX B for a range of first-aid competencies which make up a basic curriculum. Organisations that are contracted to train first-aid personnel may be notified of any particular risks or hazards in a workplace so that the first-aid course that is provided can be tailored to include the risks specific to that workplace.

Additional special training may be undertaken to deal with unusual risks and hazards. This will enable the first-aider to be competent in dealing with such risks. Such special training may be separate from the basic course or an extension to it but does not need approval by the HSE. The first-aid certificate awarded may be endorsed to verify that special training has been received.

It is important for employers to understand that first-aid at work certificates are valid for three years. Employers wishing to arrange for re-testing of competence training must do so before the original first-aid certificate expires. If the certificate of any first-aider expires, he or she will have to attend a full course of training to regain their first-aid at work certificate. Employers can arrange for re-testing of competence training up to three months before the expiry date of the certificate – the new first-aid at work certificate will then run from the date of the expiry of the old certificate. It is advisable for employers to keep a record of first-aiders in the company, together with their certification dates, in order to assist them in organising refresher training. Employers should develop a programme of knowledge and skills training for their first-aiders to enable them to be updated on new skills and to make them aware of suitable sources of first-aid information, such as occupational health services and training organisations qualified by the HSE to conduct first-aid at work training.

Appointed persons

F7019 An appointed person is an individual who takes charge of first-aid arrangements for the company including looking after the facilities and equipment and calling the emergency services when required. The appointed person is allocated these duties when it is found through the first-aid needs assessment that a first-aider is not necessary. The appointed person is the minimum requirement an employer can have in the workplace. Clearly, even if the company is considered a low health and safety risk and, in the opinion of the employer, a first-aider is unnecessary, an accident or illness still may occur – therefore somebody should be nominated to call the emergency services, if required.

Appointed persons are not first-aiders – and therefore they should not be called upon to administer first-aid if they have not received the relevant training. However, employers may consider it prudent to send appointed persons on first-aid training courses. Such a course normally last for four hours and includes the following topics:

- action necessary in an emergency;
- cardio-pulmonary resuscitation;

- first-aid treatment for the unconscious casualty;
- first-aid treatment for the bleeding or the wounded.
- HSE approval is not required for this training.

The only time an appointed person can replace a first-aider is when the first-aider is absent, due to circumstances which are temporary, unforeseen and exceptional. Appointed persons cannot replace first-aiders who are on annual leave. If the first-aid assessment has identified a requirement for first-aiders, they should be available whenever there is a need for them in the place of work.

Records and record keeping

F7020 It is considered good practice to keep records of incidents which required the attendance of a first-aider and treatment of an injured person. It is advisable for smaller companies to have one record book but for larger organisations this may not be practicable.

The data entered in the record book should include the following:

- the date, time and place of the incident;
- the injured person's name and job title;
- a description of the injury or illness and of the first-aid treatment administered;
- details of where the injured person went after the incident, i.e. hospital, home or back to work;
- the name and signature of the first-aider or person who dealt with the incident.

This information may be collated, to help the employer improve the environment with regard to health and safety in the workplace. It could be used to help determine future first-aid needs assessment and will be helpful for insurance and investigative purposes. The statutory accident book is not the same as the record book, but they may be combined.

First-aid resources

F7021 Having completed the first-aid needs requirements assessment, the employer must provide the resources, i.e. the equipment, facilities and materials, that will be needed to ensure that an appropriate level of cover is available to the employees at all relevant times. First-aid equipment, suitably marked and obtainable, must be made available at specific sites in the workplace.

First-aid containers

F7022 First-aid equipment must be suitably stocked and contained in a properly identifiable container. One first-aid container with sufficient quantity of first-aid materials must be made available for each worksite – this is the minimum level of first-aid equipment. Larger premises, for example, will require the provision of more than one container.

First-aid containers should be easily accessible and, where possible, near hand-washing facilities. The containers should be used only for first-aid equipment. Tablets and medications should not be kept in them. The first-aid materials within the containers should be protected from damp and dust.

Having completed the first-aid needs assessment, the employer will have a good idea as to what first-aid materials should be stocked in the first-aid containers. If there is no specific risk in the workplace, a minimum stock of first-aid materials would normally comprise the following (there is no mandatory list):

- a leaflet giving guidance on first-aid (for example, HSE leaflet *Basic advice on first aid at work*);
- 20 individually wrapped sterile adhesive dressings (assorted sizes), appropriate to the type of work;
- two sterile eye pads;
- four individually wrapped triangular bandages (preferably sterile);
- six safety pins;
- six medium-sized individually wrapped sterile unmedicated wound dressings – approximately 12cm x 12cm;
- two large sterile individually wrapped unmedicated wound dressings – approximately 18cm x 18cm;
- one pair of disposable gloves.

This list is a suggestion only – other equivalent materials will be deemed acceptable.

An examination of the first-aid kits should be conducted frequently. Stocks should be replenished as soon as possible after use, and ample back-up supplies should be kept on the company premises. Any first-aid materials found to be out of date should be carefully discarded.

All first-aid containers should have a white cross on a green background as identification.

Additional first-aid resources

F7023 If the results of the assessment suggest a need for additional resources such as scissors, adhesive tape, disposable aprons, or individually wrapped moist wipes, they can be kept in the first-aid container if space allows. Otherwise they may be kept in a different container as long as they are ready for use if required.

If the assessment highlights the need for such items as protective equipment, they must be securely stored next to first-aid containers or in first-aid rooms or in the hazard area itself. Only persons who have been trained to use these items may be allowed to use them.

If there is a need for eye irrigation, and mains tap water is unavailable, at least a litre of sterile water or sterile normal saline solution (0.9%) in sealed, disposable containers should be provided. If the seal is broken, the containers should be disposed of and not reused. Such containers should also be disposed of when their expiry date has been passed.

First-aid kits for travelling

F7024 First-aid kits for travelling may contain the following items:

- a leaflet giving general guidance on first-aid (for example, HSE leaflet *Basic advice on first aid at work*);
- six individually wrapped sterile adhesive dressings;
- one large sterile unmedicated dressing – approximately 18cm x 18cm;

- two triangular bandages;
- two safety pins;
- individually wrapped moist cleansing wipes;
- one pair of disposable gloves.

This list is a suggestion only – it is not mandatory. However, the kit must be regularly inspected and topped up from a back-up store at the home site.

Rooms designated as first-aid areas

F7025 A suitable room should be made available for first-aid purposes where the first-aid needs assessment found such a room to be necessary. Such room(s) should have sufficient first-aid resources, be easily accessible to stretchers and be easily identifiable, and where possible should be used only for administering first-aid.

First-aid rooms are normally necessary in organisations operating within high-risk industries. Therefore they would be deemed to be necessary on, chemical, ship building and large construction sites or on large sites remote from medical services. A person should be made responsible for the first-aid room.

On the door of the first-aid room a list of the names and telephone extensions of all the first-aiders should be displayed, together with details as to how and where they may be contacted on site.

First-aid rooms should:

- have enough space to hold a couch with space in the room for people to work, a desk, a chair and any other resources found necessary (see below);
- where possible, be near an access point in the event that a person needs to be taken to hospital;
- have heating, lighting, and ventilation;
- have surfaces that can be easily washed;
- be kept clean and tidy; and
- be available and ready for use whenever employees are in the workplace.

The following is a list of resources that may be found in a first-aid room:

- a record book for logging incidents where first-aid has been administered;
- a telephone;
- a storage area for storing first-aid materials;
- a bed/couch with waterproof protection and clean pillows and blankets;
- a chair;
- a foot operated refuse bin with disposable yellow clinical waste bags or some receptacle suitable for the safe disposal of clinical waste;
- a sink that has hot and cold running water – also drinking water and disposable cups;
- soap and some form of disposable paper towel.

If the designated first-aid room has to be shared with the working processes of the company, the employer must consider the implications of the room being needed in an emergency and whether the working processes in that room could be stopped immediately. Can the equipment in the room be removed in an emergency so as not to interfere with any administration of first-aid? Can the first-aid resources and equipment be stored in such a place as to be available quickly when necessary? Lastly, the room must be appropriately identified and, where necessary, be sign-posted by white lettering or symbols on a green background.

Appendix A
Assessment of first-aid needs – checklist

The minimum first-aid provision for each worksite is:

- a suitably stocked first-aid container;
- an appointed person to take charge of first-aid arrangements;
- information for employees on first-aid arrangements.

This checklist will help you assess whether you need to make any additional provisions.

	Aspects to consider	Impact on first-aid provision
1.	What are the risks of injury and ill health arising from the work as identified in your risk assessment?	If the risks are significant you may need to employ first-aiders.
2.	Are there any specific risks, such as:	You will need to consider:
	• hazardous substances;	• specific training for first-aiders;
	• dangerous tools;	• extra first-aid equipment;
	• dangerous machinery;	• precise siting of first-aid equipment;
	• dangerous loads or animals?	• informing emergency services;
		• a first-aid room.
3.	Are there parts of your establishment where different levels of risk can be identified (e.g. a university with research labs)?	You will probably need to make different levels of provision in different parts of the establishment.
4.	Are large numbers of people employed on site?	You may need to employ first-aiders to deal with the higher probability of an accident.
5.	What is your record of accidents and cases of ill-	You may need to:
	health? What type are they and where did they happen?	• locate your provision in certain areas;
		• review the contents of the first-aid box.
6.	Are there inexperienced workers on site, or	You will need to consider:
	employees with disabilities or special health problems?	• special equipment;
		• local siting of equipment.

7.	Are the premises spread out, e.g. are there several buildings on the site or multi-floor buildings?	You will need to consider provision in each building or on several floors.
8.	Is there shiftwork or out-of-hours working?	Remember that there needs to be first-aid provision at all times people are at work.
9.	Is your workplace remote from emergency medical services?	You will need to: ● inform local medical services of your location; ● consider special arrangements with the emergency services.
10.	Do you have employees who travel a lot or work alone?	You will need to: ● consider issuing personal first-aid kits and training staff in their use; ● consider issuing personal communicators to employees.
11.	Do any of your employees work at sites occupied by other employers?	You will need to make arrangements with the other site occupiers.
12.	Do you have any work experience trainees?	Remember that your first-aid provision must cover them.
13.	Do members of the public visit your premises?	You have no legal responsibilities for non-employees, but HSE strongly recommends that you include them in your first-aid provision.
14.	Do you have employees with reading or language difficulties?	You will need to make special arrangements to give them first-aid information.

Do not forget to allow for leave or absences of first-aiders and appointed persons. First-aid personnel must be available at all times when people are at work.

Appendix B
First-aid competencies

Employees who have successfully completed first-aid training must be able to apply the following competencies:

(*a*) to act safely, promptly and effectively when an incident occurs at work;

(*b*) to administer cardio-pulmonary resuscitation (CPR) promptly and effectively;

(*c*) to administer first-aid safely, promptly and effectively to a casualty who is unconscious;

(*d*) to administer first-aid safely, promptly and effectively to a casualty who is wounded or bleeding;

(*e*) to administer first-aid safely, promptly and effectively to a casualty who:

- is burnt or scalded;
- has an injury to bones, muscles or joints;
- is in shock;
- has an eye injury;
- is suffering from poisoning;
- has been overcome by gas or fumes;

(*f*) the transportation of the casualty safely in the workplace;

(*g*) the recognition and management of common major illnesses;

(*h*) the recognition and management of minor illnesses;

(*i*) the management of records and the provision of written information to medical staff if required.

First-aiders must also know and understand the following elements of first-aid at work:

(*a*) the legal requirements relating to the provision of first-aid at work;

(*b*) first-aider responsibilities and procedures in an emergency;

(*c*) safety and hygiene in first-aid procedures;

(*d*) how to use first-aid equipment provided in the workplace.

Food Safety and Standards

Introduction

F9001 Legislation has governed the sale of food for centuries. In Europe we can trace it back at least to the Middle Ages, and the ancient Hebrew food laws, found notably in the Book of Deuteronomy, show that it goes back even further. Two themes have existed from the start:

- *Food Safety* – the protection of the health and well being of anyone eating food; and
- *Food Standards* – the control of composition and adulteration of food for the prevention of fraud.

UK legislation is structured principally around the *Food Safety Act 1990* supported by a whole raft of more detailed regulations. Increasingly over the past thirty years directives and regulations from the European Union have determined UK legislation. More recently, devolution within the UK has seen the responsibility for food policy shifted to the regional assemblies in Scotland, Wales and Northern Ireland. This may allow greater diversity of regulation to develop within the UK as far as this is possible within the EU framework.

Food safety

F9002 The principal legislation is the *Food Safety Act 1990*. Since the inception of the European Economic Community (EEC), now the European Union (EU), European 'hygiene' regulations have developed on the basis of specific measures for specific food sectors, the so-called 'vertical' regulations. A full complement of these was in place by the 1990s covering the processing of most foods of animal origin including meat, fish, eggs and dairy products. The vertical regulations do not in general cover businesses selling food direct to the ultimate consumer, for example catering or retail businesses.

Only in 1993 did the EU publish the 'horizontal' Food Hygiene Directive 93/43. This establishes general principles for food hygiene and sets hygiene standards for those businesses not covered by 'vertical' legislation. These include food factories processing non-animal products and also retail or catering outlets. This directive passed into UK law as the *Food Safety* (*General Food Hygiene*) *Regulations 1995* (*SI 1995 No 1763*) ('the General Food Hygiene Regulations'). These Regulations were supplemented by the *Food Safety* (*Temperature Control*) *Regulations 1995* (*SI 1995 No 2200*) ('the Temperature Control Regulations').

The EU is now planning a radical overhaul of the 'vertical' directives and regulations, which will be consolidated around the framework of a revised version of Directive 93/43. The final draft was agreed in April 2004 and, if the amending legislation is passed in time, will take affect from January 2006. The following account will be largely restricted to the provisions of the Food Safety Act, the 1995

General Food Hygiene Regulations and the Temperature Control Regulations. Food processors who may be covered by the 'vertical' legislation will need to take more specific advice.

Food standards

F9003 Food standards is governed by a complex range of regulations. Despite its name, the *Food Safety Act 1990* is also the primary legislation covering most aspects of food standards and a number of regulations have been made under it. Over the recent past there has been a significant change in emphasis. Previously there was a tendency to prescribed compositional standards for foods. For example, the meat content of certain products, the fruit content of drinks and so on. A little of this legacy remains, but the approach proved inflexible especially in a climate of rapid innovation and product development. The emphasis has now shifted to allow more diversity of composition, but supported by informative labelling. The *Food Labelling Regulations 1996* (*SI 1996 No 1499*) play an important role. The *Food Labelling* (*Amendment*) *Regulations 1998* (*SI 1998 No 1398*) took this even further with a specific requirement that pre-packed foods must carry a 'quantitative ingredient declaration' or 'Quid'. This dictates that the label must carry a declaration of the percentage quantity of the main ingredient or ingredients of any food. Food standards and labelling may also be enforced through the more general *Trades Descriptions Act 1968*.

Food labelling can also have a bearing on safety or issues thought to impact on safety. Subsequent, actual, or proposed amendments to labelling regulations have dealt with topics such as the labelling of ingredients that may cause severe allergenic in some consumers and the labelling of Genetically Modified foods.

Administrative regulations

F9004 Finally, there are more general regulations made under the *Food Safety Act 1990*. These may regulate the law enforcement process, setting criteria for inspectors, food analysts or examiners. Or they may be more general requirements for food businesses such as the *Food Premises* (*Registration*) *Regulations 1991* (*SI 1991 No 2825*). Under these Regulations all food premises must register with the local authority. The Regulations have been amended twice (by *SI 1993 No 2022* and *SI 1997 No 723*).

Of course many other pieces of consumer protection legislation can apply; notably the *Weights and Measures Act 1985*, which controls the way in which food is sold by weight or other measures, and the *Prices Act 1974*. The *Price Indications* (*Food and Drink on Premises*) *Order 1079* requires prices for food and drinks to be displayed by catering premises following a particular format.

Policy and enforcement

F9005 Until the year 2000, government policy on food came largely from the Ministry of Agriculture, Fisheries and Food (MAFF) and the Department of Health (DH). MAFF had particular responsibility for the 'upstream' parts of the food chain and for food standards. DH was naturally more involved in questions of food safety, and they took a role in liaison with enforcement authorities.

April 2000 saw the establishment of the Food Standards Agency (FSA). This is a non-ministerial government department designed to take all of the policy functions from MAFF and DH into an organisation at arms length from the government. This should allow it to act more independently and primarily in the consumers' interest. The FSA is controlled by a board of directors and operates through a chief executive and officials, the majority of whom joined the FSA from similar roles in MAFF or DH. Complementary FSA structures exist in Scotland, Wales and Northern Ireland. The FSA is created and empowered by the *Food Standards Act 1999*.

Complex arrangements are in place for the enforcement of food legislation. Traditionally local authorities have had the responsibility although central government officials have particular responsibilities in 'upstream' parts of the food chain. This is particularly true of 'meat hygiene' in abattoirs and cutting plants. This role was taken away from local authorities to a central Meat Hygiene Service (MHS) with the MHS itself becoming a branch of the FSA from April 2000. The FSA has also developed a new role of local authority monitoring.

Most food enforcement at the production and retail level falls to two groups of local government officers:

- Environmental Health Officers (EHOs) have responsibility for food safety legislation. They are generally employed at district council (town hall) level.
- Trading Standards Officers (TSOs) take control of food standards issues, including weights and measures. Generally they are employed at county council level.

The distinction between district and county councils has become increasingly blurred with much local government organised on a unitary basis. This includes most English metropolitan areas and all authorities in Scotland. In these cases TSO and EHO functions are usually combined in the same department.

Food Safety Act 1990

F9006 The *Food Safety Act 1990* describes the offences that may be committed if food legislation is contravened and also explains the defences that may be used if charges are instigated. It sets up the mechanisms to enforce the laws and fixes penalties that may be levied. The Act also enables the government to make regulations that include more detailed food safety and consumer protection measures.

The Act is divided into four principle sections:

(i) Part I: Definitions and responsibilities for enforcement;

(ii) Part II: Main provisions;

(iii) Part III: Administrative and enforcement issues such as powers of entry; and

(iv) Part IV: Miscellaneous arrangements notably the power to issue codes of practice.

The following is a brief summary of the main provisions contained within the Act.

Section 1: definitions

F9007 Under the *Food Safety Act 1990* 'food' has a wide meaning and includes:

- drink;
- articles and substances of no nutritional value which are used for human consumption;
- chewing gum and similar products;
- articles and substances used as ingredients in any of the above foods.

Food does not include live animals, birds or fish (although shellfish that are eaten raw and alive such as oysters and similar are classed as 'food'). Neither does 'food' include animal feed, nor controlled drugs that might be taken orally.

The scope of the Act is very wide ranging and covers all commercial businesses. A business may be a canteen, club, school, hospital, care home and so on, whether or not it is run for profit. Government establishments that once had 'crown immunity' are treated no differently to other food businesses. Charity events that sell food are also subject to the Act.

The Act covers any premises from abattoir to retail superstore, from street vendor to a five star hotel. It includes any place (including premises used only occasionally for a food business such as a village hall), any vehicle, mobile stall and temporary structure.

The *Food Safety Act 1990, s 2* provides an extended meaning to the word 'sale' of food, and *s 3* establishes a presumption that food is intended for human consumption. The meaning of 'sale' of food is very wide. The Act does not apply only to sales where money changes hands and the business is run for profit. It also covers food given as a prize or as a reward by way of business promotion or entertainment. Entertainment includes social gatherings, exhibitions, games, sport and so on. Thus, if food, which turns out to be unfit is given as a prize for a darts competition, it could be subject to an action under the 1990 Act. If someone has food in their possession the Act presumes that the intention is to sell it unless it can be proved otherwise. For example, food past its 'use by' date must be segregated from food that is for sale and should be marked clearly otherwise an EHO or TSO could presume that it was for sale.

The *Food Safety Act 1990, ss 4–6* describe who does what under food law and defines the various responsibilities of local and central government.

Part II: main provisions of the Act

F9008 The main provisions of the *Food Safety Act 1990* fall within *ss 7* to *22*. The offences are divided into two types: food safety and consumer protection which each breakdown into two separate offences:

(*i*) *Food Safety*

- *Section 7*: Rendering food injurious to health.
- *Section 8*: Selling food not complying with food safety requirements.

(*ii*) *Consumer Protection*

- *Section 14*: Food not of the nature, substance or quality demanded.
- *Section 15*: Falsely presenting or describing food.

Section 7: rendering food injurious to health

F9009 It is an offence to do anything intentional that would make food harmful to anyone that eats it. Even if one did not know that it would have that effect, an offence would still have been committed. For example subjecting food to poor temperature control that allows bacteria to multiply could render food injurious to health.

Food is considered injurious to health if it would harm most people who ate it. It would not be considered 'injurious to health' if it contained an ingredient (like peanuts) that may only affect a small proportion of people who have a specific allergy. It would be injurious to health if it were likely to cause immediate harm to anyone that ate it, for example botulism in yoghurt. It would also be injurious to health if the harm were cumulative over a long period, for example fungal toxins in cereal products.

Section 8: selling food not complying with food safety requirements

F9010 Whilst *s 7* deals with the person who renders food injurious to health, the *Food Safety Act 1990, s 8* goes on to prevent him or anyone else from selling the food. This can apply to anyone in the food chain. The predecessor to the 1990 Act only intercepted food as it was sold to the consumer. The 1990 Act allows action to be taken at any stage in the chain.

Food would fail to comply with food safety requirements if:

(*a*) it has been rendered injurious to health as described above;

(*b*) it is unfit for human consumption;

(*c*) it is so contaminated that it would be unreasonable to expect it to be eaten.

Food would be unfit for consumption if it was putrid or toxic or it contained serious foreign material, for example a dead mouse.

The third part of *s 8* was a new provision in 1990 and gave wide scope to prosecute someone for selling food that fails to meet a customer's expectations. For example the food is mouldy or contains a rusty nail or there are excessive antibiotic residues in meat.

(The *Food Safety Act 1990, ss 9–13* provide enforcement procedures for food safety offences.) (These sections are dealt with from F9014–F9019.)

Section 14: selling food not of the nature, substance or quality demanded

F9011 It is an offence to supply to the prejudice of the purchaser food which is not of the nature, substance or quality demanded. The purchaser in this case does not have to be a customer in a retail store or catering outlet. One company, large or small may purchase from another and an offence is committed if the food is inferior in nature or substance or quality to that which they demanded. Once again this allows the law to be applied at any point of the food chain.

The purchaser does not have to buy the food for his or her own use to be prejudiced. They may intend to give it to someone else, for example members of their family.

The three parts of the offence are separate and a charge will be brought under one of the three according to the circumstances.

If a purchaser asks for cod and gets coley, or beef mince contains a mixture of lamb or chicken, it would be not of the 'nature' demanded.

If a purchaser asked for diet cola and was served regular cola, or expected a sheep's milk cheese and it was made from cow's milk, it would be not of the 'quality' demanded.

Food that was not of the quality or substance demanded often formed the basis for complaints of mouldy food or foreign material. Such complaints are likely to be taken under the *Food Safety Act 1990, s 8(2)(c)* (i.e. that the food is so contaminated that it would be unreasonable to expect it to be eaten).

More recently, the above provision has been used when a susceptible customer has requested information about peanuts or similar ingredients to which they may have a severe allergic reaction. If, notwithstanding such a request, the consumer is sold food that triggers a reaction, cases have been brought on the grounds that the food was not of the 'substance' demanded.

Section 15: falsely presenting or describing food

F9012 It would be an offence to make a statement on the food label that was untrue, or even if there was a misleading pictorial representation. It is also an offence if material that is technically accurate is presented in such a way as to mislead the consumer.

The *Food Safety Act 1990, s 15* is backed up with very detailed requirements on labelling in the *Food Labelling Regulations 1996* (*SI 1996 No 1499*) *as amended.* For some offences, especially labelling by caterers on menus or chalkboards, TSOs may invoke the more general requirements under the *Trades Descriptions Act 1968.*

Enforcement

F9013 The *Food Safety Act 1990* gives enforcement officers strong powers, namely:

- To enter food premises to investigate possible offences.
- To inspect food.
- To detain or seize suspect food.
- To take action that requires a business to put things right.
- If all else fails:
  - to prosecute,
  - to prohibit the use of premises or equipment,
  - to bar an individual from working in a food business.

Under the *Food Safety Act 1990, ss 32* and *33*, enforcement officers have rights of access at any reasonable time and an offence will be committed if they are obstructed. They must be given reasonable information and assistance. They can inspect premises, processes and records. If records are kept on a computer, they are entitled to have access to them. They can copy records, take samples, take photographs and even videos. If they are refused access to any of these, they can apply to a magistrate for a warrant. They must give 24 hours notice to enter private houses used in connection with a food business.

Of course none of these requirements to co-operate with officers and give them access to information cancel the basic right to avoid self-incrimination. In the extreme situation one has the right to remain silent and to consult a solicitor.

Officers themselves commit an offence if they reveal any trade secrets learned in the course of official duties.

The *Food Safety Act 1990, ss 29–31* allow officers to procure samples and control the sampling and testing procedures. A section 40 code of practice (no. 7) covers 'Sampling for Analysis or Examination'. Samples that are taken carelessly may give unreliable results and may be not be acceptable as evidence in court.

The *Food Safety Act 1990*, ss *9–13* provide enforcement officers with a series of enforcement measures to control food or food businesses that present a health risk.

Section 9: detention and seizure of food

F9014 If an officer suspects that food is unfit or fails to meet food safety requirements, he can issue a detention notice. This requires the person in charge of the food to keep it in a specified place and not to use it for human consumption pending investigation. The detention notice must be written on a prescribed form. It is an offence to ignore or contravene a detention notice even if you may believe that it is unjustified. The officer has 21 days to complete his investigations. If he concludes that the food was actually safe, he must withdraw the notice (using another prescribed form) and restore the food to the person in charge.

On the other hand, the EHO may conclude that the food is unsafe. He can seize it and take it before a justice of the peace (JP). The EHO may reach this conclusion as a result of tests performed during the 21 days after serving the detention notice. But in some circumstances, he may be confident that it is unsafe from the beginning and seize the food immediately without using a detention notice. Once again there is a form that must be used and the EHO must state exactly what is happening and when. After seizure the matter should be taken before the JP within 2 days, quicker in the case of perishable food.

The hearing before the JP may not only decide the fate of the batch of food but may also be a forerunner to other legal proceedings. One is allowed to make representations at this hearing and to call witnesses. It is probably advisable to take specific legal advice from a solicitor when the first detention notice is issued. It is wise to have legal representation at a hearing following seizure.

If the JP decides that the food is unsafe he can order it to be destroyed and he may order the offender to cover the costs of its disposal. If the seized food is only part of a larger batch the court will presume that the whole batch fails unless evidence is provided to the contrary.

If it turns out that the EHO was wrong, the offender may be entitled to compensation. If the food has deteriorated as a result they may be entitled to compensation equal to the loss in value. If a reasonable sum cannot be agreed with the authority there is an arbitration procedure.

Section 10: improvement notices

F9015 EHOs have a range of powers to deal with unsatisfactory premises from informal advice through to prosecution and emergency prohibition. If an authorised officer believes that a business does not comply with hygiene or processing regulations, he can issue a formal improvement notice demanding that it puts matters right. This provides a quicker and cheaper remedy than taking the business to court. An improvement notice should not be ignored. It is an offence to fail to comply with a notice and either matters must be put right as indicated on the notice or an appeal lodged against it.

An improvement notice must contain four elements:

(i) the reasons for believing that the requirements are not being complied with;

(ii) the ways in which regulations are being breached;

(iii) the measure which should be taken to put things right;

(iv) the time allowed for putting things right (this cannot be less than 14 days).

One can choose to put things right in a different way to that suggested by the EHO provided that the EHO is satisfied that it will have the required effect.

Authority FORM 1

Food Safety Act 1990 - Section 9

DETENTION OF FOOD NOTICE

Reference Number:

1. To: --

   Of: --

2. Food to which this notice applies:

   Description: --

   Quantity: --

   Identification marks: --

3. *THIS FOOD IS NOT TO BE USED FOR HUMAN CONSUMPTION.*

   In my opinion, the food does not comply with food safety requirements because: ------------------

   ----

   ----

4. The food must not be removed from:

   ----

   *unless it is moved to:

   (*Officer to delete if not applicable)

5. Within 21 days, either this notice will be withdrawn and the food released, or the food will be seized to be dealt with by a justice of the peace, or in Scotland a sheriff or magistrate, who may condemn it.

Signed: --- Authorised Officer

Name in capitals: --

Date: --

Address: --

Tel: ------------------------- Fax:--

Please read the notes overleaf carefully. If you are not sure of your rights or the implications of this notice, you may want to seek legal advice.

Authority: ______________________________ FORM 2

Food Safety Act 1990—Section 9

WITHDRAWAL OF DETENTION OF FOOD NOTICE

1. To: ....
   Of: ....
   ....

2. Detention Notice Number .... dated .... and served on you on .... (date) is now withdrawn. The food described in paragraph 3 below can now be used for human consumption.

3. Food released for human consumption:
   Description: ....
   Quantity: ....
   Identification marks: ....

   Signed: .... Authorised Officer
   Name in capitals: ....
   Date: ....
   Address: ....
   ....
   ....
   Tel: .... Fax ....

Please read the notes overleaf carefully. If you are not sure of your rights or the implications of this notice, you may want to seek legal advice.

Authority:____ FORM 3

Food Safety Act 1990—Section 9

FOOD CONDEMNATION WARNING NOTICE

Reference Number:

1. To: ....
   Of: ....
   ....

2. This Notice applies to the following food which has been seized by an officer of this authority:
   Description ....
   Quantity ....
   Identification marks ....

3. *IT IS MY INTENTION TO APPLY TO A JUSTICE OF THE PEACE, OR IN SCOTLAND A SHERIFF OR MAGISTRATE, AT*
   ....
   ON .... (DATE) AT .... AM/PM FOR THE ABOVE FOOD TO BE CONDEMNED.
   because ....

4. As the person in charge of the food, you are entitled to attend and to bring witnesses.

5. A copy of this notice has also been given to:
   ....
   ....
   who may also attend and bring witnesses.

Signed: .... Authorised Officer
Name in capitals: ....
Date: ....
Address: ....
....
....
Tel: .... Fax: ....

Please read the notes overleaf carefully. If you are not sure of your rights or the implications of this notice, you may want to seek legal advice.

Authority: .... FORM 1

Food Safety Act 1990—Section 10

IMPROVEMENT NOTICE

Reference Number:

1. To: .... (Proprietor of the food business)

   At: ....

   ....

   .... (Address of proprietor)

2. In my opinion the:

   ....

   ....

   [Office to insert matters which do not comply with the Regulations]

   in connection with your food business ....

   .... (Name of business)

   at ....

   .... (Address of business)

   do/does* not meet the requirements of ....

   of the .... Regulations

   because:

   ....

   ....

   [* Officer to delete as appropriate]

3. In my opinion, the following measures are needed for you to comply with these Regulations: ....

   ....

   ....

4. These measures or measures that will achieve the same effect must be taken by: .... (date)

5. *It is an offence not to comply with this improvement notice by the date stated.*

   Signed: .... Authorised Officer

   Name in capitals: ....

   Date: ....

   Address: ....

   ....

   Tel: .... Fax: ....

> *Please read the notes overleaf carefully.*
> *If you are not sure of your rights or the implications of this notice, you may want to seek legal advice.*

If a party does not agree with an improvement notice the *Food Safety Act 1990, s 37* gives the right to appeal to a magistrate within one month of its issue or within the time specified on the notice if shorter than one month. The EHO must inform the party about their rights to appeal and provide the name and address of the court when he serves the notice.

The notice may contain a number of points but the whole notice does not have to be appealed. One may appeal just one point or simply ask for a time extension. The notice is effectively suspended during the appeal, but it is best to let the EHO, as well as the court, know about the appeal. The *Food Safety Act 1990, s 39* allows the magistrate to modify or cancel any part of the notice.

(Until April 2001 there was an additional procedure involved in the issue of improvement notices. Any improvement notice had to be preceded with a 'minded to' notice. This was an informal notification from the EHO that he or she intended to issue an improvement notice. It gave the opportunity to put things right without a formal improvement notice. The 'minded to' procedure was ended by the *Regulatory Reform Act 2001*.)

Section 11: prohibition orders

F9016 If one is successfully prosecuted for food safety offences, the EHO can also apply for a prohibition order. He or she must establish that there is a danger to the public.

The first step the authority must take is to get a successful prosecution for a breach of food hygiene or food safety regulations. The offence may be failing to comply with an improvement notice. If the court decides that there is an ongoing risk to public health it must issue a prohibition order. The order can deal with any of one three issues depending upon the actual risk:

- to prohibit a process or treatment,
- to deal with the *structure* of the premises or the use of equipment,
- to deal with a problem that relates to the *condition* of the premises or equipment.

In either of the last two cases, the order may prohibit the use of the equipment or the premises.

Any prohibition order must be served on the proprietor of the business. If it relates to equipment it will also be fixed to the equipment and if it relates to the premises it will be fixed prominently to the premises. It is automatically an offence if one knowingly breaches a prohibition order.

As soon as a party believes that they have put matters right and that the health risk no longer exists they can apply to the EHO to have the prohibition lifted. The EHO must respond by reaching a decision within a fortnight and if they believe that things are satisfactory, issue a certificate within another three days. The certificate will state that enough has been done to ensure that there is no longer an unacceptable risk to public health. If the authority disagrees and refuses to issue a certificate they must give their reasons and one can appeal to a magistrate to have the order lifted. The authority must give information on the right to appeal, and who to contact as well as the time scale (in this case one must appeal within one month).

PROHIBITION ORDER

(Food Safety Act 1990, s. 11)

Magistrates' Court

(Code)

Date :

Accused : WHEREAS

of :

being the proprietor of a food business carried on at premises at

has today been convicted by this court of an offence under regulations to which section 11 of the Food Safety Act 1990 applies [by virtue of section 10(3)(b) of the said Act], namely:-

Decision : AND WHEREAS the Court is satisfied that there exists a risk of injury to health by reason of

[the use for the purposes of the said food business of a certain [process] [treatment], namely

]

[the [construction] [state or condition] of the premises at

used for the purposes of the said food business]

[the [use for the purposes of the said food business] the [state or condition] of certain equipment [used for the purposes of the said food business], namely

]

and it is ORDERED that

Order : [the use of the said [process][treatment] for the purposes of the business] [the use of the said [premises] [equipment] for the purposes of [the business] [any other food business of the same class or description, namely

]

[any food business]]
[and that the participation by the accused in the management of any food business [of the said class or description]
is prohibited.

[By Order of the Court] [*Justice of the Peace*] [*Justices' Clerk*]

Delete any words within square brackets which do not apply.

FS 46

Personal prohibition under section 11

F9017 The court can also prohibit a person from running or managing a food business. One can apply to the court to lift a personal prohibition but cannot do so within 6 months of its imposition and if an appeal is unsuccessful it cannot be appealed again for three months.

Section 12: emergency prohibition

F9018 *Section 11* prohibitions are imposed only by a court following conviction. But in an emergency, an EHO can act on his or her own authority and with immediate effect. They must be satisfied that a business presents an imminent risk to health. In such circumstances they can serve an emergency prohibition notice without prior reference to a court. Emergency prohibition can apply to the whole premises or a specific part. The notice will be fixed in a prominent place and anyone removing it or deliberately ignoring it is guilty of an offence.

The officer must take the matter before a magistrates' court within three days of serving the notice. At least one day before it goes to court the EHO must serve a notice on the proprietor of the business telling him of the court hearing. Once again, if one finds themself faced with action of this kind it is advisable to seek specific legal advice.

If the court agrees that the EHO's actions were justified it will make an emergency prohibition order which replaces the emergency prohibition notice. The arrangements for lifting the order are the same as a *section 11* order. If the court does not uphold an emergency prohibition, the proprietor may seek compensation for loss or damages. An emergency prohibition order cannot be made against an individual. Prohibition of a person can only be made under *section 11* following a conviction.

Section 13: emergency control orders

F9019 This power would normally be exercised by the government. It may be that an unsafe batch of food has already been distributed around the country. Emergency prohibition of the manufacturing site would stop further production but it would not control the risk from food already in the system. The *Food Safety Act 1990, s 13* allows an emergency control order to require all steps to be taken that will remove the threat. This is a wide-ranging power that will only be used in exceptional circumstances. Of course in most cases, any business told that it has received unsafe food from a supplier would stop selling it simply to protect their own reputation. If an emergency control order is invoked, there is no right of appeal and no compensation arrangements.

Voluntary arrangements are in place to co-ordinate food hazards and issue warnings and organise recalls when food is distributed more widely than the local area in which it is produced. These warnings are circulated by electronic mail to EHOs around the country. Where possible, trade organisations and the media are also informed. The Food Standards Agency is demonstrating a greater commitment to freedom of information than previous departments. Nowadays food hazard warnings and recalls are better publicised, not least through the FSA website at: www.foodstandards.gov.uk

Sections 16 to 18: power to make regulations

F9020 The *Food Safety Act 1990, ss 16–18,* include wide-ranging enabling powers that allow ministers to make regulations. For example:

Authority: .... FORM 2

Food Safety Act 1990–Section 12

EMERGENCY PROHIBITION NOTICE

Reference Number:

1. To: .... (Proprietor of the food business)

   At: ....

   ....

   .... (Address of proprietor)

2* I am satisfied that: ....

   ....

   ....

   at ....

   -- (Address of business)

POSES AN IMMINENT RISK OF INJURY TO HEALTH because:

....

....

....

*(*See Note 1 overleaf)*

3. *YOU MUST NOT USE IT FOR THE PURPOSE OF THIS/ANY/THIS OR ANY SIMILAR* FOOD BUSINESS.*

[* Officer to delete as appropriate]

Signed: .... Authorised Officer

Name in capitals: ....

Date: ....

Address: ....

....

Tel: .... Fax: ....

> *Please read the notes overleaf carefully. If you are not sure of your rights or the implications of this notice, you may want to seek legal advice.*

- To prohibit specified substances in foods.
- To prohibit a certain food process.
- To require hygienic conditions in commercial food premises.
- To dictate microbiological standards for foods.
- To make standards on food composition.
- To specify requirements for food labelling.
- To restrict and control claims made in the advertising of food.

They also provide the power to ban from sale food originating from potentially diseased sources. For example these powers were used to prohibit the sale of certain parts of beef cattle for human food on the basis that they may have been suffering from BSE. Much of the detail of food law is in the regulations made under *s 16*.

Section 17 allows the government to bring EU provisions into UK law. *Section 18* regulations will cover very particular circumstances such as genetically modified foods.

Section 19: registration and licensing

F9021 The *Food Safety Act 1990, s 19* permits the government to bring in regulations demanding licensing or registration of certain food businesses.

Registration and licensing confer significantly different powers on enforcement authorities. Registration is simply an administrative exercise. It is designed primarily to let enforcement agencies know what food businesses are operating in their area. It also provides some basic information on the size of the business and the type of food that they produce. No conditions are attached to registration, the local authority cannot refuse to register a business, there is no charge, and it does not have to be renewed periodically.

Licensing (or 'prior approval') is quite different. It is a control measure. Precise licensing criteria will be specified and there would normally be prior inspection and approval before a new business is allowed to open. There is often a charge for a licence, which must be renewed periodically. The enforcement authority will usually have the right to withdraw a licence if standards fall. The business would have to close.

Amongst enforcement officers there is generally a preference for licensing rather than registration. But at present, licensing only applies to a handful of different types of food business mostly meat or dairy operations. More recently, licensing was imposed on retail butchers' shops following the recommendation of the Pennington Group after the North Lanarkshire E coli outbreak. Shops selling both raw and cooked meat must be licensed. The licensing conditions include compliance with the General Food Hygiene Regulations supplemented by enhanced requirements for Hazard Analysis Critical Control Points ('HACCP') and staff training. At the outset the annual licence fee was £100.

Most other types of food businesses are subject only to registration. The *Food Premises (Registration) Regulations 1991* (*SI 1991 No 2828*) (as amended by *SI 1993 No 2022* and *SI 1997 No 723*) were enacted under *s 19* and described later.

Defences against charges under the Food Safety Act 1990

F9022 The following paragraphs explain the defences available under the *Food Safety Act 1990.*

Section 21: defence of due diligence

F9023 The so called 'due diligence' defence first appeared within food safety legislation when the *Food Safety Act 1990* was passed. Previously, the provision had been tried and tested within a number of other pieces of legislation after it first appeared just over 100 years before in the *Merchandise Marks Act 1887.* Indeed legislation that included a due diligence defence, such as the *Trades Descriptions Act 1968*, already overlapped with food legislation. So although the defence was new to food law there were numerous precedents to signal the implications of the due diligence provision.

The *Food Safety Act 1990* creates offences of strict liability. The prosecution is not required to show that the offence was committed intentionally. However the due diligence defence is intended to balance the protection of the consumer against the right of traders not to be convicted for an offence that they have taken all reasonable care to avoid committing. The intention of the due diligence defence is to encourage traders to take proper responsibility for their products and processes.

The *Food Safety Act 1990, s 21* states: 'in any proceedings for an offence under any of the preceding paragraphs of this Part … it shall be … a defence for the person charged to prove that he took all reasonable precautions and exercised all due diligence to avoid the commission of the offence by him or any person under his control.'

The onus is on the business to establish its defence on the balance of probabilities. Part of the due diligence defence may be to establish that someone else was at fault. If that is what is intended, the prosecutor must be informed at least seven days before the hearing, or if one has been in court already, within one month of their appearance. Under the 1990 Act, one can no longer rely upon a supplier's warranty alone, although a warranty may be part of due diligence defence.

It is impossible to give precise advice on what any business would need to do to have a due diligence defence. Cases that have been decided by the courts serve only to demonstrate that each case is decided on its own facts. What is certain is that doing nothing will never provide a due diligence defence. One must take some positive steps.

The defence has two distinct parts:

- Taking *reasonable precautions* involves setting up a system of control. One must consider the risks that threaten their operation and take *all reasonable precautions* that may be expected of a business of the size and type.
- Exercising *due diligence* means that one continues to take those precautions on a continuous basis.

All reasonable precautions and *all* due diligence are needed. The courts have decided consistently that if a precaution could reasonably have been taken but was not taken then the defence would not succeed.

The test is 'what is reasonable?' One will be expected to take greater precautions with high risk ready to eat foods than are needed for boiled sweets or biscuits. What is reasonable for a large-scale food business may not be reasonable for a smaller enterprise.

One must be careful that the scope of 'due diligence' covers all aspects of food law requirements. It is no good having excellent control of food safety hazards if there are no precautions in place to ensure that the composition and labelling of food complies with the regulations.

Documentation is important. Unless the precautions are written down, it will be difficult to persuade the EHO or the court that a proper system is in place. The same is true of records of any checks made to demonstrate that the system is followed diligently. The burden of proof lies with you to establish that the defence is satisfied.

A good 'HACCP' plan (see below) may be helpful in showing that a systematic approach to identifying the food safety precautions needed in the operation has been adopted. Cross-reference to industry guidelines or codes of practice may show that the system has a sound basis. Documentation from the hazard analysis will add to the evidence that precautions have been taken. Records such as specifications, cleaning schedules, training programmes and correspondence with suppliers or customers may all play a part in establishing the defence. An organisation chart and job descriptions that identify roles and responsibilities will also be useful evidence of the precautions taken.

Not all food businesses have the same defence. Some traders within the retail and catering end of the food supply chain may use a simpler defence of 'deemed due diligence' from the *Food Safety Act 1990, s 21 (3), (4)*. Even then 'deemed due diligence ' is not available against the *s 7* offence of rendering food injurious to health. Manufacturers and importers must satisfy the full defence. Larger businesses may be expected to take more precautions than smaller businesses. Those selling 'own label' packs may have greater responsibility for upstream activity than those selling manufacturers' 'branded' products.

Section 35: penalties under the Food Safety Act 1990

F9024 The courts decide penalties on the merits of individual cases but they are limited by maximum figures included in the *Food Safety Act 1990, s 35*. For most offences the crown court can send offenders to prison for up to 2 years and/or impose an unlimited fine. Even the magistrates' court, where most cases are heard, can set fines of up to £5,000 per offence and up to 6 months imprisonment. For the two food safety offences in *sections 7* and *8* and the first of the consumer protection offences (*s 14*), the fine may be up to £20,000. In Scotland the sheriff may impose equivalent penalties.

Each set of regulations made under the Act will have its own level of penalties that will not exceed the levels mentioned here.

Section 40: codes of practice

F9025 The *Food Safety Act 1990, s 40* allows for the development of codes of practice. Section 40 codes of practice are primarily intended to instruct and inform enforcement officers and related bodies or personnel, for example food analysts or examiners. In carrying out their duties, enforcement professionals must have regard for section 40 codes of practice. Any failure to follow a code requirement might seriously jeopardise their ability to bring a successful prosecution. By the middle of the year 2001 there were eighteen codes made under section 40. Some had already undergone several revisions. The subjects of early codes are general matters, such as the demarcation of responsibility for enforcing different parts of the Act or general inspection procedures. Later ones provide more specific information on the enforcement of particular regulations, for example the Dairy Hygiene Regulations.

The most significant of the general codes is CoP number 9 on food hygiene inspections. It contains detailed advice to environmental health officers on how to conduct such inspections. It includes a 'risk rating' system whereby inspectors can prioritise different food businesses and determine the necessary inspection frequency. Codes also give guidance on such issues as the powers of inspecting officers and the communication of findings to the business. The FSA has indicated its intention to radically revise the section 40 codes and probably to consolidate them in fewer separate codes.

Section 40 codes should not be confused with 'industry guides' that are promoted by the EU General Food Hygiene Directive and the General Food Hygiene Regulations. These 'guides' provide more detailed practical explanation of the implications of the particular regulations in a food business sector. There are also numerous other guides and codes of practice published by industry, government, and other organisations.

Food premises regulations

F9026 The following paragraphs deal with the main Regulation that covers food premises.

The Food Premises (Registration) Regulations 1991 (as amended by SI 1993 No 2022 and SI 1997 No 723)

F9027 All food premises including retail and catering premises must register with the local authority unless they are already required by other legislation to be licensed. The Regulations demand the registration of individual food *premises* and not food businesses. So premises that are used only occasionally by food businesses, and maybe even by different food businesses, must be registered by the owner of the premises. For example a church hall, village hall or scout hut used from time to time for the sale of food must be registered. The criterion is that they are used by commercial food businesses for five or more days (which do not have to be consecutive) in any period of five weeks.

Similarly, markets or other premises used by more than one food business must also be registered. The market operator or the person in charge of the premises is responsible. Staff restaurants must be registered. Contractors and clients should have a clear agreement as to who will action this. All premises should have been registered by May 1992. New premises must apply to be registered at least 28 days before opening. This is designed to allow an EHO the opportunity to look at a business before it opens. However, if the EHO does not take that opportunity, one does not have to wait for a visit or for permission from the EHO before opening provided that the necessary registration form has been sent off.

If mobile food premises are used they may have to be registered. These can range from a handcart in a market to a forty-foot trailer used for hospitality. If one uses their own moveable premises in a market, they must be registered even though the market is registered by the operator. (If stalls provided by the market are used, registration is not necessary). If mobile food premises are operated, the premises at which they are normally kept must be registered, with the food authority local to the base.

To register, one must supply a range of information on a registration form obtained from the local EHO department, which is normally contacted via the Town Hall. The local authority may prefer for their version of the form to be used instead. Registration must take place 28 days before the business opens. In case of doubt, it is best to keep a note of when the form was sent. If possible check that it has been received. Sending the form by fax may be helpful in establishing the transmission date.

Of course, some of the information may change over time. The Regulations only demand that the authorities are notified about certain fundamental points. If the proprietor or the nature of the business changes they should be notified of this, e.g. changing from selling fruit and vegetables only to become a deli. The authority must be notified of the change within 28 days. Many of the other pieces of information on the application form may also change, for example the name of the manager, the phone number, the number of employees and so on but these changes do not need to be notified. By failing to register, giving false or incomplete information or failing to notify any of the changes mentioned above, can trigger prosecution proceedings.

Food premises exempted from registration fall into a number of categories as follows.

Already licensed or registered under other measures

F9028

- Slaughterhouses.
- Poultry slaughterhouses and cutting premises.
- Meat export cutting premises and stores.
- Meat product plants.
- Butchers' shops selling both raw and ready to eat meat.
- Dairies or farm dairies.
- Milk distribution premises.

Premises used only occasionally

F9029

- Premises used for less than five days (not necessarily consecutive) in five consecutive weeks.

Low risk activities exempt unless used for retail sales

F9030

- Where game is killed in sport (grouse moors).
- Where fish is taken for food (but not processed).
- Where crops are harvested, cleaned, stored or packed. However, if the crops are put into the final consumer pack they must register.
- Where honey is harvested.
- Where eggs are produced or packed.
- Livestock farms and markets.
- Shellfish harvesting areas.
- Places where no food is kept, for example an administrative office or head quarters of a food business.

Some domestic premises

F9031 Some domestic premises that might technically be deemed to be food premises are also exempt in the following situations:

FORM OF APPLICATION FOR REGISTRATION OF FOOD PREMISES

1. Address of premises ....
(or address at which movable premises are kept) Post code ....

2. Name of food business.. Telephone no: ....
(trading name)

3. Type of premises Please tick ALL the boxes that apply

Farm/smallholding ☐
Food/manufacturing/processing ☐
Slaughterer ☐
Packer ☐
Importer ☐
wholesale/cash and carry ☐
Distribution/warehousing ☐
Retailer ☐
Market ☐
Restaurant/cafe/snack bar ☐
Staff restaurant/canteen/kitchen ☐
Catering ☐
Hospital/residential home/school ☐
Hotel/pub/guest house ☐
Private house used for a food business ☐
Premises used by a number of businesses ☐
Moveable premises ☐
Other: please give details ....
....

4. Does your business handle or involve any of the following? Please tick ALL the boxes that apply

Chilled foods ☐
Frozen foods ☐
Fruit and vegetables ☐
Fish/fish products ☐
Fresh/frozen meat ☐
Fresh/frozen poultry ☐
Meat products or delicatessen ☐
Dairy products ☐
Eggs ☐
Bakery ☐
Sandwiches ☐
Confectionery ☐
Ice cream ☐
Alcoholic drinks ☐
Canning ☐
Vacuum packing ☐
Bottling and other packing ☐
Table meals/snacks ☐
Takeaway food ☐
Accommodation ☐
Delivery service ☐
Chilled food storage ☐
Bulk storage ☐
Use of private water supply ☐
Other: please give details ....
....

5. Are vehicles or ships used for transporting food kept at or used from the premises? Yes/No

Are vehicles, stalls or ships used for preparing or selling food, kept at or used from the premises? Yes/No

Number of vehicles/stalls/ships kept at or used from the premises, and used for preparing, selling or transporting food.
5 or less ☐ 6–10 ☐
11-50 ☐ 51 plus ☐

6. Name(s) of proprietor(s) of food business ....
Address of business head office or registered office ....
if different from address of premises
.... Post code ....

7. Name of manager if different from proprietor ....

8. If this is a new business.......................
Date you intend to open

9. If this is a seasonal business ....
Period during which you intend to be open each year

10. Number of people engaged in food business
Count part-timer(s) (25 hrs per week or less) as one-half

0–10 11–50 51 plus (Please tick one box)
☐ ☐ ☐

It is an offence to give false or incomplete information

The completed form should be sent to:

[ ]

[ ]

Signature ....
Date ....
Name ....
(BLOCK CAPITALS)
Position in company/business.........................

- Childminders caring for no more than six children (*SI 1993 No 2022*).
- People preparing foods for sale in WI Country Markets Ltd. (*SI 1997 No 723*).
- Where the resident is not the owner of the food business (for example a mobile food vehicle is sometimes kept at the premises of the driver) If the premises are used for the purpose of for example peeling shrimps or prawns they must be registered.
- Premises used for the production and sale of honey.
- Where crops are produced, cleaned, packed and sold whether wholesale or retail.
- Premises which provide bed and breakfast accommodation in not more than three bedrooms.

Some vehicles

F9032

- Private cars.
- Aircraft.
- Ships unless permanently moored or used for pleasure excursions in inland or coastal waters.
- Food vehicles normally based outside the UK.
- Vehicles or stalls kept at premises that are themselves registered or exempt.
- Market stalls owned by the market controller.
- Tents, marquees and awnings.

Other exemptions

F9033

- Places where the main activity has nothing to do with food but where light refreshments such as biscuits and drinks are served to customers without charge (for example hairdressers salons).
- Where food is sold only through vending machines.
- Places run by voluntary or charitable organisations and used only by those organisations provided that no food is stored on the premises except tea, coffee, dry biscuits etc. For example some village or church halls, but not if they are used more than five days in five weeks by commercial caterers.
- Crown premises exempted for security reasons.
- Places supplying food and drink in religious ceremonies.
- Stores of food kept for an emergency or national disaster.

Food hygiene regulations

F9034 The following account will concentrate on the *Food Safety* (*General Food Hygiene*) *Regulations 1995* (*SI 1995 No 1763*) that stem from the 'horizontal' food hygiene Directive 93/43. It will also cover the *Food Safety* (*Temperature Control*) *Regulations 1995* (*SI 1995 No 2200*). Both of these Regulations apply to the retail and catering sale of all types of food and also to the production and preparation of non-animal products at earlier points of the food supply chain. More specific 'vertical' regulations apply to manufacture and processing of animal products including fish and dairy products. A detailed account of the vertical regulations is beyond the scope of this publication.

Food Safety (General Food Hygiene) Regulations 1995

F9035 The structure of the *Food Safety* (*General Food Hygiene*) *Regulations 1995* (*SI 1995 No 1763*) is closely aligned to the principles established by the Codex Alimentarius of the World Health Organisation. Food safety controls must be based on a HACCP system, supported by a range of hygiene pre-requisites. The only difference is that the hazard analysis requirement within the 1995 Regulations is not quite so full and formal as HACCP. Opinions differ as to exactly how far short it is of full HACCP, but the most obvious omission is that there is no overt requirement for documentation or records.

Hygiene has a wider meaning than just the safety of food. It is defined in the Regulations as 'all measures necessary to ensure the safety and wholesomeness of food during preparation, processing, manufacturing, packaging, storing, transportation, distribution, handling and offering for sale or supply to the consumer.' Wholesomeness means 'fitness for human consumption as far as hygiene is concerned.'

Early parts of the Regulations, specify some broad hygiene objectives. Latter parts specify some particular goals that must be achieved to help meet those objectives.

Regulation 4(*1*) demands that 'A proprietor of a food business shall ensure that any of the following operations, namely, the preparation, processing, manufacturing, packaging, storing, transportation, distribution, handling and offering for sale or supply, of food are carried out in a hygienic way'.

The so-called 'hazard analysis' provision is contained within *Reg 4*(*3*). This effectively demands a systematic approach to food safety controls appropriate to the business. The historical tendency towards very prescriptive hygiene regulations is seen to be inflexible with the very wide diversity of the food industry. HACCP is a formalised system that allows effective controls to be developed that suit the products and production methods of the individual business.

Hazard analysis

F9036 The *Food Safety* (*General Food Hygiene*) *Regulations 1995* (*SI 1995 No 1763*) specify five elements that must be included in the process of identifying steps in the activities of the business that are 'critical' to ensuring that food is safe:

- Analyse the food hazards that may be encountered.
- Identify the points in the process where these may occur.
- Decide which of these points may be 'critical' to the safety of the food.
- Identify effective controls that can be applied at critical points. In every case, identify a system of monitoring those controls, and implement both control and monitoring into the process.
- Review the system periodically and make changes if necessary.

HACCP is a powerful and sophisticated management tool that can be used by businesses to develop a food control plan. Full and formal HACCP requires quite detailed expert knowledge and training. A significant amount of literature has built up over the years. It serves well those larger businesses that employ staff with technical training. Smaller food businesses tend to find HACCP literature to be too full of jargon and inaccessible.

Clear guidance has been developed by central government for some specific food sectors. The most notable recent example was a guide for butchers' shops in England that was drafted in preparation for the introduction of licensing. Other guidance aimed at the catering industry was published much earlier at the start of the 1990s, notably *Assured Safe Catering*. Although serving a useful purpose at the time, this document is now seriously in need of updating. Without up to date central guidance, ideally from the FSA, interpretation and implementation of hazard analysis is likely to remain confused and inconsistent. The new EU regulations will enhance the hazard analysis provisions to include other elements of the HACCP system, notably verification, validation and documentation. In advance of this the FSA has been developing new guidance for the implementation of HACCP in food service businesses. But it appears that radically different directions are being followed in the various devolved regions of the UK.

Critical control points (CCP) for microbiological food poisoning hazards will tend to focus on a small number of areas.

COOKING: In most processes there is likely to be a risk of pathogens in raw ingredients, so cooking is frequently a CCP. Adequate cooking temperatures will be the control measure, and target temperatures should be established. These can be monitored either by checking the food temperature directly or monitoring the process. For example using a cooking time and temperature that you know will achieve a satisfactory temperature.

CONTAMINATION: Process steps at which ready to eat food may become contaminated are also likely to be CCPs. Contamination could come from contact with raw food ('cross contamination'), or from contact with contaminated equipment or personnel. Separation of raw and ready to eat foods, and cleaning and disinfection routines will be appropriate control measures. In a good hazard analysis system these procedures will be specified objectively to allow them to be monitored. Written cleaning schedules will help. Nowadays there are also rapid test systems that will assess effectiveness of cleaning.

TIME AND TEMPERATURE CONTROLS: Cooling of food after cooking is often a CCP, as are storage times and temperatures of ready to eat foods. In most cases, controls will be expressed as a combination of time and temperature. The critical limits may be just a few hours if the food is kept at warm room temperature, or many days if it is kept in good refrigeration. In this area technology is providing increasingly sophisticated monitoring options include automatic alarms and real-time data monitoring. The temperature control regulations also intervene by prescribing certain targets.

Documentation and Records

F9037 A proper HACCP system would be fully documented. Any monitoring or verification procedures would be recorded and records kept. There is no explicit requirement for either of these in the *Food Safety* (*General Food Hygiene*) *Regulations 1995* (*SI 1995 No 1763*) except in butchers' shops covered by the licensing amendment. Most businesses would be advised to keep some concise and succinct documentation. Without it, it may not be easy to demonstrate compliance with the legislation. It will certainly be difficulty to muster a convincing due diligence defence if that should become necessary.

Hygiene 'pre-requisites'

F9038 Codex Alimentarius, the definitive guide to HACCP, says that any HACCP system must be supported by basic hygiene controls. These have become known as the pre-requisites to HACCP. The *Food Safety* (*General Food Hygiene*) *Regulations 1995* (*SI 1995 No 1763*), *Sch 1* (*'The Rules of Hygiene'*) take a similar approach by establishing certain requirements for structures and services. They cover the following subjects within ten chapters:

(i) Chapter I: General requirements.

(ii) Chapter II: Rooms where food is prepared.

(iii) Chapter III: Movable or temporary premises, etc.

(iv) Chapter IV: Transport.

(v) Chapter V: Equipment.

(vi) Chapter VI: Food waste.

(vii) Chapter VII: Water Supply.

(viii) Chapter VIII: Personal hygiene.

(iv) Chapter IX: Protection of food from contamination.

(v) Chapter X: Training

Chapter I: General requirements for food premises, equipment and facilities

F9039 Food premises must be kept clean, and in good repair and condition. The layout, design, construction, and size must permit good hygiene practice and be easy to clean and/or disinfect and should protect food against external sources of contamination such as pests.

Adequate sanitary and handwashing facilities must be available and lavatories must not lead directly into food handling rooms. Washbasins must have hot and cold (or better still mixed) running water and materials for cleaning and drying hands. Where necessary there must be separate facilities for washing food and hands. Drainage must be suitable and there must be adequate changing facilities.

Premises must have suitable natural or mechanical ventilation. Ventilation systems must be accessible for cleaning, for example to give easy access to filters, and adequate natural and/or artificial lighting.

Chapter II: Specific requirements in food rooms

F9040 Food rooms should have surface finishes that are in good repair, easy to clean and, where necessary, disinfected. This would, for instance, apply to wall, floor and equipment finishes. The rooms should also have adequate facilities for the storage and removal of food waste. Of course, every food premises must be kept clean. How they are cleaned and how often, will vary. For example it will be different for a manufacturer of ready-to-eat meals than for a bakery selling bread.

There must be facilities, including hot and cold water, for cleaning and where necessary disinfecting tools and equipment. There must also be facilities for washing food wherever this is needed.

Chapter III: Temporary and occasional food businesses and vending machines

F9041 Most of the requirements apply equally to food businesses trading from temporary or occasional locations like marquees or stalls. For reasons of practicability, some requirements are slightly modified in this chapter.

Chapter IV: Transport of food

F9042 The design of containers and vehicles must allow cleaning and disinfection. Businesses must keep them clean and in good order to prevent contamination and place food in them so as to minimise risk of contamination. Precautions must be taken if containers or vehicles are used for different foods or for both food and non-food products. You should also separate different products to protect against the risk of contamination, and clean vehicles or containers effectively between loads.

Chapter V: Equipment

F9043 All equipment and surfaces that come into contact with food must be well constructed and kept clean. One interpretation of this requirement is that wooden cutting boards are not deemed suitable for use with ready to eat foods.

Chapter VI: Food waste

F9044 Food and other waste must not accumulate in food rooms any more than is necessary for the proper functioning of the food business. Containers for waste must be kept in good condition and easy to clean and disinfect. Waste storage containers should be lidded to keep out pests. The waste storage area must be designed so that it can be easily cleaned and prevent pests gaining access. There should be arrangements for the frequent removal of refuse and the area should be kept clean.

Chapter VII: Water supply

F9045 There must be an adequate supply of potable (drinking) water. Ice must be made from potable water.

Chapter VIII: Personal hygiene

F9046 Anyone who works in a food handling area must maintain a high degree of personal cleanliness. The way in which they work must also be clean and hygienic. Food handlers must wear clean and, where appropriate, protective clothes. Anyone whose work involves handling food should:

- follow good personal hygiene practices;
- wash their hands routinely when handling food;
- never smoke in food handling areas;
- be excluded from food handling if carrying an infection that may be transmitted through food if there is a risk of contamination of the food.

In this case, the obligation on the business proprietor to take action is supplemented by an obligation on every employee contained within the *Food Safety* (*General Food Hygiene*) *Regulations 1995* (*SI 1995 No 1763*), *Reg 5*. Personnel working in food handling areas are obliged to report any illness (like diarrhoea or vomiting, infected wounds, skin infections) immediately to the proprietor of the business. If a manager receives such notification he or she may have to exclude them from food handling areas. Such action should be taken urgently. If there is any doubt about the need to exclude, it is best to seek urgent medical advice or consult the local council EHO.

These personal hygiene provisions are wide ranging. They apply not only to food handlers, but anyone working in food handling areas. There have been well-documented cases of cleaners contaminating working surfaces or equipment with pathogens that are later transmitted to food.

Chapter IX: Preventing food contamination

F9047 Food and ingredients must be protected against contamination that may make them unfit for human consumption or a health hazard. For example, raw poultry must not be allowed to contaminate ready-to-eat foods.

Chapter X: Training and supervising food handlers

F9048 Whereas nine of the chapters deal with structural and physical pre-requisites, the final chapter deals with the higher-level issue of competence of personnel to do their job. This provision is more subtle than it is often given credit for. Prior to the publication of the Regulations, many training providers hoped that the legislation would simply prescribe certain levels of food hygiene training, ideally with a requirement for regular refresher courses. The provision is still subject to occasional criticism that it is too lenient on training requirements. In fact the provision recognises that there is more to it than that. It refers to three complementary elements, instruction, training and supervision.

Food handlers must receive proper supervision, instruction and/or training in food hygiene that is commensurate with their work. Of course they may need training in the principles of food hygiene. They must also be instructed on how to do their particular jobs properly and supervised to make sure that they follow instructions. The requirements will differ according to the nature of the business and the job of the individual staff member.

Industry guides to good hygiene practice

F9049 The *Food Safety* (*General Food Hygiene*) *Regulations 1995* (*SI 1995 No 1763*) introduce a new concept of voluntary industry guides to good hygiene practice. These provide more detailed guidance on complying with the Regulations as they relate to specific industry sectors. They are usually produced by trade associations and recognised by the government (formerly the Department of Health latterly the FSA). Importantly, enforcement officers are obliged to have regard for them when examining how businesses are operating. The publication of a guide for the vending industry in 2001 brought the total number of guides published in the UK to eight since the publication of the Catering Guide in 1995. Sectors covered are:

- Vending.
- Catering.
- Retail.
- Baking.
- Wholesale.
- Markets and Fairs.
- Fresh Produce.
- Flour Milling.

Food Safety (Temperature Control) Regulations 1995

F9050 Growth of pathogenic bacteria in food will significantly increase the risk of food poisoning. Indeed growth of any micro-organisms in food will compromise its wholesomeness. Temperature controls at certain food holding or processing steps will be CCPs in most HACCP plans. The importance of good control of food temperatures is emphasised by the fact that some controls are prescribed in regulations.

The *Food Safety* (*Temperature Control*) *Regulations 1995* (*SI 1995 No 2200*) require food business proprietors to observe certain temperatures during the holding of food if this is necessary to prevent a risk to health. The Regulations only cover safety issues. For example hard cheese that may go mouldy if kept at room temperature is not covered because it would not support the growth of pathogens. These Regulations impact on similar businesses to those covered by the General Food Hygiene Regulations. Again, the vertical regulations set more specific rules, including temperature controls, for most foods of animal origin during processing and handling prior to the retail or catering outlet.

The Temperature Control Regulations make a fairly simple issue extraordinarily complex. We do not even have the same regulations in Scotland as the rest of the UK. (The EU Directive 93/43 does not recommend specific temperature controls and for the time being, decisions are made nationally.)

The following table contains a very brief summary of the requirements. The Regulations prescribe temperatures for some but not all steps. The Regulations in Scotland also have similar gaps but not always at the same step. The industry guides, particularly the catering guide, fill in the gaps with recommendations of good practice.

	Rest of the UK	*Scotland*	*Recommended good practice targets*
Chill store	8°C maximum	Not specified	5°C
Cook	Not specified	Not specified	70°C for 2 minutes or 75°C minimum
Hot hold	63°C	63°C	63°C min
Cool	Not specified	Not specified	Below 10°C in 4 hours
Reheat	Not specified	82°C	70°C for 2 minutes or 75°C minimum

Frozen storage is not covered by the General Food Hygiene Regulations and will not be a food safety CCP. Other legislation and best practice indicate a frozen storage temperature of –18°C

Relationship between time and temperature

F9051 Almost invariably the control of micro-organisms in food depends upon a combination of time and temperature. Temperature alone does not have an absolute effect. For example, destruction of micro-organisms can be effected in a very short time (seconds) at temperatures above 100°C or in a couple of minutes at around 70°C. A similar thermal destruction can be achieved even at temperatures as low as 60°C but exposure for around 45 minutes will be necessary. Similarly in chilled storage, food can remain safe and wholesome for many days if the temperature can be kept close to freezing point at around –1°C. However, it will have a much shorter life at higher storage temperatures around 10°C.

Of course one should note that the above remarks are generalisations and not all micro-organisms will react in the same way. Some organisms show greater resistance to heating; some are able to grow at storage temperatures as low as 3°C whereas others will show little growth a temperatures cooler than 15°C. But the general point that control is a function of time and temperature remains true.

Broadly speaking, this relationship is recognised in the Regulations. In some instances, time is actually prescribed in the temperature control regulations. An example is the so called 'four hour rule' for display of food at a temperature warmer than 8°C. In other cases times are implicit in the temperature control regulations, such as the necessity to cool food rapidly after heating. Actual parameters for compliance are outlined in the industry guides. For other situations there is an inter-relationship with other regulations. For example the storage life of pre-packed, microbiologically perishable food must be controlled under requirements of the Food Labelling Regulations. It must be labelled with an indication of its 'minimum durability' that takes the form of a 'use by' date. There is an onus on the producer to ascertain a safe shelf life having regard for the nature of the food and its likely storage conditions and to use this as the basis of the date mark.

Temperature controls – England and Wales

F9052 The regulations governing temperature controls were largely new provisions introduced in the early 1990s and refined during the first half of the decade culminating in the *Food Safety* (*Temperature Control*) *Regulations 1995* (*SI 1995 No 2200*). The structure of the Regulations in England and Wales has the main requirement that chilled storage should be at 8°C or cooler, followed by a series of exemptions for particular circumstances. The general requirement under *Reg 10* of these Regulations comes from the EU directive and it is generally assumed that compliance with the other more specific regulations will deliver compliance with *Reg 10*. The requirements are outlined in the table below.

Provision within the Temperature Control Regulations	*Requirement*	*Comment*
Reg 10	A general requirement for all food to be kept under temperature control if that is needed to keep it safe.	Applies to raw materials and foods in preparation. Limited periods outside temperature control are permitted for certain practicalities. No temperature is specified. A combination of time and temperature will be important.

Reg 4	Chilled food must be kept at 8°C or cooler.	Applies only to foods that would become unsafe.
Reg 5	Various cold foods are exempt: shelf stable, canned foods, raw materials, cheeses during ripening, and others where there is no risk to health.	Soft cheeses once ripe, and perishable food from opened cans must be kept below 8°C.
Reg 6	Manufacturers may recommend higher storage temperature/shorter storage life (provided that safety is verified).	Caterers must use the food within the 'use by' date indicated.
Reg 7(1)	Cold food on display or for service can be warmer than 8°C. A maximum of 4 hours is allowed.	Any item of food can be displayed outside temperature control only once. The burden of proof is on the caterer.
Reg 7(2)	Tolerances outside of temperature control are also allowed during transfer or preparation of food, and defrost or breakdown of equipment.	No time/temperature limits specified. Both should be minimized consistently with food safety.
Regs 8, 9	Hot food should be kept at 63°C or hotter.	Food may be kept at a temperature cooler than 63°C for maximum 2 hours if it is for service or on display.
Reg 11	Food must be cooled quickly after heating or preparation.	No limits are specified – must be consistent with food safety.

Note that all temperatures specified are food temperatures not the air temperature of refrigerators, vehicles or hot cabinets.

Chilled storage and foods included in the scope of the regulations

F9053 The principal rule for chilled storage is contained within the *Food Safety (Temperature Control) Regulations 1995 (SI 1995 No 2200), Reg 4(1)*: 'Subject to paragraph (2) and regulation 5, no person shall keep any food — (a) which is likely to support the growth of pathogenic micro-organisms or the formation of toxins; and (b) with respect to which any commercial operation is being carried out, at or in food premises at a temperature above 8 degrees C.'

Regulation 4(1) only applies to food that will support the growth of pathogenic micro-organisms. Such foods must be kept at 8°C or cooler. The types of food that will be subject to temperature control are indicated in the following table.

It is often good practice to keep foods at temperatures cooler than 8°C either to preserve quality or to allow longer-term storage and to allow a margin of error below the legal standard. Industry guides suggest that one should aim for a target food temperature of 5°C. This is especially important in cabinet fridges. There can be significant temperature rises during frequent door opening.

Food type	*Comments*
Cooked meats and fish, meat and fish products.	Includes prepared meals, meat pies, pates, potted meats, quiches and similar dishes based on fish.
Cooked meats in cans that have been pasteurised rather than fully sterilised.	Typically large catering packs of ham or cured shoulder.
Cooked vegetable dishes.	Includes cereals, rice and pulses. Some cooked vegetables or dessert recipes may have sufficiently high sugar content* (possibly combined with other factors like acidity) to prevent the growth of pathogenic bacteria. These will not be subject to mandatory temperature control.
Any cooked dish containing egg or cheese.	Includes flans, pastries etc.
Prepared salads and dressings.	Includes mayonnaise and prepared salads with mayonnaise or any other style of dressing. Some salads or dressings may have a formulation (especially the level of acidity**) that is adequate to prevent growth of pathogens.
Soft cheeses/mould ripened cheeses (after ripening).	Cheeses will include Camembert, Brie, Stilton, Roquefort, Danish Blue and any similar style of cheese.
Smoked or cured fish, and raw scombroid fish.	For example smoked salmon, smoked trout, smoked mackerel etc. Also raw tuna, mackerel and other scombroid fish.
Any sandwiches whose fillings include any of the foods listed in this Table.	

Low acid** desserts and cream products.	Includes dairy desserts, fromage frais and cream cakes. Some artificial cream may be 'ambient stable' due to low water activity and/or high sugar. Any product that does not support the growth of pathogenic micro-organisms does not have to be kept below 8°C. It may be necessary to get clarification from suppliers.
Fresh pasta and uncooked or partly cooked pasta and dough products.	Includes unbaked pies and sausage rolls, unbaked pizzas and fresh pasta.
Smoked or cured meats which are not ambient stable.	Salami, parma hams and other fermented meats will not be subject to temperature controls if they are ambient shelf stable.

*Technically Aw (water activity) is the key criterion.

**Technically, pH 4.5 (or more acid) is the critical limit.

Mail order foods

F9054 There is a controversial exemption from the specific 8°C requirement if the food is being conveyed by post or by a private or common carrier to the ultimate consumer. The sender still has a responsibility for the safety of the food but the specific 8°C temperature does not apply. This exemption for mail order foods cannot apply to supplies to caterers or retailers neither of whom meet the definition of 'ultimate consumers'.

Short 'shelf life' products

F9055 Some perishable foods are allowed to be kept at ambient temperatures for the duration of their shelf life with no risk to health. This may be because they are intended to be kept for only very short periods (sandwiches). Bakery products like fresh pies, pasties, custard tarts can also be kept without refrigeration for limited periods.

Canned foods and similar

F9056 One does not have to keep sterilised cans or similar packs under temperature control until the hermetically sealed pack is opened. After that perishable food must be kept chilled (for example corned beef, beans, canned fish, and dairy products). Some canned meats are not fully sterilised (e.g. large catering packs of ham or pork shoulder). They must be kept chilled even before the can is opened.

High acid canned or preserved foods (some fruit, tomatoes, etc.) do not have to be kept chilled after opening for safety reasons. However it is advisable to remove these from the can for chilled storage after opening. It will inhibit mould growth and avoid any chemical reaction with the metal can body.

Raw food for further processing

F9057 Raw food for further processing does not have to be kept at 8°C or cooler. Thus it is not against the Regulations to keep raw meat, poultry and fish out of the fridge. Of course for quality reasons it is best kept in the fridge. Processed foods must be kept at 8°C or cooler even if they are to be heated again.

Raw meat or fish that is intended to be eaten without further processing (for example beef for steak tartare, or fish for sushi) will not be exempted by this provision. It must be kept chilled. Raw scombroid fish (tuna, mackerel, etc.) will also not be exempted by this provision. The 'scombrotoxin' is heat stable and processing will not render contaminated food fit for consumption. Scombroid fish must be kept at 8°C or cooler.

Variations from 8°C storage

F9058 The *Food Safety* (*Temperature Control*) *Regulations 1995* (*SI 1995 No 2200*), *Reg 4*(*1*), allow the producer of a food to recommend that it can be kept safely at a temperature above 8°C. He must label the food clearly with the recommended storage temperature and the safe shelf life at that temperature. He must also have good scientific evidence that the food is safe at that temperature. Little use has been made of this provision.

The 4-hour rule

F9059 The '4-hour rule' allows cold food to be kept above 8°C when it is on display. This is crucial. Without it you could not serve food on a buffet without refrigeration. The time that the food is on display must be controlled. The maximum time allowed is four hours. Most catering operations and many retailers depend upon this exemption for at least some of their operation.

Only one such period of display is allowed no matter how short. For example, if you put a dish of food on display for 1 hour at the end of a service period, you cannot have another 3 hours above 8°C at the next service. Food uneaten at the end of a display period does not have to be discarded provided that it is still fit for consumption. The food must be cooled to 8°C or cooler and kept at that temperature until it can be used safely.

The burden of proof is on the business. One must be able to demonstrate that the time limit is observed. Good management of food displays will be important. The amount of food on display must be kept to a minimum consistent with the pattern of trade.

There needs to be systems to help keep to the time limit and to demonstrate that they are being adhered to. These may include the labelling of dishes to indicate when they went onto display. Avoid topping-up of bulk displays of food. Food at the bottom of the dish will remain on display for much longer than 4 hours if topping-up is allowed.

Exemptions for other contingencies

F9060 The Regulations allow for the fact that food may rise above 8°C for limited periods of time in unavoidable circumstances such as:

- transfers to or from vehicles;
- during handling or preparation;
- during defrost of equipment;
- during temporary breakdown of equipment.

The Regulations do not put specific figures on the length of time allowed or how warm the food might become. This tolerance is allowed as a defence and the burden of proof will be on the business to show that:

- the food was unavoidably above 8°C for one of the reasons allowed;
- that it was above 8°C for only a limited period;
- that the break in temperature control was consistent with food safety.

The acceptable limits will obviously depend upon the combination of time and temperature. Under normal circumstances, a single period of up to two hours is unlikely to be questioned.

All transfers of food must be organised so that exposure to warm ambient temperatures is reduced and that rises in food temperature are kept to a minimum. For example, put deliveries away quickly and move chilled food first, then frozen, then grocery

If food from a cash and carry warehouse is being collected, insulated bags or boxes should be used for any chilled foods to which the Regulations apply. The chilled food should be taken straight back to the outlet and quickly put them into chilled storage.

Equipment breakdown should be avoided by ensuring planned and regular maintenance.

Hot food

F9061 Hot food must be kept at 63°C or hotter if this is necessary for safety. This will apply to the same types of food described in the earlier table (see F9053 above). For example the rule does not apply to hot bread or doughnuts.

Food must be kept at 63°C or hotter, whether:

- it is in the kitchen or bakery awaiting service or dispatch;
- or in transit to a serving point, no matter how near or far;
- or actually on display in the serving area.

Some tolerance or limited exemptions are allowed for practical handling reasons. As similar to the 4-hour rule for chilled food, hot food can be kept for service or display at less than 63°C for one period of up to 2 hours.

Again one must be able to show that:

- the food was for service or on display for sale;
- it had not been kept for more than 2 hours;
- it had only had one such period.

Generally there should be less problems with meeting the 63°C target in hot display equipment than encountered in achieving 8°C in 'chilled' display units. Hot display units are not designed to heat or cook foods from cold. They should be used for holding only not heating.

Cooling

F9062 Food that becomes warmer than 8°C during processing must be cooled again to 8°C 'as quickly as possible'. In another of the more controversial aspects of the Regulations, no time and temperatures are specified. Over the years a wide variety of cooling times have been recommended by different sources. Cooling times as short as 1.5 hours are specified in UK Department of Health guidelines on cooking chilled foods and published in the early 1980s. Many EHOs have regarded this figure as definitive. More recent literature from the USA recommends a cooling time of exactly ten times that long. The UK industry guides recommend that cooling between 60°C and 10°C should be accomplished in a maximum of 4 hours. Recent UK research suggests that this may be conservative.

Note that rapid cooling is not only necessary for food that has been heated. Food that has become warm during processing must also be returned to chilled storage below 8°C as quickly as possible after the process is completed.

Temperature controls – Scotland

F9063 The Scottish Regulations include at *Reg 16* of the *Food Safety* (*Temperature Control Regulations*) *1995* (*SI 1995 No 2200*) the same general requirement from the EU directive found at *Reg 10*. Otherwise the Scottish Regulations have remained largely unchanged for many years. Some would say that they are now anachronistic. No chilled storage temperature is specified, yet the 82°C requirement for re-heating pre-cooked food appears unnecessarily high, unsupported by the science and potentially detrimental to the quality of many foods.

Provision within the Temperature Control Regulations	*Requirement*	*Comment*
Reg 16	A general requirement similar to that which applies to England and Wales under *Reg 10* for all food to be kept under temperature control if that is needed to keep it safe. Additionally it includes the need for rapid cooling.	Applies to raw materials and foods in preparation. Limited periods outside temperature control are permitted for certain practicalities. No temperature is specified. A combination of time and temperature will be important.
Reg 13	Cold food must be kept in a cool place or refrigerator.	No temperature is specified.
Reg 13	Hot food must be kept at 63°C or hotter.	There are exemptions for food on display, during preparation, etc.
Reg 14	Food that is reheated must reach 82°C.	Exemption in place if 82°C is detrimental to food quality.
Reg 15	Gelatin must be boiled, or held at 71°C for at least 30 minutes.	Unused glaze must be chilled quickly and kept in a refrigerator.

Again, note that all temperatures specified are food temperatures not the air temperature of refrigerators, vehicles or hot cabinets.

Cool storage

F9064 The *Food Safety (Temperature Control Regulations) 1995 (SI 1995 No 2200), Reg 13* will apply to any food that may support the growth of food poisoning organisms within the 'shelf life' for which the food will be kept.

The types of food that will be subject to the provision are the same as those that fall under *Reg 4* of the Regulations (applicable to England and Wales).

If cold, this food must be kept either in a refrigerator or a cool well-ventilated place. The Scottish Regulations do not specify a temperature but in practice, it is advisable to follow the same rules and recommendations that apply in the rest of the UK. In any case, the hazard analysis requirement of the General Food Hygiene Regulations does apply in Scotland, and one is obliged to establish controls at CCPs.

Hot food

F9065 The target for hot food is the same as England and Wales, 63°C or hotter.

Cooling

F9066 The Scottish Regulations demand rapid cooling and the requirement is considered to be identical to *Reg 11* of the Regulations (applicable to England and Wales).

Exemptions during practical handling

F9067 In certain circumstances food does not have to be kept cold or above 63°C:

- if it is undergoing preparation for sale;
- if it is exposed for sale or it has already been sold;
- if it is being cooled;
- if it is available to consumers for sale;
- if it is shelf stable.

Again these Regulations have remained unchanged for many years and are less detailed than those that apply in the rest of the UK.

Reheating of food

F9068 In Scotland only, food that has been heated and is being reheated must be raised to a temperature of 82°C or hotter. Fortunately this is qualified by a clause to say that this is not necessary if it is detrimental to the quality of the food. This should provide the opportunity for most businesses to reheat food to the more sensible, perfectly adequate and scientifically verified temperature of around 70°C.

Gelatine

F9069 Scotland also has specific rules about the use of gelatine:

1. Immediately before it must be boiled it or held at at 71°C or hotter for at least 30 minutes
2. Any left over gelatine must be discarded or cooled quickly and stored in a refrigerator or cool larder.
3. Leftover gelatine must be pasteurised again before re-use, using either of the treatments described in point 1 above.

Food Safety Manual – food safety policy and procedures

F9070 Both the *Food Safety Act 1990* and the Regulations made under it provide an obligation on proprietors of food businesses to operate within the law. In addition, the *Food Safety Act 1990, s 21* provides the defence of 'due diligence'. One way for a business to demonstrate its commitment to operating safely and within the law is to document its management system, including the relevant policies and procedures. This documentation may form a significant part of a due diligence defence. The policy and procedures will illustrate the precautions that should be taken by the business and its staff in their day-to-day operations. The procedures will normally require the keeping of various records, and these will in turn help to demonstrate that the '*reasonable precautions*' are being followed '*with all due diligence*'.

It is often convenient to keep these policies and procedures together in a 'Food Safety Manual'. The following provides a brief outline of the contents of a Food Safety Manual for a typical food business.

Policy and procedures

F9071 The 'policy' will be a statement of intent by the business. A Food Safety Manual may begin with a fairly broad statement of policy. Later there may be more specific policy statements linked to particular outcomes, for example a policy to take all reasonable precautions to minimise the contamination of food with foreign material.

Procedures will detail the ways in which the business intends to put the policies into effect. A typical layout of any 'Quality Manual' is to begin each section or subject heading with a statement of policy, and to follow that with the procedures intended to deliver it. Procedures are often referred to by the acronym SOPs – 'Standard Operating Procedures'.

General policy and organisation

F9072 A Food Safety Manual would normally begin with a top-level statement of the business's policy with regard to food safety. Recognising the wider scope of 'food safety' within the meaning of the 1990 Act, it may be advisable to add a statement about commitment to true and accurate labelling of food and adherence to compositional requirements.

The statement should be signed by the proprietor of the business, who is the person with ultimate legal responsibility. In a large group this may be the group chief executive.

The early part of the manual should also illustrate the organisational structures and management hierarchy that has responsibility for implementing the policy and procedures. There may be technical support services, (e.g. pest control, testing laboratories) either from inside or outside the business. Their roles and responsibilities can also be identified within the organisational structures in this part of the manual.

Hazard Analysis (HACCP)

F9073 Today, any food safety policy must be centred on a hazard analysis approach. The Food Safety Manual should include a policy statement to that effect. This will be followed by procedures to describe how you will go about the hazard analysis and record its outcomes and implementation.

Pre-requisites to Hazard Analysis

F9074 A modern approach to HACCP will focus on a very small number of process steps that are truly critical to the safety of the food. These are the steps that *must* be controlled (and monitored) to ensure that the food is safe. But this control must operate against a background of good hygiene practice. The so-called 'pre-requisites' to HACCP must all be in place. The Food Safety Manual should include policies and procedures for all of these.

- *Design – structures and layout*

  The Food Safety Manual should include a policy that the design, structure and layout will promote hygienic operation. The procedures should detail specific arrangements to achieve that. This section should also cover points such as the services, especially water and ventilation; provision for washing of hands, equipment and food; and removal of waste.

- *Equipment specification*

  The policy should be that the business will only acquire and use equipment that is fit for its purpose. The focus is often on cleanability alone. However in many cases, the ability of equipment to do the job can be even more important. For example, can refrigeration equipment keep food at a specified temperature in operational conditions?

- *Maintenance*

  A policy for maintenance should address both the premises in general and the equipment. Procedures should detail the arrangements for routine and emergency maintenance. There should be a record of any maintenance whether routine or emergency.

- *Cleaning and disinfection*

  Detailed cleaning schedules for sections of the building or particular pieces of equipment should be available. Cleaning should be recorded.

- *Pest Control*

  Employing a pest control contractor is only one small part of the pest control procedures. The prevention of pest access, the integrity of pest proofing, cleaning and the need for vigilance should also be included in the procedures. Records would normally include any sightings of or evidence of pests, positions of bait stations, visits by the contractor together with observations and action taken, and even details such as the changing of UV tubes on Electronic Flying Insect Killers.

- *Personal hygiene*

  The business should have a very clear policy on personal hygiene. The arrangements will include a very clear statement of the dress code and personal hygiene requirements for all staff working in or passing through food areas. There must also be very particular policies and procedures to deal with staff suffering illness that may potentially be foodborne.

- *Food purchasing*

  It may be easy to state a policy that the business should purchase only from reputable suppliers. It may be more difficult to deliver such a policy, especially for smaller businesses that do not have the resources to inspect their suppliers. One may be able to rely on third party inspection systems that have developed during the late 1990s; especially those operated by accredited inspection or certification bodies.

Supply specifications may also be drafted. Such specifications will usually cover food 'quality' in its broadest sense. They are the ideal route through which to state your food safety conditions such as delivery temperature, shelf life remaining after delivery, packaging and so on.

- *Storage*

  Most businesses will have documented procedures for the storage of various categories of food as well as for stock checks to ensure proper storage temperature and stock rotation.

- *Food handling and preparation*

  The general policy should be to manage all steps in food handling and preparation so as to minimise contamination with any hazardous material or pathogenic micro-organisms. And there should be precautions to minimise the opportunities for growth of any bugs that might get into the food. The procedures will be unique to the business, its particular range of foods and preparation methods.

- *Foreign material*

  You should have a policy to take all precautions to minimise the risk of foreign material contamination including chemical hazards such as cleaning materials. In general terms, procedures will cover three points:

  (*a*) Excluding or controlling potential sources of contamination.

  (*b*) Protecting food from contamination.

  (*c*) Taking any complaints seriously, tracking down what caused them and stopping them from happening again.

Allergy (Anaphylaxis)

F9075 A responsible business will have a policy to protect susceptible customers from exposure to foods to which they may be allergic. In general terms procedures will rely upon:

(*a*) Awareness amongst staff of the types of food that may cause reaction.

(*b*) Labelling of the foods that include such ingredients.

(*c*) Careful use of these ingredients in the kitchen to avoid 'cross-contamination' to other dishes.

(*d*) An ability to supply reliable information about all ingredients on request from guests.

Vegetarians and other dietary preferences

F9076 Similarly, most food businesses today recognise the need to cater for guests with particular dietary preferences, notably vegetarians. You should have a policy to be able to supply vegetarian dishes properly formulated and accurately labelled.

Consumer concerns

F9077 From time to time particular issues will come to the forefront of consumer consciousness. For example, the irradiation of food, genetically modified ingredients or animal welfare issues such as veal production, foie gras and intensive rearing of poultry. The Food Safety Manual could include a statement of the policy on whichever of these issues that one thinks might be relevant. One would also need procedures to deliver them.

Labelling and composition of foods and 'fair-trading'

F9078 In view of the fact that the *Food Safety Act 1990* deals not only with health and hygiene one may also wish to include policies and procedures on fair-trade topics such as food composition and labelling.

Product recall

F9079 Businesses involved in central production and distribution of food that will have several days 'shelf life' if not more should have contingency plans for product recall. Do you label the product in such a way that you can identify batches? Do you have records of where it might have been delivered?

Complaints

F9080 Complaints are an important source of information. A business should have a policy and procedures to deal with them. These will normally include keeping a record of every complaint and making every attempt to track down its cause.

Training, instruction and supervision

F9081 Cutting across all of the other policies and procedures will be a policy on staff training, instruction and supervision. Typically a business will keep a personal training record for all members of staff.

Summary

F9082 Modern businesses recognise the value of a properly documented management system. A Food Safety Manual can provide the focus of such a system. The design of a Food Safety Manual must be a fine balance between conflicting objectives. On the one hand, it must be reasonably comprehensive if it is to stand up as part of a 'due diligence' defence. On the other hand, it must be sufficiently concise to be usable as an effective working tool within the business.

Gas Safety

Introduction

G1001 Not surprisingly, explosions caused by escaping gas have been responsible for large-scale damage to property and considerable personal injury, including death. Although escape of metered gas or gas in bulk holders might possibly have attracted civil liability under the rule in *Rylands v Fletcher*, statutory authority to perform a public utility (i.e. supply gas) used to constitute a defence in such proceedings, coupled with the fact that negligence had to be proved in order to establish liability for personal injury (*Read v J Lyons & Co Ltd [1947] AC 156*). This situation has now changed in favour of imposition of strict product liability for injury/damage caused by escape of gas and incomplete or inefficient combustion causing carbon monoxide poisoning.

According to research undertaken by the Consumer's Association, over half of the appliances surveyed in dwellings had not been serviced in accordance with the requirements of the Gas Safety Regulations. As some well-publicised court cases have demonstrated, landlords and those they employ to deal with gas installations face heavy fines and possibly imprisonment for breach of the regulations (see G1026 below).

Changes have also been made to the gas industry itself, stemming from the Government's policy of creating competition in the gas supply field. These changes have resulted in the introduction of new and revised legislation, namely:

- *Gas Act 1986*

  This has been considerably modified and extended by the *Gas Act 1995*.

- *Gas Safety (Installation and Use) Regulations 1998 (SI 1998 No 2451)*

  These Regulations were made under the authority of the *Health and Safety at Work etc Act 1974*.

- *Gas Safety (Management) Regulations 1996 (SI 1996 No 551)*

  These Regulations were made under the authority of the *Health and Safety at Work etc Act 1974*.

- *Pipelines Safety Regulations 1996 (SI 1996 No 825)*

  These Regulations were made under the authority of the *Health and Safety at Work etc Act 1974*.

- *Gas Safety (Rights of Entry) Regulations 1996 (SI 1996 No 2535)*

  These Regulations were made under the authority of the *Gas Act 1986* (as amended).

In practice, many of the legislative changes were introduced for the purposes of increasing responsibilities for safety of gas processors, gas transporters and gas suppliers and are therefore outside the scope of this handbook. This chapter concentrates on the legislation which deals with the knowledge required by landlords, managing agents, health and safety managers, employers and other responsible persons who must ensure that gas appliances and fittings are installed safely and checked by a competent person every twelve months.

It should be noted that although the legislation discussed in this chapter is current at the date of publication, the Health and Safety Executive (HSE) is at present considering changes that could affect the *Gas Act 1986* (as amended) and some of the regulations referred to in this section. The proposed changes are outlined in a HSE document entitled *'Fundamental Review of Gas Safety Regime Proposals for Change'*, which is the result of consultation with interested parties throughout the gas industry.

Gas is defined in the *Gas Act 1986* (as amended) and the *Gas Safety* (*Installation and Use*) *Regulations 1998* (*SI 1998 No 2451*) to include methane, ethane, propane, butane, hydrogen and carbon monoxide mixtures of any two or more of these gases together with inert gases or other non-flammable gases; or combustible mixtures of one or more of these gases and air. However, the definition does not include gas consisting wholly or mainly of hydrogen when used in non-domestic premises.

Gas supply – the Gas Act 1986, as amended

G1002 The *Gas Act 1986* (as amended) requires the Secretary of State for the Environment to establish a Gas Consumers' Council and appoint an officer – the Director General of Gas Supply – to perform the functions relating to the supply of gas as set out in *Part I* of the Act.

As far as safety in the supply of gas is concerned, the Secretary of State and the Director each have a principal duty to exercise the functions assigned to them under Part I for the protection of the public from dangers arising from the conveyance or from the use of gas conveyed through pipes.

The following persons are defined in the *Gas Act 1986* as having duties and responsibilities in respect of the safe supply of gas:

- *Domestic customer*

  A domestic customer is a person who is supplied by a gas supplier with gas conveyed to particular premises at a rate which is reasonably expected not to exceed 2,500 therms a year.

- *Owner*

  In relation to any premises or other property, 'owner' includes a lessee, and cognate expressions are construed accordingly.

- *Relevant authority*

— in relation to dangers arising from the conveyance of gas by a public gas transporter, or from the use of gas conveyed by such a transporter, 'relevant authority' means that transporter; and

— in relation to dangers arising from the conveyance of gas by a person other than a public gas transporter, or from the use of gas conveyed by such a person, 'relevant authority' means the Secretary of State for the Environment.

With regard to safety, the *Gas Act 1986* (as amended) provides for:

- licensing and general duties [*Gas Act 1986, ss 4A, 7–10* and *23*];

- security [*Gas Act 1986, s 11*];
- safety regulations regarding rights of entry for inspection of connected systems and equipment and the making safe and investigation of gas escapes [*Gas Act 1986, ss 18* and *18A*];
- pipeline capacity [*Gas Act 1986, ss 21, 22, 22A*];
- standards of performance [*Gas Act 1986, s 33* as amended by the *Competition and Service (Utilities) Act 1992*].

Gas supply management – the Gas Safety (Management) Regulations 1996 (SI 1996 No 551)

G1003 Principally, the *Gas Safety (Management) Regulations 1996* (*SI 1996 No 551*) require an appointed person – the 'network emergency co-ordinator' – to prepare a safety case for submission to and acceptance by HSE before a gas supplier can convey gas in a specified gas network [*SI 1996 No 551, Reg 3*].

The safety case

G1004 The safety case document must contain:

- the name and address of the person preparing the safety case (the duty holder);
- a description of the operation intended to be undertaken by the duty holder;
- a general description of the plant, premises and interconnecting pipes;
- technical specifications;
- operation and maintenance procedures;
- a statement of the significant findings of the risk assessment carried out pursuant to the *Management of Health and Safety at Work Regulations 1999* (*SI 1999 No 3242*), *Reg 3*;
- particulars to demonstrate the adequacy of the duty holder's management system to ensure the health and safety of his employees and of others (in respect of matters within his control);
- particulars to demonstrate adequacy in the dissemination of safety information;
- particulars to demonstrate the adequacy of audit and reporting arrangements;
- particulars to demonstrate the adequacy of arrangements for compliance with the duty of co-operation [*Gas Safety (Management) Regulations 1996* (*SI 1996 No 551*), *Reg 6*];
- particulars to demonstrate the adequacy of arrangements for dealing with gas escapes and investigation [*Gas Safety (Management) Regulations 1996* (*SI 1996 No 551*), *Reg 7*];
- particulars to demonstrate compliance with the content and characteristics of the gas to be conveyed in the network [*Gas Safety (Management) Regulations 1996* (*SI 1996 No 551*), *Reg 8*];

- particulars to demonstrate the adequacy of the arrangements to minimise the risk of supply emergency;
- particulars of the emergency procedures [*Gas Safety* (*Management*) *Regulations 1996* (*SI 1996 No 551*), *Reg 3*(*1*) and *Sch 1*].

Duties of compliance and co-operation

G1005 The duty holder must ensure that the procedures and arrangements described in the safety case and any revision of it are followed. In addition, a duty of co-operation is placed upon specified persons including:

- a person conveying gas in the network;
- an emergency service provider;
- the network emergency co-ordinator in relation to a person conveying gas;
- a person conveying gas in pipes which are not part of a network;
- the holder of a licence issued under the *Gas Act 1986, s 7*;
- the person in control of a gas production or processing facility [*Gas Safety* (*Management*) *Regulations 1996* (*SI 1996 No 551*), *Regs 5 and 6*].

The *Gas Safety* (*Management*) *Regulations 1996* (*SI 1996 No 551*), *Reg 7* deals with the duties and responsibilities of gas suppliers, gas transporters and 'responsible persons' regarding actions to be taken in the event of gas escape incidents. All incidents must be investigated including those resulting in an accumulation of carbon monoxide gas from incomplete combustion of a gas fitting.

Anyone discovering or suspecting a gas leak must notify British Gas plc immediately by telephone. British Gas are obliged to provide a continuously manned telephone service in Great Britain. The person appointed 'responsible person' for the premises must take all reasonable steps to shut off the gas supply.

The reporting of gas incidents to the HSE is laid down in the *Reporting of Injuries, Diseases and Dangerous Occurrences Regulations 1995* (*SI 1995 No 3163*) (*RIDDOR*) in conjunction with HSE Form F2508G.

For further information regarding compliance with these Regulations see the HSE publication L80: '*A guide to the Gas Safety* (*Management*) *Regulations 1996*' (ISBN 0 7176 1159 0).

Rights of entry – the Gas Safety (Rights of Entry) Regulations 1996 (SI 1996 No 2535)

G1006 Where an officer authorised by a public gas transporter has reasonable cause to suspect an escape of gas into or from premises supplied with gas, he is empowered to enter the premises and to take any steps necessary to avert danger to life or property. [*Gas Safety* (*Rights of Entry*) *Regulations* (*SI 1996 No 2535*), *Reg 4*].

Inspection, testing and disconnection

G1007 On production of an authenticated document, an authorised person must be allowed to enter premises in which there is a service pipe connected to a gas main for the purpose of inspecting any gas fitting, flue or means of ventilation. Fittings and any part of a gas system may be disconnected and sealed off when it is necessary to do so for the purpose of averting danger to life or property.

It is incumbent on the authorised officer carrying out any disconnection or sealing off activities to give a written statement to the consumer within five days. The notice must contain:

- the nature of the defect;
- the nature of the danger in question;
- the grounds and the manner regarding the appeal procedure available to the consumer.

Prominent and conspicuous notices must also be affixed at appropriate points of the gas system regarding the consequences of any unauthorised reconnection to the gas supply. [*Gas Safety* (*Rights of Entry*) *Regulations* (*SI 1996 No 2535*), *Regs 5–8*].

Prohibition of reconnection

G1008 It is an offence for any person, except with the consent of the relevant authority, to reconnect any fitting or part of a gas system. This provision is qualified by the term 'knows or has reason to believe that it has been so disconnected'. [*Gas Safety* (*Rights of Entry*) *Regulations* (*SI 1996 No 2535, Reg 9*].

Pipelines – the Pipelines Safety Regulations 1996 (SI 1996 No 825)

G1009 The *Pipelines Safety Regulations 1996* (*SI 1996 No 825* as amended by *SI 2003 No 2563*) apply to all pipelines in Great Britain, both on and offshore, with the following exceptions:

- pipelines wholly within premises;
- pipelines contained wholly within caravan sites;
- pipelines used as part of a railway infrastructure;
- pipelines which convey water.

For the purposes of the Regulations, a pipeline for supplying gas to premises is deemed not to include anything downstream of an emergency control valve, that is, a valve for shutting off the supply of gas in an emergency, being a valve intended for use by a consumer of gas. [*SI 1996 No 825, Reg 3*(*4*)].

The Regulations complement the *Gas Safety* (*Management*) *Regulations 1996* (*SI 1996 No 551*) and include:

- the definition of a pipeline [*SI 1996 No 825, Reg 3*];
- the general duties for all pipelines [*SI 1996 No 825, Regs 5–14*];
- the need for co-operation among pipeline operators [*SI 1996 No 825, Reg 17*];
- arrangements to prevent damage to pipelines [*SI 1996 No 825, Reg 16*];
- the description of a dangerous fluid [*SI 1996 No 825, Reg 18*];
- notification requirements [*SI 1996 No 825, Regs 20–22*];
- the major accident prevention document [*SI 1996 No 825, Reg 23*];
- the arrangements for emergency plans and procedures [*SI 1996 No 825, Regs 24–26*].

Generally, the Regulations place emphasis on 'major accidents' and the preparation of emergency plans by local authorities. As with the *Gas Safety (Management) Regulations 1996 (SI 1996 No 551)*, the detail relates to the specification and characteristics of gas and the design of the pipes to convey it. The design of gas service pipelines is specifically addressed in the HSE Approved Code of Practice, L81: '*Design, construction and installation of gas service pipes*' (1996) (ISBN 0 7176 1172 8). For further information on and guidance to the *Pipelines Safety Regulations 1996*, see the HSE Publication L82: '*A guide to the Pipelines Safety Regulations 1996*' (ISBN 0 7176 1182 5).

Gas systems and appliances – the Gas Safety (Installation and Use) Regulations 1998 (SI 1998 No 2451)

G1010 The *Gas Safety (Installation and Use) Regulations 1998 (SI 1998 No 2451)* are supported by the HSE Approved Code of Practice and Guide, L56: '*The Gas Safety (Installation and Use) Regulations 1998*'.

The Regulations aim at protecting the gas-consuming public and cover natural gas, liquefied petroleum gas (LPG), landfill gas, coke, oven gas, and methane from coal mines when these products are 'used' by means of a gas appliance. Such appliances must be designed for use by a gas-consumer for heating, lighting and cooking. However, a gas appliance does not include a portable or mobile appliance supplied with gas from a cylinder except when such an appliance is under the control of an employer or self-employed person at a place of work. Similarly, the Regulations do not cover gas appliances in domestic premises where a tenant is entitled to remove such appliances from the premises.

In addition to the general provisions governing the safe installation of gas appliances and associated equipment by HSE approved and competent persons, the 1998 Regulations lay down specific duties for landlords.

Except in the case of 'escape of gas' (*SI 1998 No 2451, Reg 37*), and certain types of valves to control pressure fluctuations, the Regulations do not apply to the supply of gas when used in connection with:

- bunsen burners in an educational establishment [*SI 1998 No 2451, Reg 2*];
- mines or quarries [see *Mines and Quarries Act 1954*];
- factories [see *Factories Act 1961 (FA61), s 175*] or electrical stations [*FA61, s 123*], institutions [*FA61, s 124*], docks [*FA61, s 125*] or ships [*FA61, s 126*];
- agricultural premises;
- temporary installations used in connection with any construction work within the meaning of the *Construction (Design and Management) Regulations 1994 (SI 1994 No 3140)*;
- premises used for the testing of gas fittings; or
- premises used for the treatment of sewage.

Note: The Regulations would apply in relation to the above premises (or parts of those premises) if used for domestic or residential purposes or as sleeping accommodation, but not to hired touring caravans. [*SI 1998 No 2451, Reg 2*].

Generally, gas must be used in order to come within the scope of the Regulations. Therefore, the venting of waste gas from coal mines and landfill sites is excepted unless the gas is collected and intended for use. Gas used as motive power or from grain drying is also not covered by the Regulations. Service pipes (that is, the pipes for distributing gas to premises from a distribution main and the outlet of the first emergency control downstream of the main) remain the property of the gas supplier or gas transporter. At least 24 hours' notice should be given to the relevant organisation before any work can be performed on a service pipe. [*L56, paras 2–11*].

'Work' in relation to a gas fitting is defined in the Regulations as including any of the following activities carried out by any person, whether an employee or not, that is to say:

- installing or reconnecting the fitting;
- maintaining, servicing, permanently adjusting, disconnecting, repairing, altering or renewing the fitting or purging it of air or gas;
- where the fitting is not readily movable, changing its position; and
- removing the fitting.

The Approved Code of Practice, L56: '*The Gas Safety* (*Installation and Use*) *Regulations 1998*', provides further guidance on the meaning of work and states that 'work' for the purposes of these Regulations also includes do-it-yourself activities, work undertaken for friends and work for which there is no expectation of reward or gain, such as charitable work. [*L56, para 12*].

Duties and responsibilities

G1011 Duties are placed on a wide range of persons associated with domestic premises and on persons concerned with the supply of gas to those premises and to industrial premises which have accommodation facilities.

Most of the legal requirements are 'absolute', i.e. they are not qualified by 'as far as is practicable' or 'so far as is reasonably practicable'; in all cases the duty must be satisfied to avoid committing an offence. Persons affected by the Regulations include:

- *individuals*. These include householders and other members of the general public;
- *responsible person*. This is the occupier of the premises or, where there is no occupier or the occupier is away, the owner of the premises or any person with authority for the time being to take appropriate action in relation to any gas fittings;
- *landlord*. In England and Wales the landlord is defined as follows:

  (i) where the relevant premises are occupied under a lease, the person for the time being entitled to the reversion expectant on that lease or who, apart from any statutory tenancy, would be entitled to possession of the premises; and

  (ii) where the relevant premises are occupied under a licence, the licensor, save where the licensor is himself a tenant in respect of those premises.

  In Scotland, the landlord is the person for the time being entitled to the landlord's interest under a lease;
- *tenant*. In England and Wales the tenant is defined as follows:

  (i) where the relevant premises are so occupied under a lease, the person for the time being entitled to the term of that lease; and

(ii) where the relevant premises are so occupied under a licence, the licensee.

In Scotland, the tenant is the person for the time being entitled to the tenant's interest under a lease.

Duties are imposed on employers and the self-employed to ensure that all persons carrying out work in relation to gas fittings are competent to do so and are members of HSE approved organisations. [*Gas Safety (Installation and Use) Regulations 1998 (SI 1998 No 2451), Reg 3*].

Persons connected with work activities include:

- *supplier.* In relation to gas, a 'supplier' means:
  - (i) a person who supplies gas to any premises through a primary meter; or
  - (ii) a person who provides a supply of gas to a consumer by means of the filling or refilling of a storage container designed to be filled with gas at the place where it is connected for use, whether or not such container is or remains the property of the supplier; or
  - (iii) a person who provides gas in refillable cylinders for use by a consumer whether or not such cylinders are filled, or refilled, directly by that person and whether or not such cylinders are or remain the property of that person.

    Note: A retailer is not a supplier when he sells a brand of gas other than his own.
- *transporter.* This is defined as meaning a person, other than a supplier, who conveys gas through a distribution main.
- *gas installer.* This is any person who installs, services, maintains, removes or repairs gas fittings whether he is an employee, self-employed or working on his own behalf (for example, a do-it-yourself activity).

Escape of gas

G1012 It is the duty of the responsible person for the premises to take immediate action to shut off the gas supply to the affected appliance or fitting and notify the gas supplier (or the nominated gas emergency call-out office if different from the supplier). All reasonable steps must be taken to prevent further escapes of gas, and the gas supplier must stop the leak within twelve hours from the time of notification. This may entail cutting off the gas supply to the premises. [*Gas Safety (Installation and Use) Regulations 1998 (SI 1998 No 2451), Reg 37(1)–(3)*].

An escape of gas also includes an emission of carbon monoxide resulting from incomplete combustion in a gas fitting. However, the legal duties regarding the action to be taken by the gas supplier are limited to making safe and advising of the need for immediate action by a competent person to examine, and if necessary carry out repairs, to the faulty fitting or appliance. [*Gas Safety (Installation and Use) Regulations 1998 (SI 1998 No 2451), Reg 37(8)*].

Competent persons and quality control

G1013 No work is permitted to be carried out on a gas fitting or a gas storage vessel except by a competent person. [*Gas Safety (Installation and Use) Regulations 1998 (SI 1998 No 2451), Reg 3(1)*].

Where work to a gas fitting is to any extent under their control – or is to be carried out at any place of work under their control – employers (and self-employed persons) must ensure that the person undertaking such work is registered with an HSE-approved body such as the Council of Registered Gas Installers (CORGI). [*Gas Safety* (*Installation and Use*) *Regulations 1998* (*SI 1998 No 2451*), *Reg 4*].

A gas fitting must not be installed unless every part of it is of good construction and sound material, of adequate strength and size to secure safety and of a type appropriate for the gas with which it is to be used. [*Gas Safety* (*Installation and Use*) *Regulations 1998* (*SI 1998 No 2451*), *Reg 5*(*1*)].

Competent persons – qualifications and supervision

G1014 To achieve competence in safe installation, a person's training must include knowledge of purging, commissioning, testing, servicing, maintenance, repair, disconnection, modification and dismantling of gas systems, fittings and appliances. In addition, a sound knowledge of combustion and its technology is essential, including:

- properties of fuel gases,
- combustion,
- flame characteristics,
- control and measurement of fuel gases,
- gas pressure and flow,
- construction and operation of burners, and
- operation of flues and ventilation.

To reach the approval standard required by *Gas Safety* (*Installation and Use*) *Regulations 1998* (*SI 1998 No 2451*), *Reg 3*, gas installers and gas fitters should know:

- where/how gas pipes/fittings (including valves, meters, governors and gas appliances) should be safely installed;
- how to site/install a gas system safely, with reference to safe ventilation and flues;
- associated electrical work (e.g. appropriate electrical power supply circuits, that is, overcurrent and shock protection from electrical circuits, earthing and bonding);
- electrical controls appropriate to the system being installed/maintained/repaired;
- when/how to check the whole system adequately before it is commissioned;
- how to commission the system, leaving it safe for use.

In addition, they should know how to recognise and test for conditions that might cause danger and what remedial action to take, as well as being able to show consumers how to use any equipment they have installed or modified, including how to shut off the gas supply in an emergency. They should also alert customers to the significance of inadequate ventilation and gas leaks and the need for regular maintenance/servicing.

To meet the criteria for HSE approval, individual gas fitting operatives must be assessed (or reassessed) by a certification body accredited by the United Kingdom

Accreditation Service. The scheme requires every registered gas fitter or gas installer to possess a certificate, which consumers can ask to see.

In 2003 HSE announced changes to the Nationally Accredited Certification Scheme (ACS) for Individual Gas Fitting Operatives.

The changes mean that a prepared and experienced operative is able to complete a 'tailored' domestic natural gas initial assessment in less than 3 days and a 'tailored' re-assessment in one day. This follows a recommendation of the Health and Safety Commission (HSC) *Fundamental Review of Gas Safety*.

Initial assessments and re-assessments take the form of 'tailored' assessments. For those entering the ACS regime for the first time the 'tailored' assessment covers basic gas safety issues plus appliance specific safety issues associated with the installation, maintenance and repair of four types of appliance – cookers, gas fires, water heaters and central heating boilers. The 'tailored' re-assessment for domestic work covers changes in legislation, standards and technology demonstrated across the range of four appliances. The re-assessment also ensures that operatives have retained their gas safety competence with regard to carrying out gas work safely and commissioning appliances, e.g. gas tightness, ventilation and flueing, and the ability to recognise and take action to remove dangerous and at risk situations.

For those operatives wishing to be assessed on fewer or more appliances than the four covered by the 'tailored' assessments, certification bodies, through their assessment centres, can offer packaged assessments which cover the required variation without unnecessary duplication of tasks successfully completed.

Assessment centres are already offering domestic natural gas and LPG in permanent dwellings 'tailored' and packaged initial assessments. Further work is being undertaken to extend the 'tailored' approach to the remainder of LPG work, and commercial catering and commercial work.

Materials and workmanship

G1015 It is incumbent on gas installers to acquaint themselves with the appropriate standards about gas fittings and to ensure that the fittings they use meet those standards. Most gas appliances are subject to the *Gas Appliances* (*Safety*) *Regulations 1992* (*SI 1992 No 711*) and therefore should carry the CE mark or an appropriate European/British Standard.

It is an offence to carry out any work in relation to a gas fitting or gas storage vessel other than in accordance with the appropriate standards and in such a way as to prevent danger to any person. [*Gas Safety* (*Installation and Use*) *Regulations 1998* (*SI 1998 No 2451*), *Reg* 5(*3*)]. Gas pipes and pipe fittings installed in a building must be metallic or of a type constructed in an encased metallic sheath and installed, so far as is reasonably practicable, to prevent the escape of gas into the building if the pipe should fail. Pipes or pipe fittings made from lead or lead alloy must not be used. [*Gas Safety* (*Installation and Use*) *Regulations 1998* (*SI 1998 No 2451*), *Reg* 5(*2*)].

General safety precautions

G1016 A general duty is imposed on all persons in connection with work associated with gas fittings to prevent a release of gas unless steps are taken which ensure the safety of any person [*SI 1998 No 2451, Reg* 6(*1*)].

The *Gas Safety* (*Installation and Use*) *Regulations 1998* (*SI 1998 No 2451*), *Regs* 6(*2*)–(*6*) provide that the following precautions must be observed when carrying out work activities:

- a gas fitting must not be left unattended unless every complete gasway has been sealed with an appropriate fitting;
- a disconnected gas fitting must be sealed at every outlet;
- smoking or the use of any source of ignition is prohibited near exposed gasways;
- it is prohibited to use any source of ignition when searching for escapes of gas;
- any work in relation to a gas fitting which might affect the tightness of the installation must be tested immediately for gas tightness.

With regard to gas storage vessels it is an offence for any person intentionally or recklessly to interfere with a vessel, or otherwise do anything which might affect it so that the subsequent use of that vessel might cause a danger to any person. [*SI 1998 No 2451*), *Reg 6*(*9*)]. In addition:

- gas storage vessels must only be installed where they can be used, filled or refilled without causing danger to any person [*SI 1998 No 2451, Reg 6*(*7*)];
- gas storage vessels, or appliances fuelled by LPG which have an automatic ignition device or a pilot light, must not be installed in cellars or basements [*SI 1998 No 2451, Reg 6*(*8*)];
- methane gas must not be stored in domestic premises [*SI 1998 No 2451, Reg 6*(*10*)].

Gas appliances

G1017 Precautions to be observed regarding gas appliances apply to everyone, not just gas installers. Generally, it is an offence for the occupier, owner or other responsible person to permit a gas appliance to be used if at any time he knows, or has reason to suspect, that the appliance is unsafe or that it cannot be used without constituting a danger to any person. [*SI 1998 No 2451, Reg 34*]. It is also an offence for anyone to carry out work in relation to a gas appliance which indicates that it no longer complies with approved safety standards. [*SI 1998 No 2451, Reg 26*(*7*), (*8*)]. The *Gas Safety* (*Installation and Use*) *Regulations 1998* (*SI 1998 No 2451*), *Reg 27* imposes similar safety requirements on persons who install or connect gas appliances to flues (see G1019 below).

All gas appliances must be installed in a manner which permits ready access for operation, inspection and maintenance. [*SI 1998 No 2451, Reg 28*]. Manufacturer's instructions regarding the appliance must be left with the owner or occupier of the premises after its installation. [*SI 1998 No 2451, Reg 29*]. If the appliance is of the type where it is designed to operate in a suspended position, the installation pipework and other associated fittings must be constructed and installed as to be able to support the weight of the appliance safely. [*SI 1998 No 2451, Reg 31*].

During installation it is essential to ensure that all gas appliances are:

- connected (in the case of a flued domestic gas appliance) to a gas supply system by a permanently fixed rigid pipe [*SI 1998 No 2451, Reg 26*(*2*)];
- installed with a means of shutting off the supply of gas to the appliance unless it is not reasonably practicable to do so [*SI 1998 No 2451, Reg 26*(*6*)].

With the exception of the direct disconnection of the gas supply from a gas appliance, or the purging of gas or air which does not adversely affect the safety of that appliance, all commissioning operations and any other work performed on a gas appliance must be immediately examined to ensure the effectiveness of flues, and checks must be made with regard to:

- the supply of combustion air;
- the operating pressure or heat input; and
- the operation of the appliance to ensure its safe functioning.

[*SI 1998 No 2451, Reg 26*(*9*)].

Any defects must be rectified and reported to the appropriate responsible person as soon as is practicable; the defect must be reported to the appropriate gas supply organisation if a responsible person or the owner is not available. [*SI 1998 No 2451, Reg 26*(*9*)].

Room-sealed appliances

G1018 A room-sealed appliance is an appliance whose combustion system is sealed from the room in which the appliance is located and which obtains air for combustion from a ventilated uninhabited space within the premises or directly from the open air outside the premises. The products of combustion must be vented safely to open air outside the premises.

A room-sealed appliance must be used in the following rooms:

- a bathroom or a shower room [*SI 1998 No 2451, Reg 30*(*1*)];
- in respect of a gas fire, gas space heater or gas water heater (including instantaneous water heaters) of more than 14 kilowatt gross heat input, in a room used or intended to be used as sleeping accommodation [*SI 1998 No 2451, Reg 30*(*2*)].

Gas heating appliances which have a gross heat input rating of less than 14 kilowatt or an instantaneous water heater may incorporate an alternative arrangement, i.e. an approved safety control designed to shut down the appliance before a build-up of dangerous combustion products can occur [*Reg 30*(*3*)].

For the purposes of the *Gas Safety* (*Installation and Use*) *Regulations 1998* (*SI 1998 No 2451*), *Reg 30*(*1*)–(*3*), a room also includes:

- a cupboard or compartment within such a room; or
- a cupboard, compartment or space adjacent to such a room if there is an air vent from the cupboard, compartment or space into such a room [*SI 1998 No 2451, Reg 30*(*4*)].

Flues and dampers

G1019 A flue is a passage for conveying the products of combustion from a gas appliance to the external atmosphere and includes the internal ducts of the appliance. Generally, it is prohibited to install a flue other than in a safe position, and, in the case of a power-operated flue, it must prevent the operation of the appliance should the draught fail [*Gas Safety* (*Installation and Use*) *Regulations 1998* (*SI 1998 No 2451*), *Reg 27*(*4*), (*5*)].

A flue must be suitable and in a proper condition for the operation of the appliance to which it is fitted. It is prohibited to install a flue pipe so that it enters a brick or masonry chimney in such a manner that the seal cannot be inspected. Similarly, an appliance must not be connected to a flue surrounded by an enclosure, unless the enclosure is sealed to prevent spillage into any room or internal space other than where the appliance is installed. [*Gas Safety (Installation and Use) Regulations 1998 (SI 1998 No 2451), SI 1998 No 2451, Reg 27(1)–(3)*].

Manually operated flue dampers must not be fitted to serve a domestic gas appliance, and, similarly, it is prohibited to install a domestic gas appliance to a flue which incorporates a manually operated damper unless the damper is permanently fixed in the open position.

In the case of automatic dampers, the damper must be interlocked with the gas supply so that the appliance cannot be operated unless the damper is open. A check must be carried out immediately after installation to verify that the appliance and the damper can be operated safely together and without danger to any person.

Gas fittings

G1020 Gas fittings are those parts of apparatus and appliances designed for domestic consumers of gas for heating, lighting, cooking or other approved purposes for which gas can be used (but not for the purpose of an industrial process occurring on industrial premises), namely:

- pipework;
- valves; and
- regulators, meters and associated fittings.

[*Gas Safety (Installation and Use) Regulations 1998 (SI 1998 No 2451), SI 1998 No 2451, Reg 2*].

Emergency controls

G1021 An emergency control is a valve for use by a consumer of gas for shutting off the supply of gas in an emergency. Emergency controls must be appropriately positioned with adequate access, and there must be a prominent notice or other indicator showing whether the control is open or shut. A notice must also be posted either next or near the emergency control indicating the procedure to be followed in the event of an escape of gas. [*Gas Safety (Installation and Use) Regulations 1998 (SI 1998 No 2451), SI 1998 No 2451, Reg 9*].

Meters and regulators

G1022 The *Gas Safety (Installation and Use) Regulations 1998 (SI 1998 No 2451), Reg 12* provides that gas meters must be installed where they are readily accessible for inspection and maintenance. A meter must not be so placed as to adversely affect a means of escape or where there is a risk of damage to it from electrical apparatus. After installation, meters and other other associated fittings should be tested for gastightness and then purged so as to remove safely all air and gas other than the gas to be supplied. [*SI 1998 No 2451, Reg 12(6)*].

Meters must be of sound construction so that, in the event of fire, gas cannot escape from them and, where a meter is housed in an outdoor meter box, the box must be so designed as to prevent gas entering the premises or any cavity wall, i.e. any escaping gas must disperse to the air outside. Combustible materials must not be kept inside meter boxes. Where a meter is housed in a box or compound that includes a lock, a suitably labelled key must be provided to the consumer. [*SI 1998 No 2451, Reg 13*].

The *Gas Safety (Installation and Use) Regulations 1998 (SI 1998 No 2451), Regs 14–17* prescribe detailed precautions to be observed during the installation of the various types of meters. Most of these are for compliance by gas installers and should be included in work procedures. In essence, a meter must not be installed in service pipework unless:

- there is a regulator to control the gas pressure [*SI 1998 No 2451, Reg 14*(*1*)(*a*),(*b*)];
- a relief valve or seal is fitted which is capable of venting safely [*SI 1998 No 2451, Reg 14*(*1*)(*c*)]; and
- the meter contains a prominent notice (in permanent form) specifying the procedure in the event of an escape of gas [*SI 1998 No 2451, Reg 15*].

Similar requirements are imposed regarding gas supplies from storage vessels and re-fillable cylinders. [*SI 1998 No 2451, Reg 14*(*2*)–(*4*)].

Where gas is supplied from a primary meter to a secondary meter, a line diagram must be provided and prominently displayed showing the configuration of all meters, installation pipework and emergency controls. It is incumbent on any person who changes the configuration to amend the diagram accordingly. [*SI 1998 No 2451, Reg 17*].

Installation pipework

G1023 For the purposes of this chapter, 'installation pipework' covers any pipework for conveying gas to the premises from a distribution main, including pipework which connects meters or emergency control valves to a gas appliance, and any shut-off devices at the inlet to the appliance. The term 'service pipe' is used to describe any pipe connecting the distribution main with the outlet of the first emergency control downstream from the distribution main. Similarly, 'service pipework' are those pipes which supply gas from a gas storage vessel. All service pipes normally remain the property of the gas supplier or transporter and their permission must be obtained before any work associated with such pipes can be performed.

Installation pipework must not be installed:

- where it cannot safely be used, having regard to other pipes, pipe supports, drains, sewers, cables, conduits and electrical apparatus;
- in or through any floor or wall, or under any building, unless adequate protection is provided against failure caused by the movement of these structures;
- in any shaft, duct or void, unless adequately ventilated;
- in a way which would impair the structure of a building or impair the fire resistance of any part of its structure;
- where deposit of liquid or solid matter is likely to occur, unless a suitable vessel for the reception and removal of the deposit is provided. (It should be noted that such clogging precautions do not normally need to be taken in respect of natural gas or LPG as these are 'dry' gases.)

[*Gas Safety* (*Installation and Use*) *Regulations 1998* (*SI 1998 No 2451*), *Regs 18–21*].

Following work on installation pipework, the pipes must immediately be tested for gastightness and, if necessary, a protective coating applied; followed, where gas is being supplied, by satisfactory purging to remove all unnecessary air and gas. Except in the case of domestic premises, the parts of installation pipework which are accessible to inspection must be permanently marked as being gas pipes. [*Gas Safety* (*Installation and Use*) *Regulations 1998* (*SI 1998 No 2451*), *Regs 22, 23*].

Testing and maintenance requirements

G1024 According to the *Gas Safety (Installation and Use) Regulations 1998 (SI 1998 No 2451), Reg 33(1)*, a gas appliance must be tested when gas is supplied to premises to verify that it is gastight and to ensure that:

(*a*) the appliance has been installed in accordance with the requirements of the 1998 Regulations;

(*b*) the operating pressure is as recommended by the manufacturer;

(*c*) the appliance has been installed with due regard to any manufacturer's instructions accompanying the appliance; and

(*d*) all gas safety controls are in proper working order.

If adjustments are necessary in order to comply with (*a*) to (*d*) above, but cannot be carried out, the appliance must be disconnected or sealed off with an appropriate fitting. [*SI 1998 No 2451, Reg 33(2)*].

A general duty is imposed on employers and self-employed persons to ensure that any gas appliance, installation pipework or flue in places of work under their control is maintained in a safe condition so as to prevent risk of injury to any person. [*SI 1998 No 2451, Reg 35*].

A similar duty is imposed on landlords of premises occupied under a lease or a licence to ensure that relevant gas fittings and associated flues are maintained in a safe condition. [*SI 1998 No 2451, Reg 36(2)*]. A relevant gas fitting includes any gas appliance (other than an appliance which the tenant is entitled to remove from the premises) or installation (not service) pipework installed in such premises.

It is the responsibility of landlords to ensure that each gas appliance and flue is checked by a competent person (HSE approved person) at least every 12 months. [*SI 1998 No 2451, Reg 36(3)*]. Such safety checks shall include, but not be limited to, the following:

- the effectiveness of any flue;
- the supply of combustion air;
- the operating pressure and/or heat input of the appliance;
- the operation of the appliance to ensure its safe functioning.

Nothing done or agreed to be done by a tenant can be considered as discharging the duties of a landlord in respect of maintenance except in so far as it relates to access to the appliance or flue for the purposes of carrying out such maintenance or checking activities. [*SI 1998 No 2451, Reg 36(10)*].

In addition, it is the duty of landlords to ensure that appliances and relevant fittings are not installed in any room occupied as sleeping accommodation such as to cause a contravention of *Reg 30(2)* or (*3*) (see G1018 above).

Inspection and maintenance records

G1025 The *Gas Safety (Installation and Use) Regulations 1998 (SI 1998 No 2451)* mark a significant increase in the obligations now placed upon landlords.

A landlord is under a duty to retain a record of each inspection of any gas appliance or flue for a period of two years from the date of the inspection. The record must include:

- the date on which the appliance or flue is checked;

- the address of the premises at which the appliance or flue is installed;
- the name and address of the landlord of the premises at which the appliance or flue is installed;
- a description of, and the location of, each appliance and flue that has been checked together with details of defects and any remedial action taken.

The person who carries out the safety check must enter on the record his name and his particulars of registration. Then he should sign the record confirming that he has examined the following:

- the effectiveness of any flue;
- the supply of combustion air;
- the operating pressure and/or heat input of the appliance;
- the operation of the appliance to ensure its safe functioning.

[*SI 1998 No 2451, Reg 36*(*3*)].

Within 28 days of the date of the check, the landlord must ensure that a copy of the record is given to each tenant of the premises to which the record relates. [*SI 1998 No 2451, Reg 36*(*6*)(*a*)]. In addition, before any new tenant moves in, the landlord must provide any such tenant with a copy of the record – however, where the tenant's right to occupation is for a period not exceeding 28 days, a copy of the record may instead be prominently displayed within the premises. [*SI 1998 No 2451, Reg 36*(*6*)(*b*)]. A copy of the inspection record given to a tenant under *Reg 36*(*6*)(*b*) need not include a copy of the signature of the person who carried out the inspection, provided that it includes a statement that the tenant is entitled to have another copy, containing a copy of such signature, on request to the landlord at an address specified in the statement. Where the tenant makes such a request, the landlord must provide the tenant with such a copy of the record as soon as is practicable. [*SI 1998 No 2451, Reg 36*(*8*)].

The *Gas Safety* (*Installation and Use*) *Regulations 1998* (*SI 1998 No 2451*), *Reg 36*(*7*) provides that where any room occupied, or about to be occupied, by a tenant does not contain any gas appliance, the landlord may – instead of giving the tenant a copy of the inspection record in accordance with *Reg 36*(*6*) (see above) – prominently display a copy of the record within the premises, together with a statement endorsed upon it that the tenant is entitled to have his own copy of the record on request to the landlord at an address specified in the statement. Where the tenant makes such a request, the landlord must provide the tenant with a copy of the record as soon as is practicable.

Interface with other legislation

G1026 Although wide-ranging in their scope the *Gas Safety* (*Installation and Use*) *Regulations 1998* (*SI 1998 No 2451*) are not concerned directly with product safety. Certain premises, gas fittings and uses are also subject to exceptions. However, in many of these instances where exceptions are made similar requirements for gas safety are to be found in the *Health and Safety at Work etc Act 1974* and supporting legislation. The most important of these are:

- *Gas Appliances* (*Safety*) *Regulations 1995* (*SI 1995 No 1629*)

  These Regulations were made under the authority of the *Consumer Protection Act 1987*. They implement European directives that require gas appliances and fittings to conform with essential safety requirements and to be safe when used normally.

- *Management of Health and Safety at Work Regulations 1999* (*SI 1999 No 3242*) (*see also* RISK ASSESSMENT)

  These Regulations were made under the *Health and Safety at Work etc Act 1974*. They require employers and the self-employed to undertake a suitable and sufficient assessment of the risks to the health and safety of their employees and others who may be affected by the work that they do. These Regulations further require employers (including those who install or supply gas or gas appliances or who undertake gas maintenance work) to ensure controls are in place and kept under review.

- *Provision and Use of Work Equipment Regulations 1998* (*SI 1998 No 2306 as amended by SI 2002 No 2174*)) (*see also* MACHINERY SAFETY)

  These Regulations were made under the authority of the *Health and Safety at Work etc Act 1974*. They impose health and safety requirements on work equipment including any gas appliance, apparatus, fitting or tool used or provided for use at work. Equipment must be suitable for its intended use, be maintained and inspected and users provided with appropriate health and safety information and where necessary written instructions.

- *Pressure Systems Safety Regulations 2000* (*SI 2000 No 128*)

  These Regulations were made under the authority of the *Health and Safety at Work etc Act 1974*. Their objective is the prevention of serious injury from the failure of a pressure system or its components, including those that contain gas in excess of 0.5 bar above atmospheric pressure and pipelines used to convey gas at 2 bar above atmospheric pressure. Under these Regulations duties are placed on those who design, manufacture, use and maintain gas systems overlapping with similar duties under the *Gas Safety* (*Installation and Use*) *Regulations 1998* (*SI 1998 No 2451*)

- *Building Regulations 1991* (*SI 1991 No 2768*) (*see also* WORKPLACES – HEALTH, SAFETY AND WELFARE)

  These Regulations (as amended in 1994) were made under the authority of the *Building Act 1984*. They are supported by non-mandatory approved documents that provide practical guidance on for example the safe installation of gas appliances used for heating – Approved Document J is concerned with air supply to gas heaters and discharge of the products of combustion to the open air.

Summaries of relevant court cases

G1027 The importance of the provisions relating to the correct installation and regular maintenance aspects of gas appliances, even in the earlier Gas Safety Regulations, is demonstrated in the following summaries of recent court cases.

Unsafe gas fittings

G1028 A Nottingham landlord narrowly escaped a charge of manslaughter when he was tried and convicted of carrying out unsafe work practices which resulted in the deaths of two people.

The court heard of makeshift repairs to a gas boiler which then leaked carbon monoxide into an adjacent flat. The gas supplier (British Gas) gave evidence of having previously made the boiler inoperable because of its unsafe condition but it was reconnected by the landlord without any authority to do so, in December 1992.

The landlord was charged under the *Gas Safety (Installation and Use) Regulations 1984* and the *Health and Safety at Work etc Act 1974, s 3(2)*. He was accused of failing in his duty as a self-employed person to conduct his undertaking so as not to put himself or others at risk and of failing to carry out work in relation to a gas fitting in a proper and workmanlike manner.

The judge made it clear that if the legislation had carried a penalty of imprisonment, he would have imposed it. The seriousness of the case could only be reflected by a heavy fine and he sentenced the landlord with a fine of £32,000 with £24,800 costs.

[*Leicester Crown Court, June 1996*].

Note: The *Gas Safety (Installation and Use) Regulations 1994* introduced a wider scope of control and increased the penalties for contravention to an unlimited fine and a maximum of two years' imprisonment.

Local authority (as landlord) guilty of bad gas safety management

G1029 A local authority found itself in court for nine breaches of the *Gas Safety (Installation and Use) Regulations 1994* (failing to carry out safety checks and to keep records) and for failing to ensure the safety of tenants, contrary to the *Health and Safety at Work etc Act 1974, s 3(1)*. The case was brought after the discovery of more than 150 flues which were left disconnected after refurbishment work on a London council estate; and on other estates, annual safety checks had not been carried out.

The magistrate commented that there had been serious risk to hundreds of householders and imposed fines on the council of £27,000 for the breaches of the *Gas Safety (Installation and Use) Regulations 1994* and £17,000 for the offences which contravened the *Health and Safety at Work etc Act 1974*.

[*Clerkenwell Magistrates' Court, January 1997*].

Note: The HSE pointed out that local authorities were the largest landlords in the country and that failure to manage in this case had put many tenants at risk of carbon monoxide poisoning and fire and explosion.

Gas explosion results in three separate prosecutions

G1030 A major gas explosion at a twenty two-storey tower block resulted in the prosecution of a borough council, an installation company and a self employed electrician. The explosion blew off the roof of a boiler house, only minor injuries were sustained but there was considerable property damage from flying debris. There had been serious risk to hundreds of householders.

An investigation revealed that new gas burners and gas pressure booster pumps had been inadequately installed and not commissioned correctly.

The borough council was fined £75,000.00 for failing to manage the project.

The installation company was fined £10,200.00 for responsibility for the inadequate installation and management of the project.

The electrician was fined £3,000.00 for failing to install and commission the boilers correctly, not being competent, and not registered with CORGI.

[*Middlesex Guildhall Crown Court, September 1999*].

Student death – landlord and gas fitter imprisoned

G1031 A landlord and an unregistered gas fitter were charged and convicted of manslaughter following the death of a student who inhaled carbon monoxide fumes from an inappropriate type of gas boiler which was not correctly ventilated.

The investigation into the incident by the HSE originated with the *Gas Safety (Installation and Use) Regulations 1994* which required gas fitters to be registered with the Council of Registered Gas Installers (CORGI). The charge of manslaughter was brought because of the seriousness of the offence: it had resulted in the death of one student and the hospitalisation of four other tenants who had been affected by poisonous fumes.

Both the landlord and the gas fitter pleaded guilty to manslaughter: the landlord was judged to be most responsible and was imprisoned for two years. The gas fitter was imprisoned for fifteen months.

[*Stafford Crown Court, December 1997*].

Note: Broadly, the law of manslaughter is part of the law of homicide. It is less serious than murder with the sentence at the discretion of the court. For health and safety purposes, manslaughter may be tentatively defined as killing without the intention to kill or to cause serious bodily harm. This case was the second health and safety manslaughter charge to result in imprisonment.

Gas fitter jailed – chimney flue capped with concrete caused death

G1032 This case involved an unregistered gas fitter who failed to notice that the flue into which he had connected a gas fire was capped with a concrete slab.

The fitter, the landlord and the landlord's son were charged with manslaughter following the death of a tenant in a bedsit. The prosecuting counsel told the court that the landlord of the bedsit, and his son who was looking after the property on behalf of his father, were also responsible for the death. He gave evidence of negligence in 20 other properties where failure to maintain gas appliances properly could have had similar tragic consequences.

In the case in question the judge said that the landlord and his son had been lulled into a false sense of security by a gas safety certificate provided by the fitter. He imposed the following sentences:

Gas fitter: 12 months' imprisonment, six of which were suspended;

Landlord and his son: 9 months' imprisonment suspended for 18 months – the landlord's son was also ordered to pay £16,000 costs.

[*Norwich Crown Court, April 1998*].

Unsafe gas appliance – landlord guilty

G1033 A landlord was charged with:

- failing to maintain gas appliances in a safe condition, contrary to the *Gas Safety (Installation and Use) Regulations 1994, Reg 35A*; and with
- providing false information in connection with his failure to comply with an improvement notice, contrary to the *Health and Safety at Work etc Act 1974, ss 20* and *21*.

Two tenants who suffered dizziness and headaches informed the gas emergency company which then inspected the gas boiler and declared it unsafe. The boiler was switched off and a label was attached, warning that it should not be used.

The landlord instructed an unregistered gas fitter to repair the boiler; he failed to check for carbon monoxide leaks after switching the boiler back on. Two days later the two tenants collapsed after inhaling poisonous fumes and were hospitalised.

The HSE found that both the boiler and a gas cooker in the same property were leaking dangerously high levels of carbon monoxide gas. The landlord gave false information about the identity of the fitter who had carried out the work and switched the boiler back on.

The Crown Court judge said that the landlord was very lucky not to be facing manslaughter charges and imposed fines and costs totalling £24,000.

[*Cardiff Crown Court, May 1998*].

Gas safety requirements in factories

G1034 Where part of plant contains explosive/flammable gas under pressure greater than atmospheric, that part must not be opened unless:

- before the fastening of any joint of any pipe, connected with that part of the plant, any flow of gas has been stopped by a stop-valve;
- before such fastening is removed, all practicable steps have been taken to reduce the gas pressure in the pipe or part of the plant, to atmospheric pressure; and
- if such fastening is loosened or removed, inflammable gas is prevented from entering the pipe/part of the plant, until the fastening has been secured or securely replaced.

No hot work must be permitted on any plant or vessel which has contained gas until all practicable steps have been taken to remove the gas and any fumes arising from it, or to render it non-explosive or non-flammable. Similarly, and if any plant or vessel has been heated, no gas shall be allowed to enter it until the metal has cooled sufficiently to prevent any risk of ignition.

[*Factories Act 1961, s 31*(*3*), (*4*)].

Pressure fluctuations

G1035 The *Gas Safety* (*Installation and Use*) *Regulations 1998* (*SI 1998 No 2451*), *Reg 38*, provides that where gas is used in plant which is liable to produce pressure fluctuations in the gas supply, such as may cause danger to other consumers, the person responsible for the plant must ensure that any directions given to him by the gas transporter to prevent such danger are complied with.

If it is intended to use compressed air or any gaseous substance in connection with the consumption of gas, at least 14 days' written notice must be given to the gas transporter.

Any device fitted to prevent pressure fluctuations or to prevent the admission of a gaseous substance into the gas supply must be adequately maintained.

Harassment in the Workplace

Introduction

H1701 It is a common misconception that there exists in English law a particular cause of action for unlawful harassment in the workplace (whether on grounds of sex, race or disability). However, the law relating to harassment is but a specific example of the general law relating to discrimination in the workplace and, as such, the main pieces of legislation dealing with the subject of harassment are the *Sex Discrimination Act 1975* (*SDA 1975*), the *Race Relations Act 1976* (*RRA 1976*) and the *Disability Discrimination Act 1995* (*DDA 1995*) (together, 'the Discrimination Acts'). This said, although harassment is a particular example of discrimination, the area is of such significant importance that in the context of sex discrimination, the European Commission has been moved to publish a Recommendation (O.J.1992, L49/1) and annexed Code of Practice on the protection of the dignity of men and women at work (see H1703).

History

H1702 The first significant decision on the issue of whether harassment amounted to discrimination for the purposes of the *SDA 1975* was *Porcelli v Strathclyde Regional Council [1986] IRLR 134*. Here, the applicant was a female laboratory assistant working at a school alongside two male colleagues. The male assistants did not like the employee and they subjected her to a campaign of harassment in the form of brushing against her in the workplace and making sexually suggestive remarks to her and about her in order to make her leave her job. The campaign was a 'success' in that it did indeed force the applicant to apply for a transfer to another school. Having done so, the applicant then brought a claim against her employer claiming that the actions of her former colleagues amounted to direct sex discrimination for the purposes of the *SDA 1975, s 1(1)(a)*, in that she had been subjected to a detriment. The employer tried to defend the claim by calling in evidence the male employees responsible for the course of harassment. During their evidence they ventured to suggest that the conduct that had been meted out to the applicant would also have been displayed towards a hypothetical male comparator employee that they did not like and, consequently, the employer could not have committed sex discrimination against the applicant since the treatment was gender neutral.

The Scottish Court of Session refused to accept the defence. Lord Emslie stated of sexual harassment that it is:

> '... a particularly degrading and unacceptable form of treatment which it must be taken to have been the intention of Parliament to restrain'.

The court likened the use of sexual harassment as a 'sexual sword' which had been 'unsheathed and used because the victim was a woman'. It added that, although an equally disliked male employee may well have been subjected to a campaign of harassment, the harassment would not have had as its cutting edge the gender of the victim. The court then went on to give general guidance as to what harassment is. It stated that sexual harassment amounted to:

> '... [u]nwelcome acts which involve physical contact which, if proved, would also amount to offences at common law, such as assault or indecent assault; and also conduct falling short of such physical acts which can be fairly described as sexual harassment'.

The law has moved on since the *Porcelli* decision, most notably in the re-statement by the Employment Appeal Tribunal (EAT) in the case of *(1) Reed (2) Bull Information Systems Ltd v Stedman [1999] IRLR 299* (see below at H1710).

As will be seen in the course of this chapter, various defences to claims of harassment have been developed, and, as stated above, the European Commission was moved to set down the Recommendation and Code of Practice following a report commissioned by it from Michael Rubenstein in 1988.

The European Commission Recommendation and Code of Practice

H1703 As stated at H1701 above, the European Commission was prompted into drafting a Recommendation and Code of Practice dealing with the issue of sexual harassment following a report undertaken for it in 1988 (*The Dignity of Women at work: A report on the problem of sexual harassment in the Member States of the European Communities, 1988*). The Recommendation was made on 27 November 1991 and the Code was published in the Official Journal of the European Communities, 4 February 1992.

Content of the Recommendation

H1704 Art 1 of the Recommendation requires Member States to:

> '... take action to promote awareness that conduct of a sexual nature, or other conduct based on sex affecting the dignity of women and men at work, including conduct of superiors and colleagues, is unacceptable if:
>
> (*a*) such conducted is unwanted, unreasonable and offensive to the recipient;
>
> (*b*) a person's rejection of or submission to, such conduct on the part of employers or workers (including superiors or colleagues) is used explicitly or implicitly as a basis for a decision which affects that person's access to vocational training, access to employment, continued employment, promotion, salary or any other employment decisions; and/or
>
> (*c*) such conduct creates an intimidating, hostile or humiliating working environment for the recipient,
>
> and that such conduct may, in certain circumstances, be contrary to the principle of equal treatment within the meaning of Articles 3, 4 and 5 of Directive 76/207/EEC [the Equal Treatment Directive]'.

The Recommendation is designed to draw to the attention of employers the nature of sexual harassment and to give broad outlines as to the ways in which harassment occurs together with providing an indication of the effect of such conduct (i.e. to highlight that it may be actionable as a form of discrimination).

Content of the Code of Practice

H1705 The Code of Practice goes on to provide clear guidance to employers as to:

(*a*) the nature of sexual harassment including the effect of harassment on victims;

(*b*) a definition of harassment. This amounts to: '... unwanted conduct of a sexual nature, or other conduct based on sex affecting the dignity of women and men at work. This can include unwelcome physical, verbal or non-verbal conduct';

(*c*) the legal implications for employers; and

(*d*) the steps employers should take to prevent the spread of harassment in the workplace.

The Code goes beyond the position as it currently stands at English law. For example, it expressly states that sexual harassment is capable of occurring to homosexuals who are harassed in the workplace on account of their sexual orientation. Yet, the Court of Appeal has stated that harassment of a person on account of their sexual orientation is not *per se* discriminatory within the meaning of the *SDA 1975* (per *Smith v Gardner Merchant Limited [1998] IRLR 520*, below at H1709).

Although the EAT in Scotland held that harassment on grounds of sexual orientation can amount to discrimination (*MacDonald v Ministry of Defence [2000] IRLR 748*), this decision was reversed by the Court of Session on appeal (*[2001] IRLR 431*).

There is no denying that employers should familiarise themselves with the contents of the Code. It amounts to a sensible extension to good human resources practice and, even though it may go beyond the requirements of national sex discrimination law, it should be remembered that harassment which is not currently actionable in terms of sex discrimination may, in any event, amount to conduct which could allow employees to claim that their contracts of employment have been repudiated by their employer thereby allowing them to claim unfair dismissal.

Legal effect of the Recommendation and Code of Practice

H1706 The Recommendation and Code are not law in this country and it still remains to be seen what the Government will do to adopt the Code on a formal basis. As regards employers, they should be aware that the Code is readily accepted by employment tribunals as evidence of good practice.

Definition

The Discrimination Acts

H1707 The starting point in connection with harassment is the Discrimination Acts. Harassment on grounds of sex or race is direct discrimination, i.e. it is treatment meted out to the recipient simply because that person is a woman or a man or because of the race of a person. Disability discrimination is discrimination that is meted out to a person on account of that person's disability and which cannot be objectively justified. It seems unlikely that there could be any justification for harassment meted out to a person on the grounds of their disability.

The provisions of the Discrimination Acts dealing with harassment are contained within the *SDA 1975, s 6(2)(b)*, the *RRA 1976, s 4(2)(b)*, and the *DDA 1995*, which provide that it is unlawful to discriminate against a person employed at an establishment in Great Britain on grounds of sex, race or disability (as the case may be).

Effectively, for a cause of action to arise in relation to harassment, a three stage test has to be satisfied, this being that:

(*a*) less favourable treatment has been shown to an applicant;

(*b*) the proximate cause of the differential treatment is the applicant's gender, race or disability ('prohibited grounds'); and

(*c*) the differential treatment has subjected the victim to a detriment.

Less favourable treatment

H1708 In dealing with the question of whether or not sexual harassment amounted to less favourable treatment, the Court of Session in *Porcelli v Strathclyde Regional Council [1986] IRLR 134* stated that:

> 'In a case of alleged sexual harassment, the correct primary question is not "was there sexual harassment?", a phrase which is not found in the statute, but "was the applicant less favourably treated on the ground of her sex than a man would have been treated?" If that question is answered in the affirmative, there is discrimination within the meaning of the Sex Discrimination Act'.

The Discrimination Acts do, however, have their limitations. In the first instance, it is entirely conceivable that situations could arise where although an employer treats an employee in a manner that seems to amount to a *prima facie* case of harassment, it may do so in circumstances that would not amount to less favourable treatment. For example, for the purposes of the *SDA 1975*, there can be circumstances arising from conduct that could equally affront a male employee. A good illustration is the case of *Stewart v Cleveland Guest (Engineering) Ltd [1994] IRLR 440*. Here, the applicant was a female engineer who frequently had to make visits to the factory floor in order to carry out her duties. She was upset at the display of posters of nude women that adorned the factory floor area and complained to the employer's management. She asked for the posters to be removed. Her requests were ignored and the employee resigned and claimed that she had been constructively dismissed. She alleged that she had been subjected to less favourable treatment on the grounds of her sex which had ultimately led to her constructive dismissal. An employment tribunal and, on appeal, the EAT, found that the display of such posters was gender neutral since a male employee might also have been equally upset at the posters. In the circumstances, it was not possible to say that the employee had been treated less favourably than a male comparator.

Another example is the case of *Balgobin v London Borough of Tower Hamlets [1987] IRLR 401*. This case concerned complaints by female employees who claimed that they had been subjected to the detriment of being required to work with their alleged harasser after an investigation into the allegations of harassment was found to be inconclusive. Again, the EAT refused to accept that the employee had been treated less favourably on grounds of sex since it was possible that if a male employee had been harassed, the employer would have treated the male in a similar way. This is not to say that all cases will be treated in this fashion. Indeed, the EAT held in *Driskel v Peninsula Business Services [2000] IRLR 151* that, where a male supervisor engages in vulgar behaviour with a junior male colleague, it does not follow that such behaviour will provide a defence if the supervisor attempts to engage in similar behaviour with a female subordinate since behaviour of this nature where directed at a female subordinate is more likely to be intimidatory.

Finally, in the context of sex discrimination, it is conceivable that a harasser may be bi-sexual and therefore a defence could exist to the effect that the harasser would indeed treat a man in a similar fashion.

This is not to say that employers can use these lines of defence to totally escape liability. It needs to be remembered that although treatment might not be less favourable for the purposes of the Discrimination Acts, it may still amount to a breach of the implied duty in a contract of employment to maintain the trust and confidence of an employee and the harassment and may therefore be a precursor to an unfair dismissal claim.

In spite of the above, it seems that an argument still exists in the context of sexual harassment to the effect that, since sexual harassment is gender specific (in much the same way as pregnancy), there is no requirement for a comparison to be made as to whether a male employee would have been treated in the same way. Indeed, the EAT accepted this view in *British Telecommunications Ltd v Williams [1997] IRLR 668*. The EAT even went so far as to hold that it would be no defence for an employer to state that a male employee would have been treated in a similar fashion. Unfortunately, some of the credibility in the *Williams* case is lost in that it made no reference to the decision in *Stewart*. There must also be some doubt as to whether the argument can survive following the decision of the European Court of Justice in *Grant v South-West Trains Ltd [1998] CMLR 993* (dealing with an employer's policy of not providing travel concessions to partners in same sex relationships) which held that, provided a policy or treatment is applied equally to men and women, there can be no discrimination.

Differential treatment is on prohibited grounds of sex

H1709 In order for unlawful discrimination to occur, it is necessary that a claimant be treated less favourably than an actual or hypothetical male comparator. To this end, in the context of sex discrimination, one of the defences that has been raised is to argue that harassment against a female employee would have occurred equally to a male employee. The case of *Porcelli v Strathclyde Regional Council [1986] IRLR 134* (above) shows the limits of this line of defence. The Court of Session stated of the comparison between male and female employees:

> 'If any material part of unfavourable treatment to which a woman is subjected includes a significant element of a sexual character to which a man would not be vulnerable, the treatment is on grounds of the woman's sex within the meaning of *s 1(1)(a)*'.

It will be recalled that the defence raised by the employer in this case was that a hypothetical male employee would have been treated in a similar fashion. The defence failed and the Court of Session commented upon the nature of the treatment that had been received by the applicant in the following terms.

> 'The [employment] tribunal … concentrated on the unpleasantness of the treatment meted out to the respondent and compared it with the unpleasantness of the treatment which they considered would have been meted out to a man whom his colleagues had disliked as much as her colleagues had disliked the respondent. That was a question of fact; but the question which had to be asked after that conclusion had been reached was "Was the treatment, or any part of it, which was meted out to the respondent less favourable, than that which would have been meted out to the notional man because the respondent was a woman?".'

An example of a decision that was resolved in a contrary fashion was the case of *Smith v Gardner Merchant Limited [1998] IRLR 520*. The facts of this case were that the applicant was a male homosexual who had been dismissed for gross misconduct. He brought a claim for sex discrimination, one of the limbs of which was an allegation that he had been harassed by a female employee who had asked a series of offensive questions about his sexuality, made various accusations to him along the lines that gay people had all sorts of diseases and stated to him that gay people who spread the HIV virus should be put on an island. Mr Smith claimed that he had been treated less favourably on account of his gender. The Court of Appeal took the view that although there may have been less favourable treatment, it was on account of Mr Smith's sexual orientation rather than his gender. Accordingly, the relevant grounds of comparison that would need to be made were whether a female homosexual with the same personal characteristics of Mr Smith would have been subjected to similar or differential treatment. Whether or not such a decision is consistent with Arts 8 and 14 contained within the *Human Rights Act 1998, Sch 1*, (which recognises a right to respect for a person's private life and prohibits discrimination in relation to that right on any ground) remains to be confirmed. The Scottish EAT in *MacDonald v Ministry of Defence [2000] IRLR 748* held that the *SDA 1975* must be construed in the light of Arts 8 and 14. Since sex discrimination for the purposes of Art 14 has been held to include discrimination on the grounds of sexual orientation (in *Salgueiro da Silva Mouta v Portugal, 33290/96, 21 December 1999*) the EAT concluded that, in line with its requirement to interpret the *SDA 1975* consistently with the *Human Rights Act 1998*, discrimination on grounds of sexual orientation must be unlawful. The Court of Session reversed the decision on appeal (*[2001] IRLR 431*) and in *Pearce v Governing Body of Mayfield Secondary School [2001] IRLR 669*, the EAT and the Court of Appeal in England accepted the orthodoxy of Smith. (Note: Ms Pearce has been given leave to appeal to the House of Lords.) From a common sense point of view, there does seem to be some contradiction in the law to the extent that it does not protect homosexuals from harassment as to their sexuality and yet does protect transsexuals against treatment meted out to them on the grounds of gender reassignment (as occurred in *Chessington World of Adventures Ltd v Reed [1997] IRLR 556* — see also now the *Sex Discrimination* (*Gender Reassignment*) *Regulations 1999* (*SI 1999 No 1102*) outlawing discrimination on the grounds of gender reassignment). It remains to be seen whether the Government will pass similar legislation in relation to homosexuals following its defeat in the European Court of Human Rights in the case of *Smith & Grady v UK [1999] IRLR 734* (involving the right of homosexuals to serve in the UK armed forces).

In the context of disability discrimination, there is no requirement of an able bodied comparator (*Clark v Novacold Limited [1999] IRLR 318*) and it is simply enough for harassment to be meted out to a complainant on grounds of the complainant's disability.

Detriment

H1710 The final hurdle to be crossed in order to bring a successful claim is to show that the employee has suffered a detriment. As to what is capable of amounting to a detriment, the starting point is again the decision in *Porcelli v Strathclyde Regional Council [1986] IRLR 134*. Here, the Court of Session stated that a detriment, in its statutory context, simply means a disadvantage. The point has also been considered in *Barclays Bank plc v Kapur [1989] ICR 753* where the Court of Appeal held that, since detriment means a disadvantage, it would cover almost any discriminatory conduct by an employer against an employee. The Court of Appeal did, however, limit the scope of this statement in *Barclays Bank plc v Kapur (No 2) [1995] IRLR 87*, where it was held that an unjustified sense of grievance did not amount to a detriment.

The concept of detriment was re-examined in the case of *Reed and Bull Information Systems v Stedman [1999] IRLR 299*. The facts of the case were that the claimant had resigned claiming that the behaviour of her manager amounted to sexual harassment and therefore discrimination. The employee cited a list of fifteen alleged incidents that she considered amounted to a course of harassment. The Employment Tribunal concentrated on four of the allegations, these being such matters as telling the employee that sex was a beneficial form of exercise, telling dirty jokes to colleagues in her presence, making a pretence of looking up the employee's skirt and then laughing when she angrily left the room and stating to her when she was listening to a trial presentation that 'You're going to love me so much for my presentation so that when I finish you will be screaming out for more and you will want to rip my clothes off'. The tribunal added that, of themselves, each of the allegations was insufficient to amount to sexual harassment. However, if the claimant were able to prove the incidents, they would amount to a course of conduct that would itself be sufficient to amount to harassment. The employment tribunal went on to find that the acts had occurred and that the employer and the manager had therefore discriminated against the applicant. The employer and manager appealed to the EAT. The EAT dismissed the appeal agreeing that the actions by the manager, although not meant to be sexually harassing, were indeed discriminatory. It added:

> 'It seems to us important to stress at the outset that 'sexual harassment' is not defined by the [*SDA 1975*]. It is a colloquial expression which describes one form of discrimination in the workplace made unlawful by *s 6* of the *Sex Discrimination Act 1975*. Because it is not a precise or defined phrase, its use, without regard to s 6, can lead to confusion. Under *s 6* it is unlawful to subject a person to a "detriment" on the grounds of their sex. Sexual harassment is a shorthand for describing a type of detriment. The word detriment is not further defined and its scope is to be defined by the fact-finding tribunal on a common-sense basis by reference to the facts of the particular case. The question in each case is whether the alleged victim has been subjected to a detriment and, second, was it on the grounds of sex'.

Consequently, in order to determine whether or not conduct of a sexual nature is sexual harassment and therefore a 'detriment' within the meaning of the *SDA 1975*, it is necessary to consider whether it is a disadvantage which is itself determined from the particular circumstances of the case.

Detriment — subjective or objective determination?

H1711 The EAT in *Reed and Bull Information Systems* (above at H1710) also discussed the question as to whether a 'detriment' was something that was to be determined subjectively from the perspective of an applicant to a tribunal or objectively by the tribunal. The EAT dealt with the issue by posing a series of questions that would typically be asked in harassment cases, these being as follows.

(*a*) If a woman regards words or conduct of a sexual nature as harassment to which many women would not take exception or regard as harassment, has her claim been made out?

(*b*) If a man does not appreciate that his words or conduct are unwelcome, has her claim been proved?

(*c*) Is a 'one-off' act (words or conduct) sufficient to constitute harassment?

The tribunal answered the questions that it posed by stating as to the first of the questions that it is up to the victim of harassment to determine what she finds unwelcome or offensive. Simply because a tribunal does not necessarily find the words or conduct offensive will not lead to a complaint being dismissed. Indeed, the EAT stated that there is a range of factual situations that are capable of amounting to harassment. Inevitably, once a victim of harassment has received unwelcome attention from a person it is possible that further attention that would not ordinarily attract comment may take on a different context in the light of previous unwelcome attentions. Consequently, it is up to the victim to determine what words or conduct she is unhappy with and for a tribunal to determine in the totality of the facts whether those words or conduct could reasonably be viewed as offensive and unwelcome.

As to the second question, the EAT acknowledged that it is trite law that the motive and intention of the discriminator are irrelevant to the question of whether or not discrimination has occurred.

Finally, as to the third question, the EAT stated that the fundamental question to be addressed by a tribunal was whether the applicant was subjected to a detriment on account of her sex. It went on to add that this was not a question that was answered by counting up the number of incidents that had occurred. In fact, it was answered by looking at whether an incident or series of them damaged the working environment and constituted a barrier to sexual equality in the workplace which constituted a detriment.

Further guidance on the issue of detriment was given by the EAT in *Driskel v Peninsula Business Services Limited [2000] IRLR 151*. The case concerned allegations of a series of acts committed by the line manager of the female applicant that was alleged to amount to harassment on grounds of sex. The EAT held that, where there are a series of acts, tribunals should focus upon the cumulative effect of the acts rather than whether the acts in themselves are trivial in nature. Further, where a male supervisor engages in vulgar behaviour with male colleagues, it does not follow that a female employee subjected to vulgar behaviour is not treated less favourably since it is more likely that a female subordinate will find such behaviour to be intimidatory and likely to undermine her dignity.

Ultimately, the view of the EAT was that the test as to whether a detriment has been suffered focuses upon an applicant's subjective determination of whether conduct to which she is subjected is unwelcome and that determination has then to be judged by a tribunal which must ask itself whether that is a reasonable position for the applicant to take.

Examples of unlawful harassment

H1712 As was stated in the case of *Reed and Bull Information Systems v Stedman [1999] IRLR 299*, sexual harassment is a shorthand way of expressing the legal requirement in the *SDA 1975, s 6(2)* of a 'detriment'. The case also recognises that the type of situations in which sexual harassment can occur are legion. However, of the cases that have been reported to date they can broadly be categorised as:

(*a*) course of conduct cases;

(*b*) single incident cases; and

(*c*) failure to investigate or protect cases.

Course of conduct cases

H1713 As the heading suggests, the hallmark of this genus of cases is that an employee is subjected to repeated harassment in the course of her employment. The following are reported examples of such cases.

(*a*) *Porcelli v Strathclyde Regional Council [1986] IRLR 134*. The applicant was a female laboratory assistant and was the victim of a campaign of harassment by fellow male employees to make her leave her job. The male employees would frequently brush past the applicant in a sexually suggestive manner and direct comments of a sexual nature at her. The Court of Session held that the applicant had suffered a detriment in having to put up with the conduct which had clear sexual overtones and to which a similarly disliked hypothetical male comparator would not have been subjected.

(*b*) *Tower Boot Company Limited v Jones [1995] IRLR 529*. The applicant was a black employee who was subjected to such acts as being whipped with a leather belt, 'branded' with a red-hot screwdriver and subjected to racist name calling. The Court of Appeal held that, although employees were not employed to harass other employees, the acts had been committed during the course of the employees' employment and therefore the employer was liable for the discrimination.

(*c*) *Chessington World of Adventures Ltd v Reed [1997] IRLR 556*. The applicant was a biological male and was undergoing gender reassignment from male to female. The applicant was subjected to a prolonged campaign of ostracism and harassment (which the employer was found to have done nothing to investigate or curb) having as its proximate cause the therapy that the applicant was undertaking. In the circumstances, the employer was held to have subjected the applicant to a detriment.

(*d*) *Reed and Bull Information Systems v Stedman [1999] IRLR 299*. The applicant was a junior secretary responsible to the marketing manager of the employer. The applicant catalogued fifteen separate incidents ranging from the telling of dirty jokes by the manager to other colleagues in her presence to the pretence of looking up her skirt in the office smoking room. The EAT upheld the findings of the employment tribunal that, although the incidents by themselves would have been insufficient to amount to a detriment, collectively they amounted to a course of conduct of innuendo and general sexist behaviour amounting to harassment and therefore a detriment for the purposes of the *SDA 1975*.

(*e*) *Driskel v Peninsula Business Systems [2000] IRLR 151*. The applicant was a female employee who complained that she had been subjected to sexual banter and comments by the head of her department and, on the day before an interview for a promotion, she was told by the manager that she should wear a short skirt and a see-through blouse showing plenty of cleavage. The applicant left work and refused to return unless the manager was moved to another position. The employee's request was refused and she was subsequently dismissed. The EAT held that the dismissal was on grounds of the applicant's sex.

Single incident cases

H1714 The hallmark of 'single incident' cases is that they are effectively a 'bolt from the blue', yet comply with the requirements in *Reed and Bull Information Systems v Stedman [1999] IRLR 299* in that they provide a barrier to sexual equality in the workplace.

(*a*) *Bracebridge Engineering Ltd v Darby [1990] IRLR 3*. The applicant was physically manhandled into the office of the works manager by both a chargehand and the works manager and then indecently assaulted. She was warned not to complain because no one would believe her. The applicant did subsequently complain but the complaint was then not properly investigated by the employer. It was held by the EAT that the assault was sufficiently serious to amount to a detriment.

(*b*) *Insitu Cleaning Co Ltd v Heads [1995] IRLR 4*. The applicant was a cleaning supervisor who attended at a company meeting. Also present was the son of one of the directors of the company who greeted her by saying 'Hiya, big tits'. The applicant found the remarks particularly distressing as the director's son was approximately half her age. The EAT held that the remark was capable of subjecting the applicant to a detriment and rejected the employer's contention that the remark was of the same effect as a statement made to a bald male employee about his head.

(*c*) *Chief Constable of the Lincolnshire Police v Stubbs [1999] IRLR 81*. The applicant was a Detective Constable in the Lincolnshire Constabulary. She attended an off duty drink with some colleagues in a public house one evening after work. During the course of the evening a male colleague pulled up a stool next to her, flicked her hair and re-arranged her collar so as to give the impression that there was a relationship between himself and the applicant. The applicant found this attention to be distressing and moved away from him. On another occasion at a colleagues leaving party that she had attended with her boyfriend, she was accosted on her way to the toilet by the same officer who stated to her 'Fucking hell, you look worth one. Maybe I shouldn't say that it would be worth some money'. The applicant found the remark to be humiliating and brought proceedings for sex discrimination. The EAT upheld the decision of the employment tribunal that the statement amounted to a detriment.

Failures to investigate or protect

H1715 Where an employee brings an allegation of sexual harassment against another employee, the employer has a duty to carry out a proper investigation into the complaint and to take such further steps as are necessary in the light of the findings of the investigation. A failure to carry out such a proper investigation is capable of amounting to a detriment in its own right. The following cases are examples of alleged failures to investigate or protect employees.

(*a*) *Balgobin v London Borough of Tower Hamlets [1987] IRLR 401*. The employees were female cleaners working in a local authority hostel. They alleged that a male cook at the hostel had sexually harassed them. The employer suspended the cook and carried out a full investigation into the allegations that resulted in a failure to find that the complaints of the applicants had been proved. Consequently, the cook was allowed to return to work and the cleaners were required to carry on working with him The employees complained that they were subjected to the detriment of having to work with the cook. The EAT held that there had been no discrimination since it was possible that a male cleaner to whom homosexual advances had been made could have been treated in a similar fashion by the employer.

(*b*) *Bracebridge Engineering Ltd v Darby [1990] IRLR 3*. (See H1714 for the facts.) The assault that took place on the employee was coupled with a failure to properly investigate the incident on the part of the employer who, through its personnel manager, simply accepted that it was the word of two employees against one and that there was insufficient evidence to warrant further investigation. The EAT held that where serious allegations of sexual harassment are made, employers have a duty to properly investigate them.

(*c*) *Burton v De Vere Hotels [1996] IRLR 596*. The case took place under the provisions of both the *SDA 1975* and the identical provisions of the *RRA 1976*. The case was brought by a waitress at a hotel and concerned the detriment of suffering sexual and racial harassment from a third party not connected to the employer but whom over the employer was able to exercise control (the case concerned the applicant being subjected to sexist and racial abuse masquerading as comedy from a well known comedian at a hotel function. The applicant was humiliated in front of the audience by the comedian). Although the employer apologised to the employee for the actions of the comedian, the employee lodged proceedings for sexual and racial harassment and succeeded in the claim. The EAT held that, where an employer was in a position to exercise control over a third party so that discrimination would either not occur, or the extent of it would be reduced, the employer would be subjecting an applicant to a detriment by failing to take the steps necessary to control the employee.

(*d*) *Chessington World of Adventures Ltd v Reed [1997] IRLR 556*. (See above at H1713.) Added to the campaign of harassment that was suffered by the employee was a failure to curb the harassment when complaints were made by the applicant. This was held to be capable of amounting to a detriment since the employer was plainly able to stop the harassment.

(*e*) *Waters v Commissioner of Police for the Metropolis (No 2) [2000] IRLR 720*. The employee was allegedly subjected to bullying and a failure to investigate her complaint of rape. The House of Lords held that an employee owes a duty of care to investigate allegations of rape, bullying and harassment.

Victimisation

H1716 Victimisation is a further category of discrimination that exists under the Discrimination Acts. Victimisation exists under the *SDA 1975, s 4*, the *RRA 1976, s 2*, and the *DDA 1995, s 55*.

Victimisation occurs where a person treats another person less favourably that he treats or would treat other persons and does so for an inadmissible reason. The inadmissible reasons are set out in the *SDA 1975, s 4*(*1*), the *RRA 1976, s 2*(*1*), and the *DDA 1995, s 55*(*2*), respectively.

The victimisation provisions are aimed at preventing an employer from curtailing the right of an employee to legitimately protect his position opposite the employer. The leading case is the House of Lords decision in *Nagarajan v London Regional Transport [1999] IRLR 572* (decided under the provisions of the *RRA 1976*). Their Lordships held, by a majority (and in the face of a powerful dissenting speech by Lord Browne-Wilkinson), that victimisation occurs where less favourable treatment is meted out to a complainant because the victimised person had done an act protected under the provisions of the Act. Their Lordships went on to add that whether the discriminator intended to engage in an act of discrimination towards the victim was irrelevant. Accordingly, it is sufficient for a victimisation claim to arise that an employer is aware of the existence of the fact that an employee has a protected right under the Discrimination Acts and then unconsciously treats the employee less favourably than he would treat another not having a protected right.

However, in *Ledeatte v Tower Hamlets London Borough Council (26 June 2000)*, the EAT confirmed that there must be a causal link between the act complained of and the protected act.

Remedies

H1717 The remedies that are available to an applicant successfully bringing a victimisation complaint are the same as those arising in relation to sex discrimination — see H1725.

Employers' liability

Vicarious liability

H1718 The legal problem that occurs in connection with harassment is that, often, it is not the employer that directly subjects a victim of harassment to the course of conduct; rather it is a fellow employee who is responsible. The employer may then be held responsible for the employee's actions on the grounds of vicarious liability.

The problem that arises in legal theory with vicarious liability is that employers do not employ staff to harass other members of the workforce. Therefore, it is necessary to determine what amounts to 'the course of employment' in order to see when an employer will be liable for the harassment of its employees by other members of staff.

The course of employment

H1719 The first case to take as a line of defence the fact that a harasser was not acting in the course of his employment was *Tower Boot Co Ltd v Jones [1997] IRLR 168*. The case concerned proceedings brought under the provisions of the *RRA 1976* in respect of racial harassment. The applicant had been subjected to a particularly vicious campaign of name-calling and physical assaults. An Employment Tribunal found that this constituted racial harassment and that the employer was vicariously liable for the acts of the employees engaged in the course of conduct. The employer appealed to the EAT which found for the employer on the ground that the common law test for vicarious liability had not been satisfied. The EAT took the view that for this test to be met the aggressor employees had to be engaged in doing the business of the employer, albeit in an improper manner, i.e. the act had to be a different mode of doing what the aggressor employees were employed to do. The EAT then held that on the facts, the actions of the aggressor employees could not 'by any stretch of the imagination' be found to be an improper method of doing what the aggressors were properly employed to do.

The EAT decision was widely condemned on the grounds that the more serious that harassment was, the less likely that an employer would be found to be vicariously liable for it. Indeed, the logical conclusion to the EAT's decision was that, since people were not employed to harass other members of staff, employers would never be liable, a surprising result indeed. Unsurprisingly, the decision was appealed to the Court of Appeal which readily found for the employee. In doing so, the court held that the words, 'in the course of his employment' in the *RRA 1976, s 32(1)*, were not subject to the common law rules in the Salmond test which would artificially restrict the natural everyday sense that every layman would understand of the wording in *s 32(1)*. Consequently, harassment by an aggressor employee did not have to be connected with acts authorised to be done as part of his work. In so deciding, the Court of Appeal attacked the reasoning of the EAT, stating that its interpretation of the section creating liability:

> '... cut across the whole legislative scheme and underlying policy which was to deter racial and sexual harassment through a widening of the net of responsibility beyond the guilty employees themselves by making all employers additionally liable for such harassment'.

Requirement of causal link with employment

H1720 It is not every act of harassment that is committed by an employee that will create a liability on the part of the employer. There must still be a causal link with employment. A good example of this point is the case of *Waters v Commissioner of Police of the Metropolis [1997] IRLR 589*. In this case, the complaint was one of sexual harassment in the form of an alleged sexual assault by a male police officer on a female police officer in her section house room after they had returned from a late night walk together. Given that the attack took place outside the normal working hours of the applicant, the employment tribunal refused to follow the decision in the *Tower Boot* case and held that the applicant was in the same position as if the alleged assailant had been a total stranger to the employer. The alleged assailant did not live in the section house, both officers were both off duty and there was nothing linking the alleged assailant to the employer other than the fact that he was a police officer. The Court of Appeal ultimately upheld the decision of the tribunal on appeal.

This said, the decision in *Waters* does not provide an absolute defence to harassment undertaken after employees leave work for the day. Whether there is a causal link between the harassment and the employment, thereby attaching liability to the employer, will be a question of fact in each particular situation. In the subsequent proceedings brought by Ms Waters in respect of bullying and failure to investigate her complaint, the House of Lords held that an employer owes a duty of care to investigate allegations of rape, bullying and harassment (*Waters v Commissioner of Police for the Metropolis (No 2) [2000] IRLR 720*). An example of this point is *Chief Constable of the Lincolnshire Police v Stubbs [1999] IRLR 81*. The complainant had been the subject of distressing personal comments of a sexual nature from a fellow police officer made at a colleague's leaving party (see above at H1714). The Employment Tribunal at first instance held that since the party was an organised leaving party, the incidents were 'connected' to work and the workplace. It stated that the incident:

> '… would not have happened but for [WPC Stubbs'] work. Work-related social functions are an extension of employment and we can see no reason to restrict the course of employment to purely what goes on in the workplace'.

The EAT emphasised, however, that it would have been different 'had the discriminatory acts occurred during a chance meeting' between the employee and the harasser outside of the workplace.

The decisions give a wide degree of discretion to tribunals on the facts of a particular case to decide whether harassment is committed in the course of employment. If a tribunal finds that it is not committed in such circumstances, the effect is that, subject to *Waters (No 2)*, whilst an employee may be able to sue the harasser in tort, the remedy that is provided may well be empty if the harasser does not have the means to pay for his wrongdoing.

Employer's defence

H1721 A defence is provided to employers in relation to allegations that they are vicariously liable under, respectively, the *SDA 1975, s 41(3)* and the *RRA 1976, s 32*. The defence provides that:

> 'In proceedings brought under this Act against any person in respect of an act alleged to have been done by an employee of his it shall be a defence for that person to prove that he took such steps as were reasonably practicable to prevent the employee from doing that act, or from doing in the course of his employment acts of that description'.

A similar, although not identical, provision exists under the *DDA 1995, s 58(5)*.

The problem for employers with the defence is that, whilst it will provide a defence in a situation where an employer has a full procedure dealing with harassment (including such matters as monitoring, complaints handling and disciplinary proceedings), the provision requires that the employer must pay more than mere lip service to the existence of these matters and must also be able to show that its employees were aware of the procedures and, where relevant, that the employer had fully implemented the procedures in practice. A good example of a case where the defence succeeded was *Balgobin v London Borough of Tower Hamlets [1987] IRLR 401* (see H1715 above). Here, the alleged harasser was suspended following the initial allegations, his conduct was fully investigated and it was found that there was insufficient evidence against the cook to justify the claims. The EAT was moved to comment that it was difficult to see what further steps the employer could have taken to avoid the alleged discrimination.

Ultimately, if an employer has gone to the expense of having policies and procedures drafted then they should not be kept in the office safe. Employees should be left under no illusions as to what will happen in cases of harassment.

Liability of employees

H1722 Although an employer may be vicariously liable for acts of harassment, primary responsibility for such action will remain with the employee who has committed the discrimination and, to this end, the *SDA 1975, s 42*, and the *RRA 1976, s 33*, provide as follows.

> '(1) A person who knowingly aids another person to do an act made unlawful by this Act shall be treated for the purposes of this Act as himself doing an unlawful act of the like description.
>
> (2) For the purposes of subsection (1) an employee or agent for whose act the employer or principal is liable under section [*SDA 1975, s 41* or *RRA 1976, s 32*] (or would be so liable but for section [*41(3)* or *32(3) respectively*]) shall be deemed to aid the doing of the act by the employer or principal.'

The scope of the provision was given a liberal interpretation in *AM v WC & SPV [1999] IRLR 410*. Here, the EAT held that an Employment Tribunal had erred in not allowing a police officer to bring a complaint of harassment against another individual officer. The EAT considered that the construction of *SDA 1975, s 41(1)*, *(3)*, and *SDA, s 42*, was to make an employer vicariously liable for acts of discrimination committed by staff (*SDA 1975, s 41(1)*), to provide conscientious employers with a defence for the actions of staff where the actions are effectively a 'bolt from the blue' (*SDA 1975, s 41(3)*) and to allow an employee who has committed acts of discrimination against another to be joined as a 'partner' to proceedings commenced against an employer (*SDA 1975, s 42*). The EAT added that even where a defence did exist for an employer under *s 41(3)*, there was nothing in the SDA 1975 that prevented the victim of discrimination from launching proceedings against a discriminator employee alone.

The defence has also been considered by the House of Lords in *Anyanwu v South Bank Student Union [2001] IRLR 305*, where it was held that the words 'knowingly aids' means that the person providing aid must have given some kind of assistance to the person committing the discriminatory act which helps the discriminator to do the act. The amount or value of the help is of no importance. All that is needed is an act of some kind, done knowingly, which helps the discriminator to do the unlawful act.

The provisions under *DDA 1995, s 58(1)* and (*5*) are similar, although not identical, to those under the *SDA 1975* and *RRA 1976*.

Narrowing the extent of employers' liability

H1723 Where an employee is able to establish that he or she has been the victim of unlawful harassment, there can be circumstances where the damages that would otherwise be awarded to a victim of such discrimination will be dramatically reduced. In the context of sexual harassment, this will be where the victim's conduct shows that she would not have suffered a serious detriment as a consequence of the harassment. For example, in *Snowball v Gardner Merchant Ltd [1987] IRLR 397*, the applicant complained that she had been sexually harassed by her manager. The allegations were denied and the applicant was subjected to questioning at the tribunal hearing as to her attitude to matters of sexual behaviour so as to show that even if she had been harassed, her feelings had not been injured. In particular, evidence was called as to the fact that she referred to her bed as her 'play pen' and that she had stated that she slept between black satin sheets. On appeal, the EAT held that the evidence had been correctly admitted so as to show that the applicant was not likely to be offended by a degree of familiarity having a sexual connotation.

Likewise, in *Wileman v Minilec Engineering Ltd [1988] IRLR 144*, the EAT refused to overturn an award of damages in the sum of £50 that had been made for injury to feelings. In this case, the applicant had made a complaint of sexual harassment against one of the directors of her employer. The award had been set at this figure on account of the fact that the victim had worn revealing clothes and it made it inevitable that comment would be passed.

With respect to the EAT, these cases, whilst reflecting that harassment can still occur in cases where a complainant has a high degree of confidence in sexual banter, unfortunately appear to send out the wrong message in that they can be construed as being as a 'harasser's charter' and do not seem to recognise that a person may be particularly upset by the unwelcome attentions of a specific person. It is worth noting that both of the decisions pre-date the EC Code of Practice and the enactment of the *Human Rights Act 1998* and it will be interesting to see if the effect of the decisions is in any way curbed at a future date by these matters.

Legal action and remedies

H1724 Where a person unlawfully harasses another, the consequences of such action can be extremely far-reaching and potentially spread across the full range of actions in both civil and criminal law.

Unlawful discrimination

Elements of the tort

H1725 The most obvious form of action that a victim of harassment may consider is direct sex and race discrimination under *SDA 1975, s 1(1)(a)*, and *RRA 1976, s 1(1)(a)*, respectively. Liability arises because a person engaged in harassing another subjects that other to a detriment (*SDA 1975, s 6(2)(c)*, and *RRA 1976, s 4(2)(c)*). This head of liability is fully covered at H1703 above and the paragraphs that follow.

Disability discrimination may also occur under *DDA 1995, s 5(1)*, where the employer treats the employee less favourably on account of the employee's disability and the employer cannot show the less favourable treatment to be justified on disability neutral grounds.

Remedies

H1726 The remedies that are open to a victim of unlawful harassment are:

(*a*) a declaration;

(*b*) recommendation; or

(*c*) compensation.

A declaration is a formal acknowledgement from an employment tribunal that discrimination in the form of harassment has occurred. It will be granted in every successful case brought before a tribunal.

Recommendations are used where an applicant to an employment tribunal remains in the employment of her employer. The purpose of a recommendation is for a tribunal to give guidance to an employer relating to its business for the purposes of eliminating discrimination.

However, the main remedy that victims of harassment will seek is compensation for the loss that has been suffered as a result of the harassment including, particularly, injury to feelings.

Unfair dismissal

H1727 Unfair dismissal can arise as a claim where an employer fails to control unlawful harassment in the workplace or fails to properly investigate complaints of unlawful harassment. Such failures may and probably will also act to undermine the employee's trust and confidence in the employer. The result of such actions is that an employee may be able to claim that the employer has constructively dismissed the employee, thereby amounting to a dismissal for the purposes of the *Employment Rights Act 1996, s 95(1)(c)*. Indeed, where a unlawful harassment claim has been brought on the back of an employee's departure from his or her employment, most cases will also include a claim for unfair dismissal as well.

Proving constructive dismissal

H1728 The first hurdle that an employee has to overcome is to prove that he or she has been dismissed for the purposes of the *Employment Rights Act 1996* (*ERA 1996*), *s 95(1)(c)*. It is trite law that for an employee to show that the employee has been constructively dismissed the acts of his or her employer must show that the employer, by its conduct, evinced an intention not to be bound by the terms and conditions of the contract of employment (see, to this effect, *Western Excavating (ECC) Limited v Sharp [1978] IRLR 27*). The problem that an employee may face in this regard is that if an employer has in place a full anti-harassment policy that is properly implemented, the employee may fail to establish that the employer has acted in breach. For an example of this occurring in practice see *Balgobin v London Borough of Tower Hamlets [1987] IRLR 401* (above at H1715).

Likewise, an employee may also fail to show that an employer has breached the implied term of trust and confidence where the act is a serious one off act of harassment that the employer was unable to predict would happen and where the employer properly deals with the matter immediately upon a complaint being made. To the extent that the employer does not properly investigate a complaint in such circumstances it is almost inevitable that the employer will breach the contract of employment (see *Bracebridge Engineering Ltd v Darby [1990] IRLR 3*).

It should be noted that conduct by an employer towards an employee that may amount to a constructive dismissal for the purposes of *ERA 1996, s 95(1)(c)*, has been held by the EAT in *Commissioner of Police for the Metropolis v Harley [2001] IRLR 190* not to entitle an employee to resign and claim that he/she has been dismissed for the purposes of *DDA 1995, s 4(2)(d)*, because there is no statutory concept of constructive dismissal defined in the *DDA 1995*. The decision runs contrary to the decision of the EAT in *Derby Special Fabrication Limited v Burton [2001] IRLR 69* on the identical wording in the *RRA 1976*. Dismissal is defined by *SDA 1975, s 82(1A)*, as to include a constructive dismissal. With respect to the EAT, it is arguable that the decision in *Harley* is wrong in law given that the dismissal provision is common to all three of the Discrimination Acts. In any event, even if the EAT in *Harley* is correct in holding that constructive dismissal does not arise under the *DDA 1995*, conduct by an employer towards an employee that would amount to a constructive dismissal for the purposes of *ERA 1996, s 95(1)(c)*, may amount to a 'detriment' for the purposes of *DDA 1995, s 4(2)(d)*.

Finally, it will be recalled, that, following, *Smith v Gardner Merchant Limited [1998] IRLR 520*, sexual harassment on the grounds of sexual orientation is difficult prove since the applicant will have to show that the person subjecting her to harassment would also treat a homosexual man in like fashion. This does not release an employer from an obligation to investigate since such conduct on the part of fellow employees, if left unchecked by the employer, may breach the implied duty of the employer to maintain the employee's trust and confidence.

Remedies

H1729 The remedies that can be provided for unfair dismissal are those set out in *ERA 1986, ss 113–118*, namely:

(*a*) an order for reinstatement;

(*b*) an order for re-engagement; or

(*c*) an order for compensation.

The practical consequence of an employee claiming that he or she has been constructively dismissed as a result of an employer's breach of the implied duty of trust and confidence is that a tribunal is likely to be unwilling to make an order of reinstatement or re-engagement and therefore the reality is that most cases will provide for compensation to be awarded. The downside to claiming unfair dismissal as opposed to bringing a claim for unlawful discrimination is that, whereas, the compensation for sex discrimination is both unlimited as regards the amount that can be awarded and can contain an award for injury to feelings, the compensation that is available for unfair dismissal is subject to a cap of £52,600 and can include an award for the manner of dismissal (following *Johnson v Unisys Ltd [2001] IRLR 279*).

Actions in tort

H1730 Broadly speaking, torts are the category of actionable civil wrongs not including claims for breach of contract or those arising in equity. Acts of unlawful harassment, depending upon the particular characteristics of the harassment, are capable of amounting to a variety of different torts.

Assault and battery

H1731 Assault and battery are forms of trespass to the person. They are separate torts and occur where a person fears that either he or she will receive or has been the subject of unlawful physical contact. As was stated of the torts in *Collins v Wilcock [1984] 3 All ER 374*:

'An assault is an act which causes another person to apprehend the infliction of immediate, unlawful, force on his person; a battery is the actual infliction of unlawful force on another person'.

For the tort of assault to be committed it is irrelevant that the victim is not in fear of harm. Consequently, it is possible for an assault to occur where a victim of harassment is threatened with the possibility that he or she will receive unwanted physical contact. All that must exist for the tort to be committed is that the victim must have a reasonable apprehension that she will imminently receive unwanted contact.

Further, for either tort, the use of the word 'force' in the formulation of the tort does not mean physical violence. It is not necessary that the victim be subjected to a full-scale attack (as occurred for example in *Bracebridge Engineering Ltd v Darby [1990] IRLR 3* and *Tower Boot Company Limited v Jones [1995] IRLR 529*). It is possible for the torts to be committed where a person brushes past another suggestively (as in *Porcelli v Strathclyde Regional Council [1986] IRLR 134*) or even kisses another without consent. The torts therefore cover the full range of possible forms of contact related harassment from physical injury to unwanted molestation and are capable of providing a cause of action even though no actual physical harm has been occasioned to the victim of the particular tort.

Negligence

H1732 While negligence may imply to the layman that an act committed against a person has been committed carelessly, it is perfectly possible for the tort to be committed intentionally by a person. The classic phrasing of the tort is derived from the case of *Donoghue v Stevenson [1932] AC 562*, where Lord Atkin stated of the elements of the tort:

> 'You must take reasonable care to avoid acts or omissions which you can reasonably foresee would be likely to injure your neighbour … [who are] persons who are so closely and directly affected by my act that I ought reasonably to have them in contemplation as being so affected when I am directing my mind to the acts or omissions which are called in question'.

The test thus framed ensures that liability will be created where there is reasonable foresight of harm being inflicted upon persons foreseeably affected by an actor's conduct. Consequently, the conduct can be intentional or reckless as well as careless.

Where it is the case that an employee harasses a fellow employee, the harasser and victim are likely to be within the relationship of proximity required by the law (it being recognised that fellow employees owe a duty of care to each other).

Once it has been established that a duty is owed, three other factors become relevant these being whether harm has been sustained by a victim, whether the magnitude of the harm suffered by the victim is far greater than the actor intended and whether an employer has a defence if an action is brought for vicarious liability.

(*a*) *Requirement for loss.* For the tort of negligence to be committed, the victim must suffer loss or damage flowing from the negligent act. An obvious form that is likely to be sustained in the case of harassment is in the form of stress related illnesses that can occur from prolonged harassment (nervous shock). It is well recognised in law that where a person suffers from medically treatable stress related illnesses that are caused by the negligence of an employer, it is possible to recover damages for the injuries so inflicted (see to this end the non-harassment case of *Walker v Northumberland County Council [1995] ICR 702*).

(*b*) *'Eggshell-skulls'*. The extent of the injury suffered, provided that a victim could foreseeably have suffered a particular form of harm (e.g. distress) from the commission of the unlawful act, it is irrelevant that the final extent of the harm would not necessarily have been foreseen. This principle is known as the 'eggshell-skull' rule. The law recognises that it is possible for 'eggshell-personalities' to exist (see, for example, *Malcolm v Broadbent [1970] 3 All ER 508*). It is irrelevant that another female would not have suffered nervous shock as a consequence of harassment being meted out to her provided that the hypothetical female would have been distressed by it.

(*c*) *Vicarious liability*. Finally, there is the question as to whether it is possible for an employer to be sued for harm that is occasioned in negligence from an act of harassment. Whilst the case of *Chief Constable of the Lincolnshire Police v Stubbs [1999] IRLR 81* recognised that employers could become vicariously liable for the acts of employees for the tort of sex discrimination (due to the provisions of *s 41(1), SDA 1975*), it will also be recalled that in the earlier case of *Tower Boot Co Ltd v Jones [1997] IRLR 168* (dealing with racial harassment), it had been recognised by the EAT that, since, at common law, employers can only be liable for the misperformance of an act that an employee is performed to do, and, since an employer does not employ people to harass others, it should not be possible at common law for an employer to be vicariously liable for the harassment of an employee. Whilst the decision was overturned on appeal, the appeal only dealt with the creation of vicarious liability under the anti-discrimination legislation. In the circumstances, it is conceivable that an employer would still have a defence to a claim for negligence. However, where the claim is framed on the grounds that the employer was aware of the harassment yet failed to act when called upon to do so and harm was occasioned by the employer's failure to act, vicarious liability is not an issue since what is being complained of is the employer's primary failure.

Given that it is possible to recover damages for injury to feelings occasioned as a consequence of unlawful discrimination, it may well be the case that an employee will not want to pursue a claim for negligence. This said, it is not possible to obtain an injunction in respect of unlawful harassment under the Discrimination Acts.

Remedies in tort

H1733 Where an employee commits an actionable tort outside of harassment (the remedies for which are provided under the *Protection from Harassment Act 1997* — see below at H1734), two remedies will commonly be available to the victim. These will be:

(*a*) damages (which is available as of right for a tort); and

(*b*) an injunction (where damages are not adequate).

In tort, the correct measure of damages is that the victim should be given such a sum by way of compensation as would put the victim in the position as if the tort had not been committed (*Livingstone v Rawyards Coal Co (1880) 5 App Cas 25*). As to how much this will be is a question of fact in each case and, potentially, in the case of a course of harassment that induces nervous shock preventing an employee from working, could run into hundreds of thousands of pounds.

However, the problem incurred with the common law solution of awarding damages to the victim of a tort in the context of unlawful harassment is that the victim may not be properly compensated for on-going acts. In the circumstances, damages may not be adequate in such a situation and the better remedy may be to sue for an injunction.

Further, the use of an injunction to stop on-going harassment may be of more practical benefit to a currently distressed employee than would a future award of damages. This is because injunctions can be applied for and obtained relatively speedily as an intermediate remedy pending a full liability trial.

The problem with pursuing this remedy is that, assuming that the injunction is applied for whilst both the harasser and the victim are still in the employment of their employer, the victim's faith in his or her employer may well have become exhausted (by the employer's failure to take action in relation to the employee). Indeed, it would be surprising if this were not the case. Consequently, in many cases, the better course of action may well be to leave employment on the grounds of constructive dismissal and sue for compensation under the relevant Act.

Protection from Harassment Act

H1734 An offence and tort of harassment was created by the *Protection From Harassment Act 1997*. The legislation was passed to protect victims of 'stalking' but there is no reason why its provisions cannot be used to protect victims of sexual harassment in the workplace. Harassment is not specifically defined by the 1997 Act, although *s 7* provides that it must amount to a course of conduct causing alarm or distress to a person.

The phrase 'a course of conduct' is defined to be conduct on two or more occasions and 'conduct' includes words (*Protection From Harassment Act 1997, s 7(4)*).

It is possible to obtain an injunction against a harasser under *s 3* of the Act and also to claim damages.

Breach of contract

H1735 Where an employee harasses a fellow employee, the harasser will almost certainly breach one or more of the terms of his or her employment contract. In the first instance, most disciplinary procedures provide that harassment of fellow employees on grounds of sex, race or disability amounts to gross misconduct. Given that many contracts of employment provide that a disciplinary procedure is contractual, harassment per se may provide grounds for summary dismissal.

Even if an employer fails to provide a contractual disciplinary procedure, an employee may well be in breach of several of the implied terms under his contract of employment if he or she is engaged in unlawfully harassing a fellow employee. This is because the employee will owe a duty of care to the employer that requires the employee to take reasonable care for fellow employees. Deliberate unlawful harassment will breach this term.

Further, employees must maintain the trust and confidence of the employer during their employment. Unlawful harassment of fellow employees will undermine a victim's confidence and would provide grounds for a conduct related dismissal of a person committing acts of harassment.

Remedies

H1736 As with remedies provided for torts, the remedies for breach of contract will be either damages or an injunction.

The law places a different emphasis on the method of computing damages for breach of contract to that provided in relation to torts. Practically though, in the context of a contract of employment, the final outcome will usually be the same. In the law of contract, the victim is to be placed so far as possible in the position as if the contract had been properly performed by the employer (*Robinson v Harman (1848) 1 Exch 850*). The amount payable will vary according to the loss that has been suffered by a victim. Ordinarily, if the employee leaves his/her employment claiming constructive dismissal, the employee will be able to claim damages for loss of contractual benefits during her notice period (subject to the employee's duty to mitigate). However, it seems that in spite of the decision of the House of Lords in *Malik v BCCI [1997] IRLR 462*, following the subsequent decision of the House of Lords in *Johnson v Unisys Ltd [2001] IRLR 279*, it is not possible to recover damages for injury to feelings occasioned as a result of the humiliating manner of a dismissal.

This said, if the claim is framed by an employee as one going to the employer's duty of care to the employee (due to stress related illness being induced by harassment and the employer's unreasonable failure to prevent the cause of the stress) it is well established that an employee will be able to claim damages for the illness so caused (see, to this effect, *Walker v Northumberland County Council [1995] ICR 702*).

See H1733 for a discussion on injunctions.

Criminal action

H1737 The conduct of a person who sexually harasses another may also amount to one or more of the following crimes:

(*a*) battery;

(*b*) harassment under the *Protection from Harassment Act 1997* (see above at H1734);

(*c*) harassment under the *Public Order Act 1986*;

(*d*) racially aggravated harassment under the *Crime and Disorder Act 1998*.

Battery

H1738 Battery is a common law offence. The elements of the offences are that the harasser intentionally and unlawfully inflicts personal violence upon the victim. In *Collins v Wilcock [1984] 3 All ER 374* the Court of Appeal held that there was a general exception to the crime which embraced all forms of physical contact generally acceptable. However, it is clear from this that if the victim makes known to his/her harasser the fact that the victim does not welcome physical contact from him, the harasser will commit the offence simply by touching the victim.

The offence can only be tried summarily.

Harassment under the Protection from Harassment Act 1997

H1739 The tortious aspects of the *Protection from Harassment Act 1997* are considered at H1734 above. The 1997 Act also criminalises a course of conduct pursued by a person who 'knows or ought to know' that the conduct amounts to harassment. *Section 1(2)* of the Act provides that a person will be deemed to have knowledge where a reasonable person in possession of the same information as the harasser would think that the course of conduct amounted to harassment.

The offence is one that is triable either way. If the offence is tried on indictment, the maximum penalty is five years imprisonment or a fine or both. If tried summarily, the maximum penalty is a term of imprisonment not exceeding six months, a fine subject to the statutory maximum or both.

Harassment under the Public Order Act 1986

H1740 The offence exists under the *Public Order Act 1986, s 4A* and criminalises threatening, abusive or insulting words of behaviour thereby causing harassment, alarm or distress. The offence can be committed in a public or private place (other than a dwelling). Consequently, sexual harassment in the workplace is capable of being covered by the Act.

The offence can only be tried summarily and a fine on level 3 of the standard scale can be levied upon conviction.

Racially aggravated crimes under the Crime and Disorder Act 1998

H1741 The offence of racially aggravated harassment arises under the *Crime and Disorder Act 1998* (*CDA 1998*), *s 32*. The offence is committed where a person commits an offence against another for the purposes of the *Protection from Harassment Act 1997* and the offence is racially aggravated (as defined by *CDA 1998, s 28*).

The offence in either case is triable either way. In the event that a person is convicted on indictment of racially aggravated harassment under *CDA 1998, s 32*(*1*)(*a*), the person can be sentenced to imprisonment for a term of up to two years, a fine or to both, or for putting people in fear of violence under *CDA 1998, s 32*(*1*)(*b*), the person can be sentenced to imprisonment for a term of up to seven years, a fine or to both.

Where a person is summarily convicted of racially aggravated harassment under *CDA 1998, s 32*(*1*)(*a*), the person can be sentenced to imprisonment for a term of up to six months, or a fine not exceeding the statutory maximum or to both, or for putting people in fear of violence under *CDA 1998, s 32*(*1*)(*b*), the person can be sentenced to imprisonment for a term of up to six months or a fine not exceeding the statutory maximum or to both.

Preventing and handling claims

Harassment policy

H1742 The Equal Opportunities Commission (EOC) recommends the creation and use an of equal opportunity policy by employers. In particular, the EOC's Code of Practice states:

> 'The primary responsibility at law rests with each employer to ensure that there is no unlawful discrimination. It is important, however, that measures to eliminate discrimination or promote equality of opportunity should be understood and supported by all employees. Employers are therefore recommended to involve their employees in each opportunity policies (para 4).
>
> An equal opportunities policy will ensure the effective use of human resources in the best interests of both the organisation and its employees. It is a commitment by an employer to the development and use of employment procedures and practices which do not discriminate on grounds of sex or marriage and which provide genuine equality of opportunity for all employees. The detail of the policy will vary according to size of the organisation (para 34)'.

In order for the policy to be effective and properly used, it must have the full support of senior managers within the business and must be used fairly and properly on each occasion that its provisions become relevant. So as to achieve this, it is important that the policy is clear and made known to all employees (para 35). A harassment policy is an essential feature of a proper equal opportunities policy and should exhibit the following characteristics:

(*a*) a complaints procedure;

(*b*) confidentiality;

(*c*) a method for informal action to be taken; and

(*d*) a method for formal action to be taken.

Complaints procedure

H1743 The hallmark of an effective policy is that provision is made for a complaints procedure that both encourages justifiable complaints of harassment to be made and protects the maker of the complaint. By way of guidance, the European Commission ('EC') Code of Practice states in this regard (at section B paragraph (iii)) that a formal complaints procedure is needed so as to:

> '... give employees confidence that the organisation will take allegations of sexual harassment seriously'.

The EC Code also states that such a procedure should be used where employees feel that informal resolution of a dispute is inappropriate or has not worked.

In order for the complaints procedure to be effective, the EC Code recommends that employees should know clearly to whom to make a complaint and should also be provided with an alternative source of person to whom to complain in the event that the primary source is inappropriate (e.g. because that person is the alleged harasser).

The EC Code adds that it is good practice to allow complaints to be made to a person of the same sex as the victim.

Confidentiality

H1744 In order for complaints to be made effectively, it is necessary that some measure of confidentiality be maintained in the complaint making procedure. The problem that arises here is that there is an inevitable trade-off between the right of the victim of sexual harassment to be protected against recrimination and the right of an alleged harasser to know the case that he/she is meeting for disciplinary purposes. Indeed, to the extent that complaints are made against a person who is not told where, when and whom he/she is alleged to have harassed and the alleged harasser is then disciplined as a result of the complaint, there is a chance that the employer will be acting in breach of the duty to maintain the alleged harasser's trust and confidence.

To this end, the requirement for confidentiality exists to the extent that although a harasser may have the right to know the full nature of the complaint that he/she has to meet, the victim should be protected from recrimination in the workplace by the employer ensuring that all details surrounding the complaint are suppressed from other employees in the workplace save to the extent that it is necessary to involve them to either prove or disprove the allegations of harassment.

Further, the requirement of confidentiality also extends to ensuring that the allegations, to the extent that they are made in good faith, are not subsequently used against the complainant for any other purposes connected with the complainant's employment (thereby tying in with the duty of the employer to ensure that there is no victimisation occasioned to the complainant — see *Nagarajan v London Regional Transport [1999] IRLR 572* dealing with the issue of victimisation).

Informal action

H1745 Most victims of harassment simply want the unwanted attention or conduct towards them to stop. Whilst the employer's policy should leave employees in no doubt as to the fact that deliberate harassment will amount to a case of gross misconduct being brought against an employee, it may be the case that the form of the harassment is innocent in nature and can be remedied by informal action. As the EC Code points out, it may be the case that the victim of harassment can simply explain to the harasser that the conduct that he/she is receiving is unwelcome and makes the victim feel uncomfortable in the workplace. The EC Code goes on to state that where the victim finds it embarrassing or uncomfortable to address the harasser, it may be possible for a fellow employee (in the case of a larger organisation possibly a confidential counsellor) to make the requests to the harasser.

This said, in some cases it simply will not be possible or appropriate for the matter to be dealt with informally and where this is the case, the only possible resolution will be through formal channels.

Formal action

H1746 In the event that formal action is required to end harassment, both harassers and their victims should be made fully aware as to the procedures that an employer will adopt in relation to a claim of harassment. The EC Code provides an illustration as to the type of features that should be included in a policy. This will include providing details as to:

(*a*) how investigations are to be carried out;

(*b*) the time frame for investigations (which should be as short as possible for the benefit of both the victim of actual harassment and the alleged harasser);

(*c*) whether there is to be a period of suspension whilst a claim is investigated; and

(*d*) the penalty for an employee who is found to have engaged in harassment

The EC Code concludes that it may be necessary to require employees to be transferred to different duties or sites (in some cases even where a case is not proven against an alleged harasser so as to prevent the employees being forced to work with each other against their will). Indeed, where they are required to do so in the face of an unproved allegation, unless the matter has been thoroughly investigated by the employer, it may be a recipe for litigation as the case of *Balgobin v London Borough of Tower Hamlets [1987] IRLR 401* illustrates.

Training

H1747 Another of the statements made by the EC Code is that training of the workforce is essential in order to ensure that harassment does not occur in the first place. The EC Code specifically focuses on training for supervisory staff and managers. The recommendations for training include:

(*a*) identifying the factors contributing to a working environment free of sexual harassment;

(*b*) familiarisation of employees with the employer's harassment policy;

(*c*) specialist training of those employees required to implement the harassment policy; and

(*d*) training at induction days for new employees as to the employers policy on harassment.

Monitoring and review

H1748 The acid test of a company's equal opportunity policy is not how it looks on paper but whether and how it is implemented and monitored in practice. As to monitoring, much will depend upon the facts relating to particular employers. In the case of smaller employers, it may simply require the employer to be aware of any recommendations that are proposed by bodies such as the EOC or the Chartered Institute of Personnel and Development. However, where employers are larger, the monitoring function may require a full comparative analysis of complaints made in order to determine what matters need to be implemented with regard to training and whether any changes need to be made to the employer's policy for the purposes of such matters as investigating complaints and tightening up the policy generally.

In short, once the policy is adopted, it must be kept up to date with regard to what is acceptable within the workplace.

The Commission for Racial Equality's 'Code of Practice for the Elimination of Racial Discrimination and the Promotion of Equality of Opportunity in Employment (1983)' and the Secretary of State for Education and Employment's 'Code of Practice for the Elimination of Discrimination in the Field of Employment Against Disabled Persons or Persons Who Have a Disability (1996)' provide evidence of good practice and guidance for employers in relation to racial and disability discrimination.

Future developments

The EOC white paper — 'Equality in the 21st Century: A New Approach'

H1749 The White Paper *Equality in the 21st Century: A New Approach* sets out what needs to be done by legislation to promote equal opportunities for men and women'. It provides a full review of the current legislation under the *SDA 1975* and the *Equal Pay Act 1970* and also examines the impact of the EOC Code of Practice in relation to harassment. The White Paper ultimately concludes that a new, unifying Act dealing with all forms of sex discrimination needs to be enacted in order for the Government to fully tackle sex discrimination.

Recommendations in relation to sexual harassment

H1750 Paragraph 22 of the White Paper specifically deals with the issue of sexual harassment. It recommends that the proposed new single Act should specifically prohibit sexual harassment and adopt the EOC Code definition of harassment for these purposes.

The White Paper goes on to add that the sexual harassment should be extended to cover all areas covered by the Act and not just employment.

It remains to be seen whether the White Paper will be adopted especially in the light of the fact that the Government has announced that it does not propose to enact any further changes to employment law in the lifetime of this current Parliament.

EU proposals

H1751 The European Union (EU) has recently proposed new legislation for outlawing sexual harassment in the workplace across Member States.

If a new law is agreed by the European Parliament the EU will for the first time have a clear definition of sexual harassment. This will be helpful for UK employers and courts who will no longer have to rely on case law and judicial interpretation to deal with the basic principles.

The law will reinforce protection for employees, both male and female, who complain about discrimination, and will acknowledge the right of women to return to their jobs after maternity leave.

The proposed new law, which amends the original 1976 directive on sexual harassment and which must be implemented by 2005, does not alter the law in the UK drastically. Much of the proposed law consolidates established UK practice. However, it is likely to have a greater impact on other EU countries. While France and Belgium have laws explicitly banning sexual harassment, most other EU Member States have only partial legislation – and Portugal and Greece have not developed this area of the law.

The proposed law defines harassment:

> 'Sexual harassment shall be deemed to be discrimination on the grounds of sex at the workplace when an unwanted conduct related to sex takes place with the purposes or effect of affecting the dignity of a person and/ or creating an intimidating, hostile or offensive or disturbing environment, in particular if a person's rejection of, or submission to, such conduct is a basis for a decision which affects that person.'

It has always been advisable for employers to have specific policies on equal opportunities and sexual harassment. However, with the proposed developments it seems that employers may now need to be more proactive in their approach and attitude. The proposed new law requires employers to take preventative measures against all forms of discrimination, especially sexual harassment, as well as introducing what are referred to as 'enterprise level equality plans' which are to be made available to workers.

Hazardous Substances in the Workplace

Introduction

H2101 For most employers, the main requirements relating to hazardous substances in the workplace are those contained in the *Control of Substances Hazardous to Health Regulations SI 2002* (*SI 2002 No 2677*) ('*COSHH Regulations*'). This chapter is therefore mainly concerned with the requirements under those Regulations and the assessments they require.

The *COSHH Regulations* were first introduced in 1988 but have been subject to many changes and amendments since. Currently the *Control of Substances Hazardous to Health Regulations 2002* are in force, but these were amended in 2003 (by *SI 2003 No 978*) and further amendments are in the pipeline.

However, there are many other legal requirements affecting what would generally be regarded as hazardous substances, even though some of these do not fall within the definition of 'substance hazardous to health' contained in the *COSHH Regulations*. Relevant regulations include:

- *Chemicals (Hazard Information and Packaging for Supply) Regulations 2002* (*SI 2002 No 1689*) (commonly known as '*CHIP*');
- *Control of Asbestos at Work Regulations 2002* (*SI 2002 No 2675*);
- *Control of Lead at Work Regulations 2002* (*SI 2002 No 2676*);
- *Control of Major Accident Hazards Regulations 1999* (*SI 1999 No 743*) (known as '*COMAH*');
- *Dangerous Substances and Explosive Atmospheres Regulations 2002* (*SI 2002 No 2776*) (known as '*DSEAR*')
- *Dangerous Substances (Notification and Marking of Sites) Regulations 1990* (*SI 1990 No 304*);
- *Ionising Radiations Regulations 1999* (*SI 1999 No 3232*).

The requirements of these regulations are summarised at the end of this chapter and some of them are dealt with more fully in other chapters.

How hazardous substances harm the body

Routes of entry

H2102 Hazardous substances can be present in the workplace in a variety of forms – solids, dusts, fumes, smoke, liquids, vapour, mists, aerosols and gases. In deciding what risks are posed by these substances, three potential entry routes into the body must be considered:

- *Inhalation* – the hazardous substance may cause damage directly to the respiratory tract or the lungs. Inhalation also allows substances to enter the bloodstream via the lungs and thus affect other parts of the body.
- *Ingestion* – accidental ingestion of hazardous substances is always possible, particularly if containers are not correctly labelled. Eating, drinking and smoking in the workplace introduce the risk of inadvertently ingesting small quantities of hazardous substances.
- *The skin* – substances can damage the skin (or eyes) through their corrosive or irritant effects. Some solvents can enter the body by absorption through the skin. Damage is more likely when there are cracks or cuts in the skin.

Occupational ill health

H2103 The adverse effects of hazardous substances are many and varied and are a field of study in their own right. Many types of occupational diseases must be reported to the enforcing authorities under the requirements of the *Reporting of Injuries, Diseases and Dangerous Occurrences Regulations 1995* (*SI 1995 No 3163*) ('*RIDDOR*') and some are eligible for social security benefits under government schemes. Employees suffering from occupational ill health can of course bring civil actions for damages against their employers, former employers or others they consider responsible for their condition.

Some examples of occupational ill health caused by hazardous substances are:

- Respiratory problems
  - *Pneumoconiosis* – e.g. asbestosis, silicosis, byssinosis, siderosis.
  - *Respiratory irritation* – caused by inhaling acid or alkali gases or mists.
  - *Asthma* – sensitisation of the respiratory system which may be caused by many substances, e.g. isocyanates, flour, grain, hay, animal fur, wood dusts.
  - *Respiratory cancers* – e.g. those caused by certain chemicals, asbestos, pitch or tar.
  - *Metal fume fever* – a flu-like condition caused by inhaling zinc fumes.
- Poisoning

  Acute (short-term) or chronic (long-term) poisoning may be caused by:
  - *Metals and their compounds* – particularly lead, manganese, mercury, beryllium, cadmium.
  - *Organic chemicals* – which may affect the nervous system, the liver, the kidneys or the gastro-intestinal system.
  - *Inorganic chemicals* – as above plus possible asphyxiant effects such as those caused by carbon monoxide.
- Skin conditions
  - *Dermatitis*:

    (i) caused by primary irritants where skin tissue is damaged by acids or alkalis or natural oils are removed by solvents, detergents etc.

    (ii) due to the effects of sensitising agents, e.g. isocyanates, solvents, some foodstuffs.

— *Skin cancer* – caused by contact with pitch, tar, soot, mineral oils etc.

- Biological problems

  From contact with animals, birds, fish (including their carcasses and products) or with micro-organisms from other sources, e.g:

  — *Livestock diseases* – e.g. anthrax or brucellosis.

  — *Allergic Alveolitis* – caused by inhaling mould or fungal spores present in grain etc.

  — *Viral Hepatitis* – usually due to contact with blood or blood products.

  — *Legionnaires' Disease* – caused by inhaling airborne water droplets containing the legionella bacteria (which can occur extensively in water at certain temperatures).

  — *Leptospirosis* – Weil's disease, caused by contact with urine from small mammals, particularly rats.

  Information on the harmful effects associated with individual products may be available on packaging but certainly should be contained in a material safety data sheet provided by the manufacturer or supplier.

The COSHH Regulations summarised

Substances hazardous to health

H2104 The definition of 'substance hazardous to health' is contained in the *COSHH Regulations 2002* (*SI 2002 No 2677*), *Reg 2*(*1*) and includes:

- substances designated as very toxic, toxic, corrosive, harmful or irritant under product labelling legislation (the *CHIP Regulations*) – these substances should be clearly identified by the standard orange and black symbols required by the *CHIP Regulations*;
- substances for which the Health and Safety Commission has approved a Maximum Exposure Limit or an Occupational Exposure Standard;
- biological agents (micro-organisms, cell cultures or human endoparasites);
- other dust of any kind, when present at a concentration in air equal or greater to 10 mg/m^3 (inhalable dust) or 4 mg/m^3 (respirable dust) as a time weighted average over an 8-hour period;
- any other substance creating a risk to health because of its chemical or toxicological properties and the way it is used or is present at the workplace;

Given this widely drawn definition, in any cases of doubt it is always advisable to treat the Regulations as applying.

Duties under the Regulations

H2105 The *COSHH Regulations* place duties on employers in relation to their employees and also on employees themselves. [*SI 2002 No 2677, Reg 8*]. The self-employed have duties as if they were both employer and employee. *Regulation 3*(*1*) extends the employer's duties 'so far as is reasonably practicable' to 'any other person, whether at work or not, who may be affected by the work carried on'. Consideration must be given to:

- Others at work in the workplace, e.g. contractors, visitors or co-tenants.

- Visitors.
- Members of the public particularly in public places and buildings.
- Customers and service users.

Some of these (e.g. children) cannot be expected to behave in the same way as employees or contractors. An early prosecution under the *COSHH Regulations* was of a doctor's surgery when a young child gained access to an insecure cleaner's cupboard and drank the contents of a bottle of carbolic acid, suffering serious ill effects.

The main requirements of the Regulations are summarised below. Some will be explained in more detail later. Full details of the Regulations together with associated Approved Codes of Practice ('ACoPs') and guidance are contained in a single Health and Safety Executive ('HSE') booklet – *Control of substances hazardous to health* (L5).

- *Prohibitions relating to certain substances* [*SI 2002 No 2677, Reg 4*] – the manufacture, use, importation and supply of certain specified substances is prohibited (full details are contained in *Schedule 2* to the Regulations).
- *Application of Regulations 6 to 13* [*SI 2002 No 2677, Reg 5*] – exceptions are made from the application of these Regulations where other more specific regulations are in place. These relate to respirable dust in coal mines, lead, asbestos, the radioactive, explosive or flammable properties of substances, substances at high or low temperatures or high pressure and substances administered in the course of medical treatment.
- *Assessment* [*SI 2002 No 2677, Reg 6*] – an assessment of health risks must be made to identify steps necessary to comply with the Regulations and these steps must be implemented.
- *Prevention or control of exposure* [*SI 2002 No 2677, Reg 7*] – this must be achieved preferably through elimination or substitution or alternatively the provision of adequate controls, e.g. enclosure, LEV, ventilation, systems of work, personal protective equipment ('PPE'). This regulation requires a 'hierarchical' approach to be taken – this is explained in H2107 below.
- *Use of control measures etc.* [*SI 2002 No 2677, Reg 8*] – employees are required to make 'full and proper use of control measures, PPE etc.', and employers must 'take all reasonable steps' to ensure they do.
- *Maintenance, examination and testing of control measures* [*SI 2002 No 2677, Reg 9*] – control measures, including PPE, must be maintained in an efficient state. Controls must also be examined and tested periodically.
- *Monitoring exposure at the workplace* [*SI 2002 No 2677, Reg 10*] – monitoring (e.g. through occupational hygiene surveys) may be necessary in order to ensure adequate control or to protect employees' health.
- *Health surveillance* [*SI 2002 No 2677, Reg 11*] – surveillance of employees' health may be required in some circumstances.
- *Information, instruction and training* [*SI 2002 No 2677, Reg 12*] – must be provided for persons who may be exposed to hazardous substances (and also for those carrying out COSHH assessments).
- *Arrangements to deal with accidents, incidents and emergencies* [*SI 2002 No 2677, Reg 13*] – this new regulation, introduced in 2002, requires emergency procedures to be established and information on emergency arrangements to be provided.

What the Regulations require on assessments

Assessing the risk to health

H2106 The wording of *Regulation 6* of the *COSHH Regulations 2002* (*SI 2002 No 2677*) was changed considerably from previous versions, with *sub-paragraph* (*2*) detailing many specific factors which must be taken into account during the assessment. The Regulation now states:

'(1) An employer shall not carry on any work which is liable to expose any employees to any substance hazardous to health unless he has—

(a) made a suitable and sufficient assessment of the risk created by that work to the health of those employees and of the steps that need to be taken to meet the requirements of these Regulations; and

(b) implemented the steps referred to in sub-paragraph (a).

(2) The risk assessment shall include consideration of—

(a) the hazardous properties of the substance;

(b) information on health effects provided by the supplier, including information contained in any relevant safety data sheet;

(c) the level, type and duration of exposure;

(d) the circumstances of the work, including the amount of the substance involved;

(e) activities, such as maintenance, where there is the potential for a high level of exposure;

(f) any relevant occupational exposure standard, maximum exposure limit or similar occupational exposure limit;

(g) the effect of preventive and control measures which have been or will be taken in accordance with regulation 7;

(h) the results of relevant health surveillance;

(i) the results of monitoring of exposure in accordance with regulation 10;

(j) in circumstances where the work will involve exposure to more than one substance hazardous to health, the risk presented by exposure to such substances in combination;

(k) the approved classification of any biological agent; and

(l) such additional information as the employer may need in order to complete the risk assessment.

(3) The risk assessment shall be reviewed regularly and forthwith if—

(a) there is reason to suspect that the risk assessment is no longer valid;

(b) there has been a significant change in the work to which the risk assessment relates; or

(c) the results of any monitoring carried out in accordance with regulation 10 show it to be necessary,

and where, as a result of the review, changes to the risk assessment are required, those changes shall be made.

(4) Where the employer employs 5 or more employees, he shall record—

(a) the significant findings of the risk assessment as soon as practicable after the risk assessment is made; and

(b) the steps which he has taken to meet the requirements of regulation 7.'

In carrying out assessments, it should be noted that *Regulation 3(1)* also requires employees to take account, so far as is reasonably practicable, of others who may be affected by the work, e.g. visitors, contractors, customers, passers-by and members of the emergency services.

Prevention or control of exposure

H2107 The *COSHH Regulations 2002* (*SI 2002 No 2677*), *Reg 6(1)(a)* state that the assessment must take into account the steps needed to meet the requirements of the Regulations. *Regulation 7* of the *COSHH Regulations 2002* sets out a clear hierarchy of measures which must be taken to prevent or control exposure to hazardous substances. It states:

'(1) Every employer shall ensure that the exposure of his employees to substances hazardous to health is either prevented or, where this is not reasonably practicable, adequately controlled.

(2) In complying with his duty of prevention under paragraph (1), substitution shall by preference be undertaken, whereby the employer shall avoid, so far as is reasonably practicable, the use of a substance hazardous to health at the workplace by replacing it with a substance or process which, under the conditions of its use, either eliminates or reduces the risk to the health of the employees.

(3) Where it is not reasonably practicable to prevent exposure to a substance hazardous to health, the employer shall comply with his duty of control under paragraph (1) by applying protection measures appropriate to the activity and consistent with the risk assessment , including in order of priority -

(a) the design and use of appropriate work processes, systems and engineering controls and the provision and use of suitable work equipment and materials;

(b) the control of exposure at source, including adequate ventilation systems and appropriate organisational measures; and

(c) where adequate control of exposure cannot be achieved by other means, the provision of suitable personal protective equipment in addition to the measures required by sub-paragraphs (a) and (b).'

Thus a clear order of preference is established:

1. Prevention of exposure (by elimination or substitution).
2. Appropriate work processes, systems and engineering controls (e.g. enclosure) and suitable work equipment and materials.
3. Adequate control at source (by ventilation or organisational measures).
4. Adequate control through PPE.

(*Sub-paragraph (4)* of *Regulation 7* contains further details of what the measures referred to in *paragraph (3)* must include – these are referred to in H2109.)

Prevention of exposure

H2108 The first preference should always be to prevent exposure to hazardous substances. This could be achieved by:

- Changing work methods, e.g. adopting ultrasonic techniques or high-pressure water jets for cleaning purposes rather than using solvents.
- Using non-hazardous (or less hazardous) alternatives, e.g. replacing solvent-based paints or inks by water-based ones (or ones utilising less hazardous solvents).
- Using the same substances but in less-hazardous forms, e.g. substituting powdered materials by granules or pellets, using more dilute concentrations of hazardous substances.
- Modifying processes, e.g. eliminating the production of hazardous by-products, emissions or waste by changing process parameters such as temperature or pressure.

Exposure must be prevented 'so far as is reasonably practicable' – a definition of this qualifying phrase is provided in the chapter on RISK ASSESSMENT. The level of risk will be determined by the type and severity of the hazards associated with a substance and its circumstances of use.

Considerations weighing against the prevention of exposure being reasonably practicable might be the costs associated with using alternative methods or materials, the detrimental effect of alternatives on the product or additional risks which may be introduced by the alternatives.

For example:

- ultrasonic cleaning may not achieve the required standards of cleanliness;
- use of high pressure water can involve additional risks;
- water-based paints may result in increased problems from corrosion;
- alternative solvents may be more flammable.

However, much progress has been made in recent years in the substitution of hazardous materials. The HSE booklet *Seven steps to successful substitution of hazardous substance*s provides useful guidance.

Control measures, other than PPE

H2109 The *COSHH Regulations 2002* (*SI 2002 No 2677*), *Reg 7*(*4*) states that the control measures required by *paragraph* (*3*) shall include:

'(a) arrangements for the safe handling, storage and transport of substances hazardous to health, and of waste containing such substances, at the workplace;

(b) the adoption of suitable maintenance procedures;

(c) reducing, to the minimum required for the work concerned—

(i) the number of employees subject to exposure,

(ii) the level and duration of exposure, and

(iii) the quantity of substances hazardous to health present at the workplace;

(d) the control of the working environment, including appropriate general ventilation; and

(e) appropriate hygiene measures including adequate washing facilities.'

Examples of such control measures include:

- Enclosed or covered process vessels.
- Enclosed transfer systems, e.g. use of pumps or conveyors, rather than open transfer.
- Use of closed and clearly labelled containers.
- Plant, processes and systems of work which minimise the generation of, or suppress and contain, spills, leaks, dust, fumes and vapours. (This often involves a combination of partial enclosure and the use of local exhaust ventilation ('LEV')).
- Adequate levels of general ventilation.
- Appropriate maintenance procedures.
- Minimising the quantities of hazardous substances stored and used in workplaces.
- Restricting the numbers of persons in areas where hazardous substances are used.
- Prohibiting eating, drinking and smoking in areas of potential contamination.
- Providing showers in addition to normal washing facilities.
- Regular cleaning of walls and other surfaces.
- Using signs to indicate areas of potential contamination.
- Safe storage, handling and disposal of hazardous substances (including hazardous waste).

Paragraphs (*5*) and (*6*) of *Regulation 7* set out measures required to control exposure to carcinogens and biological agents – these are dealt with later in H2113.

Control using PPE

H2110 Control of exposure using PPE should be the last option, where prevention or adequate control by other means are not reasonably practicable. This might be the case where:

- the extent of use of hazardous substances is very small;
- adequate control by other means is impracticable from a technical viewpoint;
- control by other means is excessively expensive or difficult in relation to the level of risk;
- PPE is used as a temporary control measure pending the implementation of adequate control by other means;
- employees are dealing with emergency situations;
- infrequent maintenance activities are being carried out.

PPE necessary to control risks from hazardous substances might be:

- respiratory protective equipment (RPE);

- protective clothing;
- hand or arm protection (usually gloves);
- eye protection;
- protective footwear.

The *COSHH Regulations 2002* (*SI 2002 No 2677*), *Reg 7*(*9*) state that:

> 'Personal Protective Equipment provided by an employer in accordance with this regulation shall be suitable for the purpose and shall –
>
> (a) comply with any provision in the Personal Protective Equipment Regulations 2000 which is applicable to that item of personal protective equipment; or
>
> (b) in the case of respiratory protective equipment, where no provision referred to in sub-paragraph (a) applies, be of a type approved or shall conform to a standard approved, in either case, by the Executive.'

HSE publishes detailed guidance on PPE generally and on RPE in particular.

Adequate control

H2111 The *COSHH Regulations 2002* (*SI 2002 No 2677*), *Reg 7*(*7*), (*8*) and (*11*) define what is meant by the term 'adequate control' in respect of inhalation risks from hazardous substances:

> '(7) Without prejudice to the generality of paragraph (1), where there is exposure to a substance for which a maximum exposure limit has been approved, control of exposure shall, so far as the inhalation of that substance is concerned, only be treated as being adequate if the level of exposure is reduced so far as is reasonably practicable and in any case below the maximum exposure limit.
>
> (8) Without prejudice to the generality of paragraph (1), where there is exposure to a substance for which an occupational exposure standard has been approved, control of exposure shall, so far as the inhalation of that substance is concerned, only be treated as being adequate if –
>
> (a) that occupational exposure standard is not exceeded; or
>
> (b) where that occupational exposure standard is exceeded, the employer identifies the reasons for the standard being exceeded and takes appropriate action to remedy the situation as soon as is reasonably practicable.
>
> (11) In this regulation, "adequate" means adequate having regard only to the nature of the substance and the nature and degree of exposure to substances hazardous to health and "adequately" shall be construed accordingly.'

MEL approved

H2112 Where a Maximum Exposure Limit ('MEL') has been approved, the level of exposure must be reduced so far as is reasonably practicable as well as always being kept below the MEL at all times. More flexibility is given in respect of substances where an Occupational Exposure Standard ('OES') has been approved. If the OES is exceeded the employer must identify the reasons and remedy the situation as soon as is reasonably practicable. Listings of MELs and OESs are published regularly by HSE.

In some cases it will be relatively easy to make a judgement as to whether exposure is controlled to levels well within the MEL or OES but in other situations it may be necessary to carry out an occupational hygiene survey or even arrange for routine or periodic monitoring of exposure.

Carcinogens and biological agents

H2113 The *COSHH Regulations 2002* (*SI 2002 No 2677*), *Reg 7*(*5*) and (*6*) set out a number of specific measures which must be applied where it is not reasonably practicable to prevent exposure to carcinogens and biological agents. (These are in addition to the measures required by *paragraph* (*3*) of the regulation.)

Measures specified for controlling carcinogens are:

- totally enclosing process and handling systems (unless not reasonably practicable);
- prohibition of eating, drinking and smoking in areas that might be contaminated;
- cleaning of floors, walls and other surfaces regularly and whenever necessary;
- designating areas of potential contamination by warning signs;
- storing, handling and disposing of carcinogens safely, using closed and clearly labelled containers.

Appendix 1 to the HSE ACoP booklet provides further detail on the control of carcinogenic substances whilst Annex 2 of the booklet contains a background note on occupational cancer.

Measures for controlling biological agents are:

- use of warning signs (a biohazard sign is shown in *Schedule 3* of the Regulations);
- specifying appropriate decontamination and disinfection procedures;
- safe collection, storage and disposal of contaminated waste;
- testing for biological agents outside confined areas;
- specifying work and transportation procedures;
- where appropriate, making available effective vaccines;
- instituting appropriate hygiene measures;
- control and containment measures where human patients or animals are known (or suspected to be) infected with a Group 3 or 4 biological agent.

The *COSHH Regulations 2002* (*SI 2002 No 2677*), *Reg 7*(*10*) require work with biological agents to be in accordance with *Schedule 3* to the Regulations which is contained in the HSE COSHH ACoP booklet (L5). This sets out several detailed requirements, including standards for containment measures.

The Advisory Committee on Dangerous Pathogens (ACDP) has published the latest version of the *Approved List of Biological Agents*.

The Approved List classifies biological agents (bacteria, viruses, parasites, fungi and prions) into one of four hazard groups (HG) according to the following criteria:

- their ability to cause infection (in otherwise healthy individuals);
- the severity of the disease that may result;
- the risk that infection will spread to the community; and
- the availability of effective vaccines and treatment.

The Approved List is an approved document that works in conjunction with the *COSHH Regulations 2002* (*SI 2002 No 2677*) to provide the regulatory framework controlling the risk from biological agents at work. Only those agents classified into Groups 2–4 are listed, however, simply because an agent is not listed this does not mean it is automatically a HG1 agent.

A limited number of agents have been reclassified, removed or added to the list, reflecting the most current scientific knowledge. This version of the Approved List has seen changes to the status of Duvenhage virus and Mobala virus (both have been changed from HG2 to HG3), the removal of several HG2 bacteria, and the inclusion of the SARS virus as a HG3 agent.

Previously, the List was a supplement to the *Categorisation of biological agents according to hazard and categories of containment* guidance. The List is now a stand-alone publication that should be read in conjunction with *COSHH* and relevant ACoP and *COSHH* guidance.

Additional guidance on work with HG3 enteric pathogens (known as Appendix 24) which was previously published with the Approved List remains current but will appear in new ACDP guidance (*Biological agents: Managing the risks*) currently under preparation, due to be published by early 2005. In the interim, Appendix 24 will be separately available on the HSE website.

The Approved List can be accessed on the HSE website at: www.hse.gov.uk/hthdir/noframes/entericpathogens.pdf.

Practical aspects of COSHH assessments

The requirements summarised

H2114 COSHH assessments must involve the following:

- identification of risks to the health of employees or others;
- consideration of whether it is reasonably practicable to prevent exposure to hazardous substances creating risks (by elimination or substitution);
- if prevention is not reasonably practicable, identification of the measures necessary to achieve adequate control of exposure, as required by the *COSHH Regulations 2002* (*SI 2002 No 2677*), *Reg 7* (these control measures are described at H2109 AND H2110);
- identification of other measures necessary to comply with the *COSHH Regulations 2002* (*SI 2002 No 2677*), *Regs 8–13*, e.g:
  - — use of control measures;
  - — maintenance, examination and test of control measures etc;
  - — monitoring of exposure at the workplace;
  - — health surveillance;
  - — provision of information, instruction and training;
  - — arrangements for accidents, incidents and emergencies.

(These are described more fully at H2127–H2135 BELOW).

The assessment must be 'suitable and sufficient', and its complexity will vary considerably dependent on the circumstances. In some cases it may only be necessary to study relevant safety data sheets and decide whether existing practices are sufficient to achieve adequate control. However, for more complex processes and variable activities involving greater numbers of hazardous substances, the assessment must be much more detailed. The planning, preparation and implementation of a COSHH assessment will be very similar to other types of risk assessment and much of the guidance given in the chapter on RISK ASSESSMENT will be valid here.

Who should carry out COSHH assessments?

H2115 *Regulation* 7 of the *Management of Health and Safety at Work Regulations 1999* (*SI 1999 No 3242*) requires employers to appoint a competent person or persons to assist them in complying with their duties, which will include carrying out COSHH assessments. *Regulation 12*(*4*) of the *COSHH Regulations 2002* (*SI 2002 No 2677*) also states:

> 'Every employer shall ensure that any person (whether or not his employee) who carries out any work in connection with the employer's duties under these Regulations has suitable and sufficient information, instruction and training.'

The degree of knowledge and experience required will depend upon the circumstances. An assessment in a workplace containing a few simple uses of hazardous substances may be carried out by someone without any specialist qualifications or experience but with the capability to understand and apply the principles contained in this chapter. (However, such a person may need access to someone with greater health and safety knowledge and a more detailed understanding of the *COSHH Regulations*.) Some qualified and experienced individuals may be capable of carrying out detailed assessments of relatively high-risk situations themselves, although they are likely to need to consult others during the assessment process.

Higher risk situations, with much more complex and significant use of hazardous substances are more likely to require multi-disciplinary teams, some of whom have specialist knowledge and experience. Possible candidates for such a team might be:

- health and safety officers or managers;
- occupational hygienists;
- occupational health nurses;
- physicians with occupational health experience;
- chemical or process engineers;
- maintenance or ventilation engineers.

How should the assessments be organised?

H2116 In larger workplaces it is often best to divide them into manageable assessment units which could be based on:

- Departments or sections.
- Buildings or rooms.
- Process lines.
- Activities or services.

Where an assessment team is used, those individuals most suited for a particular situation can be selected to assess that unit.

Gathering information together

H2117 Prior to the assessment, information should be gathered together, particularly about which hazardous substances are present and what hazards are associated with them. Substances to be taken into account include:

- Raw materials used in processes.
- The contents of stores and cupboards.
- Substances produced by processes:
  - intermediate compounds;
  - products;
  - waste and by-products;
  - emissions from the process.
- Substances used in maintenance and cleaning work.
- Buildings and the work environment:
  - surface treatments;
  - pollution or contaminants;
  - bird droppings, animal faeces etc;
  - possible legionella sources.
- Substances involved in the activities of others, e.g. contractors.

Potential sources of information about the hazards associated with these substances include:

- Manufacturers' or suppliers' safety data sheets.
- Information provided on containers or packaging (including the symbols required by the *CHIP Regulations* (*SI 2002 No 1689*).
- HSE publications and the HSE website.
- Other reference books and technical literature.
- Direct enquiries to specialists or suppliers.

This information will primarily concern the hazards associated with the substance but is also likely to contain recommendations on control methods. However, the substance's actual circumstances of use may constitute a greater or lesser degree of risk than the manufacturer or supplier anticipated – it is for the assessor(s) to determine what control measures are necessary in each situation.

Other useful information may be available from previous assessments or surveys. Even if previous COSHH assessments were inadequate or are now out of date, they may still contain information of value in the assessment process. The results of previous occupational hygiene surveys (e.g. dust, gas or vapour concentrations) are also likely to be useful as will the collective results of any health surveillance work which has been carried out.

Details of existing control measures may be contained in operating procedures, health and safety rule books or listings of PPE requirements or may be referred to within training programmes. Specification information may be available for some control measures, e.g. design flow rates for ventilation equipment or performance standards for PPE such as respiratory protection or gloves. (Some of this information may also be sought during the assessment itself or prior to the preparation of the assessment record.)

Observations in the workplace

H2118 Time spent in making COSHH assessments can be reduced and made more productive by good planning and preparation. Nevertheless, observations of workplaces and work activities must still be an integral part of the COSHH assessment process.

Sufficient observations must be made to arrive at a conclusion on the level of risk and the adequacy of control measures for all the hazardous substances in the workplace – be they raw materials, products, by-products, waste, emissions etc. This should include the manner and extent to which employees (and others) are exposed and an evaluation of the equipment and working practices intended to achieve their control.

Aspects to be considered include:

- Are specified procedures being followed?
- Does local exhaust ventilation or general ventilation appear effective?
- Is PPE being used correctly?
- Is there evidence of leakage, spillage or dust accumulation?
- What equipment is available for cleaning?
- Are dusts, fumes or strong smells evident in the atmosphere?
- Are there any restrictions relating to eating, drinking or smoking?
- What arrangements are there for storage, cleaning or maintenance of PPE?
- Are there suitable arrangements for washing, showering or changing clothing?
- Are special arrangements for laundering clothing required?
- What arrangements are in place for storage and disposal of waste?
- Is there potential for major spillages or leaks?
- What emergency containment equipment or PPE is available?

Discussions with staff

H2119 It is also important to talk to those working with hazardous substances, their safety representatives and those managing or supervising their work. Questions might relate to working practices, awareness of risks or precautions, the effectiveness of precautions or experience of problems. These might include:

- Why tasks are done in particular ways?
- Are the present work practices typical?
- How effective is the LEV/general ventilation?
- What happens if it ever breaks down?
- What types of PPE are required for the work?

- Are there any problems with the PPE?
- Have workers experienced any health problems?
- How often is the workplace cleaned?
- What methods are used for cleaning?
- How is waste disposed of?
- Is there any possibility of leaks occurring or emergencies arising?
- What would happen in such situations?
- What are the arrangements for washing/cleaning/maintaining PPE?

Further assistance

H2120 Guidance on COSHH assessments is available in the HSE booklet *A step by step guide to COSHH assessment*. The HSE's Advisory Committee on Toxic Substances has also developed *COSHH essentials* which provides generic risk assessments for a wide range of substances. It is available in printed form via HSE Books but can also be accessed electronically via the HSE website. It contains advice on control measures for common tasks, e.g. mixing, weighing, spray painting.

Further investigations

H2121 Assessment in the workplace may reveal the need for further tests on the effectiveness of control measures before the assessment can be concluded. The tests may be in the form of occupational hygiene surveys – usually to measure the airborne concentrations of dust, fume, vapour or gas and compare them with the relevant MEL or OES. Alternatively there may be a need to measure the effectiveness of local exhaust ventilation or general ventilation.

Further investigations may also be necessary into matters such as:

- hazards associated with substances which were not identified during the preparatory phase of the assessment;
- whether the specifications for PPE found in use are adequate for the exposure involved;
- the feasibility of using alternative work methods (e.g. preventing exposure) or alternative methods of control to those currently in place.

COSHH assessment records

Preparation of assessment records

H2122 Much of the guidance on note taking and the preparation of assessment records contained in the chapter on RISK ASSESSMENT is relevant to COSHH assessments. Illustrative examples of completed assessment records are provided at H2124 below, although it should be stressed that no single record format will automatically cater for all types of workplaces or work activities. The essential components which must be included in COSHH assessment records are:

- the work activities involving risks from hazardous substances;
- information as to the hazardous substances involved (and their form, e.g. liquid, powder, dust etc.);
- control measures which are (or should be) in place;

- improvements identified as being necessary.

The COSHH ACoP does state that assessments need not be recorded in the simplest and most obvious cases which can be easily repeated and explained at any time. However, it suggests that in most cases assessments will need to be recorded and kept readily accessible to those who may need to know the results. The amount of information should be proportionate to the risks posed by the work. In many cases it may be acceptable to state that in the circumstances of use there is little or no risk and that no more detailed assessment is necessary. The two illustrative COSHH assessment records (below) demonstrate the contrasting levels of detail appropriate for different situations. The COSHH ACoP provides further details of what might need to be included in more comprehensive assessment records. Employees or their representatives should be informed of the results of COSHH assessments.

Review and implementation of recommendations

H2123 Once the assessment has been completed and the record has been prepared it is important that recommendations are reviewed, implemented and followed up. The review process is likely to involve people who were not involved in the original assessments – senior managers, or staff with relevant technical expertise. Whilst the assessment team may be prepared to make changes to their assessment findings or recommendations, they should not allow themselves to be pressurised into making changes they find unacceptable. If senior management choose not to implement COSHH assessment recommendations then they must accept the responsibility for their actions.

Responsibilities for implementing each recommendation within a designated timescale must be clearly allocated to individuals, possibly as part of an overall action plan. Each of these recommendations must then be followed up to ensure that they have actually been carried out and also to check that additional risks have not been created inadvertently. The assessment record should then be annotated to indicate that the recommendation has been completed (alternatively a new record could be prepared reflecting the improved situation).

Example assessment records

H2124 See the example assessment records below.

Acorn Estate Agents, Newtown

Risk Assessment		
Reference number: 6	Risk topic/issue: Hazardous substances	Sheet 1 of 1
Risks identified	**Precautions in place**	**Recommended improvements**
The substances listed below could present risks to both staff and clients:		
Office materials	Office supplies are kept in a cupboard in a part of the office not normally accessible to clients.	
With warning symbols	Generally substances are only used in very small quantities for short periods presenting no significant risk.	
Old correction fluid (harmful)		
New correction fluid (flammable)		
Spray adhesive (harmful)	If the spray adhesive is used for more than a couple of minutes, a nearby window is opened which provides adequate ventilation.	
Without warning symbols		
Photocopier and printer toners	The photocopier is used in well-ventilated areas and there are no noticeable ozone smells, even on long copying runs.	Provide disposable gloves for cleaning significant spillages of photocopier or printer toner.
Various felt tip pens (solvent based)		
Cleaning materials	The cleaner's cupboard is kept locked except when substances are being removed.	
With warning symbols		
Thick bleach (Irritant)	Suitable gloves are provided (and worn) for handling the irritant and corrosive substances in concentrated form.	Investigate replacing the thick bleach by a more dilute solution.
Polish stripper (Irritant)		
Acid descaler (Corrosive)	The cleaner is aware of the risks of mixing bleach with other substances e.g. the acid descaler.	Ensure the relief cleaner is also made aware of these risks.
Without warning symbols		
Furniture polish		
Floor polish		

Window and glass cleaner		
Signature(s) K Stephenson, R Lewis	**Name(s)** K Stephenson, R Lewis	**Date** 4/11/98
Dates for	**Recommendation follow up**: December 1998	**Next routine review**: November 2003

Rainbow Products

COSHH Assessment	
Assessment unit	Mixing Hall
Activities	Manufacture of solvent-based paints and other surface treatments.
	Solvents are pumped into the mixing vessels from an external tank farm.
	Solid constituents and some liquid components are charged into the mixing vessels from platforms above.
	After mixing, the products are pumped directly to filling stations in a neighbouring building.
Substances used (Data sheet file references)	4, 7, 14, 23, 49, 50, 51, 60, 73, 80, 92, 106, 117.
	Detailed formulations for each product are available from the Quality Control Department
Main risks	Dust from solid constituents during charging.
	Solvent vapours: – escaping from the charging hatches;
	– from minor leaks at valves and pipe connections;
	– from liquid components during charging.
	Entry into mixing vessels for cleaning or maintenance purposes.
Those at risk	Mixing Hall production employees.
	Maintenance staff when working in the area.
	Contractors involved in vessel cleaning or maintenance work
Controls and other precautions in place	Good general ventilation (specification 20 air changes per hour).
	Annual surveys of solvent vapours show levels well below all OESs.
	Hoods with LEV over each vessel charging position. These are examined and tested annually by the Maintenance Department.
	Exposure Monitoring Surveys (Aug 1996, May 1999) show production employees as well within OESs for all dusts and solvents.
	Disposable dust masks are available for vessel charging, although their use is not compulsory.
	A portable vacuum cleaning unit with suitable filter is available when required.
	Maintenance carry out an annual physical inspection of all mixer vessels and pipelines.

Observations	A weigh station is provided in a fume cupboard for weighing out smaller quantities of solid constituents (this also is examined and tested annually). Entry into vessels is controlled by the permit to work system. All employees are subject to the company's annual health screening programme. Discarded empty paper sacks were strewn on several loading platforms. Liquid component transfer containers had been left open on platforms 1 and 4 (still containing residual materials). The ventilation at charging stations 2 and 4 appeared inadequate. There were dust accumulations on all loading platforms. A brush appears to have been used to sweep up dust on some platforms. A pipe flange below mixer 3 had developed a small leak.
Recommendations	1. Improve the sack disposal containers on all loading platforms. 2. Remind staff of the importance of disposing of sacks correctly. 3. Remind staff that liquid transfer containers should have their lids replaced after use. 4. Rectify the ventilation at charging stations 2 and 4. 5. Introduce simply weekly checks on the ventilation at all charging stations. 6. Introduce weekly cleaning for all loading platforms. 7. Remind staff of the importance of using vacuum methods for cleaning. 8. Investigate obtaining vacuum cleaning units which can be lifted onto the platforms more easily. 9. Repair the leaking flange below mixer 3. 10. Increase the frequency of pipeline inspections to six monthly. 11. Develop a standard procedure for entering vessels for cleaning or maintenance purposes (linked to the permit to work system). 12. Investigate alternative cleaning methods avoiding the need for entry, e.g. immersion in solvent over weekend periods.
Signature(s) A Storm, P Gold	**Date** 2 July 2001
Recommendation follow up: October 2001	
Assessment Review: July 2003	

Ongoing reviews of assessments

H2125 The *COSHH Regulations 2002* (*SI 2002 No 2677*), *Reg 6*(*3*) require COSHH assessments to:

'be reviewed regularly and forthwith if—

(a) there is reason to suspect the risk assessment is no longer valid;

(b) there has been a significant change in the work to which the risk assessment relates; or

(c) the results of any monitoring carried out in accordance with regulation 10 show it to be necessary.'

- *Assessments no longer valid*

  Assessments might be shown to be no longer valid because of:

  — new information received about health risks, e.g. information from suppliers, revised HSE guidance, changes in the MEL or OES;

  — results from inspections or thorough examinations or tests (*Regulation 9*), e.g. indicating fundamental flaws in engineering controls;

  — regular reports or complaints about defects in control arrangements;

  — the results from workplace exposure monitoring (*Regulation 10*), e.g. showing the OES or MEL is regularly being exceeded (or approached);

  — results from health surveillance (*Regulation 11*), e.g. demonstrating an unsatisfactory or deteriorating position;

  — a confirmed case of an occupational disease.

- *Significant changes*

  Changes necessitating a review of an assessment might involve:

  — the types of substances used, their form or their source;

  — equipment used in the process or activity (including control measures);

  — altered methods of work or operational procedures;

  — variations in volume, rate or type of production;

  — staffing levels and related practical difficulties or pressures.

- *Regular review*

  The periods elapsing between reviews should relate to the degree of risk involved and the nature of the work itself. The COSHH ACoP previously stated that assessments should be reviewed at least every five years, but no longer contains such a recommendation. However, this would seem to be a reasonable maximum period to use for practical purposes.

  A review would not necessarily require a revision of the assessment – it may conclude that existing controls are still adequate despite changed circumstances. However, where changes are shown to be required, *paragraph* (*3*) of *Regulation 6* requires that these be implemented.

Further requirements of the COSHH Regulations

H2126 Some recommendations from COSHH assessments could relate to the ongoing maintenance of effective control measures and many of these aspects are covered by further specific requirements of the *COSHH Regulations*.

Use of control measures

H2127 The *COSHH Regulations 2002* (*SI 2002 No 2677*), *Reg 8* place duties on both employers and employees in respect of the proper use of control measures, including PPE.

> '(1) Every employer who provides any control measure, other thing or facility in accordance with these Regulations shall take all reasonable steps to ensure that it is properly used or applied as the case may be.
>
> (2) Every employee shall make full and proper use of any control measure, other thing or facility provided in accordance with these Regulations and where relevant shall—
>
> (a) take all reasonable steps to ensure it is returned after use to any accommodation provided for it; and,
>
> (b) if he discovers any defect therein, report it forthwith to his employer.'

Employers have duties under the *Management of Health and Safety at Work Regulations 1999* (*SI 1999 No 3242*) to monitor all of their health and safety arrangements. Areas where monitoring of COSHH control measures are likely to be necessary are:

- correct use of LEV equipment;
- compliance with specified systems of work;
- compliance with PPE requirements;
- storage and maintenance of PPE;
- compliance with requirements relating to eating, drinking or smoking;
- condition of washing and showering facilities and personal hygiene standards;
- whether defects are being reported by employees.

Where employees are unwilling to use control measures properly, employers should consider the use of disciplinary action, particularly in relation to persistent offenders.

Maintenance, examination and testing of control measures

H2128 The *COSHH Regulations 2002* (*SI 2002 No 2677*), *Reg 9* contain both general and specific requirements for the maintenance of control measures:

> '(1) Every employer who provides any control measure to meet the requirements of regulation 7 shall ensure that, where relevant, it is maintained in an efficient state, in efficient working order, in good repair and in a clean condition.

(2) Where engineering controls are provided to meet the requirements of regulation 7, the employer shall ensure that thorough examination and testing of those controls is carried out–

(a) in the case of local exhaust ventilation plant, at least once every 14 months, or for local exhaust ventilation plant used in conjunction with a process specified in Column 1 of Schedule 4, at not more that the interval specified in the corresponding entry in Column 2 of that Schedule; or

(b) in any other case, at suitable intervals.

(3) Where respiratory protective equipment (other than disposable respiratory protective equipment) is provided to meet the requirements of regulation 7, the employer shall ensure that thorough examination and, where appropriate, testing of that equipment is carried out at suitable intervals.

(4) Every employer shall keep a suitable record of the examinations and tests carried out in pursuance of paragraphs (2) and (3) and of repairs carried out as a result of those examinations and tests, and that record or a suitable summary thereof shall be kept available for at least 5 years from the date on which it was made.'

Paragraphs (*5*), (*6*) and (*7*) contain requirements specifically relating to PPE – these are examined later.

General maintenance of controls

H2129 The COSHH ACoP states that, where possible, all engineering control measures should receive a visual check at least once every week. Such checks may simply confirm that there are no apparent leaks from vessels or pipes and that LEV or cleaning equipment appear to be in working order. No records of such checks need to be kept, although it is good practice (and prudent) to do so.

The *COSHH Regulations 2002* (*SI 2002 No 2677*), *Reg 9*(*2*) require thorough examinations and tests of engineering controls. Requirements relating to LEV are reviewed below at H2130 but for other engineering controls such examinations and tests must be 'at suitable intervals', and suitable records must be kept for at least five years. The nature of examinations and tests will depend upon the engineering control involved and the potential consequences of its deterioration or failure. Examples might involve:

- detailed visual inspections of tanks and pipelines;
- inspections and non-destructive testing of critical process vessels;
- testing of detectors and alarm systems;
- planned maintenance of general ventilation equipment;
- checks on filters in vacuum cleaning equipment.

Persons carrying out maintenance, examinations and testing must be competent for the purpose in accordance with the *COSHH Regulations 2002* (*SI 2002 No 2677*), *Reg 12*(*4*).

Local exhaust ventilation (LEV) plant

H2130 Most LEV systems must be thoroughly examined and tested at least once every 14 months, although *Schedule 4* of the *COSHH Regulations 2002* (*SI 2002 No 2677*), requires increased frequencies for LEV used in conjunction with a handful of specified processes. Dependent upon the design and purpose of the LEV concerned, the examination and test might involve visual inspection, air flow or static pressure measurements or visual checks of efficiency using smoke generators or dust lamps. In some cases air sampling to confirm efficiency levels, filter integrity tests or checks on air flow sensors, may be appropriate. Detailed guidance is available in an HSE booklet *The maintenance, examination and testing of local exhaust ventilation.*

Personal Protective Equipment (PPE)

H2131 All types of PPE are subject to the general maintenance requirements contained in the *COSHH Regulations 2002* (*SI 2002 No 2677*), *Reg 9*(*1*) whilst *Regulation 9*(*3*) contains specific requirements relating to non-disposable respiratory protective equipment ('RPE').

Further requirements in respect of PPE are contained in *Regulation 9*(*5*), (*6*) and (*7*) of the 2002 Regulations that state:

'(5) Every employer shall ensure that personal protective equipment, including protective clothing , is:

(a) properly stored in a well-defined place;

(b) checked at suitable intervals; and

(c) when discovered to be defective, repaired or replaced before further use.

(6) Personal protective equipment which may be contaminated by a substance hazardous to health shall be removed on leaving the working area and kept apart from uncontaminated clothing and equipment.

(7) The employer shall ensure that the equipment referred to in paragraph (6) is subsequently decontaminated and cleaned or, if necessary, destroyed.'

Some types of PPE can easily be seen to be defective by the user whilst in other cases (e.g. for gloves or clothing providing protection against strongly corrosive chemicals) it may be appropriate to introduce more formalised inspection systems.

For non-disposable RPE the COSHH ACoP states that thorough examinations and, where appropriate, tests should be made *at least* once every month, although it suggests that for RPE used less frequently, periods up to three months are acceptable. Alternatively, examination and testing prior to next use may be more appropriate. An HSE booklet *The selection, use and maintenance of respiratory protective equipment* provides detailed guidance on the subject.

Monitoring exposure at the workplace

H2132 Reference was made earlier in the chapter at H2111 and H2121 to the possible need to carry out air testing in the workplace as part of the COSHH assessment process in order to determine the adequacy of control measures. Such testing may also be necessary in order to ensure that adequate control continues to be maintained and a requirement for this is contained in the *COSHH Regulations 2002* (*SI 2002 No 2677*), *Reg 10*:

'(1) Where the risk assessment indicates that—

(a) it is requisite for ensuring the maintenance of adequate control of the exposure of employees to substances hazardous to health; or

(b) it is otherwise requisite for protecting the health of employees,

the employer shall ensure that the exposure of employees to substances hazardous to health is monitored in accordance with a suitable procedure.

(2) Paragraph (1) shall not apply where the employer is able to demonstrate by another method of evaluation that the requirements of regulation 7(1) have been complied with.

(3) The monitoring referred to in paragraph (1) shall take place—

(a) at regular intervals; and

(b) when any change occurs which may affect that exposure.

(4) Where a substance or process is specified in Column 1 of Schedule 5, monitoring shall be carried out at least at the frequency specified in the corresponding entry in Column 2 of that Schedule.

(5) The employer shall ensure that a suitable record of any monitoring carried out for the purpose of this regulation is made and maintained and that record or a suitable summary thereof is kept available—

(a) where the record is representative of the personal exposures of identifiable employees, for at least 40 years; or

(b) in any other case, for at least 5 years,

from the date of the last entry made in it.

(6) Where an employee is required by regulation 11 to be under health surveillance, an individual record of any monitoring carried out in accordance with this regulation shall be made, maintained and kept in respect of that employee.

(7) The employer shall—

(a) on reasonable notice being given, allow an employee access to his personal monitoring record;

(b) provide the Executive with copies of such monitoring records as the Executive may require; and

(c) if he ceases to trade, notify the Executive forthwith in writing and make available to the Executive all monitoring records kept by him.'

Schedule 5 to the *COSHH Regulations 2002* (*SI 2002 No 2677*) automatically requires monitoring to be carried out in processes involving vinyl chloride monomer and electrolytic chromium plating.

A number of techniques are available for monitoring air quality in the workplace. Most of these involve the use of chemical indicator tubes, direct reading instruments, or sampling pumps and filter heads. An HSE booklet *Monitoring strategies for toxic substances* provides further guidance on the subject.

Health surveillance

H2133 The *COSHH Regulations 2002* (*SI 2002 No 2677*), *Reg 11*(*1*) and (*2*) contain the main requirements relating to the need for health surveillance.

'(1) Where it is appropriate for the protection of the health of his employees who are, or are liable to be, exposed to a substance hazardous to health, the employer shall ensure that such employees are under suitable health surveillance.

(2) Health surveillance shall be treated as being appropriate where—

(a) the employee is exposed to one of the substances specified in Column 1 of Schedule 6 and is engaged in a process specified in Column 2 of that Schedule, and there is a reasonable likelihood that an identifiable disease or adverse health effect will result from that exposure; or

(b) the exposure of the employee to a substance hazardous to health is such that—

(i) an identifiable disease or adverse health effect may be related to the exposure,

(ii) there is a reasonable likelihood that the disease or effect may occur under the particular conditions of his work, and

(iii) there are valid techniques for detecting indications of the disease or the effect,

and the technique of investigation is of low risk to the employee.

The remaining parts of *Regulation 10* (*paragraphs* (*3*)–(*11*)) relate to the manner in which health surveillance is conducted and used, together with the maintenance of and access to surveillance records.

The decision as to whether health surveillance is appropriate to protect the health of employees is one that would normally be taken at the time of a COSHH assessment or during its subsequent review. However, for those processes and substances specified in *Schedule 6* to the Regulations, surveillance must be carried out.

Normally health surveillance programmes would be initiated and carried out under the overall supervision of a registered medical practitioner, and preferably one with relevant occupational health experience. However, the surveillance itself may be carried out by an occupational health nurse, a technician or a responsible member of staff, providing that individual was competent for the purpose.

There are many different procedures available for health surveillance including:

- Biological monitoring, e.g. tests of blood, urine or exhaled air.
- Biological effect monitoring, e.g. through lung function testing.
- Medical surveillance, e.g. through physical examinations.
- Interviews about possible symptoms, e.g. skin abnormalities or shortness of breath.

For health surveillance to be 'appropriate' there must firstly be a significant enough risk to justify it and there must also be a valid technique for detecting indications of related occupational diseases or ill-health effects. HSE provide specialist guidance in their booklet *Health surveillance at work*. Records of health surveillance must contain information specified in the COSHH ACoP and must be retained for at least 40 years.

Information, instruction and training

H2134 The *COSHH Regulations 2002* (*SI 2002 No 2677*), *Reg 12* contain requirements relating to information, instruction and training for persons who may be exposed to substances hazardous to health.

'(1) Every employer who undertakes work which is liable to expose an employee to a substances hazardous to health shall provide that employee with suitable and sufficient information, instruction and training.

(2) Without prejudice to the generality of paragraph (1), the information, instruction and training provided under that paragraph shall include—

(a) details of the substances hazardous to health to which the employee is liable to be exposed including—

(i) the names of those substances and the risk which they present to health,

(ii) any relevant occupational exposure standard, maximum exposure limit or similar occupational exposure limit,

(iii) access to any relevant safety data sheet, and

(iv) other legislative provisions which concern the hazardous properties of those substances;

(b) the significant findings of the risk assessment;

(c) the appropriate precautions and actions to be taken by the employee in order to safeguard himself and other employees at the workplace;

(d) the results of any monitoring of exposure in accordance with regulation 10 and, in particular, in the case of any substance hazardous to health for which a maximum exposure limit has been approved, the employee or his representatives shall be informed forthwith, if the results of such monitoring show that the maximum exposure limit is exceeded;

(e) the collective results of any health surveillance undertaken in accordance with regulation 11 in a form calculated to prevent those results from being identified as relating to a particular person; and

(f) where employees are working with a Group 4 biological agent or material that may contain such an agent, the provision of written instructions and, if appropriate, the display of notices which outline the procedures for handling such an agent or material.

(3) The information, instruction and training required by paragraph (1) shall be—

(a) adapted to take account of significant changes in the type of work carried out or methods of work used by the employer; and

(b) provided in a manner appropriate to the level, type and duration of exposure identified by the risk assessment.

(4) Every employer shall ensure that any person (whether or not his employee) who carries out work in connection with the employer's duties under these Regulations has suitable and sufficient information, instruction and training.

(5) Where containers and pipes for substances hazardous to health used at work are not marked in accordance with any relevant legislation listed in Schedule 7, the employer shall, without prejudice to any derogations provided for in that legislation, ensure that the contents of those containers and pipes, together with the nature of those contents and any associated hazards, are clearly identifiable.'

The general requirements of *paragraph* (*1*) match those found in various other regulations but the contents of *paragraph* (*2*) are much more prescriptive than the requirements of previous versions of the *COSHH Regulations*. Whilst it is important that workers are aware of the substances they are exposed to and the risks that they present, only a minority are likely to comprehend fully the significance of OESs and MELs, or understand all of the contents of a typical safety data sheet. However, a good awareness of the risks will mean employees are more likely to take the appropriate precautions. Instruction and training will be particularly relevant in relation to:

- awareness of safe systems of work;
- correct use of LEV equipment;
- use, adjustment and maintenance of PPE (especially RPE);
- the importance of good personal hygiene standards;
- requirements relating to eating, drinking and smoking;
- emergency procedures;
- arrangements for cleaning and the disposal of waste;
- contents of containers and pipes (see *Regulation 12*(*5*) above).

Employers must also inform employees or their representatives of the results of workplace exposure monitoring forthwith if the MEL is shown to have been exceeded. Employees must also be informed of the *collective* results of health surveillance, e.g. the average results within a department or on a particular shift from biological monitoring, or the numbers of employees referred for further investigation following a skin inspection.

The type of information, instruction and training provided must be appropriate for the level, type and exposure involved and adapted to take account of significant changes.

Any person carrying out work on the employer's behalf (whether or not an employee) is required by *paragraph* (*4*) to have the necessary information, instruction and training. Thus a consultant carrying out a COSHH assessment or an occupational hygienist conducting exposure monitoring must be verified by the employer as being competent for the purpose and be provided with the information necessary to carry out their work effectively.

Accidents, incidents and emergencies

H2135 *Regulation 13* was introduced in the *COSHH Regulations 2002* (*SI 2002 No 2677*) and requires employers to ensure that arrangements are in place to deal with accidents, incidents and emergencies related to hazardous substances. (This is in addition to the general duty to have procedures 'in the event of serious and imminent danger' contained in *Regulation 8* of the *Management of Health and Safety at Work Regulations 1999* (*SI 1999 No 3242*.) Based on HSE guidance in the COSHH ACoP booklet (L5) such emergencies might include:

- a serious process fire posing a serious risk to health;
- a serious spillage or leakage of a hazardous substance;
- a failure to contain biological or carcinogenic agents;
- a failure that could lead (or has led) to a sudden release of chemicals;
- a threatened significant exposure over an OES or MEL, e.g. due to a failure of LEV or other controls.

The regulation is rather lengthy and prescriptive in its requirements which are summarised below.

Arrangements should include:

- emergency procedures, such as:
  - appropriate first aid facilities;
  - relevant safety drills (tested at regular intervals);
- providing information on emergency arrangements:
  - including details of work hazards and emergencies likely to arise;
  - made available to relevant accident and emergency services;
  - displayed at the workplace, if appropriate;
- suitable warning and other communications:
  - to enable an appropriate response, including remedial actions and rescue operations.

In the event of an accident, incident or emergency, the employer must:

- take immediate steps to:
  - mitigate its effects;
  - restore the situation to normal;
  - inform employees who may be affected;
- ensure only essential persons are permitted in the affected area and that they are provided with:
  - appropriate PPE;
  - any necessary safety equipment and plant;
- in the case of a serious biological incident, inform employees or their representatives, as soon as practicable of:
  - the causes of the incident or accident;
  - the measures taken or being taken to rectify the situation.

Regulation 13(4) states that such emergency arrangements are not required where the risk assessment shows that because of the quantities of hazardous substances the risks are slight and control measures are sufficient to control that risk. (However, this 'exception' does not apply in the case of carcinogens or biological agents.)

Regulation 13(5) requires employees to report possible releases of a biological agent forthwith to their employer (or another employee with specific responsibility).

HSE guidance in the ACoP booklet states that whether or not arrangements are required under *Regulation 13* is a matter for judgement, based on the potential size and severity of accidents and emergencies which may occur. Many incidents will be capable of being dealt with by the control measures required by the *COSHH Regulations 2002* (*SI 2002 No 2677*), *Reg 7*. Even if an emergency does occur, the response should be proportionate – a small leak would not necessarily justify an evacuation of the workplace. The ACoP booklet provides further details of what emergency procedures might need to include, such as:

- the identity, location and quantities of hazardous substances present;
- foreseeable types of accidents, incidents or emergencies;
- special arrangements for emergencies not covered by general procedures;
- emergency equipment and PPE, and who is authorised to use it;
- first-aid facilities;
- emergency management responsibilities e.g. emergency controllers;
- how employees should respond to incidents;
- clear up and disposal arrangements;
- arrangements for regular drills or practice;
- dealing with special needs of disabled employees.

Other important regulations

H2136 The introduction to this chapter referred to several other legal requirements relating to hazardous substances in the workplace and these are summarised below.

Chemicals (Hazard Information and Packaging for Supply) Regulations 2002

H2137 The *Chemicals* (*Hazard Information and Packaging for Supply*) *Regulations 2002* (*SI 2002 No 1689*) (commonly known as 'CHIP') impose duties on suppliers (and importers into the EU) of substances classified as dangerous for supply. They must provide recipients with safety data sheets containing a number of specified types of information, including details on hazards; first aid, fire fighting and accidental release measures; handling and storage; exposure controls and personal protection.

Substances must be labelled according to the categories of danger they present, i.e:

- Physico-chemical, e.g. explosive, oxidising, highly flammable.
- Health, e.g.very toxic, toxic, harmful, corrosive, irritant.
- Environmental, e.g.dangerous to the environment.

This will involve the use of the standard orange and black symbols, usually accompanied by one of a number of specified risk phrases or safety phrases.

Suitable packaging must be provided for substances which:

- is designed and constructed to prevent escape or spillage of the controls;
- will not be adversely affected by its contents;
- where fitted with a replaceable closure, can be re-closed without contents escaping.

Like the *COSHH Regulations*, the *CHIP Regulations* are constantly subject to amendment and updating and employers directly affected by their detailed technical requirements should consult HSE either directly or via their publications or website.

Control of Asbestos at Work Regulations 2002

H2138 Earlier regulations on asbestos predated the original *COSHH Regulations* and thus provided a framework on which the *COSHH Regulations* were based. They continue to follow a similar pattern, involving an assessment of work which exposes employees to asbestos and the preparation of a suitable written plan of work which must prevent or reduce exposure to asbestos. Many other requirements mirror COSHH requirements, e.g. information, instruction and training; use of control measures; maintenance of control measures; requirements for air monitoring (plus standards for air testing and analysis); and health records and medical surveillance.

There are also a number of other specific requirements including those relating to the notification of work with asbestos (to HSE), to the storage, distribution and labelling of raw asbestos and asbestos waste and to the supply of products containing asbestos (which in both cases must carry specified labels).

Significant work with asbestos insulation, asbestos coating and asbestos insulating board usually requires a licence from HSE under the *Asbestos (Licensing) Regulations 1983 (SI 1983 No 1649)*. However, minor work with such materials, or work with lower risk asbestos-containing materials (asbestos cement or materials of bitumen, plastic, resins or rubber which contain asbestos) can be carried out without requiring a licence. Both licensed and unlicensed work must comply with the requirements of the *Control of Asbestos at Work Regulations 2002 (SI 2002 No 2675)* and HSE provide detailed practical guidance on how this should be achieved.

'Duty to manage asbestos in non-domestic premises'

H2139 This extremely important new requirement is contained in *Regulation 4* of the *Control of Asbestos at Work Regulations 2002 (SI 2002 No 2675)*. It will have a major impact on many employers as asbestos-containing materials ('ACMs'). (See also ASBESTOS.)

ACMs were commonly used as building materials in the 1950s, 1960s and 1970s and the use of asbestos-containing building products was not finally banned until 1999.

Regulation 4 does not come into operation until 21 May 2004 but employers should prepare early to meet its requirements. HSE has already published two key booklets – an ACoP booklet *The management of asbestos in non-domestic premises* and *A comprehensive guide to managing asbestos in premises* which contains much practical guidance. They also have a free guidance leaflet on the subject.

The 'duty holder' under this regulation will depend on the terms of any lease or contract or upon who is in 'control' of the premises. This may be the owner or leaseholder, or the employer occupying the premises, or it could be a combined responsibility (the HSE ACoP booklet provides guidance on the subject). Owners have responsibility for non-domestic premises which are unoccupied.

Other persons are required to co-operate with duty holders under the regulation. This would mean occupiers providing owners with relevant information when they vacate premises but would also require contractors, suppliers and architects involved in constructing or refurbishing premises, to provide duty holders with information about materials used.

Duty holders under *Regulation 4* must:

- take reasonable steps to find ACMs and check their condition;
- presume materials contain asbestos unless there is strong evidence to the contrary;
- keep a written record of the location and condition of actual and presumed ACMs and keep it up to date;
- assess the risk of people being exposed to asbestos;
- prepare and put into effect a plan to manage that risk.

The duty holder must ensure that anyone carrying out work as a result of the regulation is competent for the task in hand. If a separate organisation or individual is appointed, they must:

- have adequate training and experience;
- demonstrate independence, impartiality and integrity;
- have an adequate quality management system;
- carry out any survey work in accordance with recommended guidance (HSE guidance booklet MDHS 100).

MDHS 100 specifies three types of survey which can be carried out. Organisations can be accredited under ISO 17020 to undertake ACM surveys and individual surveyors under EN 45013.

Finding ACMs and assessing their condition

H2140 Information about possible ACMs should be gathered together from sources such as:

- plans and specifications;
- information from builders, architects and material suppliers;
- previous asbestos survey work.

A survey of the premises must then be carried out and materials categorised as:

- not containing asbestos materials (some are obviously wood, glass, metal, brick, stone etc.);
- materials known to be ACMs;
- materials presumed to be ACMs, until there is evidence to the contrary.

Some materials may be inaccessible, e.g. covered by wallpaper, contained within walls or fire doors or in positions which could be dangerous to access such as roof voids. In some cases it may be appropriate to make further investigations as to their composition or it may be acceptable just to presume that they are ACMs for the time being (see H2141 below). Some areas may only require a relatively straightforward inspection whilst others may justify a detailed survey including sampling of suspected ACM.

The locations and condition of actual and presumed ACMs must be clearly recorded on drawings or sketches as part of the assessment record. The use of photographs may be particularly useful.

HSE guidance sets out in detail what should be included in the assessment record.

Risk management plan

H2141 The duty holder must prepare a written plan for managing the risks from ACMs and this must include measures such as:

- clear identification of actual and presumed ACMs (by signs, colour coding etc.);
- restriction of access to or maintenance work in ACM areas (through locked doors, use of permits to work or other control systems);
- analysis of suspect material prior to its disturbance;
- treatment of ACMs which may release fibres (sealing or safe removal);
- protection of ACMs from damage (e.g. from vehicles, trolleys);
- regular monitoring of ACMs to check for deterioration, disturbance or damage (including damage by vandalism);
- ensuring information about actual and presumed ACMs is readily available to those who might disturb it (and the emergency services).

Both the risk management plan and the ACM assessment must be reviewed if there is reason to suspect they are no longer valid or there are significant changes.

Control of Major Accident Hazard Regulations 1999 (COMAH)

H2142 The *Control of Major Accident Hazard Regulations 1999* (*SI 1999 No 743*) only apply to sites containing specified quantities of dangerous substances. They require the preparation of both on-site and off-site emergency plans with the objectives of:

- containing and controlling incidents so as to minimise their effects, and to limit damage to persons, the environment and property;
- implementing the measures necessary to protect persons and the environment from the effects of major accidents;
- communicating the necessary information to the public and to the emergency services and authorities concerned in the area;
- providing for the restoration and clean-up of the environment following a major accident.

Such plans must be prepared utilising risk assessment techniques. HSE have a number of booklets providing guidance on the methodology to be followed and the parameters to be taken into account.

Control of Lead at Work Regulations 2002

H2143 Like asbestos, previous regulations on lead preceded the original *COSHH Regulations* and the *Control of Lead at Work Regulations 2002* (*SI 2002 No 2676*) continue to follow similar principles. The Regulations together with an ACoP and related guidance are contained in an HSE booklet *The control of lead at work* which is supported by other HSE guidance material.

Dangerous Substances and Explosive Atmospheres Regulations 2002 (DSEAR)

H2144 The *DSEAR Regulations* (*SI 2002 No 2776*) are concerned with protecting against risks from fire and explosion. They replaced several other regulations including the *Highly Flammable Liquids and Liquified Petroleum Gases Regulations 1972* (*SI 1972 No 917*) and (apart from the dispensing of petrol from pumps) the requirement for licensing under the *Petroleum* (*Consolidation*) *Act 1928*.

Dangerous substances

H2145 The *DSEAR Regulations* (*SI 2002 No 2776*) contain a detailed definition of 'dangerous substances' including:

- explosive, oxidising, extremely flammable, highly flammable or flammable substances or preparations (whether or not classified under *CHIP Regulations*), e.g. petrol, solvents, paints, LPG;
- other substances or preparations creating risks due to their physico-chemical or chemical properties;
- potentially explosive dusts e.g. flour, sugar, custard, pitch, wood.

Explosive atmosphere

H2146 This is defined as 'a mixture, under atmospheric conditions of air and one or more dangerous substances in the form of gases, vapours, mists or dusts, in which, after ignition has occurred, combustion spreads to the entire unburned mixture'.

The *DSEAR Regulations* (*SI 2002 No 2776*) require employers to:

- Carry out a risk assessment of work activities involving dangerous substances [*SI 2002 No 2776, Reg 5*].
- Eliminate or reduce risks as far as is reasonably practicable [*SI 2002 No 2776, Reg 6*].
- Classify places where explosive atmospheres may occur in zones, and ensure that equipment and protective systems in these places meet appropriate standards, marking zones with signs, where necessary [*SI 2002 No 2776, Reg 7*] (this duty is being phased in)
- Provide equipment and procedures to deal with accidents and emergencies [*SI 2002 No 2776, Reg 8*].
- Provide employees with suitable and sufficient information, instruction and training [*SI 2002 No 2776, Reg 9*].

Regulation 6 requires control measures to be implemented according to a specified order of priority which can be summarised as:

- Reduce the quantity of dangerous substances to a minimum.
- Avoid or minimise releases of dangerous substances.
- Control releases at source.
- Prevent the formation of an explosive atmosphere (including providing appropriate ventilation).
- Collect, contain and remove any releases to a safe place or otherwise render them safe.

- Avoid the presence of ignition sources (including electrostatic discharges).
- Avoid adverse conditions that could lead to danger (e.g. by temperature or other controls).
- Segregation of incompatible dangerous substances.

The regulation also requires measures to mitigate the detrimental effects of a fire or explosion (or other harmful physical effects). These include minimising the number of employees exposed and the provision of explosion pressure relief arrangements or explosion suppression equipment.

Dangerous Substances (Notification and Marking of Sites) Regulations 1990

H2147 The *Dangerous Substances* (*Notification and Marking of Sites*) *Regulations 1990* (*SI 1990 No 304*) are primarily for the benefit of the fire service and other emergency services. Both the fire authority and HSE must be provided with notification of dates when it is anticipated that a total quantity of 25 tonnes of 'dangerous substances' will be present on the site. (The definition of 'dangerous substances' is contained in the Regulations.) Further notifications must be made where significant changes occur in substances present, including cessation or reductions. Sites must also be marked with standard signs as specified in *Schedule 3* to the Regulations. A booklet providing detailed guidance on the regulations is available from HSE.

Ionising Radiation Regulations 1999

H2148 *Regulation 7* of the *Ionising Radiation Regulations 1999* (*SI 1999 No 3232*) require employers to carry out a risk assessment before commencing any new activity involving work with ionising radiation. The assessment must be:

> ' ... sufficient to demonstrate that—
>
> (a) all hazards with the potential to cause a radiation accident have been identified; and
>
> (b) the nature and magnitude of the risks to employers and other persons arising from those hazards have been evaluated.'

Where such radiation risks are identified, all reasonably practicable steps must be taken to:

- prevent any such accident;
- limit its consequences should such an accident occur;
- provide employees with necessary information, instruction, training and equipment necessary to restrict their exposure.

The Regulations contain many detailed requirements on the prevention of radiation accidents and related control measures. An HSE booklet contains the Regulations and an ACoP and guidance on the regulations. Much more guidance is available from HSE on specific aspects of radiation safety.

Joint Consultation in Safety – Safety Representatives, Safety Committees, Collective Agreements and Works Councils

Introduction

J3001 The provision of 'information, consultation and participation' and of 'health protection and safety at the workplace' are outlined as two of the 'fundamental social rights of workers' in accordance with the Community Charter of Fundamental Rights of 1989. Combined with Article 138 of the Consolidated Version of the Treaty Establishing the European Community (i.e. Treaty of Rome 1957), the Health and Safety Directive 1989 (resulting in 'the six pack' collection of regulations) and the United Kingdom's signing of the Social Chapter in 1998, the EU has provided great stimuli for the introduction of extensive obligations to consult on health and safety issues.

At national level, there has already been change to the obligations owed by employers to their workers. The *Working Time Regulations 1998* (*SI 1998 No 1833*) ('*Working Time Regulations*') were introduced in October 1998 and have subsequently been amended.

Furthermore, the obligations owed under the European Works Councils Directive 1994 have been transposed into national law under the *Transnational Information and Consultation of Employees Regulations 1999* (*SI 1999 No 3323*) ('*Works Councils Regulations*').

Most recently, there has been the 'Mini Works Councils Directive' (Directive 2002/19/EC). This is to be introduced on a phased basis into all workplaces with 50 or more workers (in this regard, the Government has issued a consultation document, '*High performance workplaces: Informing and consulting employees*' and draft regulations, and consultation ends on 7 November 2003). Consultation with the workforce is likely to become a central issue for industrial relations and health and safety over the next few years.

The Health and Safety Commission ('HSC') is also currently seeking to promote change. In 1999, it published the consultative document '*Employee consultation and involvement in health and safety*', which made a number of recommendations in relation to health and safety consultation. In particular, it recommended:

- giving trade union safety representatives the legal power to issue their employer with provisional improvement notices requiring the employer to rectify stated defects within a fixed period of time;
- extending the unfair dismissal protection provided to 'whistleblowers' to health and safety representatives where the representatives recommend that

workers leave their place of work due to circumstances giving rise to a serious, imminent and unavoidable danger at the workplace;

- extending access to specific information to all employees of an employer;
- the introduction of specially trained union safety representatives whose function is to undertake workplace inspections at the workplaces of different employers where their members work; and
- extending protection against victimisation to safety representatives.

HSC is proposing to publish a further consultation document on proposed legislation in this area in the early part of 2004.

This chapter looks at the existing domestic legislation that places an employer under an obligation to consult with its workforce.

Regulatory framework

J3002 The legal requirements for consultation are more convoluted than might be expected, principally because there are two groups of workers who are treated as distinct for consultation purposes. Until 1996 only those employers who recognised a trade union for any collective bargaining purpose were obliged to consult the workforce through safety representatives. The *Safety Representatives and Safety Committees Regulations 1977* (*SI 1977 No 500*) (*as amended*) ('*Safety Representatives Regulations*'), which were made under the provisions of the *Health and Safety at Work etc Act 1974* ('*HSWA 1974*'), *s 2*(*6*), came into effect in 1978 and introduced the right for recognised trade unions to appoint safety representatives. The *Safety Representatives Regulations* were amended in 1993 – by the *Management of Health and Safety at Work Regulations 1992* (*SI 1992 No 2051*) (now superceded by the *Management of Health and Safety at Work Regulations 1999* (*SI 1999 No 3242*)) – to extend employers' duties to consult and provide facilities for safety representatives.

The duty to consult was extended on 1 October 1996 by virtue of the *Health and Safety* (*Consultation with Employees*) *Regulations 1996* (*SI 1996 No 1513*) ('*Consultation with Employees Regulations*'), which have subsequently been amended. The Regulations were introduced as a 'top up' to the *Safety Representatives Regulations*, extending the obligation upon employers to consult all of their employees about health and safety measures. The *Consultation with Employees Regulations* expanded the obligation beyond just trade union appointed representatives. When introduced, the Regulations addressed the general reduction of union recognition over the preceding five years. However, the more recent trend in workforces, driven by the twin forces of Europe and the current Labour Government, is for greater worker participation in the employer's undertaking and the current framework facilitates this whether or not a union is the conduit for worker consultation.

The obligation of consultation, its enforcement and the details of the role and functions of safety representatives and representatives of employee safety, whether under the *Safety Representatives Regulations* or the *Consultation with Employees Regulations*, are almost identical but, for the sake of clarity, are dealt with separately below. The primary distinction is that different obligations apply depending upon whether the affected workers are unionised or not. The respective Regulations also cover persons working in host employers' undertakings.

There are specific supplemental provisions which apply to offshore installations, which are governed by the *Offshore Installations* (*Safety Representatives and Safety Committees*) *Regulations 1989* (*SI 1989 No 971*). Likewise, in the case of the education sector, HSC has published the 'Safety Representatives' Charter' ('the Charter'). The Charter has been developed to:

- Promote and emphasise the rights, roles and functions of safety representatives and to actively promote the involvement of safety representatives in the education sector's efforts to improve health and safety.
- Motivate employers, safety representatives and employees to work in partnership to develop a positive safety culture throughout the education sector.
- Raise awareness amongst employers of the important contribution of safety representatives towards the development of such a culture.
- Encourage employers to demonstrate their full commitment towards consulting and involving safety representatives in matters of health, safety and welfare.
- Increase the participation of education sector employees and their safety representatives in health and safety activities.
- Contribute towards an improved health and safety performance which aims to reduce accidents and ill health in the education sector.

In particular, the Charter sets out the legal obligations that employers in the education sector owe under the *Safety Representatives Regulations* and the *Consultation with Employees Regulations*.

Whilst it is not legislation itself, the Charter is evidence of best practice within the education sector.

Although the *Consultation with Employees Regulations* and the *Safety Representatives Regulations* are based upon good industrial relations practice, there is still the possibility of a prosecution by the Health and Safety Executive ('HSE') inspectors of employers who fail to consult their workforce on health and safety issues. There is substantial overlap, however, with employment protection legislation and the obligations in respect of consultation – as can be seen from the protective rights which are conferred upon safety representatives, such as the right not to be victimised or subjected to detriment for health and safety activities, together with the consequential right to present a complaint to an employment tribunal if they are dismissed or suffer a detriment as a result of carrying out their duties.

The *Consultation with Employees Regulations* also extend these rights to the armed forces. However, armed forces representatives are to be appointed rather than elected, and no paid time off is available. The Regulations do not apply to sea-going ships.

The Working Time Directive 1994 has resulted in the *Working Time Regulations 1998* (*SI 1998 No 1833*) (subsequently amended) under national law. These Regulations allow collective modification of the night working requirements in *Regulation 6*, the daily rest provisions in *Regulation 10*, the weekly rest provisions in *Regulation 11*, the rest break provisions in *Regulation 12* and also allow modification of the averaging period for calculating weekly working time under *Regulation 4* where the same is either for objective, technical or organisational reasons. The method of modification is either by collective agreement in the case of a unionised workforce or by workforce agreement (as defined in the *Working Time Regulations 1998* (*SI 1998 No 1833*), *Sch 1*) in the case of non-unionised employees.

In relation to pan-European employers, obligations are owed to workers under the European Works Councils Directive 1994. The United Kingdom adopted the Directive on 15 December 1997 and provisions were enacted into national law by the *Works Councils Regulations* (*SI 1999 No 3323*) and came into force on 15 January

2000. Whilst the *Works Councils Regulations* do not specifically encompass health and safety obligations, these issues are within the remit of a European Works Council.

Consultation obligations for unionised employers

J3003 Under the *Safety Representatives Regulations* (*SI 1977 No 500*), a trade union has the right to appoint an individual to represent the workforce in consultations with the employer on all matters concerning health and safety at work, and to carry out periodic inspections of the workplace for hazards. Every employer has a duty to consult such union-appointed safety representatives on health and safety arrangements (and, if they so request him, to establish a safety committee to review the arrangements – see J3017 below). [*HSWA 1974, s 21*].

General duty

J3004 The general duty of the employer under the *Safety Representatives Regulations* (*SI 1977 No 500*) is to 'consult with safety representatives with regard to both the making and maintaining of arrangements that will enable the employer and its workforce to co-operate in promoting and developing health and safety at work, and monitoring its effectiveness'. General guidance has been issued by HSC in the form of the Codes of Practice '*Safety Representatives and Safety Committees*' and '*Time Off for the Training of Safety Representatives*' to which regard should be had generally.

Appointment of safety representatives

J3005 The right of appointment of safety representatives was, until 1 October 1996, restricted to independent trade unions who are recognised by employers for collective bargaining purposes. The *Consultation with Employees Regulations* (*SI 1996 No 1513*) extended this to duly elected representatives of employee safety, as detailed above.

The terms 'independent' and 'recognised' are defined in the *Safety Representatives Regulations* (*SI 1977 No 500*) and follow the definitions laid down in the *Trade Union and Labour Relations* (*Consolidation*) *Act 1992, ss 5* and *178*(*3*) respectively. The *Safety Representatives Regulations* make no provision for dealing with disputes which may arise over questions of independence or recognition (this is dealt with in the *Trade Union and Labour Relations* (*Consolidation*) *Act 1992, ss 6* and *8* – the amendments that have been made to the *Trade Union and Labour Relations* (*Consolidation*) *Act 1992* by the *Employment Relations Act 1999* in relation to recognition of unions do not help as *Sch A1* is confined to recognition disputes concerning pay, hours and holiday).

Safety representatives must be representatives of recognised independent trade unions, and it is up to each union to decide on its arrangements for the appointment or election of its safety representatives. [*Safety Representatives Regulations* (*SI 1977 No 500*), *Reg 3*]. Employers are not involved in this matter, except that they must be informed in writing of the names of the safety representatives appointed and of the group(s) of employees they represent. [*Safety Representatives Regulations* (*SI 1977 No 500*), *Reg 3*(*2*)].

The *Safety Representatives Regulations* (*SI 1977 No 500*), *Reg 8*, state that safety representatives must be employees except in the cases of members of the Musicians'

Union and actors' Equity. In addition, where reasonably practicable, safety representatives should have at least two years' employment with their present employer or two years' experience in similar employment. The HSC guidance notes advise that it is not reasonably practicable for safety representatives to have two years' experience, or employment elsewhere, where:

- the employer is newly established,
- the workplace is newly established,
- the work is of short duration, or
- there is high labour turnover.

The same general guidance is followed for employee safety representatives under the *Consultation with Employees Regulations* (*SI 1996 No 1513*).

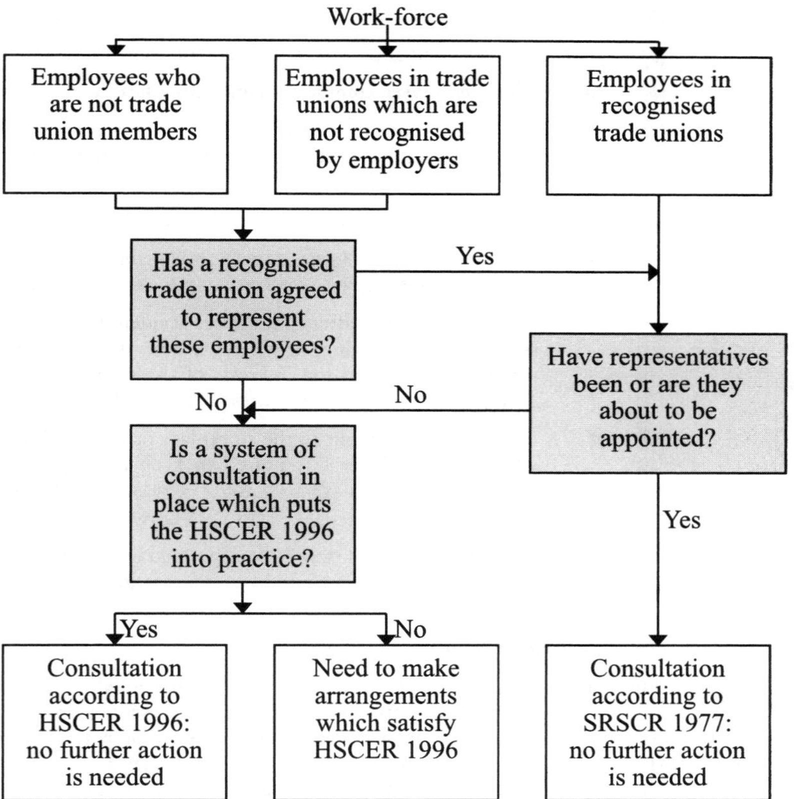

Number of representatives for workforce

J3006 The *Safety Representatives Regulations* (*SI 1977 No 500*) do not lay down the number of safety representatives that unions are permitted to appoint for each workplace. This is a matter for unions themselves to decide, having regard to the number of workers involved and the hazards to which they are exposed. HSE's view is that each safety representative should be regarded as responsible for the interests of a defined group of workers. This approach has not been found to conflict with existing workplace trade union organisation based on defined groups of workers. The size of these groups varies from union to union and from workplace to workplace. While normally each workplace area or constituency would need only one safety representative, additional safety representatives are sometimes required where workers are exposed to numerous or particularly severe hazards; where workers are distributed over a wide geographical area or over a variety of workplace locations; and where workers are employed on shiftwork.

Role of safety representatives

J3007 The *Safety Representatives Regulations* (*SI 1977 No 500*), *Reg 4*(*1*) (as amended by the *Management of Health and Safety at Work Regulations 1992* (*SI 1992 No 2051*) – now superseded by the *Management of Health and Safety at Work Regulations 1999* (*SI 1999 No 3242*)) lists a number of detailed functions for safety representatives:

(*a*) to investigate potential hazards and causes of accidents at the workplace;

(*b*) to investigate employee complaints concerning health etc. at work;

(*c*) to make representations to the employer on matters arising out of (*a*) and (*b*) and on general matters affecting the health etc. of the employees at the workplace;

(*d*) to carry out the following inspections (and see J3008 below):

- of the workplace (after giving reasonable written notice to the employer – see *Safety Representatives Regulations, Reg 5*);
- of the relevant area after a reportable accident or dangerous occurrence (see accident reporting) or if a reportable disease is contracted, if it is safe to do so and in the interests of the employees represented (see *Safety Representatives Regulations* (*SI 1977 No 500*), *Reg 6*);
- of documents relevant to the workplace or the employees represented which the employer is required to keep (see *Safety Representatives Regulations* (*SI 1977 No 500*), *Reg 7*) – reasonable notice must be given to the employer;

(*e*) to represent the employees they were appointed to represent in consultations with HSE inspectors, and to receive information from them (see J3016 below);

(*f*) to attend meetings of safety committees.

These functions are interrelated and are to be implemented proactively rather than reactively. Safety representatives should not just represent their members' interests when accidents or near-misses occur or at the time of periodic inspections, but should carry out their obligations on a continuing day-to-day basis. The *Safety Representatives Regulations* make this obligation clear by stating that safety representatives have the functions of investigating potential hazards and members' complaints *before* accidents, as well as investigating dangerous occurrences and the causes of accidents *after* they have occurred. These functions are only assumed when the employer has been notified in writing by the trade union or workforce of the identity of the representative.

Thus safety representatives may possibly be closely involved not only in the technical aspects of health, safety and welfare matters at work, but also in those areas which could be described as quasi-legal. In other words, they may become involved in the interpretation and clarification of terminology in the *Safety Representatives Regulations*, as well as in discussion and negotiation with employers as to how and when the regulations may be applied. This would often happen in committee meetings.

Under the *Safety Representatives Regulations* (*SI 1977 No 500*), *Reg 4A*(*1*) (introduced by the *Management of Health and Safety at Work Regulations 1992*), the subjects on which consultation 'in good time' between employers and safety representatives should take place are:

- the introduction of any new measure at a workplace which may substantially affect health and safety;
- arrangements for appointing competent persons to assist the employer with health and safety and implementing procedures for serious and imminent risk;
- any health and safety information the employer is required to provide; and
- the planning and organisation of health and safety training and health and safety implications of the introduction (or planning) of new technology.

The safety representative's terms of reference are, therefore, broad, and exceed the traditional 'accident prevention' area. For example, the *Safety Representatives Regulations* (*SI 1977 No 500*), *Reg 4*(*1*) empowers safety representatives to investigate 'potential hazards' and to take up issues which affect standards of health, safety and welfare at work. In practice it is becoming clear that four broad areas are now engaging the attention of safety representatives and safety committees:

- health;
- safety;
- environment;
- welfare.

These four broad areas effectively mean that safety representatives can, and indeed often do, examine standards relating, for example, to noise, dust, heating, lighting, cleanliness, lifting and carrying, machine guarding, toxic substances, radiation, cloakrooms, toilets and canteens. The protective standards that are operating in the workplace, or the lack of them, are now coming under much closer scrutiny than hitherto.

Workplace inspections

J3008 Arrangements for three-monthly and other more frequent inspections and reinspections should be by joint arrangement. The TUC advises that the issues to be discussed with the employer can include:

- more frequent inspections of high risk or rapidly changing areas of work activity;
- the precise timing and notice to be given for formal inspections by safety representatives;
- the number of safety representatives taking part in any one formal inspection;
- the breaking-up of plant-wide formal inspections into smaller, more manageable inspections;
- provision for different groups of safety representatives to carry out inspections of different parts of the workplace;

- the kind of inspections to be carried out, e.g. safety tours, safety sampling or safety surveys;
- the calling in of independent technical advisers by the safety representatives.

Although formal inspections are not intended to be a substitute for day-to-day observation, they have on a number of occasions provided an opportunity to carry out a full-scale examination of all or part of the workplace and for discussion with employers' representatives about remedial action. They can also provide an opportunity to inspect documents required under health and safety legislation, e.g. certificates concerning the testing of equipment. It should be emphasised that, during inspections following reportable accidents or dangerous occurrences, employers are not required to be present when the safety representative talks with its members. In workplaces where more than one union is recognised, agreements with employers about inspections should involve all the unions concerned. It is generally agreed that safety representatives are also allowed under the Regulations to investigate the following:

- potential hazards;
- dangerous occurrences;
- the causes of accidents;
- complaints from their members.

This means that imminent risks, or hazards which may affect their members, can be investigated right away by safety representatives without waiting for formal joint inspections. Following an investigation of a serious mishap, safety representatives are advised to complete a hazard report form, one copy being sent to the employer and one copy retained by the safety representative.

Rights and duties of safety representatives

Legal immunity

J3009 Ever since the *Trade Disputes Act 1906*, trade unions (and employers' associations) have enjoyed immunity from liability in tort for industrial action, taken or threatened, in contemplation or furtherance of a trade dispute (although such freedom of action was subsequently curtailed by the *Trade Union and Labour Relations* (*Consolidation*) *Act 1992, s 20*). Not surprisingly, perhaps, this immunity extends to their representatives acting in a lawful capacity. Thus, the *Safety Representatives Regulations* (*SI 1977 No 500*) state that none of the functions of a safety representative confers legal duties or responsibilities [*Safety Representatives Regulations* (*SI 1977 No 500*), *Reg 4*(*1*)]. As safety representatives are not legally responsible for health, safety or welfare at work, they cannot be liable under either the criminal or civil law for anything they may do, or fail to do, as a safety representative under the *Safety Representatives Regulations*. This protection against criminal or civil liability does not, however, remove a safety representative's legal responsibility as an employee. Safety representatives must carry out their responsibilities under the *HSWA 1974, s 7* if they are not to be liable for criminal prosecution by an HSE inspector. These duties as an employee are to take reasonable care for the health and safety of one's self and others, and to co-operate with one's employer as far as is necessary to enable him to carry out his statutory duties on health and safety.

Time off with pay

General right

J3010 Under the *Safety Representatives Regulations* (*SI 1977 No 500*), safety representatives are entitled to take such paid time off during working hours as is necessary to perform their statutory functions, and reasonable time to undergo training in accordance with a Code of Practice approved by HSC.

Definition of 'time off'

J3011 The *Safety Representatives Regulations* (*SI 1977 No 500*), *Reg 4*(*2*) provides that the employer must provide the safety representative with such time off with pay during the employee's working hours as shall be necessary for the purposes of:

- performing his statutory functions; and
- undergoing such training in aspects of those functions as may be reasonable in all the circumstances.

Further details of these requirements are outlined in the Code of Practice attached to the *Safety Representatives Regulations* and the HSC Approved Code of Practice on time off for training. The Code of Practice is for guidance purposes only – its contents are recommendations rather than requirements. However, it is guidance that an employment tribunal can and will take into account if a complaint is lodged in relation to an employer's unreasonable failure to allow time off.

The combined effect of the ACAS Code No 3: '*Time Off for Trade Union Duties and Activities*' (1991) carried over into the *Trade Union and Labour Relations* (*Consolidation*) *Act 1992* (and reissued in 1998) and the HSC Approved Code of Practice on time off is that shop stewards who have also been appointed as safety representatives are to be given time off by their employers to carry out both their industrial relations duties and their safety functions, and also paid leave to attend separate training courses on industrial relations and on health and safety at work – this includes a TUC course on *COSHH* (*Gallagher v The Drum Engineering Co Ltd, COIT 1330/89*).

An employee is not entitled to be paid for time taken off in lieu of the time he had spent on a course. This was held in *Hairsine v Hull City Council [1992] IRLR 211* when a shift worker, whose shift ran from 3 pm to 11 pm, attended a trade union course from 9 am to 4 pm and then carried out his duties until 7 pm. He was paid from 3 pm to 7 pm and he could claim no more. However, where more safety representatives have been appointed than there are sections of the workforce for which safety representatives could be responsible, it is not unreasonable for an employer to deny some safety representatives time off for fulfilling safety functions (*Howard and Peet v Volex plc (HSIB 181)*).

A decision of the EAT seems to favour jointly sponsored in-house courses, except as regards the representational aspects of the functions of safety representatives, where the training is to be provided exclusively by the union (*White v Pressed Steel Fisher [1980] IRLR 176*). Moreover, one course per union per year is too rigid an approach (*Waugh v London Borough of Sutton (1983) HSIB 86*).

Definition of 'pay'

J3012 The amount of pay to which the safety representative is entitled is contained in the *Safety Representatives Regulations* (*SI 1977 No 500*), *Sch 2*.

Recourse for the safety representative

J3013 Where the employer's refusal to allow paid time off is unreasonable, he must reimburse the employee for the time taken to attend. [*Safety Representatives Regulations* (*SI 1977 No 500*), *Reg 4*(*2*), *Sch 2*]. In the case of *Scarth v East Herts DC (HSIB 181)*: the test of reasonableness is to be judged at the time of the decision to refuse training.

Safety representatives who are refused time off to perform their functions or who are not paid for such time off are able to make a complaint to an employment tribunal. [*Safety Representatives Regulations* (*SI 1977 No 500*), *Reg 11*].

Facilities to be provided by employer

J3014 The type and number of facilities that employers are obliged to provide for safety representatives are not spelled out in the Regulations, Code of Practice or guidance notes, other than a general requirement in the *Safety Representatives Regulations* (*SI 1977 No 500*), *Reg 5*(*3*) which states, *inter alia*, that 'the employer shall provide such facilities and assistance as the safety representatives shall require for the purposes of carrying out their functions'. Formerly, the requirement to provide facilities and assistance related only to inspections.

Trade unions consider that the phrase 'facilities and assistance' includes the right to request the presence of an independent technical adviser or trade union official during an inspection, and for safety representatives to take samples of substances used at work for analysis outside the workplace. The TUC has recommended that the following facilities be made available to safety representatives:

- a room and desk at the workplace;
- facilities for storing correspondence;
- inspection reports and other papers;
- ready access to internal and external telephones;
- access to typing and duplicating facilities;
- provision of notice boards;
- use of a suitable room for reporting back to and consulting with members;
- other facilities should include copies of all relevant statutes, regulations, Approved Codes of Practice and HSC guidance notes; and copies of all legal or international standards which are relevant to the workplace.

Disclosure of information

J3015 Employers are required by the *Safety Representatives Regulations* to disclose information to safety representatives which is necessary for them to carry out their functions. [*Safety Representatives Regulations* (*SI 1977 No 500*), *Reg 7*(*1*)]. A parallel provision exists under the *Management of Health and Safety at Work Regulations 1999*, *Reg 10*(*2*) in relation to the information that is to be provided to the parents of a child to be employed by an employer.

Regulation 7 is consolidated by paragraph 6 of the Code of Practice which details the health and safety information 'within the employer's knowledge' that should be made available to safety representatives. This should include:

- plans and performance and any changes proposed which may affect health and safety;

- technical information about hazards and precautions necessary, including information provided by manufacturers, suppliers and so on;
- information and statistical records on accidents, dangerous occurrences and notifiable industrial diseases; and
- other information such as measures to check the effectiveness of health and safety arrangements and information on articles and substances issued to homeworkers.

The exceptions to this requirement are where disclosure of such information would be 'against the interests of national security'; where it would contravene a prohibition imposed by law; any information relating to an individual (unless consent has been given); information that would damage the employer's undertaking; and information obtained for the sole purpose of bringing, prosecuting or defending legal proceedings. [*Safety Representatives Regulations* (*SI 1977 No 500*), *Reg 7*(*2*)].

However, the decision in *Waugh v British Railways Board [1979] 2 All ER 1169* established that where an employer seeks, on grounds of privilege, to withhold a report made following an accident, he can only do so if its dominant purpose is related to actual or potential hostile legal proceedings. In this particular case, a report was commissioned for two purposes following the death of an employee: (*a*) to recommend improvements in safety measures, and (*b*) to gather material for the employer's defence. It was held that the report was not privileged.

This was followed in *Lask v Gloucester Health Authority [1986] HSIB 123* where a circular '*Reporting Accidents in Hospitals*' had to be discovered by order after an injury to an employee whilst he was walking along a path.

Where differences of opinion arise as to the evaluation or interpretation of technical aspects of safety information or health data, unions are advised to contact the local offices of HSE, because of HSE expertise and access to research.

Technical information

J3016 HSE inspectors are also obliged under the *HSWA 1974, s 28*(*8*), to supply safety representatives with technical information – factual information obtained during their visits (i.e. any measurements, testing and results of sampling and monitoring), notices of prosecution, copies of correspondence and copies of any improvement or prohibition notices issued to their employer. The latter places an absolute duty on an inspector to disclose specific kinds of information to workers or their representatives concerning health, safety and welfare at work. This can also involve personal discussions between the HSE inspector and the safety representative. The inspector must also tell the representative what action he proposes to take as a result of his visit. Where local authority health inspectors are acting under powers granted by the *HSWA 1974* (see ENFORCEMENT), they are also required to provide appropriate information to safety representatives.

Safety committees

J3017 There is a duty on every employer, in cases where it is prescribed (see below), to establish a safety committee if requested to do so by safety representatives. The committee's purpose is to monitor health and safety measures at work. [*HSWA 1974, s 2*(*7*)]. Such cases are prescribed by the *Safety Representatives Regulations* (*SI 1977 No 500*) and limit the duty to appoint a committee to requests made by trade union safety representatives.

Establishment of a safety committee

J3018 If requested by at least two safety representatives in writing, the employer must establish a safety committee. [*Safety Representatives Regulations* (*SI 1977 No 500*), *Reg 9*(*1*)].

When setting up a safety committee, the employer must:

- consult with both:
  - the safety representatives who make the request; and
  - the representatives of recognised trade unions whose members work in any workplace where it is proposed that the committee will function;
- post a notice, stating the composition of the committee and the workplace(s) to be covered by it, in a place where it can easily be read by employees;
- establish the committee within three months after the request for it was made.

[*Safety Representatives Regulations* (*SI 1977 No 500*), *Reg 9*(*2*)].

Function of safety committees

J3019 In practical terms, trade union appointed safety representatives are now using the medium of safety committees to examine the implications of hazard report forms arising from inspections, and the results of investigations into accidents and dangerous occurrences, together with the remedial action required. A similar procedure exists with respect to representatives for tests and measurements of noise, toxic substances or other harmful effects on the working environment.

Trade unions regard the function of safety committees as a forum for the discussion and resolution of problems that have failed to be solved initially through the intervention of the safety representative in discussion with line management. There is, therefore, from the trade unions' viewpoint, a large measure of negotiation with its consequent effect on collective bargaining agreements.

If safety representatives are unable to resolve a problem with management through the safety committee, or with HSE, they can approach their own union for assistance – a number of unions have their own health and safety officers who can, and do, provide an extensive range of information on occupational health and safety matters. The unions, in turn, can refer to the TUC for further advice.

The 'Brown Book', which contains the *Safety Representatives Regulations*, Code of Practice and guidance, was revised in 1996 to include the amendments made in 1993 by the *Management of Health and Safety at Work Regulations 1992* (now superceded by the *Management of Health and Safety at Work Regulations 1999* (*SI 1999 No 3242*)) (see above) and the *Consultation with Employees Regulations* (see J3021).

Non-unionised workforce – consultation obligations

J3020 A representative of employee safety is an elected representative of a non-unionised workforce who is assigned with broadly the same rights and obligations as a safety representative in a unionised workforce.

General duty

J3021 The *Consultation with Employees Regulations* (*SI 1996 No 1513*) introduced a new duty to consult any employees who are not members of a group covered by safety representatives (appointed under the *Safety Representatives Regulations*). Employers therefore have the choice of consulting their employees either directly or by way of an appointed representative of employee safety. The obligation is to consult those employees in good time on matters relating to their health and safety at work.

Number of representatives for workforce

J3022 Guidance notes on the *Consultation with Employees Regulations* (*SI 1996 No 1513*) state that the number of safety representatives who can be appointed depends on the size of the workforce and workplace, whether there are different sites, the variety of different occupations, the operation of shift systems and the type and risks of work activity. A DTI Workplace Survey has concluded that in a non-unionised workplace which has appointed worker representatives, it is usual for there to be several representatives, with the median being three. The survey estimated that there are approximately 218,000 representatives across all British workplaces with 25 or more employees.

Role of safety representatives

J3023 The functions of representatives of employee safety are:

- to make representations to the employer of potential hazards and dangerous occurrences at the workplace which affect or could affect the group of employees he represents;
- to make representations to the employer on general matters affecting the health and safety at work of the group of employees he represents, and in particular on such matters as he has been consulted about by the employer under the *Consultation with Employees Regulations*; and
- to represent that group of employees in consultations at the workplace with inspectors appointed under the *HSWA 1974*.

Rights and duties of representatives of employee safety

Time off with pay

J3024 Under the *Consultation with Employees Regulations* (*SI 1996 No 1513*), *Reg 7(1)(b)*, the right that a representative of employee safety has to take time off with pay is generally the same as that for safety representatives.

Definition of 'time off'

J3025 An employer is under an obligation to permit a representative of employee safety to take such time off with pay during working hours as shall be necessary for:

- performing his functions; and
- undergoing such training as is reasonable in all the circumstances.

A candidate standing for election as a representative of employee safety is also allowed reasonable time off with pay during working hours in order to perform his functions as a candidate *SI 1996 No 1513, Reg 7(2)*].

Definition of 'pay'

J3026 The *Consultation with Employees Regulations* (*SI 1996 No 1513*), *Sch 1* deal with the definition of pay, and generally the definition is the same as that for union safety representatives.

Provision of information

J3027 The employer must provide such information as is necessary to enable the employees or representatives of employee safety to participate fully and effectively in the consultation. In the case of representatives of employee safety, the information must also be sufficient to enable them to carry out their functions under the *Consultation with Employees Regulations* (*SI 1996 No 1513*).

Information provided to representatives must also include information which is contained in any record which the employer is required to keep under *RIDDOR 1995* (*SI 1995 No 3163*) and which relates to the workplace or the group of employees represented by the representatives. Note that there are exceptions to the requirement to disclose information similar to those under the *Safety Representatives Regulations* (*SI 1977 No 500*), *Reg 7* (see J3015 above).

Relevant training

J3028 Representatives of employee safety must be provided with reasonable training in respect of their functions under the *Consultation with Employees Regulations* (*SI 1996 No 1513*), for which the employer must pay.

Remedies for failure to provide time off or pay for time off

J3029 A representative of employee safety, or candidate standing for election as such, who is denied time off or who fails to receive payment for time off, may make an application to an employment tribunal for a declaration and/or compensation. As in the case of safety representatives, the remedies obtainable (set out in the *Consultation with Employees Regulations* (*SI 1996 No 1513*), *Sch 2* are similar to those granted to complainants under the *Trade Union and Labour Relations* (*Consolidation*) *Act 1992, s 168* (time off for union duties).

Recourse for safety representatives

J3030 Safety representatives (whether they are appointed under the *Safety Representatives Regulations* (*SI 1977 No 500*) or the *Consultation with Employees Regulations* (*SI 1996 No 1513*) are provided with statutory protection for the proper execution of their duties.

An employee who is:

- designated by his employer to carry out a health and safety related function,
- a representative of employee safety, or
- a candidate standing for election as such,

has the right not to be subjected to any detriment or unfairly dismissed on the grounds that:

— having been designated by the employer to carry out a health and safety related function, he carried out, or proposed to carry out, the function;

— he undertook, or proposed to undertake, any function(s) consistent with being a safety representative or member of a safety committee;

— he took part in, or proposed to take part in, consultation with the employer; or

— he took part in an election of representatives of employee safety.

In relation to a detriment claim, where the employer infringes any of these rights, the employee has the right to make a complaint to an employment tribunal under the *Employment Rights Act 1996, s 44(1)* (as amended by the *Consultation with Employees Regulations* (*SI 1996 No 1513*), *Reg 8*. An employment tribunal can make a declaration and also award compensation. [*Employment Rights Act 1996, ss 48, 49*]. There is no minimum qualifying period of service nor upper age limit for bringing such a claim.

If the employer unfairly dismisses such an employee for one of the above reasons, or where it is the principal reason for the dismissal, that dismissal shall be deemed automatically unfair. [*Employment Rights Act 1996, s 100,* as amended by the *Consultation with Employees Regulations* (*SI 1996 No 1513*), *Reg 8*]. It will also be an automatically unfair dismissal to select a representative or candidate for redundancy for such a reason. [*Employment Rights Act 1996, s 105*]. The normal minimum qualifying period of service, the normal upper age limit and the cap on the compensatory award for unfair dismissal claims do not apply. [*Employment Rights Act 1996, ss 108, 109 and 124(1A)* as amended by the *Employment Relations Act 1999, s 37(1)*)].

Under the *Employment Rights Act 1996, s 103A* a right to automatically claim unfair dismissal would exist for a worker who is dismissed for making a protected disclosure. A right also exists under the *Employment Rights Act 1996, s 47B* for workers to claim that they have been subjected to a detriment for making a protected disclosure. 'Protected disclosures' in this regard are capable of covering the situation where the health and safety of an individual has been or is likely to be endangered. [*Employment Rights Act 1996, s 43B(1)(d)*].

European developments in health and safety

Introduction

J3031 Health and safety law will continue to be subject to change in the future with the implementation of further European directives and HSC programme of modifying and simplifying health and safety law. At a European level the most important legislation is the directives made under Art 138 of the Treaty of Rome (as amended). The Framework Directive, from which the *Consultation with Employees Regulations* (*SI 1996 No 1513*) were derived, will continue to drive forward developments in UK health and safety law.

Further developments have occurred under European law which are having an impact at national level these being in the form of the Working Time Directive (93/104) and the European Works Councils Directive (94/45). These Directives have been implemented into national law under the *Working Time Regulations 1998* (*SI 1998 No 1833*) (as amended) and the *Works Councils Regulations* (*SI 1999 No 3323*) respectively.

The Working Time Regulations 1998 (as amended)

J3032 A full discussion of the impact of the *Working Time Regulations* (*SI 1998 No 1833*) is beyond the scope of this chapter – readers are referred to WORKING TIME for a discussion of the impact of the *Working Time Regulations* generally.

That said, the *Working Time Regulations* have introduced a joint consultation function into the operation of the Regulations by use of collective, workforce and relevant agreements. The Regulations provide that it is possible to vary the extent to which the Regulations must be strictly complied with through the use of these devices. The various types of agreements can be described as follows:

- Collective agreements – these are defined by *s 178 of the Trade Union and Labour Relations* (*Consolidation*) *Act 1992* as being agreements between independent trade unions and employers;
- Workforce agreements – these were created by the Regulations and are defined in *the Working Time Regulations* (*SI 1998 No 1833*), *Reg 2* and *Sch 1*. They amount to agreements between an employer and either duly elected worker representatives of the employer or, in the case of an employer employing less than 20 workers, a majority of the individual workers themselves, where the agreement concluded:
  - is in writing;
  - has effect for a specified period not exceeding five years;
  - applies to either:
    - (i) all of the relevant members of the workforce, or
    - (ii) all of the relevant members of the workforce who belong to a particular sub-group;
  - is signed:
    - (i) by the worker representatives or by the particular group of workers, or
    - (ii) in the case of an employer having less than 20 employees on the date on which the agreement is first concluded, either by appropriate representatives or by a majority of the workers working for the employer; and
  - before being made available for signature, copies of the agreement were provided to all of the workers to whom the agreement was intended to apply, together with such guidance as the workers might reasonably require in order to understand the draft agreement;
- Relevant agreements – these are workforce agreements that cover a worker, any provision of a collective agreement that is individually incorporated into the contract of employment of a worker, or any other agreement in writing between a worker and his employer that is legally enforceable (e.g. a staff handbook).

By *Regulation 23*(*a*) of the *Working Time Regulations* (*SI 1998 No 1833*), it is possible to modify the provisions relating to:

- the length of night work (see *Working Time Regulations* (*SI 1998 No 1833*), *Reg 6*); and
- the minimum daily and weekly rest periods and rest breaks (see *Working Time Regulations* (*SI 1998 No 1833*), *Regs 10–12* respectively).

Modification must be by way of a collective or workforce agreement and such an agreement must make provision for a compensatory rest period of equivalent length. [*Working Time Regulations* (*SI 1998 No 1833*), *Reg 24*].

Further, by *Regulation 23(b)* of the *Working Time Regulations* (*SI 1998 No 1833*), it is possible for an employer and its workers to agree by collective or workforce agreement to vary the reference period for calculating the maximum working week from the usual 17 weeks to a 52-week period if there are objective or technical reasons relating to the organisation which justify such a change.

The variation provisions allow a degree of flexibility where it is necessary for the interests of an employer's business to effect such change for operational reasons whilst still ensuring the protection of the health and safety of the workforce. The provisions also ensure that any change that is to be made must survive collective scrutiny of the employer's workforce.

European Works Councils and Information and Consultation Procedures

J3033 The provisions relating to European Works Councils are at present confined to large pan-European entities.

As stated at J3001, Directive 2002/14/EC will make 'mini works councils' obligatory for any employer employing more than 50 workers. Transitional provisions will apply initially limiting the impact of Directive 2002/14/EC to employers employing more than 150 employees. All employers in the United Kingdom employing 50 or more employees will have to comply with the Directive by 23 March 2007. National legislation to implement Directive 2002/14/EC is not expected until 2004.

The European Work Council Directive (94/45) was incorporated into national law by the *Works Councils Regulations* (*SI 1999 No 3323*). A full discussion of the operation of the *Works Councils Regulations* is beyond the scope of this chapter, which instead focuses upon the health and safety aspect of the Regulations.

The *Works Councils Regulations* govern employers employing a total of 1,000 or more workers where at least 150 workers are so employed in each of two or more member states.

Their main purpose is procedural. *Part IV* of the *Works Council Regulations* (*SI 1999 No 3323*) creates machinery between workers and their employer for the purpose of establishing either a European Works Council ('EWC') or an Information and Consultation Procedure ('ICP'). *Regulation 17(1)* of the *Works Councils Regulations* (*SI 1999 No 3323*) provides that the central management of the employer and a special negotiating body (defined in *Part III* of the Regulations) are bound to:

> '... negotiate in a spirit of co-operation with a view to reaching a written agreement on the detailed arrangements for the information and consultation of employees in a Community-scale undertaking or Community-scale group of undertakings'.

Regulation 17(3) of the *Works Councils Regulations* (*SI 1999 No 3323*) leaves the choice of whether to proceed with an EWC or an ICP to the parties.

The parties are free to include in the agreement reference to whatever matters are likely to affect the workers of the employer at a trans-national level. Health and safety is clearly such an issue.

If:

- the parties fail to agree the content of the agreement; or

- within six months of a valid request being made to the central management of an employer the employer fails to negotiate so as to create either an EWC or ICP; or
- after three years of negotiation to produce an agreement for an EWC or ICP the parties cannot agree as to the constitution,

default machinery is provided by the *Schedule* to the *Works Councils Regulations* [*Works Councils Regulations* (*SI 1999 No 3323*), *Reg 18*(*1*)]. *Paragraph 6* of the *Schedule* provides in relation to EWCs that:

> 'The competence of the European Works Council shall be limited to information and consultation on the matters which concern the Community-scale undertaking or Community-scale group of undertakings as a whole or at least two of its establishments or group undertakings situated in different Member States'.

In relation to ICPs, *para 7*(*3*) of the *Schedule* to the *Works Councils Regulations* (*SI 1999 No 3323*) provides that meetings of ICPs:

> '… shall relate in particular to the structure, economic and financial situation, the probable development of the business and of production and sales, the situation and probable trend of employment, investments, and substantial changes concerning organisation, [and] introduction of new working methods or production processes …'

Although not expressly providing for discussion of health and safety issues, the provisions relating to both EWCs and ICPs will, by implication, include debate of health and safety matters.

Lifting Machinery and Equipment

Introduction

L3001 There are several distinct categories or groups of machinery which relate to lifting. Each group is dealt with within the regulatory framework in a slightly different way.

The legislation separates lifting machinery from lifts which are permanently installed in buildings. The latter are covered by the EU's Lifts Directive which is implemented into UK law by the *Lifts Regulations 1997* (*SI 1997 No 831*). The *Lifts Regulations 1997* make reference to the Machinery Directive and the Construction Products Directive.

The lifting machinery which is not covered by the Lifts Directive is covered by the Machinery Directive which is implemented into UK law by the *Supply of Machinery* (*Safety*) *Regulations 1992* (*SI 1992 No 3073* as amended by *SI 1994 No 2063*). The Machinery Directive deals directly with safety for suppliers of lifting machines and lifting accessories.

There are specific requirements for lifting machinery in addition to the common requirements for all machinery. These extra requirements deal with matters such as lifting coefficients for tests, whether machines lift goods and/or people, and what control systems are appropriate (see the *Supply of Machinery* (*Safety*) *Regulations 1992, Schedule 3 para 4* 'Essential health and safety requirements to offset the particular hazards due to a lifting operation').

In addition, specific types of lifting equipment including lifting devices which raise people to a height of more than three metres, and vehicle servicing lifts, are within the *Supply of Machinery* (*Safety*) *Regulations 1992, Schedule 4*, and require the conformity to be subject to scrutiny by a notified body within the Machinery Directive. 'State-of-the-art' for safety is established in some cases in the relevant transposed harmonised standards, e.g. BS EN 1443: 1999 – Vehicle lifts.

Lifts controlled by electrical equipment are also subject to the relevant parts of the *Electrical Equipment* (*Safety*) *Regulations 1994* (*SI 1994 No 3260*) implementing the Low Voltage Directive, and the *Electromagnetic Compatibility Regulations 1992* (*SI 1992 No 2372* as amended by *SI 1994 No 3080*). Lifts installed in buildings are also subject to the IEE Wiring Regulations (BS 7671: 1992 – Requirements for electrical installations).

These above rules and regulations apply to the supply of lift equipment. In addition the *Lifting Operations and Lifting Equipment Regulations 1998* (*SI 1998 No 2307* as amended by *SI 2002 No 2174*) apply to their use and upkeep.

This chapter examines:

- hoists and lifts;
- forklift trucks;
- lifting machinery other than hoists and lifts – particularly cranes and the safe use of mobile cranes;

- lifting tackle (ropes, rings, hooks and slings);
- lifting operations on construction sites; and
- statutory requirements relating to lifting equipment, namely:
  - (i) the *Lifts Regulations 1997* (*SI 1997 No 831*);
  - (ii) the *Lifting Operations and Lifting Equipment Regulations 1998* (*SI 1998 No 2307* as amended by *SI 2002 No 2174*).

Hoists and lifts

L3002 The safe use and maintenance of hoists and lifts is governed primarily by the *Lifting Operations and Lifting Equipment Regulations 1998* (*SI 1998 No 2307* as amended by *SI 2002 No 2174*).

Examples include: goods lifts; man hoists (for example those found in flour mills), paternoster lifts (for transporting passengers vertically – usually up to six); scissors lifts and passenger lifts.

Powered working platforms are commonly used for fast and safe access to overhead machinery/plant, stored products, lighting equipment and electrical installations as well as for enabling maintenance operations to be carried out on high-rise buildings. Their height, reach and mobility give them distinct advantages over scaffolding, boatswain's chairs and platforms attached to fork lift trucks. Typical operations are characterised by self-propelled hydraulic booms, semi-mechanised articulated booms and self-propelled scissors lifts.

Platforms should always be sited on firm level working surfaces and their presence indicated by traffic cones and barriers. Location should be away from overhead power lines – but, if this is not practicable, a permit to work system should be instituted. A key danger arises from overturning as a consequence of overloading the platform. Maximum lifting capacity should, therefore, be clearly indicated on the platform as well as in the manufacturer's instructions, and it is inadvisable to use working platforms in high winds (i.e. above Force 4 or 16 mph). Powered working platforms should be regularly maintained and only operated by trained personnel.

A simple guide on the thorough examination and testing of hoists and lifts aimed at helping small businesses is published by HSE Books. *'A thorough examination and testing of lifts: simple guidance for lift owners'* explains in a free leaflet what duty holders need to do to comply with the law. It includes a summary of the legal requirements, an explanation of the purpose of thorough examination and what it should include and advice on how to select a competent person to carry out the examination.

Fork lift trucks

L3003 Fork lift trucks are the most widely used item of mobile mechanical handling equipment. There are several varieties which are as follows:

(*1*) *Pedestrian-operated stackers – manually-operated and power-operated*

Manually-operated stackers are usually limited in operation, for example, for moving post pallets, and cannot pick up directly from the floor. Whereas power-operated stackers are pedestrian-operated or rider-controlled, operate vertically and horizontally and can lift pallets directly from the floor.

(*2*) *Reach trucks*

Reach trucks enable loads to be retracted within their wheel base. There are two kinds, namely, (*a*) moving mast reach trucks, and (*b*) pantograph reach trucks. Moving mast reach trucks are rider-operated, with forward-mounted load wheels enabling carriage to move within the wheel base – mast, forks and load moving together. Pantograph reach trucks are also rider-operated, reach movement being by pantograph mechanism, with forks and load moving away from static mast.

(*3*) *Counterbalance trucks*

Counterbalance trucks carry loads in front counterbalanced to the weight of the vehicle over the rear wheels. Such trucks are lightweight pedestrian-controlled, lightweight rider-controlled or heavyweight rider-controlled.

(*4*) *Narrow aisle trucks*

With narrow aisle trucks the base of the truck does not turn within the aisle in order to deposit/retrieve load. There are two types, namely, side loaders for use on long runs down narrow aisles, and counterbalance rotating load turret trucks, having a rigid mast with telescopic sections, which can move sideways in order to collect/deposit loads.

(*5*) *Order pickers*

Order pickers have a protected working platform attached to the lift forks, enabling the driver to deposit/retrieve objects in or from a racking system. Conventional or purpose-designed, they are commonly used in racked storage areas and operate well in narrow aisles.

Lifting machinery (other than hoists/lifts)

L3004 All lifting equipment is covered by the *Lifting Operations and Lifting Equipment Regulations 1998* (*SI 1998 No 2307* as amended by *SI 2002 No 2174*) which includes cranes, lifts as well as components such as chains, ropes, slings, shackles and eyebolts.

Cranes

L3005 Cranes are widely used in lifting/lowering operations in construction, dock and shipbuilding works. The main hazard, generally associated with overloading or incorrect slewing, is collapse or overturning, the latter in consequence of the crane driver exceeding the 'maximum permitted moment' (mpm). Contact with overhead power lines is also a danger. In such cases, the operator should normally remain inside the cab and not allow anyone to touch the crane or load; the superintending engineer should immediately be informed. (For reportable dangerous occurrences in connection with cranes etc., see ACCIDENT REPORTING AND INVESTIGATION at A3035.)

There are several varieties of crane in frequent use, namely, fixed cranes used at docks and railway sidings; tower cranes on construction sites; mobile cranes used for lifting/lowering loads onto particular locations; overhead travelling cranes – these last operating along a fixed railtrack; in addition, on construction sites there are rough-terrain cranes as well as crawler and wheeled cranes all carrying suspended loads. Persons in the foreseeable impact area of an overhead travelling crane are especially at risk. In order to avoid accidents, electrical supply to the crane should be isolated and a permit to work system instituted. Trained signallers (or banksmen) should be on hand to direct movement of the crane and in a position to see the load clearly and be clearly seen by the driver.

Safe lifting operations by mobile cranes

L3006 A mobile crane is a crane which is capable of travelling under its own power.

Safe lifting operations – as per BS 7121 'Safe use of cranes' – depend on co-operation between supervisor (or appointed person), slinger (and/or signaller) and crane driver.

- *Appointed person*

  Overall control of lifting operations rests with an 'appointed person', who can, where necessary, stop the operation. Failing this, control of operations will be in the hands of the supervisor (who, in some cases, may be the slinger). The appointed (and competent) person must ensure that:

  (i) lifting operations are carefully planned and executed;

  (ii) weights and heights are accurate;

  (iii) suitable cranes are provided;

  (iv) the ground is suitable;

  (v) suitable precautions are taken, if necessary, regarding gas, water and electricity either above or below ground;

  (vi) personnel involved in lifting/lowering are trained and competent.

  (vii) access within the vicinity at where the lifting operation is undertaking is minimised.

- *Supervisor*

  Supervisors must:

  (i) direct the crane driver where to position the crane;

  (ii) provide sufficient personnel to carry out the operation;

  (iii) check the site conditions;

  (iv) report back to the appointed person in the event of problems;

  (v) supervise and direct the slinger, signaller and crane driver;

  (vi) stop the operation if there is a safety risk.

- *Crane driver*

  Crane drivers must:

  (i) erect/dismantle and operate the crane as per manufacturer's instructions;

  (ii) set the crane level before lifting and ensure that it remains level;

  (iii) decide which signalling system is to apply;

  (iv) inform the superviser in the event of problems;

  (v) carry out inspections/weekly maintenance relating to

  — defects in crane structure, fittings, jibs, ropes, hooks, shackles,

— correct functioning of automatic safe load indicator, over hoist and derrick limit switches.

Drivers should always carry out operations as per speeds, weights, heights and wind speeds specified by the manufacturer, mindful that the weight of slings/lifting gear is part of the load. Such information should be clearly displayed in the cab and not obscured or removed. Windows and windscreens should be kept clear and free from stickers containing operational data. Any handrails, stops, machinery guards fitted to the crane for safe access should always be replaced following removal for maintenance; and tools, jib sections and lifting tackle properly secured when not in use.

After a load has been attached to the crane hook by the lifting hook, tension should be taken up slowly, as per the slinger's instructions, the latter being in continuous communication with the driver. In the case of unbalanced loads, drivers/slingers should be familiar with a load's centre of gravity – particularly if the load is irregularly shaped. Once in operation, crane hooks should be positioned directly over the load, the latter not remaining suspended for longer than necessary. Moreover, suspended loads should not be directed over people or occupied buildings.

- *Slingers*

  Slingers must:

  (i) attach/detach a load to/from the crane;

  (ii) use correct lifting appliances;

  (iii) direct movement of a load by correct signals. Any part of a load likely to shift during lifting/lowering must be adequately secured by the slinger beforehand. Spillage or discharge of loose loads (e.g. scaffolding) can be a problem, and such loads must be properly secured/fastened. Nets are useful for covering palletised loads (e.g. bricks).

- *Signallers*

  Quite frequently, signallers are responsible for signalling in lieu of slingers. Failing this, their remit is to transmit instructions from slinger to crane driver, when the former cannot see the load.

Lifting tackle

L3007 Lifting tackle refers to:

- chain slings;
- rope slings;
- rings;
- hooks;
- shackles;
- swivels.

[*Factories Act 1961, s 26*(*3*)].

Materials used for the manufacture of lifting equipment

L3008 All materials have unique physical properties and will behave in different ways depending on the conditions to which they are exposed. For example some materials are more likely to suffer the effects of exposure to high temperatures whilst others are more vulnerable to low temperatures.

Chains

L3009 A chain is a classic example of lifting tackle, in spite of an increase in use of wire ropes. There are several varieties, including mild high-tensile, and alloy steel. The principal risk is breakage, usually occurring in consequence of a production defect in a link, or through application of excessive loads.

Ropes

L3010 Ropes are used quite widely throughout industry in lifting/lowering operations, the main hazard being breakage through overloading and/or natural wear and tear. Ropes are either of natural (e.g. cotton, hemp) or man-made fibre (e.g. nylon, terylene). Natural fibre ropes, if they become wet or damp, should be allowed to dry naturally and kept in a well-ventilated room. They do not require certification prior to service but six months' examination thereafter is compulsory (all lifting equipment accessories require a six monthly examination to meet the requirements under the *Lifting Operations and Lifting Equipment Regulations 1998*). If the specified safe working load (SWL) cannot be maintained by a rope, it should forthwith be withdrawn from service. Of the two, man-made fibre ropes have greater tensile strength, are not subject so much to risk from wear and tear, are more acid/corrosion-resistant and can absorb shock loading better.

Wire ropes are much in use in cranes, lifts, hoists, elevators etc. and are normally lubricant-impregnated to minimise corrosion and reduce wear and tear.

Slings, hooks, eyebolts, pulley blocks etc.

L3011 Slings are chains or ropes (either fibre or wire) and should be of adequate strength. Wire rope slings should always be well lubricated and, where multiple slings are in use, the load should be evenly distributed. Hooks are of forged steel and should be fitted with a safety catch to ensure that the load does not slip off. Eyebolts are mainly for lifting heavy concentrated loads; there being several types, namely, dynamo, collar and eyebolt incorporating link. Pulley blocks, ideally made from shock-resistant metal, are also widely used in lifting and lowering operations. In particular, however, blocks designed for use with fibre rope should not be operated with wire rope.

Lifting operations on construction sites

L3012 Lifting/lowering activities are typical on construction sites. Hazards peculiar to this environment, which crane drivers should beware of, are:

- uneven floor surface;
- adverse weather, e.g. heavy rain, causing ground sinkage, thereby necessitating levelling checks or mat packing;
- underground services, e.g. drains, sewers, mains;

- overhead wires and obstructions;
- uncompacted landfill;
- movement and circulation of other heavy mobile vehicles;
- excavations. Crane operators should always check with supervisors before travelling close to excavations, and, in any case, regularly check the excavation face, especially after heavy rainfall. Moreover, crane travel, whether with or without loads on site, should proceed only when accompanied by a slinger to keep an eye open for obstructions and hazards and warn other people in the vicinity. When travelling minus load, chains/slings should be detached from the hook and the hookblock secured. Similarly, cranes should not be left unattended unless:
  - (i) loads have been detached,
  - (ii) the lifting device has been secured,
  - (iii) the engine has been turned off,
  - (iv) brakes and locks have been applied,
  - (v) the ignition key has been removed.

Cranes travelling on public highways are governed by the same rules and regulations as apply to other road users; drivers must be in possession of a valid current driving licence and over 21. They should be aware of the crane's overall clearance height and see that slings, shackles etc. are either removed or secured.

Statutory requirements relating to lifting equipment

L3013 The *Lifts Regulations 1997* (*SI 1997 No 831*) came into force on 1 July 1997 and the *Lifting Operations and Lifting Equipment Regulations 1998* (*SI 1998 No 2307* as amended by *SI 2002 No 2174*) came into force on 5 December 1998. These two regulations replaced the previous provisions contained in the *Factories Act 1961*, the *Offices, Shops and Railway Premises* (*Hoists and Lifts*) *Regulations 1968*, the *Construction* (*Lifting Operations*) *Regulations 1961* and the *Lifting Plant and Equipment* (*Records of Test and Examination etc.*) *Regulations 1992*.

This section examines:

- the *Lifts Regulations 1997* (*SI 1997 No 831*); and
- the *Lifting Operations and Lifting Equipment Regulations* (*SI 1998 No 2307* as amended by *SI 2002 No 2174*).

In this context certain words need explanation:

'Lift' – this means an appliance serving specific levels, having a car moving along guides which are rigid – or along a fixed course even where it does not move along guides which are rigid (for example, a scissor lift) – and inclined at an angle of more than 15 degrees to the horizontal and intended for the transport of:

- persons,
- persons and goods, or
- goods alone if the car is accessible, that is to say, a person may enter it without difficulty, and fitted with controls situated inside the car or within reach of a person inside;

'Accessory for lifting' – this means 'work equipment for attaching loads to machinery for lifting';

'Lifting equipment' – this means 'work equipment for lifting or lowering loads and includes its attachments used for anchoring, fixing or supporting it'.

The Lifts Regulations 1997

L3014 These Regulations apply to lifts permanently serving buildings or constructions; and safety components for use in such lifts. [*Reg 3*].

These Regulations do not apply to:

- the following lifts – and safety components for such lifts:
  - — cableways, including funicular railways, for the public or private transportation of persons,
  - — lifts specially designed and constructed for military or police purposes,
  - — mine winding gear,
  - — theatre elevators,
  - — lifts fitted in means of transport,
  - — lifts connected to machinery and intended exclusively for access to the workplace,
  - — rack and pinion trains,
  - — construction-site hoists intended for lifting persons or persons and goods.

[*Reg 4, Schedule 14*].

- any lift or safety component which is placed on the market (i.e. when the installer first makes the lift available to the user, but see L3019 below) and put into service before 1 July 1997 [*Reg 5*].
- any lift or safety component placed on the market and put into service on or before 30 June 1999 which complies with any health and safety provisions with which it would have been required to comply if it was to have been placed on the market and put into service in the United Kingdom on 29 June 1995.

  This exclusion does not apply in the case of a lift or a safety component which;
  - — unless required to bear the CE marking pursuant to any other Community obligation, bears the CE marking or an inscription liable to be confused with it; or
  - — bears or is accompanied by any other indication, howsoever expressed, that it complies with the Lifts Directive.

[*Reg 6*].

- any lift insofar as and to the extent that the relevant essential health and safety requirements relate to risks wholly or partly covered by other Community directives applicable to that lift [*Reg 7*].

General requirements

General duty relating to the placing on the market and putting into service of lifts

L3015

- Subject to *Reg 12* (see L3019 below), no person who is a responsible person shall place on the market and put into service any lift unless the requirements of paragraph (ii) below have been complied with in relation to it [*Reg 8*(*1*)].
- *Regulation 8*(*2*) provides that the requirements in respect of any lift are that:
  - — it satisfies the relevant essential health and safety requirements and, for the purpose of satisfying those requirements:
    - — where a transposed harmonised standard covers one or more of the relevant essential health and safety requirements, any lift constructed in accordance with that transposed harmonised standard shall be presumed to comply with that (or those) essential health and safety requirement(s); and
    - — by calculation, or on the basis of design plans, it is permitted to demonstrate the similarity of a range of equipment to satisfy the essential safety requirements;
  - — the appropriate conformity assessment procedure in respect of the lift has been carried out in accordance with *Reg 13*(*1*) (see L3020 below);
  - — the CE marking has been affixed to it by the installer of the lift in accordance with *Schedule 3*;
  - — a declaration of conformity has been drawn up in respect of it by the installer of the lift; and
  - — it is in fact safe.

  Note: Reg 18(*1*) provides that a lift which bears the CE marking, and is accompanied by an EC declaration of conformity in accordance with *Reg 8*(*2*), is taken to conform with all the requirements of the *Lifts Regulations 1997*, including *Reg 13* (see L3020 below), unless reasonable grounds exist for suspecting that it does not so conform.
- Any technical documentation or other information in relation to a lift required to be retained under the conformity assessment procedure used must be retained by the person specified in that respect in that conformity assessment procedure for any period specified in that procedure [*Reg 8*(*3*)].

In these Regulations, 'responsible person' means:

- in the case of a lift, the installer of the lift;
- in the case of a safety component, the manufacturer of the component or his authorised representative established in the Community; or
- where neither the installer of the lift nor the manufacturer of the safety component nor the latter's authorised representative established in the Community, as the case may be, have fulfilled the requirements of *Reg 8*(*2*) (see above) or *Reg 9*(*2*) (see L3016 below), the person who places the lift or safety component on the market.

General duty relating to the placing on the market and putting into service of safety components

L3016 Safety components include devices for locking landing doors; devices to prevent falls; overspeed limitation devices; and electric safety switches.

- Subject to *Reg 12* (see L3020 below), no person who is a responsible person shall place on the market and put into service any safety component unless the requirements of paragraph (ii) below have been complied with in relation to it [*Reg 9*(*1*)].
- *Regulation 9*(*2*) provides that the requirements in respect of any safety component are that:
  - it satisfies the relevant essential health and safety requirements and for the purpose of satisfying those requirements where a transposed harmonised standard covers one or more of the relevant essential health and safety requirements, any safety component constructed in accordance with that transposed harmonised standard shall be presumed to be suitable to enable a lift on which it is correctly installed to comply with that (or those) essential health and safety requirement(s);
  - the appropriate conformity assessment procedure in respect of the safety component has been carried out in accordance with *Reg 13*(*1*) (see L3021 below);
  - the CE marking has been affixed to it, or on a label inseparably attached to the safety component, by the manufacturer of that safety component, or his authorised representative established in the Community, in accordance with *Schedule 3* (which specifies the requirements for CE marking);
  - a declaration of conformity has been drawn up in respect of it by the manufacturer of the safety component or his authorised representative established in the Community; and
  - it is in fact safe.

  Note: Reg 18(*1*) provides that a safety component – or its label – which bears the CE marking, and is accompanied by an EC declaration of conformity in accordance with *Reg 9*(*2*), is taken to conform with all the requirements of the *Lifts Regulations 1997*, including *Reg 13* (see L3021 below), unless reasonable grounds exist for suspecting that it does not so conform.
- Any technical documentation or other information in relation to a safety component required to be retained under the conformity assessment procedure used must be retained by the person specified in that respect in that conformity assessment procedure for any period specified in that procedure [*Reg 9*(*3*)].

General duty relating to the supply of a lift or safety component

L3017 Subject to *Reg 12* (see L3020 below) any person who supplies any lift or safety component but who is not a person to whom *Reg 8* or *9* applies (see L3015 and L3016 above) must ensure that that lift or safety component is safe [*Reg 10*].

Penalties for breach of Regs 8, 9 or 10

L3018 A person who is convicted of an offence under *Reg 8, 9* or *10* above is liable to imprisonment (not exceeding three months) or to a fine not exceeding level 5 on the standard scale, or both. *Reg 22*, however, provides a defence of due diligence: i.e. the defendant must show that he took all reasonable steps and exercised all due diligence to avoid committing the offence.

Specific duties relating to the supply of information, freedom from obstruction of lift shafts and retention of documents

L3019

- The person responsible for work on the building or construction where a lift is to be installed and the installer of the lift must keep each other informed of the facts necessary for, and take the appropriate steps to ensure, the proper operation and safe use of the lift. Shafts intended for lifts must not contain any piping or wiring or fittings other than that which is necessary for the operation and safety of that lift.
- Where, in the case of a lift, for the purposes of *Reg 8(2)* (see L3015 above) the appropriate conformity assessment procedure is one of the procedures set out in *Reg 13(2)(a), (b)* or *(c)* (see L3021 below), the person responsible for the design of the lift must supply to the person responsible for the construction, installation and testing all necessary documents and information for the latter person to be able to operate in absolute security.

  A person who is convicted of an offence under (i) or (ii) above is liable to imprisonment (not exceeding three months) or to a fine not exceeding level 5 on the standard scale, or both. *Reg 22*, however, provides a defence of due diligence: i.e. the defendant must show that he took all reasonable steps and exercised all due diligence to avoid committing the offence.
- A copy of the declaration of conformity mentioned in *Reg 8(2)* or *9(2)* must:
  - in the case of a lift, be supplied to the EC Commission, the member states and any other notified bodies, on request, by the installer of the lift together with a copy of the reports of the tests involved in the final inspection to be carried out as part of the appropriate conformity assessment procedure referred to in *Reg 8(2)*; and
  - be retained, by the person who draws up that declaration, for a period of ten years – in the case of a lift, from the date on which the lift was placed on the market; and in the case of a safety component, from the date on which safety components of that type were last manufactured by that person.

  A person who fails to supply or keep a copy of the declaration of conformity, as required above, is liable on summary conviction to a fine not exceeding level 5 on the standard scale. *Reg 22*, however, provides a defence of due diligence: i.e. the defendant must show that he took all reasonable steps and exercised all due diligence to avoid committing the offence.

[*Reg 11*].

Exceptions to placing on the market or supply in respect of certain lifts and safety components

L3020 For the purposes of *Reg 8, 9* or *10*, a lift or a safety component is not regarded as being placed on the market or supplied:

- where that lift or safety component will be put into service in a country outside the Community; or is imported into the Community for re-export to a country outside the Community – but this paragraph does not apply if the CE marking, or any inscription liable to be confused with such a marking, is affixed to the lift or safety component or, in the case of a safety component, to its label; or
- by the exhibition at trade fairs and exhibitions of that lift or safety component, in respect of which the provisions of these Regulations are not satisfied, if:
  - a notice is displayed in relation to the lift or safety component in question to the effect that it does not satisfy those provisions; and that it may not be placed on the market or supplied until those provisions are satisfied; and
  - adequate safety measures are taken to ensure the safety of persons.

[*Reg 12*].

Conformity assessment procedures

L3021 For the purposes of *Reg 8(2)* or *9(2)* (see L3015 and L3016 above), the appropriate conformity assessment procedure is as follows:

For lifts – one of the following procedures:

- if the lift was designed in accordance with a lift having undergone an EC type-examination as referred to in *Schedule 5*, it must be constructed, installed and tested by implementing:
  - the final inspection referred to in *Schedule 6*; or
  - the quality assurance system referred to in *Schedule 11* or *Schedule 13*; and

  the procedures for the design and construction stages, on the one hand, and the installation and testing stages, on the other, may be carried out on the same lift;
- if the lift was designed in accordance with a model lift having undergone an EC type-examination as referred to in *Schedule 5*, it must be constructed, installed and tested by implementing:
  - the final inspection referred to in *Schedule 6*; or
  - one of the quality assurance systems referred to in *Schedule 11* or *Schedule 13*; and

  all permitted variations between a model lift and the lifts forming part of the lifts derived from that model lift must be clearly specified (with maximum and minimum values) in the technical dossier required as part of the appropriate conformity assessment procedure;
- if the lift was designed in accordance with a lift for which a quality assurance system pursuant to *Schedule 12* was implemented, supplemented by an examination of the design if the latter is not wholly in accordance with the harmonised standards, it must be installed and constructed and tested by implementing, in addition, the final inspection referred to in *Schedule 6* or one of the quality assurance systems referred to in *Schedule 11* or *Schedule 13*;

- the unit verification procedure, referred to in *Schedule 9*, by a notified body (see L3023 below); or
- the quality assurance system in accordance with *Schedule 12*, supplemented by an examination of the design if the latter is not wholly in accordance with the transposed harmonised standards.

For safety components – one of the following procedures:

- submit the model of the safety component for EC type-examination in accordance with *Schedule 5* and for production checks by a notified body (see L3023 below) in accordance with *Schedule 10*;
- submit the model of the safety component for EC type-examination in accordance with *Schedule 5* and operate a quality assurance system in accordance with *Schedule 7* for checking production; or
- operate a full quality assurance system in accordance with *Schedule 8*.

[*Reg 13*].

Requirements fulfilled by the person who places a lift or safety component on the market

L3022 Where in the case of a lift or a safety component, any of the requirements of *Regs 8, 9, 11* and *13* to be fulfilled by the installer of the lift or the manufacturer of the safety component or, in the case of the latter, his authorised representative established in the Community, have not been so fulfilled such requirements may be fulfilled by the person who places that lift or safety component on the market [*Reg 14*].

This provision, however, does not affect the power of an enforcement authority to take action in respect of the installer of the lift, the manufacturer of the safety component or, in the case of the latter, his authorised representative established in the Community in respect of a contravention of or a failure to comply with any of those requirements.

Notified bodies

L3023 For the purposes of these Regulations, a notified body is a body which has been appointed to carry out one or more of the conformity assessment procedures referred to in *Reg 13* which has been appointed as a notified body by the Secretary of State in the United Kingdom or by a member State.

[*Reg 15*].

The Lifting Operations and Lifting Equipment Regulations 1998 (as amended by SI 2002 No 2174)

L3024 The requirements imposed by these Regulations on an employer in relation to lifting equipment apply in respect of lifting equipment provided for use or used by an employee of his at work. They also apply to a self-employed person, with regard to lifting equipment he uses at work, and to anyone with control, to any extent, of:

- lifting equipment;
- a person at work who uses or supervises or manages the use of lifting equipment; or
- the way in which lifting equipment is used,

and to the extent of his control.

General requirements

Strength and stability

L3025 Every employer must ensure that lifting equipment is of adequate strength and stability for each load, having regard in particular to the stress induced at its mounting or fixing point – and that every part of a load and anything attached to it and used in lifting it is of adequate strength [*Reg 4*].

Lifting equipment for lifting persons

L3026 Every employer must ensure that lifting equipment for lifting persons:

- is such as to prevent a person using it being crushed, trapped or struck or falling from the carrier;
- is such as to prevent so far as is reasonably practicable a person using it, while carrying out activities from the carrier, being crushed, trapped or struck or falling from the carrier;
- has suitable devices to prevent the risk of a carrier falling, and, if the risk cannot be prevented for reasons inherent in the site and height differences, the employer must ensure that the carrier has an enhanced safety coefficient suspension rope or chain which is inspected by a competent person every working day;
- is such that a person trapped in any carrier is not thereby exposed to danger and can be freed.

[*Reg 5*].

Positioning and installation

L3027 Employers must ensure that lifting equipment is positioned or installed in such a way as to reduce to as low as is reasonably practicable the risk of the lifting equipment or a load striking a person, or the risk from a load:

- drifting;
- falling freely; or
- being released unintentionally,

and that otherwise it is safe.

Employers must also ensure that there are suitable devices for preventing anyone from falling down a shaft or hoistway [*Reg 6*].

Marking of lifting equipment

L3028 Employers must ensure that:

- machinery and accessories for lifting loads are clearly marked to indicate their safe working loads;
- where the safe working load of machinery for lifting loads depends on its configuration, either the machinery is clearly marked to indicate its safe working load for each configuration, or information which clearly indicates its safe working load for each configuration is kept with the machinery;

- accessories for lifting are also marked in such a way that it is possible to identify the characteristics necessary for their safe use;
- lifting equipment which is designed for lifting persons is appropriately and clearly marked to this effect; and
- lifting equipment which is not designed for lifting persons but which might be mistakenly so used is appropriately and clearly marked to the effect that it is not designed for lifting persons.

[*Reg 7*].

Organisation of lifting operations

L3029 Every employer must ensure that any lifting or lowering of a load which involves lifting equipment is properly planned by a competent person, appropriately supervised and carried out in a safe manner [*Reg 8*].

Thorough examination and inspection

L3030 An employer is under a duty to ensure:

- before lifting equipment is put into service for the first time by him, that it is thoroughly examined for any defect unless:
  - the lifting equipment has not been used before; and
  - in the case of lifting equipment for which an EC declaration of conformity could or (in the case of a declaration under the *Lifts Regulations 1997*) should have been drawn up, the employer has received such declaration made not more than twelve months before the lifting equipment is put into service; or
  - if obtained from the undertaking of another person, it is accompanied by physical evidence referred to in (iv) below;

[*Reg 9*(*1*)].

- where the safety of lifting equipment depends on the installation conditions, that it is thoroughly examined – after installation and before being put into service for the first time and after assembly and before being put into service at a new site or in a new location, to ensure that it has been correctly installed and is safe to operate [*Reg 9*(*2*)];
- that lifting equipment which is exposed to conditions causing deterioration which is liable to result in dangerous situations is:

  thoroughly examined;

  - at least every six months, in the case of lifting equipment for lifting persons or an accessory for lifting;
  - at least every twelve months, in the case of other lifting equipment; or
  - in either case, in accordance with an examination scheme; and
  - whenever exceptional circumstances which are liable to jeopardise the safety of the lifting equipment have occurred, and

  if appropriate for the purpose, is inspected by a competent person at suitable intervals between thorough examinations, to ensure that health and safety conditions are maintained and that any deterioration can be detected and remedied in good time;

[*Reg 9(3)*].

(iv) that no lifting equipment leaves his undertaking; or, if obtained from the undertaking of another person, is used in his undertaking, unless it is accompanied by physical evidence that the last thorough examination required to be carried out under this regulation has been carried out.

[*Reg 9(4)*].

Reports and defects

L3031

- A person making a thorough examination for an employer under *Reg 9* must:
  - immediately notify the employer of any defect in the lifting equipment which in his opinion is or could become a danger to anyone;
  - write a report of the thorough examination (see L3031 below) to the employer and any person from whom the lifting equipment has been hired or leased;
  - where there is in his opinion a defect in the lifting equipment involving an existing or imminent risk of serious personal injury send a copy of the report to the relevant enforcing authority ('relevant enforcing authority' means, where the defective lifting equipment has been hired or leased by the employer, the Health and Safety Executive – and otherwise, the enforcing authority for the premises in which the defective lifting equipment was thoroughly examined).
- A person making an inspection for an employer under *Reg 9* must immediately notify the employer of any defect in the lifting equipment which in his opinion is or could become a danger to anyone, and make a written record of the inspection.
- Every employer who has been notified under (i) above must ensure that the lifting equipment is not used before the defect is rectified; or, in the case of a defect which is not yet but could become a danger to persons, after it could become such a danger.

[*Reg 10*].

Prescribed information

L3032 *Schedule 1* specifies the information that must be contained in a report of a thorough examination, made under *Reg 10* (see L3031 above). The information must include the following:

(i) The name and address of the employer for whom the thorough examination was made.

(ii) The address of the premises at which the thorough examination was made.

(iii) Particulars sufficient to identify the lifting equipment including its date of manufacture, if known.

(iv) The date of the last thorough examination.

(v) The safe working load of the lifting equipment or, where its safe working load depends on the configuration of the lifting equipment, its safe working load for the last configuration in which it was thoroughly examined.

(vi) In respect of the first thorough examination of lifting equipment after installation or after assembly at a new site or in a new location:

— that it is such thorough examination; and

— if in fact this is so, that it has been installed correctly and would be safe to operate.

(vii) In respect of all thorough examinations of lifting equipment which do not fall within (vi) above:

(*a*) whether it is a thorough examination under *Reg 9*(*3*):

— within an interval of six months;

— within an interval of twelve months;

— in accordance with an examination scheme; or

— after the occurrence of exceptional circumstances;

(*b*) if in fact this is so, that the lifting equipment would be safe to operate.

(viii) In respect of every thorough examination of lifting equipment:

(*a*) identification of any part found to have a defect which is or could become a danger to anyone, and a description of the defect;

(*b*) particulars of any repair, renewal or alteration required to remedy a defect found to be a danger to anyone;

(*c*) in the case of a defect which is not yet but could become a danger to anyone:

— the time by which it could become such a danger;

— particulars of any repair, renewal or alteration required to remedy the defect;

(*d*) the latest date by which the next thorough examination must be carried out;

(*e*) particulars of any test, if applicable;

(*f*) the date of the thorough examination.

(ix) The name, address and qualifications of the person making the report; that he is self-employed or, if employed, the name and address of his employer.

(x) The name and address of a person signing or authenticating the report on behalf of its author.

(xi) The date of the report.

Keeping of information

L3033 An employer who obtains lifting equipment to which the 1998 Regulations apply, and who receives an EC declaration of conformity relating to it, must keep the declaration for so long as he operates the lifting equipment [*Reg 11*(*1*)].

Regulation 11(*2*)(*a*) provides that the employer must ensure that the information contained in every report made to him under *Reg 10*(*1*) (see L3031 above) is kept available for inspection:

- in the case of a thorough examination under *Reg 9(1)* (see L3030 above) of lifting equipment other than an accessory for lifting, until he stops using the lifting equipment;
- in the case of a thorough examination under *Reg 9(1)* (see L3030 above) of an accessory for lifting, for two years after the report is made;
- in the case of a thorough examination under *Reg 9(2)* (see L3030 above), until he stops using the lifting equipment at the place it was installed or assembled;
- in the case of a thorough examination under *Reg 9(3)* (see L3030 above), until the next report is made under that paragraph or the expiration of two years, whichever is later.

The employer must ensure that every record made under *Reg 10(2)* (see L3031 above) is kept available until the next such record is made [*Reg 11(2)(b)*].

Cranes, hoists and lifting equipment – common causes of failure

General

L3034 The principal cause of failure in all forms of lifting equipment is that of overloading, i.e. exceeding the specified safe working load (SWL) of the crane, forklift truck, hoist, chain, etc. in use for a specific lifting job. Every year there are numerous accidents and scheduled dangerous occurrences reported which are caused as a result of overloading. A second common cause of failure, and one which is inexcusable, is associated with neglect of the equipment while not in use. Neglect may be associated with poor or inadequate maintenance of the fabric of a crane or its safety devices, or simply a failure to store rope slings properly while not in use. One of the results of neglect is corrosion of metal surfaces resulting in weakened crane structures, wire ropes and slings. This is why lifting equipment is subject to statutory inspection.

Examples of specific causes of failure are outlined below.

Cranes

L3035

- Failure to lift vertically, e.g. dragging a load sideways along the ground before lifting.
- 'Snatching' loads, i.e. not lifting slowly and smoothly.
- Exceeding the maximum permitted moment, i.e. the product of the load and the radius of operation.
- Excessive wind loading, resulting in crane instability.
- Defects in the fabrication of the crane, e.g. badly welded joints.
- Incorrect crane assembly in the case of tower cranes.
- Brake failure (rail-mounted cranes).
- In the case of mobile cranes:
  - (i) failure to use outriggers;
  - (ii) lifting on soft or uneven ground; and
  - (iii) incorrect tyre pressures.

Hoists and lifts

L3036
- Excessive wear in wire ropes.
- Excessive broken wires in ropes.
- Failure of the overload protection device.
- Failure of the overrun device.

Ropes

Fibre ropes

L3037
- Bad storage in wet or damp conditions resulting in rot and mildew.
- Inadequate protection of the rope when lifting loads with sharp edges.
- Exposure to direct heat to dry, as opposed to gradual drying in air.
- Chemical reaction.

Wire ropes

L3038
- Excessive broken wires.
- Failure to lubricate regularly.
- Frequent knotting or kinking of the rope.
- Bad storage in wet or damp conditions which promotes rust.

Chains

L3039
- Mechanical defects in individual links.
- Application of a static in excess of the breaking load.
- Snatch loading.

Slings

L3040 Slings are manufactured in natural or man-made fibre or chain. The safe working load of any sling varies according to the angle formed between the legs of the sling.

The relationship between sling angle and the distance between the legs of the sling is also important. (See Table 1 below.)

Table 1

Safe working load for slings

Sling Angle	*Distance between legs*
30°	1/2 leg length
60°	1 leg length
90°	11/3 leg length
120°	12/3 leg length

For a one tonne load, the tension in the leg increases as shown in Table 2 below.

Table 2

Safe working load for slings – increased tension

Sling leg angle	*Tension in leg* (*tonnes*)
90°	0.7
120°	1.0
151°	2.0
171°	6.0

Other causes of failure in slings are:

- Cuts, excessive wear, kinking and general distortion of the sling legs.
- Failure to lubricate wire slings.
- Failure to pack sharp corners of a load, resulting in sharp bends in the sling and the possibility of cuts or damage to it.
- Unequal distribution of the load between the legs of a multi-leg sling.

Hooks

L3041
- Distortion of the hook due to overloading.
- Use of a hook without a safety catch.
- Stripping of the thread connecting the hook to the chain fixture.

Forklift trucks

L3042
- Uneven floors, steeply inclined ramps or gradients, i.e. in excess of 1:10 gradient.
- Inadequate room to manoeuvre.
- Inadequate or poor maintenance of lifting gear.
- The practice of driving forwards down a gradient with the load preceding the truck.
- Load movement in transit.
- Sudden or fast braking.
- Poor stacking of goods being moved.
- Speeding.
- Turning corners too sharply.
- Not securing load sufficiently e.g. pallets stacked poorly.
- Hidden obstructions in the path of the truck.
- Use of the forward tilt mechanism with a raised load.
- Generally bad driving, including driving too fast, taking corners too fast, striking overhead obstructions, particularly when reversing and excessive use of the brakes.

Mobile lifting equipment

L3043 Mobile work equipment is any equipment which carries out work whilst travelling. It may be self-propelled, towed or remotely controlled and it may incorporate attachments. An example of mobile lifting equipment is an excavator involved in digging tasks.

The risk of mobile lifting equipment overturning is another cause for concern, and the problem was addressed by the additional requirements introduced in the *Provision and Use of Work Equipment Regulations 1998* (*SI 1998 No 2306* as amended by *SI 2002 No 2174*). These implement into UK legislation the amending directive to the *Use of Work Equipment Directive* (*AUWED*). Under this legislation, workers must be protected from falling out of the equipment and from unexpected movement e.g. overturning.

To prevent such equipment overturning, the following action should be taken:

- fit stabilisers (e.g. outriggers or counterbalance weights);
- have in place a structure which ensures that it does no more than fall on its side;
- ensure the structure gives sufficient clearance to anyone being carried if it overturns more than on its side;
- installation of roll-over protective structure;
- provision of harnesses or a suitable restraining system to protect against being crushed;
- use only on firm ground;
- avoid excessive gradients.

HSE has published guidance on the fitting and use of restraining systems on lift trucks. The guidance explains when operator restraint should be fitted to a lift truck, when it should be used, and what to do if it cannot be fitted. It also describes the type of lift trucks most at risk of overturning and gives advice on how to prevent overturning accidents. The guidance is aimed at employers, drivers and others with a responsibility for managing the safe operation of lift trucks. Copies of *'Fitting and use of restraining systems on lift trucks'*, MISC 241, can be ordered free of charge from HSE Books.

Safe stacking of materials

L3044 Once materials, containers etc. have been lifted, they must remain stable until it is necessary to bring them down again. Whether 'block stacked', i.e. self-supporting, or stored in shelving or racks, there is a hazard and attention to basic principles is important.

Methods of storage depend on the shape and fragility of the material or package. Cylinders stored 'on the roll' are one of the more hazardous materials. The bottom layer must be properly secured to prevent movement; subsequent layers can rest on the one below, or be laid on battens and wedged. Progressively, as forklift trucks – and particularly reach trucks and narrow aisle stackers – lift to greater heights, so storage racking has been constructed higher. The Storage Equipment Manufacturers' Association has published a code of practice, which covers pallet racks, drive-in and drive-through racks and cantilever racks, setting out guidance to users. ('*Code of practice for the use of static racking*', Storage Equipment Manufacturers' Association, McLaren Building, 35 Dale End, Birmingham B4 7LN. Tel: 0121 200 2100).

Racking

L3045 A manufacturer will install racking to customer's requirements in terms of loading, pallet height and aisle width depending on the type of handling equipment used. Heavier loads must not be used; if there is to be a change of load stored, the manufacturer must be consulted. Racking has collapsed through either overloading or impact damage to an upright member.

Where trucks are used the following are to be recommended:

- bolting to the floor; and
- a column guard or guide rail to protect corners at the ends of aisles.

Pallets

L3046 Many types of load are carried on pallets and, whilst there are many metal pallets in use, the vast majority are of timber for economy of cost and weight. When damaged they are hazardous – and they are easily damaged by the dangerous forks of a truck. The pallet should, therefore, be designed according to the load it is to carry and, if it is to be stored in racking, the type of rack.

There is a useful HSE Guidance Note which indicates some of the considerations of design, and makes recommendations for inspection of new and used pallets. ('*Safety in the use of pallets*', HSE Guidance Note PM 15). See also British Standard BS ISO 6780: '*General purpose flat pallets for through transit of goods. Principal dimensions and tolerances*' (1988).

Forklift trucks

L3047 Forklift trucks (for legal requirements, see L3003 above) are potentially hazardous for a variety of reasons:

- they usually work in fairly congested areas;
- when elevating a heavy load, stability is bound to be reduced;
- there is always the possibility of a load falling down;
- when travelling with a load in the lowered position, the driver's visibility is often impaired;
- even when travelling unladen, the forks projecting at the front are dangerous.

Truck population in the UK is now reckoned to be more than a quarter of a million; there are, therefore, a great many older machines in use which can exhibit many faults. In recent years many manufacturers have been fitting load guards, overhead guards, and warning horns as standard. Some now include a transparent window in the overhead guard (a wise extra where there is a possibility of small items falling down and being able to pass through the guard members) and masts with improved forward visibility. There are British and International Standards Organisation (ISO) standards for stability testing and practically all manufacturers test accordingly. (British Standards BS 3726, BS 5777). Another safety move has been the introduction of standard control symbols, particularly useful now that more trucks are being imported. (BS 5829 '*Specification for control symbols for powered industrial trucks*').

Requirements for safe forklift truck operation

L3048 An analysis of safe truck operation identifies three principal aspects as the potential cause of truck accidents: the driver, the truck and the system of work.

The driver

L3049 Drivers should be in good health, with sound vision and hearing. They should be over 18 years of age and trained within an approved training scheme. Drivers should observe the following precautions:

- regulate speed with visibility;
- use the horn whenever turning a blind corner;
- be constantly aware of pedestrians and vehicles on roadways, loading bays, storage areas and transfer points (the use of convex mirrors located at strategic points greatly reduces the risk of collision);
- drive in reverse when the load obscures vision;
- travel with the forks down, and not operate the forks when in motion;
- use prescribed lanes/routes; *no short cuts*;
- stick to factory speed limits, e.g. 10 mph;
- slow down on wet or uneven surfaces;
- use the handbrake and tilt mechanism correctly;
- take care on ramps (max 1:10);
- when leaving the truck at any time, put the controls in neutral position, switch the power off, apply the brakes, and ensure the key or connector plug is removed.

Drivers should not:

- carry passengers;
- park in front of fire appliances or fire exits;
- turn around on ramps;
- permit unauthorised use, e.g. by contractors.

The truck

L3050 On no account should trucks in a defective or dangerous condition be used.

A daily check system should be operated, prior to starting or on handover to another driver, which covers brakes, lights, steering, horn, battery, hydraulics and speed controls.

The system of work

L3051 On no account should the maximum rated load capacity be exceeded. Loads should always be placed dead centre on the forks.

The truck should be driven with the forks well under the load, with the load located firmly against the fork carriage and the mast tilted to suit the stability of the load being carried.

The following general points should be observed:

- slinging should be undertaken only at designated slinging points;
- a load which looks unsafe should never be moved;
- broken, defective or inadequate strength pallets should never be used;

- care must be taken at overhead openings, pipework, ducting, conduits, etc.;
- the stability of a stack should always be checked before moving the forks.

Training of truck operators

L3052 Although training in safe handling, storage and transportation is necessary under the general legal duties imposed by *HSWA* (see above), special hazards exist with respect to forklift trucks (see further L3003 above). Demonstrably, proper training of operators is the only way to safer use of lift trucks. RTITB Ltd publish '*Lift Truck Operator and Instructor Training Recommendations*'.

Although there are no specific regulations on this area, the HSE ACoP requires, following general principles under *HSWA s 2*, that employers ensure their employees are sufficiently trained for the job. Competence, of course, can be demonstrated by the employees having completed a training course provided by a training provider accredited by one of the four accrediting bodies. The HSE recognises four accrediting bodies. These are:

(*a*) RTITB Ltd,
Ercall House,
8 Pearson Road,
Central Park,
Telford
TF2 9TX
Telephone: (01952) 520 200

(*b*) Lantra National Training,
National Agricultural Centre,
Stoneleigh,
Near Kenilworth,
Warwickshire
CV8 2LG
Telephone: (024) 7669 6996

(*c*) Association of Industrial Truck Trainers,
Independent Training Standards Scheme and Register,
Scammell House,
High Street,
Ascot,
Berkshire
SL5 7JF
Telephone: (01530) 417 234

(*d*) Construction Industry Training Board (CITB),
Bircham Newton Training Centre,
King's Lynn,
Norfolk
PE31 6RH
Telephone: (01485) 577 577

After completing a course and receiving a certificate of competence, an operator needs proper supervision by a qualified person. Lamentably, many trainees (and so-called operators, in many cases) fail to understand the theory of counterbalancing. Most appreciate the effect of load weight, but not many grasp the effect of *load centre* which is equally essential to safe truck operation (e.g. the overturning moment: load weight × distance from pivot point, which is the front wheels).

Pedestrian-controlled lift trucks

L3053 Useful information for operators of pedestrian-controlled lift trucks is published by the British Industrial Truck Association in pocket book format. ('*Operator's safety code for powered industrial trucks*', British Industrial Truck Association, Scammell House, 9 High Street, Ascot, Berks SL5 7JF. Tel: 01344 623 800). Features which are increasingly being built in by manufacturers are:

- toe guards around the wheels; and
- a safety button in the head of the control handle to reverse the machine in the event of the operator being trapped between the machine and some other object.

Lighting

Introduction

L5001 Increasingly, over the last decade or so, employers have come to appreciate that indifferent lighting is both bad economics and bad ergonomics, not to mention potentially bad industrial relations; and conversely, good lighting uses energy efficiently and contributes to general workforce morale and profitability – that is, operating costs fall whilst productivity and quality improve. Alternatively, poor lighting reduces efficiency, thereby increasing the risk of stress, denting workforce morale, promoting absenteeism and leading to accidents, injuries and even deaths at work.

Ideally, good lighting should 'guarantee' (*a*) employee safety, (*b*) acceptable job performance and (*c*) good workplace atmosphere, comfort and appearance. This is not just a matter of maintenance of correct lighting levels. Ergonomically relevant are:

- horizontal illuminance;
- uniformity of illuminance over the job area;
- colour appearance;
- colour rendering;
- glare and discomfort;
- ceiling, wall, floor reflectances;
- job/environment illuminance ratios;
- job and environment reflectances;
- vertical illuminance.

Statutory lighting requirements

General

L5002 Adequate standards of lighting in all workplaces can be enforced under the general duties of the *Health and Safety at Work etc Act 1974*, which require provision by an employer of a safe and healthy working environment. *Regulation 8* of the *Workplace (Health, Safety and Welfare) Regulations 1992* (*SI 1992 No 3004* as amended by *SI 2002 No 2174*), *req*uires that all workplaces have suitable and sufficient lighting and that, so far is reasonably practicable, the lighting should be natural light.

More particularly, however, people should be able to work and move about without suffering eye strain and having to avoid shadows. Local lighting may be necessary at individual workstations and places of particular risk. Outdoor traffic routes used by pedestrians should be adequately lit after dark. Lights and light fittings should avoid dazzle and glare and be so positioned that they do not cause hazards, whether fire, radiation or electrical. Switches should be easily accessible and lights should be replaced, repaired or cleaned before lighting becomes insufficient. Moreover, where persons are particularly exposed to danger in the event of failure of artificial lighting, emergency lighting must be provided. [*Workplace* (*Health, Safety and Welfare*) *Regulations 1992* (*SI 1992 No 3004* as amended by *SI 2002 No 2174*), *Reg 8*(*3*)] (and ACOP).

To accommodate these 'requirements', refresh rates (flickering), a combination of general and localised lighting (not to be confused with 'local' lighting, e.g. desk light) is often necessary. Moreover, state of the art visual display units have thrown up some occupational health problems addressed by the *Health and Safety* (*Display Screen Equipment*) *Regulations 1992* (*SI 1992 No 2792* as amended by *SI 2002 No 2174*).

Specific processes

Lighting and VDUs

L5003 Annex A to BS EN 924–6–1999, provides guidance on illuminance levels, specifically in relation to use with display screen equipment (BS EN 9241 – Ergonomic requirements for office work with visual display terminals (VDTs); Part 6 – Guidance on work environment).

Machinery and work equipment – Provision and Use of Work Equipment Regulations 1998 (SI 1998 No 2306 as amended by SI 2002 No 2174)

L5004 Lighting requirements, applicable to woodworking machines in the *Woodworking Machines Regulations 1974* (*SI 1974 No 903*), *Reg 43* have now been revoked. This now comes under the *Provision and Use of Work Equipment Regulations 1998* (*SI 1998 No 2306* as amended by *SI 2002 No 2174*), *Reg 21*. This Regulation is very broad and merely requires that suitable and sufficient lighting is provided, taking into account the operations to be carried out. To ascertain what is required to meet this duty the HSE guidance has been produced. This refers to occasions when additional or localised lighting may be needed such as:

- when the task requires a high perception of detail;
- when there is a dangerous process;
- to reduce visual fatigue;
- during maintenance operations.

Additional guidance is provided in the HSE Guidance Note HS(G) 38.*'Lighting at Work'*

Lighting requirements on construction sites – Construction (Health, Safety and Welfare) Regulations 1996 (SI 1996 No 1592)

L5005 Every workplace; approach to the workplace; and traffic route in and around the workplace, shall have suitable and sufficient lighting. Where reasonably practicable, this lighting shall be by natural light. The colour of every artificial light must not adversely affect or change the perception of any sign or signal provided for health and safety reasons. Suitable and sufficient secondary lighting must be provided in places where there is risk to health and safety in the event of the failure of the primary lighting. [*SI 1996 No 1592*, *Reg 25*]

Lighting requirements for electrical equipment – Electricity at Work Regulations 1989 (SI 1989 No 635)

L5006 So as to prevent injury, adequate lighting must be provided at all electrical equipment on which or near which work is being done in circumstances that may give rise to danger. [*Electricity at Work Regulations 1989* (*SI 1989 No 635*), *Reg 15*].

Sources of light

Natural lighting

L5007 Daylight is the natural, and cheapest, form of lighting, but it has only limited application to places of work where production is required beyond the hours of daylight, at all seasons, and where day-time visibility is restricted by climatic conditions. But, however good the outside daylight, windows can rarely provide adequate lighting alone for the interior of large floor areas. Single storey buildings can, of course, make use of insulated opaque roofing materials, but the most common provision of daylight is by side windows. There is also the fact that the larger the glazing area of the building, the more other factors such as noise, heat loss in winter, and unsatisfactory thermal conditions in summer must be considered.

Modern conditions, where the creation of pleasant building interior environment requires the balanced integration of lighting, heating, air conditioning, acoustic treatment, etc., are such that lighting cannot be considered in isolation. At the very least, natural lighting will have to be supplemented for most of the time with artificial lighting, the most common source for which is electric lighting.

Electric lighting

L5008 Capital costs, running costs and replacement costs of various types of electric lighting have a direct bearing on the selection of the sources of electric lighting for particular application. Such costs are as important considerations as the size, heat and colour effects required of the lighting. The efficiency of any type of lamp used for lighting is measured as light output, in lumens, per watt of electricity. Typical values for various types of lamp are as follows (the term 'lumen' is explained in the discussion of standards of illuminance in L5007 above).

Type of lamp	*Lumens per watt*
Incandescent lamps	10 to 18
Tungsten halogen	22
High pressure mercury	25 to 55
Tubular fluorescent	30 to 80 (depending on colour)
Mercury halide	60 to 80
High pressure sodium	100

In general, the common incandescent lamps (coiled filament lamps, the temperature of which is raised to white heat by the passage of current, thus giving out light) are relatively cheap to install but have relatively expensive running costs. A discharge or fluorescent lighting scheme (which works on the principle of electric current passing through certain gases and thereby producing an emission of light) has higher capital costs but higher running efficiency, lower running costs and longer lamp life. In larger places of work the choice is often between discharge and fluorescent lamps. The normal mercury discharge lamp and the low pressure sodium discharge lamp have restricted colour performance, although newly developed high pressure sodium discharge lamps and colour corrected mercury lamps do not suffer from this disadvantage.

Standards of lighting or illuminance

The technical measurement of illuminance

L5009 The standard of illuminance (i.e. the amount of light) required for a given location or activity depends on a number of variables, including general comfort considerations and the visual efficiency required. The unit of illuminance is the 'lux' which equals one lumen per square metre: this unit has now replaced the 'foot candle' which was the number of lumens per square foot. The term 'lumen' is the unit of luminous flux, describing the quantity of light received by a surface or emitted by a source of light.

Light measuring instruments

L5010 For accurate measurement of the degree of illuminance at a particular working point, a reliable instrument is required. Such an instrument, suitable for most measurements, is a pocket lightmeter which incorporates the principle of the photo-electric cell, which generates a tiny electric current in proportion to the light at the point of measurement. This current deflects a pointer on a graduated scale measured in lux. Manufacturers' instructions should, of course, be followed in the care and use of such instruments.

Average illuminance and minimum measured illuminance

L5011 HSE Guidance Note HS(G) 38 'Lighting at work' (1997) relates illuminance levels to the degree or extent of detail which needs to be seen in a particular task or situation. Recommended illuminances are shown in Table 1 at L5010 below.

This guidance note makes recommendations both for average illuminance for the work area as a whole and for minimum measured illuminance at any position within it. As the illuminance produced by any lighting installation is rarely uniform, the use of the average illuminance figure alone could result in the presence of a few positions with much lower illuminance which pose a threat to health and safety. The minimum measured illuminance is therefore the lowest illuminance permitted in the work area taking health and safety requirements into account.

The planes on which the illuminances should be provided depend on the layout of the task. If predominantly on one plane, e.g. horizontal, as with an office desk, or vertical, as in a warehouse, the recommended illuminances are recommended for that plane. Where there is either no well defined plane or more than one, the recommended illuminances should be provided on the horizontal plane and care taken to ensure that the reflectances of surfaces in working areas are high.

Illuminance ratios

L5012 The relationship between the lighting of the work area and adjacent areas is significant. Large differences in illuminance between these areas may cause visual discomfort or even affect safety levels where there is frequent movement, e.g. forklift trucks. This problem arises most often where local or localised lighting in an interior exposes a person to a range of illuminance for a long period, or where there is movement between interior and exterior working areas exposing a person to a sudden change of illuminance. To reduce hazards and possible discomfort specific recommendations shown in Table 1 below should be followed.

Where there is conflict between the recommended average illuminances shown in Table 1 and maximum illuminance ratios shown in Table 2, the higher value should be taken.

Table 1

Average illuminances and minimum measured illuminances for different types of work

General activity	*Typical locations/types of work*	*Average illuminance (Lx)*	*Minimum measured illuminance (Lx)*
Movement of people, machines and vehicles[1]	Lorry parks, corridors, circulation routes	20	5
Movement of people, machines and vehicles in hazardous areas; rough work not requiring any perception of detail	Construction site clearance, excavation and soil work, docks, loading bays, bottling and canning plants	50	20
Work requiring limited perception of detail[2]	Kitchens, factories, assembling large components, potteries	100	50
Work requiring perception of detail	Offices, sheet metal work, bookbinding	200	100
Work requiring perception of fine detail	Drawing offices, factories assembling electronic components, textile production	500	200

Notes

[1] Only safety has been considered, because no perception of detail is needed and visual fatigue is unlikely. However, where it is necessary to see detail to recognise a hazard or where error in performing the task could put someone else at risk, for safety purposes as well as to avoid visual fatigue, the figure should be increased to that for work requiring the perception of detail.

[2] The purpose is to avoid visual fatigue: the illuminances will be adequate for safety purposes.

Table 2

Maximum ratios of illuminance for adjacent areas

Situations to which recommendation applies	*Typical location*	*Maximum ratio of illuminances*		
		Working area		*Adjacent area*
Where each task is individually lit and the area around the task is lit to a lower illuminance	Local lighting in an office	5	:	1
Where two working areas are adjacent, but one is lit to a lower illuminance than the other	Localised lighting in a works store	5	:	1
Where two working areas are lit to different illuminances and are separated by a barrier but there is frequent movement between them	A storage area inside a factory and a loading bay outside	10	:	1

Maintenance of light fitments

L5013 The lighting output of a given lamp will reduce gradually in the course of its life but an improvement can be obtained by regular cleaning and maintenance, not only of the lamp itself but also of the reflectors, diffusers and other parts of the luminaire. A sensible and economic lamp replacement policy is called for (e.g. it may be more economical, in labour cost terms to change a batch of lamps than deal with them singly as they wear out).

Qualitative aspects of lighting and lighting design

L5014 While the quantity of lighting afforded to a particular location or task in terms of standard service illuminance is an important feature of lighting design, it is also necessary to consider the qualitative aspects of lighting, which have both direct and indirect effects on the way people perceive their work activities and dangers that may be present. The quality of lighting is affected by the presence or absence of glare, the distribution of the light, brightness, diffusion and colour rendition.

Glare

L5015 This is the effect of light which causes impaired vision or discomfort experienced when parts of the visual field are excessively bright compared with the general surroundings. It may be experienced in three different forms:

- disability glare – the visually disabling effect caused by bright bare lamps directly in the line of vision;
- discomfort glare – caused by too much contrast of brightness between an object and its background, and frequently associated with poor lighting design. It can cause discomfort without necessarily impairing the ability to see detail. Over a period it can cause visual fatigue, headaches and general fatigue;
- reflected glare – is the reflection of bright light sources on shiny or wet work surfaces, such as plated metal or glass, which can almost entirely conceal the detail in or behind the object which is glinting.

  N.B. The Illuminating Engineering Society (IES) publishes a Limiting Glare Index for each of the effects in (*a*) and (*b*) above. This is an index representing the degree of discomfort glare which will be just tolerable in the process or location under consideration. If exceeded, occupants may suffer eye strain or headaches or both.

Distribution

L5016 Distribution is concerned with the way light is spread. The British Zonal Method classifies luminaires (light-fittings) according to the way they distribute light from BZ1 (all light downwards in a narrow column) to BZ10 (light in all directions). A fitting with a low BZ number does not necessarily imply less glare, however. Its positioning, the shape of the room and the reflective surfaces present are also significant.

The actual spacing of luminaires is also important when considering good lighting distribution. To ensure evenness of illuminance at operating positions, the ratio between the height of the luminaire and the spacing of it must be considered. The IES spacing: height ratio provides a basic guide to such arrangements. Under normal circumstances, e.g. offices, workshops and stores, this ratio should be between 11/2:1 and 1:1 according to the type of luminaire.

Brightness

L5017 Brightness or 'luminosity' is very much a subjective sensation and, therefore, cannot be measured. However, it is possible to consider a brightness ratio, which is the ratio of apparent luminosity between a task object and its surroundings. To ensure the correct brightness ratio, the reflectance (i.e. the ability of a surface to reflect light) of all surfaces in the working area should be well maintained and consideration given to reflectance values in the design of interiors. Given a task illuminance factor (i.e. the recommended illuminance level for a particular task) of 1, the effective reflectance values should be ceilings – 0.6, walls – 0.3 to 0.8, and floors – 0.2 to 0.3.

Diffusion

L5018 This is the projection of light in all directions with no predominant direction. The directional flow of light can often determine the density of shadows, which may prejudice safety standards or reduce lighting efficiency. Diffused lighting will reduce the amount of glare experienced from bare luminaires.

Colour rendition

L5019 Colour rendition refers to the appearance of an object under a specific light source, compared to its colour under a reference illuminant, e.g. natural light. Good standards of colour rendition allow the colour appearance of an object to be properly perceived. Generally, the colour rendering properties of luminaires should not clash with those of natural light, and should be just as effective at night when there is no daylight contribution to the total illumination of the working area.

Stroboscopic effect

L5020 One aspect of lighting quality that formerly gave trouble was the stroboscopic effect of fluorescent tubes which gave the illusion of motion or even the illusion that a rotating part of machinery was stationary. With modern designs of fluorescent tubes, this effect has largely been eliminated.

Machinery Safety

Introduction

M1001 The introduction of the harmonised European Union ('EU') regulations and standards in health and safety has changed the approach to the prevention of accidents and ill health at work. In the place of a prescriptive set of standards and regulations, the harmonised regulations and standards represent a remarkable breakthrough in the identification, assessment and control of machinery and work equipment risks.

As far as machinery safety is concerned, the introduction of the EU concept for the proactive 'goal-setting' approach has fundamentally changed the approach to the prevention of accidents and ill health at work. In the place of prescriptive legislation and standards, the new risk-based approach to machinery and work equipment safety has fundamentally changed the way for the identification, assessment and control of machinery hazards. The new approach is essentially based on structured and systematic assessment of risks.

This chapter aims to review this approach, which is based on the EU Machinery and Use of Work Equipment Directives as well as the Transposed Harmonised EU Machinery Safety Standards. It aims to give guidance and examples on the link between the structured approach for hazard identification and analysis as well as risk evaluation. This helps in developing action (health and safety) plans for the implementation of a package of measures for machinery and work equipment accident/ill health prevention in a practical way.

Currently, statutory duties are dually (but separately) laid on both manufacturers of machinery for use at work and employers and users of such machinery and equipment, prior compliance, on the part of manufacturers leading to compliance on the part of employers/users. General and specific duties are imposed on manufacturers by the *Supply of Machinery* (*Safety*) *Regulations 1992* (*SI 1992 No 3073*) as amended by the *Supply of Machinery* (*Safety*) (*Amendment*) *Regulations 1994* (*SI 1994 No 2063*).

Duties of employers are incorporated into the *Provision and Use of Work Equipment Regulations 1998* (*SI 1998 No 2306*).

Legal requirements relating to machinery

M1002 The introduction of the *Health and Safety at Work etc. Act* ('*HSWA 1974*') in 1974, in addition to placing general duties on employers under *section 2* of the Act, has placed the responsibility of machinery safety on designers, manufacturers and suppliers. This is still applicable only in the UK, in addition to the EU requirements.

The original version of the text in *section 6*(*1*) of *HSWA 1974* placed the following duties on designers, manufacturers and suppliers:

> 'It shall be the duty of any person who designs/supplies … any article for use at work to ensure, so far as is reasonably practicable, that the article is so designed … as to be safe and without risks to health … when properly used'.

The original *section 6* concentrated on protection against physical injuries during normal use of machinery and placed obligations on the user to follow the manufacturer instructions. The term 'so far as is reasonably practicable' has been interpreted by the courts as to indicate that in deciding an adequate level of safety, both the risk and cost of dealing with must be taken into account. This is a difficult concept for the majority of machine designers who understandably are cost driven.

The introduction of the European Directive on product liability, which was implemented in the UK under the *Consumer Protection Act* in 1987, in addition to the deficiencies shown by several court cases, have resulted in the revision of *section 6* in 1988.

The current *section 6(1)* of the 1974 Act requires that:

> 'It shall be the duty of any person who designs/ supplies … any article for use at work to ensure, so far as is reasonably practicable, that the article is so designed … that it will be safe without risk to health at all times: when being set, used, cleaned or maintained by a person at work'.

It should be noted that the term 'when properly used' had disappeared, as it is expected that designers should take account of normal behaviour of operators and maintenance personnel and the machine design should accommodate foreseeable misuse.

Machinery safety – the risk based approach

M1003 In May 1985 the EU Ministers agreed to a New Approach to Technical Harmonisation and Standards to overcome the problem of trade between European partners. The Machinery Safety Directive (89/392/EEC), subsequently amended, is one of the Product Safety Directives and sets out the essential health and safety requirements ('EH&SR's') for machinery which must be met before machinery is placed on the market anywhere within the EU.

EH&SRs are expressed in general terms and it is intended that the European Harmonised Standards should fill in the detail so that machinery designers and suppliers have clear guidance on how to achieve conformity with the Directive.

The Use of Work Equipment Directive (89/655/EEC) ('UWED') was also introduced to outline the responsibilities of management for the protection of workers from machinery and work equipment. In the UK this became the *Provision and Use of Work Equipment Regulations 1992* (*SI 1992 No 2932*) ('*PUWER'92*). The amending EU Directive on the Use of Work Equipment Directive (95/63/EC) has been implemented as the *Provision and Use of Work Equipment Regulations 1998* (*SI 1998 No 2306*) ('*PUWER '98*').

The Provision and Use of Work Equipment Regulations 1998 (PUWER'98)

M1004 The first EU Directive aimed at the use of work equipment was the Use of Work Equipment by Workers at Work Directive (89/655/EEC) ('UWED'). This Directive required all member states of the EC to have the same minimum requirements for the selection and use of work equipment, and was implemented in the UK by *PUWER'92* (*SI 1992 No* 2932). The Directive was subsequently amended and the non-lifting aspects of the amendments by Directive 95/63/EC ('AUWED') were implemented in the UK by *PUWER'98* (*SI 1998 No 2306*). These Regulations came into force on 5 December 1998 and on the same date *PUWER'92* was revoked. The lifting aspects of AUWED have been implemented by the *Lifting Operations and Lifting Equipment Regulations 1998* (*SI 1998 No 2307*) ('*LOLER*').

The 1998 Regulations are aimed at employers, and are designed to implement EU Directive 89/655/EEC, made under Article 118A of the *Single European Act*. This is aimed specifically at protecting the health and safety of workers in the Community. PUWER is also part of the so-called '6 pack Regulations'. They came into force on 1 January 1993 and applied immediately to equipment supplied after that date. *Regulations 1–10* apply to existing items of equipment but employers have until 1 January 1997 to ensure existing equipment meets the 'hardware' requirements in *Regulations 11–24*. Until such time as they do, existing equipment in factories will remain subject to the requirements of *sections 12–17* of the *Factories Act 1961*, which *PUWER* repealed.

PUWER'98, was made under the *HSWA 1974*. HSC states that the primary objective of *PUWER'98* is to ensure that work equipment should not result in health and safety risks, regardless of its age, condition or origin.

Structure of PUWER'98

M1005 The Regulations are structured in five main parts. These are set out as follows:

Part I: Regulations 1–3

This includes interpretation, definitions and application of *PUWER'98*.

Part II: Regulations 4–10 (*General –software*)

This sets out the 'management' duties of *PUWER'98* covering: selection of suitable equipment, maintenance, inspection, specific risks, information, instruction and training. *Regulation 10* covers the conformity of work equipment with the requirements of EU Directives on product safety.

Part II: Regulations 11–24 (*General – hardware*)

This deals with the hardware aspects of *PUWER'98*. It covers the guarding of dangerous parts of work equipment, the provision of appropriate stop and emergency stop controls, stability, lighting and suitable warning or devices.

Part III: Regulations 25–30 (*Mobile work equipment*)

This deals with specific risks associated with the use of self-propelled, towed and remote controlled mobile work equipment.

Part IV: Regulations 31–35 (*Power presses*)

This deals with the management requirements for the safe use and thorough examination of power presses. (Readers are advised to refer to a HSE publication 'Power presses: Maintenance and thorough examination' (HSG236), which provides practical advice on what, when and how to maintain presses.)

Part V: Regulations 36–39

Cover transitional provisions, repeal of Acts and revocation of instruments.

Risk assessment and PUWER'98

M1006 There is no requirement in *PUWER'98* (*SI 1998 No 2306*) to carry out a risk assessment. The general risk assessment requirement is made under *Regulation 3*(*1*) of the *Management of Health and Safety at Work Regulations 1999* (*SI 1999 No 3242*), which is explained in general terms in the associated HSC publications.

The HSC publication, 'Safe use of work equipment: Approved Code of Practice and Guidance' (L22), contains more specific guidance on risk assessment in relation to work equipment. This publication indicates that the factors to be considered in a risk assessment to meet the requirements of *PUWER'98* should include: type of work equipment, substances and electrical or mechanical hazards to which people may be exposed. Action to eliminate/control any risk might include, for example, during maintenance: disconnection of power supply, supporting parts of the work equipment which could fall, securing mobile equipment so that it cannot move and depressurising pressurised systems. Reference is also made to the use of the HSE publication '5 steps to risk assessment' (INDG163) (Revised 1998).

In dealing with risk assessment the HSC publication L22 refers to 'significant risk' which is defined by in that document as 'one which could foreseeably result in a major injury or worse'. This definition implies that if the consequences of exposure to a hazard could result in a minor injury, there is no need to comply with many of *PUWER'98* requirements.

'Risk' is defined as the combination of the probability (or chance) of a harm being realised coupled with the consequences (severity) as a result of exposure to the harm. Depending on which criteria is used to evaluate risks, if the chance of a minor injury is likely, then this would constitute a significant risk. The lack of guidance on agreed criteria for the evaluation of health and safety risks makes the task of deciding whether risks are tolerable somewhat subjective and arbitrary. The HSE's '5 steps to risk assessment' may be too generic to result in adequate hazards identification and analysis or in the evaluation of risks, particularly in relation to machinery.

Summary of main requirements of PUWER'98

Part II: General (Regulations 4–24)

M1007 *Part II* of *PUWER'98* (*SI 1998 No 2306*) applies to all work equipment, including machinery; mobile plant and work equipment.

Regulation 4 – Suitability

Equipment must be suitable, by design, construction or adaptation, for the actual work it is provided to do. In selecting the work equipment, the employer must take account of the environment in which the equipment is to be used and must ensure that the equipment is used only for the operations, and under the conditions, for which it is suitable.

Regulation 5 – Maintenance

Equipment must be maintained in an efficient state, in efficient working order and in good repair. (The term 'efficient' relates to how the condition of the equipment might affect health and safety, not productivity). There is no requirement to keep a maintenance log, but where one exists (and this is good management practice) it must be kept up to date.

Regulation 6 – Inspection

Regulation 6 of *PUWER'98* (*SI 1998 No 2306*) includes a requirement for the inspection of work equipment. The purpose of an inspection is to identify whether the work equipment can be operated, adjusted and maintained safely and that any deterioration, e.g. defects, damage or wear can be detected and repaired before it results in unacceptable risks.

The guidance to *PUWER'98* states that inspection is only necessary where there is a significant risk resulting from incorrect installation or re-installation, deterioration or as a result of exceptional circumstances which could affect the safe operation of the work equipment. Examples given in the guidance of work equipment unlikely to need an inspection include hand tools, non-powered machinery and powered machinery such as a reciprocating fixed blade metal cutting saw!

The extent of the inspection required will depend on the level of risk associated with the work equipment, and should include, where appropriate visual and functional checks to the more comprehensive inspection which may require some dismantling and/or testing. The frequency of inspections made under *Regulation 6* is to be decided by a 'competent person'.

A fundamental requirement under *Regulation 6* is that every employer must ensure that the result of an inspection is recorded and kept until the next inspection is recorded. Although records do not have to be kept in a particular form, the example shown in FIGURE 1 below identifies the main items to be recorded on an inspection record sheet.

Components, which must receive particular attention as part of an inspection, are the safety-related and safety-critical parts, for example interlocking devices, safeguards, protection devices, controls and control systems.

Regulation 6 also states that no work equipment should leave an undertaking, or be obtained from an undertaking of another person (for example by hiring), unless it is accompanied by physical evidence that the last inspection has been carried out.

Regulation 7 – Specific risks

Regulation 7 of *PUWER'98* (*SI 1998 No 2306*) requires employers to ensure that, where the use of work equipment is likely to involve a specific risk, the use, repairs, modifications, maintenance or servicing of such equipment is restricted to those persons who have been specifically designated to perform operations of that description. It is further required that employers must provide adequate training to perform such tasks safely.

The Approved Code of Practice ('ACoP') and Guidance to *PUWER'98* did not define what is meant by 'specific risks', or what makes such risks 'specific'. The examples given in the Guidance L22 on machinery which poses specific risks, e.g. a platen printing machine or a drop-forging machine did not make reference to which specific risks are associated with these machines.

The hazards associated with the use, maintenance, repair etc. of the above example of machines should be identified if a suitable and sufficient risk assessment is carried out. For example, hazards associated with a drop forge might include: noise, vibration, crushing, shearing, heat, splashes of molten metal, toxic fumes, etc. It may not be impractical to provide adequate protection against all these hazards by means of physical safeguards. As a result, significant 'residual' risks remain which may require the use of adequate personal protective equipment and safe systems of work. Are these meant to be 'specific' risks?

Figure 1: Example of an inspection record sheet

INSPECTION RECORD

Type and model of equipment: Press Brake-Komato ....
Serial number/Identification mark: KP/012 ....
Normal location of equipment: Press Shop ....
Date inspection carried out: 08/09/2002 ....

Type of Inspection:
Visual check:
Functional check:
Dismantling/testing: ✓

Who carried out inspection: John Smith ....

Items and safety-related parts inspected

1. Guard-Interlocking Device ....
2. Photo-electric Trip Device ....
3. Fixed and Interlocking Safeguards ....
4. Pneumatic Clutch and Brakes Systems ....
5. Machine Control and Electrical Systems ....

Faults and defects found:
1. Small air leak near clutch ....
2. Interlocking device wiring loose ....

....

Action/corrective measures taken:
1. Seals replaced for pneumatic system ....
2. Re-wiring of interlocking system ....

....

To whom faults have been reported: Kevin Taylor ....

Date repairs/actions were carried out: 12/09/2003 ....

The ACoP relating to *Regulation 7* requires employers to ensure that risks are always controlled by (in the order given):

(*a*) eliminating the risks, or if that is not possible;

(*b*) taking 'hardware' (physical) measures to control risks such as the provision of guards; but if risks cannot be adequately controlled;

(*c*) taking appropriate 'software' measures to deal with the residual risks, such as following safe systems of work and the provision of information, instruction and training.

Regulation 8 – Information and instructions

Managers, supervisors and users of work equipment must be provided with adequate health and safety information and, where appropriate, written instructions relating to work equipment. This must include the conditions in which, and the methods by which, the equipment may be used, the limitations on its use together with any foreseeable difficulties that might arise and the required actions.

Regulation 9 –Training

Managers, supervisors and all users of work equipment must receive adequate health and safety training in the methods, which may be adopted when using equipment, any risks, which may arise from such use and precautions to be taken. The level of training will depend on the circumstances of use and the competence of the employee. Other regulations may contain specific training requirements. Particular attention must be paid to the training and supervision of young persons.

Regulation 10 – Conformity with Community requirements

Regulation 10 of *PUWER'98* (*SI 1998 No 2306*) (amended by the *Health and Safety* (*Miscellaneous Amendments*) *Regulations 2002* (*SI 2002 No 2174*)), applies to items of work equipment provided for use in the premises or undertaking of the employer for the first time after 31 December 1992. The amended *Regulation 10*(*1*) requires employers to ensure that work equipment complies at all times with any EH&SR's that applied to that kind of work equipment at the time of its *first supply* or when it is *first put into service*. It places a duty on the employer that complements those on manufacturers and suppliers in other jurisdictions regarding the initial integrity of equipment.

The amended *Regulation 10*(*2*) defines the term 'essential requirements' – which are the essential health and safety requirements listed in the Regulations implementing the relevant directives in the UK (which are listed in *Schedule 1* of the 1998 Regulations).

The amended 1998 Regulations no longer disapply *Regulations 11–19* and *22–29*. This means that inspectors are now free to use these particular provisions if they feel that these Regulations are more appropriate than using the more generally worded.

The amendment is designed to deal with the situation where a piece of work equipment is supplied in compliance with the relevant supply legislation when it is first supplied/put into service, but the employer then alters the equipment such that it no longer complies. In this case *Regulations 11–19* and *22–29* can be used to ensure that the employer returns the equipment to a state of safety and compliance.

The amended Regulations also allow the use of *Regulations 11–19* and *22–29* for equipment that was not in compliance with the supply legislation at the time of supply/putting into service. In this situation inspectors can use *Regulations 11–19* and *22–29* to ensure that employers take the necessary action to bring this equipment into compliance, if appropriate.

The employer can demonstrate compliance with *Regulation 10* if the following conditions are satisfied:

- that the employer where appropriate checks to see that the equipment bears a CE marking and asks for a copy of the EU Declaration of Conformity;

- the employer also needs to check that adequate operating instructions have been provided and that there is information about residual hazards such as noise and vibration;
- more importantly, the employer should check the equipment for obvious faults.

The guidance offers additional advice in the form of HSE booklet 'Buying new machinery'.

If however, work equipment does not comply with relevant product legislation, for example a new machine has been supplied without suitable guards, *Regulation 11* of *PUWER'98* (*SI 1998 No 2306*) could be used to ensure the user does provide adequate guards.

The HSE booklet 'Buying new machinery' (INDG271) is written in easy to understand language and contains two checklists: Checklist A – What should I talk to a supplier about? and Checklist B – What do I do when I have bought new machinery? In both checklists, the employer relies on guidance given by the machine designers/suppliers. This guidance is aimed at small/medium-size enterprises and the checklists are somewhat superficial and over simplified. Basic requirements such as the format of Declarations of Conformity/Incorporation, or the structure/relevance of the Transposed Harmonised EU Standards are not discussed.

Regulation 11 – Dangerous parts of machinery

Regulation 11(*1*) of *PUWER'98* (*SI 1998 No 2306*) requires employers to ensure that access to dangerous parts of the machine is prevented, or if access is needed to ensure that the machine is stopped before any part of the body reaches the danger zone. *Regulation 11*(*2*) sets out the following hierarchy of preventive measures (see also M1027 onwards):

(*a*) fixed enclosing guards;

(*b*) other guards or protective devices;

(*c*) protection appliances, e.g. jigs, holders, push sticks, etc.

(*d*) the provision of information, instruction, training and supervision.

This hierarchy follows the traditional preference of fixed guards and is open to criticism. If an individual needs frequent access into the machine danger zone, a fixed guard could be the worst option, as the machine can operate without it.

The European Commission did express concern that the original *Regulation 11*(*2*)(*d*) – the provision of information, instruction, training and supervision, could be used as the sole solution to working with the unguarded parts of dangerous machinery.

To address this and other concerns, *Regulation 11* was amended so that the provision of information, instruction, training and supervision is now seen as an additional requirement in each case and not as part of the hierarchy of measures.

Regulation 11(*3*) describes the need for the provision of safeguards/safety devices which are robust, difficult to defeat, well maintained and meet the requirements of relevant EN standards. Illustrated examples of the options given in *Regulation 11* are shown in the guidance to *PUWER'98*.

Regulation 12 – Protection against specified hazards

Appropriate measures must be taken to prevent or adequately control exposure to 'specified hazards' arising from the use of work equipment. These measures should be other than the provision of personal protective equipment or of information, instruction, training and supervision, so far as is reasonably practicable. The 'specified hazards' are: falling or ejected articles or substances; rupture or disintegration of parts of equipment; work equipment catching fire or overheating; and unintended or premature discharges or explosions.

Regulation 13 – High or very low temperature

Work equipment, parts of work equipment and any article or substance produced, used or stored in work equipment must be protected so as to prevent burns, scalds or sears.

Regulations 14 to 18 – Controls and control systems

Work equipment must be provided with:

- One or more controls for starting the equipment or for controlling the operating conditions (speed, pressure etc.), where the risk after the change is greater than or of a different nature from the risks beforehand.
- One or more readily accessible stop controls that will bring the equipment to a safe condition in a safe manner and which operate in priority to any control that starts or changes the operating conditions of the equipment.
- One or more readily accessible emergency stop controls – unless by the nature of the hazard, or by the adequacy of the normal stop controls (above), this is unnecessary. Emergency stop controls must operate in priority to other stop controls.
- Controls should be clearly visible and identifiable (including appropriate marking). The position of controls should be such that the operator can establish that no person is at risk as a result of the operation of the controls. Where this is not possible, safe systems of work must be established to ensure that no person is in danger as a result of equipment starting or where this is not possible, there should be an audible, visible or other suitable warning whenever work equipment is about to start. Sufficient time and means shall be given to a person to avoid any risks due to the starting or stopping of equipment.
- Control systems should not create any increased risk, which ensure that any faults or damage in the control systems or losses of energy supply do not result in additional or increased risk, and which do not impede the operation of any stop or emergency stop control.

Regulation 18 – Control systems

Regulation 18 of *PUWER'98* (*SI 1998 No 2306*) relates to work equipment control systems. This requires that:

'(1) Every employer shall—

(a) ensure, so far as is reasonably practicable, that all control systems of work equipment are safe; and

(b) are chosen making due allowance for failures, faults and constraints to be expected in the planned circumstances of use.'

A control system is defined as: a system which responds to input signals and generates an output signal which causes the equipment to perform controlled functions.

The input signals may be made by the operator (manual control), or automatically controlled by the equipment through sensors or protection devices, e.g. guard interlocking devices, photoelectric device, emergency stop or speed limiters.

The objective of *Regulation 18* is to prevent or reduce the likelihood that a single failure in the interlocking device, wiring, contactors, hardware, software could cause injury or ill health.

The subject of control systems generally and the safety related functions of control systems in particular are complex, even for engineers. Although the Harmonised Standard BS EN 954–1: 1996 'Control systems with safety-related functions' provides guidance for the design and integrity of control systems based on risk assessment.

The complexity of this subject is further compounded by the use of programmable electronic systems in the control system. Guidance on the use of programmable electronic systems for safety related applications are not covered by the EU Harmonised Standards. However, the International Electro-technical Standard IEC 61508 is generally used to identify, based on risk assessment the relevant Safety Integrity Level (SIL) required for the control system.

Regulation 19 – Isolation from sources of energy

Equipment must be provided, where appropriate, with identifiable and readily accessible means of isolating it from all its sources of energy. Reconnection of any source of energy must not expose anyone to any health and safety risk.

Regulation 20 – Stability

All work equipment must be stabilised where necessary – by clamping, tying or fastening.

Regulation 21 – Lighting

Suitable and sufficient lighting, taking account of the operations carried out, must be provided. This may entail additional lighting where the ambient lighting is insufficient for the required tasks.

Regulation 22 – Maintenance operations

Work equipment must be constructed or adapted in such a way that maintenance operations, which involve a risk to health or safety, can be carried out while the equipment is shut down. If this is not possible, the work should be carried out in such a way as to prevent exposure to risk and appropriate measures taken to protect those carrying out the maintenance work. Such measures might include, for instance, functions designed into the equipment that limit the power, speed or range of movement of dangerous parts during maintenance.

Regulation 23 – Markings

Equipment must be marked for the purposes of ensuring health and safety with clearly visible markings.

Regulation 24 – Warnings

Equipment must incorporate appropriate warnings or warning devices, which must be unambiguous, easily perceived and easily understood. These requirements are nothing new since they already exist in some other part of legislation or at least constitute accepted good practice. However, the terminology of *PUWER'98* (*SI 1998 No 2306*) is very prescriptive when compared to many of the other harmonised 'goal-setting' regulations.

Second-hand machinery

Second-hand machinery supplied within the EU is not required to be 'CE' marked, nor issued with a Declaration of Conformity/Incorporation. These however should subject to a suitable and sufficient risk assessment and should comply with *Regulations 11* to *24* of *PUWER'98* (*SI 1998 No 2306*).

Second-hand machinery supplied from outside the EU should be treated as new machinery and will be subject to CE marking requirements.

It is the management responsibility to ensure that second-hand machinery and modification work do comply with the requirements of the *PUWER'98*.

Part III: Mobile work equipment

M1008 *Part III* of *PUWER'98* (*SI 1998 No 2306*) added new requirements for the management of mobile equipment.

Mobile equipment is defined as any work equipment, which carries out work while it is travelling or which travels between different locations where it is used to carry out work. Mobile work equipment may be self-propelled, towed or remote controlled and may incorporate attachments. *Regulations 25* to *30*, in addition to other requirements of *PUWER'98* (e.g. training, guarding and inspection), apply to all work equipment including mobile work equipment. New mobile equipment taken into use from 5th December 1998 should comply with all requirements of *PUWER'98*.

Regulation 25 – Employees carried on mobile work equipment

Regulation 25 of *PUWER'98* (*SI 1998 No 2306*) contains a general requirement relating to the risks to people (drivers, operators and passengers) carried by mobile equipment, when it is travelling. This includes risks of people falling from the equipment or from unexpected movement while it is in motion or stopping.

This Regulation requires every employer to ensure that no employee is carried by mobile work equipment unless:

(*a*) it is suitable for carrying persons; and

(*b*) it incorporates features for reducing to as low as is reasonably practicable risks to their safety, including risks from wheels or tracks.

Mobile work equipment can be made suitable for carrying persons by means of operator stations, seats or work platforms to provide a secure place.

Features for reducing risks may include the following:

- Falling object protective structures ('FOPS'). This may be achieved by a strong safety cab or protective cage.
- Restraining systems – this can be full-body seat belts, lap belts or purpose-designed restraining system. This should provide adequate protection against injury through contact with or being flung from the mobile equipment if it comes to a sudden stop moves unexpectedly or rolls over.
- Speed adjustment – when carrying people, mobile equipment should be driven within safe speed limits for stability.
- Guards and barriers fitted to mobile work equipment to prevent contact with wheels and tracks.

Regulation 26 – Rolling over of mobile work equipment

In addition to the main general requirements of *Regulation 25*, *Regulation 26* of *PUWER'98* (*SI 1998 No 2306*) covers the measures necessary to protect employees where there are risks from roll-over while travelling. If this risk is significant, then the following measures should be considered:

- Stabilisation – this might include fitting appropriate counterbalance weights, wider wheels and locking devices.
- Structures to prevent rolling over by more than 90°, e.g. boom of a hydraulic excavator, when positioned in its recommended travel position.
- Roll-over protective structures ('ROPS') – normally fitted on mobile equipment to withstand the force for roll over through 180°.

Regulation 27 – Overturning of forklift trucks

Regulation 27 of *PUWER'98* (*SI 1998 No 2306*) applies to FLT's fitted with vertical masts, which effectively protect seated operators from being crushed between the FLT and the ground in the event of roll-over. FLT's capable of rolling over 180° should be fitted with ROPS.

FLT's with a seated ride-on operator can roll-over in use. There is also a history of accidents on counterbalanced, centre control, high lift trucks. Restraining systems will normally be required on these trucks.

Regulation 28 – Self-propelled work equipment

- Preventing unauthorised start-up – this could be achieved by a starter key or device, which is issued only to authorised personnel.
- Minimise consequences of a collision of rail-mounted work equipment – this can be achieved by introducing safe systems of work, buffers or automatic means to prevent contact.
- Devices for stopping and braking – all self-propelled work equipment should have brakes to enable it to slow down and stop in a safe distance and park safely.
- Emergency braking and stopping facilities – a secondary braking system may be required if risk is significant.

- Driver's field of vision – where the driver's direct field of vision is impaired then mirrors or more sophisticated visual or sensing facilities may be necessary.
- Lighting for use in the dark – to be fitted to equipment if outside lighting is inadequate.
- Carriage of appropriate fire-fighting equipment – where escape from self propelled work equipment in the case of fire is not easy.

Regulation 29 – Remote-controlled, self-propelled work equipment

- This is self-propelled work equipment that is operated by controls which have no physical link with it, for example radio control.
- Alarms or flashing lights – to be considered for the prevention of collision with others.
- Sensing or contact devices – to prevent or reduce risk of injury following contact.
- Hold-to-run control devices – so that any hazardous movement can stop when the controls are released.
- Control range – once equipment leaves its control range should stop and remain in a safe state.

Regulation 30 – Drive shafts

- A 'drive-shaft' is a device, which conveys the power from the mobile work equipment to any work equipment connected to it. In agriculture these devices are known as power take-off shafts.
- Shaft seizure – if seizure can result in significant risk, for example the ejection of parts due to equipment break-up, guards should be fitted in accordance with *Regulation 12* of *PUWER'98* (*SI 1998 No 2306*).
- Isolation from source of energy – to reduce the risk associated with drive shaft stalling.
- Support on cradle – when drive shaft and its guards are not in use, the drive shaft should be supported on cradle or equivalent for protection against damage.

The Supply of Machinery (Safety) Regulations 1992 (as amended)

M1009 General and specific duties relating to machinery safety are laid on designers and manufacturers of machinery for use at work by the *Supply of Machinery* (*Safety*) *Regulations 1992* (*SI 1992 No 3073*), as amended by the *Supply of Machinery* (*Safety*) (*Amendment*) *Regulations 1994* (*SI 1994 No 2063*).

Supply of safe machinery (including safety components, roll-over protective structures and industrial trucks) is governed by these Regulations. The 1994 Regulations extend to machinery for lifting people (e.g. elevating work platforms). [*SI 1994 No 2063, Reg 4, Sch 2, para 10*]. The combined effect of the two sets of Regulations requires that, when machinery or components are properly installed, maintained and used for their intended purposes, there is no risk (except a minimal one) of their being a cause or occasion of death or injury to persons, or damage to property. [*SI 1992 No 3073*, *Reg 2*(*2*), as amended by *SI 1994 No 2063, Reg 4, Sch 2, para 5*].

The EU Machinery Directive (89/392/EEC) as amended is aimed at machinery designers and suppliers. This is a so-called 'Product Directive' made under Article 100A of the *Single European Act* to ensure fair competition by eliminating barriers to trade. The relevant products have to meet certain essential health and safety requirements ('EH&SRs') before they can be placed on the market within the Community.

The manufacturer affixing a CE-Mark to the machinery claims compliance with all relevant EU Directives. The EU Machinery Directive was implemented in the UK under the *Supply of Machinery* (*Safety*) *Regulations 1992*.

EU Machinery Directive

M1010 The EU Machinery Directive requires:

- The clear allocation of responsibilities to machinery suppliers.
- Type approval: 'attestation' via a technical construction file and a Declaration of Conformity before the CE mark is affixed to the machine.
- Intended use of machinery has to be clearly defined.
- Principles of Safety Integration. Manufacturers must demonstrate that the following order of options for machinery safety are followed:
  - (i) elimination or reduction of risks by seeking to design inherently safe machinery;
  - (ii) necessary protection measures taken (e.g. safeguarding or safety devices) to deal with the risks which cannot be eliminated;
  - (iii) users of residual risks are informed.

The majority of the requirements of the Directive are couched as 'objects to be achieved'.

The Directive and EN standards show that while traditional mechanical hazards continue to merit attention, manufacturers must also consider: discomfort, fatigue and psychological stress; electricity, including static; sources of energy supply other than electricity; errors of fitting; temperature extremes; fire and explosion; noise and vibration; radiation; laser equipment; and emissions of gases, etc.

Definition of a machine

M1011 A 'machine' is defined by the EU Machinery Directive as 'an assembly of linked parts or components, at least one of which moves, with the appropriate actuators, control and power circuits, joined together for a specific application, in particular for the processing, treatment, moving or packaging of a material'. The definition includes an assembly of machines which functions as an integral whole as well as interchangeable equipment (not being a spare part or tool) which can be assembled with a machine by the operator – for instance, an item of agricultural equipment attached to a tractor.

The main elements in the framework for demonstrating compliance are summarised as follows.

Essential health and safety requirements ('EH&SR')

M1012 The route to demonstrating compliance starts with the identification of relevant EH&SRs relating to the design and construction of machinery. These are set out in Annex I of the EU Machinery Directive, and include the following:

- Principles of safety integration.
- Materials and products health and safety.
- Lighting.
- Design of the machine to facilitate its handling.
- Safety and reliability of the control system.
- Control devices.
- Starting, change-over and stopping devices.
- Emergency stops.
- Mode selection.
- Power supply failure.
- Failure of the control system, including software.
- Stability.
- Risk of break-up during operation.
- Risks due to falling or ejected objects.
- Sharp surfaces, edges or angles.
- Hazards associated with moving parts.
- Required characteristics of guards and protection devices.
- Fire, explosion, noise, vibration, radiation, laser equipment, etc.

Suppliers should identify relevant EH&SRs applicable to the machine and show how these have been considered.

Conformity assessment procedure carried out

M1013 The responsibility for demonstrating that the machinery satisfies EH&SR's rests with the manufacturer, supplier or the importer into the EU ('the responsible person').

The EU Machinery Directive sets out various ways in which a supplier can demonstrate compliance with EH&SR's. This is known as 'conformity assessment procedure'. There are three different conformity assessment procedures, detailed in *Regulations 13*, *14* and *15* of the *Supply of Machinery* (*Safety*) *Regulations 1992* (*SI 1992 No 3073*). These are:

- for most machinery other than that listed in *Schedule 4* of *SI 1992 No 3073* (Annex IV of the Directive);
- Schedule 4 (Annex IV) machinery manufactured in accordance with the EU Transposed Harmonised Standards; or
- Schedule 4 machinery not manufactured in accordance with the EU Harmonised Standards or where they do not exist.

Schedule 4 machinery lists categories of machines that pose special risks. These include mechanical presses, press brakes, injection and compression plastics moulding machines.

For machinery listed in *Schedule* 4 of the *Supply of Machinery* (*Safety*) *Regulations 1992* (*SI 1992 No 3073*) that is manufactured in conformity with Harmonised Transposed Standards, the machinery technical file is submitted to an approved body for EC type-examination. The approved body may simply verify that the Transposed Harmonised Standards have been correctly applied and draw up a certificate of adequacy.

The technical file

M1014 The conformity procedure requires the supplier to assemble records, which technically describe the rationale behind the approach to ensuring that the machine satisfies EH&SR's. These records are referred to as the 'technical file'. The technical file should include a risk assessment (if applicable) and other technical information which includes: technical drawings, test results, a list of the EH&SR's, Harmonised Standards, other standards and technical specifications that were used in the design of the machinery. It should also include a description of the methods adopted to eliminate hazards, other relevant technical information and a copy of the instructions for the machinery.

For machinery posing special hazards (listed in Annex IV of the EU Machinery Directive), further action is necessary which usually involves verification by an approved body that either recognised standards have been correctly applied or an EU type-examination of the machine shows that it satisfies the relevant provisions of the Directive.

The supplier must retain the technical file and all documentation relating to particular machinery for a minimum of 10 years after the production of the last unit of that machinery.

EU Declaration of Conformity/Incorporation

M1015 An EU Declaration of Conformity must be drawn up which states that the relevant machinery complies with all the EH&SR's that apply to it.

Such a declaration must include a description of the machinery (with make, type and serial number), indicate all relevant provisions with which the machinery complies, give details of any EU-type examination and specify any standards and technical specifications which have been used. Reference should also be made to all other relevant EU Product Directives, e.g. Low Voltage, EMC and Simple Vessel Directives.

The machine supplier prior to the commissioning or operation of the machine should issue the Declaration of Conformity. A typical example of a Declaration of Conformity is shown in FIGURE 2 below.

The three most relevant sections on this form are:

1. Relevant EU Directives applicable to the product;
2. The Transposed Harmonised Standards used; and
3. Other standards and technical specifications relevant to the machinery.

The declaration must identify European Transposed Harmonised Standards applicable to the machine design and construction. Other national and international standards relevant to the machine, which are not covered by relevant European

Figure 2: Example of a Declaration of Conformity

EU DECLARATION OF CONFORMITY

In accordance with:

The Supply of Machinery (Safety) Regulations 1992

WE DECLARE THAT THIS MACHINE CONFORMS WITH THE ESSENTIAL HEALTH AND SAFETY REQUIREMENTS

Description of the machine, type and serial number:

Business name and full address of manufacturer:

All relevant product directives complied with:
Machinery Directive (89/392/EEC) and all amendments thereto.
Low Voltage, CE marking and EMC Directives, ...

The machine complies with the following transposed harmonised standards:
EN 292-1, EN 292-2, EN 294, EN 60204-1, EN 954-1, EN 1050, EN 1088

Other standards and technical specifications used:
IEC 1508

IDENTIFICATION of the person empowered to sign on behalf of the Company: Usually Managing Director

I certify that on 14th of January 1996, the above machine conformed with the EH&SR's of all the relevant Product Directives.

Signature: .... Date: ....

standards should also be identified on the certificate. For example IEC 61508, which is an international electro-technical standard for the design and selection of programmable electronic systems with safety related applications. The Declaration of Conformity should also make reference to all relevant EU product directives, e.g. Low Voltage and Electro-magnetic Compatibility Directives.

A Declaration of Incorporation is required where the machinery is intended for incorporation into another machinery or assembly with other machinery to constitute machinery covered by the Directive. The Declaration of Incorporation is similar to the declaration of conformity where relevant Harmonised Standards are listed. The only difference is that a 'CE' mark is not required, and a condition included in the declaration, which states that: 'the equipment must not be put into service until the machinery into which it is incorporated has been declared in conformity with the provision of the Machinery Directive'. See Annex IIB of the Directive for more detail.

'CE' marking

M1016 A 'CE' marking (meaning European Conformity) on a product or system indicates a legal declaration by the supplier that the product complies with EH&SRs of the EU Machinery Directive and all other EU directives relevant to it.

The CE mark is regarded as properly affixed only if: (a) an EU declaration of conformity has been issued, (b) it is affixed in a distinct, visible, legible and indelible manner, and (c) the machinery complies with the requirements of other relevant directives. It is an offence to affix the CE mark to machinery, which does not satisfy the EH&SRs, or to machinery that is not safe. The CE mark is described in the Machinery Directive and consists of a symbol 'CE' and the last two figures of the year in which the mark is affixed.

Requirements of other EU directives

M1017 Machinery suppliers need to identify, apart from the Machinery Directive, all other EU Directives relevant to their products. It is then necessary to identify specific legislation, which implement them. Usually more than one EU directive may apply to particular machinery. For example, in addition to the Machinery Directive, an item of electrically driven machine would be subject to the following Directives:

- Electromagnetic Compatibility Directive (89/336/EEC) as amended. This is implemented in the UK as the *Electromagnetic Compatibility Regulations 1992* (*SI 1992 No 2732*).
- Low Voltage Directive (72/23/EEC) implemented in the UK under the *Electrical Equipment* (*Safety*) *Regulations 1994* (*SI 1994 No 3260*).
- Other EU directives, which may be relevant to some machinery include Simple Pressure Vessel Directive (87/404/EEC) as amended, Potential Explosive Atmosphere Directive (94/9/EEC) and Telecom Terminal Equipment Directive (91/263/EEC).

The machine is in fact 'safe'

M1018 The key objective of the EU Machinery Directive is that the onus is placed on suppliers to demonstrate that a machine complies with relevant EH&SRs before it is placed for sale within the market place of the EU. The requirement for the machine to be in fact safe is an UK rather than EU requirement.

'Safe' is defined by the *Supply of Machinery* (*Safety*) *Regulations 1992* (*SI 1992 No 3073*) as, 'when the machinery is properly installed and maintained and used for the purpose for which it is intended, there is no risk (apart from one reduced to a minimum) of it being the cause or occasion of death or injury to persons or, where appropriate, to domestic animals or damage to property. The practicability at the time of manufacture, of reducing the risk may be taken into account when considering whether or not the risk has been reduced to a minimum'.

Failure to comply with relevant EU requirements will mean that the machinery cannot legally be supplied in the UK and could result in prosecution and penalties on conviction of a fine up to £5,000 or in some cases, of imprisonment for up to three months, or of both.

The same rules apply everywhere in the EU, so machinery complying with the EU regime may be supplied in any Member State.

The Transposed Harmonised European Machinery Safety Standards

M1019 Application of the European Transposed Harmonised Machinery Safety Standards is fundamental to the risk-based approach. These standards have now replaced corresponding British Standards relating to machinery safety, as well as other EU member countries national standards.

A process for incorporating the European Machinery Safety Standards into international standards has been adopted by ISO/IEC since 1990. The American National Standards Institute (ANSI) is also considering the EU risk-based approach to machinery safety.

Status and scope of the Harmonised Standards

M1020 The role and status of the European Harmonised Standards has been frequently misunderstood, mainly due to the previously prescriptive nature of the health and safety legislation and standards. The 'new approach' is based on the legal status of the EH&SRs as an objective, but the Harmonised Standards represent ways for fulfilling the details of the individual EH&SRs. The Harmonised Standards are not of the same legal status as the EH&SRs, but do reflect recognised practice and should be the first consideration.

EH&SRs are expressed in general terms and it is intended that the European Harmonised Standards should fill in the detail so that designers have clear guidance on how to achieve conformity with the directive".

The Machinery Directive states 'objectives to be achieved' where the ES&HRs constitute the legal objective requiring to be complied with and the Harmonised Standards are the current recognised methods, and therefore do not constitute the legal duty as to the method of compliance.

Structure of the Harmonised Standards

M1021 The European Standards start by listing the machine or process hazards that are considered relevant within its scope and use. In addition, each health and safety requirement should include an explicit safety aim, and to assess whether a risk has been reduced to an acceptable level.

An overview of the structure for the Harmonised European Standards is shown in FIGURE 3 below.

These Standards are introduced in three levels: A, B and C. 'A' type Standards provide general requirements applicable to all machines. The key standard relating to the EU Machinery Directive is EN 292: 1991 Machinery safety: Basic concepts, general principles for design. This Standard is introduced in two parts:

- BS EN 292: 1991 – Part 1: Machinery safety Basic concepts and methodology.
- BS EN 292: 1991 – Part 2 and Part 2/A1: Technical principles and specifications.

BS EN 1050: 1997 – Machinery safety: Principle or risk assessment is largely regarded as an 'A' type Standard. BS EN 1050 provides a framework for risk assessment and the range of hazards to be considered in the assessment.

There is no standard risk assessment technique, as this will depend on the nature of machine hazards and the degree of interactions with humans. Risk assessment however should be suitable and sufficient. In other words, the degree of detail and techniques used are a function of the risk level and the complexity of the situation under study.

'B' type Standards are standards on techniques or components, which could be applicable to a large number of machines. 'B1' Standards provide generic guidance and information, e.g. EN 294: 1992 'minimum height and distance from the hazard'; whereas 'B2' Standards outline generic safety standards for 'hardware', e.g. electro-sensitive safety systems, two-hand controls, control systems.). The B type or horizontal group safety standards therefore deal with only one safety aspect or one safety related device at the time.

Figure 3: Structure of the Harmonised European Standards

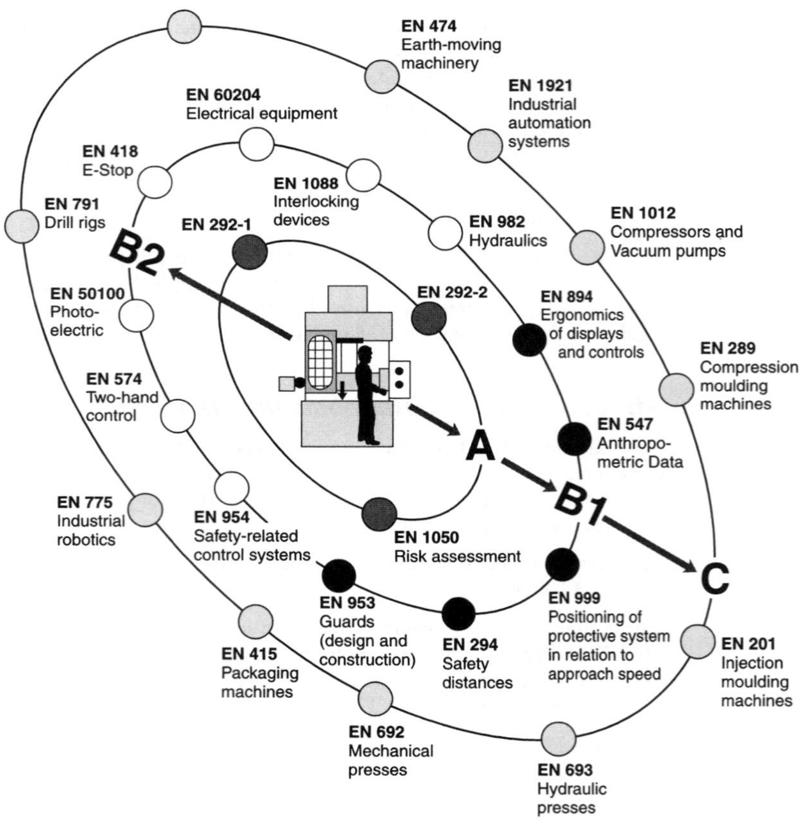

'C' type Standards are drawn up to cover a particular type or class of machines. Examples of C Standards include EN 289 1994 – Compression moulding machines and EN 692 for mechanical presses. A large number of 'C' Standards are yet to be produced or introduced in provisional form. These may be classed as vertical standards. In following a particular 'C' Standard, cross-reference is extensively made to the relevant sections of 'A' and 'B' type Standards.

It is important to stress that the use of European Harmonised Standards is not mandatory. However, compliance with relevant EN standards is one of the most effective means for demonstrating that relevant EH&SRs are met. The other alternative would be the application of a robust risk assessment, which considers all relevant hazards and hazardous situations listed in EN 292 and EN 1050.

Machinery risk assessment

M1022 The risk-based approach involves a structured and systematic risk assessment. Risk assessment is essentially a proactive means to foresee how accidents can happen and to decide what corrective and preventive actions are needed in advance to prevent such accidents.

Risk assessment is defined as ' ... A structured and systematic procedure for *identifying hazards* and *evaluating risks* in order to prioritise decisions to reduce risks to a *tolerable* level'.

This section will provide guidance on the following key elements of a structured and systematic risk assessment:

- techniques for hazard identification and analysis;
- criteria for the estimation and evaluation of risks;
- the concept of 'ALARP' and risk tolerability, and
- the preferred hierarchy for risk control options.

In addition to demonstrating compliance with relevant legislation/standards, risk assessment is now crucial as it aids in the consistency of decision making and the cost effectiveness in allocation of resources for health, safety and environmental issues.

Risk assessment is vital for two reasons: the level of risk determines priority which should be accorded to and the selection of appropriate health, safety and environmental control measures to deal with different hazards, and the standard of integrity of corrective/preventive measures will depend on the level of risk.

Risk assessment techniques are complementary to the more pragmatic ways of problem identification and assessment. They highlight systematically how hazards can occur and provide a clearer understanding of their nature and possible consequences, thereby improving the decision-making process for the most effective way to prevent injury and damage to health/environment.

The techniques range from relatively simple qualitative methods of hazard identification and analysis to the advanced quantitative methods for risk assessment in which numerical values of risk frequency or probability are derived.

Hazard and risk – definitions

M1023 A *hazard* is defined as 'a potential to cause harm'.

Harm is defined as:

- injury or damage to health;
- damage to the environment;
- economic losses (interruption to production/asset damage).

Risk is defined as:

- 'the chance (probability) of the harm being realised, combined with its consequences; or
- Risk = Chance of exposure to the hazard x Consequences (severity).

FIGURE 4 below illustrates the concept of risk. If for example it was estimated that the chance of an accident involving the collision between a fork-lift truck and an operator as 'remote', and the resulting injury severity as fatal 'catastrophic', then the Risk Matrix shows this as risk level 'A' or high, which should warrant some urgent attention. If on the other hand it was estimated that the chance of slipping on a wet surface in part of the workplace as 'occasional', which would result in 'minor' injury. The risk matrix shows this to be risk level 'C' or low, which is broadly acceptable.

Figure 4: The Risk Matrix

	Probability				
	A	B	C	D	E
Severity	**Improbable** **1 in 100,000** years	**Remote** **1 in 10,000** years	**Occasional** **1 in 1,000** years	**Probable** **1 in 100** years	**Frequent** **1 in 10** years
5. Catastrophic					
4. Severe				**HIGH RISK A**	
3. Critical			**MEDIUM RISK (ALARP) B**		
2. Marginal					
1. Negligible	**LOW RISK C**				

Risk assessment framework

M1024 Hazard analysis is an essential part of the overall risk assessment framework. The flow-chart shown in FIGURE 5 (based on EN 1050) outlines elements of the overall framework. The effectiveness of hazard analysis and risk evaluation processes involves a number of practical considerations.

Before commencing the study, its objectives and scope must be defined explicitly. The purpose of the study and the fundamental assumptions made must be clearly stated. In order to limit the analysis, the human/system boundaries must also be defined, as well as its intended design, use, operation and layout.

The risk assessment procedure contains two essential elements:

Figure 5: Risk assessment framework

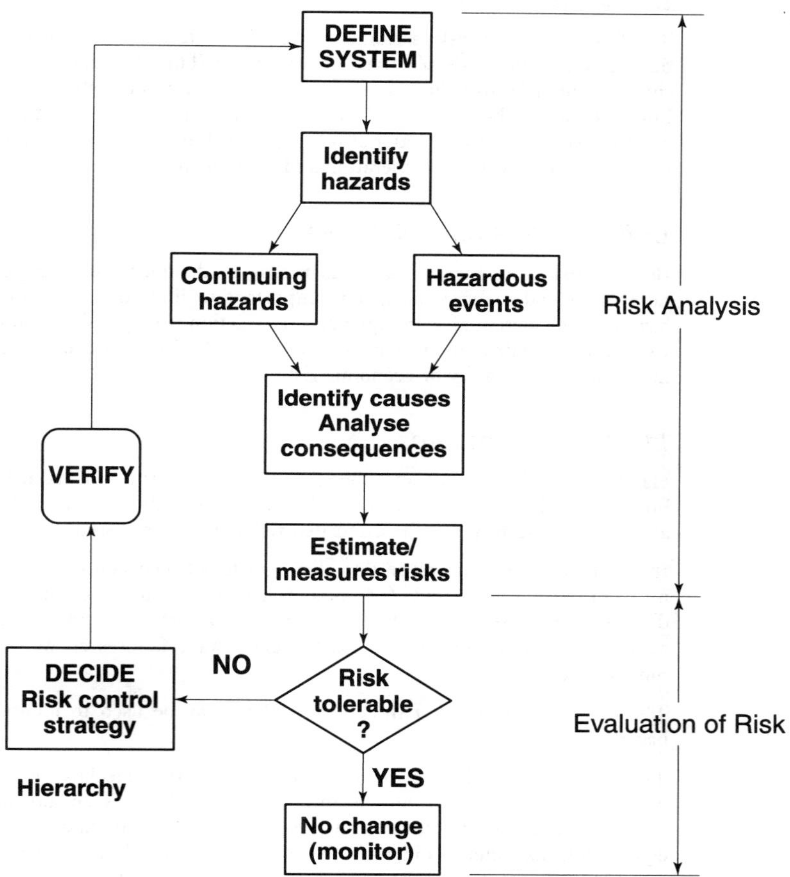

1. *Risk analysis.* Where hazards and hazardous situations are systematically identified and their consequences are analysed. The level of risk associated with each hazardous situation is estimated or measured.

2. *Risk evaluation.* In which a judgement is made as to whether the level of risk is acceptable/tolerable or whether some corrective/preventive actions are needed.

Following any modifications in the plant/process design or operation/maintenance procedures, there is a need to verify that the original hazards/hazardous situations are controlled, and that new hazards are not introduced as a result of these modifications. It is crucial that the risk assessment is updated as a result of design/procedural evolution. The safety management system developed by the employer as a result of risk assessment is also a living system, capable of learning from experience as the people who run it, and should reflect what is actually taking place, and not what should happen.

The main elements of risk assessment, ideally, should include the following:

Define the system/activity

The first step in any risk assessment is to provide terms of reference to the detailed description of the process/activity being studied. This clear description is needed at this stage in order to avoid confusion at a later stage and to define the boundary and interfaces with other processes/activities. This step also involves the identification of the design objectives, material being handled/processed and exposure patterns for individuals involved in operation and maintenance.

Define the study objective

Because risk assessment is used as means to demonstrating compliance with a number of codes/regulations, it is important to limit the study to the range of hazards and potential consequences in relation to specific requirements, e.g. COSHH i.e. control of substances hazardous to health', noise at work, the environment, machinery safety or economic risks.

Identify the hazards

Having defined the scope and objectives, the identification of hazards is the third, but the most important step in any risk assessment study because any hazard omitted at this stage will result in the associated risk not to be assessed.

It is important in this respect to distinguish between continuing hazards (those inherent in the work activity/equipment/machinery/substances under normal conditions), e.g. noise, toxic substances, mechanical hazards, and hazards which can result from failures/error, e.g. hardware/software failures as well as foreseeable human error.

The checklist shown in TABLE 1 at M1025 may be used to identify continuing hazards.

The range of continuing hazards outlined in the checklist is based on the EU-Machinery Directive and is intended to be used as an aid-memoire. This approach to hazard identification will focus attention on parts of the work activity/equipment/machinery to identify the sources of each relevant hazard, who may be exposed to them, e.g. the operator, maintenance, and the task involved, e.g. loading, removing, tool setting, cleaning, adjusting.

Hazards resulting from equipment failure and human error require an open-ended type of analysis, based on brain storming. This approach normally involves the following:

1. A detailed hazard identification to be carried out for all stages of the process life-cycle. This ideally should include the identification of contributory causes for each hazards.
2. Analysis of systems of work and established procedures to identify, who might be exposed to a hazard and when, e.g. operator under normal conditions, maintenance/tool fitter/cleaner, etc.

Structured and systematic techniques which could assist in the identification of hazardous events identified by BS EN 1050: 1997 include the following:

- *Hazard and Operability Study* ('HAZOP') – a qualitative technique to identify hazards resulting from hardware failures and human error. This involves potential causes and consequences

- *Failure Modes and Effects Analysis* ('FM&EA') – an inductive technique to identify and analyse hardware failures. This technique can be used to quantify risks.
- *Task Analysis* – an inductive technique to identify the opportunity for human error. An approach, based on task-based hazards identification and analysis is introduced in this document.

Machinery hazards identification

M1025 A person may be injured at machinery as a result of one or more of the following:

(*a*) contact or entanglement with the machinery;

(*b*) crushing between a moving part of the machine and fixed structures;

(*c*) being struck by ejected parts of the machinery;

(*d*) being struck by material ejected from the machinery.

These are regarded as mechanical hazards. The British Standard 5304: 1988 'Safety of machinery' classified mechanical hazards in the form of type of injury caused by exposure to the hazard.

The main types of mechanical hazards shown in FIGURE 6 are described as:

- **Crushing**: Occurs when part of the body is caught between a moving part of a machine against a fixed object, e.g. underneath scissor lift or between the tools of a press.
- **Shearing**: Parts of the body may be sheared by scissor action caused by parts of the machine, e.g. mechanism of scissor lift or oscillating pendulum.
- **Cutting and severing**: Cutting hazards include contact with circular saws, guillotine knife, rotary knives or moving sheet metal.
- **Entanglement**: Occurs as a result of clothing or hair contact with rotating objects or catching on projections or in gaps, e.g. drills, rotating workpiece or belt fasteners.
- **Drawing-in or trapping**: Occurs when part of the body is caught between two counter-rotating parts, e.g. gears, mixing mills or between belt and pulley or chain and chain wheel.
- **Impact**: Impact hazards are caused by a moving object striking the body without penetrating it. Examples include: being hit by a robot arm or by moving traffic.
- **Stabbing and puncture**: flying objects, swarf or rapid moving parts, e.g. sewing machine.
- **Friction and abrasion**: contact with moving rough or abrasive surfaces, e.g. abrasive wheels.
- **High pressure fluid injection**: sudden release of fluid under pressure can cause tissue damage similar to crushing. Examples include: water jetting, compressed air jets and high pressure hydraulic systems.

Since a hazard is defined as 'a potential to cause injury or damage to health', it is therefore important to include health hazards in addition to the mechanical hazards which are aimed at physical injury. Other hazards considered by BS EN 1050 include the following:

Figure 6: Examples of mechanical hazards

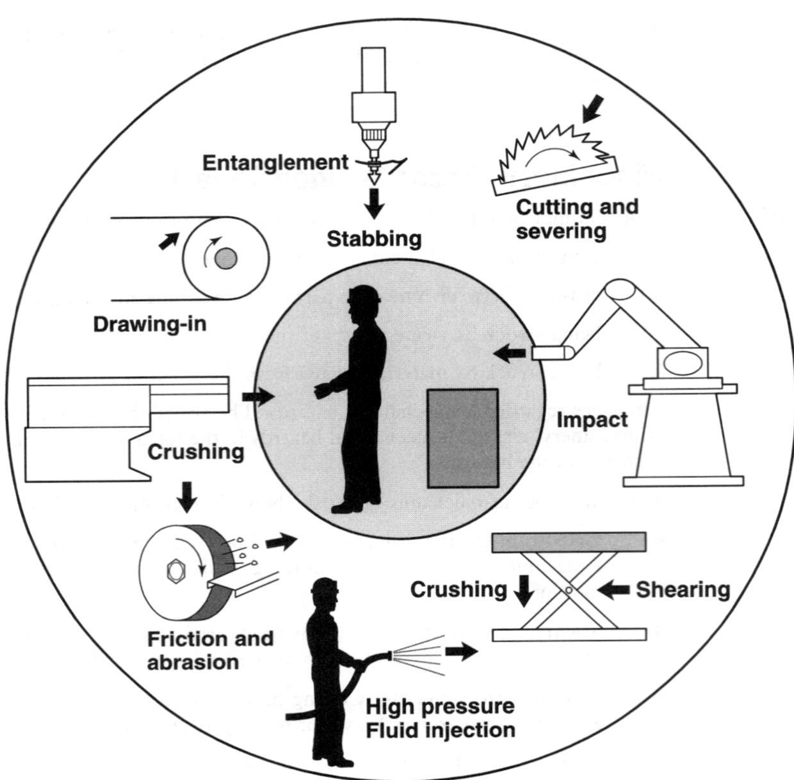

- Electrical hazards – to include direct and indirect contact with live electricity and electro-static phenomena.
- Radiation hazards – to include ionising and non-ionising sources as well as electro-magnetic effect.
- Work environment hazards – to include hot/cold ambient conditions, noise, vibration, humidity and poor lighting.
- Hazardous materials and substances – to include toxic, flammable and explosive substances.
- Work activity hazards – to include highly repetitive actions, stressful posture, lifting/handling heavy items, visual fatigue and poor workplace design.
- The neglect of ergonomic principles.

Table 1: Hazards identification checklist

Type of hazard	*Source*	*Task involved* (Who is exposed and when?)
1. Mechanical hazards 1.1 Crushing 1.2 Shearing 1.3 Cutting/severing 1.4 Entanglement 1.5 Drawing-in/trapping 1.6 Impact 1.7 Stabbing/puncture 1.8 Friction/abrasion 1.9 High pressure fluid injection 1.10 Slips/trips/falls 1.11 Falling objects 1.12 Other mechanical hazards		
2. Electrical hazards 2.1 Direct contact 2.2 Indirect contact 2.3 Electrostatic phenomena 2.4 Short circuit/overload 2.5 Source of ignition 2.6 Other electrical hazards		
3. Radiation hazards 3.1 Lasers 3.2 Electro-magnetic effects 3.3 Ionising/non-ion radiation 3.4 Other radiation hazards		
4. Hazardous substances 4.1 Toxic fluids 4.2 Toxic gas/mist/fumes/dust 4.3 Flammable fluids 4.4 Flammable gas/mist/fumes/dust 4.5 Explosive substances 4.6 Biological substances 4.7 Other hazardous substances		
5. Work activities hazards		

5.1 Highly repetitive actions 5.2 Stressful posture 5.3 Lifting/handling heavy items 5.4 Mental overload/stress 5.5 Visual fatigue 5.6 Poor workplace design 5.7 Other workplace hazards		
6. Work environment hazards 6.1 Localised hot surfaces 6.2 Localised cold surfaces 6.3 Significant noise 6.4 Significant vibration 6.5 Poor lighting 6.6 Hot/cold ambient temperature 6.7 Other work environment hazards		

Options for machinery risk reduction

M1026 The most effective risk control measures are those implemented at the machine/work equipment design stage. It is a legal requirement in the EU that all machinery supplied after January 1995 should carry a 'CE' mark. This demonstrates that the machine complies with all relevant EU machinery directives, which include meeting the ES&HRs. Fundamental to this approach, manufacturers and suppliers need to carry out a risk assessment and demonstrate that all risks are adequately controlled by design, rather than by procedures. The following risk control options reflect the preferred order of priority:

Technical measures to be taken at the design stage

1. To make the machinery inherently safer (design out hazards) as a priority:
   - to eliminate hazards at the design stage;
   - to substitute hazardous substances used by the machine or process by less hazardous ones.
2. To provide protection from the hazards by the machine design, e.g.to avoid the need for access to hazardous parts of the machine.
3. To improve machinery reliability by reducing the chance for fail-to-danger of critical components/systems.
4. To provide protection from the hazards by adequate safeguarding.
5. To reduce ease of access to danger zones (e.g. safeguards or safety devices which allow safe access, but prevent access at other times).
6. To reduce the chance of potential human error:

— by improving the ergonomics of machine control layout to reduce the chance for unintended errors;

— by reducing the chance for violations by making maintenance, cleaning, adjustments, etc. tasks convenient, easy and safe.

Procedural measures (make work tasks safer)

These may be achieved by the following measures:

- Planned preventive maintenance and regular inspection of machines and safeguards/safety devices.
- Safe systems of work (which would minimise the needs for access into the danger zone).
- Permit-to-work procedures (to formalise precautions in the face of a hazard).
- Adequate personal protection equipment ('PPE') and planning for emergencies.

Behavioural measures (develop safer people)

This involves the following:

- Selection and certification of personnel against specific work tasks.
- Training: basic skills; systems and procedures as well as knowledge of hazards.
- Adequate instructions, warnings and supervision.
- Improve safety culture within the business.
- Ask: why should safeguards/safety devices be violated/defeated? *

  * Note that violations are prevented if the perceived inconvenience of defeating a safeguard/safety device exceeds the benefit of so doing.

Types of safeguards and safety devices

M1027 The main types of safeguards and safety devices can be classified as follow:

1. Fixed guards.
2. Fixed guards with adjustable element.
3. Automatic guards.
4. Interlocked guards.
5. Safety devices.

   These are described as:

   5.1 Trip devices, e.g. photoelectric light curtains, pressure sensitive devices.

   5.2 Two-hand control devices.

Machinery safeguards – general considerations

M1028 It is a general requirement that all machinery safeguards and safety devices must have the following characteristics (adapted from BS EN 292: 1991 Part 2 and EH&SR's) outlined below.

Figure 7: Design/selection of safeguards

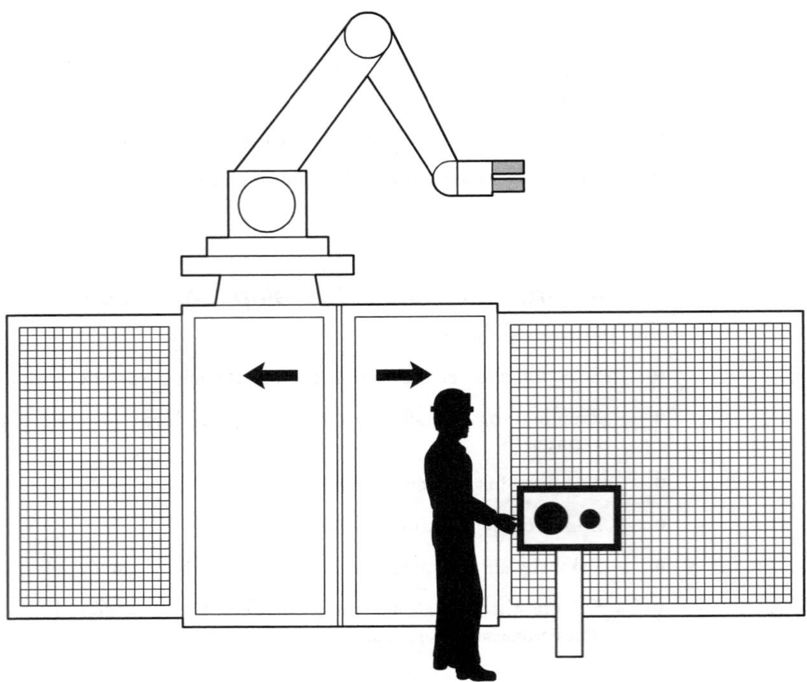

Safeguards and safety devices should:

- Be of robust construction: strength, stiffness and durability to prevent ejected parts of the machine/components or material penetrating the guard.
- Not give rise to additional hazards.
- Not be easy to bypass or render non-operational.
- Be located at an adequate distance from the danger zone.
- Cause minimum obstruction of view for machine operators.
- Enable essential work (e.g. maintenance, cleaning) to be done without safeguard removal.

The two most applicable EU Standards relevant to safeguards are BS EN 953: 1994 (design and construction), and BS EN 294: 1994 (safety distances).

Fixed guards

M1029 BS EN 953: 1998 'Safety of machinery: Guards: General requirements for the design and construction of fixed and movable guards' defines a fixed guard as 'a guard kept in place (i.e. closed)', either permanently or by means of fasteners (screws, nuts) making removal/opening impossible without using tools.

A fixed guard is a guard attached to a machine by a fixing method, which is not linked to, or associated with, the controls, motion, or hazardous condition of a machine. Therefore, a machine can operate normally with the guard removed.

Figure 8: Example of a fixed guard

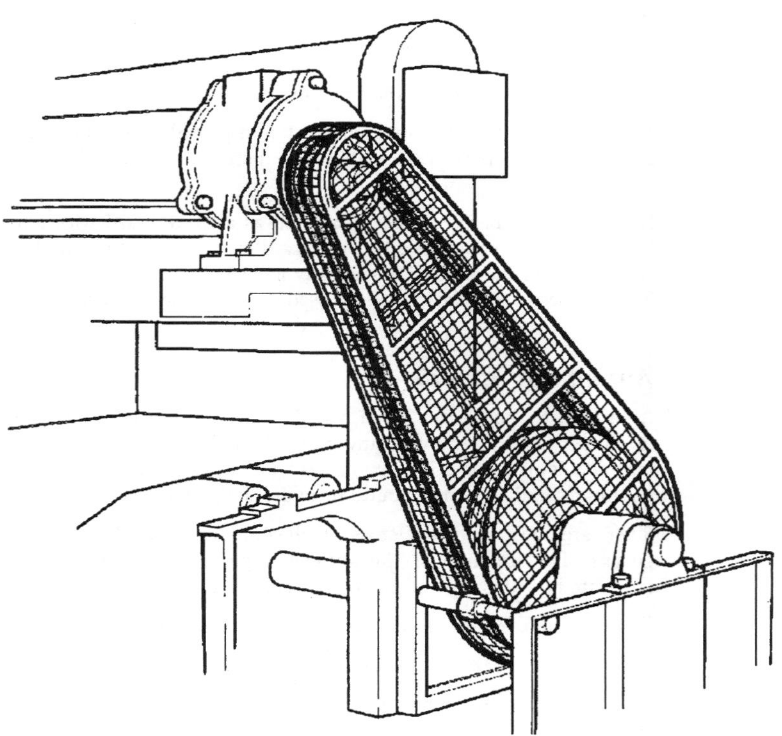

A fixed guard should when fitted be incapable of being displaced casually, so the method of fixing is important. The ideal method of fixing the guard would be of a captive type, which would make unofficial access difficult.

It is clear from experience that if the need for access into hazardous parts of a machine fitted with a fixed guard, there will be a tendency not to replace the guard back onto the machine. Therefore an interlocked guard is more suited in this situation.

Adjustable guards

M1030 Adjustable guards comprise a fixed guard with adjustable elements that the setter or operator has to position to suit the job being worked on. They are widely used for woodworking and tool-room machines. Where adjustable guards are used, the operators should be familiar and trained in how to adjust them so that full protection can be obtained.

FIGURE 9 shows the application of a telescopic guard to a heavy duty-drilling machine. The idea here is that as the drill descends into the workpiece, the fixed guard will always enclose the drill. The approach to selecting the most appropriate safeguard should start by hazard identification. Hazards associated with the use of the drilling machine include the following:

- entanglement with the drill;
- stabbing/puncture by the swarf;
- ejection of broken drill bit;
- stabbing by the drill bit; and
- possible ejection of the work-piece.

It is noted that the telescopic guard should provide protection against most of the above hazards apart from stabbing/puncture by the drill and ejection of the work-piece.

Self-adjusting guards are also classified as a fixed guard with adjustable element. This guard prevents access to the hazard except when the guard is forced open by the passage of the work. They usually incorporate a spring-loaded pivoted element. These types of guards are mainly used on woodworking machinery.

Automatic guards

M1031 An automatic guard is a guard, which is moved into position automatically by the machine, thereby removing any part of a person from the hazardous area of the machine. This is sometimes known as 'sweep away' guard and is used on some paper guillotines and large mechanical or hydraulic presses.

There are however many factors which could limit the effectiveness of these types of guards. The example of the automatic guard shown in FIGURE 10 is used to illustrate how this type of guard can cause injury.

Some possible hazards associated with this automatic guard include the following:

- shearing hazard between the moving and fixed parts of the guard;
- impact with the moving part of the guard;
- crushing hazard against fixed structures nearby as well as the ergonomics of the guard, e.g. height of guard in relation to operators and the size of guard gaps.

Interlocking guards

M1032 Interlocking guards are usually movable, e.g. they could be hinged, sliding or removable. These guards are used, where frequent access to hazardous parts of a machine may be needed, and are connected to the machine controls by means of 'position sensors'. These interlocking elements could be electrical, mechanical, magnetic, hydraulic or pneumatic. The position sensors interlock the guard with the

Figure 9: Fixed guard with adjustable element

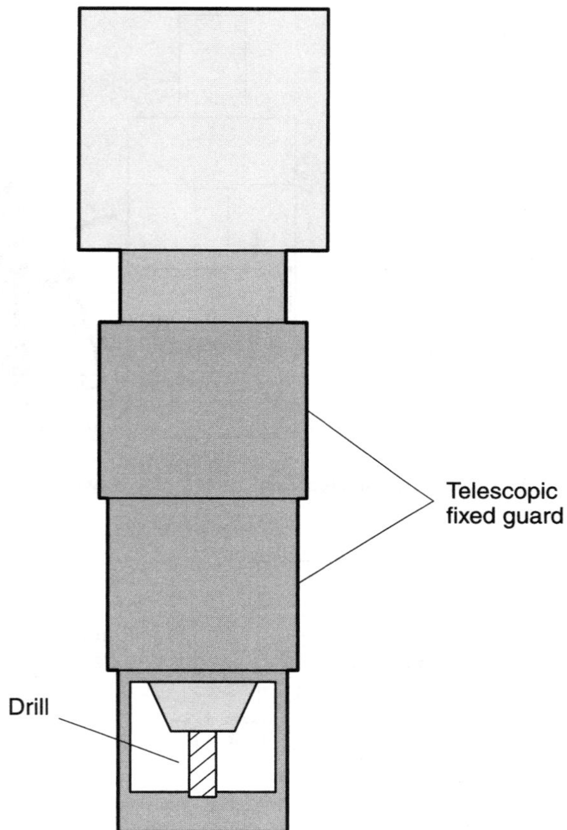

power source of the hazard. When the guard is open, the power is isolated thus allowing safe access into the relevant part of the machine.

The British Standard BS EN 1088: 1997 'Interlocking devices associated with guards' provides detailed guidance on the design and selection of the appropriate type of devices. Risk assessment plays a vital role in the selection of type and level of integrity on interlocking devices.

The integrity of the interlocking mechanism must ensure that the device is reliable, capable of resisting interference, difficult to defeat and the system should not, as far as possible, fail-to-danger. The term 'fail-safe' is no longer used, as it can give the impression that if a component fails in a certain mode this will always result in the machine stopping even when the safeguard is closed. Fail-to-danger of an interlocking device means that the machine can be operated when the guard is open. Therefore to minimise the probability of failure-to-danger is a preferred option.

Interlocking guards are convenient, as they give ready access into the danger zone, while ensuring the safety of the operator. However, some of these guards may be

Figure 10: Example of automatic guard

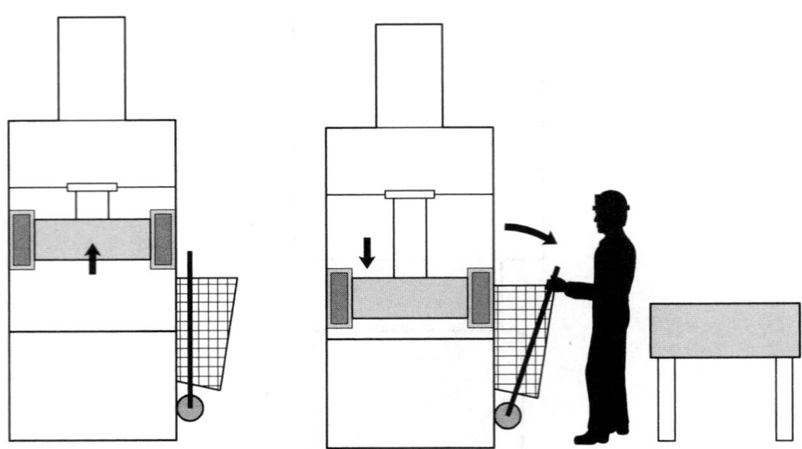

Figure 11: Main types of electrical interlocking switches

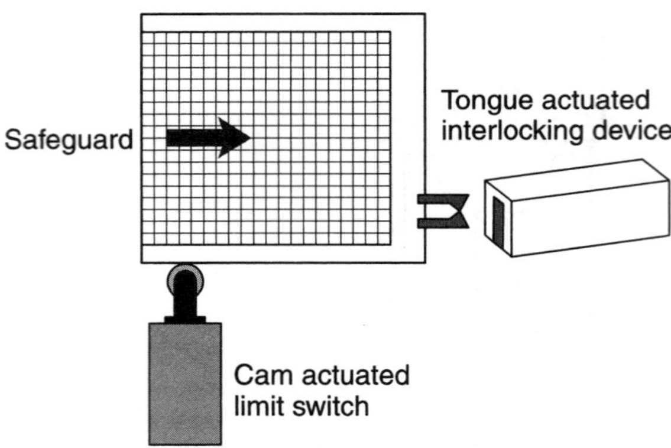

defeated to carry out necessary tasks, e.g. maintenance, setting, fault finding etc. Therefore, additional measures such as locking off or isolation may need to be introduced under a safe system of work. It is however a requirement that the machine designer should consider how these tasks can be safely performed at the design stage.

The choice of interlocking method must be compatible with the level of risk and should take into account the probability of failure of the interlocking element(s), proportion of time a person is exposed to the hazard and the potential consequences (e.g. injury severity and/or ill health).

BS EN 1088: 1997 classifies interlocking guards in two broad categories:

- Guard is interlocked with the source of power.

- Guard is interlocked with respect to the motion itself.

There are clear advantages for interlocking the guard with the moving parts rather than the source of power, as the power might fail-to-danger due to wiring/earth faults or failures of the interlocking switch. The main requirements for the movable interlocking guards are:

(*a*) the hazardous machine functions covered by the guard cannot operate until the guard is closed;

(*b*) if the guard is opened while hazardous machine functions are operating, a stop instruction is given;

(*c*) when the guard is closed, the hazardous machine functions covered by the guard can operate, but the closure of the guard does not by itself initiate their operation.

Interlocking guards may be fitted with a locking device so that the guard remains closed and locked until any risk of injury from the hazardous machine functions has passed.

Types of interlocking devices

M1033 Some common types of interlocking elements include the following:

- *Cam-operated limit switch interlocks*: are one of most popular types of interlocks used generally as they are versatile, effective and easy to install and use. Another type of mechanically actuated device is the 'tongue' actuator, which opens or closes the contacts of a switch. Both types of interlocks are shown in FIGURE 14.
- *Trapped key interlocks*: a master key, which controls the power supply through a switch at the master key box, has to be turned OFF before the keys for individual guards can be released. The master switch cannot be turned to ON until all the individual keys are replaced in the master box. This type of interlock is commonly used in electrical isolation.
- *Captive-key interlocking*: involves a combination of an electrical switch and a mechanical lock in a single assembly.
- *Direct manual switch or valve interlocks*: where a switch or valve controlling power source cannot be operated until a guard is closed, and the guard cannot be opened at any time the switch is in run position (position sensors).
- *Mechanical interlocks*: provide a direct linkage from guard to power or transmission control. In other words the switch cannot be reached when the guard is open.
- *Magnetic switches* with the actuating 'coded' magnet attached to the guard: have the disadvantage that they can be defeated easily. However, types using a shaped magnet have been used with removal guards on screw conveyors. Normal reed switches are not acceptable unless they incorporate special current limiting features. Reed switches are often encapsulated and have application in flammable atmospheres. A similar type of switch relies on inductive circuits.
- *Electro-mechanical device*: can provide a time delay where the first movement of the bolt trips the machine, but the bolt has to be unscrewed a considerable distance before the guard is released. Time delay arrangements are necessary when the machine being guarded has a long run down time. A solenoid-operated bolt can also be used in conjunction with a time delay circuit.

- *Mechanical restraints* (*scotches*): are used as a back-up to other forms of interlocks on certain types of presses to protect the operator against crushing injury between platens.

Trapped key interlocking (Power isolation)

M1034 The electrical interlocking switches interrupt the power source of the hazard by switching of a circuit, which controls the power-switching device. This is known as control interlocking.

Power interlocking on the other hand involves the direct switching of the power supply to the hazard, e.g. isolation of power. The most practical means of power interlocking is a trapped key system.

An example of such a system is illustrated in FIGURE 12. The power supply isolation switch is operated by key 'A' which is trapped in position while the isolation switch is in the ON position. When the key is turned and released, the isolation switch contacts are locked open thus isolating the power supply. The machine safeguard door in the meantime is locked closed. To open the guard, the isolation key 'A' is inserted into the guard-locking unit and turned to release the door. The key is then trapped in position and cannot be removed until the guard is closed and locked again.

One of the drawbacks of this system is that unknown to others, a person may still be inside the machine while another person decides to start-up the machine by closing and locking the guard. To overcome this foreseeable problem, the use of a personal key 'B' which is also trapped in the guard locking unit and is released when the guard is open. Key 'A' cannot be released from the guard unless key 'B' is inserted and turned into the guard locking unit. Key 'B' can also be used for other tasks, e.g. inch mode for presses or programming mode for robotics.

Figure 12: Trapped key system for power interlocking

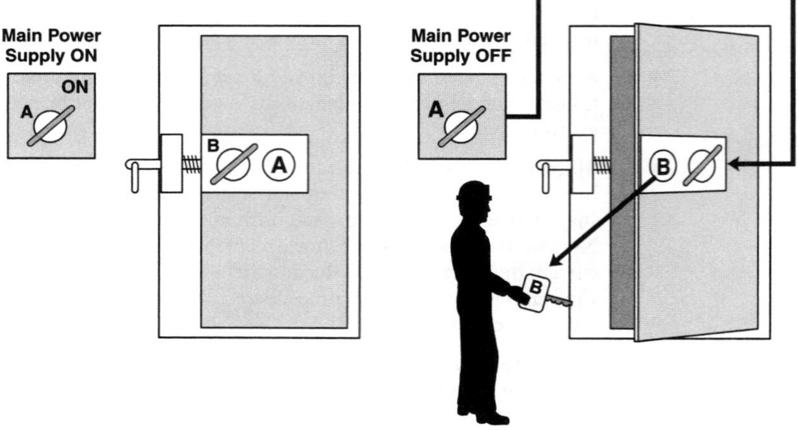

This type of system is reliable and does not require electrical wiring to the guard. The main disadvantage is that because if requires the exchange of the keys every time, it may not be suitable if access is frequently required. Strict control of the keys is essential.

Mechanical interlocking

M1035

The system shown in FIGURE 13 is known as guard inhibited interlocking. This is

Figure 13: Example of mechanical interlocking system

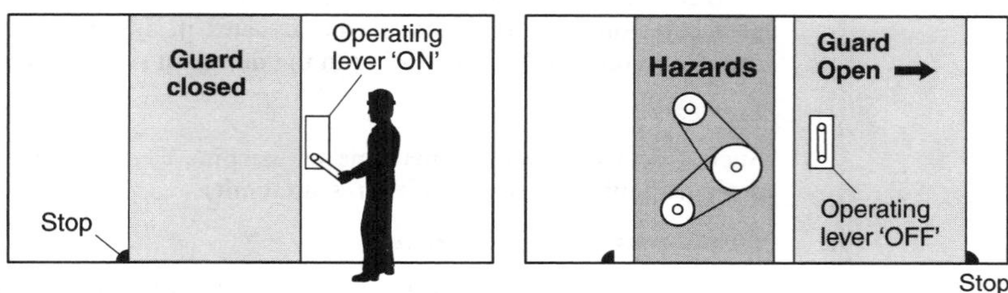

achieved in the above example by a sliding guard, when open the guard retains the lever of a switch or a control valve in the power OFF position. When the guard has been closed, the lever or switch can be moved to the ON position, thus inhibiting the guard from being open.

This system however is easy to defeat by removing one of the door-stops, or even by removing the guard.

Other forms of mechanical interlocks include mechanical restraints (scotches) to prevent gravity fall of vertical machinery, e.g. scissors lifts and presses.

Safety devices

M1036 A safety device is a protective appliance, other than a guard, which eliminates or reduces risk, alone or associated with a guard. There are many forms of safety device available.

(*a*) *Trip device*

A trip device is one which causes a machine or machine elements to stop (or ensures an otherwise safe condition) when a person or a part of his body goes beyond a safe limit (EN 292–1: 1991). Trip devices take a number of forms – for example, mechanical, electro-sensitive safety systems and pressure-sensitive mat systems.

Trip devices may be:

(i) mechanically actuated – e.g. trip wires, telescopic probes, pressure-sensitive devices; or

(ii) non-mechanically actuated – e.g. photo-electric devices, or devices using capacitive or ultrasonic means to achieve detection.

(*b*) *Enabling (control) device*

This is an additional manually operated control device used in conjunction with a start control and which, when continuously actuated, allows a machine to function.

(*c*) *Hold-to-run control device*

This device initiates and maintains the operation of machine elements for only as long as the manual control (actuator) is actuated. The manual control returns automatically to the stop position when released.

(*d*) *Two-hand control device*

A hold-to-run control device which requires at least simultaneous actuation by the use of both hands in order to initiate and to maintain, whilst hazardous condition exists, any operation of a machine thus affording a measure of protection only for the person who actuates it. EN 574: 1996 lays down specific recommendations relating to the design of such devices.

(*e*) *Limiting device*

Such a device prevents a machine or machine elements from exceeding a designed limit (e.g. space limit, pressure limit).

(*f*) *Limited movement control device*

The actuation of this sort of device permits only a limited amount of travel of a machine element, thus minimising risk as much as possible; further movement is precluded until there is a subsequent and separate actuation of the control.

(*g*) *Deterring/impeding device*

This comprises any physical obstacle which, without totally preventing access to a danger zone, reduces the probability of access to this zone by preventing free access.

Managing Health and Safety

Introduction

M2001 There are many reasons why employers need to manage health and safety well. These are usually grouped under three main headings:

- *Humanitarian*

  In a typical year in the UK around 300 people are killed in work-related accidents, 30,000 people suffer major injuries (fractured limbs, amputations etc), over 2,000 people die as a result of occupational diseases and at least 10,000 other deaths are partly attributable to occupational ill health. Employers would obviously not wish to be involved in these accidents and cases of ill health. The psychological effects of individual incidents on employers, employees and others are sometimes quite traumatic.

- *Legal*

  Society has been increasingly unwilling to accept such a toll of accidents and ill health. In many respects the UK has led the world in health and safety law and now European directives result in a steady flow of health and safety legislation. Many regulations (particularly the *Management of Health and Safety at Work Regulations1999* (*SI 1999 No 3242*) – see M2005 below), specifically require the application of appropriate management techniques.

- *Financial*

  Accidents and ill-health can have a direct financial impact on a business through fines and compensation claims. However, there may also be significant indirect costs associated with the unavailability of injured employees or damaged equipment, or due to the impact of legal sanctions such as prohibition notices. Loss of reputation following serious accidents has had a significant adverse impact on a number of companies, particularly in the rail sector, and supply-chain pressure can result in those whose health and safety standards are found lacking, finding it increasingly difficult to gain business.

  Insurance will only protect employers against a limited range of direct and indirect costs. In 1993 the HSE published the results of a detailed study of 'The costs of accidents at work'. This was based on five individual case studies and demonstrated that the ratio of 'insured' costs to 'uninsured' costs could be as high as 1:36. The ways in which such costs may be incurred are described in FIGURE 1.

(Expanded version of diagram available in HSE booklet HSG 96.)

Legal requirements

M2002 Almost all recent health and safety legislation contains provisions relating to effective management, often requiring the application of risk assessment and control techniques. Set out below are the requirements of *sections 2* and *3* of the *Health and*

Figure 1: The costs of accidents

INSURED

	DIRECT	INDIRECT
INSURED	• Employers' liability claims* • Public liability claims* • Major damage* • External legal costs*	• Major business interruption* • Product liability*
UNINSURED	• Fines* • Sick Pay* • Minor damage • Lost or damaged product • Clear-up costs	• Minor business interruption • Prohibition notice interruption • Customer dissatisfaction (product delay) • First aid and medical attention • Investigation time and internal administration • Diversion of attention/spectating time • Loss of key staff • Hiring/training replacement staff • Hiring replacement equipment • Lowered employee morale • Damage to external image

UNINSURED

* Only these costs can be quantified without considerable effort

Safety at Work etc. Act 1974 ('*HSWA 1974*'), together with those contained in the *Management of Health and Safety at Work Regulations 1999* (*SI 1999 No 3242*) (the '*Management Regulations*').

Duties of employers to their employees

M2003 *Section 2* of *HSWA 1974* sets out a number of general duties that employers hold towards their employees. *Subsection (1)* of *section 2* contains an all-embracing requirement:

> 'It shall be the duty of every employer to ensure, so far as is reasonably practicable, the health, safety and welfare of all his employees'.

Subsection (2) of *section 2* contains a number of more specific requirements (again qualified by the phrase 'so far as is reasonably practicable') relating to:

- provision and maintenance of plant and systems of work;
- use, handling, storage and transport of articles and substances;
- provision of information, instruction, training and supervision;
- places of work and means of access and egress;
- the working environment, facilities and welfare arrangements.

It is inconceivable that an employer could achieve compliance with these extremely wide-ranging duties without the application of effective management principles.

In addition, *subsection (3)* of *section 2* requires employers to prepare and revise a written statement of their policy with respect to the health and safety at work of his employees and the organisation and arrangements for carrying out that policy. This duty is examined in greater detail at M2025 below and in the chapter entitled STATEMENTS OF HEALTH AND SAFETY POLICY.

Duties of employers to others

M2004 *Section 3(1)* of *HSWA 1974* places a general duty on employers in respect of persons other than their employees:

> 'It shall be the duty of every employer to conduct his undertaking in such a way as to ensure, so far as is reasonably practicable, that persons not in his employment who may be affected thereby are not exposed to risks to their health or safety'.

Persons protected by this requirement include contractors (and their employees), visitors, customers, members of the emergency services, neighbours, passers-by and other members of the general public – including trespassers, although the 'so far as is reasonably practicable' qualification is of particular relevance to them. (*Subsection (2)* of *section 3* places a similar duty on the self-employed.)

Once again, effective health and safety management provides the only way of complying with this duty.

The Management Regulations

M2005 The *Management Regulations* (*SI 1999 No 3242*) came into operation on 29 December 1999. They modified the 1992 Management Regulations which in turn had developed many of the principles already established by the *HSWA* in 1974. The Regulations make it quite clear that health and safety must be managed systematically, like any other aspect of an organisation's affairs. Some of the main requirements of the Regulations relate to:

- *Risk assessments* – to identify risks and related precautions (*Regulation 3*).
- *Management systems* – to ensure that precautions are implemented (*Regulation 5*).
- *A competent source of health and safety advice* being appointed (*Regulation 7*).
- *Emergency procedures* being developed (*Regulation 8*).

HSE booklet L21 contains the Regulations in full together with an Approved Code of Practice (ACOP) and related guidance. The 1999 Regulations incorporated previous amendments to the 1992 Regulations, relating to young persons, new or expectant mothers and fire risks.

The requirements of a number of the regulations (e.g. risk assessment) are dealt with more fully elsewhere in this publication (see RISK ASSESSMENT) and therefore receive only limited attention here.

Risk assessment (Regulation 3)

M2006 Employers (and the self-employed) must make 'suitable and sufficient' assessments of the risks to their employees and others who may be affected by their work activities, in order to identify the measures necessary to comply with the law. The 'significant findings' of the assessment must be recorded by those with five or more employees.

Given the widely drawn nature of the requirements of *HSWA 1974* and other legislation, all risks must be included in this exercise. There is no need to repeat other 'assessments' required under more specific regulations, provided these are still valid. An assessment of fire risks and precautions must be carried out, whether or not the workplace already has a fire certificate under the *Fire Precautions Act 1971*. Both the risks and the related precautions must be identified.

The *Management Regulations* (*SI 1999 No 3242*) require special consideration to be given to risks to young persons (under 18s) (*Reg 3*(*4*) and (*5*)) and to new and expectant mothers and their babies by virtue of *Regulation 16*.

Principles of prevention to be applied (Regulation 4)

M2007 Preventive and protective measures must be implemented on the basis of principles specified in *Schedule 1* to the *Management Regulations* (*SI 1999 No 3242*). These include:

- avoiding risks;
- evaluating risks which cannot be avoided;
- combating risks at source;
- replacing the dangerous by the non-dangerous or less dangerous;
- giving collective protective measures priority over individual measures, i.e. following good principles of health and safety management.

Health and safety arrangements (Regulation 5)

M2008 Employers must make and give effect to appropriate 'arrangements' for the effective planning, organisation, control, monitoring and review of preventive and protective measures. Records should be kept of these arrangements by those with 5 or more employees.

While *Regulation 3* of the *Management Regulations* (*SI 1999 No 3242*) requires precautions to be identified, *Regulation 5* requires them to be put into practice effectively, utilising a systematic management cycle (see FIGURE 2).

Figure 2

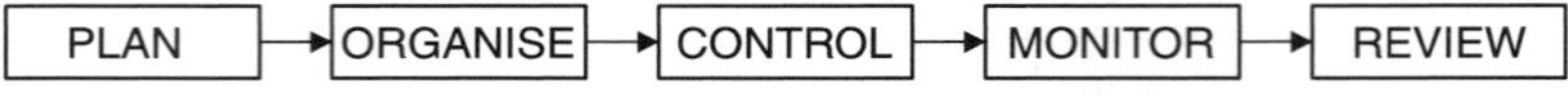

The risk assessment cannot remain a meaningless piece of paper – *Regulation 5* of the *Management Regulations* (*SI 1999 No 3242*) requires it to be put into effect. Much of the remainder of this chapter is concerned with the practical application of these management principles.

Health surveillance (Regulation 6)

M2009 Employers must ensure that employees are provided with appropriate health surveillance where this is identified by the risk assessment as being necessary. In most cases surveillance should already have been identified as being necessary to comply with other specific regulations (e.g. COSHH, asbestos). However, there may be other types of health risk not adequately covered by other legislation, e.g. colour blindness for train drivers or electricians; risks of vibration white finger.

Health and safety assistance (Regulation 7)

M2010 Employers must appoint one or more 'competent' persons to assist them in complying with the law. The numbers of such persons and the time and resources available to them must be adequate for the size of and the risks present in the undertaking.

Competence in relation to this requirement might involve:

- knowledge and understanding of the work activities involved;
- knowledge of the principles of risk assessment and prevention;
- understanding of relevant legislation and current best practice;
- the capability to apply this in a practical environment;
- an awareness of one's own limitations (in experience or knowledge);
- the willingness and ability to learn.

Although formal health and safety qualifications give a good guide to competence, they are not an automatic requirement. Small employers are allowed to appoint themselves under this Regulation, providing they are competent. The *Management Regulations* (*SI 1999 No 3242*) state (in *Regulation 7(8)*) that there is a preference for the competent person to be an employee rather than someone from outside, e.g. a consultant.

Procedures for serious and imminent danger and for danger areas (Regulation 8)

M2011 Procedures must be established 'to be followed in the event of serious and imminent danger to persons at work'. These procedures must be implemented where appropriate. Access must also be restricted to areas which are known to be dangerous.

Most employers are familiar with the process of preparing a fire evacuation procedure. Similar procedures must be established for dealing with other types of emergency which may have been identified through the process of risk assessment as being foreseeable. Such emergencies might include bomb threats, leaks or discharges of hazardous substances, runaway processes, power failure, violence, animal escape, severe weather etc. Competent persons must be appointed to implement evacuation where necessary.

Areas identified as being dangerous, e.g. because of the presence of electrical risks, chemical risks, potentially dangerous animals (or humans), must have access to them restricted to those who have received adequate instruction on the risks present and the precautions to be taken.

Contacts with external services (Regulation 9)

M2012 Necessary contacts must be arranged with emergency services (and others if appropriate) regarding first aid, emergency medical care and rescue work, particularly in respect of the types of emergencies referred to in M2011 above.

Information for employees (Regulation 10)

M2013 Employers must provide employees with comprehensible and relevant information on:

- risks to their health and safety identified by risk assessment;
- the related preventive and protective measures;
- emergency procedures (and evacuation co-ordinators);
- risks notified to them by other employers in shared workplaces.

Regulation 10(2) of the *Management Regulations* (*SI 1999 No 3242*) requires parents or guardians of children (under the minimum school leaving age) to be provided with information on risks the child may be exposed to and the precautions identified as necessary by the risk assessment. This is of relevance in relation to school work experience programmes and part-time work by schoolchildren.

Co-operation and co-ordination (Regulation 11)

M2014 Where two or more employers share a workplace (whether temporarily or permanently) each employer must:

- co-operate with the other employers where necessary for them to comply with the law;
- take all reasonable steps to co-ordinate their precautions with others;
- take all reasonable steps to inform the others of risks arising out of their work.

(Self-employed persons are also covered by this requirement.)

This type of co-operation and co-ordination was in effect already required under *sections 2* and *3* of *HSWA 1974*. In many cases a main employer may already control the worksite with the other employers assisting. However, in some circumstances it may be necessary to identify an individual or organisation to carry out a co-ordinating role. Even those without employees in the workplace (e.g. landlords) may still be required to co-operate in order to meet their obligations under *section 4* of *HSWA 1974*.

Persons working in host employers' or self-employed persons' undertakings (Regulation 12)

M2015 Host employers must ensure that employers whose employees are working in their undertaking are provided with comprehensible information on:

- risks arising out of or in connection with the host's activities; and
- measures being taken by the host to comply with the law.

Hosts have an additional duty to ensure that visiting employees are given instructions and information on the risks present and any emergency or evacuation procedures.

(Under *Regulation 12* of the *Management Regulations* (*SI 1999 No 3242*) the self-employed are treated as both employers and employees).

Over and above the requirements of *Regulation 11* of the *Management Regulations* (*SI 1999 No 3242*) and *HSWA 1974*, hosts are given twin duties of informing visiting employers and of *ensuring* that visiting employees are instructed and informed. Whether they do this themselves or require others to do it on their behalf will depend on local circumstances. *Regulation 12* and *Regulation 11* inter-relate closely with the requirements of the *Construction, Design and Management Regulations 1994* (*SI 1994 No 3140*) (*CDM*) .

Capabilities and training (Regulation 13)

M2016 Employers must take into account employees' capabilities as regards health and safety in entrusting tasks to them. Employees' intellectual and physical capabilities must be taken into account as well as the demands of the task and the knowledge and experience required to perform it. Adequate health and safety training must be provided to employees:

- on recruitment into the undertaking;
- on exposure to new or increased risks because of:
  - transfer or a change in responsibilities;
  - new or changed work equipment;
  - new technology;
  - new or changed systems of work.

Training must:

- be repeated periodically where appropriate;
- be adapted to take account of new or changed risks;
- take place during working hours.

The requirements on training develop on the base provided by *section 2* of *HSWA 1974*. Many other sets of regulations also have specific training requirements. Even where employees have been trained in the same way their capabilities may differ and *Regulation 13* of the *Management Regulations* (*SI 1999 No 3242*) requires employers to take ability into account. It also identifies the possible need for refresher training.

Employees' duties (Regulation 14)

M2017 Employees must use machinery, equipment, dangerous substances etc in accordance with the training and instructions they have been provided with. They must also inform their employer (or his representative) of situations representing serious and immediate danger and of any shortcomings in the employer's protection arrangements. To a large extent, this requirement duplicates those already contained in *section 7* of *HSWA 1974* and other legislation. A failure to report a dangerous situation is likely to be considered a serious 'omission' under *section 7*.

Temporary workers (Regulation 15)

M2018 *Regulation 15* of the *Management Regulations* (*SI 1999 No 3242*) creates several detailed requirements in relation to the use of fixed-term contract or agency staff. It should always be remembered that temporary workers are still affected by most of the other requirements of these Regulations.

Risk assessment in respect of new or expectant mothers (Regulations 16, 17 and 18)

M2019 Where there are:

- women workers of childbearing age, and
- the work could involve risk to the mother or baby, e.g. processes, working conditions, physical, biological or chemical agents (see Annexes I and II of EC Directive 92/85/EEC),

the risk assessments made under *Regulation 3(1)* of the *Management Regulations* (*SI 1999 No 3242*) must take account of such risks.

For individual employees the employer must, *if necessary*:

- alter working conditions or hours of work, or *if necessary*,
- suspend the employee from work,

in order to avoid such risks.

Regulation 17 of the *Management Regulations* (*SI 1999 No 3242*) allows doctors (or midwives) to require employers to suspend women from night work where necessary for health and safety reasons. *Regulation 18* of the 1999 Regulations states that employers need not take action under *Regulations 16* and *17* in respect of individual employees unless the employee has notified them in writing of her pregnancy or that she has given birth within the previous 6 months or is breastfeeding. (The employer may request confirmation of pregnancy through a certificate from a doctor or midwife.) Issues relating to new and expectant mothers are dealt with in more detail in the chapter entitled VULNERABLE PERSONS.

Protection of young persons (Regulation 19)

M2020 Employers must ensure young persons are protected from risks which are a consequence of their:

- lack of experience;
- absence of awareness of risks;
- lack of maturity,

e.g. by appropriate training, supervision, restrictions (see also *Regulation 3* of the *Management Regulations* (*SI 1999 No 3242*)).

Young persons may not be employed for work:

- beyond their physical or psychological capacity;
- involving harmful exposure to toxic, carcinogenic, genetically damaging agents etc;
- involving harmful exposure to radiation;
- involving risks they cannot recognise or avoid (due to insufficient attention, lack of experience or training);
- in which there are health risks from extreme cold or heat, noise or vibration.

Such work may be carried out if it is necessary for the training of young persons who are not children, providing they will be supervised by a competent person and risks are reduced to the lowest level reasonably practicable.

This topic is dealt with more fully in the chapter VULNERABLE PERSONS.

Other requirements

M2021 The remaining requirements under the *Management Regulations 1999* (*SI 1999 No 3242*) cover the following:

- *Regulation 20 – Exemption certificates*

  (For the armed forces in the interests of national security)
- *Regulation 21 – Provisions as to liability*

  (Of employers in relation to criminal proceedings)
- *Regulation 22 – Restriction of civil liability for breach of statutory duty*

  (A breach of a duty imposed on an employer by these Regulations does not confer a right of action in civil proceedings insofar as that duty applies for the protection of persons not in his employment. The previous exclusion of right of action under the Regulations by employees was removed in 2003.)
- *Regulation 23 – Extension outside Great Britain*

  (The Regulations apply to various offshore activities)
- *Regulations 23–27*

  (Contain minor amendments in respect of first aid, offshore installations and pipeline works, mines and construction)
- *Regulations 28–30*

  (Deal with administrative matters, revocations and transitional arrangements)

Accident causation

M2022 The domino model of accident or loss causation has been used by many to demonstrate how accidents result from failures in health and safety management. The model is based on five dominos standing on their edges in a line:

- The *fifth and final domino* in the line represents all the *potential losses* which can result from an accident or incident – personal injury, damage to equipment, materials, premises, disruption of business etc;
- The *fourth domino* in the line represents the specific *accident or incident* which occurred – for example a failure of an item of lifting equipment;
- This failure could have a number of *immediate causes* which are represented by the *third domino* – overloading or misuse of the lifting equipment, an undetected defect etc;
- Behind these immediate causes are a variety of *basic or underlying causes*, represented by the *second domino* – an absence of training, a lack of supervision, failure to conduct regular inspections and thorough examinations of the lifting equipment;
- These in turn are indicative of a *lack of management control* which is represented by the *first domino* – in a well-managed workplace appropriate training is provided, there is effective supervision, and inspection and thorough examination procedures are in place.

If there is effective management control then all the dominos stay in position. However, if management control breaks down then the other dominos will topple eventually causing an accident or incident and the resultant losses.

Health and safety management systems

M2023 The vast majority of occupational safety and health ('OSH') management systems (and environmental management systems) follow the guidance given in HSE's publication 'Successful health and safety management', (HSG65) and embellished in BS 8800:1996, Guide to Occupational Health and Safety Management Systems, which outlines a system based on:

- establishing a policy with targets and goals;
- organising to implement it,
- setting forth practical plans to achieve the targets/goals;
- measuring performance against targets/goals; and
- reviewing performance.

The whole process is overlaid by auditing. This outline may be remembered via the use of the mnemonic: POPIMAR:

Policy

Organising

Planning

Implementing

Measuring/monitoring

Auditing

Reviewing

These key elements are linked in the form of a continual improvement loop (see FIGURE 3).

The more times organisations go round the loop, the better will be their OSH management performance.

Within BS 8800, two alternative approaches are suggested:

- the HSG65 approach;
- the ISO 14001 approach (BSI BS EN ISO 14001: 1996, Environmental Management Systems – Specification with Guidance for Use).

BS 8800 itself is merely a guide; it is *not* a certifiable standard; whereas ISO 14001 *is* a certifiable standard. To date, there is no equivalent in health and safety to the ISO 9000 standard for quality systems or the ISO 14001 standard for environmental management systems. OHSAS 18001 is not a recognised certifiable ISO standard.

BS 8800 builds on the HSG65 approach (outlined above) but starts off the loop with an initial status review in order to establish 'Where are we now?' BS 8800 also includes an alternative approach aimed at those organisations wishing to base their OSH management system not on HSG65 but on BS EN ISO 14001, the environmental management systems standards which again incorporates an initial status review.

A review checklist

M2024 In order to undertake an initial status review, use may be made of the following checklist:

Figure 3: Continual improvement loop

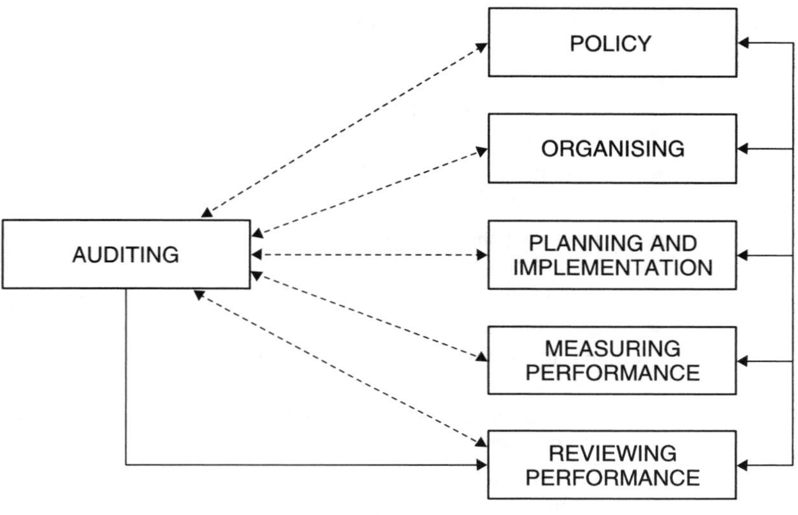

- Does a current OSH policy exist for the organisation/location?
- Is the policy up to date (i.e. not more the three years old)?
- Has the policy been signed (and dated) by a director/senior manager who has (site) responsibility for OSH?
- Does the policy recognise that OSH is an integral and critical part of the business performance?
- Does the policy commit the organisation/location to achieve a high level of OSH performance with legal compliance seen as a minimum standard?
- Does this commitment include the concept of continual improvement?
- Does the policy state that adequate and appropriate resources, i.e. time, money, people, will be provided to ensure effective policy implementation?
- Does the policy allow for the setting and publishing of OSH objectives for the organisation/location and/or for individual directors, managers and supervisors?
- Does the policy clearly place the prime responsibility for the management of OSH on to line management, from the most senior executive to first line supervision?
- Has the policy been effectively brought to the attention of *all* employees/agency staff/temporary employees?
- Are copies of the policy on display throughout the organisation/location?
- Is the policy understood, implemented and maintained at all levels within the organisation/location via the use of suitable arrangements?

- Does the policy ensure that employee involvement and consultation takes place in order to gain their commitment to it and its implementation?
- Does the policy require that it gets periodically reviewed (at least every three years) in order to ensure that management and compliance audit systems are in place?
- Does the policy require that all employees at all levels, including agency staff and temporary employees, receive appropriate training to ensure that they are competent to carry out their duties?
- Does the policy contain a section dealing with the organisational framework – people and their duties – so as to facilitate effective implementation?
- Does the policy contain a section dealing with the need for risk assessments to be undertaken – both general and specific – their significant findings acted upon, and an assessment record-keeping system maintained?
- Does the policy contain a section dealing with the arrangements – systems and procedures – by which the policy will be implemented on a day-to-day basis?
- Does the policy contain a section dealing with the monitoring/measurement of OSH performance?
- Does the policy contain a section dealing with planning for and reviewing the organisation's policy implementation and overall OSH performance?

  (N.B. Again this may/should be in the arrangements section.)

Policy

Requirements under HSWA 1974

M2025 *Section 2(3)* of *HSWA 1974* states:

> 'Except in such cases as may be prescribed, it shall be the duty of every employer to prepare and as often as may be appropriate revise a written statement of his general policy with respect to the health and safety at work of his employees and the organisation and arrangements for the time being in force for carrying out that policy, and to bring the statement and any revision of it to the notice of his employees.'

The *Employers' Health and Safety Policy Statements (Exception) Regulations 1975 (SI 1975 No 1584)* except from these provisions any employer who carried on an undertaking in which for the time being he employs less than five employees. In this respect regard is to be given only to employees present on the premises at the same time (*Osborne v Bill Taylor of Huyton Ltd [1982] IRLR 17*).

Contents of policy statements

M2026 The policy is the first stage in the POPIMAR management system and the statement must reflect that. Reference to its contents was made in the checklist provided in M2024 above. Some contents will be common to all organisations, whatever their size or work activity. Typical of these are the following:

- A commitment to achieving high standards of health and safety (not only in respect of employees but also others who may be affected by the organisation's activities).

- Reference to the importance of people to the organisation and the organisation's moral obligations.
- A statement that compliance with legal obligations must be achieved (although these should be regarded as a minimum standard).
- Effective management of health and safety must be a key management objective (it must continue right through the line of management responsibility).
- Reference to the provision of resources necessary to implement the policy.
- A commitment to consulting employees on health and safety matters.

Where the organisation is fully committed to the type of management system represented by POPIMAR this should be reflected in the policy statement by:

- A reference to achieving continuous improvement in health and safety standards.
- Reference to the setting of health and safety objectives and the allocation of responsibility for their implementation (whether on an annual basis or otherwise).

The statement must be signed by an appropriate director or senior manager to reflect a high level commitment to the policy and must be dated. Review of the policy as a whole, where appropriate, is a legal requirement, as well as a key component of the POPIMAR system. An out of date policy statement signed by someone who has long since left the organisation is not indicative of a real commitment to effective health and safety management.

Organisation

M2027 The HSE booklet 'Successful health and safety management' (HSG65) refers to the need to establish a 'health and safety culture' which it defines as 'the product of individual and group values, attitudes, perceptions, competencies and patterns of behaviour that determine the commitment to, and the style and proficiency of, an organisation's health and safety management.' Put in other words it is 'the way we do things here'. In order to achieve such a culture, activities are necessary on four fronts (often called the four Cs):

- establishing and maintaining *control* in respect of health and safety matters;
- securing *co-operation* between individuals and groups;
- achieving effective *communication*;
- ensuring the *competence* of individuals to make their contribution;

In effect the four Cs encompass both the organisation and arrangements which are required to implement the health and safety policy statement.

Control

M2028 Management must take responsibility and provide clear direction in respect of all matters relating to health and safety. Key elements of control include:

- *Clear definitions of responsibilities*

  Responsibilities must be allocated to all levels of line management as well as persons with specialist health and safety roles. These responsibilities should be set out in the health and safety policy (see the chapter STATEMENTS OF HEALTH AND SAFETY POLICY) and also included in individuals' job descriptions.

- *Nomination of a senior person to control and monitor policy implementation*

  This should ensure that health and safety is seen to be important to the organisation. The senior figure must be involved in the development of plans and particularly in monitoring their implementation. They may also be involved in formal means of consultation with the workforce such as health and safety committees.

- *Establishment of performance standards*

  The old adage of 'what gets measured gets done' applies just as much to health and safety as to other aspects of management. Performance standards must be incorporated into health and safety plans and also into health and safety related procedures and arrangements. Standards must specify:

  — what needs to be done;

  — who is responsible for carrying it out;

  — when (or how often) the task must be carried out;

  — what is the expected result,

  For example, an investigation must be made into all accidents requiring first aid treatment. This must be carried out by the injured employee's supervisor and a report submitted on the company report form within 24 hours.

- *Adequate levels of supervision*

  The precise level of supervision will depend on the risks involved in the work and the competence of those carrying it out. Supervisors are key players in informing, instructing and, to an extent, training the members of their teams. They also act as an important point of reference on health and safety matters, particularly on whether it is safe to continue with a task.

Co-operation

M2029 Health and safety must become everyone's business – there must be real commitment to health and safety throughout the organisation. Commitment comes where there is a genuine consultation and involvement. This can partially be achieved through formal means of consultation, such as safety representatives and health and safety committees. Hazard report books and suggestion schemes can also play their part. However, much can also be achieved informally by line managers and health and safety specialists by simply 'walking the job' and talking to those at work about the health and safety aspects of their job.

Communication

M2030 Good communication is a key element of all aspects of running any organisation. Employees need to be made aware of what is being done to improve health and safety standards (and why), and also of their own role in achieving this improvement. Information needs to move in various directions.

- *Incoming information*

  The organisation must keep abreast of developments in the outside world, e.g:

— potential and actual changes in legal requirements;

— technical developments in respect of control measures;

— new risks coming to light;

— good practices in health and safety management.

- *Information circulation within the organisation*

  Information will need to be passed down from management levels but there must also be means of passing information up the line and across the organisation. Important information to circulate includes:

  — the health and safety policy document itself;

  — senior management's commitment to implementing it;

  — current plans for achieving improvement;

  — progress reports on previous plans;

  — other performance reports, e.g. accident frequency rates;

  — proposed changes to health and safety procedures and arrangements;

  — employees' views on such proposals;

  — comments and ideas for improvement;

  — details of accidents and incidents, together with lessons learned as a result.

  Information can be communicated in written form (e.g. policy documents, booklets, newsletters, notices, e-mails etc.) but face to face communication is also important. This can include health and safety committee meetings, health and safety discussions or briefings at other meetings, tool box talks. Once again, 'walking the job' by management can be invaluable in achieving two way communication in an informal setting.

- *Outgoing information*

  Health and safety related information may also need to be communicated outside the organisation, for example:

  — formal reports to enforcing authorities (e.g. as required by RIDDOR);

  — information about products supplied or manufactured;

  — information to other organisations involved in the same business (about potential risks or successful control measures);

  — emergency planning information (to the emergency services, local authorities and neighbours who may be involved).

Competence

M2031 All employees need to be capable of carrying out their jobs safely and effectively. What is required to achieve this will vary depending upon the individual's role in the organisation. Important elements include:

- *Recruitment and appointment arrangements*

  Employees must have appropriate levels of intellectual and physical abilities to carry out their jobs. Whilst knowledge and experience can be acquired subsequently, certain levels may be a pre-requisite for some posts.

- *Training*

  Systems should be in place to ensure that staff receive adequate training for their work, particularly following their recruitment, transfer or promotion. Arrangements will also be necessary to train those given particular responsibilities, e.g. first aiders, fire wardens. Other training needs may also be identified as a result of audits, job appraisals, changes in technology, legislative changes etc. Refresher training may also be necessary for some staff.

- *Succession and contingency planning*

  Account must be taken of the need to have sufficient competent staff available at all times in some safety-critical roles. Staff may be absent for short periods, e.g. holidays or sickness, for longer periods due to serious injury or illness, or leave permanently, whether for other jobs or on retirement.

The chapter TRAINING AND COMPETENCE IN OCCUPATIONAL SAFETY AND HEALTH provides more information on this subject.

Planning

M2032 'Failure to plan is planning to fail'.

Effective planning results in an occupational safety and health management system ('OSHMS') which controls risks, reacts to changing demands and sustains a positive OSH culture. Such planning involves designing, developing and installing risk control systems ('RCSs') and workplace precautions ('WPs') commensurate to the risk profile of the organisation. It should also involve the operation, maintenance and improvements to the OSHMS to suit changing needs, hazards and risks.

Planning is the first stage of the PDCA cycle: Plan, Do, Check, Act, which is the cornerstone of the vast majority of management systems.

Plan:	Policies, organisation, hazard identification, risk assessments,change management
Do (Implement):	The implementation of the management plans and processes Worker consultation is a crucial aspect of planning and doing which helps to create a positive culture
Check (Monitor):	Inspections, audits, exposure monitoring, health surveillance, attendance/absence monitoring, accident reporting and investigation, statistical/causal analyses, testing of emergency arrangements/procedures
Act (Audit/Review):	Take control action based on monitoring activities, keep track of developments/corrective measures, continually improve processes and plans

The planning process

M2033 The planning process involves:

- Setting objectives.
- Risk control systems.
- Legal and other requirements.
- OSH management arrangements.
- Emergency plans and procedures.

Setting objectives

This relies on having an annual plan for the organisation with clearly outlined individual and collective 'SMART' OSH targets. SMART refers to the need for the targets to be: specific, measurable, achievable, realistic, trackable.

The annual OSH plan should:

- outline individual and collective action plans for the current year;
- target continual improvement in OSH performance;
- fix accountabilities via SMART targets;
- be linked to job descriptions/performance appraisals;
- be adequately resourced – time, money, people.

Risk control systems

These should ideally evolve from an ongoing review of all general and specific risk assessments in place within the organisation. Indeed, as part of the annual plan, all current risk assessments should be reviewed, updated, and recommunicated to all concerned.

Risk should be reduced to tolerable or acceptable levels via an ongoing process of hazard identification and risk assessment within the working environment. Suitable RCSs should be developed and implemented with the goal of hazard elimination and risk minimisation. As part of the process all significant findings/risk assessments should be recorded in writing.

Legal and other requirements

These need to be incorporated into the annual plan in order to establish and maintain arrangements which ensure that all current and emerging OSH legal and other requirements are complied with and understood throughout the organisation.

Bench-marking within the organisation's sector can ensure that best practice and performance measures are adopted and communicated.

OSH management arrangements

These should reflect both the needs of the business and the overall risk profile and should cover all elements of the POPIMAR model. Specific arrangements should include:

- plans, objectives, personnel, resources;
- operational plans to control risks;
- contingency plans for emergencies;
- planning for organisational arrangements;
- plans covering change management (positive OSH culture);
- plans for interaction with third parties (e.g. contractors);
- planning for performance measurement, audits and reviews;
- implementing corrective actions.

Emergency plans and procedures

These should be in place and reviewed as part of the annual planning processes. Specific serious and imminent dangers should have been flagged up via previous risk assessments in accordance with the *Management Regulations* (*SI 1999 No 3242*), *Reg 8*.

These emergency plans should ensure that controls are in place, are in writing and have been communicated to all concerned in connection with all foreseeable disasters, including fire, flood, structural collapse/subsidence, chemical spillage, transport accident, bomb scare, explosion, plane crash etc.

Implementation

M2034 In order to ensure smooth implementation of the OSH policy and management system, it is imperative to have sufficient written records and documentation as this is the key element in organisational communication and continual improvement.

It is therefore important to assemble and retain OSH knowledge and competence within the organisation. However, in order to ensure efficiency and effectiveness of the OSHMS, documentation should be kept to the minimum and should be tailored to suit organisational needs. The detail should be in proportion to the level and complexity of the associated hazards and risks.

Specifically, the key risk areas requiring implementation of commensurate risk control systems may be grouped as follows:

- selection and training;
- general/specific risk assessments;
- safe systems of work/permit to work systems;
- management of third parties on site;
- workplace/work equipment safety and health;
- fire precautions;
- first aid requirements;
- use of personal protective equipment.

These are generally referred to as 'arrangements' in most organisational OSH policies.

As outlined in the section on Planning above at M2032, this 'doing' section requires the use of action plans and SMART targets in order to ensure that the written down standards and procedures are actually achieved in practice on a continual basis. The phrase 'Mind the gap(s)' springs to mind!

Most organisations split their arrangements into general and specific. General arrangements require implementation throughout the organisation whereas specific will only operate at certain locations or for certain tasks.

Within the implementation section of the OSHMS, the risk control systems relevant to the organisation should be briefly described, together with who is responsible for their implementation and who should be involved in the implementation process.

In some organisations the detail may well be consigned to an OSH manual which is referred to but is not part of the OSH policy.

Monitoring

'What gets measured gets done!

M2035 Monitoring of the OSHMS is vital to ensure continual improvement. Monitoring procedures fall into two distinct categories:

- proactive;
- reactive.

Proactive monitoring takes place before the event (accident/disease) and includes audits, inspections and specific checks.

Reactive monitoring is always after the event and includes accident reporting and investigation, numerical/causal analysis of accidents/diseases, compilation of accident statistics, and adherence to the RIDDOR '95 requirements on accident notification.

- *OSH inspections – proactive monitoring*

  This should involve a competent inspection team going around the workplace on a regular (ideally monthly) basis in order to identify hazards – unsafe acts as well as unsafe conditions – with a view to their rectification.

  Such inspections should be joint – both management and workforce representatives – and should always involve talking to the people at risk in the workplace being inspected. This may help to flag up unsafe practices, near misses and shortcomings in the agreed systems and procedures. In most cases use will be made of a specific inspection checklist which ideally has been compiled by the people undertaking the inspection.

  In all cases there needs to be a post-inspection action plan which clearly indicates who is going to take control action and by when. There should also be full communication of the findings and recommendations from all such inspections.

- *Specific checks – proactive monitoring*

  Such checks in the workplace may involve: housekeeping inspections; checks on adherence to agreed safe systems of work, permit to work systems, site safety rules; utilisation of personal protective equipment; documentary checks – risk assessments, statutory inspection records, maintenance records and training records.

- *Accident reporting and investigation systems – reactive monitoring*

  This should encompass all injury, damage and near-miss accidents and occupational ill-health in order to ensure RIDDOR'95 compliance and an effective accident database with which to plan further accident reduction strategies.

  All reported accidents, including occupational diseases, should be promptly investigated in order to identify causes and to implement commensurate control measures so as to prevent a recurrence. Most organisations will have set up an internal accident reporting and investigation system to capture these accident data. Any investigation culture should be positive, i.e. *not* blame-apportioning or fault-finding.

  The main purposes of accident investigation are therefore to:

  — discover the immediate and underlying causes (note: plural!);

- — prevent recurrences;
- — minimise legal liability;
- — collect sufficient data on which to base future plans;
- — identify accident/illness/absence trends;
- — ensure RIDDOR'95 compliance;
- — maintain records for organisational purposes;
- — continually improve the OSHMS performance.

Audit

M2036 Formal OSH audits should be undertaken at regular intervals, ideally annually, by trained and competent auditors, sometimes external to the organisation.

All audit findings (successes as well as shortcomings) should be highlighted and communicated to all concerned. Any recommendations for improvement should be prioritised and allocated to named individuals for action within agreed, finite time-scales. The progress on action completion should also be followed up and communicated.

The main targets of any audit are to reduce risks, with avoidance being the best strategy, and to ensure continual improvement of the OSHMS.

The audit needs to compare actual OSH performance in the workplace against agreed standards – mind the gaps! – such as:

- legislative compliance;
- best practice (bench-marking);
- national/international standards.

Essentially the OSH audit, as with financial audits, is a deep and critical appraisal of all elements (POPIMAR) of the OSHMS and should be seen as being additional to the routine monitoring and reviews. Ideally, the auditors need to be independent of the area/activity/location being audited and the audit checklist should be tailored to suit the needs/hazards/risks of the organisation.

The audit scope should address key areas such as:

- is the overall OSHMS capable of achieving and maintaining the required, agreed standards?
- is the organisation fulfilling all OSH obligations?
- what are the strengths and weaknesses of the OSHMS?
- is the organisation actually doing and achieving what it claims?
- are the audit results acted upon, communicated and followed up?
- are the audit reports used in management reviews?

Review

M2037 Reviews of the performance of the OSHMS generally fall into two categories:

- periodic status review ('PSR');
- management review.

The main purposes of the PSR are to make judgements on the adequacy of performance and to ensure that the right decisions are made and taken in connection with the nature and timings of the remedial actions.

Specifically, the PSR should consider:

- overall OSHMS performance;
- performance of individual elements;
- audit findings;
- internal and external factors such as changes in organisational structure, production cycles, pending legislation, introduction of new technology;
- anticipated future changes;
- information gathered from proactive and reactive monitoring.

The PSR should be a continual process which includes management and supervisory responses to system shortcomings such as non-implementation of agreed risk control systems, substandard performances, or failure to assess the impact of forward plans and objectives on their part of the organisation. Successes should be highlighted.

The effectiveness of the PSR can be enhanced by clearly establishing responsibilities for implementing the remedial actions identified and by setting deadlines for action completion – by whom? by when? The PSR needs to be auditable, trackable, fully communicated and followed up.

The management review should be undertaken annually by the board/senior management. It should ensure that the OSHMS is suitable, adequate, effective and efficient, and that it satisfies and achieves the aims and objectives as stated in the OSH policy.

The review should be based on the audit results and on the periodic status review reports. Specifically it should:

- establish new OSH objectives for continual improvement;
- develop OSH annual plans;
- consider changes to OSHMS arrangements;
- review findings which should then be documented, communicated to stakeholders via annual reports and followed up to ensure implementation and continual improvement.

Continual improvement

M2038 Continual improvement is at the heart of the OSHMS. It essentially is a fundamental commitment to better manage OSH risks proactively so that accidents and ill-health are reduced (system effectiveness) and/or the system achieves its desired aims using less resource (system efficiency).

The International Labour Office Guidelines on Occupational Safety and Health Management Systems (ILO, 2001) includes the following model which clearly illustrates the importance of the goal of continual improvement.

The route to continual improvement has been highlighted throughout the course of this chapter. It involves the use of SMART targets, the setting of OSH key performance indicators ('KPIs'), the use of performance appraisals which include

Figure 4

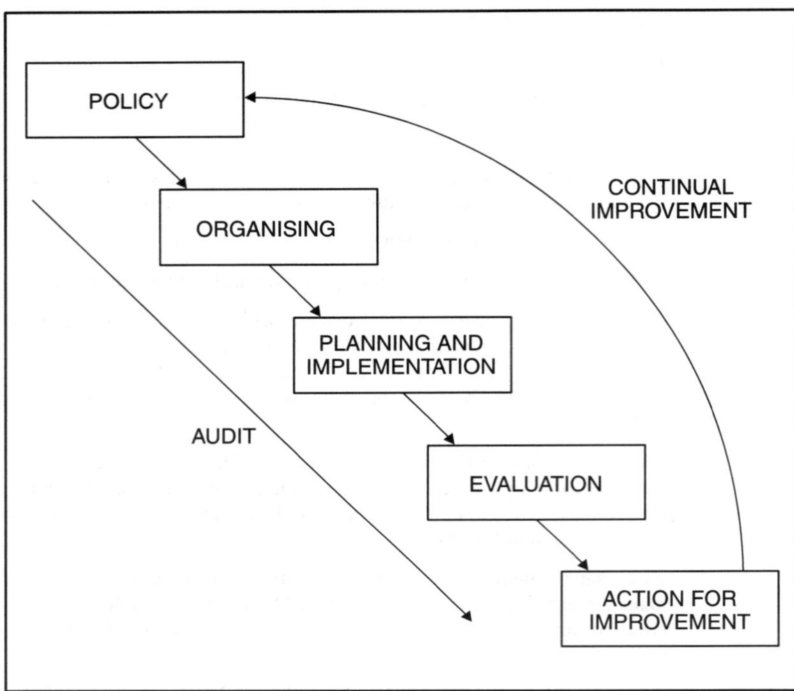

individual and collective OSH performance reviews, and accountability fixing for OSH responsibilities at all levels within the organisation.

The goal of continual improvement should manifest itself in a number of ways:

- better year on year results – less injuries, diseases, damage, near misses;
- steady, improved results using less resources;
- better targeted and more efficient and effective OSHMS;
- indications of breakthrough performances;
- more positive, proactive OSH culture;
- OSHMS improvements: more comprehensive, easier to understand, better all round, i.e. 'we improve as we get better!'

In order to achieve continual improvement, the following aspects should be taken into account:

- regular OSH audits;
- statistical improvements;
- bench-marking;
- industry/sector guidelines;

- ownership of work processes:
  - managers, supervisors;
  - workers representatives;
- proactive monitoring and targets;
- effective and efficient operation via workforce involvement;
- capture and evaluate ideas for improvement, ensure regular feedback to originators;
- resourced, implemented and monitored action plans;
- SMART improvement action plans;
- OSHMS training;
- OSHMS documentation/reference manual.

POPIMAR in action

M2039 The example below shows how the POPIMAR principles can be followed in practice. It relates to the introduction into an organisation of a permit to work ('PTW') system to control certain high risk activities.

Policy

The PTW system will become part of the 'arrangements' for implementing the health and safety policy statement. The PTW procedure (see below) is likely to contain a general statement about the importance of ensuring that appropriate controls are in place to minimise the risks associated with potentially high risk activities.

Organisation

This will primarily centre around the development of a formal PTW procedure which will involve:

- *Control*

  The procedure must identify:
  - what activities will be controlled by the PTW;
  - who will issue (and cancel) PTWs;
  - the timing sequence for issuing and cancelling individual PTWs;
  - the format of the PTW form.
- *Co-operation*

  The draft procedure must be reviewed with those likely to be involved in issuing and working under PTWs. This could involve an ad hoc consultative group in addition to safety representatives and the health and safety committee.
- *Communication*

  A wider section of the workforce must be informed about the proposed procedure and the reasons for its introduction.

- *Competence*

  Those who are to issue the PTWs must receive formal training in the procedure and use of the PTW form. There should also be a formal appointment process to ensure that they have successfully assimilated the training and can relate it to their practical work situations. Arrangements for this should be referred to in the procedure.

Planning and implementation

Once the procedure has been finalised the timescales and responsibilities must be identified for achieving the various component parts of its implementation. This will include:

- delivery of training to PTW issuers;
- formal appointment of PTW issuers (after an appropriate test process);
- printing and distributing the PTW forms;
- obtaining any necessary equipment (e.g. isolation locks, testing equipment);
- announcement of the formal introduction of the PTW procedure.

Measuring performance

Monitoring of the PTW can be carried out in various ways:

- a survey soon after the procedure's introduction;
- ongoing monitoring during formal health and safety inspections;
- periodic mini-audits of the procedure (e.g. by health and safety specialists or managers).

Auditing

The operation of the PTW procedure would be included in any general OSH auditing activities.

Reviewing performance

Both monitoring and audit findings must be reviewed by appropriate individuals and groups such as:

- health and safety specialists;
- the health and safety committee;
- senior management;
- ad hoc review teams.

Where satisfactory standards are not being achieved, the reviewers must consider how the POPIMAR sequence can best be applied to taking remedial action.

Manual Handling

Introduction

M3001 The *Manual Handling Operations Regulations 1992* (*SI 1992 No 2973* as amended by *SI 2002 No 2174*) came into operation on 1 January 1993 as part of the so-called 'six pack' of new Regulations. Although the Regulations implemented European Directive 90/269/EEC, the HSE had previously attempted to introduce Regulations on manual handling in an attempt to reduce the huge toll of accidents from this source. At the time of the introduction of the Regulations more than a quarter of the accidents reported each year to the enforcing authorities (the HSE and local authorities) were associated with manual handling. In 1990/91 65 per cent of handling accidents resulted in sprains and strains, whilst 5 per cent were fractures. Many manual handling injuries have cumulative effects eventually resulting in physical impairment or even permanent disability.

There is already evidence of the beneficial effects of the Regulations, particularly in respect of the increasing use of mechanical handling aids and improved manual handling training. By 1996/97 the proportion of handling accidents causing fractures had been reduced to 3.3 per cent. However, the battle was far from over with 54 per cent of RIDDOR reportable accidents in the health and social work sector still attributable to handling. Several other sectors had over 30 per cent of accidents involving handling, including food and drink, textiles, construction, wholesale and retail, transport/storage and education.

What the Regulations require

M3002 Various terms used in the *Manual Handling Operations Regulations 1992* (*SI 1992 No 2973*) are defined in *Regulation 2* which states:

> 'manual handling operations' means any transporting or supporting of a load (including the lifting, putting down, pushing, pulling, carrying or moving thereof) by hand or by bodily force.
>
> 'Injury' does not include injury caused by any toxic or corrosive substance which:
>
> (a) has leaked or spilled from a load;
>
> (b) is present on the surface of a load but has not leaked or spilled from it; or
>
> (c) is a constituent part of a load;
>
> and 'injured' shall be construed accordingly;'

(This in effect establishes a demarcation with the *COSHH Regulations* (*SI 2002 No 2677*) – where there is oil on the surface of a load making it difficult to handle that is a matter for the *Manual Handling Operations Regulations*, whereas any risk of dermatitis is covered by the *COSHH Regulations*).

> 'Load includes any person and any animal'.

(This is of great importance to a number of work sectors including health and social care, agriculture and veterinary work).

HSE guidance accompanying the Regulations states that an implement, tool or machinery being used for its intended purpose is not considered to constitute a load – presumably establishing a demarcation with the requirements of PUWER 98 (*SI 1998 No 2306*). However, when such equipment is being moved before or after use it will undoubtedly be a load for the purpose of these Regulations.

Regulation 2(2) imposes the duties of an employer to his employees under the Regulations on a self-employed person in respect of himself. *Regulation 3* excludes the normal ship-board activities of a ship's crew under the direction of the master from the application of the Regulations (these are subject to separate merchant shipping legislation).

The duties of employers under the Regulations are all contained in *Regulation 4* which (as in the *COSHH Regulations*) establishes a hierarchy of measures that employers must take. *Regulation 4* states:

> '(1) Each employer shall –
>
> (a) so far as is reasonably practicable, avoid the need for his employees to undertake any manual handling operations at work which involve a risk of their being injured; or
>
> (b) where it is not reasonably practicable to avoid the need for his employees to undertake any manual handling operations at work which involve a risk of their being injured–
>
> (i) make a suitable and sufficient assessment of all such manual handling operations to be undertaken by them, having regard to the factors which are specified in column 1 of Schedule 1 to these Regulations and considering the questions which are specified in the corresponding entry in column 2 of that Schedule,
>
> (ii) take appropriate steps to reduce the risk of injury to those employees arising out of their undertaking any such manual handling operations to the lowest level reasonably practicable, and
>
> (iii) take appropriate steps to provide any of those employees who are undertaking any such manual handling operations with general indications and, where it is reasonably practicable to do so, precise information on –
>
> (aa) the weight of each load, and
>
> (bb) the heaviest side of any load whose centre of gravity is not positioned centrally.
>
> (2) Any assessment such as is referred to in paragraph (1)(b)(i) of this Regulation shall be reviewed by the employer who made it if –
>
> (a) there is reason to suspect that it is no longer valid; or
>
> (b) there has been a significant change in the manual handling operations to which it relates,
>
> and where as a result of any such review changes to an assessment are required, the relevant employer shall make them.'

As a result of the *Health and Safety* (*Miscellaneous Amendments*) *Regulations 2002* (*SI 2002 No 2174*), a further paragraph was added to *Regulation 4*. This states:

> '(3) In determining for the purpose of this regulation whether manual handling operations at work involve a risk of injury and in determining the appropriate steps to reduce that risk regard shall be had in particular to—
>
> (a) the physical suitability of the employee to carry out the operations;
>
> (b) the clothing, footwear or other personal effects he is wearing;
>
> (c) his knowledge and training;
>
> (d) the results of any relevant risk assessment carried out pursuant to regulation 3 of the Management of Health and Safety at Work Regulations 1999;
>
> (e) whether the employee is within a group of employees identified by that assessment as being especially at risk; and
>
> (f) the results of any health surveillance provided pursuant to regulation 6 of the Management of Health and Safety Regulations 1999.'

The hierarchy of measures which must be taken by employers where there is a risk of injury can be summarised as follows:

- Avoid the operation (if reasonably practicable);
- Assess the remaining operations;
- Reduce the risk (to the lowest level reasonably practicable);
- Inform employees about weights.

Reference will be made later in the chapter to a variety of measures which can be taken to avoid or reduce risks. The factors to which regard must be given during the assessment (as specified in *Schedule 1* to the Regulations) are the *tasks, loads, working environment* and *individual capability* together with other factors referred to in *Regulation 4*(*3*). These factors and the questions specified by *Schedule 1* are also covered later in the chapter.

Regulation 5 places a duty on employees. It states:

> 'Each employee while at work shall make full and proper use of any system of work provided for his use by his employer in compliance with Regulation 4(1)(b)(ii) of the Regulations.'

Regulation 6 provides for the Secretary of State for Defence to make exemptions from the requirements of the Regulations in respect of home forces and visiting forces and their headquarters.

Regulation 7 extends the Regulations to apply to offshore activities such as oil and gas installations, including diving and other support vessels. Previous provisions relating to manual handling were replaced or revoked by *Regulation 8*. The Regulations themselves are extremely brief but they are accompanied by considerable HSE guidance, much of which will be referred to during this chapter.

Risk of injury from manual handling

Assessment guidelines

M3003 Appendix 1 of the guidance accompanying the *Manual Handling Operations Regulations 1992* (*SI 1992 No 2973*) (HSE booklet L23 M*anual Handling Operations 1992*:

Guidance on the Regulations) provides assessment guidelines which can be used to filter out manual handling operations involving little or no risk and help to identify where a detailed risk assessment is necessary. However, it must be stressed that *these are not weight limits* – they must not be regarded as safe weights nor must they be treated as thresholds which must not be exceeded. The HSE guidance states that application of the guidelines will provide a reasonable level of protection to around 95 per cent of working men and women. Some workpeople will require additional protection (e.g. pregnant women, frail or elderly workers) and this should be considered in relation to the assessment of individual capability.

In the latter part of 2003 the HSE published a free leaflet, *Manual Handling Assessment Charts* (INDG 383), to provide a tool to assist in identifying high risk manual handling operations during assessments. It uses numerical methods and a green, amber, red, purple colour coding system to categorise risks as low, medium, high and very high respectively. Risks from lifting, carrying and team handling are covered, although not from pushing and pulling. The leaflet also contains guidance on assessment technique.

Lifting and lowering

The guidelines for lifting and lowering operations are shown in the accompanying diagram (below). They assume that the load can be easily grasped with both hands and that the operation takes place in reasonable working conditions, using a stable body position. Where the hands enter more than one box, the lowest weights apply. A detailed assessment should be made if the weight guidelines are exceeded or if the hands move beyond the box zones.

These guideline weights apply up to approximately 30 operations per hour. They should be reduced for more frequent operations:

- by 30 per cent for 1 or 2 operations per minute;
- by 50 per cent for 5 to 8 operations per minute;
- by 80 per cent for more than about 12 operations per minute.

Other factors likely to require a more detailed assessment are where:

- workers do not control the pace of work, e.g. on assembly lines;
- pauses for rest are inadequate;
- there is no change of activity allowing different muscles to be used;
- the load must be supported for any length of time.

Carrying

Similar guidelines figures can be applied for carrying where the load is held against the body and carried up to 10 metres without resting. For longer distances or loads held below knuckle height a more detailed assessment should be made. The guideline figures can be applied for carrying loads on the shoulder for distances in excess of 10 metres but an assessment may be required for the lifting of the load onto and off the shoulder.

Pushing and pulling

Where loads are slid, rolled or supported on wheels the guideline figures assume that force is applied with the hands between knuckle and shoulder height. The guideline figure for starting or stopping the load is a force of about 25kg (about 250

Newtons) for men and 10kg (about 100 Newtons) for keeping the load in motion (these figures reduce to 16kg and 7kg respectively for women).

In practice it is extremely difficult to make a judgement as to whether forces of these magnitude are being applied. Many employers are likely to make assessments of significant pushing or pulling activities anyway. There are no guideline limits on distances over which loads may be pushed or pulled, providing there is sufficient opportunity for rest.

Handling when seated

The accompanying diagram illustrates the guideline figures for handling whilst seated. Any handling activities outside the specified box zones should be assessed in any case.

Fig. 1: Lifting and lowering

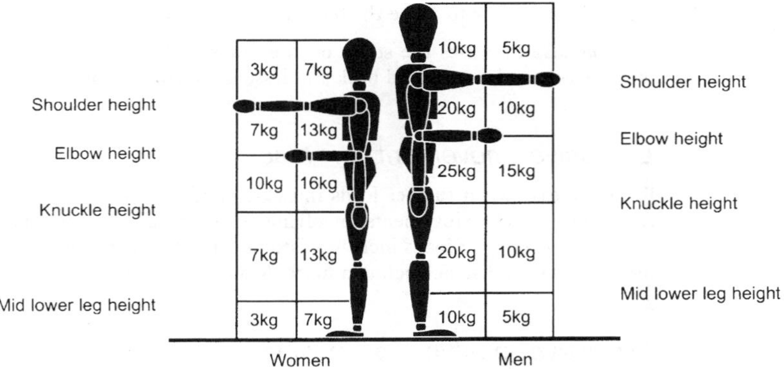

Fig. 2: Handling while seated

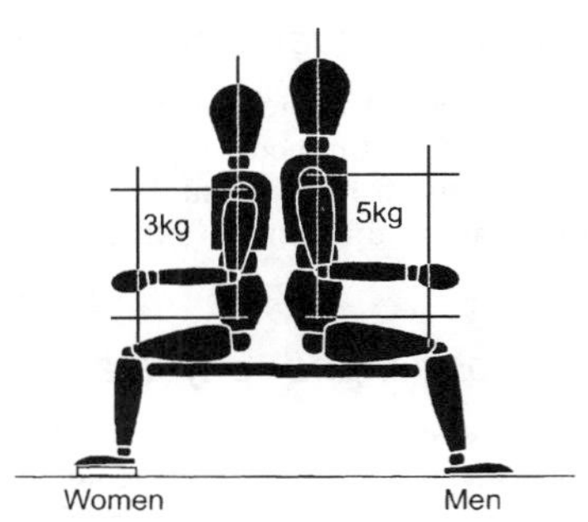

Schedule 1 to the *Manual Handling Operations Regulations 1992* (*SI 1992 No 2973*) contains a list of questions which must be considered during assessments of manual handling operations. All of these indicate factors which may increase the risk of injury. (Most of these factors have been included in the checklist for assessments provided later in the chapter).

Task related factors

Holding or manipulating loads at a distance from the trunk

M3004 Stress on the lower back increases as the load is moved away from the trunk. Holding a load at arms length imposes five times the stress as the same load very close to the trunk.

Unsatisfactory body movement or posture

The risk of injury increases with poor feet or hand placings, e.g. feet too close together, body weight forward on the toes, heels off the ground.

Twisting the trunk whilst supporting a load increases stress on the lower back – the principle should be to move the feet, not twist the body.

Stooping also increases the stress on the lower back whilst *reaching upwards* places more stress on the arms and back and lessens control on the load.

Excessive movement of loads

Risks increase when further loads must be *lifted or lowered*, especially when lifting from floor level. Movements involving a *change of grip* are particularly risky. Excessive *carrying* of loads increases fatigue – hence the guidance figure of about 10 metres carrying distance before a more detailed assessment is made.

Excessive pushing or pulling

Here the risks relate to the forces which must be exerted, particularly when starting to move the load, and the quality of the grip of the handler's feet on the floor. Risks also increase when pushing or pulling below knuckle height or above shoulder height.

Risk of sudden movement

Freeing a box jammed on a shelf or a jammed machine component can impose unpredictable stresses on the body, especially if the handler's posture is unstable. Unexpected movement from a client in the health care sector or an animal in agricultural work can produce similar effects.

Frequent or prolonged physical effort

Risk of injury increases when the body is allowed to become tired, particularly where the work involves constant repetition or a relatively fixed posture. The periodic inter-changing of tasks within a work group can do much to reduce such risks.

Clothing, footwear etc

The effects of any PPE or other types of clothing, footwear etc. required for the task may make manual handling more difficult or increase the possibilities of fatigue, dehydration etc.

Insufficient rest and recovery

The opportunity for rest and recovery will also help to reduce risks, especially in relation to tasks which are physically demanding.

Rate of work imposed by a process

In such situations workers are often not able to take even short breaks, or to stretch or exercise different muscle groups. More detailed assessments should be made of assembly line or production line activities. Again periodic role changes within a work team may provide a solution.

Load related factors

Is the load too heavy?

M3005 The guideline figures provided earlier are relevant here but only form part of the picture. Task-related factors (e.g. twisting or repetition) may also come into play or the load may be handled by two or more people. Size, shape or rigidity of the load may also be important.

Is the load bulky or unwieldy?

Large loads will hinder close approach for lifting, be more difficult to get a good grip of or restrict vision whilst moving. The HSE guidance suggests that any dimension exceeding 75cm will increase risks and the risks will be even greater if it is exceeded in more than one dimension. The positioning of suitable handholds on the load may somewhat reduce the risks.

Other factors to consider are the effects of wind on a large load, the possibility of the load hitting obstructions or loads with offset centres of gravity (not immediately apparent if they are in sealed packages).

Is the load difficult to grasp?

Where loads are large, rounded, smooth, wet or greasy, inefficient grip positions are likely to be necessary, requiring additional strength of grip. Apart from the possibility of the grip slipping there are likely to be inadvertent changes of grip posture, both of which could result in loss of control of the load.

Is the load unstable?

The handling of people or animals was referred to earlier. Not only do such loads lack rigidity, but they may also be unpredictable and protection of the load as well as the handler is of course an important consideration. Other potentially unstable loads are those where the load itself or its packaging may disintegrate under its own weight or where the contents may shift suddenly, e.g. a stack of books inside a partially full box.

Is the load sharp, hot etc?

Loads may have sharp edges or rough surfaces or be extremely hot or cold. Direct injury may be prevented by the use of protective gloves or clothing but the possibility of indirect risks (e.g. where sharp edges encourage an unsuitable grip position or cold objects are held away from the body) must not be overlooked.

Factors relating to the working environment

Space constraints

M3006 Areas of restricted headroom (often the case in maintenance work) or low work surfaces will force stooping during manual handling and obstructions (e.g. in front of shelves) will result in other unsatisfactory postures. Restricted working areas or the lack of clear gangways will increase risks in moving loads, especially heavier or bulkier items.

Uneven, slippery or unstable floors

All of the above can increase the risks of slips, trips and falls. The availability of a firm footing is a major factor in good handling technique – such a footing will be a rarity on a muddy construction site. Moving workplaces (e.g. trains, boats, elevating work platforms) will also introduce unpredictability in footing. (The guideline weights should be reduced in such situations.)

Variation in levels of floors or work surfaces

Risks will increase where loads have to be handled on slopes, steps or ladders, particularly if these are steep. Any slipperiness of their surface, e.g. due to ice, rain, mud will create further risks. The need to maintain a firm handhold on ladders or steep stairways is another factor to consider. Movement of loads between surfaces or shelving may also need to be considered, especially if there is considerable height change, e.g. floor level to above shoulder height.

Extremes of temperature or humidity

High temperatures or humidity will increase the risk of fatigue, with perspiration possibly affecting the grip. Work at low temperatures (e.g. in cold stores or cold weather) is likely to result in reduced flexibility and dexterity. The need for gloves and bad weather clothing is another factor to consider.

Ventilation problems or gusts of wind

Inadequate ventilation may increase fatigue whilst strong gusts of wind may cause considerable danger with larger loads, e.g. panels being moved on building sites.

Poor lighting

Lack of adequate lighting may cause poor posture or an increased risk of tripping. It can also prevent workers identifying risks associated with individual loads, e.g. sharp edges or corners, offset centres of gravity.

Capability of the individual

M3007 Manual handling capabilities will vary significantly between individual workers. Reference was made earlier in the chapter to the assessment guideline weights providing a reasonable level of protection to around 95 per cent of workers and those guideline figures differed for men and women. *Regulation 4(3)* of the *Manual Handling Operations Regulations 1992* (*SI 1992 No 2973*) (as amended) now specifically refers to factors which must be taken into account in respect of individual workers. Individual factors include:

Gender

Whilst there is overlap between the capabilities of men and women, generally the lifting strength of women is less than men.

Age

The bodies of younger workers will not have matured to full strength, whilst with older workers there will be a gradual decline in their strength and stamina, particularly from the mid 40's onwards. Such persons will be the types of groups of employees referred to in *Regulation 4(3)(e)* of the *Manual Handling Operations Regulations 1992* (*SI 1992 No 2973*) (as amended).

Experience

More mature workers may be better able to recognise their own capabilities and pace themselves accordingly, whilst also having acquired better handling techniques.

Pregnancy

The need to protect new or expectant mothers is emphasised in the chapter RISK ASSESSMENT.

Previous injury or ill-health

Workers with long term musculoskeletal problems, a history of injuries or ill-health or short term injuries (including those from non-work related causes) will all justify additional protection. Such problems should be detected by the health surveillance required by *Regulation 6* of the *Management of the Health and Safety at Work Regulations 1999* (*SI 1999 No 3242*). (See M3015 later in the chapter.)

Physique and stature

These factors will vary considerably within any workforce. Whilst bigger will often mean stronger, it must not be overlooked that some tasks or work locations may force taller workers to stoop or adopt other unsuitable handling postures.

Generally the objective should be to ensure that all manual handling operations can be performed satisfactorily by most reasonably fit and healthy employees, although a minority of workers may justify restrictions on their handling activities (see later in the chapter).

During the assessment process it is important that special note is taken of any manual handling operations which:

- Require unusual strength, height etc, e.g. reels of wrapping paper can only be safely positioned in a machine by persons above a certain height.
- Create additional risk to workers who are pregnant, disabled or have health problems.

  Such tasks may need to be reassessed to reduce risks or it may be necessary to restrict some individuals from carrying them out.
- Require special information or training.

  The health care sector provides many examples of situations where both the worker and the client (the load) may be at risk if correct technique is not used. Many other tasks are likely to require specific techniques to be adopted – whether in manual handling itself or the use of handling aids.

Clothing, footwear etc

Some individuals may be affected more than others by the use of PPE or clothing, e.g. gloves, protective suits, breathing apparatus. These may significantly affect the user's mobility or dexterity in respect of manual handling. Similarly uniforms or costumes (e.g. in the entertainment industry) may also be a factor to be taken into account.

Avoiding or reducing risks

M3008 The *Manual Handling Operations Regulations 1992* (*SI 1992 No 2973*) require employers to avoid the need for employees to undertake manual handling operations involving a risk of injury, so far as is reasonably practicable. Measures for avoiding such risks can be categorised as:

- Elimination of handling.
- Automation or mechanisation.
- Reduction of the load.

Elimination of handling

M3009 It may be possible for employers to eliminate manual handling altogether (or to eliminate handling by their own employees) by means such as:

Redesigning processes or activities

For example, so that activities such as machining or wrapping are carried out in situ, rather than a product being manually handled to a position where the activity take place; a treatment is taken to a patient rather than vice versa (this may have other benefits to the patient).

Using transport better

For example, allowing maintenance staff to drive vehicles carrying tools or equipment up to where they are working, rather than manually handling them into place.

Requiring direct deliveries

For example, requiring suppliers to make deliveries directly into a store rather than leaving items in a reception area (such suppliers may be better equipped in terms of handling aids and their staff may be better trained in respect of manual handling).

Automation or mechanisation

M3010 The automation or mechanisation of a work activity is best considered at the design stage although there is no reason in principle why such changes cannot be made later, e.g. as the result of a risk assessment. In making changes it is important to avoid creating additional risks, e.g. the use of fork lift trucks for a task in an already cramped workplace may not be the best solution. However, as employers have become more aware of manual handling risks (many prompted by the advent of the Regulations) a much greater range of mechanical handling solutions to manual handling problems have become available.

These include:

Mechanical lifting devices

Fork lift trucks, mobile cranes, lorry mounted cranes, vacuum devices and other powered handling equipment are commonly used to eliminate or greatly reduce the manual handling of loads.

Manually-operated lifting devices

Pallet and stacker trucks, manually operated chain blocks or lever hoists can all be used to move heavy loads using very limited physical force.

Powered conveyors

As well as fixed conveyors (both belt and driven roller types), mobile conveyors are being used increasingly, particularly for the loading and unloading of vehicles, e.g. stacking of bundles of newspapers in delivery vans.

Non-powered conveyors, chutes etc

Free-running roller conveyors, chutes or floor-mounted trolleys allow loads to move under the effects of gravity or to be moved manually with little effort. Such devices may operate between different levels or be set into the workplace floor.

Trolleys and trucks

Trolleys and trucks can be used to greatly reduce the manual handling effort required in transporting loads. Some types also incorporate manual or mechanically-powered lifting mechanisms. Specialist trolleys are designed for carrying drums or other containers whilst an ingenious triple wheel system can be lifted to some trucks to aid them in climbing or descending stairs.

Lifting tools

Special lifting tools are available to reduce the manual handling effort required for lifting or lowering certain loads, e.g. paving slabs, manhole covers, drums or logs.

All of these and many other types of devices are illustrated in HSE booklet HSG 115 *Manual handling: Solutions you can handle*'.

Reduction of the load

M3011 There is considerable scope to avoid the risk of injury by reducing the sizes of the loads which have to be handled. In some cases the initiative for this has come from suppliers of a particular product whereas in others the impetus has come from customers or from users of equipment, often as a result of their manual handling assessments. Examples include:

- Packaged building materials (e.g. cement) reduced to 25kg.
- Photocopying paper now supplied in 5 ream boxes (approximately 12kg) – 10 ream boxes were previously commonplace.
- Newspaper bundle sizes reduced below 20kg (printers reduce the number of copies in a bundle as the number of pages in the paper increases).
- Customers specifying maximum container weights they will accept from suppliers, e.g. weights of brochures from commercial printers.
- Equipment being separated into component parts and assembled where it is to be used, e.g. emergency equipment used by fire services.

Where it is not reasonably practicable to avoid a risk of injury from an manual handling operation, the employer must carry out an assessment of such operations with the aim of reducing the risks to the lowest level reasonably practicable . The four main factors to be taken into account during the assessment are *the task, the load, the working environment* and *individual capability*. Possible risk reduction measures related to each of these factors are described below.

(Later in the chapter these are summarised in an assessment checklist – see M3020 below).

Task-related measures

Mechanical assistance

M3012 The types of mechanical assistance described earlier can be used to reduce the risk of injury, even if such risks cannot be avoided entirely. The use of roller conveyors, trolleys, trucks or levers can reduce considerably the amount of force required in manual handling tasks.

Task layout and design

The layout of tasks, storage areas etc should be designed so that the body can be used in its more efficient modes, e.g:

- heavier items stored and moved between shoulder and mid-lower leg height;
- allowing loads to be held close to the body;
- avoiding reaching movements, e.g. over obstacles or into deep bins;
- avoiding twisting movements or handling in stooped positions, e.g. by repositioning work surfaces, storage areas or machinery;
- providing resting places to aid grip changes whilst handling;

- reducing lifting and carrying of loads, e.g. by pushing, pulling, sliding or rolling techniques;
- utilising the powerful leg muscles rather than arms or shoulders;
- avoiding the need for sustaining fixed postures, e.g. in holding or supporting a load;
- avoiding the need for handling in seated positions.

Work routines

Work routines may need to be altered to reduce the risk of injury using such measures as:

- limiting the frequency of handling loads (especially those that are heavier or bulkier);
- ensuring that workers are able to take rest breaks (either formal breaks or informal rest periods as and when needed);
- introducing job rotation within work teams (allowing muscle groups to recover whilst other muscles are used or lighter tasks undertaken).

Team handling

Tasks which might be unsafe for one person might be successfully carried out by two or more. HSE guidance indicates that the capability of a two-person team is approximately two thirds of the total of their individual capabilities; and of a three-person team approximately one half of their total. Other factors to take into account in team handling are:

- the availability of suitable handholds for all the team;
- whether steps or slopes have to be negotiated;
- possibilities of the team impeding each other's vision or movement;
- availability of sufficient space for all to operate effectively;
- the relative sizes and capabilities of team members.

Load-related measures

Weight or size reduction

M3013 The types of measures described earlier in the chapter might be adopted to reduce the risk of injury even if it is not possible to avoid the risk entirely. However, reducing the individual load size will increase the number of movements necessary to handle a large total load. This could result in a different type of fatigue and also the possibility of corners being cut to save time. Sizes of loads may also be reduced to make them easier to hold or bring them closer to the handler's body.

Making the load easier to grasp

Handles, grips or indents can all make loads easier to handle (as demonstrated by the handholds provided in office record storage boxes). It will often be easier handling a load placed in a container with good handholds than handling it alone without any secure grip points. The positioning of handles or handholds on a load can also influence the handling technique used, e.g. help avoid stooping. Handholds should be wide enough to accommodate the palm and deep enough to accommodate the knuckles (including gloves where these are worn).

The use of slings, carrying harnesses or bags can all assist in gaining a secure grip on loads and carrying them in a more comfortable and efficient position.

Making the load more stable

Loads which lack rigidity themselves or are in insecure packaging materials may need to be stabilised for handling purposes. Use of carrying slings, supporting boards or trays may be appropriate. Partially full containers may need to have their contents wedged into position to prevent them moving during handling.

Reducing other risks

Loads may need to be cleaned of dirt, oil, water or corrosive materials for safe handling to take place. Sharp corners or edges or rough surfaces may have to be removed or covered over. There may also be risks from very hot or very cold loads. In all cases the use of containers (possibly insulated) other handling aids or suitable gloves or other PPE may need to be considered.

Improving the work environment

Providing clear handling space

M3014 The provision of adequate gangways and sufficient space for handling activities to take place will be linked with general workplace safety issues. Low headroom, narrow doorways and congestion caused by equipment and stored materials are all to be avoided. Good housekeeping standards are essential to safe manual handling.

Floor condition and design

Even in temporary workplaces such as construction sites, every effort should be made to ensure that that firm, even floors are provided where manual handling is to take place. Special measures may be necessary to allow water to drain away or to clear promptly any potentially slippery materials (e.g. food scraps in kitchens). Where risks are greater special slip-resistant surfaces may need to be considered. In outdoor workplaces the routine application of salt or sand to surfaces made slippery by ice or snow may need to be introduced.

Differing work levels

Measures may be necessary to reduce the risks caused by different work levels. Slopes may need to be made more gradual, steps may need to be provided or existing steps or stairways made wider to accommodate handling activities. Work benches may need to be modified to provide a uniform and convenient height.

Thermal environment and ventilation

Unsatisfactory temperatures, high humidity or poor ventilation may need to be overcome by improved environmental control measures or transferring work to a more suitable area. For work close to hot processes or equipment, in very cold conditions (e.g. refrigerated storage areas) or carried on out of doors the use of suitable PPE may be necessary.

Gusts of wind

Where gusts of wind (or powerful ventilation systems) could affect larger loads extra precautions, in addition to those normally taken, may be necessary e.g:

- using handling aids (e.g. trolleys);
- utilising team handling techniques;
- adopting a different work position;
- following an alternative transportation route.

Lighting

Suitable lighting must be provided to permit handlers to see the load and the layout of the workplace and to make accurate judgements about distance, position and the condition of the load.

Individual capability

M3015 In ensuring that individuals are capable of carrying out manual handling activities in the course of their work, account must be taken both of factors relating to the individual and those relating to the activities they are to perform. Steps which may be appropriate include:

Medical screening

Workers may be screened both prior to being offered employment and periodically thereafter to check whether they are physically capable of carrying out the full range of manual handling tasks in the workplace. In some extreme situations (e.g. the emergency services) an individual may be deemed unsuitable for employment if they are not sufficiently fit. However, legislation prevents employers discriminating against those with disabilities *without good reason*. In most workplaces the results of such screening may simply result in restrictions on the range of tasks individuals are allowed to perform.

Long-term restrictions

Long-term restrictions may be appropriate where workers are identified through screening programmes as not being capable of carrying out certain manual handling activities without undue risk, or they become incapable due to some long term injury or the effects of the ageing process. It may be necessary to place long term restrictions on young workers until they attain an appropriate age at which their capabilities can be re-assessed.

Short-term restrictions

Relatively short-term restrictions on manual handling activities may be necessary because of:

- pregnancy (or a particular stage of pregnancy)
- recently having given birth
- injury or illness of short duration.

Fitness programmes

An increasing number of employers actively encourage employers to be physically fit, thus reducing their risk of injury during manual handling activities. A number of Fire Services have particularly pro-active programmes.

General manual handling training

Many employers offer general training in manual handling, either to all employees or to all those employees expected to engage in significant manual handling work. Aspects to cover in such training include:

- recognition of potentially hazardous operations;
- avoiding manual handling hazards;
- dealing with unfamiliar handling operations;
- correct use of handling aids;
- proper use of PPE;
- working environment factors;
- importance of good housekeeping;
- factors affecting individual capability (knowing one's limitations);
- good handling technique.

Task-specific training and instruction

Training and instruction will often need to relate to the safety of specific manual handling tasks, e.g:

- how to use specific handling aids;
- specific PPE requirements;
- specific handling techniques.

General principles of handling technique are summarised in HSE leaflet INDG 143 *Getting to grips with manual handling: A short guide for employers.* Many other HSE publications deal with specific occupational areas and often contain guidance on training (see 'HSE Guidance' at M3019). *Regulation 4(3)(c)* of the *Manual Handling Operations Regulations 1992* (*SI 1992 No 2973*) (as amended) now requires employees' level of knowledge and training to be considered during the risk assessment.

Planning and preparation

M3016 The principles of planning and preparation for manual handling assessments are no different from those for any other types of risk assessments.

Who will carry out the assessments?

M3017 HSE guidance contained in the main HSE booklet on the *Manual Handling Operations Regulations 1992* (L 23) states that while one individual may be able to carry out an assessment in relatively straightforward cases, in others it may be more appropriate to establish an assessment team. It also refers to employers and managers with a practical understanding of the manual handling tasks to be performed, the loads to be handled, and the working environment being better able to conduct assessments than someone outside the organisation (although it is not stated in the guidance that in-house personnel will also usually be more aware of the capabilities of employees carrying out manual handling tasks).

The guidance refers to the individual or team performing the assessments needing to possess the following knowledge and expertise:

- the requirements of the Regulations;
- the nature of the handling operations;
- a basic understanding of human capabilities;
- awareness of high risk activities;
- practical means of reducing risk.

The HSE also acknowledges that there will be situations where external assistance may be required. These might include:

- training of in-house assessors;
- assessing risks which are unusual or difficult to assess;
- re-designing equipment or layouts to reduce risks;
- training of staff in handling techniques.

How will the assessments be arranged?

M3018 As for general risk assessments, the workplace or work activities will often need to be divided into manageable assessment units. Depending on how work is organised, these might be based on:

- Departments or sections.
- Buildings or rooms.
- Parts of processes.
- Product lines.
- Work stations.
- Services provided.
- Job titles.

Members of the assessment team can then be allocated to the assessment units where they are best able to make a contribution.

Gathering information

M3019 It is important to gather information prior to the assessment in order to identify potentially high risk manual handling activities and the precautions which ought to be in place to reduce the risks. Relevant sources might include:

Accident investigation reports

Dependent upon the requirements of the organisation's accident investigation procedure, these may only include the more serious accidents resulting from manual handling operations.

Ill health records

These may reveal short absences or other incidents due to manual handling which may have escaped the accident reporting system. Enquiries may need to be made in respect of the health surveillance records of individuals (see M3015 above).

Accident treatment records

Accident books or other treatment records may identify regularly recurring minor accidents which do not necessarily result in any absences from work, eg cuts or abrasions from sharp or rough loads.

Operating procedures etc

Reference to manual handling activities and precautions which should be adopted may be contained in operating procedures, safety handbooks etc.

Work sector information

Trade associations and similar bodies publish information on manual handling risks and how to overcome them.

HSE Guidance

Much guidance is available from the HSE – some of the more important publications were referred to earlier in the chapter. The table below gives details of additional guidance relating to specific work sectors.

HSE publications on manual handling in specific work sectors

	Manual handling in the health service (1998)
	Getting to grips with handling problems: Worked examples of assessment and reduction of risk in the health service (1993)
HSG 119	Manual handling in drinks delivery (1994)
HSG 196	Moving food and drink: Manual handling solutions for the food and drink industries (2000)
HSG 171	Well handled: Offshore manual handling solutions (1997)
	Manual handling in paper mills (1998)
	Picking up the pieces: Prevention of musculoskeletal disorders in the ceramics industry (1996)
AS 23	Manual handling solutions for farms (2000)
INDG 125	Handling and stacking bales in agriculture (1998) – free leaflet.
IACL 105	Handling the news: Advice for employers on manual handling of bundles (1999) – free leaflet.
IACL 106	Handling the news: Advice for newsagents and employees on safe handling of bundles (1999) – free leaflet.
IACL 103	Manual handling in the textile industry (1998)

	Manual handling in the rubber industry (1999)
APIS 2	Manual handling operations: Baggage handling at airports
INDG 332	Manual packing in the brick industry (2000) – free leaflet.
INDG 326	Manual handling in the railway industry: Advice for employers (2000) – free leaflet.
HSG 149	Backs for the future: Safe manual handling in construction
CIS 37	Handling heavy building blocks
CAIS	Manual handling in the catering industry
INDG 318	Manual handling solutions in woodworking

Making the assessment

M3020 As with other types of assessments, visits to the workplace are an essential part of manual handling assessments. Manual handling operations need to be observed (sometimes in some detail) and discussions with those carrying out manual handling work, their safety representatives and their supervisors or managers need to take place.

The assessment checklist provided on the following pages provides guidance on the factors which may need to be taken into account during the observations and discussions and also on possible measures to reduce the risk. The checklist utilises the four main factors specified in the Regulations – task, load, working environment and individual capability. In practice each of these will have greater or lesser importance, depending upon the manual handling activity.

Observations

M3021 Sufficient time should be spent in the workplace to observe a sufficient range of the manual handling activities taking place and to take account of possible variations in the loads handled (raw materials, finished product, equipment etc). Aspects to particularly be considered include:

- comparison of actual methods used with those specified in operating procedures, industry standards etc;
- the level and manner of use of handling aids and their effectiveness;
- handling techniques adopted;
- physical conditions, e.g. access, housekeeping, floor surfaces, lighting;
- variations between employees, e.g. size, strength, technique adopted.

Discussions

M3022 In discussing manual handling issues with workpeople an enquiring approach should be taken, utilising open-ended questions wherever possible. Aspects which might be discussed include:

- whether normal conditions are being observed;
- their awareness of procedures, rules etc;
- their training in the use of handling aids or handling techniques;

MANUAL HANDLING ASSESSMENT CHECKLIST

THE TASK

ASSESSMENT FACTORS	REDUCING THE RISK
DISTANCE OF THE LOAD FROM THE TRUNK	TASK LAYOUT
BODY MOVEMENT / POSTURE • Twisting • Reaching • Stooping • Sitting	USE THE BODY MORE EFFECTIVELY eg SLIDING OR ROLLING THE LOAD SPECIAL SEATS
DISTANCE OF MOVEMENT • Height (? grip change) • Carrying (? over 10m) • Pushing or pulling	RESTING PLACE / TECHNIQUE USE OF TROLLEYS etc TECHNIQUE / FLOOR SURFACE
RISK OF SUDDEN MOVEMENT	AWARENESS / TRAINING
FREQUENT / PROLONGED EFFORT	IMPROVED WORK ROUTINE eg Job rotation
CLOTHING / FOOTWEAR ISSUES RATE OF WORK IMPOSED BY A PROCESS	ADEQUATE REST OR RECOVERY PERIODS

THE LOAD

ASSESSMENT FACTORS	REDUCING THE RISK
HEAVY	MAKE THE LOAD LIGHTER TEAM LIFTING MECHANICAL AIDS
BULKY (Any dimensions above 75cm)	MAKE THE LOAD SMALLER
UNWIELDY (Offset centre of gravity)	INFORMATION / MARKING
DIFFICULT TO GRASP (Large, rounded, smooth, wet, greasy)	MAKE IT EASIER TO GRASP • Handles • Handgrips • Indents • Slings • Carrying devices • Clean the load
UNSTABLE • Contents liable to shift • Lacking rigidity	WELL-FILLED CONTAINERS USE OF PACKING MATERIALS SLINGS / CARRYING AIDS
SHARP EDGES / ROUGH SURFACES	AVOID OR REDUCE THEM GLOVES OR OTHER PPE
HOT OR COLD	CONTAINERS (? insulated)

THE WORKING ENVIRONMENT

ASSESSMENT FACTORS	**REDUCING THE RISK**
SPACE CONSTRAINTS AFFECTING POSTURE	ADEQUATE GANGWAYS, FLOORSPACE, HEADROOM
UNEVEN, SLIPPERY OR UNSTABLE FLOORS	WELL MAINTAINED SURFACES PROVISION OF DRAINAGE SLIP-RESISTANT SURFACING GOOD HOUSEKEEPING PROMPT SPILLAGE CLEARANCE
VARIATION IN WORK SURFACE LEVEL eg steps, slopes, benches	STEPS NOT TOO STEEP GENTLE SLOPES UNIFORM BENCH HEIGHT
HIGH OR LOW TEMPERATURES HIGH HUMIDITY POOR VENTILATION	BETTER ENVIRONMENTAL CONTROL RELOCATING THE WORK SUITABLE CLOTHING
STRONG WINDS OR OTHER AIR MOVEMENT	RELOCATING THE WORK ALTERNATIVE ROUTE HANDLING AIDS TEAM HANDLING
LIGHTING	SUFFICIENT WELL-DIRECTED LIGHT
MOVING WORKPLACE • Boat • Train • Vehicle	

INDIVIDUAL CAPABILITY

ASSESSMENT FACTORS	**REDUCING THE RISK**
INDIVIDUAL FACTORS STRENGTH HEIGHT FLEXIBILITY STAMINA	BALANCED WORK TEAMS 'SELF SELECTION' MAKE ASSISTANCE AVAILABLE ENCOURAGE FITNESS
PROBLEMS WITH CLOTHING ETC. PREGNANCY INJURY / HEALTH PROBLEM • Back trouble • Hernia • Temporary injury	SPECIAL ASSESSMENTS FORMAL RESTRICTIONS CLEAR INSTRUCTIONS

TASK FACTORS OPERATIONS OR SITUATIONS REQUIRING PARTICULAR	FORMALISED WORK PROCEDURES CLEAR INSTRUCTIONS ADEQUATE TRAINING
• Awareness of risks • Knowledge • Method of approach • Technique	• Recognising potentially hazardous operations • Dealing with unfamiliar operations • Correct use of handling aids • Proper use of PPE • Environmental factors • Importance of housekeeping • Knowing one's own limitations • Good technique

- possible variations in work activities, e.g. product changes, seasonal variations, differences between day and night shift;
- what happens if handling equipment or handling aids break down or are unavailable;
- whether assistance is available if required;
- what manual handling problems have been experienced?;
- have there been any injuries with manual handling?;
- reasons precautions are not being taken.

Notes

M3023 Notes should be taken during the assessment of anything which might to be of relevance such as:

- further detail about manual handling risks found;
- handling risks which are discounted as insignificant;
- handling aids available;
- other control measures in place and appearing effective;
- problems identified or concerns expressed;
- alternative precautions worthy of consideration;
- related procedures, records or other documents.

These rough notes will eventually form the basis of the assessment.

After the assessment

Assessment records

M3024 HSE guidance states that the significant findings of the assessment must be recorded unless:

- the assessment could very easily be repeated and explained at any time because it is so simple and obvious; or

- the manual handling operations are quite straightforward, of low risk, are going to last only a very short time, and the time taken to record them would be disproportionate.

As should be the case in relation to other types of assessment, it is better to make a simple record if there is any doubt.

There is no standard format for recording an assessment. The main HSE Guidance booklet (L 23) includes a sample record format together with a useful worked example. This record format includes a checklist of assessment factors under the four main headings (task, load, working environment, capability) in a similar listing to the assessment checklist we have provided. Alongside each checklist item the assessor has space to:

- identify whether the risk from this factor is low, medium or high;
- provide a more detailed description of the problems;
- identify possible remedial action.

Some will no doubt find this checklist format meets their needs. However, many assessment factors often have little or no relevance to a specific manual handling operation. As a result, the eventual record contains a lot of blank paper, with only occasional comments. Consequently it is often preferable to keep the checklist separate and only include within the record a reference to the risks, precautions and recommended improvements which are relevant to the operation being assessed. Two worked examples of this type of record are included below.

MANUAL HANDLING ASSESSMENT		*BROADACRES DEVELOPMENT*	DATE: *27 January 2004*	ASSESSMENT BY: A Backworth
No.	ACTIVITY	RISKS	EXISTING PRECAUTIONS	RECOMMENDATIONS
1	HANDLING OF STATIONERY AND PRINTED MATERIALS	Heaviest stationery items are 5 ream boxes of paper (approximately 12 kg). Some packs of printed promotional material weigh over 20 kg.	Suppliers deliver items directly to main store.	Require printing suppliers to supply materials in packs weighing no more than 12 kg.
		Items are moved from main store on the ground floor to smaller stores on other floors.	Trolley and lift used for transporting between main store and smaller stores.	Ensure paper and printed materials are stored on shelves between knee and shoulder height.
2	HANDLING OF ARCHIVED RECORDS	Records are normally stored in file boxes weighing no more than 20 kg. They are moved into and out of the record store on the third floor.	File boxes are kept below shoulder height on substantial racking. They are moved around using the trolley and lift.	Access to the Accounts section of the archive store racking was obstructed by various items which must be removed.
		(Older records are removed for storage off-site by a specialist company.)	Movement to off-site storage is carried out by Facilities staff who have had manual handling training.	Some records were stored in oversize boxes — this practice must cease.
3	MOVEMENT OF EQUIPMENT PC monitors, TV sets, projection equipment etc.	These items (of varying weights) are moved between the equipment store on the first floor and locations where they are required.	The trolley and lift are used where appropriate. IT staff have been trained in handling techniques.	Training section assistants should also receive handling technique training. Provide dust coats (so that staff can carry dirty items close to their bodies).
			Some racking is provided in the equipment store but this is inadequate and many items are stored on the floor.	Install additional racking in the equipment store and keep heavier items on more accessible shelves.
4	MOVEMENT OF FURNITURE Desks, seating etc.	Desks and seating are moved as required for individual workstations.	Facilities section staff have had manual handling training.	Training section assistants require training (see above).
		(Major moves are carried out by contractors.)	Team lifting is used for larger items, e.g. desk tops. The lift is used as appropriate.	Suitable racking must be provided in the equipment store for desk and table tops.
		Seating and tables are moved when required for major meetings and training events.	Desk and table tops are kept in the equipment store.	

GOLD STAR ENGINEERING	MANUAL HANDLING ASSESSMENT	
MAINTENANCE STORES	Assessment by: *T Wrightson*	Date: *28 January 2004*
ACTIVITIES/ RISKS	EXISTING PRECAUTIONS	RECOMMENDATIONS
Handling activities involve the movement of equipment and materials within the stores building and the yard outside, together with two storage containers and a flammable liquid store situated within the yard. Both equipment and materials must also be moved to locations within the factory.	A fork lift truck is used to handle heavier items within the yard. A hand-operated pallet truck and a lightweight lifting truck are available. Several trucks and trolleys are provided including a stair-climbing truck, gas cylinder and fire extinguisher trolleys and a drum transporter. Skates, bogie units and dollies are also available for transporting heavy items.	
There is considerable variation in the sizes and weights of items handled. Some items are palletised, some are supplied in boxes and some are kept in bins and cages.	Most internal and external floor surfaces are in good condition.	A section of the floor just inside the main stores entrance requires repair.
	There is a gradual ramp into the main stores building but not into the storage containers or the flammable liquid store (the latter having a sill at its entrance).	Suitable permanent or removable ramps should be provided to allow handling aids to be used in the storage containers and flammable liquid store.
Drums (up to 100 litres capacity), gas cylinders and large fire extinguishers must also be moved.	Lighting standards are good in the main stores building, the flammable liquid store and the yard, but there are no lights in the storage containers.	Lighting must be provided within the storage containers.
Racking and cupboards are provided within the building and inside the outside storage containers.	Heavier items are generally stored either on the floor or on accessible levels on racking or in cupboards.	Ensure walkways between racking are kept clear.
Some items to be handled have sharp or rough edges.	Latex coated knitted gloves (for grip) and gloves with chrome leather palms are available (as are gloves for chemical risks).	
	Most stores' staff have attended a one-day course in manual handling techniques tailored to their needs.	Develop a formal training programme to ensure that all stores staff receive training in manual handling technique and the correct use of the various handling aids available.

Review and implementation of recommendations

M3025 As for other types of risk assessments, there will be a need to review recommendations and prepare action plans for their implementation. Recommendations will need to be followed up in order to ensure that they have actually been implemented and also to assess any new risks which may have been introduced. This latter point is particularly relevant to manual handling as changes in working practice may have resulted in different techniques being adopted (with their own attendant risks) or the introduction of handling aids may have created additional training needs.

Assessment review

M3026 *Paragraph 4(2)* of the *Manual Handling Operations Regulations 1992* requires a manual handling assessment to be reviewed if:

'(a) there is reason to suspect that it is no longer valid; or

(b) there has been a significant change in the manual handling operations to which it relates;'

As is the case for other assessments, a review would not necessarily result in the assessment being revised, although a revision may be deemed to be necessary. A significant manual handling related injury should automatically result in a review of the relevant assessment. It is recommended that assessments are reviewed periodically in any case in order to take account of small changes in work practices which will eventually have a cumulative effect. The periods for such reviews might vary between two and five years – depending upon the degree of risk involved and the potential for gradual changes to take place.

Training

M3027 The need for specific manual handling technique training may frequently be one of the recommendations made as a result of a manual handling assessment. With manual handling activities featuring in a wide range of work sectors many employers have chosen to provide all or a significant proportion of their workforce with general training in good handling technique. Such training is often provided as part of a more comprehensive health and safety induction programme. Guidance on the content of such general training is contained in paragraphs 173 and 174 of HSE Guidance booklet (L 23) and also in other HSE publications dealing with specific work sectors (see the table earlier in the chapter).

Noise at Work

Introduction

N3001 This chapter is aimed at company directors, legal advisors, health and safety officers, staff representatives and others who need a comprehensive, concise reference source on the issue of noise at work.

It will be helpful to anyone who needs to commission a workplace noise assessment or to undertake a noise control project. It will also assist those who wish to devise long-term noise reduction and equipment-buying strategies.

Despite decades of warnings and legislation, noise-induced hearing loss is still a huge but underrated problem, with a slow, cumulative and irreversible effect. As a result, legislation is in the process of being tightened, so that many more workplaces are about to fall into the net (N3002).

This chapter discusses the scale of problems arising from workplace noise exposure, the legal obligations on employers to reduce the risk of noise-induced hearing damage both in the short term and as a long-term obligation (N3012); and the risks in terms of compensation claims (N3022). An overview of social security benefits is also provided (N3018).

The chapter gives a short introduction to the perception of sound and the various indexes used to measure noise (N3004). It provides general guidance on methods of assessing and reducing noise (see N3029 and N3035), and suggests sources of more detailed guidance on these matters (see N3023).

Scale of the problem and future developments

N3002 A Medical Research Council survey in 1997–98 estimated that there were 507,000 people in Great Britain suffering from hearing difficulties as a result of exposure to noise at work. This is much higher than HSE's own estimate of 87,000 based on its 2001/02 self-reported work-related Illness (SWI) survey. The large difference is probably due to the survey methods, and suggests that even people who suffer hearing difficulties are often unaware that their noisy workplace has contributed to their problems.

The Department of Social Security (DSS) pays disablement benefit to people who suffer at least 50 decibels (dB) of noise-induced hearing loss in both ears. This level of hearing loss is roughly equivalent to listening to the television or a conversation through a substantial brick wall. Moreover, claimants must have been employed for at least ten years in a specified noisy occupation. Despite these restrictions, the number of new awards appears to have stuck at just over 260 per year, after a significant fall during the 1980s and 1990s from around 1,500 a year in 1984–85 to a low of 226 in 2000. There are currently about 14,000 people – 99 per cent of them men, receiving the benefit from DSS.

The number of claimants with between 35 and 49 decibels of hearing loss (still a severe disability, equivalent to listening through a substantial partition wall in a house) shows an improving trend. According to statistics from audiological examinations, the number of claimants has fallen from 1200 in 1995 to 710 in 1999.

It is likely that some of this improvement is a result of the contraction of traditionally noisy industries such as ship-building and coal mining, with the armed forces now showing some of the highest rates of incidence.

However, there is concern about noise exposure in some of the newer occupations, such as workers in pubs and clubs where noise levels often exceed the second action level of the *Noise at Work Regulations 1989* (*SI 1989 No 1790*).

The EU's Physical Agents (Noise) Directive (2003/10/EC) requires changes to be made to the existing UK regulations (i.e. Noise At Work Regulations) and HSE are currently consulting on new Control of Noise at Work Regulations. The new regulations will reduce the action levels by 5 dB, requiring exposure to be assessed when it exceeds 80 dB(A) averaged over 8 hours (in some cases, averaged over 5 days each of 8 hours). Hearing protection must be made available at this exposure, as well as information and training. Health surveillance must be made available where exposure reaches 85 dB(A) and hearing protectors must be worn. There will be a limit on exposure of 87 dB(A), the exposure to be calculated to include the effect of any hearing protection that is provided. Corresponding changes are made to action levels in relation to peak sound pressures. This proposal is far more practicable and realistic than in the original EU proposals, following representations from Britain (amongst others) that it was preferable to ensure greater compliance with current limits than to introduce more stringent limits that are not rigorously enforced.

Hearing damage

N3003 We all lose hearing acuity with age, and this loss is accelerated and worsened by exposure to excessive noise. Fortunately, in early life, we have much greater hearing acuity than modern life demands, and a small loss is readily compensated by turning up the volume of the radio or television.

Loud noises cause permanent damage to the nerve cells of the inner ear in such a way that a hearing aid is ineffective. At first, the damage occurs at frequencies above normal speech range, so that the sufferer may have no inkling of the problem, although it could be identified easily by an audiogram, which measures the sensitivity of the ear at a number of frequencies across its normal range. As exposure continues, the region of damage progresses to higher and lower frequencies. Damage starts to extend into the speech range, making it difficult to distinguish consonants, so that words start to sound the same. Eventually, speech becomes a muffled jumble of sounds.

Some people are much more susceptible to hearing damage than others. A temporary dullness of hearing or tinnitus (a ringing or whistling sound in the ears) when emerging from a noisy place are both indicative of damage to the nerve cells of the inner ear. Because these symptoms tend to disappear with continued exposure, they are often misinterpreted as the ear becoming 'hardened', whereas in fact it is losing its ability to respond to the noise. Even when hearing has apparently returned to normal, a little of the sensitivity is likely to have been lost. Therefore, anyone who experiences dullness of hearing or tinnitus after noise exposure must take special care to avoid further exposure. They may also be advised to seek medical advice and an audiogram.

Apart from catastrophic exposures (usually to explosive sounds), the risk of hearing loss is closely dependent on the 'noise dose' received, and especially on the cumulative effect over a period of time. British Standard BS5330 provides a procedure for estimating the risk of 'hearing handicap' due to noise exposure.

Handicap is there defined as a hearing loss of 30 dB, which is sufficiently severe to impair the understanding of conversational speech or the appreciation of music. This may be compared with the loss of 50 dB required for DSS disability benefit.

The perception of sound

N3004 Sound is caused by a rapid fluctuation in air pressure. The human ear can hear a sensation of sound when the fluctuations occur between 20 times a second and 20,000 times a second. The rate at which the air pressure fluctuates is called the frequency of the sound and is measured in Hertz (Hz). The loudness of the sound depends on the amount of fluctuation in the air pressure. Typically, the quietest sound that can be heard (the threshold of hearing) is zero decibels (0 dB) and the sound becomes painful at 120 dB. Surprisingly perhaps, zero decibels is not zero sound. The sensitivity and frequency range of the ear vary somewhat from person to person and deteriorate with age and exposure to loud sounds.

The human ear is not equally sensitive to sounds of different frequencies (i.e. pitch): it tends to be more sensitive in the frequency range of the human voice than at higher or lower frequencies (peaking at about 4 kHz). When measuring sound, compensation for these effects can be made by applying a frequency weighting, usually the so-called 'A' weighting, although other weightings are sometimes used for special purposes.

The ear has an approximately logarithmic response to sound: for example, every doubling or halving in sound pressure gives an apparently equal step increase or decrease in loudness. In measuring environmental noise, sound pressure levels are therefore usually quoted in terms of a logarithmic unit known as a decibel (dB). To signify that the 'A' weighting has been applied, the symbol of dB(A) is often used. However, current practice tends to prefer the weighting letter to be included in the name of the measurement index. For example, dB(A) L_{eq} and dB L_{Aeq} both refer to the 'A' weighted equivalent sound level in decibels.

Depending upon the method of presentation of two sounds, the human ear may detect differences as small as 0.5 dB(A). However, for general environmental noise the detectable difference is usually taken to be between 1 and 3 dB(A), depending on how quickly the change occurs. A 10 dB(A) change in sound pressure level corresponds, subjectively, to an approximate doubling or halving in loudness. Similarly, a subjective quadrupling of loudness corresponds to a 20 dB(A) increase in sound pressure level (SPL). When two sounds of the same SPL are added together, the resultant SPL is approximately 3 dB(A) higher than each of the individual sounds. It would require approximately *nine* equal sources to be added to an original source before the subjective loudness is doubled.

Noise indices

N3005 The sound pressure level of industrial and environmental sound fluctuates continuously. A number of measurement indices have been proposed to describe the human response to these varying sounds. It is possible to measure the physical characteristics of sound with considerable accuracy and to predict the physical human response to characteristics such as loudness, pitch and audibility. However, it is not possible to predict *subjective* characteristics such as annoyance with certainty. This should not be surprising: one would not expect the light meter on a camera to be able to indicate whether one was taking a good or a bad photograph, although one would expect it to get the physical exposure correct. Strictly speaking, therefore, a meter can only measure sound and not noise (which is often defined as sound unwanted by the recipient): nevertheless, in practice, the terms are usually interchangeable.

Equivalent continuous 'A' weighted sound pressure level, L_{Aeq}

N3006 This unit takes into account fluctuations in sound pressure levels. It can be applied to all types of noise, whether continuous, intermittent or impulsive. L_{Aeq} is defined as the steady, continuous sound pressure level which contains the same energy as the actual, fluctuating sound pressure level. In effect, it is the energy-average of the sound pressure level over a period which must be stated, e.g. $L_{Aeq,\ (8\text{–hour})}$.

This unit is now being put forward as a universal noise index, because it can be used to measure all types of noise, although it has yet to supplant older units in certain cases, particularly for the assessment of road traffic noise where calculation techniques and regulatory criteria have not been updated.

Levels measured in L_{Aeq} can be added using the rules mentioned earlier. There is also a time trade-off: if a sound is made for half the measurement period, followed by silence, the L_{Aeq} over the whole measurement period will be 3 dB less than during the noisy half of the period. If the sound is present for one-tenth of the measurement period, the L_{Aeq} over the whole measurement period will be 10 dB less than during the noisy tenth of the measurement period. This is a cause of some criticism of L_{Aeq}: for discontinuous noise, such as may arise in industry, it does not limit the maximum noise level, so it may be necessary to specify this as well.

Daily personal noise exposure, $L_{EP,d}$

N3007 This is used in the *Noise at Work Regulations 1989* (*SI 1989 No 1790*) as a measure of the total sound exposure a person receives during the day. It is formally defined in a *Schedule* to the Regulations.

$L_{EP,d}$ is the energy-average sound level (L_{Aeq}) to which a person is exposed over a working day, disregarding the effect of any ear protection which may be worn, adjusted to an 8–hour period. Thus, if a person is exposed to 90 dB L_{Aeq} for a 4-hour working day, their $L_{EP,d}$ is 87 dB, but if they work a 12–hour shift, the same sound level would give them an $L_{EP,d}$ of 91.8 dB.

Maximum 'A' weighted sound pressure level, $L_{Amax,T}$

N3008 The maximum 'A'-weighted root-mean-square (rms) sound pressure level during the measurement period is designated L_{Amax}. Sound level meters indicate the rms sound pressure level averaged over a finite period of time. Two averaging periods (T) are defined, S (slow) and F (fast), having averaging times of 1 second and 1/8th second. It is necessary to state the averaging period. Maximum SPL should not be confused with peak SPL which is a measure of the instantaneous peak pressure.

Peak sound pressure

N3009 Instantaneous peak sound pressure can only be measured with specialist instruments. It is used in the assessment of explosive sounds, such as from gunshots and blasting, and in hearing damage assessments of this type of sound. Where the maximum sound level exceeds 125 dB $L_{Amax,F}$, then an accurate measurement of peak pressure is advisable to check for compliance with the peak action level of the *Noise at Work Regulations 1989* (*SI 1989 No 1790*).

Background noise level, L_{A90}

N3010 L_{A90} is the level of sound exceeded for 90 per cent of the measurement period. It is therefore a measure of the background noise level, in other words, the sound drops below this level only infrequently. (The term 'background' should not be confused with 'ambient', which refers to *all* the sound present in a given situation at a given time).

Sound power level, L_{WA}

N3011 The sound output of an item of plant or equipment is frequently specified in terms of its sound power level. This is measured in decibels relative to a reference power of 1 pico-Watt (dB re 10^{-12} W), but it must not be confused with sound pressure level. The sound pressure level at a particular position can be calculated from a knowledge of the sound power level of the source, provided the acoustical characteristics of the surrounding and intervening space are known. As a crude analogy, the power of a lamp bulb gives an indication of its light output, but the illumination of a surface depends on its distance and orientation from the source, and the presence of reflecting and obstructing objects in the surroundings.

To give a rough idea of the relationship between sound power level and sound pressure level, then for a noise source which is emitting sound uniformly in all directions close above a hard surface in an open space, the sound pressure level 10m from the source would be 28 dB below its sound power level.

General legal requirements

N3012 Statutory requirements relating to noise at work *generally* are contained in the *Noise at Work Regulations 1989* (*SI 1989 No 1790*). Replacing the previous Department of Employment (voluntary) Code of Practice on Noise (1972), these Regulations require employers to take reasonably practicable measures, on a long-term on-going basis, to reduce employees' exposure to noise at work to the lowest possible level, and to lower noise exposure where employees are exposed to levels of 90 dB(A) or above, or to peak action level or above (200 Pascals). In addition, ear protectors must be provided and worn and ear protection zones designated. Estimates throughout industry overall suggest that about 1.3 million workers may be exposed above 85 dB(A), the first action level.

There are now specific provisions relating to vibration exposure which are covered in the VIBRATION chapter. This is addition to the provision contained in the *Social Security* (*Industrial Injuries*) (*Prescribed Diseases*) *Regulations 1985* (*SI 1985 No 967*) and the *Reporting of Injuries, Diseases and Dangerous Occurrences Regulations 1985* (*SI 1985 No 2023*). Occupational deafness and certain forms of vibration-induced conditions, i.e. vibration-induced white finger, are prescribed industrial diseases for which disablement benefit is payable (though the 14 per cent disablement rule will obviously limit the number of successful claimants (see C6011 COMPENSATION FOR WORK INJURIES/DISEASES). Damages may also be awarded against the employer (see further N3022 below).

Noise At Work Regulations 1989 (SI 1989 No 1790)

N3013 The following duties are laid on employers.

1. To make (and update where necessary) a formal noise assessment, where employees are likely to be exposed to:

   (*a*) first action level or above (85 dB(A)),

   (*b*) peak action level or above (200 Pascals).

   [*SI 1989 No 1790, Regs 2*(*1*), *4*(*1*)].

Such assessment should be made by a competent person and adequately:

(i) identify which employees are exposed, and

(ii) provide the employer with such information as will enable him to carry out his statutory duties, and

(iii) when there is reason to suppose that the assessment is no longer valid, or when there has been a significant change in the work to which the assessment relates, review noise levels and make any changes recommended by the review.

[*SI 1989 No 1790, Reg 4*(*2*)].

2. To keep an adequate record of such assessment until a further assessment is made. [*SI 1989 No 1790, Reg 5*].

3. (As a long-term strategy, and on an on-going basis), to reduce the risk of damage to the hearing of their employees from exposure to noise to the lowest level reasonably practicable. [*SI 1989 No 1790, Reg 6*].

4. To reduce, so far as is reasonably practicable, the exposure to noise of employees (other than by provision of personal ear protectors), where employees are likely to be exposed to (*a*) 90 dB(A) or above or (*b*) peak action level (200 Pascals) or above. [*SI 1989 No 1790, Reg 7*].

5. To provide, at the request of an employee, suitable and efficient personal ear protectors where employees are likely to be exposed to 85 dB(A) or above but less than 90 dB(A). [*SI 1989 No 1790, Reg 8*(*1*)].

6. To designate ear protection zones, indicating:

(*a*) that it is an ear protection zone, and

(*b*) the need for employees to wear personal ear protectors whilst in such zone where any employee is likely to be exposed to 90 dB(A) or above, or to peak action level or above. Moreover, no employee should enter such zone unless he is wearing personal ear protectors.

[*SI 1989 No 1790, Reg 9*].

Ear protection so provided must be maintained in an efficient state and employees must report any defects in it to the employer and see that it is fully and properly used. [*SI 1989 No 1790, Reg 10*].

7. To provide employees, likely to be exposed to 85 dB(A) or above, or to peak action level or above, with adequate information, instruction and training with regard to:

(*a*) risk of damage to that employee's hearing,

(*b*) steps the employee can take to minimise the risk,

(*c*) the requirement on employees to obtain personal ear protectors from the employer, and

(*d*) the employee's duties under the regulations.

[*SI 1989 No 1790, Reg 11*].

In addition, there are specific legal requirements applying to tractor cabs and offshore installations and construction sites (see below).

(From the *Health and Safety* (*Safety Signs and Signals*) *Regulations 1996* (*SI 1996 No 341*)

Figure 1: Sign for informing that ear protectors must be worn (white on a circular blue background)

EAR PROTECTION ZONE
EAR PROTECTORS MUST BE WORN

Legal obligations of designers, manufacturers, importers and suppliers of plant and machinery

N3014 The *Supply of Machinery* (*Safety*) *Regulations 1992* (*SI 1992 No 3073* as amended) require manufacturers and suppliers of noisy machinery to design and construct such machinery so that the risks from noise emissions are reduced to the lowest level taking account of technical progress. Information on noise emissions must be provided when specified levels are reached.

If a machine is likely to cause people at work to receive a daily personal noise exposure exceeding the first or peak action levels, adequate information on noise must be provided. If the second or peak action levels are likely to be exceeded, this should include a permanent sign or label, or if the machine may be noisy in certain types of use, an instruction label which could be removed following noise testing.

Specific legal requirements

Agriculture

N3015 The *Agriculture (Tractor Cabs) Regulations 1974* (*SI 1974 No 2034*) (as amended by *SI 1990 No 1075*) provide that noise levels in tractor cabs must not exceed 90 dB(A) or 86 dB depending on which annex is relevant in the certificate under Directive 77/311/EEC. [*SI 1974 No 2034, Reg 3*(*3*)].

Equipment for construction sites and other outdoor uses

N3016 An EC Directive relating to the noise emission in the environment by equipment for use outdoors (Directive 2000/14/EC) was issued in 2000 to consolidate and update 23 earlier directives relating to the noise emission of construction plant and certain other equipment. The Directive was implemented by the *Noise Emission in the Environment by Equipment for Use Outdoors Regulations 2001* (*SI 2001 No 1701*). The earlier statutory instruments were revoked from 3 January 2002 and replaced by the requirements of the new statutory instrument.

The 2001 Regulations apply to a very wide range of powered equipment used outdoors. They do not apply to most non-powered equipment, equipment for the transport of goods or persons by road, rail, air or on waterways, or equipment for use by military, police or emergency services.

Each type of equipment is subject to a maximum permissible sound power level, dependent on the power of the equipment's engine or other drive system. The permissible sound power level will be reduced in January 2006 by 2 or 3 dB, relative to the level permitted in 2002.

Examples of the type of equipment covered include: cranes, hoists, excavators, dozers, dump trucks, compressors, generators, pumps, power saws, concrete breakers, compactors, glass recycling containers, lawnmowers and other gardening equipment.

The Regulations define the noise testing methods in great detail and also require a standardised marking of conformity indicating the guaranteed sound power level. Interestingly, it has been noted that many manufacturers simply use the maximum permitted level as the guaranteed level, which would seem to indicate that they see no marketing advantage in promoting quieter products.

The Supply of Machinery (*Safety*) *Regulations, 1992* (*SI 1992 No 3073* as amended) are also relevant to noise emissions from machinery.

Code of practice for noise and vibration control

N3017 The code of practice for basic information and procedures for noise and vibration control, BS 5228: Parts 1 to 5, gives detailed guidance on the assessment of noise and vibration from construction sites, open-cast coal extraction, piling operations, and surface mineral extraction. Part 1, revised in 1997, gives detailed noise calculation and assessment procedures. Its predecessor (published in 1984) was an approved code of practice under the *Control of Pollution Act, 1974*. Part 1 of the code of practice is principally concerned with environmental noise and vibration, but also briefly recites the *Noise at Work Regulations 1989* (*SI 1989 No 1790*).

Compensation for occupational deafness

Social Security

N3018 The most common condition associated with exposure to noise is occupational deafness. Deafness is prescribed occupational disease A10 (see OCCUPATIONAL HEALTH AND DISEASES). Prescription rules for occupational deafness have been extended three times, in 1980, 1983 and 1994. It is defined as: 'sensorineural hearing loss amounting to at least 50 dB in each ear being the average of hearing losses are 1, 2 and 3 kHz frequencies, and being due, in the case of at least one ear, to occupational noise'. [*Social Security* (*Industrial Injuries*) (*Prescribed Diseases*) *Amendment Regulations 1989* (*SI 1989 No 1207*), *Reg 4*(*5*)]. Thus, the former requirement for hearing loss to be measured by pure tone audiometry no longer applies. Extensions of benefit criteria relating to occupational deafness are contained in the *Social Security* (*Industrial Injuries*) (*Prescribed Diseases*) *Regulations 1985* (*SI 1985 No* 967), which are amended, as regards assessment of disablement for benefit purposes, by the *Social Security* (*Industrial Injuries*) (*Prescribed Diseases*) *Amendment Regulations 1994* (*SI 1994 No 2343*).

Conditions for which deafness is prescribed

N3019 Occupational deafness is prescribed for a wide range of occupations involving working with, or in the immediate vicinity of:

- mechanically-powered grinding and percussive tools used on metal; on rock and stone in quarries or in mines, in sinking shafts and in tunnelling;
- sawing masonry blocks, jet channelling of stone and masonry, vibrating metal moulding boxes in the concrete products industry;
- forging, cutting, burning and casting metal;
- plasma spray guns engaged in the deposition of metal;
- air arc gouging;
- high pressure jets of water or a mixture of water and abrasive material in the water jetting industry (including work under water);
- textile manufacturing where the work is undertaken in rooms or sheds in which there are machines engaged in weaving man-made or natural (including mineral) fibres or in the high speed false twisting or fibres; or mechanical bobbin cleaning;
- machines for automatic moulding, pressing or forming of glass hollow ware and continuous glass toughening furnaces; and spinning machines using compressed air to produce glass wool or mineral wool;
- a wide range of woodworking and metalworking machines;
- chain saws in forestry;
- work in ships' engine rooms;
- work on gas turbines in connection with performance testing on a test bed, installation testing of replacement engines in aircraft, and acceptance testing in armed service fixed wing combat planes.

The Regulations are detailed and should be consulted for precise definitions of these occupations. [*Social Security* (*Industrial Injuries*) (*Prescribed Diseases*) *Regulations 1985* (*SI 1985 No 967*); *Social Security* (*Industrial Injuries*) (*Prescribed Diseases*)

Amendment No 2 Regulations 1987 (*SI 1987 No 2112*); *Social Security* (*Industrial Injuries*) (*Prescribed Diseases*) *Amendment Regulations 1994* (*SI 1994 No 2343*)].

'Any occupation' covers activities in which an employee is engaged under his contract of employment. The fact that the work-force is designated, classified of graded by reference to function, training or skills (e.g. labourer, hot examiner, salvage and forge examiner) does not of itself justify a conclusion that each separate designation, classification or grading involves a separate occupation (*Decision of the Commissioner No R*(*I*) *3/78*).

'Assistance in the use' of tools qualifies the actual use of tools, not the process in the course of which tools are employed. Thus, a crane driver who positions bogies to enable riveters to do work on them and then goes away, assists in the process of getting bogies repaired, which requires use of pneumatic tools – is not assistance in the actual use of tools, for the purposes of disablement benefit. The position is otherwise when a crane holds a bogie in suspension to enable riveters to work safely on them. Here the crane driver assists in the actual use of pneumatic percussive tools (*Decision of the Commissioner No R*(*I*) *4/82*).

Conditions under which benefit is payable

N3020 For a claimant to be entitled to disablement benefit for occupational deafness, the following conditions under the *Social Security* (*Industrial Injuries*) (*Prescribed Diseases*) *Regulations 1985* (*SI 1985 No 967*) currently apply:

- they must have been employed:
  - (i) at any time on or after 5 July 1948, and
  - (ii) for a period or periods amounting (in the aggregate) to at least ten years.
- there must be permanent sensorineural hearing loss, and loss in each ear must be at least 50 dB; and
- at least loss of 50 dB in one ear must be attributable to noise at work [*SI 1985 No 967, Sch 1, Pt 1*].

  (There is a presumption that occupational deafness is due to the nature of employment [*SI 1985 No 967, Reg 4*(*5*)]);
- the claim must be made within five years of the last date when the claimant worked in an occupation prescribed for deafness [*SI 1985 No 967, Reg 25*(*2*)];
- any assessment of disablement at less than 20 per cent is final [*SI 1985 No 967,Reg 33*].

A person, whose claim for benefit is turned down because he/she had not worked during the five years before the claim in one of the listed occupations, may claim if he/she continued to work and later met the time conditions. If a claim is turned down as the disability is less than 20 per cent, the claimant must wait three years before re-applying. If, by waiting three years, it would be more than five years since the applicant worked in one of the listed occupations, the three-year limit is waived.

Assessment of disablement benefit for social security purposes

N3021 The extent of disablement is the percentage calculated by:

(*a*) determining the average total hearing loss due to all causes for each ear at 1, 2 and 3 kHz frequencies; and (*b*) determining the percentage degree of disablement for each ear; and then

(*c*) determining the average percentage degree of binaural disablement.

[Social Security (Industrial Injuries) (Prescribed Diseases) Amendment Regulations 1989, Reg 4(2)].

The following chart (TABLE 1 below), shows the scale for all claims made on/after 3 September 1979.

Table 1

Percentage degree of disablement in relation to hearing loss

Hearing loss	Percentage degree of disablement
50–53 dB	20
54–60 dB	30
61–66 dB	40
67–72 dB	50
73–79 dB	60
80–86 dB	70
87–95 dB	80
96–105 dB	90
106 dB or more	100

[*Social Security* (*Industrial Injuries*) (*Prescribed Diseases*) *Regulations 1985, Reg 34, Sch 3, Pt II,* as amended]

Any degree of disablement, due to deafness at work, assessed at less than 20 per cent, must be disregarded for benefit purposes. [*Social Security* (*Industrial Injuries*) (*Prescribed Diseases*) *Amendment Regulations 1990* (*SI 1990 No 2269*)].

Action against employer at common law

N3022 There is no separate action for noise at common law; liability comes under the general heading of negligence. Indeed, it was not until as late as 1972 that employers were made liable for deafness negligently caused to employees (*Berry v Stone Manganese Marine Ltd [1972] 1 Lloyd's Rep 182*). Absence of a previous general statutory requirement on employers regarding exposure of employees to noise sometimes led to the law being strained to meet the facts (*Carragher v Singer Manufacturing Co Ltd 1974 SLT (Notes) 28* relating to the *Factories Act 1961, s 29*: 'every place of work must, so far as is reasonably practicable, be made and kept safe for any person working there', to the effect that this is wide enough to provide protection against noise). Admittedly, it is proper to regard noise as an aspect of the working environment (*McCafferty v Metropolitan Police District Receiver [1977] 2 All ER 756* where an employee, the plaintiff, who was a ballistics expert, suffered ringing in the ears as a result of the sounds of ammunition being fired from different guns in the course of his work. When he complained about ringing in the ears – the ballistics room had no sound-absorbent material on the walls and he had not been supplied with ear protectors – he was advised to use cotton wool, which was useless. It was held that his employer was liable since it was highly foreseeable that the employee would suffer hearing injury if no steps were taken to protect his ears, cotton wool being useless). Moreover, although there are specific statutory requirements to minimise exposure to noise (in agriculture and offshore operations and construction operations, see N3015, N3016 above), these have generated little or no case law.

The main points established at common law are as follows.

(*a*) As from 1963, the publication date by the (then) Factory Inspectorate of 'Noise and the Worker', employers have been 'on notice' of the dangers to hearing of their employees arising from over-exposure to noise (*McGuinness v Kirkstall Forge Engineering Ltd 1979, unreported*). Hence, consistent with their common law duty to take reasonable care for the health and safety of their employees, employers should 'provide and maintain' a sufficient stock of ear muffs.

This was confirmed in *Thompson v Smiths, etc.* (see (*d*) below). However, more recently, an employer was held liable for an employee's noise-induced deafness, even though the latter's exposure to noise, working in shipbuilding, had occurred *entirely before* 1963. The grounds were that the employer had done virtually nothing to combat the ***known*** noise hazard from 1954–1963 (apart from making earplugs available) (*Baxter v Harland & Wolff plc, Northern Ireland Court of Appeal 1990 (unreported)*).

This means that, as far as Northern Ireland is concerned, employers are liable at common law for noise-induced deafness as from 1 January 1954 – the earliest actionable date. Limitation statutes preclude employees suing prior to that date (*Arnold v Central Electricity Generating Board [1988] AC 228*).

(*b*) Because the true nature of deafness as a disability has not always been appreciated, damages have traditionally not been high (*Berry v Stone Manganese Marine Ltd [1972]* – £2,500 (halved because of time limitation obstacles); *Heslop v Metalock (Great Britain) Ltd (1981)* – £7,750; *O'Shea v Kimberley-Clark Ltd (1982)* – £7,490 (tinnitus); *Tripp v Ministry of Defence [1982] CLY 1017* – £7,500).

(*c*) Damages will be awarded for exposure to noise, even though the resultant deafness is not great, as in tinnitus (*O'Shea v Kimberley-Clark Ltd, The Guardian, 8 October 1982*).

(*d*) Originally the last employer of a succession of employers (for whom an employee had worked in noisy occupations) was exclusively liable for damages for deafness, even though damage (i.e. actual hearing loss) occurs in the early years of exposure (for which earlier employers would have been responsible) (*Heslop v Metalock (Great Britain) Ltd, The Observer, 29 November 1981*). More recently, however, the tendency is to *apportion* liability between offending employers (*Thompson, Gray, Nicholson v Smiths Ship Repairers (North Shields) Ltd*; *Blacklock, Waggott v Swan Hunter Shipbuilders Ltd*; *Mitchell v Vickers Armstrong Ltd [1984] IRLR 93*). This is patently fairer because some blame is then shared by the original employer(s), whose negligence would have been responsible for the actual hearing loss.

(*e*) Because of the current tendency to apportion liability, even in the case of pre-1963 employers (see *McGuinness v Kirkstall Forge Engineering Ltd* above), contribution will take place between earlier and later insurers.

(*f*) Although judges are generally reluctant to be swayed by scientific/statistical evidence, the trio of shipbuilding cases (see (*d*) above) demonstrates, at least in the case of occupational deafness, that this trend is being reversed (see Table 2 below, the 'Coles-Worgan classification'); in particular, it is relevant to consider the 'dose response' relationship published by the National Physical Laboratory (NPL), which relates long-term continuous noise exposure to expected resultant hearing loss. This graph always shows a rapid increase in the early years of noise exposure, followed by a trailing off.

(*g*) Current judicial wisdom identifies three separate evolutionary aspects of deafness, i.e. (i) hearing loss (measured in decibels at various frequencies); (ii) disability (i.e. difficulty/inability to receive everyday sounds); (iii) social handicap (attending musical concerts/meetings etc.). That social handicap is a genuine basis on which damages can be (*inter alia*) awarded, was reaffirmed in the case of *Bixby, Case, Fry and Elliott v Ford Motor Group (1990, unreported)*.

General guidance on noise at work

Guidance on the Noise at Work Regulations 1989

N3023 HSE has revised and updated its Noise Guides which are regarded as providing authoritative and detailed guidance on the Noise at Work Regulations. These are now published in a single volume called *Reducing noise at work Guidance on the Noise at Work Regulations 1989* (ISBN 0 7176 1511 1). It covers the following areas:

- legal duties of employers to prevent damage to hearing;
- duties of designers, manufacturers, importers and suppliers;
- how to choose a competent person – advice for employers;
- how to carry out a noise assessment – advice for the competent person;
- control of noise exposure – advice for employers and engineers;
- selection and use of personal ear protection – advice for employers.

Guidance on specific types of equipment

N3024 HSE has also issued a number of guidance documents for specific types of machine or activity, including:

- Hazards arising from ultrasonic processes (Local Authority Circular 59/1, 2000).
- Metal cutting circular saws (Noise Reduction (Local Authority Circular 59/2, 2000).
- The reduction of noise from pneumatic breakers, hammers, drills, etc (Local Authority Circular 59/4, 2000).
- Reducing noise from CNC presses (Engineering Information Sheet No. 39, 2002).

Other guidance

N3025 Some working environments carry particular difficulties in the assessment and control of noise exposure, mainly because of the variability of the noise levels and the length of exposure. The following specific guidance may be of assistance:

- *A guide to reducing the exposure of construction workers to noise*, R A Waller, Construction Industry Research and Information Association, CIRIA Report 120 (1990).

- *Offshore installations: Guidance on design and construction, Part II, Section 5*, Department of Energy (1977).
- *Guide to health, safety and welfare at pop concerts and similar events*, Health and Safety Commission, Home Office and the Scottish Office, HMSO (1993).
- *Practical solutions to noise problems in agriculture*, HSE Research Report 212 (2004).

Reducing noise in specific working environments

N3026 Although many noisy industries are in decline, noise is now becoming an issue in some newer industries, as described below.

Discos and night clubs

N3027 These venues can be a problem because operators have two conflicting needs: they need to provide loud music for their clients, but they must provide a safe working environment for staff. The music levels on the dance floor are often over 100 dB(A) to provide the buzz that clients expect. It is, of course, an interesting point as to whether clients will in the future be looking for compensation for noise-induced hearing loss.

A number of different techniques may need to be used in order to achieve adequate control of the noise exposure of staff. Loudspeakers should be kept away from the bar area. The bar itself can be designed to be partially screened from the loudspeakers and with sound absorbent to reduce noise levels in the locality. A high-quality sound system with low distortion levels can produce the bass frequencies that clients want, with less high-frequency distortion. This results in lower overall sound pressure levels than a poorer system would produce. Such a system can have the additional advantage of reducing environmental noise levels, which are often problematic. Some clubs provide a quiet 'chill-out' room to which staff and members can retreat.

Ear plugs can be used by staff patrolling the dance floor and seating areas, and jobs can be rotated between staff so that the average noise exposure is reduced. Most ear plugs will reduce sound at speech frequencies by more than at the bass frequencies usual in disco music and this may reduce the ability of staff to understand speech. With the anticipated new Control of Noise at Work Regulations, it will be important to ensure that ear plugs give adequate protection at the important bass frequencies.

Call centres

N3028 Call centres can be very noisy simply because of the large number of people talking on the telephone. If poorly-designed, each operator will inevitably raise their voice so as to be heard above the noise produced by their neighbours. This becomes self-defeating as others then raise their voices further. They may also need to raise the volume of their earpiece, and this could lead to excessive sound exposure. There have been successful cases of compensation for deafness (sometimes only in one ear) suffered by people using telephone and radio headsets, and employers should ensure that staff do not adjust them to be excessively loud.

A number of techniques must be used to achieve a satisfactory acoustic environment. Firstly the basic acoustics of the call centre must be satisfactory. Working areas should be as sound-absorbent as feasible, with acoustic ceiling tiles and a thick, sound absorbent carpet. Acoustic screens should be used between workstations unless they are widely-spaced. The screens need to have thick absorbent surfaces, and should be high enough to screen operators when they are sitting at their workstations.

Staff training must include discussion of appropriate voice and earpiece levels, and the risks of excessive levels. They need to be encouraged to keep their voices down: it should be explained that it is not necessary for them to raise their voices in order for the customer to hear them clearly.

Workplace noise assessments

Competent person

N3029 The *Noise at Work Regulations 1989* (*SI 1989 No 1790*) require surveys to be made by a 'competent person'. This is someone who understands the requirements of the Regulations, is experienced in using sound level meters and in making workplace noise assessments. The Regulations do not stipulate a particular form of certification for such people, although the Institute of Acoustics and other organisations do provide formal training and certification that is intended to meet these requirements.

Sound level meters

N3030 It is usually more economical to employ an acoustics consultant for workplace noise assessments than to send a member of staff on a training course and to hire a sound level meter to do the measurements in-house. It may be helpful if the report is clearly independent should an employee make a claim. Suitable sound level meters are expensive. A precision meter is required and these cost some thousands of pounds, depending on their capabilities, and they must be sent away for laboratory calibration at regular intervals. Those meters that can be bought cheaply from electronics suppliers are not suitable: they could be misleading even as a 'screening' tool. This is because cheap meters do not have an accurate response to rapid or impulsive sounds, and usually do not contain an L_{Aeq} function as this requires an on-board computer chip for the complex averaging process.

Assessing the daily personal noise exposure

N3031 It is important to appreciate that noise assessments must determine the daily personal noise exposure of each individual member of staff – it is not sufficient simply to measure the noise level in various parts of the workplace and to mark the noisy areas as 'hearing protection zones'.

In some workplaces, this is not particularly difficult to do: if the employee has a fixed workstation and fixed working routine, it will only be necessary to measure at a location representative of their head position, for a sufficient period of time to cover a few work cycles – maybe five minutes or so.

However, it is quite common for a worker to use a variety of machines or to move between different parts of the works during a normal day. Some workers, such as crane or lorry drivers, may have a work environment where noise levels vary a great deal and it is not practicable for the noise surveyor to accompany them throughout a representative period.

These considerations mean that a workplace noise assessment needs planning before it is made. The first item is to establish the different types of work that are being done on the site: this will need a meeting with the site manager to determine how many staff there are, the jobs that they do, shift-working patterns, and the hours of work, including when and where meal-breaks are taken. It is also important to consider the sensitivities of the workforce: it is helpful to explain in brief terms what you are doing and to confirm what they do. They will also want to know the results, but the surveyor should explain that this cannot be provided immediately as the daily personal noise exposure needs to be calculated. In some workplaces, it may be necessary to ensure that union representatives are consulted.

A scale plan of the workplace will be needed so that measurement locations, noise sources and noisy areas can be recorded for the report. The surveyor will wish to ensure that all machinery and plant is tested, and will want to note anything that is out of use during the survey.

Dosimeters

N3032 Where there are people such as lorry drivers, whom the surveyor cannot accompany, then it will probably be best to ask them to wear a dosimeter. This is a small electronic device about the size of a mobile phone that clips into the top pocket, with a microphone that clips to the lapel or some convenient part of the clothing near to the ear. The dosimeter does not record the actual sound, only the sound level. (It is not a 'spy in the cab'.) Dosimeters usually contain some protection to deter tampering or falsification and to prevent the worker obtaining the readout, although the surveyor will need to be alert to the risks.

Where workers move between various workstations during their working day, then it is usually acceptable to measure the noise level at each workstation and to calculate the daily personal noise exposure from an estimate of the proportion of the day that each worker typically spends at each one.

Although it may seem that this is less accurate than using a dosimeter, it is important to recognise that each of the work-cycles will usually produce a slightly different noise exposure, as there are so many things that affect the noise levels. Moreover, it is necessary to obtain actual workstation noise levels in order to identify any ear protection zones that need to be marked out. In some borderline cases, maybe dosimetry might settle the matter, but it would be better to look at ways of reducing noise levels, since the effect on any one individual's hearing is difficult to determine, even though the Regulations define action levels precisely.

Hazardous environments

N3033 Special care is required in hazardous working environments, especially those with an explosion risk (such as gas and petroleum processing plants). Many sound level meters use a high 'polarisation voltage' for the microphone, which could produce a spark. 'Intrinsically safe' meters can be obtained for such instances, although they can be less convenient for ordinary use.

Requirements for the report

N3034 A written report of the survey must be produced and kept as a record of the survey at least until the next survey is made. The report should state all the information used in the survey, and should contain a plan showing the areas surveyed and hearing protection zones that were identified. It must identify individual staff or jobs where noise exposure exceeds any of the action levels. The report should also contain a statement of the employer's duties resulting from the findings of the survey. However, it is not usual for the report to include detailed specifications for any noise remediation that might be needed. This is usually a separate study, as the issues cannot be known prior to the survey and there will usually be a number of options which will require detailed discussion with the employer.

Noise reduction

N3035 It is the duty of every employer to reduce the risk of hearing damage from noise at work as far as reasonably practicable. This is an on-going and long-term obligation. Unfortunately, it is often difficult to reduce the noise of many items of equipment, particularly hand-held tools. Remarkably, it is still the case that quietness does not seem to be a primary selling point for tools and equipment, and it should not be assumed that newer equipment will necessarily be quieter. Moreover, although many items of equipment must be labelled with their sound power, the label often shows only the permitted maximum level rather than the actual level, so this is of little help in choosing the quietest items. Furthermore, it should not be assumed that smaller items of plant will be quieter than larger items. This is because larger items often have more scope for noise control in the form of extra casings, insulation and silencers, as size and weight are less of a limitation.

With the advent of the proposed new Control of Noise At Work Regulations, the buying policy will become a very important aspect of noise reduction, since it is often difficult, inconvenient and expensive to retrospectively fit noise control equipment. A noise target level should be set and then equipment bought to comply with this, taking into account the combined effect of many items operating simultaneously, and the effect of reverberant build-up of sound in the workplace.

When designing noise reduction measures, it is essential to identify the major sources of noise first, since no amount of work in reducing minor sources will have any effect on the overall noise level. Noise is usually created by a moving or vibrating part of the equipment inducing vibrations in the surrounding air. Noise can be reduced by preventing such vibrations from reaching the operator of the equipment: the most effective solution is to block 'air-paths' with a solid, heavy casing. Even tiny air-holes can allow the sound to escape, which is a problem when a flow of cooling air is needed, or where there must be continuous access to feed work-pieces.

A different approach is to fit 'damping material' to vibrating surfaces, to reduce the amplitude of vibration. Damping material is especially effective on thin sheet metal, such as equipment casings, hoppers, containers and ductwork. This is because even a small amount of vibration can cause their surfaces to 'radiate' a large amount of noise energy. The vibrational energy is absorbed in the damping material, rather than being radiated. There are a number of ways of applying damping material: a bitumastic or rubber sheet can be glued to the surface, a second sheet of metal can be glued by an elastic adhesive to the surface, or the item can be made from a laminate composed of two sheets of metal glued with an elastic adhesive.

It should be noted that sound absorptive material (mineral fibre, etc) is not used as a damping material. However, it can be used to provide a vibration break when cladding a vibrating surface. For example, if noise is breaking out from a duct, it can be wrapped in mineral fibre and then clad with a thin metal sheet. The mineral fibre separates the inner and outer cladding, thereby preventing much of the vibration in the wall of the duct from reaching the cladding, which therefore radiates much less noise.

Sometimes, a less 'springy' material than steel can be used. At one time lead was popular for being both heavy and floppy, but is now considered too hazardous for general use. However, plasterboard is a popular acoustic material, as it is reasonably heavy, has good natural damping and is inexpensive. It can often be used to prevent noise breakout, as long as there is a vibration break between it and the radiating surface.

Ventilation and cooling is often a problem, since enclosing a piece of equipment can prevent 'natural' cooling. Moreover, fans used to provide a flow of cooling air can be very noisy in themselves, not counting the noise that escapes along the ductwork. The best solution in such cases is to place the fan on the inside of the casing and to fit an attenuator ('silencer') on the external side of the fan. Attenuators are essentially rectangular or circular ducts with an acoustic absorbent lining.

There are many different types of fan and it is important to select one that is appropriate for the task in hand, taking account of the volume of air that needs to be moved and the pressure (resistance) that the fan must work against. Small, fast-running fans tend to be noisier than larger slower-running fans for a given duty. The amount of noise produced by a fan and the amount of attenuation provided by a silencer are well-characterised and so ventilation systems can be designed with confidence over their performance, although this is a specialist job.

Sometimes, developments intended for another purpose can have the additional benefit of reducing noise. For example, there was a need to develop a circular saw blade that gave better cutting accuracy. It was found that vibration of the saw blade caused irregularities in the cut, so slots were cut in the surface of the blade and these were filled with a resin to maintain the strength of the blade. This extra damping was effective in reducing vibration, thereby improving both the cut and the noise levels.

Despite the technology, it is often difficult to obtain adequate noise reduction. This is because of the huge range of the hearing mechanism. A 10 dB reduction requires the escaping noise energy to be reduced by 90 per cent, a 20 dB reduction requires the escaping noise energy to be reduced by 99 per cent, and a 30 dB reduction requires 99.9 per cent of the energy to be removed. Because of these considerations, it may be advisable to employ a specialist who can design a solution with a minimum of trial and error.

Ear protection

N3036 There are two main categories of ear protectors:

(i) circumaural protectors (ear muffs) which fit over and surround the ears, and which seal to the head by cushions filled with soft plastic foam or a viscous liquid; and

(ii) ear plugs, which fit into the ear canal.

Ear protectors will only be effective if they are in good condition, suit the individual and are worn properly. Ear protectors can be uncomfortable, especially if they press too firmly on the head or ear canal, or cause too much sweating. Some users may be tempted to bend the head-band so that the muffs do not press so tightly, which reduces the effectiveness of the seal. Ear muffs are also less effective for people with thick spectacle frames, long hair or beards that prevent the muff from sealing fully against the head.

The amount of protection differs between different designs, and usually ear protectors give more protection at middle frequencies and less protection at low frequencies. Where the noise has strong tones (often described by words such as rumble, drone, hum, whine, screech) then particular care is needed to ensure that the protectors give adequate attenuation at those frequencies. However, it can be counterproductive to choose ear protectors that give far more attenuation than necessary, since these will be heavier, with greater head-band or insert pressure and so less comfortable to wear. And an ear protector that is not worn does not give any protection at all. Ear plugs should not be used by people with ear infections and certain other ear conditions, and it may be desirable for such people to obtain medical advice.

Because individuals differ greatly in their preference, the employer should select more than one type of suitable protector and offer the user a personal choice where possible. British Standard BS EN 485: 1994 gives guidance and recommendations for the selection, use, care and maintenance of hearing protectors.

Ear protectors need to be maintained in a clean and hygienic condition. Insert types should not be shared between people, and muffs should not usually be shared. With ear muffs, it is necessary to check that the sealing cushion is not damaged, as this will seriously reduce its effectiveness. On some makes it is possible to change a damaged cushion. The ear cups have a foam lining which must not be discarded. If it gets dirty, it can usually be taken out and washed in detergent, then dried and replaced.

If the headband tension becomes weak, the muffs should be discarded.

Some earplugs are designed for one-off use, but others are reusable. These need to be kept in a protective container when not in use, and regularly washed according to the manufacturer's instructions. If the seals become damaged or hardened, the plugs must be replaced.

The employer needs to ensure that staff know how to check the condition of their ear protection and how to obtain replacements as soon as the need occurs.

Acoustic booths

N3037 In some workplaces, it may be possible for operators to be located in a control booth from which they can monitor the plant, only occasionally entering the noisy area. They may still need put on ear defenders when leaving the acoustic booth, since it may require only a few seconds or minutes to exceed the allowable daily noise dose.

Even where it is not practical to have a separate control booth, it may be possible to use an acoustic refuge – a small cabin with an open side – that will give some respite from the noise. However, this should not be expected to give more than a nominal sound reduction in most cases.

Work rotation

N3038 Work rotation can sometimes be a method of reducing the daily personal noise exposure. The staff work part of the day in a noisy area and the rest of the day in a quiet area. If a person works for half their day in a place with a noise level of 93 dB L_{Aeq} (4 hours) and the other half of their day in a place with a noise level of 70 dB L_{Aeq} (4 hours) they would just meet a noise exposure of 90 dB. However, this is not a very satisfactory method as it can easily go wrong. They may spend longer than expected in the noisy workplace, or there may be a higher than expected noise level in the quiet area, for example. And again, whilst this approach could satisfy the legal requirements, some people may suffer hearing loss from shorter exposure to loud noise.

Noise barriers

N3039 Noise barriers often give very limited noise reduction within buildings. This is because sound can be reflected over the barrier from the ceiling, pipe-work, and other overhead fittings. It is usually necessary to use noise barriers in conjunction with sound baffles suspended from the ceiling. Suspended sound baffles on their own have very limited application. They can sometimes be helpful in very reverberant work-spaces where the staff are not exposed to the direct sound of noisy machines.

PA and music systems

N3040 Some workplaces use public address and music systems. One study found that they were a major source of excessive noise in lorry cabs. Care must be taken to ensure that in noisy workplaces these system are not so loud that they create excessive noise in themselves. This also applies to people who wear radio or telephone headsets.

Sound systems for emergency purposes

N3041 Voice messages can be superior to bells or sirens to convey warnings and instructions in emergencies. However, these must be properly audible in noisy places. British Standard BS 7443: 1991 gives specifications for sound systems for emergency purposes.

Occupational Health and Diseases

Introduction

O1001 Occupational disease as a topic is often in the public eye. There has been a significant amount of public and legal debate about asbestos-related disease and work-related stress conditions over recent years. The label 'occupational' causes much difficulty often due to the conflicting views over what causes disease. The debate tends to be polarised and highly emotive.

There are about 70 or so prescribed occupational diseases or conditions recognised under the *Social Security (Industrial Injuries) (Prescribed Diseases) Regulations 1985* (*SI 1985 No 967*) (as amended) for which benefit may be claimed subject to certain qualifications (see Table 2 in O1053). There are also many other conditions which may have an occupational cause and which are not on the prescribed list. Some conditions may have more than one cause(e.g. hearing loss in the inner ear may be due to prolonged exposure to excessive noise, but may additionally be due to the effects of ageing). Some conditions may have a non-occupational origin which is then exacerbated by work. Establishing occupational causation is therefore a complicated exercise that involves the careful elimination of other possible causes

The compensation factor cannot be overlooked as the confirmation of an occupational cause often opens the door to a possible claim for benefit from the Department of Social Security or for civil liability for an employer or other negligent party. Those who suffer from a non-occupational condition will usually visit their GP or hospital in the first instance, but those who believe there is an occupational cause sometimes consult their trade union or a solicitor before seeking medical advice.

Diagnosis of some occupational diseases is difficult: in some cases there are no universally recognised clinical tests and the examiner must rely on the subjective history provided by the patient. For example, in the early stages of vibration white finger, the symptoms may be transient and not present at the time of the examination

The extent of occupational disease in the United Kingdom is unknown. It certainly runs into millions of sufferers but numbers depend on the extent of reporting, the gathering of this information and in some cases, the accuracy of the diagnosis. Department of Social Security statistics suggest that there are at least 60,000 new claims each year, but this is only the tip of the iceberg as many do not qualify for benefit. Insurance statistics of civil claims made may be a more reliable indicator but reporting is spasmodic and there may be overlap where one person is claiming damages from several parties.

Occupational disease has a clear impact, not just in terms of sickness absence but also in relation to the additional costs associated with managing lost time/productivity/human resources at employer level and paying/administering benefits at state level.

Occupational diseases are widespread amongst those who have worked in hazardous industries such as mining, construction and certain heavy industries. We are still left with the heavy burden and effect of industrial practices in the 1950's, 1960's and 1970's. There is, however, an increasing occurrence of 'white collar' disease: new technology has brought new problems. The widespread use of visual display units has brought potential risks of musculo-skeletal disorders and eyesight problems (see also DISPLAY SCREEN EQUIPMENT). Increasing work pressures and the emphasis on individual quality of life has seen the increased incidence of stress-related conditions, especially in the public sector.

This pattern is reflected by the changing emphasis of health and safety legislation. Previous legislation based on the *Factories Act 1961* and the *Offices, Shops and Railway Premises Act 1963* (together with detailed regulations applying to specific trades and processes), has been replaced by new legislation which is more widely based and which applies generally to people 'at work'. The first radical change was the *Health and Safety at Work etc Act 1974* (*HSWA 1974*), but subsequent legislation has largely been due to the implementation of various EU health and safety at work directives.

Some conditions are known as short-tail diseases because the symptoms often become manifest within a short period of time from the relevant exposure (e.g. dermatitis). Others (like asbestos-related diseases) are long-tail as the symptoms may not arise for many years after the relevant exposure. One particular asbestos disease, mesothelioma, may not become manifest until 40 or 50 years later. This means that there could be generations of workers whose exposure ceased in the 1960s and 1970s but whose symptoms may not appear until the first part of the 21st century. When such conditions do arise, investigation of exposure will be exceedingly difficult with the passage of time. One should also remember that some diseases are suffered by those who do not have direct contact. This so-called 'neighbourhood exposure' could affect others working nearby, those washing a worker's dusty overalls or, in one case, even children playing in a dusty factory yard.

Many diseases are incurable although there may be treatment that alleviates the symptoms. The progression of some diseases may be halted by prompt intervention, but the removal of the worker from the harmful exposure is often necessary and this may mean the loss of a job or the transfer to lower-paid work. Regrettably, some diseases, especially lung conditions, are often fatal. Most long-tail diseases are caused by past neglect of proper preventative action, even when viewed in the light of knowledge at the time. Strict observance of current legislation and codes of practice should go a long way to ensure that the present epidemic of occupational disease does not continue.

Long-tail diseases

O1002 Most of these diseases originate from the former manufacturing industries and, because of the latent period between the first exposure and the onset of the condition, we are now faced with the legacy of those workers who are suffering because of past neglect. Many of these workers are no longer in the jobs where their exposure occurred and many of their former workplaces have either closed or have been drastically changed. This makes investigation into the cause of the disease somewhat difficult.

Modern health and safety legislation should help to eradicate many of the old problems that were the root cause of these conditions but it would be naive to think that no-one remains at risk from long-tail hazards. There are also those working today who are already suffering from one or more of these diseases due to past exposures, so it is essential that any present employer is made aware of their pre-existing problems and takes adequate steps to prevent further harm.

Three typical long-tail diseases have been selected for detailed comment:

- Asbestos-related diseases;
- Noise-induced hearing loss (NIHL); and
- Hand arm vibration syndrome (HAVS).

Asbestos-related diseases

O1003 Asbestos is a name given to a group of minerals whose common feature is their fibrous nature. It is found in rock fissures in certain parts of the world, including Canada and Southern Africa. It is a very versatile mineral which, when broken down into fibres, may be spun, woven and incorporated into many compounds. There are three main kinds of asbestos:

(*a*) *Chrysotile* (*white*). This is the most commonly used, and has strong, silky, flexible fibres which are easily used in the making of asbestos textiles.

(*b*) *Crocidolite* (*blue*). This has brittle fibres with high tensile strength which are highly resistant to chemicals and sea water.

(*c*) *Amosite* (*brown*). This has long, fairly strong fibres with good insulation properties.

It is commonly believed that blue asbestos is the most dangerous of the three types because of its fibre length and diameter, but all three types should be regarded as being dangerous to health because of their indestructible nature. The fibres have the propensity to penetrate all the body's natural defence mechanisms and to lodge permanently in the lungs.

The use of asbestos grew considerably through the first part of the 20th century and probably reached a peak in the 1960s and early 1970s. Its uses are many and varied and include:

(*a*) manufacture of asbestos textiles;

(*b*) thermal insulation (lagging), especially in shipyards and power stations;

(*c*) electrical insulation;

(*d*) clutch and brake linings;

(*e*) asbestos cement products (like roofing sheets and pipework).

Dockers who unloaded sacks of raw asbestos may also have been exposed, and exposure may be an ongoing problem where asbestos lagging has to be removed.

From the early part of the 20th century, some of the health problems associated with asbestos were beginning to become apparent. The first statutory controls were contained in the *Asbestos Industry Regulations 1931* but this legislation was limited to the asbestos manufacturing processes. After World War II it was suspected that certain groups of shipyard workers may have been at risk and some asbestos controls were incorporated in the *Shipbuilding and Ship-repairing Regulations 1960* (*SI 1960 No 1932*). There were also more general requirements in the *Factories Acts of 1937 and 1961* relating to ventilation and the removal of dust. It is probably fair to say that most of these statutory requirements went unheeded by the majority of employers until the mid to late 1960s, when the extent of the asbestos problem began to be fully realised.

Firstly, an investigation into conditions in a mining area of South Africa, published in 1960 found a high incidence of a rare cancer, mesothelioma, amongst those who lived and worked there. Secondly, an investigation in 1965 discovered a high incidence of this disease in the London area. Public concern grew and the *Asbestos Regulations 1970* (of much wider application than the 1931 Regulations), were introduced. Further stringent measures have since been taken, culminating in the *Control of Asbestos at Work Regulations 1987* (*SI 1987 No 2115*) (*now revoked and replaced by SI 2002 No 2675*) and the *Asbestos* (*Licensing*) *Regulations 1983* (*SI 1983 No 1649*) (*as amended by SI 1998 No 3233*) and the *Asbestos* (*Prohibitions*) *Regulations 1992* (*SI 1992 No 3067*). (See also amending regulations *SI 1999 No 2373, SI 1999 No 2977 and SI 2003 No 1889.*)

The *Control of Asbestos at Work Regulations 2002* (*SI 2002 No 2675*) came into force on 21 May 2004 and are causing a considerable degree of concern for those with responsibility for the repair and maintenance of non-domestic premises. *Regulation 4* of the Regulations imposes a duty on such persons to manage and control any asbestos in the building. (See ASBESTOS.)

The conditions which may arise from asbestos exposure are:

(*a*) mesothelioma;

(*b*) asbestosis;

(*c*) bronchial carcinoma (lung cancer);

(*d*) diffuse pleural thickening; and

(*e*) pleural plaques.

Mesothelioma

O1004 This is a cancer of the pleura, a thin lining which surrounds the lungs. Sometimes it may also affect the peritoneum, the lining of the abdominal cavity. It is believed that only a relatively small degree of exposure is necessary for the disease to be initiated. Blue asbestos is thought to be the culprit but other types cannot be excluded.

There is usually an extremely long latent period between exposure and onset during which time the cells are slowly multiplying but remain undetected. Patients may first suffer chest pains and breathlessness and begin to lose weight, but by the time the condition is diagnosed, it is too late. Progression is rapid, the cancer invades the lungs and the condition becomes extremely painful and distressing. This type of cancer does not currently respond to treatment or surgical intervention and most patients die within 12 to 18 months of the initial diagnosis. New gene therapy treatments are on the horizon and have proved successful in animals with the condition. Mesothelioma is quite rare in the normal population, so where there is a history of asbestos exposure, causation is seldom in dispute (but see *Fairchild* in O1009). The condition may be detected on X-ray or CT scan.

Because the uses of asbestos peaked in the 1960s and 1970s and because the latency period is typically 40 to 50 years, the incidence of mesothelioma is likely to peak in the first quarter of the 21st century.

Asbestosis

O1005 This is a type of scarring or fibrosis of the lungs which is caused by asbestos bodies penetrating through to the alveoli – the tiny vessels at the end of the airways where oxygen is taken into the bloodstream and carbon dioxide removed. The lung's defences cause the fibrosis to occur and this gradually spreads so as to block the actions of the alveoli. Unlike mesothelioma, asbestosis is a dose-related condition: it is believed that causation depends upon a significant degree of exposure over several year for it to take hold, and the severity of the disease may be related to the extent of the exposure and the period of time over which the exposure takes place. The latency period is thought to be at least ten years from first exposure.

Symptoms may include breathlessness on exertion, crackling sounds in the base of the lungs and finger clubbing. The fibrosis may sometimes be seen on X-ray or CT scan and a lung function test may reveal reduced gas transfer. Asbestosis is not invariably fatal but may cause severe respiratory disability. Complications, particularly lung cancer, may arise which will have a significant effect on life expectancy.

Lung cancer

O1006 This disease is of course widely associated with cigarette smoking but it is believed that it can arise directly as a result of asbestos exposure in non-smokers. However, where there is a combination of cigarette smoking and asbestos exposure, the likelihood of lung cancer developing is extremely high.

Diffuse pleural thickening

O1007 This is a fairly uncommon condition where asbestos bodies cause a thickening of the lining of the pleura, resulting in breathlessness. Unlike mesothelioma, it is a benign condition but those suffering from it could develop the malignant condition at a later stage.

Pleural plaques

O1008 These are small areas of fibrous thickening on the walls of the pleura which usually cause no respiratory disability but which may be detected on X-ray examination. They are probably an indication that the individual has had asbestos exposure.

Although there is no disability, the patient who is told about the finding may be understandably anxious about the future. In a small percentage of cases the condition may progress into one of the disabling conditions listed above. For this reason, those who have been advised that they have pleural plaques may seek an award of provisional damages, with a proviso that they may return to the court for a more substantial award at a later date if they succumb to one of the more serious conditions.

Civil liability

O1009 Historically there had been few disputes about liability to the individual claimant. If there had been exposure and if the *Factories Acts* applied, the stringent requirements of the sections relating to ventilation and the removal of dust made it difficult for most employers to defend. Problems sometimes arose over apportionment where there was more than one defendant. Sharing and handling arrangements between insurers helped to resolve this type of dispute. For example it was usual (because of the latent period) to agree that any asbestos exposure less than ten years before diagnosis of the disease was not causally relevant.

However, the last few years have seen significant legal developments in this field, including a challenge to the status quo which culminated in the landmark House of Lords decision in *Fairchild v Glenhaven Funeral Services Ltd and Others [2002] UKHL 22, [2002] All ER (D) 139 (Jun).*

In *Holtby v Brigham & Cowan (Hull) Ltd [2000] 3 All ER 421* the court was concerned with the liability of one employer in an asbestos-related claim involving asbestosis, where the claimant had been potentially exposed to asbestos during other periods of employment. The Court of Appeal made a distinction between 'divisible' (i.e. dose-related conditions like asbestosis/pleural thickening) and 'non-divisible' conditions (like mesothelioma). Where the condition is divisible, the court will do its best to apportion the damages to reflect the defendant's responsibility for the injury suffered in the course of that employment. Where the condition is not divisible, the claimant will recover in full. The court was split 2–1 on the question of burden of proof. The majority said that the burden stayed with claimant throughout, but the dissenting opinion of Clarke LJ has great force.

Fairchild v Glenhaven Funeral Services Ltd and Others concerned six appeals in cases of asbestos-related mesothelioma arising from multiple periods of employment and exposure. In December 2001 the Court of Appeal decided three central issues:

(*a*) *Causation* (*Fairchild v Glenhaven Funeral Services Ltd and Others; Fox v Spousal (Midlands) Ltd; Edwin Matthews v Associated Portland Cement Manufacturers (1978) Ltd and British Uralite plc*)

The court ruled that the claimants could not recover damages for their asbestos-related mesothelioma where the exposure took place with more than one employer – even if those employers were in breach of duty because it was impossible to say which employer was responsible for the guilty asbestos fibre which triggered the disease. A material increase in the risk of developing mesothelioma was, the court said, insufficient to establish causation.

(*b*) *Occupiers Liability* (*Dyson v John Watson Field and Others; Babcock Int Ltd v National Grid Co plc; and Fairchild*)

In these cases occupiers had instructed competent contractors to work at their premises. The court ruled that any exposure to asbestos dust by their visitors did not arise from the static state of the premises but the way in which contractors did their work. The occupiers' duty of care did not extend beyond ensuring that visitors were reasonably safe from the use of the premises themselves (and did not include risks arising from dangerous activities carried out on those premises.

(*c*) *Provisional Damages* (*Pendleton Stone v Webster Eng Ltd and Others*)

The court held that arguments over causation should be considered at the time a claim for additional damages is made (based on scientific knowledge at that time). A claimant is not prevented (by the ruling on causation) from claiming provisional damages should he or she contract mesothelioma in the future.

This judgment caused a public outcry and was appealed to the House of Lords. On the 16 May 2002 the House of Lords allowed the appeal on causation without handing down written reasons. On 20 June 2002 the Lords gave written reasons:

(*a*) The usual 'but for' causation test would be relaxed in mesothelioma claims involving exposure to asbestos by more than one employer for policy reasons. In these cases, it is sufficient to demonstrate that an employer's breach of duty materially increased the risk of the claimant developing mesothelioma. The claim would not fail simply because the claimant could not establish (on current scientific knowledge) which employment had been responsible for the culpable asbestos fibre.

(*b*) That this 'relaxed test' on causation should not have general application and should be extended with significant restraint and certainly not without 'good reason". The Lords restricted their findings presently to cases where:

(i) science could not prove how and who caused injury;

(ii) a breach of duty has materially increased the risk of injury and was capable of causing the relevant injury; and

(iii) the agent arising from the defendant's breach was the cause of injury and not just one of a number of potential causes.

(*c*) Each employer in these type of claims will be held jointly and severally liable to the claimant – the claimant being entitled to recover damages in full against each defendant, leaving the defendant's having to pursue contribution from other employers as appropriate.

What has been the fall out to this judgement? We have had the Court of Appeal Judgement in *Barker v Sr Gobain Pipelines plc* in May 2004. In this case the claimant had suffered asbestos exposure during a period of self employment. The defendants argued that he could not be described as an 'innocent victim' – a prerequisite for the modified test in *Fairchild.* Further they argued that there should be an apportionment to reflect the claimant's own exposure whilst self employed and employed elsewhere. Both arguments were rejected by the Court of Appeal. Attempts are now being made to appeal this case to the House of Lords. (See also *Philips v Syndicate 922 (14 May 2003).*)

We do now have the ABI Guidelines for the apportioning and handling of employer's liability mesolthelioma claims. These have been produced by the insurance industry as a response to *Fairchild* and in an attempt to avoid future conflict and difficulties in the settlement of these claims.

In *Teresa Maguire v Harland & Wolff plc and Others (2004) EWHC 577*, the High Court decided that it was reasonably foreseeable to the employer of a boilermaker in the period 1961–65 that the boilermaker's wife was at risk of serious injury to her health as result of exposure to asbestos dust carried home each day on her husband's work clothes.

Finally we have the pleural plaques test case litigation listed for hearing in the High Court in November 2004. The court will consider:

- whether pleural plaques are a compensatable injury;
- if so, what are the appropriate general damages;
- is it appropriate to award damages for future risks given that that the 'risks' do not manifest or flow from the pleural plaques themselves,
- if so, how should these awards be assessed?

Noise-induced hearing loss (NIHL)

O1010 The condition is more commonly called 'occupational deafness' but noise-induced hearing loss (NIHL) is now a more acceptable term because the word 'deafness' implies almost total loss of hearing whereas noise exposure at work often causes only a partial hearing loss.

The general view is that NIHL is the most common occupational disease, although work-related stress conditions are on the increase. Although no reliable figures are available, it is believed that at least 2 million workers in the UK may have been exposed to excessive noise for at least a significant period of their employment. The DSS statistics are of little assistance because benefit is not paid until the claimant has a high level of hearing loss. Although the disease was extremely common in the traditional heavy industries, like shipbuilding, boilermaking, steel manufacture and mining, most factories have some noisy areas and there are many parts of the service industry where workers are at risk.

Noise

O1011 Noise is simply unwanted sound. This is sometimes subjective: a rock band may give one person much pleasure but cause annoyance to another. However, it is generally accepted that excessive noise at work is not only unwanted but potentially harmful.

There are three important factors when considering the effects of noise:

- intensity (or loudness);
- frequency (or pitch); and
- daily dose (or duration of exposure).

Intensity

O1012 The louder the sound, the greater the likelihood of hearing damage. The range of intensity levels which may be heard by the human ear varies enormously from the slightest whisper to the sound of a jet engine on take-off. The latter may be 100 million, million (10^{14}) times as loud as the former so a logarithmic scale, known as the decibel (dB) scale, is used to simplify the numbering of the ratios. 0 dB is the threshold of normal hearing, 140 dB would be the sound of a jet engine, whereas factory noise or a heavy goods vehicle would be typically in the 80 to 90 dB range. As the scale is logarithmic, the following examples should be noted:

- 90dB is 10 times as loud as 80dB.
- ± 1dB represents an increase/decrease of 1.26 ($10^{0.1}$) in the intensity level.
- 93 dB is twice as loud as 90dB (3dB = 1.26^3).
- 96 dB is four times as loud as 90dB.
- 87dB is half as loud as 90dB.

Noise intensity levels may be measured in the workplace by a noise level meter. Personal dose meters may also be used to assess an individual's noise dose over a period of time.

Frequency

O1013 Sounds may occur at various frequencies which range from a low rumble to a high-pitched whine or whistle. The human ear responds differently to sounds of the same intensity but at different frequencies – some high frequency sounds will appear to be louder. Frequencies are banded together in octaves (like the notes on a piano) but when measuring the effects of noise at work, it is usual to average out the bands and to apply a weighting factor which allows for the differing responses of the ear. This is known as the A-weighted scale and a measurement of intensity would be expressed as 90 dB(A), for example.

Frequencies are usually measured at their octave band centres. The unit of measurement is the Hertz (Hz). 1 Hz = 1 cycle per second; 1000Hz = 1 kiloHertz (1 kHz). The range of hearing in a normal healthy young adult is between about 20 Hz and 20 kHz but the range for the understanding of most speech is between about 1 and 3 kHz.

Daily dose

O1014 The risk of hearing damage from noise at work is dose related. The unprotected ear may be exposed to a moderately loud noise for a short period of time without harm but if the exposure continues for several hours, there may be a risk of temporary hearing loss. If this daily pattern of exposure continues over a number of years, the temporary hearing loss will inevitably become permanent. The UK control standard is based on daily personal exposure to noise over an 8 hour working day ($L_{EP,d.}$).

Control measures

O1015 The first official attempt at assessing the risk to hearing from noise at work and advising on preventative steps was a booklet called *'Noise and the Worker'*, published by the Ministry of Labour in 1963. The booklet advised employers to measure noise levels and, if they exceeded a level which would now be expressed at approximately 89 dB(A), to take preventative measures. This would include the issue of personal hearing protection, which by that time was reasonably effective in attenuating high noise levels. *'Noise and the Worker'* was followed by more detailed and updated guidance in a Code of Practice, published in 1972 by the Department of Employment.

With a few minor exceptions, there was no specific legislation aimed at noise control at work until the *Noise at Work Regulations 1989* (*SI 1989 No 1790*) (implementing an EU Noise at Work Directive) were introduced. These regulations had the following main objectives:

(*a*) General duty to reduce the risk of hearing damage to the lowest level reasonably practicable.

(*b*) Assessments of noise exposure to be made.

(*c*) Reduce noise exposure as far as reasonably practicable.

(*d*) Provision of information to workers.

(*e*) Two main action levels at or above 85 dB(A) and 90 dB(A):

  (i) At the first level (85dB(A)) and above, ear protection must be provided to workers on request;

  (ii) At the second level (90dB(A)), the protection must be issued to all those exposed and ear protection zones established where all those entering must wear protection.

HSE has published updated guidance on ear protection for employers to refer to entitled '*Ear protection: Employers' duties explained*', and for employees '*Protect your hearing*'. These update and compliment an earlier HSE guidance '*Reducing noise at work: Guidance on the Noise at Work Regulations 1989*'.

The HSE like to emphasize that ear protection should be a last resort and that reduction of noise at source is always preferable.

See also the Physical Agents (Noise) Directive which looks to reduce the two action levels to 80dB(A) and 85dB(A) respectively. The Directive provides that health surveillance must be instituted where noise levels exceed 80dB(A). It is due to be implemented by 15 February 2006.

Damage to hearing by noise

O1016 The human ear is divided into three sections:

(*a*) The outer ear, which extends along the ear canal to the ear drum;

(*b*) The middle ear, containing a set of three bones, the ossicles, which link the ear drum to the inner ear; and (*c*) The inner ear which contains the cochlea, the organ of hearing.

The cochlea is connected to the brain by the auditory nerve. Inside the cochlea are receptor cells which have hair-like tufts and groups of these cells respond to different frequency ranges. Continuous exposure to loud noise will tend to permanently damage these cells so that the efficiency of the hearing organ is gradually diminished.

The cells which respond to the high frequency ranges are situated close to the oval window in the cochlea, a membrane which connects the cochlea to the ossicles in the middle ear. These cells tend to suffer from the greatest damage by noise and those exposed typically have their greatest hearing loss at or about the 4 kHz frequency. The effect of this is that they have difficulty in distinguishing the high-pitched consonant sounds like sh, th, p etc., especially in crowded places where there others are talking. They may also have problems hearing the doorbell or telephone ringing.

Other typical indicators of NIHL are:

(*a*) It causes damage only to the inner ear (sensori-neural loss);

(*b*) It is in both ears (bilateral);

(*c*) The extent of the loss is about the same in each ear (symmetrical); and

(*d*) It develops gradually over a period of years of exposure to high noise levels.

There are many other causes of hearing loss, some of which, like presbyacusis (senile deafness) also affect the inner ear. Correct diagnosis will involve the consideration of all the other possible causes. The most common method of measuring hearing loss is by pure tone audiometry, whereby tones are delivered to the subject through an earphone and responses are recorded on a graph. The object is to establish the threshold of hearing for a given frequency, this threshold being measured in dB (i.e. the intensity level at which the subject begins to hear the sound). An audiogram of a typical NIHL subject will show that the hearing threshold is greatest at or around the 4 kHz frequency and this will be shown by a dip or notch in the graph.

Tinnitus

O1017 This is a symptom which consists of a sensation of noises in the ear without external stimulus. Typically, it consists of ringing noises but whistling, buzzing, roaring or clicking sounds are reported by some. Its cause is not well established but it can arise in connection with several conditions of the inner ear, including NIHL but also with presbyacusis and menière's disease, for example.

The effect of tinnitus varies enormously from mild to very severe. Most people have experienced a ringing sensation from time to time, for example after visiting a disco or perhaps during a respiratory infection. These transient symptoms cause few problems, but there are some who suffer from continuous, loud noises in the ear which often disturb sleep, and in rare cases, may even cause them to become suicidal.

Civil liability

O1018 As with asbestos claims, the investigation of noise exposure is often difficult to investigate because of the passage of time. Some employers may have done noise surveys in the past, but these frequently provide only a snapshot of conditions at a particular time, and it is often argued by claimants that these surveys are unrepresentative. Defendants also have the problem of co-ordinating the claim where there are multiple defendants.

Because specific statutory control only started in 1990, many NIHL claims are based on common law. In this context, the actions of a reasonable and prudent employer in the light of current knowledge and invention is relevant (following *Stokes v GKN [1968] 1 WLR 1776*). The so-called guilty knowledge began, for most defendants, only in 1963 when *'Noise and the Worker'* was first published, so, where there was pre-1963 exposure, liability and damages should be apportioned between the 'innocent' and 'guilty' periods (see *Thompson v Smiths Ship Repairers [1984] 1 All ER 881*).

We are beginning to see a growth of claims particularly from telephone operators and those within the leisure industry (e.g. workers in pubs and clubs where noise levels can exceed 90dB(A)).

Hand arm vibration syndrome (HAVS)

O1019 This is the name given to a group of diseases relating to vibration induced conditions in the upper limbs. The most common and widely known condition is vibration white finger (VWF). This is an occupationally induced form of a condition known as Raynaud's Phenomenon and arises after prolonged and excessive use of vibration hand tools.

Recent HSE funded research suggests that over a million workers are potentially exposed to harmful levels of hand arm vibration. The construction industry was particularly affected with around 460,000 individuals involved.

Vibration

O1020 Like NIHL, VWF is dose related and many who suffer from an occupational hearing loss may also have this condition as hand-held tools which are noisy may also vibrate excessively, but VWF is confined to those who actually use the equipment. The measurement factors are:

- magnitude of vibration;
- frequency of the vibration; and
- the daily dose.

Magnitude

O1021 Measurement of magnitude is complex and needs to be done by a skilled person. The aim is to measure the acceleration of the tool, the unit of measurement being in meters per second squared (m/s^2). An accelerometer is fixed to the handle of the tool and measurements are taken along three axes (x, y and z). Variations in results are likely due to the condition of the tool, the material being worked upon and the way in which the operator grips the tool.

Frequency

O1022 As with noise measurement, the frequency range is an important factor, but with vibration exposure the harmful effects are most likely to be in the low frequency range from about 2 Hz to 1250 Hz. A weighted average is normally used.

Daily dose

O1023 The current action level given in the HSE guidance (HSG88) applies when the daily dose reaches 2.8 m/s^2 over 8 hours. Because of the problems with variable measurement it is suggested that the following guidelines may be appropriate:

(*a*) any tool causing tingling and numbness after 10 minutes' use should be suspect;

(*b*) almost any exposure which exceeds 2 hours a day is likely to cause harm; and

(*c*) exposure which exceeds 30 minutes a day with tools like caulking or chipping hammers, rock drills, pneumatic road breakers, or heavy duty portable grinders is likely to cause harm.

Effects of exposure to excessive vibration

O1024 There are two main symptoms of the VWFcondition:

(*a*) *Vascular*: blanching of one or more parts of the fingers; and

(*b*) *Neurological*: tingling and (sometimes) numbness in the fingers.

In the early stages, the attacks are transient, perhaps lasting up to one hour. As the patient recovers from the attack, there is aching and redness. Initial attacks occur in cold conditions, usually in winter. It is believed that early vascular damage may be subject to spontaneous recovery but by the time the neurological damage occurs the condition will be permanent.

It should be emphasised that these symptoms are the same whether the subject suffers from the non occupational Raynaud's Phenomenon or from VWF. The former is a naturally occurring condition in the general population and is particularly common in women where it is believed that about 10 per cent suffer from it. Raynaud's Phenomenon may arise as a complication of various conditions, including rheumatoid arthritis, frostbite, and vascular disease, and may also arise as a result of cigarette smoking or from trauma or surgery to the hands.

Where the condition is work-related, continuous exposure may cause damage to the blood supply to the fingers and to the nerve endings. With time, the attacks become more prolonged and the effects more serious and in a very few cases, gangrene may occur. The condition starts at the tips of the fingers, but as it progresses, more parts of the fingers are affected. It is rare for the thumbs to be involved. As cold conditions may bring on attacks, workers may be advised to wear gloves and perhaps avoid working in exposed locations.

If the disease is diagnosed at an early stage, the worker may be advised to cease work with vibratory tools. If this is done, it is unlikely that the condition will progress and in some cases, a complete recovery may be possible.

Classification of symptoms

O1025 There are two recognised scales for assessing the severity of symptoms. The Taylor-Pelmear scale is the oldest and uses four stages of severity but is perhaps too crude a measure for medico-legal purposes and may rely on some non-clinical factors, such as a change of job. The later Stockholm scale grades symptoms separately for vascular and sensori-neural components and for the number of affected fingers on each hand. This more detailed approach is now favoured by many examiners although they will often refer to both scales in a medico-legal context.

Preventative steps

O1026 These may be summarised as follows:

(*a*) Risk assessment – identifying the processes and individuals at risk.

(*b*) Warnings – advising those at risk of the hazards and action to be taken.

(*c*) Training and adequate supervision.

(*d*) Minimising daily exposure time – by job rotation etc.

(*e*) Proper tool control and maintenance.

(*f*) Medical surveillance to ensure that symptoms are reported promptly and that harmful exposure does not continue.

(*g*) Automation of process where necessary.

It is questionable whether the wearing of gloves will reduce the magnitude of vibration and their use should be confined to the role of keeping the hands warm and/or protection from sharp edges etc.

Civil liability

O1027 There is no specific statutory requirement relating to VWF so most of the claims are based on duty at common law. As with deafness claims, there is a date when guilty knowledge commences and, in most cases, this will be 1 January 1976 or thereabouts. This date is used following the publication of a book – *Vibration white finger in industry* by Taylor and Pelmear (1975) – and by the introduction of a draft British Standard (DD 43) in the same year. (See also *Armstrong v British Coal, The Times, 6 December 1996*, Kemp & Kemp at H6A-002 and *White v Holbrook Castings [1985] IRLR 215.*)

In *Doherty and Others v Rugby Joinery (UK) ltd (2004) EWCA Civ 147* the Court of Appeal considered the date of guilty knowledge within the woodworking industry. Although the court was keen not to create a general date of knowledge for the woodworking industry (and said that future cases should be decided on their own facts) the date adopted in this case was 1992. This case serves a useful reminder that in some limited instances the relevant date of guilty knowledge may be more recent than 1 January 1976.

The Court of Appeal has confirmed that an employer in a VWF claim will only be liable for the excess damage which can be attributed to that employer's negligence or breach of duty – thus confirming the dose related nature of the condition (See *Allen and Others v British Rail Engineering Ltd and Another [2001] All ER (D) 291 (Feb)*. Defendants can use this decision to argue for broad brush reductions to reflect inevitable VWF that can be proved during any period of negligence.

In the *Armstrong* case, the court made reference to the HSE Guidance *'Hand Arm Vibration'* (1994) (HSG 88) which provides the starting point when looking at current standards. Will the courts ever find an employer in breach of duty where the daily dose does not exceed the action levels in HSG 88? In *Allen*, the court ruled that an employer could be liable even where HSG 88 was not exceeded, and where that employer knew that other employees were suffering from VWF but failed to take additional preventative measures in the form of health surveillance and vibration surveys.

Following agreement of the EU Physical Agents (Vibration) Directive in March 2002 (2002/44/EC) there will be changes introduced into UK law, most notably a reduction in current action levels. The Directive requires employers, where there is likely to be a risk from exposure to vibration:

- To reduce exposure to a minimum.
- To provide information and training.
- To assess exposure levels.
- When exposure reaches the exposure action level (to be reduced to 2.5 m/s2 for hand-arm vibration over an 8-hour working day) to carry out a programme of measures to reduce exposure and provide appropriate health surveillance.
- To keep exposure below the exposure limit value (to be 5.0 m/s2 for hand-arm vibration over an 8-hour working day)

Exposure action and limit values are also given by the Directive for whole-body vibration.

This Directive is due to be implemented by 6 July 2005. Transitional provisions apply for existing, new and agricultural/forestry equipment.

The changing pattern of occupational diseases

O1028 Technological and socio-economic changes have continued to alter the pattern and nature of occupational health problems in the UK and abroad. The gradual shift from manufacturing to service-related employment has resulted in the development of many so-called 'white collar' diseases. Obviously, there is a geographic pattern to the development of these new diseases.

The 1992 'Six Pack' Regulations reflect a partial response to these changes and a move away from the factory based legislation of yesteryear.

Stress

O1029 Two growth areas include:

(*a*) Work-related stress.

(*b*) Work-related upper limb disorders (WRULDs).

Work-related stress

O1030 Stress is hardly a new problem restricted to modern man. People have suffered from stress-related conditions, one suspects back to our very earliest stages of evolution. We all suffer stress in one form or another during our lifetime whether it be due to work pressures or difficult life events related to personal tragedy.

What is stress?

O1031 The word 'stress' has many connotations and meanings but is frequently used to describe the psychological, physiological and behavioural responses to external stimuli. Not all stimuli causes excessive stress levels and not all stress is bad.

The HSE's definition of stress ascribes a negative connotation:

> 'the reaction people have to excessive pressures or the types of demands placed upon them arises when they worry that they cannot cope.'

The late 20th Century saw the development of a new breed of claim – work-related stress. These claims stem not only from a changed perception of what is now acceptable in the workplace but also the courts, or perhaps more importantly, lawyers' willingness to expand the sphere of employers' liability.

It is important to emphasise that stress is not a separate diagnosable illness nor a discrete medical condition – the word is commonly used to describe symptomology including depression, anxiety, guilt, apathy, sleeplessness, behavioural changes (psychological) and tension, high blood pressure, weight loss, loss of appetite (physical). Media presentation of work-related stress concentrates very much on the psychological aspects of work-related stress.

We should not underestimate the socio-economic impact of work-related stress. Research undertaken by the International Stress Management Association UK and Royal Sun Alliance revealed that 70 per cent of UK adults had experienced stress at work – 49 per cent indicating that stress levels had increased over the last twelve months.

The Health and Safety Commission (HSC) concluded following an extensive public consultation exercise (which included the publication *'Managing Stress at Work'*) that work-related stress was a serious health and safety problem which could only be tackled in part by the existing health and safety legislation. At present an approved code of practice in the field is deemed unenforceable but HSE has this under review.

What causes stress?

O1032 When looking for work-related trends there are a number of factors or 'stressors' which can contribute towards an individuals level of stress:

(*a*) Time pressures.

(*b*) Overload/underload.

(*c*) Interpersonal relationships.

(*d*) Working hours.

(*e*) Working environment – physical and systemic.

(*f*) Personality.

(*g*) Changes – technological, procedural and systemic.

Stress-related illness often arises as a result of a combination of the above as well as contribution from external stressors (e.g. death, family problems or physical illness). It is the multi-factorial nature of stress-related illness that makes it difficult to identify the cause and to prevent occurrence. It is often impossible to unravel the occupational causes from the non-occupational elements and this creates particular difficulties in the medico-legal context.

Although it is dangerous to generalise in this area, research has shown that some employees especially nurses, professionals, education/welfare/security workers and senior management are particularly affected by work-related stress. See HSE report *'The scale of occupational stress: A further analysis of the impact of demographic factors and type of job'*. Men and women appear to be similarly afflicted although this report suggests a higher incidence within non-white ethnic groups.

Civil liability

O1033 The following sections explain civil liability under both statute and common law.

Statutory

O1034 The Government has sought to address the number of hours worked by any one individual through the enactment of the *Working Time Regulations 1998* (*SI 1998 1833*) (as amended by *SI 1999 No 3372*). These Regulations implement Council Directive 93/104 (1992 OJL 307/18) and Council Directive 94/93 (1994 OJL 216/12). Obligations are imposed on employers concerning the maximum average weekly working time of workers, the average normal hours of night workers, the provision of health assessments for night workers, rest breaks to be given to workers engaged in certain kinds of work and keeping records of workers' hours of work. Whilst these Regulations do not confer a civil right of action they may be relevant in 'overload' claims.

Under the *Management of Health and Safety at Work Regulations 1999* (*SI 1999 No 3242*) employers are required to make a suitable and sufficient assessment of the risks to the health and safety of their employees to which they are exposed whilst at work. This includes excessive stress levels whether it be caused by difficult time pressures or bullying by co-workers.

There have been claims brought under the *Protection from Harrassment 1997* (e.g. *Majrowski v Guy's & St Thomas NHS Trust (2004)* where the claimant failed).

Common law

O1035 Until a few years ago *Walker v Northumberland County Council [1995] 1 All ER 737* was the most frequently cited authority – seen by many as a watershed decision in this field despite following well established principles. Mr Walker managed four teams of social service field workers in a depressed area of Northumberland. He found the strain of the job too much and at the end of November 1986 suffered a nervous breakdown. His symptoms included anxiety, headaches, sleeplessness and an inability to cope with any levels of stress. On medical advice he was absent from work for three months. He returned to work but in September 1987 stress induced symptoms returned and he was subsequently diagnosed as suffering from stress-related anxiety and was advised to take sick leave. In February 1988 he had a second mental breakdown and was later dismissed on the grounds of permanent ill health. The local authority was held liable. It was found that an employer does owe a duty of care to an employee not to cause him psychiatric injury as a result of the volume or character of work which the employees is required to do. Although the first breakdown was found not to be reasonably foreseeable to the defendant, the second one was. It was reasonably foreseeable that there was a real risk of repetition of his illness if he was exposed to the same workload and if his duties were not alleviated by effective additional assistance. In continuing to employ him but providing no effective help the council had acted unreasonably and was in breach of it's duty of care.

On the issue of foreseeability, Colman J in *Walker* said:

> 'The question is whether it ought to have foreseen that Mr Walker was exposed to a risk of mental illness materially higher than that which would ordinarily affect (an employee) in his position with a really heavy workload.
>
> For if the foreseeable risk were not materially greater than that, there would not, as a matter of reasonable conduct, be any basis upon which the council's duty to act arose."

In saying this he acknowledged that not all work stress can be removed nor would it be reasonable to impose such a duty on employers.

On the issue of breach and standard of care, Colman J said:

> 'It is clear law that an employer has a duty to provide his employee with a reasonably safe system of work or to take reasonable steps to protect him from

risks which are reasonably foreseeable … the standard of care required for the performance of that duty must be measured against a yardstick of reasonable conduct on the part of a person in the position of that person who owes the duty … It calls for no more than a reasonable response, what is reasonable being measured by the nature of the employer-employee relationship, the magnitude of the risk of injury which was reasonably foreseeable, the seriousness of the consequence for the person to whom the duty is owed of a risk of eventuating and the cost in practicality of preventing the risk … the practicality of remedial measures must clearly take into account the resources and facilities at the disposal of the personal body owing the duty of care … and the purpose of the activity which has given rise to the risk of injury.'

Walker was expressly approved by the Court of Appeal in *Garrett v London Borough of Camden [2001] EWCA Civ 395.*

This case was followed by four occupational stress appeals which came before the Court of Appeal in November 2001 under the lead case name *Hatton v Sutherland and Others [2002] 2 All ER 1.* Judgment was handed down on 5 February 2002 and the following specific guidance was given to practitioners and the lower courts:

(*a*) There are no special control mechanisms applying to claims for psychiatric or physical illness arising from stress of work. Ordinary principles of employer's liability apply.

(*b*) The threshold test is whether this *'kind of harm to this particular employee was reasonably foreseeable i.e.* an *injury to health which is attributable to stress at work'*.

(*c*) Foreseeability depends upon what the employer knows (or ought reasonably to know) about the individual employee. Psychiatric injury is usually harder to foresee than physical injury. An employer is entitled to assume that an employee can withstand the normal pressures of work unless he knows of a particular vulnerability.

(*d*) There are no intrinsically dangerous occupations – the test is the same for all.

(*e*) Factors relevant to the threshold test:

— The nature and extent of the work done by the employee including signs by other employees.

— Signs from the employee of impending harm to health.

(*f*) An employer is generally entitled to take what he is told by the employee at face value unless he has good reason to think to the contrary. There is no need for searching enquiries.

(*g*) Before there is a duty to take steps, the indications of impending harm to health must be plain enough for any reasonable employer to realise that he should do something about it.

(*h*) The employer is only in breach of duty if he has failed to take steps which are reasonable in the circumstances bearing in mind the factors cited by Colman J in *Walker* above.

(*i*) The size and scope of the employer's operation, its resources and the demands it faces are relevant to what is reasonable.

(*j*) An employer can only be reasonably expected to take steps which are likely to do some good – expert evidence is likely to be required.

(*k*) An employer who offers a confidential advice service (including referral to counsellors) is unlikely to be found in breach of duty.

(*l*) If the only reasonable and effective step would be to dismiss or demote the employee, the employer will not be in breach of duty in allowing a willing employee to continue in his or her job. (But see also *Wayne Coxall v Goodyear GB Ltd [2003] 1 WLR* below.)

(*m*) The court must be able to identify which steps both could and should have been taken before finding any breach of duty.

(*n*) The claimant must show that the breach of duty has caused or materially contributed to the harm suffered.

(*o*) Where there is more than one cause – the employer should only pay for the proportion of the harm attributable to his/her wrongdoing unless the harm is truly indivisible. The defendant must raise the issue of apportionment.

(*p*) Assessment of damage will take into account any pre-existing disorder and any chance of inevitable breakdown.

One of the *Hatton* appeals, *Barber v Somerset County Council [2004] UKHL 13*, was appealed to the House of Lords.

Mr Barber had been appointed head of the mathematics department of one of the defendant county council's comprehensive schools in 1983. Staff restructuring took place in 1995 and he had become 'mathematical area of expertise co-ordinator'. To maintain his salary level he had also taken on the post of project manager for public and media relations. He had worked long hours (between 61 and 70 hours a week) in discharging his new responsibilities.

Toward the end of 1995 he had begun to feel the strain. In February 1996 he spoke to one of the two deputy heads about 'work overload'. In May and June of that year he had been off work for three weeks with depression, with sick notes showing 'overstressed/depression' and 'stress'. On his return to work, he had completed the council's form of sickness declaration and had given his reasons for absence as 'overstressed/depression'. That form had been countersigned by one of the deputy heads.

Toward the end of June 1996 Mr Barber decided to take the initiative by arranging a meeting with the headmistress about his problems. She had however treated him unsympathetically, observing that all the staff were under stress. During July of that year, Mr Barber had had separate meetings with each of the deputy heads about his workload and his health, but no positive steps had been taken to help him.

In the Autumn term of 1996 Mr Barber had found himself with the same or even possibly a slightly heavier workload. In November 1996, he suffered a nervous breakdown when at school. He never returned to work and in March 1997, when he was 52 years old, he took early retirement on the grounds of ill health.

Mr Barber commenced civil proceedings against his employers, the council, claiming damages for psychiatric injury caused by work-related stress (WRS). The county court found that the responses of the headmistress and her two deputies to Mr Barber's difficulties had been inadequate:

> 'In my view a prudent employer faced with the knowledge of work overload dating back to the autumn of 1995 and increasing into 1996 such that the employee had to take time off work for stress, would have investigated the employee's situation to see how his difficulties might be improved … it must have been apparent … that the risk of injury to [Mr Barber's] mental health was significant and higher than that which would have related to a teacher in a similar position with a heavy workload.'

The county court accordingly found that the council had been in breach of its common law duty of care to Mr Barber and awarded him £101,547 damages, with interest. The council appealed.

The council's appeal was allowed by the Court of Appeal in *Hatton*, which observed:

> 'Unless he knows of some particular problem or vulnerability, an employer is usually entitled to assume that his employee is up to the normal pressures of the job. It is only if there is something specific about the job or the employee or the combination of the two that he has to think harder. But thinking harder does not necessarily mean that he has to make searching or intrusive inquiries. Generally he is entitled to take what he is told by or on behalf of the employee at face value.'

Mr Barber appealed.

Mr Barber's appeal was allowed by the House of Lords by a majority of four to one. In his leading judgment Lord Walker, referring to the passage cited above from the Court of Appeal judgement, observed that though it was 'useful practical guidance', it had to be read as that and not as having anything like statutory force. Every case depended on its own particular facts and the best statement of general principle remained that Mr Justice Swanwick in *Stokes v Guess, Keen and Nettlefold (Bolts and Nuts) Ltd [1968] 1 WLR 1776*, where he said: '[T]he overall test is still the conduct of the reasonable and prudent employer, taking positive thought for the safety of his workers in the light of what he knows or ought to know … [W]here he has in fact greater than average knowledge of the risks, he may be thereby obliged to take more than the average of standard precautions. He must weigh up the risk in terms of the likelihood of injury occurring and the potential consequences if it does, and he must balance against this the probable effectiveness of the precautions that can be taken to meet it and the expense and inconvenience they involve.'

Lord Walker then turned to another passage from the Court of Appeal judgement in *Hatton*, in which that court noted that: '[T]his was a classic case in which it was essential to consider at what point the school's duty to take some action was triggered, what that action should have been, and whether it would have done some good … it is difficult … to identify a point at which the school had a duty to take the positive steps identified by the judge. It might have been different if Mr Barber had gone to [one of the deputy heads] at the beginning of the Autumn term [in 1996] and told him that things had not improved over the holiday. But it is expecting far too much to expect the school authorities to pick up the fact that the problems were continuing without some such indication.'

Looking at that, Lord Walker said that the issue of the council's duty of care to Mr Barber was very close to the borderline. It had not been, he emphasised, a clear case of a flagrant breach of duty any more than it had been an obviously hopeless case. The county court had however concluded that the council had been in breach of its duty and he thought that, on the facts, there had been insufficient reasons for the Court of Appeal to set aside this finding.

At the very least, in the circumstances, the school management team – the headmistress and the two deputy heads – should have taken the initiative in making sympathetic inquiries about Mr Barber when he returned to work in June 1996 and making some reduction in his workload to ease his return. Even a small reduction in his duties, coupled with the feeling that the team was on his side, might by itself have made a real difference.

In any event, his condition should have been monitored and, if it did not improve, some more drastic action would have had to be taken. Supply teachers cost money, but not as much as the cost of the permanent loss through psychiatric illness of a valued member of the school's staff.

Lord Walker concluded that the employer's duty to take some action arose in June and July 1996, when Mr Barber saw separately each member of the school's management team. It continued as long as nothing was done to help him.

There has been a mixed reaction to the decision. Despite the result of the individual case, insurers and employers breathed a sigh of relief given their lordships endorsement of the Court of Appeal guidance. Employee groups have received the result with less enthusiasm.

Other cases to note in the field are:

Young v The Post Office [2002] EWCA Civ 661 – where the claimant successfully sued his employer following a second breakdown. The Court of Appeal said it would be unusual (although theoretically possible) for a finding of contributory negligence to be made in a stress claim.

In *Wayne Coxall v Goodyear GB Ltd [2003] 1 WLR 536*, the Court of Appeal considered whether an employer had a duty to dismiss an employee in circumstances where that employment continued to expose the employee to a risk of injury. Although this concerned an occupational asthma claim, it is interesting because the Court of Appeal upheld the lower court's decision to distinguish *Hatton* and found that there was a positive duty to dismiss the employee in the circumstances. The two decisions do no sit comfortably together.

Practitioners should have regard to the following HSE guidance when considering foreseeability and the standard of care in this area:

- 'Stress at work: Guide for employers' (1995) HSG 116
- 'Help on work-related stress: A short guide' (1998) INDG281
- 'Managing occupational stress: A guide for managers and teachers in the schools sector' (1990) (ISBN 011885559X)
- 'Tackling work related stress: A manager's guide to improving and maintaining employee health and well-being' (1999) HSG 218
- *'Tackling work related stress: A guide for employees'* INDG 341
- See also the HSE's 6 draft management standards (Demands, Control, Support, Role, Change and Relationships) which introduce benchmarks for achievements by employers. Many feel that these management standards will form the plinth for any new civil actions in stress-related actions.

Causation and the nature of injury are important features of these claims. The claimant must establish that he or she has suffered a definable psychiatric illness (see *Page v Smith [1995] 2 All ER 736* and *Fraser v State Hospital Board of Scotland, The Times, 11 September 2000*). The illness must have been materially contributed to by the breach of duty and there will be no compensatable claim if the same extent of psychiatric illness would have occurred in any event and but for the work stressors.

Overload claims continue to feature prominently in this field but harassment/victimisation claims should not be overlooked. In the four cases heard by the Court of Appeal in *Hatton*, only the case of *Olwen Jones v Sandwell Metropolitan BC* contained any element of treatment which could be categorised as harassment. This was the only case to succeed out of the four appeals in the Court of Appeal (*Barber* succeeding in the Lords) – and involved the situation where the employer knew that the employee was being badly treated by another employee and could have done something about it.

In *Waters v Commissioner of the Police of the Metropolis [2000] IRLR 720* the House of Lords refused to strike out a claim against the metropolitan police where a police officer alleged that her employer had breached their duty of care to prevent assaults by other officers. Lord Slynn said:

> 'If an employer knows that acts being done by employees during their employment may cause physical or mental harm to a particular employee and he does nothing to supervise or prevent such acts, when it is in his power to do so, it is clearly arguable that he may be in breach of duty to that employee he may also be in breach of that duty if he can foresee that such acts may happen and if they do, that physical or mental harm may be caused to an individual.'

Lord Hutton agreed but said that employees may have to accept some unpleasantness from co-workers and employers will not be liable without actual or constructive knowledge of the harassment.

Finally we should not ignore the importance of the decisions in *Johnson v Unisys [2001] 2 WLR 1076*, *Eastwood v Magnox Electric plc [2002] All ER (D) 366 (Mar)*, *Robert McCabe v Cornwall County Council and The Governing Body of Mounts Bay School (18 December 2002)* and *Christopher Dunnachie v Kingston Upon Hull City Council CA (17 February 2004)* concerning the appropriate jurisdiction for psychiatric claims connected with dismissal. In the lead judgment in *Eastwood* and *McCabe*, Lord Nicholls said that if, before dismissal, an employee has an accrued right of action, that cause of action remains unimpaired by his subsequent dismissal. In cases like *Eastwood* and *McCabe*, where the losses flow from conduct that precedes the dismissal, that cause of action is independent of any claim to the employment tribunal for unfair dismissal. In essence the Lords have confined the effect of the *Johnson* decision to the actual act of dismissal. Those employees where the dismissal was the last straw that made the claimant ill, will probably be caught by *Johnson* and still have to bring any claim before the employment tribunal.

Preventative steps

O1036 These can be summarised as follows:

(*a*) Risk assessment.

(*b*) Recognition of tell tale signs of excess stress e.g absences, lateness, staff turnover, poor quality or quantity of work.

(*c*) Act on identified stressors/reports of excess stress.

(*d*) Monitor stress-related absences, ensure any return is monitored and refer to medical profession for clarification of capacity to work.

Work-related upper limb disorders (WRULDs)

Definition

O1037 Such disorders were previously referred to as RSI (this terminology as a separate form of injury was rejected by Prosser J in *Mughal v Reuters [1993] IRLR 571*).

Claims arise out of a wide variety of jobs. In relation to white-collar workers, claims are commonly made by VDU operators and typists and data processors. A recent IOM study funded by HSE found that secretarial staff and temps were significantly over-represented amongst female workers reporting upper limb disorders(ULD's). As regards blue-collar workers, claims can be seen from heavy repetitive manual or routine repetitive line workers (for example assembly work).

Upper limb disorders (ULDs) include a wide range of conditions affecting the fingers, hands, wrists and forearms. Some are prescribed diseases, some are not. Some are known to be work-related, some not. Medical opinion remains fairly divided.

See Table 1 below for a list of common upper limb disorders.

Table 1

Common upper limb disorders

Medical condition	*Description*	*Clinical signs and symptoms*
Peritendonitis Crepitans PD A8	Inflammation of tendon of hand or forearm.	Swelling, heat, redness, pain and crepitus (creaking) of wrist. Can be caused by rapid repeated movement and thus caused by work.
Carpal Tunnel Syndrome	Compression of the medial nerve of the wrist as it passes through the carpal tunnel.	Pain, numbness in palm and fingers on the lateral side of the hand and wrist, difficulty in moving fingers, decreased hand strength. Electro-conductivity diagnostic tests. It is a prescribed disease where associated with vibration.
Tendonitis PD A8	Inflammation of tendons.	Pain and swelling. Tendons can become locked in their sheaths so that the fingers become locked in flexion. Repeated extension and flexion of fingers thought to be causative.
Tenosynovitis (trigger digit/trigger thumb) PD A8	Inflammation of the tendon sheath.	Swelling, tenderness and pain in the tendon sheath of fingers or thumbs, accompanied by a characteristic 'locking'. Can be caused by repetitive motions of the digits.
De Quervains Syndrome (De Quervains Stenosing Tenovaginitis) PD A8	Thickening of the tendon sheath over the radial borders of the wrist at the junction of the thumb/wrist.	Reduction in the grip strength, tenderness and swelling, pain on radial side of wrist. Caused by repeated pinching and gripping with thumb or repeated ulnar deviation of wrist. Can develop spontaneously.
Tennis Elbow (Lateral Epicondylitis)	Inflammation of the tendons that attach the forearm muscles to the bony knob on the outer elbow.	Tenderness of outer or inner aspect of the elbow, extending into the forearm along a palpable band in line with the extensor muscles. Accompanied by pain and swelling. No clear medical evidence that the conditions are related to work although they are exacerbated by heavy repetitive manual labour.

Golfer's Elbow (Medial Epicondylitis)	Inflammation of the tendons of the forearm muscles that attach to the inner aspect of the bony knob of the elbow.	
PDA4 (Writer's cramp)	Cramp of the hand or forearm.	Can be caused by prolonged periods of handwriting, typing or other repetitive movements of the fingers, hand or arms.
Ganglions	Cysts containing viscous, mucinous fluid found in vicinity of joints and tendons in various parts of the body, although the majority are on the wrist.	Not believed to be work-related but can be exacerbated by work.

Civil liability

O1038 To bring a successful WRULD claim the claimant will need to establish:

(*a*) That it was foreseeable to a reasonable and prudent employer that the work engaged in could give rise to an upper limb disorder.

(*b*) That he or she has suffered an injury (but see *Alexander v Midland Bank plc* below).

(*c*) That this injury has been caused or exacerbated by his work and as a result of any breach of duty owed to him either in common law or pursuant to statute by his employers.

Problems arise because a claimant has difficulty in satisfying one of the definable medical conditions listed in Table 1 above. In *Mughal v Reuters [1993] IRLR 571* the plaintiff was unable to establish one of the specific upper limb conditions and was suffering from a diffuse range of symptoms. Prosser J could not accept RSI as a separate form of injury and in the absence of a specific organic condition rejected the claim.

It is not enough for a claimant to say that he or she has suffered 'passing minimal discomfort' (see *Griffiths v British Coal Corporation (Unreported)*).

Many of the recent cases have turned on the establishment of an organic condition. Following *Mughal, some* claimants suffering from diffuse symptoms attempted to identify their problem as a specific upper limb condition (see for example *Pickford*). Alternatively, a number of alternative labels such as fibro-myalgia or reflex sympathetic dystrophy were adopted on their behalf by the treating or expert medical practitioner.

In *Alexander and Others v Midland Bank [1999] IRLR 723* five female bank employees claimed that they were suffering from regional fibro-myalgia arising from their work as data processor operators. The claimants all worked in the nine district centres set up by Midland Bank to process cheques and vouchers. Their job involved rapid repetitive keying in work on encoding machines over long periods of time, often one-handed. By 1991 some of the claimants began to report symptoms in the upper limbs. The pain and discomfort became more persistent and intense. It was held at first instance that the plaintiffs' symptoms were more than passing minimal discomfort and were sufficient to form an actionable claim for personal injury notwithstanding the lack of precision in the description of the pathology of their pain. HH Byrt QC found that the pressure of the decoders' work contributed towards their injuries, and the bank knew or ought to have reasonably foreseen that these factors invited the risk of injury and were held in breach of their duty.

The bank appealed. The Court of Appeal found that the lack of precise pathological/physiological explanation for the claimants' symptoms did not rule out recovery providing there was medical evidence that those symptoms were caused by the work/breach of duty.

In *Amosu and Others v The Financial Times, 31 July 1998 (Unreported)* the court considered a claim by five journalists at the *Financial Times* who each claimed to have suffered specific musculo-skeletal disorders following the introduction of a computerised system for the writing, editing and printing of the paper. The allegations included the provision of bad and poorly ergonomically designed work stations and the exposure to excessive pressures of work with inadequate rest breaks. On this occasion the court held that each of the claimants had failed to demonstrate or prove on the balance of probabilities that they had suffered from the various physical problems as alleged.

In the majority decision by the House Lords in *Pickford v Imperial Chemical Industries plc [1998] 1 WLR 1189* the claimant failed to satisfy the court that her symptoms were organic and caused by typing work. Mrs Pickford alleged that she had developed prescribed disease PD A4 commonly known as writer's cramp because of the large amount of typing work carried out at speed for long periods without breaks. The House of Lords found that it was not reasonably foreseeable that a secretary working in the same work regime as the plaintiff would be likely to suffer from the condition. There was no specific duty on ICI to provide rest breaks as her work was sufficiently varied to allow rotation between typing and other known repetitive work. Further there was no specific duty to warn of the specific risks of developing PD A4 as this was not practice in the industry at the time and such warnings would have been counter productive precipitating the very condition it was intended to avoid.

Employers' duties and prevention

O1039 The primary source is the *Health and Safety (Display Screen Equipment) Regulations 1992 (SI 1992 No 2792)* as amended by the *Health and Safety (Miscellaneous Amendments) Regulations 2002 (SI 2002 No 2174)*. See also the *Management of Health and Safety At Work Regulations 1999 (SI 1999 No 3242)* and the *Workplace (Health, Safety and Welfare) Regulations 1992 (SI 1992 No 3004)*.

Original guidance came from the HSE publication *'Work-related upper limb disorders: A guide to prevention'* (1990) and stated that prudent employers should look to:

(*a*) Provide suitable and adequate health and safety training/information relating to the use of equipment, posture and breaks;

(*b*) Provide a suitable work station;

(*c*) Provide suitable rotation and breaks. Employers should carry out risk assessments looking at the necessary application of force, the speed of work, the repetition and the awkwardness of any specific task;

(*d*) Provide warnings where appropriate; and

(*e*) Undertake health surveillance.

On 28 February 2002 the HSE published revised guidance (*'Upper limb disorders in the workplace'*, HSG 60(rev) (2002) which is much more specific and specifically addresses the need to consider both keyboard and mouse use.

There guidance contains more detailed identification/definition of risk factors including:

- Repetition
- Working posture
- Force
- Duration of exposure
- Working environment
- Psychosocial factors
- Individual differences

The guidance gives 7 key stages:

- Understand the issues and commit to action
- Create the right organisational environment
- Assess the risk of Upper Limb Disorders (ULD's) in the workplace
- Reduce the risks of ULD's
- Educate and inform your workforce.
- Manage any episode of ULD's
- Carry out regular checks on programme effectiveness

See also the HSE publication *'Working with VDUs'* (INDG36(rev1)) (1998); HSE Press release E199:02, HSE Press Release E030:03 'New guidance on using computers and preventing RSI'; HSE publication *'The law on VDUs: An easy guide'* (HSG 90).

Preventative management

Safety policy

O1040 *Section 2(3)* of *HSWA 1974* requires those employing five or more people to have a written statement of general policy on matters of health and safety and to revise it as necessary. The document should detail the organisation and arrangements for carrying out the policy and it should be brought to the notice of all employees.

This is an important document which should be signed by a senior person within the organisation (usually the chief executive). It should clearly set out the obligations of line managers in carrying out the objectives of the organisation. Where there are specific hazards (e.g. asbestos) the policy should deal with the measures to be taken. This may be in the form of an internal code of practice.

Risk assessment

O1041 This is the cornerstone of modern occupational health and safety management and stems from EU Directives on health and safety which have been subsequently implemented by UK legislation.

Regulations which contain the requirement to carry out risk assessments include:

(*a*) *Control of Asbestos at Work Regulations 2002* (*SI 2002 No 2675*) (as amended);

(*b*) *Control of Lead at Work Regulations 2002* (*SI 2002 No 2676*);

(*c*) *Control of Substances Hazardous to Health Regulations 2002* (*COSHH*) (*SI 2002 No 2677*);

(*d*) *Health and Safety* (*Display Screen Equipment*) *Regulations 1992* (*SI 1992 No 2792* (as amended);

(*e*) *Management of Health and Safety at Work Regulations 1999* (*SI 1999 No 3242*);

(*f*) *Manual Handling Operations Regulations 1992* (*SI 1992 No 2793*)(as amended);

(*g*) *Noise at Work Regulations 1989* (*SI 1989 No 1790*);

(*h*) *Personal Protective Equipment at Work Regulations 1992* (*SI 1992 No 2966*);

(*i*) *Workplace* (*Health, Safety and Welfare*) *Regulations 1992* (*SI 1992 No 3004*).

An employer is required to carry out a suitable and sufficient risk assessment where there may be a risk of injury or disease. Having identified the areas of risk, protective measures which either eliminate or minimise the risk must be indicated.

The assessments may be carried out by outside agencies, but are often best done internally where there are staff familiar with the processes and potential hazards. Management should consider seeking the help and advice of employees in this task, particularly where there are employee health and safety representatives. Assessments are not necessarily confined to the risks to employees, but others who may be affected by their undertaking: this may include contractors, customers, delivery drivers and others likely to visit the premises. Where employees are working away from the normal workplace, they need to be given special consideration.

Risk assessments are not static: they must be reviewed constantly in the light of changing circumstances. They will normally be in writing and these records should be retained for as long as they are relevant. It is particularly important that where there is a risk of occupational disease, assessments should deal specifically with the problem. For example, where there is a problem with excessive noise, there should be a detailed survey of noisy areas and risks to individuals, but this may also need to be supported by health surveillance in the form of audiometry to test the effectiveness of the control measures. Regular safety audits are also an effective way of checking the effectiveness of the risk assessments.

Ceasing exposure

O1042 In some cases, the risk of harm from a substance or process is so great that consideration should be given to ceasing the exposure entirely. This may mean closing the workplace or discontinuing the hazardous process. Such radical steps may be necessary where the process carries a risk of serious injury and the employer is not satisfied that control measures will eliminate the risk. An example of this might be the total ban on the use or handling of asbestos and the substitution by another insulating material.

Minimising exposure

O1043 A less radical solution may be to reduce the level of exposure to the lowest level that is reasonably practicable. The aim should be to reduce the exposure not only to the occupational exposure standard (which should be regarded as the maximum allowable in any event), but to aim for the lowest level that is reasonably achievable. For example, in a noisy workplace, an employer may aim to reduce exposure to the first action level of 85 dB(A) but cannot be confident that this level may always be achieved. Further engineering measures could bring the level to well below 85 dB(A), ensuring that none of the workforce is at risk and avoiding the necessity to have hearing protection available.

Enclosing the process

O1044 Risks arising from a hazardous process may be substantially reduced if it is enclosed to ensure that there is no escape of the substance. However, there must be provision for access by personnel to carry out essential maintenance and the employer must ensure that such persons are adequately trained and are provided with the necessary personal protection when working in the enclosure.

Enclosing the operator

O1045 This is the reverse of the above system of control. It is especially suitable where the process is on a scale that makes enclosure uneconomic (e.g. a noisy bottling plant). The operator and the plant controls are situated in a sealed control room and access on to the floor of the plant is carefully controlled with the necessary personal protection worn.

Automating or mechanising the process

O1046 The aim of this system is to remove the human element from some or all of the process. Where there is a high incidence of musculo-skeletal problems in the workplace, this may be an appropriate measure to consider. However, the different risks involved with the introduction of new machinery may need to be assessed.

Minimising exposure time

O1047 Many occupational diseases are dose-related so every effort should be made to ensure that workers are exposed to hazards for as little time as possible during their working day. In theory, this is a very effective control measure and should be cost-effective, but it does need proper supervision and worker co-operation to ensure that it is effective. A simple example is where VDU operators are given adequate rest breaks coupled with other tasks (e.g. filing) which are performed away from the display screen.

Monitoring exposure

O1048 Monitoring is an important part of the process for ensuring that exposure is maintained at or below the maximum safe level. For example, dust levels may be monitored by regular environmental samples and supplemented by personal dose sampling on workers who are at risk. Continuous automatic sampling may be necessary in some cases. Monitoring should be done regularly and proper records maintained.

Personal protective equipment (PPE)

O1049 The use of PPE is now governed by the *Personal Protective Equipment at Work Regulations 1992* (*SI 1992 No 2966*). In the guidance notes to the Regulations it is stressed that in the hierarchy of control measures, PPE must be regarded as a last resort: engineering controls and safe systems of work should come first although there may be a need for personal protection whilst such measures are being implemented. The reason for the 'last resort' status is because there are so many problems involved in the reliance upon PPE as a safe method of prevention, and in some workplaces there is reluctance by the workforce to wear the protection provided.

There are however circumstances where the employer has no alternative but to use PPE. If this is so, there are many points to consider to ensure that the protection is effective, including:

(*a*) risk assessment to ensure that if PPE is the only means of controlling exposure, the equipment is suitable and effective;

(*b*) ensuring that the necessary equipment is issued to all workers who may be at risk;

(*c*) ensuring that the equipment fits correctly and is compatible with other PPE that may be worn;

(*d*) arrangements for the maintenance and replacement of the equipment;

(*e*) suitable accommodation for the equipment (e.g. lockers);

(*f*) information, instruction and training to advise workers about the risks of exposure, the need for the protection, the manner of its use, and how to maintain the equipment;

(*g*) information given needs to be comprehensible to the worker – a point to consider where there may be language difficulties.

If a control system relies on PPE, it is essential that there is adequate supervision to ensure that equipment is being properly worn at all relevant times. Having regard to all the problems, an employer may well be advised to look again at other control measures.

Information, instruction, training and supervision

O1050 These have been discussed in relation to PPE but they are essential ingredients of the total programme of preventative management.

(*a*) *Information.* Those who are at risk must be told the precise nature of the risks that they face and what they need to do to avoid or minimise these risks. For example, if there is a problem with noise, they should be told how noise can damage hearing, what areas of the workplace are hazardous, and where they must wear hearing protection. Instructional videos are often available for this sort of education.

(*b*) *Instruction.* In addition to giving advice on the health risks, a worker should also receive clear instructions as to any steps which need to be taken. This instruction may be verbal but is best supplemented by an instruction booklet and perhaps warning notices. Instructions may need to be in foreign languages where appropriate.

(*c*) *Training.* Following instruction there must be adequate training to ensure that the worker understands what needs to be done. This may include, for example, a demonstration on the correct method of wearing PPE.

(*d*) *Supervision.* Adequate supervision is the key to effective preventative management. If this does not exist then all the assessments, plans and procedures are simply waste paper. First line supervisors are on the spot and can immediately see what is right or wrong. They can ensure for example that protection is being worn or that hazardous dust is not allowed to escape. Employers should ensure that health and safety is an essential part of the supervisor's role and that they are given the time and resources to perform that function.

Medical surveillance

O1051 The purpose of medical surveillance is to identify at an early stage health risks and control measures. There are certain statutory requirements for this (eg. in the COSHH Regulations) but in addition, employers may choose to implement their own programme. The surveillance may take several forms:

(*a*) pre-employment questionnaire and/or examination to ensure that new entrants are screened for suitability;

(*b*) new starters may have initial tests, such as audiometry, which will establish a benchmark against which further progress can be checked;

(*c*) regular examinations, say every year, to ensure that health is maintained;

(*d*) providing a service where workers with symptoms may report them immediately (this is particularly beneficial with work-related upper limb disorders, for example); and

(*e*) a referral procedure where those returning from sickness absence may be examined as to their continuing suitability for certain work.

Future trends

O1052 Increasing technological development will undoubtedly change the risks to which workers are exposed and as a consequence the nature of occupational health problems.

Asthma and other environmental conditions are likely to be at the forefront of new claims along with WRULDs and stress.

As individuals become ever more litigious we can expect lawyers to attempt to push open new causes of action founded on new occupational diseases.

Although law and practice change to reflect each new risk there is inevitably a time gap and whilst the incidence of many traditional industrial diseases have been halted or reduced we should not expect an eradication of all occupational health diseases. The very technological advancements which have reduced many of the traditional risks in industry have brought a whole new set of occupational health issues and problems.

Current list of prescribed occupational diseases

O1053 The current list of prescribed occupational diseases is contained in the *Social Security (Industrial Injuries) (Prescribed Diseases) Regulations 1985 (SI 1985 No 967)* (the main regulations), the *Social Security (Industrial Injuries) (Prescribed Diseases) Amendment Regulations 1987 (SI 1987 No 335)*, the *Social Security (Industrial Injuries) (Prescribed Diseases) (Amendment No 2) Regulations 1987 (SI 1987 No 2112)*, the *Social Security (Industrial Diseases) (Prescribed Diseases) Amendment Regulations 1989 (SI 1989 No 1207)*, the *Social Security (Industrial Injuries) (Prescribed Diseases) Amendment Regulations 1990 (SI 1990 No 2269)*, the *Social Security (Industrial Injuries) (Prescribed*

Diseases) *Amendment Regulations 1991* (*SI 1991 No 1938*), the *Social Security* (*Industrial Injuries*) (*Prescribed Diseases*) *Amendment Regulations 1993* (*SI 1993 No 862*), the *Social Security* (*Industrial Injuries*) (*Prescribed Diseases*) *Amendment* (*No 2*) *Regulations 1993* (*SI 1993 No 1985*), the *Social Security* (*Industrial Injuries*) (*Prescribed Diseases*) *Amendment Regulations 1994* (*SI 1994 No 2343*), *Social Security* (*Industrial Injuries and Diseases*) (*Miscellaneous Amendments*) *Regulations 1996* (*SI 1996 No 425*), *Social Security* (*Industrial Injuries*) (*Prescribed Diseases*) *Amendment Regulations 2000* (*SI 2000 No 1588*) and the *Social Security* (*Industrial Injuries*) (*Prescribed Diseases*) *Amendment Regulations 2002* (*SI 2002 No 1717*). *Schedule 1* to the main regulations and occupations for which they are prescribed are reproduced in Table 2 below. Pneumoconiosis and the occupations for which it is prescribed are set out in *Schedule 2* to the main regulations and is also reproduced below. Pneumoconiosis and the occupations for which it is prescribed are set out in Part II of the Table 2 below. Compensation for pneumoconiosis etc. is dealt with in COMPENSATION FOR WORK INJURIES AND DISEASES and below.

The question whether a claimant is suffering from a prescribed disease is a diagnosis one, falling within the remit of the Medical Appeal Tribunal; the question whether a disease is prescribed is for the Adjudicating Officer, the Social Security Appeal Tribunal and the Commissioner. Where a claimant is suffering from a prescribed disease, it is open to the Adjudicating Officer to seek to show that the particular disease was not due to the nature of the claimant's employment (R(I) 4/91 – primary neoplasm C 23, invoking *Reg 4* of the *Social Security* (*Industrial Injuries*) (*Prescribed Diseases*) *Regulations 1985*). In addition, a Medical Appeal Tribunal can decide disablement issues (R(I) 2/91 – disablement assessed at 7% in respect of vibration white finger, A 11).

Moreover, medical practitioners and specially qualified medical practitioners can now adjudicate upon matters relating to industrial injuries and prescribed diseases (hitherto the exclusive province of medical boards or special medical boards). [*Social Security* (*Industrial Injuries and Adjudication*) *Regulations 1993* (*SI 1993 No 861*); *Social Security* (*Industrial Injuries*) (*Prescribed Diseases*) *Amendment* (*No 2*) *Regulations 1993* (*SI 1993 No 1985*)].

Table 2

Part I

List of prescribed diseases and the occupations for which they are prescribed

	Occupation
A Conditions due to physical agents	Any occupation involving:
[A1 Leukaemia (other than chronic lymphatic leukaemia) or cancer of the bone, female breast, testis or thyroid.	Exposure to electro-magnetic radiations (other than radiant heat) or to ionising particles where the dose is sufficient to double the risk of the occurrence of the condition.]
A2 <…> cataract.	[Frequent or prolonged exposure to radiation from red-hot or white-hot material.]
A3 Dysbarism, including decompression sickness, barotrauma and osteonecrosis.	Subjection to compressed or rarefied air or other respirable gases or gaseous mixtures.

A4 Cramp of the hand or forearm due to repetitive movements.	Prolonged periods of handwriting, typing or other repetitive movements of the fingers, hand or arm.
A5 Subcutaneous cellulitis of the hand (beat hand).	Manual labour causing severe or prolonged friction or pressure on the hand.
A6 Bursitis or subcutaneous cellulitis arising at or about the knee due to severe or prolonged external friction or pressure at or about the knee (beat knee).	Manual labour causing severe or prolonged external friction or pressure at or about the knee.
A7 Bursitis or subcutaneous cellulitis arising at or about the elbow due to severe or prolonged external friction or pressure at or about the elbow (beat elbow).	Manual labour causing severe or prolonged external friction or pressure at or about the elbow.
A8 Traumatic inflammation of the tendons of the hand or forearm, or of the associated tendon sheaths.	Manual labour, or frequent or repeated movements of the hand or wrist.
A9 Miner's nystagmus.	Work in or about a mine.
[A10 Sensorineural hearing loss amounting to at least 50 dB in each ear, being the average of hearing losses at 1, 2 and 3 kHz frequencies, and being due in the case of at least one ear to occupational noise (occupational deafness).]	[The use of, or work wholly or mainly in the immediate vicinity of the use of, a—
	(a) band saw, circular saw or cutting disc to cut metal in the metal founding or forging industries, circular saw to cut products in the manufacture of steel, powered (other than hand powered) grinding tool on metal (other than sheet metal or plate metal), pneumatic percussive tool on metal, pressurised air arc tool to gouge metal, burner or torch to cut or dress steel based products, skid transfer bank, knock out and shake out grid in a foundry, machine (other than a power press machine) to forge metal including a machine used to drop stamp metal by means of closed or open dies or drop hammers, machine to cut or shape or clean metal nails, or plasma spray gun to spray molten metal;

	(b) pneumatic percussive tool:- to drill rock in a quarry, on stone in a quarry works, underground, for mining coal, for sinking a shaft, or for tunnelling in civil engineering works;
	(c) vibrating metal moulding box in the concrete products industry, or circular saw to cut concrete masonry blocks;
	(d) machine in the manufacture of textiles for:- weaving man-made or natural fibres (including mineral fibres), high speed false twisting of fibres, or the mechanical cleaning of bobbins;
	(e) multi-cutter moulding machine on wood, planing machine on wood, automatic or semi-automatic lathe on wood, multiple cross-cut machine on wood, automatic shaping machine on wood, double-end tenoning machine on wood, vertical spindle moulding machine (including a high speed routing machine) on wood, edge banding machine on wood, bandsawing machine (with a blade width of not less than 75 millimetres) on wood, circular sawing machine on wood including one operated by moving the blade towards the material being cut, or chain saw on wood;
	(f) jet of water (or a mixture of water and abrasive material) at a pressure above 680 bar, or jet channelling process to burn stone in a quarry;
	(g) machine in a ship's engine room, or gas turbine for: performance testing on a test bed, installation testing of a replacement engine in an aircraft, or acceptance testing of an Armed Service fixed wing combat aircraft;

	(h) machine in the manufacture of glass containers or hollow ware for:- automatic moulding, automatic blow moulding, or automatic glass pressing and forming;
	(i) spinning machine using compressed air to produce glass wool or mineral wool;
	(j) continuous glass toughening furnace;
	(k) firearm by a police firearms training officer; or
	(l) shot-blaster to carry abrasives in air for cleaning.]
A11 Episodic blanching, occurring throughout the year, affecting the middle or proximal phalanges or in the case of a thumb the proximal phalanx, of—	(a) the use of hand-held chain saws in forestry; or
(a) in the case of a person with 5 fingers (including thumb) on one hand, any 3 of those fingers, or	(b) the use of hand-held rotary tools in grinding or in the sanding or polishing of metal, or the holding of material being ground, or metal being sanded or polished, by rotary tools; or
(b) in the case of a person with only 4 such fingers, any 2 of those fingers, or	(c) the use of hand-held percussive metal-working tools, or the holding of metal being worked upon by percussive tools, in riveting, caulking, chipping, hammering, fettling or swaging; or
(c) in the case of a person with less than 4 such fingers, any one of those fingers or, as the case may be, the one remaining finger (vibration white finger).	(d) the use of hand-held powered percussive drills or hand-held powered percussive hammers in mining, quarrying, demolition, or on roads or footpaths, including road construction; or
	(e) the holding of material being worked upon by pounding machines in shoe manufacture.
[A12 Carpal tunnel syndrome.	The use of hand-held powered tools whose internal parts vibrate so as to transmit that vibration to the hand, but excluding those which are solely powered by hand.]
B. Conditions due to biological agents	Any occupation involving

B1 Anthrax.	Contact with animals infected with anthrax or the handling (including the loading or unloading or transport) of animal products or residues.
B2 Glanders.	Contact with equine animals or their carcases.
B3 Infection by leptospira.	(a) Work in places which are, or are liable to be, infested by rats, field mice or voles, or other small mammals; or
	(b) work at dog kennels or the care or handling of dogs; or
	(c) contact with bovine animals or their meat products or pigs or their meat products.
B4 Ankylostomiasis.	Work in or about a mine.
B5 Tuberculosis.	Contact with a source of tuberculous infection.
B6 Extrinsic allergic alveolitis (including farmer's lung).	Exposure to moulds or fungal spores or heterologous proteins by reason of employment in:—
	(a) agriculture, horticulture, forestry, cultivation of edible fungi or malt-working; or
	(b) loading or unloading or handling in storage mouldy vegetable matter or edible fungi; or
	(c) caring for or handling birds; or
	(d) handling bagasse.
B7 Infection by organisms of the genus brucella.	Contact with—
	(a) animals infected by brucella, or their carcases or parts thereof, or their untreated products; or
	(b) laboratory specimens or vaccines of, or containing, brucella.
B8 Viral hepatitis.	Any occupation involving: Contact with—
	(a) human blood or human blood products; or
	(b) a source of viral hepatitis.

B9 Infection by Streptococcus suis.	Contact with pigs infected by Streptococus suis, or with the carcases, products or residues of pigs so infected.
[B10	
(a) Avian chlamydiosis	Contact with birds infected with chlamydia psittaci, or with the remains or untreated products of such birds.
B10	
(b) Ovine chlamydiosis	Contact with sheep infected with chlamydia psittaci, or with the remains or untreated products of such sheep.
B11 Q fever	Contact with animals, their remains or their untreated products.]
[B12 Orf.	Contact with sheep, goats or with the carcasses of sheep or goats.
B13 Hydatidosis.	Contact with dogs.].
C. Conditions due to chemical agents	
C1	
[(a) Amaemia with a haemoglobin concentration of 9g/dL or less, and a blood film showing punctate basophilia; (b) peripheral neuropathy; (c) central nervous system toxicity].	The use or handling of, or exposure to the fumes, dust or vapour of, lead or a compound of lead, or a substance containing lead.
C2 [Central nervous system toxicity characterised by parkinsonism].	The use or handling of, or exposure to the fumes, dust or vapour of, manganese or a compound of manganese, or a substance containing manganese.
C3 Poisoning by phosphorus or an inorganic compound of phosphorus or poisoning due to the anti-cholinesterase or pseudo anti-cholinesterase action of organic phosphorus compounds.	The use or handling of, or exposure to the fumes, dust or vapour of, phosphorus or a compound of phosphorus, or a substance containing phosphorus.
[C4 Primary carcinoma of the bronchus or lung.	Exposure to the fumes, dust or vapour of arsenic, a compound of arsenic or a substance containing arsenic.]
[C5A Central nervous system toxicity characterised by tremor and neuropsychiatric disease.	Exposure to mercury or inorganic compounds of mercury for a period of, or periods which amount in aggregate to, 10 years or more.

C5B Central nervous system toxicity characterised by combined cerebellar and cortical degeneration.	Exposure to methylmercury.]
[C6 Peripheral neuropathy.	The use or handling of, or exposure to, carbon disulphide (also called carbon disulfide).]
[C7 Acute non-lymphatic leukaemia.	Exposure to benzene.]
C8 <…>	<…>
C9 <…>	<…>
C10 <…>	<…>
C11 <…>	<…>
[C12	
(a) Peripheral neuropathy;	Exposure to methyl bromide (also called bromomethane).]
(b) central nervous system toxicity.	
[C13 Cirrhosis of the liver.	Exposure to chlorinated naphthalenes.]
C14 <…>	<…>
C15 <…>	<…>
[C16	
(a) Neurotoxicity;	Exposure to the dust of gonioma kamassi.]
(b) cardiotoxicity.	
[C17 Chronic beryllium disease.	Inhalation of beryllium or a beryllium compound.]
[C18 Emphysema.	Inhalation of cadmium fumes for a period of, or periods which amount in aggregate to, 20 years or more.]
[C19	
(a) Peripheral neuropathy;	Exposure to acrylamide.]
(b) central nervous system toxicity.	
C20 Dystrophy of the cornea (including ulceration of the corneal surface) of the eye.	[Exposure to quinone or hydroquinone.]
[C21 Primary carcinoma of the skin.	Exposure to arsenic or arsenic compounds, tar, pitch, bitumen, mineral oil (including paraffin) or soot.]

[C22	
(a) Primary carcinoma of the mucous membrane of the nose or paranasal sinuses;	Work before 1950 in the refining of nickel involving exposure to oxides, sulphides or water-soluble compounds of nickel.]
(b) primary carcinoma of the bronchus or lung.	
[C23 Primary neoplasm of the epithelial lining of the urinary tract.	(a) the manufacture of 1-naphthylamine, 2-naphthylamine, benzidine, auramine, magenta or 4-aminobiphenyl (also called biphenyl-4-ylamine); (b) work in the process of manufacturing methylene-bis-orthochloroaniline (also called MbOCA) for a period of, or periods which amount in aggregate to, 12 months or more; (c) exposure to 2-naphthylamine, benzidine, 4-aminobiphenyl (also called biphenyl-4-ylamine) or salts of those compounds otherwise than in the manufacture of those compounds; (d) exposure to orthotoluidine, 4-chloro-2-methylaniline or salts of those compounds; or (e) exposure for a period of, or periods which amount in aggregate to, 5 years or more, to coal tar pitch volatiles produced in aluminium smelting involving the Soderberg process (that is to say, the method of producing aluminium by electrolysis in which the anode consists of a paste of petroleum coke and mineral oil which is baked *in situ*).]
[C24	
(a) Angiosarcoma of the liver;	Exposure to vinyl chloride monomer in the manufacture of polyvinyl chloride.]
(b) acro-osteolysis characterised by	
(i) lytic destruction of the terminal phalanges,	

(ii) in Raynaud's phenomenon, the exaggerated vasomotor response to cold causing intense blanching of the digits, and	
(iii) sclerodermatous thickening of the skin;	
(c) liver fibrosis.	
[C25 Vitiligo.	The use or handling of, or exposure to, paratertiary-butylphenol (also called 4-*tert* butylphenol), paratertiary-butylcatechol (also called 4-*tert*-butylcatechol), para-amylphenol (also called *p*-pentyl phenol isomers), hydroquinone, monobenzyl ether of hydroquinone (also called 4-benzyloxyphenol) or mono-butyl ether of hydroquinone (also called 4-butoxyphenol).]
[[C26	
(a) Liver toxicity;	The use or handling of, or exposure to, carbon tetrachloride (also called tetrachloromethane).]
(b) kidney toxicity.	
[C27 Liver toxicity.	The use or handling of, or exposure to, trichloromethane (also called chloroform).]
C28 <…>	<…>
[C29 Peripheral neuropathy.	The use or handling of, or exposure to, n-hexane or n-butyl methyl ketone.]]
[C30	
(a) Dermatitis; (b) ulceration of the mucous membrane or the epidermis.	The use or handling of, or exposure to, chromic acid, chromates or dichromates.]
D. Miscellaneous Conditions	
D1 Pneumoconiosis.	Any occupation—
	(a) set out in Part II of this Schedule;
	(b) specified in regulation 2(b)(ii).

D2 Byssinosis.	Any occupation involving: Work in any room where any process up to and including the weaving process is performed in a factory in which the spinning or manipulation of raw or waste cotton or of flax, or the weaving of cotton or flax, is carried on.
D3 Diffuse mesothelioma (primary neoplasm of the mesothelium of the pleura or of the pericardium or of the peritoneum).	[Exposure to asbestos, asbestos dust or any admixture of asbestos at a level above that commonly found in the environment at large.]
[D4 Allergic rhinitis which is due to exposure to any of the following agents— (a) isocyanates; (b) platinum salts; (c) fumes or dusts arising from the manufacture, transport or use of hardening agents (including epoxy resin curing agents) based on phthalic anhydride, tetrachlorophthalic anhydride, trimellitic anhydride or triethylenetetramine; (d) fumes arising from the use of rosin as a soldering flux; (e) proteolytic enzymes; (f) animals including insects and other arthropods used for the purposes of research or education or in laboratories; (g) dusts arising from the sowing, cultivation, harvesting, drying, handling, milling, transport or storage of barley, oats, rye, wheat or maize, or the handling, milling, transport or storage of meal or flour made therefrom; (h) antibiotics; (i) cimetidine; (j) wood dust; (k) ispaghula; (l) castor bean dust; (m) ipecacuanha; (n) azodicarbonamide;	Exposure to any of the agents set out in column 1 of this paragraph.]

(o) animals including insects and other arthropods or their larval forms, used for the purposes of pest control or fruit cultivation, or the larval forms of animals used for the purposes of research or education or in laboratories;	
(p) glutaraldehyde;	
(q) persulphate salts or henna;	
(r) crustaceans or fish or products arising from these in the food processing industry;	
(s) reactive dyes;	
(t) soya bean;	
(u) tea dust;	
(v) green coffee bean dust;	
(w) fumes from stainless steel welding.]	
D5 Non-infective dermatitis of external origin (<...> excluding dermatitis due to ionising particles or electro-magnetic radiations other than radiant heat).	Exposure to dust, liquid or vapour or any other external agent [except chromic acid, chromates or bi-chromates,] capable of irritating the skin (including friction or heat but excluding ionising particles or electro-magnetic radiations other than radiant heat).
D6 Carcinoma of the nasal cavity or associated air sinuses (nasal carcinoma).	(a) Attendance for work in or about a building where wooden goods are manufactured or repaired; or (b) attendance for work in a building used for the manufacture of footwear or components of footwear made wholly or partly of leather or fibre board; or (c) attendance for work at a place used wholly or mainly for the repair of footwear made wholly or partly of leather or fibre board.
	Any occupation involving:
D7 Asthma which is due to exposure to any of the following agents:—	Exposure to any of the agents set out in column 1 of this paragraph.
(a) isocyanates;	
(b) platinum salts;	

(c) fumes or dusts arising from the manufacture, transport or use of hardening agents (including epoxy resin curing agents) based on phthalic anhydride, tetrachlorophthalic anhydride, trimellitic anhydride or triethylenetetramine; (d) fumes arising from the use of rosin as a soldering flux; (e) proteolytic enzymes; [(f) animals including insects and other arthropods used for the purposes of research or education or in laboratories] (g) dusts arising from the sowing, cultivation, harvesting, drying, handling, milling, transport or storage of barley, oats, rye, wheat or maize, or the handling, milling, transport or storage of meal or flour made therefrom (occupational asthma [(h) antibiotics; (i) cimetidine; (j) wood dust; (k) ispaghula; (l) castor bean dust; (m) ipecacuanha; (n) azodicarbonamide] [(o) animals including insects and other arthropods or their larval forms, used for the purposes of pest control or fruit cultivation, or the larval forms of animals used for the purposes of research, education or in laboratories; (p) glutaraldehyde; (q) persulphate salts or henna; (r) crustaceans or fish or products arising from these in the food processing industry; (s) reactive dyes; (t) soya bean; (u) tea dust; (v) green coffee bean dust;

(w) fumes from stainless steel welding;	
(x) any other sensitising agent.]	
D8 Primary carcinoma of the lung where there is accompanying evidence of one or both of the following—	(a) The working or handling of asbestos or any admixture of asbestos; or
(a) asbestosis;	(b) the manufacture or repair of asbestos textiles or other articles containing or composed of asbestos; or
[(b) unilateral or bilateral diffuse pleural thickening extending to a thickness of 5mm or more at any point within the area affected as measured by a plain chest radiograph (not being a computerised tomography scan or other form of imaging) which—	(c) the cleaning of any machinery or plant used in any of the foregoing operations and of any chambers, fixtures and appliances for the collection of asbestos dust; or
(i) in the case of unilateral diffuse pleural thickening, covers 50% or more of the area of the chest wall of the lung affected; or	
(ii) in the case of bilateral diffuse pleural thickening, covers 25% or more of the combined area of the chest wall of both lungs.]	
	(d) substantial exposure to the dust arising from any of the foregoing operations.
[D9 Unilateral or bilateral diffuse pleural thickening extending to a thickness of 5mm or more at any point within the area affected as measured by a plain chest radiograph (not being a computerised tomography scan or other form of imaging) which—	(a) The working or handling of asbestos or any admixture of asbestos; or
(i) in the case of unilateral diffuse pleural thickening, covers 50% or more of the area of the chest wall of the lung affected; or	(b) the manufacture or repair of asbestos textiles or other articles containing or composed of asbestos; or
(ii) in the case of bilateral diffuse pleural thickening, covers 25% or more of the combined area of the chest wall of both lungs.]	(c) the cleaning of any machinery or plant used in any of the foregoing operations and of any chambers, fixtures and appliances for the collection of asbestos dust; or Any occupation involving:
	(d) substantial exposure to the dust arising from any of the foregoing operations.

[D10 [Primary carcinoma of the lung]	(a) Work underground in a tin mine; or (b) exposure to bis(chloromethyl)ether produced during the manufacture of chloromethyl methyl ether; or (c) exposure to zinc chromate calcium chromate or strontium chromate in their pure forms.]
[D11 Primary carcinoma of the lung where there is accompanying evidence of silicosis.	Exposure to silica dust in the course of— (a) the manufacture of glass or pottery; (b) tunnelling in or quarrying sandstone or granite; (c) mining metal ores; (d) slate quarrying or the manufacture of artefacts from slate; (e) mining clay; (f) using siliceous materials as abrasives; (g) cutting stone; (h) stonemasonry; or (i) work in a foundry.]
[D12 Except in the circumstances specified in regulation 2(d)— (a) chronic bronchitis; or (b) emphysema; or (c) both, where there is accompanying evidence of a forced expiratory volume in one second (measured from the position of maximum inspiration with the claimant making maximum effort) which is—	Exposure to coal dust by reason of working underground in a coal mine for a period or periods amounting in aggregate to at least 20 years (whether before or after 5th July 1948) and any such period or periods shall include a period or periods of incapacity while engaged in such an occupation.]

[(i) at least one litre below the appropriate mean value predicted, obtained from the following prediction formulae which give the mean values predicted in litres—

For a man, where the measurement is made without back-extrapolation, (3.62 × Height in metres) – (0.031 × Age in years) – 1.41; or, where the measurement is made with back-extrapolation, (3.71 × Height in metres) – (0.032 × Age in years) – 1.44;

For a woman, where the measurement is made without back-extrapolation, (3.29 × Height in metres) – (0.029 × Age in years) – 1.42; or, where the measurement is made with back-extrapolation, (3.37 × Height in metres) – (0.030 × Age in years) – 1.46; or]

(ii) less than one litre.

Part II

Occupations for which Pneumoconiosis is prescribed

1. Any occupation involving—

(a) the mining, quarrying or working of silica rock or the working of dried quartzose sand or any dry deposit or dry residue of silica or any dry admixture containing such materials (including any occupation in which any of the aforesaid operations are carried out incidentally to the mining or quarrying of other minerals or to the manufacture of articles containing crushed or ground silica rock);

(b) the handling of any of the materials specified in the foregoing sub-paragraph in or incidental to any of the operations mentioned therein, or substantial exposure to the dust arising from such operations.

2

Any occupation involving the breaking, crushing or grinding of flint or the working or handling of broken, crushed or ground flint or materials containing such flint, or substantial exposure to the dust arising from any of such operations.

3

Any occupation involving sand blasting by means of compressed air with the use of quartzose sand or crushed silica rock or flint, or substantial exposure to the dust arising from sand and blasting.

4

Any occupation involving work in a foundry or the performance of, or substantial exposure to the dust arising from, any of the following operations:—

(a) the freeing of steel castings from adherent siliceous substance;

(b) the freeing of metal castings from adherent siliceous substance—

   (i) by blasting with an abrasive propelled by compressed air, by steam or by a wheel; or

   (ii) by the use of power-driven tools.

5

Any occupation in or incidental to the manufacture of china or earthenware (including sanitary earthenware, electrical earthenware and earthenware tiles), and any occupation involving substantial exposure to the dust arising therefrom.

6

Any occupation involving the grinding of mineral graphite, or substantial exposure to the dust arising from such grinding.

7

Any occupation involving the dressing of granite or any igneous rock by masons or the crushing of such materials, or substantial exposure to the dust arising from such operations.

8

Any occupation involving the use, or preparation for use, of a grindstone, or substantial exposure to the dust arising therefrom.

9

Any occupation involving—

(a) the working or handling of asbestos or any admixture of asbestos;

(b) the manufacture or repair of asbestos textiles or other articles containing or composed of asbestos;

(c) the cleaning of any machinery or plant used in any foregoing operations and of any chambers, fixtures and appliances for the collection of asbestos dust;

(d) substantial exposure to the dust arising from any of the foregoing operations.

10

Any occupation involving—

(a) work underground in any mine in which one of the objects of the mining operations is the getting of any mineral;

(b) the working or handling above ground at any coal or tin mine of any minerals extracted therefrom, or any operation incidental thereto;

(c) the trimming of coal in any ship, barge, or lighter, or in any dock or harbour or at any wharf or quay;

(d) the sawing, splitting or dressing of slate, or any operation incidental thereto.

11

Any occupation in or incidental to the manufacture of carbon electrodes by an industrial undertaking for use in the electrolytic extraction of aluminium from aluminium oxide, and any occupation involving substantial exposure to the dust arising therefrom.

12

Any occupation involving boiler scaling or substantial exposure to the dust arising therefrom.

Occupiers' Liability

Introduction

O3001 Inevitably, by far the greater part of this work details the duties of employers (and employees) both at common law, and by virtue of the *Health and Safety at Work etc Act 1974* (*HSWA 1974*) and other kindred legislation. This notwithstanding, duties are additionally laid on persons (including companies and local authorities) who merely *occupy* premises which other persons either visit or carry out work activities upon, e.g. repair work or servicing. More particularly, persons in control of premises (see O3004 below for meaning of 'control'), that is, occupiers of premises and employers, where others work (*although not their employees*), have a duty under *HSWA 1974, s 4* to take reasonable care towards such persons working on the premises. Failure to comply with this duty can lead to prosecution and a fine on conviction (see further ENFORCEMENT). This duty applies to premises not *exclusively* used for private residence, e.g. lifts/electrical installations in the *common parts* of a block of flats, and exists for the benefit of workmen repairing/servicing them (*Westminster City Council v Select Managements Ltd [1985] 1 All ER 897*). Moreover, a person who is injured while working on or visiting premises, may be able to sue the occupier for damages, even though the injured person is not an employee. Statute law relating to this branch of civil liability (i.e. occupiers' liability) is to be found in the *Occupiers' Liability Act 1957* (*OLA 1957*) and, as far as trespassers are concerned, in the *Occupiers' Liability Act 1984* (*OLA 1984*). (In Scotland, the law is to be found in the *Occupiers' Liability* (*Scotland*) *Act 1960*.)

Duties owed under the Occupiers' Liability Act 1957

O3002 An occupier of premises owes the same duty, the 'common duty of care', to all his lawful visitors. [*OLA 1957, s 2*(*1*)]. 'The common duty of care is a duty to take such care as in all the circumstances of the case is reasonable to see that the visitor will be reasonably safe in using the premises for the purposes for which he is invited or permitted by the occupier to be there.' [*OLA 1957, s 2*(*2*)]. Thus, a local authority which failed, in severe winter weather, to see that a path in school grounds was swept free of snow and treated with salt and was not in a slippery condition, was in breach of *s 2*, when a schoolteacher fell at 8.30 a.m. and was injured (*Murphy v Bradford Metropolitan Council [1992] 2 All ER 908*).

The duty extends only to requiring the occupier to do what is 'reasonable' in all the circumstances to ensure that the visitor will be safe. Some injuries occur where no one is obviously to blame. For example, in *Graney v Liverpool County Council (1995) (unreported)* an employee who slipped on icy paving slabs could not recover damages against his employer. The Court of Appeal held that many people slip on icy surfaces in cold weather – it did not follow that in this particular case the employer was to blame.

Nature of the duty

O3003 The *OLA 1957* is concerned only with civil liability. 'The rules so enacted in relation to an occupier of premises and his visitors shall also apply, in like manner and to like extent as the principles applicable at common law to an occupier of premises and his invitees or licensees would apply … '. [*OLA 1957, s 1(3)*]. This contrasts with the *HSWA 1974*, in which the obligations are predominantly penal measures enforced by the HSE (or some other 'enforcing authority'). The *OLA 1957* cannot be enforced by a state agency – nor does it give rise to criminal liability. Action under the *OLA 1957* must be brought in a private suit between parties.

The liability of an occupier towards lawful visitors at common law was generally based on negligence; so, too, is the liability under the *OLA 1957*. It is never strict. In *Neame v Johnson [1993] PIQR 100*, a case concerning an ambulance man who was injured in the defendant's house when carrying him unconscious in a chair in poor lighting conditions, the ambulance man knocked over a pile of books stacked by a wall on a landing, and consequently slipped on a book, injuring himself. It was held that the pile of books did not create a 'reasonably foreseeable risk of injury', and so there was no liability. Nor does the duty of the occupier extend to replacing glass panels in an ageing building. In *McGivney v Golderslea Ltd (1997) (unreported)* the defendant owned a block of flats which had been built in 1955 – the glass in the door at the foot of the communal stairs had been installed at the same time. When the plaintiff, who was visiting a friend on the third floor, came down the stairs, he slipped and his hand went through the glass panel on the front door resulting in personal injuries. The plaintiff claimed that the defendant was in breach of the *OLA 1957, s 2(2)*. However, at the time the flats were built the glass satisfied the building regulations then in force. Since 1955 building regulations have been updated and, in such a building being constructed today, stronger glass must be used such that on the balance of probabilities the plaintiff's accident would not have occurred. The Court of Appeal found that there was no duty on the defendant to replace the glass and the plaintiff's claim therefore failed.

The relationship of occupier and visitor is less immediate than that of employer and employee, and is not, as far as occupiers' liability is concerned, underwritten by compulsory insurance (but by public liability insurance, which is not obligatory but advisable).

Who is an 'occupier'?

O3004 'Occupation' is not defined in the Act, which merely states that the 'rules regulate the nature of the duty imposed by law in consequence of a person's occupation or control of premises … but they (shall) not alter the rules of the common law as to the persons on whom a duty is so imposed or to whom it is owed … .'. [*OLA 1957, s 1(2)*]. The meaning of 'occupation' must, therefore, be gleaned from the rules of common law. Where premises, including factory premises, are leased or subleased, control may be shared by lessor and lessee or by sublessor and sublessee. 'Wherever a person has a sufficient degree of control over premises that he ought to realise that any failure on his part to use care may result in injury to a person coming lawfully there, then he is an "occupier" and the person coming lawfully there is his "visitor" and the "occupier" is under a duty to his "visitor" to use reasonable care. In order to be an occupier it is not necessary for a person to have entire control over the premises. He need not have exclusive occupation. Suffice it that he has some degree of control with others' (per Lord Denning in *Wheat v E Lacon & Co Ltd [1965] 2 All ER 700*). In *Jordan v Achara (1988) 20 HLR 607*, the plaintiff, who was a meter reader, was injured when he fell down stairs in a basement of a house. The defendant landlord, who was the owner of the house, had divided it into flats.

Because of arrears of payment, the electricity supply had been disconnected. The local authority, having arranged for it to be reconnected for the tenants, recovered payment by way of rents paid directly to the authority. It was held that the landlord was liable for the injury, under the *OLA 1957*, since he was the occupier of the staircase and the passageway (where the injury occurred) and his duties continued in spite of the local authority being in receipt of rents.

Liability associated with dangers arising from maintenance and repair of premises will be that of the person responsible, under the lease or sublease, for maintenance and/or repair. 'The duty of the defendants here arose not out of contract, but because they, as the requisitioning authority, were in law in possession of the house and were in practice responsible for repairs ... and this control imposed upon them a duty to every person lawfully on the premises to take reasonable care to prevent damage through want of repair' (per Denning LJ in *Greene v Chelsea BC [1954] 2 All ER 318*, concerning a defective ceiling which collapsed, injuring the appellant, a licensee). Significantly also, managerial control constitutes 'occupation' (*Wheat v Lacon*, see above, concerning an injury to a customer at a public house owned by a brewery and managed by a manager – both were held to be 'in control'). Moreover, if a landlord leases part of a building but retains other parts, e.g. roof, common staircase, lifts, he remains liable for that part of the premises (*Moloney v Lambeth BC (1966) 64 LGR 440*, concerning a guest injured on a defective common staircase in a block of council flats). The Court of Appeal has now held (*Ribee v Norrie, The Times 22 November 2000*), that the proper question was not what actual control the landlord did or did not exercise but what power of control he had, i.e. whether or not he had the authority to act and was in a position where he could have taken steps to prevent certain risks.

However, it is not the occupier's duty to make sure that the visitor is *completely safe*, but only reasonably so. For example, in *Berryman v London Borough of Hounslow, The Times, 18 December 1996,* the plaintiff injured her back when she was forced to carry heavy shopping up several flights of stairs because the lift in the block of flats maintained by the landlord was broken. The landlord's obligation to take care to ensure that the lift he provided was reasonably safe was discharged by employing competent contractors. There was no causal connection between the plaintiff's injury on the stairs and the absence of a lift service. In other words, the injury to the plaintiff was not a 'foreseeable consequence' of the broken lift.

Premises

O3005 The *OLA* regulates the nature of the duty imposed by law in consequence of a person's occupation of premises. [*OLA 1957, s 1(2)*]. This means that the duties are not *personal* duties but depend on occupation of *premises*; and extend to a 'person occupying, or having control over, any fixed or movable structure, including any vessel, vehicle or aircraft' [*OLA 1957, s 1(3)(a)*] (e.g. a car, *Houweling v Wesseler [1963] 40 DLR(2d) 956-Canada* or, more recently, a sea wall in *Staples v West Dorset District Council (1995) 93 LGR 536* (see O3006 below)). In *Bunker v Charles Brand & Son Ltd [1969] 2 All ER 59* the defendants were contractors digging a tunnel for the construction of the Victoria Line of London Underground. To this end they used a large digging machine which moved forward on rollers. The plaintiff was injured when he slipped on the rollers. The defendants were held to be occupiers of the tunnel, even though it was owned by London Transport.

To whom is the duty owed?

Visitors

O3006 Visitors to premises entitled to protection under the Act are both (*a*) invitees and (*b*) licensees. This means that protection is afforded to all lawful visitors, whether the visitors enter for the occupier's benefit (clients or customers) or for their own benefit (factory inspectors, policemen), though not to persons exercising a public or private way over premises. [*OLA 1957, s 2*(6)]. (See below O3007 *Countryside and Rights of Way Act 2000*.)

Nevertheless, occupiers are under a duty to erect a notice warning visitors of the immediacy of a danger (*Rae (Geoffrey) v Mars (UK) [1990] 3 EG 80* where a deep pit was situated very close to the entrance of a dark shed, in which there was no artificial lighting, into which a visiting surveyor fell, sustaining injury. It was held that the occupier should have erected a warning). In some instances it may be necessary not only to provide proper warning notices but also to arrange for the supervision of visitors at the occupier's premises (*Farrant v Thanet District Council (1996) (unreported)*). However, there is no duty on an employer to light premises at night, which are infrequently used by day and not occupied at night. It is sufficient to provide a torch (*Capitano v Leeds Eastern Health Authority, Current Law, October 1989* where a security officer was injured when he fell down a flight of stairs at night, while checking the premises following the sounding of a burglar alarm. The steps were formerly part of a fire escape route but were now infrequently used). However, landowners need not warn against obvious dangers (see below O3017).

Trespassers

O3007 '… a trespasser is not necessarily a bad man. A burglar is a trespasser; but so too is a law-abiding citizen who unhindered strolls across an open field. The statement that a trespasser comes upon land at his own risk has been treated as applying to all who trespass, to those who come for nefarious purposes and those who merely bruise the grass, to those who know their presence is resented and those who have no reason to think so.' (*Commissioner for Railways (NSW) v Cardy [1960] 34 ALJR 134*).

Common law defines a trespasser as a person who:

(*a*) goes onto premises without invitation or permission; or

(*b*) although invited or permitted to be on premises, goes to a part of the premises to which the invitation or permission does not extend; or

(*c*) remains on premises after the invitation or permission to be there has expired; or

(*d*) deposits goods on premises when not authorised to do so.

The *Countryside and Rights of Way Act 2000* (*CRWA 2000*) now gives the public the right to roam across the open countryside in England and Wales. *Section 13* of *CRWA 2000* amends *s 1*(*4*) of the *OLA 1957* so that: *'A person entering any premises in exercise of [the right to roam] is not, for the purposes of this Act, a visitor of the occupier of the premises.'* Therefore, ramblers under the *CRWA 2000* are also classed as 'trespassers'. Indeed, *CRWA 2000* amends *s 1* of the *OLA 1984* so that no duty of care is owed to a person exercising the right to roam under the Act in respect of: *'a risk resulting from the existence of any natural feature of the landscape, or any river, stream, ditch or pond whether or not a natural feature … or a risk of that person suffering injury when passing over, under or through any wall, fence or gate, except by proper use of the gate of or a style.'* The 2000 Act therefore explicitly states what the position is in law – whilst 'visitors' falling under the 1957 Act must refer to case law.

Duty owed to trespassers, at common law and under the Occupiers' Liability Act 1984

Common law

O3008 The position under common law used to be that an occupier was not liable for injury caused to a trespasser, unless the injury was either intentional or done with reckless disregard for the trespasser's presence (*R Addie & Sons (Collieries) Ltd v Dumbreck [1929] AC 358*). In more recent times, however, the common law has adopted an attitude of humane conscientiousness towards simple (as distinct from aggravated) trespassers. '… the question whether an occupier is liable in respect of an accident to a trespasser on his land would depend on whether a conscientious, humane man with his knowledge, skill and resources could reasonably have been expected to have done, or refrained from doing, before the accident, something which would have avoided it. If he knew before the accident that there was a substantial probability that trespassers would come I think that most people would regard as culpable failure to give any thought to their safety' (*Herrington v British Railways Board [1972] 1 All ER 749*). The effect of this House of Lords case was that it became possible for a trespasser to bring a successful action in negligence under the common law. However, the precise nature of the occupier's duty to the trespasser remained unclear, leading to Parliamentary intervention culminating in the *OLA 1984*.

Occupiers' Liability Act 1984

O3009 The *OLA 1984* introduced a duty on an occupier in respect of trespassers, that is persons whether they have 'lawful authority to be in the vicinity or not' who may be at risk of injury on his premises. [*OLA 1984, s 1*]. Thus, an occupier owes a duty of care to a trespasser in respect of any injury suffered on the premises (either because of any danger due to the state of the premises, or things done or omitted to be done) [*OLA 1984, s 1*(*3*)(*a*)], in the following circumstances:

(*a*) if he was aware of the danger or had reasonable grounds to believe that it exists;

(*b*) if he knows or has reasonable grounds to believe that a trespasser is in the vicinity of the danger concerned, or that he may come into the vicinity of the danger; and

(*c*) if the risk is one against which, in all the circumstances of the case, he may reasonably be expected to offer some protection.

Where under the *OLA 1984*, the occupier is under a duty of care to the trespasser in respect of the risk, the standard of care is 'to take such care as is reasonable in all the circumstances of the case to see that he does not suffer injury on the premises by reason of the danger concerned'. [*OLA 1984, s 1*(*4*)]. The duty can be discharged by issuing a warning, e.g. posting notices warning of hazards; these, however, must be explicit and not merely vague. Thus, 'Danger' might not be sufficient whereas 'Highly Flammable Liquid Vapours – No Smoking' would be [*OLA 1984, s 1*(*5*)] (see O3015 below). Moreover, under *OLA 1984* there is no duty to persons who willingly accept risks (see O3016 below). However, the fact that an occupier has taken precautions to prevent persons going on to his land, where there is a danger, does not mean that the occupier has reason to believe that someone would be likely to come into the vicinity of the danger, thereby owing a duty to the trespasser, under the *OLA 1984, s 1*(*4*) (*White v St Albans City and District Council, The Times, 12 March 1990*). In *Revil v Newbury, The Times, 3 November 1995,* the court considered the duty of an occupier to an intruder. It found that the intruder could recover damages despite being engaged in unlawful activities at the time the injury occurred. The defendant was held to have injured the intruder by using greater force than was justified (by shooting the intruder through a locked door) in protecting himself and his property. The doctrine of *ex turpi causa non oritur actio* (i.e. no cause of action may be founded on an immoral or illegal act) cannot, without more, be relied upon by the defendant to escape liability. Violence may be returned with necessary violence, but the force used must not exceed the limits of what is reasonable in all the circumstances. For the extent of the intruder's contributory negligence, see O3011 below.

Children

O3010 'An occupier must be prepared for children to be less careful than adults.' [*OLA 1957, s 2*(*3*)(*a*)]. Where an adult would be regarded as a trespasser, a child is likely to qualify as an implied licensee, and this in spite of the stricture that 'it is hard to see how infantile temptations can give rights however much they excuse peccadilloes' (per Hamilton LJ in *Latham v Johnson & Nephew Ltd [1913] 1 KB 398*). If there is something or some state of affairs on the premises (e.g. machinery, a boat, a pond, bright berries, a motor car, forklift truck, scaffolding), this may constitute a 'trap' to a child. If the child is then injured by the 'trap', the occupier will often be liable. Though sometimes the presence of a parent may be treated as an implied condition of the permission to enter premises (e.g. when children go on to premises at dusk – *Phipps v Rochester Corporation [1955] 1 All ER 129*). Perhaps the current common law position regarding 'child-trespass' was best put as follows: 'The doctrine that a trespasser, however innocent, enters land at his own risk, that in no circumstances is he owed a duty of reasonable or any care by the owners or occupiers of the land, however conscious they may be of the likelihood of his presence and of the grave risk of terrible injury to which he will probably be exposed, may have been all very well when rights of property, particularly in land, were regarded as more sacrosanct than any other human right … It is difficult to see why today this doctrine should not be buried' (per Salmon LJ in *Herrington v British Railways Board*, concerning a six-year-old boy electrocuted on the defendant's electrified line) (but see O3017 below, *Titchener v British Railways Board*). Even though the children had been warned of the danger, the defendant company was still held liable in *Southern Portland Cement v Cooper [1974] AC 623*.

The case of *Jolley v London Borough of Sutton, The Times, May 24 2000*, concerned an old wooden boat that had been dumped on a grassed area belonging to a block of local authority flats. It became rotten and decayed, but was not covered or fenced off. The council placed a danger sticker on the boat, warning people not to touch it. This did not deter two teenage boys, who decided to repair it so they could take it sailing. Using a car jack and some wood from home, they jacked up the front of the boat so that they could repair the hull. One of the boys was lying on the ground underneath the boat when the prop gave way. The boat fell on him and caused very serious spinal injuries.

The Court of Appeal ruled that although it was foreseeable that children who were attracted by the boat might climb on it and be injured by the rotten planking giving way, it was not reasonably foreseeable that an accident could occur as a result of the boys deciding to work under a propped-up boat. However, the House of Lords decided this was wrong and that the council was liable; the trial judge was right in saying that the wider risk was that children would meddle with the boat at the risk of some physical injury, and what the boys did was not so different from normal play. The council should have removed the boat.

Contributory negligence

O3011 In a number of cases brought under the *OLA 1957*, the question of contributory negligence is often considered and impacts considerably on the amount of damages eventually awarded. This may be because the duty on the occupier is only to take such care as is reasonable to see that the visitor will be reasonably safe in using his premises. Therefore, it is possible to imagine that, even where an occupier is found to have breached his duty of care, the injured party may also, by his own conduct, have contributed to the circumstances giving rise to the injury and may have exacerbated the severity of the injuries sustained. The *Law Reform (Contributory Negligence) Act 1945, s 1(1)* provides that 'where any person suffers damage as the result partly of his own fault and partly of the fault of another person, a claim in respect of that damage shall not be defeated by reason of the fault of the person suffering the damage, but the damage recoverable in respect thereof shall be reduced to such extent as the court thinks just having regard to the claimant's share in the responsibility of the damage'. There is nothing to suggest that the above provision does not apply to children or trespassers. However, Lord Denning in *Gough v Thorn [1966] 3 All ER 398* held that 'a very young child cannot be guilty of contributory negligence. An older child may be. But it depends on the circumstances'. In *Revil v Newbury* (see O3009 above), although the defendant's conduct was not reasonable, the plaintiff intruder was found to be two-thirds to blame for the injuries he suffered and his damages were reduced accordingly. A claimant cannot be guilty of 100 per cent contributory negligence (*Anderson v Newham College of Further Education [2002] EWCA Civ 505*) if evidence showed that the entire fault was to lie with the claimant then there could be no liability on the defendant.

Dangers to guard against

O3012 The duty owed by an occupier to his lawful visitors is a 'common duty of care', so called since the duty is owed to both invitees and licensees, i.e. those having an interest in common with the occupier (e.g. business associates, customers, clients, salesmen) and those permitted by regulation/statute to be on the premises, e.g. factory inspectors/policemen. That duty requires that the dangers against which the occupier must guard are twofold: (*a*) structural defects in the premises; and (*b*) dangers associated with works/operations carried out for the occupier on the premises.

Structural defects

O3013 As regards structural defects in premises, the occupier will only be liable if either he actually knew of a defect or foreseeably had reason to believe that there was a defect in the premises. Simply put, the occupier would not incur liability for the existence of latent defects causing injury or damage, unless he had *special* knowledge in that regard, e.g. a faulty electrical circuit, unless he were an electrician; whereas, he would be liable for patent (i.e. obvious) structural defects, e.g. an unlit hole in the road. This duty now extends to 'uninvited entrants', e.g. trespassers, under the *OLA 1984, s 1*.

Workmen on occupier's premises

O3014 'An occupier may expect that a person, in the exercise of his calling, will appreciate and guard against any special risks ordinarily incident to it, so far as the occupier leaves him free to do so.' [*OLA 1957, s 2(3)(b)*]. This has generally been taken to imply that risks associated with system or method of work on customer premises are the exclusive responsibility of the employer (not the customer or occupier) (*General*

Cleaning Contractors Ltd v Christmas (see W9040 WORK AT HEIGHTS)), though how far this rule now applies specifically to window cleaners themselves is doubtful (see *King v Smith [1995] ICR 339* at W9040 WORK AT HEIGHTS). This means that risks associated with the system or method of work on third party premises are the responsibility of the employer not the occupier (*General Cleaning Contractors Ltd v Christmas [1952] 2 All ER 1110*, concerning a window cleaner who failed to take proper precautions in respect of a defective sash window: it was held that there was no liability on the part of the occupier, but liability on the part of the employer). The occupier will only incur liability for a structural defect in premises which the oncoming workman would not normally guard against as part of a safe system of doing his job, i.e. against 'unusual' dangers. 'And with respect to such a visitor at least, we consider it settled law that he, using reasonable care on his part for his own safety, is entitled to expect that the occupier shall on his part use reasonable care to prevent damage from unusual danger, which he knows or ought to know' (per Willes J in *Indermaur v Dames [1866] LR 1 CP 274*, concerning a gasfitter testing gas burners in a sugar refinery who fell into an unfenced shaft and was injured).

The law is not entirely settled in the case of firemen (*Sibbald v Sher Bros, The Times, 1 February 1981*): 'It (is) … very unlikely that the duty of care owed by the occupier to workers was the same as that owed to firemen,' (per Lord Fraser of Tullybelton), (but) it is arguable that a 'fireman (is) a "neighbour" of the occupier in the sense of Lord Atkin's famous dictum in *Donoghue v Stevenson [1932] AC 562*, so that the occupier owes him some duty of care, as for instance, to warn firemen of an unexpected danger or trap of which he knew or ought to know' (per Waller LJ in *Hartley v British Railways Board (1981) SJ 125*). The recent case of *Simpson v A I Dairies Farms Limited [2001] LTL January 12 2001* reinforces this as a farmer who failed to warn a fireman of the presence of an uneven drain in a yard which was currently concealed by water was deemed liable when the fireman was injured.

One's neighbour was defined in law as follows: 'persons who are so closely and directly affected by my act that I ought reasonably to have them in contemplation as being so affected when I am directing my mind to the acts or omissions which are called in question' (per Lord Atkin in *Donoghue v Stevenson [1932] AC 562*).

In summary, an occupier will be liable:

(*a*) if he exposes a fireman to the risk of injury/death over and above the normal risks; and

(*b*) in accordance with the general principles of negligence, in non-emergency situations (as per *Ogwo v Taylor [1987] 3 All ER 961*, applying the 'neighbour' principle of *Donoghue v Stevenson*).

In both cases, the basis of action is negligence, either at common law or under the *OLA 1957* or both.

Dangers associated with works being done on premises

O3015 Where work is being done on premises by a contractor, the occupier is not liable if he:

(*a*) took care in selecting a competent contractor; and

(*b*) satisfied himself that the work was being properly done by the contractor.

[*OLA 1957, s 2(4)(b)*].

Liability for a person injured as the result of work being carried out by an independent contractor tends also to remain the responsibility of the occupier as it is s/he who has 'control' of the land. The occupier must (1) act reasonably in selecting and entrusting work to the independent contractor; (2) take reasonable steps to supervise the work; and (3) use reasonable care to check that the work has been properly done. In *Jackson v Goodwood Estates [1999] LTL April 5 1999* a visitor to Goodwood Park tripped over a cable which was lying across the gateway. The owner had given instructions to the contractor that all cables be buried. However, the court found that the owner had not made sure that those instructions were implemented and had no system in place for checking, and was therefore liable.

As regards the checking of works, it may be highly desirable (indeed necessary) for an occupier to delegate the 'duty of satisfaction', especially where complicated building/engineering operations are being carried out, to a specialist, e.g. an architect, geotechnical engineer. Not to do so, in the interests of safety of visitors, is probably negligent. 'In the case of the construction of a substantial building, or of a ship, I should have thought that the building owner, if he is to escape subsequent tortious liability for faulty construction, should not only take care to contract with a competent contractor ... but also cause that work to be supervised by a properly qualified professional ... such as an architect or surveyor ... I cannot think that different principles can apply to precautions during the course of construction, if the building owner is going to invite a third party to bring valuable property on to the site during construction' (per Mocatta J in *AMF International Ltd v Magnet Bowling Ltd [1968] 2 All ER 789*).

Waiver of duty and the Unfair Contract Terms Act 1977 (UCTA)

O3016 Where damage was caused to a visitor by a danger of which he had been warned by the occupier (e.g. by notice), an explicit notice, e.g. 'Highly Flammable Liquid Vapours – No Smoking' (as distinct from a vague notice such as 'Fire Hazards') used to absolve an occupier from liability. Now, however, such notices are ineffective (except in the case of trespassers, see O3008 above) and do not exonerate occupiers. Thus: 'a person cannot by reference to any contract term or to a notice given to persons generally or to particular persons exclude or restrict his liability for death or personal injury resulting from negligence'. [*UCTA 1977, s 2*]. Such explicit notices are, however, a defence to an occupier when sued for negligent injury by a simple trespasser under the *OLA 1984*. As regards negligent damage to property, a person can restrict or exclude his liability by a notice or contract term, but such notice or contract term must be 'reasonable', and it is incumbent on the occupier to prove that it is in fact reasonable. [*UCTA 1977, ss 2(2), 11*].

The question of whether the defendant did what was reasonable fairly to bring to the notice of the plaintiff the existence of a particularly onerous condition, regarding his statutory rights, was considered in *Interfoto Picture Library Ltd v Stiletto Visual Programmes Ltd [1989] QB 433*. It was held that for a condition seeking to restrict statutory rights (e.g. under the *OLA 1957*) to be effective it must have been fairly brought to the attention of a party to the contract by way of some clear indication which would lead an ordinary sensible person to realise, at or before the time of making the contract, that such a term relating to personal injury was to be included in the contract.

Risks willingly accepted – 'volenti non fit injuria'

O3017 'The common duty of care does not impose on an occupier any obligation to a visitor in respect of risks willingly accepted as his by the visitor'. [*OLA 1957, s 2(5)*].

However, this 'defence' has generally not succeeded in industrial injuries claims. This is similarly the case with occupiers' liability claims (*Burnett v British Waterways Board [1973] 2 All ER 631* where a lighterman was held not to be bound by the terms of a notice erected by the respondent, even though he had seen it many times and understood it). This decision has since been reinforced by the *Unfair Contract Terms Act 1977*. 'Where a contract term or notice purports to exclude or restrict liability for negligence a person's agreement to or awareness of it is not of itself to be taken as indicating his voluntary acceptance of any risk.' [*UCTA 1977, s 2(3)*]. The *OLA 1957* makes clear that there may be circumstances in which even an explicit warning will not absolve the occupier from liability. Nevertheless, the occupier is entitled to expect a reasonable person to appreciate certain obvious dangers and take appropriate action to ensure his safety. In those circumstances no warning would appear to be required. It was not necessary, for example, to warn an adult of sound mind that it was dangerous to go near the edge of a cliff in *Cotton v Derbyshire Dales District Council, The Times, 20 June 1994*. In *Darby v The National Trust [2000] EWCA Civ 189* it was held there was no duty on the defendant to warn against the risk of swimming in a pond as the dangers of drowning were no more then an adult should have foreseen. The risk to a competer swimmer was an obvious one (*Staples v West Dorset District Council [1995] PIQR P439*). There was therefore no relative causative risk; a sign would have told the deceased no more than he already knew.

This is not, however, the position with respect to trespassers. The *OLA 1984, s 1(6)* states, 'No duty is owed … to any person in respect of risks willingly accepted as his by that person … '. This applies even if the trespasser is a child (see O3010 above) (*Titchener v British Railways Board 1984 SLT 192* where a 15-year-old girl and her boyfriend aged 16 had been struck by a train. The boy was killed and the girl suffered serious injuries. They had squeezed through a gap in a fence to cross the line as a short-cut to a disused brickworks which was regularly used. It was held by the House of Lords that the respondent did not owe a duty to the girl to do more than they had done to maintain the fence. It would have been 'quite unreasonable' for the respondent to maintain an impenetrable and unclimbable fence. But the 'duty (to maintain fences) will tend to be higher with a very young or a very old person than with a normally active and intelligent adult or adolescent' (per Lord Fraser)). *Scott and Swainger v Associated British Ports and British Railways Board [2000] All ER (D) 1937*, concerned schoolboys who on separate occasions were both severely injured whilst attempting to 'train surf' i.e. hitch rides on slow passing freight trains. The case was decided on the narrow issue of causation. The judge found as a question of fact that the provision of a fence would not have deterred them. In this situation, the court seemed to conclude that defendants may not do more than fence in dangers in order to protect children and that by scaling a fence the claimants would have willingly consented to the risk of injury.

This position might well be otherwise if anti-trespasser devices were not satisfactorily maintained (*Adams v Southern Electricity Board, The Times, 21 October 1993* where the electricity board was held to be liable to a teenage trespasser for injuries sustained when he climbed on to apparatus by means of a defective anti-climbing device). Also, see *Tomlinson v Congleton Borough Council [2002] EWCA Civ 309 CA* in which a lake had warning notices of danger and prohibiting swimming. It was held that by ignoring these notices the claimant entered the lake as a trespasser. The claimant was aware of the danger of swimming and *Darby v The National Trust* was referred to. However as the borough council were aware that notices were being ignored and that injuries had already been caused, and were also aware of further steps which could alleviate the danger but which had not been implemented, the Court of Appeal held by a majority judgement that the council were liable with 2/3 contributory negligence for the claimant.

In the curious case of *Arthur v Anka [1996] 2 WLR 602* a car was parked without permission in a private car park. It was clamped but the driver removed his vehicle with the clamp still attached. The owner of the car park sued the driver for the return of the wheel clamp and payment of the fine for illegal parking. The question of *volenti* was considered in detail. The judge held that the owner parked his car in full knowledge that he was not entitled to do so and that he was therefore consenting to the consequences of his action (payment of the fine); he could not complain after the event. The effect of this consent was to render conduct lawful which would otherwise have been tortious. By voluntarily accepting the risk that his car might be clamped, the driver accepted the risk that the car would remain clamped until he paid the reasonable cost of clamping and de-clamping.

Actions against factory occupiers

O3018 The *Workplace* (*Health, Safety and Welfare*) *Regulations 1992* (*SI 1992 No 3004* as amended by *SI 2002 No 2174*) which came into force, in so far as existing workplaces are concerned, on 1 January 1996 now apply with respect to the health, safety and welfare of persons in a defined workplace. The application of the Regulations is not confined to factories and offices – they also affect public buildings such as hospitals and schools, but they do not apply to, for example, construction sites and quarries. Employers, persons who have control of the workplace and occupiers are all subject to the Regulations, the provisions of which address the environmental management and condition of buildings, such as the day-to-day considerations of ventilation, temperature control and lighting and potentially dangerous activities such as window cleaning. In many ways the Regulations reflect the obligations imposed by the old *Factories Act 1961*. For example, *Regulation 12*(*3*) requires floors and traffic routes, 'so far as is reasonably practicable', to be kept free from obstacles or substances which may cause a person to 'slip, trip or fall'. These Regulations also closely mirror the general obligation placed on employers by *HSWA 1974, s 2* to ensure so far as is reasonably practicable, the health, safety and welfare of their employees whilst at work. Breach of these Regulations gives rise to civil liability by virtue of *HSWA 1974, s 47*(*2*) (for criminal liability, see O3019 below and ENFORCEMENT).

An approved Code of Practice and Guidance to the Regulations is available from the Health and Safety Commission.

Occupier's duties under HSWA 1974

O3019 In addition to civil liabilities under the *OLA 1957* and at common law, occupiers of buildings also have duties, the failure of which to carry out can lead to criminal liability under the *HSWA 1974*. More particularly, 'each person who has, to any extent, control of premises (i.e. "non-domestic" premises) or the means of access thereto or egress therefrom or of any plant or substance in such premises' must do what is reasonably practicable to see that the premises, means of access and egress and plant/substances on the premises, are safe and without health risks. [*HSWA 1974, s 4*(*2*)]. This section applies in the case of (*a*) non-employees and (*b*) non-domestic premises. [*HSWA 1974, s 4*(*1*)]. In other words, it places health and safety duties on persons and companies letting or sub-letting premises for work purposes, even though the persons working in those premises are not employees of the lessor/sublessor. The duty also extends to non-working persons, e.g. children at a play centre (*Moualem v Carlisle City Council (1994) 158 JPN 786*). As with most other forms of leasehold tenure, the person who is responsible for maintenance and repairs of the leased premises is the person who has 'control' (see, by way of analogy, O3004 above). [*HSWA 1974, s 4*(*3*)]. Included as 'premises' are common parts of a block of flats. These are 'non-domestic' premises (*Westminster City Council v Select Managements Ltd [1985] 1 All ER 897* which held that being a 'place' or 'installation on land', such areas are 'premises'; and they are not 'domestic', since they are in common use by the occupants of more than one private dwelling).

Moreover, the reasonableness of the measures which a person is required to take to ensure the safety of those premises is to be determined in the light of his knowledge of the expected use for which the premises have been made available and of the extent of his control and knowledge, if any, of the use thereafter. More particularly, if premises were not a reasonably foreseeable cause of danger to anyone acting in a way a person might reasonably be expected to act, in circumstances that might reasonably be expected to occur during the carrying out of the work, further measures would not be required against unknown and unexpected events (*Mailer v Austin Rover Group [1989] 2 All ER 1087* where an employee of a firm of cleaning contractors, whilst cleaning one of the appellant's paint spray booths and the sump underneath it, was killed by escaping fumes. The contractors had been instructed by the appellants not to use paint thinners from a pipe in the booth (which the appellants had turned off but not capped) and only to enter the sump (where the ventilator would have been turned off) with an approved safety lamp and when no one was working above. Contrary to those instructions, an employee used thinners from the pipe, which had then entered the sump below, where the deceased was working with a non-approved lamp, and an explosion occurred. It was held that the appellant was not liable for breach of the *HSWA 1974, s 4(2)*).

When considering criminal liability of an occupier under the *HSWA 1974* in relation to risks to outside contractors, *s 4* is likely to be of less significance in the future because of the decision of the House of Lords in *R v Associated Octel [1994] ICR 281*, a decision which in effect enables the *HSWA 1974, s 3* to be used against occupiers. *Section 3(1)* provides that 'It shall be the duty of every employer to conduct his undertaking in such a way as to ensure, so far as is reasonably practicable, that persons not in his employment who may be affected thereby are not thereby exposed to risks to their health or safety'. The decision establishes that an occupier can still be conducting his undertaking by engaging contractors to do work, even where the activities of the contractor are separate and not under the occupier's control. *Section 3* therefore requires the occupier to take steps to ensure the protection of contractors' employees from risks not merely arising from the physical state of the premises but also from the process of carrying out the work itself.

However, the judgment of *Makepiece v Evans Brothers (Reading) (A firm) (Court of Appeal, 23 May 2000)* should be noted in that it was held that an occupier would not usually be liable to an employee of a contractor employed to carry out work at the occupiers' premises if the employee was injured as a result of any unsafe system of work used by the employee and the contractor. It was not generally reasonable to expect an occupier of premises, having engaged a contractor whom he had reasonable grounds to regard as competent, to supervise the contractor's activities in order to ensure that the was discharging his duties to ensure a safe systems of work for his employees. Even if the occupier may have reason to suspect that the contractors may have been using an unsafe system of work, this would not, in itself, be enough to impose upon him liability under the act or in negligence.

Personal Protective Equipment

Introduction

P3001 Conventional wisdom suggests that 'safe place strategies' are more effective in combating health and safety risks than 'safe person strategies'. Safe systems of work, and control/prevention measures serve to protect everyone at work, whilst the advantages of personal protective equipment are limited to the individual(s) concerned. Given, however, the fallibility of any state of the art technology in endeavouring to achieve total protection, some level of personal protective equipment is inevitable in view of the obvious (and not so obvious) risks to head, face, neck, eyes, ears, lungs, skin, arms, hands and feet.

Current statutory requirements for employers to provide and maintain suitable personal protective equipment are contained in the *Personal Protective Equipment at Work Regulations 1992* (*SI 1992 No 2966* as amended by *SI 2002 No 2174*), enjoining them to carry out an assessment to determine what personal protective equipment is needed and to consider employee needs in the selection process. Adjunct to this, common law insists, not only that employers have requisite safety equipment at hand, or available in an accessible place, but also that management ensure that operators use it (see *Bux v Slough Metals Ltd [1974] 1 All ER 262*). That the hallowed duty to 'provide and maintain' has been getting progressively stricter is evidenced by *Crouch v British Rail Engineering Ltd [1988] IRLR 404* to the extent that employers could be in breach of either statutory or common law duty (or both) in the case of injury/disease to a member of the employee's immediate family involved, say, in cleaning protective clothing – but the injury/disease must have been foreseeable (*Hewett v Alf Brown's Transport Ltd [1992] ICR 530*).

This section deals with the requirements of the *Personal Protective Equipment at Work Regulations 1992*, the *Personal Protective Equipment* (*EC Directive*) *Regulations 1992* (*SI 1992 No 3139*) (now revoked and consolidated by *SI 2002 No 1144*), the duties imposed on manufacturers and suppliers of personal protective equipment, the common law duty on employers to provide suitable personal protective equipment, and the main types of personal protective equipment and clothing and some relevant British Standards.

The Personal Protective Equipment at Work Regulations 1992 (SI 1992 No 2966 as amended by SI 2002 No 2174)

P3002 With the advent of the *Personal Protective Equipment at Work Regulations 1992*, all employers must make a formal assessment of the personal protective equipment needs of employees and provide ergonomically suitable equipment in relation to foreseeable risks at work (see further P3006 below). However, employers cannot be expected to comply with their new statutory duties, unless manufacturers of personal protective equipment have complied with theirs under the requirements of the *Personal Protective Equipment* (*EC Directive*) *Regulations 1992* (*SI 1992 No 3139*) and *SI 1994 No 2326* (now revoked and consolidated by *SI 2002 No 1144*) – that is, had their products independently certified for EU accreditation purposes. Although imposing a considerable remit on manufacturers, EU accreditation is an indispensable condition precedent to sale and commercial circulation (see P3012 below).

It is important that personal protective clothing and equipment should be seen as 'last resort' protection – its use should only be prescribed when engineering and management solutions and other safe systems of work do not effectively protect the worker from the danger.

Employees must be made aware of the purpose of personal protective equipment, its limitations and the need for on-going maintenance. Thus, when assessing the need for, say, eye protection, employers should first identify the existence of workplace hazards (e.g. airborne dust, projectiles, liquid splashes, slippery floors, inclement weather in the case of outside work) and then the extent of danger (e.g. frequency/velocity of projectiles, frequency/severity of splashes). Selection can (and, indeed, should) then be made from the variety of CE-marked equipment available, in respect of which manufacturers must ensure that such equipment provides protection, and suppliers ascertain that it meets such requirements/standards [*HSWA s 6*] (see further PRODUCT SAFETY). Typically, most of the risks will have already been logged, located and quantified in a routine risk/safety audit (see RISK ASSESSMENT), and classified according to whether they are physical/chemical/biological in relation to the part(s) of the body affected (e.g. eyes, ears, skin).

Selection of personal protective equipment is a first stage in an on-going routine, followed by proper use and maintenance of equipment (on the part of both employers and employees) as well as training and supervision in personal protection techniques. Maintenance presupposes a stock of renewable spare parts coupled with regular inspection, testing, examination, repair, cleaning and disinfection schedules as well as keeping appropriate records. Depending on the particular equipment, some will require regular testing and examination (e.g. respiratory equipment), whilst others merely inspection (e.g. gloves, goggles). Generally, manufacturers' maintenance schedules should be followed.

Suitable accommodation must be provided for protective equipment in order to minimise loss or damage and prevent exposure to cold, damp or bright sunlight, e.g. pegs for helmets, pegs and lockers for clothing, spectacle cases for safety glasses.

On-going safety training, often carried out by manufacturers for the benefit of users should combine both theory and practice.

Work activities/processes requiring personal protective equipment

P3003 Examples abound of processes/activities of which personal protective equipment is a prerequisite, from construction work and mining, through work with ionising radiations, to work with lifting plant, cranes, as well as handling chemicals, tree felling and working from heights. Similarly, blasting operations, work in furnaces and drop forging all require a degree of personal protection.

Statutory requirements in connection with personal protective equipment

P3004 General statutory requirements relating to personal protective equipment are contained in of the *Health and Safety at Work etc Act 1974, ss 2, 9*, and in the *Personal Protective Equipment at Work Regulations 1992*, 'Every employer shall ensure that suitable personal protective equipment is provided to his employees who may be exposed to a risk to their health or safety at work except where, and to the extent that such risk has been adequately controlled by other means which are equally or more effective'. [*Reg 4*(*1*)]. More specific statutory requirements concerning personal protective equipment exist in the *Personal Protective Equipment at Work Regulations 1992* (in tandem with the *Personal Protective Equipment* (*EC Directive*) *Regulations 1992* (*SI 1992 No 3139*) and *SI 1994 No 2326*) (now revoked and consolidated by *SI 2002 No 1144*)), and sundry other recent regulations applicable to particular industries/processes, e.g. asbestos, noise, construction (see P3008 below). Where personal protective equipment is a necessary control measure to meet a specific statutory requirement or following a risk assessment, this must be provided free of charge [*HWSA* s 9]. However, where it is optional or non essential, there is nothing to stop the employer from seeking a financial contribution.

General statutory duties – HSWA 1974

P3005 Employers are under a general duty to ensure, so far as reasonably practicable, the health, safety and welfare at work of their employees – a duty which clearly implies provision/maintenance of personal protective equipment. [*HSWA s 2*(*1*)].

Specific statutory duties – Personal Protective Equipment at Work Regulations 1992 (SI 1992 No 2966 as amended by SI 2002 No 2174)

Duties of employers

P3006 Employers – and self-employed persons in cases (*a*), (*b*), (*c*), (*d*) and (*e*) below, must undertake the following:

(*a*) Formally assess (and review periodically) provision and suitability of personal protective equipment. [*Reg 6*(*1*)(*3*)].

The assessment should include:

(i) risks to health and safety not avoided by other means;

(ii) reference to characteristics which personal protective equipment must have in relation to risks identified in (i); and

(iii) a comparison of the characteristics of personal protective equipment having the characteristics identified in (ii).

[*Reg 6*(*2*)].

The aim of the assessment is to ensure that an employer knows which personal protective equipment to choose; it constitutes the first stage in a continuing programme, concerned also with proper use and maintenance of personal protective equipment and training and supervision of employees.

(*b*) Provide suitable personal protective equipment to his employees, who may be exposed to health and safety risks while at work, except where the risk has either been adequately controlled by other equally or more effective means. [*Reg 4*(*1*)].

Personal protective equipment is not suitable unless:

(i) it is appropriate for risks involved and conditions at the place of exposure;

(ii) it takes account of ergonomic requirements and the state of health of the person who wears it;

(iii) it is capable of fitting the wearer correctly; and

(iv) so far as reasonably practicable (for meaning, see ENFORCEMENT), it is effective to prevent or adequately control risks involved without increasing the overall risk.

[*Reg 4*(*3*)].

(*c*) Provide compatible personal protective equipment – that is, that the use of more than one item of personal protective equipment is compatible with other personal protective equipment. [*Reg 5*].

(*d*) Maintain (as well as replace and clean) any personal protective equipment in an efficient state, efficient working order and in good repair. [*Reg 7*].

(*e*) Provide suitable accommodation for personal protective equipment when not being used. [*Reg 8*].

(*f*) Provide employees with information, instruction and training to enable them to know:

(i) the risks which personal protective equipment will avoid or minimise;

(ii) the purpose for which and manner in which personal protective equipment is to be used;

(iii) any action which the employee might take to ensure that personal protective equipment remains efficient.

[*Reg 9*].

(*g*) Ensure, taking all reasonable steps, that personal protective equipment is properly used. [*Reg 10*].

Summary of employer's duties

- duty of assessment;
- duty to provide suitable PPE;
- duty to provide compatible PPE;
- duty to maintain and replace PPE;
- duty to provide suitable accommodation for PPE;
- duty to provide information and training;
- duty to see that PPE is correctly used.

Duties of employees

P3007 Every employee must:

- use personal protective equipment in accordance with training and instructions [*Reg 10*(*2*)];
- return all personal protective equipment to the appropriate accommodation provided after use [*Reg 10*(*4*)]; and
- report forthwith any defect or loss in the equipment to the employer [*Reg 11*].

Specific requirements for particular industries and processes

P3008 In addition to the general remit of the *Personal Protective Equipment at Work Regulations 1992*, the following regulations impose specific requirements on employers to provide personal protective equipment up to the EU standard.

- *Control of Lead at Work Regulations 2002* (*SI 2002 No 2676*), *Regs 6, 8* – supply of suitable respiratory equipment and where exposure is significant, suitable protective clothing and equipment (see APPENDIX 4);
- *Control of Asbestos at Work Regulations 2002* (*SI 2002 No 2675*), *Reg 10* – supply of respiratory protective equipment/suitable protective clothing (see APPENDIX 4);
- *Control of Substances Hazardous to Health Regulations 2002* (*SI 2002 No 2677* as amended by *SI 2003 No 978*) (*COSHH*), *Regs 7, 8, 9* – supply of suitable protective equipment where employees are foreseeably exposed to substances hazardous to health;
- *Noise at Work Regulations 1989* (*SI 1989 No 1790*), *Reg 8* – personal ear protectors;
- *Construction* (*Head Protection*) *Regulations 1989* (*SI 1989 No 2209*), *Reg 3* – suitable head protection (for the position of Sikhs, see C8115 CONSTRUCTION AND BUILDING OPERATIONS);
- *Ionising Radiations Regulations 1999* (*SI 1999 No 3232*), *Reg 9* – supply of suitable personal protective equipment and respiratory protective equipment;
- *Shipbuilding and Ship-repairing Regulations 1960* (*SI 1960 No 1932*), *Regs 50, 51* – supply of suitable breathing apparatus, belts, eye protectors, gloves and gauntlets.

[*Miscellaneous Factories* (*Transitional Provisions*) *Regulations 1993* (*SI 1993 No 2482*)].

Increased importance of uniform European standards

P3009 To date, manufacturers of products, including personal protective equipment (PPE), have sought endorsement or approval for products, prior to commercial circulation, through reference to British Standards (BS), HSE or European standards, an example of the former being Kitemark and an example of the latter being CEN or CENELEC. Both are examples of *voluntary* national and international schemes that can be entered into by manufacturers and customers, under which both sides are 'advantaged' by conformity with such standards. Conformity is normally achieved following a level of testing appropriate to the level of protection offered by the product. With the introduction of the *Personal Protective Equipment* (*EC Directive*) *Regulations 1992* as amended by *SI 1994 No 2326* (now revoked and consolidated by *SI 2002 No 1144*) (see P3012 below), this voluntary system of approval was replaced by a statutory certification procedure.

Towards certification

P3010 Certification as a condition of sale is probably some time away. Transition from adherence based on a wide range of voluntary national (and international) standards to a compulsory universal EU standard takes time and specialist expertise. Already, however, British Standards Institution (BSI) has adopted the form BS EN as a British version of a uniform European standard. But, generally speaking, uniformity of approach to design and testing, on the part of manufacturers, represents a considerable remit and so statutory compliance with tougher new European standards is probably not likely to happen immediately. Here, too, however, there is evidence of some progress, e.g. the EN 45000 series of standards on how test houses and certification bodies are to be established and independently accredited. In addition, a range of European standards on PPE have been implemented in BS EN format, for example, on eye protection, fall arrest systems and safety footwear.

Interim measures

P3011 Most manufacturers have elected to comply with the current CEN standard when seeking product certification. Alternatively, should a manufacturer so wish, the inspection body can verify by way of another route to certification – this latter might well be the case with an innovative product where standards did not exist. Generally, however, compliance with current CEN (European Standardisation Committee) or, alternatively, ISO 9000: Quality Assurance has been the well-trodden route to certification. To date, several UK organisations, including manufacturers and independent bodies, have established PPE test houses, which are independently accredited by the National Measurement Accreditation Service (NAMAS). Such test houses will be open to all comers for verification of performance levels.

Personal Protective Equipment (EC Directive) Regulations 1992 (SI 1992 No 3139 as amended by SI 1994 No 2326)

P3012 The *Personal Protective Equipment (EC Directive) Regulations 1992* (as amended by *SI 1994 No 2326*) required most types of PPE (for exceptions, see P3019 below) to satisfy specified certification procedures, pass EU type-examination (i.e. official inspection by an approved inspection body) and carry a 'CE' mark both on the product itself and its packaging before being put into commercial circulation. Indeed, failure on the part of a manufacturer to obtain affixation of a 'CE' mark on his product before putting it into circulation, is a criminal offence under the *Health and Safety at Work etc Act 1974*, s 6 carrying a maximum fine, on summary conviction, of £20,000 (see ENFORCEMENT). Enforcement is through trading standards officers.

The *Personal Protective Equipment Regulations 2002* (*SI 2002 No 1144*) now revoke and consolidate the *Personal Protective Equipment (EC Directive) Regulations 1992*. The 2002 Regulations implement the Product Safety Directive 89/696/EEC, on the approximation of the laws of the Member States relating to personal protective equipment. The new Regulations came into force on 15 May 2002.

The 2002 Regulations place a duty on responsible persons who put personal protective equipment (PPE) on the market to comply with the following requirements:

- the PPE must satisfy the basic health and safety requirements that are applicable to that type or class of PPE;
- the appropriate conformity assessment procedure must be carried out;
- a CE mark must be affixed on the PPE.

The Regulations will be enforced by the Weights and Measures Authorities (District Councils in Northern Ireland).

CE mark of conformity

P3013 The *Personal Protective Equipment (EC Directive) Regulations 1992* as amended by the *Personal Protective Equipment (EC Directive) (Amendment) Regulations 1994* (*SI 1994 No 2326*) (now revoked and consolidated by *SI 2002 No 1144*) specify procedures and criteria with which manufacturers must comply in order to be able to obtain certification. Essentially this involves incorporation into design and production basic health and safety requirements. Hence, by compliance with health and safety criteria, affixation of the 'CE' mark becomes a condition of sale and approval (see *fig. 1* below). Moreover, compliance, on the part of manufacturers, with these regulations, will enable employers to comply with their duties under the *Personal Protective Equipment at Work Regulations 1992. Reg 4(3)(e)* of the *Personal Protective Equipment at Work Regulations 1992* requires the PPE provided by the employer to comply with this and any later EU directive requirements.

CE accreditation is not a synonym for compliance with the CEN standard, since a manufacturer could opt for certification via an alternative route. Neither is it an approvals mark. It is rather a quality mark, since all products covered by the regulations are required to meet formal quality control or quality of production criteria in their certification procedures. [*Arts 8.4, 11, 89/686/EEC*].

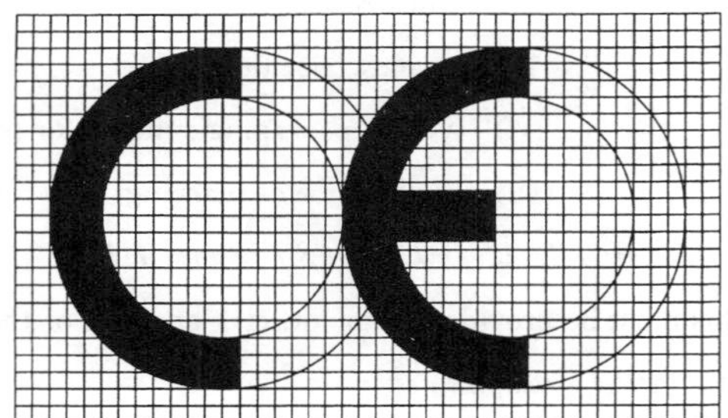

Certification procedures

P3014 The *Personal Protective Equipment (EC Directive) Regulations 1992* as amended by *SI 1994 No 2326* (now revoked and consolidated by *SI 2002 No 1144*), identify three categories of PPE:

- PPE of simple design;
- PPE of non-simple design;
- PPE of complex design.

Each of these have different requirements for manufacturers to follow.

In order to obtain certification, manufacturers must submit to an approved body the following documentation (except in the case of PPE of simple design, where risks are minimal).

(*1*) *PPE – simple design*

This includes use of PPE where risks are minimal e.g. gardening gloves, helmets, aprons, thimbles. This type of equipment must be subject to a manufacturers declaration of conformity.

(*2*) *PPE – non-simple design*

In order to obtain certification, manufacturers must submit to an approved body the following documentation:

(*a*) Technical file, i.e.

(i) overall and detailed plans, accompanied by calculation notes/ results of prototype tests; and

(ii) an exhaustive list of basic health and safety requirements and harmonised standards taken into account in the model's design.

(*b*) Description of control and test facilities used to check compliance with harmonised standards.

(*c*) Copy of information relating to:

(i) storage, use, cleaning, maintenance, servicing and disinfection;

(ii) performance recorded during technical tests to monitor levels of protection;

(iii) suitable PPE accessories;

(iv) classes of protection appropriate to different levels of risk and limits of use;

(v) obsolescence deadline;

(vi) type of packaging suitable for transport;

(vii) significance of markings.

[*Art 1.4, Annex II; Art 8.1, Annex III*].

This must be provided in the official language of the member state of destination.

(*3*) *PPE – complex design*

This includes use of PPE for protection against mortal/serious dangers and requires compliance with a quality control system.

This category of PPE, including:

(*a*) filtering, respiratory devices for protection against solid and liquid aerosols/ irritant, dangerous, toxic or radiotoxic gases;

(*b*) respiratory protection devices providing full insulation from the atmosphere, and for use in diving;

(*c*) protection against chemical attack or ionising radiation;

(*d*) emergency equipment for use in high temperatures, whether with or without infrared radiation, flames or large amounts of molten metal (100°C or more);

(*e*) emergency equipment for use in low temperatures (–50°C or less);

(*f*) protection against falls from heights;

(*g*) protection against electrical risks; and

(*h*) motor cycle helmets and visors;

must satisfy either

(i) — an EU quality control system for the final product, or

— a system for ensuring EU quality of production by means of monitoring;

and in either case,

(ii) an EU declaration of conformity.

[*Art 8.4*].

Basic health and safety requirements applicable to all PPE

P3015 All PPE must:

- be ergonomically suitable, that is, be able to perform a risk-related activity whilst providing the user with the highest possible level of protection;
- preclude risks and inherent nuisance factors, such as roughness, sharp edges and projections and must not cause movements endangering the user;
- provide comfort and efficiency by facilitating correct positioning on the user and remaining in place for the foreseeable period of use; and by being as light as possible without undermining design strength and efficiency; and
- be accompanied by necessary information – that is, the name and address of the manufacturer and technical file (see P3014 above).

[*Annex II*].

General purpose PPE – specific to several types of PPE

P3016

- PPE incorporating adjustment systems must not become incorrectly adjusted without the user's knowledge;
- PPE enclosing parts of the body must be sufficiently ventilated to limit perspiration or to absorb perspiration;
- PPE for the face, eyes and respiratory tracts must minimise risks to same and, if necessary, contain facilities to prevent moisture formation and be compatible with wearing spectacles or contact lenses;
- PPE subject to ageing must contain date of manufacture, the date of obsolescence and be indelibly inscribed if possible. If the useful life of a product is not known, accompanying information must enable the user to establish a reasonable obsolescence date. In addition, the number of times it can be cleaned before being inspected or discarded must (if possible) be affixed to the product; or, failing this, be indicated in accompanying literature;
- PPE which may be caught up during use by a moving object must have a suitable resistance threshold above which a constituent part will break and eliminate danger;
- PPE for use in explosive atmospheres must not be likely to cause an explosive mixture to ignite;
- PPE for emergency use or rapid installation/removal must minimise the time required for attachment and/or removal;

- PPE for use in very dangerous situations (see P3015 above) must be accompanied with data for exclusive use of competent trained individuals, and describe procedure to be followed to ensure that it is correctly adjusted and functional when worn;
- PPE incorporating components which are adjustable or removable by user, must facilitate adjustment, attachment and removal without tools;
- PPE for connection to an external complementary device must be mountable only on appropriate equipment;
- PPE incorporating a fluid circulation system must permit adequate fluid renewal in the vicinity of the entire part of the body to be protected;
- PPE bearing one or more identification or recognition marks relating to health and safety must preferably carry harmonised pictograms/ideograms which remain perfectly legible throughout the foreseeable useful life of the product;
- PPE in the form of clothing capable of signalling the user's presence visually must have one or more means of emitting direct or reflected visible radiation;
- multi-risk PPE must satisfy basic requirements specific to each risk.

[*Art 2, Annex II*].

Additional requirements for specific PPE for particular risks

P3017 There are additional requirements for PPE designed for certain particular risks as follows.

PPE protection against

- *Mechanical impact risks*
  - (i) impact caused by falling/projecting objects must be sufficiently shock-absorbent to prevent injury from crushing or penetration of the protected part of the body;
  - (ii) falls. In the case of falls due to slipping, outsoles for footwear must ensure satisfactory adhesion by grip and friction, given the state and nature of the surface. In the case of falls from a height, PPE must incorporate a body harness and attachment system connectable to a reliable anchorage point. The vertical drop of the user must be minimised to prevent collision with obstacles and the braking force injuring, tearing or causing the operator to fall;
  - (iii) mechanical vibration. PPE must be capable of ensuring adequate attenuation of harmful vibration components for the part of the body at risk.
- (*Static*) *compression of part of the body* – must be able to attenuate its effects so as to prevent serious injury of chronic complaints.
- *Physical injury* – must be able to protect all or part of the body against superficial injury by machinery, e.g. abrasion, perforation, cuts or bites.
- *Prevention of drowning* – (lifejackets, armbands etc.) must be capable of returning to the surface a user who is exhausted or unconscious, without danger to his health. Such PPE can be wholly or partially inherently buoyant

or inflatable either by gas or orally. It should be able to withstand impact with liquid, and, if inflatable, able to inflate rapidly and fully.

- *Harmful effects of noise* – must protect against exposure levels of the *Noise at Work Regulations 1989* (i.e. 85 and 90 dB(A)) and must indicate noise attenuation level.
- *Heat and/or fire* – must possess sufficient thermal insulation capacity to retain most of the stored heat until after the user has left the danger area and removed PPE. Moreover, constituent materials which could be splashed by large amounts of hot product must possess sufficient mechanical-impact absorbency. In addition, materials which might accidentally come into contact with flame as well as being used in the manufacture of fire-fighting equipment, must possess a degree of non-flammability proportionate to the risk foreseeably arising during use and must not melt when exposed to flame or contribute to flame spread. When ready for use, PPE must be such that the quantity of heat transmitted by PPE to the user must not cause pain or health impairment. Second, ready-to-use PPE must prevent liquid or steam penetration and not cause burns. If PPE incorporates a breathing device, it must adequately protect the user. Accompanying manufacturers' notes must provide all relevant data for determination of maximum permissible user exposure to heat transmitted by equipment.
- *Cold* – must possess sufficient thermal insulating capacity and retain the necessary flexibility for gestures and postures. In particular, PPE must protect tips of fingers and toes from pain or health impairment and prevent penetration by rain water. Manufacturers' accompanying notes must provide all relevant data concerning maximum permissible user exposure to cold transmitted by equipment.
- *Electric shock* – must be sufficiently insulated against voltages to which the user is likely to be exposed under the most foreseeably adverse conditions. In particular, PPE for use during work with electrical installations (together with packaging), which may be under tension, must carry markings indicating either protection class and/or corresponding operating voltage, serial number and date of manufacture; in addition, date of entry into service must be inscribed as well as of periodic tests or inspections.
- *Radiation*

  (*a*) Non-ionising radiation – must prevent acute or chronic eye damage from non-ionising radiation and be able to absorb/reflect the majority of energy radiated in harmful wavelengths without unduly affecting transmission of the innocuous part of the visible spectrum, perception of contrasts and distinguishment of colours. Thus, protective glasses must possess a spectral transmission factor so as to minimise radiant-energy illumination density capable of reaching the user's eye through the filter and ensure that it does not exceed permissible exposure value. Accompanying notes must indicate transmission curves, making selection of the most suitable PPE possible. The relevant protection-factor number must be marked on all specimens of filtering glasses.

  (*b*) Ionising radiation – must protect against external radioactive contamination. Thus, PPE should prevent penetration of radioactive dust, gases, liquids or mixtures under foreseeable use conditions. Moreover, PPE designed to provide complete user protection against external irradiation must be able to counter only weak electron or weak photon radiation.

- *Dangerous substances and infective agents*

  (*a*) respiratory protection – must be able to supply the user with breathable air when exposed to polluted atmosphere or an atmosphere with inadequate oxygen concentration. Leak-tightness of the facepiece and pressure drop on inspiration (breath intake), as well as in the case of filtering devices purification capacity, must keep the contaminant penetration from the polluted atmosphere sufficiently low to avoid endangering health of the user. Instructions for use must enable a trained user to use the equipment correctly;

  (*b*) cutaneous and ocular contact (skin and eyes) – must be able to prevent penetration or diffusion of dangerous substances and infective agents. Therefore, PPE must be completely leak-tight but allow prolonged daily use or, failing that, of limited leak-tightness restricting the period of wear. In the case of certain dangerous substances/infective agents possessing high penetrative power and which limit duration of protection, such PPE must be tested for classification on the basis of efficiency. PPE conforming with test specifications must carry a mark indicating names or codes of substances used in tests and standard period of protection. Manufacturers' notes must contain an explanation of codes, detailed description of standard tests and refer to the maximum permissible period of wear under different foreseeable conditions of use.

- *Safety devices for diving equipment* – must be able to supply the user with a breathable gaseous mixture, taking into account the maximum depth of immersion. If necessary, equipment must consist of

  (*a*) a suit to protect the user against pressure;

  (*b*) an alarm to give the user prompt warning of approaching failure in supply of breathable gaseous mixture;

  (*c*) life-saving suit to enable the user to return to the surface.

[*Art 3, Annex II*].

Excluded PPE

P3018

- PPE designed and manufactured for use specifically by the armed forces or in the maintenance of law and order (helmets, shields);
- PPE for self-defence (aerosol canisters, personal deterrent weapons);
- PPE designed and manufactured for private use against:

  (i) adverse weather (headgear, seasonal clothing, footwear, umbrellas),

  (ii) damp and water (dish-washing gloves),

  (iii) heat (e.g. gloves);

- PPE intended for the protection or rescue of persons on vessels or aircraft, not worn all the time.

[*Annex I*].

Main types of personal protection

The main types of personal protection covered by the *Personal Protective Equipment at Work Regulations 1992* are (*a*) head protection, (*b*) eye protection, (*c*) hand/arm protection, (*d*) foot protection and (*e*) whole body protection.

In particular, these regulations do not cover ear protectors and most respiratory protective equipment – these areas are covered by other existing regulations and guidance, for example the *Noise at Work Regulations 1989* and HSE's guidance booklet HS(G) 53, 'Respiratory protective equipment: a practical guide for users'.

Head protection

P3019 This takes the form of:

- crash, cycling, riding and climbing helmets;
- industrial safety helmets – to protect against falling objects;
- scalp protectors (bump caps) – to protect against striking fixed obstacles;
- caps/hairnets – to protect against scalping,

and is particularly suitable for the following activities:

(i) building work – particularly on scaffolds;

(ii) civil engineering projects;

(iii) blasting operations;

(iv) work in pits and trenches;

(v) work near hoists/lifting plant;

(vi) work in blast furnaces;

(vii) work in industrial furnaces;

(viii) ship-repairing;

(ix) railway shunting;

(x) slaughterhouses;

(xi) tree-felling;

(xii) suspended access work, e.g. window cleaning.

Eye protection

P3020 This takes the following forms:

- safety spectacles – these are the same as prescription spectacles but incorporating optional sideshields; lenses are made from tough plastic, such as polycarbonate – provide lateral protection;
- eyeshields – these are heavier than safety spectacles and designed with a frameless one-piece moulded lens; can be worn over prescription spectacles;
- safety goggles – these are heavier than spectacles or eye shields; they are made with flexible plastic frames and one-piece lens and have an elastic headband – they afford total eye protection; and

- faceshields – these are heavier than other eye protectors but comfortable if fitted with an adjustable head harness – faceshields protect the face but not fully the eyes and so are no protection against dusts, mist or gases.

Eye protectors are suitable for working with:

(i) chemicals;

(ii) power driven tools;

(iii) molten metal;

(iv) welding;

(v) radiation;

(vi) gases/vapours under pressure;

(vii) woodworking.

Hand/arm protection

P3021 Gloves provide protection against:

- cuts and abrasions;
- extremes of temperature;
- skin irritation/dermatitis;
- contact with toxic/corrosive liquids,

and are particularly useful in connection with the following activities/processes:

(i) manual handling;

(ii) vibration;

(iii) construction;

(iv) hot and cold materials;

(v) electricity;

(vi) chemicals;

(vii) radioactivity.

Foot protection

P3022 This takes the form of:

- safety shoes/boots;
- foundry boots;
- clogs;
- wellington boots;
- anti-static footwear;
- conductive footwear,

and is particularly useful for the following activities:

(i) construction;

(ii) mechanical/manual handling;

(iii) electrical processes;

(iv) working in cold conditions (thermal footwear);

(v) chemical processes;

(vi) forestry;

(vii) molten substances.

Respiratory protective equipment

P3023 This takes the form of:

- half-face respirators,
- full-face respirators,
- air-supply respirators,
- self-contained breathing apparatus,

and is particularly useful for protecting against harmful:

(i) gases;

(ii) vapours;

(iii) dusts;

(iv) fumes;

(v) smoke;

(vi) aerosols.

Whole body protection

P3024 This takes the form of:

- coveralls, overalls, aprons;
- outfits to protect against cold and heat;
- protection against machinery and chainsaws;
- high visibility clothing;
- life-jackets,

and is particularly useful in connection with the following activities:

(i) laboratory work;

(ii) construction;

(iii) forestry;

(iv) work in cold-stores;

(v) highway and road works;

(vi) food processing;

(vii) welding;

(viii) fire-fighting;

(ix) foundry work;

(x) spraying pesticides.

Some relevant British Standards for protective clothing and equipment

P3025 Although not, strictly speaking, a condition of sale or approval, for the purposes of the *Personal Protective Equipment (EC Directive) (Amendment) Regulations 1994* (*SI 1994 No 2326*) (now revoked and consolidated by *SI 2002 No 1144*), compliance with British Standards (e.g. ISO 9000: 'Quality Assurance') may well become one of the well-tried verification routes to obtaining a 'CE' mark. For that reason, if for no other, the following non-exhaustive list of British Standards on protective clothing and equipment should be of practical value to both manufacturers and users.

Table 1

Protective clothing

BSEN 60903: 1986	Specification for rubber gloves for electrical purposes
BS 3314: (1998)	Specification for protective aprons for wet work
BS 5426: 1993	Specification for workwear and career wear
BS 5438: 1976	Methods of test for flammability of vertically oriented textile fabrics and fabric assemblies subjected to a small igniting flame
1989 (1995)	Methods of test for flammability of textile fabrics when subjected to a small igniting flame applied to the face or bottom edge of vertically oriented specimens
BS 6408: (1990)	Specification for clothing made from coated fabrics for protection against wet weather
BS EN 340: 1993	Protective clothing. General Requirements
BS EN 341: (1999)	Personal Protective Equipment against falls from a height. Descender devices
BS EN 348: 1992	Protective clothing. Determination of behaviour of materials on impact of small splashes of molten metal
BS EN 353: 1993	Personal protective equipment against falls from a height. Guided type of fall arresters
BS EN 354: 1993	Personal protective equipment against falls from a height. Lanyards
BS EN 355: 1993	Personal protective equipment against falls from a height. Energy Absorbers
BS EN 360: 1993	Personal protective equipment against falls from a height. Retractable type fall arresters
BS EN 361: 1993	Personal protective equipment against falls from a height. Full body harnesses
BS EN 362: 1993	Personal protective equipment against falls from a height. Connectors

BS EN 363: 1993	Personal protective equipment against falls from a height. Fall arrest systems
BS EN 364: 1993 (1999)	Personal protective equipment against falls from a height. Test methods
BS EN 365: 1993	Personal protective equipment against falls from a height. General Requirements for instructions for use and for marking
BS EN 366: 1993	Protective clothing. Protection against heat and fire
BS EN 367: 1992	Protective clothing. Protection against heat and fires
BS EN 368: 1993	Protective clothing. Protection against liquid chemicals
BS EN 369: 1993	Protective clothing. Protection against liquid chemicals
BS EN 373: 1993	Protective clothing: Assessment of resistance of materials to molten metal splash
BS EN 381 Part 1: 1993	Protective clothing for users of hand-held chain saws
BS EN 463: 1995	Protective clothing. Protection against liquid chemicals. Test method
BS EN 464: 1994	Protective clothing. Protection against liquid and gaseous chemicals including liquid aerosols and solid particles. Test method
BS EN 465: 1995	Protective clothing: Protection against liquid chemicals. Performance requirements for chemical protective clothing
BS EN 4661: 1995	Protective clothing: Protection against liquid chemicals. Performance requirements for chemical protective clothing
BS EN 467: 1995	Protective Clothing: protection against liquid chemicals.
BS EN 468: 1995	Protective clothing for use against liquid chemicals. Test method
BS EN 469: 1995	Protective clothing for firefighters
BS EN 470	Protective clothing for use in welding and allied processes
Part 1: 1995	General requirements
BS EN 471: 1994	Specification for high visibility warning clothing
BS EN 510: 1993	Specification for protective clothing for use where there is a risk of entanglement with moving parts
BS EN 530: 1995	Abrasion resistance of protective clothing material
BS EN 531: 1995	Protective clothing for workers exposed to heat
BS EN 532: 1995	Protective clothing. Protection against heat and flame

BS EN 533: 1997	Clothing for protection against heat or flame
Protective footwear	
BS 2723: 1956 (1995)	Specification for fireman's leather boots
BS 4676: 1983 (1996)	Specification for gaiters and footwear for protection against burns and impact risks in foundries
BS 5145: 1989	Specification for lined industrial vulcanized rubber boots
BS 6159:	Polyvinyl chloride boots
Part 1: 1987	Specification for general and industrial lined or unlined boots
BS EN	Safety, protective and occupational footwear for professional use
BS EN 345	Safety footwear for professional use
BS EN 346	Protective footwear for professional use
BS EN 347	Occupational footwear for professional use
BS EN 381	Protective clothing for users of hand-held chain saws
Head protection	
BS 6658: 1985 (1995)	Specification for protective helmets for vehicle users
BS 6473: 1984	Specification for protective hats for horse and pony riders
BS 3864: 1989	Specification for protective helmets for firefighters
BS 4033: 1978	Specification for industrial scalp protectors (light duty)
BS 4423: 1969	Specification for climbers helmets
BS 4472: 1988	Specification for protective skull caps for jockeys
Face and eye protection	
BS 679 : 1989	Specification for filters, cover lenses and backing lenses for use during welding and similar industrial operations
BS 1542: 1982 (1995)	Specification for eye, face and neck protection against non-ionising radiation arising during welding and similar operations
BS 2092: 1987	Specification for eye protectors for industrial and non-industrial users
BS 2724: 1987 (1995)	Specification for sunglare eye protectors for general use
BS 4110: 1999	Specification for eye-protectors for vehicle users
BS 7028: 1999	
BS EN 167: 1995	Personal eye protection. Optical test methods

BS EN 168: 1995	Personal eye protection. Non-optical test methods
BS EN 169: 1992	Specification for filters for personal eye-protection equipment used in welding and similar operations
BS EN 170: 1992	Specification for filters for ultraviolet filters used in personal eye-protection equipment
BS EN 171: 1992	Specification for infra-red filters used in personal eye protective equipment
BS EN 172: (2000)	Specification for sunglare filters used in personal eye-protectors equipment for industrial use
Respiratory protection	
BS 4001	Care and maintenance of underwater breathing apparatus
Part 1 (1998)	
Part 2 (1995)	Standard diving equipment
BS 4275: 1997	Guide to implementing an effective respiratory protective device programme
BS 4400: 1969 (1995)	Method for sodium chloride particulate test for respirator filters
BS EN 132: 1999	Respiratory protective devices. Definitions of terms and pictograms
BS EN 133: 1991	Respiratory protective devices. Classification
BS EN 134: 1998	Respiratory protective devices. Nomenclature of compounds
BS EN 135: 1999	Respiratory protective devices. List of equivalent terms
BS EN 136: 1998	Respiratory protective devices. Full face mask requirements for testing and marking
BS EN 137: 1993	Specification for respiratory protective devices self-contained open-circuit compressed air breathing apparatus
BS EN 138: 2000	Respiratory protective devices. Specification for fresh air hose breathing apparatus for use with full face mask, half mask or mouthpiece assembly
BS EN 139: 1995	Respiratory protective devices. Compressed air line breathing apparatus for use with a full face mask, half mask or a mouthpiece assembly. Requirements, testing, marking
BS EN 141: 1991	Specification for gas filters and combined filters used in respiratory protective equipment
BS EN 143: 1991	Specification for particle filters used in respiratory protective equipment
BS EN 144	Respiratory protective devices. Gas cylinder valves

BS EN 145: 1998	Respiratory protective devices. Self-contained closed-circuit compressed oxygen breathing apparatus
BS EN 146: 1992	Respiratory protection devices. Specification for powered particle filtering devices incorporating helmets or hoods
BS EN 147: 1992	Respiratory protection devices. Specification for power assisted particle filtering devices incorporating full face masks, half masks or quarter masks
BS EN 148	Specification for thread connection
BS EN 149: 1992	Specification for filtering half masks to protect against particles
BS EN 371: 1992	Specification for AX gas filters and combined filters against low boiling organic compounds used in respiratory protective equipment
BS EN 372: 1992	Specification for SX gas filters and combined filters against specific named compounds used in respiratory protective equipment
Radiation protection	
BS 1542: 1982 (1995)	Specification for equipment for eye, face and neck protection against non-ionising radiation arising during welding and similar operations
BS 3664: 1963	Specification for film badges for personnel radiation monitoring
BS EN 269: 1995(2000)	Respiratory protective devices. Specification for powered fresh air hose breathing apparatus incorporating a hood.
BS EN 270: 1995	Respiratory protective devices. Compressed air line breathing apparatus incorporating a hood. Requirements, testing, marking
Hearing protection	
BS EN 352:	Hearing protectors. Safety requirements and testing
Part 1 1993	Ear muffs
Part 2 1993	Ear plugs
Part 3 1997	Ear muffs attached to an industrial safety helmet
Hand protection	
BS EN 374:	Protective gloves against chemicals and microorganisms
Part 1: 1994	Terminology and performance requirements
Part 2: 1994	Determination of resistance to penetration
Part 3: 1994	Determination of resistance to permeation by chemicals

BS EN 511: 1994	Specification for protective gloves against cold
BS EN 421: 1994	Protective gloves against ionising radiation and radioactive contamination
BS EN 407: 1994	Protective gloves against thermal risks (heat and/or fire)
BS EN 388: 1994	Protective gloves against mechanical risks

Common law requirements

P3026 In addition to general and/or specific statutory requirements, the residual combined duty, on employers at common law, to provide and maintain a safe system of work, including appropriate supervision of safety duties, still obtains, extending, where necessary, to protection against foreseeable risk of eye injury (*Bux v Slough Metals Ltd [1974] 1 All ER 262*) and dermatitis/facial eczema (*Pape v Cumbria County Council [1991] IRLR 463*). *Pape* concerned an office cleaner who developed dermatitis and facial eczema after using Vim, Flash and polish. The employer was held liable for damages at common law, since he had not instructed her in the dangers of using chemical materials with unprotected hands and had not made her wear rubber gloves.

Consequences of breach

P3027 Employers who fail to provide suitable personal protective equipment, commit a criminal offence, under the *Health and Safety at Work etc Act 1974* and the *Personal Protective Equipment at Work Regulations 1992* and sundry other regulations (see P3006–P3008 above). In addition, if, as a result of failure to provide suitable equipment, an employee suffers foreseeable injury and/or disease, the employer will be liable to the employee for damages for negligence. Conversely, if, after instruction and, where necessary, training, an employee fails or refuses to wear and maintain suitable personal protective equipment, he too can be prosecuted and/or dismissed and, if he is injured or suffers a disease in consequence, will probably lose all or, certainly, part of his damages.

Product Safety

Introduction

P9001 With the expansion of EU directives and other legislation aimed at manufacturers, designers, importers and suppliers, product safety emerges as a fast growth area – a development expedited by the introduction of the *General Product Safety Regulations 1994* (*SI 1994 No 2328*). Indeed, there is an identifiable trend towards placing responsibilities on producers and those involved in commercial circulation as well as on employers and occupiers (see MACHINERY SAFETY and PERSONAL PROTECTIVE EQUIPMENT). Although there is no question of removal of duties from employers and occupiers and users of industrial plant, machinery and products, there is a realisation that the *sine qua non* of compliance on the part of employers and users with their safety duties is compliance with essential health and safety requirements on the part of designers and manufacturers. This trend is likely to continue with more products being brought within the scope of the EU's 'new approach' regime which promotes supply of products throughout the single market provided they meet essential requirements which can be demonstrated by compliance with harmonised European standards. As far as civil liability is concerned, this trend may well result in a long-term shift in the balance of liability for injury away from employers and occupiers towards designers, manufacturers and other suppliers.

It was thalidomide that first focused serious attention on the legal control of product safety, leading to the Medicinal Products Directive (65/65/EEC) and the *Medicines Act 1968*, the first comprehensive regulatory system of product testing, licensing and vigilance. Since then a host of regulations has appeared, covering areas as diverse as pencils, aerosols, cosmetics and electrical products. The General Product Safety Directive (92/59/EEC), implemented by the *General Product Safety Regulations 1994* (*SI 1994 No 2328*), requires in a general way that producers of consumer goods place only safe products on the market and undertake appropriate post-marketing surveillance. The significance of these Regulations lies not in their substantive requirements (which are not at all detailed or especially onerous). Their importance lies more in the completion of an all-embracing consumer product safety regime across the European Union. Whilst specific regulatory requirements will take precedence (e.g. duties specified in sectoral directives or directives dealing with particular risks such as the Electromagnetic Compatibility Directive (89/336/EEC)) aspects of the *General Product Safety Regulations 1994* are meant to cover any gaps in the consumer protection network left by existing regulatory requirements.

The General Product Safety Directive was recently amended and re-issued as Directive 2001/95/EC to incorporate even more post-marketing obligations and powers for enforcing authorities which will apply to consumer products, even those already covered by sector-specific regulatory requirements. At the time of going to print the UK implementing legislation had not been adopted. Member States have until 15 January 2004 to implement the Directive.

The regulation of products for use in the workplace has developed along separate lines (although there is an increasing overlap by virtue of the 'new approach' directives, such as those for machinery or electromagnetic compatibility, which apply uniform requirements for consumer and workplace products albeit with separate enforcement regimes). Under the *Health and Safety at Work etc. Act 1974* (*HSWA 1974*), *s 6* all articles and substances for use at work are required to be as safe as is reasonably practicable, and manufacturers and other suppliers are under obligation to provide information and warnings for safety in use. Various product-specific regulations have also been made under *HSWA 1974*, for example the *Chemicals* (*Hazard Information and Packaging for Supply*) *Regulations 1994* (*SI 1994 No 3247*) (now revoked and re-enacted with amendments by *SI 2002 No 1689*) and the *Asbestos* (*Prohibitions*) *Regulations 1992* (*SI 1992 No 3067*)).

As far as commercial considerations are concerned, quality and fitness for purpose of products has been an implied statutory term in contracts for sale of goods in legislation stemming from the late nineteenth century and which is now found in the *Sale of Goods Act 1979*, as amended by the *Sale and Supply of Goods Act 1994* and the *Sale of Goods* (*Amendment*) *Act 1995*. These 1994 amendments included a more focused definition of the standard of *satisfactory quality* which a supplier is obliged by contract to provide, the criteria for which now explicitly includes safety, durability, and freedom from minor defects. The ability to exclude these statutory terms, or insist that buyers waive the rights they have, is now strictly curtailed by the *Unfair Contract Terms Act 1977* (banning exclusion of the most significant rights in consumer contracts, and subjecting exclusions in standard form commercial contracts to a statutory test of 'reasonableness'). In addition, and based on the Directive on Unfair Terms and Consumer Contracts (93/13/EEC), there are now other restrictions contained in the *Unfair Terms in Consumer Contracts Regulations 1999* (*SI 1999 No 2083*) imposing a requirement of fairness on most terms in consumer contracts which are not individually negotiated. These Regulations also lay down a requirement for suppliers to ensure that terms are expressed in plain, intelligible language. The most important restriction of all for present purposes in these two sets of Regulations is the effective ban on notices and contract terms seeking to exclude liability for death or personal injury resulting from negligence, a prohibition which is reinforced in relation to defective products by the *Consumer Protection Act 1987, s 7*.

This section deals with criminal and civil liabilities for unsafe products for consumer and industrial use as well as the duties imposed on sellers and suppliers of consumer products and after-sales personnel as far as contract law is concerned.

Consumer products

General Product Safety Regulations 1994

Duties of producers and distributors

P9002 The *General Product Safety Regulations 1994* (*SI 1994 No 2328*) lay down general requirements concerning the safety of products intended for or likely to be used by consumers. The Regulations will be superseded by UK legislation implementing the General Product Safety Directive 2001/95/EC which is required by the Directive to be brought into force by 15 January 2004. Directive 2001/95/EC replaces in its entirety the 1992 General Product Safety Directive 92/59/EEC and significantly increases the obligations on producers and distributors and has other important implications. In the meantime, the *General Product Safety Regulations 1994* remain in full force until that date. See P9008 for details of the revised Directive.

Products used solely in the workplace are not covered by the 1994 Regulations, but the fact that a product has a wider industrial application will not prevent it being a consumer product for the purposes of these Regulations if it is supplied to consumers. All manner of products are potentially caught, including clothing, primary agricultural products, DIY equipment and motor vehicles. What can be considered to be a consumer product will be extended even further by the new Directive (see P9008).

The Regulations do not apply to:

- second-hand products which are antiques;
- products supplied for repair or reconditioning before use, provided that the purchaser is clearly informed accordingly;
- products exported direct by the UK manufacturer to a country outside the European Union.

The Regulations are disapplied in relation to any product where there are specific provisions in European law governing *all aspects* of the safety of a product. Where there are product-specific provisions in European law which do not cover all aspects, the 1994 Regulations apply to the extent that specific provision is not made by European law. [*General Product Safety Regulations 1994* (*SI 1994 No 2328*), *Reg 3*(*c*) and *Reg 4*].

The main requirements of the Regulations are that:

(*a*) no producer may place a product on the market unless it is safe [*General Product Safety Regulations 1994* (*SI 1994 No 2328*), *Reg 7*];

(*b*) producers provide consumers with relevant information, so as to enable them (i.e. consumers) to assess risks inherent in products throughout the normal or foreseeable period of use, where such risks are not immediately obvious without adequate warnings, and to take precautions against those risks [*General Product Safety Regulations 1994* (*SI 1994 No 2328*), *Reg 8*(*1*)(*a*)];

(*c*) producers update themselves regarding risks presented by their products and take appropriate action (if necessary recall and withdrawal), for example by:

  (i) identifying products/batches of products by marking,

  (ii) sample testing,

  (iii) investigating and following up *bona fide* complaints,

  (iv) keeping distributors informed accordingly

  [*General Product Safety Regulations 1994* (*SI 1994 No 2328*), *Reg 8*(*1*)(*b*), (*2*)];

(*d*) in order to enable producers to comply with their duties, distributors must act with due care and in particular:

  (i) must not supply products to any person which are known or presumed (on the basis of information in their possession and as professionals) to be dangerous products, and

  (ii) must participate in monitoring safety of products placed on the market – in particular, by passing on information on product risks and co-operating in action to avoid risks.

  [*General Product Safety Regulations 1994* (*SI 1994 No 2328*), *Reg 9*(*a*)(*b*)].

Presumption of conformity with safety requirements

P9003 A 'safe' product is a product which, under normal or reasonably foreseeable conditions of use (as well as duration), does not present any risk (or only minimal risks) compatible with the product's use, considered as acceptable and consistent with a high level of protection for the safety and health of consumers, with reference to:

(*a*) product characteristics (e.g. composition, packaging, assembly and maintenance instructions);

(*b*) the effect on other products, where use with other products is reasonably foreseeable;

(*c*) product presentation (i.e. labelling, instructions for use and disposal); and

(*d*) categories of consumers at serious risk – in particular, children.

[*General Product Safety Regulations 1994* (*SI 1994 No 2328*), *Reg 2*(*1*)].

In this connection, where a product conforms with specific UK safety requirements (e.g. the *Plugs and Sockets etc* (*Safety*) *Regulations 1994* (*SI 1994 No 1768*)) there is a presumption that the product is safe, until the contrary is proved. In the absence of specific regulations governing the health and safety aspects of a product, the assessment of safety is to be made by taking into account:

(i) voluntary national standards implementing European standards;

(ii) EU technical specifications, or, failing these,

— standards drawn up in the UK, or

— codes of practice relating to the product, or

— state of the art and technology

and the safety that consumers may reasonably expect.

[*General Product Safety Regulations 1994* (*SI 1994 No 2328*), *Reg 10*].

Offences, penalties and defences

P9004 Contravention of *Regs 7–9*(*a*) (above) is an offence, carrying, on summary conviction, a maximum penalty of:

(*a*) imprisonment for up to three months, or

(*b*) a fine of £5,000, or

(*c*) both.

[*General Product Safety Regulations 1994* (*SI 1994 No 2328*), *Reg 12*].

Preparatory acts by producers and distributors

P9005 No producer or distributor may:

(*a*) offer or agree to place on the market any dangerous product, or expose or possess such product for placing on the market, or

(*b*) offer or agree to supply any dangerous product, or expose or possess same for supply.

Contravention is an offence, carrying the same maximum penalties as breach of *Reg 12* (see P9004 above).

[*General Product Safety Regulations 1994* (*SI 1994 No 2328*), *Reg 13*].

It is worth noting that it may be easier for an enforcement officer to bring a prosecution for the above offences than for the offence under *Reg 9*(*a*) for supplying products which are known or presumed (on the basis of information in their possession and as professionals) to be dangerous products. In the latter case, the prosecution has the burden of proving knowledge of the defect and also that the distributor did not act with due care. Trading Standards Officers may bring a prosecution on the basis of these offences even in a case where there has been an actual supply and there is evidence to suggest that this is already occurring.

Offence by another person

P9006 As with breaches of *HSWA* and similarly-oriented legislation, where an offence is committed by 'another person' in the course of his commercial activity, that person can be charged, whether or not the principal offender is prosecuted. Similarly, where commission of an offence is consented to or connived at, or attributable to neglect, on the part of a director, manager or secretary, such persons can be charged in addition to or in lieu of the body corporate. [*General Product Safety Regulations 1994* (*SI 1994 No 2328*), *Reg 15*].

Defence of 'due diligence' – and exceptions

P9007 It is a defence for a person charged under these Regulations to show that he took all reasonable steps and exercised all due diligence to avoid committing the offence. [*General Product Safety Regulations 1994* (*SI 1994 No 2328*), *Reg 14*(*1*)]. (See further ENFORCEMENT.)

The exceptions to the above are:

(*a*) Where, allegedly, commission of the offence was due to:

- (i) the act or default of another, or
- (ii) reliance on information given by another,

a person so charged cannot, without leave of the court, rely on the defence of 'due diligence', unless he has served a notice, within, at least, seven days before the hearing, identifying the person responsible for the commission or default. [*General Product Safety Regulations 1994* (*SI 1994 No 2328*), *Reg 14*(*2*)(*3*)].

(*b*) A person so charged cannot rely on the defence of 'information supplied by another', unless he shows that it was reasonable in all the circumstances for him to have relied on the information. In particular, did he take steps to verify the information or did he have any reason to disbelieve the information? [*General Product Safety Regulations 1994* (*SI 1994 No 2328*), *Reg 14*(*4*)].

(*c*) A distributor charged with an offence cannot rely on the defence where he has contravened *Reg 9*(*b*) concerning passing on information and co-operating in action to avoid product risks. [*General Product Safety Regulations 1994* (*SI 1994 No 2328*), *Reg 14*(*5*)].

A mere recommendation on the part of an importer that labels should be attached to boxes by retailers does not constitute 'due diligence' for the purposes of *Reg 14*(*1*). (In *Coventry County Council v Ackerman Group plc [1995] Crim LR 140*, an egg boiler imported by the defendant failed to contain instructions that eggs should be broken into the container and yolks pricked before being microwaved. The defendant had learned of the problem and had printed instructions which were sent to all retailers, recommending that they be fixed to the boxes.)

Compliance with recognised standards (such as British Standards) will not amount to the defence of due diligence if a product is nevertheless unsafe for the user. In

Whirlpool (UK) Ltd and Magnet Ltd v Gloucestershire County Council (1993, unreported) cooker hoods which were intrinsically safe and which met applicable standards for the purposes of the *Low Voltage Electrical Equipment* (*Safety*) *Regulations 1989* failed to meet the general safety requirement contained in the then applicable consumer protection legislation because they were liable to result in fires when used in conjunction with certain gas hobs.

General Product Safety Directive

P9008 On 15 January 2004, UK legislation implementing the General Product Safety Directive 2001/95/EC will replace the *General Product Safety Regulations 1994* (*SI 1994 No 2328*). The following paragraphs explain the new obligations and other changes that the new Directive will bring.

Duties of producers and distributors

P9009 The Directive increases the scope of consumer products covered by it. The new definition (with new wording in italics) is as follows:

> '… any product – *including in the context of providing a service* – which is intended for consumers or likely, *under reasonably foreseeable conditions*, to be used by consumers even if not intended for them, and is supplied or made available, whether for consideration or not, in the course of a commercial activity, and whether new, used or reconditioned.
>
> This definition shall not apply to second-hand products supplied as antiques or as products to be repaired or reconditioned prior to being used, provided that the supplier clearly informs the person to whom he supplies the product to that effect'.

The scope of the Directive is therefore extended as follows:

1. It encompasses products which are made available to consumers, as well as supplied to them. This would include many products provided in beauty centres, theme parks or playgrounds or in connection with transport, sporting events or health treatments.
2. It includes products supplied or made available in the context of providing a service. This would include repairs and after-sales service, maintenance and cleaning work, hotel and restaurant services, health treatment, provision of gas and electricity etc or a tyre inflation gauge and compressed air pump at a garage.

In addition, new recital 10 makes clear that, although business products are not included *per se*, the new Directive does include products which are designed exclusively for professional use but have subsequently migrated to the consumer market, such as laser pens.

The new Directive includes a provision to prohibit the export from the European Community any products which have been the subject of a European Commission-initiative decision (after consulting the Member States and, whenever it proves necessary also a Community scientific committee) that requires Member States to take measures in relation to that particular product on the grounds that it presents a serious risk (unless the decision specifically provides that the products may still be exported outside the Community).

A number of significant obligations are added to those referred to above in P9002 and these are that:

(*a*) producers keep a register of complaints, if necessary.

[*General Product Safety Directive 2001/95/EC, Article 5.1*]

(*b*) producers and distributors notify the competent authorities immediately if they know or ought to know, on the basis of information in their possession and as professionals, that a product they have placed on the market poses risks to the consumer that are compatible with the general safety requirement, i.e. that it is dangerous, and of any action taken to prevent such risks [*General Product Safety Directive 2001/95/EC, Article 5.3*];

(*c*) producers and distributors co-operate with the competent authorities, at the request of the latter, on action taken to avoid risks posed by products that they supply or have supplied.

[*General Product Safety Directive 2001/95/EC, Article 5.4*]

(*d*) producers (as part of their duty to update themselves regarding risks and take appropriate action) adequately and effectively warn consumers or recall from consumers, if necessary (thus the recall obligation is extended beyond the chain of distribution and includes products already supplied or made available to the consumer/user).

[*General Product Safety Directive 2001/95/EC, Article 5.1*]

(*e*) distributors keep and provide the documentation necessary for tracing the origin of products.

[*General Product Safety Directive 2001/95/EC, Article 5.2*]

There has been no change in the policy that specific General Product Safety Directive provisions do apply to products regulated by other directives where such provisions are not covered by the particular sectoral directive applicable to the product. These new post-marketing obligations would, therefore, generally apply to most consumer products even if subject also to regulation under sectoral directives, but guidance is awaited from the European Commission to clarify exactly which provisions do not apply to specific-product sectors. It is intended that this guidance will be made available before the date for implementation of the new Directive into national law, so that Member States can incorporate its principles clearly into their national legislation. It may be that sectoral directives will in due course be amended to provide for similar obligations.

The European Commission is to produce a guide containing simple and clear criteria on notifications required in (*b*) above, together with the content and a standard form for notifications. These notification provisions constitute a major increase in regulation for producers and distributors. They are based on reporting requirements which apply in the United States of America and have arisen out of some frustration that Member States and the Commission have not been informed about product withdrawals in Europe until US authorities inform them that the product has been withdrawn from the US market.

The notification provisions will require producers and distributors to put in place sophisticated procedures to capture and assess safety information. Deciding whether new information means that a product is no longer legally safe can be a major issue in practice and is best undertaken against a background of previous safety data and an assessment that the product's safety is acceptable. Companies' systems may need to provide for obtaining independent expert advice on these issues from scientists and lawyers. For certain products, at least, it will be necessary to undertake a documented risk analysis before a product is placed on the market, and subsequently to keep it up to date, so as to define the level of anticipated risks with a product's use, to determine that these are acceptable and that the general safety requirement is met, and to provide a background and baseline statement of acceptable risks against which any increase in risk may be assessed.

Recalls in (*d*) above are only required as a last resort where other measures would not suffice to prevent the risks involved, where the producer considers it necessary or where they are obliged to do so by a Member State. Practical issues which may arise include: the ability of a system adequately to contact users over safety issues; to measure the number and percentage of products returned in a recall; and decide what level of returns would be acceptable. UK guidance may assist to an extent (see *Consumer Product Recall: A Good Practice Guide* (1999) Department of Trade and Industry).

Complementary to the obligation to recall products from consumers in (*d*) above is a new power for Member States to order or organise the immediate withdrawal of any dangerous product already on the market and alert consumers to the risks it presents and to order or co-ordinate or, if appropriate, to organise with producers and distributors its recall from consumers and its destruction in suitable conditions. [*General Product Safety Directive 2001/95/EC, Article 8.1*(*f*)].

Presumption of conformity with safety requirements

P9010 What will be considered to be a safe product is slightly different in the new Directive. In the factors to be taken into account when considering the safety of a product, the new Directive requires that: the instructions for assembly and, where applicable, for installation and maintenance are taken into account in the characteristics of the product; and also that the elderly are a particular category of consumer to consider when taking into account the categories of consumers at risk when using the product, along with children. The factors to be taken into account are in particular:

(*a*) product characteristics (e.g. composition, packaging, instructions for assembly and, where applicable, for installation and maintenance);

(*b*) the effect on other products, where use with other products is reasonably foreseeable;

(*c*) product presentation (i.e. labeling, warnings, instructions for use and disposal, any other indication of information regarding the product); and

(*d*) categories of consumers at serious risk – in particular, children and the elderly.

[*General Product Safety Directive 2001/95/EC, Article 2*(*b*)].

There is a new provision in the new Directive introducing a system whereby European harmonised standards can be mandated under the Directive and will confer a presumption of conformity with the Directive's safety obligations, in the same way as applies under many other product regulatory directives. Reference to such standards will be published in the *Official Journal of the European Communities* with particular reference to this Directive.

A product will still be deemed safe if it conforms with specific national safety requirements, unless the contrary is proved. In the absence of specific harmonised standards mandated and listed in the Official Journal or specific regulations governing the health and safety aspects of a product, the assessment of safety is to be made by taking into account:

(i) voluntary national standards implementing European standards other than those listed in the Official Journal with reference to the General Product Safety Directive; or

(ii) standards drawn up in the Member State in which the product is marketed; or

(iii) European Commission recommendations setting guidelines on product safety assessment; or

(iv) product safety codes of good practice in the relevant sector; or

(v) state of the art and technology; or

(vi) the safety that consumers may reasonably expect.

[*General Product Safety Directive 2001/95/EC, Article 3.3*].

Market surveillance and enforcement

P9011 The Directive's market surveillance and enforcement provisions have been strengthened and there are increased obligations on Member States. The European Commission also has an obligation to promote and take part in the operation of a European network of authorities of the Member States.

The obligations and enforcement powers of Member States have been expanded to include the following elements:

(*a*) a detailed definition of the tasks, powers, organisation and cooperation arrangements for market surveillance of competent authorities;

(*b*) establishment of sectoral surveillance programmes by categories of products or risks and the monitoring of surveillance activities, findings and results;

(*c*) the follow-up and updating of scientific and technical knowledge concerning the safety of products;

(*d*) periodic review and assessment of the functioning of the control activities and, if necessary, revision of the approach and organisation of surveillance;

(*e*) procedures to receive and follow-up complaints from consumers and others on product safety, surveillance or control activities;

(*f*) exchange of information between Member States on risk assessment, dangerous products, test methods and test results, recent scientific developments and other aspects relevant for control activities;

(*g*) joint surveillance and testing projects between surveillance authorities;

(*h*) exchange of expertise and best practices, and cooperation in training activities;

(*i*) improved collaboration at Community level on tracing, withdrawal and recall of dangerous products.

[*General Product Safety Directive 2001/95/EC, Articles 6–10*]

An extended list of enforcement powers is included in the new Directive and includes the power for a Member State to order or organise the issuance of warnings about, or recall of, dangerous products. [*General Product Safety Directive 2001/95/EC, Article 8.1*].

Other significant changes

P9012 Other significant changes can be summarised as follows:

1. Member States are required to take due account of the precautionary principle when they are taking enforcement measures. The precautionary principle is that Member States should use a precautionary approach to risks where there is scientific uncertainty as to the level of risk.

   [*General Product Safety Directive 2001/95/EC, Article 8.2*]

2. Information which is available to the authorities relating to consumer health and safety, in particular on product identification, the nature of any risk and on measures taken, shall in general be available to the public, in accordance with the requirements of transparency. However, information which is obtained by the authorities shall not be disclosed if the information, by its nature, is covered by professional secrecy in duly justified cases.

   [*General Product Safety Directive 2001/95/EC, Article 16.1*].

Industrial products

P9013 The *HSWA 1974* was the first Act to place a *general* duty on designers and manufacturers of industrial products to design and produce articles and substances that are safe and without health risks when used at work. Prior to this date legislation had tended to avoid this approach, e.g. the *Factories Act 1961*, on the premise that machinery could not be made design safe (see M1001 MACHINERY SAFETY). Statutory requirements had tended to concentrate on the duty to guard and fence machinery, and with the placement of a duty upon the user/employer to inspect and test inward products for safety. There is a general residual duty on the employers/users of industrial products to inspect and test them for safety under *HSWA 1974, s 2*.

These duties notwithstanding, the trend of legislation in recent years has been towards safer design and manufacture of products for industrial and domestic use. Thus, *HSWA 1974, s 6* as updated by the *Consumer Protection Act 1987, Sch 3* imposes general duties on designers, manufacturers, importers and suppliers of products ('articles and substances') for use at work, to their immediate users. Contravention of s 6 carries with it a maximum fine, on summary conviction, of £20,000 or an unlimited fine in the Crown Court (see ENFORCEMENT). In addition, a separate duty is laid on installers of industrial plant and machinery. However, because of their involvement in the key areas of design and manufacture, more onerous duties are placed upon designers and manufacturers of articles and substances for use at work than upon importers and suppliers, who are essentially concerned with distribution and retail of industrial products. However, under *HSWA 1974, s 6(8A)* importers are made liable for the first time for the faults of foreign designers/manufacturers.

There has been a recent interesting development in case law regarding the interaction between *HSWA 1974*, European directives concerning the placing on the market and putting into service of certain product types, and the UK legislation which implements the European directives. In the case of *R (on the application of Junttan OY) v Bristol Magistrates' Court [2002] EWHC 566 (Admin)*, concerning a judicial review of a decision of the Bristol Magistrates' Court involving a prosecution brought by the HSE based on offences under *sections 3* and *6* of *HSWA 1974* regarding machinery, the court held that it was not open to the HSE to bring proceedings under *HSWA 1974, s 6*. The court ruled that the HSE should have brought proceedings under the UK Regulations implementing the European directive applicable to such machinery. The basis of the decision is that, where there is a specific statutory offence under regulations implementing a European directive applicable to the product in question and where the offence under the regulations covers exactly the same ground as *HSWA 1974, s 6*, the offence under the regulations is the offence which gives effect to the directive and proceedings should be brought under the regulations and not *HSWA 1974, s 6*. Fines are lower under the regulations.

Regulations made under the Health and Safety at Work etc Act 1974

P9014 Various safety regulations are made under the umbrella of *HSWA 1974* and these often implement European health and safety directives. The *Provision and Use of Work Equipment Regulations 1998* (*SI 1998 No 2306* as amended by *SI 2002 No 2174*) are such Regulations which implemented European Directive 89/655/EEC.

The 1998 Regulations place obligations on employers and on certain persons having control of work equipment, or of persons who use or supervise or manage the use of work equipment or of the way in which the equipment is used. The key requirement is to ensure that work equipment is maintained in an efficient state, in efficient working order and in good repair. [*Provision and Use of Work Equipment Regulations 1998* (*SI 1998 No 2306* as amended by *SI 2002 No 2174*), *Reg 5* (*1*)]. It was recently confirmed that this imposes strict liability on the employer (*Stark v The Post Office, The Times, 29 March 2000*) and that, where a workplace product injures someone, the employer will be liable even in situations where through wear and tear the product has become dangerous. The employer's duty to maintain translates into a duty to ensure that there are no latent defects.

Criminal liability for breach of statutory duties

P9015 This refers to duties laid down in *HSWA 1974* as revised by the *Consumer Protection Act 1987, Sch 3*. The duties exist in relation to articles and substances for use at work and fairground equipment. There are no civil claims rights available to employees under these provisions except for breach of safety regulations. (For the specific requirements now applicable to machinery for use at work, see MACHINERY SAFETY.)

Definition of articles and substances

P9016 An article for use at work means:

(*a*) any plant designed for use or operation (whether exclusively or not) by persons at work; and

(*b*) any article designed for use as a component in any such plant.

[*HSWA 1974, s 53*(*1*)].

A substance for use at work means 'any natural or artificial substance (including micro-organisms), whether in solid or liquid form or in the form of a gas or vapour'. [*HSWA 1974, s 53*(*1*)].

An article upon which first trials/demonstrations are carried out is not an article for use at work, but rather an article which *might* be used at work. The purpose of trial/demonstration was to determine whether the article could safely be later used at work (*McKay v Unwin Pyrotechnics Ltd, The Times, 5 March 1991* where a dummy mine exploded, causing the operator injury, when being tested to see if it would explode when hit by a flail attached to a vehicle. It was held that there was no breach of *HSWA 1974, s 6(1)(a)*).

Duties in respect of articles and substances for use at work

P9017 *HSWA 1974, s 6* (as amended by the *Consumer Protection Act 1987, Sch 3*) places duties upon manufacturers and designers, as well as importers and suppliers, of (*a*) articles and (*b*) substances for use at work, whether used exclusively at work or not (e.g. lawnmower, hair dryer).

Articles for use at work

P9018 Any person who designs, manufactures, imports or supplies any article for use at work (or any article of fairground equipment) must:

(*a*) ensure, so far as is reasonably practicable (for the meaning of this expression see E15031 ENFORCEMENT), that the article is so designed and constructed that it will be safe and without risks to health at all times when it is being (i) set, (ii) used, (iii) cleaned or (iv) maintained by a person at work;

(*b*) carry out or arrange for the carrying out of such testing and examination as may be necessary for the performance of the above duty;

(*c*) take such steps as are necessary to secure that persons supplied by that person with the article are provided with adequate information about the use for which the article is designed or has been tested and about any conditions necessary to ensure that it will be safe and without risks to health at all such times of (i) setting, (ii) using, (iii) cleaning, (iv) maintaining *and* when being (v) dismantled, or (vi) disposed of; and

(*d*) take such steps as are necessary to secure, so far as is reasonably practicable, that persons so supplied are provided with all such revisions of information as are necessary by reason of it becoming known that anything gives rise to a serious risk to health or safety.

[*HSWA 1974, s 6(1)(a)–(d) as amended by the Consumer Protection Act 1987, Sch 3*].

(See P9020 and P9021 below for further duties relevant to articles for use at work.)

In the case of an article for use at work which is likely to cause an employee to be exposed to 85 dB(A) or above, or to peak action level (200 pascals) or above, adequate information must be provided about noise likely to be generated by that article. [*Noise at Work Regulations 1989* (*SI 1989 No 1790*), *Reg 12*] (see further NOISE AT WORK and VIBRATION).

Substances for use at work

P9019 Every person who manufactures, imports or supplies any substance must:

(*a*) ensure, so far as is reasonably practicable, that the substance will be safe and without risks to health at all times when it is being (i) used, (ii) handled, (iii) processed, (iv) stored, or (v) transported by any person at work or in premises where substances are being installed;

(*b*) carry out or arrange for the carrying out of such testing and examination as may be necessary for the performance of the duty in (*a*);

(*c*) take such steps as are necessary to secure that persons supplied by that person with the substance are provided with adequate information about:

(i) any risks to health or safety to which the inherent properties of the substance may give rise;

(ii) the results of any relevant tests which have been carried out on or in connection with the substance; and

(iii) any conditions necessary to ensure that the substance will be safe and without risks to health at all times when it is being (*a*) used, (*b*) handled, (*c*) processed, (*d*) stored, (*e*) transported and (*f*) disposed of; and

(*d*) take such steps as are necessary to secure, so far as is reasonably practicable, that persons so supplied are provided with all such revisions of information as are necessary by reason of it becoming known that anything gives rise to a serious risk to health or safety.

[*HSWA 1974, s 6(4)* as amended by the *Consumer Protection Act 1987, Sch 3*].

Additional duty on designers and manufacturers to carry out research

P9020 Any person who undertakes the design or manufacture of any article for use at work must carry out, or arrange for the carrying out, of any necessary research with a view to the discovery and, so far as is reasonably practicable, the elimination or minimisation of any health or safety risks to which the design or article may give rise. [*HSWA 1974, s 6(2)*].

Duties on installers of articles for use at work

P9021 Any person who erects or installs any article for use at work in any premises where the article is to be used by persons at work, must ensure, so far as is reasonably practicable, that nothing about the way in which the article is erected or installed makes it unsafe or a risk to health when it is being (*a*) set, (*b*) used, (*c*) cleaned, or (*d*) maintained by someone at work. [*HSWA 1974, s 6(3)* as amended by the *Consumer Protection Act 1987, Sch 3*].

Additional duty on manufacturers of substances to carry out research

P9022 Any person who manufactures any substance must carry out, or arrange for the carrying out, of any necessary research with a view to the discovery and, so far as is reasonably practicable, the elimination or minimisation of any health/safety risks at all times when the substance is being (*a*) used, (*b*) handled, (*c*) processed, (*d*) stored, or (*e*) transported by someone at work. [*HSWA 1974, s 6(5)* as amended by the *Consumer Protection Act 1987, Sch 3*].

No duty on suppliers of industrial articles and substances to research

P9023 It is not necessary to repeat any testing, examination or research which has been carried out by designers and manufacturers of industrial products, on the part of importers and suppliers, in so far as it is reasonable to rely on the results. [*HSWA 1974, s 6(6)*].

Custom built articles

P9024 Where a person designs, manufactures, imports or supplies an article for or to another person on the basis of a written undertaking by that other to ensure that the article will be safe and without health risks when being (*a*) set, (*b*) used, (*c*) cleaned, or (*d*) maintained by a person at work, the undertaking will relieve the designer/manufacturer etc. from the duty specified in *HSWA 1974, s 6*(*1*)(*a*) (see P9017 above), to such extent as is reasonable, having regard to the terms of the undertaking. [*HSWA 1974, s 6*(*8*) as amended by the *Consumer Protection Act 1987, Sch 3*].

Importers liable for offences of foreign manufacturers/designers

P9025 In order to give added protection to industrial users from unsafe imported products, the *Consumer Protection Act 1987, Sch 3* has introduced a new subsection (*HSWA 1974, s 6*(*8A*)) which, in effect makes importers of unsafe products liable for the acts/omissions of foreign designers and manufacturers. *Section 6*(*8A*) states that nothing in (*inter alia*) *s 6*(*8*) is to relieve an importer of an article/substance from any of his duties, as regards anything done (or not done) or within the control of:

(*a*) a foreign designer; or

(*b*) a foreign manufacturer of an article/substance.

[*HSWA 1974, s 6*(*8A*)].

Proper use

P9026 The original wording of *HSWA 1974, s 6*(*1*)(*a*), *6*(*4*)(*a*) and *6*(*10*) concerning 'proper use' excluded 'foreseeable user error' as a defence, which had the consequence of favouring the supplier. Thus, if a supplier could demonstrate a degree of operator misuse or error, however reasonably foreseeable, the question of initial product safety was side stepped. Moreover, 'when properly used' implied, as construed, that there could only be a breach of *s 6* once a product had actually been *used*. This was contrary to the principle that safety should be built into design/production, rather than relying on warnings and disclaimers. The wording in *s 6* is now amended so that only *unforeseeable* user/operator error will relieve the supplier from liability; he will no longer be able to rely on the strict letter of his operating instructions. [*Consumer Protection Act 1987, Sch 3*].

Powers to deal with unsafe imported goods

P9027 *HSWA 1974* does not empower enforcing authorities to stop the supply of unsafe products at source or prevent the sale of products by foreign producers after they have been found to be unsafe, but enforcement officers have the power to act at the point of entry or anywhere else along the distribution chain to stop unsafe articles/substances being imported by serving prohibition notices (see ENFORCEMENT).

Customs officers can seize any imported article/substance, which is considered to be unsafe, and detain it for up to two (working) days. [*HSWA s 25A* (incorporated by *Schedule 3* to the *Consumer Protection Act 1987*)].

In addition, customs officers can transmit information relating to unsafe imported products to HSE inspectors. [*HSWA 1974, s 27A* (incorporated by the *Consumer Protection Act 1987, Sch 3*)].

Civil liability for unsafe products – historical background

P9028 Originally at common law where defective products caused injury, damage and/or death, redress depended on whether the injured user had a contract with the seller or hirer of the product. This was often not the case, and in consequence many persons, including employees repairing and/or servicing products, were without remedy. This rule, emanating from the decision of *Winterbottom v Wright (1842) 10 M & W 109*, remained unchanged until 1932, when *Donoghue v Stevenson [1932] AC 562* was decided by the House of Lords. This case was important because it established that manufacturers were liable in negligence (i.e. tort) if they failed to take reasonable care in the manufacturing and marketing of their products, and in consequence a user suffered injury when using the product in a reasonably foreseeable way. More particularly, a 'manufacturer of products, which he sells in such a form as to show that he intends them to reach the ultimate consumer in the form in which they left him with no reasonable possibility of intermediate inspection, and with the knowledge that the absence of reasonable care in the preparation or putting up of the products will result in an injury to the consumer's life or property, owes a duty to the consumer to take reasonable care' (per Lord Atkin). In this way, manufacturers of products which were defective were liable in negligence to users and consumers of their products, including those who as intermediaries, repair, maintain and service industrial products, it being irrelevant whether there was a contract between manufacturer and user (which normally there was not).

Donoghue v Stevenson is a case of enormous historical importance in the context of liability of manufacturers, but although the principle has become well established and is still widely applied it did not give a reliable remedy to injured persons, who still had to satisfy the legal burden of proving that the manufacturer had not exercised reasonable care, a serious obstacle to overcome in cases involving technically complex products. In some instances it became possible to avoid this obstacle, by establishing liability on other bases.

Defective equipment supplied to employees

P9029 At common law an employer obtaining equipment from a reputable supplier was unlikely to be found liable to an employee if the equipment turned out to be defective and injured him (*Davie v New Merton Board Mills Ltd [1959] AC 604*). Although it represented no bar to employees suing manufacturers direct for negligence, the effect of this decision was reversed by the *Employers' Liability (Defective Equipment) Act 1969* rendering employers strictly liable (irrespective of negligence) for any defects in equipment causing injury. In such circumstances the employer would be able to claim indemnity for breach of contract by the supplier of the equipment.

Breach of statutory duty

P9030 Whilst legislation and regulations dealing with domestic and industrial safety are principally penal and enforceable by state agencies (trading standards officers and health and safety inspectors) if injury, death or damage occurs in consequence of a breach of such statutory duty, it may be possible to use this breach as the basis of a civil liability claim. Here there may be strict liability if there are absolute requirements, or the duty may be defined in terms of what it is practicable or reasonably practicable to do (see E15031 ENFORCEMENT).

As far as workplace products are concerned, *HSWA 1974, s 47* bars a right of action in civil proceedings in respect of any failure to comply with *HSWA 1974, s 6* (or the other general duties under the Act). However a breach of a duty imposed by health and safety regulations will generally be actionable in this way, unless the particular regulations in question contain a proviso to the contrary.

The position is the same in relation to consumer safety regulations made under the *Consumer Protection Act 1987* (*Consumer Protection Act 1987, s 41*(*1*)). No provision is made in the *General Product Safety Regulations 1994* (*SI 1994 No 2328*) for any breach thereof to give rise to civil liability for the benefit of an injured person. Given that the Regulations stem from European law it is unlikely that they would be construed in such a way as to give more extensive rights than those contained in the Product Liability Directive (85/374/EEC) (see P9031).

Consumer Protection Act 1987

P9031 At European level it was deemed necessary to introduce a degree of harmonisation of product liability principles between Member States, and at the same time to reduce the importance of fault and negligence concepts in favour of liability being determined by reference to 'defects' in a product. Thus the nature of the product itself would become the key issue, not the conduct of the manufacturer. After protracted debate the Product Liability Directive (85/374/EEC) was adopted.

Introduction of strict product liability

P9032 The introduction of strict product liability is enshrined in Britain within the *Consumer Protection Act 1987, s 2*(*1*). Thus, where any damage is *caused* wholly or partly by a defect (see P9035 below) in a product (e.g. goods, electricity, a component product or raw materials), the following may be liable for damages (irrespective of negligence):

(*a*) the producer;

(*b*) any person who, by putting his name on the product or using a trade mark (or other distinguishing mark) has held himself out as the producer;

(*c*) any person who has imported the product into a Member State from outside the EU, in the course of trade/business.

[*Consumer Protection Act 1987, s 2*(*1*)(*2*)].

Producers

P9033 Producers are variously defined as:

(*a*) the person who manufactured a product;

(*b*) in the case of a substance which has not been manufactured, but rather won or abstracted, the person who won or abstracted it;

(*c*) in the case of a product not manufactured, won or abstracted, but whose essential characteristics are attributable to an industrial process or agricultural process, the person who carried out the process.

[*Consumer Protection Act 1987, s 1*(*2*)(*c*)].

Liability of suppliers

P9034 Although producers are principally liable, intermediate suppliers can also be liable in certain circumstances. Thus, any person who supplied the product is liable for damages if:

(*a*) the injured person requests the supplier to identify one (or more) of the following:

  (i) the producer,

  (ii) the person who put his trade mark on the product,

  (iii) the importer of the product into the EU; and

(*b*) the request is made within a reasonable time after damage/injury has occurred *and* it is not reasonably practicable for the requestor to identify the above three persons; and

(*c*) within a reasonable time after receiving the request, the supplier fails either:

  (i) to comply with the request, or

  (ii) identify his supplier.

[*Consumer Protection Act 1987, s 2*(*3*)].

Importers of products into the EU and persons applying their name, brand or trade mark will also be directly liable as if they were original manufacturers. [*Consumer Protection Act 1987, s 2*(*2*)].

Defect – key to liability

P9035 Liability presupposes that there is a defect in the product, and indeed, existence of a defect is the key to liability. Defect is defined in terms of the absence of safety in the product. More particularly, there is a 'defect in a product ... if the safety of the product is not such as persons generally are entitled to expect' (including products comprised in that product). [*Consumer Protection Act 1987, s 3*(*1*)].

Defect can arise in one of three ways and is related to:

(*a*) construction, manufacture, sub-manufacture, assembly;

(*b*) absence or inadequacy of suitable warnings, or existence of misleading warnings or precautions;

(*c*) design.

The definition of 'defect' implies an entitlement to an expectation of safety on the part of the consumer, judged by reference to *general* consumer expectations not individual subjective ones. (The American case of *Webster v Blue Ship Tea Room Inc 347 Mass 421, 198 NE 2d 309 (1964)* is particularly instructive here. The claimant sued in a product liability action for a bone which had stuck in her throat, as a result of eating a fish chowder in the defendant's restaurant. It was held that there was no liability. Whatever her own expectations may have been, fish chowder would not be fish chowder without some bones and this is a general expectation.)

The court will require a claimant to be specific and be particular about exactly what the alleged defect is (*Paul Sayers and Others v SmithKline Beecham plc and Others (MMR/MR litigation) [1999] MLC 0117*).

Consumer expectation of safety – criteria

P9036 The general consumer expectation of safety must be judged in relation to:

(*a*) the marketing of the product, i.e.:

  (i) the manner in which; and

  (ii) the purposes for which the product has been marketed;

(iii) any instructions/warnings against doing anything with the product; and

(*b*) by what might reasonably be expected to be done with or in relation to the product (e.g. the expectation that a sharp knife will be handled with care); and

(*c*) the time when the product was supplied (e.g. a product's shelf-life).

A defect cannot arise retrospectively by virtue of the fact that, subsequently, a safer product is made and put into circulation. [*Consumer Protection Act 1987, s 3*(*2*)].

Time of supply

P9037 Liability attaches to the *supply* of a product (see P9041 below). More particularly, the producer will be liable for any defects in the product existing at the time of supply (see P9045(*d*) below); and where two or more persons collaborate in the manufacture of a product, say by submanufacture, either and both may be liable, that is, severally and jointly (see P9043 below).

Contributory negligence

P9038 A person who is careless for his own safety is probably guilty of contributory negligence and, thus, will risk a reduction in damages. [*Consumer Protection Act 1987, s 6*(*4*)]. However, in the product liability context, carelessness of the user may mean that there is no liability at all on the part of the producer. If, for example, clear instructions and warnings provided with the product had been disregarded, when compliance would have avoided the accident, it is highly unlikely that the product would be found to be defective for the purposes of the Act. No off-setting of damages for contributory fault of the claimant would arise.

Absence or inadequacy of suitable/misleading warnings

P9039 The common law required that the vendor of a product should point out any latent dangers in a product which he either knew about or ought to have known about. Misleading terminology/labelling on a product or product container could result in liability for negligence (*Vacwell Engineering Ltd v BDH Chemicals Ltd [1969] 3 All ER 1681* where ampoules containing boron tribromide, which carried the warning 'Harmful Vapours', exploded on contact with water, killing two scientists. It was held that this consequence was reasonably foreseeable and accordingly the defendants should have researched their product more thoroughly). In a similar product liability action today, the manufacturers would be strictly liable (subject to statutory defences) and, if injury/damage followed the failure to issue a written/pictorial warning, as required by law (e.g. the *Chemicals* (*Hazard Information and Packaging for Supply*) *Regulations 2002* (*SI 2002 No 1689*)), there would be liability.

The duty, on manufacturers, to research the safety of their products, before putting them into circulation (see *Vacwell Engineering Ltd v BDH Chemicals Ltd*) is even more necessary and compulsory now, given the introduction of strict product liability. This includes safety in connection with directions for use on a product. A warning refers to something that can go wrong with the product; directions for use relate to the best results that can be obtained from products, if the directions are followed. In the absence of case law on the point it is reasonable to assume that in order to avoid actions for product liability, manufacturers should provide both warnings, indicating the worst results and dangers, and directions for use, indicating the best results; the warning, in effect, identifying the worst consequences that could follow if directions for use were not complied with.

In the case of *Worsley v Tambrands Ltd (No 2) [2000] MCR 0280*, it was held that a tampon manufacturer had done what was reasonable in all the circumstances to warn a woman about the risk of toxic shock syndrome from tampon use. They had placed a clear legible warning on the outside of the box directing the user to the leaflet. The leaflet was legible, literate and unambiguous and contained all the material necessary to convey both the warning signs and the action required if any risk were present. The manufacturer could not cater for lost leaflets or for those who chose not to replace them. This give valuable guidance as to the extent that manufacturers are expected to warn users of their products of the risks associated with their products.

Defect must exist when the product left the producer's possession

P9040 This situation tends to be spotlighted by alteration of, modification to or interference with a product on the part of an intermediary, for instance, a dealer or agent. If a product leaves an assembly line in accordance with its intended design, but is subsequently altered, modified or generally interfered with by an intermediary, in a manner outside the product's specification, the manufacturer is probably not liable for any injury so caused. In *Sabloff v Yamaha Motor Co 113 NJ Super 279, 273 A 2d 606 (1971)*, the claimant was injured when the wheel of his motor-cycle locked, causing it to skid, then crash. The manufacturer's specification stipulated that the dealer attach the wheel of the motor-cycle to the front fork with a nut and bolt, and this had not been done properly. It was held that the dealer was liable for the motor-cyclist's injury (as well as the assembler, since the latter had delegated the function of tightening the nut to the dealer and it had not been properly carried out).

Role of intermediaries

P9041 If a defect in a product is foreseeably detectable by a legitimate intermediary (e.g. a retailer in the case of a domestic product or an employer in the case of an industrial product), liability used to rest with the intermediary rather than the manufacturer, when liability was referable to negligence (*Donoghue v Stevenson [1932] AC 562*). This position does not duplicate under the *Consumer Protection Act 1987*, since the main object of the legislation is to fix producers with strict liability for injury-causing product defects to users. Nevertheless, there are common law and statutory duties on employers to inspect/test inward plant and machinery for use at the workplace, and failure to comply with these duties may make an employer liable. In addition, employers, in such circumstances, can incur liability under the *Employers' Liability* (*Defective Equipment*) *Act 1969* and so may seek to exercise contractual indemnity against manufacturers.

Comparison with negligence

P9042 The similarity between product liability and negligence lies in causation. Defect must be the material *cause* of injury. The main arguments against this are likely to be along the lines of misuse of a product by a user (e.g. knowingly driving a car with defective brakes), or ignoring warnings (a two-pronged defence, since it also denies there was a 'defect'), or – as is common in chemicals and pesticides cases – a defence based on alternative theories of causation of the claimant's injuries.

Product liability differs from negligence in that it is no longer necessary for injured users to prove absence of reasonable care on the part of manufacturers. All that is now necessary is proof of (*a*) defect (see P9035 above) and (*b*) that the defect caused the injury. It will, therefore, be no good for manufacturers to point to an unblemished safety record and/or excellent quality assurance programmes, or the lack of foreseeability of the accident, since the user is not trying to establish negligence. How or why a defect arose is immaterial; what is important is the fact that it exists.

Liability in negligence still has a role to play in cases where liability under the *Consumer Protection Act 1987* cannot be established because, for example, the defendant is not a 'producer' as defined, or because the statutory defence or time limit would bar a strict liability claim (see P9044 and P9048). There is greater scope under the law of negligence for liability to be established against distributors and retailers who may be held responsible for certain defects, especially where inadequate warnings and instructions have been provided (e.g. *Goodchild v Vaclight, The Times, 22 May 1965*).

Joint and several liability

P9043 If two or more persons/companies are liable for the same damage, the liability is joint and several. This can, for example, refer to the situation where a product (e.g. an aircraft) is made partly in one country (e.g. England) and partly in another (e.g. France). Here both partners are liable (joint liability) but in the event of one party not being able to pay, the other can be made to pay all the compensation (several liability). [*Consumer Protection Act 1987, s 2(5)*].

Parameters of liability

P9044 For certain types of damage including (*a*) death, (*b*) personal injury and (*c*) loss or damage to property liability is included and relevant to private use, occupation and consumption by consumers. [*Consumer Protection Act 1987,s 5(1)*]. However, producers and others will not be liable for:

(*a*) damage/loss to the defective product itself, or any product supplied with the defective product [*Consumer Protection Act 1987, s 5(2)*];

(*b*) damage to property not 'ordinarily intended for private use, occupation or consumption' e.g. car/van used for business purposes [*Consumer Protection Act 1987, s 5(3)*]; and

(*c*) damage amounting to less than £275 (to be determined as early as possible after loss) [*Consumer Protection Act 1987, s 5(4)*].

Defences

P9045 The following statutory defences are open to producers:

(*a*) the defect was attributable to compliance with any requirement imposed by law/regulation or a European Union rule/regulation [*Consumer Protection Act 1987, s 4(1)(a)*];

(*b*) the defendant did not supply the product to another (i.e. did not sell/hire/lend/exchange for money/give goods as a prize etc. (see below 'supply')) [*Consumer Protection Act 1987, s 4(1)(b)*];

(*c*) the supply to another person was not in the course of that supplier's business [*Consumer Protection Act 1987, s 4(1)(c)*];

(*d*) the defect did not exist in the product at the relevant time (i.e. it came into existence after the product had left the possession of the defendant). This principally refers to the situation where for example a retailer fails to follow the instructions of the manufacturer for storage or assembly [*Consumer Protection Act 1987, s 4(1)(d)*];

(*e*) that the state of scientific and technical knowledge at the relevant time was not such that a producer 'might be expected to have discovered the defect if it had existed in his products while they were under his control' (i.e. development risk) [*Consumer Protection Act 1987, s 4(1)(e)*] (see P9049 below);

(*f*) that the defect was:

(i) a defect in a subsequent product (in which the product in question was comprised); and

(ii) was wholly attributable to:

(A) design of the subsequent product; or

(B) compliance by the producer with the instructions of the producer of the subsequent product.

[*Consumer Protection Act 1987, s 4(1)(f)*]. This is known as the component manufacturer's defence.

The meaning of 'supply'

P9046 Before strict liability can be established under the *Consumer Protection Act 1987*, a product must have been '*supplied*'. This is defined as follows:

(*a*) selling, hiring out or lending goods;

(*b*) entering into a hire-purchase agreement to furnish goods;

(*c*) performance of any contract for work and materials to furnish goods (e.g. making/repairing teeth);

(*d*) providing goods in exchange for a consideration other than money (e.g. trading stamps);

(*e*) providing goods in or in connection with the performance of any statutory function/duty (e.g. supply of gas/electricity by public utilities);

(*f*) giving the goods as a prize or otherwise making a gift of the goods.

[*Consumer Protection Act 1987, s 46(1)*].

Moreover, in the case of hire-purchase agreements/credit sales the effective supplier (i.e. the dealer), and not the ostensible supplier (i.e. the finance company),is the 'supplier' for the purposes of strict liability. [*Consumer Protection Act 1987, s 46(2)*].

Building work is only to be treated as a supply of goods in so far as it involves provision of any goods to any person by means of their incorporation into the building/structure, e.g. glass for windows. [*Consumer Protection Act 1987, s 46(3)*].

No contracting out of strict liability

P9047 The liability to person who has suffered injury/damage under the *Consumer Protection Act 1987*, cannot be (*a*) limited or (*b*) excluded:

(*a*) by any contract term; or

(*b*) by any notice or other provision.

[*Consumer Protection Act 1987, s 7*].

Time limits for bringing product liability actions

P9048 No action can be brought under the *Consumer Protection Act 1987, Part I* (i.e. product liability actions) after the expiry of ten years from the time when the product was first put into circulation (i.e. the particular item in question was supplied in the course of business/trade etc.). [*Limitation Act 1980, s 11A*(*3*); *Consumer Protection Act 1987, Sch 1*]. In other words, ten years is the cut-off point for liability. An action can still be brought for common law negligence after this time.

A recent case has serious implications for this cut-off point. In *SmithKline Beecham plc and Another v Horne-Roberts [2001] EWCA CIV 2006*, the court allowed a claimant to substitute a new defendant for an existing one in a strict liability claim despite the fact that the substitution was outside the ten year cut-off period.

However all actions for personal injury caused by product defects must be initiated within three years of whichever event occurs later, namely:

(*a*) the date when the cause of action accrued (i.e. injury occurred); or

(*b*) the date when the injured person had the requisite knowledge of his injury/damage to property.

[*Limitation Act 1980, s 11A*(*4*); *Consumer Protection Act 1987, Sch 1*].

But if during that period the injured person died, his personal representative has a further three years from his death to bring the action. (This coincides with actions for personal injuries against employers, except of course in that case there is no overall cut-off period of ten years.) [*Limitation Act 1980, s 11A*(*5*)].

Development risk

P9049 This will probably emerge as the most important defence to product liability actions. Manufacturers have argued that it would be wrong to hold them responsible for the consequences of defects which they could not reasonably have known about or discovered. The absence of this defence would have the effect of increasing the cost of product liability insurance and stifle the development of new products. On the other hand, consumers maintain that the existence of this defence threatens the whole basis of strict liability and allows manufacturers to escape liability by, in effect, pleading a defence associated with the lack of negligence. For this reason, not all EU states have allowed this defence; the states in favour of its retention are the United Kingdom, Germany, Denmark, Italy and the Netherlands. The burden of proving development risk lies on the producer and it seems likely that he will have to show that no producer of a product of that sort could be expected to have discovered the existence of the defect. 'It will not necessarily be enough to show that he (the producer) has done as many tests as his competitor, nor that he did all the tests required of him by a government regulation setting a minimum standard.' (Explanatory memorandum of EC Directive on Product Liability, Department of Trade and Industry, November 1985).

Additionally, the fact that judgments in product liability cases are 'transportable' could have serious implications for the retention of development risk in the United Kingdom (see P9050 below).

Transportability of judgments

P9050 The so-called 'Brussels Regulation' (Regulation 44/2001 of 22 December 2000 on *Jurisdiction and the Recognition and Enforcement of Judgments in Civil and Commercial Matters*) requires judgments given in one of the Member States (except Denmark which will continue to apply the 1968 Brussels Convention) to be enforced in another. The United Kingdom is bound by this Regulation as of 1 March 2002, which, by virtue of the *Civil Jurisdiction and Judgments Order 2001* (*SI 2001 No 3929*) (which amends the *Civil Jurisdiction and Judgments Act 1982*), is part of UK law. Where product liability actions are concerned, litigation can be initiated in the state where the defendant is based or where injury occurred and the judgment of that court 'transported' to another Member State. This could pose a threat to retention of development risk in the United Kingdom and other states in favour of it from a state against it, e.g. France, Belgium, Luxembourg.

Contractual liability for substandard products

P9051 Contractual liability is concerned with defective products which are substandard (though not necessarily dangerous) regarding quality, reliability and/or durability. Liability is predominantly determined by contractual terms implied by the *Sale of Goods Act 1979* and the *Sale and Supply of Goods Act 1994*, in the case of goods sold; the *Supply of Goods* (*Implied Terms*) *Act 1973*, where goods are the subject of hire purchase and conditional and/or credit sale; and the *Supply of Goods and Services Act 1982*, where goods are supplied but not sold as such, primarily as a supply of goods with services; hire and leasing contracts are subject to the 1982 Act as well.

Exemption or exclusion clauses in such contracts may be invalid by virtue of the *Unfair Contract Terms Act 1977*, which also applies to such transactions. Moreover, 'standard form' contracts with consumers where the terms have not been individually negotiated, which contain 'unfair terms' – that is, terms detrimental to the consumer – will have such terms excised, if necessary, by the Director General of Fair Trading, under the *Unfair Terms in Consumer Contracts Regulations 1999* (*SI 1999 No 2083*). These Regulations extend to most consumer contracts between a seller or supplier and a consumer. [*Unfair Terms in Consumer Contracts Regulations 1999,* (*SI 1999 No 2083*)*, Reg 4*(*1*)]. (See P9057 below.)

Contractual liability is strict. It is not necessary that negligence be established (*Frost v Aylesbury Dairy Co Ltd [1905] 1 KB 608* where the defendant supplied typhoid-infected milk to the claimant, who, after its consumption, became ill and required medical treatment. It was held that the defendant was liable, irrespective of the absence of negligence on his part).

Sale of Goods Acts

P9052 Conditions and warranties as to fitness for purpose, quality and merchantability were originally implied into contracts for the sale of goods at common law. Those terms were then codified in the *Sale of Goods Act 1893*. However, this legislation did not provide a blanket consumer protection measure, since sellers were still allowed to exclude liability by suitably worded exemption clauses in the contract. This practice was finally outlawed, at least as far as consumer contracts were concerned, by the *Supply of Goods* (*Implied Terms*) *Act 1973* and later still by the *Unfair Contract Terms Act 1977*, the current statute prohibiting contracting out of contractual liability and negligence. Indeed, consumer protection has reached a height with the *Unfair Terms in Consumer Contracts Regulations 1999* (*SI 1999 No 2083*), invalidating 'unfair terms' in most consumer contracts of a standard form nature (see P9062 below).

More recently, the law relating to sale and supply of goods has been updated by the *Sale of Goods Act 1979* and the *Sale and Supply of Goods Act 1994*, the latter replacing the condition of 'merchantable quality' with 'satisfactory quality'. The difference between the 'merchantability' requirement, under the *1979 Act*, and its replacement 'satisfactory quality', under the *1994 Act*, is that, under the former Act, products were 'usable' (or, in the case of food, 'edible'), even if they had defects which ruined their appearance; now they must be free from minor defects as well as being safe and durable. Further, under the previous law, a right of refund disappeared after goods had been kept for a reasonable time; under current law, there is a right of examination for a reasonable time after buying.

Sale of Goods Act 1979

P9053 In 1979 a consolidated *Sale of Goods Act* was passed and current law on quality and fitness of products is contained in that Act. Another equally important development has been the extension of implied terms, relating to quality and fitness of products, to contracts other than those for the sale of goods, that is, to hire purchase contracts by the *Supply of Goods (Implied Terms) Act 1973*, and to straight hire contracts by the *Supply of Goods and Services Act 1982*. In addition, where services are performed under a contract, that is, a contract for work and materials, there is a statutory duty on the contractor to perform them with reasonable care and skill. In other words, in the case of services liability is not strict, but it is strict for the supply of products. This is laid down in the *Supply of Goods and Services Act 1982, s 4*. This applies whether products are simultaneously but separately supplied under any contract, e.g. after-sales service, say, on a car or a contract to repair a window by a carpenter, in which latter case service is rendered irrespective of product supplied.

Products to be of satisfactory quality – sellers/suppliers

P9054 The *Sale of Goods Act 1979* (as amended) writes two quality conditions into all contracts for the sale of products, the first with regard to satisfactory quality, the second with regard to fitness for purpose.

Where a seller sells goods in the course of business, there is an implied term that the goods supplied under the contract are of satisfactory quality, according to the standards of the reasonable person, by reference to description, price etc. [*Sale of Goods Act 1979, s 14(2)* as substituted by the *Sale and Supply of Goods Act 1994, s 1(2)*]. The 'satisfactory' (or otherwise) quality of goods can be determined from:

(*a*) their state and condition;

(*b*) their fitness for purpose (see P9057 below);

(*c*) their appearance and finish;

(*d*) their freedom from minor defects;

(*e*) their safety; and

(*f*) their durability.

However, the implied term of 'satisfactory quality' does not apply to situations where:

(i) the unsatisfactory nature of goods is specifically drawn to the buyer's attention prior to contract; or

(ii) the buyer examined the goods prior to contract and the matter in question ought to have been revealed by that examination.

[*Sale of Goods Act 1979, s 14(2)* as substituted by the *Sale and Supply of Goods Act 1994, s 1(2A), (2B)* and *(2C)*].

Sale by sample

P9055 In the case of a contract for sale by sample, there is an implied condition that the goods will be free from any defect making their quality unsatisfactory, which would not be apparent on reasonable examination of the sample. [*Sale of Goods Act 1979, s 15*(*2*) as substituted by the *Sale and Supply of Goods Act 1994, s 1*(*2*)].

Conditions implied into sale – sales by a dealer

P9056 The condition of satisfactory quality only arises in the case of sales by a dealer to a consumer, not in the case of private sales. The *Sale of Goods Act 1979, s 14*(*2*) also applies to second-hand as well as new products. It is not necessary, as it is with the 'fitness for purpose' condition (see P9058 below), for the buyer in any way to rely on the skill and judgment of the seller in selecting his stock, in order to invoke *s 14*(*2*). However, if the buyer has examined the products, then the seller will not be liable for any defects which the examination should have disclosed. Originally this applied if the buyer had been given opportunity to examine but had not, or only partially, exercised it. In *Thornett & Fehr v Beers & Son [1919] 1 KB 486*, a buyer of glue examined only the outside of some barrels of glue. The glue was defective. It was held that he had examined the glue and so was without redress.

Products to be reasonably fit for purpose

P9057 'Where the seller sells goods in the course of a business and the buyer, expressly or by implication, makes known

(*a*) to the seller, or

(*b*) where the purchase price or part of it is payable by instalments and the goods were previously sold by a credit-broker to the seller, to that credit-broker,

any particular purpose for which the goods are being bought, there is an implied condition that the goods supplied under the contract are reasonably fit for that purpose, whether or not that is a purpose for which such goods are commonly supplied, except where the circumstances show that the buyer does not rely or that it is unreasonable for him to rely, on the skill or judgment of the seller or credit-broker.' [*Sale of Goods Act 1979, s 14*(*3*)].

Reliance on the skill/judgment of the seller will generally be inferred from the buyer's conduct. The reliance will seldom be express: it will usually arise by implication from the circumstances; thus to take a case of a purchase from a retailer, the reliance will be in general inferred from the fact that a buyer goes to the shop in the confidence that the tradesman has selected his stock with skill and judgment (*Grant v Australian Knitting Mills Ltd [1936] AC 85*). Moreover, it is enough if the buyer relies partially on the seller's skill and judgment. However, there may be no reliance where the seller can only sell goods of a particular brand. A claimant bought beer in a public house which he knew was a tied house. He later became ill as a result of drinking it. It was held that there was no reliance on the seller's skill and so no liability on the part of the seller (*Wren v Holt [1903] 1 KB 610*).

Even though products can only be used normally for one purpose, they will have to be reasonably fit for that particular purpose. A claimant bought a hot water bottle and was later scalded when using it because of its defective condition. It was held that the seller was liable because the hot water bottle was not suitable for its normal purpose (*Priest v Last [1903] 2 KB 148*). But, on the other hand, the buyer must not be hypersensitive to the effects of the product. A claimant bought a Harris Tweed coat from the defendants. She later contracted dermatitis from wearing it. Evidence showed that she had an exceptionally sensitive skin. It was held that the coat was reasonably fit for the purpose when worn by a person with an average skin (*Griffiths v Peter Conway Ltd [1939] 1 All ER 685*).

Like *s 14*(*2*), *s 14*(*3*) extends beyond the actual products themselves to their containers and labelling. A claimant was injured by a defective bottle containing mineral water, which she had purchased from the defendant, a retailer. The bottle remained the property of the seller because the claimant had paid the seller a deposit on the bottle, which would be returned to her, on return of the empty bottle. It was held that, although the bottle was the property of the seller, the seller was liable for the injury caused to the claimant by the defective container (*Geddling v Marsh [1920] 1 KB 668*).

Like *s 14*(*2*) (above), *s 14*(*3*) does not apply to private sales.

Strict liability under the Sale of Goods Act 1979, s 14

P9058 Liability arising under the *Sale of Goods Act, s 14* is strict and does not depend on proof of negligence by the purchaser against the seller (*Frost v Aylesbury Dairy Co Ltd* (see P9051 above)). This fact was stressed as follows in the case of *Kendall v Lillico [1969] 2 AC 31*: 'If the law were always logical one would suppose that a buyer who has obtained a right to rely on the seller's skill and judgment, would only obtain thereby an assurance that proper skill and judgment had been exercised, and would only be entitled to a remedy if a defect in the goods was due to failure to exercise such skill and judgment. But the law has always gone further than that. By getting the seller to undertake his skill and judgment the buyer gets ... an assurance that the goods will be reasonably fit for his purpose and that covers not only defects which the seller ought to have detected but also defects which are latent in the sense that even the utmost skill and judgment on the part of the seller would not have detected them' (per Lord Reid).

Dangerous products

P9059 As distinct from applying to merely substandard products, both *s 14*(*2*) and (*3*) of the *Sale of Goods Act 1979* can be invoked where a product is so defective as to be unsafe, but the injury/damage must be a reasonably foreseeable consequence of breach of the implied condition. If, for instance, therefore, the chain of causation is broken by negligence on the part of the user, in using a product knowing it to be defective, there will be no liability. In *Lambert v Lewis [1982] AC 225*, manufacturers had made a defective towing coupling which was sold by retailers to a farmer. The farmer continued to use the coupling knowing that it was unsafe. As a result, an employee was injured and the farmer had to pay damages. He sought to recover these against the retailer for breach of *s 14*(*3*). It was held that he could not do so.

Credit sale and supply of products

P9060 Broadly similar terms to those under the *Sale of Goods Act 1979, s 14* exist, in the case of hire purchase, credit sale, conditional sale and hire or lease contracts, by virtue of the Acts described above at P9051 having been modified by the *Sale and Supply of Goods Act 1994*.

Unfair Contract Terms Act 1977 (UCTA)

P9061 In spite of its name, this Act is not directly concerned with 'unfairness' in contracts, nor is it confined to the regulation of contractual relations. The main provisions are as follows:

(*a*) Liability for death or personal injury resulting from negligence cannot be excluded by warning notices or contractual terms. [*UCTA 1977, s 2*]. (This does not necessarily prevent an indemnity of any liability to an injured person being agreed between two other contracting parties: if A hires plant to B on terms that B indemnifies A in respect of any claims by any person for injury, the clause may be enforceable (see *Thompson v T Lohan (Plant Hire) Ltd [1987] 2 All ER 631).*)

(*b*) In the case of other loss or damage, liability for negligence cannot be excluded by contract terms or warning notices unless these satisfy the requirement of reasonableness.

(*c*) Where one contracting party is a consumer, or when one of the contracting parties is using written standard terms of business, exclusions or restrictions of liability for breach of contract will be permissible only in so far as the term satisfies the requirement of reasonableness [*UCTA 1977, s 3*]; as far as contracts with consumers are concerned, it is not possible to exclude or restrict liability for the implied undertakings as to quality and fitness for purpose contained in the *Sale of Goods Act 1979* (as amended) or in the equivalent provisions relating to hire purchase and other forms of supply of goods [*UCTA 1977, s 6*(*2*)]. Separate provisions make it an offence to include this type of exclusion clause in a consumer contract [*Consumer Transactions* (*Restrictions on Statements*) *Order 1976* (*SI 1976 No 1813*)].

(*d*) In all cases where the reasonableness test applies, the burden of establishing that a clause or other provision is reasonable will lie with the person who is trying to rely on the term in question. The consequence of contravention of the Act is, however, limited to the offending term being treated as being ineffective; the courts will give effect to the remainder of the contract so far as it is possible to do so.

Unfair Terms in Consumer Contracts Regulations 1999 (SI 1999 No 2083)

P9062 Consumers faced with standard form contracts are given more ammunition to combat terminological obscurity, legalese and inequality of bargaining power, by the *Unfair Terms in Consumer Contracts Regulations 1999* (*SI 1999 No 2083*). These Regulations have the effect of invalidating any 'unfair terms' in consumer contracts involving products and services, e.g. sale, supply, servicing agreements, insurance and, as a final resort, empowering the Director General of Fair Trading to scrutinise 'standard form' terms, with a view to recommending, where necessary, their discontinued use. Contracts affected by the Act will remain enforceable minus the unfair terms (which will be struck out), in so far as this result is possible. [*Unfair Terms in Consumer Contracts Regulations 1999* (*SI 1999 No 2083*), *Reg 8*(*2*)].

The contracts affected are only 'standard form' ones, that is, contracts whose terms have not been 'individually negotiated' between seller/supplier and consumer [*Unfair Terms in Consumer Contracts Regulations 1999* (*SI 1999 No 2083*), *Reg 5*]; and a term is taken not to have been 'individually negotiated', where it has been drafted in advance and the consumer has not been able to influence the substance of the term [*Unfair Terms in Consumer Contracts Regulations 1999* (*SI 1999 No 2083*), *Reg 5*(*2*)]. Significantly, it is incumbent on the seller/supplier, who claims that a term was 'individually negotiated' (i.e. not unfair) to prove that it was so. [*Unfair Terms in Consumer Contracts Regulations 1999* (*SI 1999 No 2083*), *Reg 5*(*4*)]. However, if the terms of the contract are in plain, intelligible language, fairness, relating to subject matter or price/remuneration, cannot be questioned [*Unfair Terms in Consumer Contracts Regulations 1999* (*SI 1999 No 2083*), *Reg 6*(*2*)].

The key provisions are:

(*a*) an 'unfair term' is one which has not been individually negotiated and causes a significant imbalance in the parties' rights and obligations, detrimentally to the consumer, contrary to the underlying tenet of good faith.

[*Unfair Terms in Consumer Contracts Regulations 1999* (*SI 1999 No 2083*), *Reg 5*(*1*)]

(*b*) an 'unfair term' is not binding on the consumer.

[*Unfair Terms in Consumer Contracts Regulations 1999* (*SI 1999 No 2083*), *Reg 8*(*1*)]

(*c*) terms of standard form contracts must be expressed in plain, intelligible language and, if there is doubt as to the meaning of a term, it must be interpreted in favour of the consumer except in proceedings brought under *Reg 12* (see (*e*) below).

[*Unfair Terms in Consumer Contracts Regulations 1999* (*SI 1999 No 2083*), *Reg 7*]

(*d*) complaints about 'unfair terms' in standard forms are to be considered by the Director General of Fair Trading, who may bring proceedings for an injunction to prevent use of the terms in future if the proponent of the unfair terms does not agree to desist from using them.

[*Unfair Terms in Consumer Contracts Regulations 1999* (*SI 1999 No 2083*), *Reg 8*];

(*e*) instead of the Director General of Fair Trading, a qualifying body named in *Sch 1* (statutory regulators, trading standards departments and the Consumers' Association) may notify the Director that it agrees to consider the complaint and may bring proceedings for an injunction to prevent the continued use of an unfair contract term.

[*Unfair Terms in Consumer Contracts Regulations 1999* (*SI 1999 No 2083*), *Regs 10 and 12*].

Examples of 'unfair terms'

P9063 Terms which have the object or effect of:

(*a*) excluding or limiting the legal liability of a seller or supplier in the event of the death of a consumer or personal injury to the latter resulting from an act or omission of that seller or supplier;

(*b*) inappropriately excluding or limiting the legal rights of the consumer vis-à-vis the seller or supplier or another party in the event of total or partial non-performance or inadequate performance by the seller or supplier of any of the contractual obligations, including the option of offsetting a debt owed to the seller or supplier against any claim which the consumer may have against him;

(*c*) making an agreement binding on the consumer whereas provision of services by the seller or supplier is subject to a condition whose realisation depends on his own will alone;

(*d*) permitting the seller or supplier to retain sums paid by the consumer where the latter decides not to conclude or perform the contract, without providing for the consumer to receive compensation of an equivalent amount from the seller or supplier where the latter is the party cancelling the contract;

(*e*) requiring any consumer who fails to fulfil his obligation to pay a disproportionately high sum in compensation;

(*f*) authorising the seller or supplier to dissolve the contract on a discretionary basis where the same facility is not granted to the consumer, or permitting the seller or supplier to retain the sums paid for services not yet supplied by him where it is the seller or supplier himself who dissolves the contract;

(*g*) enabling the seller or supplier to terminate a contract of indeterminate duration without reasonable notice except where there are serious grounds for doing so;

(*h*) automatically extending a contract of fixed duration where the consumer does not indicate otherwise, when the deadline fixed for the consumer to express this desire not to extend the contract is unreasonably early;

(*i*) irrevocably binding the consumer to terms with which he had no real opportunity of becoming acquainted before the conclusion of the contract;

(*j*) enabling the seller or supplier to alter the terms of the contract unilaterally without a valid reason which is specified in the contract;

(*k*) enabling the seller or supplier to alter unilaterally without a valid reason any characteristics of the product or service to be provided;

(*l*) providing for the price of goods to be determined at the time of delivery or allowing a seller of goods or supplier of services to increase their price without in both cases giving the consumer the corresponding right to cancel the contract if the final price is too high in relation to the price agreed when the contract was concluded;

(*m*) giving the seller or supplier the right to determine whether the goods or services supplied are in conformity with the contract, or giving him the exclusive right to interpret any term of the contract;

(*n*) limiting the seller's or supplier's obligation to respect commitments undertaken by his agents or making his commitments subject to compliance with a particular formality;

(*o*) obliging the consumer to fulfil all his obligations where the seller or supplier does not perform his;

(*p*) giving the seller or supplier the possibility of transferring his rights and obligations under the contract, where this may serve to reduce the guarantees for the consumer, without the latter's agreement;

(*q*) excluding or hindering the consumer's right to take legal action or exercise any other legal remedy, particularly by requiring the consumer to take disputes exclusively to arbitration not covered by legal provisions, unduly restricting the evidence available to him or imposing on him a burden of proof which, according to the applicable law, should lie with another party to the contract.

[*Unfair Terms in Consumer Contracts Regulations 1999* (*SI 1999 No 2083*), *Sch 2*].

Risk Assessment

Introduction

R3001 Although the term 'risk assessment' probably first came into common use as a result of the *Control of Substances Hazardous to Health Regulations 1988* (commonly known as the '*COSHH Regulations*' and revised several times since), similar requirements had actually previously been contained in both the *Control of Lead at Work Regulations 1980* and the *Control of Asbestos at Work Regulations 1987*.

In practice a type of risk assessment had already been necessary for some years particularly as a result of the use of the qualifying clause 'so far as is reasonably practicable' in a number of the sections of the *Health and Safety at Work etc Act 1974* ('*HSWA 1974*').

HSWA 1974 requirements

R3002 *HSWA 1974, s 2* contains the general duties of employers to their employees with the most general contained within *s 2(1)*:

> 'It shall be the duty of every employer to ensure, so far as is reasonably practicable, the health, safety and welfare at work of all employees.'

Other more specific requirements are contained in *s 2(2)* and these are also qualified by the term 'reasonably practicable'.

HSWA 1974, s 3 places general duties on both employers and the self-employed in respect of persons other than their employees. *Section 3(1)* states:

> 'It shall be the duty of every employer to conduct his undertaking in such a way as to ensure, so far as is reasonably practicable, that persons not in his employment who may be affected thereby are not exposed to risks to their health or safety.'

Employers thus have duties to contractors (and their employees), visitors, customers, members of the emergency services, neighbours, passers-by and the public at large. This may to a certain point extend to include trespassers. Self-employed persons are put under a similar duty by virtue of *HSWA 1974, s 3(2)* and must also take care of themselves. In each case these duties are subject to the 'reasonably practicable' qualification.

HSWA 1974, s 4 places duties on each person who has, to any extent, control of non-domestic premises used for work purposes in respect of those who are not their employees. *HSWA 1974, s 6* places a number of duties on those who design, manufacture, import or supply articles for use at work, or articles of fairground equipment and those who manufacture, import or supply substances. Many of these obligations also contain the 'reasonably practicable' qualification.

What is reasonably practicable?

R3003 The phrase 'reasonably practicable' is not just included within the key sections of *HSWA 1974* but is also contained in a wide variety of regulations. Lord Justice Asquith provided a definition in his judgment in the case of *Edwards v National Coal Board [1949] 1 All ER 743* in which he stated:

> 'Reasonably practicable' is a narrower term than "physically possible" and seems to me to imply that a computation must be made by the owner in which the quantum of risk placed on one scale and the sacrifice involved in the measures necessary for averting risk (whether in money, time or trouble) is placed in the other, and that, if it be shown that there is a gross disproportion between them – the risk being insignificant in relation to the sacrifice – the defendants discharge the onus on them. Moreover, this computation falls to be made by the owner at a point in time anterior to the accident.'

HSWA 1974, s 40 places the burden of proof in respect of what was or was not 'reasonably practicable' (or 'practicable', see R3004 below) on the person charged with failure to comply with a duty or requirement. Employers and other duty holders must establish the level of risk involved in their activities and consider the various precautions available in order to determine what is reasonably practicable i.e. they must carry out a form of risk assessment.

Practicable and absolute requirements

R3004 Not all health and safety law is qualified by the phrase 'reasonably practicable'. Some requirements must be carried out 'so far as is practicable'. 'Practicable' is a tougher standard to meet than 'reasonably practicable' – the precautions must be possible in the light of current knowledge and invention (*Adsett v K and L Steelfounders & Engineers Ltd [1953] 1 WLR 773*). Once a precaution is practicable it must be taken even if it is inconvenient or expensive. However, it is not practicable to take precautions against a danger which is not yet known to exist (*Edwards v National Coal Board [1949] 1 All ER 743*), although it may be practicable once the danger is recognised.

Many health and safety duties are subject to neither 'practicable' nor 'reasonably practicable' qualifications. These absolute requirements usually state that something 'shall' or 'shall not' be done. However, such duties often contain other words which are subject to a certain amount of interpretation e.g. 'suitable', sufficient', 'adequate', 'efficient', 'appropriate' etc.

In order to determine whether requirements have been met 'so far as is practicable' or whether the precise wording of an absolute requirement has been complied with, a proper evaluation of the risks and the effectiveness of the precautions must be made.

The Management of Health and Safety at Work Regulations 1999

R3005 The Management Regulations were introduced in 1992 and revised by the 1999 Regulations. They were intended to implement the European Framework Directive (89/391) on the introduction of measures to encourage improvements in the safety and health of workers at work. The *Management of Health and Safety at Work Regulations 1999* (*SI 1999 No 3242*), *Reg 3* require employers and the self-employed to make a suitable and sufficient assessment of the risks to both employees and persons not in their employment. The purpose of the assessment is to identify the measures needed 'to comply with the requirements and prohibitions imposed ... by or under the relevant statutory provisions ... ' i.e. identifying what is needed to comply with the law.

Given the extremely broad obligations contained in *HSWA 1974, ss 2, 3, 4* and *6*, all risks arising from work activities should be considered as part of the risk assessment process (although some risks may be dismissed as being insignificant). Compliance with more specific requirements of regulations must also be assessed, whether these are absolute obligations or subject to 'practicable' or 'reasonably practicable' qualifications. The requirement for risk assessment introduced in the Management Regulations simply formalised what employers (and others) should have been doing all along i.e. identifying what precautions they needed to take to comply with the law. More detailed requirements of the Management Regulations are covered later in the chapter.

Any such additional precautions must then be implemented (the *Management of Health and Safety at Work Regulations 1999* (*SI 1999 No 3242*), *Reg 5* contains requirements relating to the effective implementation of precautions).

Common regulations requiring risk assessment

R3006 An increasing number of codes of regulations contain requirements for risk assessments. Several of these regulations are of significance to a wide range of work activities and are dealt with in more detail elsewhere in the looseleaf. The main regulations are as follows.

Control of Substances Hazardous to Health Regulations 2002 (COSHH)

R3007 The *Control of Substances Hazardous to Health Regulations 2002* (*SI 2002 No 2677*), *Reg 6* require employers to make a suitable and sufficient assessment of the risks created by work liable to expose any employees to any substance hazardous to health and of the steps that need to be taken to meet the requirements of the regulations. See HAZARDOUS SUBSTANCES IN THE WORKPLACE, which provides more details of the COSHH requirements.

The Health and Safety Executive ('HSE') booklet L5, contains both the regulations and the associated Approved Code of Practice. There are many other relevant HSE publications including HSG97 '*A step by step guide to COSHH assessment*'.

Noise at Work Regulations 1989

R3008 The *Noise at Work Regulations 1989* (*1989 No 1790*), *Reg 4* require employers to make a noise assessment which is adequate for the purposes of:

- identifying which employees are exposed to noise above defined action levels;
- providing information to comply with other duties under the Regulations (e.g. reduction of noise exposure, provision of ear protection, establishment of ear protection zones and informing employees).

See NOISE AND VIBRATION which provides more details of the requirements of the *Noise at Work Regulations 1989*. HSE booklet L108 '*Reducing Noise at Work*' contains guidance both on the regulations and on the assessment process.

Manual Handling Operations Regulations 1992

R3009 Employers are required by the *Manual Handling Operations Regulations 1992* (*SI 1992 No 2793*), *Reg 4* to make a suitable and sufficient assessment of all manual handling operations at work which involve a risk of employees being injured, and to take appropriate steps to reduce the risk to the lowest level reasonably practicable (They must avoid such manual handling operations if it is reasonably practicable to do so).

See MANUAL HANDLING which provides further details of the Regulations and the carrying out of assessments. HSE booklet L23, '*Manual Handling: Manual Handling Operations Regulations 1992 – guidance on regulations*' provides detailed guidance on the Regulations and manual handling assessments.

Health and Safety (Display Screen Equipment) Regulations 1992

R3010 The *Health and Safety* (*Display Screen Equipment*) *Regulations 1992* (*SI 1992 No 2792*), *Reg 2* require employers to perform a suitable and sufficient analysis of display screen equipment (DSE) workstations for the purpose of assessing risks to 'users' or 'operators' as defined in the Regulations. Risks identified in the assessment must be reduced to the lowest extent reasonably practicable.

The HSE booklet L26 '*Work with display screen equipment*' contains guidance on the Regulations and on workstation assessments.

Personal Protective Equipment at Work Regulations 1992

R3011 Under the *Personal Protective Equipment at Work Regulations 1992* (*SI 1992 No 2966*), *Reg 6* employers must ensure that an assessment is made to determine risks which have not been avoided by other means and identify personal protective equipment ('PPE') which will be effective against these risks. The Regulations also contain other requirements relating to the provision of PPE; its maintenance and replacement; information, instruction and training; and the steps which must be taken to ensure its proper use.

See PERSONAL PROTECTIVE EQUIPMENT which provides further details of the requirements under the Regulations. HSE booklet L25, '*Personal protective equipment at work*' provides detailed guidance on the Regulations and the assessment of PPE needs.

Fire Precautions (Workplace) Regulations 1997

R3012 The *Fire Precautions* (*Workplace*) *Regulations 1997* (*SI 1997 No 1840*) together with the *Management of Health and Safety at work Regulations 1999* (*SI 1999 No 3242*), make it quite explicit that employers must carry out an assessment of fire risks and fire precautions. *Part II* of the 1997 Regulations contains specific requirements relating to fire safety which must be included in the assessment.

Amendments to the 1997 Regulations in 1999 removed the exemption from risk assessment requirements originally given to holders of fire certificates (under the *Fire Precautions Act 1971* and other legislation). They too are required to carry out fire risk assessments.

See FIRE PREVENTION AND CONTROL which contains further details both on the Regulations and on fire precautions generally. The HSE/Home Office publication *'Fire Safety: An employer's guide'* provides extensive guidance on factors to be taken into account during fire risk assessments.

Control of Asbestos at Work Regulations 2002

R3013 The *Control of Asbestos at Work Regulations 2002* (*SI 2002 No 2675*) continue the requirement contained in the 1987 Regulations of the same title, for employers to make a risk assessment before carrying out work which is liable to expose employees to asbestos. *Regulation 6* contains several specific requirements on what the risk assessment must involve and other regulations detail many precautions which must be taken in asbestos work. Some types of asbestos work can only be done by companies licensed by the HSE. There are several important HSE reference publications on work with asbestos:

- L11 A guide to the Asbestos (Licensing) Regulations 1983 as amended;
- L27 Work with asbestos which does not normally require a licence;
- L28 Work with asbestos insulation, asbestos coating and asbestos insulating board.

The 2002 Regulations introduced new requirements for 'The management of asbestos in non-domestic premises'. These are contained in *Regulation 4* which does not come into operation until 21 May 2004. Duty holders under this regulation are defined within it and will normally be the owner or leaseholder, the employer occupying the premises or a combination of these (depending upon the terms of any lease or contract or who controls the premises).

Duty holders must carry out an assessment of their premises in order to identify any asbestos-containing material ('ACM') and assess its condition. They must then prepare a written plan for managing the risks from ACM, including such measures as regular inspections of confirmed and presumed ACM, restricting access to or maintenance in ACM areas and, where necessary, protection, sealing or removal of ACM.

Further details on these requirements are contained in ASBESTOS and also in HAZARDOUS SUBSTANCES IN THE WORKPLACE.

Important HSE references worth referring to are:

- L127 The management of asbestos in non-domestic premises.
- HSG227 A comprehensive guide to managing asbestos in premises.

Dangerous Substances and Explosive Atmospheres Regulations 2002

R3014 These Regulations (known sometimes as 'DSEAR') replace many older pieces of legislation including the *Highly Flammable Liquids and Liquefied Petroleum Gases Regulations 1972* and many of the licensing requirements of the *Petroleum* (*Consolidation*) *Act 1928*. They are concerned with protecting against risks from fire and explosion. The definition of 'dangerous substances' contained within the regulations includes such substances as petrol, many solvents, some types of paint, LPG etc. and also potentially explosive dusts.

Employers are required by the *DSEAR* (*SI 2002 No 2776*), *Reg 5* to carry out a risk assessment of work involving dangerous substances. The *DSEAR* (*SI 2002 No 2776*), *Reg 6* require a range of specified control measures to be implemented in order to eliminate or reduce risks as far as reasonably practicable. Places where explosive atmospheres may occur must be classified into zones and equipment and protective systems in these zones must meet appropriate standards. [*DSEAR* (*SI 2002 No 2776*), *Reg 7*]. The *DSEAR* (*SI 2002 No 2776*), *Reg 8* require equipment and procedures to deal with accidents and emergencies whilst the *DSEAR* (*SI 2002 No 2776*), *Reg 9* states that employees must be provided with information, instruction and training. Further details are contained in HAZARDOUS SUBSTANCES IN THE WORKPLACE.

Specialist regulations requiring risk assessment

R3015 Some codes of regulations requiring risk assessments are of rather more specialist application, although some are still covered in some detail elsewhere in the publication. References to relevant chapters of the looseleaf and to key HSE publications of relevance are included below. These regulations include:

- *Control of Lead at Work Regulations 2002* (*SI 2002 No 2676*)

  See HAZARDOUS SUBSTANCES IN THE WORKPLACE.

  L132 Control of lead at work.

- *Supply of Machinery* (*Safety*) *Regulations 1992* (*SI 1992 No 3073*)

  The Regulations include a variety of procedures which must be followed in assessing conformity of machinery with essential health and safety requirements set out in the Machinery Directive.

  See MACHINERY SAFETY.

  INDG270 Supplying new machinery: Advice to suppliers, free leaflet.

  INDG271 Buying new machinery: A short guide to the law, free leaflet.

- *Control of Major Accident Hazard Regulations 1999* (*SI 1999 No 743*) (*COMAH*)

  The Regulations only apply to sites containing specified quantities of dangerous substances.

  See DISASTER AND EMERGENCY MANAGEMENT SYSTEMS (DEMS).

  L111 A guide to the Control of Major Accident Hazard Regulations 1999.

  HSG 190 Preparing safety reports: Control of Major Accident Hazard Regulations 1999.

  HSG 191 Emergency planning for major accidents: Control of Major Accident Hazard Regulations.

- *Ionising Radiations Regulations 1999* (*SI 1999 No 3232*)

  See HAZARDOUS SUBSTANCES IN THE WORKPLACE.

  L121 Work With Ionising Radiation: Ionising Radiations Regulations 1999 – approved code of practice and guidance.

Related health and safety concepts

R3016 Risk assessment techniques are an essential part of other health and safety management concepts.

- *Safe systems of work*

  Employers are required under *HSWA 1974, s 2(2)(a)* to provide and maintain 'systems of work that are, so far as is reasonably practicable, safe and without risks to health'.

  Both the *Confined Spaces Regulations 1997* (*SI 1997 No 1713*) and the *Lifting Operations and Lifting Equipment Regulations 1998* (*LOLER*) (*SI 1998 No 2307*) contain similar requirements for safe systems of work. A safe system of work can only established through a process of risk assessment.

- *'Dynamic risk assessment'*

  The term 'dynamic risk assessment' is often used to describe the day to day judgements that employees are expected to make in respect of health and safety. However, employers must ensure that employees have the necessary knowledge and experience to make such judgements. The employer's 'generic risk assessments' must have identified the types of risks which might be present in the work activities, established a framework of precautions (procedures, equipment etc.) which are likely to be necessary and provided guidance on which precautions are appropriate for which situations.

- *Permits to work*

  A permit to work system is a formalised method for identifying a safe system of work (usually for a high risk activity) and ensuring that this system is followed. The permit issuer is expected to carry out a dynamic risk assessment of the work activity and should be more competent in identifying the risks and the relevant precautions than those carrying out the work.

- *CDM health and safety plans*

  A key component of the *Construction (Design and Management) Regulations 1994* (*CDM*) (*SI 1994 No 3140*) is the requirement for a health and safety plan. Essentially this process requires an assessment of risks involved in the project and the identification and eventual implementation of appropriate precautions. (The requirements of the CDM Regulations are covered in more detail in CONSTRUCTION AND BUILDING OPERATIONS.)

- *Method statements*

  Method statements usually involve a description of how a particular task or operation is to be carried out and should identify all the components of a safe system of work arrived at through a process of risk assessment.

Management Regulations requirements

R3017 The general requirement concerning risk assessment is contained in the *Management of Health and Safety at Work Regulations 1999* (*'the Management Regulations'*) (*SI 1999 No 3242*), *Reg 3*. Changes to the original regulations passed in 1992 mean that the risk assessment must now include fire risks and precautions (see R3012 above) and also risks to both young persons (under 18's) (see R3024 below) and new and expectant mothers (see R3025 below). Other regulations require more specific types of risk assessment e.g. of hazardous substances (COSHH), noise, manual handling operations, display screen equipment and personal protective equipment (see R3006–R3015 above).

Regulation 3(1) of the Management Regulations states:

> 'every employer shall make a suitable and sufficient assessment of:

(a) the risks to the health and safety of his employees to which they are exposed whilst they are at work; and

(b) the risks to the health and safety of persons not in his employment arising out of or in connection with the conduct by him of his undertaking,

for the purpose of identifying the measures he needs to take to comply with the requirements or prohibitions imposed upon him by or under the relevant statutory provisions and by Part II of the Fire Precautions (Workplace) Regulations 1997.'

Regulation 3(2) imposes similar requirements on self-employed persons.

Regulation 3(3) requires a risk assessment to be reviewed if:

- there is reason to suspect that it is no longer valid; or
- there has been a significant change in the matters to which it relates.

Regulation 3(4) requires a risk assessment to be made or reviewed before an employer employs a young person, and *Regulation 3(5)* identifies particular issues which must be taken into account in respect of young persons (especially their inexperience, lack of awareness of risks and immaturity). Further requirements in respect of young persons are contained in *Regulation 19*. *Regulation 16* contains specific requirements concerning the factors which must be taken into account in risk assessments in relation to new and expectant mothers. These relate to processes, working conditions and physical, biological or chemical agents. Assessments in respect of young persons and new or expectant mothers are dealt within more detail later in the chapter.

Regulation 3(6) requires employers who employ five or more employees to record:

— the significant findings of their risk assessments; and

— any group of employees identified as being especially at risk.

Methods of recording assessments are described later in this chapter (see R3041 below).

HSE booklet L21, *'Management of health and safety at work'*, contains the Management Regulations in full, the associated Approved Code of Practice ('ACoP') and Guidance on the Regulations.

Hazards and risks

R3018 The ACoP to the Management Regulations (*SI 1999 No 3242*) provides definitions of both hazard and risk.

A *hazard* is something with the potential to cause harm.

A *risk* is the likelihood of potential harm from that hazard being realised.

The *extent of the risk* will depend on:

- the likelihood of that harm occurring;
- the potential severity of that harm (resultant injury or adverse health effect);
- the population which might be affected by the hazard i.e. the number of people who might be exposed.

As an illustration, work at heights involves a hazard of those below being struck by falling objects. The extent of the risk might depend on factors such as the nature of the work being carried out, the weight of objects which might fall, the distance to the ground below, the numbers of people in the area etc.

Evaluation of precautions

R3019 The ACoP to the Management Regulations (*SI 1999 No 3242*) clearly states that risk assessment involves 'identifying the hazards present and evaluating the extent of the risks involved, taking into account existing precautions and their effectiveness.'

The evaluation of the effectiveness of precautions is an integral part of the risk assessment process. This is overlooked by some organisations who concentrate on the identification (and often the quantification) of risks without checking whether the intended precautions are actually being taken in the workplace and whether these precautions are proving effective.

Using the illustration in the previous paragraph, the risk assessment must take account of what precautions are required, such as:

- use of barriers and/or warning signs at ground level;
- use of tool belts by those working at heights;
- provision of edge protection on working platforms;
- use of head protection by those at ground level.

'Suitable and sufficient'

R3020 Risk assessments under the Management Regulations (*SI 1999 No 3242*) (and several other regulations) must be 'suitable and sufficient', but the phrase is not defined in the Regulations themselves. However, the ACoP to the Regulations states that 'The level of risk arising from the work activity should determine the degree of sophistication of the risk assessment.' The ACoP also states that insignificant risks can usually be ignored, as can risks arising from routine activities associated with life in general ('unless the work activity compounds or significantly alters those risks').

In practice, a risk can only be concluded to be insignificant if some attention is paid to it during the risk assessment process and, if there is any scope for doubt, it is prudent to state in the risk assessment record which risks are considered insignificant.

Winter weather (with its attendant rain, ice, snow or wind) may be considered to pose a routine risk to life. However, driving a fork lift truck in an icy yard or carrying out agricultural or construction work in a remote location may involve a far greater level of risk than normal and require additional precautions to be taken.

The risk assessment must take into account both workers and other persons who might be affected by the undertaking.

For example, a construction company would need to consider risks to (and from) their employees, sub contractors, visitors to their sites, delivery drivers, passers by and even possible trespassers on their sites.

Similarly a residential care home should take into account risks to (and from) their staff, visiting medical specialists, visiting contractors, residents and visitors to residents.

Reviewing risk assessments

R3021 The *Management of Health and Safety at Work Regulations 1999* (*SI 1999 No 3242*), *Reg 3*(*3*) require a risk assessment to be reviewed if:

'(a) there is reason to suspect it is no longer valid; or

(b) there has been a significant change in the matters to which it relates ; and where as a result of any such review changes to an assessment are required, the employer or self-employed person concerned shall make them.'

The ACoP to the Management Regulations states that those carrying out risk assessments 'would not be expected to anticipate risks that were not foreseeable.' However, what is foreseeable can be changed by subsequent events. An accident, a non-injury incident or a case of ill-health may highlight the need for a risk assessment to be reviewed because:

- a previously unforeseen possibility has now occurred;
- the risk of something happening (or the extent of its consequences) is greater than previously thought;
- precautions prove to be less effective than anticipated.

A review of the risk assessment may also be required because of significant changes in the work activity e.g. changes to the equipment or materials used, the environment where the activity takes place, the system of work used or to the numbers or types of people carrying out the activity. The need for a risk assessment review may be identified from elsewhere e.g. from others involved in the same work activity, through trade or specialist health and safety journals, from the suppliers of equipment or materials or from the HSE or other specialist bodies. Routine monitoring activity (inspections, audits etc.) or consultation with employees may also identify the need for an assessment to be reviewed. A review of the risk assessment does not necessarily require a repeat of the whole risk assessment process but it is quite likely to identify the need for increased or changed precautions.

In practice workplaces and the activities within them are constantly subject to gradual changes and the ACoP states 'it is prudent to plan to review risk assessments at regular intervals'. The frequency of such reviews should depend on the extent and nature of the risks involved and the degree of change likely. It is advisable that all risk assessments should be reviewed at least every five years.

There are many activities where the nature of the work or the workplace itself changes constantly. Examples of such situations are construction work or peripatetic maintenance or repair work. Here it is possible to carry out 'generic' assessments of the types of risks involved and the types of precautions which should be taken. However, some reliance must be placed upon workers themselves to identify what precautions are appropriate for a given set of circumstances or to deal with unexpected situations. Such workers must be well informed and well trained in order for them to be competent to make what are often called 'dynamic' risk assessments (see R3016 above).

Related requirements of the Management Regulations

R3022 A number of other requirements within the Management Regulations (*SI 1999 No 3242*) are closely related to the risk assessment process.

- *Regulation 4: Principles of prevention to be applied*

The *Management of Health and Safety at Work Regulations 1999* (*SI 1999 No 3242*), *Reg 4* and *Sch 1* require preventive and protective measures to be implemented on the basis of specified principles.

- *Regulation 5: Health and safety arrangements*

  Under the *Management of Health and Safety at Work Regulations 1999* (*SI 1999 No 3242*), *Reg 5*, employers are required to have appropriate arrangements for the effective planning, organisation, control, monitoring and review of preventive and protective measures. Those employers with five or more employees must record these arrangements. This application of the 'management cycle' to health and safety matters is dealt with in greater detail in R3023 below.

- *Regulation 6: Health surveillance*

  Under the *Management of Health and Safety at Work Regulations 1999* (*SI 1999 No 3242*), *Reg 6*, where risks to employees are identified through a risk assessment, they must be 'provided with such health surveillance as is appropriate.' (Surveillance is also likely to be necessary to comply with the requirements of more specific regulations e.g. *COSHH* (*SI 2002 No 2677*), *Control of Asbestos at Work Regulations 2002* (*SI 2002 No 2675*), *Ionising Radiations Regulations 1999* (*SI 1999 No 3242*)). Surveillance may be appropriate to deal with risks such as colour blindness and other vision defects (e.g. in electricians and train or vehicle drivers) or blackouts or epilepsy (e.g. for drivers, operators of machinery and those working at heights).

  HSE guidance is available on this important area (see HSG61 *'Health surveillance at work'*).

- *Regulation 8: Procedures for serious and imminent danger and for danger areas*

  The *Management of Health and Safety at Work Regulations 1999* (*SI 1999 No 3242*), *Reg 8* require every employer to 'establish and where necessary give effect to appropriate procedures to be followed in the event of serious and imminent danger to persons at work in his undertaking.' It also refers to the possible need to restrict access to areas 'on grounds of health and safety unless the employee concerned has received adequate health and safety instruction.'

  The need for emergency procedures or restricted areas should of course be identified through the process of risk assessment. Situations 'of serious and imminent danger' might be due to fires, bomb threats, escape or release of hazardous substances, out of control processes, personal attack, escape of animals etc. Areas may justify access to them being restricted because of the presence of hazardous substances, unprotected electrical conductors (particularly high voltage), potentially dangerous animals or people etc.

Application of the management cycle

R3023 The time spent in carrying out risk assessments will be wasted unless those precautions identified as being necessary are actually implemented. The *Management of Health and Safety at Work Regulations 1999* (*SI 1999 No 3242*), *Reg 5* require the application of a five stage 'management cycle' to this process.

- Planning

  The employer must have a planned approach to health and safety involving such measures as:

- a health and safety policy statement;
- annual health and safety plans;
- development of performance standards.

- Organisation

  An organisation must be put in place to deliver these good intentions e.g:

  - individuals allocated responsibility for implementing the health and safety policy or achieving elements of the health and safety plan;
  - arrangements for communicating with and consulting employees (using such measures as health and safety committees, newsletters, noticeboards);
  - the provision of competent advice and information.

- Control

  Detailed health and safety control arrangements must be established. The extent of such arrangements will reflect the size of the organisation and the degree of risk involved in its activities but they are likely to involve:

  - formal procedures and systems e.g. for accident and incident investigation, fire and other emergencies, the selection and management of contractors;
  - the provision of health and safety related training e.g. at induction, for the operation of equipment or certain activities, for supervisors and managers;
  - provision of appropriate levels of supervision.

- Monitoring

  The effectiveness of health and safety arrangements must be monitored using such methods as:

  - health and safety inspections (both formal and informal);
  - health and safety audits;
  - accident and incident investigations.

- Review

  The findings of monitoring activity must be reviewed and, where necessary, the management cycle applied once again to rectify shortcomings in health and safety arrangements.

  The review process might involve:

  - joint health and safety committees;
  - management health and safety meetings;
  - activities of health and safety specialists.

The management cycle can also be applied to ensure the effective implementation of individual health and safety management procedures (e.g. for the investigation and reporting of accidents and non-injury incidents) or to specific types of health and safety precautions (e.g. guarding of machinery or precautions for working at heights).

Children and young persons

R3024 Previous Acts and regulations identified many types of equipment which children and young persons were not allowed to use or processes or activities that they must not be involved in. Many of these 'prohibitions' were revoked by the *Health and Safety (Young Persons) Regulations 1997 (SI 1997 No 135)*, since incorporated into the *Management of Health and Safety at Work Regulations 1999 (SI 1999 No 3242)*. (A few 'prohibitions' still remain). The emphasis has now changed to restrictions on the work which children and young persons are allowed to do, based upon the employer's risk assessment.

The *Health and Safety (Training for Employment) Regulations 1990 (SI 1990 No 1380)* have the effect of giving students on work experience training programmes and trainees on training for employment programmes the status of 'employees'. The immediate provider of their training is treated as the 'employer'. (There are exceptions for courses at educational establishments, i.e. universities, colleges, schools etc.). Therefore employers have duties in respect of all children and young persons at work in their undertaking: full-time employees, part-time and temporary employees and also students or trainees on work placement with them.

The term 'child' and 'young person' are defined in the *Management of Health and Safety at Work Regulations 1999 (SI 1999 No 3242), Reg 1(2)*.

- *Child* is defined as a person not over compulsory school age in accordance with:
  - the *Education Act 1996, s 8* (for England and Wales); and
  - the *Education (Scotland) Act 1980, s 31* (for Scotland).

  (In practice this is just under or just over the age of sixteen)
- *Young Person* is defined as 'any person who has not attained the age of eighteen'.

Some of the prohibitions remaining from older health and safety regulations use different cut off ages.

See VULNERABLE PERSONS which provides much more detail on the requirements of the Management Regulations and the factors to be taken into account when carrying out risk assessments in respect of children and young persons. These factors centre around their:

- lack of experience;
- lack of awareness of existing or potential risks; and
- immaturity (in both the physical and psychological sense).

That chapter also contains in APPENDIX C a checklist of 'Work presenting increased risks for children and young persons'.

New or expectant mothers

R3025 Amendments made in 1994 to the previous Management Regulations implemented the European Directive on Pregnant Workers, requiring employers in their risk assessments to consider risks to new or expectant mothers. These amendments were subsequently incorporated into the 1999 Management Regulations. The *Management of Health and Safety at Work Regulations 1999 (SI 1999 No 3242), Reg 1* contains two relevant definitions:

- '*New or expectant mother*' means an employee who is pregnant; who has given birth within the previous six months; or who is breastfeeding.
- '*Given birth*' means 'delivered a living child or, after twenty-four weeks of pregnancy, a stillborn child.'

The requirements for 'risk assessment in respect of new and expectant mothers' are contained in the *Management of Health and Safety at Work Regulations 1999* (*SI 1999 No 3242*), *Reg 16* which states in *paragraph* (*1*):

> '(1) Where —
>
> (a) the persons working in an undertaking include women of child-bearing age; and
>
> (b) the work is of a kind which could involve risk, by reason of her condition, to the health and safety of a new or expectant mothers, or to that of her baby, from any processes or working conditions, or physical, biological or chemical agents, including those specified in Annexes I and II of Council Directive 92/85/EEC on the introduction of measures to encourage improvements in the safety and health at work of pregnant workers and workers who have recently given birth or are breastfeeding,
>
> the assessment required by regulation 3(1) shall also include an assessment of such risk.'

Regulation 16(*4*) states that in relation to risks from infectious or contagious diseases an assessment must only be made if the level of risk is in addition to the level of exposure outside the workplace. (The types of risk which are more likely to affect new or expectant mothers are described in R3026 below). *Regulation 16*(*2*) and (*3*) set out the actions employers are required to take if these risks cannot be avoided. *Paragraph* (*2*) states:

> 'Where, in the case of an individual employee, the taking of any other action the employer is required to take under the relevant statutory provisions would not avoid the risk referred to in paragraph (1) the employer shall, if it is reasonable to do so, and would avoid such risks, alter her working conditions or hours of work.'

Consequently where the risk assessment required under *Reg 16*(*1*) shows that control measures would not sufficiently avoid the risks to new or expectant mothers or their babies, the employer must make reasonable alterations to their working conditions or hours of work.

In some cases restrictions may still allow the employee to substantially continue with her normal work but in others it may be more appropriate to offer her suitable alternative work.

Any alternative work must be:

— suitable and appropriate for the employee to do in the circumstances;

— on terms and conditions which are no less favourable.

Regulation 16(*3*) states:

> 'If it is not reasonable to alter the working conditions or hours of work, or if it would not avoid such risk, the employer shall, subject to section 67 of the 1996 Act, suspend the employee from work for so long as is necessary to avoid such risk.'

(The 1996 Act referred to is the *Employment Rights Act 1996* which provides that any such suspension from work on the above grounds is on full pay. However, payment might not be made if the employee has unreasonably refused an offer of suitable alternative work.)

The *Management of Health and Safety at Work Regulations 1999* (*SI 1999 No 3242*), *Reg 17* deals specifically with night work by new or expectant mothers and states:

> 'Where —
>
> (a) a new or expectant mother works at night; and
>
> (b) a certificate from a registered medical practitioner or a registered midwife shows that it is necessary for her health or safety that she should not be at work for any period of such work identified in the certificate,
>
> the employer shall, subject to section 46 of the 1978 Act, suspend her from work for so long as is necessary for her health or safety.'

Such suspension (on the same basis as described above) is only necessary if there are risks arising from work. The HSE do not consider there are any risks to pregnant or breastfeeding workers or their children working at night per se. They suggest that any claim from an employee that she cannot work nights should be referred to an occupational health specialist. The HSE's own Employment Medical Advisory Service are likely to have a role to play in such cases.

The requirements placed on employers in respect of altered working conditions or hours of work and suspensions from work only take effect when the employee has formally notified the employer of her condition. The *Management of Health and Safety at Work Regulations 1999* (*SI 1999 No 3242*), *Reg 18* states:

> 'Nothing in paragraph (2) or (3) of regulation 16 shall require the employer to take any action in relation to an employee until she has notified the employer in writing that she is pregnant, has given birth within the previous six months, or is breastfeeding.'

Regulation 18(*2*) states that the employer is not required to maintain action taken in relation to an employee once the employer knows that she is no longer a new or expectant mother; or cannot establish whether this is the case.

Risks to new or expectant mothers

R3026 The HSE booklet HSG122 *'New and expectant mothers at work: A guide for employers'* provides considerable guidance on those risks which may be of particular relevance to new or expectant mothers, including those listed in the EC Directive on Pregnant Workers (92/85/EEC). These risks might include:

Physical agents

- Manual handling.
- Ionising radiation.
- Work in compressed air.
- Diving work.
- Shock, vibration etc.
- Movement and posture.
- Physical and mental pressure.
- Extreme heat.

Biological agents

Many biological agents in hazard groups 2, 3 and 4 (as categorised by the Advisory Committee on Dangerous Pathogens) can affect the unborn child should the mother be infected during pregnancy.

Chemical agents

- Substances labelled with certain risk phrases.
- Mercury and mercury derivatives.
- Antimitotic (cytotoxic) drugs.
- Agents absorbed through the skin.
- Carbon Monoxide.
- Lead and lead derivatives.

Working conditions

The HSE's frequently stated position in respect of work with display screen equipment (DSE) is that radiation from DSE is well below the levels set out in international recommendations, and that scientific studies taken as a whole do not demonstrate any link between this work and miscarriages or birth defects. There is no need for pregnant women to cease working with DSE. However, the HSE recommend that, to avoid problems from stress or anxiety, women are given the opportunity to discuss any concerns with someone who is well informed on the subject.

Other factors associated with pregnancy may also need to be taken into account e.g. morning sickness, backache, difficulty in standing for extended periods or increasing size. These may necessitate changes to the working environment or work patterns.

Other vulnerable persons

R3027 As well as children and young persons and new or expectant mothers, there are other vulnerable members of the workforce who merit special consideration during the risk assessment process. Such persons include:

- *Lone workers*

  Some staff members may work alone continuously or only part of the time. They may be working on the employer's own premises or outside, within the community. The level of risk may relate to the worker's location, the activities they are involved in or the type of people they might encounter.

  Amongst lone workers will be:

  - security staff;
  - cleaners;
  - isolated receptionists or enquiry staff;
  - some retail staff;
  - some maintenance workers;
  - staff working late, at nights or at weekends;
  - keyholders (particularly at risk if called to suspected break-ins);
  - delivery staff;

— sales and technical representatives;

— property surveyors etc.

- *Disabled persons*

  Legislation concerning disability discrimination will affect general access requirements within work premises. The *Health and Safety* (*Miscellaneous Amendments*) *Regulations 2002* (*SI 2002 No 2174*) introduced an additional requirement into *Regulation 25* of the *Workplace* (*Health, Safety and Welfare*) *Regulations 1992* (*SI 1992 No 3004*):

'Where necessary, those parts of a workplace (including in particular doors, passageways, stairs, showers, washbasins, lavatories and workstations) used or occupied directly by disabled persons at work shall be organised to take account of such persons.'

  In considering access in general and also specific risks which may affect disabled persons, account must be taken of all possible types of disability:

  — persons lacking mobility (including wheelchair users);

  — those lacking physical strength (particularly in relation to manual handling);

  — persons with relevant health conditions (especially on COSHH-related matters);

  — blind or visually impaired persons;

  — those who have hearing difficulties;

  — persons with special educational needs.

  Whilst employers have particular duties towards their own employees, the risk assessment must also consider risks to disabled persons who are:

  — on work placements;

  — visitors;

  — service users;

  — neighbours or passers-by.

  All types of risks must be considered but particular note must be taken of:

  — access posing particular dangers to disabled persons;

  — the presence of hazardous substances;

  — potentially dangerous work equipment;

  — hot items or surfaces;

  — the need for special emergency arrangements for disabled persons.

- *Inexperienced workers*

  Some older persons may (like young people) be lacking in experience of some types of work or lack awareness of existing or potential risks of particular work activities. Some may also lack a suitable degree of maturity, particularly those with special needs. This must be taken into account during the risk assessment process, especially in situations such as:

  — training or retraining programmes;

— work activities with rapid staff turnover;

— activities involving people with special needs.

Who should carry out the risk assessment?

R3028 The *Management of Health and Safety at Work Regulations 1999* (*SI 1999 No 3242*), *Reg 7* require employers to appoint competent health and safety assistance. Such persons must have sufficient training and experience or knowledge together with other qualities, in order to be able to identify risks and evaluate the effectiveness of precautions to control those risks. (*Regulation 7*(*8*) expresses a preference for such persons to be employees as opposed to others e.g. consultants).

The HSE guidance on risk assessment states that in small businesses the employer or a senior manager may be quite capable of carrying out the risk assessment. Larger employers may create risk assessment teams who might be drawn from managers, engineers and other specialists, supervisors or team leaders, health and safety specialists, safety representatives and other employees. Many employers continue to utilise consultants to co-ordinate or carry out their risk assessments. Whether the assessments are to be carried out by an individual or by a team, others will need to be involved during the process. Managers, supervisors, employees, specialists etc. will all need to be consulted about the risks involved in their work and the precautions that are (or should be) taken.

Assessment units

R3029 In a small workplace it may be possible to carry out a risk assessment as a single exercise but in larger organisations it will usually be necessary to split the assessment up into manageable units. If done correctly this should mean that assessment of each unit should not take an inordinate amount of time and also allows the selection of the people best able to assess an individual unit. Division of work activities into assessment units might be by department or sections, buildings or rooms, processes or product lines, or services provided.

As an illustration a garage might be divided into:

- Servicing and repair workshop.
- Body repair shop.
- Parts department.
- Car sales and administration.
- Petrol and retail sales.

Relevant sources of information

R3030 There is a wide range of documents which might be of value during the risk assessment process including:

- *Previous risk assessments*

  (including risk assessments carried out to comply with specific regulations e.g. COSHH or Noise regulations).

- *Operating procedures*

  (where these include health and safety information).

- *Safety handbooks etc.*
- *Training programmes and record*s
- *Accident and Incident records*
- *Health and safety inspection or audit reports*
- *Relevant Regulations and Approved Codes of Practice*

  (Risk assessment involves an evaluation of compliance with legal requirements and therefore an awareness of the legislation applying to the workplace in question is essential.)

- *Relevant HSE publications*

  The HSE publishes a wide range of booklets and leaflets providing guidance on health and safety topics. Some of these relate to the specific requirements of regulations, others deal with specific types of risks whilst some publications deal with sectors of work activity.

  All of these can be of considerable value in identifying which risks the HSE regard as significant and in providing benchmarks against which precautions can be measured. The guidance on specific types of workplaces (which includes engineering workshops, motor vehicle repair, warehousing, kitchens and food preparation, golf courses, horse riding establishments and many others) should form an essential basis for those carrying out assessments in those sectors.

  HSE Books regularly publishes a detailed catalogue of HSE publications. Their booklet '*Essentials of Health and Safety at Work*' provides an excellent starting point for those in small businesses needing guidance or carrying out risk assessment. The booklet also contains a useful reference section to other HSE publications which may be of relevance. To obtain these publications contact HSE Books, PO Box 1999, Sudbury, Suffolk CO10 2WA (tel: 01787 881165; fax: 01787 313995; website: www.hsebooks.co.uk).

- *HSE website*

  The HSE website at www.hse.gov.uk contains an ever-growing resource of guidance material. Some of this is available for specific work areas, e.g. agriculture, construction, engineering, food manufacture, haulage, motor vehicle repair, offshore oil and gas, railways, textiles and footwear. It is also possible to search for information on a wide range of particular risk topics. Much of this information (particularly free HSE publications) can be downloaded directly from the website. A subscription service (HSE Direct) also allows access to and downloading of priced HSE publications.

- *Trade Association codes of practice and guidance*
- *Information from manufacturers and suppliers*

  (Equipment handbooks or substance data sheets).

- *General health and safety reference books*

Consider who might be at risk

R3031 The risk assessment process must take account of all those who may be at risk from the work activities i.e. both employees and others. It is important that all of these are identified.

- *Employees*

  Different categories of employees to take into account might include:

  - production workers;
  - maintenance workers;
  - administrative staff;
  - security officers;
  - cleaners;
  - delivery drivers;
  - sales representatives;
  - others working away from the premises;
  - temporary employees.

  Some of these employees may merit special considerations:

  - children and young persons (see R3024);
  - women of childbearing age (i.e. potential new or expectant mothers) (see R3025 and R3026);
  - other vulnerable persons (see R3027);

- *Contractors and their staff*

  Contracted services might involve:

  - construction or engineering projects;
  - routine maintenance or repair;
  - hire of plant and operators;
  - support services e.g. catering, cleaning, security, transport;
  - professional services e.g. architects, engineers, trainers;
  - supply of temporary staff.

  (Contractors also have duties to carry out risk assessments in respect of their own staff.)

- *Others at risk*

  The types of people might be put at risk by the organisation's activities will depend upon the nature and location of those activities. Groups of people to be considered include:

  - volunteer workers;
  - co-occupants of premises;
  - occupants of neighbouring premises;
  - drivers making deliveries;
  - visitors (both individuals and groups);

- — residents e.g. in the care or hospitality sectors;
- — passers-by;
- — users of neighbouring roads;
- — trespassers;
- — customers or service users.

Identify the issues to be addressed

R3032 The issues which will need to be addressed during the risk assessment process should be identified. This may be done in respect of the assessment overall or separately for each of the assessment units and will be based upon the knowledge and experience of those carrying out the assessment and the information gathered together from the sources described earlier. This will create an initial list of headings and sub-headings for the eventual record of the risk assessment, although in practice this list is likely to be amended along the way. The sample assessment records contained later in the chapter (see R3049 and R3050) will demonstrate how such a list can be built up for typical workplaces.

These headings are likely to consist of the more common types of risk e.g. fire, vehicles, work at heights, together with some specialised types of risk associated with the work activities such as violence, lasers or working in remote locations. A checklist of possible risks to be considered during risk assessments is provided in R3034 below. This includes some of the risks which have specific regulations associated with them.

In some situations particular notes may also be made to check on the effectiveness of the precautions which should be in place to control the risks e.g. standards of machine guarding, compliance with personal protective equipment (PPE) requirements or the quality and extent of training.

Variations in work practices

R3033 Consideration should also be given at this stage to possible variations in work activities which may create new risks or increase existing risks. Such variations might involve:

- Fluctuations in production or workload demands.
- Reallocation of staff to meet changing workloads.
- Seasonal variations in work activities.
- Abnormal weather conditions.
- Alternative work practices forced by equipment breakdown/unavailability.
- Urgent or 'one-off' repair work.
- Work carried out in unusual locations.
- Difference in work between days, nights or weekends.

Further variations may emerge later on in the assessment process.

Checklist of possible risks

R3034 This checklist is intended to assist in identifying which issues need to be addressed during risk assessments. In some cases the headings and/or sub-headings might be used in the form shown, in other cases it may be more appropriate to combine them or modify the titles.

Work Equipment

- Process machinery
- Other machines
- Powered tools
- Handtools
- Knives/blades
- Fork-lift trucks
- Cranes
- Lifts
- Hoists
- Lifting equipment
- Vehicles

Services/power sources

- Electrical installation
- Compressed air
- Steam
- Hydraulics
- Other pressure systems
- Buried services
- Overhead services

Access

- Vehicle routes
- Rail traffic
- Pedestrian access
- Work at heights
- Ladders and stepladders
- Scaffolding
- Mobile elevating work platforms
- Falling objects
- Glazing

Storage

- Shelving and racking
- Stacking
- Silos and tanks
- Waste

Work activity

- Burning or welding
- Entry into confined spaces
- Electrical work
- Excessive fatigue or stress
- Handling cash/valuables
- Use of compressed gases
- Molten metal

Fire and explosion prevention

- Flammable liquids *
- Flammable gases *
- Storage of flammables *
- Hot work

* DSEAR requirements

External factors

- Violence or aggression
- Robbery
- Large crowds
- Animals
- Clients' activities

Work locations

- Heat
- Cold
- Severe weather
- Deep water
- Tides

Other factors

- Vibration
- Lasers
- Ultra violet/infra red radiation
- Work related upper limb disorder
- Remote locations
- Work alone
- Homeworking
- Poor hygiene
- Infestations
- Work abroad
- Work in domestic property
- Clients' premises
- Site security

Risks/issues subject to separate assessment requirements	
Hazardous substances (COSHH)	Lead
Noise	Asbestos work
Manual handling	Asbestos in premises
Display screen equipment workstations	Conformity of machinery
PPE needs	Major accident hazards (COMAH)
Fire precautions	Ionising radiation
Dangerous Substances and Explosive Atmospheres (DSEAR)	

Making the risk assessment

R3035 Good planning and preparation can reduce the time spent in actually making the risk assessment as well as enabling that time to be used much more productively. However, it is essential that time is spent in work locations, seeing how work is actually carried out (as opposed to how it should be carried out).

Observation

R3036 Observation of the work location, work equipment and work practices is an essential part of the risk assessment process. Where there are known to be variations in work activities a sufficient range of these should be observed to be able to form a judgement on the extent of the risks and the adequacy of precautions. Evaluations can be made of the effectiveness of fixed guards, the suitability and condition of access equipment, compliance with PPE requirements, the observance of specified operating procedures or working practices and many other aspects. It should also be borne in mind that work practices may change once workers realise they are under observation. Initial or undetected observation of working practices may be the most revealing.

Discussions

R3037 Discussions with people carrying out work activities, their safety representatives and those responsible for supervising or managing them are also an essential part of risk assessment. Amongst aspects of the work that might be discussed are possible variations in the work activities, problems that workers encounter, workers' views of the effectiveness of the precautions available, the reasons some precautions are not utilised and their suggestions for improving health and safety standards.

Tests

R3038 In some situations it may be appropriate to test the effectiveness of safety precautions e.g. the efficiency of interlocked guards or trip devices, the audibility of alarms or warning devices or the suitability of access to remote workplaces e.g. crane cabs, roofs. When carrying out such tests care must be taken by those carrying out the assessment not to endanger themselves or others, nor to disrupt normal activities.

Further investigations

R3039 Frequently further investigations will need to be made before the assessment can be concluded. Such investigations may involve detailed checks on standards or records, or enquiries into how non-routine situations are dealt with. Examples of further checks or enquiries which might be appropriate are:

- Detailed requirements of published standards e.g. design of guards, thickness of glass.
- Contents of operating procedures.
- Maintenance or test records.
- Contents of training programmes or training records.

Once again the potential list is endless although lines of further enquiry should be indicated by the observations and discussions during the initial phase of the assessment.

Notes

R3040 Rough notes should be made throughout the assessment process. It will be on these notes that the eventual assessment record will be based. It will seldom be possible to complete an assessment record 'on the run' during the assessment itself. The notes should relate to anything likely to be of relevance, such as:

- Risks discounted as insignificant.
- Further detail on risks or additional risks identified during the assessment.
- Risks which are being controlled effectively.
- Descriptions of precautions which are in place and effective.
- Precautions which do not appear to be effective.
- Alternative precautions which might be considered.
- Problems identified or concerns expressed by others.
- Related procedures, records or other documents.

Assessment records

R3041 This phase of the risk assessment process concludes with the preparation of the assessment records. It may be preferable to record the findings in draft form initially, with a revised version being produced once the assessment has been reviewed more widely and/or recommended actions have been completed.

- *Decide on the record format*

  Details of the content of assessment records and examples of one assessment record format are provided later in the chapter (see R3049 and R3050) but many alternatives are available. Different types of records may be appropriate for different departments, sections or activities. Use of a 'model' assessment might be relevant for similar workplaces or activities. Some (or all) of the assessment record might be integrated into documented operating procedures.

- *Identify the section headings to be used*

  During the preparatory phase of the assessment a list of risks and other sources to be addressed in each assessment unit was prepared as an aide memoire. This list will now need to be converted into section headings for the assessment records. As a result of the assessment some of the headings may have been sub-divided into different headings whilst others may have been merged.

- *Prepare the assessment records*

  The rough notes made during assessment must then be converted into formal assessment records. Preparing the records will normally require at least half of the time that was spent in the workplace carrying out the assessment and sometimes might even take longer. It is wise not to allow too long to elapse between assessing in the workplace and preparation of the assessment record. Notes will seem much more intelligible and memory will often be able to 'colour in' between the notes.

- *Identify the recommendations*

  The assessment will almost inevitably result in recommendations for improvements and these must be identified. As can be seen later, some assessment record formats incorporate sections in which recommendations can be included but in other cases separate lists will need to be prepared. The use of risk rating matrices (see below) may assist in the prioritisation of recommendations.

Risk rating matrices

R3042 The ACoP accompanying the *Management of Health and Safety at Work Regulations 1999* (*SI 1999 No 3242*) describes how the risk assessment process needs to be more sophisticated in larger and more hazardous sites. Nevertheless it suggests that quantification of risk will only be appropriate in a minority of situations. Some organisations use risk-rating systems to assist in the identification of priorities. Most involve matrices, utilising a simple combination of the likelihood of a hazard having an adverse effect and the severity of the consequences if it did. Some use numbers to produce a risk rating, as in the example below:

Risk Assessment Matrix			*Likelihood of adverse effect*		
			Unlikely	Possible	Frequent
			1	2	3
Severity of consequences	Minor	1	1	2	3
	Moderate	2	2	4	6
	Severe	3	3	6	9

The numbers can be replaced by descriptions of the level of risk as shown in the next example:

Risk Assessment Matrix		*Likelihood of adverse effect*		
		Unlikely	Possible	Frequent
Severity of consequences	Minor	Low	Low	Medium
	Moderate	Low	Medium	High
	Severe	Medium	High	Very high

The author's view is generally against the use of such matrices on the basis that time is often spent considering risk values at the expense of evaluating the effectiveness of the controls which ought to be in place.

After the assessment

R3043 Whilst the recording of the risk assessment is an important legal requirement, it is even more important that the recommendations for improvement identified during the assessment are actually implemented. This is likely to involve several stages.

Review and implementation of the recommendations

R3044 There may be a need to involve others outside (and probably senior to) the risk assessment team. The reasoning behind the recommendations can be explained and various alternative ways of controlling risks can be evaluated. Changes to the assessment findings or recommendations may be made at this stage but the risk assessment team should not allow themselves to be browbeaten into making alterations that they do not feel can be justified. Similarly they should not hold back on making recommendations they consider are necessary just because they believe that senior management will not implement them. The assessment team should carry out their duties to the best of their abilities in identifying what precautions are necessary in order to comply with the law – the responsibility for achieving compliance rests with their employer.

Some recommendations may need to be costed in respect of the capital expenditure or staff time required to implement them. It is unlikely that all recommendations will be able to be implemented immediately; there may be a significant lead time for the delivery of materials or the provision of specialist services from external sources. Once costings and prioritisation have been agreed, the recommendations should be converted into an action plan with individuals clearly allocated responsibility for each element of the plan, within a defined timescale. Some members of the risk assessment team (particularly health and safety specialists) may have responsibility for implementing parts of the plan or providing guidance to others.

Recommendation follow-up

R3045 Even in well-intentioned organisations, recommendations for improvement that have been fully justified and accepted are often still not implemented. It is essential that the risk assessment process includes a follow-up of the recommendations made. The recording format provided later in the chapter includes reference to this. As well as establishing that the improvements have actually been carried out, consideration should also be given to whether any unexpected risks have inadvertently been created.

Once the follow-up has been carried out, the assessment record should be annotated or revised to take account of the changes made. If recommendations have not been implemented there is a clear need for the situation to be referred back to senior management for them to take action to overcome whatever are the obstacles to progress.

Assessment review

R3046 The *Management of Health and Safety at Work Regulations 1999* (*SI 1999 No 3242*) and the associated ACoP state that assessments must be reviewed in certain circumstances (see R3021 above). The ACoP also states that 'it is prudent to plan to review risk assessments at regular intervals.' The frequency for reviews should be established which relates to the extent and nature of the risks involved and the likelihood of creeping changes (as opposed to a major change which would automatically justify a review).

The review may, however, conclude that the risks are unchanged, the precautions are still effective and that no revision of the assessments is necessary. The process of conducting regular reviews of assessments and, where appropriate, making revisions may be aided by the application of document control systems of the type used to achieve compliance with ISO 9000 and similar standards.

Content of assessment records

R3047 The *Management of Health and Safety at Work Regulations 1999* (*SI 1999 No 3242*), *Reg 3*(*6*) states:

> 'Where the employer employs five or more employees, he shall record —
>
> (a) the significant findings of the assessment; and
>
> (b) any group of his employees identified by it as being especially at risk.'

The accompanying ACoP refers to the record as representing 'an effective statement of hazards and risks which then leads management to take the relevant actions to protect health and safety.' It goes on to state that the record must be retrievable for use by management, safety representatives, other employee representatives or visiting inspectors. The need for linkages between the risk assessment, the record of health and safety arrangements (required by *Reg 5* of the Management Regulations) and the health and safety policy is also identified. The ACoP allows for assessment records to be kept electronically as an alternative to being in written form.

The essential content of any risk assessment record should be:

- Hazards or risks associated with the work activity.
- Any employees identified as especially at risk.
- Precautions which are (or should be) in place to control the risks (with comments on their effectiveness).
- Improvements identified as being necessary to comply with the law.

Other important details to include are:

- Name of the employer.
- Address of the work location or base.
- Names and signatures of those carrying out the assessment.
- Date of the assessment.
- Date for next review of the assessment.

(These might be provided as a introductory sheet.)

Illustrative assessment records

R3048 In the final pages of this chapter the risk assessment methodology described earlier is used to provide illustrations of how the process can be applied in two different types of workplace. In each case, relevant risks or issues to be addressed are listed, one of those risks or issues is selected, relevant regulations, references and other key assessment points relating to that risk or issue are identified and an illustration of how the completed assessment record might look is provided. The content of the form used is similar to that provided in the HSE's *'Five steps to risk assessment'* leaflet but the layout is felt to be more user-friendly.

The illustrations used are for a supermarket and a newspaper publisher.

A supermarket

R3049 Relevant risk topics are likely to include:

- Food processing machinery – in the delicatessen.
- Knives – delicatessen and butchery.
- Fork-lift trucks – warehousing areas.
- Vehicle traffic – delivery vehicles, customer and staff vehicles.
- Vehicle unloading – in the goods inward area.
- Pedestrian access – external to and inside the store.
- Glazed areas – store windows, display cabinets.
- Electrical installation – the power system within the store.
- Electrical equipment – including tills, cleaning equipment, office equipment.
- Shelving – in the store.
- Racking – in warehouse areas.
- Refrigerators – in the store and warehouse.
- Fire – general risks only, precautions must take account of customers.
- Aggression – e.g. from unhappy customers.
- Possible robbery – the supermarket will hold large quantities of cash.
- WRULD – for checkout operators.
- Hazardous substances – cleaning materials, office supplies.
- Manual handling – e.g. shelf stacking, movement of trolleys.
- Display screen equipment – used at workstations in the office.
- PPE requirements – throughout the supermarket.
- Asbestos – the possible presence of asbestos-containing materials within the premises.
- Use of contractors – for maintenance and repair work.

A sample risk assessment record for **vehicle traffic** is provided below.

In conducting this risk assessment:

- The requirements of the *Workplace (Health, Safety and Welfare) Regulations 1992* must be complied with.

- HSE booklet HS(G)136 *'Workplace Transport Safety'* is likely to be relevant.
- Particular attention should be paid to:
  - signage and road markings;
  - observation of vehicle movements and speeds;
  - conditions during busy periods and hours of darkness.

ABC Supermarkets, Newtown

Risk Assessment		
Reference number: 4	Risk topic/issue: Vehicle traffic	Sheet 1 of 1
Cross references: HSE booklet IND(G) 136 Risk assessments 5 (vehicle unloading) and 6 (pedestrian access)		
Risks identified	**Precautions in place**	**Recommended improvements**
Vehicles making deliveries to the 'Goods inward' bay present risks to each other and to any pedestrians on the access road	Prominent 10 mph are in place on the roadway. The speed limit is enforced effectively. Signs prohibit use of the road by pedestrians and customer or staff vehicles.	
There are also risks to other road users as they leave and rejoin the main road.	There are give way signs and road markings at the junction with the main road.	Reposition the advertising sign which partly blocks visibility on rejoining the main road
	Vehicles are parked in a holding area prior to backing up to the 'Goods Inward' bay.	
	Movement of vehicles is controlled by a designated member of the supermarket staff.	Provide this designated staff member with a high visibility waterproof jacket.
	The area is well lit by roadside lamps and floodlights on the side of the building.	
Customer and staff vehicles circulating in the car park area present risks to each other and to pedestrians in the area.	The access road around the car park is one way and well indicated by signs. The entrance and exit are well separated from each other and the good access road.	
There are also risks to other road users at the entrance from and access back into the main road.	There are prominent 15 mph signs around the roadways (some vehicles exceed this speed). There are also some give way markings and signs at all roadway junctions.	Provide clearly marked speed ramps at suitable locations.
	Parking bays are well marked.	The surface of some bays in the south-west corner of the car park should be repaired.
	Pedestrian crossing points are clearly marked and signed. The area is well lit by lighting towers which are protected at the base.	
	Security staff inspect, and where necessary, salt roadways in icy or snowy weather	Include the goods access road, Goods inward bay and car park in routine safety inspections.
Signature(s) A Smith, B Jones	**Name(s)** A Smith, B Jones	**Date** 11/12/00
Dates for	**Recommendation follow up March 2001**	**Next routine review December 2003**

A newspaper publisher

R3050 A large workplace like this would need to be divided into assessment units (see R3029 above), which might consist of:

- Common facilities, services etc. – e.g. fire, electrical supply, lifts, vehicle traffic, presence of asbestos.
- Reel handling and stands – e.g. supply of reels of newsprint to the press area.
- Platemaking – e.g. equipment and chemicals used to produce printing plates.
- Printing press – e.g. press machinery, solvents, noise etc.
- Despatch – e.g. inserting equipment, newspaper stacking, strapping and loading.
- Circulation and transport – e.g. distribution and other vehicles, fuel, waste disposal.
- Maintenance – e.g. workshops, garage, maintenance activities.
- Offices – e.g. editorial and administrative areas.

Selecting the **reel handling and stands** unit, risk topics are likely to include:

- Fork-lift trucks – used to unload, transport and stack reels.
- Reel storage – stability of stacks, access issues.
- Reel handling equipment – hoists, conveyors and the reel stands feeding the press.
- Wrappings and waste – removal of wrappings, storage and disposal of waste paper etc.
- Noise and dust – from the operation of the reel stands and nearby press.

A sample assessment record for **reel handling equipment** is provided below.

In conducting this risk assessment:

- The requirements of the *Provision and Use of Work Equipment Regulations 1998* (*PUWER*), the *Lifting Operation and Lifting Equipment Regulations 1998* and the *Manual Handling Operations Regulations 1992* must be complied with.
- Reference may need to be made to BS 5304:1988 'Safety of machinery' or to other standards for conveyors or specialist handling equipment.
- Particular attention should be paid to:
  - guarding standards and the possible presence of unguarded dangerous parts;
  - any need for manual handling of the reels;
  - training issues relating to the above;
  - statutory examination records (for the hoist).

Newtown News

Risk Assessment		
Reference number: B3	Risk topic/issue: **Reel handling and stands – reel handling equipment**	Sheet 1 of 1
Cross references: Risk assessments B1 (Fork-lift trucks), B2 (Reel storage), B5 (Noise and dust)		
Risks identified	**Precautions in place**	**Recommended improvements**
The equipment below presents risks to all staff working in the area.		
Reel hoist – Carries reel down from the reels store to the reel stand basement. It is fed by fork-lift trucks and feeds onto the roller conveyor system.	Slow moving hoist protected by a substantial mesh guard.	
	No need for access within the hoist enclosure.	
	Sign states 'Do not ride on hoist'.	Replace this sign by one complying with the *Safety Signs Regulations*.
	Gap between base of hoist and roller conveyor (no shear trap).	
	Statutory examinations by Insurance Engineers (kept by Work Engineer).	
Roller conveyor – This is in a T formation and consists of powered and free running rollers with a turntable at the junction. Reels are transferred to holding bays or floor-based trolleys by fork-lift truck.	All drives for the conveyor are fully enclosed and there are no in-running nips. The conveyor is protected by kerbs from fork-lift damage.	
	Signs prohibit climbing on the conveyor system.	Replace these signs as above.
Floor-based trolleys – This system carries reels right up to the transfer carriages which load them onto the reel stands.	Reels can be easily moved by one person using the trolley system. Reels can be safely rolled across the floor onto the trolleys if the correct technique is used.	Include formal training on correct techniques in the induction programme for new staff in the area.
Some reels must be transferred manually from the holding bays onto the trolleys.	Two persons are required to move the transfer carriages up to the reel stands.	
Reel stands – The reel stands rotate mechanically ensuring a constant web feed to the press.	There are no accessible in-running nips. Operatives do not need to approach the rotating reels.	
The rotating drive shaft for the reel stands is approximately 4 metres above the ground level.	This shaft is considered to be 'safe by position' for normal operating purposes.	Ensure a safe system of work for maintenance work near the shaft, e.g. by using a permit to work procedure.
Signature(s) J Young, K Old	**Names(s)** J Young, K Old	**Date** 26/3/00
Dates for	**Recommendation follow up June 2000**	**Next routine review March 2002**

Safe Systems of Work

Introduction

S3001 This chapter examines the need for safe systems of work ('SSWs') within the framework of an occupational safety and health management system ('OSHMS').

In essence SSWs outline the control measures needed to avoid or reduce risk in the workplace and are usually one of the main results of a suitable and sufficient risk assessment.

This chapter considers:

- the components of a safe system of work – both legal and others (S3002–S3006);
- appropriate safe system of work required for level of risk involved (S3007);
- development of SSWs, including job safety analysis; preparation of job safety instruction and safe operating procedures (S3008–S3010);
- permit to work systems (S3011–S3022);
- isolation procedures (S3023).

This particular chapter will be of use to guide those managers and supervisors who have the responsibility to develop and utilise SSWs, including permit to work systems, within their areas of control. It will also be useful to occupational safety and health ('OSH') practitioners who have to advise line management on practical development and implementation of such systems.

Essentially, a safe system of work is written down, and chronologically lists the methods of doing a particular job in such a way so as to avoid or minimise risk within the workplace. The safe system of work is therefore a very important workplace precaution or control measure which should be known, understood, adhered to and rigorously enforced within the workplace.

All employers should therefore utilise suitable and sufficient safe systems of work commensurate with the degree of risk highlighted in a risk assessment, especially when the legal requirement for SSWs are taken into account (see S3003 and S3004).

The pitfalls of not utilising SSWs within the risk control system include:

- an increase in the number of accidents/injuries/diseases;
- the increased risk of prosecution;
- increased employers' liability insurance premiums;
- a more dangerous workplace.

Remember – SAFE SYSTEMS OF WORK MEAN FEWER ACCIDENTS!

Components of a safe system of work

S3002 The components of a safe system of work comprise of legal elements – common law (S3003) and statute law (S3004) – and other components which explain where SSWs fit into the overall framework of an occupational safety and health management system ('OSHMS') (S3006). Indeed SSWs are considered to be the cornerstone of all successful OSHMSs.

Legal components – common law

S3003 The need for SSWs can be traced back to 1905 when Lord McLaren in *Bett v Dalmey Oil Co (1905)7F (Ct of Sess) 787* judged that:

> 'the obligation is threefold: the provision of a competent staff of men, adequate material, and a proper system and effective supervision.'

In 1938, the common law duty of care was highlighted in the case of *Wilsons and Clyde Coal Co Ltd v English [1938 AC 57]* where Lord Wright said that:

> 'the whole course of authority consistently recognises a duty which rests on the employer, and which is personal to the employer, to take reasonable care for the safety of his workmen, whether the employer be an individual, a firm or a company, and whether or not the employer takes any share in the conduct of the operations.'

This duty of care implies the need for SSWs and also that delegation to an employee does not remove the employer's duty to provide a safe system of work.

The question of what constitutes a safe system of work was further considered by the Court of Appeal in 1943 by Lord Greene, Master of the Rolls, in the case of *Speed v Thomas Swift Co Ltd [1943] 1 All ER 539*, who said:

> 'I do not venture to suggest a definition of what is meant by a system. But it includes, or may include according to circumstances, such matters as the physical layout of the job, the setting of the stage, the sequence in which the work is to be carried out, the provision of warnings and notices, and the issue of special instructions.'

He also said that: 'a system may be adequate for the whole course of the job, or it may have to be modified or improved to meet circumstances which arise: such modifications or improvements appear to me equally to fall under the heading of system.'

He added that: 'the safety of a system must be considered in relation to the particular circumstances of each particular job.'

The conclusion here is that the safe system of work must be tailored to the job to which it applies by co-ordinating the work, planning the arrangements and layout, stating the agreed method of work, providing the appropriate equipment, and giving employees the necessary information, instruction, training and supervision. All the above are designed to prevent harm to persons at work.

Legal components – statute law

S3004 The above common law precedents (at S3003) were taken into account during the drafting of the *Health and Safety at Work etc. Act 1974* ('*HSWA 1974*').

Section 2(2)(a) of *HSWA 1974* requires 'the provision and maintenance of plant *and systems of work* that are, so far as is reasonably practicable, safe and without risks to health'.

Furthermore, *section 2(3)* of *HSWA 1974* requires employers having five or more employees to prepare a *written* health and safety policy statement, together with the organisation and *arrangements* for carrying it out.

These 'written arrangements' should include a safe system of work.

This general duty has been tightened up and more clearly defined since the enactment of the *Management of Health and Safety at Work Regulations1999* (*SI 1999 No 3242*) ('*MHSWR*').

Specifically, *Regulation 3* of *MHSWR* (*SI 1999 No 3242*) requires 'suitable and sufficient' risk assessments. A safe system of work should be the end result of the vast majority of risk assessments.

Regulation 4 of *MHSWR* (*SI 1999 No 3242*) requires the 'Principles of Protection' to be applied in connection with the introduction of 'any preventive and protective measures'. These measures are control measures and are 'arrangements' within the written health and safety policy. These principles of protection are listed in *Schedule 1* to the Regulations and should be referred to when developing SSWs.

Further clarification of 'arrangements' is given in *Regulation 5* of *MHSWR* (*SI 1999 No 3242*): 'arrangements for the effective planning, organisation, control, monitoring and review of the preventive and protective measures'. This infers that as SSWs are part of the preventive and protective (i.e. control) measures, then they should be planned, monitored and reviewed.

Summary of legal components of SSWs

S3005 From the common and statute law references above (S3004), it may be seen that the key elements of SSWs are:

- adequate plant and equipment:
  - the right tools for the job;
  - plant designed for safe access and isolation.
- competent staff:
  - properly trained and experienced in the work they do;
  - instructed clearly in the work to be done;
  - provided with the necessary information on substances, safe use of equipment etc.
- proper supervision:
  - to monitor the work as it progresses;
  - to ensure the safe system of work is adhered to.

Other components

S3006 The bulk of the SSWs component parts are derived from the common and statute law considerations above.

The safe system of work may be therefore seen as the centre piece of a jigsaw which links together all the other component parts including:

- competent staff;
- premises;

- safe design;
- plant safeguards;
- safe installation;
- safe access/egress;
- tools and equipment;
- instruction;
- supervision;
- training;
- information;
- safety rules;
- planned maintenance;
- monitoring;
- PPE;
- environment.

Hence SSWs need to take account of all of the above components, and they need to be in writing.

The relative importance of each of the components will be task-dependent. Some tasks will be heavily dependent on the intrinsic safety built into the plant at the design stage. Other jobs will rely on the skills and expertise of the competent employee(s).

It may be of use therefore to consider the SSWs components under the sub-headings: hardware and software.

The 'hardware' includes: design, installation, premises/plant, tools and equipment, and working environment – heating, lighting, ventilation, noise control.

The 'software' comprises: planned maintenance, competent employees, adequate supervision, safety rules, training and information, correct use of PPE, and correct use of tools and equipment.

Which type of safe system of work is appropriate for the level of risk?

S3007 Different jobs will require different systems, depending on the levels of risk highlighted via the risk assessment process.

A very low risk job may require adherence to safety rules or a previously agreed guide, whereas a very high risk job may well require a formal written permit to work system. Both types qualify as safe systems of work.

The following matrix may prove useful:

Risk level	**Type of SSW**
Very high	Permit to Work
High	Permit (or written SSW)
Medium	Written SSW
Low	Written SSW
Very low	Verbal (with written back-up, such as safety rules)

Once the SSWs are in place, there is a need to ensure that they are reviewed at least annually, so as to ensure:

- continued legislative compliance (N.B. any new legislation);
- continued compliance with most recent risk assessment;
- SSWs still work in practice;
- any plant modifications are incorporated;
- substituted (safer) substances are allowed for;
- new work methods are incorporated;
- advances in technology are exploited;
- control measures are improved in the light of accident experience; and
- continued involvement in, and awareness of the importance of, all relevant SSWs.

It is vital that regular feedback is needed to all concerned following any updates of existing SSWs.

In essence, SSWs are one of the most important foundations of any OSHMS. They need to be:

- known, i.e. brought to the attention of all relevant employees and third parties;
- well understood by all concerned;
- adhered to by all concerned; and
- enforced by management and supervision.

Development of safe systems of work

S3008 The steps in the development of SSWs of all types are as follows:

- Hazard identification.
- Risk assessment.
- Define safety operating procedures/methods ('SOP's).
- Implement agreed system.
- Monitor system performance.
- Review system (at least annually).
- Provide feedback on system updates.

Job safety analysis ('JSA')

S3009 JSA is also known as job hazard analysis and/or task analysis. It is a technique that has evolved from the work study techniques of method study and work measurement which, in turn, formed part of the scientific management approach in the early twentieth century.

The work study engineer analysed jobs in order to improve methods of production by eliminating unnecessary job steps or changing those steps that were inefficient.

From an occupational safety and health ('OSH') viewpoint, JSA assists in the elimination of hazardous steps, thus eliminating or avoiding risky operations.

In essence, JSA is a useful risk assessment tool as it identifies hazards, describes risks and assists in the development of commensurate control measures.

The original work study approach made use of the 'SREDIM' principle:

Select	=	work to be studied
Record	=	how work is done
Examine	=	the total job
Develop	=	the best method for doing the work
Install	=	the preferred method into the company's way of working
Maintain	=	the agreed method via training, supervision etc.

Work study is used to break the total job down into its component parts and, by measuring (work measurement/time and motion) the quantity of work done in each of the component parts, the total job time could be established. If any of the component parts could be modified to make them more efficient or productive usually by saving time these improvements would be made as part of the overall study.

From measuring a number of similar tasks undertaken by different employees, standard times for each job component – and hence the total job-were derived. These were subsequently used in the development of payment and bonus systems.

JSA uses the SREDIM principle, but measures/assesses the risk content, as opposed to the work content, in each of the component parts of the job. From this detailed breakdown a safe system of work or safe operating procedure can then be compiled.

The basic JSA procedure is as follows:

1. Select the job to be analysed (Select).
2. Break the job down into its component parts in an orderly and chronological sequence of job steps (Record).
3. Critically observe and examine each job step/component part to determine whether there is any risk involved (Examine).
4. Develop control measures to eliminate or reduce the risk (Develop).
5. Formulate written SSWs and job safety instructions ('JSIs') for the job (Install).
6. Review SSWs at regular intervals to ensure continued utilisation (Maintain)

Use may be made of the following three-column layout in order to record the JSA:

Sequence of job steps	Risk factors	Controls advised

It is important to work vertically down the columns in their entirety, rather than to list one or two job steps and then work horizontally across the rows. Experience has shown that the former approach is much less time-consuming!

Once the job to be analysed has been selected, the next stage is to break the job down into its component parts or job steps. On average there should be approximately ten to twenty job steps. If there are more than twenty, too large a job for analysis has probably been chosen and one should therefore split it into two separate jobs. If there are less than ten job steps, then combine two or more jobs together.

The job steps should follow an ordered and chronological approach. Consider the following job – *changing a car wheel*:

The scenario is that you have come out of work and gone to your car where you left it on the works car park. It has a flat tyre. You are not allowed to call for assistance but you have to replace the wheel yourself! The sequence of job steps will be as follows:

1. Check handbrake on and check wheels.
2. Remove spare tyre from boot, check tyre pressure.
3. Remove hub cap/wheel trim.
4. Ensure jack is suitable and located on solid ground.
5. Ensure jacking point is sound.
6. Jack car up part way, but not so wheels leave ground.
7. Loosen wheel nuts.
8. Jack car up fully.
9. Remove nuts, then wheel.
10. Fit spare.
11. Replace nuts and tighten up.
12. Lower car.
13. Remove jack and store in boot with replaced wheel.
14. Re-tighten wheel nuts.
15. Replace trim/hub cap.
16. Ensure wheel is secure before driving off.

Once column 1 has been agreed and completed, then commence listing the associated risk factors by completing column 2. Thereafter, assign commensurate control measures in column 3.

The complete JSA should look like this:

Sequence of job steps	*Risk factors*	*Controls advised*
1. Check handbrake, check wheels	Strain to wrist/arm	Avoid snatching, rapid movements
2. Remove spare tyre, check tyre pressure	Strain to back	Use kinetic handling techniques
3. Remove hub cap/trim	Strain, abrasion to hand	Ensure correct lever used
4. Ensure jack on solid ground	Vehicle slipping, jack sinking into ground	Check jack and ground

5. Ensure jacking point is sound	Vehicle collapse	Place spare wheel under car floor as secondary support
6. Jack car up part way, wheels still on ground	Strain, bumping hands on car/jack	Avoid snatching, rapid movements. Use gloves
7. Loosen wheel nuts	Hands slipping-bruised knuckles, strain	Ensure spanner/brace in good order; avoid snatching, rapid movements. Use gloves
8. Jack car up fully	Strain, bumping hands on car/jack	Avoid snatching, rapid movements
9. Remove wheel	Strain to back, may drop onto foot	Use kinetic handling techniques. Use gloves to improve grip
10. Fit spare	Strain to back	Use kinetic handling techniques
11.Tighten nuts	Hands slipping-bruised knuckles, strain	Use gloves, avoid snatching, rapid movements
12.Lower car	Strain, bumping hands on jack/car	Avoid snatching, rapid movements
13.Remove wheel and store with jack in boot	Strain to back	Use kinetic handling techniques
14. Re-tighten nuts	Hands slipping-bruised knuckles	Use gloves, avoid snatching rapid movements
15. Replace hub cap	Abrasion to hand	Use gloves
16. Ensure wheel secure	Vehicle collapse, loose wheel	Check wheel and area around car before driving off.

The above example serves to illustrate what a typical JSA should look like. Any job/task within the workplace should be able to be recorded in this way and, ideally, displayed in close proximity to where the job is being undertaken.

The ultimate aim must be to undertake JSA on all jobs within the workplace and to communicate the findings and resultant control measures to all concerned. However, to assist in prioritising which jobs to analyse initially, the following criteria should be taken into account:

- past accident/loss experience;
- worst possible outcome/maximum potential loss ('MPL');
- probability of recurrence;
- specific legal requirements;
- newness of the job;
- number of employees at risk.

JSA may be activity-based or job-based.

Activity-based JSAs may cover:

- all work carried out above two metres;
- all driving activities:
  - internal (works transport);
  - external (driving on public highways);
- loading and unloading of vehicles.

Job-based JSAs may cover:

- the activities of a maintenance engineer repairing a roof;
- the activities of a chemical process worker obtaining samples for analysis.

The third column 'Controls advised' is in essence the job safety instructions ('JSI') and forms the basis of the written safe system of work.

Preparation of job safety instruction and safe operating procedures

S3010 The purpose of job safety instructions ('JSIs') and/or safe operating procedures ('SOPs') is to communicate the safe system of work to all concerned – supervisors, employees and contractors.

For each job step (column 1 in table in S3009) there should be a corresponding control action (column 3) designed to reduce or eliminate the risk factor (column 2) associated with the job step. The chronological ordering of the control measures listed in column 3 becomes the JSI for that particular job.

Such JSIs should be utilised in OSH training and communication, both formal (classroom style) and informal (on the job contact sessions).

All managers and supervisors should be fully knowledgeable and aware of all JSIs and SSWs that are operational within their areas of control and they should ensure that they are adhered to at all times.

JSIs may be:

- verbal (for very low risk tasks) – a word in the ear from a supervisor reminding the employee of the do's and don'ts;
- written (for low, medium and some high risk tasks) – as part of an agreed safe system of work;
- written and incorporated into safe operating procedures (usually for medium and some high risk tasks); or
- as part of an all singing, all dancing formal permit to work system (usually for 'high' or 'very high' risk tasks).

From a practical viewpoint, to aid communication and use of JSIs and SSWs, they should be listed on encapsulated cards and posted in the areas where the jobs are to be carried out. They should also be individually issued to all relevant employees, who should receive training in their application and use. All such training and communication – whether formal or informal should be recorded on individual training records.

The use of the three-column format to communicate JSIs and SSWs is well proven and is an excellent method of revitalising what in most organisations has become a rather moribund set of documents. As stated above, all jobs/tasks should be able to be fitted onto one or two sheets of A4 and use of the technique is a good way to regenerate interest in SSWs via a review.

Most organisations only review SSWs when things have gone wrong, e.g. as part of an accident investigation. This is negative and a form of reactive monitoring.

It is much better to be proactive and to review SSWs at least annually. This can be achieved by setting targets for managers and supervisors to review one safe system of work per month, together with relevant safety representatives, employees and/or contractors. Any findings or changes to the existing SSWs must rapidly be fed back to all concerned, so as to ensure continual improvement in OSH performance.

Permit to work systems

S3011 'A permit to work system (PTW) is a formal written system of work used to control certain types of work that are potentially hazardous. A permit to work is a document which specifies the work to be done and the precautions to be taken. Permits to work form an essential part of a safe system of work for many maintenance activities. They allow work to start only after safe procedures have been defined and they provide a clear record that all foreseeable hazards have been considered. A permit is needed when maintenance work can only be carried out if normal safeguards are dropped or when new hazards are introduced by the work. Examples are: entry into vessels, hot work, and pipeline breaking.' (HSE, 1997)

When may PTW systems be required?

S3012 A PTW system may be required in the following instances:

- where the risk assessment indicates 'very high' or 'high' risk;
- entry into confined spaces;
- working at heights;
- working over water;
- asbestos removal;
- high voltage electrical work;
- complex maintenance work – usually involving mechanical/electrical/chemical isolation;
- demolition work;
- work in environments which present considerable health hazards, such as:
  - radiation work;
  - thermal stress (work in hot or cold environments);
  - toxic dusts, gases, vapours;
  - oxygen enrichment or deficiency;
  - flammable atmospheres;
- lone working;
- work on or near overhead travelling cranes (7 metre rule);
- work involving contractors/third parties on site.

Essential features of PTW systems

S3013 HSE Guidance (1997) states that employers should incorporate the following essential features into bespoke PTW systems:

- provision of information;
- selection and training;
- competency-use of competent persons;
- description of work to be undertaken;
- hazards and precautions;
- procedures.

Provision of information in PTW systems

S3014 The PTW system should clearly state how the system works in practice, the job(s) it is to be used for, the responsibilities and training of those involved, and how to check that it is working effectively.

There must be clear identification in the system of those persons who may authorise particular jobs and also a note as to the limits of their competency.

There should also be a clear identification of who is responsible and competent for specifying the necessary precautions, e.g. isolation, use of RPE/PPE, emergency arrangements etc.

The PTW system paperwork should be clear, unambiguous and not misleading.

The system should be so designed that it can be used in unusual circumstances, especially to cover contractors.

Selection and training for PTW systems

S3015 Those persons authorised to issue permits must be sufficiently knowledgeable concerning the hazards and precautions associated with the plant/equipment/substances and the proposed work. They should have sufficient imagination and experience to ask enough 'what if' questions to enable them to identify all potential hazards. Lateral thinking is required here, especially if more than one trade is involved; or when own employees are working alongside contractors.

Ideally formal training should be required for all:

- permit issuers/authorisers;
- permit receivers/users;
- contractors' supervision;
- line managers/supervisors.

All staff and contractors should fully understand the importance and assurance of safety associated with PTW systems and should be 'competent persons' within the framework of the system.

What is a 'competent person'?

S3016 Recent legislation, codes of practice and guidance refer to a competent person in connection with permit to work systems in particular, and the OSHMS in general.

Regulation 3 of the *MHSWR* (*SI 1999 No 3242*) requires employers to undertake suitable and sufficient risk assessments and *Regulation 7* of the same Regulations requires employers to appoint one or more competent persons to assist them in undertaking such assessments.

MHSWR '99 defines 'competent' as follows: 'a person shall be regarded as competent for the purposes of these Regulations where he has sufficient training and experience or knowledge and other qualities to enable him properly to assist in undertaking the measures referred to in these Regulations.'

This is especially important in connection with the authorisation and receipt of permit to work documentation. Specifically, permit authorisers and receivers need:

- a knowledge and understanding of:
  - the work involved;
  - the principles of risk assessments;
  - the practical operation of commensurate control measures;
  - current OSH best practice.
- the ability to:
  - identify OSH problems;
  - implement prompt solutions;
  - evaluate the PTW system effectiveness;
  - promote and communicate the workings of the PTWS to all involved/concerned.

PTW system: work description

S3017 The permit documentation should clearly identify the work to be done, together with the associated hazards. Plans and line diagrams may be used to assist in the description, location and limitations of the work to be done.

All plant and equipment should be clearly tagged/identified/numbered, so as to assist permit issuers and users in selecting and/or isolating the correct piece of kit. All isolation switchgear should be similarly numbered or tagged. An asset register listing all discrete numbering should be maintained.

The permit should also include reference to a detailed method statement for the more complicated tasks. This should form part of the overall PTW system.

Hazards and precautions

S3018 The PTW system should require the removal of hazards – risk avoidance being the best strategy. Where this cannot be achieved then effective control measures designed to minimise the risk should be incorporated.

Any control measures should ensure legislative compliance is achieved as an absolute minimum standard. Specific regulations that may apply include: *COSHH 2002* (*SI 2002 No 2677*), *Confined Spaces Regulations 1997* (*SI 1997 No 1713*), *Control of Asbestos at Work Regulations 2002* (*SI 2002 No 2675*). Permit authorisers and users should be aware of all relevant legislation.

The permit should state those precautions that have already been taken before work commences, as well as those that are needed whilst work is in progress. This may include what physical, chemical, mechanical and electrical isolations are required and how these may be achieved and tested. Also, the need for different types of PPE and/or RPE should be specified.

Reference should also be made to any residual hazards that might remain, together with any hazards that might be introduced during the work – e.g. welding fume, vapour from cleaning solvents.

PTW system procedures

S3019 The permit should contain clear rules about how the work should be controlled or abandoned in the case of an emergency.

The permit should have a hand-back procedure which incorporates statements that the maintenance work has finished and that the plant has been made safe prior to being returned to production staff.

Time limitations and shift changeovers should be built into the PTW system.

There should be clear procedures to be followed if the work has to be suspended for any reason.

It is vitally important that there is a method for cross-referencing when two or more jobs (at least one of which is subject to a PTW system) may adversely affect each other.

A copy of the permit must be displayed as close as possible to where the work is being undertaken.

All jobs subject to a PTW system should be checked at regular intervals during the duration of the job, so as to ensure that the system is being followed, is still relevant and is working properly.

General application of PTW systems

S3020 In order for the PTW system to be applied to risk reduction in the workplace there needs to be a well understood communication system in place that everyone concerned is aware of and becomes involved in.

Most organisations make use of PTW system forms which have been designed to take account of individual site conditions and requirements. In some cases there will be general permits; in other cases special permits will be required for specific jobs, e.g. hot work, entry into confined spaces. This enables sufficient emphasis to be given to the particular hazards present and the precautions advised.

Outline of a PTW system

S3021 There are many and varied PTW system documents available in the OSH literature. The key headings/subject areas on PTW system documentation should include the following as a minimum requirement:

- permit title;
- permit number;
- date and time of issue;
- reference to other permits;
- job location;
- plant/equipment identification;

- description of work to be done;
- hazard identification;
- precautions necessary – before, during and after job/work;
- protective equipment;
- authorisation;
- acceptance;
- shift handover procedures;
- hand back;
- permit cancellation – work satisfactorily and safety completed.

Operating principles

S3022 In operating a PTW system the following principles should be observed:

- Title of permit.
- Permit number (including reference to other relevant permits or isolation certificates).
- Precise, detailed and accurate information concerning:
  - job location;
  - plant identification;
- Clear description of work to be done and its limitations.
- Hazard identification – including residual hazards and hazards introduced by the work.
- Specification of plant already made safe, together with an outline of what precautions have been taken, e.g. isolation.
- Specification of what further precautions are necessary prior to the commencing of work–e.g. rpe/ppe
- Specification of what control measures should be in place for the duration of the work.
- Clear statement of when the PTW system comes into effect, and for how long it remains in effect. A re-issue or extension should take place if the work is not completed within the allocated time. Generally, the PTW system should only be valid for the time the authoriser is on site.
- The permit document should be recognised as the master instruction which, unless it is cancelled, overrides all other instructions.
- The authorised/competent person issuing the permit must assure himself/herself – and the persons undertaking the work – that all the precautions/controls specified as necessary to make the job, plant and working environment safe and healthy have in fact been taken. This inevitably should involve an inspection of the job location by the authoriser. The authoriser's signature on the permit document gives this assurance of safety to the people doing the work.
- The person who accepts the permit becomes responsible for ensuring that all specified safety and health precautions continue in force and that only permitted work is undertaken within the location specified on the permit. At

this time, confirmation that all permit information has been communicated to all concerned should be obtained. A signature of the accepting person should also be on the permit document.

- A copy of the permit should be clearly displayed in the work area.
- If the permit needs to be extended beyond the initially agreed duration, e.g. because of shift changeover, then signatures of oncoming authorised/competent persons – together with that of the original authoriser – should be added to the appropriate section of the permit to confirm that checks have been made – again involving a visual walk-round inspection to ensure that the plant/equipment remains safe to be worked on. The new acceptor and all involved are then made fully aware of all hazards and associated precautions. A new expiry time is agreed. This should be built into relevant shift hand-over procedures.
- The hand-back procedure should involve the acceptor signing that the work has been satisfactorily completed. The authoriser of the permit also signs off certifying that the work is complete and the plant/equipment is ready for testing and recommissioning.
- The permit is then cancelled, certifying that the work has been tested and the plant/equipment has been satisfactorily completed. The copy permit should also be removed from the work location once the work has been completed.
- It is advisable to keep copies of each PTW system for a period of five years as these constitute maintenance records.

Isolation procedures

S3023 SSWs and PTW systems frequently require the effective isolation of plant and associated services, including:

- electrical power;
- pneumatic/hydraulic pressure;
- mechanical power;
- steam;
- chemical lines.

The principles of effective isolation require four steps to be taken:

1. Switch off — at the appropriate operating control panels. It is a good idea to post warning signs on the panel to alert other workers that it is switched off.

2. Isolate — by creating a physical break or gap between the source of energy and the part to be worked on. This can be achieved by operating an electrical isolator, removing a section of pipe or blanking off with a plate. N.B. Removing pipe sections is preferable to blanking off.

   Reliance on control valves alone has been a frequent cause of accidents and is not considered to be effective isolation. From a mechanical isolation viewpoint, drive belts and linkages should be physically removed.

3. Secure	the means of isolation by locking off. Multiple hasp padlock devices (e.g. Isolok) can be used when several people are working on the same piece of kit. Each puts his own padlock on the isolator so that the power cannot be reinstated until all personal padlocks are removed. It is therefore imperative to have a suite of different keys/padlocks.
4. Check	the isolation is effective by trying all available operating controls.

Many maintenance procedures will justify either a written safe systems of work or a formal PTW system; both can have the isolation procedure incorporated into the documentation.

Further reading

S3024 The following publications are all available from HSE Books.

- Permit to work systems, INDG98 (rev) (1997).
- Safe work in confined spaces: Confined Spaces Regulations 1997: Approved Code of Practice, Regulations and guidance, L101 (1997).
- Safe work in confined spaces, INDG258 (1999).
- Management of health and safety at work: Management of Health and Safety at Work Regulations 1999: Approved Code of Practice and guidance, L21 (rev) (2000).
- Emergency isolation of process plant in the chemical industry, CHIS 2 (1999).

Statements of Health and Safety Policy

Introduction

S7001 Employers have a statutory duty under *s 2(3)* of the *Health and Safety at Work etc. Act 1974* (*HSWA*) to prepare a written statement of their health and safety policy, including the organisation and arrangements for carrying it out, and to bring the statement to the attention of their employees. Employers with less than five employees are excepted from this requirement.

This section sets out the legal requirements for a policy, explains the essential ingredients that a policy should contain and provides checklists to assist in the preparation of policies.

The legal requirements

S7002 *Section 2(3)* of the *HSWA 1974* states:

> 'Except in such cases as may be prescribed, it shall be the duty of every employer to prepare and as often as may be appropriate revise a written statement of his general policy with respect to the health and safety at work of his employees and the organisation and arrangements for the time being in force for carrying out that policy, and to bring the statement and any revision of it to the notice of his employees.'

The *Employers' Health and Safety Policy Statements* (*Exception*) *Regulations 1975* (*SI 1975 No 1584*) except from these provisions any employer who carries on an undertaking in which for the time being he employs less than five employees. In this respect regard is to be given only to employees present on the premises at the same time *(Osborne v Bill Taylor of Huyton Ltd [1982] IRLR 17)*.

Directors, managers and company secretaries can be personally liable for a failure to prepare or to implement a health and safety policy. (see further *Armour v Skeen* at E15040 ENFORCEMENT).

Content of the policy statement

S7003 The policy statement should have three main parts:

The Statement of Intent

> This should involve a statement of the organisation's overall commitment to good standards of health and safety and usually includes a reference to compliance with relevant legislation. Whilst *s 2(3)* of the *HSWA 1974* only requires the statement to relate to employees, many organisations also make reference to others who may be affected by their activities e.g. contractors, clients, members of the public (This may be of particular relevance if the

policy is being reviewed by other parties such as clients). In order to demonstrate that there is commitment at a high level, the statement should preferably be signed by the chairman, chief executive or someone in a similar position of seniority.

Organisation

It is vitally important that the responsibilities for putting the good intentions into practice are clearly identified. This may be relatively simple in a small organisation but larger employers are likely to need to identify the responsibilities held by those at different levels in the management structure as well as those for staff in more specialised roles e.g. health and safety officers. In all cases the competent person who is to assist in complying with health and safety requirements should be identified. *[Management of Health and Safety at Work Regulations 1999* (*SI 1999 No 3242*), *Reg 7*].

Arrangements

The practical arrangements for implementing the policy (e.g. emergency procedures, consultation mechanisms, health and safety inspections) should be identified in this section. It may not be practicable to detail all of the arrangements in the policy document itself but the policy should identify where they can be found e.g. in a separate health and safety manual or within procedure systems.

More detailed guidance on what should be included in each of these parts is contained later together with guidance on two other important issues:

- Communication of the policy to employees.
- Revision of the policy.

The statement of intent

S7004 The statement should emphasise the organisation's commitment to health and safety. The detailed content must reflect the philosophical approach within the organisation and also the context of its work activities. Typically it might include:

- a commitment to achieving high standards of health and safety in respect of its employees;
- a similar commitment to others involved or affected by its activities (possibly specifying them e.g. contractors, visitors, clients, tenants, members of the public);
- a recognition of its legal obligations under the *HSWA 1974* and related legislation and a commitment to complying with these obligations,
- references to specific obligations e.g. to provide safe and healthy working conditions, equipment and systems of work;
- reference to the importance of health and safety as a management objective;
- a commitment to consultation with its employees.

Most employers keep the statement of intent relatively short – less than a single sheet of paper. This allows it to be prominently displayed (e.g. in reception areas or on noticeboards), or to be inserted easily into other documents such as employee handbooks.

Example:

The Midshires Development Agency (MDA) is committed to achieving high standards of health and safety not only in respect of its own employees but also in relation to tenants of the Agency's property, contractors working on the Agency's property or projects, visitors and members of the community who may be affected by the Agency's activities.

The Agency is well aware of its obligations under the Health and Safety at Work Act and related legislation and is fully committed to meeting those obligations. The successful management of health and safety is a key management objective.

MDA supports the concept of consultation with its staff on health and safety matters and has established a Health and Safety Committee to provide a forum for such consultation. This policy will be distributed to all staff and will be reviewed on an annual basis.

A Champion

Chief Executive

August 2000

Organisation

S7005 The statement of intent is of little value unless the organisation for implementing these good intentions is clearly established. This usually involves identifying the responsibilities of people at different levels in the management organisation and also those with specific functional roles e.g. the health and safety officer. How these responsibilities are allocated will depend upon the structure and size of the organisation. The responsibilities of the managing director of a major company will be quite different from those of a managing director in a small business. The latter will quite often have responsibilities held by a supervisor in a larger organisation e.g. ensuring staff comply with PPE requirements.

Most organisations prefer to identify the responsibilities through job titles rather than by name as this avoids the need to revise the document when individuals change jobs or leave. Responsibilities might be allocated to levels in the organisation e.g.

- Managing Director/Chief Executive
- Senior Managers
- Junior Managers
- Supervisors/Foremen/Chargehands
- Employees

and also to those with specialist roles e.g.

- Health and Safety Officer
- Occupational Hygienist
- Occupational Health staff
- Personnel staff
- Maintenance Engineer

Where health and safety advice is provided from outside (e.g. a consultant or a specialist at a different location), this should be stated and the means of contacting this person for advice should be identified.

The following examples are provided as an illustration of how responsibilities might be allocated in a medium sized organisation. The allocation will undoubtedly be different for businesses of greater or lesser size.

Examples:

- *Managing Director*

  The Managing Director has overall responsibility for health and safety and in particular for:

  — ensuring that adequate resources are available to implement the health and safety policy;

  — ensuring health and safety performance is regularly reviewed at board level;

  — monitoring the effectiveness of the health and safety policy;

  — reviewing the policy annually.

- *Works Manager*

  The Works Manager is primarily responsible for the effective management of health and safety within the factory. In particular this includes;

  — delegating specific health and safety responsibilities to others;

  — monitoring their effectiveness in carrying out those responsibilities;

  — ensuring the company has access to adequate competent health and safety advice;

  — ensuring that safe systems of work are established within the factory;

  — ensuring that premises and equipment are adequately maintained;

  — ensuring that risk assessments are carried out;

  — ensuring that adequate health and safety training is provided;

  — chairing the health and safety committee.

- *Supervisors*

  Each Supervisor is responsible for the effective management of health and safety within his or her own area or function. In particular this includes:

  — ensuring that safe systems of work are implemented;

  — enforcing PPE requirements;

  — ensuring that staff are adequately trained for the tasks they perform;

  — monitoring premises and work equipment, reporting faults where necessary;

  — identifying and reporting health and safety related problems and issues;

  — identifying training needs;

  — investigating and reporting on accidents and incidents;

— participating in the risk assessment programme;

— setting a good example on health and safety matters.

- *All employees*

  All employees have a legal obligation to take reasonable care for their own health and safety and for that of others who may be affected by their actions e.g. colleagues, contractors, visitors, delivery staff. Employees are responsible for:

— complying with company procedures and health and safety rules;

— complying with PPE requirements;

— behaving in a responsible manner;

— identifying and reporting defects and other health and safety concerns;

— reporting accidents and near miss incidents to their supervisor;

— suggesting improvements to procedures or systems of work;

— co-operating with the company on health and safety matters.

- *Health and Safety Officer*

  The Health and Safety Officer is responsible for co-ordinating many health and safety activities and for acting as the primary source of health and safety advice within the company. These responsibilities specifically include:

— co-ordinating the company's risk assessment programme;

— acting as secretary to the Health and Safety Committee;

— administering the accident investigation and reporting procedure;

— liaising with the HSE, the company's insurers and other external bodies;

— submitting reports as required by RIDDOR;

— co-ordinating the health and safety inspection programme;

— identifying health and safety training needs;

— providing or sourcing health and safety training;

— providing health and safety induction training to new staff;

— identifying the implications of changes in legislation or HSE guidance;

— preparing and submitting progress reports on an annual health and safety action programme;

— sourcing additional specialist health and safety assistance when necessary.

Arrangements

S7006 The duty to provide information on the arrangements for carrying out the health and safety policy overlaps with the duty imposed by *Regs 3* and *5* of the *Management of Health and Safety at Work Regulations* (*SI 1999 No 3242*) to record the results of risk assessments and of arrangements 'for the effective planning, organisation, control, monitoring and review of the preventive and protective measures'. Rather than providing all of the detail on arrangements within the health and safety policy many employers prefer to provide information on where these details can be found, for example in:

- risk assessment records;
- health and safety manuals or handbooks;
- formal health and safety procedures (e.g. within an ISO 9000 controlled system).

This has the benefit of keeping the policy document itself relatively short – an aid in the effective communication of the policy to employees. Some key aspects of health and safety arrangements which should be included or referred to in this section of the policy are those for:

- conducting and recording risk assessments;
- specialised risk assessments e.g. COSHH, Noise, Manual Handling, DSE workstations;
- assessing and controlling risks to young persons and new or expectant mothers;
- establishing PPE standards and the provision of PPE;
- operational procedures (where these relate to health and safety);
- permit to work systems;
- routine training e.g. induction, operator training;
- specialised training e.g. first aiders, fork lift drivers;
- health screening e.g. pre-employment medicals, regular monitoring, DSE user eye tests;
- statutory inspections;
- consultation with employees e.g. health and safety committees, staff meetings, shift briefings;
- health and safety inspections;
- auditing health and safety arrangements;
- investigation and reporting of accidents and incidents;
- selection and management of contractors;
- control of visitors;
- fire prevention, control and evacuation;
- dealing with other emergencies e.g. bomb threats, chemical leaks;
- first-aid;
- maintaining and repairing premises and work equipment.

Communication of the policy

S7007 Employers must bring the policy (and any revision of it) to the notice of their employees. The most important time for doing this is when new employees join the organisation and communication of the policy should be an integral part of any induction programme. Young people in particular need to be reminded of their personal responsibilities at this stage, as well as any restrictions on what they can and can't do. Even experienced workers may need to be made aware that they are joining an organisation that takes health and safety seriously, as well as knowing what the PPE rules are and where they can obtain the PPE from.

Many employers provide their employees with a personal copy of the policy whilst others include it in a general employee handbook or display it prominently e.g. on noticeboards. In all cases communication will be aided by keeping the policy itself relatively brief and providing the detail (particularly in respect of 'arrangements') elsewhere.

Review

S7008 The policy must be revised 'as often as may be appropriate' but the need for revision can only be identified through a review. In their free guidance leaflet for small firms on preparing a health and safety policy document (IND(G) 324 *'Stating your business'*) the HSE recommend an annual review and this should not be too onerous even for small employers, for whom it should be a fairly simple affair.

In larger organisations the policy could be subjected to an annual review by incorporating it within an ISO 9000 document control system or making the review an annual task of the health and safety committee. Alternatively the review could be made the responsibility of an individual (e.g. the Managing Director or the Health and Safety Officer), either formally through specifying this within the organisational responsibilities or informally through a simple diary entry.

The review may well conclude that there is no need for change but it may identify the need for revision such as:

- A new Managing Director should make his personal commitment to the statement of intent by signing and re-issuing it.
- Changes in the management structure necessitate a reallocation of responsibilities for health and safety.
- New 'arrangements' for health and safety have been established (or existing ones have been altered) and the policy needs to be amended to match this.

A policy which is clearly well out of date does not reflect well on the organisation's commitment to health and safety, nor on the effectiveness of its health and safety management.

Stress and Violence in the Workplace

Introduction

S11001 Stress is becoming an increasing concern for employers and employees alike. More seems to be demanded by organisations in less time. At work, life is becoming more competitive, with more people chasing the same goals.

People are very complex and the way in which they react to events within their lives varies considerably. Sometimes, a single event may cause a crisis in someone's life. How they respond will depend upon a number of factors, including other circumstances that may exist within their life, their own reaction to the event and the length of time it takes to deal with the event. There are people who, when stressed, will make sure that everyone around them knows it. Others will become withdrawn, and internalise their stress. Physically retreating from stressful situations, such as by taking frequent smoking breaks, is one way that some people cope. Others will obviously be angry, they care little about anyone except themselves, and their tension is obvious. This may result in behaviour that appears to be aggressive or threatening. Indeed, in some cases the stress may build to an extent that someone does resort to expressing their feelings by using violent behaviour. Conversely, for others, stress will be a result of them being exposed to violent or threatening situations. Examples where this may occur include: care workers, retailers, police and other emergency support workers, and call centre staff.

The Health and Safety Executive (HSE) estimates that up to 13.14 million days are lost each year due to stress at work. This costs society between £3.7 billion and £3.8 billion every year (based on 1995/96 prices), and the problem is growing. Research commissioned by the HSE has indicated that about half a million people experienced work-related stress at a level they believe was making them ill, and up to 5 million people in the UK feel 'very' or 'extremely' stressed by their work. Occupational stress is one of the key target areas that the HSE is concentrating upon improving over the next few years.

Definition of stress

S11002 Taken in its simplest form, the Oxford English dictionary defines stress as 'demand upon physical or mental energy'. The HSE's definition of stress is given as 'the adverse reaction people have to excessive pressures or other types of demand placed upon them'. This has been seen by some as a rather negative view, as it implies that stress produces only adverse effects. The HSE recognises that pressure in itself is not necessarily bad and some people even thrive on it. However, heath problems can arise when the pressure being experienced is seen as excessive by the individual, and the pressure is ongoing.

Stress often occurs when the pressures placed upon a person exceed their capacity to cope. This can be negative or positive. For some people, the pressure of a tight deadline will 'positively stress' them, with the end result being that it enhances their performance. This may be due to their fear of failure or the consequences of not meeting the deadline. This is particularly true of people working in competitive sport and those people who set themselves challenges e.g. to increase sales figures by 50 per cent in the next 12 months. In these cases, the pressures placed upon them have a positive effect in that it produces drive, challenge and excitement.

Understanding the stress response

S11003 People's ability to respond to difficult or stressful situations has evolved over many million years. The mind recognises a particular need to respond to an external source (both physical and mental). The alert message is sent to the brain which then initiates dramatic changes in the hypothalamus. This is the control centre which integrates our reflex actions and coordinates the different activities within our bodies. When the hypothalamus is activated, the muscles, the brain lungs and the heart are given priority over any other bodily activity. The chemical adrenaline is produced and the message of a threat is passed on to other parts of the body which prepares itself for vigorous physical action – known as the fight or flight instinct. When the threat has abated, the parasympathetic nervous system takes over again and the body returns to a state of equilibrium, (known as homeostasis) where all our functions are in balance. This is demonstrated in the stress response curve.

Figure 1: Stress Response Curve

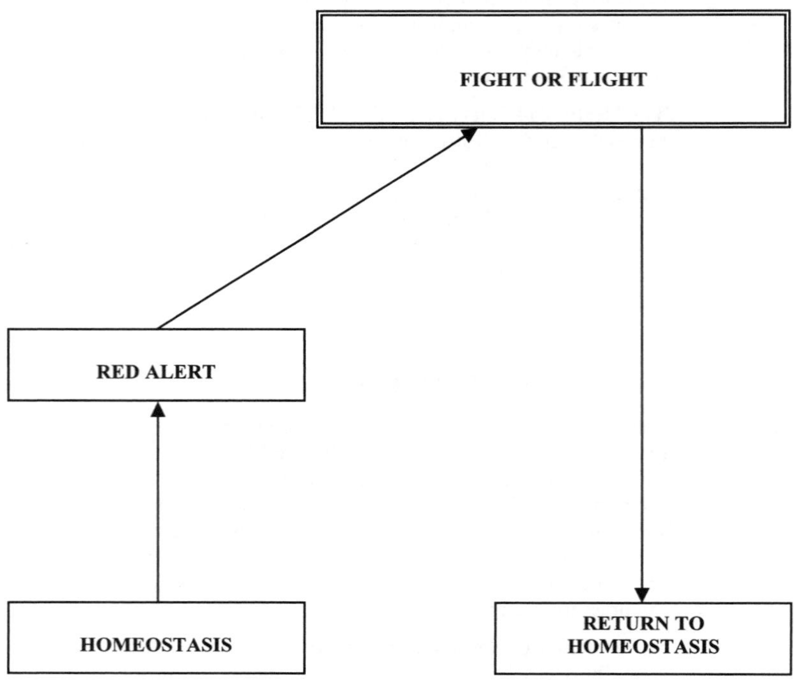

To summarise, given below are the physical responses to the 'fight or flight' mechanism.

- brain goes on red alert and stimulates hormonal changes including the production of adrenalin and noradrenaline (stress hormones).
- muscles tense, ready for action.
- eye pupils dilate to enable the danger to be seen more clearly.
- the heart beat increases to provide more blood to the muscles, therefore blood pressure increases.
- breathing increases to provide extra oxygen for the tensed muscles.
- liver releases glucose to provide extra energy for the muscles.
- digestive system shut down, making the mouth go dry and the sphincters close.
- sweating occurs in anticipation of expending extra energy.
- immune system slows down.

These reactions are highly complex, but demonstrate the increased demands that are placed upon the body when feelings of stress occur. It is easy to see, therefore, how this response can have a detrimental effect upon well being and health if the stress is prolonged.

Why manage stress?

S11004 The following paragraphs explain the importance of managing stress.

Legal issues

S11005 There is no legislation that relates specifically to stress but the following Regulations touch upon the issue.

Health and Safety at Work etc Act 1974

S11006 Under the *Health and Safety at Work etc Act 1974, s 2(1)* employers have a general duty to ensure, so far as is reasonably practicable, the health, safety and welfare of employees at work. This implies that measures should be in place to ensure that employees do not suffer stress-related illness as a result of their work.

Management of Health and Safety at Work Regulations 1999

S11007 Under the *Management of Health and Safety at Work Regulations 1999* (*SI 1999 No 3242*), *Reg 3* employers must also take into consideration the risk of stress-related ill health or unsafe working practices that may be caused or exacerbated by stress. Stress needs to be considered when undertaking general risk assessments (e.g. a risk assessment of the work environment, or of a particular task) as people's responses may differ when they are feeling the symptoms of stress, or overload. Some organisations may decide to undertake a specific assessment of occupational stress within their workplace, i.e. factors associated with the work that are known to affect the way people feel and respond.

Working Time Regulations 1998

S11008 The *Working Time Regulations 1998* (*SI 1998 No 1833*) seek to reduce the effects on employees who are subjected to working long hours, or who have insufficient rest periods during and between work shifts (these issues are known to be occupational stressors). The Regulations are detailed, and cover minimum rest times during a shift (20 minutes every six hours), minimum rest time between shifts (11 hours) and maximum number of hours that can be worked within a weekly period (48 hours). The maximum number of hours may be calculated over a 17-week reference period.

There are a number of exclusions that apply to certain categories of personnel under the Regulations, and these are under review in the European Courts at present.

Detailed requirements apply to employers to maintain a record of hours worked for each employee who is covered by this legislation.

For further information see working time.

HSE guidance

S11009 The HSE has issued two guidance booklets on managing stress, one of which is targeted at employers and the other at employees. The booklet '*Tackling Work-related stress: A manager's guide*' provides practical advice to managers on areas such as:

- Developing a business case.
- Gaining workers and trade unions to buy in.
- Identifying and targeting the key stress issues within the organisation.
- Reviewing progress and success.

The other booklet is targeted at employees. '*Tackling work-related stress: A guide for employees*' seeks to provide information to employees on issues such as:

- What stress is (and its definition).
- Symptoms experienced when suffering from stress.
- What to do if an employee feels they are suffering from stress.
- Methods of managing stress.

Employers cannot be prosecuted for not following HSE guidance, but failure to follow guidance is often seen as persuasive in the lower courts.

On-going research into stress

S11010 The HSE has been working with partners to develop standards of good management practice that employers can use to measure their performance in tackling a range of key occupational stressors. In 2002 an evaluation of scientific evidence to support the need for standards of control, demand and support for occupational stress occurred. Following this, the first draft of a standard has been prepared, and this has been updated again to make it simpler to understand and apply. The Management Standards are being developed with partners who are piloting them between March and December 2003, and organisations who wish to try the pilot study may obtain the Standards and supporting material via the HSE's website.

Criminal and civil liability

S11011 Whilst there is potential for prosecution in criminal law for failure to meet the above duties, the likelihood is that compliance will be driven by civil litigation. To date, there have been no prosecutions under criminal law for failure to manage stress at work. Enforcement of the general duty of care under health and safety legislation has been due to successful claims being made for compensation following allegations by an employee that they have suffered illness or ill-health as a result of being exposed to unacceptable levels of stress at work.

Indeed, there have been some large insurance settlements for employees who have successfully claimed for stress at work. The most notable was *Walker v Northumberland County Council [1995] IRLR 35*. In this case, Mr Walker, a senior social worker, had been employed by the council for 17 years. During the 1980s his workload had increased, and as a result, he suffered a nervous breakdown. Upon his return to work he told his employers that his workload must be reduced. This was not done, and Mr. Walker suffered another nervous breakdown which resulted in him being dismissed for permanent ill health. He claimed compensation from his employer, and was successful on the grounds that the employer had been aware that he was under extreme pressure at work. A key question in the case was whether Mr Walker's risk of mental illness was materially higher than that which would normally affect a senior social worker with a heavy workload.

As regards the second breakdown, the employer should have foreseen a risk that Mr Walker's career would end and they should have appreciated that Mr Walker was more vulnerable to damage than he had been before his first breakdown. Because of this, it was deemed reasonable for Mr Walker to have received assistance, but in not providing such assistance, the council was in breach of their common law duty of care.

In February 2002, a landmark decision by the Court of Appeal in *Hatton v Sutherland and Others [2002] 2 All ER 1* overturned three out of four stress claims and quashed the awards for damages initially given to the workers in the High Court. The Appeal Court ruling stated that for a compensation claim to succeed it must first be 'reasonably foreseeable' to the employer that the employee concerned would suffer a psychiatric injury as a result of stress in the workplace. When deciding whether a risk is 'foreseeable' the following needs to be taken into account:

- nature and extent of the work undertaken by the employee (e.g. is the workload more than is normal for that particular job?).
- signs from the employee themselves of impending harm to their health as a result of work-related stress (e.g. has an employee already suffered harm to their health from work-related stress?).

The Appeal Court ruling further stated that employers are entitled to assume that an employee can withstand the normal pressures of a job, except where they are aware of a particular problem or vulnerability (e.g. the employee has warned the employer that they are suffering from work-related stress). In addition, employers are entitled to take at face value what they have been told by their employees about the effects of work-related stress on their mental health. For example, if an employee has had sickness absence due to stress but does not inform their employer of the cause of the absence, then it would be reasonable for the employer to expect that the worker would be fit to return to their normal work.

The Court of Appeal set out a number of factors that must be taken into account by the courts once it has been established that it was reasonably foreseeable that an employee would suffer harm as a result of work-related stress. These include:

- an employer is only in breach of their duty of care if they fail to take reasonable steps to prevent the employee from suffering as a result of stress.
- indications of impending harm to health must be plain enough for any reasonable employer to realise.
- when considering what actions are reasonable for an employer to take to reduce risk of harm from stress, the size and resources of the employer's operation may be taken into account.
- it is not enough for a claimant to show they are suffering harm from a result of occupational stress – they must demonstrate that their 'injury' was caused by the employer's breach of duty of care by failing to take reasonable steps to protect the worker.
- an employer will not be deemed to be in breach of their duty of care if they allow a willing employee to continue with their job, despite them stating they have been suffering from stress, if the only reasonable alternative is to dismiss the employee.
- if an employer offers a confidential staff counselling service for those who experience stress at work with referral to treatment services, they will be unlikely to be found in breach of their duty of care.

The court ruling places more responsibility on the employee to let an employer know if they are suffering from symptoms of occupational stress, rather than leaving it totally to the employer to identify a 'stressed' employee. In practice, therefore, an employee is unlikely to have a valid claim unless they can demonstrate that they have told their employer that they are suffering from work-related stress. Note that this does not take away the employer's duty to be proactive and seek to identify stress risk factors in the workplace, and to implement suitable precautions to minimise these risks.

See also OCCUPATIONAL HEALTH AND DISEASES O1035.

Occupational stress risk factors

S11012 The following paragraphs discuss in detail the most common occupational stress risk factors.

The physical environment

S11013 The physical environment can be described as the general conditions in which people have to work. The work environment is often seen as a contributory cause for stress at work, i.e. something that compounds a problem that is being experienced at work. For others the work environment may be the main cause of stress amongst the workforce, especially if the conditions prevent work targets from being achieved, or detract workers from undertaking their work effectively.

Our general work conditions have improved drastically over the past century, but so have our expectations. The environments in which people work will vary considerably, and will be dependant upon the type of work that is being undertaken. For example, a die casting factory is traditionally a hot, dirty, and malodorous environment. Any investments are usually spent upon the equipment to produce the product rather than to enhance the work environment, and managers may be reluctant to spend a heavy outlay on a workplace that is likely to become dusty and dirty within a short time. These types of environment often have less time and effort spent on them in maintaining high standards of cleanliness, good housekeeping or ensuring that lighting remains in good order throughout the premises.

This can, however, have a knock-on effect for staff, both in their mental attitude towards the workplace, in their general conduct and in their general well being. It has been found, following several studies of the workplace, that an environment that is poorly maintained produces a negative attitude amongst the staff – 'if management don't care then why should we' type approach. If the workplace is allowed to become dirty, untidy or cramped, then workers will respond accordingly with the 'why should we tidy up – management don't seem to mind' type attitude. A workplace that is dull, untidy, badly designed or scruffy is likely to make employees feel negative about coming to work, and they are more likely to respond negatively to issues that arise at work.

The condition of the workplace

S11014 The condition of the workplace can be described as its general standard of maintenance and decoration, both internal and external. Housekeeping standards within the workplace could be included under this heading. To understand how a workplace condition can affect how a person feels and responds, consider the following examples.

- *Scenario 1*

  An old hospital building was first constructed during the early 1900s. The hospital management has struggled to keep up with the increase in usage over the years and the structural changes that have been necessary to accommodate more patients and more services. The hospital is crowded and cluttered. Government financial cutbacks have been a common setback and funding has focussed on provision of medical care, rather than on maintenance.

  The walls are grey and flooring is mainly grey concrete. It is cracked in many areas, causing ruts and trip hazards. Lighting is poor and several overhead tubes are not working. Paint on the wall is dirty, marked in places and chipped in others. Some of the windows are cracked.

  Doors are damaged due to the constant throughfare of hospital trolleys being pushed against them.

  Sacks of dirty laundry, and black sacks containing rubbish, seem to be constantly accumulating in corners awaiting collection.

- *Scenario 2*

  This scene depicts another hospital that was constructed in the early 1900s. Similar challenges have been faced within the hospital to provide increased services and a higher level of care. Budgets restraints also present a challenge.

  The walls are still grey but have been painted within the last few years. However, different parts of the hospital have received a splash of colour by way of painting the support pillars, door surrounds etc with bright contrasting colours. The doors are less damaged as they have metal plates fitted to reduce damage from the trolleys. Windows are cracked in some places, but have been cleaned, which has resulted in more light coming into the hospital.

  Rubbish bags and dirty laundry bags are out of sight. The floor looks cleaner and there are less pot holes and cracks in the flooring.

- *Scenario 3*

  This scene depicts a hospital that was also built in the early 1900s. However, despite the hospital facing the same challenges as the hospitals in the first two scenarios, the management team have adopted a much more proactive approach to general decoration. Staff have been encouraged to participate in choosing colour schemes when new services have been introduced and the patients' views have been taken into account when planning any refurbishment. This has resulted in local colleges being invited to practise their artistic skills in main areas at little or no cost to the hospital. Waiting rooms have brightly coloured murals along the walls. Corridors are coloured differently for each separate unit (e.g. cardiology, rehabilitation) and collages and other pictures have been displayed along the walls. These have resulted in interesting schemes for people to enjoy as they walk through the corridors.

  Some extra money has been made available each year for ongoing repairs which has resulted in a higher standard of maintenance. A proactive maintenance schedule is in place whereby all lighting tubes are replaced annually, thereby significantly minimising lighting failure. Staff have been encouraged to share their requirements or preferences for lighting in different areas, and for reporting to management when problems are occurring. Windows are kept clean.

  Wards are brightly coloured, and curtains have been fitted at windows which are pleasant to look at. Plants are situated around the hospital and a water feature has been installed in the main reception waiting area.

  Waste collection points have been clearly defined. Laundry collection/ delivery areas are clearly marked and waste is not allowed to accumulate.

An employee working in any one of these environments, doing work which is known to be emotionally and physically demanding, would obviously prefer to be working in an environment depicted in scenario 3.

People feel better and therefore respond more positively when their environments are pleasant. This positive factor is increased when staff feel that they have had some level of control over their own environment, even if it is something as basic as deciding on the best areas for storage or waste collection points.

The workplace layout

S11015 The workplace layout can make a significant impact upon a worker's efficiency, safety and general well being. This is where the study of ergonomics can have a major input into reducing stress levels at work. Ergonomics is a specialist area, which looks at the interaction of the worker, the work equipment and the work environment. The use of an expert may be needed at some stage in workplace layout, but the first step should always be to discuss employee requirements and to see how these can be accommodated within current restraints. See ERGONOMICS E17021.

Simple changes to a work layout can make a drastic improvement in a person's efficiency, safety and general well being, especially when the idea has emanated from the workforce itself. The worker is the expert in the particular task that is being undertaken, and whilst an employer does not want to provide a 'wish list' to the workforce, valuable information can be ascertained merely by finding out if and what causes frustration within a work layout.

An employer might consider asking its staff the following questions.

Question	Yes/ No	Comments
Are worksurfaces large enough for the work and equipment that are needed?		
Are worksurfaces of a suitable height (too high or too low) for the worker(s)?		
Is there sufficient space for the worker to get to, from and around their own work area?		
Are storage facilities in close proximity and easy to access?		
Is there sufficient storage?		
Is storage being used to full effect?		
Do workers have somewhere to store their own personal items?		
Are seats or suitable rest areas provided, at or near to the workstations?		
Are walkways wide enough to enable people and equipment to pass through (e.g. trolleys)?		
Are work areas in large areas clearly identifiable, signed and labelled?		
Is the layout designed to minimise environmental hazards (e.g. glare from windows, noise from nearby machines)?		

Poorly designed layouts can induce other health conditions such as musculo-skeletal disorders e.g. back problems from having to stoop, stretch or crouch to undertake a task. Stress creates physical tension which can contribute to these health conditions.

Staff facilities

S11016 Everyone occasionally feels a need to take a break from their work, and this may be more of a need than a desire in some work situations that are known to incur high levels of stress e.g. dealing with the public, working in the care profession. Having nowhere to go to take a break or let off steam may result in stress levels increasing to a point where a worker becomes distressed. This will be compounded when other aspects of the work are problematic, for example, the shift patterns are badly designed or working hours are excessive.

Staff facilities should include basic facilities such as sanitary conveniences and washing facilities, and somewhere from where drinks can be prepared or obtained. Employers should ensure that such facilities:

- accommodate the numbers of people using them.
- take into consideration the need for privacy (especially if workers need to change clothes or shower before or after work).
- are maintained to a good state of hygiene and cleanliness.
- provide basic equipment such as soap, drying equipment, toilet paper and sanitary bins.
- enable workers to obtain both hot and cold drinks, including provision to clean spillages or clean utensils if drinks are being prepared by the workers.
- provide a ready supply of hot/warm and cold water, and potable water for drinking purposes.

The *Workplace (Health, Safety and Welfare) Regulations 1992* (*SI 1992 No 3004*) and supporting guidance contain detailed minimum standards for sanitary conveniences, washing facilities, drinking water facilities for changing clothes.

These Regulations also contain provisions for rest facilities, and require that somewhere that is suitable and sufficient is made available for people to eat meals where meals are regularly eaten in the workplace. Resting and eating facilities must ensure that people who do not smoke are protected from the effects of tobacco smoke. Therefore the onus of protection is for the non-smoker rather than the rights of the smoker. Implementation of these requirements will vary considerably, according to the company's activities and the resources available. Separate smoking and non-smoking facilities would be ideal but are not practicable in many workplaces. Some workplaces offer rest rooms for smokers, whilst others require that smokers take their 'smoke break' outside of the workplace.

This legislation also requires the provision of rest facilities for workers who are pregnant, or who are nursing mothers. Some employers have expressed concern at the practicalities of meeting this requirement. However, the female workforce is increasing as more mothers are returning to work earlier and employers have a legal as well as moral duty to ensure that these provisions are available. These types of issue do have an impact upon the stress levels of employees.

For further information see WORKPLACES – HEALTH, SAFETY AND WELFARE W11001.

Environmental conditions

S11017 When considering stress induced by environmental conditions, an employer should consider issues such as lighting, heating, ventilation, humidity and space.

In some cases, the environmental conditions may be the sole or primary cause of stress, in others it may be a factor that exacerbates other stressful conditions within the workplace. Poor environmental conditions are the cause of other workplace issues such as sick building syndrome.

Several regulations contain specific duties for provision of environmental standards. These include the *Workplace (Health, Safety and Welfare) Regulations 1992* (*SI 1992 No 3004*), the *Schedule* to the *Health and Safety (Display Screen Equipment) Regulations 1992* (*SI 1992 No 2792*) and the *Provision and Use of Workplace Regulations 1998* (*SI 1998 No 2306*).

An employer should take the following into account when deciding on the standards that should be achieved in the workplace.

- legal requirements contained in various statutes.

- official guidance issued by the HSE, trade associations or other recognised bodies such as the Chartered Institute of Building Service Engineers.
- the nature of the work (work requiring intricate detail will need a higher level of lighting than one undertaking general tasks).
- the amount of physical effort in a task (a cooler background temperature will be needed where workers exert physical effort).
- the need to combine workers' needs with other stakeholders (e.g. customers, patients).
- the level of humidity in the environment (work in a humid environment will reduce a worker's ability to control their own internal temperature by sweating, thus increasing the risk of heat exhaustion).
- the amount of space for the work components, the work equipment, and the worker's need for movement.

There are several sources of guidance which will help an employer to interpret conditions that will be deemed as 'reasonable' to meet statutory requirements for a work environment. A good starting point, other than a specific trade association, is the Chartered Institute of Building Service Engineers. This institute can provide detailed guidance on recommended environmental standards, and should be referred to when deciding optimum work conditions for different types of work activity.

Employers should take into account, however, that any such guidance should not be used in isolation. The most effective way of reducing physical and emotional stress from work environments is to involve the staff. It is not possible to please everyone, but by obtaining a general consensus an employer has an opportunity of ascertaining if any problem exists, where the problem might be, and the extent that the environment may be cause or adding to a problem. Where such involvement has occurred, it will be necessary to ensure that staff do receive feedback on the decisions made, including reasons why suggestions made by staff have not been adopted. People are likely to be more accepting of a negative decision when they can understand the thought process behind it.

Another important key to reducing stress induced by environmental conditions is to provide as much control as possible over the local environment. For example, it is usual in large open workspaces for the lights to be controlled by only a few switches. The lighting circuits could be changed so that a smaller area is controlled by each switch, enabling more control over an employee's immediate working vicinity. The strength and colour of the tubes could also be altered to further meet individual preferences.

The work

S11018 There are many aspects of work that may induce pressure. Pressure at work is not necessarily bad as it helps to generate enthusiasm and can produce energy. However, prolonged pressure is likely to create a state of nervous tension as the person has no respite from the constant rush of adrenaline that occurs when stress is induced. Unless this is relieved, the end result can lead to health problems.

The trick is to try to achieve a balance between providing a challenge to staff, and overloading them.

The following areas are known to be occupational stressors, and should be reviewed when any stress management strategy is being considered.

Targets deadlines and standards

S11019 Targets are necessary at work. This is because an employee who is not given a target will set their own target which may not be in line with organisational expectations. However, problems arise when the standard that has been set by the employer exceeds the individual's ability to meet it.

When a deadline or target is unachievable, a sense of helplessness may be experienced by the person. Similar emotions may be experienced when each time the target is met the employer increases the next target. Whilst for some employees this will represent a sense of excitement, there will inevitably be a time when an employee begins to feel defeated, undermined or useless. This problem becomes more serious when there is no mechanism for the employee to let the employer know how they are feeling, or when there is no system for monitoring staff behaviour and performance once the target has been set. If the employee feels that their job depends upon meeting the target then the feeling of pressure is increased, as the consequences of failure are more significant to that individual.

There are various reasons why a deadline or target may be unachievable. Managers should consider the following questions, and perhaps agree with the employee what is reasonable when all the restraints have been taken into consideration. This will help the employee to perceive a sense of involvement and control over the amount of pressure being placed upon them.

Assessment of targets			
Question	**Yes/ No**	**Comments**	**Agreed action**
Is the employee sufficiently trained/experienced to meet the standard?			
Are there adequate tools/equipment to meet the deadline/target?			
What other demands are being made which may affect the standard/deadline being met?			
Is there any mechanism for the employee to report early if he/she feels they are not likely to meet the standard?			
To what extent can machine failure etc affect the target being met?			
Is the time frame for the amount of work being set reasonable (i.e. is this being achieved reasonably by others)?			

What are the consequences of meeting or failing the target?			
Are there any other factors that the employee feels may detrimentally affect the target/deadline being achieved?			

People who have experienced continual pressure from unrealistic deadlines or targets often explain their feelings with comments such as not being able to see the light at the end of the tunnel and being on a treadmill that they cannot get off. When these feelings become excessive, a common response is for the employee to take time off for sickness absence, or even to leave the company. Early warning signs that an employee may be feeling excessive pressure may include erratic behaviour or mood swings, reduced or erratic work performance, looking worn out or tired or even compulsive behaviour (increased smoking, drinking or eating). Some people, of course, will respond with the opposite effects, i.e. by losing their appetite, becoming very quiet or seeming to become particularly conscientious.

Time demands

S11020 Stress from conflicting demands on time may be a sign of 'role conflict' or 'role ambiguity', i.e. which task is more important. It may be that the time available is insufficient to deal with the extent of work being required. If time demands being made upon the employer are such that everything is expected to be undertaken at the same time, then the worker will suffer.

Different people have different ways of handling their work. Some people prefer to complete one task before giving any attention to another, whilst others appear to thrive on dealing with numerous matters all at the same time. Problems arise when the manager operates to one of these principles whilst the subordinates preferred style is totally different. The end results can be disastrous, with neither party feeling that a job is being completed satisfactorily and both parties feeling that they are being misunderstood. Two-way dialogue between management and employees is needed to ensure that each party understands the other's needs, and to agree parameters for working.

A commonly reported cause of stress is knowing that several other tasks are building up whilst only one task is being addressed. The employee will either rush through all the work, to try to reduce the mountain that is piling, or select one task that they feel is the most important at the expense of others which are perceived to be less important. Neither response is likely to result in staff feeling satisfied that they are coping and doing a good job. A key to managing this type of problem is to:

- review work loads.
- take into account deadlines.
- agree priorities for each task that is building up.
- ensure the most suitable type of person is allocated (some people deal better with a high throughput that requires low levels of accuracy whilst others prefer work that require less output but a higher level of attention).
- encourage and develop self-organisation for the individual to manage their work load.

Working hours and shift patterns

S11021 Management of working hours is a key feature in reducing stress levels at work. When considering working hours, the employer should take into consideration:

- the length of the shift (how many hours worked each day).
- the shift pattern, especially if rotating shifts are used.
- rest times between shifts.
- rest and break times during the shift.

As mentioned previously (see S11008) the *Working Time Regulations 1998* (*SI 1998 No 1833*) contain requirements on the number of hours that should be worked each week, the minimum rest time between shifts and the minimum rest breaks during shifts. Regardless of this whether an employee is excluded from the Regulations or chooses to 'opt out', the employer has a duty of care to ensure that health, safety and welfare are not detrimentally affected by the work.

Problems with working hours seem to crystalize when the time spent at work impinges on employees' home and social activities. Someone who habitually works long hours often feels too tired to enjoy their private time. An employee will be prepared to 'trade off' a well-designed shift pattern in order meet social demands being made at home. It is clear, therefore, that a balance must be struck between shift patterns that have been deemed to be the most effective, against the social needs of the workers.

The most frequently reported source of stress among shift workers who work night shifts is the interruption in their sleep, or difficulty in being able to sleep during the day. Possible sources of interruptions are numerous and include doorbells, telephones, increased traffic noise, family, pets and neighbours. Workers who operate to rotating shifts are more likely to feel the effects of sleep disturbance.

The need for variety during night shifts in terms of the range of tasks, and the mental and physical output necessary, is more important for night workers. The attention span will reduce, typically between 3.00 a.m. – 4.00 a.m. and this will be exacerbated if the work is monotonous, requiring high levels of concentration or is physically demanding. Fatigue is unavoidable during night work, but will be reduced if the interest of the shift worker is maximised.

Travel during work

S11022 Many employees enjoy travelling as part of their work as it provides a sense of variety, change in environment and a level of autonomy about their work pattern (e.g. by not feeling that they are being watched). Problems arise when the amount of time travelling exceeds the amount of time that can be spent in the workplace, or impinges upon the time available for the remainder of work that needs to be completed. Travelling is also exhausting, both physically and mentally, and can reduce a person's work performance. Other problems associated with travelling include feelings of isolation from colleagues, family and friends, especially when extended stays away from home occur.

Apart from drivers of heavy goods or public services vehicles, there are no maximum times set in legislation that a worker may travel on the road without taking a break. Therefore an employer should undertake a risk assessment, and agree with the workers what travel arrangements are considered reasonable. Factors that need to be taken into account include:

- preferred method of travel.

- manual handling issues of the traveller (cases, files etc).
- the length of the journey (including times for delays).
- how much time the employee has been given to reach their destination (i.e. work time versus private travelling).
- travel routes (especially if using the road).
- the need for regular two-way communication with the worker whilst away from the workplace.
- suitability of overnight accommodation, including location of accommodation.

Organisations such as ROSPA (Royal Society for the Prevention of Accidents) have produced detailed guidance and literature detailing results of their research into occupational road risk, and this may be a useful source of reference when considering the physical and psychological risks associated with travelling.

Technology and work equipment

S11023 There is nothing more frustrating that having a mismatch between the tools necessary to complete a task and the task itself. If the problem is simply being given the wrong tool to complete the task, then this can quickly be rectified (provided the employer is informed). However, stress levels increase when equipment constantly fails to meet its design criteria, breaks down due to overload or poor maintenance, or when there is a skill's shortage amongst the workers.

To overcome the problem the employer should undertake a task analysis to identify what tools and equipment are needed, and to ensure that they are provided. Ongoing maintenance may be needed, including an emergency call-out service where the job is dependant upon the work being completed within a certain time. This needs to be followed by a training needs' analysis, to identify if and what type of training is required to ensure workers can use the equipment effectively and safely.

Most people are instinctively sociable, and respond to both verbal and non-verbal behaviour and communication whenever they meet. Someone making a light-hearted comment can lighten a dull or pressurised meeting, a smile will indicate whether an ambiguous comment is meant as a joke. This type of behaviour is missing from computerised communication i.e. e-mail or even telephone. Messages received without the additional benefit of non-verbal communication may come across as blunt, aggressive or hostile and may make the recipient feel threatened when the opposite was the intention. Stress will result from the perception of aggression or threat, especially if that person feels helpless to respond or defend themselves. This needs to be taken into account when transmitting difficult messages, to avoid inducing unnecessary anxiety. In extreme cases, e-mail communication has been known to cause severe disruption to workplace harmony and cause individuals to become insecure or even sick.

Role conflict or ambiguity

S11024 People can work much more effectively if they know what they are supposed to do, how it is supposed to be undertaken, why it needs to be undertaken in a particular way and how their role fits into the corporate objective.

When job roles conflict or are ambiguous the end result is confusion and frustration for everyone involved. If this becomes a frequent event then those feelings of frustration may increase into anger and resentment. This may be compounded when the person's level of authority is limited and they have no access or authority to change what is causing the problem. The following scenarios may be indicative of conflicting or ambiguous roles.

- *Scenario 1*

  A secretary is told by her manager to manage the diary and to keep all the clients satisfied when they ring. Several potential clients have telephoned to book meetings which the secretary has entered into the book, but her boss is very displeased when he finds out and tells her to cancel them. Nearly all other bookings made into the diary by the secretary get changed or altered, causing the secretary additional work and upset. When challenged by the manager about why all these bookings have been made, the secretary defends herself by stating that she is doing what she was told to do. The boss responds by saying that nothing should be booked without checking with him first.

- *Scenario 2*

  A departmental manager has responsibility for reducing expenditure in his area of authority. However, he has also been given responsibility for managing health and safety. The health and safety standards in the department are poor. The manager identifies a need for health and safety training which will improve standards but this will take him over the agreed spending threshold which will directly affect profitability and his requirement to reduce expenditure.

Clerical/administration support

S11025 Many employees now undertake their own basic administration tasks, using basic computer software packages. This is acceptable provided they have the basic skills to use the equipment, and the time allocated within their work patterns to complete the administration. Common indicators that completion of paperwork is becoming problematic may include:

- workers regularly taking work home in the evenings/weekends.
- employees not using their holiday entitlement.
- increase in errors within written correspondence (typographical, grammar or content).
- tasks being half completed.
- paperwork piling up on workstations.
- messages not being answered.
- comments from workers about bureaucracy.

Employers should ensure that time is made available to enable workers to complete any necessary paperwork, and that the equipment is accessible. This is particularly relevant where the worker is based in a non-office environment, and would not automatically have access to a desk, paper, filing system etc readily to hand.

Most people do not like excessive paper-work, and some do not like any paper-work. Employers should therefore seek to streamline their systems so that administration is minimised, and simplified.

Cover during staff absence

S11026 Administration often becomes a stress issue when staff are constantly required to cover for colleagues. The additional pressures this causes are unlikely to be problematic if it is for only short duration, e.g. sickness or holiday. However, staff who feel that they are frequently undertaking additional work to cover for shortages will begin to feel the effects of the burden. Early symptoms to look for may include reduced efficiency, increased error rates, lower patience or attention thresh-hold, or increased sickness absence.

Managers should seek to avoid relying on employee goodwill as the sole method for dealing with the workload when a colleague is absent. Use of temporary staff should only be used as a stopgap. Often their effectiveness is limited as they are unfamiliar with the work and work systems. It also takes time to train a temporary worker, and if this becomes a regular exercise then the use of temporary workers will be seen as a hindrance rather than a help.

Employers should therefore adopt a more proactive stance, and consider what alternatives would be suitable to deal with staff shortages. Examples may include use of multi-skilled workers who could be moved around, or 'floating' staff who are employed specifically to cover during other staff absences.

Company structure and job organisation

S11027 The following paragraphs discuss the relevance of company structure and job organisation on stress levels.

Hierarchy of control

S11028 Ambiguity over job roles and the level of control over a task is a major stressor at work.

Problems occur when employees are given responsibility for a task but have no authority to implement necessary systems, training etc to effectively deal with the task. Colleagues and fellow team members will judge both the corporate reaction and the reaction of the particular manager to any decisions that are made within an organisation, and will build their own attitudes as a result when showing commitment back to the organisation. It is therefore very important that everyone knows what their level of responsibility is, and the authority that they have to deal with a given situation. When a challenge is presented at work the employee's physiological state will alter to deal with the event, (increased blood pressure, etc). Any nervousness or heightened state of alertness to deal with the situation will be better controlled by the individual if they know exactly how they are supposed to respond, and the parameters in which they can deal with the given situation.

The interface between two overlapping parts of an organisation can also be a cause of stress – each will have their own priorities and their own way of working. Good communication and clearly defined lines of authority are vital to identify the essential patterns of overlap and to enhance the organisation's efficiency.

To overcome such problems the employer should review where any overlap in processes or responsibilities may occur, and where any gaps or areas that have not been clearly defined may be present. These should be cross-checked against the recipients (i.e. employee's) perceptions and interpretations of the responsibilities to ensure that misunderstandings do not occur.

Organisational changes and arrangements

S11029 People generally do not like change. It upsets the person's comfort factor and invariably leads to a period of uncertainty whilst they are adapting to the changes. Some people adapt better to changes in their lives than others. Having an awareness that any change is likely to cause stress is the first step to managing the problem.

It is necessary to communicate as clearly as possible the nature of any changes, and the consequences of these to the workers who are affected. Even bad news can be dealt with positively if it is communicated honestly, openly and with sensitively.

Responsibility

S11030 Once a responsibility has been allocated to an individual, there should be sufficient scope allocated to enable the person to carry out the task. This does not mean that the employee is then left to deal with a given responsibility without any mechanism for support or assistance if the responsibility becomes more extensive or complex than first anticipated. Being given a responsibility without any access to support when needed will inevitably lead to stress.

The feeling of a lack of appreciation, through non-realisation of an individual's potential, often arises from being given insufficient responsibility. Many people, especially those in jobs which do not enjoy high status in their organisation, feel frustration through lack of challenge, or feel frustrated when they can see things going on around them when they do not have an opportunity to be involved in making a contribution. Giving people autonomy, i.e. increasing the degree of self-control over their work, will help to counteract this.

Developing a system for employee recognition and staff development will help to identify where an employee needs more or less responsibility within their work. This will also provide a mechanism for a manager to explain why more responsibility has not been given to an individual who feels they want or need it, and for managers and employees to work together to reach a level which is mutually agreeable.

Individual perception of company

S11031 If staff have a negative view of the company, they will respond accordingly. Employees who feel that the company is going nowhere, or does not care about its image, will not care about the company's future or its image. This will give them no sense of achievement or pride in their employment. The end result will be a higher staff turnover and the associated problems that go with this. Output will not be as effective as a motivated team of workers, and standards of work are likely to be lower.

Involving staff in the company's vision, or business plan will help employees to see what the company is hoping to achieve and the goals that have been set. Even if the employees do not fully understand or agree with the vision they will at least have an understanding of what is happening when changes begin to occur.

Perception of one's self-esteem and role within the company

S11032 Everyone needs to feel that they have a valid role to play, in their life and in their work; everyone needs to feel valued and needed. When this does not occur in the workplace, the result is an employee who feels worthless, undervalued and misunderstood. The problem will become compounded if the individual concerned is someone who is ambitious or is trying to achieve promotion or a more fulfilling role.

Previous experience

S11033 Someone who has experience of the environment will be more aware of the penalties and rewards associated with the work. They are likely to have been exposed to any associated work stressors before and may be more adept at managing them. However, this can sometimes have the opposite effect, in that if someone has been exposed to high levels of stress in a similar type job previously then this may well have reduced their tolerance to pressures.

Personal factors

S11034 Everyone responds differently to events that occur in their lives. This very much depends upon their basic personality traits, their upbringing and their experiences in life.

Various studies have been undertaken in behaviour patterns. The research undertaken by Friedman and Rosenman *Type 'A' behaviour and your heart* (1975) (ISBN 0704501589) gave some insight into why some people are more prone to stress-related disease. In these studies, two main risk behaviours were identified, which were called Type A for high risk, and Type B for low risk. In general, Type B behaviours are the opposite of Type A behaviours. These categories should only be considered as a general yardstick as there are no absolutes, and most people will fall somewhere between the two extremes identified.

Type A people were said to be:

- autonomous, dominant and self-confident.
- very self-driven, busy and self-disciplined.
- aggressive, hostile and impatient.
- ambitious and striving for upward social mobility.
- preoccupied with competitive activities, enjoying the challenge of responsibility.
- believing that their behaviour pattern is responsible for their own success.

Type A people were said to work longer hours and spend less time in relaxation and recreational activities. They received less sleep, communicated less with their partners and tended to derive less pleasure from socialising.

In contrast, type B people tend to:

- find it easier and are more effective at delegating.
- work well in a team.
- allow subordinates some flexibility in how a task is undertaken.
- avoid, or do not set, unrealistic targets.
- keep a sense of balance in the events that occur in their lives.
- be more accepting of a given situation.

Type B people are often more secure within themselves and more accepting of their strengths and limitations. A very simplistic summary of this type of character could be that they are simply more 'laid back'.

The level of responsibility and control over work

S11035 These are major factors as to whether someone is likely to suffer the effects of stress. Karasek and Theorell *Healthy Work* (1996), a comprehensive study of job demand and autonomy, showed that stress is a function of the combination of psychological demand and 'decision latitude', or autonomy.

The term 'executive stress' has been widely used, insinuating that perhaps people with high levels of responsibility are more likely to suffer from occupational stress. Research has shown, in fact, that this is not true. Whilst the level of responsibility increases with 'white collar' roles, so does the level of autonomy about how the work is to be carried out. Material rewards are usually higher, which helps to enhance self-esteem within the worker.

Karasek and Theorell's studies showed that the greatest risks are among those occupations with least autonomy and of a lower status. Examples include bus drivers, production line workers and care staff. These are classic examples of the conclusion reached by Karasek and Theorell that it 'is the bossed, and not the bosses' who are most likely to suffer from stress. They often have little or no control over how the work is to be undertaken but have to deal with the problems and difficulties that arise when a job is badly designed or managed. They also have less material reward for the work undertaken.

Qualifications and competencies

S11036 There is a rather cynical saying that people are promoted to their level of incompetence. Whilst this may be an exaggerated view there is some truth in the concept that people who are newly promoted may not immediately be competent. The same could apply to almost any role at work, if there is a mismatch between the task that is required and the skills that are possessed by the people undertaking the work. People need to grow and develop into their roles at work, to a point that they feel comfortable and competent in what they do without becoming bored or complacent. An employee who works beyond their competence is likely to cause problems for the organisation in that work standards may reduce and production may be affected. From the workers' point of view they are not likely to feel confident or able that they can achieve what is required, and this will have a knock-on effect upon their self-esteem. The end result is likely to be that such an employee would feel frustrated and dissatisfied, and maybe very unhappy. The manager is likely to feel equally frustrated and dissatisfied because the work is not being completed satisfactorily.

Personal issues

S11037 Separating personal problems from work issues is difficult for some people, and impossible for others. Both work and personal issues have an indirect effect upon each other and if an employee is experiencing a stressful situation in their home life then this may impact upon work performance.

An employer cannot be responsible for personal issues in an employee's life, but an awareness that a problem exists for the individual is useful knowledge for the employer. Systems should be developed that enable a manager to offer support and sympathy whilst the difficulty exists. For example, if a family member is in hospital perhaps working hours could be adapted slightly to fit in with the hospital's visiting times. The reaction to an individual who is going through a crisis, in terms of the level of support and flexibility shown by the manager, can stay with that person for a long time. Genuine help being offered by the manager can be an investment which promotes commitment from that person at a later stage.

A pre-agreed workload level may be acceptable in normal circumstances, but intolerable when a long term problem is occurring in someone's personal life. Therefore, the events going on in the employees' lives outside work will, at some stage, have an effect on their ability to cope with work. If these events are positive then the ability to cope will improve, with the opposite effect happening when the personal events are negative.

Managers need to take a balanced approach when responding to an individual's personal issues. An employee who is considered to have a good 'credit' balance in terms of their commitment and enthusiasm to the organisation may be treated more sympathetically by the organisation if a crisis develops in their lives. However, this could in itself have a significant knock-on effect to other employees who feel that they have not been treated so considerately, and this can induce feelings of being undervalued, misinterpreted or simply overlooked by those colleagues.

Stress management techniques

S11038 The following paragraphs provide detailed guidance and advice on managing stress.

Stress auditing/surveys

S11039 Stress audits or surveys are a good starting point, as this will give the employer an indication of whether a stress issue exists in the workplace, and the areas where this may be most problematic.

There are several approaches to audits and surveys, ranging from a simple checklist that employees are asked to complete, to detailed analyses with scoring systems that are undertaken by specialists. The type of audit that is used will depend upon the organisational culture and the extent to which an employer feels that a stress problem exists.

A survey in its simplest form may merely be a short questionnaire that asks employees to list their three most positive and three most negative aspects about their work. Guided headings may be included, perhaps with a tick box and a space for the employee to write comments explaining why it is seen as a positive/negative aspect of work. Headings that are used could include areas that the employer suspects are causing pressures upon the employees. This type of approach is often used as a pre-audit review, to enable employers to gain a feel as to any patterns which might be obvious to the workers.

Stress assessments are more detailed, and require planning, thought and sensitivity. They usually take the form of a guided list of questions or areas that the assessor wishes to cover about the work. They can include all members of staff, or selected members of staff from certain job titles to act as a representation. The interview needs to be undertaken by someone who is sufficiently competent to complete such a task. Competencies necessary to achieve this may include:

- counselling skills.
- polished interview techniques.
- knowledge of the company and its management structure.
- knowledge of procedures and practices.

- understanding of health and safety issues, including stressors at work.
- understanding knowledge of human behaviour (i.e. human factors at work).
- the ability to gain the confidence of the interviewee.
- the ability to listen.

> Points to remember:
>
> - Reassure staff that all comments will be confidential.
> - Respect the confidentiality of your staff.
> - Tell your staff what you are planning to do, what you hope to achieve, why you are doing it, and the benefits that may be in it for them.
> - Provide a mechanism by which anyone can discuss any queries or concerns about the audit process beforehand.
> - Involve all levels of employment, from senior board members to junior staff.
> - Consider issuing a pre-interview questionnaire which you can review before your interview (e.g. three best and three worst aspects of their job).
> - Design a simple framework around which you can develop your audit/survey format.
> - Ask simple, open questions.
> - Give people time to consider questions asked, or points that have been raised.
> - Allow plenty of time for the interview, and sufficient time afterwards to write up the key findings.
> - Tell your staff what you plan to do with any information collected.
> - Involve them as much as possible in any further decisions or developments arising from the survey.

After the audit:

- Inform staff and/or representatives about the key findings and what the company plans to do in response.
- Provide (where possible) an indication of time frames as to when and how these changes can expect to occur.
- Involve staff and/or representatives.
- Explain why any suggestions may not be taken forward by the company, don't just refuse.
- Follow up any changes you make to ensure they are having the effect you intend.
- Keep staff and representatives involved at all times during the design and implementation of any systems or procedures resulting from the audit.
- Lead by example – managers can communicate important signals about the importance of managing stress.

Stress management policy

S11040 A proactive management strategy for occupational stress should include development of a policy. This should highlight the company's philosophy and recognition of the issue, and clearly state the procedures in place to eliminate or manage stress associated with work.

The policy should include:

- the company's definition of occupational stress.
- recognition that stress is a work issue that will be taken seriously and sympathetically by management.
- acknowledgement that stress is not an illness or a weakness.
- explanation of the procedures the company will implement to identify occupational stressors and implement risk reduction control systems.
- details of personnel who have been allocated with responsibilities for implementing and monitoring risk reduction measures for stress.
- explanation of employees' responsibilities and procedures for reporting to management when they feel they are suffering from stress.
- training and support packages, including any counselling services that may be available.

On-going monitoring

S11041 Like any other health and safety system it is necessary to monitor systems and procedures to ensure they are effective. Monitoring will have the added benefit of helping to identify a stressed employee who is unwilling to report they have a problem. Monitoring can be undertaken in many ways including:

- monitoring during team meetings, by encouraging a general discussion about positive and negative aspects of work that have occurred.
- monitoring during appraisals and other one-to-one meetings with staff.
- implementing an anonymous stress report box in which people who are finding a work situation to be continuously problematic can report this without any fear of recrimination.
- circulating staff attitude or opinion surveys which will encourage feedback from workers.
- observation of workers' behaviour patterns and relationships with colleagues (looking for any obvious differences from the norm, or spotting relationships/atmospheres that appear to be tense).
- monitoring accident and incident statistics, and sickness absence records.

Information and training

S11042 Provision of information and training is a general duty under the Health and Safety at Work etc Act 1974, s 2(2)(c). This will include information and training in occupational stress. Information should be given to employees to enable them to understand the organisation's views on occupational stress, its policies and its procedures for dealing with stress and the part that employees can play in managing occupational stress for themselves.

Training can be used to underpin any findings from surveys or assessments that may have been undertaken, and to help the policy on stress to be understood and effectively implemented. Training should be provided to managers and to staff in general. Managers will need to know:

- What stress is and it contrasts with or effects occupational stress.
- Symptoms of stress and how to recognise someone who may be suffering from stress.
- Health risks associated with stress.
- Occupational factors which cause or contribute towards stress.
- Identifying, separating and managing personal and management stressors.
- Outline of company policy and procedures on stress.
- Identifying measures to identify stress.
- Understanding of measures to manage/reduce occupational stress.

Employees will need to understand for themselves exactly what stress is, and how to recognise for themselves when they or a colleague may be experiencing symptoms. The training will enable the organisation to demonstrate that it is a caring employer.

The Chartered Institute of Environmental Health (CIEH) has developed a one day accredited course called 'Stress Awareness'. This provides a detailed programme that covers the key issues associated with occupational stress. The fact that is accredited by a recognised body will help to increase its profile to employees who may be more cynical about the subject itself, or the organisation's commitment to employee welfare.

Violence and stress

S11043 The risk of a violent attack at work is a serious occupational hazard for many people, particularly for those whose jobs bring them into regular contact with the public.

What is violence?

S11044 Violence is defined by the HSE as 'any incident in which an employee is abused, threatened or assaulted by a member of the public in circumstances arising out of the course of his/her employment'.

This definition clearly includes verbal abuse as a violence issue, and this is probably the most common type of violence that most people at work experience. People will have different perceptions about the types of behaviours that they find threatening. For some it will merely cause annoyance, but others may become very distressed if exposed to verbal abuse or threatening behaviour.

The legal issue

S11045 The employer's duty to manage risks from violence at work is enshrined within the general duty of care under the *Health and Safety at Work etc Act 1974, s 2.* This requires employers to ensure, so far as is reasonably practicable, the health, safety and welfare at work of its employees. Threats of violence can affect a person's mental health and their general welfare, whereas physical violence will detriment a person's safety. Employers therefore need to ensure they have sufficient systems in place to reduce risks from violence at work. Any such systems will need to be supported by adequate information and training to enable the employees to understand the hazards, be familiar with the processes in place to reduce the risks, and know how to deal with a situation in which they may be exposed to a violent situation.

Under the *Management of Health and Safety at Work Regulations 1999* (*SI 1999 No 3242*), *Reg 3*, an employer must undertake general risk assessments. In practice, this means any general risks that may apply at work, and would include risks associated with actual or potential exposure to violence. The assessment should consider the physical, emotional and psychological effects on an employee who is exposed to a threatening situation at work. Some violent incidents cannot be predicted but many are foreseeable, especially where an employee has to deal with the public. Therefore employers have a responsibility to identify these and seek to prevent them. Where a real risk of violence exists, employers must:

- identify the hazards arising from the jobs that they ask employees to do.
- evaluate the risks (by looking at the frequency of exposure, duration of exposure, number of people likely to be affected and the possible effects or outcome).
- plan measures to remove hazards and reduce the risks.
- implement and monitor these measures.
- train and inform all workers who may be affected by exposure to violence.

A record must be kept of the assessment.

The HSE has produced a number of guidance documents that set out procedures to enable employers to manage the problem effectively, and provide a framework for investigating the risk of violence at work. These include: '*Preventing violence to staff*' (which involved both HSE and Tavistock Institute of Human Relations, and includes case studies); '*Violence at work: A guide for employers*' (IND(G)69L.

Recognising the problem

S11046 The first step is to ascertain if staff are being exposed to any type of violence, and then to ascertain the:

- type of violence (e.g. verbal abuse, physical threats, actual assault).
- frequency of exposure.
- duration that each incident occurred.
- any consequences as a result (sickness absence, injury resulting, police action etc).

To do this, the employer is likely to need to agree a definition of violence, and to spell this out in very simple and clear language. The information required above can be ascertained in many formats, including questionnaires (which could be anonymous), incident report books, e-mail reports, verbal feedback or during meetings and training courses.

Near-miss situations, where no one was actually hurt but the potential for harm to be there is as stressful as when someone is actually injured. The threat that this could recur and escalate will always be in the employee's mind.

Processing the information

S11047 Finding out the actual and potential experiences that employees are faced with will provide a benchmark from which an employer can begin to consider the extent of concern or stress that is being experienced by those exposed.

Any information obtained about violence issues should be categorised to identify whether there is a general problem, or a specific issue with a particular category of workers or specific job. It is important to ascertain any outcomes to violent events as these will have a bearing on the level of stress that a victim or the victim's colleagues may be experiencing as a result.

Risk management procedures

S11048 These will vary, according to the existence or extent of the problem. An employer should, however, have as a minimum a policy which recognises violence as a workplace issue and outlines the arrangements in place to eliminate or control the problem. The policy should include a commitment to introduce measures for:

- reducing the risk of violence at work.
- offering full support for people who have been affected by a violent incident.
- encouraging employees to report incidents in which they felt threatened.
- assurance that all reported incidents will be investigated and taken seriously.

Procedures for managing the risk may include the following.

Risk assessment

S11049 This is a proactive management strategy in that it seeks to pre-empt the potential for violence and associated stress, in order that management can implement controls to minimise the chances of a violent incident occurring, or at least reduce the impact if it does occur. In practice, a risk assessment is only effective if it has the full understanding and involvement of the people who are exposed to the risk, and the company is prepared to spend time collecting and analysing data of past events. This will include actual violence, verbal threats and abuse, attempts at violence and general threatening behaviour. These will need to be correlated with any time lost from work (e.g. to attend hospital) or for subsequent sickness absence. Note that sickness absence may occur a few weeks after the event, as some people do not respond to situations immediately.

Emergency/initial response procedure

S11050 A system should be in place to guide employees in the event that they are exposed to a stressful or violent situation. The procedure will give them some comfort and security to help them maintain control of the situation. It should include guidance on how to prevent a situation from escalating, and also what to do in the event that help may be required.

The procedure should also cater for actions that should be taken as soon as possible after the incident has happened. This may include notification of senior management (depending on the event that occurred), contacting support teams, holding informal group meetings with others involved or affected by the incident. The idea should be to respond to the victim's immediate needs and to help the person(s) to feel that what they are experiencing is a normal reaction to a disturbing event, rather than them feeling that they have somehow failed. (This is a very common response for an individual who has been exposed to a violent incident.)

Post incident support

S11051 This is a reactive management strategy, but very important in its impact upon management of the situation. Providing support for staff should be part of the overall policy on preventing and controlling violence at work. Support measures will help to minimise and control any impact on staff and assist in their recovery from the incident. Any post incident support must be readily available and easily accessible, so as to provide support quickly after the event has been reported.

Victim support groups are available from a number of organisations, including the police and the Suzy Lamplugh Trust. Alternatively, the company may consider using the services of their own counsellor to deal with any post-traumatic situations.

A preventative strategy

S11052 The issue of violence should be taken into account in all aspects of running an effective and safe organisation. Employers should therefore consider the following:

- design and layout of buildings.
- staffing levels.
- job design, systems of work and working practices.
- policy development, with dissemination of procedures.
- communication channels.
- incident and near miss recording, and other statistical monitoring.
- provision of training and information for workers.
- risk assessment programmes, including review of assessments.
- spot checks and other monitoring strategies to check if systems are being adhered to.
- proper reporting procedures.
- counselling and support for the victim and their colleagues.

Useful publications

S11053 Various publications produced by the HSE provide useful guidance on management of violence to staff.

- HSG 133: *Preventing violence to retail staff*
- *Violence to staff in the education sector* (ISBN 0 7176 1293 7)
- HS(G)100: *Prevention of violence to staff in banks and building societies*
- *Violence and aggression to staff in the health services* (ISBN 0717614662)

Videos:

- 'All stressed Up' (particularly aimed at women), available from Leeds Animation Workshop, 45 Bayswater Row, Leeds, West Yorkshire LS8 5LF (tel: 01532 484997).
- 'Personal safety at work' (1994), video and training pack produced by Cinegrade Production, and distributed by the Suzy Lamplugh Trust.

Training and Competence in Occupational Safety and Health

Introduction

T7001 Although the term 'incompetent' is widely recognised by people in the world at large and can quite easily be defined as not qualified or able to undertake a defined action or task, when looking at the definition of competent or competence, the boundaries are less easy to define. Someone may be qualified or able to undertake a task to a certain level or in a specified area, so they could be considered competent either generally at a lower level or in part. This particular dilemma is one that has faced most professions and occupations during their development stages and has led to the formation of professional bodies and trade associations, who as one of their major reasons for existence have set standards of practice for their own particular disciplines. This is perhaps where competent performance related to health and safety is very different to other professions and trades in so much that everyone, whether at work or not, needs to have knowledge of how to remain safe and healthy and from many different angles and perspectives.

This becomes particularly relevant in workplaces where both employers and employees have moral and legal responsibilities both for themselves and others in terms of maintaining a safe working environment. Everyone needs to have a level of competence in terms of occupational health and safety, from employees who need to ensure that their actions do not endanger either themselves or their colleagues, to directors of the company who are responsible for the health and safety of the workforce and need to know what this entails. Additionally specialist advisers in health and safety will need to have much higher and broader levels of competence in occupational health and safety issues, so that the advice they give to both the management and the workforce in organisations is clear and of a standard expected from a professional practitioner.

The Health and Safety Commission ('HSC'), made a clear statement regarding health and safety competence in their 'Strategy on Health and Safety Training' which has the vision: 'Everyone at work should be competent to fulfil their roles in controlling risk'. It also makes a statement about the training required to do this:

> 'Health and safety training covers all training and developmental activities aimed at providing worker, including safety representatives, and managers with:
>
> (a) greater awareness of health and safety issues;
>
> (b) specific skills in risk assessment and risk management;
>
> (c) skills relating to the hazards of particular tasks and occupations; and
>
> (d) a range of other skills, including those relating to job specification and design, contract management, ergonomics, occupational health etc.
>
> The specific aims of the strategy are to:
>
> - raise awareness of the importance of health and safety training;

- bring about a substantial improvement in the quality and quantity of health and safety training;
- promote an awareness of the importance of competence in controlling risk;
- influence providers of the education system to provide the necessary framework of basic knowledge and skills.

With the objectives of the strategy being to ensure competence by:

- encouraging employers and trade unions to recognise the need to provide health and safety training of good quality;
- engaging HSE and Local Authority (LA) inspectors to assess the competence of workers and managers and the adequacy of training provided by employers. This forms an important part of inspection, investigation and enforcement activities;
- continuing to seek partnerships elsewhere to provide training of appropriate quality and quantity;
- influencing other government departments to reflect the need for health and safety training provision in their responsibility; and
- ensuring that all parts of the education system provide a foundation of knowledge upon which health and safety training can be built.'

This chapter focuses on:

- the components of competence and the acquisition of this;
- how it may be attained and assessed for all the sectors of the workforce;
- the legal and managerial requirements for competence; and
- the various organisations that fulfil roles in the process.

In all cases, the chapter recognises that any level of competence is time bounded. Skills and knowledge become out dated very quickly in the light of new technologies and practices, and is a particular aspect that needs to be considered in maintaining the health and safety of the workforce.

Recognising competence

T7002 It is worth having a more in depth look at some of the more common definitions of competence, before exploring the ways and means of attaining specific and relevant levels of health and safety competence. These definitions, although not in actual conflict with each other do have slightly different slants on the concept, which very much depends on the source of the definition. Clearly people from outside the health and safety world may well view competence in a slightly different way to those more familiar with the legal concepts. This may well lead to the position where employers believe they have an appropriate level of competent advice (as described in the next section at T7002 below), but may in fact be leaving themselves exposed to both criminal and civil actions in the event of an accident, where it can be proved that the appropriate level of advice was not available. When either selecting individuals or training programmes for employees, employers should be advised to seek appropriate advice to ensure that they are legally and morally covered in terms of health and safety; good advice can also have a considerable effect on a company's profit margins!

Looking at the concept of competence from a plain English perspective *The New Oxford Dictionary of English* defines competence as the ability to do something successfully or efficiently and competent as having the necessary ability, knowledge, or skill to do something successfully.

The National Vocational Standards define competence as 'the ability to perform to the standards required in employment across a range of circumstances and to meet changing demands'.

In case law a competent person is viewed as 'one who is a practical and reasonable man who knows what to look for and how to recognise it when he sees it' (*Gibson v Skibs A/S Marina and Orkla Grobe A/B and Smith and Coggins Ltd [1966] 2 All ER 476*). It could also be added that they should also know what to do with the acquired knowledge once they have found it.

All of these definitions basically amount to the same thing, but with a slightly different slant, that competence is a combination of knowledge, experience and skills. It should be recognised that competence itself is based on the outcomes of this combination of requirements and it is a demonstration that an individual can perform to specified standards rather than simply a record of a person's qualifications and experience, with which it is often confused. It is this simple definition of knowledge, skills and experience that is used throughout this chapter, recognising that the actual determination of an individual's level of competence does require a further stage of assessment by an employer. Only the training and education leading to the development of competence can be identified here.

Legal obligations for training and competence

Health and Safety at Work etc Act 1974

T7003 The *Health and Safety at Work etc Act 1974* ('*HSWA 1974*') is the principal enabling legislation for the UK. It imposes duties on employers, employees, bodies corporate, manufactures and others, as well as establishing the HSC which recommends policy regarding health and safety, and the Health and Safety Executive ('HSE') which enforces this policy and to serve a range of legal notices to employers.

Training is specifically mentioned within *section 2* of the *HSWA 1974* which prescribes the general duties of employers. *Section 2(2)* of the 1974 Act states that in addition to other requirements it is an employer's duty to ensure 'the provision of such information, instruction, training and supervision as is necessary to ensure, so far as is reasonably practicable, the health and safety at work of his employees'. This in fact does not place an absolute duty on an employer to provide training but recognises that having carried out generic and specific risk assessments of the workplace an employer may well believe that training would be an effective measure of control. Many employers take this approach in addition to other risk control methods.

Competence is not specifically identified within the *HSWA 1974* other than an allusion in *section 19(1)* which refers to the appointment of inspectors. Here it relates to 'such persons having suitable qualifications as it thinks necessary for carrying into effect the relevant statutory provisions within its field of responsibility'.

Perhaps the most important part of the *HSWA 1974* in relation to training and competence is in *section 15* which relates to the power to make health and safety regulations. The first set of these regulations as relevant to training and competence are the *Safety Representatives and Safety Committees Regulations 1977* (*SI 1977*

No 500) which require an employer to allow a trade union appointed safety representative to take time off 'to undergo such training as may be reasonable in the circumstances'.

Management Regulations

T7004 Major changes to the requirement for competence on a general basis came about as a result of the UK's enactment of a series of European directives on health and safety via a series of regulations. The *Management of Health and Safety at Work Regulations 1992* (*SI 1992 No 2051*) (now revoked) was a set of generic regulations relating to the overall management of health and safety and worked in conjunction with already existing regulations such as the Control of Substances Hazardous to Health Regulations ('COSHH') and the Noise at Work Regulations. The 1992 Management Regulations actually introduced the concept of 'competent person'. The Regulations were amended in 1999 with *Regulation 6* becoming *Regulation 7* in the *Management of Health and Safety at Work Regulations 1999* (*SI 1999 No 3242*) ('*MHSWR 1999*'). These Regulations are the main focus when considering competence in health and safety.

The *MHSWR 1999* (*SI 1999 No 3242*), *Reg 7*(*1*) states that:

> 'Every employer shall, appoint one or more competent persons to assist him in undertaking the measures he needs to take to comply with the requirements and prohibitions imposed upon him by or under the relevant statutory provisions'.

The *MHSWR 1999* (*SI 1999 No 3242*), *Reg 7*(*5*) states that:

> 'A person shall be regarded as competent for the purposes of paragraph (1) where he has sufficient training and experience or knowledge and other qualities to enable him to assist in undertaking the measures referred to in that paragraph'.

T7005 The Approved Code of Practice that accompanies the *MHSWR 1999* (*SI 1999 No 3242*) makes some specific conditions regarding the competent person role in these Regulations. In summary these are:

- Employers are solely responsible for ensuring that the person appointed to fulfil the 'competent person' role is capable of applying the principles of risk assessment and prevention together with a current understanding of legislation and health and safety standards.
- Where there is more than one person appointed to fulfil the role, or parts of it, then these people must co-operate.
- Competence in the Regulations does not necessarily depend on the particular possession of particular skills or qualifications and simple situations may require and understanding of relevant current best practices and an awareness of the limitations of one's own knowledge but have an ability to seek external help where necessary.
- More complex situations will require a person with a higher level of knowledge and experience and in these cases employers are advised to check the appropriateness of health and safety qualifications and /or membership of professional bodies. The competence-based qualifications accredited by the Qualification Curriculum Authority (QCA) can act as a guide.

The *MHSWR 1999* (*SI 1999 No 3242*), *Reg 8*(*1*)(*b*) imposes a duty on employers to nominate competent persons who have the ability to carry out evacuations of an establishment in the event of serious or imminent danger. *Regulation 8*(*3*) goes on to say that such a person will have sufficient training and experience or knowledge and other qualities, defining competence in a similar way to the *MHSWR 1999* (*SI 1999 No 3242*), *Reg 7*. Also, in *Regulation 15* of the 1999 Regulations which relates to temporary workers, an employer is again required to determine if a temporary employee has the competence to carry out the job in a healthy and safe manner.

T7006 The *MHSWR 1999* (*SI 1999 No 3242*) make several statements particularly in relation to the requirement for training. *Regulation 3*(*5*) relates specifically to young people and *Regulation 13* specifically relates to capabilities and training for all employees. *Regulation 13*(*2*) particularly states that every employer shall ensure that his employees are provided with adequate health and safety training, and very specifically relates to the various times that this should be given, i.e:

- on recruitment;
- on being exposed to new or increased risks because of being transferred or having a change of responsibility;
- on the introduction of new work equipment or a significant change to it;
- on the introduction of new work equipment;
- on the introduction of a new system of work or a change to an existing system.

The *MHSWR 1999* (*SI 1999 No 3242*), *Reg 13* also says that this training should be repeated periodically and be adapted to take account of any changes or new risks and should also take place during working hours.

T7007 Overall the *MHSWR 1999* (*SI 1999 No 3242*) impose a very comprehensive requirement for both training and the recognition of competent advice in the workplace. They have highlighted and made a legal requirement of what has been good practice in workplaces for many years. In addition to these Regulations there are several more hazard-specific pieces of legislation, which also require training to be delivered and competence to be demonstrated.

The policy unit of HSE has published an outline map on competence, training and certification. The map gives an overview for competence and training as described in legislation and can be found on the HSE website. The mapping process identifies that there are four different groups in which training and competence requirements can be identified:

- *Group 1*

  General goal-setting requirements where there is no precise detail of how the requirement should be met – there are 37 pieces of legislation that have this approach;

- *Group 2*

  General goal-setting requirements qualified by some specific requirements for people or bodies with particular responsibilities – there are 8 instances of this in legislation;

- *Group 3*

  General goal-setting requirements qualified by particularly strong specific guidance, sometimes agreed with the industry/sector – there are 6 instances of this in legislation;

- *Group 4*

  Specific requirements for competence or training, certification/qualifications and approval, including statutory requirements to meet a performance standard – there are 40 specific requirements of this nature in legislation.

Within group 1 there are five high-risk areas where an employer is required to have arrangements for training and levels of competence written into plans or reports. These are:

- construction (site supervisor's plan);
- control of major hazards (in accordance with the *Control of Major Accident Hazards Regulations 1999* (*SI 1999 No 3242*);
- offshore safety cases;
- railway safety cases (in accordance with the *Railways* (*Safety Case*) *Regulations 2000* (*SI 2000 No 2688*);
- nuclear site licences.

There are specific requirements for people or bodies within group 2. This includes medical personnel responsible for medical surveillance under the *Control of Substances Hazardous to Health Regulations 2002* (*SI 2002 No 2677* as amended by *SI 2003 No 978*), and the requirement of specific training within the *Provision and Use of Work Equipment Regulations 1998* (*SI 1998 No 2306* as amended by *SI 2002 No 2174*).

In group 3 there is strong specific guidance advocating the use of particular programmes of training. Regulations included in this group are: the *Noise at Work Regulations 1989* (*SI 1989 No 1790*), the *Manual Handling Operations Regulations 1992* (*SI 1992 No 2793* as amended by *SI 2002 No 2174*), and the *Health and Safety* (*Display Screen Equipment*) *Regulations 1992* (*SI 1992 No 2792* as amended by *SI 2002 No 2174*).

Group 4 lists those areas where there are specific legal requirements for named training programmes. Activities worth particular noting in this area are radiation protection advisors, gas fitting (CORGI), adventure activity licensing, mining, the carriage of dangerous goods and diving.

Employers should make themselves aware of any specifically cited requirements in their sphere of business in addition to the more general management requirements described in the *MHSWR 1999* (*SI 1999 No 3242*). They should give the necessary training required that will lead their workforce to perform competently with regards to health and safety.

Setting national standards in health and safety

T7008 During the 1980s it was identified by a Government Working Party, that there was no effective system of vocational qualifications within the UK. What did exist had evolved from individual employment sectors and as such there was no commonality of level and the standards varied. This was a very difficult situation for employers who were unable to determine the validity of qualifications; whilst some sectors had highly respected qualifications others had none. A system was needed that would recognise the skills people needed and was reliable, consistent and well structured. In 1986 the Government established the National Council for Vocational Qualifications ('NCVQ') to set up a comprehensive framework of vocational standards covering all occupations and industries. One notable fact was that all the standards were required to have a unit of health and safety within their structure.

The standards were developed by organisation know as 'Lead Bodies'. These were independent groups of professional representing employers, trade unions, government departments, local authorities and other relevant groups. In the health and safety area the organisation setting the standards was known as the Occupational Health and Safety Lead Body ('OHSLB'). HSE formed the secretariat for the Lead Body and included representatives from the CBI, TUC, and Department for Employment and Local Authorities. They developed the national standards for Occupational Health and Safety Practice, Regulation and Radiation Protection. These standards were launched in 1995 and became available as the National and Scottish Vocational Qualifications ('NVQ/SVQ').

HSE adopted the OHSLB Enforcement Standards, in addition to its established academic training programme, in the Guidance Note attached to the *Management of Health and Safety at Work Regulations 1992* – recognising the Practice Standards, as representing levels of competence for those giving advice on health and safety. The Institution of Occupational Safety and Health ('IOSH'), the chartered professional body for practitioners in health and safety also recognised the standards as underpinning their competence categories of membership; level 4 at their full member category and level 3 at technician level. (See T7010 below for clarification of levels.)

The ownership of the standards passed from the OHSLB to a new organisation, the Employment National Training Organisation ('ENTO'), in 1998, following a government rethink of the structure of how the national vocational standards should be organised. ENTO's immediate priority was to cross-functionally map the constituent standards it had inherited. This included, as well as occupational health and safety, the personnel, training and development and trade union standards. Once this was completed its next task was to revise the existing standards, which commenced in 2000 with a consultation process.

T7009 The new standards in Occupational Health and Safety Practice ('OHS') were launched on the 13 August 2002 in conjunction with the leading players in the health and safety world: IOSH, the Royal Society for the Prevention of Accidents (RoSPA), the National Examining Board for Occupational Safety and Health ('NEBOSH') and the British Safety Council. The OHS standards themselves had undergone a fundamental revision during this process and now became available at level 4 and 5 – the latter reflecting the increasing requirement for higher standards of competence for those who strategically manage or advise on health and safety issues. HSE also adopted the revised level 5 standards for enforcement.

An earlier initiative by ENTO had identified that the mandatory health and safety units included in the wider range of vocational standards were numerous (well over 400) and variable. It was its belief, as the training organisation for people who work with people at work, that a collection of stand-alone units for people who needed the competence to fulfil certain health and safety responsibilities, but who were not health and safety practitioners, was necessary to bring a level of consistency in the competence requirements across the whole range of vocational standards. These were produced and launched in 1999, and as such they have no particular level and can be fitted into other sets of vocational standards as appropriate. The eight separate units known as 'Health and Safety for People at Work' can be used to demonstrate a level of health and safety competence for the whole range of employees from workers to managers and directors. They cover the range of knowledge, skills and practical experience, which a person has to have to be able to do a particular task in a safe manner and are designed to allow people to show that they are competent in the safety issues described.

In conclusion, there are now standards available within all the UK national frameworks to recognise the level of competence required by all people in the workforce. These standards, as well being capable of being used as the standards against which the N/SVQ qualifications are assessed, also form the basis for other types of qualification in the health and safety area.

Education and training towards national standards of competence

T7010 This section will focus in more detail on the required competence levels in health and safety across the whole spectrum of the workforce. This will include strategic health and safety practitioners or managers, health and safety advisers and practitioners, health and safety technicians, managers and supervisors with line responsibilities for health and safety, and last but by no means least, the workforce itself. Some of the qualifications that have been traditionally associated with health and safety such as those from the British Safety Council or NEBOSH will be equated to the national frameworks to give an indication of the parity of the different routes available.

The framework of competence standards and associated vocational-related qualifications ('VRQs') is organised in the constituent bodies by the nationally associated bodies highlighted in T7008 above. For brevity they will be referred to using the English body QCA. However, this should be taken as meaning all the bodies, which do in fact have a working association.

In addition to these frameworks there is another UK qualification framework specifically for the higher education sector. This is known as the Quality Assurance Agency for Higher Education and is referred to as 'QAA'. For a clearer understanding, the relationship of qualifications on the QAA framework to the QCA framework are shown in the following chart:

QCA	QAA
Level 5	Post-graduate Diploma, Masters Degree
Level 4	Higher Education Diploma, Ordinary Degree and Honours Degree
Level 3	'A' Level (access to higher education)
Level 2	—
Level 1	—

This chart is a rough guide only for the purposes of this chapter. It has no legal standing as to the equivalence of level.

Competence requirements for occupational health and safety practitioners

T7011 IOSH is the leading professional body in the UK for those involved in occupational health and safety practice. It was founded in 1945 and incorporated by Royal Charter in 2003 and currently has over 27,000 subscribers including over 12,000 in competent practice categories of membership. IOSH, as the chartered professional body sets the standards for those in the practice of health and safety. However, since their inception, IOSH has taken an active part in the development of the national vocational standards for OHS, and Health and Safety for People at Work, and now uses these national vocational standards as part of the definition of its membership criteria.

T7012 The national vocational standards at level 4 are recognised by the profession as representing the core competence requirements for those in health and safety practice. There are, however, many ways of achieving these standards through the various routes of qualification together coupled with the development of skills and experience. The recently revised standards, which are loosely based on the safety management model developed in the HSE publication *Successful health and safety management* (HSG65), have moved away from the approach taken by the original level 4 standards. These original standards focused on the level of risk faced by an organisation and the competence a person would need to develop to ensure that the advice given to control these risks was given in an effective manner. For the level 4 standards this was deemed to be for all industry sectors which included complex, co-existing high risks where failure to control them could be life or organisation threatening. The level 3 standards introduced at the same time were for those in lower risk workplaces, where control is well documented or for those who formed part of a team in higher risk areas reporting to a level 4 standard competent person. Level 3 Practice Standards do not appear in the revised suite but have been replaced by a different level 3 in health and safety based on the generic standards Health and Safety for People at Work. The new standards, which are all at level 4, form the basis of competent practice, irrespective of the organisation in which a person operates.

The titles of the revised OHS Standards, which are available on a CD-ROM from ENTO, are as follows:

G3 Evaluate and develop own practice

H2 Promote a positive health and safety culture

H3 Develop and implement a health and safety policy

H4 Develop and implement effective communication systems for health and safety information

H5 Develop and maintain individual and organisational competence in health and safety matters

H6 Identify and evaluate health and safety hazards

H7 Assess health and safety risks

H8 Determine and implement health and safety risk control measures

H9 Develop and implement active monitoring systems for health and safety

H10 Develop and implement reactive monitoring systems for health and safety

H11 Develop, implement and test health and safety emergency response systems and procedures

H12 Develop and implement health and safety review systems

H13 Develop and implement health and safety audit systems

The NVQ qualifications are actually assessed against these standards. This type of qualification is an assessment of the ability of a person to perform to the laid down standards and involves an assessor who will monitor and lead the candidate through the process of developing a practice portfolio. For an NVQ at this higher level it is also necessary for the assessor backed up by the verification processes required by the awarding bodies, to determine that a candidate has the requisite domain knowledge to carry out the competence based tasks required.

This type of qualification has proved very popular in occupational health and safety practice as it is particularly suited to those who have many years of experience in giving health and safety advice, based on sound experience but who are not happy or comfortable with the traditional exam-based type of qualifications available. In fact in IOSH's subscription year ending on 31 March 2003, 25 per cent of all those entering the Member (corporate) category of IOSH held a qualification of this type.

T7013 The national vocational standards are also used by the QCA to assess the relevance of vocationally-related qualifications to the overall national framework. Organisations applying to the QCA to be placed on this national framework must show that they have robust systems of quality control in place and have an examination process totally divorced from the delivery of their courses. This is known as part A accreditation. They must also show that their syllabus delivers the underpinning knowledge requirements of the national vocational standards, known as part B accreditation. NEBOSH is one organisation that follows this route and the long established NEBOSH Diploma (Dip2OSH) is currently being revised to meet the new OHS Standards at level 4. The current NEBOSH Diploma, which is divided into part 1 and part 2, was structured to meet the original Practice Standards at level 3 and 4. However, as the level 3 Practice Standards have been withdrawn this has lead to the reversion to a one-part NEBOSH Diploma. It is also expected that the British Safety Council will shortly also have QCA accreditation for its long established Diploma in Safety Management (DipSM) following revisions which are currently taking place. Although other organisations such as the Chartered Institute of Environmental Health (CIEH) and the Royal Society for the Promotion of Health (RSPH) have qualifications on the framework, as yet they have nothing above level 3.

The QCA accreditation verifies that the qualifications on its framework are robust and have fair levels of achievement recognised at the specified levels. Universities offering qualifications in occupational health and safety also have to undergo similar processes. Universities are themselves responsible for awarding qualifications within the QAA frameworks and have introduced stringent and searching quality control procedures internally, by a system of internal validation and external examination by appointed examiners. The QAA holds the overarching quality control on the qualifications offered. Also it is normal for universities offering vocationally-related degrees to seek external accreditation from the professional body in that area. IOSH undertakes this accreditation process for occupational health and safety degrees and post-graduate qualifications and published its own criteria for the conditions of accreditation and syllabus content in 1997. These have currently been revised to fall in line with the national vocational standards of competence and to maintain the IOSH accreditation universities will be expected to adjust their syllabuses to meet the knowledge requirements within the next few years.

There are a range of different levels of qualification available from the higher education ('HE') sector which form the Higher Education Diploma and Foundation degrees which roughly equate to level 4 on the QCA framework to the more academically challenging Bachelors and Post-Graduate degree programmes. There are currently 28 universities and related colleges in the UK offering this type of programme, which are proving very popular. In fact this is now by far the biggest route of entry into the professional membership category of IOSH amounting to 52 per cent of all entrants in the subscription year ending on 31 March 2003.

It should be clarified that academic qualifications such as those from the HE sector, NEBOSH or other similar bodies, are not assessments of competence in the same way as an NVQ qualification. They are a demonstration that a programme of academic work leading to the development of knowledge and some skills has been undertaken and formally assessed by examination and/or other suitable form of academic assessment. This by itself would not be sufficient to demonstrate competence but with suitable experience and further development of skills in a practical setting, the knowledge gained can form the vital knowledge part of the knowledge, skills and experience equation. However, it should also be recognised that although the achievement of an NVQ actually demonstrates competence, this is only in the workplace setting in which the person undertaking the qualification is employed. This means that the broader knowledge base required of a competent practitioner has not necessarily been covered or assessed to the same extent to a person who has taken the academic qualifications.

T7014 There are clearly pros and cons for each type of qualification and this is a debate, which continues throughout the safety profession and the deliverers of training and education. This is not helpful to employers and those needing to employ competent health and safety practitioners who need to be able to assess the similarities and differences between the varying types of initial qualification. Health and safety is a diverse profession and attracts people from many different backgrounds, so it is advantageous and necessary to have the variety of different ways to qualification currently available. IOSH as the professional body currently accepts all the different qualification routes but recognises that all have particular strengths and weaknesses. It is proposed that in a future re-definition of the IOSH membership structure that further assessment in addition to initial qualifications will be necessary to achieve this. This will be tailored to the initial qualification so that those with an NVQ will need to undertake some further formal assessment of knowledge whilst those with knowledge based qualifications will need to demonstrate the acquisition of skills. It is expected that this type of approach will commence within the next two years, which should be some clarification to the position. Potential employers will be able to use the IOSH membership structure and particularly the requirements for its full category of membership as a yardstick of competence for safety practitioners irrespective of whether they are a member of the professional body

Competence in health and safety practice at strategic management level

T7015 In the current set of national vocational standards, a level 5 set of competencies in OHS was introduced. This reflects the fact that many more health and safety professional are now moving into more strategic management roles within organisations and also the fact that many, who would not consider themselves to be health and safety professionals also operate at this strategic management level role. Although there are a plethora of health and safety training courses specifically targeted at mangers, very few of them are actually pitched at this higher level of management expertise. This is perhaps not surprising, as most managers would incorporate health and safety within their workplace role on day-to-day basis, implementing the rules and policies that have already been developed, using their personal knowledge of the processes of the organisation for which they work.

It has been a long-standing request from the health and safety community that health and safety should be included in MBA programmes so that potential future leaders of industry should have health and safety included within their academic programmes. As yet this has not borne much fruit but one or two organisations do now do this. Perhaps of more potential is the fact that HSC included in its *Revitalising Health and Safety* document the fact that the education for those in safety-critical professions should include health and safety specifically.

The titles of standards for OHS at level 5 are listed as follows:

G3	Evaluate and develop own practice
H1	Develop and review the organisation's health and safety policy
H2	Promote a positive health and safety culture
H3	Develop and implement the health and safety policy
P11	Develop a strategy and plan for people resourcing change management
H15	Influence and keep pace with improvements in health safety practice
L2	Identify organisational learning and development needs
B3 (METO)	Manage and use financial resources
B4 (METO)	Secure financial resources for your organisation

These standards are reflective of health and safety strategies but are in reality management standards. Some of the standards are in fact taken from other sets of standards, such as those from the management-training organisation (METO). How much of these standards will be used as NVQs is as yet undetermined, as they are still relatively new and form part of the new suite of standards launched by ENTO during 2002.

Competence in health and safety for managers and supervisors

T7016 Although those in safety practice can set the standards and form the systems to ensure that health and safety risks are controlled, it is actually those in everyday contact with the people and work interface who can actually make the most difference to the overall health and safety performance. The competence of this group can be achieved by a variety of ways and means and it is this group who have the knowledge of the day-to-day workings of an organisation and therefore who will be able to have the most effect in the workplace.

The ENTO suite of standards Health and Safety for People at Work were designed as competence standards for all people in the workplace and many of them were targeted at the managerial and supervisory groups. Initially it was anticipated that managers and supervisors would make their own choice from the suite depending on their own role in the organisation for which they are employed. However, at the recent revision it was suggested that the complete suite, with the exception of Unit D which is for placement review officers, would form a level 3 NVQ qualification in its own right. The titles of the standards are given below:

H&SA	Ensure your own actions reduce risks to health and safety
H&SB	Monitor procedures to control risks to health and safety
H&SC	Develop procedures to control risks to health and safety
H&SE	Promote a health and safety culture within the workplace
H&SF	Investigate and evaluate incidents and complaints in the workplace
H&SG	Conduct an assessment of risks within the workplace
H&SH	Ensure your own actions aim to protect the environment.

These standards can be assessed as competence via the NVQ route but at this level the majority of industries still favour the more traditional training approach to providing knowledge and skills. There are numerous programmes, which take this approach. Some are assessments of both knowledge and also have an element of skills assessment, whilst others programmes take a more traditional knowledge approach assuming that the skills can be gained in the actual workplace. This section highlights a few of these programmes but it is not exhaustive and normal training selection criteria should be applied when selecting a programme.

T7017 The NEBOSH National General Certificate in Occupational Safety and Health is an examined-basis qualification designed to help non-specialists to discharge their duties or functions in workplace health and safety. Such people include managers, supervisors and employee representatives. This certificate covers much of the underpinning knowledge requirements for the ENTO level 3 Health and Safety for People at Work standards. It is normally taught over 80–100 hours by all modes of delivery including distance learning. NEBOSH makes it clear in its literature that being awarded a certificate does not imply competence but the underpinning knowledge towards competence. Although not specifically designed as an introduction to safety practice this qualification is often uses as a starter for people moving into the profession, who would then progress to the NEBOSH Diploma.

T7018 The Chartered Institute of Environmental Health ('CIEH') Advanced Health and Safety Certificate is also classified as a level 3 qualification on the QCA framework. However, it is a much shorter programme than the NEBOSH Certificate. It is at a similar level but does not have the full breadth of syllabus found in the NEBOSH Certificate.

T7019 The Royal Society for the Promotion of Health ('RSPH') Advanced Diploma in Health and Safety in the Workplace is again a QCA level 3 course. It has 36 guided learning hours and assessments are carried out by coursework assignments. The title is a little misleading, as this is a basic health and safety qualification. The QCA framework does not specify the titles of programmes only the levels.

T7020 The British Safety Council Certificate in Safety Management is a 5-day programme at a similar level 3. It is targeted specifically at managers with the Diploma in Safety Management being aimed at practice level.

T7021 The IOSH Managing Safety course has been specifically written for managers and focuses on two to three units of the ENTO Health and Safety for People at Work standards. The course particularly focuses on the risk management process and how to use it in day-to-day workplace activities. This course does not appear on the QCA framework as it is run as a collaboration between IOSH and its training delivering members, and as such does not have the separation of marking from delivery that is required for this accreditation. The course is flexible and can be tailored to meet the specific requirements of organisations.

T7022 In addition to these qualifications run by health and safety bodies most of the large awarding bodies also offer health and safety programmes. Both OCR and City and Guilds have qualifications at QCA level 2 as well as offering the level 3 NVQ Health and Safety for People at Work.

IOSH also offer a Safety for Senior Executives briefing programme which is tailored to give executives, directors and the most senior managers in organisations the knowledge and information they require to run their organisations in a safe and healthy manner. This group is vital to the facilitation of good health and safety standards in the workplace, and they must show a commitment towards making sure standards are maintained and where possible improved on a continuing and regular basis. Training at this level needs to specifically recognise the requirements of this group of people, and routine training hammering out day-to-day hazards and specific details of the legislative requirements is not appropriate. More in depth consideration and exploration of the financial and moral aspects of health and safety is required, together with an understanding of the legal responsibilities and possible penalties for failure to ensure health and safety in the organisations for which they are responsible.

Most of the organisations also offer courses of a slightly shorter nature for those in supervisory roles. These are shortened versions of the manager programmes. It can be argued, however, that supervisors are the real front-line for safety and their training should be exactly the same as for managers but the reality is that financial constraints often mean that these shorter programmes are used.

Competence in health and safety for safety representatives

T7023 Many people who perform the function of safety representatives in the workplace, as defined by the *Safety Representatives and Safety Committees Regulations 1977* (*SI 1977 No 500*), actually train in health and safety in a similar way to their colleagues undertaking similar courses to those described for managers and supervisors. There is, however a specific programme run by the TUC under the auspices of the National Open College Network (NOCN), which has been specifically designed to meet the requirements of this group. This programme, the TUC Certificate in Occupational Health and Safety, is a level 3 programme that builds on the TUC basic stage 1 and stage 2 health and safety courses or other union's equivalent courses. The assessment of the course is by a competence-based type of assessment similar to the NVQ qualifications with the addition of workplace-biased projects of practical value to the candidate's workplace and union.

The course is available for union representatives on either a part-time day release basis or by distance learning. The course additionally gives access to higher education for those who wish to progress further.

The TUC Education Service normally pays for the course, provided that the candidate is nominated by a trade union and is a valuable contribution towards competent safety advice in the workplace. This is recognised by IOSH who currently offer TechSP membership for those with this qualification and able to demonstrate at least two years pro-rata full-time equivalent experience.

Competent performance by the workforce

T7024 Last but by no means least in the range of the competence chain that leads to better health and safety performance in the workplace is the workforce itself. The *HSWA 1974* makes it quite clear that employers have a legal duty to ensure, as far as is reasonably practicable, the health, safety and welfare at work of the people for whom they are responsible and the people who may be affected by the work that they do. The 1974 Act further requires employers to provide information, instruction training and supervision as is necessary to ensure, as far as reasonably practicable, the health and safety at work of employees.

It is the training aspect of these requirements, which has a huge variance across employers, on which this section focuses.

The ENTO recognised the fact there should be a standard available for everyone in the different suites of occupational standards, i.e. the Health and Safety for People at Work suite of standards. Within this suite there is a unit specifically targeted at everyone in a working environment, regardless of position or the number of hours worked. The unit A in the suite titled 'Ensure that your own actions reduce risks to health and safety' is about making sure that risks to health and safety are not created or ignored. It is divided into two elements:

A1 Identify the hazards and evaluate the risks in your workplace

A2 Reduce the risks to health and safety in your workplace

The elements then go on to identify what a person should be able to do to act in a competent manner towards health and safety issues in the workplace. Clearly this is competence which is developed whilst undertaking an employment role but it will be necessary to include training that delivers the knowledge that is required to perform to this level. This is normally carried out at an early stage in a person's employment, normally induction training on starting a job.

These standards are straightforward and form a good series of competence requirements for a workforce. The series of performance criteria for identifying hazards and evaluating risks includes such things as:

- correctly naming and locating the persons responsible for health and safety in the workplace;
- identifying workplace policies which are relevant to working practices;
- identifying working practices in any part of a job role which could harm yourself or other people;
- identifying those aspects of the workplace which could harm yourself or other people;
- evaluating which of the potentially harmful working practices and the potentially harmful aspects of the workplace are those with the highest risk to yourself or to others;
- reporting those hazards which present a high risk to the persons responsible for health and safety in the workplace;
- dealing with hazards with low risks in accordance with workplace policies and legal requirements

The attainment of this level of competence would be assessed for an NVQ across a wide range of workplace policies including methods of working, equipment, hazardous substances, smoking, eating, drinking and drugs and what to do in the event of an emergency. Similar sets of competence requirements are also required for A2 but these are based on reducing the risks to health and safety in the workplace in addition to the recognition required in the unit A1.

These standards can be used to plan health and safety training and they can also be used to analyse where training is actually needed by comparing existing skills and competence of the workforce with the standards and then identifying if there are any gaps, which can be covered by training. This can be done on an individual or group basis and the training offered as a result of this analysis can be offered in a specifically targeted way.

T7025 There are a variety of training programmes at this primary level, which develop the knowledge requirements for health and safety in the workplace. Many are quality controlled by either awarding bodies or bodies with a health and safety remit. Some of the well-known programmes are IOSH Working Safely, CIEH Basic Health and Safety Certificate, RSPH Foundation Certificate in Health and Safety, the Workplace and awards from City and Guilds, OCR and Edexcel. Many organisations seek this external verification to show that they have carried out their training requirements with regard to their duties under the *HSWA 1974*. However, there are many perfectly good induction programmes run by organisations themselves which allow this knowledge part of the competence requirements to be achieved. It is the fact that good training is available that is of vital importance, and the national standards of competence give an indication of what needs to be achieved from any training programme.

Competence for contractors and safety passports

T7026 Leading on from the requirement of the development of a workforce that is competent to perform its role in a healthy and safe manner, is the requirement to look at all those people who also may be present in a workplace.

Although organisations have the ability to control the training and competence of individuals that they have directly employed, by developing programmes which meet their specific requirements, there is one area that they need to control but do not necessarily have complete control over – that is outside contractors coming into their workplaces. This is particularly relevant in some industrial sectors such as construction and related industries where the use of contracted labour is extensive. However, at some point all employing organisations may have people other than their own employees within their workplaces. Many organisations control this situation by having 'permit to work' systems that ensure that anyone entering a workplace has sufficient knowledge of the health and safety arrangements that are in place.

However within the last few years there has been a move to a more transferable skills type of scheme that can allow those who do move between workplaces to show that they have sufficient underpinning knowledge of health and safety matters to allow briefer induction programmes to be given when moving between different workplaces and sites. These are the safety passport schemes.

These passports available from several industry-related schemes, allow employers to determine who may be given access to their workplaces. The employing organisation makes their workplace a passport-controlled environment. The passports themselves generally take the form of a credit-card sized plastic card, containing the name, photograph and other relevant details of an individual, they also verify what training a person has undertaken towards receiving this passport. The passport usually also has an expiry date.

The advantages of passport schemes are:

- they can save time and money through reduced induction training;
- they demonstrate an organisation's commitment to a safe and health working environment;
- they can create good relationships with the supply chain between organisations and their suppliers;
- they can be used by employees of companies as well as sub-contractors;
- contractors can move between companies and demonstrate that they have the necessary health and safety awareness;
- probably most importantly, they can increase a positive safety culture within organisations, which can lead to a reduction in accidents or ill health.

However, it does need to be borne in mind that a passport cannot identify that a worker is competent or be an effective substitute for workplace management including risk assessment and site-specific information.

Example of safety passport

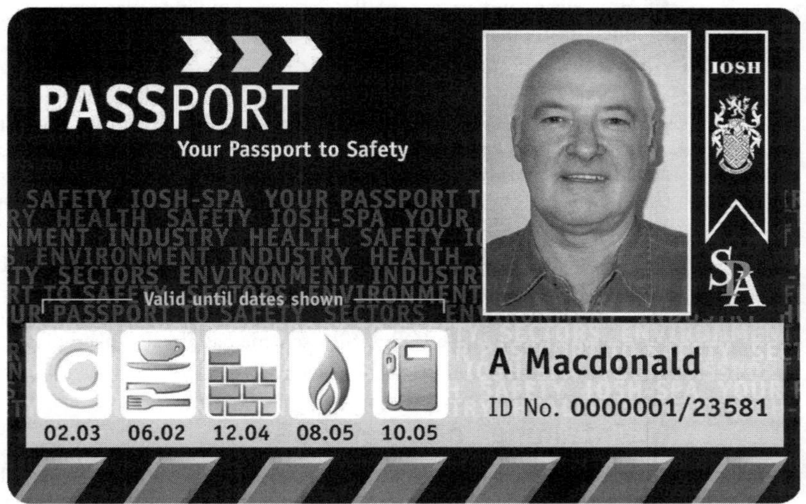

T7027 Many large organisations are now taking up this approach to as part of their safety management systems, this in turn will lead to smaller organisations, who are often suppliers to or sub-contractors for the larger organisations to also adopt the schemes as a business decision.

The largest of the schemes currently available is the Construction Skills Certification Scheme ('CSCS') which to date has issues over a quarter of a million passports. This organisation has a management board that has members from the Construction Confederation; Federation of Master Builders; GMB Trade Union; National Specialist Contractors Council; Transport and General Workers Union and the Union of Construction Allied Trades and Technicians. It also includes observer members from relevant government departments and the Confederation of Construction Clients.

The Construction Industry Training Board (CITB) manages the operation of the CSCS and delivers training through its series of regional offices. To obtain a passport the candidates have to sit a 40-question multiple-choice test through the same centres as used for driving tests, although this is expected to be expanded to other centres using an on-line technique in the near future.

The Client/Contractor National Safety Group ('CCNSG') is one of the oldest schemes and was developed for contractor site personnel. Representatives from major clients are nominated by local safety groups to serve on the CCNSG and trade unions; contractors and safety groups also have places. The training is a basic two-day training programme. The Engineering Construction Industry Training Board (ECITB) manages the scheme. Around 160,000 people currently hold this passport.

A fairly recent addition to the market is the IOSH/Safety Pass Alliance scheme. This was launched in 1991 and consists of a core-training day and an industry-specific day. This particular scheme has a high level of card security, which gives a plastic tamper-proof card, which uses foil holograms and a' holocote' finish, as well as a tailored signature strip. All details are verifiable on the IOSH/Safety Pass Alliance database. The first sector day launched was for the food and drink industry. Client driven initiatives in the water; building maintenance; engineering/ manufacturing; ports and motor industries are following this.

Other passport initiatives are: Airport Construction Training Alliance (ACTA); Gas and Water Industry; Register of Energy Sector Engineers, Technologist and Support Staff (RESETS); Engineering Services SKILL Card; Sentinel for Railway Track Workers; Federation of Bakers' Contractors Passport (FBCP); National Micro-electronics Institute (NMI); Metal Industry Skills and Performance Institute (MetSkill); Safety Certification for Contractors (SCC) and the Port Skills and Safety Limited (PSSL).

In addition to this list there are some other initiatives within specific industries. Clearly some rationalisation of the schemes would be useful to improve the currency of the passports and this may well be a factor in this market over the next few years. One such move with the IOSH/SPA scheme has already taken place.

Maintaining competence

T7028 As we have seen competence is the development of sufficient knowledge, skills and experience to be able to perform a task to the laid down standards in a workplace. These standards may be prescribed by a legislative requirement, national standards or by best practice in the workplace. However competence is prescribed or developed – there is a cycle that needs to be recognised in the attainment and then in the maintenance of the desired level.

Competence is actually a developed attribute and passes through several stages, these can generally be recognised as distinct stages:

- *unconscious incompetence* – a person is unaware of the requirements to perform in a competent manner;
- *conscious incompetence* – a person begins to undertake training and becomes aware of what they do not know;
- *conscious competence* – a person has undergone sufficient training to be able to complete a task in a competent manner and is aware of this;
- *unconscious competence* – a person now performs tasks in a fully competent manner and has now developed to become unaware of this as it is part of their behavioural patterns.

  At this point in the cycle there can be two pathways, a person can continue to perform in a competent manner or they can unconsciously develop bad habits, which can lead to a diminishment in their performance.

The final bullet point indicates that there is a need for those responsible for the workforce to revisit both their own current knowledge and that of their employees on a regular basis to ensure that performance relating to health and safety is in fact to the level they believe it to be.

There can be a number of triggers that can identify that there may be a need for refresher or further training. This is not necessarily simply because of a time lapse since the initial training took place – other issues such as the introduction of new equipment, changes to methods of use, the re-site of equipment, changes of shift patterns etc. should be considered. Also safety audits may identify that there have been behavioural changes that need addressing.

Many of the external agencies, which play a role in the quality control of training programmes, will specify an expiry date for their accreditation. The safety passport schemes have particular expiry dates usually 3–5 years built into the actual passport, whilst organisations such as IOSH have a three-year expiry term on their Managing and Working series of programmes.

For professionals who practice in health and safety there is a particular need for the maintenance and development of their skills. The area of health and safety is constantly changing both in terms of legislative requirements and in the knowledge of best practice in health and safety issues. For those on the IOSH Register of Safety Practitioners ('RSP'), which is the register of those who are acknowledged by the professional body as competent in the practice of health and safety, there is a mandatory requirement to complete Continuing Professional Development ('CPD') to a prescribed level, roughly equating to 5 days of activity per year. All those on the RSP and the considerable number of IOSH members who undertake CPD on a voluntary basis produce a log book with details of the activities that show how both the maintenance and development have been achieved. Employers looking to recruit either an employee or consultant at this level are advised that they should ask to view this log as an indication of continuing competence. This is one particular aspect that IOSH are currently reviewing to incorporate a modern more reflective approach to this activity, in which the professional develops plans for both development and maintenance of skills and after a predetermined period, currently two years, reflects back to determine their achievement or sometimes lack or deficiency of achievement over this period. This is an extension of the concept of life-long learning which is now being incorporated into the initial training programmes for those in occupational health and safety practice and in fact features within the national vocational standards.

To sum up it is critical for good performance in health and safety in the workplace, that the initially developed knowledge and skills are constantly maintained. This can only be achieved by regular checking of the competence levels of the entire workforce in organisations. Some industry sectors have formalised this approach within their safety management systems and here competence frameworks are developed and monitored on a formal basis. This is good practice and should be considered by those who have not yet taken this approach.

Useful contacts

T7029 The following is a list of useful contacts for organisations that have been mentioned in this chapter:

Airport Construction Training Alliance
Website: www.acta.easitrack.com

British Safety Council
70 Chancellors Road
London
W6 9RS
Tel: 020 8741 1231
Fax: 020 8741 4555
Website: www.britsafe.org

Chartered Institute of Environmental Health (CIEH)
Chadwick Court
15 Hatfields
London
SE1 8DJ
Tel: 020 7928 6006
Fax: 020 7828 5866
Website: www.cieh.org

City and Guilds
1 Giltspur Street
London
EC1A 9DD
Tel: 020 7294 2468
Fax: 020 7294 2400
Website: www.city-and-guilds.co.uk

Client Contractor National Safety Group (CCNSG)/
Engineering Construction Industry Training Board
Blue Court
Church Lane
Kings Langley
Hertfordshire
WD4 8JP
Tel: 01502 712 329
Website: www.ecitb.org.ik/safetypassport/safetypassport.html

Construction Skills Certification Scheme (CSCS)
Tel: 01485 578 777
Website: www.cscs.uk.com

Employment National Training Organisation (ENTO)
Kimberley House
47 Vaughan Way
Leicester
LE1 4SG
Tel: 0116 251 7979
Fax: 0116 251 1464
Website: www.empnto.co.uk

Engineering Services Skill Card
Old Mansion House
Eamont Bridge
Penrith
Cumbria
CA10 2BX
Tel: 01768 860 406
Fax: 01768 860 401
Website: www.skillcard.org.uk

Federation of Bakers (FBCP)
6 Catherine Street
London
WC2B 5JW
Tel: 020 7420 7190
Website: www.bakersfederation.org.uk

Gas and Water Industries National Training Organisation (GWINTO)
The Business Centre
Edward Street
Redditch
Worcestershire
B97 6HA
Tel: 01527 584 848
Website: www.gwinto.co.uk

Health and Safety Commission (HSC) (Contact HSE)

Health and Safety Executive (HSE)
Rose Court
2 Southwark Bridge
London
SE1 9HS
Tel: 020 7717 6000
Fax: 020 7717 6996
Website: www.hse.gov.uk

Institution of Occupational Safety and Health (IOSH)
The Grange
Highfield Drive
Wigston
Leicester
LE18 1NN
Tel: 0116 257 3100
Fax: 0116 257 3101
Website: www.iosh.co.uk

Metal Industries Skills and Performance (MetSkill)
5–6 Meadowcroft
Amos Road
Sheffield
S9 1BX
Website: www.sinto.co.uk

National Examination Board in Occupational Safety and Health (NEBOSH)
5 Dominus Way
Meridian Business Park
Leicester
LE19 1QW
Tel: 0116 263 4700
Fax: 0116 282 4000
Website: www.nebosh.org.uk

National Microelectronics Institute (NMI)
1 Michaelson Square
Kirkton Campus
Livingston
EH54 7DP
Tel: 0131 449 8507
Website: www.nmi.org.uk

OCR Examinations
Westwood Way
Coventry
CV4 8JQ
Tel: 0247 647 0033
Fax: 0247 642 1944
Website: www.meg.org.uk

Port Skills and Safety Ltd
Africa House
64–78 Kingsway
London
WC2B 6AH
Tel: 020 7242 3538
Website: www.bpit.co.uk

Qualification Assurance Agency for Higher Education (QAA)
Southgate House
Southgate Street
Gloucester
GL1 1UB
Tel: 01452557000
Fax: 01452 557 070
Website: www.qca.ac.uk

Qualification Curriculum Authority (QCA)
83 Piccadilly
London
W1J 8QA
Tel: 020 7509 5555
Fax: 020 7509 6666
Website: www.qca.org.uk

Royal Society for the Promotion of Health
38A St George's Drive
London
SW1V 4BH
Tel: 020 7630 0121
Fax: 020 7976 6847
Website: www.rsph.org

Register of Energy Sector Engineers, Technologists and Support Staff (RESETS)
Operations Enterprise House
Cherry Orchard Lane
Salisbury
SP2 7LD
Tel: 01722 427 226
Fax: 01722 414 165
Website: www.resets.org

Trade Union Congress (TUC)
Congress House
Great Russell Street
London
WC1B 3LS
Tel: 020 7636 4030
Website: www.tuc.org.uk

Ventilation

Introduction

V3001 Air being invisible and intangible, there is a dangerous tendency to take it for granted. As breathing is predominantly an involuntary activity, it underlines the need to be selective about what we breathe, with whom we do so, and in what conditions. The twin objectives of good ventilation are the elimination or dispersal of particulate matter (see V3010 below) and the replacement of stale, hot or humid air (often associated with production processes or equipment) with fresh air free from impurity, the aim being to induce a sense of individual comfort in employees. The *Workplace* (*Health, Safety and Welfare*) *Regulations 1992* (*SI 1992 No 3004* as amended by *SI 2002 No 2174*) guidance states that the fresh air supply rate to workplaces should not normally fall below 5 to 8 litres per second, per occupant. When considering the appropriate fresh air supply rate, the factors that should be considered include: the amount of floor space available for each occupant; the work actives being carried out; whether there is smoking allowed within the work areas; and whether there are other sources of airborne contamination, such as that arising from work activities, process machinery, heaters, furniture, furnishings, etc. In addition to air changes, the movement of air across the work area/occupant should also be taken into consideration. The CIBSE Guide *'Volume A: Environmental design'* recommends that 'At normal temperatures the air flow velocity should be between 0.1 to 0.15 metres per second and up to 0.25 metres per second during summer.' Draughts should be kept to a minimum. The composition of pure dry air is 20.94% oxygen, 0.03% carbon dioxide and 79.03% nitrogen and inert gases.

Satisfactory ventilation is normally achievable by means of windows (or other similar apertures), as well as by recycled air, particularly in warm weather or hot workplaces. Air recirculated via mechanical ventilation or air conditioning systems should be adequately filtered and impurities removed, and be impregnated with fresh air prior to recirculation. To that end, ventilation systems should be designed with fresh air inlets and kept open, regularly and properly cleaned, tested and maintained. In order to avoid uncomfortable draughts the velocity of ventilation systems may have to be controlled. Employers who fail to ventilate effectively and follow relevant official guidance on medical surveillance of workers can be liable at common law if an employee becomes sensitised to a respiratory sensitiser in the workplace, even in the absence of evidence that the relevant occupational exposure limits have been exceeded (*Douglas Reilly v Robert Kellie & Sons Ltd (1989) HSIB 166* where an employee contracted occupational asthma whilst working with low levels of isocyanate fumes. It was held that the employer was liable at common law for breach of the *Factories Act 1961, s 63* (now repealed), since the hazards associated with isocyanate fumes had been known about since the early 1980s).

This section deals with current statutory requirements, contained in the *Health and Safety at Work etc Act 1974* (general) (see V3004 below), the *Workplace* (*Health, Safety and Welfare*) *Regulations 1992* (*SI 1992 No 3004*) (specific) (see V3005 below), and certain specific processes where adherence to ventilation requirements is paramount (see V3006 below). (For requirements relating to ventilation on construction sites see C8057 CONSTRUCTION AND BUILDING OPERATIONS.) Origins of pollution of the working atmosphere, specific pollutants and their insidious effect on health, along with strategies for controlling them, are also covered.

Pollution of the working environment and dangers of exposure to dust and fumes

V3002 Pollution of the working environment can occur through the generation of airborne particulates, such as dusts, fumes, mists and vapours (as defined in V3003 below). Of these, probably dust and fumes pose the principal hazard to workers, some of the situations being well documented, e.g. tungsten and silicon processes resulting in pneumonia followed by progressive fibrosis, milling cotton resulting in byssinosis, weaving hemp in chronic bronchitis, underground mining in emphysema and copper welding/electroplating with cadmium also causing emphysema (see further OCCUPATIONAL HEALTH AND DISEASES). Dusts can be fibrogenic, that is they can lead to fibrotic changes to lung tissue, or, alternatively be toxic, eventually leading to poisoning of body systems. Examples of fibrogenic dusts are: silica, cement dust and certain metals such as tungsten; toxic dusts include: arsenic, lead, mercury and beryllium. More particularly, dusts are classified according to the response on the worker, as follows:

- benign pneumoconiosis, such as siderosis, associated with work with iron particles; here there is no permanent lung disorder;
- pneumonitis or acute inflammation of lung tissue, caused by inhalation of metallic fumes, e.g. zinc oxide fumes; this can result in death;
- extrinsic allergic alveolitis, e.g. farmer's lung; this can eventuate into a disabling condition;
- tumour-forming, e.g. asbestosis, plural mesothelioma (caused by work with crocidolite); characterised by a high mortality rate;
- nuisance particulates, e.g. dust from combustion of solid fuels; however causing no permanent lung damage.

Dust at the workplace has a variety of origins ranging from:

- dust arising in connection with cleaning and treatment of raw materials, e.g. sandblasting in foundries;
- dust emitted in operations such as refining, grinding, milling, cutting, sanding; and
- dusts manufactured for specific treatments or dressings; to
- general environmental dust, e.g. caused by sweeping factory floors or by fuel combustion.

Fumes are formed through vaporisation or oxidation of metals, e.g. typically lead and welding fumes. Lead processes emitting dust and fume should be enclosed and maintained under negative pressure by an enclosing hood. Moreover, the fume should be treated before any escape into the atmosphere. Welding fume, caused by action of heat and ultraviolet light, produces carbon monoxide and ozone and is potentially harmful, leading to 'welder's lung'. For that reason, welding should only be carried out in ventilated areas. Welding workshops should be equipped with adequate mechanical ventilation, and local exhaust ventilation should be incorporated at the point of fume emission as a supplement to general ventilation. If welding is carried out in a confined space (see the *Confined Spaces Regulations 1997*

(*SI 1997 No 1713*)), a permit to work system should be in force. Welders should of necessity be familiar with various relevant forms of respiratory protection (see further PERSONAL PROTECTIVE EQUIPMENT).

Polluting agents

V3003 Pollution of the workplace environment may take place through the generation of airborne particulates, such as dusts, fumes, mists and vapours. These are defined below.

Particulate: a collection of solid particles, each of which is an aggregation of many molecules.

Dust: an aerosol composed of solid inanimate particles. (Standard ILO definition).

Fumes: airborne fine solid particulates formed from a gaseous state, usually by vaporisation or oxidation of metals.

Mist: airborne liquid droplets.

Vapour: a substance in the form of a mist, fume or smoke emitted from a liquid.

Statutory requirements

General – HSWA 1974

V3004 The general duty upon employers under the *Health and Safety at Work etc Act 1974* ('*HSWA 1974*') to 'ensure, so far as reasonably practicable, the health, safety and welfare at work of all their employees' [*HSWA 1974, s 2(1)*], must be regarded as including a duty to provide employees with an adequate and renewable supply of pure and uncontaminated air. Employers must also provide and maintain a safe working environment. [*HSWA 1974, s 2(2)(e)*]. In consequence, employers may be in breach if the workplace is not adequately ventilated and specific dust and fume hazards not controlled. In addition, all employers must inform, instruct and train their employees in health and safety procedures. This means that they, or their representatives, must know how to use, test and maintain equipment for ensuring air purity and ventilation as well as dust control (*R v Swan Hunter Shipbuilders Ltd [1981] ICR 831*, where the deaths of the subcontractor employees might have been avoided, had they been properly instructed and trained in the use of respirators). (See further C8145 CONSTRUCTION AND BUILDING OPERATIONS.)

Facilities managers (i.e. those responsible for the well-being of occupants and visitors) are required to maintain a clean and healthy ventilation system in the workplace. This may mean the procurement of ventilation hygiene services from a specialist contractor which often demands a level of technical expertise outside the scope of the majority.

'General ventilation guidance for employers', HSG202, is published by HSE Books.

The Heating and Ventilating Contractors' Association (HVCA) have published a Guide to Good Practice, '*Cleanliness of Ventilation Systems – TR/17*'. This Guide and others on ventilation are available from HVCA Publications (tel: 01768 864771).

Responding to industry needs, the Building Services Research and Information Association has developed a Standard Specification for Ventilation Hygiene* to allow competitive tenders to be provided by contractors on an equal basis. The client thereby achieves optimum value for money by securing complete but not excessive action by the contractor. Correspondingly, contractors face less ambiguity in performance expectations.

The Standard Specification is in two parts, the basic specification with accompanying detailed guidance. Supported by the Department of Environment Transport and Regions, this is the first of a range of specifications written for facilities managers. It provides straightforward, easy to use clauses of industry-standard performance requirements. The guidance document explains in detail the scope of work required to fulfil a ventilation hygiene contract.

* '*Guidance and Standard Specification for Ventilation Hygiene*', price £40, is available from BSRIA publications sales, tel: 01344 426511.

Specific – Workplace (Health, Safety and Welfare) Regulations 1992 (SI 1992 No 3004)

V3005 Effective and suitable provision must be made to ensure that every enclosed workplace is ventilated by a sufficient quantity of fresh or purified air. Plant designed and used for such purposes must be accompanied with visible and/or audible means of warning of any failure, which might affect health or safety. [*Workplace* (*Health, Safety and Welfare*) *Regulations 1992* (*SI 1992 No 3004* as amended by *SI 2002 No 2174*), *Reg 6*].

Specific processes or industries

Dangerous substances and explosive atmospheres

V3006 The *Dangerous Substances and Explosive Atmosphere Regulations 2002* (*SI 2002 No 2776*) ('*DSEAR*') which came into force on 9 December 2002, govern the control of fire and explosion risks from dangerous substances and potentially explosive atmospheres. The Regulations apply to any workplace where there is present any substance or mixture of substances with the potential to create a risk from energetic (energy-releasing) events such as fire, explosions, thermal runaway from exothermic reactions etc. Such substances include: petrol, liquefied petroleum gas (LPG), paints, varnishes, solvents and certain types of dust that are explosive (e.g. wood dust).

The main requirements of *DSEAR* are that employers and the self-employed must:

- carry out an assessment of the fire and explosion risks of any work activities involving dangerous substances;
- provide measures to eliminate, or reduce as far as is reasonably practicable, the identified fire and explosion risks;
- apply measures, so far as is reasonably practicable, to control risks and to mitigate the detrimental effects of a fire or explosion;
- provide equipment and procedures to deal with accidents and emergencies; and
- provide employees with information and precautionary training.

Where explosive atmospheres may occur:

- the workplaces should be classified into hazardous and non-hazardous places; and any hazardous places classified into zones on the basis of the frequency and duration of an explosive atmosphere, and where necessary marked with a sign;
- equipment in classified zones should be safe and satisfy the requirements of the *Equipment and Protective Systems Intended for Use in Potentially Explosive Atmospheres 1996* (*SI 1996 No 192*); and,

- the workplaces should be verified as meeting the requirements of DSEAR, by a competent person.

Lead processes

V3007 The *Control of Lead at Work Regulations 1998* (*SI 1998 No 543*) (now revoked) came into force on 1 April 1998. They were introduced to ensure that work involving lead might be covered by a single set of regulations. The Regulation have now been superseded by the *Control of Lead at Work Regulations 2002* (*SI 2002 No 2676*).

The aims of the Regulations and Approved Code of Practice, Regulations and Guidance, (COP2) are to:

- protect the health of people at work by preventing or, where this is not practicable, adequately controlling their exposure to lead;
- monitor the amount of lead that employees absorb so that individuals whose work involves significant exposure (as defined by *Control of Lead at Work Regulations 2002* (*SI 2002 No 2676*), *Reg 2*) to lead at work can be taken off such work before their health is affected.

The Regulations apply to all work which exposes employees to lead in any form in which it may be inhaled, ingested or absorbed through the skin, for example as dust, fume or vapour.

The duties imposed by these Regulations extend not only to employers but also to self-employed persons. [*Control of Lead at Work Regulations 2002* (*SI 2002 No 2676*), *Reg 3*(*2*)].

Control of Lead at Work Regulations 2002 (SI 2002 No 2676)

V3008 An employer who is working with lead, or a substance or material containing it has a duty to ensure that his employees are not exposed to lead, or, where this is not reasonably practicable (see E15039 ENFORCEMENT), to ensure that appropriate control measures are adequately controlling the exposure to lead of his employees and of anyone else who may be affected by their work with lead and/or lead-based products – such persons may include:

- persons working for other employers, such as maintenance staff or cleaners;
- visitors to the work area; and
- families of those exposed to lead at work who may be affected by lead carried home unintentionally on clothing and/or footwear.

Exposure to lead covers all routes of possible exposure, i.e. inhalation, absorption through the skin and ingestion.

Without prejudice to the *Management of Health and Safety at Work Regulations 1999* (*SI 1999 No 3242*), *Reg 3*, employers must not carry on any work which is liable to expose any employees to lead at work unless they have first made an assessment as to whether the exposure of any employees to lead is liable to be significant. [*Control of Lead at Work Regulations 2002* (*SI 2002 No 2676*), *Reg 5*].

Measures for controlling exposure to lead may include one or more of the following:

- using substitutes, i.e. lead-free or low-solubility lead compounds;
- using lead or lead compounds in emulsion or paste form to minimise the formation of dust;

- using temperature controls to keep the temperature of molten lead below 500°C, the level above which fume emission becomes significant, though the formation of lead oxide and the emission of dust is still possible below this temperature;
- the containment of lead, lead materials, compounds, fumes or dust in totally enclosed plant and in enclosed containers such as drums and bags. The container must be so designed that no lead is allowed to leak out. Where it is necessary to open such containers, this should be carried out under exhaust ventilation conditions, if reasonably practicable;
- where total enclosure is not reasonably practicable, an effective ventilation system must be in operation before work is allowed to commence. This may consist of:

  (i) partial enclosure such as booths designed to prevent the lead escaping – these should be fitted with exhaust ventilation;

  (ii) various types of exhaust hoods which should be as close to the lead source as is reasonably practicable so that they may take the lead dust, fume or vapour away from the employee's breathing zone;

  (iii) an extract ductwork system that is adequate to remove the dust, fume or vapour from the source area;

  (iv) a dust and/or fume collection unit with a sound and adequate filtration system that will both remove the lead source from the workplace and prevent it from re-entering;

  (v) fans of a suitable type placed in the system after the collection and filtration units so that the units are kept under negative pressure, thus ensuring that any escape of lead is minimised;
- wet methods which include:

  (i) the wetting of lead and lead materials, e.g. wet grinding and pasting processes. Wet methods should be used during rubbing and/or scraping lead-painted surfaces;

  (ii) the wetting of floors and work benches whilst certain types of work are being carried out, e.g. work with dry lead compounds and pasting processes in the manufacture of batteries.

  Wetting should be sufficiently thorough to prevent dust forming, and the wetted materials or surfaces should not be allowed to dry out since this can create dry lead dust which then is liable to be hazardous if it becomes airborne. Water sprays are not a fully effective method of controlling airborne dust.

  Wetting methods should not be used when they are liable to be unsafe, such as:

  (i) at furnaces where they could cause an explosion;

  (ii) when lead materials containing arsenides or antimonides could, on contact with water, produce highly toxic arsine or stibine gases;
- providing and maintaining a high standard of cleanliness.

Special care and attention should be given to the design of plant and systems to eliminate possible areas which might increase the risk to the employee of exposure from lead.

Before any plant, equipment or systems are used in lead or lead compound work, there should be an adequate checking procedure in place to ensure that such plant, equipment or systems are as designed and that they meet the standard required under the *Control of Lead at Work Regulations 2002* (*SI 2002 No 2676*) for the safety of all those who could be affected by lead.

Asbestos

V3009 No process must be carried on in any factory unless exhaust equipment is provided, maintained and used, which prevents the entry into air of asbestos dust. [*Control of Asbestos at Work Regulations 2002* (*SI 2002 No 2675*), *Reg 12*(*2*)].

In addition, where it is not reasonably practicable to reduce exposure of employees to below the 'control limits', employers must provide employees concerned with suitable respiratory protective equipment. [*Control of Asbestos at Work Regulations 2002* (*SI 2002 No 2675*), *Reg 10*(*4*)].

Moreover, where the concentration of asbestos is likely to exceed any 'control limit', the employer must designate that area a 'respirator zone' and ensure that only permitted employees enter/remain in that zone. [*Control of Asbestos at Work Regulations 2002* (*SI 2002 No 2675*), *Reg 17*(*1*)(*b*)]. (See further ASBESTOS.)

Control of airborne particulates

V3010 Airborne particulates may be controlled in the following ways.

(*a*) *Substitution replacement*

The use of a less harmful toxic substance or modification to a process may minimise or totally eliminate the hazard. For example, substitution of soap solutions for organic solvents is sometimes possible in cleaning operations.

(*b*) *Suppression*

The use of a wet process for handling powders or other particulates is an effective form of control. In the cleaning process, it may be possible to damp down floors prior to removal of dust sooner than resorting to dry sweeping.

(*c*) *Isolation*

This form of control entails enclosure of all or part of a process or the actual point of dust production and may be incorporated in machinery/plant, in which case it is necessary to ensure seals are maintained. Total enclosure of large processes involving grinding of metals or other materials, linked to an extract ventilation system, is an effective method of isolating dust from the operator's breathing zone.

(*d*) *Extract/exhaust ventilation*

Removal of dust or fume at the point of emission by entraining it in a path of fresh air and taking it to an extract hood or other collection device is the role of extract/ exhaust ventilation (see the HSE's guidance booklet HSG54: '*The maintenance, examination and testing of local exhaust ventilation*'). The air velocity required to provide this movement depends upon the type of material, varying from about 0.5 metres per second for gases to 10 metres per second or above for some dusts. In fact for large dense dust particles from grinding or cutting operations, it may be necessary to arrange for the trajectory of the emitted particles to be encompassed by the exhaust hood so that the material is literally thrown into the hood by its own energy.

Extract ventilation systems generally take three distinct forms:

(i) *Receptor systems*: the contaminant enters the system without inducement, and is transported from the hood through ducting to a collection point by the use of a fan.

(ii) *Captor systems*: in this system moving air captures the contaminant at some point outside the hood and induces its flow into it. The rate of air flow must be sufficient to capture the contaminant at the furthest point of origin, and the air velocity induced at this point must be high enough to overcome the effects of cross currents created by open doors, windows or moving parts of machinery.

(iii) *High velocity low volume systems*: dusts from high speed grinding machines in particular require very high capture velocities. With an HVLV system high velocities at the source are created by extracting from small apertures very close to the source of the contaminant. These high velocities can be achieved with quite low air flow rates.

(*e*) *Dilution ventilation*

In certain cases, it may not be possible to extract particulates close to their point of origin. Where the quantity of contaminant is small, uniformly evolved and of low toxicity, it may be possible to dilute it by inducing large volumes of air to flow through the contaminated region. Dilution ventilation is most successfully used to control vapours from low toxicity solvents, but is seldom satisfactory in the control of dust and fumes.

(*f*) *Cleaning procedures*

Whilst the above methods may be effective in preventing environmental contamination of the workplace, there will inevitably be a need for efficient cleaning procedures wherever dusty processes are operated. Hand sweeping should be replaced by the use of industrial vacuum cleaning equipment, or in situ (ring main) systems which incorporate hand-held suction devices connected via ducting to a central collection point.

(*g*) *Air cleaning*

Air cleaning is often employed, either to prevent emission of noxious substances into the atmosphere or to enable some of the air to be recirculated during winter months, thus reducing heated air costs and lowering fuel bills. For particulates such as dusts and grit, air cleaning may involve some form of inertial separator, such as settling chambers or cyclones, usually followed by bag filters. For very fine dusts and fumes, bag filters are often used as the pre-filter, followed by absolute or electrostatic filters. For gases and vapours, the cleaning is usually achieved by wet scrubbers (device where the air is passed in close contact with a liquid) to take the gas into solution or react chemically with it. Scrubbers can also be used for particulate material. The selection of air cleaning types will depend upon the properties of the materials emitted and the size ranges of the particulates.

To generate air flow in the duct, various types of fans are available. The type of fan must be suitable for the system in which it is installed. This requires a knowledge of the system resistance and the fan characteristics in order to generate the desired air flow with minimum noise and power consumption.

For systems with little ducting and airflow resistance, axial fans may be used to generate high air flows. For ducted systems, centrifugal fans are often used, creating generally lower air flows per size, but at the higher air pressures required to overcome the resistances imposed by ducting and air cleaners. The fan should not be used outside its duty range. It should be sited either outside the building or as near to the discharge as possible, so that air cleaners and all ducting within the building are on the negative static pressure side of the fan, thus preventing leakage of the contaminant through cracks or defects in the duct.

New guidance* on the use of the dust lamp for observing the presence of airborne particles in the workplace has been published by the Health & Safety Executive (HSE) in the series 'Method for the Determination of Hazardous Substances' (MDHS). The guidance has been written for occupational hygienists, ventilation engineers, and health and safety practitioners. It briefly explains the principle of the dust lamp, its use in observing the presence of airborne particles and identifies its advantages and limitations. Despite its name, the dust lamp can be used to reveal the presence of many different types of airborne particulates both solids (dusts, fumes, fibres) and liquids (organic or inorganic mists). The dust lamp can be used to gauge the size and direction of movement of a particle cloud, but it does not give a quantitative measure of either concentration or particle size.

* '*MDHS 82: The dust lamp: A simple tool for observing the presence of airborne particles*', price £15.00, is available from HSE Books.

Respiratory protection

V3011 Respiratory protection implies the provision and use of equipment such as dust masks, general purpose dust respirators, positive pressure powered dust respirators, self-contained breathing apparatus or other forms of such protection. Where dust emission is intermittent or other controls are not available (as above), then resort may be made to respiratory protection. The work and care required in the selection and establishment of a respiratory protection programme, including the training of operators in the correct use of the equipment, is likely to be at least as great as any other control system. The protection must be selected to give adequate cover and minimum discomfort and the need for other personal protective devices should not be ignored. (See also PERSONAL PROTECTIVE EQUIPMENT.)

Vibration

The scope of this chapter

V5001 This chapter deals with the effects of vibration on people and the obligations of employers to assess, monitor and reduce the risks of adverse health effects of hand-arm and whole-body vibration on their workforce. This area has emerged from relative obscurity following an EC Directive (2002/44/EC) called the Physical Agents (Vibration) Directive. The Health and Safety Commission (HSC) is currently preparing new regulations on the Control of Vibration at Work, which are expected to come into operation in July 2005.

This chapter is intended to provide some basic information on vibration and to help companies to get a feel for the likely scope and requirements of the new regulations, and to consider the implications for their businesses.

An introduction to the effects of vibration on people

V5002 The effects of vibration on people can be divided into three broad classes:

- building vibration perceptible to occupiers;
- whole-body vibration of people in vehicles etc, and industrial situations;
- hand-arm vibration of people operating certain tools or machines.

It is convenient to consider these classes separately, as the cause, effect and method of control is different in each case.

Hand-arm vibration and vibration-induced white finger (VWF)

V5003 The effects of industrial vibration on people received little attention until an award of damages to seven ex-British Coal miners in July 1998, in a landmark Court of Appeal decision (*Armstrong and Others v British Coal Corporation (1998) (unreported)*. The miners received compensation ranging from £5,000 to £50,000 for the effects of vibration white finger ('VWF'), a form of hand-arm vibration syndrome characterised by the fingers becoming numb and turning white. In its early stages, the disease is reversible, but continued exposure leads to permanent damage and even gangrene, resulting in the loss of fingers or even a complete hand. The decision was important not only for the size of the awards, but also in setting out the terms in which British Coal were negligent and the standards of exposure that would be reasonable.

The Government set up a compensation scheme for miners (now closed to new claimants) to avoid the need for individual litigation, which is eventually expected to award 40,000 miners a total of £500 million. However, by January 2001, only £212 million had been paid.

This landmark award is inevitably being followed by many others. For example, in August 2000, eight workers of North-West Water received a £1.2 million settlement. They contracted VWF whilst using jackhammers and whackers when breaking and re-instating concrete surfaces.

A Medical Research Council survey in 1997–98 gave an estimate of over 300,000 sufferers from VWF in Great Britain – more than eight times the number of self-reported sufferers identified in HSE's self-reported work-related illness surveys in 1995.

The number of new cases of VWF assessed for disablement benefit dropped to 2,428 in 2001/02 from a peak of 5,403 in 1990/91. The majority of claims are from current or former coal-miners and water supply workers, probably because of the higher awareness in these industries, due to the legal actions.

In recent years, VWF has become the most common prescribed disease under the Industrial Injuries Scheme. Approximately 40 per cent of those reporting VWF in the 1995 survey also reported work-related deafness or other ear problems, indicating that work which exposes people to hand-arm vibration is often noisy.

Carpal Tunnel Syndrome

V5004 Use, either frequent or intermittent, of hand-held vibratory tools, can result in injury to the wrist: Carpal tunnel syndrome ('CTS') is one example of this. It is thought to arise in part from trapping or compression of nerves in the wrist. It can arise from repetitive twisting or gripping movements of the hand as well as from the use of vibrating tools. CTS arising from the use of vibrating tools was prescribed occupational disease A12 in April 1993. [*Social Security* (*Industrial Injuries*) (*Prescribed Diseases*) *Amendment Regulations 1993* (*SI 1993 No 862*), *Reg 6*(*2*)]. The number of new cases assessed for benefit continues to rise, with 797 cases in 2001/02.

Whole-body vibration

V5005 The four principal effects of whole-body vibration ('WBV') are considered to be:

- degraded health;
- impaired ability to perform activities;
- impaired comfort;
- motion sickness;

Exposure to whole-body vibration causes a complex distribution of oscillatory motions and forces within the body. These may cause unpleasant sensations giving rise to discomfort or annoyance, resulting in impaired performance (e.g. loss of balance, degraded vision) or present a health risk (e.g. tissue damage or deleterious physiological change). However, there is little evidence that vibration directly affects thought processes. Many factors influence human response to vibration, including the type and direction of vibration and the type of person involved. The present state of knowledge does not permit a definitive dose-effect relationship between whole-body vibration and injury or health.

People can overcome moderate effects of vibration on task performance by making greater effort, and this may initially improve performance, although at the cost of greater fatigue, with ultimate degradation of performance.

Effects on comfort depend greatly on the circumstances, for example whether the person expects to be able to read or write in the prevailing conditions.

Low frequency oscillation of the body can cause the motion sickness syndrome (kinetosis) characterised principally by pallor, sweating, nausea and vomiting.

Musculoskeletal disorders may result from WBV, a common example being back pain in drivers of lorries, tractors and other off-road vehicles. However, it is not easy to assess the degree to which WBV is the cause of back problems, as these can also be caused by lifting goods, climbing when entering or leaving vehicles, and even from prolonged sitting in the driving position.

Buildings and structures

V5006 Vibration can affect buildings and structures, but a detailed treatment of this is beyond the scope of the present chapter. People are more sensitive to vibration than are buildings or structures, so vibration within a building is likely to become unacceptable to the occupants at values well below those which pose a threat to a structurally sound building. For detailed technical guidance on the measurement and evaluation of the effects of vibration on buildings, British Standard BS 7385: Part 1: 1990 [ISO 4866: 1990] and BS 7385: Part 2: 1993 may be consulted.

Vibration measurement

Measurement units

V5007 Vibration is the oscillatory motion of an object about a given position. The rate at which the object vibrates (i.e. the number of complete oscillations per second) is called the frequency of the vibration and is measured in Hertz (Hz). The frequency range of principal interest in vibration is from about 0.5 Hz to 100 Hz, i.e. below the range of principal interest in noise control.

The magnitude of the vibration is now generally measured in terms of the acceleration of the object, in metres per second squared (i.e. metres per second, per second) denoted $m.s^{-2}$ or m/s^2. (In mathematical notation, the minus sign in front of the index, such as in s^{-2}, means that s^2 is on the bottom line. Thus in $m.s^{-2}$, m is 'divided by' s^2. In vibration work, other values of index may be encountered, such as 1 and 1.75.)

Vibration magnitude can also be measured in terms of peak particle velocity, in metres per second (denoted $m.s^{-1}$ or m/s), or in terms of maximum displacement (in metres or millimetres).

For simple vibratory motion, it is possible to convert a measurement made in any one of these terms to either of the other terms, provided the frequency of the vibration is known, so the choice of measurement term is to some extent arbitrary. However, *acceleration* is now the preferred measurement term because modern electronic instruments generally employ an *accelerometer* to detect the vibration. As the name suggests, this responds to the acceleration of the vibrating object and hence this characteristic can be measured directly.

Direction and frequency

V5008 The human body has different sensitivity to vibration in the head to foot direction, the side to side direction, and in the front to back direction, and this sensitivity varies according to whether the person is standing, sitting or lying down. In order to assess the effect of vibration, it is necessary to measure its characteristics in each of the three directions and to take account of the recipient's posture. The sensitivity to vibration is also highly frequency-dependent. BS 6841: 1987 'Measurement and evaluation of human exposure to whole-body mechanical vibration and repeated shock' provides a set of six frequency-weighting curves for use in a variety of situations. The curves may be considered analogous to the 'A'-weighting used for noise measurement. However, few instruments have these weightings built into them.

Vibration dose value

V5009 In most situations, vibration magnitudes do not remain constant, and the concept of vibration dose value ('VDV') has been developed to deal with such cases. This is analogous to the concept of noise dose used in the *Noise at Work Regulations 1989* (*SI 1989 No 1790*). Again, few instruments are capable of measuring VDV. By applying the frequency-weighting system, it is possible to introduce a single-number vibration rating. The proposed Control of Vibration at Work Regulations are likely to set an exposure limit value for whole-body vibration of $9.1\ m.s^{-1.75}$.

Structural vibration of buildings can be felt by the occupants at low vibration magnitudes. It can affect their comfort, quality of life and working efficiency. Low levels of vibration may provoke adverse comments, but certain types of highly sensitive equipment (e.g. electron microscopes) or delicate tasks may require even more stringent criteria.

Adverse comment regarding vibration is likely when the vibration magnitude is only slightly in excess of the threshold of perception and, in general, criteria for the acceptability of vibration in buildings are dependent on the degree of adverse comment rather than other considerations such as short-term health hazard or working efficiency.

Vibration levels greater than the usual threshold may be tolerable for temporary or infrequent events of short duration, especially when the risk of a startle effect is reduced by a warning signal and a proper programme of public information.

Detailed guidance on this subject may be found in British Standard BS 6472: 1992 'Evaluation of human exposure to vibration in buildings (1 Hz to 80 Hz)'.

Causes and effects of hand-arm vibration exposure

V5010 Intense vibration can be transmitted to the hands and arms of operators from vibrating tools, machinery or workpieces. Examples include the use of pneumatic, electric, hydraulic or engine-driven chain-saws, percussive tools or grinders. The vibration may affect one or both arms, and may be transmitted through the hand and arm to the shoulder.

The vibration may be a source of discomfort, and possibly reduced proficiency. Habitual exposure to hand-arm vibration has been found to be linked with various diseases affecting the blood vessels, nerves, bones, joints, muscles and connective tissues of the hand and forearm, most commonly VWF.

VWF arises from progressive loss of blood circulation in the hand and fingers, sometimes resulting in necrosis (death of tissue) and gangrene for which the only solution may be amputation of the affected areas or complete hand. Initial signs are mild tingling and numbness of the fingers. Further exposure results in blanching of the fingers, particularly in cold weather and early in the morning. The condition is progressive to the base of the fingers, sensitivity to attacks is reduced and the fingers take on a blue-black appearance. The development of the condition may take up to five years according to the degree of exposure to vibration and the duration of such exposure.

Vibration-induced white finger as an occupational disease

V5011 Vibration white finger ('VWF') is reportable under the *Reporting of Injuries, Diseases and Dangerous Occurrences Regulations 1995* (*RIDDOR*) (*SI 1995 No 3163*).

VWF is prescribed occupational disease A11. It is described as: 'episodic blanching, occurring throughout the year, affecting the middle or proximate phalanges or in the case of a thumb the proximal phalanx, of:

(*a*) in the case of a person with five fingers (including thumb) on one hand, any three of those fingers;

(*b*) in the case of a person with only four such fingers, any two of those fingers; or

(*c*) in the case of a person with less than four such fingers, any one of those fingers or ... the remaining one finger.'

[*Social Security* (*Industrial Injuries*) (*Prescribed Diseases*) *Regulations 1985* (*SI 1985 No 967*)].

Action at common law

V5012 In July 1998, the Court of Appeal upheld an award of damages to seven employees of British Coal claiming damages for VWF. The court determined that after January 1976, British Coal should have implemented a range of precautions, including training, warnings, surveillance and job rotation where exposure to vibration was significant.

The court went on to consider what degree of exposure would be reasonable. They supported the standards set out in the HSE Guidance booklet *Hand-arm vibration* published in 1994, which establishes limits for vibration dose measured as an eight-hour average or A8. The court considered that an exposure of 2.8 m/sec^2 A8 would be an appropriate level for prudent employers to use. This level of exposure would lead to a 10 per cent risk of developing finger blanching (the first reversible stage of VWF) within eight years.

The court also suggested a form of wording to warn employees: 'If you are working with vibrating tools and you notice that you are getting some whitening or discolouration of any of your fingers then, in your own interests, you should report this as quickly as possible. If you do nothing, you could end up with some very nasty problems in both hands.'

Stages of vibration white finger

V5013 Damages are awarded according to the stage of the disease at the time of the action from the date the employer should have known of the risk. The Taylor-Pelmear Scale is usually used to describe these stages as follows.

Taylor-Pelmear Scale System		
Stage	*Grade*	*Description*
0	—	No attacks
1	Mild	Occasional attacks affecting the tips of one or more fingers

2	Moderate	Occasional attacks affecting the tips and the middle of the fingers (rarely the base of the fingers) on one or more fingers
3	Severe	Frequent attacks affecting the entire length of most fingers
4	Very severe	As in stage 3, with damaged skin and possible gangrene in finger tips

Prescription of hand-arm vibration syndrome

V5014 Foreseeably, prescription may extend to hand-arm vibration syndrome ('HAVS') instead of VWF. The former would cover recognised neurological effects as well as the currently recognised vascular effects of vibration. Neurological effects will include numbness, tingling in the fingers and reduced sensibility. Moreover, the current list of occupations, for which VWF is prescribed, may be replaced by a comprehensive list of tools/rigid materials against which such tools are held, including:

- percussive metal-working tools (e.g. fettling tools, riveting tools, drilling tools, pneumatic hammers, impact screwdrivers);
- grinders/rotary tools;
- stone working, mining, road construction and road repair tools;
- forest, garden and wood-working machinery (e.g. chain saws, electrical screwdrivers, mowers/shears, hedge trimmers, circular saws); and
- miscellaneous process tools (e.g. drain suction machines, jigsaws, pounding-up machines, vibratory rollers, concrete levelling vibratibles).

Physical Agents (Vibration) Directive (2002/44/EC) and proposed Control of Vibration at Work Regulations

V5015 After ten years of debate, in 2002 the European Commission issued the Physical Agents (Vibration) Directive 2002/44/EC on the protection of workers from exposure to vibration. Member states have until July 2005 to transpose the directive into national law and accordingly, HSC have produced and consulted on draft Control of Vibration at Work Regulations. These recognise that the health risks and methods of control of hand-arm vibration and of whole-body vibration are very different. The Regulations will apply to work activities on mainland Britain, on offshore workplaces and in aircraft in flight over Britain, although exemption certificates may be granted to emergency services, air transport and the Ministry of Defence. Separate Vibration Regulations will be introduced in Northern Ireland and for sea transport.

For *hand-arm vibration*, there will be an exposure action value of 2.5 m/s^2, at which specified action is to be taken to reduce risk, and an exposure limit value at 5 m/s^2 daily personal exposure which must not be exceeded except in certain pre-permitted circumstances. The daily personal exposure values are to be adjusted (normalised) to an 8-hour reference period.

For *whole-body vibration*, there will be an exposure action value at 0.5 m/s^2 and an exposure limit value of 1.15 m/s^2 daily personal exposure, offending work activities must be notified. (HSC are consulting on whether vibration dose value should be used instead of acceleration for the action value.)

A long transitional period is proposed. The limit values will not come into effect until July 2007 for new equipment or July 2010 for existing equipment, or until July 2014 for existing equipment used by employees in the agricultural and forestry sectors.

Employers must undertake a health risk assessment if any of their employees are likely to be exposed to a vibration risk. This includes consideration of people who may be particularly susceptible to vibration, the maintenance of equipment and the availability of replacement equipment that reduces vibration exposure.

Consideration must be given to aggravating circumstances, such as exposure to cold, awkward working conditions and workplace exposure extending beyond normal working hours, such as when rest areas subject to vibration, as often encountered in offshore installations.

Health surveillance is required in circumstances where there is a link between the level of exposure and known adverse health effects, and there are a range of obligations when an adverse effect is detected.

Information, instruction and training is to be provided to persons who may be exposed to a risk from vibration.

Impact on employers

V5016 The new Regulations will impose a new discipline on employers in an unfamiliar area. Few employers or health and safety officers will have any feeling for the degree of vibration risk that their processes entail, even though there have been some specific obligations on employers for some years, in addition to the general duty of care under the *Health and Safety at Work etc. Act 1974*.

Such obligations include, amongst others, those under:

- the *Reporting of Injuries, Diseases and Dangerous Occurrences Regulations 1995* (*SI 1995 No 3163*);
- the *Provision and Use of Work Equipment Regulations 1998* (*SI 1998 No 2306*);
- the *Supply of Machinery* (*Safety*) *Regulations 1992* (*SI 1992 No 3073* as amended by *SI 2004 No 2063*).

Advice and obligations in respect of hand-arm vibration

Vibration magnitudes

V5017 The energy level of the hand tool is significant. Percussive action tools, such as compressed air pneumatic hammers, operate within a frequency range of 33–50 Hz. These cause considerable damage whereas rotary hand tools, which operate within the frequency range 40–125 Hz, are less dangerous.

The Health and Safety Executive (HSE) intends to publish guidance on the vibration magnitudes of a variety of tools in a selection of typical working conditions. Vibration can vary not only with the type and design of the tool, but also with other things such as the task, the operator's technique and the material being worked.

For example, electric impact drills (hammer drills) create vibration of around 10 m/s^2 on average, but can go as high as 25 m/s^2. At 10 m/s^2, the exposure action value would be reached after about 30 minutes use and the exposure limit value would be reached after about 2 hours use.

Angle grinders produce a similar range of exposure values. Older models of needle scalers produce average vibration magnitudes in the range of 8 to 18 m/s^2, but newer vibration-reduced models produce vibration magnitudes between 4 and 7 m/s^2. It is clear from this data that most of these tools could not be used for a whole working day without exceeding the vibration limit value, and none could be used without exceeding the exposure action value. Indeed, it appears that most currently-available vibrating hand tools will exceed the vibration exposure action value, and are very likely to exceed the exposure limit value if used for a whole working day.

It would seem that all employers whose work processes use vibratory tools will be advised to undertake a vibration risk assessment.

Managing vibration exposure

V5018 Where a tool is used for a well-defined process, it might be helpful to control the vibration exposure by limiting the number of at-risk tasks in a day, rather than the number of hours use. For example, it may be found that an operator can drill 50 to 60 holes 100 mm deep in concrete before reaching the exposure action value, and 200 to 230 such before reaching the exposure limit (this is a theoretical example).

Many work processes that involve hand-held power tools will require employees to undertake a variety of tasks, possibly using different tools, each with different vibration exposures. In this case, it could be more difficult to set a limit on the number of operations that are allowable. An alternative is to use a 'points' system.

For example, an employer might allocate points to vibration exposure such that 100 points per day is equal to the exposure action value. The employer assesses the vibration 'dose' of an angle grinder as 30 points per hour, and of an impact drill as 120 points per hour. This means that the grinder gives an exposure of 5 points every 10 minutes and the impact drill gives 20 points for every 10 minutes use. Thus 20 minutes use of the drill plus 120 minutes use of the grinder would cause the exposure action value to be reached.

The allocation of points to each tool will have to be calculated using a formula which depends on the vibration magnitude of the tool and process. Note that the system described here is only intended to be indicative, as vibration dose does not strictly obey the rules of simple arithmetic.

Personal protection against hand-arm vibration

V5019 Workers can reduce the risk of injury from hand-arm vibration by keeping the blood flowing while working, for example by keeping warm, not smoking especially whilst working, and exercising hands and fingers. Gloves and other warm clothing should be provided when working in cold conditions.

Tools should be designed for the job, both to lessen vibration and to reduce the strength of grip and amount of force needed. The equipment should be used in short bursts rather than long sessions. Symptoms should not be ignored

There are a number of 'anti-vibration' gloves on the market that conform with the relevant standard BS EN ISO 101819:1997. However, these may seriously limit dexterity and can in some circumstances actually increase the risk of vibration injury. The standard is under revision, but at present, anti-vibration gloves should not be relied upon to reduce vibration exposure significantly. Very often, good quality industrial gloves will provide equal levels of protection.

Manufacturers' warning of risk

V5020 The *Supply of Machinery* (*Safety*) *Regulations 1992* (*SI 1992 No 3073* as amended by *SI 2004 No 2063*), establish essential health and safety requirements for machinery supplied in the European Economic Area. This includes information on vibration from hand-held, hand guided and mobile machinery. Machinery manufacturers and suppliers have a duty to warn whenever a machine carries a vibration risk, and to provide information for hand-held or hand-guided machines where vibration emissions exceed 2.5 m/s^2. This is the vibration emission of the machine, not the daily exposure of the user, which will depend on the amount of use of this and other machines. It should be noted that 2.5 m/s^2 is not to be regarded as an acceptable target for the amount of vibration a machine may emit – there is an overriding duty to reduce vibration emission as far as possible within the current state of the art.

Manufacturers are required to test vibration emissions in accordance with test procedures aimed at obtaining a reproducible value, and they must declare both the typical measured value and the amount of uncertainty (variation) of measurements from the typical value. Thus, the typical value plus the uncertainty value should give the 'worst case', but it may still not represent the total amount of vibration in real use. This can complicate the choice of tool, to the extent that sometimes a tool that gives slightly more vibration than another tool may still be better if it is more comfortable, appropriate or quicker for the task.

Moreover, tools that are incorrectly assembled, used wrongly, with worn or loose parts can give many times the amount of vibration than normal.

An action plan

V5021 Although there is a long transitional period, employers should commence an action plan as soon as possible, starting with a risk assessment. All hand-held vibrating tools, and especially those using some sort of hammer action, have a risk: rotary tools usually have lower vibration emissions, but can still present a risk when used for long periods.

Employers should also check any warnings that the equipment suppliers provide, and whether there is any generally-known vibration problem in their industry. They should also ask their employees whether any of them have any problems that could be symptomatic of HAVS. If any employee is diagnosed with HAVS or vibration-related carpal tunnel syndrome, the employer must report this to the local health and safety authority (either HSE or the local authority depending on the industry.)

Formal assessments

V5022 It is quite difficult to measure the vibration emission of equipment in actual use, as the measuring device must be attached to the equipment in such a place and in such a manner as to represent the vibration being transmitted to the operator. The measuring system must be capable of determining the amount of vibration over a range of frequencies and in different directions. Hence, it is usually necessary to rely on general advice.

The first part of the plan should therefore be to try to find ways to eliminate the use of vibrating tools, to use the most appropriate equipment to get the job done quickly and safely, and to make sure that the tool is correctly assembled and used, with the correct cutting bit kept sharp.

Advice and obligations in respect of whole-body vibration

Incidence of whole-body vibration

V5023 Although some millions of people are exposed to whole-body vibration ('WBV') at work, most of them are users of road vehicles and are not likely to be at risk of injury from WBV. Nevertheless, more than 1.3 million workers are thought to be exposed to vibration above the exposure action value and more than 20,000 are thought to be exposed to more than the WBV limit value.

WBV can cause or worsen back pain, and the effect is aggravated when the vibration contains severe shocks and jolts. These conditions occur particularly in off-road driving, such as in farming, construction and quarrying, but can also occur in small, fast boats at sea and in some helicopters. Old railway vehicles running on trackwork in poor condition can also give rise to high levels of WBV.

Back pain can arise from many causes other than WBV, and can be aggravated by poor driving posture, sitting for long periods without being able to change position, awkwardly placed controls that require the operator to stretch or twist to reach them, manual lifting, and repeatedly climbing into or jumping out of a high cab.

Jobs which combine two or more of these factors can increase the risk further, for example, driving over rough ground whilst twisting round to check on the operation of equipment being operated behind the vehicle.

Assessing the risk of WBV problems

V5024 Road transport drivers are not usually at risk unless there are other aggravating factors such as long hours of driving, or use of unmade or poor road surfaces. However, if there is a history of back pain in the job, this could be indicative of a problem, and it may be wise to consider the regulations and guidance on manual handling of materials in addition to the vibration regulations.

Operators of off-road vehicles are at greater risk. Indicators of risk include the operator being thrown around in the cab, or receiving shocks and jolts through the seat, or if the manufacturer warns of a risk. A history of back pain is again an indicative factor.

Keeping records

V5025 Risk assessments should be recorded. These should show those jobs or processes identified as being at risk from WBV; an estimate of the amount of exposure; comparison of exposure with the action and limit values; the available risk controls; the plans to control and monitor risks; and the effectiveness of the control measures.

Reducing the risk of WBV problems

V5026 When a risk is identified the obvious initial actions are to ensure that the machine is suitable for the job, is used correctly and driven at safe speeds for the ground conditions, and that seats are correctly adjusted for the driver's size and weight. Access routes on work sites should be maintained in good condition, for example by using a scalping blade and roller.

Exposure assessments

V5027 An exposure assessment should be done by a competent person, but an initial assessment will usually be made from published data and does not need to be highly accurate, so it can often be done by company staff who have had some training. If the initial assessment shows that exposure is approaching or in excess of the exposure limit value, a measurement of vibration levels may be needed: this is more complex and is likely to require a specialist.

Health surveillance

V5028 HSE currently considers that because there is no way in which specific occurrences of back pain can be attributed to WBV, it is not appropriate to undertake health surveillance. However, it suggest that 'health monitoring' is advisable, though not obligatory. Health monitoring is a system for reporting and recording symptoms that could be attributable to WBV, manual handling or posture, so that action can be taken if there is a high incidence of problems.

HSE guidance on hand-arm and whole-body vibration

V5029 HSE publishes a full guide on *Hand-arm vibration* (HS(G)88), ISBN 0 7176 0743 7. It also publishes a book of 51 case studies dealing with vibration, titled *Vibration solutions: Practical ways to reduce the risk of hand-arm vibration injury'*, ISBN 0 7176 0954 5. An HSE video pack *Hard to handle* includes a book and a leaflet, along with a 15-minute video giving an introduction to the risks of hand-arm vibration and how to manage them. A CD-ROM *The successful management of hand-arm vibration* (1999), for managers and safety specialists, has also been published by HSE.

In addition to the above, HSE publishes a number of leaflets, including:

- *Hand-arm vibration: Advice for employees and the self-employed* (1999), ISBN 0 7176 1554 5;
- *Health risks from hand-arm vibration: Advice for employers* (1998), ISBN 0 7176 1553 7;
- *Reducing the risk of hand-arm vibration injury among stonemasons* (1998), MISC 112.

A whole new series of guidance documents are in preparation for use with the proposed new regulations.

Vulnerable Persons

Introduction

V12001 This chapter deals with groups of people who, for one reason or another, are more vulnerable to health and safety risks. Such people include:

- children and young persons;
- new or expectant mothers;
- lone workers;
- disabled persons; and
- inexperienced workers.

Where these persons are at work they are subject to the protection of the full range of health and safety legislation – the *Health and Safety at Work etc. Act 1974* ('*HSWA 1974*') and the various regulations made under it. Of particular relevance are the *Management of Health and Safety at Work Regulations 1999* (*SI 1999 No 3242*) ('the *Management Regulations*') and their requirements for risk assessments. The potential vulnerability of some members of the workforce must be taken into account during the risk assessment process. In addition, the *Management Regulations* contain specific requirements in respect of children and young persons and also new or expectant mothers. These will be explained later in the chapter.

However, *HSWA 1974, s 3* places duties on employers and the self-employed in relation to persons not in their employment such as visitors, occupants of premises, service users, neighbours or passers-by. Risks to such persons must also be considered during risk assessments with appropriate attention given to those who are particularly vulnerable. The general principles of risk assessment are described in the chapter dealing with that subject.

This chapter concentrates on legislation of particular relevance to vulnerable persons and many of the practical considerations to be taken into account when carrying out risk assessments relating to such people.

Relevant legislation

V12002 Whilst the *Management Regulations* (*SI 1999 No 3242*) impose specific requirements in relation to children and young persons, together with new or expectant mothers, there are other regulations where the vulnerability of some people is either recognised directly or must be taken into account indirectly through some form of risk assessment.

Management Regulations requirements

V12003 The terms 'child' and 'young person' are defined in *Regulation 1(2)* of the *Management Regulations* (*SI 1999 No 3242*), whilst *subsections* (*4*) and (*5*) of

Regulation 3 contain requirements relating to risk assessments and young people. *Regulation 19* of the Regulations deals with precautions which must be taken to ensure the 'Protection of Young Persons'.

'New or expectant mother' is also defined in *Regulation 1(2)* whilst there are three regulations relating directly to them:

- *Regulation 16* – deals with risk assessments;
- *Regulation 17* – covers possible suspension from night work;
- *Regulation 18* – contains requirements in respect of notification to the employer of pregnancy or having given birth.

All these definitions and requirements are explained in full later in the chapter.

Other regulations directly affecting vulnerable persons

V12004

- *Children and young persons*

  Several sets of regulations contain age-related prohibitions or restrictions on specified activities. These are set out and explained in TABLE A later in the chapter (see V12015 below).

  Also of relevance are the *Health and Safety (Training for Employment) Regulations 1990 (SI 1990 No 1380)* which give students on work experience programmes and trainees on training for employment programmes the status of 'employees'. The immediate provider of the training is treated as their 'employer'. Whilst this particularly affects children and young persons, it also applies to adult participants in training programmes. (Courses at schools, colleges and universities etc. are not covered by these requirements.)

- *New or expectant mothers*

  Some regulations contain requirements for employees engaged in certain activities to be free from medical or physical conditions which make them unfit for that activity. Typical of these are the *Work in Compressed Air Regulations 1996 (SI 1996 No 1656)* and the *Diving at Work Regulations 1997 (SI 1997 No 2776)*.

  Both the *Ionising Radiations Regulations 1999 (SI 1999 No 3232)* and the *Control of Lead at Work Regulations 2002 (SI 2002 No 2676)* impose tighter exposure standards on *all* women of reproductive capacity whilst the latter Regulations would almost certainly result in the suspension from lead work of a woman once her pregnancy was notified. (These requirements are explained in more detail later in the chapter – see V12018 below.)

Regulations with indirect implications relating to vulnerable persons

V12005

- *Construction (Design and Management) Regulations 1994 (SI 1994 No 3140) (CDM)*

  When construction work is to be carried out in or close to premises occupied by children (or those with special educational needs) the CDM Health and Safety Plan must take particular account of their lack of awareness of risks and general curiosity. If normal access routes in premises are affected this may create risks or problems for pregnant women, disabled persons or the elderly.

- *Manual Handling Operations Regulations 1992* (*SI 1992 No 2793*)

  The limited manual handling capabilities of, and potential risks to, both children and young persons, and new or expectant mothers must be taken into account during risk assessments. However, risks to lone workers (with no source of assistance), persons with temporary or permanent physical disabilities and inexperienced workers (who may be unaware of relevant handling techniques) must also be considered.

- *Control of Substances Hazardous to Health Regulations 2002* (*SI 2002 No 2677*) (*COSHH*)

  Children, persons with special educational needs or those with fading mental faculties may all be at extra risk from hazardous substances as they may not be able to read or understand warnings or instructions. COSHH assessments should take account of the potential presence of such vulnerable persons and ensure the secure storage of hazardous substances where this is appropriate, e.g. by locking cleaners' cupboards.

- *Personal Protective Equipment at Work Regulations 1992* (*SI 1992 No 2966*)

  Use of certain types of personal protective equipment ('PPE') may present difficulties for pregnant women or workers with physical difficulties. Young and inexperienced workers are likely to need more detailed explanations of the purposes of different types of PPE and how and when they are to be used, whilst those with limited intellectual capabilities will need even closer attention.

- *Health and Safety* (*Display Screen Equipment*) *Regulations 1992* (*SI 1992 No 2792*)

  The suitability of a display screen equipment ('DSE') workstation for a pregnant woman will need to be assessed, particularly as her size increases. Disabled workers or other staff with physical problems (e.g. past injuries) are likely to require special attention during workstation assessments, especially in relation to the suitability of chairs and other ergonomic aspects.

Children and young persons

V12006 For many years work done by children and young persons was subject to an often bewildering array of prohibitions – equipment that children and young persons were not allowed to use or processes or activities that they must not be involved in. However, since 1997 the emphasis has switched from such prohibitions towards restrictions, based upon a process of risk assessment.

Many of the previous age-related prohibitions were revoked but some still remain. The more important of these are listed in TABLE A at V12015 below. However, the ongoing programme of modernisation of health and safety legislation is continually sweeping away many of these residual prohibitions. (Any cases of doubt should be referred to the HSE.)

Definitions

V12007 The terms 'child' and 'young person' are defined in *Regulation 1*(*2*) of the *Management Regulations* (*SI 1999 No 3242*).

Child is defined as under the minimum school leaving age ('MSLA') in accordance with:

- *section 8* of the *Education Act 1996* (for England and Wales);
- *section 31* of the *Education (Scotland) Act 1980* (for Scotland).

(The MSLA is just before or just after the age of sixteen)

Young Person is defined as 'any person who has not attained the age of eighteen'.

Some of the prohibitions remaining from older health and safety regulations use different cut off ages – where relevant these are referred to in TABLE A at V12015 below. Further detail is also provided in the HSE guidance booklet HSG 165 'Young people at work'.

Risk assessment requirements

V12008 *Regulation (3)* of the *Management Regulations (SI 1999 No 3242)* contains important requirements (first introduced in 1997) in relation to risk assessments and young persons. These are:

'(4) An employer shall not employ a young person unless he has, in relation to risks to the health and safety of young persons, made or reviewed an assessment in accordance with paragraphs (1) and (5).

(5) In making or reviewing the assessment, an employer who employs or is to employ a young person shall take particular account of—

(a) the inexperience, lack of awareness of risks and immaturity of young persons;

(b) the fitting-out and layout of the workplace and the workstation;

(c) the nature, degree and duration of exposure to physical, biological and chemical agents;

(d) the form, range and use of work equipment and the way in which it is handled;

(e) the organisation of processes and activities;

(f) the extent of the health and safety training provided or to be provided to young persons; and

(g) risks from agents, processes and work listed in the Annex to Council Directive 94/33/EC(b) on the protection of young persons at work.'

(Note: *Paragraph (1)* of the regulation contains the basic requirement for employers to carry out risk assessments.)

Protection of young persons

V12009 In assessing the risks to young persons, the employer must take particular note of *Regulation 19* of the *Management Regulations (SI 1999 No 3242)* which states:

'(1) Every employer shall ensure that young persons employed by him are protected at work from any risks to their health or safety which are a consequence of their lack of experience, of absence of awareness of existing or potential risks or the fact that young persons have not yet fully matured.

(2) Subject to paragraph (3), no employer shall employ a young person for work—

(a) which is beyond his physical or psychological capacity;

(b) involving harmful exposure to agents which are toxic or carcinogenic, cause heritable genetic damage or harm to the unborn child or which in any way chronically affect human health;

(c) involving harmful exposure to radiation;

(d) involving the risk of accidents which it may reasonably be assumed cannot be recognised or avoided by young persons owing to their insufficient attention to safety or lack of experience or training; or

(e) in which there is a risk to health from—

(i) extreme cold or heat;

(ii) noise; or

(iii) vibration,

and in determining whether work will involve harm or risk for the purposes of this paragraph, regard shall be had to the results of the assessment.

(3) Nothing in paragraph (2) shall prevent the employment of a young person who is no longer a child for work—

(a) where it is necessary for his training;

(b) where the young person will be supervised by a competent person; and

(c) where any risk will be reduced to the lowest level that is reasonably practicable.'

Some of the practical implications of these provisions are considered later in the chapter.

Information on risks to children

V12010 *Regulation 10* of the *Management Regulations* (*SI 1999 No 3242*) deals with 'information for employees' and contains two paragraphs relating to the employment of children. These state:

'(2) Every employer shall, before employing a child, provide a parent of the child with comprehensible and relevant information on—

(a) the risks to his health and safety identified by the assessment;

(b) the preventive and protective measures; and

(c) the risks notified to him in accordance with Regulation 11(1)(c).

(3) The reference in paragraph (2) to a parent of the child includes—

(a) in England and Wales, a person who has parental responsibility, within the meaning of section 3 of the Children Act 1989, for him; and

(b) in Scotland, a person who has parental responsibility, within the meaning of section 8 of the Law Reform (Parent and Child) (Scotland) Act 1986, for him.'

This requirement to provide information to parents includes situations where children are on work experience programmes (where they have the status of employees by virtue of the *Health and Safety* (*Training for Employment*) *Regulations 1990* (*SI 1990 No 1380*)) and also includes part time or temporary work.

Some of the practical implications of this requirement are considered later in this chapter.

Disapplication of requirements

V12011 The wide-ranging impact of these provisions is lessened to a limited extent by *paragraph* (*2*) in R*egulation 2* of the *Management Regulations* (*SI 1999 No 3242*) – 'Disapplication of these Regulations'.

> '(2) Regulations 3(4), (5), and 10(2) and 19 shall not apply to occasional work or short-term work involving—
>
> (a) domestic service in a private household; or
>
> (b) work regarded as not being harmful, damaging or dangerous to young people in a family undertaking.'

However, the term 'family undertaking' is not defined in the Regulations. HSE guidance indicates that this should be interpreted as meaning a firm, owned by, and employing members of the same family, i.e. husbands, wives, fathers, mothers, grandfathers, grandmothers, stepfathers, stepmothers, sons, daughters, grandsons, granddaughters, stepsons, stepdaughters, brothers, sisters, half-brothers and half-sisters.

As far as is known this narrow interpretation of 'family undertaking' has not been tested in the courts. Common usage of the term would suggest small and medium-sized businesses controlled and managed by members of the same family but not necessarily only employing family members, as implied by the words used by the HSE.

Capabilities and training

V12012 *Regulation 13* of the *Management Regulations* (*SI 1999 No 3242*) states in *paragraph* (*1*) that 'Every employer shall, in entrusting tasks to his employees, take into account their capabilities as regards health and safety'. Quite clearly the capabilities of children and young persons will be somewhat different from those of more experienced and mature employees. (This requirement is also of relevance to other vulnerable persons e.g. disabled or inexperienced workers.)

Risk assessment in practice

V12013 The requirements of *Regulations 3*(*4*), (*5*) and *19* of the *Management Regulations* (*SI 1999 No 3242*) are based upon the contents of a European Directive (94/33/EC) and it is not particularly easy for the employer to identify exactly what he or she must (or must not) do.

A good starting point is to consider the three characteristics associated with young people which are mentioned in both *Regulations 3*(*5*) and *19*(*1*):

- lack of experience;
- lack of awareness of existing or potential risks; and
- immaturity (in both the physical and psychological sense).

All young people share these characteristics but to differing extents – for example, one would have different expectations of a school leaver who had already been playing a prominent role in a family business, e.g. a farm, as opposed to a work experience student with no previous exposure to the world of work. Employers must also be aware that the degrees of physical and psychological maturity of young people vary hugely.

These three characteristics must then be considered in respect of the risks involved in the employer's work activities and particularly those identified in *Regulation 3(5)(b)–(g)* and *Regulation 19(2)*.

The contents of those Regulations (including those risks listed in the Annex to Council Directive 94/33/EC) have been consolidated into a single checklist for employers which is contained in TABLE B.

The purpose of all risk assessments is to identify what measures the employer needs to take in order to comply with the law. Additional measures which the employer must consider in order to provide adequate protection for young persons are:

- not exposing the young person to the risk at all;
- providing additional training;
- providing close supervision by a competent person;
- carrying out additional health surveillance (as required by *Regulation 6* of the *Management Regulations* (*SI 1999 No 3242*) or other regulations e.g. *COSHH* (*SI 2002 No 2677*));
- taking other additional precautions.

Regulation 19(3) allows more latitude in respect of young persons who are no longer children in relation to the requirement to completely exclude them from the types of risks specified in *Regulation 19(2)*.

In deciding what precautions are required, the employer must consider both young people generally and the characteristics of individual young persons. Additional training and/or supervision may be necessary in respect of young people with 'special needs'.

Young people should gradually acquire more experience, awareness of risks and maturity, particularly as they pass through formal training programmes within the NVQ system. As this occurs, restrictions on their activities may be progressively removed.

Provision of information

V12014 *Regulation 10(1)* of the *Management Regulations* (*SI 1999 No 3242*) requires employers to provide all employees with comprehensible and relevant information on:

- risks to their health and safety identified by risk assessments (or notified by other employers);
- preventive and protective measures (i.e. appropriate precautions);
- emergency procedures and arrangements.

This requirement is of particular importance in relation to young persons. *Regulation 10(2)* requires employers *also* to provide information to parents of children (under the MSLA) on the risks the children will be exposed to and the precautions which are in place.

The information may be provided either orally or in writing or possibly both. Key items (e.g. critical restrictions or prohibitions) should be recorded. Means of providing information might include:

- induction training programmes;
- employee handbooks or rulebooks;
- job descriptions;
- formal operating procedures;
- trainee agreement forms (increasingly common for young persons on formal training programmes e.g. modern apprenticeships);
- information forms for parents of work experience students.*

  * Further guidance on work experience is provided in the HSE booklet HSG 199 'Managing health and safety on work experience: A guide for organisers'.

The information must be comprehensible – special arrangements may be necessary for young people whose command of English is poor or for those with special needs. The type of information required to be provided will obviously relate to the work activities and the risks involved. The content must be relevant – both to the workplace and the young person. It might include the following types of information:

- general risks present in the workplace

  e.g. fork lift trucks are widely used in the warehouse.
- general precautions taken in respect of those risks

  e.g. all fork lift drivers are trained to the standard required by the Approved Code of Practice ('ACoP').
- specific precautions in respect of the young person

  e.g. the induction tour includes identification of areas where fork lift trucks operate and indication of warning signs.
- restrictions or prohibitions on the young person

  e.g. X will not be allowed to drive fork lift trucks or any other vehicles (he will be considered for fork lift truck training after attaining the age of 17).
- supervision arrangements

  e.g. X will be supervised by the warehouse foreman (or other persons designated by him).
- PPE requirements

  e.g. safety footwear must be worn by all employees working in the warehouse (this is supplied by the company).

Where restrictions or prohibitions are removed (e.g. after successful completion of training programmes) an appropriate record should be made, either on the original restriction/prohibition or within the individual's training record.

V12015

Table A *Prohibitions on children and young persons*
Outright prohibitions on work by children and young persons continue to be revoked, and to be replaced by a risk assessment approach. Prohibitions known still to be in place at the time of publication in 2000 of the HSE booklet 'Young people at work: A guide for employers' (HSG 165) were:
Carriage of explosives and dangerous goods
Under-18s may not be employed as a driver or attendant of an explosives vehicle, be responsible for the security of the explosives and may only enter the vehicle under the direct supervision of someone over the age of 18. (There are some exceptions where the risks are low). [*Carriage of Explosives by Road Regulations 1996* (*SI 1996 No 2093*]. Under-18s may not supervise road tankers or vehicles carrying dangerous goods nor supervise the unloading of petrol from a road tanker at a petrol filling station. [*Carriage of Dangerous Goods by Road Regulations 1996* (*SI 1996 No 2095* as amended].
Agriculture
Under-13s may not ride on vehicles and machines including tractors, trailers etc. [*Prevention of Accidents to Children in Agriculture Regulations 1998* (*SI 1998 No 3262*]. HSE guidance also states that children (under the minimum school leaving age) should not operate certain machines and tractors carrying out certain operations.
Ionising radiation
Under-18s may not be designated as 'classified persons'. Dose exposure limits are lower for under-18s. [*Ionising Radiations Regulations 1999 SI 1999 No 3232*].
Lead
Under-18s may not be employed in certain lead processes – lead smelting and refining, lead-acid battery manufacturing – or to clean places where such processes are carried out.
Mines and quarries
Various restrictions exist for under-18s and under-16s relating to use of winding and rope haulage equipment, conveyors at work faces, locomotives, shunting and shot firing. Some restrictions also apply to under-21s and under-22s.
Shipbuilding and shiprepairing
Under-18s, until they have been employed in a shipyard for six months, may not be employed on staging or in any part of a ship where they are liable to fall more than two metres or into water where there is a risk of drowning. [*Shipbuilding and Shiprepairing Regulations 1960* (*SI 1960 No 1932*].
Docks
Under-18s may not operate powered lifting appliances in dock operations unless undergoing a suitable course of training under proper supervision of a competent person (serving members of HM Forces are exempt). [*Docks Regulations 1988* (*SI 1988 No 1655*)].

Work presenting increased risks for children and young persons

V12016 The checklist below (TABLE B) is based upon *Regulations 3(5)* and *19(2)* of the *Management Regulations* (*SI 1999 No 3242*) and the Annex to the European Council Directive 94/33/EC. It is intended to assist employers conducting risk assessments in respect of work by children and young persons.

These types of work or situations of exposure to risk are not necessarily prohibited, although the requirements of *Regulation 19(2)* must be taken into account. However, restrictions could be required for young persons (particularly children) and additional precautions may be required to provide them with adequate protection from risk. Many situations present no greater risk to young persons than adults and restrictions (e.g. close supervision) may only be necessary until the employer is sure that the young person is fully aware of the risks and is capable of taking the necessary precautions.

Table B
Excessively physically demanding work • Manual handling operations where the force required or the repetitive nature could injure someone whose body is still developing (including production line work); • Certain types of piece work (particularly if peer pressure may result in them tackling tasks or working at speeds that are too much for them).
Excessively psychologically demanding work • Work with difficult clients or situations where there is a possibility of violence or aggression; • Difficult emotional situations e.g. dealing with death, serious illness or injury; • Decision making under stress.
Harmful exposure to physical agents • Ionising radiation (separate exposure limits apply to young persons); • Non-ionising electromagnetic radiation, e.g. lasers, UV from electric arc welding or lengthy exposure to sunlight, infra red from furnaces or burning/welding; • Risks to health from extreme cold or heat; • Excessive noise; • Hand-arm vibration, e.g. from portable tools; • Whole-body vibration, e.g. from off-road vehicles; • Work in pressurised atmospheres and diving work.
Harmful exposure to biological or chemical agents • Very toxic, toxic, harmful, corrosive and irritant substances; • Substances causing heritable genetic damage or harming the unborn child; • Carcinogenic substances;

- Asbestos (including asbestos-containing materials);
- Lead and lead compounds;
- Biological risks, e.g. legionella, leptospirosis (Weil's disease), zoonoses.

Work equipment

Where there is an increased risk of injury due to the complexity of precautions required or the level of skill required for safe operation e.g:

- Woodworking machines (particularly saws, surface planing and vertical spindle moulding machines);
- Food slicers and other food processing machinery;
- Certain types of portable tools such as chainsaws;
- Power presses;
- Vehicles such as fork lift trucks, mobile cranes, construction vehicles;
- Other cranes and lifting hoists;
- Firearms.

Some young people may not have the physical size or strength necessary to operate equipment which has been designed for adults.

Dangerous materials or activities

- Work with explosives, including fireworks;
- Work with fierce or poisonous animals, e.g. on farms, in zoos or veterinary work;
- Certain types of electrical work, e.g. exposure to high voltage or live electrical equipment;
- Handling of flammable liquids or gases, e.g. petrol, acetylene, butane, propane;
- Work with pressurised gases;
- Work in large slaughterhouses (risks from animal handling, use of stunning equipment);
- Holding large quantities of cash or valuables.

Dangerous workplaces or workstations

- Work at heights, e.g. on high ladders or other unprotected forms of access
- Work in confined spaces, particularly where the risks specified in the *Confined Spaces Regulations 1997* (*SI 1997 No 1713*) are present;
- Work where there is a risk of structural collapse, e.g. in construction or demolition activities or inside old buildings.

Further guidance on risks to young people and appropriate precautions is contained in HSE booklet HSG 165 'Young people at work. A guide for employers.'

New or expectant mothers

V12017 Amendments made in 1994 to the previous *Management Regulations* (*SI 1992 No 2051* now revoked) implemented the European Directive on Pregnant Workers

(92/85/EEC), requiring employers in their risk assessments to consider risks to new or expectant mothers. These amendments were subsequently incorporated into the *Management Regulations 1999* (*SI 1999 No 3242*). *Regulation 1* of the *Management Regulations* contains two relevant definitions:

'New or expectant mother' means an employee who is pregnant; who has given birth within the previous six months; or who is breastfeeding.

'Given birth' means 'delivered a living child or, after twenty-four weeks of pregnancy, a stillborn child'.

Requirements of the Management Regulations

V12018 The requirements for 'Risk assessment in respect of new and expectant mothers' are contained in *Regulation 16* of the *Management Regulations 1999* (*SI 1999 No 3242*) which states in *paragraph* (*1*):

> 'Where—
>
> (a) the persons working in an undertaking include women of child-bearing age; and
>
> (b) the work is of a kind which could involve risk, by reason of her condition, to the health and safety of a new or expectant mothers, or to that of her baby, from any processes or working conditions, or physical, biological or chemical agents, including those specified in Annexes I and II of Council Directive 92/85/EEC on the introduction of measures to encourage improvements in the safety and health at work of pregnant workers and workers who have recently given birth or are breastfeeding,
>
> the assessment required by Regulation 3(1) shall also include an assessment of such risk.'

Paragraph (*4*) of the Regulation states that in relation to risks from infectious or contagious diseases an assessment must only be made if the level of risk is in addition to the level of exposure outside the workplace.

The types of risk which are more likely to affect new or expectant mothers are described later in the chapter. *Paragraphs* (*2*) and (*3*) of *Regulation 16* set out the actions employers are required to take if these risks cannot be avoided. *Paragraph* (*2*) states :

> 'Where, in the case of an individual employee, the taking of any other action the employer is required to take under the relevant statutory provisions would not avoid the risk referred to in paragraph (1) the employer shall, if it is reasonable to do so, and would avoid such risks, alter her working conditions or hours of work.'

Consequently where the risk assessment required under *Regulation 16*(*1*) shows that control measures would not sufficiently avoid the risks to new or expectant mothers or their babies, the employer must make reasonable alterations to their working conditions or hours of work.

In some cases restrictions may still allow the employee to substantially continue with her normal work but in others it may be more appropriate to offer her suitable alternative work.

Any alternative work must be:

- suitable and appropriate for the employee to do in the circumstances;
- on terms and conditions which are no less favourable.

Paragraph (*3*) of *Regulation 16* states:

> 'If it is not reasonable to alter the working conditions or hours of work, or if it would not avoid such risk, the employer shall, subject to section 67 of the 1996 Act, suspend the employee from work for so long as is necessary to avoid such risk.'

The 1996 Act referred to here is the *Employment Rights Act 1996* which provides that any such suspension from work on the above grounds is on full pay. However, payment might not be made if the employee has unreasonably refused an offer of suitable alternative work. Employment continues during such a suspension, counting as continuous employment in respect of seniority, pension rights etc. Contractual benefits other than pay do not necessarily continue during the suspension. These are a matter for negotiation and agreement between the employer and employee, although employers should not act unlawfully under the *Equal Pay Act 1970* and the *Sex Discrimination Act 1975*. Enforcement of employment rights is through employment tribunals.

Regulation 17 of the *Management Regulations 1999* (*SI 1999 No 3242*) deals specifically with night work by new or expectant mothers and states:

> 'Where—
>
> (a) a new or expectant mother works at night; and
>
> (b) a certificate from a registered medical practitioner or a registered midwife shows that it is necessary for her health or safety that she should not be at work for any period of such work identified in the certificate,
>
> the employer shall, subject to section 46 of the 1978 Act, suspend her from work for so long as is necessary for her health or safety.'

Such suspension (on the same basis as described above) is only necessary if there are risks arising from work. The HSE does not consider there are any risks to pregnant or breastfeeding workers or their children from working at night per se. It suggests that any claim from an employee that she cannot work nights should be referred to an occupational health specialist. The HSE's own Employment Medical Advisory Service are likely to have a role to play in such cases.

The requirements placed on employers in respect of altered working conditions or hours of work and suspensions from work only take effect when the employee has formally notified the employer of her condition. *Regulation 18* of the *Management Regulations 1999* (*SI 1999 No 3242*) states:

> '(1) Nothing in paragraph (2) or (3) of regulation 16 shall require the employer to take any action in relation to an employee until she has notified the employer in writing that she is pregnant, has given birth within the previous six months, or is breastfeeding.
>
> (2) Nothing in paragraph (2) or (3) of regulation 16 or in regulation 17 shall require the employer to maintain action taken in relation to an employee—
>
> > (a) in a case—
> >
> > > (i) to which regulation 16(2) or (3) relates; and
> > >
> > > (ii) where the employee has notified her employer that she is pregnant, where she has failed, within a reasonable time of being requested to do so in writing by her employer, to

produce for the employer's inspection a certificate from a registered medical practitioner or a registered midwife showing that she is pregnant;

(b) once the employer knows that she is no longer a new or expectant mother; or

(c) if the employer cannot establish whether she remains a new or expectant mother.'

Risks to new or expectant mothers

V12019 The HSE booklet HSG 122 'New and expectant mothers at work: A guide for employers' provides considerable guidance on those risks which may be of particular relevance to new or expectant mothers, including those listed in the EC Directive on Pregnant Workers (92/85/EEC). This guidance is both summarised and augmented below.

Physical agents

V12020 *Manual handling*

Pregnant women are particularly susceptible to risk from manual handling activities as also are those who have recently given birth, especially after a caesarean section. Manual handling assessments (as required by the *Manual Handling Operations Regulations 1992* (*SI 1992 No 2793*) are dealt with in the chapter MANUAL HANDLING in this publication.

Noise

The HSE does not consider that there are any specific risks from noise for new or expectant mothers. Compliance with the requirements of the *Noise at Work Regulations 1989* (*SI 1989 No 1790*) should provide them with sufficient protection.

Ionising radiation

The foetus may be harmed by exposure to ionising radiation, including that from radioactive materials inhaled or ingested by the mother. The *Ionising Radiations Regulations 1999* (*SI 1999 No 3232*) set an external radiation dose limit for the abdomen of any woman of reproductive capacity and also contain a specific requirement to provide information to female employees who may become pregnant or start breast-feeding. Systems of work should be such as to keep exposure of pregnant women to radiation from all sources as low as reasonably practicable. Contamination of a nursing mother's skin with radioactive substances can create risks for the child and special precautions may be necessary to avoid such a possibility.

Several HSE publications provide detailed guidance on work involving ionising radiation. (See L121 'Work with ionising radiation: Ionising Radiations Regulations 1999: Approved Code of Practice and Guidance' (2000); and INDG334 'Working safely with radiation: Guidelines for expectant and breast-feeding mothers' (2001), in particular.)

Other electromagnetic radiation

The HSE does not consider that new or expectant mothers are at any greater risk from other types of radiation, with the possible exception of over exposure to radio-frequency radiation which could raise the body temperature to harmful levels. Compliance with the exposure standards for electric and magnetic fields published by the National Radiological Protection Board should provide adequate protection.

Work in compressed air

Although pregnant women may not be at greater risk of developing the 'bends', potentially the foetus could be seriously harmed by gas bubbles in the circulation should this condition arise. There is also evidence that women who have recently given birth have an increased risk of the bends.

The *Work in Compressed Air Regulations 1996* (*SI 1996 No 1656*), *Reg 16*(*2*) states:

'... the compressed air contractor shall ensure that no person works in compressed air where the compressed air contractor has reason to believe that person to be subject to any medical of physical condition which is likely to render that person unfit or unsuitable for such work.'

This would appear to prohibit such work by pregnant women or those who have recently given birth. In its booklet HS(G)122 the HSE states that there is no physiological reason why a breastfeeding mother should not work in compressed air although they point out that practical difficulties would exist. HSE booklet L96 'A guide to the Work in Compressed Air Regulations 1996' (1996) provides detailed guidance on the requirements of the *Work in Compressed Air Regulations*.

Diving work

The HSE draws attention to the possible effects of pressure on the foetus during underwater diving by pregnant women and states that they should not dive at all.

Under the *Diving at Work Regulations 1997* (*SI 1997 No 2776*), *Reg 15* divers must have a certificate of medical fitness to dive and the HSE guidance to doctors issuing such certificates advises that pregnant workers should not dive.

Regulation 13(*1*) of the *Diving at Work Regulations 1997* (*SI 1997 No 2776*) states that 'No person shall dive in a diving project ... if he knows of anything (including any illness or medical condition) which makes him unfit to dive.'

A series of HSE booklets provide general guidance on different types of diving projects.

Shock, vibration etc.

Major physical shocks or regular exposure to lesser shocks or low frequency vibration may increase the risk of a miscarriage. Activities involving such risks (e.g. the use of vehicles off road) should be avoided by pregnant women.

Movement and posture

Fatigue from standing and other physical work has been associated with miscarriage, premature birth and low birth weight. Ergonomic considerations will increase as the pregnancy advances and these could affect display screen equipment workstations, work in restricted spaces (e.g. for some maintenance or cleaning activities) or work on ladders or platforms. Underground mining work is likely to involve movement and posture problems and will also be subject to some of the other 'physical agents' described in this section. Driving for extended periods or travel by air may also present postural problems.

Pregnant women should be allowed to pace their work appropriately, taking longer and more frequent breaks. They may need to be restricted from carrying out certain tasks. Seating may need to be provided for work that is normally done standing and adjustments to display screen equipment and other workstations may be necessary.

Physical and mental pressure

Excessive physical or mental pressure could cause stress and lead to anxiety and raised blood pressure. Workplace stress is a complex issue which has recently been receiving increased attention. Fatigue issues are referred to above but other possible causes of stress may also need to be considered. These might be associated with the workload of individual pregnant employees (e.g. for those in management or administrative roles), the pressure of decision-making (e.g. in the health care or financial sectors) or the trauma of potential work situations (e.g. serious accidents dealt with by the emergency services).

Extreme cold or heat

Pregnant women are less tolerant of heat and may be more prone to fainting or heat stress. Although the risk is likely to reduce after birth, dehydration may impair breastfeeding. Exposure to prolonged heat at work, e.g. at furnaces or ovens, should be avoided.

Maintenance or cleaning work in hot situations should also be avoided, particularly if this involves use of less secure forms of access such as ladders, where fainting could result in a serious fall.

The HSE does not consider that there are any specific problems from working in extreme cold although obviously appropriate precautions should be taken as for other workers e.g. the provision of warm clothing.

Biological agents

V12021 Many biological agents in hazard groups 2, 3 and 4 (as categorised by the Advisory Committee on Dangerous Pathogens) can affect the unborn child should the mother be infected during pregnancy. Some agents can cause abortion of the foetus and others can cause physical or neurological damage. Infections may also be passed on to the child during or after birth, e.g. while breastfeeding.

Agents presenting risks to children include hepatitis B, HIV, herpes, TB, syphilis, chickenpox, typhoid, rubella (german measles), cytomegalovirus and chlamydia in sheep. Most women will be at no more risk from these agents at work than living within the community but the risks are likely to be higher in some work sectors, e.g. laboratories, health care, the emergency services and those working with animals or animal products.

Details of appropriate control measures for biological agents are contained in *Schedule 3* and *Appendix 2* of the ACoP booklet to the *COSHH Regulations 2002* (*SI 2002 No 2677*). There is a separate HSE publication on 'Infections in the workplace to new and expectant mothers' (1997). When carrying out a risk assessment in respect of new or expectant mothers, normal containment or hygiene measures may be considered sufficient, but there may be the need for special precautions such as use of vaccines. Where there is a high risk of exposure to a highly infectious biological agent it may be necessary to remove the worker entirely from the high risk environment.

Chemical agents

V12022 *Substances labelled with certain risk phrases*

The *Chemicals (Hazard Information and Packaging for Supply) Regulations 2002 (SI 2002 No 1689)* ('*CHIP*') require many types of hazardous substances to be labelled with specified risk phrases. Several of those are of relevance to new or expectant mothers:

R40: possible risk of irreversible effects;

R45: may cause cancer;

R46: may cause heritable genetic damage;

R49: may cause cancer by inhalation;

R61: may cause harm to the unborn child;

R63: possible risk of harm to the unborn child;

R64: may cause harm to breastfed babies;

R68: possible risk of irreversible effects.

The *CHIP Regulations 2002* are regularly subject to changes and employers should be alert for other risk phrases which indicate risks in respect of new or expectant mothers.

Control of such substances at work is already required by the *COSHH Regulations 2002 (SI 2002 No 2677)* or the separate regulations governing lead and asbestos. Risk assessments in relation to pregnant women or those who have recently given birth may indicate that normal control measures are adequate to protect them also. However, additional precautions may be necessary, e.g. improved hygiene procedures, additional PPE or even restriction from work involving certain substances.

The ACoPs relating to the *COSHH Regulations 2002 (SI 2002 No 2677)* (contained in HSE booklet L5) together with HSG 193 'COSHH Essentials: Easy steps to control chemicals' provide further details of the types of precautions which may be appropriate.

Mercury and mercury derivatives

The HSE states that exposure to organic mercury compounds can slow the growth of the unborn baby, disrupt the nervous system and cause the mother to be poisoned. It considers that there is no clear evidence of adverse effects on the foetus from mercury itself and inorganic mercury compounds. Mercury and its derivatives are subject to the *COSHH Regulations 2002 (SI 2002 No 2677)* and normal control measures (see above) may be adequate to protect new or expectant mothers.

(See HSE publications L5 and HSG193 as mentioned above; EH17 'Mercury and its inorganic divalent compounds' (1996); and MS12 'Mercury: Medical guidance notes' (1996).)

Antimitotic (cytotoxic) drugs

These drugs (which may be inhaled or absorbed through the skin) can cause genetic damage to sperm and eggs and some can cause cancer. Those workers involved in preparation or administration of such drugs and disposal of chemical or human waste are at greatest risk, e.g. pharmacists, nurses and other health care workers. Antimitotic drugs can present a significant risk to those of either gender who are trying to conceive a child as well as to new or expectant mothers, and all those working with them should be made aware of the hazards.

The *COSHH Regulations 2002* (*SI 2002 No 2677*) again apply to the control of these substances (see above). Since there is no known threshold limit for them, exposure must be reduced to as low a level as is reasonably practicable.

Agents absorbed through the skin

Various chemicals including some pesticides may be absorbed through the skin causing adverse effects. These substances are identified in the tables of occupational exposure limits in HSE booklet EH40 'Occupational exposure limits' (an updated version is published annually), in which some substances are accompanied by an annotation 'Sk'. Many such agents (particularly pesticides which are subject to the *Control of Pesticides Regulations 1986* (*SI 1986 No 1510*)) will also be identified by product labels.

Effective control of these chemicals is obviously important in respect of all employees (the *COSHH Regulations 2002* (*SI 2002 No 2677*) applying once again) although risk assessments may reveal the need for additional precautions in respect of new or expectant mothers, e.g. modified handling methods, additional PPE, or even their restriction from activities involving exposure to such substances.

HSE booklet L9 'Safe use of pesticides for non-agricultural purposes' contains a further ACoP in relation to COSHH obligations.

Carbon monoxide

Exposure of pregnant women to carbon monoxide can cause the foetus to be starved of oxygen with the level and duration of exposure both being important factors. There are not felt to be any additional risks from carbon monoxide to mothers who have recently given birth or to breast-fed babies.

Once again it is important to protect all members of the workforce from high levels of carbon monoxide (the *COSHH Regulations 2002* (*SI 2002 No 2677*) require it) but it may also be necessary to ensure that pregnant women are not regularly exposed to carbon monoxide at lower levels, e.g. from use of gas-fired equipment or other processes or activities. HSE Guidance Note EH43 provides general guidance on carbon monoxide.

Lead and lead derivatives

High occupational exposure to lead has historically been linked with high incidence of spontaneous abortion, stillbirth and infertility. Decreases in the intellectual performance of children have more recently been attributed to exposure of their mothers to lead. Since lead can enter breast milk there are potential risks to the child if breastfeeding mothers are exposed to lead.

All women of reproductive capacity are prohibited from working in many lead processing activities. All work involving exposure to lead is subject to the *Control of Lead at Work Regulations 2002* (*SI 2002 No 2676*). Even where women of reproductive capacity are allowed to work with lead or its compounds the blood-lead concentrations contained within the Regulations as an 'action level' and a 'suspension level' are set at half the figures for adult males. This is intended to ensure that women who may become pregnant already have low blood lead levels.

Once pregnancy is confirmed, the doctor carrying out medical surveillance (as required by *Regulation 10* of *SI 2002 No 2676*) would normally be expected to suspend the woman from work involving significant exposure to lead. There is, however, no specific requirement in the Regulations themselves that this must happen. Detailed guidance on the Regulations is contained in HSE booklet L132 'Control of lead at work' (2002).

Working conditions

V12023 HSE guidance in HS(G)122 in respect of the 'working conditions' referred to in the *Management Regulations* (*SI 1999 No 3242*), *Reg 16*(*1*) repeats the HSE's stated position on work with DSE contained in Appendix 2 of the HSE booklet L26 'Work with display screen equipment'. Its view is that radiation from DSE is well below the levels set out in international recommendations and that scientific studies taken as a whole do not demonstrate any link between this work and miscarriages or birth defects.

There is no need for pregnant women to cease working with DSE. However, the HSE recommends that, to avoid problems from stress or anxiety, women are given the opportunity to discuss any concerns with someone who is well informed on the subject.

Other aspects of pregnancy

V12024 Although the HSE booklet on 'New and expectant mothers at work' (HSG 122) draws attention to other features of pregnancy which employers may wish to take into account, it suggests that employers have no legal obligation to do so. However, this view has not been tested by the courts and an argument could be advanced that some of these represented 'working conditions' which involve risk to the mother or her baby within the terms of the *Management Regulations* (*SI 1999 No 3242*), *Reg 16*(*1*)(*b*).

The impact of these aspects will vary during the course of the pregnancy and employers are likely to need to keep the situation under review. A modified version of the appendix from the HSE booklet dealing with these 'other aspects' is provided in TABLE C below.

Aspects of pregnancy that may affect work

Table C	
Aspects of pregnancy	*Factors in work*
Morning sickness	Early shift work Exposure to nauseating smells Availability for early meetings Difficulty in leaving job
Frequent visits to toilet	Difficulty in leaving job/site of work

Breastfeeding	Access to private area Use of secure, clean refrigerators to store milk
Tiredness	Overtime/long working hours Evening work More frequent breaks Private area to sit or lie down
Increasing size	Difficulty in using protective clothing Work in confined areas Manual handling difficulties Posture at DSE workstations (Dexterity, agility, co-ordination, reach and speed of movement may also be impaired)
Comfort	Problems working in confined or congested workspaces
Backache	Standing for extended periods Posture for some activities Manual handling
Travel i.e. fatigue, stress, static posture	Reduce need for lengthy journeys Change travel methods
Varicose veins	Standing/sitting for long periods
Haemorrhoids	Working in hot conditions/sitting
Vulnerability to passive smoking	Effective smoking policy (priority to the needs of non-smokers)
Vulnerability to stress	Modified duties Adjustments to working conditions or hours
Vulnerability to falls	Restrictions on work at heights or on potentially slippery surfaces
Vulnerability to violence (e.g. from contact with the public)	Changes to workplace layout or staffing Modified duties
Lone working	Effective communication and supervision Possible need for improved emergency arrangements

A practical approach to assessing risks

V12025 Essentially risk assessment in respect of new or expectant mothers must be carried out by an employer if:

- there are women of childbearing age; and
- their work could involve risks to new or expectant mothers or their babies.

Most organisations employ women of childbearing age and, particularly in the case of large businesses, there may be a number of work activities that could create relevant risks. In some cases (e.g. those involving exposure to hazardous substances or radiation) it may be appropriate to stipulate that women are not allowed to work in certain activities, processes or departments once their pregnancy has been notified and/or for a finite period after they return to work after giving birth.

However, in many workplaces the issues will be far from clear cut – it would be neither necessary nor practical to prohibit pregnant workers from carrying out *any* manual handling or from standing up, sitting down or climbing up ladders. Also it is difficult for any employer to identify in advance exactly what steps need to be taken to alter working conditions or hours of work so that risks can be avoided for every female employee who may become pregnant.

A much more practical approach is for the employer to develop a checklist similar to the sample provided at the end of this section. Such a checklist can be prepared by carrying out a review of the organisation's activities in order to identify risks which are present which *may* be of relevance in relation to new or expectant mothers. The risks described in the previous part of this chapter provide a good starting point from which a workplace-specific checklist can be developed. In some workplaces it may be appropriate to develop more than one such checklist (e.g. for production, maintenance and administration) because the profiles of risks are different.

Such a checklist should be completed by a suitable employer's representative (a personnel or health and safety specialist or the pregnant woman's manager perhaps) together with the pregnant woman herself. The checklist is intended to provoke discussion about the employee's possible exposure to the risks that the employer has identified so that any necessary additional precautions (or changes to work practices) can be agreed. There is the opportunity to identify whether any further review of the situation is necessary. The process would be repeated once the mother returned to work after giving birth.

The sample checklist provided below is for a bakery. Female staff are likely to be involved in production work where there is significant potential for risk to new or expectant mothers, and they may also be employed in laboratory or maintenance work. Risks for staff working in administrative activities must also be considered, e.g. those associated with DSE workstations. However, the greatest problems could be for female staff delivering to retail outlets. As well as risks from manual handling, falls and possibly excessive driving, the difficulties for such staff in accessing suitable rest and toilet facilities must also be taken into account.

BUNN THE BAKERS	NEW OR EXPECTANT MOTHERS	RISK ASSESSMENT CHECKLIST
Name of Employee:		Work location:
	Possible risks to consider	**Precautions/changes agreed**
Bakery and delivery work	Manual handling (production, maintenance, deliveries) Excessive standing (production lines) Awkward sitting (production lines) Difficult access (cleaning, maintenance) Excessive heat (near to ovens) Hazardous substances (e.g. cleaning materials,maintenance, laboratory) Significant risk of falls (e.g. slippy floors, maintenance work, trips on delivery work) Excessive driving (deliveries) Other	
Administrative activities	DSE workstation layout Unsuitable meeting times Manual handling (e.g. stationery, records) Significant travel (especially by car or air) Other	
All activities	Ability to leave workplace temporarily Availability of rest area/toilets Major pressure or stress situations Possible effects of fatigue Other	
Signature (for Bunn the Bakers):		Signature (new/expectant mother):
Date:		Date for further review (if any):
This form should be completed when the pregnancy is first notified *and* when the new mother returns to work.		

Risks to non-employees

V12026 As part of their general risk assessments, employers must also consider risks to new or expectant mothers who are not part of their own workforce. Such persons may be employed by another organisation or be self-employed but they may also be service users or members of the public. Even though other employers have duties to assess risks to their staff, 'host' employers still have a statutory duty (and a 'duty of care') to other people's employees, particularly in respect of risks of which they may be aware, but which might not be immediately apparent to others. In this context, risks to pregnant women especially could include:

- slips, trips and falls, e.g. due to slippy or uneven surfaces;
- accidental impact from persons, e.g. running children or accidental contacts related to sports activities;
- accidental collision with mobile equipment;
- use of off-road transport;
- assault, e.g. on police or prison premises;
- ionising radiation;
- hazardous substances (see the earlier reference to risk phrases at V12022 above).

Lone workers

V12027 There are no general legal restrictions on working alone – indeed many work activities would be extremely impractical if there were. However, the activities of lone workers must be subject to the same sort of risk assessment process as any other type of work. Even those regulations imposing the greatest restrictions on lone working (i.e. *Electricity at Work Regulations 1989* (*SI 1989 No 635*) and the *Confined Spaces Regulations 1997* (*SI 1997 No 1713*) still require the element of risk assessment to be applied in identifying suitable safe systems of work.

The various types of lone workers

V12028 There are many different types of lone working. These can be divided into three categories:

People working alone in fixed workplaces

This category includes:

- workers in small shops and kiosks;
- people working in more remote parts of larger premises, e.g. isolated reception or enquiry desks, private interview rooms, security gatehouses;
- laboratories, stores;
- staff working late, at nights or at weekends;
- staff responding to alarm calls etc, e.g. 'keyholders';
- staff working at home.

Staff working on a variety of premises belonging to other employers

Such people will be working independently of their colleagues but may have others working around them for at least part of the time. They include:

- persons carrying out construction or plant installation work;
- maintenance and repair workers;
- cleaners, painters, decorators etc;
- other types of contracting staff.

Persons working within the wider community

These staff may be working in urban, rural or remote locations, sometimes in places which are accessible to the public but also on private property (often in domestic premises). This includes:

- mobile security staff;
- domestic installation, maintenance and repair staff;
- delivery staff (including postal workers);
- sales and technical representatives;
- architects, property surveyors and estate agents;
- rent collectors and meter readers;
- social workers, home carers and district nurses;
- police, traffic wardens and probation staff;
- agricultural and forestry workers;
- utility workers (electricity, gas, water, sewage, telecommunications etc.);
- vehicle recovery staff;

and many other similar types of work.

Risks for lone workers

V12029 In some situations lone workers may be at risk because they are not physically capable of carrying out certain activities safely on their own. These may relate to:

- *Manual handling activities*

  – where team handling is identified as being necessary.

- *Use of certain types of equipment*

  – particularly heavy and bulky equipment, or where another person needs to be available to carry out adjustments.

- *Operation of controls etc.*

  – controls may be physically separate from each other or from related instrumentation requiring a second person to simultaneously activate controls or relay information (e.g. where a banksman may be required for lifting operations).

- *Achieving safe access*

  – a second person may be necessary to foot a ladder or ensure safe use of other forms of temporary access equipment.

Some work activities may involve such a high degree of risk that another person needs to be present to assist in the identification of risks or act as a deterrent to such risks. These situations include:

- *Some types of maintenance and repair work*

  – ensuring that equipment has been correctly isolated and other precautions taken (two heads are better than one!).

- *Live electrical work*

  – ensuring that the person carrying out the work is fully aware of what is live and what is not.

- *Work involving potential risks of aggression or even violence*

  – handling complaints, passing on unwelcome news, dealing with persons with aggressive or violent histories, working in high risk areas of the community, handling cash or valuables.

- *Activities where staff feel highly vulnerable*

  – one to one encounters, possibly with unknown persons, particularly in remote locations, empty premises or domestic property (female staff often feel more vulnerable but males should also be made aware of risks of unwarranted accusations against them)

There may also be risks to lone workers where a second person may be needed to take appropriate emergency actions such as:

- *Isolating electrical supplies*

  – particularly where live work is being carried out.

- *Rescuing a worker from inside a confined space*

  – e.g. by use of a rescue line or wearing suitable respiratory protective equipment ('RPE') and PPE.

- *Giving immediate first aid attention*

  – particularly important for electrical accidents or persons seriously affected by hazardous substances, or asphyxiated.

- *Safely recovering a worker who has fallen whilst wearing a safety harness*

The general potential for accidents and illness to lone workers from less foreseeable causes must also be considered, even though these may not justify the immediate presence of another person.

The above examples are of the more common types of risks that lone workers may encounter. Employers will no doubt be able to add to the list, particularly after consulting those who carry out lone work and their representatives.

Controlling risks to lone workers

V12030 The previous section identified the types of risks that lone workers may be subject to. Set out below is a variety of measures which may be necessary to control risks.

Whilst some measures will be relevant to most, if not all risk situations, others may only have limited application. It is in effect a menu from which to choose the appropriate precautions.

- *General precautions*

  Lone workers must be made aware of the types of risks that they may encounter and the precautions which must be taken to control those risks. This will require information, instruction and training, supported by effective ongoing supervision.

  Some types of lone working may justify the creation of formal written procedures or rules whilst in others the lone workers themselves may be expected to carry out a degree of dynamic risk assessment (see R3016 in RISK ASSESSMENT).

  As a result of the employer's risk assessment some activities may be identified as being ones which workers must never carry out alone. These must be communicated effectively to the lone worker, who must also be given guidance on where the necessary assistance can be obtained. This may be achieved by the temporary presence of work colleagues or assistance may be forthcoming from the customer's personnel or from others working in the vicinity.

- *Communication*

  Good communications are essential to ensuring the safety of lone workers. This is likely to involve:

  — provision of suitable communication equipment, e.g.mobile phones, portable radios, direct radio links, monitored CCTV surveillance;

  — availability of emergency alarms, e.g. fixed alarm buttons, portable attack alarms (some types of alarms detect an absence of movement by the alarm user or require a regular response from the user);

  — availability of information on lone workers' whereabouts to others, e.g.work colleagues, control centres, security staff, partners, friends etc. (this may involve the use of location boards, computerised or desk diaries, itinerary sheets or just word of mouth);

  — ongoing communication from the lone worker, e.g. about delays, problems, changes in plans;

  — ongoing checks on the lone worker's wellbeing;

  — pre-emptive contacts from lone workers encountering unexpected risks (explaining where they are, who they're with or what they're doing and requesting a return call in an agreed period of time);

  — checks to ensure that the lone worker has safely returned to base (or home).

- *Specific precautions*

  There are many specific precautions which can be taken to reduce the risks to lone workers. The nature of these will depend upon the location of the lone worker and the type of activity involved. Examples of these include:

  — physical separation of the lone worker from people they must deal with, e.g. by use of screens or wide desks;

- ensuring that assistance is readily available to the lone worker when required, e.g. for certain manual handling operations or live electrical work;
- placing lone workers where assistance from members of the public is at hand, e.g. arranging encounters, particularly with higher risk contacts, in public buildings or areas, including hotels, cafes etc.

- *Emergency arrangements*

  Here too, the nature of the arrangements will vary according to the location and activity of the lone worker. Arrangements are likely to include:

  - response to alarms – who is expected to respond and what actions should they take;
  - follow-up of lone workers who do not return when expected or do not respond to calls;
  - provision of basic first aid equipment for lone workers, so that they can carry out simple first aid for themselves or receive treatment from others coming to their aid.

Monitoring of lone workers

V12031 Those responsible for managing and supervising lone workers must ensure that their activities are monitored effectively. Since only a limited amount of observation of their work in the field is likely to be practicable, regular consultation with the staff involved will be important. Aspects to monitor for are:

- new risks which have become apparent;
- existing risks which may have increased;
- the effectiveness of existing precautions;
- workers' views on new precautions which may be appropriate.

It will also be necessary to actively monitor the effectiveness of precautions through actions such as:

- checking the effectiveness of communication equipment;
- testing alarms;
- enquiring as to the current whereabouts of lone workers;
- testing the awareness of staff on how they should react to emergency situations.

Disabled persons

Types of disability

V12032 Disabilities come in many different forms. HSE guidance to local authorities on the *Disability Discrimination Act 1995* refers to impairments affecting the ability to perform normal day to day activities affecting at least one of the following:

- mobility;
- manual dexterity;
- physical co-ordination;

- continence;
- ability to lift, carry or otherwise move everyday objects;
- speech, hearing or eyesight;
- memory or ability to concentrate, learn or understand; or
- perception of the risk of physical danger.

The last two on this list have a close relationship with the consideration of risks to young persons referred to earlier in the chapter.

Although not included in the list above, account must also be taken of those with relevant health conditions and allergies, particularly when assessing risks relating to hazardous substances.

Disabled persons at risk

V12033 Employers in their risk assessments must take account of all disabled persons who may be affected by their activities, including:

- permanent and temporary employees;
- persons on training or work experience placements (who have the status of employees);
- visitors;
- customers or service users;
- neighbours and passers-by.

Access for disabled persons

V12034 The *Health and Safety* (*Miscellaneous Amendments*) *Regulations 2002* (*SI 2002 No 2174*) introduced an additional requirement into *Regulation 25* of the *Workplace* (*Health, Safety and Welfare*) *Regulations 1992* (*SI 1992 No 3004*):

> 'Where necessary, those parts of a workplace (including in particular doors, passageways, stairs, showers, washbasins, lavatories and workstations) used or occupied directly by disabled persons at work shall be organised to take account of such persons.'

Factors which must be taken into account in complying with the above requirement and in preventing accidents to disabled persons are:

- the locations where disabled persons work or must make access to;
- the width of access routes;
- maintaining access routes free from obstructions and slipping or tripping hazards;
- standards of guardrails and handrails;
- heights of door handles, light swtiches, alarms, controls etc.

(It is not the intention of this chapter to consider the general accessibility of work premises to disabled members of the public.)

Other issues of relevance

V12035 There are a number of other issues which must be considered when carrying out risk assessments in respect of disabled persons, whether members of the workforce or not, including:

- hazardous substances:
  - their potential effects on those with relevant health conditions or allergies;
  - the abilities of exposed persons to understand the risks or implement precautions.
- work equipment:
  - physical ability to handle the equipment;
  - intellectual capability to understand risks and implement necessary precautions.
- hot (or very cold) items or surfaces:
  - the ability to identify such items or surfaces;
  - a perception of the risks involved;
  - the ability to keep clear of such risks.
- workstation layout:
  - ergonomically suited to the needs of the disabled worker;
  - provision of suitable seating is of particular importance.
- emergency arrangements:
  - the ability to hear or see alarm signals;
  - the physical abilty to react to the alarm with the necessary speed;
  - the potential for panic amongst those with some types of disability.

Controlling risks to disabled persons

V12036 As a result of the risk assessment process employers must control risks to the standards required by legislation – either the general standards required by the *HSWA 1974* or the more specific requirements of individual sets of regulations. In respect of employees they must also comply with the requirements of *Regulation 13*(*1*) of the *Management Regulations* (*SI 1999 No 3242*) which states:

> 'Every employer shall, in entrusting tasks to his employees, take into account their capabilities as regards health and safety.'

This will require employers to:

- Identify the disabilities of individual workers (and others) via:
  - application forms and questionnaires;
  - interviews;
  - medical examinations.
- Identify situations where persons with such disabilities may be at risk:
  - the types of issues referred to above.

In order to control risks to the standards required, it may be necessary for the employer to:

- modify existing equipment;
- provide alternative equipment;
- make modifications to the workplace;
- provide a work location suited to the individual's disability;
- provide additional or modified training (making special arrangements for those with learning difficulties);
- ensure emergency arrangements take account of disabled persons;
- restrict the activities and/or locations of disabled persons.

Employers have a duty under the *Disability Discrimination Act 1995* ('DDA') to make 'reasonable adjustment' to the workplace or working arrangements of disabled persons.

Avoiding discrimination

V12037 Employers must find a balance between avoiding exposing disabled persons to risk and discriminating against them unfairly. HSE guidance (Local Authority Circular 63/3) states that:

> 'The DDA does not prevent employers from continuing to stipulate essential health requirements for particular types of employment but they must be able to justify these and show that it would not be reasonable to waive them in any individual case.'

Employers may enquire about disability or ask a disabled person to undergo a medical examination but this must be justified by the relevance of the disability to the job in question.

There are concerns that employers may use health and safety reasons as an excuse for discriminating against disabled persons. Some of these concerns were found to be justified by research carried out on behalf of the HSE entitled 'Extent of use of health and safety requirements as a false excuse for not employing sick or disabled persons'. Although this showed that health and safety was frequently being used as the rationale for non-recruitment or dismissal, it found considerably more evidence of employers overcoming health and safety difficulties to retain workers with a disability, health condition or injury.

There are many practical actions which can be taken to accommodate disabled workers safely in the workplace, as described in the section above. Even where restrictions must be placed on the disabled, these can often relate to a very limited range of activities or locations, rather than preventing their employment altogether.

Inexperienced workers

V12038 Workers of all ages may (like young persons) be lacking in experience of some types of work or lack awareness of the risks associated with particular work activities. *Regulation 13* of the *Management Regulations* (*SI 1999 No 3242*) as well as requiring employers to take into account the capabilities of employees in entrusting tasks to them (see above), also contains several requirements about training.

Paragraph (2) of the *Regulation 13* states:

> 'Every employer shall ensure that his employees are provided with adequate health and safety training – on their being recruited into the employer's undertaking; and on their being exposed to new or increased risks ... '

Such new or increased risks are specified as:

- transfer or change of responsibilities;
- due to new or changed work equipment;
- because of new or changed systems of work.

Training and supervision of inexperienced workers

V12039 General health and safety induction training and job specific training is necessary in all workplaces but particular account of how to cope with these training needs must be taken in:

- work activities with rapid staff turnover;
- employment utilising large numbers of temporary workers, e.g. to meet seasonal needs or special orders (special arrangements will be necessary for workers without a good command of English);
- workplaces providing placements for those on government-funded training and re-training schemes;
- work activities involving workers with special educational needs.

Even when inexperienced workers have received appropriate training they will still require ongoing supervision, in many respects similar to that required for young persons. The imposition of restrictions as described earlier in this section (in relation to children and young persons) may also be appropriate.

Work at Heights

Introduction

W9001 There are many circumstances in working life where people are substantially above ground or floor level (or near the edge of an opening or excavation) and at risk, however remote, of falling. Such risk also arises where access is necessary to roofs clad in material which is either inherently fragile, or has become weakened through age. Then there are situations where those working at a height may cause articles to fall on those below. Two-fifths of all reported major injuries are caused by falls from a height; as are half the fatal injuries to construction workers. The management of such risks is clearly a priority in many situations, particularly in construction, facilities management and plant maintenance.

People may fall, or cause things to fall, while going to or leaving an elevated workplace; or such an accident may happen while they are there and preoccupied with what they are doing. In the one case there will be a need to provide safe stairways or ladders etc, and in the other, the proper protection of edges from which persons or articles might fall, or where appropriate, the provision of nets or harnesses. But it must be realised that since the introduction of the *Management of Health and Safety at Work Regulations*, and particularly since the amended Regulations of 1999 (*SI 1999 No 3242*), such customary approaches to the management of the risk are not sufficient – once, that is, a risk requiring a remedy has been identified. The strategy (principles of prevention and protection) set out in *Regulation 4* and *Schedule 1* of the *Management of Health and Safety at Work Regulations 1999* requires that the actual need to be at such heights, or to be there so often, must be questioned *before* the appropriate precautions are considered.

This is not to say, of course, that where well recognised and *reliable* means of access and workplace protection have been provided, there is any pressing need to redesign the job to avoid work at heights – but when changes are contemplated (and change occurs all the time in construction, for instance) this approach must be properly considered. Indeed where, say, the existing precautions entail the use of portable ladders, there will be an immediate need to consider if this is appropriate.

W9002 In construction, *Regulation 6* of the *Construction* (*Health, Safety and Welfare*) *Regulations* (*SI 1996 No 1592*) supplements the general requirement referred to above (W9001) in the case of ladders used in that industry by limiting the circumstances in which they may be used. The approach of Health and Safety Executive (HSE) inspectors when encountering portable ladders on site is not now just to check the angle at which the ladder is set and how well it is footed or secured etc, but to ask whether its use can be justified at all in the circumstances found. A specific enforcement campaign on this is planned for 2004/05 and one can expect this sharpening of attitude to become more general, particularly when the new all-industry regulations on working at heights (see W9008 below) are in force. There is perhaps at this point a danger that some readers may lose patience with the argument (which in their view is an example of nanny state extremism). What needs to be borne in mind is that health and safety law is driven by national statistics, and what might appear locally as a remote possibility when repeated daily across the country, may produce a very objectionable number of deaths at the end of the year. Furthermore, the particular strategic drive to reduce risk in the first place rather than manage it later rests on the sound perception that human failure is a fact of existence: precautions that rely on the constant repetition of correct actions by people will simply not be as reliable as the consequences of reduced inherent risk. If a job can be re arranged so that it can be done from the ground nobody can fall; and a permanent ladder will not slip.

W9003 This strategic approach to the management of risk extends to designers within the meaning of the *Construction (Design and Management) Regulations 1994 (SI 1994 No 3140 as amended by SI 2000 No 2380)*, (*'CDM Regulations'*), *Reg 13* (see C8029). This duty, to have *adequate regard* to the need to avoid or reduce risk applies not just to risks to those constructing a building, but to risks to those maintaining it and specifically (through inclusion of the defined term *cleaning work*) to the protection of window cleaners. Further, work at heights in and on the completed structure is one of the matters to be covered in the health and safety file required under these Regulations (see C8038). It is no longer the case that an architect or engineer can design a structure without careful thought as to how it is to be maintained and how the risks of that can be minimised. So increasingly in the case of new buildings, one can expect there to be somewhat less need to be above floor level for foreseeable maintenance purposes and that where that remains necessary, as it often will be, there should be properly thought-through provision for the work to be done safely.

W9004 Before going on to describe and comment on the legal duties concerning the management of the risks of working at heights, it will be helpful to consider how employers may incur those duties The following cases are not exhaustive but cover the more common situations:

- as the employer of those working at height in that employer's premises;
- as an employer bringing in contractors to those premises while work continues there;
- as a contractor whether or not the client's workforce is present;
- as an employer who brings in a contractor to carry out construction work (as defined in the *CDM Regulations* for the purposes of those and other regulations) while normal work continues;
- as an employer who engages a contractor to carry out construction work on empty premises or a green or brownfield site.

The first three of these cases are either clear or fairly straightforward. For instance the general duties of an employer to contractors carrying out work in the employer's premises are clear – specific requirements about organisation and the provision of information are in the *Management of Health and Safety at Work Regulations 1999* (*SI 1999 No 3242*). All these duties, insofar as they relate to work at heights, will be discussed in the following paragraphs of this chapter. In the case of contractors engaged to undertake construction work, as defined in the *CDM Regulations* (*SI 1994 No 3140* as amended by *SI 2000 No 2380*) (see C8031), however, the situation is more complicated and not without doubt. Since construction contributes so largely to the to the risk of falling, it is necessary to refer to a significant change in the official view of the extent of the duties of the construction client.

There is no need to repeat here all the considerations set out in C8028, which can readily be referred to, but it should noted that the in the current edition of the

Approved Code of Practice to the *CDM Regulations* the Health and Safety Commission (HSC) expresses the view that construction clients should take a more active role in relation in the health and safety management of the construction work itself than just following the duties of a client as set out in the *CDM Regulations*.

HSC apparently believes that the *Management of Health and Safety at Work Regulations 1999* apply directly to the employer's acts as a client over and above any duties he might have in relation to the interaction of his own staff and those of the contractor. This would mean, for instance that an employer having work done to the roof of an empty building, with none of his own workers even remotely present, would have to monitor the observance of the necessary precautions by the contractor. According to HSC, if the employer lacks the technical knowledge to do this effectively he must engage advice under *Regulation 7* of the 1999 Regulations; whereas, previously it was thought that the need for such advice would relate only to the specifics of his own operations. This might very well mark a very substantial increase in an employer's responsibilities in relation to working at heights. As stated in C8028, it is arguable that HSC are not correct in their opinion, but employers having construction work done must be aware of this official view and it would seem wise for occupiers of premises etc to pay rather more attention in what follows to matters that ordinarily are the duty of contractors.

Should, however, such an occupier extend his interest to the point of controlling the way that construction work is done he could well acquire typical and specific contractor duties, including those relating to the prevention of falls. [*Construction (Health, Safety and Welfare) Regulations 1996* (*SI 1996 No 1592*), *Reg 4*].

W9005 Sometimes work at heights is undertaken in a crisis situation. For instance, a fault may have developed on an item of plant which needs immediate attention for production to continue, or the factory roof may be leaking resulting in damage to stored products below. In both cases, there may be insufficient time to properly consider the precautions necessary to ensure safe working at height. Procedures should be planned and in force to deal with this sort of contingency. The danger a while at a height when an emergency occurs will be addressed in the new regulations referred to in W9008 below.

W9006 Another aspect of work at heights is window cleaning. Windows in factories, workplaces and high-rise commercial and office blocks must be regularly and periodically cleaned. [*Workplace (Health, Safety and Welfare) Regulations 1992* (*SI 1992 No 3004* as amended by *SI 2002 No 2174*), *Reg 16*]. Frequency of cleaning depends on the type of establishment, whether office, shop or factory etc (see W9044 below) – but in any case the method selected should ensure protection for window cleaners (see W9042, W9043 and W9045 below). Window cleaning, particularly in high-rise commercial properties has posed serious safety problems for contract window cleaning companies and self-employed window cleaners. On the one hand, contract cleaning companies need to use the best available window cleaning equipment and methods, and, on the other, building owners/occupiers need to make safety anchorage points, in so far as they are permitted to do so, conveniently accessible on the building (see further BS 8213 for guidance). Crucially, in this respect, buildings, windows and skylights on buildings may need to be fitted with suitable devices (see W9046–W9048 below) to allow windows and skylights to be securely cleaned to satisfy the safety requirements. [*Workplace (Health, Safety and Welfare) Regulations 1992* (*SI 1992 No 3004* as amended by *SI 2002 No 2174*), *Reg 16*(*1*)].

More generally, the building designer has a duty to pay adequate attention to the avoidance or reduction of risk to those cleaning windows and any transparent or translucent wall, ceiling or roof where they might fall more than two metres. [*CDM Regulations* (*SI 1994 No 3140* as amended by *SI 2000 No 2380*), *Regs 2*(*1*), *13*(*2*)(*a*)]. As falls from heights account for by far the greater majority of major injuries, the need to comply with fall-preventive statutory requirements and window cleaning regulations at all times is paramount.

Current statutory requirements

W9007 While significant change is pending (see W9008 below), current statutory requirements relating to falls from heights are contained in the *Workplace* (*Health, Safety and Welfare*) *Regulations 1992* (*SI 1992 No 3004* as amended by *SI 2002 No 2174*). There are also specific requirements relating to prevention of falls on construction sites, in the form of the *Construction* (*Health, Safety and Welfare*) *Regulations 1996* (*SI 1996 No 1592*), *Regs 5–8*. As noted above, the *Management of Health and Safety at Work Regulations 1999* (*SI 1999 No 3242*), *Reg 3*, specifies that there must be a proper risk assessment. Therefore individuals cannot be required to work at heights from which they may be injured or killed until this assessment has taken place. There is a clear duty, according to the degree of risk revealed, to avoid the risk altogether or to reduce the extent or potentially dangerous nature of the work, wherever reasonable, before adopting the precautions in the specific regulations referred to. [*Management of Health and Safety at Work Regulations 1999* (*SI 1999 No 3242*), *Reg 4 and Sch 1*]. (See also paragraphs 30 and 31 of the Approved Code of Practice (ACoP) to the 1999 Regulations.) As we have already seen, in practical terms this means asking at the outset 'do we have to do it at all?', 'do we only have to do so much of it?' or 'do we have to do it in that way?'.

Specific requirements relating to windows and window cleaning are contained in the *Workplace* (*Health, Safety and Welfare*) *Regulations 1992* (*SI 1992 No 3004* as amended by *SI 2002 No 2174*), *Regs 14–16*. We have already referred to the designers' duty to take into consideration window cleaning along with the risk implications of their designs for construction and maintenance workers. [*CDM Regulations, Reg 13* (*SI 1994 No 3140* as amended by *SI 2000 No 2380*)].

Recent developments

W9008 Following the adoption of the second amendment to the Use of Work Equipment Directive 89/655/EEC in 2001, new regulations on working at heights were required to be in force by July 2004, but because of concern, principally among outdoor activity interests, they have been delayed. The intention is to produce a single code of regulations applying to all industries. Essentially they will make general the kind of requirements that have applied to construction and shipbuilding for many years. But they will embody the detail of this recent Directive.

There is little that is really new for the construction industry within the draft regulations published in the consultative document (CD192) (still available from HSE's website). Furthermore, while specific detailed legislation has not previously applied to most other industries, current good health and safety practice, (in compliance with the general duties under *HSWA 1974* and related regulations – as sharpened by the Management Regulations), should already have led to action substantially close to what the new regulations will require. For instance, *Regulation 13* of the *Workplace* (*Health, Safety and Welfare*) *Regulations 1992* (*SI 1992 No 3004* as amended by *SI 2002 No 2174*) is very relevant (the ACoP to those Regulations already provides some detailed guidance). Much has been made of draft Regulation 6: 'every employer shall ensure that work is not carried out at a height where it is reasonably practicable to carry out the work safely otherwise than at height' – but that is essentially a specific way of stating a general requirement that already applies to work at heights, as well as other work, in *Regulation 4* and *Schedule 1* of the *Management of Health and Safety at Work Regulations 1999* (*SI 1999 No 3242*). There appear to be, however, issues of principle in relation to certain work activities supporting recreation in which risk is an inherent part of the enjoyment, and although managed, is freely accepted by responsible persons at a level that all agree would be quite wrong at, or in relation to, an ordinary workplace. A clear example would be commercial instruction of climbers on an actual mountain etc. This point is discussed further at W9051 below. The point of general interest, however, is that after media notice and an unprecedented Early Day Motion in Parliament, the Government is looking into the possibility of some exemption. It is understood that the regulations are expected to be made by Christmas 2004.

The following is a summary of what the Directive requires to be implemented. Employers should:

- select the most suitable and safe equipment based on their risk assessment. Selection must take account of how often the equipment will be used, at what height and for how long;
- take steps to prevent or arrest falls from height, giving priority to collective measures, such as working platforms netting, over personal protection measures, such as harnesses;
- use ladders and rope access, such as abseiling, only where other safer equipment, such as scaffolding is not justified;
- ensure that the bearing components of scaffolding are prevented from slipping or from moving accidentally during work at height, and that the dimensions and layout of the decks are suitable for work and allow safe passage and use and;
- ensure ladders are used safely and positioned and secured to ensure that they are stable and do not slip in use.

There are in fact rather more detailed requirements in relation to working at heights than originally appeared in Annexe 4 of the Temporary or Mobile Construction Sites Directive, which was implemented in the *Construction* (*Health, Safety and Welfare*) *Regulations 1996* (*SI 1996 No 1592*) – and this detail appears in the draft regulations. The corresponding requirements of the construction regulations will be repealed and replaced by the new regulations to provide a set of common regulations applicable to all activities already mentioned.

It should be noted that the order of preference of the preventative measures now outlined in the *Construction* (*Health Safety and Welfare*) *Regulations 1996* (*SI 1996 No 1592*), *Reg 6* will have to be changed. Where working platforms with guard rails and toe boards are not practicable, or where because of the short duration of the work they are not reasonably practicable, means of personal suspension (safety harnesses) are currently preferred over means of arresting falls (safety nets) (see W9020 and W9021). This however contradicts the fundamental strategy of preventative measures of the Framework Directive set out in the *Management of Health and Safety at Work Regulations 1999* (*SI 1999 No 3242*), *Reg 4* and *Sch 1*, where collective measures, which include nets, must be considered before personal ones. The new Directive follows the Framework Directive in this respect.

A further point of interest to those familiar with existing legislation is a reference to fragile *surfaces* rather than fragile materials. This is intended to make it clear that action is required in relation to materials that may originally have been load bearing but have been weakened by damage or weathering. It seems that this terminology has caused unforeseen grief in the climbing world.

HSE are preparing guidance for all industries, but HSE documents for the construction industry itself will be revised accordingly.

W9009 As mentioned in W9001 above, the risks from work at heights must be assessed and considered in light of the fundamental principles of prevention and protection set out in the *Management of Health and Safety at Work Regulations 1999* (*SI 1999 No 3242*) (*MHSWR*), *Sch 1*. Before deciding the type of precautions necessary, the approach explained in paragraphs 30 and 31 of the *MHSWR* ACoP must be followed. Once it is established what extent of the work cannot be reduced or avoided altogether (the higher the risk the greater the case for elimination, or reduction at source), the precautions specified in the Regulations can be applied. Care must also be taken to give preference to collective precautions: fixed means of access, working platforms etc. over personal measures such as providing individuals with safety harnesses etc. Training cannot be a substitute for the physical precautions required but must complement them.

So far as reasonably practicable (for meaning, see E15039 ENFORCEMENT), suitable and effective measures must be taken (other than by provision of personal protective equipment, training, information, supervision etc.) to prevent:

- any person falling a distance likely to cause personal injury; or
- any person being struck by a falling object likely to cause personal injury.

Any area from which this might happen must be clearly indicated – particularly in the case of pits or tanks. [*Workplace* (*Health, Safety and Welfare*) *Regulations 1992* (*SI 1992 No 3004* as amended by *SI 2002 No 2174*), *Reg 13*].

Suitable and effective measures are:

- fencing;
- covering; and
- fixed ladders.

Fencing

W9010 This should primarily prevent people falling from edges and objects falling onto people, and where it has to be removed temporarily (as where goods/materials are admitted), temporary measures should be put in train (e.g. provision of handholds). Except in the case of roof edges or other points to which there is no general legitimate access, secure fencing should be provided where a person might fall:

- 2 metres or more, or
- less than 2 metres, where the risk of injury is otherwise greater (e.g. internal traffic route below).

Fencing should be sufficiently high to prevent falls, both of people and objects, over or through it. Minimally, it should consist of two guard-rails at suitable heights. The top of the fencing should generally be at least 1,100 millimetres above the surface from which a person might fall, be of adequate strength and stability to restrain any person or object liable to fall against it.

Covers

W9011 Pits, tanks, vats, sumps, kiers etc. can be securely covered (instead of fenced). Covers must be strong and able to support loads imposed on them as well as passing traffic; nor should they be easily detachable and removable. They should be kept securely *in situ* except for purposes of inspection or access. Uncovered tanks, pits or structures must be fenced if there is a traffic route over them.

Fixed ladders

W9012 Assuming a staircase is impractical, fixed ladders (sloping ones are safer than vertical ones) (which should be provided in pits etc.) should be:

- of sound construction,
- properly maintained,
- securely fixed.

Rungs should be horizontal and provide adequate foothold without reliance on nails or screws. In the absence of an adequate handhold, stiles should extend to at least 1,100 millimetres above any landing (or the highest rung) to step or stand on. Fixed ladders with a vertical distance of more than 6 metres should normally have a landing at every 6 metre point. Where possible, each run should be out of line with the previous run to reduce falling distance. And where a ladder passes through a floor, the opening should be as small as possible, fenced (if possible) and a gate provided to prevent falls.

(For ladders on construction sites, see W9023 below and C8066 CONSTRUCTION AND BUILDING OPERATIONS.)

Pits, tanks, vats, kiers etc. – new workplaces

W9013 Serious accidents have occurred as a result of workers, often maintenance workers, falling into or overreaching or overbalancing (and falling) into pits etc. containing dangerous corrosive or scalding substances. Thus, where there is a risk of a person or employee falling into a dangerous or corrosive or scalding substance, (so far as is reasonably practicable) pits, tanks (or similar vessels) must be either:

- fenced, or
- covered.

[*Workplace* (*Health, Safety and Welfare*) *Regulations 1992* (*SI 1992 No 3004* as amended by *SI 2002 No 2174*), *Reg 13*(*5*)]

and means of access/egress into pits/tanks – in the form (preferably) of fixed ladders (or the equivalent) – should be provided. [*Workplace* (*Health, Safety and Welfare*) *Regulations 1992* (*SI 1992 No 3004* as amended by *SI 2002 No 2174*), *Reg 13*(*6*)].

Construction sites – the Construction (Health, Safety and Welfare) Regulations 1996 (SI 1996 No 1592)

W9014 Construction work is associated with a large number of falls, some of which are fatal. A fair proportion of serious accidents occur as a result of workers falling off (or through) working platforms, scaffolds, toe boards, fragile roofs and personal suspension equipment. This situation is addressed by the *Construction* (*Health, Safety and*

Welfare) Regulations 1996 (*SI 1996 No 1592*), *Regs 6–8*. The *Management of Health and Safety at Work Regulations 1999* (*SI 1999 No 3242*), *Regs 3, 4,* and *Sch 1* also address the situation whereby employers have a duty to assess risks and to avoid or reduce them at source. (See particularly the advice to *Reg 4* in paragraphs 30 and 31 of the ACoP). Designers too must pay adequate regard to the avoidance or reduction of risk. [*CDM Regulations* (*SI 1994 No 3140* as amended by *SI 2000 No 2380*), *Reg 13*(*2*)(*a*)].

As we have seen above, these duties will entail such matters as reviewing the need firstly to have the work – or so much of it – done at heights; and where the work cannot be avoided or reduced, giving preference to collective measures such as secure working platforms over individual ones such as safety harnesses.

W9015 Suitable and sufficient steps must be taken to prevent any person (direct or indirect employees) from falling, so far as is reasonably practicable. [*Construction* (*Health, Safety and Welfare*) *Regulations 1996* (*SI 1996 No 1592*), *Reg 6*(*1*)]. This will usually involve the provision of:

- suitable working platforms;
- guard-rails, toe-boards (not required in respect of stairway or rest platforms of a scaffold used as a means of access or egress to or from a place of work) [*Construction* (*Health, Safety and Welfare*) *Regulations 1996* (*SI 1996 No 1592*), *Reg 6*(*9*)]; and
- barriers.

These can be removed for movement of materials, but must be replaced as soon as practicable. [*Construction* (*Health, Safety and Welfare*) *Regulations 1996* (*SI 1996 No 1592*), *Reg 6*(*4*)].

Where working platforms etc. are not practicable, or are not reasonably practicable because the duration of the work is short, means of personal suspension should be used (see W9019 below). Where it is not practicable to provide neither platforms nor personal suspension, then other precautions should be taken as well as means of arresting falls (nets) (see W9021). It should be noted that this preference for personal suspension over safety nets – which are a collective means of preventing injury from falls – is contrary to the hierarchy within the *Management of Health and Safety at Work Regulations 1999* (*SI 1999 No 3242*), *Reg 4* and *Sch 1*. This approach will be reversed as soon as the new regulations are made in accordance with the second amendment to the Use of Work Equipment Directive (see W9008 above). The preference for collective over personal precautions stems from experience that the use of personal precautions depends upon constant human compliance, which is often absent. Collective precautions are more easily managed because they depend less on human behaviour and thus are more reliable.

Mobile elevating work platforms (MEWPs) provide safe and effective access to working positions above ground level. HSE Guidance in Operational Circular 314/19 gives information on the matters inspectors take into account when considering this equipment. This covers safe systems of work; planning; training and experience.

High-risk situations occur where:

- there are protruding features which could catch or trap the carrier;
- nearby vehicles or mobile plant could foreseeably collide with, or make accidental contact with the MEWP;

- the nature of the work being done from the basket may mean operators are more likely to lean out or are handling awkward work pieces which may move unexpectedly; and
- unexpected or rapid movement of the machine or overturning is possible.

These assessments will be of particular interest to architects and engineers considering MEWPS for window cleaners etc. since they are indications of the unsuitability of this means of access.

Definition

W9016 A working platform refers to any platform used (*a*) as a place of work, or (*b*) a means of access or egress (to or from) a place of work, including:

- a scaffold;
- a suspended scaffold;
- a cradle;
- a mobile platform;
- a trestle;
- a gangway;
- a run;
- a gantry;
- a stairway; and
- a crawling ladder.

[*Construction* (*Health, Safety and Welfare*) *Regulations 1996* (*SI 1996 No 1592*), *Reg 2*(*1*)].

Requirements relating to guard-rails and toe boards

W9017 The Regulations require that:

- guard-rails, toe-boards and barriers and other similar means of protection must be:
  - (i) suitable and of sufficient strength and rigidity for the purpose for which they are being used, and
  - (ii) so placed, secured and used as to ensure, so far as is reasonably practicable, the avoidance of accidental displacement;
- structures supporting guard-rails, toe-boards, barriers and other means of protection, or a structure to which these are attached, must be of a suitable and sufficient strength for the purpose for which they are used;
- main guard-rails must be at least 910 millimetres above the edge from which a person is liable to fall (an increase to 950mm is one of the changes proposed in the coming regulations);
- there must not be an unprotected gap of more than 470 millimetres between any guard-rail, toe-board or barrier and the walking surface;
- toe-boards must be not less than 150 millimetres high; and

- guard-rails, toe-boards, barriers and other similar means of protection must be so placed as to prevent, so far as is reasonably practicable, the fall of a person, material or objects from a place of work.

[*Construction* (*Health, Safety and Welfare*) *Regulations 1996* (*SI 1996 No 1592*), *Sch 1*].

Requirements relating to working platforms

W9018 The Regulations require work platforms to conform to the following requirements.

- *Stability* – they must be:
  - (i) suitable and of sufficient strength and rigidity for the intended use,
  - (ii) erected and used to ensure, so far as is reasonably practicable, the avoidance of accidental displacement,
  - (iii) remain stable and if modified or altered, remain stable after modification or alteration, and
  - (iv) dismantled so as to avoid accidental displacement.
- *Safety* – they must be:
  - (i) of sufficient dimensions to permit free passage of persons and safe use of equipment and materials, and, so far as is reasonably practicable, be a safe working area,
  - (ii) at least 600 millimetres wide,
  - (iii) so reasonably constructed that the surface has no gap likely to cause injury, or from which there is any risk of any person below the platform being struck by falling objects,
  - (iv) so erected, used and maintained to prevent, so far as is reasonably practicable, any slipping or tripping, or any person being caught between a working platform and an adjacent structure, and
  - (v) provided with such handholds and footholds as are necessary to prevent, so far as is reasonably practicable, any person from slipping or falling from a working platform.
- *Load* – they must not be so loaded as to give rise to danger of collapse or deformation which could affect its safe use.
- *Supporting structures*– must be:
  - (i) of suitable and sufficient strength and rigidity for the intended purpose,
  - (ii) so erected and, where necessary, securely attached to another structure as to ensure stability,
  - (iii) so altered or modified as to ensure stability when altered or modified, and
  - (iv) erected on surfaces which are stable and of a sufficient strength and suitable composition to ensure the safe support of the structure, working platform and any load intended to be placed upon the working platform.

[*Construction* (*Health, Safety and Welfare*) *Regulations 1996* (*SI 1996 No 1592*), *Sch 2*].

Personal suspension equipment – fall preventative requirements

W9019 Personal suspension equipment refers to suspended access (other than a working platform) for use by an individual, including a boatswain's chair and abseiling equipment, but does not include a suspended scaffold or cradle. [*Construction* (*Health, Safety and Welfare*) *Regulations 1996* (*SI 1996 No 1592*), *Reg 2*(*1*)].

Where compliance with guard-rail, toe-board or working platform duties (see above) is not reasonably practicable, (e.g. owing to short-term work), suitable personal suspension equipment must be provided and used. [*Construction* (*Health, Safety and Welfare*) *Regulations 1996* (*SI 1996 No 1592*), *Reg 6*(*3*)(*c*)].

See also the discussion on the preference for personal suspension over means of arresting falls in W9008 and W9015 above.

It should be noted that new requirements will be introduced within the new working at heights regulations. Double rope working will be the normal standard, though single rope working would be allowed where a second rope would entail greater risk. None the less, this coming duty is causing difficulties in recreational climbing.

Requirements relating to personal suspension equipment

W9020 Personal suspension equipment must be:

- of suitable and sufficient strength, having regard to the work being carried out and the load, including any person it is intended to bear;
- securely attached to a structure or plant, and the structure or plant must be suitable and of sufficient strength and stability to support the equipment and load;
- installed or attached so as to prevent uncontrolled movement of equipment; and
- suitable and sufficient steps must be taken to prevent any person from falling or slipping from personal suspension equipment.

[*Construction* (*Health, Safety and Welfare*) *Regulations 1996* (*SI 1996 No 1592*), *Sch 3*].

Means of arresting falls

W9021 Where compliance with guard-rail, toe-board, working platform or personal suspension equipment requirements is not reasonably practicable, suitable and sufficient means for arresting falls must be: (i) provided, and (ii) used. [*Construction* (*Health, Safety and Welfare*) *Regulations 1996* (*SI 1996 No 1592*), *Reg 6*(*3*)(*d*)].

Equipment provided for arresting falls must adhere to the following requirements:

- The equipment must be suitable and of sufficient strength to safely arrest the fall of any person liable to fall.
- The equipment must be securely attached to a structure or to a plant, and the structure or plant (and the means of attachment) must be suitable and of sufficient strength and stability to safely support the equipment and any person liable to fall.

- Suitable and sufficient steps must be taken to ensure, so far as is reasonably practicable, that in the event of a fall, equipment does not cause injury to a person.

[*Construction* (*Health, Safety and Welfare*) *Regulations 1996* (*SI 1996 No 1592*), *Sch 4*].

See also the discussion on the preference for personal suspension over means of arresting falls in W9008 and W9015 above.

Installation and erection of scaffolds, personal suspension equipment and fall-arresting devices

W9022 Installation and erection of any scaffold, personal suspension equipment or fall-arresting devices, or substantial addition or alteration thereto (not including personal attachment of any equipment or fall-arresting devices to a person for whose safety such equipment or device is provided) must be carried out under the supervision of a competent person. [*Construction* (*Health, Safety and Welfare*) *Regulations 1996* (*SI 1996 No 1592*), *Reg 6*(*8*)]. Training to CITB standards in fall arrest safety equipment is now available. Guidance on the use of fall arrest equipment during the erection, dismantling and alteration of scaffolding is available from Construction Industry Publications, 60 New Coventry Road, Sheldon, Birmingham B26 3AY (tel: 0121 722 8200).

Ladders

W9023 Owing to a growing number of fatalities and serious injuries to workers involving the use of ladders, ladders can no longer be used as:

- a place of work, or
- a means of access or egress, to or from a place of work, unless it is reasonable to do so, having regard to:
  - (i) the nature of work and the duration of work, and
  - (ii) the risk to the safety of any person at work arising from the use of a ladder.

[*Construction* (*Health, Safety and Welfare*) *Regulations 1996* (*SI 1996 No 1592*), *Reg 6*(*5*)].

Ladders must be:

- of suitable and sufficient strength for their intended purpose;
- so erected as to avoid displacement;
- (where they are of three metres in height or more), secured, so far as is practicable, and where not practicable, a person must be positioned at the foot of the ladder to prevent it slipping during use (see further the case of *Boyle v Kodak Ltd [1969] 2 All ER 439*, where an experienced painter failed to secure a ladder, contrary to the *Building* (*Safety, Health and Welfare*) *Regulations 1948, Reg 29* and was injured);
- (where used as a means of access between places of work), sufficiently secured to prevent slipping and/or falling.

In addition, ladders must also:

- (where the top of the ladder is used as a means of access to another level), extend to a sufficient height above the level to which it gives access, so as to provide a safe handhold (unless a suitable alternative handhold is provided);
- (where a ladder or run of ladders rises vertically 9 metres or more above its base), be provided, where practicable, with a safe landing area or rest platform at suitable intervals;
- be erected on surfaces which are stable and firm, of sufficient strength, and of suitable composition safely to support the ladder and its load.

[*Construction* (*Health, Safety and Welfare*) *Regulations 1996* (*SI 1996 No 1592*), *Sch 5*].

(For practical safety criteria in the use of ladders, see also C8066 CONSTRUCTION AND BUILDING OPERATIONS.)

Maintenance of fall-preventative equipment

W9024 All scaffolds, working platforms, toe-boards, guard-rails, personal suspension equipment, fall-arresting devices and ladders must be properly maintained. [*Construction* (*Health, Safety and Welfare*) *Regulations 1996* (*SI 1996 No 1592*), *Reg 6*(*7*)].

Fragile materials

W9025 Suitable and sufficient steps must be taken to prevent any person falling through any fragile material. [*Construction* (*Health, Safety and Welfare*) *Regulations 1996* (*SI 1996 No 1592*), *Reg 7*(*1*)]. In particular:

- no person must pass or work on or from fragile material, through which he would be liable to fall 2 metres (or more), unless suitable and sufficient platforms, coverings or other similar means of support are provided and used so that the weight of any person so passing or working is supported by such supports;
- no person must pass or work near fragile materials through which he would be liable to fall 2 metres (or more), unless suitable and sufficient guard-rails or coverings are provided and used to prevent any person working or passing from falling; and
- in the above cases, prominent warning notices must be affixed to the place where the material is located.

[*Construction* (*Health, Safety and Welfare*) *Regulations 1996* (*SI 1996 No 1592*), *Reg 7*(*2*)].

For the possible effect of such notices, see OCCUPIERS' LIABILITY.)

HSE are particularly concerned about reducing the risk from fragile materials and consider that designers, having regard to their duty to avoid or reduce risk under the *CDM Regulations*, should not specify them. To assist designers they have agreed standards for strength and durability with manufacturers' representatives. (See Advisory Committee for Roofwork Material Standards: ACR (M) 001: 2000, available from The Fibre Cement Manufacturers' Association, Ghyll House Cock Road, Colton, Stowmarket IP14 4QA, tel: 01449 781 577.)

Falling objects

W9026 With relation to this hazard the Regulations require:

- that in order to prevent danger to any person, suitable and sufficient steps must be taken to prevent, so far as is reasonably practicable, the fall of any material or object (e.g. the provision of working platforms or guard-rails etc.) [*Construction* (*Health, Safety and Welfare*) *Regulations 1996* (*SI 1996 No 1592*), *Reg 8*(*1*)];
- where compliance is not reasonably practicable, suitable and sufficient steps must be taken to prevent any person being struck by any falling material or object likely to cause injury (e.g. the provision of a covered traffic route or roadway) [*Construction* (*Health, Safety and Welfare*) *Regulations 1996* (*SI 1996 No 1592*), *Reg 8*(*3*)];
- that no material or object must be thrown or tipped from a height, where it is likely to cause injury [*Construction* (*Health, Safety and Welfare*) *Regulations 1996* (*SI 1996 No 1592*), *Reg 8*(*4*)]; and
- that materials or equipment must be so stored as to prevent danger to any person from:
  (i) collapse,
  (ii) overturning, or
  (iii) unintentional movement of such materials and equipment [*Construction* (*Health, Safety and Welfare*) *Regulations 1996* (*SI 1996 No 1592*), *Reg 8*(*5*)].

Inspection of working platforms etc.

W9027 Working platforms and personal suspension equipment must be inspected by a competent person:

- before being taken into use for the first time;
- after any substantial addition, dismantling or other alteration;
- after any event likely to have affected its strength or stability; and
- at regular intervals not exceeding seven days.

[*Construction* (*Health, Safety and Welfare*) *Regulations 1996* (*SI 1996 No 1592*), *Reg 29*(*1*), *Sch 7*].

Also, the employer or person who controls the way that construction work is done by anyone using a scaffold must ensure that it is stable and complies with the safeguards required by the Regulations before it is used by such persons for the first time. [*Construction* (*Health, Safety and Welfare*) *Regulations 1996* (*SI 1996 No 1592*), *Reg 29*(*2*)].

Where a person may fall more than two metres, the competent person must make a report within the working period in which the inspection is completed. The report must contain the following:

- the name and address of the person for whom the inspection was carried out;
- the location of the place of work inspected;
- a description of what was inspected;
- date and time of the inspection;

- details of anything that could cause risk;
- details of any consequent action;
- details of further action considered necessary;
- name and position of the person making the report.

[*Construction (Health, Safety and Welfare) Regulations 1996 (SI 1996 No 1592), Reg 30(1), (5), Sch 8*].

The report must be provided within 24 hours to the person for whom it was prepared. It must be kept on site until completion and for a further three months at the office of the person for whom it was prepared; in addition, it must be available to an inspector at all reasonable times and copies or extracts must be sent to an inspector if required. [*Construction (Health, Safety and Welfare) Regulations 1996 (SI 1996 No 1592), Reg 30(2)–(4)*].

No report is required on a mobile tower scaffold unless it has remained in the same place for seven days or more [*Construction (Health, Safety and Welfare) Regulations 1996 (SI 1996 No 1592), Reg 30(6)(a)*] and only report of an addition, dismantling or alteration (see (*b*) above) is required in any period of 24 hours. [*Construction (Health, Safety and Welfare) Regulations 1996 (SI 1996 No 1592), Reg 30(6)(b)*].

Practical guidelines

W9028 Construction or extensive maintenance work will probably entail risks both to those carrying it out and to the ordinary worker. These risks must be assessed and any existing precautions reviewed [*Management of Health and Safety at Work Regulations 1999 (SI 1999 No 3242), Reg 3*]. The appropriate action should also be taken according to *Schedule 1* of the 1999 Regulations. [*Management of Health and Safety at Work Regulations 1999 (SI 1999 No 3242), Reg 4*].

Where work must be done at heights, suitable access equipment or other safeguards must be provided and attention paid to the following:

- Who is in control and what is the amount of co-ordination necessary, i.e. have the responsibilities of the occupier, employer and contractor been clearly defined?
- What level of supervision and training is required with regard to the type of hazard and level of risk? (including selection of the appropriate method, and the use of approved equipment).
- What other considerations are necessary, e.g. is there any danger to works employees from the contractors' operations, or vice-versa?

Many of these matters may seem obvious, but lives have been lost because the contractor or maintenance worker did not appreciate the fragile nature of a roof or the fact that it had been weakened owing to the nature of the process. These considerations point to the need for careful preparation and planning of such work to ensure that there is proper co-ordination and co-operation. [*Management of Health and Safety at Work Regulations 1999 (SI 1999 No 3242), Regs 11 and 12*]. The contractor must be aware of any risks from the operations on the premises that might affect his workforce and of the precautions that the occupier is taking to control them. Furthermore, the occupier must ensure that comprehensible information about this has been passed on to the construction workers. [*Management of Health and Safety at Work Regulations 1999 (SI 1999 No 3242), Reg 12(1), (3)*]. The contractor must be provided with the names of the relevant persons who have been appointed to ensure evacuation in an emergency, while the occupier must take all reasonable steps to ensure that the contractor's employees themselves know who is to implement their own evacuation. [*Management of Health and Safety at Work Regulations 1999* (*SI 1999 No 3242*), *Reg 12*(*4*)].

Where the *CDM Regulations* (*SI 1994 No 3140* as amended by *SI 2000 No 2380*) apply, the occupier (as client) must give the planning supervisor information about the premises. This information should be taken into account by the designers in the avoidance or reduction of risk and in the subsequent preparation of the health and safety plan. It is important to note that this information must include anything that the client could find out by making reasonable enquiries. For instance, the obvious risks to construction workers of falling should not obscure the possible danger that work at heights may disturb any asbestos that might be there – it should not be forgotten that asbestos is a greater cause of death in the construction trades than falling.

Since the inception of the *Management of Health and Safety at Work Regulations 1999* and *CDM Regulations*, larger firms should have well-established systems to deal with these matters, including the appointment of competent principal and other contractors. Smaller firms, or those that do not regularly commission construction work, while they are entitled to the advice of the planning supervisor where the *CDM Regulations* apply, may care to consider the following as appropriate when selecting contractors or in subsequent dealings with them (the views of HSC on the extent of clients' duties are mentioned in W9004 above):

- the time for the tendering contractor to respond to the initial (pre-tender) stage of the health and safety plan;
- the time for the contractor's own investigation of health and safety matters (e.g. a survey) where the Regulations do not apply;
- the response to the significant risks – knowledge of precautions; possession of advisory booklets etc; any safety training of managers and supervisors; signs that the impact of risks arising from the operations has been appreciated;
- how the contractor checks his health and safety performance;
- how staff (including casuals) engaged for this job are or will be trained;
- the protective clothing and equipment that will be provided (see PERSONAL PROTECTIVE EQUIPMENT);
- the resources to be devoted to health and safety management (ask to see the person who will actually do the managing);
- the arrangements for necessary liaison throughout the job (not forgetting the contractor's arrangements for delivering file information to the planning supervisor);
- potential risks associated with adverse weather conditions, such as a sudden downpour, dense fog, snow lying on a roof or high winds; in particular, the responsibility for calling operators off a roof or high-level position given adverse weather conditions;
- the provision of safety harnesses and belts, including the necessary anchorage points for them, or safety nets in certain situations (see W9042 below).

In addition the occupier/employer should ensure that his maintenance supervisor or clerk of works, etc. has access to the information contained in the health and safety at work guidance/advisory booklets. He may also provide simple checklists to cover his own operations.

HSE has published an important guidance booklet on safety in roof work.

The guidance is relevant not only to contractors and others who actually carry out work on roofs, but also to those who have an influence on workers' health and safety, such as planning supervisors, clients, designers and manufacturers.

It carries forward existing, well-known messages on basic safety precautions, such as the fundamental need for edge protection. Effective fall arrest equipment is essential during industrial roofing projects, HSE's preferred method being properly installed and maintained safety netting. The guidance also contains important new material on:

- *the use of safety nets, particularly in industrial roofing*

  HSE is convinced that there is a substantial role for a greater use of safety nets in industrial roofing. Although they are not the only way of arresting falls, it is inspectors' consistent experience that harnesses, for example, are frequently not provided or worn. Where they are used, more often than not they are used incorrectly.

- *the role of designers in creating roofs that are safe to build and subsequently to maintain*

  Designers can often eliminate risks by designing them out at source, but they need to understand the problems faced by contractors to realise these benefits. If not, they could be breaking the law, as well as helping to perpetuate one of the biggest workplace killers. The *Construction (Design and Management) Regulations 1994 (SI 1994 No 3140 as amended by SI 2000 No 238)* require designers to have regard to health and safety risks for construction and maintenance workers in their designs. Their role in reducing the presence of fragile materials in roofs is clearly highlighted in the guidance as the most effective area in which they can contribute to improving health and safety.

- *the role of clients in providing relevant information ensuring that adequate resources are allocated for safety*
- *the risks from fragile roofing material and how to manage them*

'*Health and safety in roof work*' (ref HSG33), price £8.50, is available from HSE Books, PO Box 1999, Sudbury, Suffolk CO10 2WA.

Windows and window cleaning – all workplaces

W9029 Window cleaning hazards occur principally in high-rise properties such as office blocks and residential tower blocks. These contain vast areas of glass which should be maintained in a clean state on both inner and outer surfaces. Cleaning the interior surfaces generally presents no difficulties, though accidents have arisen from the unsafe practice of window cleaners reaching up to clean windows from step ladders which have not been securely placed on highly-polished office floors and, in consequence, falling through the glass windows and injuring themselves. Moreover, if windows can be cleaned from inside, company safety rules and procedures should prevent employees going out onto sills in cases where a window/windows, if in proper working order, can be cleaned from inside. Failure to operate this rule can involve the employer in both criminal and civil liability. Equally important, if not more so, the customer or client must ensure that his windows are in proper working order and regularly inspected and maintained, if they are capable of being cleaned from inside (*King v Smith [1995] ICR 339* (see W9040 below)).

It is the cleaning of outer surfaces that can present serious hazards if proper precautions are not taken, and this part of the section is concerned with the duties of workplace occupiers in relation to windows and window-cleaning on their premises (see W9037 below). In addition, liability of employers at common law in relation to window cleaning personnel is also covered (see W9040 below), and methods of protection for window cleaners as well as methods of cleaning (see W9042–W9048 below).

Statutory requirements relevant to windows in all workplaces (except construction sites) – the Workplace (Health, Safety and Welfare) Regulations 1992 (SI 1992 No 3004 as amended by SI 2002 No 2174)

Composition of windows

W9030 Every window (and every transparent/translucent surface in a door or gate) where necessary for reasons of health or safety must be:

- of safety material (e.g. polycarbonates, glass blocks, glass which breaks safely (laminated glass) or ordinary annealed glass meeting certain minimal criteria);
- protected against breakage of transparent/translucent material (e.g. by a screen or barrier); and
- appropriately (and conspicuously) marked so as to make it apparent (e.g. with coloured lines/patterns).

[*Workplace* (*Health, Safety and Welfare*) *Regulations 1992* (*SI 1992 No 3004* as amended by *SI 2002 No 2174*), *Reg 14*].

HSC in October 1996 approved a change to its publication '*Workplace health, safety and welfare*', the ACoP which supports the *Workplace* (*Health, Safety and Welfare*) *Regulations 1992*. The change became necessary as HSC acknowledged that there was confusion in the business world on what the law says about glazing in existing workplaces that became subject to the *Workplace Regulations* on 1 January 1996. The change relates to workplace glazing and now reads:

> 'In assessing whether it is necessary, for reasons of health and safety, for transparent or translucent surfaces in doors, gates, walls and partitions to be of a safety material or be adequately protected against breakage, particular attention should be paid to the following cases:
>
> (*a*) in doors and gates, and door and gate side panels, where any part of the transparent or translucent surface is at shoulder height or below;
>
> (*b*) in windows, walls and partitions, where any part of the transparent or translucent surface is at waist level or below, except in glasshouses where people there will be likely to be aware of the presence of glazing and avoid contact.
>
> This paragraph does not apply to narrow panes up to 250mm wide measured between glazing beads.'

Position and use of windows

W9031 No window, skylight or ventilator, capable of being opened, must be likely to be:

- opened,

- closed,
- adjusted, or
- in a position when open,

so as to expose the operator to risk of injury. [*Workplace* (*Health, Safety and Welfare*) *Regulations 1992* (*SI 1992 No 3004* as amended by *SI 2002 No 2174*), *Reg 15*].

In other words, it must be possible to reach openable windows safely; and window poles or a stable platform should be kept ready nearby. Window controls should be positioned so that people are not likely to fall out of or through the window. And where there is a danger of falling from a height, devices should prevent the window opening too far. In order to prevent people colliding with them, the bottom edge of opening windows should generally be 800 millimetres above floor level (unless there is a barrier to stop or cushion falls). Staircase windows should have controls accessible from a safe foothold, and window controls beyond normal reach should be gear-operated or accessible by pole (see BS 8213 and GS 25 1983). Manually operated window controls should not be higher than 2 metres above floor level.

Cleaning windows

W9032 All windows and skylights in workplaces must be designed and constructed to allow safe cleaning. [*Workplace* (*Health, Safety and Welfare*) *Regulations 1992* (*SI 1992 No 3004* as amended by *SI 2002 No 2174*), *Reg 16*(*1*)]. If they cannot be cleaned from the ground (or similar surface), the building should be fitted with suitable safety devices to be able to comply with the general safety requirement (see W9031). [*Workplace* (*Health, Safety and Welfare*) *Regulations 1992* (*SI 1992 No 3004* as amended by *SI 2002 No 2174*), *Reg 16*(*2*)].

Civil liability in connection with fall-preventive regulations

W9033 Because regulations give rise to civil liability when breached, even if silent (which the 1992 Regulations are), if breach leads to injury/damage, an action for breach of statutory duty would lie (see further INTRODUCTION).

Window cleaners as employees or self-employed contractors

W9034 Where window cleaners are employees of a contract window cleaning company, the company as employer owes a duty to its employees, under *HSWA 1974, s 2*(*1*) and (*2*), to provide a safe method of work, and instruction and training in job safety, e.g. provision of information as to how to tackle the job and the use of safety harnesses etc. All workplaces, buildings, windows and skylights should be fitted with suitable devices to allow the window or skylight to be cleaned safely, to comply with the *Workplace* (*Health, Safety and Welfare*) *Regulations 1992* (*SI 1992 No 3004* as amended by *SI 2002 No 2174*), *Reg 16* (see W9032 above).

Alternatively, window cleaners may be self-employed. In such cases they themselves commit an offence if they fail to provide themselves with adequate protection, e.g. safety harnesses, because of the requirement of *HSWA 1974, s 3*(*2*) (see INTRODUCTION).

Who provides protection to window cleaners on high-rise properties?

W9035 Where multi-storey properties are concerned, the duty of protection may fall upon one of three parties (or, at least, there may be division of responsibility between them):

- the contract window cleaning firm or the self-employed window cleaner;
- the building owner;
- the building occupier or business operator (i.e. the building occupier who is not the owner of the building).

Contract window cleaning firms and self-employed window cleaners

W9036 Both the contract window cleaning firm and the self-employed window cleaner have a statutory duty to provide protection (under *HSWA 1974, ss 2(1)(2), 3(2)*) and the employer/occupier, under the *Workplace (Health, Safety and Welfare) Regulations 1992* (*SI 1992 No 3004* as amended by *SI 2002 No 2174*), *Regs 15 and 16* (see W9031, W9032 above). This applies to contract cleaning companies (not just of windows) in respect of equipment left on the occupier's premises and used by the occupier's employees, even though the contract cleaning company is not actually working (*R v Mara [1987] 1 WLR 87* where one of the occupier's employees was electrocuted when using polisher/scrubber, which had a defective cable, to clean loading bay on a Saturday afternoon, when cleaning company did not operate its undertaking. It was held that the director of the cleaning company was in breach of *HSWA 1974, s 3(1)*). In addition, if an accident arose from the work, the window cleaning company could, as employer, be liable at common law, if it failed to take reasonable care and exercise control. There is no action for damages for breach of statutory duty under *HSWA 1974*. In this connection, any employer, when prosecuted under the 1974 Act or sued at common law would have to show that he had clearly instructed employees not to clean windows where no proper safety precautions had been taken, in order to avoid liability. Also, window cleaning companies would have to satisfy themselves, before instructing employees to clean windows, as to what safety precautions (if any) were provided, and that, if necessary, employees were told to test for defective sashes (see further *King v Smith* at W9040 below).

Division of responsibility between window cleaning firm or self-employed window cleaner and building owner or occupier (i.e. employer)

W9037 Although a window cleaner, whether an employee or self-employed, would be expected to have his own safety harness, it would be up to the contractor to provide fixing points for harness attachment. The responsibility for provision of harness anchorage points (e.g. safe rings, i.e. eyebolts) will be with either the building owner or building occupier (business operator) whose windows are being cleaned. Two situations are possible here:

- the building occupier is the owner of the building; or
- the building occupier (as is usual) is not the owner of the building.

Where the building occupier is the building owner

W9038 In this situation, the window cleaning contract is placed by the building occupier/owner and the latter, having control of the building, is therefore responsible under *HSWA 1974, s 3(1)* and the *Workplace (Health, Safety and Welfare) Regulations 1992* (*SI 1992 No 3004* as amended by *SI 2002 No 2174*), *Regs 15 and 16,* and at common law (and under the contract) to protect window cleaning personnel on his premises. Courts may also be prepared to imply terms into such contracts that such premises be reasonably safe to work on.

This position (that is, where the building owner and the building occupier are the same person) is not, however, the norm. On the contrary, the normal position is that the building owner and business operator are separate persons or companies.

Where the building occupier (that is, the employer) is not the building owner

W9039 Given that the employee or self-employed window cleaner must provide his own safety harness (see W9037 above), the responsibility of providing a safe ring for harness attachment lies with the building owner or occupier. More precisely, where the building owner employs a contract window cleaning firm, then it is up to the building owner (*inter alia*, as an implied contractual term) to make safety anchorage provision. He cannot escape from this contractual obligation owing to the strictures of the *Unfair Contract Terms Act 1977* (see OCCUPIERS' LIABILITY). He would also be required to do so by virtue of *HSWA 1974, s 3(1)*– the duty towards persons working on premises who are not one's employees.

This is the exception, not the norm. Generally, window cleaning contracts are placed by building occupiers (business operators), and so in the great majority of cases it is the *building occupier* who is responsible under *HSWA 1974, s 3(1)* and under the *Workplace (Health, Safety and Welfare) Regulations 1992* (*SI 1992 No 3004* as amended by *SI 2002 No 2174*), under the express or implied terms of a contract, and/or at common law (should an accident occur) for ensuring the safety of window cleaning personnel. Thus, the building occupier would have to provide a safety ring for harness attachment. A problem could arise here if the building owner refused permission for such attachment – resort to legal action would be, it is thought, the only ultimate solution. And occupiers should be under no illusion that failure to provide safety anchorage points can (and will, in all probability) result in the imposition of heavy fines! (By way of consolation, however, the cost of making available certain types of protection to window cleaners may often be less than the penalty incurred for failing to provide it.)

Common law liability to window cleaners

W9040 At common law (and under *HSWA 1974, s 2(2)*), an employer must provide and maintain a safe system of work. This extends to provision of safe appliances, tools etc. and giving information/training on how to carry out tasks safely. Nevertheless, there is a division of responsibility between employer and occupier of premises where a window cleaner is working. The former, the employer, must see that the employee is provided with a safe method of work; but this did not extend to defects in the premises (e.g. window sills) of the occupier causing injury to employees (*General Cleaning Contractors Ltd v Christmas [1952] 2 All ER 1110* where an experienced employee window cleaner was injured whilst cleaning windows at a club. There were no fittings on the building to which safety belts could have been attached. A defective sash dropped onto the employee's hand, causing him to lose his handhold and fall. It was held that the employer was liable; he should have provided wedges to prevent sashes from falling and also instructed employees to test for dangerous sashes). Modern conventional wisdom suggests, however, that an employer does not provide a safe system of work if a window cleaner has to clean windows from an outside window sill, which are capable of being cleaned from inside if they are in proper working order. It is incumbent on the customer-occupier to ensure that such windows are in a good state of maintenance and proper working order. (In the case of *King v Smith [1995] ICR 339*, a window cleaner was injured when he fell from a sill on a local authority building. The court held both employer and customer liable, the former 70 per cent, the latter 30 per cent.)

Much of the earlier common law liability will, in all probability, be replaced in due course by case law arising in connection with breach of duty under the *Workplace (Health, Safety and Welfare) Regulations 1992* (*SI 1992 No 3004* as amended by *SI 2002 No 2174*), where such duties are strict.

In practice it is important that owners and occupiers of buildings assure themselves that the proposed method of cleaning the windows by the cleaning contractor is safe and in compliance with *Reg 16* of the 1992 Regulations, and the owners/occupiers should verify that the agreed system of work is being followed by the contractor's employees, or by a self-employed contractor, as the case may be.

Accidents to window cleaners – typical causes

W9041

- The most common fatal accident is that of a cleaner falling from an external window sill, ledge, or similar part of a building, due to loss of balance as the result of a slip, or the breakage of part of a sill, or the failure of a pull handle or part of a building being used as a handhold.
- Other fatalities have been due to falls through fragile roofing where cleaners have relied upon the roof for support when cleaning or glazing windows, or gaining access for such work, and falls from suspended scaffolds or boatswain's chairs due to failure of the equipment.
- Falls from ladders, which account for a substantial proportion of injuries, are occasionally fatal; they include falls due to the unexpected movement of a ladder such as the top sliding sideways or the foot slipping outwards, failure of part of the ladder, and falls when stepping on or off it.

Methods of protection for window cleaners

W9042

- If the building was designed with totally self-pivoting windows, this is probably the ideal situation in terms of safeguarding the window cleaner. Note the qualified duty in this regard on the building designer in the *CDM Regulations* (*SI 1994 No 3140* as amended by *SI 2000 No 2380*), *Reg 13*(*2*)(*a*). However, a wide variety of window designs are used in buildings – sliding sashes, hinged casements and fixed lights. All require a slightly different technique for cleaning.
- Ordinary ladders are not suitable (and should not be used) for multi-storey properties of more than two storeys.
- Hydraulic platforms are effective but work can only safely proceed at a relatively slow rate; moreover, their maximum reach is sometimes limited, and they cause nuisance to passers-by.

- Gondola cages are used on high-rise buildings with good results, but are of little use for low and medium-rise blocks. They can also be dangerous to window cleaning personnel, as they are inclined to sway about in high winds. Moreover, this method of window cleaning is expensive.

Probably the most practical and economical method of complying with legal requirements for the safety of window cleaning personnel is provision of safe ring/harness combinations. However, note the presumption in favour of collective over individual precautions that is contained within the *Management of Health and Safety at Work Regulations 1999* (*SI 1999 No 3242*), *Reg 4* and *Sch 1* and the similar duty imposed upon building designers by the *CDM Regulations* (*SI 1994 No 3140* as amended by *SI 2000 No 2380*), *Reg 13*(*2*)(*a*)(*iii*). These bolts can be installed rapidly, without damaging the building structurally, and without impairing its aesthetic appearance. Such bolts must, however, comply with British Standard BS 970: 1980 Part 1 (high tensile carbon steel) or Part 4 (premium stainless steel); also the whole anchorage system must comply with BS 5845: 1980.

Safety rings should be installed by a specialist company operating in the field. In this way a window cleaner can clip his harness onto the ring and step onto the window ledge and clean the window safely. After proper installation, anchorage points should be professionally examined from time to time to ensure that they remain firm and safe.

Another (relatively inexpensive) device for use in cleaning outer surfaces on high-rise office blocks and industrial multi-storey properties is the new mobile safety anchor. The cost is often shared between the contract window cleaning company (or self-employed window cleaner) and the building owner or occupier.

Suspended scaffolds are commonly used for window cleaning in high-rise buildings. Guidance on the design, construction and use of suspended scaffolds is given in:

- HSE guidance PM Machinery 30 'Suspended access equipment' (1983);
- BS 6037: 1981 'Code of Practice for permanently installed suspended access equipment'; and
- BS 5974: 1982 'Code of practice for temporarily installed suspended scaffolds and access equipment'; and
- GS 25 'Prevention of falls to window cleaners' (HSE Books).

Whether a suspended scaffold is used, either permanently or temporarily installed, it is essential to ensure that:

- safe means of access to and egress from the cradle are provided;
- properly planned inspection and maintenance procedures for each installation are carried out;
- there are instructions that work shall be carried out only from the cradle; and
- operatives are properly trained in the use of the cradle.

Where power-driven equipment is used, operatives should be familiar with:

- relevant instructions from the manufacturer or supplier;
- any limitations on use, for example, due to wind conditions or length of suspension rope;
- the correct operation of the controls, particularly those affecting the raising or lowering of the cradle;
- the safety devices fitted to the equipment; and

- the procedure if the equipment does not work properly.

Precautions against the failure of a cradle having a single suspension rope at each end are described in BS 6037 (Clause 14.3), for example, the provision at each suspension point of a second rope (safety rope) and an automatic device to support the platform. In some cases, protection against suspension rope failure at the cradle end can be obtained by fitting a manually operated clamping device.

Travelling ladders are permanently installed in some buildings. In other cases a ladder can be suspended from a specially designed frame on the roof. In such cases, the cleaner should wear a safety harness or belt, attached to an automatic fall arresting device on the side of the ladder, and have a safe place at which he can step on or off the ladder.

Cleaning methods

W9043 Window glass can normally be cleaned satisfactorily using plain water, liberally applied, followed by leather off and polishing with a scrim. The use of squeegees is on the increase and, although they cannot safely be used by a person standing on a window sill, they can be safely operated from the ground. Very dirty glazing in factories, foundries or railway premises may need treatment with ammonia or strong soda solution; and hydrofluoric acid in diluted solution may be necessary when cleaning skylights or roof glazing which has remained uncleaned for a long time. Here the working area should be adequately sheeted to protect passers-by, and, in the case of roof glazing, the interior of the building should be protected against penetrating drops. Stringent personal precautions are necessary when handling acids (e.g. eye shields, rubber gloves, boots (see further PERSONAL PROTECTIVE EQUIPMENT), which should only be used as a last resort) and such work should only be carried out by specialist firms. Hydrofluoric acid should never be used on vertical glazing.

Frequency of cleaning

W9044 The recommended frequencies of external and internal cleaning, as dictated by current good practice, are as follows:

Shops	weekly
Banks	twice a month
Offices/hotels	monthly
Hospitals	monthly
Factories – light industry	monthly
Heavy industry	every two months
Schools	every two months

Access for cleaning

W9045 There are three ways in which windows can be cleaned, affecting access:

- external cleaning (with access exclusively from the outside);
- internal cleaning;

- a mixed system, whereby windows accessible from the inside are cleaned internally, and the rest are cleaned externally, access being through the opening lights of the facade.

Moreover, in factories and other industrial concerns windows are, not infrequently, obstructed by machines, reflecting in-plant lay-out at initial design – a matter for safety officers and safety representatives when carrying out safety audits (see JOINT CONSULTATION – SAFETY REPRESENTATIVES AND SAFETY COMMITTEES).

Cleaning from the outside

W9046 Assuming windows cannot be cleaned from inside the workplace or building, as a matter of design, there should be safe external access in the form of permanent walkways, with guard-rails or other protective devices to prevent cleaners (and other users) from falling down. Such walkways should be at least 400 millimetres wide and guard-rails at least 900 millimetres above the walkway, with a knee rail. Safety of window cleaners is best guaranteed by installation of totally self-pivoting windows (see W9042 above). However, a wide variety of window designs is currently in use in buildings – sliding sashes, fixed lights and hinged casements, all requiring different cleaning techniques.

Cleaning from ladders

W9047 Portable ladders, aluminium or timber, should rest on a secure base and reasonably practicable precautions, to avoid sliding outwards at the base and sideways at the top, should be taken to:

- secure fixing at the top to preclude lateral or outward movement, e.g. by fastening to an eyebolt/ringbolt;
- fasten rung to eyebolt/ringbolt or other anchorage point (see further W9042 above) at a height of 2 metres;
- failing this, a person should be stationed at the base in order to steady the ladder or an approved base anchoring device should be used.

The '1 out 4 up' rule (for ladders) suggests that ladders are safest when placed at an angle of 75° to the horizontal; lesser angles indicate that the ladder is more likely to slide outwards at the base. Moreover, window cleaners (and other users) should avoid:

- overreaching (as this can unbalance the ladder);
- proximity with moving objects either above or below (e.g. overhead travelling crane or vehicles operating in the workplace or delivering to the workplace);
- positioning near to vats or tanks containing dangerous fluids/substances or near to unguarded machinery or exposed electrical equipment. (As for *in situ* travelling ladders, see W9042 above.)

Safety harnesses

W9048 In the absence of other, more satisfactory means of cleaning windows, e.g. suspended scaffolds, hydraulic platforms etc., the other reasonably practicable alternative is a safety harness. However, this has the shortcoming that, if a window cleaner falls, he is still likely to be injured. For this reason, harnesses and safety belts must be up to 'free fall' distance (that is, the distance preceding arrest of fall) of, at least,

- 2 metres – safety harness, or
- 0.6 metres – safety belt.

[*BS 1397: 'Specification for industrial safety belts, harnesses etc.'*].

Because some walls may not be strong enough or otherwise suitable, inspection by a competent person should always precede selection of permanent fixed anchorage points. Permanent fixed anchorages should comply with BS 5845 (see W9042 above) and be periodically inspected and tested for exposure to elements. Failing this, temporary anchorage or even mobile anchorage, given the same built-in safeguards, may suffice (see W9042 above).

Cleaning from the inside

W9049 Outer surfaces by design should be able to be cleaned from inside the workplace without use of steps or stepladders. Also, size of aperture and weight of window is relevant. Different types of windows present different dangers; for example, in the case of reversible pivoted/projecting windows, a safety catch is necessary to maintain the window in a fully reversed position; in the case of louvres, there should be sufficient space for a cleaner's hand to pass between the blades, which should incorporate a positive hold-open position to avoid the danger of blowing shut (see BS 8213, Part 1 1991: 'Safety in use and during cleaning of windows').

Maximum safe reach to clean glass immediately beneath an open window is 610 millimetres downwards (that is, 2' 0"), 510 millimetres upwards (1' 8") and 560 millimetres sideways (1'10"). Horizontally or vertically pivoted windows, reversible for cleaning purposes, are probably the best safety option (see W9042 above). Given that windows should be accessible without resort to a ladder, short-of-stature window cleaners should invariably make use of cleaning aids to reach further up glazing panels. Built-in furniture should never be placed near windows to obstruct access; nor should blinds or pelmets inhibit the operation of windows and window controls.

Mixed system of cleaning

W9050 Where window cleaners clean windows externally without facilities for ladders or cradles, the building designer should appreciate that the cleaner's safety depends on good foothold and good handhold. The practice of cleaners having to balance, like trapeze artists, on narrow sills or transoms, is patently dangerous. Owing to the frequency of failure of apparently safe and adequate footholds and handholds, it is imperative that suitable and convenient safety bolts or fixings, to which the cleaner may fix his safety belt, should be provided. Where possible, the safety eyebolt should be fitted on the inside of the wall. Moreover, internal bolts are not weakened by the weather. Where safety eyebolts are fitted to the window frame, architects should pay particular regard at design stage to the fixing of the frame to the building structure, so as to ensure that fixings can withstand the extra load, imposed on the frame, in the event of the cleaner falling.

Where glazing areas are incorporated in roofs, provision should be made for cleaning both sides of the glass, and, where possible, walkways should be provided, both externally and internally. Internal walkways can also be designed to serve for maintenance of electric lighting installations. Where walkways are not feasible, access by permanent travelling ladders should be considered.

The *CDM Regulations* (*SI 1994 No 3140* as amended by *SI 2000 No 2380*) place specific responsibility upon designers to ensure that their designs avoid or reduce health and safety risks in maintenance (including window cleaning) (see C8005

CONSTRUCTION AND BUILDING OPERATIONS). Practical advice for designers on this subject, and on cleaning buildings in general, can be found in CIRIA report 166, *CDM Regulations: Work sector guidance for designers*, at section D6.

W9051 Before concluding this section it is necessary to say something more about the controversy generated by the official view that the Use of Work Equipment Directive is intended to cover the work activities involved in recreational climbing and hill walking etc. The first point to note is that according to climbing sources, the Government has undertaken to establish the position elsewhere in the EU. It is no doubt the case, as HSE officials have said, that the European *Commission* takes the view that the Directive applies; but are they carrying their point with member states? The controversy has probably gone too far to be settled in this country by just a relaying of what has been said in Brussels.

However, there may be a deeper question of policy manifest here. It is understood that those most concerned have subscribed to a declaration by the British Mountaineering Council that *'the BMC recognises that climbing and mountaineering are activities with a danger of personal injury or death. Participants in these activities should be aware of this and accept these risks and be responsible for their own actions and involvement'*.

This does not sit well with the principles of prevention and protection as covered in *Schedule 1* of *MHSWR* (*SI 1999 No 3242*), though the wording of draft Regulation 6 quoted in W9008 above fully allows for the acceptance of risk where avoidance is not, as in the present case, reasonably practicable. It is rather the extent of the measures that then have to be observed which is the trouble. They are clearly aimed at a low level of risk indeed and if adopted, would render the commercial provision of climbing instruction on actual mountainsides impracticable – even ridiculous – the plastering of Snowdonia with warning signs has been feared! That is not to say that much might not be done in the controlled environment of a training centre, but at some stage instruction must extend to where the techniques being taught are to be practised. It seems therefore that a serious case can be made for both workers in a particular activity (the instructors) and those protected via *section 3* of *HSWA 1974* (the would be mountaineers) to be exposed to what ordinarily would be unacceptable risk. Furthermore the argument is being advanced that there is a human right to at least somewhat risky recreation, together with, one deduces, a practical corollary that commercial support is necessary, if this is not to be stultified.

There is perhaps a further point here. It is hard not to be struck, when reading the consultative document, with the thought that, when the Directive was being negotiated, climbers on mountainsides were not often in mind. The extension of orthodox (industrially derived) health and safety thinking to other areas has recently received hostile comment in relation to the HSE's role in the regulation of the railways, and was endorsed by the House of Commons Select Committee on Transport. The Government's intention to remove safety regulation from the HSE to a body with a *railway industry* focus has since been announced.

Working Time

Introduction

W10001 The *Working Time Regulations 1998* (*SI 1998 No 1833*) ('the Regulations') came into force on 1 October 1998 and were the first piece of English law which set out specific rules governing working hours, rest breaks and holiday entitlement for the majority of workers (as opposed to those in specialised industries). The Regulations were introduced in order to implement the provisions of the Working Time Directive (93/104/EC) and the Young Workers Directive (94/33/EC). This chapter deals only with the Regulations as they govern adult workers, i.e. those aged 18 or over.

The Working Time Directive ('the Directive') was adopted on 23 November 1993 as a health and safety measure under Article 138 (formerly Article 118a) of the Treaty of Rome 1957, thereby requiring only a qualified majority, rather than a unanimous, vote. As such, the UK Government, which was opposed to the Directive, was unable to avoid its impact. The Government challenged the Directive on the basis that it was not truly a health and safety measure and should, in fact, have been introduced under Article 100 as a social measure which would have required a unanimous vote. The European Court of Justice (ECJ) has nevertheless upheld the status of the Directive as a health and safety measure.

The stated purpose of the Working Time Directive (acknowledged by the ECJ) is to lay down a minimum requirement for health and safety as regards the organisation of working time. This is, of course, significant to the extent that the English courts and tribunals will interpret the Regulations in accordance with the Directive by adopting the 'purposive' approach, which is now accepted under English law.

When they were first introduced, the Regulations (and accompanying guidance produced by the DTI) were met with vociferous criticism from the business world. Just a year later the Government sought to address these initial concerns. Following a limited consultation in July of 1999, amendments were tabled to two key areas of the Regulations (see W10013 and W10033 below). In addition in March 2000, the DTI issued a second (more concise) version of its guidance, which contains many changes from the original. It is still unclear how much weight tribunals will attach to this publication, which does not have the force of law.

Despite all the controversy regarding the Regulations and the implementation of the Directive it would appear that the Regulations have had little impact on the long-hours culture within the UK. This may of course change if the European Commission removes the 48-hour limit 'opt out' when it is reviewed towards the end of 2003 (see W10013).

The structure of the Regulations is to prescribe various limits and entitlements for workers and then to set out a sequence of exceptions and derogations from these provisions. The structure of this chapter will roughly follow that format.

Definitions

W10002 The *Working Time Regulations 1998* (*SI 1998 No 1833*), *Reg 2* sets out a number of basic concepts (some more familiar than others), which recur consistently and which underpin the legislation. These are as follows.

Working time

W10003 Working time is defined, in relation to a worker, as:

(*a*) any period during which he is working, *at his employer's disposal* and carrying out his activity or duties;

(*b*) any period during which he is receiving relevant training; and

(*c*) any *additional* period which is to be treated as working time for the purposes of the Regulations under a relevant agreement (see W10006 below).

This definition initially raised numerous uncertainties, in particular, in relation to workers who are 'on call' or who operate under flexible working arrangements.

Despite this initial confusion, the ECJ decision in the case of *SIMAP v Conselleria de Sandidad y Consumo de la Generalitat Valencia (C303/98)* confirmed that time spent on call (in this case by doctors at health centres) is 'working time' provided the doctors are available and physically present. In addition, periods during which doctors are subject to the 'on-call' system (namely being available to come in to work while being outside the health centre is working time to the extent that the doctor spends it effectively exercising his professional responsibility). A similar approach may now be taken in other sectors.

The DTI has given examples of what it considers constitutes working time. In its revised guidance it states that 'time spent travelling outside normal working time' is not working time. This seems to address the difficult question of non-routine travel. For example, it has never been clear whether someone who is required to travel from London to Edinburgh on a Sunday, to be at a meeting in Edinburgh at 9 a.m. on Monday morning, can count that time as working time. The DTI's guidance would suggest that it would not be working time. In practice, however, it seems that it would depend on the facts.

It should be noted that the definition of working time can be clarified in a relevant agreement between the parties, but only by specifying any *additional* period which can be treated as working time. The relevant agreement cannot alter, and in particular cannot narrow, the absolute definition of working time.

Worker

W10004 A worker is any individual who has entered into or works under (or where the employment has ceased), worked under:

(*a*) a contract of employment; or

(*b*) any other contract, whether express or implied and (if express) whether oral or in writing, whereby the individual undertakes to do or perform personally any work or services for another party to the contract whose status is not by virtue of the contract that of a client or customer of any profession or business undertaking carried on by the individual.

This definition is wider than just employees and covers any individuals who are carrying out work for an employer, unless they are genuinely self-employed, in that the work amounts to a business activity carried out on their own account (see also the reference to agency workers at W10005). When the Regulations were first introduced, the definition of 'worker' was considered to be a novel concept. It has now become a familiar one and appears in several other contexts such as the *Part Time Workers (Prevention of Less Favourable Treatment) Regulations 2000 (SI 2000 No 1551)* and the *National Minimum Wage Act 1998*.

In *Byrne Bros (Foamwork) Ltd v Baird & Others [2002] ICR 667* the Employment Appeal Tribunal considered whether carpenters and labourers who had been offered work on a building site were 'workers' for the purposes of the Regulations. One significant factor was that the tradesmen were entitled to appoint a substitute to carry out the work but the EAT concluded that a limited power to appoint substitutes was inconsistent with an obligation to provide work on a personal basis. On the evidence, as these individuals worked at the site for a significant and indefinite period and were under the close control of the company, they were more akin to workers than to individuals carrying on a business undertaking.

Agency workers

W10005 The *Working Time Regulations 1998 (SI 1998 No 1833), Reg 36* makes specific provision in relation to agency workers who do not otherwise fall into the general definition of workers. This provides that where (i) any individual is employed to work for a principal under an arrangement made between an agent and that principal, and (ii) the individual is not a worker because of the absence of a contract between the individual and the agent or principal, then the Regulations will apply as if that individual were a worker employed by whichever of the agent or principal is responsible for paying or actually pays the worker in respect of the work. Again, individuals who are genuinely self-employed are excluded from this definition.

Collective, workforce and relevant agreements

W10006 These play a significant role in the Regulations as employers and employees can, by entering into such agreements (where the Regulations so allow), effectively supplement or derogate from the strict application of the Regulations. Employers and employees who need a certain amount of flexibility in their working arrangements may well find one or other of these agreements will facilitate compliance with the Regulations.

Collective agreement

W10007 This is an agreement with an independent trade union within the meaning of the *Trade Union and Labour Relations (Consolidation) Act 1992, s 178*. The definition in *s 178* is wide enough to cover agreements reached between trade unions and employers under the statutory recognition process.

Workforce agreement

W10008 This is a new concept under English law and since its introduction by the Regulations has not been widely utilised. It is an agreement between an employer and the duly elected representatives of its employees or, in the case of small employers (i.e. those with 20 or fewer employees), potentially the employees themselves. In order for a workforce agreement to be valid, it must comply with the conditions set out in the *Working Time Regulations 1998 (SI 1998 No 1833), Sch 1*, which provides that a workforce agreement must:

(*a*) be in writing;

(*b*) have effect for a specified period not exceeding five years;

(*c*) apply either to

(i) all of the relevant members of the workforce, or

(ii) all of the relevant members of the workforce who belong to a particular group;

(*d*) be signed by

(i) the representatives of the workforce or of the particular group of workers, or

(ii) where an employer employs 20 or fewer workers on the date on which the agreement is first made available for signature, either appropriate representatives or by a majority of the workers employed by him;

(*e*) before the agreement was made available for signature, the employer must have provided all the workers to whom it was intended to apply with copies of the text of the agreement and such guidance as they might reasonably require in order to understand it fully.

'Relevant members of the workforce' are defined as all of the workers employed by a particular employer (excluding any worker whose terms and conditions of employment are provided for wholly or in part in a collective agreement). Therefore, as soon as a collective agreement is in force in respect of any worker, the provisions of any workforce agreement in respect of that worker would cease to apply.

Paragraph 3 of *Schedule 1* to the Regulations sets out the requirements relating to the election of workforce representatives. These are as follows:

(*a*) the number of representatives to be elected shall be determined by the employer;

(*b*) candidates for election as representatives for the workforce must be relevant members of the workforce, and the candidates for election as representatives of a particular group must be members of that group;

(*c*) no worker who is eligible to be a candidate can be unreasonably excluded from standing for election;

(*d*) all the relevant members of the workforce must be entitled to vote for representatives of the workforce and all the members of a particular group must be entitled to vote for representatives of that group;

(*e*) the workers must be entitled to vote for as many candidates as there are representatives to be elected;

(*f*) the election must be conducted so as to secure that (i) so far as is reasonably practicable those voting do so in secret, and (ii) the votes given at the election are fairly and accurately counted.

Relevant agreement

W10009 This is an 'umbrella provision' and means a workforce agreement which applies to a worker, any provision of a collective agreement which forms part of a contract between a worker and his employer, or any other agreement in writing which is legally enforceable as between the worker and his employer. A relevant agreement could, of course, include the written terms of a contract of employment. It would only include the provisions of staff handbooks, policies etc where it could be shown that these were 'legally enforceable' as between the parties.

Maximum weekly working time

48-hour working week

W10010 It is, of course, the 48-hour working week which initially caused so much controversy. The *Working Time Regulations 1998* (*SI 1998 No 1833*), *Reg 4* provides that an employer shall take all reasonable steps, in keeping with the need to protect the health and safety of workers, to ensure that a worker's *average* working time (including overtime) shall not exceed 48 hours for each 7-day period.

Regulation 4(*6*) provides a specific formula for calculating a worker's average working time over a reference period, which the Regulations have determined as 17 weeks (for exceptions, see W10012). This is as follows:

$$\frac{a+b}{c}$$

where

a is the total number of hours worked during the reference period;

b is the total number of hours worked during the period which

- (i) begins immediately after the reference period, and
- (ii) consists of the number of working days equivalent to the number of 'excluded days' during the reference period; and

c is the number of weeks in the reference period.

'Excluded days' are days comprised of annual leave, sick leave or maternity leave and any period in respect of which an individual has opted out of the 48-hour limit (see W10013 below).

When the case of *Barber and Others v RJB Mining (UK) Ltd [1999] IRLR 308* was first reported it was initially thought that this decision would add considerably to the importance of this provision. The facts of this case were as follows. For the 17 weeks from 1 October 1998 (the date the Regulations came into force), the plaintiffs worked more than the average of 48 hours per week. Letters seeking the workers' agreement to opt out of the 48-hour limit were sent out on 7 December 1998. Each of the plaintiffs refused to sign the opt-out. On 25 January 1999, each plaintiff refused to carry out further work until his average working hours fell to within the specified limit. Each was required to continue working and did so 'under protest' and without prejudice to their rights in the proceedings.

In his judgement, Mr Justice Gage said:

> 'It seems to me clear that Parliament intended that all contracts of employment should be read so as to provide that an employee should work no more than an average of 48 hours in any week during the reference period. In my judgement, this is a mandatory requirement which must apply to all contracts of employment.'

He stated that although *Regulation 4*(*1*) does not prohibit an employer from requiring his employees to work longer hours, it does not preclude this interpretation as the obligation is in keeping with the stated objective of the Directive, of providing for the health and safety of employees.

The plaintiffs had clearly worked more than their 48-hour average. The judge continued:

> 'Having held that para(1) of Reg 4 provides free-standing legal rights and obligations under their contract of employment, it must follow that to require the plaintiffs to continue to work before sufficient time has elapsed to bring the weekly average below 48 hours is a breach of Reg 4(1).'

In theory, the ruling could add an extra string to the worker's bow in the form of a claim for breach of contract if their employer fails to comply with their obligations under *Regulation 4(1)*. Workers may be able to claim damages, a declaration or possibly an injunction in the event of a breach. However, it should be remembered that the case is a first instance decision and unfortunately, due to the 'politics' of the circumstances surrounding the cases will not be appealed. Indeed, there have been no other reported decisions where a worker has succeeded in bringing a claim for breach of contract where their employer has failed to comply with its obligations under *Regulation 4(1)*. It is, of course, worth bearing in mind that the steps being taken by the European Commission in relation to the potential infringement proceedings against the UK Government for failure to implement the Directive correctly may give further guidance on this point in due course.

The issue of an employer's obligations under the Regulations where they are not the only employer of the worker is problematic. *Regulation 4(2)* states 'an employer shall take all reasonable steps, in keeping with the need to protect the health and safety of workers, to ensure that the limit specified in *paragraph (1)* is complied with in the case of each worker employed by him in relation to whom it applies'. The DTI, in its guidance, recommend that an employer should agree an opt-out with a worker (see below) if the employer knows that the worker has a second job, if the total time worked is in excess of 48 hours a week. They also suggest that employers may wish to make an enquiry of their workforce about any additional employment. If a worker does not tell an employer about other employment and the employer has no reason to suspect that the worker has another job, it is extremely unlikely, the DTI suggest, that the employer would be found not to have complied.

In *Brown v Controlled Packaging Services Ltd (1999) (unreported)*, a worker who had another job in a bar on Tuesday and Thursday evenings was told at his interview that his basic working week would be 40 hours but that on occasion he would be required to work overtime. He soon discovered that the amount of overtime expected of him conflicted with the bar job, as on a number of occasions he was required to start work at 6 a.m. instead of 8 a.m. and he did not want to start work early on Wednesdays or Fridays. After the Regulations came into force, the worker was asked to sign an opt-out agreement which was viewed by his manager as a formality to ensure that existing overtime arrangements complied with the Regulations. Brown refused to sign the agreement because long hours would be incompatible with his evening job. Brown left and claimed constructive dismissal. The Employment Tribunal found that the dismissal was automatically unfair under the *Employment Rights Act 1996, s 101A* – i.e. for a reason connected with the employee exercising his rights under the Regulations.

Reference period

W10011 The crucial issue in calculating the average number of hours worked is the question of when the reference period starts. The reference period is any period of 17 weeks in the course of a worker's employment, unless a relevant agreement provides for the application of successive 17-week periods. [*Working Time Regulations 1998 (SI 1998 No 1833), Reg 4(3)*]. This means that unless the parties specify that the reference period is a defined period of 17 weeks, followed by a successive period of

17 weeks, then the reference period will become a 'rolling' 17-week period. This could be significant where there is a marked variation in the hours that an individual works in a particular period, from week to week. In such a situation an employer would be recommended to include provision in a relevant agreement for successive 17-week reference periods to ensure that it can comply with the maximum weekly working requirements. Such a provision may be desirable in any event, for ease of administration.

For the first 17 weeks of employment, the average is calculated by reference to the number of weeks actually worked. [*Working Time Regulations 1998* (*SI 1998 No 1833*), *Reg 4*(*4*)].

Exceptions and derogations to Regulation 4

W10012 There are currently various exceptions to the maximum working week.

(*a*) *Regulation 5* provides that an individual can agree with his employer to opt out of the 48-hour week (see W10013 below).

(*b*) The 48-hour limit does not apply at all in the case of certain workers whose working time is unmeasured; in respect of other workers, part of whose working time is unmeasured, it only applies to that part of their working time which is measured. [*Working Time Regulations 1998* (*SI 1998 No 1833*), *Reg 20* – see W10033 below].

(c) In special cases (as described in *Reg 21* – see W10034 below), the 17-week reference period over which the 48 hours are averaged will be automatically extended to 26 weeks. [*Working Time Regulations 1998* (*SI 1998 No 1833*), *Reg 4*(*5*)].

(*d*) A collective or workforce agreement can extend the 17-week reference period to a period not exceeding 52 weeks, if there are objective or technical reasons concerning the organisation of work. [*Working Time Regulations 1998* (*SI 1998 No 1833*), *Reg 23*].

Agreement to exclude 48-hour working week

W10013 The 48-hour limit on weekly working time will not apply where a worker agrees with his employer in writing that the maximum 48-hour working week should not apply in the individual's case, provided that certain requirements are satisfied. These are:

(*a*) the agreement must be in writing;

(*b*) the agreement may specify its duration or be of an indefinite length – however, it is always open to the worker to terminate the agreement by giving notice, which will be the length specified in the agreement (subject to a maximum of 3 months), or, if not specified, 7 days; and

(*c*) the employer must keep up-to-date records of all workers who have signed such an agreement.

Until December 1999, employers were under controversial and onerous record keeping requirements in relation to workers who had signed opt out agreements. Since December 1999, employers only need to keep records of who has opted out.

A worker cannot be forced to sign an opt-out agreement. Moreover, the Regulations provide protection where an employee has suffered any detriment on the grounds that he has refused to waive any benefit conferred on him/her by the Regulations.

[*Employment Rights Act 1996, s 45A(1)(b)*, inserted by the *Working Time Regulations 1998* (*SI 1998 No 1833*), *Reg 31(1)(b)* – see W10041 below]. Furthermore, a worker will always have the right to terminate the opt-out agreement so that the 48-hour limit applies, on giving, at most, three months' notice.

On a practical note, the agreement by a worker to exclude the 48-hour working week must be a written agreement between the employer and the individual worker. It cannot take the form of or be incorporated in a collective or workforce agreement. However, it could, of course, be part of the contract of employment, (although it would be advisable for an employer to clearly delineate the agreement to work more than 48 hours from the rest of the contract of employment). It seems that employers' record keeping obligations in this regard would be satisfied by keeping an up-to-date list of workers who have signed opt-out agreements.

The ability of employers to seek the agreement of workers to 'opt out' of the 48-hour week (which is specifically permitted by the Directive) arguably runs a 'coach and horses' through the spirit of the legislation. Severe criticism has been levelled at the UK which is the only country which allows workers generally to opt-out of the maximum working week. The use of the opt-out is due to be reviewed by the EU Council by 23 November 2003 and it will be interesting to see if the long-hours culture which seems to prevail in the UK will be affected should a decision be taken to remove the 'opt-out'.

Rest periods and breaks

Daily rest period

W10014 It should be noted that currently an employer's only obligation is to ensure that its employees are allowed to take their rest breaks – they are not required to compel employees to take them. This is one area which is currently being reviewed by the European Commission as part of the first steps in a potential infringement action against the UK Government for failing to implement the Directive properly.Adult workers are entitled to an uninterrupted rest period of not less than 11 consecutive hours in each 24-hour period. [*Working Time Regulations 1998* (*SI 1998 No 1833*), *Reg 10*].

Exceptions and derogations

W10015 The provisions regarding daily rest periods do not apply where a worker's working time is unmeasured. [*Working Time Regulations 1998* (*SI 1998 No 1833*), *Reg 20(1)*].

Derogations from the rule may be made with regard to shift work [*Reg 22*], or by means of collective or workforce agreements [*Reg 23*] or where there are special categories of workers [*Reg 21*]. However, in all cases (except for workers with unmeasured working time under *Reg 20(1)* – see W10033 below) compensatory rest must be provided [*Reg 24*].

Weekly rest period

W10016 Adult workers are entitled to an uninterrupted rest period of not less than 24 hours in each seven-day period. [*Working Time Regulations 1998* (*SI 1998 No 1833*), *Reg 11*]. However, if his employer so determines, an adult worker will be entitled to either two uninterrupted rest periods (each of not less than 24 hours) in each 14-day period or one uninterrupted rest period of not less than 48 hours in each 14-day period.

The entitlement to weekly rest is in addition to the 11-hour daily rest entitlement which must be provided by virtue of *Regulation 10*, except where objective or technical reasons concerning the organisation of work would justify incorporating all or part of that daily rest into the weekly rest period.

For the purposes of calculating the entitlement, the 7 or 14-day period will start immediately after midnight between Sunday and Monday, unless a relevant agreement provides otherwise. The Regulations do not require Sunday to be included in the minimum weekly rest period.

Derogations from the entitlement to weekly rest periods are the same as those for daily rest.

Rest breaks

W10017 By virtue of the *Working Time Regulations 1998* (*SI 1998 No 1833*), *Reg 12*, adult workers are entitled to a rest break where their daily working time is more than six hours. Details of this rest break, including duration and the terms on which it is granted, can be regulated by a collective or workforce agreement. If no such agreement is in place, the rest break will be for an uninterrupted period of not less than 20 minutes and the worker will be entitled to spend that break away from his workstation, if he has one. In its latest guidance the DTI states that the break should be taken during the six-hour period and not at the beginning or end of it, at a time to be determined by the employer.

Derogations from the entitlement to rest breaks include cases specified in the *Working Time Regulations 1998* (*SI 1998 No 1833*), *Reg 21* (see above), where working time is unmeasured and cannot be predetermined [*Reg 20*(*1*)], and also by means of a collective or workforce agreement [*Reg 23*]. As before, compensatory rest must be provided other than where *Regulation 20* applies [*Reg 24*].

Monotonous work

W10018 The *Working Time Regulations 1998* (*SI 1998 No 1833*), *Reg 8* provides that where the pattern according to which an employer organises work is such as to put the health and safety of a worker employed by him at risk, in particular because the work is monotonous or the work rate is predetermined, the employer shall ensure that the employee is given adequate rest breaks. This Regulation is phrased in virtually identical terms to Article 13 of the Working Time Directive and unfortunately, its incorporation into the Regulations has not clarified its meaning in any way!

It is not clear how rest breaks in *Regulation 8* would differ from those under *Regulation 12*. It may be that in the case of monotonous work, an employer might have to consider giving employees shorter breaks more frequently as opposed to one longer continuous break. This was the suggestion in the Government's consultative document.

No derogations from these provisions apply except in the case of domestic workers.

Night work

Definitions

W10019 Before considering the detailed provisions in relation to night work contained in *Regulation 6*, an understanding of the definitions of 'night time' and 'night worker' contained in the *Working Time Regulations 1998* (*SI 1998 No 1833*), *Reg 2*, is necessary. These are as follows.

Night time in relation to a worker means a period:

(*a*) the duration of which is not less than 7 hours; and

(*b*) which includes the period between midnight and 5 a.m.,

which is determined for the purposes of the Regulations by a relevant agreement or, in the absence of such an agreement, the period between 11 p.m. and 6 a.m.

Night worker means a worker:

(i) who *as a normal course* works at least three hours of his daily working time during night time (for the purpose of this definition, it is stated in the Regulations that (without prejudice to the generality of that expression) a person works hours 'as a normal course' if he works such hours on a majority of days on which he works – a person who performs night work as part of a rotating shift pattern may also be covered); or

(ii) who is likely during night time to work at least such proportion of his annual working time as may be specified for the purposes of the Regulations in a collective or workforce agreement.

The question of when someone works at least 3 hours of his daily working time 'as a normal course' was addressed in the case of *R v Attorney General for Northern Ireland ex parte Burns [1999] IRLR 315*. In this case the Northern Ireland High Court held that a worker who spent one week in each 3-week cycle working at least 3 hours during the night was a 'night worker'. The High Court said that 'as a normal course' meant nothing more than as a regular feature. The DTI in its guidance refers to this case and says that it expects further clarification from the European Court in due course. It states that occasional and or ad hoc work at night does not make you a 'night worker'.

The question is certainly open, following the *Burns* case, of how frequent something has to be in order to become 'regular'.

Length of night work

W10020 If a worker falls within these definitions, then he or she is a night worker for the purposes of the Regulations. The *Working Time Regulations 1998* (*SI 1998 No 1833*), *Reg 6* then goes on to provide that an employer shall take all reasonable steps to ensure that the normal working hours of a night worker do not exceed an average of 8 hours in any 24-hour period. This is averaged over a 17-week reference period which is calculated in the same way as in *Regulation 4* (see W10011 above).

As with the maximum working week, there is a formula for calculating a night worker's average normal hours for each 24-hour period as follows:

$$\frac{a}{b-c}$$

where:

a is the normal (not actual) working hours during the reference period;

b is the number of 24-hour periods during the applicable reference period; and

c is the number of hours during that period which comprise or are included in weekly rest periods under *Regulation 11*, which is then divided by 24.

The Regulations exclude night shift overtime hours from those which count towards 'normal' hours. This is currently being considered by the European Commission as part of the preliminary steps in a potential infringement procedure against the UK Government for not implementing the Directive correctly.

If the night work involves 'special hazards or heavy physical or mental strain', then a strict eight-hour time limit is imposed on working time in each 24-hour period and no averaging is allowed over a reference period. The identification of night work with such characteristics is by means of either a collective or workforce agreement which takes account of the specific effects and hazards of night work, or by the risk assessment which all employers are required to carry out under the *Management of Health and Safety at Work Regulations 1999* (*SI 1999 No 3242*) (see RISK ASSESSMENT).

The derogations in the *Working Time Regulations 1998* (*SI 1998 No 1833*), *Reg 21* (special categories of workers) apply to the provisions on length of night work. Workers whose working time is unmeasured or cannot be predetermined (and where it is only partly unmeasured or predetermined, to the extent that it is unmeasured) are also excluded. [*Working Time Regulations 1998* (*SI 1998 No 1833*), *Reg 20*]. Other exemptions may be made by means of collective or workforce agreement. [*Working Time Regulations 1998* (*SI 1998 No 1833*), *Reg 23*].

Health assessment and transfer of night workers to day work

W10021 An employer must, before assigning a worker to night work, provide him with the opportunity to have a free health assessment. [*Working Time Regulations 1998* (*SI 1998 No 1833*), *Reg 7*]. The purpose of the assessment is to determine whether the worker is fit to undertake the night work. While there is no reliable evidence as to any specific health factor which rules out night work, a number of medical conditions could arise or could be made worse by working at night, such as diabetes, cardiovascular conditions or gastric intestinal disorders.

Employers are under a further duty to ensure that each night worker has the opportunity to have such health assessments 'at regular intervals of whatever duration may be appropriate in his case'. [*Working Time Regulations 1998* (*SI 1998 No 1833*), *Reg 7*(*1*)(*b*)].

The Regulations do not specify the way in which the health assessment must be carried out, nor is there specific reference to medical assessments, so that strictly speaking, such assessments could be carried out by qualified health professionals rather than by a medical practitioner. The latest DTI guidance contains a sample health questionnaire.

Contrast this with the position as regards the transfer from night to day work. If a night worker is found to be suffering from health problems that are recognised as being connected with night work, he or she is entitled to be transferred to suitable day work 'where it is possible'. In such a situation, the Regulations provide that a 'registered medical practitioner' must have advised the employer that the worker is suffering from health problems which the practitioner considers to be connected with the performance by that night worker of night work.

Annual leave

Entitlement to annual leave

W10022 The right under the *Working Time Regulations 1998* (*SI 1998 No 1833*), *Reg 13* to 4 weeks' paid holiday each year is the first time under English law that workers have had a statutory right to paid holiday. The right was originally available to all workers within the scope of the Regulations, providing that they had been continuously employed for 13 weeks (this requirement is satisfied if the worker's relations with his employer have been governed by a contract during the whole or part of each of those weeks). The 13-week qualifying period was challenged by BECTU in the case *R v Secretary of State for Trade and Industry ex parte BECTU Case–173/99 (25 June 2001) (ECJ)* and the ECJ concluded that the 13 week qualifying period was unlawful.

The Regulations were amended by the *Working Time (Amendment) Regulations 2001 (SI 2001 No 3256)* and now provide for the entitlement to paid annual leave to begin on the first day of employment.

Where a worker began work after 25 October 2001, the employer now has the option to use an accrual system during the first year of employment. The amount of leave which the worker may take builds up monthly in advance at the rate of one twelfth of the annual entitlement each month. If the calculation does not produce a whole number the accrued leave entitlement is rounded up either to a half day (if the calculation is less than a half day) or a whole day (if the calculation exceeds a half day). Any rounded up element of leave taken would, of course, be deducted from the worker's annual entitlement to paid leave.

For the purposes of *Regulation 13*, a worker's leave year begins on any day provided for in a relevant agreement, or, where there is no such provision in a relevant agreement, 1 October 1998 or the anniversary of the date on which the worker began employment (whichever is the later). In the majority of cases, written statements of terms and conditions contain a reference to the holiday or leave year. If they do not, employers are recommended to ensure that they do so. Failure in this regard could result in administrative confusion as each employee could have a different holiday year for the purposes of calculations under the Regulations.

The statutory leave entitlement may be taken in instalments, but it can only be taken in the leave year to which it relates and a payment in lieu of the statutory entitlement cannot be made except where the worker's employment is terminated (but see comments at W10024 below). It should be noted that this relates only to the entitlement under the Regulations, so that if an employer provides for annual leave over and above the statutory entitlement, the enhanced element of the holiday can be carried forward or paid for in lieu as agreed between the parties. In *Miah v La Gondola Ltd (1999) (unreported)*, the Employment Tribunal found that a worker who was paid an additional week's pay instead of taking a week's holiday (due under the Regulations) during his employment, was entitled, on the termination of that contract, to payment for that proportion of annual leave which was due to him in accordance with *Regulation 14(3)(b)* and the payment paid to him during his employment in lieu of the week's statutory holiday had to be disregarded.

For many workers (and employers), the introduction of rights to paid annual leave was significant. It is understood that before the Regulations came into force some 2.5 million workers had no right to a holiday at all. Employment Tribunals have had to resolve many disputes arising from this right.

Compensation related to entitlement to annual leave

W10023 Where an employee has outstanding leave due to him when the employment relationship ends, the *Working Time Regulations 1998 (SI 1998 No 1833), Reg 14* specifically provides that an allowance is payable in lieu. It states that in the absence of any relevant agreement to the contrary, the amount of such allowance will be determined by the formula

$(a \times b) - c$

where:

a is the period of leave to which the worker is entitled under the Regulations;

b is the proportion of the worker's leave year which expired before the effective date of termination;

c is the period of leave taken by the worker between the start of the leave year and the effective date of termination (*Reg 14(3)*).

Regulation 14(4) provides that where there is a relevant agreement which so provides, an employee shall compensate his employer – whether by way of payment, additional work or otherwise – in relation to any holiday entitlement taken in excess of the statutory entitlement.

Interestingly in the case of *Witley & District Mens Club v Mackay [2001] IRLR 595* the EAT held that where a relevant agreement (in this case a collective agreement) purported to allow an employer to dismiss an employee for gross misconduct without payment for any accrued holiday pay this provision would be void as the *Working Time Regulations 1998* (*SI 1998 No 1833*), *Reg 35* specifically provides for a sum to be paid. It is open to debate whether a provision of a relevant agreement which permitted the payment of a notional sum (say, £1) on termination in the event of gross misconduct would be valid.

Payment, and notice requirements, for annual leave

W10024 The *Working Time Regulations 1998* (*SI 1998 No 1833*), *Reg 16* specifies the way in which a worker is paid in respect of any period of annual leave: this is at the rate of a week's pay in respect of each week of leave.

In order to calculate the amount of a 'week's pay', employers are referred to *ss 221–224* of the *Employment Rights Act 1996* (although the calculation date is to be treated as the first day of the period of leave in question, and the relevant references to *ss 227 and 228* do not apply). The application of these provisions can lead to a disproportionately high level of holiday pay for workers paid on a commission only basis. This is because *s 223(2)* stipulates that if, during any of the 12 weeks preceding the worker's holiday, no remuneration was payable by the employer to the worker concerned, account should be taken of remuneration in earlier weeks so as to bring the number of weeks of which account is taken up to 12. In other words, it seems that the calculation of a 'commission only' worker's holiday entitlement will be based solely on the worker's earnings for any weeks (up to a maximum of 12) in which they received commission payments-any slow, unproductive weeks will be disregarded.

In the case of *Smith & Others v Chubb Security Personnel Ltd (1999) (unreported)*, employees who were obliged to work such hours as were detailed in their duty roster were paid holiday pay on the basis of a 40-hour week when in fact they worked a 3-week cycle of fourteen 12 hour days which averaged out at 56 hours per week. The Tribunal decided that their pay should be calculated on the basis of a 56-hour week.

However, note also *Bamsey v Albon Engineering and Manufacturing plc (Judgment 25 March 2004)*, in which the Court of Appeal were called upon to decide whether the overtime that the employee regularly worked should have been taken into account when calculating the amount of his holiday pay entitlement.

Mr Bamsey's contract of employment with his employers Albon Engineering entitled him to a basic working week of 39 hours, with a substantial compulsory, but non-guaranteed, overtime. The employer had in fact required him to work a regular shift pattern of four 12-hour days and one 10-hour day – a 58-hour working week.

Over the 12 weeks before the holiday period in question the employee had averaged 60 hours work, earning nearly £330 a week.

However, the employer paid him for the period of annual leave on the basis of his 39-hour week, just short of £200 a week.

Mr Bamsey complained to an employment tribunal, contending that he should have been paid over the period of annual leave at the same rate as averaged, with overtime, while at work. His complaint was rejected by the employment tribunal and, on appeal, by the Employment Appeal Tribunal. He appealed to the Court of Appeal.

The *Working Time Regulations 1998* (*SI 1998 No 1833*), *Reg 16* provides that a worker 'is entitled to be paid in respect of any period of annual leave ... at the rate of a week's pay in respect of each week of leave.' That same Regulation goes on to say that *sections 221* to *224* of the *Employment Rights Act 1996*, are concerned with the calculation of a week's pay in respect of 'employments with normal working hours.'

The critical question confronting the Court of Appeal was whether *Regulation 16* of the 1998 Regulations incorporated, along with *sections 221* to *224* of the 1996 Act, the definition of 'normal working hours' in *section 234* of that Act. If it did, the appeal failed. *Section 234* of the 1996 Act provides that overtime will not be incorporated within 'normal working hours' either where the overtime is voluntary or where, as in the instant case, the employee is obliged to and does overtime but there is no corresponding obligation on the employer to provide overtime.

The Court of Appeal concluded that *Regulation 16* of the 1998 Regulations *did* incorporate *section 234* of the 1996 Act. This meant that in the particular case Mr Bamsey's overtime was not to be taken into account calculating his holiday pay, because he was not contractually entitled to it. His 'normal working hours' for the purposes of *Regulation 16* of the 1998 Regulations were those which comprised his basic working week – 39 hours.

The *Working Time Regulations 1998* (*SI 1998 No 1833*), *Reg 15* deals with the dates on which the leave entitlement can be taken, and sets out detailed requirements as to notice. This Regulation is extremely complicated and before going into the details, it should be noted that alternative provisions concerning the notice requirements can be contained in a relevant agreement. Bearing in mind the detailed nature of the provisions, it is recommended that employers consider specifying such notice requirements in their contracts of employment, thereby avoiding the need for confusion at a later stage.

Regulation 15 provides that a worker must give his employer notice equivalent to twice the amount of leave he is proposing to take. An employer can then prevent the worker from taking the leave on a particular date by giving notice equivalent to the number of days' leave which the employer wishes to prohibit. An employer can also, by giving notice equivalent to twice the number of days' leave in question, require an employee to take all or part of his leave on certain dates. This would, of course, be useful with regard to seasonal shutdowns over summer and Christmas holidays etc. An employee has no right to serve a counter notice and so, provided the employer has given proper notice, the employee must take the holiday as required. The Regulations do not require any notice to be given in writing. However, given the level of detail that must be included in any notice, employers would be advised to give the notice in writing and require their employees to do the same. This would also ensure that employers have accurate records of required holiday periods.

Employers who give more than 4 weeks' holiday and who are contractually obliged to pay their workers more for the statutory 4 weeks than they pay for the remaining leave will be interested in *Barton & Others v SCC Ltd (1999) (unreported)*. In this case, an employer sought to argue that monies paid by way of holiday pay for leave in excess of the statutory minimum leave under the Regulations can go towards the employer's liability to pay a week's pay for each week of leave taken under the Regulations (calculated under the *Working Time Regulations 1998* (*SI 1998 No 1833*), *Reg 16*). In addition, the employer argued that they had the right to chose which days off would be paid at the statutory rate and which at the contractual rate. The Employment Tribunal found in favour of the workers on both issues. *Regulations 16*(*1*) and (*5*) make it clear that the Employment Tribunal have to focus on the individual holiday week in question and the payment made for that individual week. With regard to which holiday is paid at which rate, the Employment Tribunal said that the right to holiday pay is an entitlement and it is a matter for the worker to choose which holiday weeks he or she seeks to take pursuant to his or her statutory entitlement.

The right to paid leave together with the relevant rate of pay have resulted in a number of controversial and sometimes conflicting decisions. In particular, there have been conflicting decisions as to whether it is possible to 'roll-up' holiday pay in an hourly rate and whether an employee is entitled to paid holiday whilst on sick leave.

There have been conflicting decisions as to whether it is permissible to roll-up holiday pay into an hourly rate. In the first case of *Johnson v Northbrook College (unreported) (Case No. 3102727/99)*, the Employment Tribunal expressly found that if an hourly rate incorporated an element of holiday pay, the employer's duties under the Regulations would have been discharged. (On the facts, as the employee had not agreed to the purported variation of her contract to include holiday pay in an hourly rate, the employer was found to be in breach of the Regulations.) This decision is, of course, not binding on subsequent Employment Tribunals and indeed, is difficult to reconcile with the case of *Barton & Others v SCC Ltd* where the Employment Tribunal considered that it was necessary to look at the week in which the leave was actually taken and to consider what payments, if any, were made during that week. In the *Johnson* case, no payment was actually made when leave was taken as the holiday pay was rolled up into the hourly rate.

The same issue arose again at EAT level in the case of *Gridquest Ltd t/a Select Employment v Blackburn [2002] IRLR 168* where the EAT concluded that it was permissible for employers to 'roll-up' an element of holiday pay into the overall hourly rate paid to workers. In direct contrast, a later decision of the EAT in a Scottish case, *MPB Structures v Munro (EAT/1257/01)* (which was argued under *Reg 35* (Restrictions on Contracting Out) of the 1998 Regulations and did not (at EAT level) refer to either *Reg 16*(*5*) or indeed the *Gridquest case),* held that adding an allowance of 8 per cent to each pay packet to cover the cost of holidays did not comply with the requirement of the Regulations to ensure that workers are entitled to paid leave. The EAT in the *Munro* case took the view that the aim of the Regulations was to ensure that workers were able to take their leave and that this could only be achieved where a worker was paid at the time when he took the leave.

The *Gridquest* case was subsequently appealed to the Court of Appeal and it was hoped that their decision would clarify the state of the law. The Court of Appeal concluded that, on the facts, there was no agreement between the workers and their employer regarding rolling up holiday pay into the hourly rate and accordingly it declined to comment on whether any such agreement would be lawful under the Regulations. The *Munro* decision was also appealed and the Scottish Court of Session concluded that rolling up holiday pay was not lawful under the Regulations. The decision of the Scottish Court of Session is not, strictly speaking, binding on English tribunals as was evidenced by the subsequent English EAT decision in *Marshalls Clay Products v Caulfield (as yet unreported)*. At the time of going to print, only a brief summary of this decision was available. However, it appears to suggest that employers can roll up holiday pay into an hourly rate provided that this is done with the express agreement of the worker. Hopefully, when the case is fully reported, employers will be given additional guidance as to when it is permissible to roll up holiday pay.

In addition, until recently there have been conflicting decisions on the question of a worker's entitlement to annual leave while on sick leave. The issue has now been clarified by the EAT decision in *Kigass Aero Components Ltd v Brown (EAT/481/00)* where the EAT held that workers are entitled to take their paid annual leave under the Regulations even if they are on sick leave during that period provided that they comply with the notification requirements under the Regulations. The EAT acknowledged that this was a somewhat surprising result which does not reflect the underlying health and safety purpose of the Regulations.

A further case which has given employers some cause for concern is the decision in *List Design Group v Catley (EAT/0966/00)*. In this particular case, the EAT held that a worker could mount a claim for a series of failures to make payments in respect of holiday pay where the claim was based on the unlawful deduction from wages provisions in the *Employment Rights Act 1996*. (See also EMPLOYMENT PROTECTION.) Normally, under the Regulations a worker would only be able to make a claim in respect of untaken holiday where the claim is brought within 3 months of the failure to allow the worker to take paid holiday. The Regulations do not permit a worker to bring a claim in respect of untaken holiday in previous holiday years. In contrast, under the provisions for unlawful deductions from wages workers can, in certain circumstances, bring a claim for serial failures to pay 'wages' (which for these purposes includes holiday pay due under the Regulations). Therefore, by bringing a claim under the deduction from wages provisions rather than the Regulations a worker may be able to bring a claim covering several holiday years where holiday was neither taken nor paid for. Accordingly as Mr Catley's case was based on a series of deductions from wages, he was able to claim 'back holiday pay' to the date when the Regulations came into force.

Records

W10025 The publicity and controversy surrounding the introduction of the Regulations related primarily to the maximum working week and the annual leave entitlement. However, in practice the provisions relating to record keeping proved very significant for employers, in particular those with workers who had signed opt-outs, where the rigorous record keeping requirements arguably undermined the benefit to an employer of the freedom given by the opt-out. Accordingly, the UK Government significantly watered down the requirements for employers *vis a vis* opted out workers, and the current position is outlined below.

The *Working Time Regulations 1998* (*SI 1998 No 1833*), *Reg 9* introduces an obligation on employers to keep specific records of working hours. All employers (except in relation to workers serving in the armed forces – *Reg 25*(*1*)) are under a duty to keep records which are adequate to show that certain specified limits are being complied with in the case of each entitled worker employed by him. These limits are:

(*a*) the maximum working week (see W10010 above);

(*b*) the length of night work (including night work which is hazardous or subject to heavy mental strain) (see W10020 above); and

(*c*) the requirement to provide health assessments for night workers (see W10021 above).

These records are required to be maintained for two years from the date on which they were made.

In addition, where a worker has agreed to exclude the maximum working week under the *Working Time Regulations 1998* (*SI 1998 No 1833*), *Reg 5* (see W10013 above), an employer is required to keep up-to-date records of all workers who have opted out. Note that, in contrast with the obligations under *Regulation 9*, no time limit is imposed for the maintenance of such records.

Note that the obligation on employers to keep records may increase as a result of the steps taken by the European Commission as part of a potential infringement action against the UK government for failing to implement the Directive correctly.

Excluded sectors

W10026 Prior to 1 August 2003, the following sectors of activities (listed below) were totally excluded from the terms of the Regulations by *Regulation 18*:

(*a*) air, rail, road, sea, inland waterway and lake transport;

(*b*) sea fishing;

(*c*) other work at sea;

(*d*) the activities of doctors in training.

Following the adoption of five Directives aimed at extending the working time provisions to previously excluded sectors, this is no longer the case and most of the above sectors have gained some protection under the Regulations. The Directives are:

- Horizontal Amending Directive.
- Road Transport Directive.
- Seafarers' Directive (outside the scope of this chapter).
- Seafarers' Enforcement Directive (outside the scope of this chapter).
- Aviation Directive (outside the scope of this chapter).

The Horizontal Amending Directive was implemented by the UK with effect from 1 August 2003 (see the *Working Time* (*Amendment*) *Regulations 2003* (*SI 2003 No 1684*).

Application of the new directives to previously excluded sectors

Mobile and non-mobile workers

W10027 A mobile worker is defined in the *Working Time* (*Amendment*) *Regulations 2003* (*SI 2003 No 1684*) as: 'any worker employed as a member of travelling or flying personnel by an undertaking which operates transport services for passengers or goods by road or air'. Non-mobile workers are undefined and are therefore any worker who is not a mobile worker, for example, all office based staff, warehouse staff etc. Unfortunately the Amendment Regulations do not address the situation where a worker carries out a combination of mobile and non-mobile activities (for example, a worker who spends part of his time based in a depot loading goods onto lorries and the remainder of his time driving an HGV delivering the goods.) It is likely that future cases will provide further guidance on this point.

Road transport sector

Non-mobile workers

W10028 Following the implementation of the *Working Time (Amendment) Regulations 2003 (SI 2003 No 1684)* on 1 August 2003, the Regulations are extended in full to non-mobile workers in the road transport sector.

Mobile workers

W10029 As of 1 August 2003, mobile workers who are *not* covered by the Road Transport Directive (such as drivers of vehicles of less than 3.5 tonnes) will benefit from the average 48-hour working week, four weeks' paid holiday, appropriate health checks if they are night workers and provision for adequate rest (unless there are exceptional circumstances as provided for in *Regulation 21(e)* of the *Working Time Regulations 1998 (SI 1998 No 1833)*. They are not entitled to the daily rest requirements set out in *Regulation 10(1)* or the weekly rest periods set out in *Regulation 11(1)* and *11(2)* of the 1998 Regulations. In addition, they are not entitled to the rest breaks set out in *Regulation 12(1)*. Further, the exemptions for special categories of workers may apply (see W10034 below).

As of 1 August 2003, mobile workers who are covered by the Road Transport Directive (mainly drivers of vehicles of over 3.5 tonnes) will be entitled to 4 weeks' paid holiday and appropriate health checks if they are night workers.

In addition, mobile workers who are covered by the Road Transport Directive, will gain additional protection once the Directive has been implemented. The Road Transport Directive specifies a maximum working week of 60 hours (provided always that the average working week over a 4-month period does not exceed 48 hours). Night work is restricted to 10 hours and self-employed drivers are excluded from the ambit of the Directive. It is due to be implemented by 23 March 2005.

Rail sector

W10030 Following the implementation of the *Working Time (Amendment) Regulations 2003 (SI 2003 No 1684)* on 1 August 2003, the Regulations are, in principle, extended in full to all workers in the rail transport sector.

It is, however, possible to derogate from the Regulations in respect of railway workers whose activities are intermittent, whose working time is spent on board trains or whose activities 'are linked to transport timetables and to ensuring the continuity and regularity of traffic'. For more information on these partial exemptions see W10034 below.

Other sectors

W10031 It should be noted that the Horizontal Amending Directive together with the other sector-specific directives detailed above provide different entitlements for mobile workers in the sea-fishing sector, aviation sector, inland waterway and lake transport sectors and also in relation to offshore working. In addition, in relation to doctors in training the Horizontal Amending Directive provides for a gradual phasing in of the

Regulations: initial implementation must be completed by 1 August 2004 and full implementation of the provisions by 1 August 2012 at the latest.

Partial exemptions

Domestic service

W10032 The following provisions of the Regulations do not apply in relation to workers employed as domestic servants in private households:

(*a*) the 48-hour working week (see W10010 above);

(*b*) length of night work (see W10020 above);

(*c*) health assessments for night workers (see W10021 above);

(*d*) monotonous work (see W10018 above).

Unmeasured working time

W10033 The requirements of the Regulations listed below are, by virtue of the *Working Time Regulations 1998* (*SI 1998 No 1833*), *Reg 20*(*1*), not applicable to workers where, on account of the specific characteristics of the activity in which they are engaged, the duration of their working time is not measured or pre-determined or can be determined by the workers themselves. The provisions excluded are:

(*a*) the 48-hour working week (see W10010 above);

(*b*) minimum daily and weekly rest periods and rest breaks (see W10014 above);

(*c*) length of night work (see W10020 above).

This would leave such workers with requirements relating to monotonous work, the requirement for health assessments for night workers, the requirement to keep records and annual leave. It should also be noted that there is no requirement to provide such workers with compensatory rest where the requirements of the Regulations are not complied with.

This provision is often thought to exclude the Regulations in relation to 'managing executives or other persons with autonomous decision-making powers'. However, it should be noted that this is simply *one* of the three examples given in *Regulation 20*(*1*), as regards the type of worker with unmeasured working time. The application of the Regulation could be much wider than this, and will depend on whether the worker in question is genuinely able to control his work, to the extent of being able to determine how many hours he works.

The UK Government made an amendment to *Regulation 20* in December 1999 so that the working time of workers does not, effectively count towards their weekly working time, to the extent that it is unmeasured, even if their time is not wholly unmeasured. Prior to the amendment, workers were either within the *Regulation 20* 'unmeasured' exemption or they were outside it. The exemption provides a 'half way house' for workers who have some part of their working time set in advance (e.g. under their contract of employment) but choose on their own account to work longer hours. It is as yet unclear how widely this provision will be interpreted i.e. when extra hours can genuinely be said not to be 'required' by the employer.

Special categories of workers

W10034 The *Working Time Regulations 1998* (*SI 1998 No 1833*), *Reg 21* provides that in the case of certain categories of employees, the following Regulations do not apply:

(*a*) length of night work (see W10020 above);

(*b*) minimum daily rest and weekly rest periods and rest breaks (see W10014 above).

This leaves the 48-hour working week, requirements relating to monotonous work, the requirement for health assessments for night workers, the requirement to keep records and annual leave. However, it should be noted that where any of the allowable derogations are utilised, an employer will have to provide compensatory rest periods (see W10037 below).

Further, for these categories of workers, the reference period over which the 48-hour working week is averaged is 26 weeks and not 17. [*Working Time Regulations 1998* (*SI 1998 No 1833*), *Reg 4*(*5*)].

The special categories of worker are as follows:

(i) where the worker's activities mean that his place of work and place of residence are distant from one another, or the worker has different places of work which are distant from one another (including cases where the worker carries out offshore work);

(ii) workers engaged in security or surveillance activities, which require a permanent presence in order to protect property and persons – examples given are security guards and caretakers;

(iii) where the worker's activities involve the need for continuity of service or production – specific examples are:

  (A) services relating to the reception, treatment or care provided by hospitals or similar establishments (including the activities of doctors in training), residential institutions and prisons;

  (B) workers at docks or airports;

  (C) press, radio, television, cinematographic production, postal and telecommunications services, and civil protection services;

  (D) gas, water and electricity production, transmission and distribution, household refuse collection and incineration;

  (E) industries in which work cannot be interrupted on technical grounds;

  (F) research and development activities;

  (G) agriculture;

(iv) any industry where there is a foreseeable surge of activity – specific cases suggested are:

  (A) agriculture;

  (B) tourism;

  (C) postal services;

(v) where the worker's activities are affected by:

  (A) unusual and unforeseeable circumstances beyond the control of the employer;

  (B) exceptional events which could not be avoided even with the exercise of all due care by the employer;

  (C) an accident or the imminent risk of an accident;

(vi) where the worker works in railway transport and:

(A) his activities are intermittent; or

(B) he works on board trains; or

(C) his activities are linked to transport timetables and to ensuring the continuity and regularity of traffic.

Shift workers

W10035 Shift workers are defined by the *Working Time Regulations 1998* (*SI 1998 No 1833*), *Reg 22* as workers who work in a system whereby they succeed each other at the same work station according to a certain pattern, including a rotating pattern which may be continuous or discontinuous, entailing the need for workers to work at different times over a given period of days or weeks.

In the case of shift workers the provisions for daily rest periods (see W10014 above) and weekly rest periods (see W10016 above) can be excluded in order to facilitate the changing of shifts, on the understanding that compensatory rest must be provided (see W10037 below). In addition, those provisions also do not apply to workers engaged in activities involving periods of work split up over the day – the example given is that of cleaning staff.

Collective and workforce agreements

W10036 By virtue of the *Working Time Regulations 1998* (*SI 1998 No 1833*), *Reg 23*(*a*), employers and employees have the power to exclude or modify the following provisions:

(*a*) length of night work (see W10020 above);

(*b*) minimum daily rest and weekly rest periods and rest breaks (see W10014 above),

by way of collective or workforce agreements. Compensatory rest must be provided (see W10032 below).

In addition, the *Working Time Regulations 1998* (*SI 1998 No 1833*), *Reg 23*(*b*) allows the reference period for calculating the maximum working week to be extended to up to 52 weeks, if there are objective or technical reasons concerning the organisation of work to justify this.

This Regulation gives the parties a good deal of flexibility (on the understanding that they are prepared to consent with each other) to opt out of significant provisions contained in the Regulations.

Compensatory rest

W10037 The *Working Time Regulations 1998* (*SI 1998 No 1833*), *Reg 24* provides that where a worker is not strictly governed by the working time rules because of:

(*a*) a derogation under *Regulation 21* (see W10034 above); or

(*b*) the application of a collective or workforce agreement under *Regulation 23*(*a*) (see W10036 above); or

(*c*) the special shift work rules under *Regulation 22* (see W10035 above);

and as a result, the worker is required by his employer to work during what would otherwise be a rest period or rest break, then the employer must wherever possible allow him to take an equivalent period of compensatory rest. In exceptional cases, in which it is not possible for objective reasons to grant such a rest period, the employer must afford the employee such protection as may be appropriate in order to safeguard his health and safety.

Enforcement

W10038 The Regulations divide the enforcement responsibilities between the Health and Safety Executive and the Employment Tribunals. In essence, the working time limits are enforced by the Health and Safety Executive and local authority environmental health departments. The entitlements to rest and holiday are enforced through Employment Tribunals.

Interestingly, despite the assertions that the Directive is a health and safety measure, there do not appear to be any successful prosecutions by the Health and Safety Executive and there has been only one successful prosecution by a local authority under the Regulations in the past two years. In contrast, there have been numerous cases lodged at the Employment Tribunal or High Court since the implementation of the Regulations.

Health and safety offences

W10039 The *Working Time Regulations 1998* (*SI 1998 No 1833*), *Reg 28* provides that certain provisions of the Regulations (referred to as 'the relevant requirements') will be enforced by the Health and Safety Executive (except to the extent that a local authority may be responsible for their enforcement by virtue of *Regulation 28*(*3*)). The relevant requirements are:

(*a*) the 48-hour working week;

(*b*) length of night work;

(*c*) health assessment and transfers from night work;

(*d*) monotonous work;

(*e*) record keeping; and

(*f*) failure to provide compensatory rest, where the provision concerning the length of night work is modified or excluded.

Any employer who fails to comply with any of the relevant requirements will be guilty of an offence and shall be liable on summary conviction (in the magistrates' court) to a fine not exceeding the statutory maximum and on conviction on indictment (in the Crown Court) to a fine.

In addition, the Health and Safety Executive can take enforcement proceedings utilising certain provisions of the *Health and Safety at Work etc Act 1974*, as set out in *Regulations 28* and *29*. Employers may face criminal liability under these provisions, the sanctions for which range, according to the offence, from a fine to two years' imprisonment. (For enforcement of health and safety legislation generally, see ENFORCEMENT.)

Employment tribunals

Enforcement of the Regulations

W10040 By virtue of the *Working Time Regulations 1998* (*SI 1998 No 1833*), *Reg 30*, certain provisions of the Regulations may be enforced by a worker presenting a claim to an Employment Tribunal where an employer has refused to permit him to exercise such rights. These are:

(*a*) daily rest period;

(*b*) weekly rest period;

(*c*) rest break;

(*d*) annual leave;

(*e*) failure to provide compensatory rest, insofar as it relates to situations where daily or weekly rest breaks are modified or excluded;

(*f*) the failure to pay the whole or any part of the amount relating to paid annual leave, or payment on termination in lieu of accrued but untaken holiday.

A complaint must be made within (i) three months (other than in the case of members of the armed forces – see below) of the act or omission complained of (or in the case of a rest period or leave extending over more than one day, of the date on which it should have been permitted to begin) or (ii) such further period as the tribunal considers reasonable, where it is satisfied that it was not reasonably practicable for the complaint to be presented within that time. (The time limit in respect of members of the armed forces is six months.)

Where an Employment Tribunal decides that a complaint is well-founded, it must make a declaration to that effect and can award compensation to be paid by the employer to the worker. This shall be such amount as the Employment Tribunal considers just and equitable in all the circumstances, having regard to (i) the employer's default in refusing to permit the worker to exercise his right and (ii) any loss sustained by the worker which is attributable to the matters complained of. With regard to complaints relating to holiday pay or payment in lieu of accrued holiday on termination, the tribunal can also order the employer to pay the worker the amount which it finds properly due.

There is no qualifying service period with regard to such complaints being presented to a tribunal.

Protection against detriment, and against unfair dismissal

W10041 The *Working Time Regulations 1998* (*SI 1998 No 1833*), *Reg 31* inserts a new *section 45A* into the *Employment Rights Act 1996*, so as to provide protection for a worker against being subjected to any detriment, where the worker has:

(*a*) refused (or proposed to refuse) to comply with a requirement which the employer imposed (or proposed to impose) in contravention of the Regulations;

(*b*) refused (or proposed to refuse) to forgo a right conferred on him by the Regulations;

(*c*) failed to sign a workforce agreement or make any other agreement provided for under the Regulations, such as an individual opt-out from the maximum weekly working time limit;

(*d*) been a candidate in an election of work place representatives or, having been elected, carries out any activities as such a representative or candidate; or

(*e*) in good faith, (i) made an allegation that the employer has contravened a right under the Regulations, or (ii) brought proceedings under the Regulations.

The right to bring these claims would (as with discrimination claims) allow an individual to pursue a claim relating to a breach of the Regulations whilst continuing in employment.

By virtue of the inserted *section 101A* of the *Employment Rights Act 1996*, the dismissal of an employee on all but ground (*e*) above is automatically unfair, although this will only apply to employees and not to the wider definition of worker (for the position of workers who are not employees, see below). The compensation available would be subject to any cap on unfair dismissal compensation (the current cap on unfair dismissal compensatory awards is £52,600). Employees whose contracts were terminated on ground (*e*) above would be protected from dismissal for assertion of a statutory right, by virtue of *section 104* of the 1996 Act.

An employee would also be protected if selected for redundancy on any of the grounds listed above. This would be automatically unfair selection, by virtue of *section 105* of the 1996 Act.

Workers (i.e. those who are not employees) whose contracts are terminated for any of the grounds set out above can claim that they have suffered a detriment, and they may claim compensation, which would be capped in the same way as an award for unfair dismissal.

Breach of contract claims

W10042 Following the *RJB Mining case* (see W10010), there is an argument that workers may also be able to claim for breach of contract, if their employer breaches the obligations in the *Working Time Regulations 1998* (*SI 1998 No 1833*), *Reg 4*(*1*). (In practice, however, this type of claim has not arisen since the determination of this case.) If the claim is for damages and is outstanding at the termination of the employment, then the worker could in principle take their case to the Employment Tribunal (within 3 months of the effective date of termination of the contract giving rise to the claim). If the claim is for a declaration or an injunction, then the worker will have to apply to the High Court, where special rules apply.

Contracting out of the Regulations

W10043 Under the *Working Time Regulations 1998* (*SI 1998 No 1833*), *Reg 35*, an agreement to contract out of the provisions of the Regulations can be made via a conciliation officer or by means of a compromise agreement, and the provisions are similar and consistent with the current provisions in the *Employment Rights Act 1996, s 203* (as amended by the *Employment Rights* (*Dispute Resolution*) *Act 1998*).

Conclusion

W10044 The *Working Time Regulations 1998* (*SI 1998 No 1833*) broke new ground in English law. Despite the adverse publicity surrounding them their introduction does not appear to have unduly changed current industrial practice and does not appear to have led to a reduction in the long hours culture which is prevalent within the UK.

The amendments to the Directive which extend the Regulations to both non-mobile and mobile workers within the previously excluded sectors are, however, likely to have a significant impact on some sectors, most notably the road transport sector. In addition, it is arguable that the impact of the Regulations will be increased if the European Commission's review of the use of opt-outs in 2003 results in the removal of the opt-out provisions.

Finally, it should be noted that the European Commission has taken the first steps which could lead to infringement proceedings against the UK government for failing to implement the Directive properly. There are three grounds of complaint: the lack of a requirement to measure time worked voluntarily over normal working time, the exclusion of night shift overtime hours from those which count towards normal hours and the lack of a duty on employers to ensure that workers take breaks and holidays. We shall have to wait and see whether this results in further changes to the Regulations.

Workplaces – Health, Safety and Welfare

Introduction

W11001 The *Workplace (Health, Safety and Welfare) Regulations 1992* (*SI 1992 No 3004* as amended by *SI 2002 No 2174*) set out a wide range of basic health, safety and welfare standards applying to most places of work. The Workplace Regulations cover not only offices, shops and factories covered by earlier legislation in this area, but apply to a much wider range of workplaces, including schools, hospitals, theatres, cinemas and hotels, for example.

The *Health and Safety (Miscellaneous) Amendments Regulations 2002* (*SI 2002 No 2174*) mean that the 1992 Regulations now apply to every workplace, as previously excluded workplaces such as factories and mines are now included. In addition, a new *Regulation 25A* states that employers must consider the needs of disabled workers.

The Health and Safety Executive (HSE) reviewed the Workplace Regulations for the European Commission in 2003 and concluded that they were working well.

The Workplace Regulations set down detailed standards for premises used as places of work in the following areas:

(*a*) maintenance [*SI 1992 No 3004, Reg 5*] (see W11003);

(*b*) ventilation [*SI 1992 No 3004, Reg 6*] (see VENTILATION);

(*c*) temperature [*SI 1992 No 3004, Reg 7*] (see W11005);

(*d*) lighting [*SI 1992 No 3004, Reg 8*] (see LIGHTING);

(*e*) cleanliness and waste storage [*SI 1992 No 3004, Reg 9*] (see W11007);

(*f*) room dimensions and space [*SI 1992 No 3004, Reg 10*] (see W11008);

(*g*) workstations and seating [*SI 1992 No 3004, Reg 11*] (see W11009);

(*h*) conditions of floors and traffic routes [*SI 1992 No 3004, Reg 12*] (see W11010);

(*j*) freedom from falls and falling objects [*SI 1992 No 3004, Reg 13*] (see WORK AT HEIGHTS);

(*k*) windows and transparent or translucent doors, gates and walls (see W11012);

(*l*) windows, skylights and ventilators [*SI 1992 No 3004, Reg 18*] (see W11013);

(*m*) window cleaning [*SI 1992 No 3004, Regs 14—16*] (see WORK AT HEIGHTS);

(*n*) organisation of traffic routes [*SI 1992 No 3004, Reg 17*] (see W11015);

(*o*) doors and gates [*SI 1992 No 3004, Reg 18*] (see W11016);

(*p*) escalators and travelators [*SI 1992 No 3004, Reg 19*] (see W11017);

(*q*) sanitary conveniences/washing facilities [*SI 1992 No 3004, Regs 20, 21*] (see W11019–W11023);

(*r*) drinking water [*SI 1992 No 3004, Reg 22*] (see W11024);

(*s*) clothing accommodation and facilities for changing clothing [*SI 1992 No 3004, Reg 23*] (see W11026–W11027); and

(*t*) rest/meal facilities [*SI 1992 No 3004, Reg 25*] (see W11027).

In addition to these Regulations, there is other legislation which contain health, safety and welfare standards for particularly high risk industries and workplaces, including construction sites, mines, quarries and the railway industry. This legislation is dealt with in other chapters.

This chapter looks at the provisions of the *Workplace* (*Health, Safety and Welfare*) *Regulations 1992*, together with guidance provided in the associated approved code of practice (ACoP). It outlines the requirements of the *Building Regulations 1991* (*SI 1991 No 2768*), which apply where workplaces are being built, extended or modified, and looks at the *Disability Discrimination Act 1995, s 6*, which requires that reasonable adjustments be made to workplaces where necessary to ensure that disabled workers are not put at a substantial disadvantage.

It also sets out the provisions of the *Health and Safety* (*Safety Signs and Signals*) *Regulations 1996* (*SI 1996 No 341*).

Workplace (Health, Safety and Welfare) Regulations 1992 (SI 1992 No 3004)

Definitions

W11002 A 'workplace' is any non-domestic premises available to any person as a place of work, including:

(*a*) canteens, toilets;

(*b*) parts of a workroom or workplace (e.g. corridor, staircase, or other means of access/egress other than a public road);

(*c*) a completed modification, extension, or conversion of an original workplace;

but excluding

(i) boats, ships, hovercraft, trains and road vehicles (although the requirements in *regulation 13*, which deal with falls and falling objects, apply when aircraft, trains and road vehicles are stationary inside a workplace);

(ii) building operations/works of engineering construction;

(iii) mining activities.

[*Workplace* (*Health, Safety and Welfare*) *Regulations 1992* (*SI 1992 No 3004*), *Regs 2, 3* and *4*].

The definition of 'work' includes work carried out by employees and self-employed people and the definition of 'premises' includes outdoor places. The Regulations do not apply to private dwellings, but they do apply to hotels, nursing homes and to parts of premises where domestic staff are employed, for example in the kitchens of hostels.

Regulations 20 to *25* (which deal with toilets, washing, changing facilities, clothing accommodation, drinking water and eating and rest facilities) apply to temporary work sites, but only so far as is reasonably practicable.

General maintenance of the workplace

W11003 All workplaces, equipment and devices should be maintained

(*a*) in an efficient state,

(*b*) in an efficient working order, and

(*c*) in a good state of repair.

[*Workplace* (*Health, Safety and Welfare*) *Regulations 1992* (*SI 1992 No 3004*), *Reg 5*].

Dangerous defects should be reported and acted on as a matter of good housekeeping (and to avoid possible subsequent civil liability). Defects resulting in equipment/plant becoming unsuitable for use, though not necessarily dangerous, should lead to decommissioning of plant until repaired – or, if this might lead to the number of facilities being less than required by statute, repaired forthwith (e.g. a defective toilet).

To this end, a suitable maintenance programme must be instituted, including:

(*a*) regular maintenance (inspection, testing, adjustment, lubrication, cleaning);

(*b*) rectification of potentially dangerous defects and the prevention of access to defective equipment;

(*c*) record of maintenance/servicing.

There are more detailed regulations dealing with plant and equipment used at work, including the *Provision and Use of Work Equipment Regulations 1998* (*SI 1998 No 2306* as amended by *SI 2002 No 2174*), which are examined in detail in MACHINERY SAFETY.

Ventilation

W11004 *Workplace* (*Health, Safety and Welfare*) *Regulations 1992* (*SI 1992 No 3004*), *Reg 6* deals with the provision of sufficient ventilation in enclosed workplaces. There must be effective and suitable ventilation in order to supply a sufficient quantity of fresh or purified air. If ventilation plant is necessary for health and safety reasons, it must give warning of failure.

The ACoP to the Regulations sets out that ventilation should not cause uncomfortable draughts, and that it should be sufficient to provide fresh air for the occupants to breathe, to dilute any contaminants and to reduce odour.

The HSE publication, *General ventilation in the workplace: Guidance for employers*, provides detailed guidance in this area. It sets out that the fresh air supply rate to workplaces should not normally fall below 5 to 8 litres per second, per occupant, and that in deciding the appropriate rate, employers should consider factors including:

- The amount of floor space available to each worker;
- The type of work being carried out;
- Whether people smoke in the workplace; and
- What contaminants are being discharged by, for example, any process machinery, heaters, furniture and furnishings.

This guidance also lists sources of further information and help, from organisations including the Chartered Institution of Building Services Engineers (CIBSE) and the Heating and Ventilating Contractors Association (HVCA) and is available from HSE Books price £4.00.

Ventilation is dealt with in detail in VENTILATION.

Temperature

W11005 The temperature in all workplaces inside buildings should be reasonable during working hours. [*Workplace (Health, Safety and Welfare) Regulations 1992 (SI 1992 No 3004), Reg 7*]. Workroom temperatures should enable people to work (and visit sanitary conveniences) in reasonable comfort, without the need for extra or special clothing. Although the Regulations themselves do not specify a maximum or minimum indoor workplace temperature, the approved code of practice (ACoP) sets out that the minimum acceptable temperature is 16°C at the workstation, except where work involves considerable physical effort, when it reduces to 13°C (dry bulb thermometer reading). Space heating of the average workplace should be 16°C, and this should be maintained throughout the remainder of the working day. However, this is a minimum temperature. The method of heating or cooling should not result in dangerous or offensive gases or fumes entering the workplace.

The following temperatures for different types of work are recommended by the Chartered Institute of Building Services Engineers (CIBSE):

(*a*) heavy work in factories 13°C;

(*b*) light work in factories 16°C;

(*c*) hospital wards and shops 18°C; and

(*d*) office and dining rooms 20°C.

Maintenance of such temperatures may not always be feasible, as, for instance, where hot/cold production/storage processes are involved, or where food has to be stored. In such cases, an approximate temperature should be maintained. With cold storage, this may be achievable by keeping a small chilling area separate or by product insulation; whereas, in the case of hot processes, insulation of hot plant or pipes, provision of cooling plant, window shading and positioning of workstations away from radiant heat should be considered in order to achieve a reasonably comfortable temperature. Moreover, where it is necessary from time to time to work in rooms normally unoccupied (e.g. storerooms), temporary heating should be installed. Thermometers must be provided so that workers can periodically check the temperatures.

Where, despite the provision of local heating or cooling, temperatures are still not reasonably comfortable, suitable protective clothing or rest facilities should be provided, or there should be systems of work in place, such as job rotation, to minimise the length of time workers are exposed to uncomfortable temperatures.

HSE guidance on the Regulations, *Workplace health, safety and welfare: A short guide for managers*, sets out how to carry out an assessment of the risk to workers' health from working in either a hot or cold environment. This advises that employers should look at personal factors, such as body activity, the amount and type of clothing and duration of exposure, together with environmental factors, including the ambient temperature and radiant heat, and if the work is outdoor, sunlight, wind velocity and the presence of rain or snow.

It sets out that any assessment needs to consider:

- Measures to control the workplace environment, particularly heat sources;
- Restriction of exposure by, for example, reorganising tasks to build in rest periods or other breaks from work;

- Medical pre-selection of employees to ensure that they are fit to work in these environments;
- Use of suitable clothing;
- Acclimatisation of workers to the working environment;
- Training in the precautions to be taken; and
- Supervision to ensure that the precautions the assessment identifies are taken.

HSE produces further guidance on temperature at work, *Thermal comfort in the workplace: Guidance for employers*. This is available from HSE Books, price £3.50.

For employers with outdoor workers there is a free leaflet, *Keep your top on*, which contains guidance on working outdoors in sunny weather, also available from HSE Books.

Lighting

W11006 Every workplace must be provided with suitable and sufficient lighting. [*Workplace (Health, Safety and Welfare) Regulations 1992 (SI 1992 No 3004), Reg 8*]. This should be natural lighting, so far as is reasonably practicable. There should also be suitable and sufficient emergency lighting where necessary. The ACoP sets out that in order to be suitable and sufficient, lighting must enable people to work and move about safely. HSE guidance on the Regulations advises that where necessary, local lighting should be provided at individual workstations, and at places of particular risk such as crossing points on traffic routes.

An HSE booklet, *Lighting at work*, gives detailed guidance in this area and is available from HSE Books, price £9.25. LIGHTING also deals with this area in greater detail.

General cleanliness

W11007 All furniture and fittings of every workplace must be kept sufficiently clean. Surfaces of floors, walls and ceilings must be capable of being kept sufficiently clean and waste materials must not accumulate other than in waste receptacles. [*Workplace (Health, Safety and Welfare) Regulations 1992 (SI 1992 No 3004), Reg 9*].

The level and frequency of cleanliness will vary according to workplace use and purpose. Obviously a factory canteen should be cleaner than a factory floor. Floors and indoor traffic routes should be cleaned at least once a week, though dirt and refuse not in suitable receptacles should be removed at least daily, particularly in hot atmospheres or hot weather. Interior walls, ceilings and work surfaces should be cleaned at suitable intervals and ceilings and interior walls painted and/or tiled so that they can be kept clean. Surface treatment should be renewed when it can no longer be cleaned properly. In addition, cleaning will be necessary to remove spillages and waste matter from drains or sanitary conveniences. Methods of cleaning, however, should not expose anyone to substantial amounts of dust, and absorbent floors likely to be contaminated by oil or other substances difficult to remove, should be sealed or coated, say, with non-slip floor paint (not covered with carpet!).

Workroom dimensions/space

W11008 Every room in which people work should have sufficient (*a*) floor area,

(*b*) height, and

(*c*) unoccupied space

for health, safety and welfare purposes. [*Workplace* (*Health, Safety and Welfare*) *Regulations 1992* (*SI 1992 No 3004*), *Reg 10*].

Workrooms should have enough uncluttered space to allow people to go to and from workstations with relative ease. The number of people who may work in any particular room at any time will depend not only on the size of the room but also on the space given over to furniture, fittings, equipment and general room layout. Workrooms should be of sufficient height to afford staff safe access to workstations. If, however, the workroom is in an old building, say, with low beams or other possible obstructions, this should be clearly marked, e.g. 'Low beams, mind your head'.

The total volume of the room (when empty), divided by the number of people normally working there, should be 11 cubic metres (minimum) per person, although this does not apply to:

(i) retail sales kiosks, attendants' shelters etc.,

(ii) lecture/meeting rooms etc.

[*Workplace* (*Health, Safety and Welfare*) *Regulations 1992* (*SI 1992 No 3004*), *Sch 1*].

In making this calculation, any part of a room which is higher than 3 metres is counted as being 3 metres high.

Where furniture occupies a considerable part of the room, 11 metres may not be sufficient space per person. Here more careful planning and general room layout is required. Similarly, rooms may need to be larger or have fewer people working in them depending on the contents and layout of the room and the nature of the work.

Workstations and seating

W11009 The Regulations stipulate the following.

(*a*) *Workstations*

Every workstation must be so arranged that:

(i) it is suitable for

(*a*) any person at work who is likely to work at the workstation, and

(*b*) any work likely to be done there;

(ii) so far as reasonably practicable, it provides protection from adverse weather;

(iii) it enables a person to leave it swiftly or to be assisted in an emergency;

(iv) it ensures any person is not likely to slip or fall.

[*Workplace* (*Health, Safety and Welfare*) *Regulations 1992* (*SI 1992 No 3004*), *Reg 11*(*1*)(*2*)].

(*b*) *Workstation seating*

A suitable seat must be provided for each person at work whose work (or a substantial part of it) can or must be done seated. The seat should be suitable for:

(i) the person doing the work, and

(ii) the work to be done.

Where necessary, a suitable footrest should be provided.

[*Workplace* (*Health, Safety and Welfare*) *Regulations 1992* (*SI 1992 No 3004*), *Reg 11*(*3*)(*4*)].

It should be possible to carry out work safely and comfortably. Work materials and equipment in frequent use (or controls) should always be within easy reach, so that people do not have to bend or stretch unduly, and the worker should be at a suitable height in relation to the work surface. Workstations, including seating and access, should be suitable for special needs, for instance, disabled workers. The workstation should allow people likely to have to do work there adequate freedom of movement and ability to stand upright, thereby avoiding the need to work in cramped conditions. More particularly, seating should be suitable, providing adequate support for the lower back and a footrest provided, if feet cannot be put comfortably flat on the floor.

Workstations with visual display units (VDUs) are subject to the *Health and Safety* (*Display Screen Equipment*) *Regulations 1992* (*SI 1992 No 2792* as amended by *SI 2002 No 2174*).

The Health and Safety Executive (HSE) revised its guidance on seating in 1998, and advises employers to use risk assessments in order to ensure that safe seating is provided. *Seating at Work*, HS(G)57 is available, price £5.95, from HSE Books.

Condition of floors and traffic routes

W11010 The principal dangers connected with industrial and commercial floors are slipping, tripping and falling. Slip, trip and fall resistance are a combination of the right floor surface and the appropriate type of footwear. Employers should ensure that level changes, multiple changes of floor surfaces, steps and ramps etc. are clearly indicated. Safety underfoot is at bottom a trade-off between slip resistance and ease of cleaning. Floors with rough surfaces tend to be more slip-resistant than floors with smooth surfaces, especially when wet; by contrast, smooth surfaces are much easier to clean but less slip-resistant. Use of vinyl flooring in public areas – basically slip-resistant – is on the increase. Vinyl floors should be periodically stripped, degreased and resealed with slip-resistant finish; linoleum floors similarly.

Apart from being safe, floors must also be hygienically clean. In this connection, quarry tiles have long been 'firm favourites' in commercial kitchens, hospital kitchens etc. but can be hygienically deceptive. In particular, grouted joints can trap bacteria as well as presenting endless practical cleaning problems. Hence the gradual transition to seamless floors in hygiene-critical areas. Whichever floor surface is appropriate and whichever treatment is suitable, underfoot safety depends on workplace activity (office or factory), variety of spillages (food, water, oil, chemicals), nature of traffic (pedestrian, cars, trucks).

Thus, floors in workplaces must:

(*a*) be constructed so as to be suitable for use. [*Workplace* (*Health, Safety and Welfare*) *Regulations 1992* (*SI 1992 No 3004*), *Reg 12*(*1*)].

They should always be of sound construction and adequate strength and stability to sustain loads and passing internal traffic; they should never be overloaded (see *Greaves v Baynham Meikle [1975] 3 All ER 99* for possible consequences in civil law).

(*b*) (i) not have holes or slopes, or

(ii) not be uneven or slippery

so as to expose a person to risk of injury. [*Workplace* (*Health, Safety and Welfare*) *Regulations 1992* (*SI 1992 No 3004*), *Reg 12*(*2*)(*a*)].

The surfaces of floors and traffic routes should be even and free from holes, bumps and slipping hazards that could cause a person to slip, trip or fall, or drop or lose control of something being lifted or carried; or cause instability or loss of control of a vehicle.

Holes, bumps or uneven surfaces or areas resulting from damage or wear and tear should be made good and, pending this, barriers should be erected or locations conspicuously marked. Temporary holes, following, say, removal of floorboards, should be adequately guarded. Special needs should be catered for, for instance, disabled walkers or those with impaired sight. (Deep holes are governed by *Workplace* (*Health, Safety and Welfare*) *Regulations 1992* (*SI 1992 No 3004*), *Reg 13* (see W9013 WORK AT HEIGHTS).) Where possible, steep slopes should be avoided, and otherwise provided with a secure handrail. Ramps used by disabled persons should also have handrails.

(*c*) be kept free from

(i) obstructions, and

(ii) articles/substances likely to cause persons to slip, trip or fall

so far as reasonably practicable. [*Workplace* (*Health, Safety and Welfare*) *Regulations 1992* (*SI 1992 No 3004*), *Reg 12*(*3*)].

Floors should be kept free of obstructions impeding access or presenting hazards, particularly near or on steps, stairs, escalators and moving walkways, on emergency routes or outlets, in or near doorways or gangways or by corners or junctions. Where temporary obstructions are unavoidable, access should be prevented and people warned of the possible hazard. Furniture being moved should not be left in a place where it can cause a hazard.

(*d*) have effective drainage. [*Workplace* (*Health, Safety and Welfare*) *Regulations 1992* (*SI 1992 No 3004*), *Reg 12*(*2*)(*b*)].

Where floors are likely to get wet, effective drainage (without drains becoming contaminated with toxic, corrosive substances) should drain it away, e.g. in laundries, potteries and food processing plants. Drains and channels should be situated so as to reduce the area of wet floor and the floor should slope slightly towards the drain and ideally have covers flush with the floor surface. Processes and plant which cause discharges or leaks of liquids should be enclosed and leaks from taps caught and drained away. In food processing and preparation plants, work surfaces should be arranged so as to minimise the likelihood of spillage. Where a leak or spillage occurs, it should be fenced off or mopped up immediately.

Staircases should be provided with a handrail. Any open side of a staircase should have minimum fencing of an upper rail at 900mm or higher, and a lower rail.

It is important also to consider the dangers posed by snow and ice upon, for example, external fire escapes.

Falls and falling objects

W11011 *Workplace (Health, Safety and Welfare) Regulations 1992 (SI 1992 No 3004), Reg 13* deals with falls and falling objects, which are covered in detail in WORK AT HEIGHTS.

It also requires that tanks, pits and other structures containing dangerous substances are securely covered or fenced where there is a risk of a person falling. Traffic routes over such open structures should also be securely fenced.

Windows and transparent or translucent doors, gates and walls

W11012 Transparent or translucent surfaces in windows, doors, gates, walls and partitions should be constructed of safety material or be adequately protected against breakage, where necessary for health and safety reasons, where:

(*a*) any part is at shoulder level or below in doors and gates; or

(*b*) any part is at waist level or below in windows, walls and partitions, with the exception of glass houses.

Screens or barriers can be used as an alternative to the use of safety materials. Narrow panels of up to 250mm width are excluded from the requirement.

Transparent or translucent surfaces should be marked to make them apparent where this is necessary for health and safety reasons.

[*Workplace (Health, Safety and Welfare) Regulations 1992 (SI 1992 No 3004), Reg 14*].

Windows, skylights and ventilators

W11013 Openable windows, skylights and ventilators must be capable of being opened, closed and adjusted safely. They must not be positioned so as to pose a risk when open.

They should be capable of being reached and operated safely, with window poles or similar equipment, or stable platforms, made available where necessary. Where there is the danger of falling from a height, devices should be provided to prevent this by ensuring the window cannot open too far. They should not cause a hazard by projecting into an area where people are likely to collide with them when open. The bottom edge of opening windows should normally be at least 800mm above floor level, unless there is a barrier to prevent falls.

[*Workplace (Health, Safety and Welfare) Regulations 1992 (SI 1992 No 3004), Reg 15*].

Ability to clean windows etc. safely

W11014 *Workplace (Health, Safety and Welfare) Regulations 1992 (SI 1992 No 3004), Reg 16* deals with the safe cleaning of windows and skylights where these cannot be cleaned from the ground or other suitable surface. This is dealt with in detail in WORK AT HEIGHTS.

Organisation of traffic routes

W11015 Traffic routes in workplaces should allow pedestrians and vehicles to circulate safely, be safely constructed, be suitably indicated where necessary for health and safety reasons, and be kept clear of obstructions.

They should be planned to give the safest route, wide enough for the safe movement of the largest vehicle permitted to use them, and they should avoid vulnerable items like fuel or chemical plants or pipes, and open and unprotected edges.

There should be safe areas for loading and unloading. Sharp or blind bends should be avoided where possible, and if they cannot be avoided, one-way systems or mirrors to improve visibility should be used. Sensible speed limits should be set and enforced. There should be prominent warning of any limited headroom or potentially dangerous obstructions such as overhead electric cables. Routes should be marked where necessary and there should be suitable and sufficient parking areas in safe locations.

Traffic routes should keep vehicles and pedestrians apart and there should be pedestrian crossing points on vehicle routes. Traffic routes and parking and loading areas should be soundly constructed on level ground. Health and Safety Executive (HSE) guidance in this area can be found in the publication, *Workplace transport safety – guidance for employers*, HS(G)136, price £7.50, available from HSE Books.

[*Workplace* (*Health, Safety and Welfare*) *Regulations 1992* (*SI 1992 No 3004*), *Reg 17*].

This area is dealt with in detail in ACCESS, TRAFFIC ROUTES AND VEHICLES.

Doors and gates

W11016 Doors and gates must be suitably constructed and fitted with safety devices. In particular,

(i) a sliding door/gate must have a device to prevent it coming off its track during use;

(ii) an upward opening door/gate must have a device to prevent its falling back;

(iii) a powered door/gate must

  (*a*) have features preventing it causing injury by trapping a person (e.g. accessible emergency stop controls),

  (*b*) be able to be operated manually unless it opens automatically if the power fails;

(iv) a door/gate capable of opening, by being pushed from either side, must provide a clear view of the space close to both sides.

[*Workplace* (*Health, Safety and Welfare*) *Regulations 1992* (*SI 1992 No 3004*), *Reg 18*].

Doors and gates that swing in both directions should have a transparent panel, unless they are low enough to see over.

Escalators and travelators

W11017 Escalators and travelators must:

(*a*) function safely;

(*b*) be equipped with safety devices;

(*c*) be fitted with emergency stop controls.

[*Workplace* (*Health, Safety and Welfare*) *Regulations 1992* (*SI 1992 No 3004*), *Reg 19*].

Welfare facilities

W11018 'Welfare facilities' is a wide term, embracing both sanitary and washing accommodation at workplaces, provision of drinking water, clothing accommodation (including facilities for changing clothes) and facilities for rest and eating meals (see W11027 below). The need for sufficient suitable hygienic lavatory and washing facilities in all workplaces is obvious. Sufficient facilities must be provided to enable everyone at work to use them without undue delay. They do not have to be in the actual workplace but ideally should be situated in the building(s) containing them and they should provide protection from the weather, be well-ventilated, well-lit and enjoy a reasonable temperature. Where disabled workers are employed, special provision should be made for their sanitary and washing requirements. Wash basins should allow washing of hands, face and forearms and, where work is particularly strenuous, dirty, or results in skin contamination (e.g. molten metal work), showers or baths should be provided. In the case of showers, they should be fed by hot and cold water and fitted with a thermostatic mixer valve. Washing facilities should ensure privacy for the user and be separate from the water closet, with a door that can be secured from the inside. It should not be possible to see urinals or the communal shower from outside the facilities when the entrance/exit door opens. Entrance/exit doors should be fitted to both washing and sanitary facilities (unless there are other means of ensuring privacy). Windows to sanitary accommodation, showers/bathrooms should be obscured either by being frosted, or by blinds or curtains (unless it is impossible to see into them from outside).

This section examines current statutory requirements in all workplaces. For requirements relating to sanitary conveniences and washing facilities on construction sites see C8056 CONSTRUCTION AND BUILDING OPERATIONS.

Sanitary conveniences in all workplaces

W11019 Suitable and sufficient sanitary conveniences must be provided at readily accessible places. In particular,

(*a*) the rooms containing them must be adequately ventilated and lit;

(*b*) they (and the rooms in which they are situated) must be kept clean and in an orderly condition;

(*c*) separate rooms containing conveniences must be provided for men and women except where the convenience is in a separate room which can be locked from the inside.

[*Workplace* (*Health, Safety and Welfare*) *Regulations 1992* (*SI 1992 No 3004*), *Reg 20*].

Washing facilities in all workplaces

W11020 Suitable and sufficient washing facilities (including showers where necessary (see W11018 above)), must be provided at readily accessible places or points. In particular, facilities must:

(*a*) be provided in the immediate vicinity of every sanitary convenience (whether or not provided elsewhere);

(*b*) be provided in the vicinity of any changing rooms – whether or not provided elsewhere;

(*c*) include a supply of clean hot and cold or warm water (if possible, running water);

(*d*) include soap (or something similar);

(*e*) include towels (or the equivalent);

(*f*) be in rooms sufficiently well-ventilated and well-lit;

(*g*) be kept clean and in an orderly condition (including rooms in which they are situate);

(*h*) be separate for men and women, except where they are provided in a lockable room intended to be used by one person at a time, or where they are provided for the purposes of washing hands, forearms and face only, where separate provision is not necessary.

[*Workplace (Health, Safety and Welfare) Regulations 1992 (SI 1992 No 3004), Reg 21*].

Minimum number of facilities – sanitary conveniences and washing facilities

(a) People at work

W11021

Number of people at work	*Number of WCs*	*Number of wash stations*
1 to 5	1	1
6 to 25	2	2
26 to 50	3	3
51 to 75	4	4
76 to 100	5	5

(b) Men at work

Number of men at work	*Number of WCs*	*Number of urinals*
1 to 15	1	1
16 to 30	2	1
31 to 45	2	2
46 to 60	3	2
61 to 75	3	3
76 to 90	4	3
91 to 100	4	4

For every 25 people above 100 an additional WC and wash station should be provided; in the case of WCs used only by *men*, an additional WC per every 50 men above 100 is sufficient (provided that at least an equal number of additional urinals is provided). [*Workplace* (*Health, Safety and Welfare*) *Regulations 1992* (*SI 1992 No 3004*), *Sch 1, Part II*].

Particularly dirty work etc.

W11022 Where work results in heavy soiling of hands, arms and forearms, there should be one wash station for every 10 people at work up to 50 people; and one extra for every additional 20 people. And where sanitary and wash facilities are also used by members of the public, the number of conveniences and facilities should be increased so that workers can use them without undue delay.

Temporary work sites

W11023 At temporary work sites suitable and sufficient sanitary conveniences and washing facilities should be provided so far as is reasonably practicable (see E15039 ENFORCEMENT for meaning). If possible, these should incorporate flushing sanitary conveniences and washing facilities with running water.

Drinking water

W11024 An adequate supply of wholesome drinking water must be provided for all persons at work in the workplace. It must be readily accessible at suitable places and conspicuously marked, unless non-drinkable cold water supplies are clearly marked. In addition, there must be provided a sufficient number of suitable cups (or other drinking vessels), unless the water supply is in a jet. [*Workplace* (*Health, Safety and Welfare*) *Regulations 1992* (*SI 1992 No 3004*), *Reg 22*].

Where water cannot be obtained from the mains supply, it should only be provided in refillable containers. The containers should be enclosed to prevent contamination and refilled at least daily. So far as reasonably practicable, drinking water taps should not be installed in sanitary accommodation, or in places where contamination is likely, for instance, in a workshop containing lead processes.

Clothing accommodation

W11025 Suitable and sufficient accommodation must be provided for:

(*a*) any person at work's own clothing which is not worn during working hours; and

(*b*) special clothing which is worn by any person at work but which is not taken home, for example, overalls, uniforms and thermal clothing.

[*Workplace* (*Health, Safety and Welfare*) *Regulations 1992* (*SI 1992 No 3004*), *Reg 23*(*1*)].

Accommodation is not suitable unless it:

(i) provides suitable security for the person's own clothing where changing facilities are required;

(ii) includes separate accommodation for clothing worn at work and for other clothing, where necessary to avoid risks to health or damage to clothing; and

(iii) is in a suitable location.

[*Workplace* (*Health, Safety and Welfare*) *Regulations 1992* (*SI 1992 No 3004*), *Reg 23*(*2*)].

Work clothing is overalls, uniforms, thermal clothing and hats worn for hygiene purposes. Workers' own clothing should be able to hang in a clean, warm, dry, well-ventilated place. If this is not possible in the workroom, then it should be put elsewhere. Accommodation should take the form of a separate hook or peg. Clothing which is dirty, damp or contaminated owing to work should be accommodated separately from the worker's own clothes.

Facilities for changing clothing

W11026 Suitable and sufficient facilities must be provided for any person at work in the workplace to change clothing where:

(*a*) the person has to wear special clothing for work, and

(*b*) the person cannot be expected to change in another room.

Facilities are not suitable unless they include:

(i) separate facilities for men and women, or

(ii) separate use of facilities by men and women.

[*Workplace* (*Health, Safety and Welfare*) *Regulations 1992* (*SI 1992 No 3004*), *Reg 24*].

Changing rooms (or room) should be provided for workers who change into special work clothing and where they remove more than outer clothing; also where it is necessary to prevent workers' own clothes being contaminated by a harmful substance. Changing facilities should be easily accessible from workrooms and eating places. They should contain adequate seating and clothing accommodation, and showers or baths if these are provided (see W11018 above). Privacy of user should be ensured. The facilities should be large enough to cater for the maximum number of persons at work expected to use them at any one time without overcrowding or undue delay.

Post Office v Footitt [2000] IRLR 243, involved an employers' appeal against an improvement notice requiring the construction of a separate changing room for women postal workers to change into and out of their uniforms, and looked at the definition of 'special clothing' and at the concept of propriety.

An environmental health officer had served an improvement notice under the *Workplace* (*Health, Safety and Welfare*) *Regulations 1992* (*SI 1992 No 3004*), *Reg 24* as she had found that any female employees wishing to change their clothing could only do so in the general area of the women's toilet facilities.

In the High Court, the judge held that the uniform worn by postal workers was 'special clothing' for the purposes of *Regulation 24*. It was held that 'special clothing' is not merely limited to clothing that is worn only at work. Therefore the fact that postal workers wear their uniform to and from work does not prevent it from being 'special clothing'. The changing facilities for women provided by the Post Office were therefore not 'suitable and sufficient' within the meaning of the Regulation.

The court also held that the fact that the changing facilities for men and women were separated was not in itself enough to satisfy the concept of propriety referred to in *Regulation 24*(*2*). There is no reason why requiring one female to undress in the presence of another cannot be said to offend against the principles of propriety. The fact that many people would have no objection to changing in the company of others of the same sex does not absolve the employer from providing facilities for those who may prefer privacy.

Rest and eating facilities

W11027 Suitable and sufficient rest facilities must be provided at readily accessible places. [*Workplace* (*Health, Safety and Welfare*) *Regulations 1992* (*SI 1992 No 3004*), *Reg 25*(*1*)].

(a) Rest facilities

A rest facility is:

(i) in the case of a new workplace, extension or conversion – a rest room (or rooms);

(ii) in other cases, a rest room (or rooms) or rest area; including

(iii) (in both cases):

- appropriate facilities for eating meals where food eaten in the workplace would otherwise be likely to become contaminated;
- suitable arrangements for protecting non-smokers from tobacco smoke. The ACoP advises that this can be achieved by providing separate areas for smokers and non-smokers, or by prohibiting smoke in rest areas;
- a facility for a pregnant or nursing mother to rest in.

Canteens or restaurants may be used as rest rooms provided that there is no obligation to buy food there (ACoP). [*Workplace* (*Health, Safety and Welfare*) *Regulations 1992* (*SI 1992 No 3004*), *Regs 25*(*2*)–(*4*)].

(b) Eating facilities

Where workers regularly eat meals at work, facilities must be provided for them to do so. [*Workplace* (*Health, Safety and Welfare*) *Regulations 1992* (*SI 1992 No 3004*), *Reg 25*(*5*)].

In offices and other workplaces where there is no risk of contamination, seats in the work area are sufficient, although workers should not be interrupted excessively during breaks, for example, by the public. In other cases, rest areas or rooms should be provided and in the case of new workplaces, this should be a separate rest room. Rest facilities should be large enough, and have enough seats with backrests and tables, for the number of workers likely to use them at one time.

Where workers regularly eat meals at work, there should be suitable and sufficient facilities. These should be provided where food would otherwise be contaminated, by dust or water for example. Seats in work areas can be suitable eating facilities, provided the work area is clean. There should be a means to prepare or obtain a hot drink, and where persons work during hours or at places where hot food cannot be readily obtained, there should be the means for heating their own food. Eating facilities should be kept clean.

Smoking

W11028 Providing protection to non-smokers from tobacco smoke in rest areas is the only specific legal requirement concerning smoking at work, although the general duty to ensure the health, safety and welfare of the employees under the *Health and Safety at Work etc. Act 1974* (*HSWA 1974*), *s 2* applies. However there have been a number of legal cases, and compensation awards have been made to employees who have claimed their health has been affected by breathing in tobacco smoke at work.

Stockport Metropolitan Council has made two out of court settlements of £25,000 and £15,000 to employees who claimed that their health was damaged as a result of passive smoking. In addition, an Employment Appeal Tribunal (EAT) case, *Dryden v Greater Glasgow Health Board [1992] IRLR 469*, held that a change to a complete smoking ban, which meant an employee who smoked had to leave her job, did not amount to constructive dismissal. In this case, the employer had consulted workers about the introduction of the ban.

In another EAT case, *Walton and Morse v Dorrington [1997]*, a tribunal decision that an employee had been constructively dismissed when her employer failed to provide her with a smoke free environment or deal with her problems relating to passive smoking, was upheld.

In the first passive smoking case to reach the courts, in May 1998, a nurse lost her action for damages against her employer. Silvia Sparrow claimed that she had developed asthma as a result of exposure to environmental tobacco smoke in a residential care home for elderly people. But the court said she had failed to prove that her former employers, St Andrew's Homes Ltd, were negligent so as to cause injury to her.

The Health and Safety Commission (HSC) has recommended that an approved code of practice (ACoP) on passive smoking should be introduced. This would give authoritative guidance on the employer's legal obligation to protect their employees from exposure to environmental tobacco smoke (ETS).

As with other ACoPs, failing to follow the code would not in itself be an offence, but the employer would have to demonstrate that equally effective methods have been adopted to signal compliance with the law.

Under the code, employers will have to determine the most reasonably practical way of controlling ETS. This could involve:

- Banning smoking in the workplace completely or partially;
- Physically segregating non-smokers from tobacco smoke;
- Providing adequate ventilation; or
- Implementing a system of work to reduce the time an employee is exposed to ETS.

There will, however, be a two-year exemption for parts of the hospitality industry, such as bars, clubs and restaurants. They will need to instead comply with the Public Places Charter during this period.

The code is currently awaiting ministerial approval.

The current HSE publication, *Passive smoking at work*, IND(G)63(L), recommends that all employers should introduce a policy to control smoking in the workplace in full consultation with employees.

The guidance sets out that environmental tobacco smoke contains carbon dioxide, hydrogen cyanide and ammonia and that passive smoking has irritant effects on the eyes, throat and respiratory tract. In addition to aggravating asthma, research indicates that it can increase the risk of lung cancer and may increase the risk of heart disease.

The HSE agrees with the advice of the Independent Scientific Committee which says that employers should regard non-smoking as the norm in enclosed workplaces and make provision for smoking, rather than vice versa, and that smokers should be segregated from non-smokers.

The guidance advises employers that they should have a specific written policy on smoking in the workplace which gives priority to the needs of non-smokers. Any policy should be introduced with proper consultation of employees and their representatives and a minimum of three months' notice of its introduction should be given. It also advises that may employers provide help to smokers to reduce or give up smoking.

It outlines that the only effective ways to achieve a smoke-free environment for non-smokers are to:

- Introduce a complete ban on indoor smoking; or
- Ban smoking in all parts of the building with the exception of designated, enclosed smoking areas.

A less effective method, which employers may have to resort to if the constraints of the workplace do not allow the above, is to segregate smokers and non-smokers in separate rooms and ban smoking in common areas.

The guidance advises employers in buildings with mechanical ventilation to consider discharging air from smoking areas separately, and if this is not reasonably practicable, decontamination systems should be used to bring the re-circulated air up to an appropriate standard.

Civil liability

W11029 There is no specific reference to civil liability in the Regulations. However, safety regulations are actionable, even if silent (as here), and, if a person suffered injury/damage as a result of breach by an employer, he could sue. Certainly, there is civil liability for breach of the *Building Regulations 1991* (*SI 1991 No 2768*) (see W11044 below).

Safety signs at work – Health and Safety (Safety Signs and Signals) Regulations 1996 (SI 1996 No 341)

W11030 Traditionally, safety signs, communications and warnings have played a residual role in reducing the risk of injury or damage at work, the need for them generally having been engineered out or accommodated in the system of work – a situation unaffected by these Regulations.

Types of signs

W11031 Safety signs and signals can be of the following types:

(*a*) permanent (e.g. signboards);

(*b*) occasional (e.g. acoustic signals or verbal communications – acoustic signals should be avoided where there is considerable ambient noise).

Interchanging and combining signs

W11032 Examples of interchanging and combining signs are:

(*a*) a safety colour (see W11033 below) or signboard to mark places where there is an obstacle;

(*b*) illuminated signs, acoustic signals or verbal communication; and (*c*) hand signals or verbal communication.

[*Health and Safety (Safety Signs and Signals) Regulations 1996 (SI 1996 No 341), Sch 1, Part I, para 3*].

Safety colours

W11033

Colour	*Meaning or purpose*	*Instructions and information*
Red	Prohibition sign	Dangerous behaviour
	Danger	Stop, shutdown, emergency cut-out services
		Evacuate
	Fire-fighting equipment	Identification and location
Yellow or Amber	Warning sign	Be careful, take precautions
		Examine
Blue	Mandatory sign	Specific behaviour or action
		Wear personal protective equipment
Green	Emergency escape, first-aid sign	Doors, exits, routes, equipment and facilities
	No danger	Return to normal

[*Health and Safety (Safety Signs and Signals) Regulations 1996 (SI 1996 No 341), Sch 1, Part I, para 4*].

Varieties of safety signs and signals

W11034 Safety signs and signals include, comprehensively:

(*a*) safety signs – providing information about health and safety at work by means of a signboard, safety colour, illuminated sign, acoustic signal, hand signal or verbal communication;

(*b*) signboards – signs giving information by way of a simple pictogram, lighting intensity providing visibility (these should be weather-resistant and easily seen);

(*c*) mandatory signs – signs prescribing behaviour (e.g. safety boots must be worn);

(*d*) prohibition signs – signs prohibiting behaviour likely to cause a health and safety risk (e.g. no smoking);

(*e*) hand signals – movement or position of arms/hands for guiding persons carrying out operations that could endanger employees;

(*f*) verbal communications – predetermined spoken messages communicated by human or artificial voice, preferably short, simple and as clear as possible.

[*Health and Safety (Safety Signs and Signals) Regulations 1996 (SI 1996 No 341), Reg 2*].

Duty of employer

W11035 It is only where a risk assessment carried out under the *Management of Health and Safety at Work Regulations 1999* (*SI 1999 No 3242*) indicates that a risk cannot be avoided, engineered out or reduced significantly by way of a system of work that resort to signs and signals becomes necessary. In these circumstances, all employers (including offshore employers) must:

(*a*) provide and maintain any appropriate safety sign(s) (see W11037–W11041 below) (including fire safety signals) but not a hand signal or verbal communication;

(*b*) so far as is reasonably practicable, ensure that correct hand signals or verbal communications are used;

(*c*) provide and maintain any necessary road traffic sign (where there is a risk to employees in connection with traffic); and

(*d*) provide employees with comprehensible and relevant information, training and instruction and measures to be taken in connection with safety signs.

[*Health and Safety* (*Safety Signs and Signals*) *Regulations 1996* (*SI 1996 No 341*), *Regs 4, 5*].

Schedule 1 to the Regulations sets out the minimum requirements concerning safety signs and signals with regard to the type of signs to be used in particular circumstances, interchanging and combining signs, signboards, signs on containers and pipes, the identification and location of fire-fighting equipment, signs used for obstacles and dangerous locations, and for marking traffic routes, illuminated signs, acoustic signals, verbal communication and hand signals.

Exclusions

W11036 Excluded from the operation of these Regulations are:

(*a*) the supply of dangerous substances or products;

(*b*) the transportation of dangerous goods;

(*c*) road traffic signs (except where there is a particular risk to employees. Where there is a risk arising from the movement of traffic and the risk is addressed by a sign stipulated in the *Road Traffic Regulations Act 1984* (e.g. speed restriction sign), these signs must be used, whether or not the Act applies to that place of work. In effect this means that where road speed and other signs are needed on a company's road, these must replicate the signs used for the purpose on public roads; and

(*d*) activities on board ship.

[*Health and Safety* (*Safety Signs and Signals*) *Regulations 1996* (*SI 1996 No 341*), *Reg 3*(*1*)].

Examples of safety signs

Prohibitory signs

W11037 Intrinsic features:

— round shape

— black pictogram on white background, red edging and diagonal line (the red part to take up at least 35% of the sign area).

fig. 1 Safety signs (prohibitory)

No smoking

Smoking and naked flames forbidden

No access for pedestrians

Do not extinguish with water

Not drinkable

No access for unauthorised persons

No access for industrial vehicles

Do not touch

Warning signs

W11038 Intrinsic features:

- triangular shape
- black pictogram on a yellow background with black edging (the yellow part to take up at least 50% of the area of the sign).

fig. 2 Safety signs (warning)

Flammable material or high temperature

Explosive material

Toxic material

Corrosive material

Radioactive material

Overhead load

Industrial vehicles

Danger: electricity

General danger

Laser beam

Oxidant material

Non-ionizing radiation

Strong magnetic field

Obstacles

Drop

Biological risk

Low temperature

Harmful or irritant material

Mandatory signs

W11039 Intrinsic features:

- round shape
- white pictogram on a blue background (the blue part to take up at least 50% of the area of the sign).

fig. 3 Safety signs (mandatory)

Eye protection must be worn
Safety helmet must be worn
Ear protection must be worn
Respiratory equipment must be worn
Safety boots must be worn
Safety gloves must be worn
Safety overalls must be worn
Face protection must be worn
Safety harness must be worn
Pedestrians must use this route
General mandatory sign (to be accompanied where necessary by another sign)

Emergency escape or first-aid signs

W11040 Intrinsic features:

- rectangular or square shape
- white pictogram on a green background (the green part to take up at least 50% of the area of the sign).

fig. 4 Safety signs (emergency escape or first-aid)

Emergency exit/escape route

This way
(supplementary information sign)

First-aid post

Stretcher

Safety shower

Eyewash

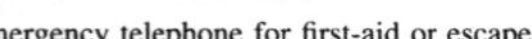

Emergency telephone for first-aid or escape

Fire-fighting signs

W11041 Intrinsic features:

- rectangular or square shape
- white pictogram on a red background (the red part to take up at least 50% of the area of the sign).

fig. 5 Safety signs (fire-fighting)

This way
(supplementary information sign)

Examples of hand signals

W11042

Meaning	*Description*	*Illustration*

A. General signals

fig. 6 Hand signals

START Attention Start of Command	both arms are extended horizontally with the palms facing forwards.	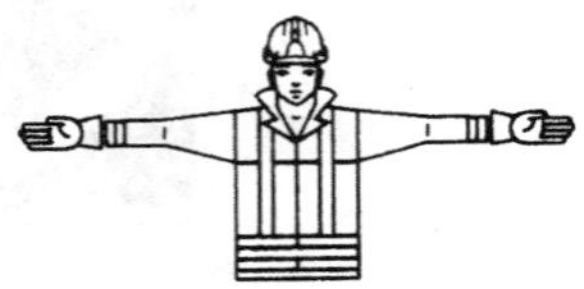

STOP Interruption End of movement	the right arm points upwards with the palm facing forwards.	
END of the operation	both hands are clasped at chest height.	

B. Vertical movements

RAISE	the right arm points upwards with the palm facing forward and slowly makes a circle.	
LOWER	the right arm points downwards with the palm facing inwards and slowly makes a circle.	
VERTICAL DISTANCE	the hands indicate the relevant distance.	

C. Horizontal movements

MOVE FORWARDS	both arms are bent with the palms facing upwards, and the forearms make slow movements towards the body.	
MOVE BACKWARDS	both arms are bent with the palms facing downwards, and the forearms make slow movements away from the body.	
RIGHT to the signalman's	the right arm is extended more or less horizontally with the palm facing downwards and slowly makes small movements to the right.	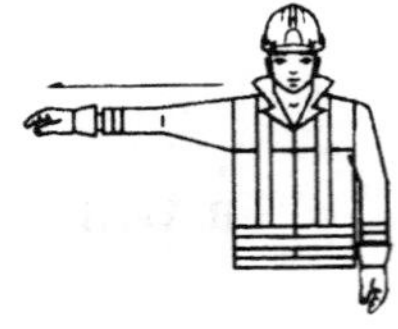
LEFT to the signalman's	the left arm is extended more or less horizontally with the palm facing downwards and slowly makes small movements to the left.	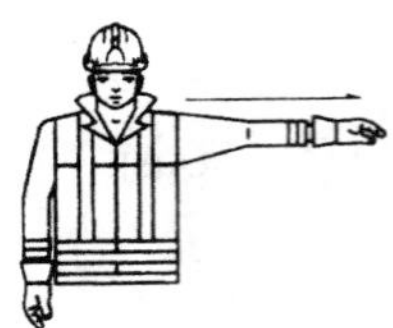
HORIZONTAL DISTANCE	the hands indicate the relevant distance.	

D. Danger

DANGER Emergency stop	both arms point upwards with the palms facing forwards.	
QUICK	all movements faster.	
SLOW	all movements slower.	

Disability Discrimination Act 1995

W11043 The *Disability Discrimination Act 1995* (*DDA 1995*) makes it unlawful for employers (who employ 15 or more people) to treat a disabled person less favourably, without a justifiable reason.

In order to gain protection under the *DDA 1995*, a person must have a substantial and long-term disability, affecting them (or which could affect them) for more than 12 months, impacting on their ability to carry out normal day-to-day activities, and affecting their mobility, dexterity, co-ordination, continence or memory. It covers mental as well as physical impairments, and applies to recruitment, selection, promotion and redundancy. It applies to all workers, including the self-employed, contract workers and agency workers, as well as employees. It does not currently apply to the police, prison and fire services nor to the armed forces.

The *DDA 1995, s 6* requires employers to make reasonable adjustments to working conditions or to the workplace to avoid putting disabled workers at a substantial disadvantage, and could therefore involve making adjustments to the physical features of workplace premises.

The Act gives examples of adjustments to the workplace that could be required, including access for wheelchairs and acquiring or modifying wheelchairs.

There is a provision in the Act to enable employers who lease their premises to obtain their landlord's permission to make alterations in order to make any necessary reasonable adjustments to the workplace.

The *Disability Discrimination (Employment Relations) Regulations 1996* (*SI 1996 No 1456*) set out the duty to make adjustments in more detail. The Regulations specify that physical features of the employer's premises are:

(a) any feature arising from the design or construction of a building on the premises or any approach to, exit from or access to such a building;

(b) any fixtures, fittings, furniture, equipment or materials in or on the premises; and

(c) any other physical element or quality of land included in the premises.

Complaints about disability discrimination are heard in employment tribunals. In addition, the Disability Rights Commission, which came into effect in April 2000, can carry out formal investigations where it believes there is discrimination and can enter into legally binding agreements with employers where the employer can choose to comply with it's recommendations. It can also assist individuals taking cases to tribunals.

In *Tawling v Wisdom Toothbrushes Ltd*, a 1997 tribunal case, a woman who had a club foot and was experiencing sciatica and other pain and discomfort as a result of standing for long periods had to take increasing amounts of sickness absence.

Her employers sought advice from an organisation providing financial support for people with disabilities, the Shaw Trust. It recommended the purchase of one of two specific types of chair costing £500 and around £1000 respectively, to which the employer would have had to fund 20 per cent of the cost. The employer did not take the advice and instead provided her with a series of ordinary chairs, which did not meet her needs and then dismissed her for poor performance.

The tribunal found that the woman had been unfairly dismissed because the employer had failed to make reasonable adjustments in not providing a suitable chair for her.

Building Regulations 1991 (SI 1991 No 2768)

W11044 Where workplaces are built, extended or altered, or where fittings such as drains, heat-producing appliances, washing and sanitary facilities or hot water storage are provided, the *Building Regulations 1991* (*SI 1991 No 2768*) normally apply.

The Regulations set down minimum standards of design and building work for the construction of buildings. They contain requirements designed to ensure the health and safety of people in and around buildings, to provide for energy conservation and to provide access for disabled people.

The requirements, which are set out in *Schedule 1* of the Regulations, concern:

- *Structure*

  Buildings must be constructed to withstand loads and ground movement. Buildings of five or more storeys must be constructed in such a way that in the event of an accident it will not suffer collapse to an extent disproportionate to the cause. Where parts of a public building, shop or shopping mall have a roof with a clear span exceeding nine metres between supports, these should be constructed so that in the event of the failure of any part of the roof, the building will not suffer collapse to an extent disproportionate to that failure;

- *Fire safety*

  Buildings must be designed and constructed with a means of escape, in the case of fire, from the building to a place of safety outside the building. Internal linings of the building must inhibit the spread of fire and the building must be designed and constructed to maintain stability for a period of time in the event of fire. Party walls should resist the spread of fire between buildings and the spread of fire within buildings should be inhibited by sub-division with fire-resisting construction as appropriate to the size and intended use of the building. The building should be designed and constructed so that the unseen spread of fire and smoke within concealed spaces in the structure and fabric of the building is inhibited. The external walls and the roof of the building should resist the spread of fire, and the building should be designed and constructed to provide facilities for fire fighters and to enable fire fighting appliances to gain access to the building. Fire safety in workplaces is dealt with in detail in FIRE PREVENTION AND CONTROL.

- *Site preparation and resistance to moisture*

  The ground to be covered by the building should be reasonably free from vegetable matter, and precautions should be taken to avoid risks to health and safety from any dangerous or offensive substances found on or in this ground, and subsoil drainage should be provided where necessary to avoid damage to the fabric of the building or the passage of ground moisture to the interior of the building. The walls, floors and roof of the building should also prevent damp penetration.

- *Toxic substances*

  If cavity wall insulation is used, precautions should be taken to prevent any subsequent release of toxic fumes into any occupied parts of the building.

- *Resistance to the passage of sound*

  This part applies only to dwellings.

- *Ventilation*

  Rooms containing sanitary conveniences must be adequately ventilated, and provision must be made to prevent excessive condensation in a roof or a roof void above an insulated ceiling.

- *Hygiene*

  Adequate sanitary conveniences shall be provided in rooms provided for that purpose and they must be separated from places where food is prepared. Adequate wash basins must be provided in or next to rooms containing wcs, and again must be separated from areas where food is being prepared. There should be suitable installation for the provision of hot and cold water to washbasins, and wcs and washbasins should be designed and installed to allow them to be effectively cleaned. There are also requirements concerning the installation of hot water storage systems.

- *Drainage and waste disposal*

  There must be a waste water drainage system which takes foul water from toilets, washbasins and sinks to a sewer, or properly constructed cesspool or septic or settlement tank. There should be adequate rainwater drainage and adequate means of storing solid waste.

- *Heat producing appliances*

  Heat producing appliances designed to burn solid fuel, oil or gas and incinerators should have adequate air supply, and adequate provision for the discharge of combustion products to the outside air. Appliances should be installed, and fireplaces and chimneys constructed to reduce the risk of the building catching fire as a consequence of their use.

- *Protection from falling, collision and impact*

  Stairs, ladders and ramps should allow users to move safely between different levels of the building, and there should be barriers where necessary to protect people from falling. Vehicle ramps and floors and roofs with vehicular access should also be guarded with barriers where necessary to protect people in and around the building.

- *Conservation of fuel and power*

  Reasonable provision should be made for the conservation of fuel and power in buildings where the floor area exceeds 30 metres squared.

- *Access and facilities for disabled people*

  Reasonable provision should be made for disabled people to gain access to and use the building; and to use any sanitary conveniences provided. This applies to people who have an impairment which limits their ability to walk or requires them to use a wheelchair for mobility, or who have impaired hearing or sight. It excludes extensions which do not include a ground storey; material alterations and parts of a building used solely for the maintenance, inspection or repair of fittings or services.

- *Glazing*

  Glazing which people are likely to come into contact with must break on impact in a way unlikely to cause injury, or resist impact without breaking or be shielded or protected from impact. Transparent glazing with which people are likely to collide should be made apparent; and

- *Materials and workmanship*

  Any building work should be carried out with proper materials and in a workmanlike manner.

There are approved documents setting out detailed practical and technical guidance on how these requirements are to be met, although the requirements can be met in alternative ways if these are available.

The *Building Regulations 1991* (*SI 1991 No 2768*), *Reg 11* requires that anyone intending to carry out building work or make a material change of use must give a building notice to, or deposit full plans with the local authority. This does not have to be done where gas appliances are being installed by, or are under the supervision of people approved under the Gas Safety Regulations. Full plans are only required where buildings are to be put to a use designated under the *Fire Precautions Act 1971*.

An approved inspector can be chosen to ensure compliance with the Regulations rather than the local authority, in which case a building notice or deposit of full plans is not required. The list of approved inspectors is available from the Association of Corporate Approved Inspectors (see website: http://www.acai.org.uk).

The Regulations and approved documents can be purchased from the Stationary Office on 0870 600 5522.

Factories Act 1961

W11045 Although large parts of the *Factories Act 1961* have now been repealed and replaced by more recent legislation, some provisions are still in force, although the Health and Safety Commission (HSC) is considering complete repeal. Additionally, civil actions for injury, relating to breach of health and safety provisions of the *Factories Act* (though not welfare) may well continue for some time, since actions for personal injury can be initiated for up to three years after injury/disease has occurred. [*Limitation Act 1980, s 11*].

Residual application of the Factories Act 1961

W11046 The *Factories Act 1961* applies to factories, as defined in *s 175*, including 'factories belonging to or in the occupation of the Crown, to building operations and works of engineering construction undertaken by or on behalf of the Crown, and to employment by or under the Crown of persons in painting buildings', e.g. hospital painters. [*Factories Act 1961, s 173*(*1*)].

Enforcement of the Factories Act 1961 and regulations

W11047 Offences under the Act are normally committed by occupiers rather than owners of factories. Unless they happen to occupy a factory as well, the owners of a factory would not normally be charged. Offences therefore relate to physical occupation or control of a factory (for an extended meaning of 'occupier', see OCCUPIERS' LIABILITY). Hence the person or persons or body corporate having managerial responsibility in respect of a factory are those who commit an offence under *s 155*(*1*). This will generally be the managing director and board of directors and/or individual executive directors. Moreover, if a company is in liquidation and the receiver is in control, he is the person who will be prosecuted and this has in fact happened (*Meigh v Wickenden [1942] 2 KB 160; Lord Advocate v Aero Technologies 1991 SLT 134* where the receiver was 'in occupation' and so under a duty to prevent 'accidents by fire or explosion', for the purposes of the *Explosives Act 1875, s 23*).

Defence of factory occupier

W11048 The main defence open to a factory occupier charged with breach of the *Factories Act 1961* is that the Act itself, or more likely regulations made under it, placed the statutory duty on some person other than the occupier. Thus, where there is a contravention by any person of any regulation or order under the *Factories Act 1961*, 'that person shall be guilty of an offence and the occupier or owner ... shall not be guilty of an offence, by reason only of the contravention of the provision ... unless it is proved that he failed to take all reasonable steps to prevent the contravention ...'. [*Factories Act 1961, s 155*(*2*)].

Before this defence can be invoked by a factory occupier or company, it is necessary to show that:

(*a*) a statutory duty had been laid on someone other than the factory occupier by a regulation or order passed under the Act;

(*b*) the factory occupier took all reasonable steps to prevent the contravention (a difficult test to satisfy).

NB. This statutory defence is not open to a building contractor (in his capacity as a notional factory occupier).

Effect on possible civil liability

W11049 Whether conviction of an employee under the *Factories Act 1961*, *s 155*(*2*) would prejudice a subsequent claim for damages by him against a factory occupier, must be regarded as an open question. Thus, in *Potts v Reid [1942] 2 All ER 161* the court said 'Criminal and civil liability are two separate things ... The legislation (the *Factories Act 1937*) might well be unwilling to convict an owner who failed to carry out a statutory duty of a crime with which he was not himself directly concerned, but still be ready to leave the civil liability untouched'. Similarly in *Boyle v Kodak Ltd [1969] 2 All ER 439* it was said, 'When considering the civil liability engrafted by judicial decision upon the criminal liability which has been imposed by statute, it is no good looking to the statute and seeing from it where the criminal liability would lie, for we are concerned only with civil liability. We must look to the cases' (per Lord Diplock). Moreover, a breach of general duties of *HSWA 1974* gives rise only to civil liability at common law and not under statute. (Though this is not the position where there is a breach of a specific regulation under *HSWA 1974*.) On the other hand, there is at least one isolated instance of an employee being denied damages where he was in breach of specific regulations (*ICI Ltd v Shatwell [1964] 2 All ER 999*). It is thought, however, that this decision would not apply in the case of breach of a *general* statutory duty, such as *s 155*(*2*).

Appendix 1

Associations, organisations and departments connected with health and safety

Advisory, Conciliation and Arbitration Service (ACAS)
Brandon House
180 Borough High Street
London
SE1 1LW
Telephone: (020) 7210 3613

Association of British Insurers (ABI)
Head Office
51 Gresham Street
London
EC2V 7HQ
Telephone: (020) 7600 3333

Association of Industrial Truck Trainers
The Springboard Centre
Mantle Lane
Coalville
Leicestershire
LE67 3DW
Telephone: (01530) 277 857

Building Research Establishment
Bucknalls Lane
Garston
Watford
Hertfordshire
WD25 9XX
Telephone: (01923) 664 000

British Chiropractic Association (BCA)
Blagrave House
17 Blagrave Street
Reading
Berkshire
RG1 1QB
Telephone: (0118) 950 5950

British Industrial Truck Association (BITA)
5–7 High Street
Sunninghill
Ascot
Berkshire
SL5 9NQ
Telephone: (01344) 623 800

British Occupational Hygiene Society
Suite 2
Georgian House
Great Northern Road
Derby
DE1 1LT
Telephone: (01332) 298 101/087

British Safety Council (BSC)
National Safety Centre
70 Chancellors Road
London
W6 9RS
Telephone: (020) 8741 1231

British Standards Institution (BSI)
389 Chiswick High Road
London
W4 4AL
Telephone: (020) 8996 9000

Chartered Institute of Environmental Health (CIEH)
Chadwick Court
15 Hatfields
London
SE1 8DJ
Telephone: (020) 7928 6006

Construction Industry Training Board (CITB)
Bircham Newton
Kings Lynn
Norfolk
PE31 6RH
Telephone: (01485) 577 577
Telephone: (0845) 609 9960

Department for Transport
Enquiry Service
76 Marsham Street
London
SW1P 4DR
Telephone: (020) 7944 9622

Department of Health (DoH)
Public Enquiries
Richmond House
79 Whitehall
London
SW1A 2NS
Telephone: (020) 7210 4850

Department for Work and Pensions (DWP)
Correspondence Unit
Room 540
The Adelphi
1–11 John Adam Street
London
WC2N 6HT
Telephone: (020) 7712 2171

Department of Trade and Industry (DTI)
Public Enquiries
1 Victoria Street
London
SW1H 0ET
Telephone: (020) 7215 5000

Environment Agency
Apollo Court
2 Bishops Square Business Park
St Albans Road West
Hatfield
Hertfordshire
AL10 9EX
Telephone: (08708) 506 506

Fire Protection Association (FPA)
Bastille Court
2 Paris Garden
London
SE1 8ND
Telephone: (020) 7902 5300

Health and Safety Commission (HSC)
Rose Court
2 Southwark Bridge
London
SE1 9HS
Telephone: (020) 7717 6000

Health and Safety Executive (HSE)
Information Centre & Public Enquiry Point
Caerphilly Business Park
Caerphilly
CF83 3GG
Telephone: (08701) 545 500

Health and Safety Publications
HSE Books
PO Box 1999
Sudbury
Suffolk
CO10 2WA
Telephone: (01787) 881 165

Independent Training Standards Scheme and Register
Armstrong House
28 Broad Street
Wokingham
RG40 1AB
Telephone: (0118) 989 3229

Institution of Occupational Safety and Health (IOSH)
The Grange
Highfield Drive
Wigston
Leicestershire
LE18 1NN
Telephone: (0116) 257 3100

Lantra National Training
Lantra House
Stoneleigh Park
Coventry
Warwickshire
CV8 2LG
Telephone: (024) 7669 6996

Loss Prevention Council
(*see* Fire Protection Association above)

Occupational & Environmental Diseases Association (OEDA)
PO Box 26
Enfield
Middlesex
EN1 2NT

Qualifications and Curriculum Authority (QCA)
83 Piccadilly
London
W1J 8QA
Telephone: (020) 7509 5555

RTITB Ltd
Access House
Halesfield 17
Telford
TF7 4PW
Telephone: (01952) 520 200

Royal Society for the Prevention of Accidents (RoSPA)
Edgbaston Park
353 Bristol Road
Edgbaston
Birmingham
B5 7ST
Telephone: (0121) 248 2000

Skills for Logistics
14 Warren Yard
Warren Farm Office Village
Stratford Road
Milton Keynes
MK12 5NW
Telephone: (01908) 313 360

Storage Equipment Manufacturers' Association (SEMA)
6th Floor
MacLaren Buildings
35 Dale End
Birmingham
B4 7LN
Telephone: (0121) 200 2100

Appendix 2

Current HSE Publications

Guidance Notes

There are 6 principal series of guidance notes available. These are: Chemical Safety (CS); Environmental Hygiene (EH); General Series (GS); Medical Series (MS); Plant and Machinery (PM), and Legal Series (L).

The following list indicates publications currently in print available from HSE Books.

Chemical safety

CS3	Storage and use of sodium chlorate and other similar strong oxidants. Revised 1998
CS15	Cleaning and gas freeing of tanks containing flammable residues. 1985
CS21	Storage and handling of organic peroxides. 1991
CS22	Fumigation. 1996
CS23	Disposal of waste explosives. 1999
CS24	The interpretation and use of flashpoint information. 1999

Environmental hygiene

EH1	Cadmium: health and safety precautions. Revised 1995
EH2	Chromium and its inorganic compounds: health hazards and precautionary measures. 1998
EH10	Asbestos: exposure limits and measurement of airborne dust concentrations. Revised 2001
EH13	Beryllium: health and safety precautions. Revised 1995
EH16	Isocyanates: health hazards and precautionary measures. Revised 1999
EH17	Mercury and its inorganic divalent compounds. Revised 1996
EH19	Antimony and its compounds: health hazards and precautionary measures. Revised 1997
EH38	Ozone: health hazards and precautionary measures. 1996
EH40/02	Occupational exposure limits. 2001
EH40/03	Occupational exposure limits. 2003
EH40	Occupational exposure limits 2002/2003 Combined pack. 2003
EH43	Carbon monoxide: health hazards and precautionary measures. Revised 1998
EH44	Dust: general principles of protection. 1997

EH47	The provision, use and maintenance of hygiene facilities for work with asbestos insulation, asbestos coating and asbestos insulating board. Revised 2002
EH50	Training operatives and supervisors for work with asbestos insulation and coatings. 1988
EH51	Enclosures provided for work with asbestos. 1999
EH57	The problems of asbestos removal at high temperatures. 1990
EH59	Respirable crystalline silica. Revised 1997
EH60	Nickel and its inorganic compounds: health hazards and precautionary measures. Revised 1997
EH63	Vinyl chloride: toxic hazards and precautions (This should be read in conjunction with approved code of practice L5). 1992
EH64	Summary criteria for occupational exposure limits. Revised 2001
EH64supp	Summary criteria for occupational exposure limits. Revised 2002
EH65/1	Trimethylbenzenes – criteria document for an OEL. 1992
EH65/2	Pulverised fuel ash – criteria document for an OEL. 1992
EH65/3	N,N-Dimethylacetamide – criteria document for an OEL. 1992
EH65/4	1,2-dichloroethane – criteria document for an OEL. 1993
EH65/5	4.4-Methylene dianiline – criteria document for an OEL. 1993
EH65/6	Epichlorohydrin – criteria document for an OEL. 1993
EH65/7	Chlorodifluoromethane – criteria document for an OEL. 1994
EH65/8	Cumene – criteria document for an OEL. 1994
EH65/9	1,4-dichlorobenzene – criteria document for an OEL. 1994
EH65/10	Carbon tetrachloride – criteria document for an OEL. 1994
EH65/11	Chloroform – criteria document for an OEL. 1994
EH65/12	Portland cement dust – criteria document for an OEL. 1994
EH65/13	Kaolin – criteria document for an OEL. 1994
EH65/14	Paracetamol – criteria document for an OEL. 1994
EH65/15	1,1,1,2-Tetrafluoroethane HFC 134a – criteria document for an OEL. 1995
EH65/16	Methyl methacrylate – criteria document for an OEL. 1995
EH65/17	p-Aramid respirable fibres – criteria document for an OEL. 1995
EH65/18	Propranolol – criteria document for an OEL. 1995
EH65/19	Mercury and its inorganic divalent compounds – criteria document for an OEL. 1995
EH65/20	Ortho-toluidine – criteria document for an OEL. 1996
EH65/21	Propylene oxide – criteria document for an OEL. 1996
EH65/22	Softwood dust – criteria document for an OEL. 1996
EH65/23	Antimony and its compounds – criteria document for an OEL. 1996
EH65/24	Platinum metal and soluble platinum salts – criteria document for an OEL. 1996
EH65/25	Iodomethane – criteria document for an OEL. 1996
EH65/26	Azodicarbonamide – criteria document for an OEL. 1996

EH65/27	Dimethyl and diethyl sulphates – criteria document for an OEL. 1996
EH65/28	Hydrazine – criteria document for an OEL. 1996
EH65/29	Acid anhydrides – criteria document for an OEL. 1996
EH65/30	Review of fibre toxicology – criteria document for an OEL. 1996
EH65/31	Rosin-based solder flux fume – criteria document for an OEL. 1997
EH65/32	Glutaraldehyde – criteria document for an OEL. 1997
EH66	Grain dust. Revised 1998
EH67	Grain dust in maltings (maximum exposure limits). 1993
EH68	Cobalt: health and safety precautions. 1995
EH70	The control of fire-water run-off from CIMAH sites to prevent environmental damage. 1995
EH72/1	Phenylhydrazine – risk assessment document. 1997
EH72/2	Dimethylaminoethanol – risk assessment document. 1997
EH72/3	Bromoethane – risk assessment document. 1997
EH72/4	3-Chloropropene – risk assessment document. 1997
EH72/5	Chlorotoluene – risk assessment document. 1997
EH72/6	2-Furaldehyde – risk assessment document. 1997
EH72/7	1,2-Diaminoethane (Ethylenediamine (EDA)) – risk assessment document. 1997
EH72/8	Aniline – risk assessment document. 1998
EH72/9	Barium sulphate – risk assessment document. 1998
EH72/10	N-Methyl-2-Pyrrolidone – risk assessment document. 1998
EH72/11	Flour dust – risk assessment document. 1999
EH72/12	Bromochlromethane – risk assessment document. 2000
EH72/13	Methyl cyanoacrylate and ethyl cyanoacrylate – risk assessment document. 2000
EH72/14	Chlorine dixide. 2000
EH72/15	Vanadium and its organic compounds – risk assessment document 2002
EH72/16	Acetic anhydride – risk assessment document 2002
EH73	Arsenic and its compounds: Health hazards and precautionary measures. 1997
EH74/1	Exposure assessment: Dichloromethane. 1998
EH72/2	Respirable crystalline silica. 1999
EH74/3	Dermal exposure to non-agricultural pesticides. 1999
EH74/4	Metalworking fluids: Exposure assessment document. 2000
EH75/1	Medium density fibreboard: (MDF) – hazard assessment document. 1999
EH75/2	Occupational exposure limits for hyperbaric conditions – hazard assessment document. 2000
EH75/3	N-propyl bromide – hazard assessment document 2002
EH75/4	Respirable crystalline silica: phase 1 – hazard assessment document 2002

EH75/5	Respirable crystalline silca: phase 2 carcinogenity – hazard assessment document 2003
EH76	Control of laboratory animal allergy 2002

General series

GS4	Safety in pressure testing. Revised 1998
GS6	Avoidance of danger from overhead electric power lines. Revised 1997
GS6W	Avoidance of danger from overhead electric power lines – Welsh version. Revised 1999
GS28/3	Safe erection of structures – part 3 working places and access 1986
GS32	Health and safety in shoe repair premises. 1984
GS38	Electrical test equipment for use by electricians. Revised 1995
GS46	In-situ timber treatment using timber preservatives: health, safety and environmental precautions. 1989
GS49	Pre-stressed concrete. 1991
GS50	Electrical safety at places of entertainment. Revised 1997
GS51	Façade retention. 1992
GS53	Single-flue steel industrial chimneys: inspection and maintenance. 1997

Medical series

MS7	Colour vision. Revised 1987
MS12	Mercury: medical guidance notes. Revised 1996
MS13	Asbestos: medical guidance notes. Revised 1999
MS17	Medical aspects of work-related exposure to organophosphates. Revised 2000
MS24	Medical aspects of occupational skin disease. Revised 1998
MS25	Medical aspects of occupational asthma. Revised 1998
MS26	A guide to audiometric testing programmes. 1995

Plant and machinery

PM4	High temperature textile dyeing machines. Revised 1997
PM5	Automatically controlled steam and hot water boilers. Revised 1989
PM15	Safety in the use of pallets. Revised 1998
PM17	Pneumatic nailing and stapling guns. 1979
PM24	Safety at rack and pinion hoists. 1981
PM29	Electrical risks from steam/water pressure cleaners. Revised 1995
PM33	Reducing bandsaw accidents in the food industry. 2000
PM38	Selection and use of electric handlamps. 1992
PM39	Hydrogen cracking of grade T(8) chain and components. Revised 1998
PM48	Safe operation of passenger carrying amusement devices: the octopus. 1985

PM55	Safe working with overhead travelling cranes. 1985
PM56	Noise from pneumatic systems. 1985
PM57	Safe operation of passenger carrying amusement devices: the big wheel. 1986
PM59	Safe operation of passenger carrying amusement devices: the paratrooper. 1986
PM60	Steam boiler blowdown systems. Revised 1998
PM61	Safe operation of passenger carrying amusement devices: the chair-o-plane. 1986
PM65	Worker protection at crocodile (alligator) shears. 1986
PM66	Scrap baling machines. 1986
PM69	Safety in the use of freight containers. 1987
PM70	Safe operation of passenger carrying amusement devices: ark/speedways. 1988
PM71	Safe operation of passenger carrying amusement devices: water chutes. 1989
PM72	Safe operation of passenger carrying amusement devices: trabant. 1990
PM73	Safety at autoclaves. Revised 1998
PM74	Forced air filtration units for agricultural vehicles. 1991
PM75	Glass reinforced plastic vessels and tanks: advice to users. 1991
PM77	Fitness of equipment used for medical exposure to ionising radiation. Revised 1998
PM78	Passenger carrying aerial ropeways. 1994
PM81	Safe management of ammonia refrigeration systems: food and other workplaces. 1995
PM82	The selection, installation and maintenance of electrical equipment for use in and around buildings containing explosives. 1997
PM83	Drilling machines: guarding of spindles and attachments. 1998
PM84	Control of safety risks at gas turbines used for power generation. Revised 2003

Health and Safety: Guidance Booklets

The purpose of this series is to provide guidance for those who have duties under *HSWA* and other relevant legislation. It gives guidance on the practical application of regulations made under *HSWA*, but should not be regarded as an authoritative interpretation of the law.

HSG6	Safety in working with lift trucks. Revised 2000
HSG17	Safety in the use of abrasive wheels. Revised 2000
HSG28	Safety advice for bulk chlorine installations. Revised 1999
HSG31	Pie and tart machines. 1986.
HSG33	Health and safety in roof work. Revised 1998
HSG37	An introduction to local exhaust ventilation. Revised 1993
HSG38	Lighting at work. Revised 1998
HSG39	Compressed air safety. Revised 1998

HSG40	Safe handling of chlorine from drums and cylinders. Revised 1999
HSG42	Safety in the use of metal cutting guillotines and shears. 1988
HSG43	Industrial robot safety: your guide to the safeguarding of industrial robots. Revised 2000
HSG45	Safety in meat preparation: guidance for butchers. 1988
HSG47	Avoiding danger from underground services. Revised 2000
HSG48	Reducing error and influencing behaviour. Revised 1999
HSG51	The storage of flammable liquids in containers. Revised 1998
HSG53	The selection, use and maintenance of respiratory protective equipment. 1998
HSG54	Maintenance, examination and testing of local exhaust ventilation. Revised 1998
HSG57	Seating at work. Revised 1998
HSG60	Upper limb disorders in the workplace: a guide to prevention. Revised 2002
HSG61	Health surveillance at work. Revised 1999
HSG62	Health and safety in tyre and exhaust fitting premises. Revised 1991
HSG65	Successful health and safety management. Revised 1997
HSG66	Protection of workers and the general public during the development of contaminated land. 1991
HSG67	Health and safety in motor vehicle repair. 1991
HSG71	Chemical warehousing: the storage of packaged dangerous substances. Revised 1998
HSG72	Control of respirable silica dust in heavy clay and refractory processes. 1992
HSG73	Control of respirable crystalline silica in quarries. 1992
HSG78	Dangerous goods in cargo transport units: packing and carriage for transport by sea. Revised 1998
HSG79	Health and safety in golf course management and maintenance. 1994
HSG85	Electricity at work: safe working practices. Revised 2003
HSG87	Safety in the remote diagnosis of manufacturing plant and equipment. 1995
HSG88	Hand-arm vibration. 1994
HSG89	Safeguarding agricultural machinery: advice for designers, manufacturers, suppliers and users. Revised 1998
HSG90	The law on VDUs: an easy guide. Revised 2003
HSG92	Safe use and storage of cellular plastics. 1996
HSG93	The assessment of pressure vessels operating at low temperature. 1993
HSG94	Safety in the design and use of gamma and electron irradiation facilities. Revised 1998
HSG95	The radiation safety of lasers used for display purposes. 1996
HSG97	A step-by-step guide to COSHH assessment. Revised 2004
HSG100	Prevention of violence to staff in banks and building societies. 1993

HSG101	The costs to Britain of workplace accidents and work-related ill health in 1995/96. Revised 1999
HSG103	Safe handling of combustible dusts: precautions against explosions. Revised 2003
HSG107	Maintaining portable and transportable electrical equipment. 1994
HSG109	Control of noise in quarries. 1993
HSG110	Seven steps to successful substitution of hazardous substances. 1994
HSG110W	Seven steps to successful substitution of hazardous substances. (Welsh version) 1996
HSG112	Health and safety at motor sports events: a guide for employers and organisers. 1999
HSG113	Lift trucks in potentially flammable atmospheres. 1996
HSG114	Conditions for the authorisation of explosives in Great Britain. 1994
HSG115	Manual handling: solutions you can handle. 1994
HSG117	Making sense of NONS: a guide to the Notification of New Substances Regulations 1993. 1994
HSG118	Electrical safety in arc welding. 1994
HSG119	Manual handling for drinks delivery. 1994
HSG120	Nuclear site licences: under the Nuclear Installations Act 1965 (as amended) Notes for applicants. 1994
HSG121	A pain in your workplace? Ergonomic problems and solutions. 1994
HSG122	New and expectant mothers at work: a guide for employers. Revised 2002
HSG123	Working together on firework displays: a guide to safety for firework display organisers and operators. Revised 1999
HSG124	Giving your own firework display: how to run and fire it safely. 1995
HSG125	A brief guide on COSHH for the offshore oil and gas industry. 1994
HSG129	Health and safety in engineering workshops. Revised 1999
HSG131	Energetic and spontaneously combustible substances: identification and safe handling. 1995
HSG132	How to deal with sick building syndrome: guidance for employers, building owners and building managers. 1995
HSG133	Preventing violence to retail staff. 1995
HSG135	Storage and handling of industrial nitrocellulose. 1995
HSG136	Workplace transport safety: guidance for employers. 1995
HSG137	Health risk management: a practical guide for managers in small and medium-sized enterprises. 1995
HSG137W	Health risk management: a practical guide for managers in small and medium-sized enterprises. (Welsh version). 1996
HSG138	Sound solutions: techniques to reduce noise at work. 1995
HSG139	The safe use of compressed gases in welding, flame cutting and allied processes. 1997
HSG140	Safe use and handling of flammable liquids. 1996
HSG141	Electrical safety on construction sites. 1995
HSG142	Dealing with offshore emergencies. Revised 2003

HSG143	Designing and operating safe chemical reaction processes. 2000
HSG144	The safe use of vehicles on construction sites: a guide for clients, designers, contractors, managers and workers involved with construction transport. 1998
HSG146	Dispensing petrol: assessing and controlling the risk of fire and explosion at sites where petrol is stored and dispensed as a fuel. 1996
HSG149	Backs for the future: safe manual handling in construction. 2000
HSG150	Health and safety in construction. Revised 2001
HSG151	Protecting the public – your next move. 1997
HSG153/1	Railway safety principles and guidance: part 1. 1996
HSG153/2	Railway safety principles and guidance: part 2 section A. Guidance on the infrastructure. 1996
HSG153/4	Railway safety principles and guidance: part 2 section C. Guidance on electric traction systems. 1996
HSG153/5	Railway safety principles and guidance: part 2 section D. Guidance on signalling. 1996
HSG153/6	Railway safety principles and guidance: part 2 section E. Guidance on level crossings. 1996
HSG153/7	Railway safety principles and guidance: part 2 section F. Guidance on trains. 1996
HSG153/8	Railway safety principles and guidance: part 2 section G. Guidance on tramways. 1997
HSG154	Managing crowds safely: a guide for organisers at events and venues. Revised 2000
HSG155	Slips and trips: guidance for employers on identifying hazards and controlling risks. 1996
HSG156	Slips and trips: guidance for the food processing industry. 1996
HSG158	Flame arresters: preventing the spread of fires and explosions in equipment that contains flammable gases and vapours. 1997
HSG159	Managing contractors: a guide for employers. 1997
HSG165	Young people at work: a guide for employers. Revised 2000
HSG166	Formula for health and safety: guidance for small and medium-sized firms in the chemical manufacturing industry. 1997
HSG167	Biological monitoring in the workplace: a guide to its practical application to chemical exposure. 1997
HSG168	Fire safety in construction: guidance for clients, designers and those managing and carrying out construction work involving significant fire risks. 1997
HSG169	Camera operations on location: guidance for managers and camera crews. 1997
HSG170	Vibration solutions: practical ways to reduce hand-arm vibration injury. 1997
HSG171	Well handled: offshore manual handling solutions. 1997
HSG172	Health and safety in sawmilling: a run-of-the-mill-business? 1997
HSG173	Monitoring strategies for toxic substances. 1997
HSG174	Anthrax: safe working and the prevention of infection. 1997

HSG175	Fairgrounds and amusement parks: guidance on safe practice: practical guidance on the management of health and safety for those involved in the fairgrounds industry. 1997
HSG176	The storage of flammable liquids in tanks. 1998
HSG177	Managing health and safety in dock work. 2002
HSG178	The spraying of flammable liquids. 1998
HSG179	Managing health and safety in swimming pools. Revised 2003
HSG180	Application of electro-sensitive protective equipment using light curtains and light beam devices to machinery. 1999
HSG181	Assessment principles for offshore safety cases. 1998
HSG182	Sound solutions offshore: practical examples of noise reduction. 1998
HSG183	Five steps to risk assessment: case studies. 1998
HSG184	Guidance on the handling, storage and transport of airbags and seat pretensioners. 1998
HSG185	Health and safety in excavations: be safe and shore. 1999
HSG186	The bulk transfer of dangerous liquids and gases between ship and shore. 1999
HSG187	Control of diesel engine exhaust emissions in the workplace. 1999
HSG188	Health risk management: a guide to working with solvents. 1999
HSG189/1	Controlled asbestos stripping techniques for work requiring a licence. 1999
HSG189/2	Working with asbestos cement. 1999
HSG190	Preparing safety reports: Control of Major Accident Hazard Regulations 1999. 1999
HSG191	Emergency planning for major accidents: Control of Major Accident Hazards Regulations 1999. 1999
HSG192	Charity and voluntary workers: a guide to health and safety at work. 1999
HSG193	COSHH essentials: easy steps to control chemicals. Revised 2003
HSG194	Thermal comfort in the workplace: guidance for employers. 1999
HSG195	The event safety guide: a guide to health, safety and welfare at music and similar events. 1999
HSG 196	Moving food and drink: Manual handling solutions for the food and drink industries. 2000
HSG197	Railway safety principles and guidance : part 3 section A. Developing and maintaining staff confidence. 2002
HSG198	The transitional arrangements for the Biocidal Products Regulations: a guide for importers and suppliers of biocides. 2001
HSG199	Managing health and safety on work experience: a guide for organisers. 2000
HSG200	Go-karts: guidance on safe operation and use. 2000
HSG201	Controlling exposure to stonemasonry dust: guidance for employers. 2001
HSG202	General ventilation in the workplace: Guidance for employers. 2000

HSG203	Controlling exposure to coating powders. 2000
HSG204	Health and safety in arc welding. 2000
HSG205	Assessing and managing risks at work from skin exposure to chemical agents: Guidance for employers and health and safety specialists. 2001
HSG206	Cost and effectiveness of chemical protective gloves for the workplace: guidance for employers and health and safety specialists. 2001
HSG207	Choice of skin care products for the workplace: guidance for employers and health and safety specialists. 2001
HSG208	A guide to the Biocidal Products Regulations for importers and suppliers of biocides. 2001
HSG209	Aircraft turnaround: A guide for airport and aerodrome operators, airlines and service providers on achieving control, co-operation and co-ordination. 2000
HSG210	Asbestos essential tasks manual: task guidance sheets for the building maintenance and allied trades. 2001
HSG212	The training of first aid at work: a guide to gaining and maintaining HSE approval. 2000
HSG213	Introduction to asbestos essentials: comprehensive guidance on working with asbestos in the building maintenance and allied trades. 2001
HSG215	A guide to the Biocidal Products Regulations for users of biocidal products. 2001
HSG216	Passenger-carrying miniature railways: guidance on safe practice. 2001
HSG217	Involving employees in health and safety: forming partnerships in the chemical industry. 2001
HSG218	Tackling work-related stress: a managers' guide to improving and maintaining employee health and well-being. 2001
HSG220	Health and safety in care homes. 2001
HSG221	Technical guidance on the safe use of lifting equipment offshore. 2002
HSG222	Effective health and safety training: a trainers' resource pack. 2001
HSG223	The regulatory requirements for medical exposure to ionising radiation. 2001
HSG224	Managing health and safety in construction: Construction (Design and Management) Regulations 1994. Approved Code of Practice and Guidance. 2001
HSG225	Handling homecare: safe efficient and positive outcomes for care workers and clients. 2002
HSG227	A comprehensive guide to managing asbestos in premises. 2002
HSG228	CHIP for everyone. 2002
HSG229	Work-related violence: Case studies managing the risk in smaller businesses. 2002
HSG230	Keeping electrical switchgear safe. 2002
HSG231	Health and safety with metalworking fluids. 2002

HSG232	Sound solutions for the food and drink industries. 2002
HSG233	A bakers dozen: thirteen essentials for a safer bakery. 2003
HSG234	Caring for cleaners: guidance and case studies on how to prevent musculoskeletal disorders. 2003
HSG235	Bulk storage of acids: guidance on the storage of hydrochloric acid and nitric acid in tanks. 2003
HSG236	Power presses: maintenance and thorough examination. 2003
HSG238	Out of control: why control systems go wrong. 2003
HSG240	Managing health and safety at recreational dive sites. 2003
HSG242	Railway safety principles and guidance: Part 3 Section B – safe movements of trains. 2003
HSG233	Fuel cells: understand the hazards, control the risks. 2004

Legal – Approved Codes of Practice (COP series), HS(R) series and new L series

Code numbers in the HS(R) series and the COP series are gradually being superseded by the L series. Publications in the Lseries contain guidance on Regulations and Approved Codes of Practice.

Approved Codes of Practice

COP6	Plastic containers with normal capacities up to 5 litres for petroleum spirit: requirements for testing and marking or labelling (in support of SI 1982 No 830) – approved code of practice. 1982
COP15	Zoos: safety, health and welfare standards for employers and persons at work – approved code of practice and guidance notes. 1985
COP20	Standards of training in safe gas installation – approved code of practice. 1987
COP25	Safety in docks: Docks Regulations 1988 – approved code of practice with regulations and guidance. 1988
COP28	Safety of exit from mines underground workings: Mines (Safety of Exit) Regulations 1988 – approved code of practice. 1988
COP35	The use of electricity at quarries: Electricity at Work Regulations 1989 – approved code of practice. 1989

HS(R) series

HSR17	A guide to the Classification and Labelling of Explosives Regulations 1983. 1983
HSR25	Memorandum of guidance on the Electricity at Work Regulations 1989. 1989
HSR27	A guide to the Dangerous Substances in Harbour Areas Regulations 1987. 1988
HSR28	Guide to the Loading and Unloading of Fishing Vessels Regulations 1988. 1988
HSR29	Notification and marking of sites: The Dangerous Substances (Notification and Marking of Sites) Regulations 1990. 1990

Lseries

L5	General COSHH ACOP, Carcinogens ACOP and Biological Agents ACOP. Control of Substances Hazardous to Health Regulations 1999 – approved code of practice . Revised 2002
L8	Legionnaires disease: The control of legionella bacteria in water systems. Revised 2000
L9	Safe use of pesticides for non-agricultural purposes. Control of Substances Hazardous to Health Regulations 1994 – approved code of practice. Revised 1991
L10	A guide to the Control of Explosives Regulations 1991: guidance on Regulations. 1991
L11	A guide to the Asbestos (Licensing) Regulations 1983 as amended: guidance on Regulations. Revised 1999
L13	A guide to the Packaging of Explosives for Carriage Regulations 1991: guidance on Regulations. 1991
L21	Management of health and safety at work. Management of Health and Safety at Work Regulations 1999 – approved code of practice. Revised 2000
L22	Safe use of work equipment: Provision and Use of Work Equipment Regulations 1998 – approved code of practice and guidance. Revised 1998
L23	Manual handling: Manual Handling Operations Regulations 1992 (as amended) – guidance on regulations. Revised 2002
L24	Workplace health, safety and welfare. Workplace (Health, Safety and Welfare) Regulations 1992 – approved code of practice and guidance. 1992
L25	Personal protective equipment at work. Personal Protective Equipment at Work Regulations 1992 – guidance on regulations. 1992
L26	Work with display screen equipment work. Health and Safety (Display Screen Equipment) Regulations 1992 as amended – guidance on regulations. Revised 2002
L27	Work with asbestos which does not normally require a licence: Control of Asbestos at Work Regulations 2002 – approved code of practice. Revised 2002
L28	Work with asbestos insulation, asbestos coating and asbestos insulating board: Control of Asbestos at Work Regulations 2002 – approved code of practice. Revised 2002
L29	A guide to the Genetically Modified Organisms (Contained Use) Regulations 1992. Revised 2000
L30	A guide to the Offshore Installations (Safety Case) Regulations 1992. Revised 1998
L31	A guide to the Public Information for Radiation Emergencies Regulations 1992. 1993
L42	Shafts and winding in mines: Mines (Shafts and Winding) Regulations 1993 – approved code of practice. 1993
L43	First-aid at mines: Health and Safety (First-Aid) Regulations 1981 – approved code of practice. 1993

L44	The management and administration of safety and health at mines. Management and Administration of Safety and Health at Mines Regulations 1993 – approved code of practice. 1993
L45	Explosives at coal and other safety-lamp mines. Coal and Other Safety-lamp Mines (Explosives) Regulations 1993 – approved code of practice. 1993
L46	The prevention of inrushes in mines – approved code of practice. 1993
L47	The Coal Mines (Owners' Operating Rules) Regulations 1993 – guidance on Regulations. 1993
L50	Railways Safety Critical Work: Railways (Safety Critical Work) Regulations 1994 – approved code of practice and guidance. 1996
L52	Railways (Safety Case) Regulations 2000 including 2001 amendments– guidance on regulations. Revised 2003
L55	Preventing asthma at work: How to control respiratory sensitisers. 1994
L56	Safety in the installation and use of gas systems and appliances. The Gas Safety (Installation and Use) Regulations 1998 – approved code of practice and guidance. Revised 1998
L59	A guide to the approval of railway works, plant and equipment. 1994
L60	Control of substances hazardous to health in the production of pottery: The Control of Substances Hazardous to Health Regulations 1994. The Control of Lead at Work Regulations 1980 – approved code of practice. 1995
L64	Safety signs and signals: The Health and Safety (Safety Signs and Signals) Regulations 1996 – guidance on regulations. 1997
L65	Prevention of fire and explosion, and emergency response on offshore installations. Offshore Installations Regulations 1995 – approved code of practice and guidance. 1997
L66	A guide to the Placing on the Market and Supervision of Transfers of Explosives Regulations 1993 (POMSTER) 1993. 1995
L70	A guide to the Offshore Installations and Pipeline Works (Management and Administration) Regulations 1995 – guidance on regulations. Revised 2002
L71	Escape and rescue from mines. Escape and Rescue from Mines Regulations 1995 – approved code of practice. 1995
L72	A guide to Borehole Sites and Operations Regulations 1995 – guidance on regulations. 1995
L73	A guide to the Reporting of Injuries, Diseases and Dangerous Occurrences Regulations 1995. Revised 1999
L74	First aid at work: The Health and Safety (First-Aid) Regulations 1981 – approved code of practice and guidance. 1997
L75	Guidance for railways, tramways, trolley vehicle systems and other guided transport systems on Reporting of Injuries, Diseases and Dangerous Occurrences Regulations 1995 – guidance on regulations. 1996
L77	Guidance to the licensing authority on the Adventure Activities Licensing Regulations 1996. The Activity Centres (Young Persons' Safety) Act 1995 – guidance on regulations. 1996

L80	A guide to the Gas Safety (Management) Regulations 1996. 1996
L81	The design, construction and installation of gas service pipes: The Pipelines Safety Regulations 1996 – approved code of practice and guidance. 1996
L82	A guide to the Pipelines Safety Regulations 1996 – guidance on regulations. 1996
L84	A guide to the well aspects of amendments of the Offshore Installations and Wells (Design and Construction etc.) Regulations 1996 – guidance on regulations. 1996
L85	A guide to the integrity, workplace environment and miscellaneous aspects of the Offshore Installations and Wells (Design and Construction etc.) Regulations 1996 – guidance on regulations. 1996
L86	Control of Substances Hazardous to health in fumigation operations: Control of Substances Hazardous to Health Regulations 1994 – approved code of practice. 1996
L87	Safety representatives and safety committees (the brown book) – approved code of practice and guidance on the regulations. Revised 1996
L88	Approved requirements and test methods for the classification and packaging of dangerous goods for carriage: Carriage of Dangerous Goods (Classification, Packaging and Labelling) and Use of Transportable Pressure Receptacles Regulations 1996. 1996
L93	Approved tank requirements: the provisions for bottom loading and vapour recovery systems of mobile containers carrying petrol. Carriage of Dangerous Goods by Road Regulations 1996. Carriage of Dangerous Goods by Rail Regulations 1996. 1996
L94	Approved requirements for the packaging, labelling and carriage of radioactive material by rail: Packaging, Labelling and Carriage of Radioactive Material by Rail Regulations 1996. 1996
L95	A guide to the Health and Safety (Consultation with Employees) Regulations 1996 – guidance on regulations. 1996
L96	A guide to the Work in Compressed Air Regulations 1996. 1996
L97	A guide to the Level Crossings Regulations 1997. 1997
L98	Railway safety miscellaneous provisions: Railway Safety (Miscellaneous Provisions) Regulations 1997 – guidance on regulations. 1997
L101	Safe work in confined spaces: Confined Spaces Regulations 1997 – approved code of practice, regulations and guidance. 1997
L102	A guide to the Construction (Head Protection) Regulations 1989. Revised 1998
L103	Commercial diving projects offshore: Diving at Work Regulations 1997 – approved code of practice. 1998
L104	Commercial diving projects inland/inshore: Diving at Work Regulations 1997 – approved code of practice. 1998
L105	Recreational diving projects: Diving at Work Regulations 1997 – approved code of practice. 1998
L106	Media diving projects. Diving at Work Regulations 1997 – approved code of practice. 1998

L107	Scientific and archaeological diving projects: Diving at Work Regulations 1997 – approved code of practice. 1998
L108	Reducing noise at work: guidance on the Noise at Work Regulations 1989. 1998
L110	A guide to the Offshore Installations (Safety Representatives and Safety Committees) Regulations 1989. 1998
L111	A guide to the Control of Major Accident Hazard Regulations. 1999
L112	Safe use of power presses: Provision and Use of Work Equipment Regulations 1998 as applied to power presses – approved code of practice. 1998
L113	Safe use of lifting equipment: Lifting Operations and Lifting Equipment Regulations 1998 – approved code of practice and guidance. 1998
L114	Safe use of woodworking machinery. 1998
L116	Preventing accidents to children in agriculture – approved code of practice. 1999
L117	Rider-operated lift trucks: operator training – approved code of practice and guidance. 1999
L118	Health and safety at quarries: Quarries Regulations 1999 – approved code of practice. 1999
L119	The control of ground movement in mines – approved code of practice and guidance. 1999
L120	Train protection systems and mark 1 rolling stock: Railway Safety Regulations 1999 – guidance on regulations. 1999
L121	Work with ionising radiation: Ionising Radiations Regulations 1999 – approved code of practice and guidance. 2000
L122	Safety of pressure systems: Pressure Systems Safety Regulations 2000 – approved code of practice. 2000
L123	Health care and first aid on offshore installations and pipeline works: Offshore Installations and Pipeline Works (First-aid) Regulations 1989 – approved code of practice and guidance. 2000
L126	A guide to the Radiation (Emergency Preparedness and Public Information) Regulations 2001: guidance on regulations. 2002
L127	The management of asbestos in non-domestic premises: regulation 4 of the Control of Asbestos at Work Regulations 2002 – approved code of practice and guidance. 2002
L128	The use of electricity in mines: Electricity at Work Regulations 1989 – approved code of practice. 2001
L129	Approved Supply List: information for the classification and labelling of substances and preparations dangerous for supply Chemicals (Hazard Information and Packaging for Supply) Regulations 2002 approved list. Revised 2002
L130	The compilation of safety data sheets: Chemical (Hazard Information and Packaging for Supply) Regulations 2002 approved code of practice. Revised 2002
L131	Approved classification and labelling guide: Chemicals (Hazard Information and Packaging for Supply) Regulations 2002 guidance on regulations. Revised 2002

L132	Control of lead at work: Control of Lead at Work Regulations 2002 approved code of practice and guidance. 2002
L133	Unloading petrol from road tankers. 2003
L134	Design of plant, equipment and workplaces
L135	Storage of dangerous substances. 2003
L136	Control and mitigation measures. 2003
L137	Safe maintenance, repair and cleaning procedures. 2003
L138	Dangerous Substances and Explosive Atmospheres. 2003

Other Publications

Accidents and emergencies

L73	A guide to the Reporting of Injuries, Diseases and Dangerous Occurrences Regulations 1995. Revised 1999
L74	First aid at work. 1997
HSE31	RIDDOR explained (available in Welsh). 1999
HSE33	RIDDOR Offshore. Revised 1999
HSG212	The training of first aid at work: a guide to gaining and maintaining HSE approval. 2001
INDG214	First aid at work – your questions answered (available in Welsh). 1996
INDG347	Basic advice on first aid at work 2002

Asbestos

INDG188	Asbestos alert for building maintenance, repair and refurbishment workers. 1995
INDG223	A short guide to managing asbestos in premises. Revised 2002
INDG255	Asbestos dust kills. Revised 1999
INDG288	Selection of suitable respiratory protective equipment for work with asbestos. Revised 2003
INDG289	Working with asbestos in buildings. 1999
L27	Work with asbestos which does not normally require a licence: Control of Asbestos at Work Regulations 2002 – approved code of practice. Revised 2002
L28	Work with asbestos insulation, asbestos coating and asbestos insulating board: Control of Asbestos at Work Regulations 2002 – approved code of practice. Revised 2002
L127	The management of asbestos in non-domestic premises: regulation 4 of the Control of Asbestos at Work Regulations 2002 – approved code of practice and guidance. 2002

Chemical Industry

–	Handle with care – assessing musculoskeletal risk in the chemical industry. 2000
HSG71	Chemical warehousing – storage of packaged dangerous substances. Revised 1998

HSG143	Designing and operating safe chemical reaction processes. 2000
HSG159	Managing contractors – a guide for employers. 1997
CHIS2	Emergency isolation of process plant in the chemical industry. 1999
CHIS3	Major accident prevention policies for lower-tier COMAH establishments. 1999
CHIS4	Use of LPG in cylinders. 1999
CHIS5	Small-scale use of LPG in cylinders. 1999
CHIS6	Better alarm handling in the chemical and allied industries. 2000
INDG98	Permit to work systems. Revised 1997
INDG243	Computer control – a question of safety. 1997
INDG245	Biological monitoring for chemicals in the workplace. 1997
INDG246L	Prepared for emergency. 1997
INDG254	Chemical reaction hazards and the risk of thermal runaway. 1997
INDG352	Read the label: how to find out if chemicals are dangerous. Revised 2002

Construction

HSG33	Health and safety in roof work. Revised 1998
HSG47	Avoiding danger from underground services. Revised 2000
HSG144	Safe use of vehicles on construction sites. 1998
HSG149	Backs for the future: safe manual handling in construction. 2000
HSG185	Health and safety in excavations: be safe and shore. 1999
HSG201	Controlling exposure to stonemasonry dust. 2001
HSG224	Managing health and safety in construction Construction (Design and Management Regulations 1994 approved code of practice. 2001
L102	Construction (Head Protection) Regulations 1989 – guidance on regulations. 1998
SIR58	Safety of construction transport. 2001
SIR59	Issues surrounding the failure of an energy absorbing lanyard. 2001
CIS18	The provision of welfare facilities at fixed construction sites. Revised 1996
CIS24	Chemical cleaners. Revised 1998
CIS27	Solvents. Revised 1998
CIS36	Silica. Revised 1999
CIS37	Handling heavy building blocks. 1993
CIS39	Construction (Design and Management) Regulations 1994: the role of the client. 1995
CIS41	Construction (Design and Management) Regulations 1994: the role of the designer. Revised 1995
CIS42	Construction (Design and Management) Regulations 1994: the pre-tender stage health and safety plan. 1995
CIS43	Construction (Design and Management) Regulations 1994: the health and safety plan during the construction phase. 1995
CIS44	CDM Regulations 1994: The Health & Safety File. 1995

CIS45	Establishing exclusion zones when using explosives in demolition. 1995
CIS46	Provision of welfare facilities at transient construction sites. 1997
CIS47	Inspections and reports. 1997
CIS49	General access scaffolds and ladders. 1997
CIS50	Personal Protective Equipment (PPE): safety helmets. 1997
CIS51	Construction fire safety. 1997
CIS52	Construction Site Transport Safety: safe use of compact dumpers. 1999
CIS53	Crossing high-speed roads on foot during temporary traffic management works. 2000
CIS54	Dust control on concrete cutting saws used in the construction industry. 2000
INDG127	Noise in construction: further guidance on the Noise at Work Regulations 1989. Revised 1994
INDG212	Workplace health and safety – Glazing: guidance on glazing for employers and people in control of workplaces. 1996.
INDG220	A guide to the Construction (Health, Safety and Welfare) Regulations 1996. 1996
INDG242	In the driving seat: advice to employers on reducing back pain in drivers and machinery operators. 1997
INDG258	Safe work in confined spaces. (Available in Welsh) 1999
INDG332	Manual packing in the brick industry. 2000
INDG344	The absolutely essential health and safety toolkit for the smaller construction contractor. 2001
INDG367	Inspecting fall arrest equipment made from webbing or rope. 2002
INDG368	Use of contractors: A joint responsibility. 2002
INDG384	The high 5: Five ways to reduce risk on site. 2003
MISC112	Reducing the risk of hand-arm vibration injury among stone masons. 1998
MISC193	Having Construction Work Done? – duties of clients under the Construction (Design and Management) Regulations 1994. 1999

Dangerous substances – general

–	The technical basis for COSHH essentials: easy steps to control chemicals. 1999
HSG188	Health risk management: a guide to working with solvents. 1998
HSG205	Assessing and managing risks at work from skin exposure to chemical agents: guidance for employees and health and safety specialists. 2001
HSG228	CHIP for everyone. 2002
L5	Control of substances hazardous to health. Control of Substances Hazardous to Health Regulations 1999. Approved Code of Practice. Revised 2002

L129	Approved Supply List: information for the classification and labelling of substances and preparations dangerous for supply Chemicals (Hazard Information and Packaging for Supply) Regulations 2002 approved list. 2002
L130	The compilation of safety data sheets: Chemical (Hazard Information and Packaging for Supply) Regulations 2002 approved code of practice. 2002
L131	Approved classification and labelling guide: Chemicals (Hazard Information and Packaging for Supply) Regulations 2002 guidance on regulations. 2002
INDG136	COSHH: a brief guide to the Regulations. Revised 2003
INDG273	Working safely with solvents: a guide to safe working practices. 1998

Dangerous substances – by type

HSG28	Safety advice for bulk chlorine installations. Revised 1999
HSG40	Safe handling of chlorine from drums and cylinders. Revised 1999
HSG187	Control of diesel engine exhaust emissions in the workplace. 1999
HSG198	The transitional arrangements for the Biocidal Products Regulations. 2001
HSG203	Controlling exposure to coating powders. 2000
HSG215	A guide to the Biocidal Products Regulations for users of biocidal products. 2001
L132	The control of lead at work: approved code of practice. Revised 2002
INDG197	Working with sewage: the health hazards – a guide for employers and employees. 1995.
INDG198	Working with sewage: the health hazards – a guide for employers. 1995
INDG230	Storage and safe handling of ammonium nitrate. 1996
INDG257	Pesticides: use them safely. 1997
INDG276	Feral honey bees. 1998
INDG286	Diesel engine exhaust emissions. 1999
INDG305	Lead and you. Revised 1998
INDG307	Hydrofluoric acid poisoning: recommendations on first aid treatment. 1999
INDG309	Safe waters: using anti fouling paints safely a guide for boat owners. 2000
INDG319	Working safely with coating powders. 2000
INDG321	A simple guide to the Biocidal Products Regulations. 2001
INDG329	Benzene and you. 2000
INDG351	Nickel and you. 1997
MSA8	Arsenic and you: arsenic is poisonous – are you at risk? 1996
MSA17	Cobalt and you: working with cobalt – are you at risk. 1995
MSA19	PCBs and you – do you know how to work safely with PCBs? 1995
MSA21	MbOCA and you: do you use MbOCA? 1996

MSB4	Skin cancer by pitch and tar. 1996

Diving

DVIS1	General hazards. 1998
DVIS2	Diving system winches. 1998
DVIS3	Breathing gas management. 1998
DVIS4	Compression chambers. 1998
DVIS5	Exposure limits for air diving operations. 1998
DVIS6	Maintenance of diving bell hoists. 1998
DVIS7	Bell run and bell lock-out times in relation to habitats. 1998
DVIS8	Diving in benign conditions and in pools, tanks, aquariums and helicopter underwater escape training. 1999

Electricity and electrical systems

INDG68	Do you use a steam/water pressure cleaner? Revised 1997
INDG139	Electric storage batteries. 1993
INDG236	Maintaining portable electrical equipment in offices and other low-risk environments. 1996
INDG237	Maintaining portable electrical equipment in hotels and tourist accommodation. 1996
INDG247	Electrical safety for entertainers. 1997
INDG354	Safety in electrical testing at work: general guidance. 2002
HSG85	Electricity at work: safe working practices. Revised 2003

Engineering

EH74/4	Metalworking fluids. 2000
HSG17	Safety in the use of abrasive wheels. 2000
HSG129	Health and safety in engineering workshops. Revised 1999
HSG204	Health and safety in arc welding. 2000
HSG231	Working safely with metalworking fluids: good practice manual. 2002
MDHS95/2	Measurement of personal exposure of metalworking machine operators to airborne water-mix metalworking fluids. 2003
EIS1	Hot work on vehicle wheels. 1992
EIS2	Accidents at metalworking lathes using emery cloth. 1993
EIS3	Monitoring requirements in the electroplating industry including electrolytic chromium processes. Revised 1998
EIS4	Workplace welfare in the electroplating industry. Revised 1998
EIS5	Health surveillance requirements in the electroplating industry. Revised 1998
EIS6	Electrical systems in the electroplating industry. Revised 1998
EIS7	Safeguarding 3 roll bending machines. Revised 1998
EIS12	Safety at manually-fed pivoting-head metal-cutting circular saws. 1998

EIS13	Safeguarding of combination metalworking machines. Revised 2000
EIS15	Control of exposure to triglycidyl isocyanurate (TGIC) in coating powders. Revised 2003
EIS16	Preventing injuries from the manual handling of sharp edges in the engineering industry. 1997
EIS18	Isocyanates: health surveillance in motor vehicle repair. 1997
EIS19	Engineering machine tools: retrofitting CNC. 1997
EIS20	Maintenance and cleaning of solvent degreasing tanks. 1998
EIS21	Immersion and cold cleaning of engineering components. 1998
EIS26	Noise in engineering. 1998
EIS27	Control of noise at metal cutting saws. 1998
EIS28	Safeguarding at horizontal boring machines. 1998
EIS29	Control of noise at power presses. 1998
EIS30	Safety in the use of hand and foot operated presses. 1999
EIS31	Cadmium in silver soldering or brazing. 1999
EIS32	Chromate primer paints. 1999
EIS33	CNC turning machines: controlling risks from ejected parts. 2001
EIS34	Surface cleaning: solvent update including the re-classification of trichloroethylene. 2002
EIS35	Safety in electrical testing: Servicing and repair of audio, TV and computer equipment. 2002
EIS36	Safety in electrical testing: Servicing and repair of audio, TV and computer equipment. 2002
EIS37	Safety in electrical testing: Switchgear and control gear. 2002
EIS38	Safety in electrical testing: Products on production lines. 2002
EIS39	Reducing noise from CNC punch presses. 2002
EIS40	Safe use of solvent degreasing plant. 2003
INDG248	Solder fume and you. 2001
INDG280	A guide to handling and storage of air bags and seatbelt pretensioners at garages and motor vehicle repair shops. 1998
INDG297	Safety in gas welding, cutting and similar processes. 1999
INDG313	Safe unloading of steel stock. 2000
INDG314	Hot work on small tanks and drums. 2000
INDG327	Take care with acetylene. 2000
INDG349	Vehicle air-conditioning systems. 2002
INDG356	Reducing ill health and accidents in motor vehicle repair. 2002
INDG365	Working safely with metalworking fluids. 2002
WGIS1	Supply of welding consumables. 1999

Flammable and explosive substances

–	Fire Safety: an employers guide. 1999
CS23	Disposal of waste explosives. 1999
HSG51	The storage of flammable liquids in containers. Revised 1998

HSG71	Chemical warehousing. 1998
HSG103	Safe handling of combustible dusts. Revised 2003
HSG114	Conditions for the authorisation of explosives in Great Britain. 1994
HSG131	Energetic and spontaneously combustible substances: identification and safe handling. 1995
HSG135	Storage and handling of industrial nitrocellulose. 1995
HSG140	Safe use and handling of flammable liquids. 1996
HSG146	Dispensing petrol. 1996
HSG158	Flame arresters. 1997
HSG176	The storage of flammable liquids in tanks. 1998
HSG178	The spraying of flammable liquids. 1998
HSR17	A guide to the Classification and Labelling of Explosive Regulations 1983. 1983
L10	A guide to the Control of Explosives Regulations 1991. 1991
L13	A guide to the Packaging of Explosives for Carriage Regulations 1991. 1991
L66	A guide to the Placing on the Market and Supervision of Transfers of Explosives Regulations (POMSTER) 1993. 1995
L138	Dangerous Substances and Explosive Atmospheres: Dangerous Substances and Explosive Atmospheres Regulations – approved code of practice and guidance. 2003
HSE8	Taking care with oxygen. 1999
INDG216	Dispensing petrol as a fuel. 1996
INDG227	Safe working with flammable substances. 1996
INDG314	Hot work on small tanks and drums. 2000
INDG331	Safe use of petrol in garages. 2000
INDG335	Is it explosive? Dangers of explosives in metal recycling. 2002
INDG370	Fire and explosion: How safe is your workplace? 2002

Foundries

FNIS2	Foundry machine guarding: introductory sheet. 1995
FNIS3	Foundry machine guarding: mould and core-making machinery. 1995
FNIS4	Foundry machine guarding: sand handling equipment. 1995
FNIS5	Foundry machine guarding: shakeouts, sand mixer and shotblasts. 1995
FNIS6	Hazards associated with foundry processes: fettling. 1995
FNIS8	Hazards associated with foundry processes: hand-arm vibration – the current picture. 1996
FNIS9	Hazards associated with foundry processes: hand-arm vibration – symptoms and solutions.
FNIS10	Hazards associated with foundry processes: hand arm vibration – assessing the need for action. 1999
FNIS11	Hand-arm vibration in foundries. 2002

FNIS12	A purchasing policy for vibration reduced tools in foundries. 2002
IACL83	Hearing protection in foundries
IACL104	Health surveillance in foundries. 1998

Health issues

L55	Preventing asthma at work. 1994
HSG122	New and expectant mothers at work: a guide for employers. Revised 2002
INDG91	Drug misuse at work: a guide for employers. 1998
INDG240	Don't mix it! A guide for employers on alcohol at work. 1996
INDG281	Work-related stress. Revised 2001
INDG304	Understanding health surveillance at work: an introduction for employers. 1999
INDG341	Tackling work-related stress: a guide for employers. 2001

Health services

HSG225	Handling homecare. 2002
HSIS1	The Reporting of Injuries, Diseases and Dangerous Occurrences Regulations 1995. 1998
MISC186	Safe use of pneumatic air tube transport systems for pathology specimens. 1999
INDG320	Latex and you. 2000

Legislation and enforcement

HSE40	Employers Liability (Compulsory Insurance) Act 1969: a guide for employers. Revised 1998
HSE5	The Employment and Medical Advisory Service and you. 2000
HSC13	Health and safety regulation: a short guide (available in Welsh). 1995
HSC14	What to expect when a health and safety inspector calls. 1998
HSC15	HSC's enforcement policy statement. 2002
HSE39	Employers Liability (Compulsory Insurance) Act 1969: a guide for employees and their representatives. 2002
INDG184	Signpost to the Safety Sign and Signals Regulations 1996: guidance on the regulations. 1996

Major hazards

–	Fire Safety: an employers guide. Revised 1999
HSG190	Preparing safety reports: Control of Major Accident Hazards (COMAH) Regulations 1999. 1999
HSG238	Out of control: why control systems go wrong and how to prevent failure. 2003
HSG191	Emergency planning for major accidents: Control of Major Accident Hazards Regulations 1999. 1999
INDG196	Safety reports: how HSE assesses these in connection with the Control of Industrial Major Accident Hazards Regulations. 1995

L111	A guide to the Control of Major Accident Hazards (COMAH) Regulations 1999. 1999

Management of health and safety

INDG163	Five steps to risk assessment. Revised 1998
INDG179	Policy statement on open government. Revised 1998
INDG218	A guide to risk assessment requirements: common provisions in health and safety law. 1996
INDG232	Consulting employees on health and safety: a guide to the law. 1996
INDG259	An introduction to health and safety: health and safety in small firms. Revised 2003
INDG275	Managing health and safety: five steps to success. 1998
INDG301	Health and safety benchmarking. 1999
INDG322	Need help on health and safety? 2000
INDG324	Stating your business. 2000
INDG343	Directors' responsibilities for health and safety. 2001
INDG355	Reduced risks, cut costs. 2002
INDG381	Passport schemes for health, safety and the environment: a good practice guide. 2003
MISC225	Securing health together. 2000
MISC227	Good Neighbour Scheme. 2000

Mines

	Safe manriding in mines. 2001
	Guidance and information on escape from mines. 2001
COP28	Safety of exit of mines underground workings: Mines (Safety of Exit) Regulations 1988 – approved code of practice. 1988
L42	Shafts and winding in mines: Mines (Shafts and Winding) Regulations 1993 – approved code of practice. 1993
L43	First aid at mines: Health and Safety (First Aid) Regulations 1981 – approved code of practice. 1993
L44	The management and administration of safety and health at mines: Management and Administration of Safety and Health at Mines Regulations 1993 – approved code of practice. 1993
L45	Explosives at coal and other safety-lamp mines: Coal and other Safety-lamp Mines (Explosives) Regulations 1993 – approved code of practice. 1993
L46	The prevention of inrushes in mines – approved code of practice. 1993
L47	The Coal Mines (Owners Operating Rules) Regulations 1993: guidance on regulations. 1993
L71	Escape and Rescue from Mines: Escape and Rescue from Mines Regulations 1995 – approved code of practice. 1995
L119	The control of ground movement in mines – approved code of practice and guidance. 1999

L128	The use of electricity in mines. 2001
QIS1	Manager's guide to safe coal cleaning and the control of pedestrians at opencast coal sites. 1994
TOP7	Improving visibility on underground free steered vehicles. 1996.

Noise

L108	Reducing noise at work: guidance on the Noise at Work Regulations 1989. 1998
HSG362	Noise at work: advice for employers. 2002
INDG363	Protect your hearing or lose it! 2002

Nuclear

INDG206	Wear your dosemeter. Revised 1995
IRIS2	Radiation doses: assessment and recording. Revised 2000
MISC434	Safety audit of Dounreay 1998–Final Report 2001. 2001
L31	A guide to the Public Information for Radiation Emergencies Regulations 1992. 1993

Oil Industry

HSG221	Technical guidance on the safe use of lifting equipment offshore. 2002
L110	A guide to the Offshore Installations (Safety Representatives and Safety Committees) Regulations 1989. Revised 1998
L123	Healthcare and first aid on offshore installations and pipeline works. Offshore Installations and Pipeline Works (First Aid) Regulations 1989. 2000
INDG119	Safety representatives and safety committees on offshore installations. Revised 1999
INDG219	How offshore helicopter travel is regulated. 1996
INDG239	Play your part! How you can help improve health and safety offshore.1996
INDG250	How HSE assesses Offshore Safety Cases. 1997
INDG277	Health and safety leadership for offshore industry. 1997
INDG361	Regulating health and safety in the UK offshore oil and gas fields: who does what? 2002

Personal Protective Equipment (PPE)

HSG53	The selection, use and maintenance of respiratory protective equipment. 1998
HSG206	Cost and effectiveness of chemical protective gloves for the workplace. 2001
HSG207	Choice of skincare products for the workplace. 2001
INDG147	Keep your top on: health risks from working in the sun – advise for outdoor workers. 1998

INDG288	Selection of suitable respiratory protective equipment for work with asbestos. Revised 2003
INDG330	Selecting protective gloves for work with chemicals. 2000
INDG337	Sun protection. 2001

Quarries

HSG73	Control of respirable crystalline silica in quarries. 1992
HSG109	Control of noise in quarries. 1993
INDG303	Do you work in a quarry? 1999
L118	Health and safety at quarries. 1999

Vibration

INDG126	Health risks from hand-arm vibration: advice for employees and the self employed. Revised 1998
INDG175	Health risks from hand-arm vibration: advice for employers. Revised 1998
INDG242	In the driving seat: advice to employers on reducing back pain in drivers and machinery operators. 1997
INDG296P	Hand-arm vibration syndrome: pocket card for employees. 1999
INDG338	Power tools: How to reduce vibration health risks. 2001

Woodworking

INDG318	Manual handling solutions in woodworking. 2000
WIS1	Wood dust. Revised 1997
WIS2	Safe stacking of sawn timber and board materials. Revised 2000
WIS6	COSHH and the woodworking industries. Revised 1997
WIS7	Accidents at woodworking machines. Revised 1999
WIS13	Noise at woodworking machines. 1991
WIS15	Safe working at woodworking machines. 1992
WIS16	Circular saw benches. Revised 1999
WIS17	Safe use of hand-fed planing machines. Revised 2000
WIS18	Safe use of vertical spindle moulding machines. Revised 2001
WIS29	Occupational hygiene and health surveillance at individual timber treatment plants. Revised 2002
WIS32	Safe collection of wood waste. 1997
WIS33	Health surveillance and wood dust. 1997
WIS34	Health and safety priorities for the woodworking industry. 1997
WIS35	Safe use of power-operated cross-cut saws. 1998
WIS36	Safe use of manually operated cross-cut saws. 1998
WIS37	PUWER 98: selection of tooling for use with hand-fed woodworking machines. 1998
WIS38	PUWER 98: rectrofitting of braking to woodworking machines. 1998
WIS39	Safe use of single-end tenoning machines. 2000

Work Equipment

HSG43	Industrial robot safety. Revised 2000
L22	Safe use of work equipment: Provision and Use of Work Equipment Regulations 1998 – approved code of practice and guidance. Revised 1998
L113	Safe use of lifting equipment. 1998
L117	Rider operated-lift trucks. 1999
INDG290	Simple guide to LOLER. 1999
INDG291	Simple guide to PUWER. 1999
INDG317	Chainsaws at work. 2000
INDG339	Thorough examination and testing of lifts. 2001
MISC156	Hiring and leasing out of plant. 1998
MISC175	Retrofitting of roll over protective structures, restraining systems and their attachment points to mobile work equipment. 1999
MISC241	Fitting and use of restraining systems on lift trucks. 2000

Workplace

HSG165	Young people at work. Revised 2000
HSG194	Thermal comfort in the workplace. 1999
HSG199	Managing health and safety on work experience: guide for organisers. 2000
HSG202	General ventilation in the workplace. 2000
HSG218	Tackling work-related stress. 2001
L8	Legionnaires' disease. Revised 2000
L26	Work with display screen equipment: Health and Safety (Display Screen Equipment) Regulations 1992 as amended by the Health and Safety (Miscellaneous Amendments) Regulations 2002 Guidance on regulations. Revised 2002
IACL27	Legionnaires' disease. Revised 2000
INDG36	Working with VDUs. Revised 1998 (Also available in Welsh)
INDG90	Understanding ergonomics at work. 2003
INDG173	Officewise. (Also available in Welsh). 1994
INDG226	Homeworking: guidance for employers and employees on health and safety. 1996
INDG244	Workplace health, safety and welfare. 1997
INDG258	Safe work in confined spaces. 1999
INDG281	Work related stress. Revised 2001
INDG293	Welfare at work. 1999
INDG333	Back in work: managing back pain in the workplace. 2000
INDG341	Tackling work-related stress. 2002
OSR1	Office, shop and railway premises. (Also available in Welsh). 1994

Appendix 3

Relevant British Standards

Access equipment	BS 6037, BS 7985
Agricultural machinery:	
combine and forage harvesters	BS EN 632
silage cutters	BS EN 703
Airborne noise emission	
earth-moving machinery	BS 6812, BS ISO 6393, 6394
hydraulic fluid power systems and components	BS 5944
portable chain saws	BS 6916–6
Ambient air: determination of asbestos fibres – direct-transfer transmission electron microscopy method	BS ISO 13794
Anchorages	
industrial safety harnesses	BS EN 361, BS EN 795
self-locking, industrial	BS EN 353, 353–1, 353–2, 354, 355, 360, 362, 365
Arc welding equipment	BS 638
Artificial daylight lamps	
colour assessment	BS 950
for sensitometry	BS ISO 7589
Artificial lighting	BS ISO 8995
Barriers, in and about buildings	BS 6180
Bromochlorodifluoromethane	
fire extinguishing systems	BS 5306–5-2, BS EN 27201–1
fire extinguishers	BS EN 27201
Carbon steel welded horizontal cylindrical storage tanks	BS 2594
Carpet cleaners, electric, industrial use	BS EN 60335–2–67, BS EN 60335–2–72
Cellulose fibres	BS 1771–2
Chain lever hoists	BS 4898
Chain pulley blocks, hand-operated	BS 3243

Chain slings	BS EN 816–2, 816–3, 816–4, 816–5, 816–6
Chairs	
office furniture, performance	BS 5459–2
office furniture, ergonomic design	BS 3044
Chemical protective clothing	
against gases and vapours	BS EN 464
liquid chemicals	BS EN 466, 467
Circular saws	
hand-held electric	BS EN 50144–2–5
woodworking	BS 411
Cleaning and surface repair of buildings	BS 6270, 8221
Closed circuit escape breathing apparatus	BS EN 13794
Clothing for protection against heat and fire	BS EN ISO 6942, BS 367, 702
Concrete cladding	BS 8297
Construction equipment	
hoists	BS 7212
suspended safety chairs, cradles	BS 2830
Control of noise (construction and open sites)	BS 5228
Cranes, safe use	BS 5744, 7121
Disabled people, means of escape	BS 5588–8
Drill Rigs	BS EN 791
Dust	
high efficiency respirators	BS EN 136, 143
particulate emission	BS 3405
Ear protectors, acoustics	BS EN 24869, BS EN ISO 4869
Earphones	
audiometry, calibration, acoustic couplers	BS EN 60318–2, 60318–3
audiometry, calibration, artificial ears	BS 4669
Earthing	BS 7430
Earth moving equipment	BS 6912
Electrical equipment	
explosive atmospheres	BS 4683, 5501, 6941, BS EN 50014, 50015, 50017, 50021, 60079

fire hazard testing	BS 6458, BS EN 60695, BS IEC 60695
guidance to wiring regulations	BS 7671
Electrical resistance materials	
bare fine resistance wires	BS 1117
earth-leakage circuit-breakers	
— AC voltage operated	BS 842
— current-operated	BS 4293
— portable RCDs	BS 7071
electric shock protection, construction of electrical equipment	BS 2754
enclosures for high-voltage cable terminations, transformers and reactors	BS 6435
fans for general purposes	BS 848
industrial electric plugs	BS EN 60309
industrial machines	BS 2771
marking for low-voltage industrial switchgear/controlgear	BS 6272
metallic	BS 115
resistivity measurement	BS 5714
static electricity	BS 5958
switchgear	BS 5486
test for resistance per unit length	BS 3466
Emergency exits	BS EN 179, 1125
Emergency lighting	BS 5266, BS EN 1838
Environmental management systems	BS EN ISO 14001
Ergonomics of the thermal environment	BS ISO 10551, BS EN ISO 7726
Ergonomic requirements for office work with VDUs	BS EN 29241
Eye protection	
equipment for eye and face protection during welding operations	BS EN 175
glossary of terms	BS EN 165
specification for sunglare filters used in personal eye protectors for industrial use	BS EN 172
Eye protectors	BS EN 165–168, 170–172, 174, 175, 207, 208, 379, 1731, 1836, 1938
Fabrics, curtains and drapes	BS 5867

Falling-object protective structures	BS EN 13627, BS EN ISO 3164, BS ISO 6055
Falsework, code of practice	BS 5975
Filters	
specification for infra-red filters used in personal eye protection equipment	BS EN 171
specification for personal eye protection equipment in welding	BS EN 169, 379
specification for protection against industrial sunglare	BS EN 172
specification for protection against laser radiation	BS EN 207
specification for ultra-violet filters used in personal eye protection equipment	BS EN 170
Fire blankets	BS EN 1869
Fire classification	BS EN 2
Fire detection/alarm systems	BS 5839, 7807, BS EN 54, BS ISO 13387–7
Fire door assemblies	BS 8214
Fire extinguishers	
disposable aerosol type	BS 6165
media	BS 6535, BS EN 1568, 27201
on premises	BS 5306
portable	BS EN 3, BS 6643, 7867, 7937
Fire hose reels	BS 5306–1, BS EN 671–1, 671–3
Fire point determination, petroleum products	
Able closed up method	BS EN ISO 13736
Equilibrium method	BS s6664–4
Pensky-Martens apparatus method	BS 2000–35, BS EN 22719
Fire precautions in design/construction and use of buildings	BS 5588
Fire protection measures, code of practice for operation	BS 7273
Fire safety signs	BS 5499
Fire terms	BS 4422
Fire tests	BS 476
Fire tests for furniture	BS 5852
Firefighters gloves	BS EN 659
First-aid reel/hoses	BS 3169
Flameproof industrial clothing	BS EN 469, 1486
Footwear	

footwear fitted with toe caps	BS EN 345, 346, 12568
methods of test for safety	BS EN 344–1, 345
occupational footwear for professional use	BS EN 347
protective clothing for users of hand-held chain saws	BS EN 381
requirements/test methods for safety protective and occupational footwear for professional use	BS EN 344
Freight containers	BS 3951, BS 1SO 1496
Gaiters and footwear for protection against burns and impact risks in foundries	BS 4676
Gas detector tubes	BS 5343
Gas-fired hot water boilers	BS 5871, 5978, 5986, 6644, 6798, BS EN 297, 677
Gas welding equipment	BS 1453, 1821, 2640, BS EN 731, 962, BS EN ISO 9692–3
Glazing	BS 6262
Gloves: medical gloves for single use	BS EN 455
Gloves: rubber gloves for electrical purposes	BS 697
Goggles, industrial/non-industrial use	BS EN 166, 167, 168
Grinding machines	
hand-held electric	BS 2769–2–3
pneumatic, portable	BS 4390
spindle noses	BS 1089
Head protection — fire fighters	BS EN 443
Headforms for use in testing protective helmets	BS EN 960
Hearing protectors	BS EN 352, 458, 24869, BS EN ISO 4869
High visibility warning clothing	BS EN 471, 1150
Hoists	
construction, safe use	BS 7212
electric, building sites	BS 4465
Hose reels with semi-rigid hose	BS EN 671–1
Hose systems with lay-flat hose cloth	BS EN 671–2
Hot environments	
estimation of heat stress on the working man	BS EN 27243
Household and similar electrical appliances	BS EN 60335

Subject	Standard
Industrial gloves	BS 697, BS EN 374, 381–7, 388, 407, 420, 60903
Industrial safety helmets, firemen's	BS EN 443
Industrial trucks	
hand-operated stillage trucks, dimensions	BS 4337
pallet trucks, dimensions	BS ISO 509
pedals, construction/layout	BS 7178
Insulating material	BS 2844, 5691, 7737, 7822, 7831, BS EN 60383–2,
Ionising radiation	
recommendation for data on shielding from ionizing radiation	BS 4094
units of measurement	BS 5775
Jib cranes	
high pedestal and portal	BS 2452
power-driven	BS 357, 1757
Ladders	
code of practice	BS 5395
permanent for chimneys, high structures	BS 4211
portable	BS 1129, 2037, BS EN 131, 1147
Life jackets	BS EN 394, 395, 396, 399
Lifting chains	BS EN 818, 5594, 6521, 7255, BS ISO 1834
Lifting slings	BS 6166
chain, welded	BS EN 818–2, 818–4
textile	BS 6668, BS EN 1492
wire rope	BS 1290, 6210
Lighting systems: automatic change-over contractors for emergency lighting	BS 764
Machine guards	
chain saws	BS ISO 6533, 6916–4
conveyors and elevators	BS 5667–18, 5667–19
earth-moving equipment	BS EN ISO 3457
Machine tools	
emergency stop equipment, functional aspects	BS EN 418
Machinery, safety of	

drafting and presentation	BS EN 414
ergonomic design principles	BS EN 614–1
hazardous substances emitted by machines	BS EN 626
indication, marking and actuation, requirements for visual, auditory and tactile signals	BS EN 61310–1
indication, marking and actuation, requirements for marking	BS EN 61310–2
minimum gaps to avoid crushing parts of body	BS EN 349
principles and specifications for machinery manufacturers	BS EN 626–1
Machines, vibration	BS 4675, 7854, BS EN 12786
Materials handling	
conveyor belts	BS 5767
freight containers	BS 3951
Mobile cranes	BS 1757
Mobile road construction machinery	BS EN 500
Mortising machines, single chain	BS 4361–16
Natural fibre ropes	BS EN 698, 701, 1261
cords, lines, twines	BS 6125
Nets, safety	BS EN 1263
Noise	
airborne, chain saws	BS 6916–6
airborne, earth-moving equipment	BS 6812, BS ISO 6393, 6394
airborne, hydraulic transmission systems	BS 5944
effects on hearing handicap	BS 5330
industrial noise, method for rating	BS 4142
sound exposure meters	BS EN 61252
Noise induced hearing loss	
effects of noise exposure	BS 5330
pure tone air conduction threshold audiometry	BS 6655
Occupational safety and health management systems	BS 8800
Office buildings, fire precautions	BS 5588–11
Office furniture, ergonomic design and selection	BS 3044
Office machines	
electrically energised, safety	BS EN 60950
glossary	BS ISO/IEC 2382–27
keyboard layout	BS 5448
noise measurement	BS 7135
Oil burning equipment	BS 799

Oil firing code of practice	BS 5410
Open bar gratings – specification	BS 4592–1
Overhead travelling cranes	
power-driven	BS 466
safe use	BS 5744
Packaging	
pictorial marking for handling of goods	BS EN ISO 780
Particulate air pollutants	
in effluent gases, measurement	BS 3405
Passenger hoists	
electric, building sites	BS 4465
vehicular	BS 6109–2
Patent glazing	BS 5516
Pedestrian guardrails (metal)	BS 7818
Performance of windows	BS 6375
Personal eye protection	
filters for welding and related techniques	BS EN 169
infrared filters	BS EN 171
non-optical test methods	BS EN 168
optical test methods	BS EN 167
specifications	BS EN 166
ultraviolet filters	BS EN 170
vocabulary	BS EN 165
Pipelines, identification marking	BS 1710
Pneumatic tools	
portable grinding machines	BS 4390
Portable fire extinguishers	BS 7863, 6643, 7867, 7937, BS EN 3
Portable tools	
pneumatic grinding machines	BS 4390
Powder fire extinguishers	
disposable, aerosol type	BS 6165
extinguishing powders for	BS EN 615
fixed fire fighting systems	BS EN 12416
on premises	BS 5306
portable, recharging	BS 6643
Power take-off	
agricultural tractors, front-mounted	BS 6818
agricultural tractors hydraulic equipment	BS 4742, 4742–8
agricultural tractors, rear-mounted	BS 5861
Powered industrial trucks	

driverless, safety	BS EN 1525
safety signs and hazard pictorials	BS ISO 15870
Process control – safety of analyser houses	BS EN 61285
Protective barriers	BS 2655–10, 6180, 6779, 7818, BS EN 1317, 1930
Protective cabs	
agricultural tractors	BS 5947
controls for external equipment	BS 5731
sound insulation performance	BS EN ISO 11957
Protective clothing	
against cold weather	DD ENV 342
against foul weather	DD ENV 343
against heat and fire	BS EN ISO 6942
against heat and flame	BS EN 702, BS ISO 15025
against molten metal splash	BS EN 373
against risk of being caught up in moving parts	BS EN 510
eye, face and neck protection, welding	BS EN 175
flameproof	BS EN 469, 531
for firefighters	BS EN 469, 1486
for industrial workers exposed to heat	BS EN 531
for use where there is risk of entanglement	BS EN 510
for users of hand-held chain saws	BS EN 381
for welders	BS EN 470
for workers exposed to heat	BS EN 531
gaiters for foundries	BS 4676
general requirements	BS EN 340
mechanical properties	BS EN 863
protection against intense heat	BS EN 367, BS EN ISO 6942
protection against liquid chemicals	BS EN 466, BS EN ISO 6529
welding	BS EN 175, 470
Protective equipment	
against falls from a height	BS EN 341
against falls from a height – guided fall type arresters	BS EN 353
Protective footwear	
vulcanised rubber	BS 7193
firemen's leather boots	BS 2723
for foundries	BS 4676

lined industrialised rubber boots	BS 7193
professional use	BS EN 346
Protective gloves	
against chemicals and micro-organisms	BS EN 374
against cold	BS EN 511
against ionising radiation	BS EN 421
against mechanical risks	BS EN 388
against thermal hazards	BS EN 407
for users of hand-held chain saws	BS EN 381–7
general requirements	BS EN 420
mechanical test methods	BS EN 388
vibration transmissibility	BS EN ISO 10819
Protective helmets	BS EN 397
Quality control	BS 5750, 7679, BS EN 295–2, 450, 12306, BS EN ISO 9000–9004
Radiation measures	
detectors, nuclear reactors	BS 5548
electroscope, exposure meters	BS 3385
film badges	BS 3664
neutron detectors	BS 5552
personal photographic dosemeters	BS 6090
portable equipment	BS 7515, 7516
test tube for	BS 5542
Radiation protection,	
area radiation monitors, X-ray and gamma radiation	BS IEC 60532
Refrigeration systems	BS 1586, 1608, 2626, 7005, BS EN 378, 1861
Resistance to ignition of upholstered furniture for non domestic seating	BS 7176, 13136, 13215
Respirators	
full masks for respiratory protective devices	BS EN 136
Respiratory protective devices	BS EN 138, 148, 269, 270, 271
Roll-over protective structures	
agricultural tractors	BS 5947
earth moving equipment	BS EN 13510, BS EN ISO 6683
industrial trucks, stacking with masted tilt forward	BS 5778
pallet stackers/high lift platform trucks	BS 5777

reach and straddle fork trucks	BS 4436
Rope pulley blocks	
gin blocks	BS 1692
synthetic fibre	BS 4344
wire	BS 4018, 4536
Rope slings	
fibre rope slings	BS 6668–1
wire rope slings	BS 1290, 6210
Rubber/plastics injection moulding machines	BS EN 201
Safety anchorages	
industrial safety harnesses	BS 6858, BS EN 795
Safety distances to prevent danger zones being reached by upper limbs	BS EN 294, 349
Safety harnesses, industrial	BS 6858, BS EN 354, 355, 358, 361–365
Safety helmets	BS EN 397, 403, 12941
Sampling methods	
airborne radioactive materials	BS 5243
particulate emissions	BS 3405
Scaffolds, code of practice	BS 5973, 5974
Scalp protectors	BS EN 812
Shaft construction and descent	BS 8008
Specification for artificial daylight for colour assessment	BS 950
Staging, portable	BS 1129, 2037
Stairs, ladders, walkways	BS 5395
Steam boilers	
electric boilers	BS 1894
oil fuelled boilers, code of practice	BS 5410–2
safety valves for	BS 6759–1
welded steel low pressure boilers	BS 855
Step ladders	
portable	BS 1129, 2037
safety of access to machinery	BS EN ISO 14122–3
Storage tanks	
carbon steel welded horizontal cylindrical	BS 2594
vertical steel welded non-refrigerated butt-welded shells	BS 2654
welded	BS 7122, BS EN 12493, 12573
Suspended access equipment	BS 6037, BS EN 1808

Suspended safety chairs	BS 2830
Suspended scaffolds, temporarily installed	BS 5974
Tables, office furniture	BS 3044, 5459–1, BS EN 527–1
Textile floor coverings	BS 3655, 5287, 5325, 5808, 5921
Textile machinery, safety requirements	BS EN ISO 11111
Transportable gas containers	
acetylene	BS EN 12863, 1800, 1801, 12755
periodic inspection, testing and maintenance	BS 5430
welded steel tanks for road transport of liquefiable gases	BS 7122
Travelling cranes	BS 357, 3037, 5744
Vertical steel welded non-refrigerated storage tanks, manufacture of	BS 2654
Vibration measurement	
chain saws	BS 6916–8
rotating shafts	BS ISO 7919, 10817
Water absorption and translucency of china or porcelain	BS 5416
Water services, installation, testing and maintenance	BS 6700
Welders, protective clothing	BS EN 470
Window cleaning	BS 8213–1
Windows, performance of	BS 6375
Woodworking noise	BS 7140
Wool and wool blends	BS 1771–1
Working platforms	
mobile, elevating	BS EN 280
permanent, suspended access	BS 6037
Workplace atmospheres	
performance of procedures for measurement of chemical agents	BS EN 482
size definitions for measurement of airborne particulates	BS EN 481
Workwear and career wear	BS 5426

Appendix 4

A Summary of Risk Assessment Requirements

Control of Asbestos at Work Regulations 2002

On whom duties placed

1. Employers.
2. Self-employed persons.

Identify of those assessed and geographical location

1. Employers.
2. Any other person, whether at work or not, who may be affected by work activity carried out by employer/self-employed person.

Hazards or risks to be assessed

Risks created by exposure of employees and other relevant persons to asbestos.

Purpose of assessment

(a) Identify type of asbestos.

(b) Determine degree and nature of exposure which may occur.

(c) Consider the effect of control measures.

(d) Set out steps to be taken to prevent that exposure or reduce it to lowest level reasonably practicable.

(e) Consider results of relevant medical surveillance.

Qualification of duty

Suitable and sufficient. In so far as the duty extends to any other person, it is also subject to the 'so far as is reasonably practicable' qualification.

When the assessment has to be made

Before work is commenced.

Record provision

Employer/self-employed person must record the significant findings of the risk assessment as soon as is practicable after it is made.

Review provision

Risk assessment to be reviewed 'regularly and forthwith' if:

(a) reason to suspect existing risk assessment is no longer valid;

(b) significant change in the work to which the risk assessment relates; or

(c) results of any monitoring carried out in accordance with Regulations show it to be necessary.

Action on review

Where, as a result of the risk assessment review, changes to it are required, such changes shall be made.

Control of Asbestos at Work Regulations 2002 (the new duty, under Reg 4, to manage the risk from asbestos in non-domestic premises, which came into force on 21 May 2004)

On whom duties placed

1. Persons having, by reason of contract or tenancy, a maintenance/repair obligation (to any extent) for non-domestic premises.

2. Persons who, where there is no such contract or tenancy, have (to any extent) control of premises or part thereof.

Identity of those assessed and geographical location

Non-domestic premises.

Hazards or risks to be assessed

Risks from asbestos.

Purpose of assessment

(a) To determine whether asbestos is, or is likely to be, present in the premises.

(b) To consider condition of any asbestos which is, or has been assumed to be, present in the premises.

Qualification of duty

Suitable and sufficient.

When assessment to be made

By 21 May 2004.

Record provision

Duty-holder must record conclusions of the risk assessment and every review.

Review provision

Risk assessment to be reviewed forthwith if:

(a) reason to suspect assessment is no longer valid; or

(b) significant change in the premises to which assessment relates.

Action on review

Where review reveals changes to risk assessment necessary, such changes shall be made.

Control of Lead at Work Regulations 2002

On whom duties placed

1. Employers.
2. Self-employed persons.

Identity of those assessed and geographical location

1. Employees
2. Any other person, whether at work or not, who may be affected by the work carried on by employer or self-employed person.

Note: Young persons and women of 'reproductive capacity' cannot be employed in any of the lead-working activities set out in *Schedule 1* of the Regulations.

Hazards or risks to be assessed

The risks to health created by working with lead.

Purpose of assessment

To allow employer/self-employed person to make a valid decision about whether the work concerned is likely to result in the exposure of any employees and other persons not employees to lead being 'significant', and to identify the measures needed to prevent or adequately control exposure.

Qualification of duty

Suitable and sufficient. In so far as the duty extends to persons not employees it is also subject to the 'so far as is reasonably practicable' qualification.

When assessment has to be made

Before work is commenced.

Record provision

Employer/self-employed person employing 5 or more employees must record:

(a) significant findings of the risk assessment as soon as is practicable after the assessment is made; and

(b) the steps taken to prevent or control exposure to lead; and

(c) the significant findings of any risk assessment review.

Review provision

The record of the assessment should be a living document, which must be revisited to ensure that it is kept up-to-date. The employer/self-employed person should make arrangements to ensure assessment is reviewed regularly. The date of the first review and the length of time between successive reviews will depend on the type of risk, the work and the judgment of the employer/self-employed person on the likelihood of changes occurring. In particular, the risk assessment should be reviewed when there is:

(a) evidence to suggest that it is no longer valid;

(b) new information on health risks;

(c) significant change in the circumstances of work.

Action on review

Implementation of any necessary changes identified by review.

Control of Substances Hazardous to Health Regulations 2002 (as amended most recently by SI 2003 No 278)

On whom duties placed

1. Employers.
2. Self-employed persons.

Identify of those assessed and geographical location

1. Employers liable to be exposed to substances hazardous to health by work.
2. Any other person, whether at work or not, who may be affected by work carried out by employer or self-employed person.

Hazards or risks to be assessed

Those likely to arise from exposure to substances hazardous to health.

Purpose of assessment

To enable employer/self-employed person to make a valid decision about the measures necessary to prevent or adequately control the exposure of their employees/other relevant persons, as above, to substances hazardous to health arising from work.

Qualification of duty

Suitable and sufficient. In so far as the duty extends to any other person, it is also subject to the 'so far as is reasonably practicable' qualification.

When the assessment has to be made

Before work is commenced.

Record provision

Where employer/self-employed person employs 5 or more employees, obligation to record:

(a) the significant findings of the risk assessment as soon as practicable after it is made; and

(b) the steps taken to prevent or control exposure to substances hazardous to health.

Review provision

Risk assessment to be reviewed 'regularly and forthwith' if:

(a) reason to suspect that it is no longer valid;

(b) there has been a significant change in the work to which the assessment relates; or

(c) results of any monitoring carried out in accordance with Regulations show it to be necessary.

Action on review

Where, as a result of the risk assessment review, changes to it are required, such changes shall be made.

Dangerous Substances and Explosive Atmospheres Regulations 2002

On whom duties placed

1. Employers.
2. Self-employed persons.

Identity of those assessed and geographical location

1. Employees.
2. Any other person, whether at work or not, who may be affected by the work of the employer.

Purpose of assessment

(a) The elimination or reduction of risks to employees and others from dangerous substances present or liable to be present in the workplace.

(b) The classification of places at the workplace where explosive atmospheres may occur into hazardous or non-hazardous places so that the appropriate steps be taken in accordance with the regulatory requirements.

(c) The putting in place of appropriate arrangements to deal with accidents, incidents or emergencies related to the presence of a dangerous substance at the workplace.

Qualification of duty

Suitable and sufficient.

When assessment has to be made

When a dangerous substance is or is liable to be present at the workplace.

Record provision

The 'significant' findings of the risk assessment, including the matters specified in the Regulations.

Review provision

Review assessment regularly and if:

(a) there is reason to suspect it is no longer valid; or

(b) there has been a significant change in the matters to which the assessment relates.

Action on review

Changes to assessment to be made where required.

Note: The record-keeping requirements only apply to those employing five or more employees.

Genetically Modified Organisms (Contained Use) Regulations 2000 (as amended by SI 2002 No 63)

On whom duties placed

Any person undertaking any activity involving genetic modification of micro-organisms and organisms other than micro-organisms.

Identity of those assessed and geographical location

Persons and the environment put at risk by the above activity.

Hazards or risks to be assessed

As above i.e. the risks created by the activity to human health and the environment.

Purpose of assessment

To identify those risks.

Qualification of duty

Suitable and Sufficient.

When assessment has to be made

Prior to undertaking any such activity.

Record provision

(a) A record of the assessment relating to the activity, and any review of that assessment.

(b) At least 10 years from the date of cessation of the activity.

Review provision

Review assessment if:

(a) reason to suspect it is no longer valid,

(b) there has been a significant change in the activity to which assessment relates.

Action on review

Risk assessment to be modified, as necessary.

Health and Safety (Display Screen Equipment) Regulations 1992 (as amended by SI 2002 No 2174)

On whom duties placed

Employers.

Identity of those assessed and geographical location

1. Users (employees who habitually use display screen equipment as a significant part of their normal work).

2. Operators (self-employed persons who habitually use display screen equipment as a significant part of their normal work).

Hazards or risks to be assessed

Workstations.

Purpose of assessment

To assess workstations for health and safety risks to which users/operators are exposed.

Qualification of duty

Suitable and sufficient.

Review provision

Review assessment if:

(a) reason to suspect it is no longer valid

(b) there has been a significant change.

Action on review

Changes to assessment to be made where required.

Details

Employers shall reduce risks identified by assessment to lowest extent reasonably practicable.

Ionising Radiations Regulations 1999

On whom duties placed

1. Employers.
2. Self-employed persons.
3. Mine managers – in so far as the duties relate to the mine or part thereof which he is manager/operator and to matters within his control.
4. Quarry operators – in so far as the duties relate to the quarry or part thereof which he is manager/operator and to matters within his control.
5. The holder of a nuclear site licence – in so far as the duties relate to the licensed site.

Identity of those assessed and geographical location

1. Employees.
2. Persons other than an employer's employees, only in so far as the exposure of these persons to ionising radiation arises from work with ionising radiation undertaken by that employer.
3. Self-employed persons.
4. Trainees, as defined.

Hazards or risks to be assessed

Any new activity involving work with ionising radiation – the 'Prior risk assessment'.

Purpose of assessment

To identify the measures needed to be taken to restrict the exposure of employees or other persons to ionising radiation.

Qualification of duty

Suitable and Sufficient.

When assessment has to be made

Before a duty holder commences a new activity involving working with ionising radiation.

Record provision

(a) No stipulation in the Regulations but ACoP and guidance says that it 'makes sense' to record the significant findings of the prior risk assessment where employers have five or more employees and/or where any group of employees is especially at risk.

(b) The significant findings.

Review provision

Review prior risk assessment if:

(a) reason to suspect it is no longer valid (or can be improved),

(b) there has been a significant change.

Action on review

Prior risk assessment to be modified, as necessary.

Management of Health and Safety at Work Regulations 1999 (as amended most recently by SI 2003 No 2457)

On whom duties placed

1. Employers.
2. Self-employed persons.

Identity of those assessed and geographical location

1. Employees (at work).
2. Persons not in the employer's employment who may face risks arising out of or in connection with the conduct by him of his undertaking.
3. New and expectant mothers.
4. Young persons.

The Management Regulations' ACoP recommends that employers should identify groups of workers particularly at risk such as night workers, homeworkers, those who work alone and disabled staff.

Hazards or risks to be assessed

(a) All risks.

(b) A risk assessment carried out by a self-employed person in circumstances where he or she does not employ others does not have to take into account duties arising under Part II of the Fire Precautions (Workplace) Regulations 1999, as amended by the Fire Precautions (Workplace) (Amendment) Regulations 1999 (SI 1999 No 1877).

(c) Physical, biological and chemical agents.

Purpose of assessment

To identify measures needed to be taken to comply with the requirements and prohibitions imposed by or under the relevant statutory provisions and by Part II of the Fire Precautions (Workplace) Regulations 1999, as amended (see above).

Qualification of duty

Suitable and Sufficient.

Record provision

(a) Five or more employees.

(b) Significant findings of the assessment and any group of employees identified by it as being especially at risk.

Review provision

Review assessment if:

(a) reason to suspect it is no longer valid (or can be improved),

(b) there has been a significant change.

Action on review

Risk assessment to be modified, as necessary.

Manual Handling Operations Regulations 1992 (as amended by SI 2002 No 2174)

On whom duties placed

1. Employers.
2. Self-employed persons.

Identity of those assessed and geographical location

Employees (at work).

The HSE's Manual Handling Regulations Guidance emphasises that employers must make allowance for those who might be pregnant or have a disability or health problem.

Hazards or risks to be assessed

All manual handling operations (with regard to factors – task, load, working environment, individual capability, other factors – listed in Schedule 1).

Purpose of assessment

To consider the questions set out in column 2 of Schedule 1 pertaining to the factors listed in that Schedule (see above).

Qualification of duty

Suitable and Sufficient.

When assessment has to be made

Make assessment where it is not reasonably practicable to avoid the need for employees to undertake any manual handling operations which involve risk of injury.

Review provision

Review assessment if:

(a) reason to suspect it is no longer valid,

(b) there has been a significant change.

Action on review

Changes to assessment to be made where required.

Details

Employers shall:

(a) take steps to reduce risk of injury to lowest level reasonably practicable,

(b) take steps to provide (to employees) general indications, and, where reasonably practicable, precise information on:

  (i) weight of load,

  (ii) heaviest side of load where centre of gravity is not central.

Noise at Work Regulations 1989 (as amended most recently by SI 1999 No 2024)

On whom duties placed

Employers and self-employed persons to ensure competent person makes assessment.

Identity of those assessed and geographical location

Employees (at work).

Hazards or risks to be assessed

Noise.

Purpose of assessment

(a) To identify which employees are exposed.

(b) To provide such information, with regard to the noise, as will facilitate compliance with Regs 7,8,9 and 11.

Qualification of duty

Adequate.

When assessment has to be made

Assessment to be made when any employee is likely to be exposed to the first action level or above or to the peak action level or above.

Record provision

Record to be kept until a further assessment is made.

Review provision

Review assessment if:

(a) reason to suspect that it is no longer valid,

(b) there has been a significant change.

Action on review

Changes to assessment to be made where required.

Personal Protective Equipment at Work Regulations 1992 (as amended most recently by SI 2002 No 2174)

On whom duties placed

1. Employers.
2. Self-employed persons.

Identity of those assessed and geographical location

Employees (at work).

Hazards or risks to be assessed

Any risks which have not been adequately controlled by other means.

Purpose of assessment

To assess risks to health and safety which have not been avoided by other means.

When assessment has to be made

Assessment to be made *before* choosing any personal protective equipment (PPE).

Review provision

Review assessment if:

(a) reason to suspect it is no longer valid,

(b) there has been a significant change.

Action on review

Changes to assessment to be made where required.

Details

Assessment to include:

(a) definition of the characteristics which PPE must have in order to be effective against the risks (taking into account any risks which the equipment itself may create),

(b) comparison of the characteristics of the PPE available to the required characteristics.

Appendix 5

A Summary of Requirements for the Provision of Information

Carriage of Dangerous Goods and Use of Transportable Pressure Equipment Regulations 2004

Carrier obliged under the Regulations to ensure that a travel document accompanies a consignment of dangerous goods by road or rail, and must keep a written record of all the information contained within that document for a period of three months after completion of journey in question.

Note: Special requirements relating to the provision of, amongst other matters, information are imposed in the specified persons where class 7 goods (i.e. special form radioactive material, etc) are to be carried by rail.

Chemicals (Hazard Information and Packaging for Supply) Regulations 2002

Employer/self-employed person to provide for other employers, self-employed persons/others

Supplier of dangerous chemicals to communicate with recipients information on the hazards presented.

Construction (Design and Management) Regulations 1994 (as amended most recently by SI 2000 No 2380)

Principal contractor

To provide:

— so far is reasonably practicable, every contractor with comprehensible information on the risks to the health or safety of that contractor or of any other employees or other persons under the control of that contractor arising out of or in connection with the construction work.

To ensure:

— so far is reasonably practicable, that every contractor who is an employer provides his employees carrying out the construction work with the information the employer must provide to those employees under the Management of Health and Safety at Work Regulations 1999, namely: the risks identified by the risk assessment; the preventative and protective measures thereby made necessary; the procedures in event of serious and imminent danger; the 'competent persons' as regards evacuation; and, the risks notified by other employers.

Control of Asbestos at Work Regulations 2002

Employer/self-employed person to provide own employees/persons not employees on the premises where the work is being carried out with adequate information

If such persons are, or are liable to be, exposed to asbestos, or if they supervise such persons who are, so that they are aware of:

(a) significant findings of the risk assessment;

(b) risks to health from asbestos;

(c) precautions which should be observed;

(d) relevant control limit and action level, in order to safeguard themselves and other relevant persons.

The above information must be:

(a) given at regular intervals;

(b) adapted to take account of significant changes in the type of work carried out by, or methods of work used by employer/self-employed person; and

(c) provided in a manner appropriate to the nature and degree of exposure identified by the risk assessment.

Note: The duty to provide adequate information to persons not employees is subject to the 'so far as is reasonably practicable' qualification.

Control of Lead at Work Regulations 2002

Employer/self-employed person to provide own employees/persons other than employees on the premises where the work is being carried out with suitable and sufficient instruction and training including:

(a) details of the form of lead to which employee/other relevant person is liable to be exposed;

(b) the significant findings of the risk assessment;

(c) the appropriate precautions and actions to be taken by employees and other persons, as above, coming within the duty, in order to safeguard both themselves and those other persons;

(d) the results of any monitoring of exposure to lead carried out in accordance with the Regulations; and

(e) the collective results of any medical surveillance undertaken in accordance with the Regulations.

Such instruction and training must be adapted to take account of significant changes in the type of work or methods of work used by the employer/self-employed person; and, provided in a manner appropriate to the level, type and duration of exposure identified by the risk assessment.

In addition, every employer/self-employed person must ensure that *any* person carrying out work in connection with the employer's/self-employed person's duties under the Regulations has sufficient instruction and training.

Note: The duty to provide suitable and sufficient instruction and training to persons other than employees is subject to the 'so far as is reasonably practicable' qualification.

Control of Substances Hazardous to Health Regulations 2002 (as amended most recently by SI 2002 No 278)

Employer/self-employed person to provide own employees/persons not employees on the premises where the work is being carried out with suitable and sufficient information, including:

(a) details, as specified, of the substances hazardous to health to which employees/other relevant persons are liable to be exposed;

(b) the significant findings of the risk assessment;

(c) the appropriate precautions and actions to be taken by employees/other relevant persons in order to safeguard both themselves and those other persons at the workplace;

(d) the results of any monitoring of exposure in accordance with the Regulations;

(e) the collective results of any health surveillance undertaken in accordance with the Regulations; and

(f) where employees and other relevant persons are working with a Group 4 biological agent (or material that may contain such an agent) the provision of written instruction and, if appropriate, the display of notices outlining procedure for handling such an agent or material.

Such information must be adapted to take account of significant changes in type of work carried out or methods of work used by employer/self-employed person; and, be provided in a manner appropriate to the level, type and duration of exposure identified by the risk assessment.

Note: The duty to provide suitable and sufficient information to persons other than employees is subject to the 'so far as is reasonably practicable' qualification.

Dangerous Substances in Harbour Areas Regulations 1987 (as amended by SI 1998 No 2885)

Employer to provide for own employees

Employer to provide information necessary to ensure his health and enable him to perform any operations in which he is involved with due regard to the health and safety of others.

Employer/self-employed person to provide for other workers

Self-employed person to ensure that he has information to ensure his and others' health and safety.

Employer/self-employed person to provide for other employers, self-employed persons/others

Operator to provide persons present on the berth with information to ensure their and others' health and safety.

Health and Safety (Display Screen Equipment) Regulations 1992 (as amended by SI 2002 No 2174)

Employer to provide for own employees

For his employees who are 'users' employer must provide information on:

(a) All aspects of health and safety relating to their workstation;

(b) Measures taken to comply with Regulations 2 (risk assessment), 3 (workstations), 4 (breaks), 5 (eyes and eyesight) and 6 (training).

Employer/self-employed person to provide for other workers

For 'users' employed by other employers, and 'operators', at work in his undertaking the employer must provide information on:

(a) All aspects of health and safety relating to their workstation;

(b) Measures taken to comply with Regulations 2 and 3 for 'users' only, Regulations 4 and 6(2) (training when workstation modified).

Health and Safety (First-aid) Regulations 1981 (as amended most recently by SI 2002 No 2174)

Employer to provide for own employees

Inform his employees of the arrangements that have been made in connection with the provision of first-aid, including the location of equipment, facilities and personnel.

Health and Safety Information for Employees Regulations 1989 (as amended by SI 1995 No 2923)

Employer to provide for own employees

Employer to ensure that the approved poster is kept displayed in a readable condition at an accessible place to the employee while he is at work and must provide the employee with the approved leaflet. Employer has to have information of EMAS office and enforcing authority on poster.

Ionising Radiations Regulations 1999

Employer/self-employed person to provide for own employees

Such information as is suitable for them to know:

(a) the health risks created by exposure to ionising radiation;

(b) the precautions which should be taken; and

(c) the importance of complying with the medical, technical and administrative requirements of the Regulations.

Employer/self-employed person to ensure

Adequate information is given to persons other than employees who are directly concerned with ionising radiation work carried on by the employer/self-employed person to ensure, so far is reasonably practicable their health and safety.

Management of Health and Safety at Work Regulations 1999 (as amended most recently by SI 2003 No 2457)

Employer/self-employed person to provide for own employees

(a) Risks identified by the assessment.

(b) Preventive and protective measures as per Schedule 1.

(c) Procedures in event of serious and imminent danger.

(d) Competent persons re evacuation.

(e) Risks notified by other employers.

Employer/self-employed person to provide for other workers

(a) Risks to health and safety arising from the conduct of the undertaking.

(b) To enable them to identify competent persons re evacuation.

(c) Any special occupational qualifications or skills needed to work safely.

(d) Any health surveillance required.

Employer/self-employed person to provide for other employers, self employed persons/others

(a) Risks to health and safety arising from the conduct of the undertaking.

(b) Measures taken in compliance.

(c) To enable them to identify competent persons re evacuation.

(d) Any special occupational qualifications or skills needed to work safely.

(e) Specific features of jobs in relation to health and safety.

Manual Handling Operations Regulations 1992 (as amended by SI 2002 No 2174)

Employer/self-employed person to provide for own employees

Those undertaking manual handling operations.

(a) General indications, and where reasonably practicable, precise information on:

(i) Weight of each load,

(ii) Heaviest side.

Noise at Work Regulations 1989 (as amended most recently by SI 1999 No 2024)

Employer/self-employed person to provide for own employees

Every employer shall provide any of his employees, who are likely to be exposed to the first action level or above, information on:

(a) Risk of damage to hearing.

(b) What steps can be taken to minimise risk.

(c) Steps that the employee must take in order to obtain personal ear protection.

(d) Employees' obligations.

Offshore Installations and Wells (Design and Construction, etc) Regulations 1996

Well-operator in installation duty holder

Provision of appropriate information so that those carrying well operations can carry them out competently.

Personal Protective Equipment at Work Regulations 1992 (as amended by SI 2002 No 2174)

Employer to provide for own employees

Such information as is adequate and appropriate on:

(a) Risks PPE will avoid/limit.

(b) Purpose/manner of use.

(c) Action to be taken by employee.

Pressure Systems Safety Regulations 2000 (as amended by SI 2004 No 568)

Employer to provide for own employees

(a) Any person who:

(i) designs for another any pressure system or any article which is intended to be a component or part thereof; or

(ii) supplies (whether as manufacturer or importer or in any other capacity) any pressure system or any such article, shall provide sufficient information concerning its design, construction, examination, operation and maintenance as may reasonably foreseeably be needed to comply with the Regulations.

(b) The employer of a person who modifies or repairs any pressure system must provide sufficient information concerning the modification or repair as may reasonably foreseeably be needed to comply with the Regulations.

Provision and Use of Work Equipment Regulations 1998 (as amended by SI 1999 No 2001) (new provisions replacing the revoked Power Press Regulations are incorporated in Part IV of these Regulations)

Employer to provide for own employees

Every employer must ensure that all employees using work equipment are provided with adequate health and safety information and, where appropriate, written instructions about the use of such equipment. The employer must also ensure that any employee supervising or managing the use of work equipment has available adequate health and safety information and, as above, where appropriate, written instructions. The information and instructions required above must include information and, where appropriate, written instructions on:

(a) All health and safety aspects arising from the use of the work equipment including conditions in which and methods by which it can be used.

(b) Any limitations on these uses.

(c) Any foreseeable difficulties that could arise.

(d) The methods to deal with them.

Safety Representatives and Safety Committees Regulations 1977 (as amended most recently by SI 1999 No 2024)

Employer to provide for own employees

To allow safety representatives to carry out inspections to allow access to relevant documents to make available all the information necessary to allow fulfilment of functions.

Employer/self-employed person to provide for other employers, self-employed persons/others

Safety representatives to receive information from inspectors.

Work in Compressed Air Regulations 1996 (as amended by SI 1997 No 2776)

Employer to provide for own employees

The 'compressed air contractor', as defined, is to ensure that adequate information etc. is given to *all* persons working in compressed air on the risks arising from such work and the precautions to be taken. The HSE Guidance to the 1996 Regulations stresses that the provision of information etc. in such circumstances may be carried out by anyone competent to do so, but the compressed air contractor is responsible for ensuring that it is carried out to standard.

Employer/self-employed person to provide for other employers, self-employed persons/others

See above.

Employer/self-employed person to provide for other workers

See above.

Appendix 6

A Summary of the Requirements for Training/Instruction and Consultation

Carriage of Dangerous Goods and Use of Transportable Pressure Equipment Regulations 2004

Any person involved in the carriage of dangerous goods by road or rail must ensure that he and those of his employees concerned with such carriage receive training which complies with, and is documented in accordance with, the requirements of the ADR/RID international agreement – whichever is applicable.

Note: A carrier carrying dangerous goods by road whose driver is obliged by the ADR agreement to receive training must ensure that that driver:

(i) has received training which complies with the ADR agreement requirements and is relevant to the goods, person and type of vehicle in question;

(ii) has received any special training required by the ADR agreement in relation to the goods in question; and

(iii) holds a certificate issued by the GB competent authority stating that the driver has participated in a training course and passed an exam in accordance with the ADR agreement requirements – in so far as they relate to the carriage of the dangerous goods in question.

Confined Spaces Regulations 1997

Employer/self-employed person to ensure

Persons likely to be involved in any emergency rescue to be trained for that purpose. The ACoP and guidance specifies the content of that training, where appropriate.

Construction (Health, Safety and Welfare) Regulations 1996 (as amended most recently by SI 1998 No 2306)

Employer provides training/instruction for employees

Adequate training of employees to avoid acts causing injury or damage.

Control of Asbestos at Work Regulations 2002

Employer/self-employed person to provide own employees/persons not employees on the premises where the work is being carried out with adequate instruction and training

If such persons are, or are liable to be, exposed to asbestos, or if they supervise such persons who are, so that they are aware of:

(a) significant findings of the risk assessment;

(b) risks to health from asbestos;

(c) precautions which should be observed;

(d) relevant control limit and action level, in order to safeguard themselves and other relevant person.

The above information must be:

(a) given at regular intervals;

(b) adapted to take account of significant changes in the type of work carried out by or methods of work used by employer/self-employed person; and

(c) provided in a manner appropriate to the nature and degree of exposure identified by the risk assessment.

Note: The duty to provide adequate information to persons not employees is subject to the 'as far as is reasonably practicable' qualification.

Control of Lead at Work Regulations 2002

Employer/self-employed person to provide own employees/persons other than employees on the premises where the work is being carried out with suitable and sufficient information, including:

(a) details of the form of lead which employee/other relevant person is liable to be exposed;

(b) the significant findings of the risk assessment;

(c) the appropriate precautions and actions to be taken by employees and other persons, as above, coming within the duty, in order to safeguard both themselves and those other persons;

(d) the results of any monitoring of exposure of lead carried out in accordance with the Regulations; and

(e) the collective results of any medical surveillance undertaken in accordance with the Regulations.

Such information must be adapted to take account of significant changes in the type of work or methods of work used by the employer/self-employer persons; and, provided in a manner appropriate to the level, type and duration of exposure identified by the risk assessment.

In addition, every employer/self-employed person must ensure that *any* person carrying out work in connection with the employer's/self-employed person's duties under the Regulations has suitable and sufficient information.

Note: The duty to provide suitable and sufficient information to persons other than employees is subject to the 'so far as is reasonably practicable' qualification.

Control of Substances Hazardous to Health Regulations 2002 (as amended most recently by SI 2003 No 278)

Employer/self-employed person to provide own employees/persons not employees on the premises where the work is being carried out with suitable and sufficient instruction and training, including:

(a) details, as specified, of the substances hazardous to health to which employees/other relevant persons are liable to be exposed;

(b) the significant findings of the risk assessment;

(c) the appropriate precautions and actions to be taken by employees/other relevant persons in order to safeguard both themselves and those other persons at the workplace;

(d) the results of any monitoring of exposure in accordance with the Regulations; and

(e) the collective results of any health surveillance undertaken in accordance with the Regulations.

Such information must be adapted to take account of significant changes in type of work carried out or methods of work used by employer/self-employed person; and, be provided in a manner appropriate to the level, type and duration of exposure identified by the risk assessment.

Note: The duty to provide such suitable and sufficient instruction and training to persons other than employers is subject to the 'so far as is reasonably practicable' qualification.

Dangerous Substances and Explosive Atmospheres Regulations 2002

Employer/self-employed person to provide own employees/persons other than employees at the workplace where the work is being carried out with suitable and sufficient information, including:

(a) details as specified of any dangerous substances present at the workplace;

(b) significant findings of the risk assessment; and

(c) suitable and sufficient information on the appropriate precautions and actions to be taken by employees and other persons, as above, coming within the duty, in order to safeguard both themselves and those other persons.

Such information must be adapted to take account of significant changes in type of work or methods of work used by employer/self-employed person; and, be provided in a manner appropriate to the risk identified in the risk assessment.

Note: The duty to provide suitable and sufficient information to persons other than employees is subject to the 'so far as is reasonably practicable' qualification.

Dangerous Substances in Harbour Areas Regulations 1987 (as amended by SI 1998 No 2885)

Employer provides training/instruction for employees

Employer and self-employed person to provide instruction, training and supervision necessary to ensure his health and safety and to enable him to perform any operations in which he is involved with due regard to the health and safety of others.

Employer provides training/instruction for specified employees/others

Operator to provide instruction, as necessary, to persons present on the berth to ensure their health and safety.

Health and Safety (Consultation with Employees) Regulations 1996 (as amended most recently by SI 2002 No 2174)

Employer provides training/instruction for employees

Employer to consult either:

(a) with employees directly, or

(b) with representatives of employee safety in non-unionised workforces,

and to allow time-off for training.

Health and Safety (Display Screen Equipment) Regulations 1992 (as amended by SI 2002 No 2174)

Employer provides training/instruction for employees

Employer shall provide adequate health and safety training in use of workstation to:

(a) his employees who are 'users' on date of coming into force of the Regulations.

(b) his employees about to become users. Employers also shall provide such training to employee 'users' when the organisation of their workstation is substantially modified.

Employer provides training/instruction for specified employees/others

Current users of DSE. Adequate health and safety training in the use of workstation, and whenever organisation of workstation modified.

Ionising Radiations Regulations 1999

Employer/self-employed person to ensure

Those of his employees engaged in work with ionising radiation are given appropriate training in the field of radiation protection and receive instruction as is suitable and sufficient for them to know:

(a) the risks to health created by exposure to ionising radiation;

(b) the precautions which should be taken; and

(c) the importance of complying with the medical, technical and administrative requirements of the Regulations.

Ionising Radiation (Medical Exposure) Regulations 2000

Employer provides training/instruction for 'practitioners or operators', as defined

Steps must be taken to ensure that every 'practitioner or operator' engaged by the employer to carry out medical exposures involving ionising radiation undertakes continuing education and training after qualification, as specified (where the employer is concurrently 'practitioner or operator' he must ensure that he undertakes such continuing education and training as may be 'appropriate').

Management of Health and Safety at Work Regulations 1999 (as amended most recently by SI 2003 No 278)

Employer provides training/instruction for employees

Adequate health and safety training: on recruitment on being exposed to new/ increased risks e.g. on transfer/change of responsibilities; new work equipment; new technology; new system of work.

Employer provides training/instruction for specified employees/others

Employees who have access to any area (where such access is subject to restriction): adequate instruction for evacuation procedures.

Noise at Work Regulations 1989 (most recently amended by SI 1999 No 2024)

Employer provides training/instruction for employees

Every employer shall provide each of his employees, who is likely to be exposed to the first action level or above, instruction and training on:

(a) risk of damage to hearing

(b) what steps can be taken to minimise risk

(c) steps that employee must take in order to obtain personal ear protectors

(d) employees' obligations.

Offshore Installations and Wells (Design and Construction, etc) Regulations 1996

Well-operator or installation duty holder to ensure

Well operations are not carried out unless persons carrying out the operation have received appropriate instruction and training.

Personal Protective Equipment at Work Regulations 1992 (as amended most recently by SI 2002 No 2174)

Employer provides training/instruction for employees

Adequate and appropriate training about risks the PPE will avoid/limit; purposes of PPE; action to be taken by employee.

Provision and Use of Work Equipment Regulations 1998 (as amended by SI 2002 No 2174)

Employer provides training/instruction for employees

All persons who use work equipment must be provided with:

adequate training for purposes of health and safety including training in the methods of use, risks and precautions. For employers' obligations regarding instruction see Provision of Information above.

Employer provides training/instruction for 'practitioners or operators', as defined

Obligations as above. The HSC's ACoP and Guidance on the Regulations emphasises that 'training and proper supervision of "young people" is particularly important because of their relative immaturity and unfamiliarity with the working environment'.

Quarries Regulations 1999 (as amended by SI 2002 No 2174)

Quarry operator must

Ensure that no person undertakes any work in a quarry unless: that person is either competent to do the work or does so under the instruction and supervision of some other person who is competent to give instruction in and to supervise the doing of that work for the purpose of training him.

Safety Representatives and Safety Committees Regulations 1977 (as amended most recently by SI 1999 No 3242)

Employer provides training/instruction for specified employees/others

Allow time off with pay for training of safety representatives (as may be reasonable); consult with safety representatives and with other trade unions representatives over establishment of safety committees; consult safety representatives on:

(a) instruction of measures which may affect health and safety of employees.

(b) arrangements for appointing those who give health and safety assistance.

(c) health and safety information provided to workforce.

Work in Compressed Air Regulations 1996 (as amended by SI 1997 No 2776)

Employer provides training/instruction for employees

The 'compressed air contractor', as defined, to provide adequate training on risks and precautions. See entry for Regulations under Provision of Information.

Table of Cases

This table is referenced to paragraph numbers in the work.

Table of Statutes

Table of Statutory Instruments

Index

References are to sections and paragraph numbers of this book.